# Lecture Notes in Computer Science     6034

*Commenced Publication in 1973*
Founding and Former Series Editors:
Gerhard Goos, Juris Hartmanis, and Jan van Leeuwen

## Advanced Research in Computing and Software Science
Subline of Lectures Notes in Computer Science

T0191671

Alejandro López-Ortiz (Ed.)

# LATIN 2010:
# Theoretical Informatics

9th Latin American Symposium
Oaxaca, Mexico, April 19-23, 2010
Proceedings

 Springer

Volume Editor

Alejandro López-Ortiz
University of Waterloo
Faculty of Mathematics
School of Computer Science
Waterloo, Ontario N2L 3G1, Canada
E-mail: alopez-o@uwaterloo.ca

Library of Congress Control Number: 2010922369

CR Subject Classification (1998): F.2, F.1, C.2, H.3, E.1, G.2

LNCS Sublibrary: SL 1 – Theoretical Computer Science and General Issues

ISSN      0302-9743
ISBN-10   3-642-12199-3 Springer Berlin Heidelberg New York
ISBN-13   978-3-642-12199-9 Springer Berlin Heidelberg New York

springer.com

© Springer-Verlag Berlin Heidelberg 2010
Printed in Germany

Typesetting: Camera-ready by author, data conversion by Scientific Publishing Services, Chennai, India
Printed on acid-free paper      06/3180

# Preface

The papers contained in this volume were presented at the 9th Latin American Theoretical Informatics Symposium held at the Benito Juárez University of Oaxaca, Oaxaca City, México, April 19-23, 2010. The LATIN series of conferences was launched in 1992 to foster the interaction between the Latin American theoretical computer science community and computer scientists around the world. LATIN 2010 was the ninth of a series, after Sao Paulo, Brazil (1992); Valparaiso, Chile (1995); Campinas, Brazil (1998); Punta del Este, Uruguay (2000); Cancun, Mexico (2002); Buenos Aires, Argentina (2004); Valdivia, Chile (2006) and Búzios, Rio de Janeiro, Brazil (2008).

From the 155 submissions, the Program Committee selected 56 papers for presentation at the conference. The selection of papers was based on originality, quality, and relevance to theoretical computer science. It is expected that most of these papers will appear in a more complete and polished form in scientific journals in the future. In addition to the contributed papers, this volume contains the abstracts of four invited plenary talks given at the conference by Cristopher Moore, Piotr Indyk, Sergio Rajsbaum, and Leslie Valiant. A special session on the life and work of the late Imre Simon was held. Prof. Simon played a key role in the development of theoretical computer science in Latin American as well as the LATIN conference. This session had contributions from Ricardo Baeza-Yates, John Brzozowski, Volker Diekert, and Jacques Sakarovitch.

We would like to thank all members of the Program Committee for their thorough work. The conference had a larger-than-usual Program Committee, which contributed to a more detailed discussion on each of the submitted papers. In addition, we thank all our sponsors, Microsoft Research, Yahoo! Research, University of Waterloo, and the Benito Juárez University of Oaxaca. Special thanks to Dante Arias, who served as the liaison for local organization arrangements.

January 2010                                      Alejandro López-Ortiz

# Organization

## Program Committee

| | |
|---|---|
| Amihood Amir | Bar-Ilan University, Israel |
| Diego Arroyuelo | Yahoo Research Latin America, Chile |
| Ricardo Baeza-Yates | Yahoo Research, Spain |
| Joan Boyar | University of Southern Denmark, Denmark |
| Gerth Brodal | Aarhus University, Denmark |
| Edgar Chavez | University of Michoacán, Mexico |
| José Correa | University of Chile, Chile |
| Irit Dinur | Weizmann Institute, Israel |
| Stephane Durocher | University of Manitoba, Canada |
| Faith Ellen | University of Toronto, Canada |
| Leah Epstein | University of Haifa, Israel |
| Cristina Fernandes | University of Sao Paolo, Brazil |
| Paolo Ferragina | University of Pisa, Italy |
| Martin Fürer | Penn State University, USA |
| Kazuo Iwama | Kyoto University, Japan |
| Valerie King | University of Victoria, Canada |
| Ravi Kumar | Yahoo Research, USA |
| Moshe Lewenstein | Bar-Ilan University, Israel |
| Alejandro López-Ortiz | University of Waterloo, Canada (Chair) |
| Jesús De Loera | UC Davis, USA |
| Kazuhisa Makino | Tokyo University, Japan |
| Dániel Marx | Budapest University, Hungary |
| Kurt Mehlhorn | Max Planck Institute for Informatics, Germany |
| Julián Mestre | Max Planck Institute for Informatics, Germany |
| Michael Mitzenmacher | Harvard University, USA |
| Ian Munro | University of Waterloo, Canada |
| S. Muthukrishnan | Google and Rutgers, USA |
| Konstantinos Panagiotou | Max Planck Institute for Informatics, Germany |
| Mike Paterson | University of Warwick, UK |
| Jorge-Luis Ramírez | Université Pierre et Marie Curie, Paris 6, France |

Iván Rapaport          University of Chile, Chile
Günter Rote            Free University of Berlin, Germany
Gelasio Salazar        San Luis Potosí University, Mexico
Hadas Shachnai         Technion, Israel
Igor Shparlinski       Macquarie University, Australia
Wojciech Szpankowski   Purdue University, USA
Subash Suri            UCSB, USA
Alfredo Viola          Universidad de la República, Uruguay
John Watrous           University of Waterloo, Canada
Renato Werneck         Microsoft Research, USA
Gerhard Woeginger      T.U. Eindhoven, The Netherlands
Norbert Zeh            Dalhousie University, Canada

## Steering Committee

Martín Farach-Colton      Rutgers Univesity, USA
Marcos Kiwi               University of Chile, Chile
Yoshiharu Kohayakawa      University of Sao Paulo, Brazil
Eduardo Laber             PUC-Rio, Brazil
Gonzalo Navarro           University of Chile, Chile
Daniel Panario            Carleton University, Canada

## Local Organization

Dante Arias
Francisco Claude
Reza Dorrigiv (Webmaster)
Robert Fraser (Publicity Chair)
Jazmín Romero
Lesvia Ruiz
Alejandro Salinger

## External Reviewers

Omran Ahmadi            Eric Bach               Ahmed Bouajjani
Mohammad Ali Abam       Boaz Ben-Moshe          Janusz Brzozowski
Bernardo Abrego         Cedric Bentz            Saverio Caminiti
Anil Ada                Piotr Berman            Erin Chambers
Louigi Addario-Berry    Sandeep Bhatt           Eric Y. Chen
Deepak Ajwani           Laurent Bienvenu        Yijia Chen
Ali Akhavi              Markus Blaser           Flavio Chierichetti
Saeed Alaei             Hans L. Bodlaender      Christian Choffrut
Noga Alon               Hans-Joachim            Yongwook Choi
Yuichi Asahiro            Boeckenhauer           Marek Chrobak
James Aspnes            Nicolas Bonichon        Jacek Cichon
Franz Baader            Glencora Borradaile     Francisco Claude

Julien Clément
Bruno Codenotti
Elad Cohen
Stephen Cohen
Robert Cori
Laszlo Csirmaz
Peter Danziger
Anirban Dasgupta
Pooya Davoodi
Erik Demaine
Jonathan Derryberry
Miriam Di Ianni
Konstantinos Drakakis
Adrian Dumitrescu
Wayne Eberly
Khaled Elbassioni
Matthias Englert
David Eppstein
Ruy Fabila
Arash Farzan
Lene Favrholdt
John Fearnley
Uriel Feige
Sandor Fekete
Silvia Fernández-
Merchant
Antonio Fernández
Jiri Fiala
Nicolas Figueroa
Rudolf Fleischer
Paola Flocchini
Guilherme D. da Fonseca
Robert Fraser
Christiane Frougny
Takuro Fukunaga
William I Gasarch
Serge Gaspers
Loukas Georgiadis
Ellen Gethner
Mark Giesbrecht
Joachim Giesen
Michael Goodrich
Zvi Gotthilf
Frederic Green
Alexander Grigoriev

Roberto Grossi
Jiong Guo
Gregory Gutin
Thomas Hackl
Xin Han
Meng He
Thomas Hildebrandt
Petr Hlineny
Markus Holzer
Han Hoogeveen
Carlos Hoppen
Xiang-dong Hou
Jing Huang
Anna Huber
Cor Hurkens
HK Hwang
Toshimasa Ishii
Allan G. Jørgensen
Maleq Kahn
Naonori Kakimura
Naoyuki Kamiyama
Juhani Karhumaki
Jun Kawahara
Idit Keidar
Hans Kellerer
Daniel Kirsten
Rolf Klasing
Stephen Kobourov
Matthias Koeppe
Christian Komusiewicz
Alexander Kononov
Tsvi Kopelowitz
Daniel Kral
Stefan Kratsch
Bala Krishnamoorthy
Danny Krizanc
Oded Lachish
Alan Lauder
Grégoire Lecerf
Asaf Levin
Erik Jan van Leeuwen
Ming Li
Mathieu Liedloff
Yury Lifshits
Daniel Lokshtanov

Sylvain Lombardy
Shachar Lovett
Vadim Lozin
Anna Lubiw
Hosam Mahmoud
Johann Makowsky
Arnaldo Mandel
Martín Matamala
Domagoj Matijevic
Nicole Megow
Jie Meng
Criel Merino
Antoine Meyer
Dimitris Michail
Avery Miller
Di Ianni Miriam
Pablo Moisset de
Espanes
Eduardo Moreno
Elena Mumford
Torsten Mütze
Viswanath Nagarajan
Lata Narayanan
Jesper Nederlof
Alantha Newman
Igor Nitto
Hirotaka Ono
Alessio Orlandi
Andrzej Pelc
Doron Peled
Alair Pereira-do-Lago
Nadia Pisanti
Valentin Polishchuk
Ely Porat
Paweł Prałat
Giuseppe Prencipe
Maurice Queyranne
Charles Rackoff
Luis Rademacher
Vijaya Ramachandran
Pedro Ramos
Dror Rawitz
Saurabh Ray
Jan Reimann
Eric Remila

Daniel Roche
Dominique Rossin
Atri Rudra
Vojtěch Rädl
Alejandro Salinger
Kai Salomaa
Davide Sangiorgi
Tamas Sarlos
Mathias Schacht
Guido Schaefer
Marcus Schaefer
Jens M. Schmidt
Eric Schost
Roy Schwartz
Siddhartha Sen
Jeffrey Shallit
Nino Shervashidze
D Sivakumar
Martin Skutella
Roberto Solís-Oba
Troels Sorensen

Francisco Soulignac
Bettina Speckmann
Heike Sperber
R. Sritharan
Rob van Stee
Damien Stehlé
Fabian Stehn
Ron Steinfeld
Nicolas Stier-Moses
Arne Storjohann
Karol Suchan
Maxim Sviridenko
Zoya Svitkina
Wojciech Szpankowski
Kenjiro Takazawa
Suguru Tamaki
Tami Tamir
Xuehou Tan
Tamir Tassa
Orestis Telelis
Alex Thomo

Alexander Tiskin
Ioan Todinca
Andrew Tomkins
Edgardo Ugalde
Johannes Uhlmann
Jorge Urrutia
Rossano Venturini
Paul Vitanyi
Jan Vondrak
Panagiotis Voulgaris
Tjark Vredeveld
Magnus Wahlström
Yoshiko Wakabayashi
Mark Ward
Mark Daniel Ward
Philipp Woelfel
Masaki Yamamoto
Amir Yehudayoff
Sheng Yu

# Table of Contents

# Continuous and Discrete Methods in Computer Science

Cristopher Moore

Department of Computer Science,
University of New Mexico
and
The Santa Fe Institute
moore@cs.unm.edu

**Abstract.** Culturally, computer scientists are generally trained in discrete mathematics; but continuous methods can give us surprising insights into many algorithms and combinatorial problems. In this pedagogical talk, I will describe two interesting places where continuous mathematics makes an entrance into computer science: proving lower bounds on the 3-colorability threshold in random graphs using differential equations, and a continuous-time version of Karmarkar's algorithm for Linear Programming, based on the so-called Newton Barrier Flow.

A. López-Ortiz (Ed.): LATIN 2010, LNCS 6034, p. 1, 2010.
© Springer-Verlag Berlin Heidelberg 2010

# Colorful Strips

Greg Aloupis[1], Jean Cardinal[1], Sébastien Collette[1,*], Shinji Imahori[2],
Matias Korman[3], Stefan Langerman[1,**], Oded Schwartz[4],
Shakhar Smorodinsky[5], and Perouz Taslakian[1]

[1] Université Libre de Bruxelles, CP212, Bld. du Triomphe, 1050 Brussels, Belgium
{galoupis,jcardin,secollet,slanger,ptaslaki}@ulb.ac.be
Supported by the Communauté française de Belgique - ARC.
[2] Graduate School of Engineering, Nagoya University, Nagoya 464-8603, Japan
imahori@na.cse.nagoya-u.ac.jp
[3] Graduate School of Information Sciences (GSIS), Tohoku University, Japan
mati@dais.is.tohoku.ac.jp
Partially supported by 21st Century Global CoE program
[4] Departments of Mathematics, Technische Universität Berlin,
10623 Berlin, Germany
odedsc@math.tu-berlin.de
[5] Ben-Gurion University, Be'er Sheva 84105, Israel
shakhar@math.bgu.ac.il

**Abstract.** We study the following geometric hypergraph coloring problem: given a planar point set and an integer $k$, we wish to color the points with $k$ colors so that any axis-aligned strip containing sufficiently many points contains all colors. We show that if the strip contains at least $2k-1$ points, such a coloring can always be found. In dimension $d$, we show that the same holds provided the strip contains at least $k(4\ln k + \ln d)$ points.

We also consider the dual problem of coloring a given set of axis-aligned strips so that any sufficiently covered point in the plane is covered by $k$ colors. We show that in dimension $d$ the required coverage is at most $d(k-1) + 1$. Lower bounds are also given for all of the above problems. This complements recent impossibility results on decomposition of strip coverings with arbitrary orientations.

From the computational point of view, we show that deciding whether a three-dimensional point set can be 2-colored so that any strip containing at least three points contains both colors is NP-complete. This shows a big contrast with the planar case, for which this decision problem is easy.

# 1   Introduction

There is a currently renewed interest in coloring problems on *geometric* hypergraphs, that is, set systems defined by geometric objects. This interest is

* Chargé de Recherches du FRS-FNRS.
** Maître de Recherches du FRS-FNRS.

A. López-Ortiz (Ed.): LATIN 2010, LNCS 6034, pp. 2–13, 2010.

motivated by applications to wireless and sensor networks [6]; conflict-free colorings [8], chromatic numbers [18], covering decompositions [17,3], or polychromatic (*colorful*) colorings of geometric hypergraphs [4] have been extensively studied in this context.

In this paper, we are interested in $k$-coloring finite point sets in $\mathbb{R}^d$ so that any region bounded by two parallel axis-aligned hyperplanes, that contains at least some fixed number of points, also contains a point of each color.

An *axis-aligned strip*, or simply a *strip* (unless otherwise specified), is the area enclosed between two parallel axis-aligned hyperplanes. A $k$-*coloring* of a finite set assigns one of $k$ colors to each element in the set. Let $S$ be a $k$-colored set of points in $\mathbb{R}^d$. A strip is said to be *polychromatic* with respect to $S$ if it contains at least one element of each color class. We define the function $p(k,d)$ as the minimum number for which there always exists a $k$-coloring of any point set in $\mathbb{R}^d$ such that every strip containing at least $p(k,d)$ points is polychromatic. This is a particular case of the general framework proposed by Aloupis, Cardinal, Collette, Langerman, and Smorodinsky in [4].

Note that the problem does not depend on whether the strips are open or closed, since the problem can be seen in a purely combinatorial fashion: an axis-aligned strip isolates a subsequence of the points in sorted order with respect to one of the axes. Therefore, the only thing that matters is the order in which the points appear along each axis. We can thus rephrase our problem, considering $d$-dimensional points sets, as finding the minimum value $p(k,d)$ such that the following holds: For $d$ permutations of a set of items $S$, it is always possible to color the items with $k$ colors, so that in all $d$ permutations every sequence of at least $p(k,d)$ contiguous items contains one item of each color.

We also study *circular* permutations, in which the first and the last elements are contiguous. We consider the problem of finding a minimum value $p'(k,d)$ such that, for any $d$ circular permutations of a set of items $S$, it is possible to $k$-color the items so that in every permutation, every sequence of $p'(k,d)$ contiguous items contains all colors.

A restricted geometric version of this problem in $\mathbb{R}^2$ consists of coloring a point set $S$ with respect to wedges. For our purposes, a wedge is any area delimited by two half-lines with common endpoint at one of $d$ given apices. Each apex induces a circular ordering of the points in $S$. This is illustrated in Figure 1. We aim at coloring $S$ so that any wedge containing at least $p'(k,d)$ points is polychromatic. In $\mathbb{R}^2$, the non-circular case corresponds to wedges with apices at infinity. In that sense, the wedge coloring problem is more difficult than the strip coloring problem.

We then study a dual version of the problem, in which a set of axis-aligned strips is to be colored so that sufficiently covered points are contained in strips from all color classes. For instance, in the planar case we study the following function $\overline{p}(k,d)$. Let $H$ be a $k$-colored set of strips in $\mathbb{R}^d$. A point is said to be polychromatic with respect to $H$ if it is contained in strips of all $k$ color classes. The function $\overline{p}(k,d)$ is the minimum number for which there always exists a

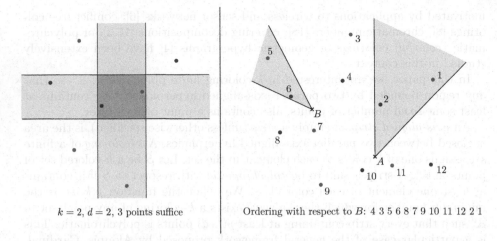

$k = 2, d = 2$, 3 points suffice | Ordering with respect to $B$: 4 3 5 6 8 7 9 10 11 12 2 1

**Fig. 1.** Illustration of the definitions of $p(k,2)$ and $p'(k,2)$. On the left, points are 2-colored so that any axis-aligned strip containing at least three points is bichromatic. On the right, two points $A$ and $B$ define two circular permutations of the point set. In this case, we wish to color the points so that there is no long monochromatic subsequence in either of the two circular orderings.

$k$-coloring of any set of strips in $\mathbb{R}^d$ such that every point of $\mathbb{R}^d$ contained in at least $\overline{p}(k,d)$ strips is polychromatic.

Note that the functions $p(k,d)$, $p'(k,d)$ and $\overline{p}(k,d)$ are monotone and non-decreasing. Since we are interested in arbitrarily large pointsets, we always consider the set that we color to be "large enough" (that is, unbounded in terms of $k$).

*Previous results.* A *hypergraph* $(S, R)$ is defined by a set $S$ (called the *ground set*) and a set $R$ of subsets of $S$. The main problem studied here is the coloring of geometric hypergraphs where the ground set $S$ is a finite set of points, and the set of ranges $R$ consists of all subsets of S that can be isolated by a single strip. In the dual case the ground set $S$ is a finite set of geometric shapes and the ranges are points contained in the common intersection of a subset of $S$. In some places in the literature, finite geometric hypergraphs are also referred to as geometric *range spaces*.

Several similar problems have been studied in this context [13,20,4], where the range space is not defined by strips, but rather by halfplanes, triangles, disks, pseudo-disks or translates of a centrally symmetric convex polygon. The problem was originally stated in terms of decomposition of *c-covers* (or *f-fold coverings*) in the plane: A *c*-cover of the plane by a convex body $Q$ ensures that every point in the plane is covered by at least *c* translated copies of $Q$. In 1980, Pach [13] asked if, given $Q$, there exists a function $f(Q)$ such that every $f(Q)$-cover of the plane can be decomposed into 2 disjoint 1-covers. A natural extension is to ask if given $Q$, there exists a function $f(k, Q)$ such that every $f(k, Q)$-cover of the plane can be decomposed into *k* disjoint 1-covers. This corresponds to a *k*-coloring of the $f(k, Q)$-cover, such that every point of the plane is polychromatic.

Partial answers to this problem are known: Pach [14] referenced an unpublished manuscript by Mani and Pach [11] showing that any 33-cover of the plane by unit disks can be decomposed into two 1-covers. This could imply that the function $f$ exists for unit disks, but could still be exponential in $k$. Recently, Tardos and Tóth [20] proved that any 43-cover by translated copies of a triangle can be decomposed into two 1-covers. For the case of centrally symmetric convex polygons, Pach [15] proved that $f$ is at most exponential in $k$. More than 20 years later, Pach and Tóth [17] improved this by showing that $f(k, Q) = O(k^2)$, and recently Aloupis et al. [3] proved that $f(k, Q) = O(k)$.

On the other hand, for the range space induced by arbitrary disks, Mani and Pach [11] (see also [16]) proved that $f(2, Q)$ is unbounded: for any constant $c$, there exists a set of points that cannot be 2-colored so that all open disks containing at least $c$ points are polychromatic. Pach, Tardos and Tóth [16] obtained a similar result for the range spaces induced by the family of either non-axis-aligned strips, or axis-aligned rectangles. Specifically, for any integer $c$ there exist $c$-fold coverings with non-aligned strips that cannot be decomposed into two coverings (i.e., cannot be 2-colored). The previous impossibilities constitute our main motivation for introducing the problem of $k$-coloring axis-aligned strips, and strips with a bounded number of orientations.

*Paper Organization.* In Section 2 we give constructive upper bounds on the functions $p$ and $p'$ for $d = 2$. In Section 3 we consider higher-dimensional cases, as well as the computational complexity of finding a valid coloring. Section 4 concerns the dual problem of coloring strips with respect to points. Our lower and upper bounds are summarized in Table 1.

**Table 1.** Bounds on $p$, $p'$ and $\bar{p}$

|  | $p(k, d)$ | $p'(k, d)$ | $\bar{p}(k, d)$ |
|---|---|---|---|
| upper bound | $k(4\ln k + \ln d)$ | $k(4\ln k + \ln d)$ | $d(k-1) + 1$ |
|  | ($2k-1$ for $d=2$) | ($2k$ for $d=2$) |  |
| lower bound | $2 \cdot \lceil \frac{(2d-1)k}{2d} \rceil - 1$ | $2 \cdot \lceil \frac{(2d-1)k}{2d} \rceil - 1$ | $\lfloor k/2 \rfloor d + 1$ |

# 2 Axis-Aligned Strips and Circular Permutations for $d = 2$

We first consider upper bounds for the functions $p(k, 2)$ and $p'(k, 2)$.

## 2.1 Axis-Aligned Strips: Upper Bound on $p(k, 2)$

We refer to a strip containing at least $i$ points as an $i$-*strip*. Our goal is to show that for any integer $k$ there is a constant $p(k, 2)$ such that any finite planar point set can be $k$-colored so that all $p(k, 2)$-strips are polychromatic.

For $d = 2$, there is a reduction to the recently studied problem of 2-coloring graphs so that monochromatic components are small. Haxell et al. [10] proved that the vertices of any graph with maximum degree 4 can be 2-colored so that every monochromatic connected component has size at most 6. For a given finite point set $S$ in the plane, let $E$ be the set of all pairs of points $u, v \in S$ such that there is a strip containing only $u$ and $v$. The graph $G = (S, E)$ has maximum degree 4, as it is the union of two paths. By the results of [10], $G$ can be 2-colored so that every monochromatic connected component has size at most 6. In particular every path of size at least 7 contains points from both color classes. To finish the reduction argument one may observe that every strip containing at least 7 points corresponds to a path (of size at least 7) in $G$. We improve and generalize this first bound in the following.

**Theorem 1.** *For any finite planar set $S$ and any integer $k$, $S$ can be $k$-colored so that any $(2k-1)$-strip is polychromatic. That is,*

$$p(k, 2) \leq 2k - 1.$$

*Proof.* Let $s_1, \ldots, s_n$ be the points of $S$ sorted by (increasing) $x$-coordinates and let $s_{\pi_1}, \ldots, s_{\pi_n}$ be the sorting by $y$-coordinates. We first assume that $k$ divides $n$, and later show how to remove the need for this assumption. Let $V_x$ be the set of $n/k$ disjoint contiguous $k$-tuples in $s_1, \ldots, s_n$. Namely, $V_x = \{\{s_1, \ldots, s_k\}, \{s_{k+1}, \ldots, s_{2k}\}, \ldots, \{s_{n-k+1}, \ldots, s_n\}\}$. Similarly, let $V_y$ be the $k$-tuples defined by $s_{\pi_1}, \ldots, s_{\pi_n}$.

We define a bipartite multigraph $G = (V_x, V_y, E)$ as follows: For every pair of $k$-tuples $A \in V_x$, $B \in V_y$, we include an edge $e_s = \{A, B\} \in E$ if there exists a point $s$ in both $A$ and $B$. Note that an edge $\{A, B\}$ has multiplicity $|A \cap B|$ and that the number of edges $|E|$ is $n$. The multigraph $G$ is $k$-regular because every $k$-tuple $A$ contains exactly $k$ points and every point $s \in A$ determines exactly one incident edge labeled $e_s$. It is well known that the chromatic index of any bipartite $k$-regular multigraph is $k$ (and can be efficiently computed, see e.g., [1,7]). Namely, the edges of such a multigraph can be partitioned into $k$ perfect matchings. Let $E_1, \ldots, E_k$ be such a partition and $S_i \subset S$ be the set of labels of the edges of $E_i$. The sets $S_1, \ldots, S_k$ form a partition (i.e., a coloring) of $S$. We assign color $i$ to the points of $S_i$.

We claim that this coloring ensures that any $(2k-1)$-strip is polychromatic. Let $h$ be a $(2k-1)$-strip and assume without loss of generality that $h$ is parallel to the $y$-axis. Then $h$ contains at least one $k$-tuple $A \in V_x$. By the properties of the above coloring, the edges incident to $A$ in $G$ are colored with $k$ distinct colors. Thus, the points that correspond to the labels of these edges are colored with $k$ distinct colors, and $h$ is polychromatic.

To complete the proof, we must handle the case where $k$ does not divide $n$. Let $i = n \pmod{k}$. Let $Q = \{q_1, \ldots, q_{k-i}\}$ be an additional set of $k-i$ points, all located to the right and above the points of $S$. We repeat our preceding construction on $S \cup Q$. Now, any $(2k-1)$-strip which is, say, parallel to the $y$-axis will also contain a $k$-tuple $A \in V_x$ disjoint from $Q$. Thus our arguments follow as before.                                                                    □

The proof of Theorem 1 is constructive and leads directly to an $O(n \log n)$-time algorithm to $k$-color $n$ points in the plane so that every $(2k-1)$-strip is polychromatic. The algorithm is simple: we sort $S$, construct $G = (V_x, V_y, E)$, and color the edges of $G$ with $k$ colors. The time analysis is as follows: sorting takes $O(n \log n)$ time. Constructing $G$ takes $O(n + |E|)$ time. As $G$ has $\frac{2n}{k}$ vertices and is $k$-regular, it has $n$ edges; so this step takes $O(n)$ time. Finding the edge-coloring of $G$ takes $O(n \log n)$ time [1]. The total running time is therefore $O(n \log n)$.

## 2.2 Circular Permutations: Upper Bound on $p'(k, 2)$

We now consider the value of $p'(k, d)$. Given $d$ circular permutations of a set $S$, we color $S$ so that every sufficiently long subsequence in any of the circular permutations is polychromatic. The previous proof for $p(k, d) \leq 2k-1$ (Theorem 1) does not hold when we consider circular permutations. However, a slight modification provides the same upper bound, up to a constant term.

**Theorem 2.** $p'(k, 2) \leq 2k$

*Proof.* If $k$ divides $n$, we separate each circular permutation into $n/k$ sets of size $k$. We define a multigraph, where the vertices represent the sets of $k$ items, and there is an edge between two vertices if two sets share the same item. Trivially, this graph is $k$-regular and bipartite, and can thus be edge-colored with $k$ colors. Each edge in this graph corresponds to one item in the permutation, thus each set of $k$ items contains points of all $k$ colors.

If $k$ does not divide $n$, let $a = \lfloor n/k \rfloor$, and $b = n \pmod{k}$. If $a$ divides $b$, we separate each of the two circular permutations into $2a$ sets, of alternating sizes $k$ and $b/a$. Otherwise, the even sets will also alternate between size $\lceil b/a \rceil$ and $\lfloor b/a \rfloor$, instead of $b/a$. We extend both permutations by adding dummy items to each set of size less than $k$, so that we finally have only sets of size $k$. Dummy items appear in the same order in both permutations. We can now define the multigraph just as before.

If we remove the dummy nodes, we deduce a coloring for our original set. As each color appears in every set of size $k$, the length of any subsequence between two items of the same color is at most $2(k - 1) + \lceil b/a \rceil$. Therefore, $p'(k, 2) \leq 2(k - 1) + \lceil b/a \rceil + 1$.

Finally, if $n \geq k(k - 1)$, then $a \geq k - 1$, and $b \leq a$, we know that $\lceil b/a \rceil \leq 1$, and thus $p'(k, 2) \leq 2k$. $\qquad\square$

# 3 Higher Dimensional Strips

In this section we study the same problem for strips in higher dimensions. We provide upper and lower bounds on $p(k, d)$. We then analyze the coloring problem from a computational viewpoint, and show that deciding whether a given instance $S \subset \mathbb{R}^d$ can be 2-colored such that every 3-strip is polychromatic is NP-complete.

## 3.1   Upper Bound on Strip Size, $p(k,d)$

**Theorem 3.** *Any finite set of points $S \subset \mathbb{R}^d$ can be $k$-colored so that every axis-aligned strip containing $k(4 \ln k + \ln d)$ points is polychromatic, that is,*

$$p(k,d) \leq k(4 \ln k + \ln d).$$

*Proof.* The proof uses the probabilistic method. Let $\{1, \ldots, k\}$ denote the set of $k$ colors. We randomly color every point in $S$ independently so that a point $s$ gets color $i$ with probability $\frac{1}{k}$ for $i = 1, \ldots, k$. For a $t$-strip $h$, let $\mathcal{B}_h$ be the "bad" event where $h$ is not polychromatic. It is easily seen that $\Pr[\mathcal{B}_h] \leq k(1 - \frac{1}{k})^t$. Moreover, $\mathcal{B}_h$ depends on at most $(d-1)t^2 + 2t - 2$ other events. Indeed, $\mathcal{B}_h$ depends only on $t$-strips that share points with $h$. Assume without loss of generality that $h$ is orthogonal to the $x_1$ axis. Then $\mathcal{B}_h$ has a non-empty intersection with at most $2(t-1)$ other $t$-strips which are orthogonal to the $x_1$ axis. For each of the other $d-1$ axes, $h$ can intersect at most $t^2$ $t$-strips since every point in $h$ can belong to at most $t$ other $t$-strips.

By the Lovász Local Lemma, (see, e.g., [2]) we have that if $t$ satisfies

$$e \cdot \left((d-1) \cdot t^2 + 2t - 1\right) \cdot k \left(1 - \frac{1}{k}\right)^t < 1$$

(where $e$ is the basis of the natural logarithm), then

$$\Pr\left[\bigwedge_{|h|=t} \overline{\mathcal{B}_h}\right] > 0.$$

In particular, this means that there exists a $k$-coloring for which every $t$-strip is polychromatic. It can be verified that $t = k(4 \ln k + \ln d)$ satisfies the condition. □

The proof of Theorem 3 is non-constructive. We can use known algorithmic versions of the Local Lemma (see for instance [12]) to obtain a constructive proof, although this yields a weaker bound. Theorem 3 holds in the more general case where the strips are not necessarily axis-aligned. In fact, one can have a total of $d$ arbitrary strip orientations in some fixed arbitrary dimension and the proof will hold verbatim.

Finally, the same proof yields the same upper bound for the case of circular permutations:

**Theorem 4.** $p'(k,d) \leq k(4 \ln k + \ln d)$

## 3.2   Lower Bound on $p(k,d)$

We first introduce a well-known result on the decomposition of complete graphs:

**Lemma 1.** *The edges of $K_{2h}$ can be decomposed into $h$ pairwise edge-disjoint Hamiltonian paths.*

This result follows from a special case of the Oberwolfach problem[5][1]. An explicit proof of this lemma can also be found in [19].

Note that if the vertices of $K_{2h}$ are labeled $V = \{1, \ldots, 2h\}$, each path can be seen as a permutation of $2h$ elements. Using Lemma 1 we obtain:

**Theorem 5.** *For any fixed dimension $d$ and number of colors $k$, let $s = \left\lceil \frac{(2d-1)k}{2d} \right\rceil - 1$. Then,*

$$p'(k, d) \geq p(k, d) \geq 2s + 1.$$

*Proof.* The first inequality comes from the fact that any polychromatic coloring with respect to circular permutations is also polychromatic with respect to strips. We now focus on showing the second inequality: let $\sigma_1, \ldots, \sigma_d$ be any decomposition of $K_{2d}$ into $d$ paths: we construct the set $P = \{p_i | 0 \leq i \leq 2d\}$, where $p_i = (\sigma_1(i), \ldots, \sigma_d(i))$. Note that the ordering of $P$, when projected to the $i$-th axis, gives permutation $\sigma_i$. Since the elements $\sigma$ decompose $K_{2d}$, in particular for any $i, j \leq 2d$ there exists a permutation in which $i$ and $j$ are adjacent.

We replace each point $p_i$ by a set $A_i$ of $s$ points arbitrarily close to $p_i$. By construction, for any $i, j \leq 2d$, there exists a $2s$-strip containing exactly $A_i \cup A_j$. Consider any possible coloring of the sets $A_i$: since $|A_i| = s$ and we are using $k$ colors, there are at least $k - s$ colors not present in any set $A_i$.

Since $\left\lceil \frac{(2d-1)k}{2d} \right\rceil - 1 < \frac{(2d-1)k}{2d}$, we conclude that $k - s > k - \frac{(2d-1)k}{2d} = k/2d$. That is, each set is missing strictly more than $k/2d$ colors. By the pigeonhole principle, there exist $i$ and $j$ such that the set $A_i \cup A_j$ is missing a color (otherwise there would be more than $k$ colors). In particular, the strip that contains set $A_i \cup A_j$ is not polychromatic, thus the theorem is shown.

We gave a set of bounded size $n = 2d$ reaching the lower bound, but we can easily create larger sets reaching the same bound: we can add as many dummy points as needed at the end of every permutation, which does not decrease the value of $p(k, d)$. □

Note that, asymptitotically speaking, the lower bound does not depend on $d$. However, by the negative result of [16], we know that $p(k, d) \to \infty$ when $d \to \infty$.

### 3.3 Computational Complexity

In Section 2, we provided an algorithm that finds a $k$-coloring such that every planar $(2k-1)$-strip is polychromatic. Thus for $d=2$ and $k=2$, this yields a 2-coloring such that every 3-strip is polychromatic.

Note that in this case $p(2, 2) = 3$, but the minimum required size of a strip for a given instance can be either 2 or 3. Testing if it is equal to 2 is easy: we can simply alternate the colors in the first permutation, and check if they also

---

[1] The authors would like to thank the anonymous referees for pointing out useful references.

alternate in the other. Hence the problem of minimizing the size of the largest monochromatic strip on a given instance is polynomial for $d = 2$ and $k = 2$. We now show that it becomes NP-hard for $d > 2$ and $k = 2$. The same problem for $k > 2$ is left open.

**Theorem 6.** *The following problem is $NP$-complete:*

**Input:** *3 permutations $\pi_1, \pi_2, \pi_3$ of an $n$-element set $S$.*
**Question:** *Is there a 2-coloring of $S$, such that every 3 elements of $S$ that are consecutive according to one of the permutations are not monochromatic?*

*Proof.* We show a reduction from NAE 3SAT (not-all-equal 3SAT) which is the following $NP$-complete problem [9]:

**Input:** A 3-CNF Boolean formula $\Phi$.
**Question:** Is there a $NAE$ assignment to $\Phi$? An assignment is called $NAE$ if every clause has at least one literal assigned True and at least one literal assigned False.

We first transform $\Phi$ into another instance $\Phi'$ in which all variables are non-negated (i.e., we make the instance monotone). We then show how to realize $\Phi'$ using three permutations $\pi_1, \pi_2, \pi_3$.

To transform $\Phi$ into $\Phi'$, for each variable $x$, we first replace the $i$th occurrence of $x$ in its positive form by a variable $x_i$, and the $i$th occurrence of $x$ in its negative form by $x'_i$. The index $i$ varies between 1 and the number of occurrences of each form (the maximum of the two). We also add the following *consistency-clauses*, for each variable $x$ and for all $i$:

$$\left(Z_i^x, x_i, x'_i\right), \left(x_i, x'_i, Z_{i+1}^x\right), \left(x_i, Z_i'^x, x'_i\right), \left(Z_i^x, Z_i'^x, Z_{i+1}^x\right)$$
$$\left(x'_i, Z_{i+1}^x, x_{i+1},\right), \left(Z_i'^x, x'_i, x_{i+1}\right), \left(x'_i, x_{i+1}, Z_{i+1}'^x,\right), \left(Z_i'^x, Z_{i+1}^x, Z_{i+1}'^x\right)$$

where $Z_i^x$ and $Z_i'^x$ are new variables. This completes the construction of $\Phi'$. Note that $\Phi'$ is monotone, as every negated variable has been replaced.

Moreover, $\Phi'$ has a $NAE$ assignment if and only if $\Phi$ has a $NAE$ assignment. To see this, note that a $NAE$ assignment for $\Phi$ can be translated to a $NAE$ assignment to $\Phi'$ as follows: for every variable $x$ of $\Phi$ and every $i$, set $x_i \equiv x$, $x'_i \equiv \overline{x}$, $Z_i^x \equiv True$, $Z_i'^x \equiv False$.

On the other hand, if $\Phi'$ has a $NAE$ assignment, then, by the consistency clauses, the variables in $\Phi'$ corresponding to any variable $x$ of $\Phi$ are assigned a consistent value. Namely, for every $i, j$ we have $x_i = x_j$ and $x_i \neq x'_i$. This assignment naturally translates to a $NAE$ assignment for $\Phi$, by setting $x \equiv x_1$.

We next show how to realize $\Phi'$ by a set $S$ and three permutations $\pi_1, \pi_2, \pi_3$. The elements of the set $S$ are the variables of $\Phi'$, together with some additional elements that are described below. Permutation $\pi_1$ realizes the clauses of $\Phi'$ corresponding to the original clauses of $\Phi$, while $\pi_2$ and $\pi_3$ realize the consistency clauses of $\Phi'$.

The additional elements in $S$ are clause elements (two elements $c_{2j-1}$ and $c_{2j}$ for every clause $j$ of $\Phi$) and *dummy elements* $\star$ (the dummy elements are not indexed for the ease of presentation, but they appear in the same order in all three permutations).

Permutation $\pi_1$ encodes the clauses of $\Phi'$ corresponding to original clauses of $\Phi$ as follows (note that all these clauses involve different variables). For each such clause $(u, v, w)$, permutation $\pi_1$ contains the following sequence:

$$c_{2j-1}, u, v, w, c_{2j}, \star, \star$$

At the end of $\pi_1$, for every variable $x$ of $\Phi'$ we have the sequence:

$$Z_1^x, Z_1'^x, Z_2^x, Z_2'^x, Z_3^x, Z_3'^x, \ldots, \star, \star$$

Permutation $\pi_2$ contains, for every variable $x$ of $\Phi$, the sequences:

$$Z_1^x, x_1, x_1', Z_2^x, x_2, x_2', Z_3^x, x_3, x_3', Z_4^x, \ldots \quad \text{and} \quad \star, \star, Z_1'^x, \star, \star, Z_2'^x, \star, \star, Z_3'^x, \ldots \star, \star$$

At the end of $\pi_2$ we have the clause-elements and remaining dummy elements:

$$\star, \star, c_1, \star, \star, c_2, \star, \star, c_3 \ldots$$

Similarly, permutation $\pi_3$ contains, for every variable $x$ of $\Phi$, the sequences:

$$x_1, Z_1'^x, x_1', x_2, Z_2'^x, x_2', x_3, Z_3'^x, x_3', \ldots \quad \text{and} \quad \star, \star, Z_1^x, \star, \star, Z_2^x, \star, \star, Z_3^x, \ldots \star, \star$$

and at the end of $\pi_3$ we have the clause-elements and remaining dummy elements:

$$\star, \star, c_1, \star, \star, c_2, \star, \star, c_3, \ldots$$

This completes the construction of $S$ and $\pi_1, \pi_2, \pi_3$. Note that for every clause of $\Phi'$ (whether it is derived from $\Phi$ or is a consistency clause), the elements corresponding to its three variables appear in sequence in one of the three permutations. Therefore, if there is a 2-coloring of $S$, such that every 3 elements of $S$ that are consecutive according to one of the permutations are not monochromatic, then there is a $NAE$ assignment to $\Phi'$: each variable of $\Phi'$ is assigned True if its corresponding element is colored '1', and False otherwise.

For the other direction, consider a $NAE$ assignment for $\Phi'$. Observe that there is always a solution where $Z_i^x$ and $Z_i'^x$ are assigned opposite values. Then assign color '1' to elements corresponding to variables assigned with True, and assign color '0' to elements corresponding to variables assigned with False. For the clause elements $c_{2j-1}$ and $c_{2j}$ appearing in the subsequence $c_{2j-1}, u, v, w, c_{2j}$, assign to $c_{2j-1}$ the color opposite to $u$, and to $c_{2j}$ the color opposite to $w$. Finally, assign colors '0' and '1' to each pair of consecutive dummy elements, respectively. It can be verified that there is no monochromatic consecutive triple in any permutation. □

*Approximability.* Note that the minimization problem (find a $k$-coloring that minimizes the number of required points) can be approximated within a constant factor: it suffices to use a constructive version of the Lovász Local Lemma (see [2]), yielding an actual coloring. The number of points that this coloring will require is bounded by a constant, provided the values of $k$ and $d$ are fixed. The approximation factor is therefore bounded by the ratio between that constant and $k$.

# 4   Coloring Strips

In this section we prove that any finite set of strips in $\mathbb{R}^d$ can be $k$-colored so that every "deep" point is polychromatic. For a given set of strips (or intervals, if $d = 1$), we say that a point is $i$-*deep* if it is contained in at least $i$ of the strips. We begin with the following easy lemma:

**Lemma 2.** *Let $\mathcal{I}$ be a finite set of intervals. Then for every $k$, $\mathcal{I}$ can be $k$-colored so that every $k$-deep point is polychromatic, while any point covered by fewer than $k$ intervals will be covered by distinct colors.*

*Proof.* We use induction on $|\mathcal{I}|$. Let $I$ be the interval with the leftmost right endpoint. By induction, the intervals in $\mathcal{I} \setminus \{I\}$ can be $k$-colored with the desired property. Sort the intervals intersecting $I$ according to their left endpoints and let $I_1, \ldots, I_{k-1}$ be the first $k-1$ intervals in this order. It is easily seen that coloring $I$ with a color distinct from the colors of those $k-1$ intervals produces a coloring with the desired property, and hence a valid coloring.    □

**Theorem 7.** *For any $d$ and $k$, one can $k$-color any set of axis-aligned strips in $\mathbb{R}^d$ so that every $d(k-1)+1$-deep point is polychromatic. That is,*

$$\bar{p}(k, d) \leq d(k-1) + 1.$$

*Proof.* We start by coloring the strips parallel to any given axis $x_i$ $(i = 1, \ldots, d)$ separately using the coloring described in Lemma 2. We claim that this procedure produces a valid polychromatic coloring for all $d(k-1)+1$-deep points. Indeed assume that a given point $s$ is $d(k-1)+1$-deep and let $H(s)$ be the set of strips covering $s$. Since there are $d$ possible orientations for the strips in $H(s)$, by the pigeonhole principle at least $k$ of the strips in $H(s)$ are parallel to the same axis. Assume without loss of generality that this is the $x_1$-axis. Then by property of the coloring of Lemma 2, $H(s)$ is polychromatic.    □

The above proof is constructive. By sorting the intervals that correspond to any of the given directions, one can easily find a coloring in $O(n \log n)$ time.

We now give a lower bound on $\bar{p}(k, d)$:

**Theorem 8.** *For any fixed dimension $d$ and integer $k$, it holds that*

$$\bar{p}(k, d) > \lfloor k/2 \rfloor d + 1$$

Proof of this claim is omitted due to space limitation.

# Acknowledgements

This research was initiated during the WAFOL'09 workshop at Université Libre de Bruxelles (U.L.B.), Brussels, Belgium. The authors want to thank all other participants, and in particular Erik D. Demaine, for helpful discussions.

# References

1. Alon, N.: A simple algorithm for edge-coloring bipartite multigraphs. Inf. Proc. Lett. 85(6), 301–302 (2003)
2. Alon, N., Spencer, J.: The Probabilistic Method, 2nd edn. John Wiley, Chichester (2000)
3. Aloupis, G., Cardinal, J., Collette, S., Langerman, S., Orden, D., Ramos, P.: Decomposition of multiple coverings into more parts. In: Proceedings of the ACM-SIAM Symposium on Discrete Algorithms, SODA 2009 (2009)
4. Aloupis, G., Cardinal, J., Collette, S., Langerman, S., Smorodinsky, S.: Coloring geometric range spaces. Discrete & Computational Geometry 41(2), 348–362 (2009)
5. Alspach, B.: The wonderful walecki construction. Bull. Inst. Combin. Appl. 52, 7–20 (2008)
6. Buchsbaum, A., Efrat, A., Jain, S., Venkatasubramanian, S., Yi, K.: Restricted strip covering and the sensor cover problem. In: ACM-SIAM Symposium on Discrete Algorithms, SODA 2007 (2007)
7. Cole, R., Ost, K., Schirra, S.: Edge-coloring bipartite multigraphs in O(E log D) time. Combinatorica 21(1), 5–12 (2001)
8. Even, G., Lotker, Z., Ron, D., Smorodinsky, S.: Conflict-free colorings of simple geometric regions with applications to frequency assignment in cellular networks. SIAM Journal on Computing 33(1), 94–136 (2004)
9. Garey, M.R., Johnson, D.S.: Computers and Intractability: A Guide to the Theory of NP-Completeness. W.H. Freeman, New York (1979)
10. Haxell, P., Szabó, T., Tardos, G.: Bounded size components: partitions and transversals. J. Comb. Theory Ser. B 88(2), 281–297 (2003)
11. Mani, P., Pach, J.: Decomposition problems for multiple coverings with unit balls (manuscript) (1986)
12. Moser, R.A.: A constructive proof of the lovász local lemma. In: Proc. of the ACM symposium on Theory of computing, New York, NY, USA, pp. 343–350 (2009)
13. Pach, J.: Decomposition of multiple packing and covering. In: 2. Kolloq. über Diskrete Geom., pp. 169–178. Inst. Math. Univ. Salzburg, Salzburg (1980)
14. Pach, J.: Decomposition of multiple packing and covering. In: 2. Kolloquium Uber Diskrete Geometrie, pp. 169–178. Inst. Math. Univ. Salzburg, Salzburg (1980)
15. Pach, J.: Covering the plane with convex polygons. Discrete & Computational Geometry 1, 73–81 (1986)
16. Pach, J., Tardos, G., Tóth, G.: Indecomposable coverings. In: Akiyama, J., Chen, W.Y.C., Kano, M., Li, X., Yu, Q. (eds.) CJCDGCGT 2005. LNCS, vol. 4381, pp. 135–148. Springer, Heidelberg (2007)
17. Pach, J., Tóth, G.: Decomposition of multiple coverings into many parts. In: Proc. of the ACM Symposium on Computational Geometry, pp. 133–137 (2007)
18. Smorodinsky, S.: On the chromatic number of some geometric hypergraphs. SIAM Journal on Discrete Mathematics 21(3), 676–687 (2007)
19. Stanton, R.G., Cowan, D.D., James, L.O.: Some results on path numbers. In: Louisiana Conference on Combin., Graph Theory and Computing (1970)
20. Tardos, G., Tóth, G.: Multiple coverings of the plane with triangles. Discrete & Computational Geometry 38(2), 443–450 (2007)

# The Mono- and Bichromatic Empty Rectangle and Square Problems in All Dimensions*

## (Extended Abstract)

Jonathan Backer and J. Mark Keil

Department of Computer Science, University of Saskatchewan, Canada

**Abstract.** The maximum empty rectangle problem is as follows: Given a set of red points in $\mathbb{R}^d$ and an axis-aligned hyperrectangle $B$, find an axis-aligned hyperrectangle $R$ of greatest volume that is contained in $B$ and contains no red points. In addition to this problem, we also consider three natural variants: where we find a hypercube instead of a hyperrectangle, where we try to contain as many blue points as possible instead of maximising volume, and where we do both. Combining the results of this paper with previous results, we now know that all four of these problems (a) are NP-complete if $d$ is part of the input, (b) have polynomial-time sweep-plane solutions for any fixed $d \geq 3$, and (c) have near linear time solutions in two dimensions.

## 1 Introduction

We use $d$ to denote the number of dimensions and $n$ to denote the total number of red and blue points. Throughout this paper, we assume that all hypercubes and hyperrectangles are axis aligned.

The maximum empty rectangle (MER) problem, defined in the abstract, was extensively studied in two and three dimensions [1,2,3,4,5,6]. Our contributions to the MER problem are in high dimensions: we show that it is NP-complete if $d$ is part of the input, and we present an algorithm that takes $O(n^d \log^{d-2} n)$ time in the worst case for any fixed dimension $d \geq 3$ (the expected run time of $O(n \log^{2d-3} n)$ is much better than the worst case).

We explore the closely related maximum empty square (MES) problem: Given a set of red points in $\mathbb{R}^d$ and a hyperrectangle $B$, find the hypercube of greatest volume that is contained in $B$ and contains no red points. We show that the MES problem is NP-complete if $d$ is part of the input, and we provide a solution that takes $O(n^{d/2} \log^2 n)$ time, for any fixed dimension $d \geq 3$.

We also investigate instances of the bichromatic shape problem: Given a set of red points and blue points in $\mathbb{R}^d$, find a figure of a certain shape that contains no red points and as many blue points as possible. Specifically, we examine the bichromatic hyperrectangle (BR) problem and the bichromatic hypercube (BS)

---

* This research is supported by the Natural Sciences and Engineering Research Council (NSERC).

A. López-Ortiz (Ed.): LATIN 2010, LNCS 6034, pp. 14–25, 2010.

problem. It was previously shown that the BR problem is NP-complete if $d$ is part of the input [7]. This reduction can be modified to show that the BS problem is also NP-complete. Our contributions to the BR and BS problems are polynomial time solutions for any fixed dimension $d \geq 3$ and near linear time solutions for $d = 2$.

Our interest in bichromatic problems was instigated by a recent paper that posed them for a variety of shapes, including squares and rectangles [8]. In that paper, Aronov and Har-Peled present an $(1 + \varepsilon)$-approximation algorithm for the bichromatic ball problem that takes $O(n^{\lceil d/2 \rceil}(\varepsilon^{-2} \log n)^{\lceil d/2+1 \rceil})$ time for any fixed dimension $d \geq 3$. They conjecture that the bichromatic circle problem in $\mathbb{R}^2$ is 3SUM-hard. This contrasts with our solution to the BS problem in $\mathbb{R}^2$, which takes $O(n \log n)$ time.

Motivated by data-analysis, Eckstein et al. explore the BR problem in high dimensions [7]. As mentioned above, they show that the BR problem is NP-hard if $d$ is part of the input. This intractability motivates the heuristic approach that they develop. In a subsequent paper, Liu and Nediak propose a $O(n^2 \log n)$ time and $O(n)$ space algorithm for the two-dimensional bichromatic rectangle problem [9]. We solve the BR problem in $O(n^d \log^{d-2} n)$ time for any fixed dimension $d \geq 3$ and in $O(n \log^3 n)$ time for $d = 2$.

All known exact solutions for the MER, MES, BS, and BR problems require at least linear storage. Motivated by potential data-mining applications, Edmonds et al. describe a heuristic solution to the MER problem that typically requires much less space [10].

For completeness, we also mention the maximum discrepancy problem, which is similar to the bichromatic shape problem: Find a figure of a certain shape that maximises the difference between the number of contained blue points and contained red points. Dobkin et al. solve the maximum discrepancy problem for rectangles in $O(n^2 \log n)$ time [11]. In their paper, they relate various discrepancy problems to problems in machine learning and computer graphics.

## 2   Preliminaries

We use $P^-$ to denote the set of red points, $P^+$ to denote the set of blue points, and $R$ to denote the shape for which we are searching (a hyperrectangle or hypercube). We look for a shape $R$ that is open so that red points can lie on $R$'s boundary without being contained in $R$. As $R$ is axis-aligned, it can be expressed as the Cartesian product of open intervals $\prod_{i=1}^{d}(a_i, b_i)$. We define the $i^{\text{th}}$ *projection* of $R$ as $(a_i, b_i)$ and the $i^{\text{th}}$ *extent* of $R$ as $b_i - a_i$. The *boundary* of $R$, denoted $\partial R$, is the difference between $\prod_{i=1}^{d}[a_i, b_i]$ and $\prod_{i=1}^{d}(a_i, b_i)$. A side of $R$ is a subset of $\partial R$ that is most extreme in some coordinate axis. Specifically, there are two sides orthogonal to the $i^{\text{th}}$ dimension: $\{x \in \partial R : \text{the } i^{\text{th}} \text{ coordinate of } x \text{ is } a_i\}$ and $\{x \in \partial R : \text{the } i^{\text{th}} \text{ coordinate of } x \text{ is } b_i\}$.

We will use $B$ to denote a hyperrectangle that bounds $P^- \cup P^+$. The hyperrectangle $B$ is given as input in the MER and MES but not in the BR and BS problems. For the BR and BS problems, we choose a sufficiently large $B$ so that there that some desired $R$ is contained in $B$: for the BR problem, any box containing $P^- \cup P^+$ suffices; for the BS problem, we can take any box containing $P^- \cup P^+$ and triple its extent in each dimension. By choosing $B$, we can more directly apply techniques developed for the MER and MES problems to the BR and BS problems.

We say that $R$ is *feasible*, if it is contained in $B$ and disjoint from $P^-$. We say that $R$ is *relevant*, if $R$ is feasible and no other feasible shape is *properly* contained within $R$. All four problems that we are considering have a relevant solution. We say that a side $S$ of $R$ is *supported*, if the interior of $S$ intersects $P^-$ or $B$. In the MER and BR problems, every side of a relevant hyperrectangle (RHR) is supported; in the MES and BS problems, the following statement holds for each relevant hypercube (RHC).

**Lemma 1.** *Two opposite sides of each RHC are supported.*

*Proof.* For contradiction, consider an RHC $R$ such that no two opposite sides are both supported. Let $S_i$ and $T_i$ denote the sides of $R$ that are orthogonal to the $i^{\text{th}}$ dimension. Assume without loss of generality that $S_i$ is unsupported. Let the vertices $u$ and $v$ of $R$ be the points $\bigcap_{i=1}^d S_i$ and $\bigcap_{i=1}^d T_i$ respectively. Note that $R$ is uniquely determined by $u$ and $v$. We can slide $u$ away from $v$ while keeping $R$ feasible, which contradicts that $R$ is relevant. □

## 3   Unbounded Dimension

Eckstein et al. show that the BR problem is NP-complete [7]. With slight modification, their construction also shows that the BS problem is NP-complete. We now prove that the MER and MES problems are NP-complete.

We first rephrase the MER and MES problems as decision problems: Given the number of dimensions $d$, a set of red points $P^-$, an enclosing hyperrectangle $B$, and a threshold $\tau$, is there a hyperrectangle (or hypercube) $R$ contained in $B$ such that $R \cap P^- = \emptyset$ and the volume of $R$ is at least $\tau$? These decision problems are in NP because it is easy to verify that a given hyperrectangle $R$ has the desired properties.

### 3.1   Maximum Empty Square

To demonstrate that the MES problem is NP-hard, we give a polynomial-time reduction from satisfiability to MES. In our reduction, each variable maps to a dimension, each clause maps to a red point, and the given formula is satisfiable if and only if there exists a sufficiently large empty hypercube.

Let $\lambda$ be the formula to be satisfied in conjunctive normal form. We denote the variables and clauses of $\lambda$ as $x_1, x_2, \ldots, x_n$ and $c_1, c_2, \ldots, c_m$ respectively. We assume without loss of generality that a variable $x_i$ and its negation $\overline{x_i}$ never occur in the same clause. For each clause $c_j$, we define a red point $p_j$ as follows: if $x_i$ occurs in $c_j$, $p_j$ is set to $\frac{1}{4}$ in the $i^{\text{th}}$ dimension; if $\overline{x_i}$ occurs in $c_j$, $p_j$ is set to $\frac{3}{4}$ in the $i^{\text{th}}$ dimension; otherwise, $p_j$ is set to $\frac{1}{2}$ in the $i^{\text{th}}$ dimension.

**Theorem 1.** *The formula $\lambda$ is satisfiable if and only if there exists a feasible hypercube $R$ contained in $[0,1]^n$ of volume at least $(\frac{3}{4})^n$.*

*Proof.* Suppose that $\lambda$ is satisfiable. Given satisfying variable assignment of $\lambda$, we construct a hypercube centre $t$ as follows: if $x_i$ is assigned true, $t$ is set to $\frac{5}{8}$ in the $i^{\text{th}}$ dimension and $\frac{3}{8}$ otherwise. Let $R$ be the hypercube of width $\frac{3}{4}$ centred at $t$. It remains to show that each point $p_j$ does not lie in $R$. At least one of the literals of $c_j$ evaluates to true. Let $x_i$ be such a literal (the case where a negation $\overline{x_i}$ is satisfied is similar). Then $t$ is equal to $\frac{5}{8}$ in the $i^{\text{th}}$ dimension and $p_j$ is equal to $\frac{1}{4}$ in the $i^{\text{th}}$ dimension. Hence $p_j$ lies on $\partial R$. Therefore $R$ is an empty hypercube of volume $(\frac{3}{4})^n$.

Suppose that there exists an empty hypercube $R$ of volume at least $(\frac{3}{4})^n$. Let $t$ be the centre of $R$. We derive a variable assignment of $\lambda$ as follows: if $t$ is at least $\frac{1}{2}$ in the $i^{\text{th}}$ dimension, $x_i$ is set to true and false otherwise. It remains to show that each clause $c_j$ is satisfied. As $R$ does not contain $p_j$, there exists a dimension $i$ such that the $i^{\text{th}}$ coordinate of $p_j$ is not in the $i^{\text{th}}$ projection of $R$. As $\frac{1}{2}$ lies inside the $i^{\text{th}}$ projection of $R$, the $i^{\text{th}}$ coordinate of $p_j$ must be either $\frac{1}{4}$ or $\frac{3}{4}$. Suppose the former (the latter case is similar). Then the $i^{\text{th}}$ coordinate of $t$ must be $\frac{5}{8}$. Hence, $x_i$ is set to true. Moreover, $x_i$ is a literal of $c_j$ because the $i^{\text{th}}$ coordinate of $p_j$ is $\frac{1}{4}$. Thus, $c_j$ evaluates true. Therefore $\lambda$ is satisfied.    $\square$

## 3.2  Maximum Empty Rectangle

To prove that the MER problem is NP-hard, we provide a reduction from independent set to MER. In our reduction, each vertex maps to a dimension, each edge maps to a point, and the given graph has an independent set of size $k$ if and only if there exists a sufficiently large empty hyperrectangle.

Let $G = (V, E)$ be a simple undirected graph. We denote the vertices and edges of $G$ as $v_1, v_2, \ldots, v_n$ and $e_1, e_2, \ldots, e_m$ respectively. Let $w \in (\frac{1}{2}, 1)$ be some constant such that $w^n > \frac{1}{2}$. For each edge $e_j$, we define a red point $p_j$ as follows: if $v_i$ is an endpoint of $e_j$, the $i^{\text{th}}$ coordinate of $p_j$ is set to $w$ and $\frac{1}{2}$ otherwise.

**Lemma 2.** *Let $R \subseteq [0,1]^n$ be a maximum-volume feasible hyperrectangle. Then $\partial R$ contains the origin and $i^{\text{th}}$ extent of $R$ is either $w$ or $1$.*

*Proof.* We first prove a weaker statement: the extent of $R$ in the $i^{\text{th}}$ dimension is either $0$, $\frac{1}{2}$, $w$, $1$, $w - \frac{1}{2}$, or $1 - w$. To see this, note that the $i^{\text{th}}$ coordinate of $p_j$ is either $\frac{1}{2}$ or $w$. If $a_i \notin \{0, \frac{1}{2}, w\}$, we can increase the volume of $R$ while keeping $R$ feasible by decreasing $a_i$, which contradicts the maximality of $R$. A similar argument holds for $b_i$. Hence, $a_i$ and $b_i$ belong to $\{0, \frac{1}{2}, w, 1\}$.

The volume of $R$ is at least $w^n$ because $\prod_{i=1}^{n}(0, w)$ is empty. Suppose for contradiction that the $i^{\text{th}}$ extent of $R$ is neither $w$ nor $1$. Then the $i^{\text{th}}$ extent is at most $\frac{1}{2}$ by the previous paragraph. Hence, the volume of $R$ is at most $\frac{1}{2}$, a contradiction because $w^n > \frac{1}{2}$. Thus, the $i^{\text{th}}$ extent of $R$ is $w$ or $1$. This implies that $a_i = 0$ for each dimension $i$. Therefore, $\partial R$ contains the origin.    $\square$

Using this lemma, we now prove the main result.

**Theorem 2.** *Let $R$ be a maximum-volume feasible hyperrectangle. The graph $G$ has an independent set of size $k$ if and only if the volume of $R$ is at least $w^{n-k}$.*

*Proof.* Suppose that $G$ has an independent set $S = \{v_{d_1}, v_{d_2}, \ldots, v_{d_k}\}$ of $k$ independent vertices. If $i \in \{d_1, d_2, \ldots, d_k\}$, set $b_i$ to 1 and $w$ otherwise. Then $R = \prod_{i=1}^n (0, b_i)$ is a hyperrectangle with volume $w^{n-k}$. It remains to show that $R$ is empty. Recall that $p_j$ corresponds to an edge $e_j$ between two vertices. One of these vertices $v_i$ is not a member of $S$. The $i^{\text{th}}$ coordinate of $p_j$ is $w$, which is not contained in the $i^{\text{th}}$ projection $(0, w)$ of $R$. Thus, $p_j$ is not contained in $R$. Therefore, $R$ is an empty rectangle with volume $w^{n-k}$.

Let $R$ be a maximum empty hyperrectangle with volume at least $w^{n-k}$. By the previous lemma, the volume of $R$ is equal to $1^i \times w^{n-i}$ for some integer $i$ between 0 and $n$. Hence, the extent of $R$ equals 1 in at least $k$ different dimensions. Let $D = \{d_1, d_2, \ldots, d_k\}$ be a set of $k$ such dimensions. Choose as a vertex set $S = \{v_{d_1}, v_{d_2}, \ldots, v_{d_k}\}$. It remains to show that $S$ is independent. Suppose for contradiction that there exists an edge $e_j$ between two vertices $v_x$, $v_y$ of $S$. Then the $x^{\text{th}}$ and $y^{\text{th}}$ projections of $R$ contains the $x^{\text{th}}$ and $y^{\text{th}}$ coordinates of $p_j$. All other coordinates of $p_j$ are $\frac{1}{2}$, which are contained in their respective projections of $R$ because $w > \frac{1}{2}$. Hence, $p_j$ is contained in $R$, which contradictions that $R$ is empty. Therefore, $S$ is independent. □

# 4    Fixed Dimension

In this section, we describe algorithms that take polynomial time for any fixed dimension $d \geq 3$. The run-time of our algorithms grows exponentially with $d$, which is consistent with the NP-completeness of these problems. Our approach to the MES and BS problems is substantially different from our approach to the MER and BR problems. However, our approaches to the MES and BS problems are similar, as are our approaches to the MER and BR problems.

## 4.1    MES and BS Problems

Finding the desired hypercube is substantially easier if we know its width $w^*$. The next lemma states that there are only $O(n^2)$ possibilities for $w^*$ for any fixed dimension $d$.

**Lemma 3.** *Let $R$ be a relevant hypercube. Its width $w$ is one of $O(dn^2)$ values and we can select the $i^{\text{th}}$ order statistic of this set in $O(dn)$ time and space.*

*Proof.* By Lemma 1, there exists two opposite supported sides of $R$. Let $i$ be the dimension orthogonal to these two sides. Let $X_i = \{x_i : x_i \text{ is the } i^{\text{th}} \text{ coordinate of some red point}\} \cup \{a_i, b_i : (a_i, b_i) \text{ is the } i^{\text{th}} \text{ projection of } B\}$. Then the $i^{\text{th}}$ projection of $R$ is $(a, b)$ where $a$ and $b$ belong to $X_i$. So given $i$, there are $O(n^2)$ possible widths. Hence, there are at most $O(dn^2)$ possible widths of $R$. Selecting the $i^{\text{th}}$ possible width is a result of Frederickson and Johnson [12]. □

Given the width $w^*$ of the desired hypercube $R$, all that remains is to position $R$. We rely on an elementary observation to do this: let $C(p, w^*)$ denote the hypercube with width $w^*$ centred at $p$; then $p \in C(q, w^*)$ if and only if $q \in C(p, w^*)$ for any two points $p$ and $q$. This observation allows us to rephrase our problem. Rather than look for a hypercube amid points, we look for the centre of $R$ amid hypercubes centred at red points.

This different perspective allows us to apply particular solution to the following problem: Given a set $S$ of axis-aligned hyperrectangles in $\mathbb{R}^d$, compute the volume of the union of $S$. To solve this problem, Overmars and Yap sweep a hyperplane $H$ perpendicular to the $d^{\text{th}}$ dimension [13]. As they sweep $H$, they maintain a compact representation of the intersection of $H$ with each hyperplane of $S$ in a tree structure. This structure allows them to quickly update the intersection as $H$ sweeps the $d^{\text{th}}$ dimension. Moreover, their structure allows efficient queries of two forms: "What is the total area of the intersection?" and "What point of $H$ is contained in the greatest (or least) number of hyperrectangles of $S$?".

**Theorem 3.** *The MES problem can be solved in $O(n^{d/2} \log n)$ time and $O(n)$ space for any fixed dimension $d \geq 3$.*

*Proof.* Let $U(S, w)$ denote the union $\bigcup_{s \in S} C(s, w)$, for an arbitrary subset $S$ of $\mathbb{R}^d$. Let $R$ be an arbitrary hypercube of width $w$. If $R$ is contained in $B$, its centre belongs to $B \setminus U(\partial B, w)$. Moreover, if $R$ contains no red points, its centre does not belong to $U(P^-, w)$. Hence, an empty hypercube of width $w$ exists if and only if $U(P^-, w)$ does not cover $B \setminus U(\partial B, w)$. To test this condition, we compare the volume of $B \cap (U(P^-, w) \cup U(\partial B, w))$ to the volume of $B$. The former increases with $w$ while the later remains constant, we can binary search for the smallest width $w^*$ such that the two volumes are equal. This takes $O(n^{d/2} \log n) \times O(\log n)$ time, for any fixed $d \geq 3$. To find the centre of $R$ with width $w^*$, we perform another plane sweep to find a point of the hyperplane $H$ that is not contained in $U(P^-, w^* - \epsilon)$, for a suitably small $\epsilon > 0$. $\quad\square$

We approach to the BS problem in a similar manner.

**Theorem 4.** *The BS problem can be solved in $O(n^{d/2+2} \log n)$ time and $O(n)$ space for any fixed dimension $d \geq 3$.*

*Proof.* A minor modification to the datastructure of Overmars and Yap allows it to answer the following query quickly: "What is the point of $H$ contained in the greatest number of hyperrectangles of $S$ but not contained in any hyperrectangle of $T$?" We can find the centre of a hypercube $R$ of width $w$ that contains as much of $P^+$ as possible but avoids $P^-$ as follows: let $\mathcal{C}(S, w) = \{C(s, w) : s \in S\}$; as we sweep $H$, we look for the point of $H$ contained in as many hyperrectangles of $\mathcal{C}(P^+, w)$ as possible but not contained in any hyperrectangle of $\mathcal{C}(P^-, w)$. Each sweep takes $O(n^{d/2} \log n)$ time and there are $O(n^2)$ possibilities for $w$. $\quad\square$

## 4.2   MER and BR Problems

To solve the MER and BR problems, we enumerate all RHRs. In the MER problem, we compute the volume of each RHR. In the BR problem, we perform an orthogonal range query to count the number of blue points in each RHR. This enumeration approach was previously applied to the MER problem in two and three dimensions [1,5,4,6]. Using a charging scheme inspired by [14,15], one can show that are at $\Theta(n^d)$ RHRs in the worst case. This complements a prior bound of $O(n \log^{d-1} n)$ RHRs on average under modest assumptions on the input [6].

In three dimensions, our enumeration is asymptotically slower than the algorithm by Datta and Soundaralakshmi by a factor of $\log n$ in the worst case [6]. However, our enumeration is much simpler, it has the same asymptotic average-case run time, and it generalises to higher dimensions. Our result is summarised in the following theorem.

**Theorem 5.** *The MER and BR problems can be solved in $O(k \log^{d-2} n)$ time and $O(k + n \log^{d-2} n)$ space where $k$ is the number of RHRs.*

The details of our enumeration are simplified by assuming that the coordinates of the red points in each dimension are unique. This can be imposed with a symbolic perturbation of the point set.

To find RHRs, we sweep $\mathbb{R}^d$ from $-\infty$ to $\infty$ with a hyperplane $H$ that is orthogonal to the $d^{\text{th}}$ dimension. The hyperplane $H$ splits $\mathbb{R}^d$ into two halves that extend towards $\infty$ and $-\infty$ in the $d^{\text{th}}$ dimension that we refer to as *above* and *below* respectively. Similarly, we refer to the sides of a hyperrectangle $R$ that are parallel to $H$ and closest to $\infty$ and $-\infty$ as top and bottom respectively.

Let $B'$ be the region of the bounding hyperrectangle $B$ that lies below $H$. As we sweep, we ensure that we have discovered all of the RHRs with respect to $B'$ (i.e. the RHRs that result from taking $B'$ as the bounding hyperrectangle and restricting our attention to the red points inside of $B'$). To do this, we maintain a list $L$ of all of the RHRs with respect to $B'$ that touch $H$. We represent each such RHR by how each of its sides is supported (i.e. by which red point or side of $B'$). Occasionally, we must add to and delete from $L$. Such events occur when $H$ passes through a red point.

Let $p_1, p_2, \ldots$ denote the red points sorted in increasing order of $d^{\text{th}}$ coordinate. We now describe how to update $L$ as $H$ passes through $p_i$. Let $H^+$ and $H^-$ be sweep planes lying just above and below $r_i$ respectively (see Figure 1). Let $L^+$ and $L^-$ be the lists associated with $P^+$ and $P^-$ respectively. The RHRs common to both $L^-$ and $L^+$ are not supported by $p_i$ on any boundary.

Let $D_i$ be the set of RHRs in $L^-$ such that $r_i$ lies directly above the top of each one. By our assumption of coordinate uniqueness, $p_i$ supports the top of each RHR in $D_i$ when the sweep plane reaches $p_i$. So $D_i$ is the set of RHRs deleted from $L^+$. To compute $D_i$, we perform an orthogonal range query whenever a new RHR $R$ is discovered. Specifically, we add $R$ to $D_j$, where $p_j$ is the red point with the lowest index that lies directly above the bottom of $R$. If there is no such $p_j$, we add $R$ to a special set $D_\infty$. We can locate $p_j$ in $O(\log^{d-1} n)$ time by using a range tree with fractional cascading. Such a tree requires $O(n \log^{d-1} n)$ time and space to construct.

Let $A_i$ be the set of RHRs in $L^+$ that are supported on some side by $p_i$. Clearly, $A_i$ is the set of RHRs added to $L^+$. Let $R_a$ be a RHR of $A_i$. Then $p_i$ does not support the top of $R_a$, which is supported by $P^+$. If $p_i$ supports the bottom of $R_a$, then $R_a$ is the region of $B$ above $p_i$ and below $H^+$. Otherwise, $R_a$ corresponds to some RHR of $D_i$ as follows (see Figure 1): Let $R'$ be the result of pushing the top of $R_a$ down so that it coincides with $H^-$. When the sweep plane is at $H^-$, the top of $R'$ is supported by the plane, but one other side is not. There is at most one such unsupported side because the co-ordinates of the red points are unique. Push this one side out until it hits an obstacle. This results in a RHR $R_d$ that belongs to $D_i$.

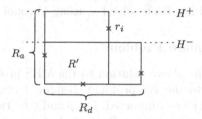

**Fig. 1.** $H_a$ corresponds to a deleted hyperrectangle $H_d$ (illustrated in 2D)

This correspondence between RHRs of $A_i$ and RHRs of $D_i$ suggests a procedure for generating $A_i$ from $D_i$. Let $R_d$ be a RHR of $D_i$. Let $S$ be one of the $(d-1)$ axis-aligned hyperplanes through $p_i$ that is perpendicular to $H$. For each such $S$, split $R_d$ along $S$, which results in two pieces. For each piece, check if it is properly supported on all all but two sides: the side coincident with $S$ (which will be supported by $p_i$) and the side coincident with $H^-$ (which will be supported by $H^+$). If a piece is properly supported add it to $A_i$. Otherwise, discard it. This process can be executed in $O(|D_i|)$ time, for any fixed $d$.

The set of all RHRs are $\bigcup_{i=1}^n D_i$. We can count the number of blue points contained in each one using a range tree with fractional cascading. Note that we do not need to compute $L$ explicitly (it can be constructed from the $D_i$). We only used it to simplify the description of our algorithm. Let $k$ be the number of RHRs. Clearly, $n \in O(k)$ because each red point supports the bottom of at least one RHR. It is straightforward to verify that this sweep-plane algorithm takes $O(k \log^{d-1} n)$ time and $O(k + n \log^{d-1} n)$ space.

This algorithm is similar to one that finds maximal negative orthants [15].

*Saving a Logarithmic Factor.* A simple observation allows us to shave a dimension off of our red-point range tree: When we query the red-point range tree for a point lying above the bottom of a RHR, we know that the desired point $p_j$ lies above the sweep plane. Hence, the height of the bottom is irrelevant, if we remove red points from the range tree as the sweep plane passes over them.

Likewise, a simple observation allows us to shave a dimension off of our blue-point range tree: The number of points in a RHR is the number of blue points directly above the bottom minus the number of blue points directly above the top. This allows us to perform a separate sweep to count the number of blue points in each RHR. Similarly, we remove points from the blue-point range tree as the sweep plane passes over them.

# 5   Two Dimensions

In this section, we *outline* efficient approaches to the two-dimensional BS and BR problems. The two-dimensional MER problem was extensively studied [1,2,3,5]. In stark contrast, the only solution to the two-dimension MES problem in the literature is described in a single line [2]: "The special case in which a largest empty square is desired has been solved ... using Voronoi diagrams."

## 5.1   Bichromatic Square Problem

We first elaborate on the above solution to the MES problem because it is the basis of our approach to the BS problem. Lemma 1 states that two opposite sides of a relevant square are supported. Let $p$ and $q$ be two points that support opposite sides of a relevant square $R$ centred at $c$ (see Figure 2a). The $L_\infty$ distance between two points $(x_1, y_1)$ and $(x_2, y_2)$ is $\max(|x_1 - x_2|, |y_1 - y_2|)$. Hence, the $L_\infty$ distance between $p$ and $c$ is exactly the $L_\infty$ distance between $q$ to $c$. Therefore, $c$ lies on the non-diagonal portion of the $L_\infty$ bisector of $p$ and $q$. We denote this bisector $E(p, q)$ because every point of $E(p, q)$ is equidistant to $p$ and $q$. As $R$ is relevant, its centre $c$ lies on an edge of the $L_\infty$-Voronoi diagram (*VD*) of the obstacles.

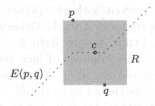

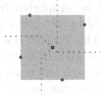

(a) Centred on horizontal segment.          (b) Centred on a Voronoi point.

**Fig. 2.** Relevant squares are centred on the edges of the Voronoi diagram, which is illustrated with dotted lines

*Remark 1.* The centre of a relevant square lies on a *non-diagonal* segment of the VD except for some degenerate cases (see Figure 2b).

Roughly speaking, the MES problem can be solved by (a) constructing the VD of the red points and (b) examining each non-diagonal segment of the VD. Step (a) can be executed in $O(n \log n)$ time and $O(n)$ space [16]. Step (b) can be

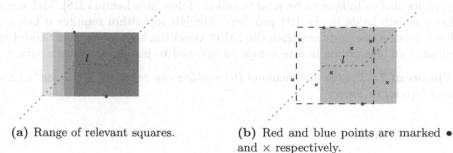

(a) Range of relevant squares.

(b) Red and blue points are marked ● and × respectively.

**Fig. 3.** Relevant squares centred on the segment $l$

executed in $O(n)$ time because the VD has $O(n)$ complexity [16]. Therefore, the MES problem can be solved in $O(n \log n)$ time and $O(n)$ space.

To solve the BS problem, we enumerate all $O(n)$ exceptions to Remark 1 and count the number of blue points in each one in $O(n \log n)$ time and $O(n)$ space. All other relevant squares are centred on non-diagonal segments of the VD. As Figure 3a illustrates, there is a continuum of relevant squares centred on a non-diagonal segment $l$. As Figure 3b illustrates, different squares centred on $l$ may contain a different number of blue points. To find a relevant square $R^*$ centred on $l$ that contains the most blue points, we sweep a square $R$ along $l$ and update the count of points contained in $R$ as blue points enter and leave it. We call the squares where a blue point either just enters or just leaves $R$ *interesting*. Clearly, we can restrict our attention to interesting squares in our search for $R^*$. To efficiently solve this problem, we rely on the following observation.

**Lemma 4.** *Each blue point enters a sweep square at most once over all horizontal segments of the VD, not just any one horizontal segment.*

This lemma implies that there are are a total of $O(n)$ interesting squares. We can enumerate all interesting squares and count the number of blue points in each one in $O(n \log n)$ time and $O(n)$ space.

**Theorem 6.** *The two-dimensional BS problem can be solved in $O(n \log n)$ time and $O(n)$ space.*

### 5.2  Bichromatic Rectangle Problem

The asymptotically fastest algorithm for the MER problem takes $O(n \log^2 n)$ time and $O(n)$ space [3]. This approach works by first considering all rectangles that are supported by the bounding box $B$. There are only $O(n)$ such boxes and they can be enumerated via a plane sweep in $O(n \log n)$ time and $O(n)$ space. The remaining relevant rectangles are only supported by red points. Via divide-and-conquer, this problem is reduced to a subproblem where all relevant rectangles contain a common point. This special case can be solved in $O(n)$ time and $O(n)$ space with a sophisticated matrix searching technique [17]. The observation that permits this technique to be used is called McKenna's Lemma [18]. McKenna's Lemma still holds in the BR problem. The BR algorithm requires a factor of $\log n$ more space and time than the MER algorithm because we must count the number of blue points in a rectangle as opposed to just computing its area.

**Theorem 7.** *The two-dimensional BR problem can be solved in $O(n \log^3 n)$ time and $O(n \log n)$ space.*

# 6    Conclusion

We have filled many gaps in what is known about the MER, MES, BR, and BS problems. It is now known that these problems are NP-complete when the dimension $d$ is part of the input, these problems have polynomial-time solutions for any fixed $d$, and these problems have very efficient solutions in two dimensions. The hardness of these problems strongly suggests that any general solution for high dimensions will have an exponential dependence in $d$. An open problem is to prove a lower bound on this dependence. This exponential dependence motivates the search for approximation algorithms. We are aware of an interesting new approximate result for the MER problem that requires $O((8ed\varepsilon^{-2})^d n \log^d n)$ time in the worst case, which is still a strong exponential dependence in $d$ [19].

We solved the MES problem in high dimensions via several plane sweeps that essentially computed the volume of a union of hypercubes. We know of an algorithm for computing the volume of the union of cubes in three dimensions that takes $O(n^{4/3} \log n)$ time [20]. Generalising this to higher dimensions remains an interesting open problem. Moreover, this algorithm for the union of cubes does not solve the depth problem, which is used in our BS algorithm.

Our solutions to the MER and BR problems have almost linear average case run times for any fixed dimension. We doubt that this is true of our solutions to the MES and BS problems. An open problem problem is to remedy this.

Our solution to the BS problem in high-dimensions is a factor of $n^2$ times slower than our solution to the MES problem. There is no such gap in two dimensions and our MER and BR problems demonstrate no such difference in run time. This suggests that the BS algorithm can be improved.

# References

1. Naamad, A., Lee, D., Hsu, W.: Maximum empty rectangle problem. Discrete Appl. Math. 8(3), 267–277 (1984)
2. Chazelle, B., Drysdale, R., Lee, D.: Computing the largest empty rectangle. SIAM J. Comput. 15, 300 (1986)
3. Aggarwal, A., Suri, S.: Fast algorithms for computing the largest empty rectangle. In: Symposium on Computational Geometry, pp. 278–290 (1987)
4. Nandy, S., Bhattacharya, B.: Maximal empty cuboids among points and blocks. Computers and Mathematics with Applications 36(3), 11–20 (1998)
5. Orlowski, M.: A new algorithm for the largest empty rectangle problem. Algorithmica 5(1), 65–73 (1990)

6. Datta, A., Soundaralakshmi, S.: An efficient algorithm for computing the maximum empty rectangle in three dimensions. Inform. Sciences 128(1-2), 43–65 (2000)
7. Eckstein, J., Hammer, P., Liu, Y., Nediak, M., Simeone, B.: The maximum box problem and its application to data analysis. Comput. Optim. Appl. 23(3), 285–298 (2002)
8. Aronov, B., Har-Peled, S.: On Approximating the Depth and Related Problems. SIAM J. Comput. 38(3), 899–921 (2008)
9. Liu, Y., Nediak, M.: Planar case of the maximum box and related problems. In: CCCG, pp. 14–18 (2003)
10. Edmonds, J., Gryz, J., Liang, D., Miller, R.J.: Mining for empty spaces in large data sets. Theor. Comput. Sci. 296(3), 435–452 (2003)
11. Dobkin, D., Gunopulos, D., Maass, W.: Computing the maximum bichromatic discrepancy, with applications to computer graphics and machine learning. J. Comput. Syst. Sci. 52(3), 453–470 (1996)
12. Frederickson, G.N., Johnson, D.B.: Generalized selection and ranking: Sorted matrices. SIAM J. Comput. 13(1), 14–30 (1984)
13. Overmars, M., Yap, C.: New upper bounds in Klee's measure problem. SIAM J. Comput. 20(6), 1034–1045 (1991)
14. Boissonnat, J.D., Sharir, M., Tagansky, B., Yvinec, M.: Voronoi diagrams in higher dimensions under certain polyhedral distance functions. Discrete & Computational Geometry 19(4), 485–519 (1998)
15. Kaplan, H., Rubin, N., Sharir, M., Verbin, E.: Counting colors in boxes. In: Bansal, N., Pruhs, K., Stein, C. (eds.) SODA, pp. 785–794. SIAM, Philadelphia (2007)
16. Lee, D.T., Wong, C.K.: Voronoi diagrams in $l_1(l_\infty)$ metrics with 2-dimensional storage applications. SIAM J. Comput. 9(1), 200–211 (1980)
17. Aggarwal, A., Klawe, M., Moran, S., Shor, P., Wilber, R.: Geometric applications of a matrix-searching algorithm. Algorithmica 2(1), 195–208 (1987)
18. McKenna, M., O'Rourke, J., Suri, S.: Finding the largest rectangle in an orthogonal polygon. In: Allerton Conference, pp. 486–495 (1985)
19. Dumitrescu, A., Jiang, M.: On the largest empty axis-parallel box amidst n points. arxiv.org reference arXiv:0909.3127 (2009) Submitted for publication
20. Agarwal, P.K., Kaplan, H., Sharir, M.: Computing the volume of the union of cubes. In: Erickson, J. (ed.) Symposium on Computational Geometry, pp. 294–301. ACM, New York (2007)

# Connectivity Is Not a Limit for Kernelization: Planar Connected Dominating Set

Qianping Gu and Navid Imani

School of Computing Science
Simon Fraser University, Burnaby BC, Canada V5A 1S6
qgu@cs.sfu.ca, navidi@sfu.ca

**Abstract.** We prove a small linear-size kernel for the connected dominating set problem in planar graphs through data reduction. Our set of rules efficiently reduce a planar graph $G$ with $n$ vertices and connected dominating number $\gamma_c(G)$ to a kernel of size at most $413\gamma_c(G)$ in $O(n^3)$ time answering the question of whether the connectivity criteria hinders the construction of small kernels, negatively (in case of the planar connected dominating set). Our result gives a fixed-parameter algorithm of time $(2^{O(\sqrt{\gamma_c(G)})} \cdot \gamma_c(G) + n^3)$ using the standard branch-decomposition based approach.

## 1 Introduction

To find an optimal solution of an NP-hard problem, one may not hope for anything better than exponential running time in the worst case. In the classical complexity theory, the size of an input instance is usually considered as the only factor for its hardness. However, many input instances consist of some parts that are relatively easy to deal with and other parts that form the real hard core of the problem. A fixed-parameter algorithm computes an optimal solution for a hard problem by restricting the *combinatorial explosion* that characterizes the exponential growth in the running time to a certain parameter. It is hoped that these parameters might take only relatively *small* values, resulting in an affordable exponential growth in which case, the fixed-parameter algorithm efficiently solves the given *parameterized problem* [22].

An efficient approach in the fixed-parameter algorithms is that before starting a cost-intensive optimal algorithm to solve a hard problem, a polynomial-time pre-processing phase is executed to shrink the input data to the hard core kernel. This is known as *data reduction to a problem kernel*. Pre-processing hard problems is not a new concept and it can be traced back to the very beginning of algorithm research. The concept of data reduction to a problem kernel was introduced by Downey and Fellows [11] for the first time to formalize reductions for parameterized complexity purposes. The vertex cover problem is probably one of the earliest problems studied in this line [21,5]. Cai et al. [6] proved that every fixed-parameter tractable problem is kernelizable. The research on finding kernels of small size for fixed-parameter tractable problems has been receiving

A. López-Ortiz (Ed.): LATIN 2010, LNCS 6034, pp. 26–37, 2010.

much attention [20,8,14,15,2,16,1]. A representative work in this line is that the dominating set problem on planar graphs is shown to have a linear size kernel [1,7,13].

Given a graph $G$ with vertex set $V(G)$, the dominating set problem asks for a minimum subset $D \subseteq V(G)$ of vertices such that every vertex in $V(G)\backslash D$ has a neighbor in $D$. The cardinality of a minimum dominating set of $G$ is known as the dominating number of $G$, denoted by $\gamma(G)$. The dominating set problem is a core NP-complete graph problem which belongs to a broader class of domination and covering problems. From applications' point of view, domination problems appear in numerous practical settings, ranging from strategic decisions such as locating radar stations or emergency services through computational biology to voting systems. The algorithmic complexity of the domination and related problems are discussed in details in the book of Haynes et al. [17]. It is known that the dominating set problem on arbitrary graphs is not fixed-parameter tractable unless $W[2] = FPT$ but when restricted to planar graphs it becomes fixed-parameter tractable [11]. The best known parameterized time complexity for the planar dominating set problem is $O(2^{11.98\sqrt{\gamma(G)}}n^{O(1)})$ [9]. Data reduction for the dominating set problem has received much attention [23,24,1,7]. Alber et al. [1] give data reduction rules which always reduce a planar graph of $n$ vertices to a problem kernel of size $O(\gamma(G))$ in $O(n^3)$ time. Recently, Guo and Niedermeier developed a generalized data reduction framework for deriving linear-size kernel on a variety of NP-hard problems on planar graphs, including the domination problems [13].

Apart from the original setting, variations of domination problems have found numerous applications and significant theoretical interests, out of which *connected dominating set* problem has probably received the most attention. The connected dominating set problem is to find a minimum dominating set $D$ of a graph $G$ such that the induced graph by $D$ is connected. Denoted by $\gamma_c(G)$ is the connected dominating number of $G$ which is calculated as the cardinality of $D$. In addition to its theoretical significance, the connected dominating set problem lies at the heart of many practical settings. An important application of this problem appears in wireless ad hoc networks [3]. It is known that the connected dominating set problem is not fixed-parameter tractable in arbitrary graphs but becomes tractable when restricted to planar graphs [10]. It has yet been open whether the planar version of the problem has a small kernel. Here we answer this question by proving that the planar connected dominating set problem admits a linear kernel of size $413\gamma_c(G)$. A recent unpublished work that implies a linear-size kernel for this problem is [4]. Although this seminal work presents meta-theorems that prove the existence of such kernels for a class of problems including planar connected dominating set, no small upperbounds on the size of the actual kernel is presented there. We also got to know about a very recent result by Lokshtanov et al. showing a linear kernel of size $3968187\gamma_c(G)$ for this problem [19] where the constant is large comparing to the best known kernelization of planar dominating set, $67\gamma(G)$ [7]. This could expose connectedness as a natural limit to deriving smaller kernels considering the difficulty of designing reduction rules to deal with the connectivity inherent in the problem. Our work falsifies this claim by obtaining an independent linear-size kernel of much smaller size for this problem[1] and positively answers the conjecture of [19] that the planar connected dominating set problem has a kernel smaller than $1000\gamma_c(G)$. Our result is achieved by developing a set of data reduction rules which reduce a planar graph $G$ of $n$ vertices to a problem kernel of size $O(\gamma_c(G))$ in $O(n^3)$ time. Considering the paramount importance of the problem in theory and practice, this brings us one step closer to efficient computation of many real life problems. From an algorithmic point of view, our linear-size kernel can be coupled with any of the previous algorithmic results to obtain an efficient fixed-parameter algorithm for the planar connected dominating set problem. In particular, using the branch-decomposition based approach of [10], our work implies a fixed-parameter algorithm of time $(2^{O(\sqrt{\gamma_c(G)})} \cdot \gamma_c(G) + n^3)$.

The organization of this paper is as follows: In Section 2, we define the terminology and notation that will be used throughout this paper. In Section 3, we design a set of data reduction rules for the planar connected dominating set problem. Section 4 deals with analysis of the size of kernel obtained after applying the reduction rules. Finally, we conclude the paper in section 5.

## 2   Preliminaries

We first introduce some definitions on fixed-parameter algorithms. Readers may refer to [22] for more details on this topic. A fixed-parameter algorithm solves a problem with an input instance of size $n$ and a parameter $k$ in $f(k) \cdot n^{O(1)}$ time for some computable function $f$ depending solely on $k$. Let $\mathcal{L}$ be a parameterized problem consisting of input pairs $(I, k)$, where $I$ is the input instance and $k$ is the parameter for $I$. Then *kernelization* or *reduction to a problem kernel* is to replace instance $(I, k)$ by a minimal reduced instance $(I', k')$ called *problem kernel* such that $k' \leq k$ and $|I'| \leq g(k)$ for some function $g$ only depending on $k$, $(I, k) \in \mathcal{L}$ if and only if $(I', k') \in \mathcal{L}$, and this reduction must be computable in polynomial time in $|I|$. Here, $g(k)$ is called the *kernel size*.

We use graphs for simple undirected graphs unless otherwise stated. We denote by $V(G)$ the vertex set and $E(G)$ the edge set of a graph $G$. Readers may refer to a textbook on graph theory for basic definitions. For a vertex $v \in V(G)$, let $N(v) = \{u|\{u, v\} \in E(G)\}$ and $N[v] = N(v) \cup \{v\}$. Given a subset $U \subseteq V(G)$, let $N(U) = \{u|v \in U, \{u, v\} \in E(G)\}$, $N[U] = N(U) \cup U$, and $G[U]$ denote the subgraph induced by the vertices of $U$. The length of a path in $G$ is the number of edges in the path. We denote by $v_1 - v_2 - \cdots - v_l$ the path $\{v_1, v_2\}, \{v_2, v_3\}, ..., \{v_{l-1}, v_l\}$. The distance between two vertices $v$ and $w$ in $G$, denoted by $d_G(v, w)$ is the length of the shortest path between $v$ and $w$ in $G$.

A vertex $u$ is *dominated* by a vertex $v$ in a graph $G$ if either $u = v$ or $\{u, v\} \in E(G)$. A vertex $v$ is dominated by a vertex set $U$ if $v$ is incident to at least one vertex from $U$. A vertex set $U'$ is dominated by a vertex set $U$ if every vertex

---

[1] The result has been obtained independently from [19] and appeared in [18].

from $U'$ is dominated by $U$. A subset $D \subseteq V(G)$ is a dominating set of $G$ if $V(G)$ is dominated by $D$. A *connected dominating set* (CDS) of $G$ is a subset $D \subseteq V(G)$ such that $D$ is a dominating set of $G$ and the subgraph $G[D]$ is connected. The CDS problem is to find a minimum CDS $D$ of $G$. The decision version of the CDS problem is to decide, given a graph $G$ and a positive integer $k$, whether $\gamma_c(G) \leq k$.

Alber et al. [1] show that the planar dominating set problem has a linear size kernel. Later, Guo et al. [13] generalize this approach to a framework for tackling a class of NP-hard problems on planar graphs. We use these works as a starting point for designing data reduction rules for the planar connected dominating set problem. The definitions in the rest of this section are adopted from [1]. For a vertex $v \in V(G)$, $N(v)$ is partitioned into:

$$N_1(v) = \{u | u \in N(v), N(u) \setminus N[v] \neq \emptyset\},$$
$$N_2(v) = \{u | u \in N(v) \setminus N_1(v), N(u) \cap N_1(v) \neq \emptyset\},$$
$$N_3(v) = N(v) \setminus (N_1(v) \cup N_2(v)).$$

For a pair of vertices $v, w \in V(G)$, let $N(v,w) = N(v) \cup N(w) \setminus \{v, w\}$ and $N[v,w] = N[v] \cup N[w]$. The neighborhood $N(v,w)$ is partitioned into:

$$N_1(v,w) = \{u | u \in N(v,w), N(u) \setminus N[v,w] \neq \emptyset\},$$
$$N_2(v,w) = \{u | u \in N(v,w) \setminus N_1(v,w), N(u) \cap N_1(v,w) \neq \emptyset\},$$
$$N_3(v,w) = N(v,w) \setminus (N_1(v,w) \cup N_2(v,w)).$$

Notice that the vertices of $N_3(v,w)$ can only be dominated by vertices of $\{v, w\} \cup N_2(v,w) \cup N_3(v,w)$. A *plane graph* is a planar graph drawn in the plane without an edge crossing. Let $G$ be a plane graph. For two vertices $v, w$ of $V(G)$ with $d_G(v,w) \leq 3$, a *region* $R(v,w)$ between $v$ and $w$ is a closed subset of the plane such that

- the *boundary* of $R(v,w)$ denoted by $\partial R$ is formed by two simple paths $P_1$ and $P_2$ that connect $v$ and $w$, and the length of each path is at most three;
- all vertices that are strictly inside the region $R(v,w)$ are from $N(v,w)$.

For a region $R = R(v,w)$, $V(R) = \{u \in V(G) | u$ is inside $R$ or on $\partial R\}$ is the set of vertices belonging to $R$. Figure 1 (a) gives an example of a region $R(v,w)$. A region $R = R(v,w)$ is called *simple* if all vertices of $V(R) \setminus \{v, w\}$ are common neighbors of $v$ and $w$ (see Figure 1 (b)). A simple region $R(v,w)$ is of *type $i$* ($i = 1, 2$) if $i$ of the vertices on its boundary except for $v, w$ have at least a neighbor outside the region. Examples of regions of types 2 and 1 are depicted in Figure 1 (b) and (c), respectively.

Given the definition of the region, for a plane graph $G$, one can envision a decomposition of $G$ into a set of non-overlapping regions. This notion is formalized as follows: Given a plane graph $G$ and a subset $D \subseteq V(G)$, a *$D$-region decomposition* of $G$ is a set $\mathcal{R}$ of regions between pairs of vertices of $D$ such that (1) for $R = R(v,w) \in \mathcal{R}$, no vertex from $D \setminus \{v, w\}$ is in $V(R)$ and (2) for two regions $R_1, R_2 \in \mathcal{R}$, $(R_1 \cap R_2) \subseteq (\partial R_1 \cup \partial R_2)$.

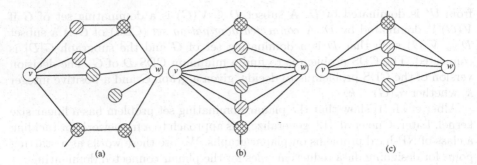

**Fig. 1.** (a) A region, (b) a simple region of type 2, and (c) a simple region of type 1

For a region decomposition $\mathcal{R}$, $V(\mathcal{R}) = \bigcup_{R \in \mathcal{R}} V(R)$ is the set of vertices in or on the boundary of a region in $\mathcal{R}$. Given a subset $D \subseteq V(G)$, $D$-region decomposition of $G$ is not necessarily unique. A $D$-region decomposition $\mathcal{R}$ is called *maximal* if no region $R$ can be added to $\mathcal{R}$ such that the resulting decomposition stays valid and more vertices are covered. In other words for a maximal $\mathcal{R}$ and $\forall R \notin \mathcal{R}$, if $V(\mathcal{R}) \subset V(\mathcal{R} \cup R)$ then $R \cup \mathcal{R}$ is not a valid region decomposition of $G$.

## 3    Reduction Rules for Connected Domination

In this section, using the inherent properties of the connected domination, we introduce five simple rules for reducing the graph. Furthermore, in order to assure the validity of our fixed-parameter reduction, for each presented rule we prove correctness and polynomial time complexity. Our first two reduction rules are similar to the two reduction rules of [1]. The intuition behind the first reduction rule is that if $N_3(v) \neq \emptyset$ then vertex $v$ is a good candidate for dominating $N_3(v)$ and the vertices in $N_2(v)$ and $N_3(v)$ can be removed.

**Rule 1:** For $v \in V(G)$, if $N_3(v) \neq \emptyset$ then remove $N_2(v) \cup N_3(v)$ from $G$ and add a new gadget vertex $v'$ with edge $\{v, v'\}$ to $G$.

Our second reduction rule is applied to the end vertices of an edge $\{v, w\}$ of $G$. The intuition behind this rule is that vertices of $\{v, w\}$ are good candidates for dominating $N_3(v, w)$ and some vertices of $N_2(v, w)$ and $N_3(v, w)$ can be removed.

**Rule 2:** For $\{v, w\} \in E(G)$, assume that $|N_3(v, w)| \geq 2$ and $N_3(v, w)$ can not be dominated by a single vertex from $N_2(v, w) \cup N_3(v, w)$.
    **Case 1:** $N_3(v, w)$ can be dominated by a single vertex of $\{v, w\}$.
    – (1.1) If $N_3(v, w) \subseteq N(v)$ and $N_3(v, w) \subseteq N(w)$ then remove $N_3(v, w)$ and $N_2(v, w) \cap N(v) \cap N(w)$ from $G$ and add a new gadget vertex $z$ with edges $\{v, z\}$ and $\{w, z\}$ to $G$.
    – (1.2) If $N_3(v, w) \subseteq N(v)$ but $N_3(v, w) \nsubseteq N(w)$ then remove $N_3(v, w)$ and $N_2(v, w) \cap N(v)$ from $G$ and add a new gadget vertex $v'$ with edge $\{v, v'\}$ to $G$.

- (1.3) If $N_3(v,w) \subseteq N(w)$ but $N_3(v,w) \nsubseteq N(v)$ then remove $N_3(v,w)$ and $N_2(v,w) \cap N(w)$ from $G$ and add a new gadget vertex $w'$ with edge $\{w,w'\}$ to $G$.

**Case 2:** If $N_3(v,w)$ can not be dominated by a single vertex from $\{v,w\}$ then remove $N_2(v,w)$ and $N_3(v,w)$ from $G$ and add new gadget vertices $v'$ and $w'$ with edges $\{v,v'\}$ and $\{w,w'\}$ to $G$.

Our next two rules are designed to be applied to a pair of vertices $v$ and $w$ of $G$ with $2 \le d_G(v,w) \le 3$. The intuition behind these rules is similar to that for Rule 2 but to remove some vertices from $N_2(v,w)$ and $N_3(v,w)$, we may need to keep some vertices which form a path between $v$ and $w$ to guarantee the connectivity of the graph induced by the dominating set while we also need to assure that there cannot be any other shorter connected path in $N_2(v,w)$ and $N_3(v,w)$ dominating $N_3(v,w)$. This makes the rules more complex than Rule 2 because there are different cases for keeping such vertices. We first introduce some notation.

A vertex $x \in N_3(v,w)$ is called a *bridge* if $x$ is dominated by a vertex from $N_2(v,w)$, $v$, and $w$, that is, $x \in N(N_2(v,w)) \cap N(v) \cap N(w)$. We denote by $B(v,w)$ the set of bridges for $v$ and $w$. Intuitively, a bridge is a good candidate for forming a path between $v$ and $w$.

**Rule 3:** For $v,w \in V(G)$ with $d_G(v,w) = 2$, assume that $|N_3(v,w)| \ge 3$. We remove some vertices from $N_3(v,w)$ but keep a path $v - p - w$ in $G$.

**Case 1:** $N_3(v,w)$ can be dominated by a single vertex of $\{v,w\}$.
- (1.1) $N_3(v,w) \subseteq N(v)$ and $N_3(v,w) \subseteq N(w)$. If $N_3(v,w)$ can not be dominated by at most two vertices of $N_2(v,w) \cup N_3(v,w)$ then:
  - If $B(v,w) \ne \emptyset$ then select a vertex from $B(v,w)$, otherwise a vertex from $N(v) \cap N(w)$, as $p$.
  - If $N_3(v,w) \backslash (B(v,w) \cup \{p\}) \ne \emptyset$ then remove $N_3(v,w) \backslash (B(v,w) \cup \{p\})$ and add a gadget vertex $z$ with edges $\{v,z\}$ and $\{z,w\}$ to $G$.
- (1.2) $N_3(v,w) \subseteq N(v)$ but $N_3(v,w) \nsubseteq N(w)$. If $N_3(v,w)$ can not be dominated by at most two vertices of $\{w\} \cup N_2(v,w) \cup N_3(v,w)$ then:
  - If $B(v,w) \ne \emptyset$ then select a vertex from $B(v,w)$, otherwise a vertex from $N(v) \cap N(w)$, as $p$.
  - If $N_3(v,w) \backslash (B(v,w) \cup \{p\}) \ne \emptyset$ then remove $N_3(v,w) \backslash (B(v,w) \cup \{p\})$ and add a gadget vertex $v'$ with edge $\{v,v'\}$ to $G$.
- (1.3) $N_3(v,w) \subseteq N(w)$ but $N_3(v,w) \nsubseteq N(v)$. If $N_3(v,w)$ can not be dominated by at most two vertices of $\{v\} \cup N_2(v,w) \cup N_3(v,w)$ then:
  - If $B(v,w) \ne \emptyset$ then select a vertex from $B(v,w)$, otherwise a vertex from $N(v) \cap N(w)$, as $p$.
  - If $N_3(v,w) \backslash (B(v,w) \cup \{p\}) \ne \emptyset$ then remove $N_3(v,w) \backslash (B(v,w) \cup \{p\})$ and add a gadget vertex $w'$ with edge $\{w,w'\}$ to $G$.

**Case 2:** $N_3(v,w)$ can not be dominated by a single vertex from $\{v,w\}$.

If $N_3(v,w)$ can not be dominated by at most two vertices of $\{v\} \cup N_2(v,w) \cup N_3(v,w)$ or at most two vertices of $\{w\} \cup N_2(v,w) \cup N_3(v,w)$ then: select a vertex from $N(v) \cap N(w)$ as $p$, remove $(N_2(v,w) \cup N_3(v,w)) \backslash \{p\}$, and add gadget vertices $v'$ and $w'$ with edges $\{v,v'\}, \{w,w'\}$ to $G$.

To introduce Rule 4, we need an additional definition. For a pair of vertices $v$ and $w$ with $d_G(v, w) = 3$, a vertex $y \in N_2(v, w) \cap N(w)$ is called a *key-neighbor* of $w$ if $y$ is dominated by a vertex $z$ of $N_3(v, w) \cap N(v)$, that is, $y \in N(w) \cap N_2(v, w) \cap N(N_3(v, w) \cap N(v))$. We call $z$ a *companion* of the key-neighbor $y$. We define similarly a key-neighbor of $v$ and a companion of the key-neighbor $y$. Intuitively, a key-neighbor $y$ and a companion $z$ of $y$ are good candidates for a shorter connected path dominating $N_3(v, w)$. As shown later, there are at most two key-neighbors of $w$ in each region $R(v, w)$ between $v$ and $w$ due to the planarity of $G$. Therefore, a companion $z$ can dominate at most two key-neighbors of $w$ and there is at most one companion which can dominate two key-neighbors of $w$ in $R(v, w)$. Similarly, a companion $z$ can dominate at most two key-neighbors of $v$ and there is at most one companion which can dominate two key-neighbors of $v$ in $R(v, w)$.

**Rule 4:** For $v, w \in V(G)$ with $d_G(v, w) = 3$, assume that $|N_3(v, w)| \geq 4$. We remove some vertices from $N_3(v, w)$ but keep a path $v - p - q - w$ in $G$.

**Case 1:** $N_3(v, w)$ can be dominated by a single vertex of $\{v, w\}$.

- (1.1) $N_3(v, w) \subseteq N(v)$. If $N_3(v, w)$ can not be dominated by at most three vertices of $\{w\} \cup N_2(v, w) \cup N_3(v, w)$ then:
  - Find a minimum subset $Z$ of companions such that $Z$ dominates every key-neighbor of $w$. If $Z \neq \emptyset$ then select a $z \in Z$ as $p$ and a key-neighbor $y$ of $w$ dominated by $z$ as $q$, otherwise select any two vertices $p$ and $q$ such that $v - p - q - w$ is a path of $G$.
  - If $N_3(v, w) \setminus (Z \cup \{p, q\}) \neq \emptyset$ then remove $N_3(v, w) \setminus (Z \cup \{p, q\})$ and add a gadget vertex $v'$ with edge $\{v, v'\}$ to $G$.
- (1.2) $N_3(v, w) \subseteq N(w)$. If $N_3(v, w)$ can not be dominated by at most three vertices of $\{v\} \cup N_2(v, w) \cup N_3(v, w)$ then
  - Find a minimum subset $Z$ of companions such that $Z$ dominates every key-neighbor of $v$. If $Z \neq \emptyset$ then select a $z \in Z$ as $p$ and a key-neighbor $y$ of $v$ dominated by $z$ as $q$, otherwise select any two vertices $p$ and $q$ such that $w - p - q - p$ is a path of $G$.
  - If $N_3(v, w) \setminus (Y \cup \{p, q\}) \neq \emptyset$ then remove $N_3(v, w) \setminus (Y \cup \{p, q\})$ and add a gadget vertex $v'$ with edge $\{v, v'\}$ to $G$.

**Case 2:** $N_3(v, w)$ can not be dominated by a single vertex of $\{v, w\}$. If $N_3(v, w)$ can not be dominated by at most three vertices of $\{v\} \cup N_2(v, w) \cup N_3(v, w)$ or at most three vertices of $\{w\} \cup N_2(v, w) \cup N_3(v, w)$ then select any two vertices $p$ and $q$ such that $v - p - q - w$ is path of $G$. If $(N_2(v, w) \cup N_3(v, w)) \setminus \{p, q\} \neq \emptyset$ then remove $(N_2(v, w) \cup N_3(v, w)) \setminus \{p, q\}$ and add gadget vertices $v'$ and $w'$ with edges $\{v, v'\}$ and $\{w, w'\}$ to $G$.

Finally, we introduce a rule which is applied to simple regions. For $v, w \in V(G)$ with $d_G(v, w) \leq 2$, let $\mathcal{R} = \mathcal{R}(v, w)$ be the union of all maximal simple regions $R(v, w)$.

**Rule 5:** For $v, w \in V(G)$, assume that $(V(\mathcal{R}) \cap N_3(v, w)) \setminus B(v, w) \neq \emptyset$. If $B(v, w) \neq \emptyset$ then select a vertex from $B(v, w)$ as $p$, otherwise select a vertex from $V(\mathcal{R}) \cap N_3(v, w)$ as $p$. Remove $(V(\mathcal{R}) \cap N_3(v, w)) \setminus (B(v, w) \cup \{p\})$. If $B(v, w) \neq \emptyset$ then add a gadget vertex $z$ with edges $\{v, z\}, \{z, w\}$ to $G$.

**Lemma 1.** *Given a graph $G$, let $G'$ be the graph obtained after applying any of the Rules 1-5 to $G$. Then a minimum CDS of $G'$ which does not contain any gadget vertices is a minimum CDS of $G$.*

**Lemma 2.** *Given a planar graph $G$, Rule 1 can be performed in $O(n)$ time for all vertices of $G$. Also Rules 2-5 can be performed in $O(n^2)$ time, for all pairs of vertices $v, w \in V(G)$.*

The sketch of the proof is almost the same as in [12], except that in order to obtain the running time of $O(n^2)$ for rules 3 and 4 ( as opposed to $O(n^3)$ and $O(n^4)$ respectively in [12]) we use a simple counting argument for checking the domination of a joint neighborhood by a path as required by rules 3 and 4.

   A graph $G$ is called *reduced* if the graph obtained after applying any of Rules 1-5 to $G$ is isomorphic to $G$.

**Theorem 1.** *Rules 1-5 convert a plane graph $G$ to a reduced plane graph $G'$ in $O(n^3)$ time such that a minimum CDS of $G'$ which does not contain any gadget vertices is a CDS of $G$ as well.*

For the proofs of the above lemmata and theorem please refer to [12].

## 4    Linear-Size Kernel

In this section, we show that the reduced graph $G'$, obtained after repetitive application of reduction Rules 1-5, has $O(\gamma_c(G))$ vertices. The proof consists of three major parts. First, by the result of [1], there is a maximal $D$-region decomposition $\mathcal{R}$ of $O(\gamma_c(G))$ regions for a plane graph $G$. Next, we prove that having applied the reduction rules repetitively to the graph, each region in $\mathcal{R}$ can have $O(1)$ vertices only. Finally, by the result of [1] we show that the number of vertices of $G$ that do not belong to any of the regions in $\mathcal{R}$ is $O(\gamma_c(G))$.

**Lemma 3.** *(Alber et al. [1]) Given a plane graph $G$ and a dominating set $D$ of $G$, a maximal $D$-decomposition $\mathcal{R}$ of at most $3|D|$ regions can be constructed.*

By this result, choosing $D$ a minimum CDS of $G$, we get a maximal $D$-region decomposition $\mathcal{R}$ of at most $3\gamma_c(G)$ regions. Next, we calculate an upperbound on the number of vertices in a region of $\mathcal{R}$.

**Proposition 1.** *Let $G'$ be the reduced graph obtained after repetitive application of Rules 1-5 to $G$. Then $G'$ has the following properties.*

1. *For every $v \in V(G')$, $N_3(v)$ does not contain any vertex of $G$.*
2. *For every pair $v, w \in V(G')$ with $d_{G'}(v, w) \leq 3$, either*
   (a) *there is a $U \subseteq (N_2(v, w) \cup N_3(v, w))$ such that $|U| \leq 3$ and $U$ dominates all vertices of $N_3(u, v)$ or*
   (b) *for every region $R(v, w)$, $N_3(v, w) \cap V(R)$ has at most two vertices from $V(G)$.*

*Proof.* (1) follows from Rule 1. For (2), if no vertex is removed from $G$ by any of Rules 2-5 then by the definition of the rules, (a) holds, otherwise we show (b). If Rule 2 successfully applied, then $N_3(v, w) = \emptyset$. If Rule 3 or Rule 5 is successfully applied, then $N_3(v, w)$ contains either bridges of $B(v, w)$ or a single vertex $p$. If Rule 4 is successfully applied, then $N_3(v, w)$ contains either some companions of $Z$ which dominate all key neighbors of $w$ (Case 1.1), or some companions of $Z$ which dominate all key neighbors of $v$ (Case 1.2), or two vertices $p$ and $q$ (all cases). Notice that the number of companions of $Z$ in a region $R(v, w)$ is at most the number of key neighbors of $v$ or $w$. Therefore, to show (b) of (2), it suffices to prove that in a region $R(v, w)$, there are at most two bridges for $d_G(v, w) = 2$, at most two key neighbors of $v$ for Case (1.2) of Rule 4, and at most two key neighbors of $w$ for Case (1.1) of Rule 4.

For $d_G(v, w) = 2$, assume that there are at least three bridges $x_1, x_2, x_3$ in $R(v, w)$. Then one bridge, say $x_3$, must be strictly inside the region $R' = R'(v, w)$ formed by the paths $v - x_1 - w$ and $v - x_2 - w$. Since $R'$ is inside $R$ and $x_1, x_2 \in N_3(v, w)$, each vertex strictly inside $R'$ is not connected to any vertex in $N_2(v, w)$, a contradiction to the fact that $x_3$ is a bridge. Thus, there are at most two bridges in $R(v, w)$.

For $d_G(v, w) = 3$, assume that there are at least three key-neighbors $x_1, x_2, x_3$ of $v$. Let $y_i \in N(w) \cap N_3(v, w), i = 1, 2, 3$ be the companions of $x_i$, respectively. Then one of the three key neighbors, $x_3$, must be strictly inside the region $R' = R'(v, w)$ formed by the paths $v - x_1 - y_1 - w$ and $v - x_2 - y_2 - w$. Since $R'$ is inside $R$ and $x_1, x_2 \in N_2(v, w)$ and $y_1, y_2 \in N_3(v, w)$, each vertex strictly inside $R'$ is not connected to any vertex in $N_1(v, w)$, a contradiction to the fact that $x_3$ is a vertex from $N_2(v, w)$. Thus, there are at most two key-neighbors of $v$ in $R(v, w)$ for Case (1.2) of Rule 4. Similarly, there are at most two key-neighbors of $w$ in $R(v, w)$ for Case (1.1) of Rule 4.                    $\square$

Next, we use Proposition 1 to upperbound the size of simple regions. We recall that for $i = 1, 2$, a simple region $R(v, w)$ is called a type-$i$ region if $V(R)$ has $i$ vertices from $N_1(v, w)$.

**Proposition 2.** *Given a reduced plane graph $G$ and a maximal D-region decomposition $\mathcal{R}$ for a CDS $D$ of $G$, a type-i region $R(v, w)$ of $\mathcal{R}$ has at most $i$ vertices from $N_1(v, w)$, $i$ vertices from $N_2(v, w)$, and $i+1$ vertices from $N_3(v, w)$.*

*Proof.* For a simple region $R = R(v, w)$ in $\mathcal{R}$, only the vertices on the boundary can have a neighbor outside $R$. By the definition of simple region, $|N_1(v, w) \cap V(R)| \leq 2$. But since $G$ is planar, every vertex in $N_1(v, w) \cap V(R)$ can contribute at most one vertex to $N_2(v, w) \cap V(R)$. Hence, we get $|N_2(v, w) \cap V(R)| \leq |N_1(v, w) \cap V(R)|$. As shown in the proof of Proposition 1, $N_3(v, w) \cap V(R)$ has at most $|N_2(v, w) \cap V(R)| \leq 2$ bridges and one gadget vertex.                    $\square$

Now we are ready to upperbound the number of vertices in a region $R(v, w)$ of $\mathcal{R}$. A key step in the proof is to upperbound the number of simple regions in a region $R(v, w)$ by Proposition 1. From this bound and Proposition 2, we can get an upper bound on the number of vertices in $R(v, w)$.

**Lemma 4.** *Given a reduced plane graph $G$ and a maximal $D$-region decomposition $\mathcal{R}$ for a CDS $D$ of $G$, every region $R = R(v, w)$ of $\mathcal{R}$ has at most 81 vertices.*

*Proof.* Let $P$ and $Q$ be the paths which form the boundary of $R$. Without loss of generality, we assume that both $P = v - p_1 - p_2 - w$ and $Q = v - q_1 - q_2 - w$ have length three (a shorter path will give a smaller number of vertices in $R$). Since only the vertices on $P$ and $Q$ can be connected to vertices outside $R$, $|N_1(v, w) \cap V(R)| \leq 4$.

Since each vertex in $N_2(v, w)$ is dominated by a vertex from $N_1(v, w)$ and a vertex from $\{v, w\}$, the vertices in $N_2(v, w)$ are included in simple regions between a vertex from $N_1(v, w)$ and one from $\{v, w\}$. From the planarity of $G$, we conclude that there are at most six such regions. In the worst case, four of the simple regions are type-1 and two of them are type-2. From Proposition 2, we get $|N_2(v, w) \cap V(R)| \leq 4 \cdot 4 + 2 \cdot 7 = 30$.

There are several cases for the size of $N_3(v, w) \cap V(R)$: (1) the condition for applying Rule 4 is not satisfied; (2) the condition for applying Rule 3 is not satisfied; (3) the condition for applying Rule 2 is not satisfied; and (4) one of Rules 2-4 has been successfully applied. For Case (1), By Proposition 1 and the definition of Rule 4, there is a subset $U \subseteq N_3(v, w)$ such that $|U| \leq 3$ and every vertex in $N_3(v, w)$ is dominated by $U$. Since each vertex in $N_3(v, w)$ is also dominated by a vertex from $\{v, w\}$, the vertices in $N_3(v, w)$ are in simple regions between a vertex from $U$ and a vertex from $\{v, w\}$. There are at most six such simple regions. Now, by Proposition 2, we know $|N_3(v, w) \cap V(R)| \leq 6 \cdot 7 + 3 = 45$. By a similar argument, we can show that $|N_3(v, w) \cap R(V)| < 45$ for Cases (2)-(4). Summarizing the above, $|V(R)| \leq |N_1(v, w) \cap V(R)| + |N_2(v, w) \cap V(R)| + |N_3(v, w) \cap V(R)| + |\{v, w\}| \leq 81$. □

Finally, we use the result of [1] to bound the number of vertices not in $V(\mathcal{R})$.

**Lemma 5.** *Given a plane graph $G$ and a CDS $D$ of $G$, if $\mathcal{R}$ is a maximal $D$-region decomposition of $G$, then $|V(G) \setminus V(\mathcal{R})| \leq 2|D| + 56|\mathcal{R}|$.*

*Proof.* Alber et al. in [1] proved that given a plane graph $G$ and a dominating set $D$ of $G$, if $\mathcal{R}$ is a maximal $D$-region decomposition, then $|V(G) \setminus V(\mathcal{R})| \leq 2|D| + 56|\mathcal{R}|$. We follow the same proof as that of Proposition 2 in [1] with the only exception that we use a separate argument for describing the simple regions in our specific reduced graph. By Proposition 2, each simple region $R(v, w)$ has at most nine vertices which is also an upper bound on the number of vertices of simple regions in [1]. Therefore, the lemma holds. □

**Theorem 2.** *For a planar graph $G$ which is reduced with respect to Rules 1 to 5, $|V(G)| \leq 413\gamma_c(G)$.*

*Proof.* From Lemma 3, we know that there are at most $3\gamma_c(G)$ regions in $\mathcal{R}$. By Lemma 4, each region has at most 81 vertices. From Lemma 5, we conclude $|V \setminus V(\mathcal{R})| \leq 2|D| + 56|\mathcal{R}|$. Therefore, $|V(G)| \leq 2\gamma_c(G) + 56 \times 3\gamma_c(G) + 81 \times 3\gamma_c(G) = 413\gamma_c(G)$. □

## 5    Concluding Remarks

This work addresses the question of finding a small linear-size kernel for the connected dominating set problem on planar graphs. Having proposed a set of simple and easy to implement reduction rules for the connected dominating set, we proved that for planar graphs, a small linear-size problem kernel can be efficiently constructed. The reduction phase has a low time complexity of only $O(n^3)$. In particular, combining our result with the branch-decomposition based approach of [10], our work implies a fixed-parameter algorithm of time $(2^{O(\sqrt{\gamma_c(G)})} \cdot \gamma_c(G) + n^3)$, for planar connected dominating set. From a methodological point of view, our constructed kernel does not deviate significantly in size from the best known kernels for the planar dominating set while the reduction time stays the same, showing that data reduction is still an efficient approach for designing small kernels even when connectivity of solution set is important. In particular, we answer the conjecture of [19] by proving that kernels of smaller than $1000\gamma_c(G)$ can be constructed for connected dominating set on planar graphs. It can be worthwhile to consider designing more sophisticated data reduction rules for further shrinking the kernel size say using the techniques in [7] and through a more detailed analysis. Developing a new framework for kernelization of a broader class of fixed-parameter tractable problems can be an important step in characterizing the hardness of FPT problems.

## References

1. Alber, J., Fellows, M.R., Niedermeier, R.: Polynomial-time data reduction for dominating set. Journal of the ACM 51(3), 363–384 (2004) (electronic)
2. Bansal, N., Blum, A., Chawla, S.: Correlation clustering, p. 238. IEEE Computer Society, Los Alamitos (2002)
3. Blum, J., Ding, M., Thaeler, A., Cheng, X.: Connected dominating set in sensor networks and MANETs. In: Handbook of combinatorial optimization, vol. B (Suppl.), pp. 329–369. Springer, New York (2005)
4. Bodlaender, H.L., Fomin, F.V., Lokshtanov, D., Penninkx, E., Saurabh, S., Thilikos, D.M.: (Meta) Kernelization, April 4 (2009), http://arxiv.org/abs/0904.0727
5. Buss, J.F., Goldsmith, J.: Nondeterminism within $P$. SIAM Journal on Computing 22(3), 560–572 (1993)
6. Cai, L., Chen, J., Downey, R.G., Fellows, M.R.: Advice classes of parameterized tractability. Annals of Pure and Applied Logic 84(1), 119–138 (1997)
7. Chen, J., Fernau, H., Kanj, I.A., Xia, G.: Parametric duality and kernelization: Lower bounds and upper bounds on kernel size. SIAM Journal on Computing 37(4), 1077–1106 (2007)
8. Chen, J., Kanj, I.: Improved exact algorithms for MAX-SAT. In: Rajsbaum, S. (ed.) LATIN 2002. LNCS, vol. 2286, pp. 341–355. Springer, Heidelberg (2002)
9. Dorn, F.: Dynamic programming and fast matrix multiplication. In: Azar, Y., Erlebach, T. (eds.) ESA 2006. LNCS, vol. 4168, pp. 280–291. Springer, Heidelberg (2006)

10. Dorn, F., Penninkx, E., Bodlaender, H.L., Fomin, F.V.: Efficient exact algorithms on planar graphs: Exploiting sphere cut branch decompositions. In: Brodal, G.S., Leonardi, S. (eds.) ESA 2005. LNCS, vol. 3669, pp. 95–106. Springer, Heidelberg (2005)
11. Downey, R.G., Fellows, M.R.: Parameterized Complexity. Monographs in Computer Science. Springer, New York (1999)
12. Gu, Q., Imani, N.: Small Kernel for Planar Connected Sominating Set. TR 2009-12, School of Computing Science, Simon Fraser University, Burnaby, BC, Canada (June 2009), ftp://fas.sfu.ca/pub/cs/TR/2009/CMPT2009-12.pdf
13. Guo, J., Niedermeier, R.: Linear Problem Kernels for NP-Hard Problems on Planar Graphs. In: Arge, L., Cachin, C., Jurdziński, T., Tarlecki, A. (eds.) ICALP 2007. LNCS, vol. 4596, pp. 375–386. Springer, Heidelberg (2007)
14. Gramm, J., Hirsch, E.A., Niedermeier, R., Rossmanith, P.: Worst-case upper bounds for MAX-2-SAT with an application to MAX-CUT. Discrete Applied Mathematics 130(2), 139–155 (2003)
15. Gramm, J., Nierhoff, T., Sharan, R., Tantau, T.: Haplotyping with missing data via perfect path phylogenies. Discrete Applied Mathematics 155(6-7), 788–805 (2007)
16. Guo, J., Niedermeier, R.: Fixed-parameter tractability and data reduction for multicut in trees. Networks 46(3), 124–135 (2005)
17. Haynes, T.W., Hedetniemi, S.T., Slater, P.J.: Fundamentals of Domination in Graphs. Monographs and Textbooks in Pure and Applied Mathematics, vol. 208. Marcel Dekker, New York (1998)
18. Imani, N.: Data Reduction for Connected Dominating Set. Master Thesis, Simon Fraser University, BC, Canada (August 2008)
19. Lokshtanov, D., Mnich, M., Saurabh, S.: Linear kernel for planar connected dominating set. To appear in Proceedings of TAMC (May 2009)
20. Mahajan, M., Raman, V.: Parameterizing above guaranteed values: MaxSat and MaxCut. Journal of Algorithms 31(2), 335–354 (1999)
21. Nemhauser, G.L., Trotter Jr., L.E.: Vertex packings: structural properties and algorithms. Mathematical Programming 8, 232–248 (1975)
22. Niedermeier, R.: Invitation to Fixed-Parameter Algorithms. Oxford Lecture Series in Mathematics and its Applications, vol. 31. Oxford University Press, Oxford (2006)
23. Weihe, K.: Covering trains by stations or the power of data reduction. In: Proceedings of Algorithms and Experiments, ALEX, pp. 1–8 (1998)
24. Weihe, K.: On the differences between "practical" and "applied". In: Näher, S., Wagner, D. (eds.) WAE 2000. LNCS, vol. 1982, pp. 1–10. Springer, Heidelberg (2000)

# Randomized Truthful Algorithms for Scheduling Selfish Tasks on Parallel Machines

Eric Angel[1], Evripidis Bampis[1], and Nicolas Thibault[2]

[1] IBISC - Université d'Évry
523 place des Terrasses
91000 Évry, France
firstname.name@ibisc.fr
[2] ERMES EAC 4441
Université Paris 2
75005 Paris, France
nicolasthibault@free.fr

**Abstract.** We study the problem of designing truthful algorithms for scheduling a set of tasks, each one owned by a selfish agent, to a set of parallel (identical or unrelated) machines in order to minimize the makespan. We consider the following process: at first the agents declare the length of their tasks, then given these bids the protocol schedules the tasks on the machines. The aim of the protocol is to minimize the makespan, i.e. the maximal completion time of the tasks, while the objective of each agent is to minimize the completion time of its task and thus an agent may lie if by doing so, his task may finish earlier. In this paper, we show the existence of randomized truthful (non-polynomial-time) algorithms with expected approximation ratio equal to $3/2$ for different scheduling settings (identical machines with and without release dates and unrelated machines) and models of execution (strong or weak). Our result improves the best previously known result [1] for the problem with identical machines ($P||C_{\max}$) in the strong model of execution and reaches, asymptotically, the lower bound of [5]. In addition, this result can be transformed to a polynomial-time truthful randomized algorithm with expected approximation ratio $3/2 + \epsilon$ (resp. $\frac{11}{6} - \frac{1}{3m}$) for $Pm||C_{\max}$ (resp. $P||C_{\max}$).

## 1 Introduction

Nowadays, there are many systems involving autonomous entities (agents). These systems are organized by protocols, trying to maximize the social welfare in the presence of private information held by the agents. In some settings the agents may try to manipulate the protocol by reporting false information in order to maximize their own profit. With false information, even the most efficient protocol may lead to unreasonable solutions if it is not designed to cope with the selfish behavior of the agents. In such a context, it is natural to study the efficiency of *truthful* protocols, i.e. protocols that are able to guarantee that no

A. López-Ortiz (Ed.): LATIN 2010, LNCS 6034, pp. 38–48, 2010.

agent has incentive to lie. This approach has been considered in many papers these last few years (see [4] for a recent survey).

In this paper, we study the problem of designing truthful algorithms for scheduling a set of tasks, each one owned by a selfish agent, to a set of parallel (identical or unrelated) machines in order to minimize the makespan. We consider the following process: before the start of the execution, the agents declare the length of their tasks, then given these bids the protocol schedules the tasks on the machines. The aim of the protocol is to minimize the makespan, i.e. the maximal completion time of the tasks, while the objective of each agent is to minimize the completion time of its task and thus an agent may lie if by doing so, his task may finish earlier. We focus on protocols without side payments that simultaneously offer a guarantee on the quality of the schedule (its makespan is not arbitrarily far from the optimum) and guarantee that the solution is truthful (no agent can lie and improve his own completion time).

## 1.1 Formal Definition

There are $n$ agents, represented by the set $\{1, 2, \cdots, n\}$ and $m$ parallel machines.

**Variants of the problem.** Depending on the type of the machines and the jobs characteristics, we consider three different variants of the problem:

- **Identical parallel machines** ($P||C_{\max}$). All the machines are identical and every task $i$ has a private value $t_i$ that represents its length. We assume that an agent cannot shrink the length of her task (otherwise he will not get his result), but if he can decrease his completion time by bidding a value larger than the real one ($b_i \geq t_i$), then he will do so.
- **Identical parallel machines with release dates** ($P|r_i|C_{\max}$). All the machines are identical and every task $i$ has now a private pair $(t_i, r_i)$, where $t_i$ is the length of task $i$ and $r_i$ its release date. Every task $i$ may bid any pair $(b_i, r_i^b)$ such that $b_i \geq t_i$ and $r_i^b \geq r_i$. A task $i$ may not bid a release date smaller than its real release date i.e. $r_i^b < r_i$, because otherwise, the task may be scheduled before $r_i$ and thus the final schedule may be infeasible.
- **Unrelated parallel machines** ($R||C_{\max}$). The machines are here unrelated. Every task $i$ has a private vector $(t_i^1, \ldots, t_i^m)$, where $t_i^j, 1 \leq j \leq m$, is the processing time of task $i$ if it is executed on machine $j$. Every task $i$ bids any vector $(b_i^1, \ldots, b_i^m)$ with $b_i^1 \geq t_i^1, \ldots, b_i^m \geq t_i^m$.

**Models of execution.** We consider two models of execution:

- *The strong model of execution:* task $i$ bids any value $b_i \geq t_i$ and its execution time is $t_i$ (i.e. task $i$ is completed $t_i$ units of time after it starts even if $i$ bids $b_i \neq t_i$).
- *The weak model of execution:* $i$ bids any value $b_i \geq t_i$ and its execution time is $b_i$ (i.e. task $i$ is completed $b_i$ units of time after it starts).

**Notation.** By $C_i$, we denote the completion time of task $i$. The objective of the protocol is to determine a schedule of the tasks minimizing the maximal completion time of the tasks or makespan, denoted in what follows by $C_{\max}$. We say that an algorithm is *truthful*, if and only if, for every task $i$, $1 \leq i \leq n$ and for every bid $b_j$, $j \neq i$, the completion time of task $i$ is minimum when $i$ bids $b_i = t_i$. In other, words, an algorithm is truthful if truth-telling is the best strategy for a player $i$ regardless of the strategy adopted by the other players.

## 1.2   Related Works

The works that are more closely related to our are those of [2], [1], [3] and [5]. In the paper by Auletta et al. [3], the authors consider the variant of the problem of $m$ related machines in which the individual function of each task is the completion time of the machine on which it is executed, while the global objective function is the makespan. They consider the strong model of execution by assuming that each task may declare an arbitrary length (smaller or greater than its real length) while the load of each machine is the sum of the true lengths of the tasks assigned to it. They provide equilibria-truthful mechanisms that use payments in order to retain truthfulness. In [1], the authors consider a different variant with $m$ identical machines in which the individual objective function of each task is its completion time and they consider the strong model of execution (but here the tasks may only report values that are greater than or equal to their real lengths). Given that for this variant the SPT (Shortest Processing Time) algorithm[1] is truthful, they focus on the design of algorithms with better approximation ratio than that of the SPT algorithm. The rough idea of their approach is a randomized algorithm in which they combine the LPT (Longest Processing Time) algorithm[2], which has a better approximation ratio than SPT but is not truthful, with a schedule (DSPT) based on the SPT algorithm where some "unnecessary" idle times are introduced between the tasks. These unnecessary idle times are introduced in the SPT schedule in order to penalize more the tasks that report false information. Indeed, in the DSPT schedule such a task is doubly penalized, since not only is its execution delayed by the other tasks but also by the introduced idle times. In such a way, it is possible to find a probability distribution over the deterministic algorithms, LPT and DSPT which produces a randomized algorithm that is proved to be truthful and with an (expected) approximation ratio of $2 - \frac{1}{m+1}(\frac{5}{3} + \frac{1}{3m})$, i.e. better than the one of SPT which is equal to $2 - \frac{1}{m}$. An optimal truthful randomized algorithm and a truthful randomized PTAS for identical parallel machines in the weak model of execution appeared in [2]. The idea of these algorithms is to introduce fake tasks in order to have the same completion time in all the machines and then to use a random order in each machine for scheduling the tasks allocated to it (including the eventual fake one). These results have been also generalized in the case of related machines and the on-line case with release

---

[1] Where the tasks are scheduled greedily following the increasing order of their lengths (its approximation ratio is $2 - 1/m$).

[2] Where the tasks are scheduled greedily following the decreasing order of their lengths (its approximation ratio is $4/3 - 1/(3m)$).

dates. Another related work, presented in [5], gives some new lower and upper bounds. More precisely, the authors proved that there is no truthful deterministic (resp. randomized) algorithm with an approximation ratio smaller than $2 - 1/m$ (resp. $3/2 - 1/2m$) for the strong model of execution. They also provide a lower bound of 1.1 for the deterministic case in the weak model (for $m \geq 3$) and a deterministic $\frac{4}{3} - \frac{1}{3m}$ truthful algorithm based the idea of bloc schedule where after inserting fake tasks in order to have the same completion time in all the machines, instead of using a random order on the tasks of each machine, the authors proposed to take the mirror of the LPT schedule.

## 1.3 Our Contribution

In the first part of the paper we consider the strong model of execution. Our contribution is a new truthful randomized non-polynomial algorithm that we call Starting Time Equalizer (STE), presented in Section 2, whose approximation ratio for the makespan is $\frac{3}{2}$ for $P||C_{\max}$. This new upper bound asymptotically closes the gap between the lower bound $\frac{3}{2} - \frac{1}{2m}$ of [5] and the previously best known upper bound of $2 - \frac{1}{m+1}\left(\frac{5}{3} + \frac{1}{3m}\right)$ for this problem [1]. We also give two polynomial-time variants of Algorithm STE, respectively with approximation ratio $\frac{3}{2} + \epsilon$ for $Pm||C_{\max}$ and $\frac{11}{6} + \frac{1}{3m}$ for $P||C_{\max}$ (we underline that both $\frac{3}{2} + \epsilon$ and $\frac{11}{6} + \frac{1}{3m}$ are better than the previous upper bound of $2 - \frac{1}{m+1}\left(\frac{5}{3} + \frac{1}{3m}\right)$). In the second part of the paper, we consider the weak model of execution. We give in Section 3.1, a new truthful randomized non-polynomial algorithm, called Mid-Time Equalizer (MTE) for the off-line problem with release dates, where the private information of each task is not only each length, but also its release date ($P|r_i|C_{\max}$). Finally, we consider the case of scheduling a set of selfish tasks on a set of unrelated parallel machines ($R||C_{\max}$) for the weak model of execution (Section 3.2) where we propose a new truthful randomized non-polynomial algorithm that we call Completion Time Equalizer (CTE). Table 1 gives a summary of the upper and lower bounds on the approximation ratio of truthful algorithms for the considered problems (with † we give the results obtained in this paper).

**Table 1.** Bounds for $m$ parallel machines

| | Deterministic | | Randomized | |
| --- | --- | --- | --- | --- |
| | Lower bound | Upper bound | Lower bound | Upper bound |
| $P||C_{\max}$ strong model | $2 - \frac{1}{m}$ [5] | $2 - \frac{1}{m}$ [6] | $\frac{3}{2} - \frac{1}{2m}$ [5] | $\frac{3}{2}$ † |
| $P||C_{\max}$ weak model | if $m = 2$ then $1 + \frac{\sqrt{105}-9}{12} > 1.1$ if $m \geq 3$ then $\frac{7}{6} > 1.16$ [5] | $\frac{4}{3} - \frac{1}{3m}$ [5] | 1 [2] | 1 [2] |
| $R||C_{\max}$ weak model | | unknown | unknown | $\frac{3}{2}$ † |
| $P|r_i|C_{\max}$ weak model | | $2 - \frac{1}{m}$ [7] | | $\frac{3}{2}$ † |

The lower bounds for truthful deterministic algorithms in the weak model for $P|r_i|C_{\max}$ and $R||C_{\max}$ are simple implications of the lower bound for truthful deterministic algorithms solving $P||C_{\max}$. Up to our knowledge, there is no interesting lower bounds for truthful randomized algorithms (resp. upper bound for truthful deterministic algorithms) for $R||C_{\max}$ and $P|r_i|C_{\max}$ (resp. $R||C_{\max}$). The upper bound $2 - \frac{1}{m}$ for $P|r_i|C_{\max}$ in the weak model holds only if we consider that each task can identified by an identification number (ID). With this assumption, we just have to consider the on-line algorithm which schedules the tasks when they become available with (for instance) the smallest ID first. This algorithm is then trivially truthful, because task i will not have incentive of bidding $(b_i > t_i, r_i^b > r_i)$ ($b_i$ has no effect on the way in which tasks are scheduled and bidding $r_i^b > r_i$ can only increase $C_i$). Moreover, as this algorithm is a particular case of Graham's list scheduling (LS) algorithm with release dates, it is $(2 - \frac{1}{m})$-competitive (because Graham's LS algorithm is $(2 - \frac{1}{m})$-competitive for $P|on\text{-}line\text{-}list|C_{\max}$, [7]).

# 2   Strong Model of Execution

## 2.1   Identical Machines

### Algorithm STE

We consider in this section the problem with identical machines ($P||C_{\max}$) in the strong model. Every task $i$ has a private value $t_i$ that represents its length and it has to bid any value $b_i \geq t_i$.

---

**Algorithm:** Starting Time Equalizer (STE)

1. Let $C_{\max}^{OPT}$ be the makespan of an optimal schedule $OPT$ for $P||C_{\max}$.
   Let $OPT_j$ be the sub-schedule of $OPT$ on machine $j$.
   Let $b_{j_1} \leq \cdots \leq b_{j_k}$ be the bids (sorted by increasing order)
   of the $k$ tasks in $OPT_j$.

2. Construct schedule $S_1$ as follows: for every machine $j$ $(1 \leq j \leq m)$,
   every task $i$ $(j_1 \leq i \leq j_k)$ in $OPT_j$ is executed on machine $j$ by
   starting at time $\sum_{l=i+1}^{k} b_{j_l}$.

3. Construct schedule $S_2$ as follows: for every machine $j$ $(1 \leq j \leq m)$,
   every task $i$ $(j_1 \leq i \leq j_k)$ in $OPT_j$ is executed on machine $j$ by
   starting at time $C_{\max}^{OPT} - \sum_{l=i+1}^{k} b_{j_l}$.

4. Choose schedule $S_1$ or $S_2$ each with probability 1/2.

---

Figure 1 illustrates the construction of schedules $S_1$ and $S_2$ in algorithm STE on machine machine $j$.

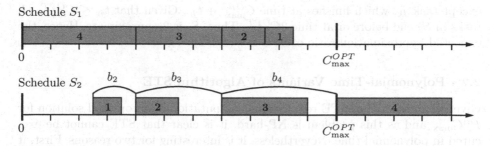

**Fig. 1.** An illustration of execution of Algorithm STE on machine $j$. We give an example of schedules $S_1$ and $S_2$ with four tasks in $OPT_j$ such that $b_{j_1} = 1$, $b_{j_2} = 1.5$, $b_{j_3} = 3$, $b_{j_4} = 4$ and $C_{\max}^{OPT} = 11$.

The main idea of the algorithm STE is to make equal the expected starting times of all the tasks. More precisely, we prove below that the expected starting time of every task in the final schedule constructed by STE, which is the average between its starting time in $S_1$ and its starting time in $S_2$, will be equal to $\frac{C_{\max}^{OPT}}{2}$ (i.e. the same value for every task). This property will be used in the proof of Theorem 1 to show that STE is truthful. In the example given in Figure 1, the expected starting time of the four tasks is $\frac{C_{\max}^{OPT}}{2}$ and it is equal to 5.5.

**Theorem 1.** *STE is a randomized, truthful and $\frac{3}{2}$-approximate algorithm in the strong model of execution for $P||C_{\max}$.*

*Proof.* As STE is a randomized algorithm, to prove it is truthful, we have to show that the expected completion time of each task is minimum when it tells the truth. By definition of STE, the expected completion time $C_i$ of any task $i$ is the average between its completion time in schedule $S_1$ and its completion time in schedule $S_2$. In the strong model of execution, every task $i$ is completed $t_i$ units of time after its starting time. Thus,

$$C_i = \frac{1}{2}\left(\left(t_i + \sum_{l=i+1}^{k} b_{j_l}\right) + \left(t_i + C_{\max}^{OPT} - \sum_{l=i+1}^{k} b_{j_l}\right)\right) = t_i + \frac{C_{\max}^{OPT}}{2}$$

For every task $i$, the completion time of task $i$ is $C_i = t_i + \frac{C_{\max}^{OPT}}{2}$ and it reaches its minimum value when $i$ tells the truth because $t_i$ does not depend on the bid $b_i$ and because $C_{\max}^{OPT}$ obviously does not decrease if $i$ bids $b_i > t_i$ instead of $b_i = t_i$. Thus, STE is truthful in the strong model of execution. Given that STE is truthful, we may consider in the following that for every $i$, we have $b_i = t_i$. Given also that STE is a randomized algorithm choosing with probability $1/2$ schedule $S_1$ and with probability $1/2$ schedule $S_2$, its approximation ratio will be the average between the approximation ratios of schedules $S_1$ and $S_2$. In $S_1$, all tasks end before or at time $C_{\max}^{OPT}$. Thus, as for every $i$, $b_i = t_i$, $C_{\max}^{OPT}$ is the makespan of an optimal solution computed with the true types of the agents, $S_1$ is optimal. In $S_2$, on every machine $j$, all tasks end before or at time $C_{\max}^{OPT}$

except task $j_k$, which finishes at time $C_{\max}^{OPT} + t_{j_k}$. Given that $t_{j_k} \leq C_{\max}^{OPT}$, all tasks in $S_2$ end before or at time $2C_{\max}^{OPT}$. Thus, $S_2$ is 2-approximate. Hence, the expected approximation ratio of STE is $\frac{1}{2}(1+2) = \frac{3}{2}$.     □

## 2.2  Polynomial-Time Variants of Algorithm STE

Given that Algorithm STE requires the computation of an optimal solution for $P||C_{\max}$ and as this problem is NP-hard, it is clear that STE cannot be executed in polynomial time. Nevertheless, it is interesting for two reasons. First, it asymptotically closes the gap between the lower bound $\frac{3}{2} - \frac{1}{2m}$ of any truthful algorithm and the previously best known upper bound of $2 - \frac{1}{m+1}\left(\frac{5}{3} + \frac{1}{3m}\right)$. Secondly, by using approximated solutions instead of the optimal one, we can obtain polynomial-time variants of STE. To precise these variants, we first need to define what we call an *increasing* algorithm.

**Definition (Increasing algorithm).** *Let $H$ and $H'$ be two sets of tasks $\{T_1, T_2, \ldots, T_n\}$ and $\{T_1', T_2', \ldots, T_n'\}$ respectively. We denote by $H \leq H'$ the fact that for every $1 \leq i \leq n$, we have $l(T_i) \leq l(T_i')$ (where $l(T)$ is the length of task $T$). An algorithm $A$ is increasing if for every pair of sets of tasks $H$ and $H'$ such that $H \leq H'$, it constructs schedules such that $C_{\max}(H) \leq C_{\max}(H')$ (where $C_{\max}(X)$ is the makespan of the solution constructed by Algorithm $A$ for the set of tasks $X$).*

As LPT (Longest Processing Time) is an increasing algorithm (See [2]) and as there exists an increasing PTAS for $Pm||C_{\max}$ (See [2]), we get the following two theorems.

**Theorem 2.** *By using LPT instead of an optimal algorithm, we obtain a polynomial-time, randomized, truthful and $(\frac{11}{6} - \frac{1}{3m})$-approximate variant of STE in the strong model of execution for $P||C_{\max}$.*

**Theorem 3.** *By using the increasing PTAS in [2] instead of an optimal algorithm, we obtain a polynomial-time, randomized, truthful and $(\frac{3}{2}+\epsilon)$-approximate variant of STE in the strong model of execution for $Pm||C_{\max}$.*

Theorem 2 (resp. Theorem 3) can be proved in a similar way as in Theorem 1. Indeed, as the completion time of each task will be $C_i = t_i + \frac{C_{\max}^{LPT}}{2}$ (resp. $C_i = t_i + \frac{C_{\max}^{PTAS}}{2}$) instead of $C_i = t_i + \frac{C_{\max}^{OPT}}{2}$ and as LPT (resp. the PTAS in [2]) is increasing, the variant of STE in Theorem 2 (resp. Theorem 3) is truthful. Moreover, as LPT is $(\frac{4}{3} - \frac{1}{3m})$-approximate for $P||C_{\max}$ (resp. the PTAS in [2] is $(1 + \epsilon)$-approximate for $Pm||C_{\max}$), we obtain that the expected approximation ratio of the variant of STE in Theorem 2 (resp. Theorem 3) is $\frac{1}{2}(\frac{4}{3} - \frac{1}{3m} + \frac{4}{3} - \frac{1}{3m} + 1) = \frac{11}{6} - \frac{1}{3m}$ (resp. $\frac{1}{2}(1 + \epsilon + 1 + \epsilon + 1) = \frac{3}{2} + \epsilon$).

## 3  Weak Model of Execution

### 3.1  Identical Machines with Release Dates

We consider in this section $P|r_i|C_{\max}$ in the weak model. Every task $i$ has now a private pair $(t_i, r_i)$ (its type), where $t_i$ is the length of task $i$ and $r_i$ its release date. Each task $i$ may bid any pair $(b_i, r_i^b)$ such that $b_i \geq t_i$ and $r_i^b \geq r_i$. Notice here that we consider that task $i$ may not bid a release date smaller than its real release date i.e. $r_i^b < r_i$, because otherwise, the task may be scheduled before $r_i$ in the final schedule and thus, the final schedule may be infeasible.

---

**Algorithm:** Mid-Time Equalizer (MTE)

1. Let $C_{\max}^{OPT}$ be the makespan of an optimal schedule $OPT$ for $P|r_i|C_{\max}$.
   Let $m_i$ be the machine where Task $i$ is executed in $OPT$.
   Let $C_i(OPT)$ be the completion time of Task $i$ in $OPT$.

2. Construct Schedule $OPT^{mirror}$ in which every task $i$ is executed on machine $m_i$ and start at Time $\max_{1 \leq j \leq n}\{r_j^b\} + C_{\max}^{OPT} - C_i(OPT)$.

3. Choose Schedule $OPT$ or $OPT^{mirror}$ each with probability $1/2$.

---

Figure 2 illustrates the construction of Schedules $OPT$ and $OPT^{mirror}$ in algorithm MTE on any machine $m_i$.

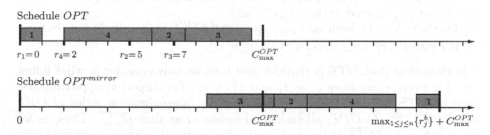

**Fig. 2.** An illustration of execution of algorithm MTE on machine $m_i$. We give an example of schedules $OPT$ and $OPT^{mirror}$ with four tasks on machine $m_i$ such that $(b_1 = 1, r_1 = 0)$, $(b_2 = 1.5, r_2 = 5)$, $(b_3 = 3, r_3 = 7)$, $(b_4 = 4, r_4 = 2)$, $\max_{1 \leq j \leq n}\{r_j^b\} = 8$ and $C_{\max}^{OPT} = 11$.

The main idea of algorithm Mid-Time Equalizer (MTE) is make equal the expected time at which every task has executed half of its total length. More precisely, we prove below that the expected mid-time of every task in the final schedule constructed by MTE is the average between its mid-time in $OPT$ and in $OPT^{mirror}$ and it is equal to $\frac{1}{2}\left(\max_{1 \leq j \leq n}\{r_j^b\} + C_{\max}^{OPT}\right)$ (i.e. the same value for every task). This property will be used in the proof of Theorem 4 in order to show that MTE is truthful in the weak model of execution. In the example given in

Figure 2, the expected mid-time of the four tasks is $\frac{1}{2} \left( \max_{1 \leq j \leq n} \{r_j^b\} + C_{\max}^{OPT} \right)$ and it is equal to $\frac{8+11}{2} = 9.5$.

Note that as we consider that for every $i$, we have $r_i^b \geq r_i$, we get $\max_{1 \leq j \leq n} \{r_i^b\} \geq \max_{1 \leq j \leq n} \{r_j\}$. Moreover, as $C_i(OPT) \leq C_{\max}^{OPT}$, every task $i$ starts in schedule $OPT^{mirror}$ at time $\max_{1 \leq j \leq n} \{r_j^b\} + C_{\max}^{OPT} - C_i(OPT) \geq \max_{1 \leq j \leq n} \{r_j\} \geq r_i$. Thus, schedule $OPT^{mirror}$ respects all the constraints of the release dates.

**Theorem 4.** *MTE is a randomized, truthful and $\frac{3}{2}$-approximate algorithm in the weak model of execution for $P|r_i|C_{\max}$.*

*Proof.* Let us prove that the expected completion time of every task is minimum when it tells the truth. By definition of MTE, the expected completion time $C_i$ of any task $i$ is the average between its completion time $C_i(OPT)$ in schedule $OPT$ and its completion time $C_i(OPT^{mirror})$ in schedule $OPT^{mirror}$. In the weak model of execution, every task $i$ is completed $b_i$ units of time after its starting time. Thus, we have

$$
\begin{aligned}
C_i &= \frac{1}{2} \left( C_i(OPT) + \max_{1 \leq j \leq n} \{r_j^b\} + C_{\max}^{OPT} - C_i(OPT) + b_i \right) \\
&= \frac{1}{2} \left( \max_{1 \leq j \leq n} \{r_j^b\} + C_{\max}^{OPT} + b_i \right)
\end{aligned}
$$

For every task $i$, its completion time $C_i = \frac{1}{2} \left( \max_{1 \leq j \leq n} \{r_j^b\} + C_{\max}^{OPT} + b_i \right)$ reaches its minimum value when $i$ tells the truth (i.e. when $i$ bids simultaneously $b_i = t_i$ and $r_i^b = r_i$), because

- for every $r_i^b \geq r_i$, both $C_{\max}^{OPT}$ and $b_i$ obviously do not decrease if $i$ bids $(b_i > t_i, r_i^b)$ instead of $(b_i = t_i, r_i^b)$, and
- for every $b_i \geq t_i$, both $\max_{1 \leq j \leq n} \{r_j^b\}$ and $C_{\max}^{OPT}$ obviously do not decrease if $i$ bids $(b_i, r_i^b > r_i)$ instead of $(b_i, r_i^b = r_i)$.

It is then clear that MTE is truthful and thus we may consider in what follow that for every $i$, we have $b_i = t_i$ and $r_i^b = r_i$. The expected approximation ratio of MTE will be the average between the approximation ratios of $OPT$ and $OPT^{mirror}$. In $OPT$, all tasks end before or at time $C_{\max}^{OPT}$. Thus, as for every $i$, $b_i = t_i$, $C_{\max}^{OPT}$ is the makespan of an optimal solution computed with the types of the agents, and thus, $OPT$ is optimal. In $OPT^{mirror}$, all tasks end before or at time $\max_{1 \leq j \leq n} \{r_j\} + C_{\max}^{OPT}$ (because for every $i$, $r_i^b = r_i$ by definition of MTE). Given that $\max_{1 \leq j \leq n} \{r_j\} \leq C_{\max}^{OPT}$, all tasks in $OPT^{mirror}$ terminate before or at time $2C_{\max}^{OPT}$. Thus, $OPT^{mirror}$ is 2-approximate. Hence the expected approximation ratio of Algorithm MTE is $\frac{1}{2}(1 + 2) = \frac{3}{2}$. □

## 3.2   Unrelated Machines

We consider in this section the case with unrelated machines ($R||C_{\max}$) in the weak model of execution. Here, every task $i$ has a private vector $(t_i^1, \ldots, t_i^m)$ (his type), where $t_i^j$ ($1 \leq j \leq m$) is the processing time of $i$ if it is executed on machine $j$. Every task $i$ bids any vector $(b_i^1, \ldots, b_i^m)$ with $b_i^1 \geq t_i^1, \ldots, b_i^m \geq t_i^m$.

---

**Algorithm:** Completion Time Equalizer (CTE)

1. Let $C_{\max}^{OPT}$ be the makespan of an optimal schedule $OPT$ for $R||C_{\max}$.
   Let $OPT_j$ be the sub-schedule of $OPT$ on Machine $j$.
   Let $b_{j_1}^j \le \cdots \le b_{j_k}^j$ be the bids (sorted by increasing order)
   of the $k$ tasks in $OPT_j$.

2. Construct schedule $S_1$ as follows: for every machine $j$ ($i \le j \le m$),
   every task $i$ ($j_1 \le i \le j_k$) in $OPT_j$ is executed on machine $j$ by
   starting at time $C_{\max}^{OPT} - \sum_{l=i}^k b_{j_l}^j$.

3. Construct schedule $S_2$ as follows: for every machine $j$ ($i \le j \le m$),
   every task $i$ ($j_1 \le i \le j_k$) in $OPT_j$ is executed on machine $j$ by
   starting at time $C_{\max}^{OPT} - b_i^j + \sum_{l=i+1}^k b_{j_l}^j$.

4. Choose schedule $S_1$ or $S_2$ each one with probability $1/2$.

---

Figure 3 illustrates the construction of schedules $S_1$ and $S_2$ in algorithm CTE on machine $j$.

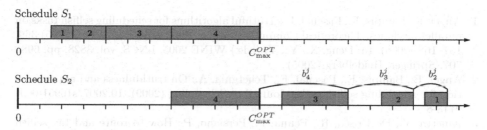

**Fig. 3.** An illustration of execution of algorithm CTE on machine $j$. An example of schedules $S_1$ and $S_2$ is given with four tasks in $OPT_j$ such that $b_{j_1}^j = 1$, $b_{j_2}^j = 1.5$, $b_{j_3}^j = 3$, $b_{j_4}^j = 4$ and $C_{\max}^{OPT} = 11$.

The intuitive idea of algorithm Completion Time Equalizer is to make equal the expected completion times of the tasks. More precisely, the expected completion time of every task in the final schedule constructed by CTE is the average between its starting time in $S_1$ and its starting time in $S_2$ and it is equal to $C_{\max}^{OPT}$ (i.e. the same for all the tasks). This property will be used in the proof of Theorem 1 to show that CTE is truthful in the weak model of execution. For instance, in the example given in Figure 1, the expected completion time of the four tasks is $C_{\max}^{OPT}$ and it is equal to 11.

**Theorem 5.** *CTE is a randomized, truthful and $\frac{3}{2}$-approximate algorithm in the weak model of execution for $R||C_{\max}$.*

*Proof.* We first show that the expected completion time of each task is minimum when it tells the truth. By definition of CTE, the expected completion time $C_i$

of any task $i$ is the average between its completion time in Schedule $S_1$ and its completion time in Schedule $S_2$. In the weak model of execution, each task $i$ is completed $b_i$ units of time after its starting time on machine $j$. Thus, we have

$$C_i = \frac{1}{2}\left(\left(b_i^j + C_{\max}^{OPT} - \sum_{l=i}^{k} b_{j_l}^j\right) + \left(b_i^j + C_{\max}^{OPT} - b_i^j + \sum_{l=i+1}^{k} b_{j_l}^j\right)\right) = C_{\max}^{OPT}$$

For every task $i$, $C_i = C_{\max}^{OPT}$ reaches its minimum value when $i$ tells the truth because $C_{\max}^{OPT}$ obviously does not decrease if for any $i,j$, task $i$ bids $b_i^j > t_i^j$ instead of $b_i^j = t_i^j$. Hence, CTE is truthful and so we can consider in the following that for every $i,j$, we have $b_i^j = t_i^j$. In schedule $S_1$, all tasks finish before or at time $C_{\max}^{OPT}$. Thus, as for every $i,j$, $b_i^j = t_i^j$, $C_{\max}^{OPT}$ is the makespan of an optimal solution computed with the types of the agents, $S_1$ is optimal. In $S_2$, on each machine $j$, all tasks end before or at time $C_{\max}^{OPT} + \sum_{l=2}^{k} b_{j_l}^j$. As $\sum_{l=2}^{k} b_{j_l}^j \leq C_{\max}^{OPT}$, all tasks in $S_2$ end before or at time $2C_{\max}^{OPT}$. Thus, $S_2$ is 2-approximate. Finally, the expected approximation ratio of algorithm CTE is $\frac{1}{2}(1+2) = \frac{3}{2}$. □

# References

1. Angel, E., Bampis, E., Pascual, F.: Truthful algorithms for scheduling selfish tasks on parallel machines. Theoretical Computer Science (short version in WINE 2005) 369, 157–168 (2006); In: Deng, X., Ye, Y. (eds.) WINE 2005. LNCS, vol. 3828, pp. 698–707. Springer, Heidelberg (2005)
2. Angel, E., Bampis, E., Pascual, F., Tchetgnia, A.: On truthfulness and approximation for scheduling selfish tasks. Journal of Scheduling (2009), 10.2007/s10951-009-0118-8
3. Auletta, V., De Prisco, R., Penna, P., Persiano, P.: How to route and tax selfish unsplittable traffic. In: 16th ACM Symposium on Parallelism in Algorithms and Architectures, pp. 196–204 (June 2004)
4. Mueller, R., Heydenreich, B., Uetz, M.: Games and mechanism design in machine scheduling - an introduction. Research Memoranda 022, Maastricht: METEOR, Maastricht Research School of Economics of Technology and Organization (2006)
5. Christodoulou, G., Gourvès, L., Pascual, F.: Scheduling selfish tasks: About the performance of truthful algorithms. In: Lin, G. (ed.) COCOON 2007. LNCS, vol. 4598, pp. 187–197. Springer, Heidelberg (2007)
6. Christodoulou, G., Koutsoupias, E., Nanavati, A.: Coordination mechanisms. In: Díaz, J., Karhumäki, J., Lepistö, A., Sannella, D. (eds.) ICALP 2004. LNCS, vol. 3142, pp. 345–357. Springer, Heidelberg (2004)
7. Graham, R.L.: Bounds for certain multiprocessing anomalies. Bell System Tech. 45, 1563 (1966)

# Almost Linear Time Computation of the Chromatic Polynomial of a Graph of Bounded Tree-Width

Martin Fürer*

Department of Computer Science and Engineering
Pennsylvania State University
University Park, PA 16802, USA
furer@cse.psu.edu
http://cse.psu.edu/~furer

**Abstract.** An $O(n \log^2 n)$ algorithm is presented to compute all coefficients of the chromatic polynomial of an $n$ vertex graph of bounded tree-width. Previously, it has been known how to evaluate the chromatic polynomial for such graphs in linear time, implying a computation of all coefficients of the chromatic polynomial in quadratic time.

**Keywords:** chromatic polynomial, counting colorings, bounded tree-width, efficient algorithms.

## 1 Introduction

The chromatic polynomial introduced by Birkhoff [1] is a well studied object in algebraic graph theory. It is a graph invariant, i.e., isomorphic graphs have the same chromatic polynomial. The value of the chromatic polynomial at a positive integer $s$ is the number of colorings with at most $s$ colors. It is not immediately obvious that such a polynomial exists, i.e., that the function with these values is a polynomial.

All graphs considered here are undirected, without multiple edges, and loop-free. Let

$$t_{(r)} = t(t-1)(t-2)\ldots(t-r+1).$$

The chromatic polynomial (see [2]) of a graph $G = (V, E)$ with $n = |V|$ vertices is defined as

$$P(G, t) = \sum_{r=1}^{n} p_r(G)\, t_{(r)}$$

where $p_r(G)$ is the number of color-partitions of $V$ into $r$ blocks, i.e., the number of partitions of $V$ into $r$ non-empty independent sets.

It is not hard to see that for any positive integer $s$, the value $P(G, s)$ is indeed the number of distinct proper colorings of $G$ with colors chosen from a set of

---

* Research supported in part by NSF Grant CCF-0728921.

A. López-Ortiz (Ed.): LATIN 2010, LNCS 6034, pp. 49–59, 2010.
© Springer-Verlag Berlin Heidelberg 2010

size $s$. This implies immediately, that the chromatic polynomial of a disconnected graph is the product of the chromatic polynomials of its connected components.

Two other cases are known to lead to a significant reduction in the time to compute the chromatic polynomial [2, pp. 66–68]. If a graph decomposes into different pieces that are either completely coupled or basically independent of each other, then it is easy to compute the chromatic polynomial of the graph from the chromatic polynomials of its pieces.

The first case occurs when $G$ is obtained from the disjoint graphs $G_1$ and $G_2$ by connecting every vertex of $G_1$ to every vertex of $G_2$ by an edge. Then every color partition of $G$ with $\ell$ color classes is composed of a color partition of $G_1$ with some number $i$ of color classes and a color partition of $G_2$ with $\ell - i$ color classes.

A trivial example of this first case is the complete graph $K_n$. There is just one color partition, and the chromatic polynomial is $t_{(n)}$.

In the second case, the intersection of two graphs $G_1$ and $G_2$ is a clique $K_\ell$, then

$$P(G,t) = \frac{1}{t_{(\ell)}} P(G_1,t) P(G_2,t).$$

A trivial example allowing the repeated application of this rule is a tree. But for a tree it is anyway easy to see that the chromatic polynomial has the simple form

$$P(T,t) = t(t-1)^{n-1}.$$

We will study a much more general situation, where the intersection of $G_1$ and $G_2$ is not restricted to be a clique, but is allowed to be any small graph.

If we are willing to move slowly and deal with an exponential explosion in the number of subgraphs to handle, there is an elegant and interesting method available. For $G = (V,E)$ and $e = \{u,v\} \in E$, define

$$G \setminus e = (V, E \setminus \{e\})$$

and

$$G/e = (V \setminus \{v\}, E \cup \{\{u,w\} : \{v,w\} \in E\} \setminus \{\{v,w\} : \{v,w\} \in E\},$$

i.e., $G \setminus e$ is obtained from $G$ by deleting the edge $e$, while $G/e$ is obtained from $G$ by identifying $u$ and $v$. Then the chromatic polynomial satisfies the simple recurrence equation

$$P(G,t) = P(G \setminus e, t) - P(G/e, t). \tag{1}$$

This is obvious from the two cases of $u$ and $v$ either having the same or different colors.

The recurrence equation immediately produces an exponential time algorithm to compute the chromatic polynomial by observing that

$$P((V, \emptyset), t) = t^n. \tag{2}$$

For graphs of bounded tree-width, we suggest not to use this algorithm. Its running time is at most of order $\phi^{|V|+|E|}$ up to polynomial factors, with $\phi$ being the golden ratio. In general, we might not expect a much better running time, because $P(G,t)$ is #P-hard for all integers $t \geq 3$. But even in the trivial case of a tree, this algorithm uses exponential time, unless the selected edges are incident or close to leaves, and the product rule for disconnected graphs is used.

Many graph polynomials have linear time algorithms for graphs of bounded tree-width. Examples are the interlace polynomials [3,4,5], as well as the MSOL-Farrell (monadic second-order logic) polynomials over a ring of unit cost [6]. For an overview of graph polynomials and polynomial time evaluation methods for graphs of bounded tree-width resulting from definability in second order logic, see the unifying article of [4].

The chromatic polynomial can be viewed as a special case of the famous Tutte polynomial, because

$$P(G,t) = (-1)^{r(E)} T(G, 1-t, 0),$$

where the rank $r(E)$ is the difference between the number of vertices and the number of connected components.

Based on this connection, the algorithms of Andrzejak [7] and Noble [8] for computing the Tutte polynomial also provide algorithms for the chromatic polynomial. Andrzejak's algorithm runs in polynomial time for graphs of bounded tree-width $k$, with the exponent being a function of $k$. Noble's algorithm runs in time $f(k)\, n$ for computing one value of the polynomial. Thus the algorithm is linear in $n$ and shows the problem to be fixed parameter tractable (FPT). The dependence of $k$ is double exponential. Makowsky et al. [9] have a polynomial time algorithm for the chromatic polynomial for the more general case of fixed clique-width.

Using the method described below in this paper, also a simple dynamic programming approach can be used to evaluate the chromatic polynomial in linear time for graphs of bounded tree-width. As an immediate application of any linear time evaluation algorithm, all coefficients of the chromatic polynomial can be computed in quadratic time for graphs of bounded tree-width. One way to obtain this corollary is to evaluate the polynomial at $n + 1$ locations and interpolate, in particular, one could evaluate at roots of unity and interpolate with an inverse FFT (fast Fourier transform). Another more direct way is to mimic the computation of a value of the chromatic polynomial. Noticing that each intermediate value is a value of some polynomial, one could always compute all coefficients of such a polynomial instead of just one value.

Until recently, no method has been known to improve the worst case complexity of such a computation below quadratic in general. Nevertheless, if a tree decomposition of a graph is nicely balanced, then the dynamic programming algorithm (which handles the subtrees of the children before the subtree of the parent) can actually be made significantly faster.

The situation is much simpler, but similar, in the case of computing the characteristic polynomial of a tree. Naturally, unlike the chromatic polynomial, the characteristic polynomial is always computable in polynomial time. On the other hand, even the characteristic polynomial of a tree is non-trivial. Traditional methods allow for the evaluation of the characteristic polynomial in linear time, and thus the computation of all its coefficients in quadratic time. A balanced tree is handled by a standard dynamic programming algorithm with fast multiplication in time $O(n \log^2 n)$. Only recently has a method been developed to compute the characteristic polynomial of any tree in time $O(n \log^2 n)$ [10].

We will apply this method for the more difficult situation of the chromatic polynomial for graphs of bounded tree-width. The method combines the advantages of the top-down recursive decomposition and the bottom-up dynamic programming. Dynamic programming fails to be efficient when the tree is not balanced. Recursive decomposition has more shortcomings. It has the advantage of being able to pick a central node as the root, implying a nice splitting of the tree. But in its traditional application, it makes many recursive calls resulting in an exponential time algorithm.

We overcome this problem, by grouping together several similar recursive calls. Each recursive call has to handle not just one, but a finite number of similar smaller subproblems. When the subproblems are split again in a balanced way by choosing a central root in each subtree, we have to deal with the difficulty of obtaining (after some recursive calls) some subtrees with an unbounded number of cases.

The problem has a surprisingly simple solution. It suffices to alternate between splitting to evenly partition the tree and splitting to bound the number of cases associated with any recursive call on a subtree.

## 2  Preliminaries

**Definition 1.** *A* tree decomposition *of a graph* $G = (V, E)$ *is a pair* $(\{X_i : i \in I\}, T)$, *where* $T = (I, F)$ *is a tree and each node* $i \in I$ *has a subset* $X_i \subseteq V$ *of vertices (called the bag of* $i$*) associated to it with the following properties.*

1. $\bigcup_{i \in I} X_i = V$, *i.e., each vertex belongs to at least one bag.*
2. *For all edges* $e = \{u, v\} \in E$ *there is at least one* $i \in I$ *with* $\{u, v\} \subseteq X_i$, *i.e., each edge is represented by at least one bag.*
3. *For every vertex* $v \in V$, *the set of indices* $i$ *of bags containing* $v$ *induces a subtree of* $T$ *(i.e., a connected subgraph).*

*If* $G$ *has a tree decomposition with largest bag size* $k + 1$, *then* $G$ *is said to have* tree-width *at most* $k$.

For every fixed $k$, there is a linear time algorithm deciding whether the tree-width is at most $k$, and if that is the case, producing a corresponding tree decomposition [11].

An [s]-*coloring* is a proper coloring (i.e., the endpoints of each edge have different colors) with the colors $[s] = \{1, 2, \ldots, s\}$.

We assume the vertices of our graph $G = (V, E)$ are enumerated. This defines a standard enumeration of the blocks of any partition of $V$ by the lowest numbered vertex in each block.

**Definition 2.** *The* standard coloring *corresponding to a partition of the vertex set $U \subseteq V$ into independent sets $U_1, U_2, \ldots$ is the coloring where the $j$th block $V_j$ obtains color $j$.*

## 2.1 Notation

For a partition $p$, we denote the number of blocks by $|p|$. For a coloring $\gamma$, we denote the number of colors by $|\gamma|$.

We use the notation $t_{(r)} = t(t-1) \ldots (t-r+1)$ and more general $p(t)_{(r)} = p(t)(p(t)-1) \ldots (p(t)-r+1)$ for any polynomial $p(t)$.

## 3 The Restricted Chromatic Polynomial

Besides the chromatic polynomial $P(G, t)$ whose value for any positive integer $s$ is the number of colorings of $G$ with at most $s$ given colors, we define the restricted chromatic polynomial counting the number of colorings of $G$ with at most $s$ given colors agreeing with a given coloring $\gamma$ on some subset $U$ of $V$.

**Definition 3.** *For a graph $G = (V, E)$, a vertex set $U \subseteq V$ and a standard coloring $\gamma$ of $U$, let $p_r(G, U, \gamma)$ be the number of those partitions of $V$ into $r$ independent sets of $G$ which agree with the color partition of $U$ produced by $\gamma$. Then the $\gamma$-restricted chromatic polynomial $P(G, U, \gamma, t)$ is defined by $P(G, U, \gamma, s)$ being (for all positive integers $s$) the number of colorings with colors from $[s] = \{1, \ldots, s\}$ agreeing with $\gamma$ on $U$.*

The following characterization of $P(G, U, \gamma, t)$ is an immediate consequence.

**Proposition 1.** $P(G, U, \gamma, t)$ *is a polynomial and*

$$P(G, U, \gamma, t) = \frac{1}{t_{(|\gamma|)}} \sum_{r=1}^{n} p_r(G, U, \gamma) t_{(r)}$$

$$= \sum_{r=1}^{n} p_r(G, U, \gamma) t_{(r-|\gamma|)}.$$

*Proof.* $\sum_{r=1}^{n} p_r(G, U, \gamma) s_{(r)}$ is the number of $[s]$-colorings of $G$ agreeing with the partition produced by the coloring $\gamma$ on $U$. Requiring the colors to agree with the standard coloring $\gamma$ on $U$ (and not just with the partition it causes on $U$) reduces the number of colorings by a factor of $s_{|\gamma|}$. □

$P(G, U, \gamma, t)$ is a polynomial, i.e., the denominator $(|\gamma|)$ in the Proposition always cancels. This division should always be done before any evaluations to avoid zero division.

Naturally, for $U$ being the empty set and $\gamma$ the empty coloring of $U$, we obtain the usual chromatic polynomial

$$P(G, t) = P(G, \emptyset, \emptyset, t).$$

# 4  The Algorithm

The essence of our algorithm is the recursive procedure Restricted-Coloring, which is called from the algorithm Chromatic-Polynomial. The auxiliary procedure Select-Root selects the tree node where the next split occurs.

An obvious initial idea to obtain an efficient algorithm is to evenly split the tree which defines the tree decomposition. A natural choice is to pick a center, i.e., the single node or one of the two nodes splitting the tree into a collection of subtrees with each subtree of the collection having at most one more than half the number of nodes of the given tree. We include the center itself in all of the subtrees.

This idea is not sufficient for our purposes, because we also have a distinguished subset $J$ of the leaves where we did previous splits. In our case, the set $J$ defines an exponential (in $|J|$) number of task to be accomplished in each of the subtrees split off. In order to bound the size of $J$ with which the procedure Restricted-Colorings is called (by the constant 3), we sometimes split not with the objective of splitting the set of all nodes evenly, but of splitting its subset $J$ evenly. For $|J| = 3$, let Splitter$(T, J)$ be the unique node of $T$ splitting the set $J$ into 3 parts. During the procedure Restricted-Colorings, $|J|$ can reach its maximum of 4 temporarily.

## Procedure Select-Root:

**Input:** A tree $T = (I, F)$ and a subset $J \subseteq I$ of at most 3 leaves.
**Output:** A node $i$ of $T$, which will be viewed as the root of $T$.
**Comment:** The first objective is to keep the intersection of $J$ small with each of the subtrees obtained by splitting the tree $T$ at $i$, even as $i$ is added to $J$. The second objective is to split the tree $T$ evenly.


   **if** $|J| = 3$ **then**
      Return Splitter$(T, J)$
   **else**
      Return Center$(T)$

**Fig. 1.** The procedure Select-Root

## Algorithm Chromatic-Polynomial:

**Input:** A graph $G = (V, E)$ with $|V| = n$ together with
   a tree decomposition $(\{X_i : i \in I\}, T = (I, F)$ of tree-width $k$.
**Output:** The coefficients $c_0, c_1, \ldots, c_n$
   of the chromatic polynomial $P(G, t) = \sum_{r=1}^{n} c_r t^r$.
   $J = \emptyset$
   $(c_0, \ldots, c_n) = $ Coefficients of Restricted-Colorings$(G, T, J)$
   Return $(c_0, \ldots, c_n)$

**Fig. 2.** The algorithm Chromatic-Polynomial

# 5    Time Complexity

As usual for such computations, we use the algebraic computation model. Each arithmetic operation is counted as one step. All binary numbers involved are at most of linear length in the number of vertices $n$. Alternatively, one could count bit operations. Then the improvement obtained by our new method are even more significant, as we would not only do more work with smaller degree polynomials, but also with smaller coefficients. We only analyze the dependence on $n$, assuming the tree-width $k$ is a constant.

Analyzing the procedure Restricted-Colorings, the first thing to note is that the sizes of $J$ and $U$ are under control. Initially $J$ and $U$ are empty. The size of $U$ is always bounded by the number of vertices in the bags of $J$, which is at most $(k+1)|J|$. The size of $J$ is at most 3 when the procedure Restricted-Colorings is called. The size of $J$ increases by 1 during the call to at most 4. All recursive calls are made with at most the current value of $J$. But if this value is 4, then three recursive calls are made with $|J| = 2$ and possibly some more with $|J| = 1$. The number of recursive calls is determined by the degree $d$ of the splitting node Splitter($T, J$).

It is important to notice that if an instantiation of the procedure Restricted-Colorings starts with $|J| = 3$, then the tree is typically not split at a center, but $|J| < 3$ for all the recursive calls. Thus every uneven split is immediately followed by even splits. $J$ never reaches size 4 twice in a row.

Let $m$ be the number of edges in the tree $T$ (not in the graph $G$). Assume $m \geq 1$, as the one vertex case is trivial and does not show up during recursive calls.

**Lemma 1.** *For $m \geq 1$ and suitable constants $c$, $c'$, and $c''$, the running time of the procedure Restricted-Colorings is at most $c\, m \lg^2 m + c'' m$ for $|J| < 3$ and at most $c\, m \lg^2 m + c'\, m \lg m + c'' m$ for $|J| = 3$.*

*Proof.* The lemma trivially holds for $m = 1$. Let $m \geq 2$ and assume the lemma is true for all trees with less than $m$ edges.

Recall that the procedure Restricted-Colorings partitions the tree $T$ edge-wise into trees $T_1, \ldots, T_d$ with $T_i = (I(i), F(i))$. Let $|F(i)| = m_i$, and for $|J| < 3$, the sizes $m_i$ are bounded by $m/2 + 1$ for all $i$. After the recursive calls for these trees $T_i$ with common root $v$, the procedure repeatedly merges pairs of approximately smallest trees into single trees until there is just one tree left, i.e., $T$ has been reassembled. Obviously, there is at most one tree $T'$ among the trees $T_i$ with three nodes contributing to its set of special vertices $U_i$. Let $m'$ be the number of edges of $T'$. Let $m_{\min} = \min_i m_i$.

**Claim:** If after some sequence of merges of pairs of trees, we have the trees $T_1, \ldots, T_{d'}$, then the time spent for the recursive calls and the merges together has been at most

$$t(d') = c \sum_{i=1}^{d'} m_i \lg^2 m_i + c'\, m_{\min} \lg m_{\min} + b\, m' \lg m' + c'' m \qquad (3)$$

## Procedure Restricted-Colorings:

**Input:** A graph $G = (V, E)$ with one of its tree decompositions $T = (I, F)$ and a subset $J \subseteq I$ of size at most 3. Let $U$ be the set of vertices in the bags of the nodes of $J$. Let $n = |V|$ and $n' = |V \setminus U|$.

**Output:** The function $\mathbf{f}$ from the set of standard colorings of $U$ into the polynomials $\mathbb{Z}[t]$ where for every standard coloring $\gamma$, $\mathbf{f}(\gamma)$ is the polynomial $P(G, U, \gamma, t)$ of degree $n'$ with $P(G, U, \gamma, s)$ being the number of $[s]$-colorings of $G$ agreeing with $\gamma$ on the subset $U$ of vertices for every positive integer $s$.

**Comment:** This procedure is only called for sets $J$ of size at most 3. Thus the size of $U$ is initially bounded by $3(k + 1)$, where $k$ is the width of the tree decomoposition. During the procedure $|J|$ increases by 1 and thus $|U|$ is always bounded by the constant $4(k + 1)$.

**if** $U = V$, i.e., all nodes of the tree $T$ are already selected **then**
    $f$ is the function with the constant value 1 for every standard coloring $\gamma$.
    Return $\mathbf{f}$
$v = $ Select-Root$(T, J)$
Consider $v$ to be the root of $T$, and let $d$ be the degree of $v$.
Let $v_1, \ldots, v_d$ be the neighbors of the root $v$.
**for** $i = 1, \ldots, d$, let $T_i = (I(i), F(i))$ be the subtree of $T$ induced by all the nodes
    reachable from $v_i$ without going through $v$ as an intermediate node.
    Let $V(I')$ be the vertices associated with the nodes in $I' \subseteq I$.
    // Thus the sets $V(I(i)) \setminus V(\{v\})$ form a partition of $V \setminus V(\{v\})$
    // and $V(\{v\}) \subseteq V(I(i))$ for all $i$.
$J = J \cup \{v\}$
$U = V(J)$
**for** $i = 1$ **to** $d$ **do**
    $\mathbf{f}_i = $ Restricted-Colorings$(G[V(I(i))], T_i, J \cap V(I(i)))$
$S = \{T_1, \ldots, T_d\}$
**while** $|S| > 1$ **do**
    Let $T_i$ and $T_j$ be two approximately smallest trees in $S$ of sizes $n_i$ and $n_j$
    respectively.
    // "Approximately smallest" means at most a constant factor bigger than the
    // smallest. Thus sorting can be avoided.
    $T_\ell = T_i \cup T_j$
    $n_\ell = n_i + n_j - 1$
    $S = S \setminus \{T_i, T_j\} \cup \{T_\ell\}$ // I.e., replace $T_i$ and $T_j$ by their union.
    **for all** standard colorings $\gamma$ of $U$ **do**
        $\mathbf{f}_\ell(\gamma) = \mathbf{f}_i(\gamma)\mathbf{f}_j(\gamma)$
        // This multiplication is done by FFT.
Now $S$ is a singleton $\{T_\ell\}$ with $T_\ell = T$.
Return $\mathbf{f}_\ell$

**Fig. 3.** The procedure Restricted-Colorings

where $c$, $c'$ and $c''$ are from the lemma and $b$ is defined by $b = c'$ if the tree $T'$ (with $|J_i| = J \cap V(I(i)) = 3$) exists and $b = 0$ otherwise.

In equation (3), the first term (the sum) is viewed as the main term covering the typical cost. The second term $(c' m_{\min} \lg m_{\min})$ has been introduced to pay for the extra cost of occasional uneven merges, i.e., merges of trees of very different sizes. The third term covers the extra cost of the subtree (if any) with $|J| = 3$. Finally, the forth term is provided to cover the base case (where all tree nodes are in $J$).

For the total time, until all merges have been done, i.e., for $d' = 1$, we will show a better bound $t'$ later.

The proof of the claim is by induction on the number of merges. The base case (just before any merges) follows directly from the inductive hypothesis of the lemma, without any need for the second term $c' m_{\min} \lg m_{\min}$. For the inductive step, we look at the difference $t(d') - t(d' + 1)$ of the allowed time after and before the merge of two trees $T_i$ and $T_j$ into $T_\ell$. We show in each case that time $t(d')l - t(d' + 1)$ is enough to perform the merge, i.e., to compute the polynomial for $T_\ell$ from the polynomials for $T_i$ and $T_j$.

First note that none of the four terms in $t(d')$ decreases during a merge (i.e., as $d'$ decreases by 1). The first term always increases by

$$c\,(m_i + m_j) \lg^2 (m_i + m_j) - c\,m_i \lg^2 m_i - c\,m_j \lg^2 m_j > 0$$

The second term increases when $m_{\min}$ increases. The last two terms clearly don't decrease.

We analyze the merges in two possible ways depending on whether the merged trees are of similar size or not. W.l.o.g., we assume $m_i \leq m_j$.

**Case "not similar":** Assume $T_i$ and $T_j$ are merged with $1 \leq m_{\min} = m_i < m_j/4$, and $\{T_i, T_j\}$ are approximately minimal, i.e., $m_h > m_j/2$ for all $h \neq i$. Here we do not assume $m_j \leq m/2$. For $|J| = 3$, it is possible to have a large tree with $m_j$ very close to $m$. Now the second term, $c' m_{\min} \lg m_{\min}$, increases from $t(d')$ to $t(d' + 1)$ by at least

$$c' \frac{m_j}{2} \lg \frac{m_j}{2} - c' m_{\min} \lg m_{\min}$$
$$> c' \frac{m_j}{2} \lg \frac{m_j}{2} - c' \frac{m_j}{4} \lg \frac{m_j}{4}$$
$$> c' \frac{m_j}{4} \lg \frac{m_j}{2}$$
$$> c' \frac{m_j}{8} \lg m_j \quad (\text{as } m_j > 4)$$

Now $c'$ is chosen large enough to make it possible to do the last merge in time $(c'/8)m_j \lg m_j$, i.e., to do the multiplications of $O(1)$ pairs of polynomials of degree $m_i$ and $m_j$ respectively using the fast Fourier transformation (FFT).

**Case "similar":** Assume $T_i$ and $T_j$ are merged with $m_i \leq m_j \leq 4m_i$. Now the first term increases from $t(d')$ to $t(d'+1)$ by

$$c(m_i + m_j)\lg^2(m_i + m_j) - cm_i\lg^2 m_i - cm_j\lg^2 m_j$$
$$\geq c(m_i + m_j)\lg^2(\tfrac{5}{4}m_j) - cm_i\lg^2 m_j - cm_j\lg^2 m_j$$
$$= c(m_i + m_j)((\lg\tfrac{5}{4} + \lg m_j)^2 - \lg^2 m_j)$$
$$= c(m_i + m_j)(2\lg\tfrac{5}{4}\lg m_j + \lg^2\tfrac{5}{4})$$
$$> C m_j\lg m_j$$

Here $C$ is first chosen large enough to make it possible to do the last merge in time $Cm_j\lg m_j$, i.e., to do the multiplications of $O(1)$ pairs of polynomials of degree $m_i$ and $m_j$ respectively using FFT. Then we make sure $c$ is chosen sufficiently large that the last inequality holds. This proves the claim.

At this point, we should notice that Claim (3) is not always strong enough to show the inductive step in the induction proof of the lemma. Indeed we can do better during the last merge. We claim a different bound $t'$ instead of just $t(1)$ to hold after the last merge, when we have just one tree $T_\ell = T$. Let

$$t' = cm\lg^2 m + am\lg m + c''m, \tag{4}$$

where $a = c'$ if $|J| = 3$ (for the set associated with the tree $T$), and $a = 0$ otherwise.

The case with $|J| = 3$ and therefore $|J_i| < 3$ for all $i$ causes no problem. Then $b = 0$, $m_{\min} = m$, and the first time bound (3) implies the second (4).

In the case $|J| < 3$, the last merge has to be handled separately. We show that this merge is always balanced and therefore significantly cheaper. We consider the stage where we are left with two trees with $m_i$ and $m_j$ edges to be merged into a tree of $m = m_\ell = m_i + m_j$ edges. We assume $m_i \leq m_j$ and claim $m_j < \frac{4}{5}m$. Otherwise, the large tree with more than $\frac{4}{5}m$ edges would have been produced by a merge involving a tree of size at least $\frac{2}{5}m$, omitting a tree of size at most $\frac{1}{5}m$, contradicting the rule of always merging approximately smallest trees.

Now the difference between the new modified bound $t'$ and the previous bound $t(2)$ is

$$t' - t(2) = cm\lg^2 m - c(m_i\lg^2 m_i + m_j\lg^2 m_j) - c'm_{\min}\lg m_{\min} - bm'\lg m'$$
$$> cm_i(\lg^2 m - \lg^2 m_i) + cm_j(\lg^2 m - \lg^2 m_j) - 2c'm\lg m$$
$$\geq cm(\lg^2 m - \lg^2 m_j) - 2c'm\lg m$$
$$> cm(\lg m + \lg m_j)(\lg m - \lg m_j) - 2c'm\lg m$$
$$= cm\lg(mm_i)\lg(m/m_j) - 2c'm\lg m$$
$$> cm\lg m\lg\tfrac{5}{4} - 2c'm\lg m$$
$$> Cm\lg m.$$

Once more, $C$ is chosen large enough to make it possible to do the last merge in time $Cm\lg m$, i.e., to do the multiplications of $O(1)$ pairs of polynomials of degree $m_i$ and $m_j$ respectively. Then we make sure $c$ is chosen sufficiently large for the last inequality to hold.    $\square$

Note that $m = |F|$ ($m$ is the number of edges in the tree $T$) and $|V| = n = \Theta(m)$. Lemma 1 immediately implies the desired complexity result for the procedure Restricted-Colorings and therefore also for the algorithm Chromatic-Polynomial.

**Theorem 1.** *For a graph $G$ with a given tree decomposition of bounded tree-width, the running time of the algorithm Chromatic-Polynomial is $O(n \log^2 n)$.*

## 6   Conclusion

We have presented a very efficient algorithm to compute all coefficients of the chromatic polynomial for a graph of bounded tree-width. It runs in almost linear time and is substantially simpler than previous algorithms, which use at least quadratic time. Also the dependence on the tree-width $k$ is as good as one could hope, namely singly exponential, whereas many other algorithms are doubly exponential in $k$.

The method is combining the advantages of recursive decomposition and dynamic programming. It has been used before for computing the characteristic polynomial of a tree [10]. It is hoped that this method has many applications for efficient computations on trees based on a clever choice of tree decompositions.

In particular, it is hoped that the method can be applied to compute several other graph polynomial efficiently. Of special interest is the Tutte polynomial. There, the situation is somewhat different, as this is a polynomial in two variables.

## References

1. Birkhoff, B.D.: A determinant formula for the number of ways of coloring a map. Ann. of Math. 14, 42–46 (1912)
2. Biggs, N.: Algebraic graph theory, 2nd edn. Cambridge Mathematical Library. Cambridge University Press, Cambridge (1993)
3. Courcelle, B.: A multivariate interlace polynomial. CoRR abs/cs/0702016 (2007)
4. Makowsky, J.A.: From a zoo to a zoology: Towards a general theory of graph polynomials. Theor. Comp. Sys. 43(3), 542–562 (2008)
5. Bläser, M., Hoffmann, C.: Fast evaluation of interlace polynomials on graphs of bounded treewidth. In: Fiat, A., Sanders, P. (eds.) ESA 2009. LNCS, vol. 5757, pp. 623–634. Springer, Heidelberg (2009)
6. Makowsky, J.A., Mariño, J.P.: Farrell polynomials on graphs of bounded tree width. Advances in Applied Mathematics 30, 160–176 (2003)
7. Andrzejak, A.: An algorithm for the Tutte polynomials of graphs of bounded treewidth. Discrete Mathematics 190(1-3), 39–54 (1998)
8. Noble, S.D.: Evaluating the Tutte polynomial for graphs of bounded tree-width. Combinatorics, Probability & Computing 7(3), 307–321 (1998)
9. Makowsky, J.A., Rotics, U., Averbouch, I., Godlin, B.: Computing graph polynomials on graphs of bounded clique-width. In: Fomin, F.V. (ed.) WG 2006. LNCS, vol. 4271, pp. 191–204. Springer, Heidelberg (2006)
10. Fürer, M.: Efficient computation of the characteristic polynomial of a tree and related tasks. In: Fiat, A., Sanders, P. (eds.) ESA 2009. LNCS, vol. 5757, pp. 11–22. Springer, Heidelberg (2009)
11. Bodlaender, H.L.: A linear time algorithm for finding tree-decompositions of small treewidth. SIAM Journal on Computing 25, 1305–1317 (1996)

# Average Parameterization and Partial Kernelization for Computing Medians

Nadja Betzler[1,*], Jiong Guo[2,**], Christian Komusiewicz[1,***],
and Rolf Niedermeier[1]

[1] Institut für Informatik, Friedrich-Schiller-Universität Jena,
Ernst-Abbe-Platz 2, D-07743 Jena, Germany
{nadja.betzler,c.komus,rolf.niedermeier}@uni-jena.de
[2] Universität des Saarlandes,
Campus E 1.4, D-66123 Saarbrücken, Germany
jguo@mmci.uni-saarland.de

**Abstract.** We propose an effective polynomial-time preprocessing strategy for intractable median problems. Developing a new methodological framework, we show that if the input instances of generally intractable problems exhibit a sufficiently high degree of similarity between each other *on average*, then there are efficient exact solving algorithms. In other words, we show that the median problems SWAP MEDIAN PERMUTATION, CONSENSUS CLUSTERING, KEMENY SCORE, and KEMENY TIE SCORE all are fixed-parameter tractable with respect to the parameter "average distance between input objects". To this end, we develop the new concept of "partial kernelization" and identify interesting polynomial-time solvable special cases for the considered problems.

## 1   Introduction

In median problems one is given a set of objects and the task is to find a "consensus object" that minimizes the sum of distances to the given input objects. Our new approach to solve in general intractable (mostly NP-hard) median problems considers an average measure for the similarity between the input objects by summing over all pairwise object distances divided by the number of these pairs. Based on this, we develop an algorithmic framework for showing that if the input objects are sufficiently "similar on average", then there are provably effective data reduction rules. In terms of parameterized algorithmics [10, 12, 18], this means that we show that the four median problems we study are fixed-parameter tractable with respect to the parameter "average distance between input objects". To the best of our knowledge, this parameter has only been studied for the KEMENY SCORE problem [5, 20] by using exponential-time dynamic programming and search tree methods. This work complements these results

---

* Supported by the DFG, research project PAWS, NI 369/10.
** Supported by the Excellence Cluster on Multimodal Computing and Interaction (MMCI). Main work was done while the author was with Friedrich-Schiller-Universität Jena.
*** Supported by a PhD fellowship of the Carl-Zeiss-Stiftung.

A. López-Ortiz (Ed.): LATIN 2010, LNCS 6034, pp. 60–71, 2010.

by polynomial-time preprocessing through data reduction. Marx [16] studies average parameterization for the CONSENSUS PATTERNS problem. He also shows fixed-parameter tractability; however in his case the parameter relates to the solution quality whereas our parameters relate to the input structure.

Let us briefly discuss the naturalness of average parameterization for two prominent median problems tackled in this paper. First, we show the fixed-parameter tractability of the NP-hard CONSENSUS CLUSTERING problem (see, e.g., [2, 6, 17]). Roughly speaking, the goal here is to find a median partition for a given set of partitions all over the same base set; this is motivated by the often occurring task to reconcile clustering information [4, 13, 17]. It is plausible that this reconciliation is only meaningful when the given input partitions have a sufficiently high degree of average similarity, because otherwise the median partition found may be meaningless since it tries to fit the demands of strongly opposing clustering proposals. Our algorithms are tailored for being efficient when there is "enough" consensus in the input.[1] Second, we also deal with the computation of Kemeny rankings (also known as rank aggregation), an NP-hard problem from the area of voting (see, e.g., [1, 2, 9, 11, 14]). As Conitzer and Sandholm [8] pointed out, one potential view of voting is that there exists a "correct" outcome (ranking), and each voter's vote corresponds to a noisy perception of this correct outcome (see [7, 9] for practical studies in this direction). Studying an average parameterization with respect to the pairwise distance between input votes reflects the view on voting proposed by Conitzer and Sandholm [8]. We develop efficient algorithms for computing Kemeny rankings in case of a reasonably small average distance between votes, again developing an effective preprocessing technique.

Within our framework, two points deserve particular attention. First, the identification of interesting *polynomial-time solvable special cases* of the underlying problems. Second, a *novel concept of kernelization* based on polynomial-time data reduction that does not yield problem kernels in the classical sense of parameterized algorithmics but only *"partial problem kernels"*. Roughly speaking, in (at least) "two-dimensional" problems as we study here (for instance, one "dimension" being the size of the base set and the other being the number of input subsets over this base set), this means that at least one dimension can be reduced such that its size only depends on the parameter value. This somewhat "weaker" concept of kernelization promises to be of wider practical use.

Due to the lack of space, most proofs are deferred to a full version of the paper.

## 2  Framework and Swap Median Permutation

In this work, we are concerned with consensus problems. Roughly speaking, the common feature of all these problems is that, given a number of combinatorial

---

[1] Indeed, a standard way of coping with too heterogeneous input partitions is to cluster the partitions and then to use CONSENSUS CLUSTERING in each "cluster of partitions", where high average similarity is to be expected [13].

objects (such as permutations, partitions etc.) over a base set $U$, to find a *median* object over $U$ that minimizes the sum of "distances" to all input objects.

The general outline of our framework reads as follows.

**Step 1.** Identify a polynomial-time solvable special case. This is done by defining a "dirtiness" concept for elements from the base set $U$ and proving that an instance of the underlying consensus problem can easily be solved when the input objects do not induce any dirty elements.

**Step 2.** Show that the number of dirty elements from $U$ is upper-bounded by a (typically polynomial) function only depending on the average distance between the given combinatorial objects.

**Step 3.** Show that the number of non-dirty elements from $U$ can be upper-bounded by a (typically polynomial) function only depending on the number of dirty elements (and, thus, also the average distance). This is achieved by developing polynomial-time data reduction rules which shrink the number of non-dirty elements (and thus $U$), generating an equivalent problem instance of smaller size.

**Step 4.** Make use of the fact that the desired median combinatorial object can be found in a running time only depending on the number of elements in $U$, and not depending on the number of combinatorial objects.

When applicable, this framework yields fixed-parameter tractability with respect to the parameter average distance. Note that a special feature of our framework is that in Step 3 we actually perform some sort of *partial kernelization*[2], a concept that may be of general interest. To illustrate our framework for efficiently solving "similar-on-average" median problems, we use the SWAP MEDIAN PERMUTATION problem (SMP for short) as a running example.[3] Herein, the combinatorial objects are permutations over the set $\{e_1, \ldots, e_m\}$, and the distance between two permutations is the *swap distance* defined as follows: A *swap* operation interchanges two elements of a permutation. For instance, swapping $e_i$ and $e_j$ in the identity permutation $e_1 \cdots e_{i-1} e_i e_{i+1} \cdots e_{j-1} e_j e_{j+1} \cdots e_m$ leads to $e_1 \cdots e_{i-1} e_j e_{i+1} \cdots e_{j-1} e_i e_{j+1} \cdots e_m$. The minimum number of swaps needed to transform a permutation $\pi_1$ into a permutation $\pi_2$ (or vice versa) is called the *swap distance* between $\pi_1$ and $\pi_2$, denoted by $d(\pi_1, \pi_2)$. Concerning notation, we follow the recent paper of Popov [19]. The formal problem definition of SMP reads as follows:

> **Input**: A set of permutations $\{\pi_1, \pi_2, \ldots, \pi_n\}$ over $\{e_1, e_2, \ldots, e_m\}$.
> **Output**: A median permutation $\pi$ with minimum distance $\sum_{i=1}^{n} d(\pi, \pi_i)$.

Now, the *average swap distance* $d$ for an input instance of SMP is defined as $d := \left( \sum_{i \neq j} d(\pi_i, \pi_j) \right) / (n \cdot (n-1))$. We present a first application of our framework using SMP as the concrete running example. After that, in the next two sections, we will provide our main results.

---

[2] The term "partial" refers to the fact that only the size of the base set is reduced, but not the number of input objects.

[3] We remark that the question of the NP-hardness of SMP seems unsettled, cf. [19].

The computation of the swap distance between two permutations can be carried out in $O(nm)$ time [3] by exploiting the tight relation between swap distances and permutation cycles of permutations. Given two permutations $\pi_1$ and $\pi_2$ of a set $U$, a *permutation cycle* of $\pi_1$ with respect to $\pi_2$ is a subset of $\pi_1$ whose elements, compared to $\pi_2$, trade positions in a circular fashion. In particular, an element $e$ having the same position in both $\pi_1$ and $\pi_2$ builds a cycle by itself. For example, with respect to permutation $e_1e_2e_3e_4e_5e_6$, permutation $e_3e_5e_1e_4e_6e_2$ has three permutation cycles $(e_1, e_3)$, $(e_4)$, and $(e_2, e_5, e_6)$. With respect to $\pi_2$, the cycle representation of $\pi_1$ as a product of disjoint permutation cycles is unique (up to the ordering of the cycles). The central observation behind the swap distance computation made by Amir et al. [3] is as follows: The swap distance between $\pi_1$ and $\pi_2$ is $m - c(\pi_1)$, where $c(\pi_1)$ is the number of permutation cycles in $\pi_1$ with respect to $\pi_2$.

First, according to Step 1, we need to define "dirty" elements. A *dominating position* of an element $e$ is a position, such that $e$ occurs at this position in more than $n/2$ input permutations. An element is called *dirty* if it has no dominating position; otherwise, it is called *non-dirty*. Lemma 1 not only shows the polynomial-time solvability of the special case but also the correctness of a data reduction rule used in Step 3.

**Lemma 1.** *Every median permutation places the non-dirty elements according to their dominating positions.*

**Lemma 2.** *SMP without dirty elements can be solved in $O(nm)$ time.*

Next, according to Step 2, we have to bound the number of dirty elements.

**Lemma 3.** *Given an SMP-instance with average swap distance $d$, there are less than $4d$ dirty elements.*

According to Step 3, the number of non-dirty elements needs to be bounded. To this end, we present the following data reduction rule.

**Reduction Rule.** In each of the input permutations, swap all non-dirty elements to their dominating positions. Remove all non-dirty elements. Record the number of the employed swap operations, which needs to be added to the distance of the median permutation of the resulting instance.

**Lemma 4.** *The above data reduction rule yields an equivalent SMP-instance with at most $4d$ elements, and it can be executed in $O(nm)$ time.*

Finally, according to Step 4, it remains to observe that for the median permutation we clearly have $O((\lceil 4d \rceil)!)$ possibilities. Hence, simply testing all of them and taking a best one yields the following theorem.

**Theorem 1.** SWAP MEDIAN PERMUTATION *is fixed-parameter tractable with respect to the parameter average swap distance.*

# 3   Consensus Clustering

Our second application of the framework deals with the NP-hard CONSENSUS CLUSTERING problem. It arises in attempts to reconcile clustering information. The goal is to find a *median partition* for a given set of partitions, which all are over the same base set. Due to its practical relevance, CONSENSUS CLUSTERING has been intensively studied in terms of the usefulness of various heuristics and accompanying experiments [4, 13]. The problem is defined as follows.

> **Input:** A set $\mathcal{C} = \{C_1, \ldots, C_n\}$ of partitions over a base set $S$.
> **Output:** A partition $C$ of $S$ with minimum distance $\sum_{C_i \in \mathcal{C}} d(C, C_i)$.

CONSENSUS CLUSTERING finds applications for example in the field of bioinformatics. Bonizzoni et al. [6] showed that CONSENSUS CLUSTERING is APX-hard even if the input consists of only three partitions. So far, the best approximation factor achievable in polynomial time is 4/3 [2].

Following Goder and Filkov [13], we call two elements $a, b \in S$ *co-clustered* with respect to a partition $C$ if $a$ and $b$ occur together in a subset of $C$ and *anti-clustered* if $a$ and $b$ occur in different subsets of $C$. Given a set $\mathcal{C}$ of partitions, we denote with $\text{co}(a, b)$ the number of partitions in $\mathcal{C}$ in which $a$ and $b$ are co-clustered and with $\text{anti}(a, b)$ we denote the number of partitions in $\mathcal{C}$ in which $a$ and $b$ are anti-clustered. Define the *distance* $d(C_i, C_j)$ between two input partitions $C_i$ and $C_j$ as the number of unordered pairs $\{a, b\}$ of elements from the base set $S$ such that $a$ and $b$ are co-clustered in one of $C_i$ and $C_j$ and anti-clustered in the other. Thus, our parameter $d$ denoting the *average distance* of a given CONSENSUS CLUSTERING instance is defined as $d := \left( \sum_{C_i, C_j \in \mathcal{C}} d(C_i, C_j) \right) / \left( n \cdot (n-1) \right)$.

Our overall goal is to show that CONSENSUS CLUSTERING is fixed-parameter tractable with respect to the average parameter $d$. To this end, we follow the approach presented in Section 2. Recall that Step 1 was to identify a polynomial-time solvable special case using a dirtiness concept.

**Definition 1.** *A pair of elements $a, b \in S$ is called a* dirty pair $a\#b$ *of a set $\mathcal{C}$ of $n$ partitions if* $\text{co}(a, b) \geq n/3$ *and* $\text{anti}(a, b) \geq n/3$. *Moreover, the predicate $(ab)$ is true iff* $\text{co}(a, b) > 2n/3$, *and the predicate $a \leftrightarrow b$ is true iff* $\text{anti}(a, b) > 2n/3$.

To show that an input instance of CONSENSUS CLUSTERING *without* dirty pairs is polynomial-time solvable, we need the following.

**Lemma 5.** *Let $\{a, b, c\}$ be a set of elements where $a$ and $c$ do not form a dirty pair. Then, $(ab) \wedge (bc) \Rightarrow (ac)$ and $a \leftrightarrow b \wedge (bc) \Rightarrow a \leftrightarrow c$.*

**Theorem 2.** CONSENSUS CLUSTERING *without dirty pairs is solvable in polynomial time.*

*Proof.* Let $C$ be an optimal solution, that is, $C$ is a partition of $S$ with minimum distance to the input partitions. It suffices to show that in $C$ the following two statements are true.

1. If $(ab)$, then $a$ and $b$ are co-clustered in $C$.
2. If $a \leftrightarrow b$, then $a$ and $b$ are anti-clustered in $C$.

Clearly, since there are no dirty pairs, any pair $a, b \in S$ must fulfill either $(ab)$ or $a \leftrightarrow b$. Hence, the two statements directly specify for each element from $S$ in which subset in $C$ it will end up.

To prove the first statement, suppose that there is an optimal solution $C$ not fulfilling the claim. Then, there must exist two subsets $S_i$ and $S_j$ in $C$ with $a \in S_i$ and $b \in S_j$. One can further partition both $S_i$ and $S_j$ into each time two subsets. More specifically, let $S_i^1 := \{x \in S_i \mid (ax)\}$ and $S_i^2 := S_i \setminus S_i^1$. The sets $S_j^1$ and $S_j^2$ are defined analogously with respect to $b$. In this way, by replacing $S_i$ and $S_j$ with $S_i^1 \cup S_j^1$, $S_i^2$, and $S_j^2$, one obtains a modified partition $C'$. Consider any $x \in S_i^1$ and any $y \in S_i^2$. Then, $x \leftrightarrow y$ follows from $(ax)$, $a \leftrightarrow y$, and Lemma 5. The same is true with respect to $S_j^1$ and $S_j^2$. Moreover, if $x \in S_i^1$ and $y \in S_j^2$, this means that $(ax)$ and $b \leftrightarrow y$, implying by Lemma 5 and using $(ab)$ that $x \leftrightarrow y$. It remains to consider $x \in S_i^1$ and $y \in S_j^1$. Then, again the application of Lemma 5 yields $(xy)$. Thus, $C'$ is a better partition than $C$ is because in $C'$ now $(ab)$ holds for all elements $a, b \in S_i^1 \cup S_j^1$ (without causing any increased cost elsewhere). This contradicts the optimality of $C$, proving the first statement. The second statement is proved analogously. $\square$

As required by Step 2 of the framework in Section 2, the next lemma upper-bounds the number of dirty pairs with the help of the average distance $d$.

**Lemma 6.** *An input instance of* CONSENSUS CLUSTERING *with average distance $d$ contains less than $9d/4$ dirty pairs.*

Step 3 of our framework now calls for a polynomial-time data reduction that reduces the number of elements that do not appear in any dirty pair. We call these elements *non-dirty elements*. To this end, we analyze the structure of an input instance. The idea is to find subsets of $S$ that contain many non-dirty elements that are all pairwisely co-clustered in more than $2n/3$ input partitions. First, we partition the input base set $S$ into two subsets $S_1$, which contains the non-dirty elements, and $S_2$, which contains the elements that appear in dirty pairs. In the following, we describe a partition of $S_1$ into equivalence classes according to the non-dirty pairs in $S_1$. Moreover, these equivalence classes also induce a partition of $S_2$. First we describe the partition $P_1 = \{S_1^1, \ldots, S_1^l\}$ of $S_1$. For each equivalence class $S_1^i \in P_1$, we demand $\forall a \in S_1^i \, \forall b \in S_1^i : (ab)$ and $\forall a \in S_1^i \, \forall b \in S \setminus S_1^i : a \leftrightarrow b$. Observe that, by Lemma 5, the partition $P_1$ of $S_1$ that fulfills these requirements is well-defined, since the predicate $(ab)$ describes a transitive relation over $S_1$. Using $P_1$, we define subsets $S_2^i$ of $S_2$. Each $S_2^i$ is defined as the set of elements $a \in S_2$ that have at least one element $b \in S_1^i$ such that $(ab)$ holds. We also define one additional set $S_2^0$ that contains all elements $a \in S_2$ such that there is no $b \in S_1$ for which $(ab)$ holds.

Finally, we obtain a set of subsets $P = \{S^0, S^1, \ldots, S^l\}$ of $S$ by setting $S^i = S_1^i \cup S_2^i$ for $1 \le i \le l$ and $S^0 = S_2^0$. The following lemma shows that $P$ is indeed a partition of $S$, and also gives some further structural property of $P$.

**Lemma 7.** *Let* $P = \{S^0, S^1, \ldots, S^l\}$ *be a set of subsets of* $S$ *constructed as described above. Then,* $P$ *is a partition of* $S$*, and for each* $S^i \in P$ *it holds that*

- $\forall a \in S^i \, \forall b \in S : (ab) \Rightarrow b \in S^i$ *and*
- $\forall a, b \in S^i, 1 \leq i \leq l : (ab) \vee a\#b$.

Informally, Lemma 7 says that inside any $S^i \in P$ we have only pairs that are *co-clustered* in more than $2n/3$ input partitions or dirty pairs; between two subsets $S^i \in P$ and $S^j \in P$ we have only dirty pairs or pairs that are *anti-clustered* in more than $2n/3$ input partitions. Clearly, the elements in $S_1^i$ then are co-clustered in more than $2n/3$ partitions with *all* elements in $S^i$ and are anti-clustered in more than $2n/3$ partitions with *all* elements in $S \setminus S^i$. This means that an $S^i$ with too many elements in $S_1^i$ is forced to become a set of an optimal partition. With the subsequent data reduction rule, we remove these sets from the input.

We introduce the following notation for subsets of $S$. For some set $E \subseteq S$, we denote with $\mathrm{dp}(E)$ the dirty pairs among the elements of $E$, that is, for a dirty pair $a\#b$ we have $a\#b \in \mathrm{dp}(E)$ if $a \in E$ and $b \in E$. Analogously, for two sets $E \subseteq S$ and $F \subseteq S$, we define $\mathrm{dp}(E, F)$ as the set of dirty pairs between $E$ and $F$, that is, for a dirty pair $a\#b$ we have $a\#b \in \mathrm{dp}(E, F)$ if $a \in E$ and $b \in F$ or vice versa.

**Reduction Rule.** Let $P$ be a partition of $S$ according to Lemma 7. If there is some $S^i \in P$ such that $|S_1^i| > |\mathrm{dp}(S^i)| + |\mathrm{dp}(S^i, S \setminus S^i)|$, then output $S^i$ as one of the sets of the solution and remove the elements of $S^i$ from all input partitions.

**Lemma 8.** *The above reduction rule is correct.*

In the following theorem, we combine Steps 3 and 4 of our framework: we show that exhaustively applying the reduction rule yields an equivalent instance whose number of elements is less than $9d$, and that this implies the fixed-parameter tractability of CONSENSUS CLUSTERING.

**Theorem 3.** CONSENSUS CLUSTERING *is fixed-parameter tractable with respect to the average distance* $d$ *between the input partitions. Each instance of* CONSENSUS CLUSTERING *can be reduced in polynomial time to an equivalent instance with less than* $9d$ *elements in the base set.*

# 4 Kemeny Rankings

In the third application of our framework, we investigate the problem of finding a "consensus ranking", that is, a so-called Kemeny ranking [11]. We first consider the NP-hard KEMENY SCORE problem and, second, the somewhat harder to attack generalization KEMENY TIE SCORE.

*Kemeny Score.* Kemeny's voting scheme can be described as follows. An *election* $(V, C)$ consists of a set $V$ of $n$ votes and a set $C$ of $m$ candidates. A *vote* is a preference list of the candidates, that is, a permutation on $C$. For instance, in the case of three candidates $a, b, c$, the order $c > b > a$ would mean that candidate $c$ is the best-liked and candidate $a$ is the least-liked for this voter. A "Kemeny consensus" is a preference list that is "closest" with respect to the so-called *Kendall-Tau distance* to the preference lists of the voters. For each pair of votes $v, w$, the Kendall-Tau distance (*KT-distance* for short) between $v$ and $w$, also known as the inversion distance between two permutations, is defined as $\text{dist}(v, w) = \sum_{\{a,b\} \subseteq C} d_{v,w}(a, b)$, where the sum is taken over all unordered pairs $\{a, b\}$ of candidates, and $d_{v,w}(a, b)$ is 0 if $v$ and $w$ rank $a$ and $b$ in the same order, and 1 otherwise. The *score* of a preference list $l$ with respect to an election $(V, C)$ is defined as $\sum_{v \in V} \text{dist}(l, v)$. A preference list $l$ with the minimum score is called a *Kemeny consensus* of $(V, C)$ and its score $\sum_{v \in V} \text{dist}(l, v)$ is the *Kemeny score* of $(V, C)$. The KEMENY SCORE problem is defined as follows:

**Input:** An election $(V, C)$.
**Output:** A Kemeny consensus $l$ with minimum score $\sum_{v \in V} \text{dist}(l, v)$.

KEMENY SCORE is NP-complete even when restricted to instances with only four votes [11]. The Kemeny score can be approximated to a factor of 8/5 by a deterministic algorithm [21] and to a factor of 11/7 by a randomized algorithm [2]. Recently, a polynomial-time approximation scheme (PTAS) for KEMENY SCORE has been developed [15]. However, its running time is impractical.

For an election $(V, C)$, the average KT-distance $d$, the parameter of this work, is defined as $d := \left( \sum_{u, v \in V, u \neq v} \text{dist}(u, v) \right) / (n(n-1))$. The KEMENY SCORE problem is known to be in FPT with respect to the parameter $d$ [5, 20]. There is a branching algorithm for KEMENY SCORE which runs in $(5.823)^d \cdot \text{poly}(n, m)$ time [20]. We extend these results by showing that the approach presented in Section 2 can be applied to KEMENY SCORE.

To identify a polynomial-time solvable special case as described in Step 1 of our framework, it is crucial to develop a concept of dirtiness. For KEMENY SCORE this is realized as follows. Let $(V, C)$ denote an election. An (unordered) pair of candidates $\{a, b\} \subseteq C$ with neither $a > b$ nor $a < b$ in more than 2/3 of the votes is called a *dirty pair* and $a$ and $b$ are called *dirty candidates*. All other pairs of candidates are called *non-dirty pairs*, and candidates that appear only in non-dirty pairs are called *non-dirty candidates*. Note that with this definition a non-dirty pair can also be formed by two dirty candidates. Let $D$ denote the set of dirty candidates and $n_d$ denote the number of dirty pairs in $(V, C)$. For two candidates $a, b$, we write $a >_{2/3} b$ if $a > b$ in more than 2/3 of the votes. Further, we say that $a$ and $b$ are *ordered according to the 2/3-majority* in a preference list $l$ if $a >_{2/3} b$ and $a > b$ in $l$. To show the following results, it will be useful to decompose the Kemeny score of a preference list into "partial scores". More precisely, for a preference list $l$ and a candidate pair $\{a, b\}$, the *partial score* of $l$ with respect to $\{a, b\}$ is $s_l(\{a, b\}) := \sum_{v \in V} d_{v,l}(a, b)$. The partial score of $l$ with respect to a subset $P$ of candidate pairs is $s_l(P) := \sum_{p \in P} s_l(p)$.

**Theorem 4.** KEMENY SCORE *without dirty pairs is solvable in polynomial time.*

*Proof.* For an input instance $(V, C)$ of KEMENY SCORE without dirty pairs, we show that the preference list "induced" by the 2/3-majority of the candidate pairs is optimal.

First, we show by contradiction that there is a preference list $l_{2/3}$ where for all candidate pairs $\{a, b\}$ with $a, b \in C$ and $a >_{2/3} b$, one has $a > b$. Assume that such a preference list does not exist. Then, there must be three candidates $a, b, c \in C$ that violate transitivity, that is, $a >_{2/3} b$, $b >_{2/3} c$, and $c >_{2/3} a$. Since $a >_{2/3} b$ and $b >_{2/3} c$, there must be at least $n/3$ votes with $a > b > c$. Since $a$ and $c$ do not form a dirty pair, it follows that $a >_{2/3} c$, a contradiction.

Second, we show by contradiction that $l_{2/3}$ is optimal. Assume that there is a Kemeny consensus $l$ with a non-empty set $P$ of candidate pairs that are not ordered according to the 2/3-majority; that is, $P := \{\{c, c'\} \mid c > c' \text{ in } l \text{ and } c' >_{2/3} c\}$. All candidate pairs that are not in $P$ are ordered equally in $l$ and $l_{2/3}$. Thus, the partial score with respect to them is the same for $l$ and $l_{2/3}$. For every candidate pair $\{c, c'\} \in P$, the partial score $s_l(\{c, c'\})$ is more than $2n/3$ and the partial score $s_{l_{2/3}}(\{c, c'\})$ is less than $n/3$. Thus, the score of $l_{2/3}$ is smaller than the score of $l$, a contradiction to the optimality of $l$.    □

Following Step 2 of our framework, the next lemma shows how the number of dirty pairs and, thus, also the number of dirty candidates, is upper-bounded by a function linear in the average KT-distance $d$.

**Lemma 9.** *Given an instance of* KEMENY SCORE *with average KT-distance $d$, there are at most $9d/2$ dirty pairs.*

The following three lemmas establish the basis for a polynomial-time data reduction rule as required in Step 3 of our framework. The basic idea is to consider the order that is induced by the 2/3-majorities of the non-dirty pairs and then to show that a dirty candidate can only "influence" the order of candidates that are not "too far away" from it in this order. Then, it is safe to remove non-dirty candidates that can be influenced by no dirty candidate.

**Lemma 10.** *For an election containing $n_d$ dirty pairs, in every Kemeny consensus at most $n_d$ non-dirty pairs are not ordered according to their 2/3-majorities.*

In the following, we show that the bound on the number of "incorrectly" ordered non-dirty pairs from Lemma 10 can be used to fix the relative order of two candidates forming a non-dirty pair. For this, it will be useful to have a concept of distance of candidates with respect to the order induced by the 2/3-majority. For $(V, C)$ and a non-dirty pair $\{c, c'\}$, define $\mathrm{dist}(c, c') := |\{b \in C : b \text{ is non-dirty and } c >_{2/3} b >_{2/3} c'\}|$ if $c >_{2/3} c'$ and $\mathrm{dist}(c, c') := |\{b \in C : b \text{ is non-dirty and } c' >_{2/3} b >_{2/3} c\}|$ if $c' >_{2/3} c$.

**Lemma 11.** *Let $(V, C)$ be an election and let $\{c, c'\}$ be a non-dirty pair. If $\mathrm{dist}(c, c') \geq n_d$, then in every Kemeny consensus $c > c'$ iff $c >_{2/3} c'$.*

Finally, the next lemma enables us to fix the position in a Kemeny consensus for a non-dirty candidate that has a large distance to all dirty candidates.

**Lemma 12.** *If for a non-dirty candidate $c$ it holds that* $\text{dist}(c, c_d) > 2n_d$ *for all dirty candidates $c_d \in D$, then in every Kemeny consensus $c$ is ordered according to the 2/3-majority with respect to all candidates from $C$.*

The correctness of the following data reduction rule follows directly from Lemma 12. It is not hard to verify that it can be carried out in $O(n \cdot m^2)$ time.

**Reduction Rule.** Let $c$ be a non-dirty candidate. If for all $c_d \in D$ one has $\text{dist}(c, c_d) > 2n_d$, then delete $c$ and record the partial score with respect to all candidate pairs that contain $c$ and are ordered according to the 2/3-majority. This score will be added to the Kemeny score of the resulting instance.

In the following, we show that after exhaustively applying the reduction rule, the number of non-dirty candidates is bounded by a function of $d$.

**Theorem 5.** *Each instance of* KEMENY SCORE *with average KT-distance $d$ can be reduced in polynomial time to an equivalent instance with at most $9d + 162 \cdot d^2$ candidates.*

*Kemeny Tie Score.* A practically relevant extension of KEMENY SCORE is KE-MENY TIE SCORE [1, 14]. Here, one additionally allows the voters to classify sets of equally liked candidates, that is, a preference list is no longer defined as a permutation of the candidates, but for two (or more) candidates $a, b$ one can have $a = b$. The term $d_{vw}(a, b)$ that denotes the contribution of the candidate pair $\{a, b\}$ to the KT-distance between two votes $v$ and $w$ is modified as follows [14]. One has $d_{v,w}(a, b) = 2$ if $a > b$ in $v$ and $b > a$ in $w$, $d_{v,w}(a, b) = 0$ if $a$ and $b$ are ordered in the same way in $v$ and $w$, and $d_{v,w}(a, b) = 1$, otherwise. Note that in the literature there are different demands for the consensus itself. For example, Hemaspaandra et al. [14] allow that the consensus list can contain ties as well whereas Ailon [1] requires the consensus list to be a "full ranking", that is, a permutation of the candidates. We consider here the more general setting used in [14]. Further, note that KEMENY TIE SCORE not only generalizes KEMENY SCORE but also includes other interesting special cases like $p$-ratings and top-$m$ lists [1].

Previous approaches [5, 20] only provide fixed-parameter tractability with respect to the average KT-distance for KEMENY SCORE. In contrast, the question of fixed-parameter tractability of KEMENY TIE SCORE with respect to the average KT-distance has been open so far. Here, we can answer this question positively by showing that the new method for partial kernelization introduced in Section 2 also applies to KEMENY TIE SCORE. This indicates that the new method can be more powerful than the dynamic programming approach in [5].

For an instance with ties, we say $a =_{2/3} b$ if $a = b$ in more than $2n/3$ votes. Then, the concept of dirtiness is adapted such that a pair of candidates $a, b$ is dirty if neither $a >_{2/3} b$ nor $a =_{2/3} b$ nor $a <_{2/3} b$. Further, we use $a \geq_{2/3} b$ to denote $(a >_{2/3} b) \vee (a =_{2/3} b)$.

**Theorem 6.** KEMENY TIE SCORE *without dirty pairs is solvable in polynomial time.*

For KEMENY TIE SCORE we can bound the number of candidates by a function only depending on the average KT-distance by proving lemmas analogous to Lemmas 9-12. For this, it is crucial to adapt the distance function between two candidates appropriately. More precisely, for two candidates $a, b$ with $a \geq_{2/3} b$, one defines $\text{dist}(a, b) := |\{c \in C : a \geq_{2/3} c \geq_{2/3} b \text{ and } c \text{ is non-dirty}\}|$.

Moreover, due to the definition of the KT-distance for the case with ties, the bound on the number of non-dirty candidates is twice as high as in the case without ties.

**Theorem 7.** KEMENY TIE SCORE *is fixed-parameter tractable with respect to the average KT-distance d. Each instance of* KEMENY TIE SCORE *with average KT-distance d can be reduced in polynomial time to an equivalent instance with at most $O(d^2)$ candidates.*

## 5   Conclusion

In applications one can easily determine the average distance parameter of the considered median problem and then decide whether the developed fixed-parameter algorithm should replace the otherwise used algorithm. Other related parameterizations which can not be computed "in advance" refer to distance measures from the input rankings to the solution. Among these, our results directly extend to the parameter "maximum distance of the input rankings from the solution" since this parameter is an upper bound for the average distance. In contrast, the "average distance of the input rankings from the solution" is clearly a lower bound for the "average distance between the input rankings". Hence, it is an interesting open question to investigate the parameterized complexity with respect to this parameter.

## References

[1] Ailon, N.: Aggregation of partial rankings, p-ratings, and top-m lists. Algorithmica (2008) (Available electronically)
[2] Ailon, N., Charikar, M., Newman, A.: Aggregating inconsistent information: ranking and clustering. Journal of the ACM 55(5), 27 pages (2008)
[3] Amir, A., Aumann, Y., Benson, G., Levy, A., Lipsky, O., Porat, E., Skiena, S., Vishne, U.: Pattern matching with address errors: rearrangement distances. Journal of Computer and System Sciences 75(6), 359–370 (2009)
[4] Bertolacci, M., Wirth, A.: Are approximation algorithms for consensus clustering worthwhile? In: Proc. 7th SDM, pp. 437–442. SIAM, Philadelphia (2007)
[5] Betzler, N., Fellows, M.R., Guo, J., Niedermeier, R., Rosamond, F.A.: Fixed-parameter algorithms for Kemeny rankings. Theoretical Computer Science 410(45), 4554–4570 (2009)

[6] Bonizzoni, P., Vedova, G.D., Dondi, R., Jiang, T.: On the approximation of correlation clustering and consensus clustering. Journal of Computer and System Sciences 74(5), 671–696 (2008)

[7] Conitzer, V.: Computing Slater rankings using similarities among candidates. In: Proc. 21st AAAI, pp. 613–619. AAAI Press, Menlo Park (2006)

[8] Conitzer, V., Sandholm, T.: Common voting rules as maximum likelihood estimators. In: Proc. 21st UAI, pp. 145–152. AUAI Press (2005)

[9] Conitzer, V., Davenport, A., Kalagnanam, J.: Improved bounds for computing Kemeny rankings. In: Proc. 21st AAAI, pp. 620–626. AAAI Press, Menlo Park (2006)

[10] Downey, R.G., Fellows, M.R.: Parameterized Complexity. Springer, Heidelberg (1999)

[11] Dwork, C., Kumar, R., Naor, M., Sivakumar, D.: Rank aggregation methods for the Web. In: Proc. 10th WWW, pp. 613–622 (2001)

[12] Flum, J., Grohe, M.: Parameterized Complexity Theory. Springer, Heidelberg (2006)

[13] Goder, A., Filkov, V.: Consensus clustering algorithms: Comparison and refinement. In: Proc. 10th ALENEX, pp. 109–117. SIAM, Philadelphia (2008)

[14] Hemaspaandra, E., Spakowski, H., Vogel, J.: The complexity of Kemeny elections. Theoretical Computer Science 349, 382–391 (2005)

[15] Kenyon-Mathieu, C., Schudy, W.: How to rank with few errors. In: Proc. 39th STOC, pp. 95–103. ACM, New York (2007)

[16] Marx, D.: Closest substring problems with small distances. SIAM Journal on Computing 38(4), 1382–1410 (2008)

[17] Monti, S., Tamayo, P., Mesirov, J.P., Golub, T.R.: Consensus clustering: A resampling-based method for class discovery and visualization of gene expression microarray data. Machine Learning 52(1-2), 91–118 (2003)

[18] Niedermeier, R.: Invitation to Fixed-Parameter Algorithms. Oxford University Press, Oxford (2006)

[19] Popov, V.Y.: Multiple genome rearrangement by swaps and by element duplications. Theoretical Computer Science 385(1-3), 115–126 (2007)

[20] Simjour, N.: Improved parameterized algorithms for the Kemeny aggregation problem. In: Chen, J., Fomin, F.V. (eds.) IWPEC 2009. LNCS, vol. 5917, pp. 312–323. Springer, Heidelberg (2009)

[21] van Zuylen, A., Williamson, D.P.: Deterministic pivoting algorithms for constrained ranking and clustering problems. Mathematics of Operations Research 34, 594–620 (2009)

# Sharp Separation and Applications to Exact and Parameterized Algorithms

Fedor V. Fomin[1], Daniel Lokshtanov[1], Fabrizio Grandoni[2], and Saket Saurabh[3]

[1] Department of Informatics, University of Bergen, N-5020 Bergen, Norway
{fedor.fomin,daniello}@ii.uib.no
[2] Dipartimento di Informatica, Sistemi e Produzione,
Università di Roma Tor Vergata,
via del Politecnico 1, 00133, Roma, Italy
grandoni@disp.uniroma2.it
[3] The Institute of Mathematical Sciences, Chennai 600113, India
saket@imsc.res.in

**Abstract.** Many divide-and-conquer algorithms employ the fact that the vertex set of a graph of bounded treewidth can be separated in two roughly balanced subsets by removing a small subset of vertices, referred to as a *separator*. In this paper we prove a trade-off between the size of the separator and the sharpness with which we can fix the size of the two sides of the partition. Our result appears to be a handy and powerful tool for the design of exact and parameterized algorithms for NP-hard problems. We illustrate that by presenting two applications.

Our first application is a parameterized algorithm with running time $O(16^{k+o(k)} + n^{O(1)})$ for the MAXIMUM INTERNAL SUBTREE problem in directed graphs. This is a significant improvement over the best previously known parameterized algorithm for the problem by [Cohen et al.'09], running in time $O(49.4^k + n^{O(1)})$.

The second application is a $O(2^{n+o(n)})$ time and space algorithm for the DEGREE CONSTRAINED SPANNING TREE problem: find a spanning tree of a graph with the maximum number of nodes satisfying given degree constraints. This problem generalizes some well-studied problems, among them HAMILTONIAN PATH, FULL DEGREE SPANNING TREE, BOUNDED DEGREE SPANNING TREE, MAXIMUM INTERNAL SPANNING TREE and their edge weighted variants.

## 1 Introduction

The aim of *parameterized* and *exact* algorithms is solving NP-hard problems exactly, with the smallest possible (exponential) worst-case running time. While exact algorithms are designed to minimize the running time as a function of the input size, parameterized algorithms seek to perform better when the instance considered has more structure than a general instance to the problem. Exact and parameterized algorithms have an old history [14,18], but they have been at the forefront in the last decade. In the last few years, many new techniques have

A. López-Ortiz (Ed.): LATIN 2010, LNCS 6034, pp. 72–83, 2010.

been developed to design and analyze exact algorithms, among them Inclusion-Exclusion, Möbius Transformation, Subset Convolution, Measure & Conquer and Iterative Compression to name a few [2,3,9,17,24].

A classical approach to solve combinatorial problems is *divide-and-conquer*: decompose the problem in two or more sub-problems, solve them independently and merge the solutions obtained. Several divide-and-conquer algorithms rely on the existence of a small *separator*, which is defined as follows. Let $G$ be an $n$-vertex graph with vertex set $V = V(G)$ and edge set $E = E(G)$. A set of vertices $S$ is called an $\alpha$-separator of $G$, $0 < \alpha \leq 1$, if the vertex set $V \setminus S$ can be partitioned into sets $V_L$ and $V_R$ of size at most $\alpha n$ such that no vertex of $V_L$ is adjacent to any vertex of $V_R$. For example, the classical result of Lipton and Tarjan that every planar graph has a $\frac{2}{3}$-separator of size $O(\sqrt{n})$ can be used to solve many NP-hard problems in planar graphs in time $O(2^{O(\sqrt{n})})$ [19].

## 1.1 Our Results

In this paper (see Section 2) we prove a trade-off between the size of the separator $S$ and the sharpness with which we can fix the size of $V_L$ and $V_R$ in the partition, for graphs of treewidth $t$. Given a function $w : X \to \mathbb{R}$, we define $w(Y) = \sum_{y \in Y} w(y)$ for any $Y \subseteq X$.

**Theorem 1 (Sharp Separation).** *Let $G = (V, E)$ be a graph of treewidth $t$ and $w : V \to \{0, 1\}$. Then for any integer $p \geq 0$ and $0 \leq x \leq w(V)$ there is a partition $(V_L, S, V_R)$ of $V$ such that $|S| \leq t\, p$, $w(V_L) \leq x + \frac{w(V)}{2^{p+1}}$, $w(V_R) \leq w(V) - x + \frac{w(V)}{2^{p+1}}$, and there is no edge in $G$ with one endpoint in $V_L$ and the other endpoint in $V_R$, that is, $S$ separates $V_L$ from $V_R$. Given a tree-decomposition of $G$ of width $t$, $S$ can be computed in polynomial time.*

Here $w$ is used to model a subset $W \subseteq V$ of vertices that we wish to separate. Theorem 1 implies for example that, with a separator of logarithmic size (for bounded treewidth graphs), we can obtain a *perfectly balanced* partition with $\max\{|V_L|, |V_R|\} \leq n/2$. In this paper we will always set $p \geq \log w(V)$, which makes the additive term $w(V)/2^{p+1}$ disappear.

Our Sharp Separation Theorem is a handy tool in the design of parameterized and exact algorithms based on the divide-and-conquer paradigm. We illustrate that by presenting two applications.

*k*-**Internal Spanning Tree.** Our first result is a parameterized algorithm for the following problem.

> $k$-INTERNAL OUT-BRANCHING: Given a digraph $D = (N, A)$ and a positive integer $k$, check whether there exists an out-branching with at least $k$ internal vertices.

The *undirected* counterpart to this problem, $k$-INTERNAL SUBTREE was first studied by Prieto and Sloper [22], who gave an algorithm with running time $2^{4k \log k} n^{O(1)}$ and a kernel of size $O(k^2)$ for the problem. Recently, Fomin et al. [10] gave an improved algorithm with running time $8^k n^{O(1)}$ and a kernel with at most $3k$ vertices. For $k$-INTERNAL OUT-BRANCHING, Gutin et al. [13] obtained an algorithm of running time $2^{O(k \log k)} n^{O(1)}$ for and gave a kernel of size of $O(k^2)$. A faster algorithm, running in time $49.4^k n^{O(1)}$ was subsequently improved by Cohen et al. [6]. In this paper we use the Sharp Separation Theorem to obtain an algorithm with running time $O(16^{k+o(k)} + n^{O(1)})$.

**Theorem 2.** *There is a one-sided-error Monte-Carlo algorithm for $k$-INTERNAL OUT-BRANCHING. The algorithm runs in polynomial-space and in time $O(16^{k+o(k)} + n^{O(1)})$, where $n$ is the size of the input digraph $D$. When an out-branching with at least $k$ internal nodes exists the algorithm fails to find one with probability at most $1/4$. This algorithm can be derandomized at the cost of an exponential $O(4^k k^{O(\log k)})$ space complexity.*

**Degree constrained spanning tree.** The second application of the Sharp Separation Theorem is an algorithm for DEGREE CONSTRAINED SPANNING TREE defined below. For a given graph $G = (V, E)$, let $d_G(v)$ denote the degree of $v \in V$ in $G$.

> DEGREE CONSTRAINED SPANNING TREE (DCST). Given a graph $G = (V, E)$ and a function $\mathcal{D} : V \to 2^{\{1,\dots,n\}}$. Find a spanning tree $T$ of $G$ maximizing $|\{v \in V : d_T(v) \in \mathcal{D}(v)\}|$.

Intuitively, $\mathcal{D}(v)$ can be seen as a set of desirable degrees for a vertex $v$ in the spanning tree. We have a *hit* each time $d_T(v) \in \mathcal{D}(v)$ for some $v$. The goal is maximizing the number of hits.

DCST naturally generalizes many NP-hard spanning tree and path problems studied in the literature. For instance we can code the famous HAMILTONIAN PATH problem, find a spanning path of a given graph, by letting $\mathcal{D}(v) = \{1, 2\}$ for all vertices; A spanning tree with $n$ hits is a Hamiltonian path. By carefully choosing the functions $\mathcal{D}(v)$ one can code many other problems as well, such as FULL DEGREE SPANNING TREE [16], BOUNDED DEGREE SPANNING TREE [12] or MAXIMUM INTERNAL SPANNING TREE [8].

For most special cases of DCST, no algorithm with running time $O(2^n n^{O(1)})$ was known until recently, and for MAXIMUM INTERNAL SPANNING TREE Fernau et al. [8] give a $O(3^n n^{O(1)})$ time algorithm, leaving the existence of a $O(2^n n^{O(1)})$ time algorithm open.

This year Nederlof [21] was able to give Inclusion-Exclusion based algorithm running in time $O(2^n n^{O(1)})$ for DCST. We use the Sharp Separation Theorem to give an alternate algorithm for the DCST problem, in particular we prove the following result.

**Theorem 3.** [⋆] [1] *The DEGREE CONSTRAINED SPANNING TREE problem can be solved in time and space $O(2^{n+o(n)})$, where $n$ is the number of nodes in the graph.*

---

[1] Proof of results labelled by ⋆ have been wholly or partially omitted due to space constraints.

Our algorithm differs from the work of Nederlof in the following ways. On one hand, his algorithm takes polynomial space and works in $2^n n^{O(1)}$ time. On the other hand, our approach is more robust. In particular our algorithm can be easily extended to find subgraphs of constant treewidth instead of trees, and also works for edge weighted variants of DEGREE CONSTRAINED SPANNING TREE.

## 1.2 Preliminaries

For basic graph terminology we refer the reader, e.g., to [7]. We just recall the definition of treewidth, and also the less standard digraph notions needed in this paper.

A *tree decomposition* of a (undirected) graph $G = (V, E)$ is a pair $(X, U)$ where $U = (W, F)$ is a tree, and $X = (\{X_i \mid i \in W\})$ is a collection of subsets of $V$ such that

1. $\bigcup_{i \in W} X_i = V$,
2. For each edge $vw \in E$, there is an $i \in W$ such that $v, w \in X_i$, and
3. For each $v \in V$ the set of vertices $\{i \mid v \in X_i\}$ forms a subtree of $U$.

The *width* of $(X, U)$ is $\max_{i \in W}\{|X_i| - 1\}$. The *treewidth* $tw(G)$ of $G$ is the minimum width over all the tree decompositions of $G$. By a classical result of Arnborg, Corneil and Proskurowski [1], a tree-decomposition of $G$ of width $t$, if any, can be computed in $O(n^{t+2})$ time. When this running time is dominated by other steps of the algorithm considered, we will just consider this decomposition as given. An *r-out-tree* in a digraph $D = (N, A)$ is a subtree $T$ of $D$ rooted at $r$, such that all arcs of $T$ are oriented away from $r$. If $T$ contains all vertices of $D$, $T$ is said to be an *r-out-branching*. For a vertex set $R$, an *R-out-forest* is a collection of $|R|$ vertex-disjoint $r$-out-trees, one out-tree for each $r \in R$.

## 2    Sharp Separation in Graphs of Bounded Treewidth

In this section we prove our Sharp Separation Theorem, which is at the heart of the algorithms described in the following sections. In order to prove that, we need the following well-known result.

**Lemma 1 ([4]).** *Given a n-vertex graph $G = (V, E)$ of treewidth $t$ and $w : V \to \mathbb{R}^+ \cup \{0\}$. There is a set $T$ of vertices of size at most $t$ such that for any connected component $G[C]$ of $G \setminus T$, $w(C) \leq w(V)/2$. Given a tree-decomposition of $G$ of width $t$, $T$ can be computed in polynomial time.*

*Proof. (Theorem 1)* We construct $V_L$, $V_R$ and $S$ iteratively, starting from empty sets, as follows. By Lemma 1 there is a set $T$ of size at most $t$ such that for any connected component $G[C]$ of $G \setminus T$, $w(C) \leq w(V)/2$. We add $T$ to $S$ and for each component $G[C]$ of $G \setminus T$, add $C$ to $V_L$ or $V_R$ if this does not violate $w(V_L) \leq x$ or $w(V_R) \leq w(V) - x$, respectively.

Let us show that at the end of the process there is at most one component $G[C]$ left. Suppose by contradiction that there are at least 2 such components, say $G[C_1]$ and $G[C_2]$. W.l.o.g. assume $w(C_1) \leq w(C_2)$. This implies that $w(V_L) + w(C_1) > x$ and $w(V_R) + w(C_1) > w(V) - x$. Consequently,

$$w(V_L) + w(V_R) + 2w(C_1) > w(V).$$

However, this contradicts the fact that

$$w(V_L) + w(V_R) + 2w(C_1) \leq w(V_L) + w(V_R) + w(C_1) + w(C_2) \leq w(V).$$

Now we iteratively reapply the construction above for $p - 1$ times, each time considering the component $G[C]$ left from previous step. Eventually we add $C$ to either $V_L$ or $V_R$.

Each time the weight of $C$ halves, so at the end of the process $w(C) \leq w(V)/2^{p+1}$. The upper bound on the weight of $V_L$ and $V_R$ follows. Since at each step we add to $S$ a set of size $t$, we eventually obtain $|S| \leq pt$. The running time claim follows immediately from Lemma 1. This concludes the proof.     □

## 3     $k$-Internal Out-Branching

In this section we use Theorem 1 to give a parameterized algorithm with running time $O(16^{k+o(k)} + n^{o(1)})$ for the $k$-INTERNAL OUT-BRANCHING problem. Our approach combines the Sharp Separation Theorem with the *divide-and-color* paradigm in [5,15] and a polynomial-sized kernel for the problem [13]. First we present a (polynomial-space) one-sided-error Monte-Carlo algorithm for $k$-INTERNAL OUT-BRANCHING with the claimed running time. We then derandomize the algorithm at the cost of an exponential space complexity.

### 3.1     A Monte-Carlo Algorithm

The first step of our algorithm is to apply the *kernelization algorithm* of Gutin et al. [13]. Given an instance $(D, k)$ of $k$-INTERNAL OUT-BRANCHING the algorithm of Gutin et al. produces a new instance $(D', k')$ with $|D'| = O(k^2)$ and $k' \leq k$ such that $D'$ has an out-branching with at least $k'$ internal vertices if and only if $D$ has an out-branching with at least $k$ internal vertices. After this step we can assume that the number $n$ of vertices in the input digraph $D$ is at most $O(k^2)$.

Now, the algorithm guesses the root $r$ of the out-branching, and verifies that there indeed is some out-branching of $D$ rooted at $r$. This guessing step, together with the following observation, allows us to search for out-trees rooted at $r$ instead of out-branchings of $D$.

**Lemma 2 ([6]).** *Let $D$ be a digraph and $r$ be a node of $D$ such that there is an $r$-out-branching of $D$. Then, for any $r$-out-tree $T$ with at least $k$ internal nodes there is an $r$-out-branching $T'$ with at least $k$ internal nodes containing $T$ as a subtree.*

When looking for $r$-out-trees with at least $k$ internal nodes, it is sufficient to restrict ourselves to $r$-out-trees with at most $2k$ nodes. The reason for this is that if some internal node sees at least two leaves of the $r$-out-tree, then one of the leaves can be removed without changing any internal nodes into leaves. We formalize this as an observation.

**Lemma 3 ([6]).** *Let $D$ be a digraph and $r$ be a node of $D$. If there is an $r$-out-tree $T$ with at least $k$ internal nodes then there is an $r$-out-tree $T'$ on at most $2k$ nodes with at least $k$ internal nodes.*

With the described preliminary steps, we have arrived at the following problem, which we call ROOTED DIRECTED $k$-INTERNAL OUT-TREE ($k$-RDIOT). Input is a digraph $D$, node $r$ and integer $k$. The digraph $D$ has $n = O(k^2)$ nodes and the objective is to decide whether there is an $r$-out-tree with at least $k$ internal nodes and at most $2k$ nodes in total.

Our algorithm splits the original problem into two smaller sub-problems by means of a proper separator, guesses the "shape" of the intersection of the out-branching with each side of the separator and solves each subproblem recursively. There are two aspects of sub-problems which do not show up in the original problem. First of all, the solution to a subproblem is not necessarily an out-tree: it is an out-forest in general. Still, the union of such forests must induce an $r$-out-tree. In order to take this fact into account, we introduce the notion of signatures.

**Definition 1.** *Let $T = (N_T, A_T)$ be an $R$-out-forest, and $Z \subseteq N_T$ be a set of nodes such that $R \subseteq Z$. The signature $\zeta_Z(T)$ of $T$ with respect to $Z$ is the $R$-out-forest $C = (Z, Q)$ where there is an arc from a vertex $u \in R$ to a vertex $v \in Z \setminus R$ if and only if there is a path from $u$ to $v$ in $T$. All vertices of $Z \setminus R$ are leaves of $C$.*

Notice that the signature of an out-forest is always a set of stars and singletons. In our recursive steps we will guess the signature of the out-forest we are looking for with respect to $Z$, where the set $Z$ includes $r$ and all the separators guessed from the original problem down to the current subproblem.

Second, in order to obtain two independent sub-problems, we need to make sure that separator nodes that are internal on both sides of the separator only get counted once. To achieve this we guess a subset $Y$ of the separator nodes, and do not count the internal nodes of the out-forest in $Y$. Altogether, a subproblem can be defined as follows.

> DIRECTED ROOTED OUT-FOREST (DROF). Input is a tuple $(D, R, C, Y, k^*, t)$ where $D = (N, A)$ is a digraph, $C = (Z, Q)$ is an $R$-out-forest with node set $Z$ for $R \subseteq Z \subseteq N$, $Y \subseteq Z$ is a node set and $k^*$ and $t$ are integers. The objective is to find an $R$-out-forest $T$ in $D$ with at least $k^*$ internal nodes outside $Y$ and at most $t$ nodes outside $Z$ such that $T$ contains $Z$ and $\zeta_Z(T) = C$.

The input instance $(D, k)$ of $k$-RDIOT is equivalent to an DROF instance $(D, R, C, Y, k, 2k)$, where $t = 2k$, $C$ is the single node $r$ and $Y = \emptyset$. Our algorithm for $k$-RDIOT initially constructs a DROF instance equivalent to the input

$k$-RDIOT instance as described above. That $k$-RDIOT instance is solved recursively in the following way. Consider a given subproblem $(D, R, C, Y, k^*, t)$. If $t \leq \log k$, that is the number of vertices outside $Z$ in the out-forest sought for is small enough, we solve the problem by brute force. In particular, we enumerate all the possible $R$-out-forests in $D$ on at most $|Z| + t$ nodes and check whether they satisfy the conditions of DROF.

Suppose now $t > \log k$, and that $(D, R, C, Y, k^*, t)$ is a "yes"-instance. Then there is an $R$-out-forest $T = (N^T, A^T)$ that satisfies the conditions of DROF. By the Sharp Separation Theorem there is a partitioning of $N^T$ into $(N_L^T, S, N_R^T)$ such that $|S| = \log k$, $|N_L^T \setminus Z| \leq t/2$, $|N_R^T \setminus Z| \leq t/2$ and there are no arcs between $N_L^T$ and $N_R^T$ in $T$. Define $Z' = Z \cup S$ and $A_{Z'}$ to be the arcs of $T[Z']$. The algorithm guesses the separator $S \subseteq N$ and for each of the $\binom{O(k^2)}{\log k}$ guesses for the separator it generates a random family of $3 \cdot 2^t \cdot |Z'|^{O(|Z'|)}$ pairs of subproblems, that is instances of DROF $\mathcal{P}_L$ and $\mathcal{P}_R$, which are solved recursively.

If for some pair $\mathcal{P}_L$ and $\mathcal{P}_R$, the algorithm returns that both $\mathcal{P}_L$ and $\mathcal{P}_R$ are "yes" instances, then the algorithm returns that $(D, C, Y, k^*, t)$ is a "yes"-instance as well. If the algorithm loops through all guesses of $S$ and all the $3 \cdot 2^t \cdot |Z'|^{O(|Z'|)}$ pairs and for each pair the algorithm returns that at least one sub-problem is a "no"-instance, the algorithm returns that $(D, C, Y, k^*, t)$ is a "no"-instance. To conclude the description of the algorithm we need to describe how the pairs $(\mathcal{P}_L, \mathcal{P}_R)$ are generated.

Before describing how the pairs are generated, define the out-forests $T_L = T[Z' \cup N_L^T]$ and $T_R = T[Z' \cup N_R^T] \setminus A_{Z'}$. Also, let $Y_R$ be $Y$ plus all the internal nodes of $T_L$ in $Z'$ and $Y_L = (Z' \setminus Y_R) \cup Y$. Now, $t_L$ and $t_R$ are the number of nodes outside $Z'$ in $T_L$ and $T_R$ respectively. Finally $k_L^*$ and $k_R^*$ are the number of internal nodes in $T_L$ outside $Y_L$ and the number of internal nodes in $T_R$ outside $Y_R$ respectively.

We next describe how a random pair $(\mathcal{P}_L, \mathcal{P}_R)$ is generated. The algorithm generates the pairs in $3 \cdot 2^t$ groups, each group with $|Z'|^{O(|Z'|)}$ pairs. For each group the algorithm partitions the node set $N \setminus Z'$ into two parts $(N_L, N_R)$ uniformly at random. For each partitioning, the algorithm guesses $C_L = \zeta_{Z'}(T_L)$, $C_R = \zeta_{Z'}(T_R)$, $Y_L$, $Y_R$, $k_L^*$, $k_R^*$, $t_L$ and $t_R$. Each set of guesses makes one pair $(\mathcal{P}_L, \mathcal{P}_R)$ of instances, where $\mathcal{P}_L = (D[N_L \cup Z'], R_L, C_L, Y_L, k_L^*, t_L)$ and $\mathcal{P}_R = (D[N_R \cup Z'], R_R, C_R, Y_R, k_R^*, t_R)$. It is easy to see that the number of possible guesses is at most $|Z'|^{O(|Z'|)}$.

The algorithm makes the guesses in a special way, making sure that if both $\mathcal{P}_L$ and $\mathcal{P}_R$ are "yes"-instances then $(D, C, Y, k^*, t)$ is a "yes"-instance as well. In particular, it makes sure that the arc sets of $C_L$ and $C_R$ are disjoint, that $C_L \cup C_R$ is an out-forest and that $\zeta(C_L \cup C_R) = C$. Also, the algorithm makes sure that $Y_L \cup Y_R = Z'$ and that $Y \subseteq Y_L$ and $Y \subseteq Y_R$. Finally, it makes sure that $t_L^* + t_R^* - |Z'| = t^*$ and that $k_L^* + k_R^* = k^*$. This concludes the description of the algorithm.

**Lemma 4.** *There is a one-sided-error Monte-Carlo algorithm for $k$-INTERNAL OUT-BRANCHING running in time $O(16^{k+o(k)} + n^{O(1)})$. When the instance is a "yes"-instance, the algorithm incorrectly returns "no" with probability at most $1/4$.*

*Proof.* Consider the algorithm above. It is enough to prove correctness and analyze the running time for the part of the algorithm that solves DROF. We first prove that when the algorithm answers yes, the answer is correct. We prove this by induction on $t$. If $t < \log k$ then the algorithm resolves the problem in a brute force manner and hence correctness follows. Suppose now that $t \geq \log k$. Since the algorithm returned yes it made a guess for $S$, a random partitioning of $N \setminus Z'$ (where $Z' = Z \cup S$) and guessed a pair $\mathcal{P}_L = (D[N_L \cup Z'], R_L, C_L, Y_L, k_L^*, t_L)$ and $\mathcal{P}_R = (D[N_R \cup Z'], R_R, C_R, Y_R, k_R^*, t_R)$ such that the algorithm returned that both $\mathcal{P}_L$ and $\mathcal{P}_R$ are "yes"-instances of DROF. By the induction hypothesis there are out-forests $T_L = (N_L^T, A_L^T)$ of $D[N_L]$ and $T_R = (N_R^T, A_R^T)$ of $D[N_R]$ with at least $k_L^*$ and $k_R^*$ inner nodes outside $Y_L$ and $Y_R$ respectively, such that $C_{Z'}(T_L) = C_L$ and $C_{Z'}(T_R) = C_R$. We prove that $T = T_L \cup T_R$ is an out-forest that satisfies the conditions of DROF.

Since the arc sets of $C_L$ and $C_R$ are disjoint, $C_L \cup C_R$ is an out-forest and $C_{Z'}(T_L) = C_L$ and $C_{Z'}(T_R) = C_R$, $T = T_L \cup T_R$ is an out-forest. Since $\zeta_Z(C_L \cup C_R) = C$ it follows that $\zeta_Z(T) = \zeta_Z(T_L \cup T_R) = C$. The number of nodes in $T$ is $t_L + t_R - Z \leq t$ and since $Y \subseteq Y_L$, $Y \subseteq Y_R$ and $Y_L \cup Y_R = Z'$ the number of inner nodes of $T$ avoiding $Y$ is at most $k_L^* + k_R^* \geq k^*$. Hence the input instance is indeed a "yes"-instance.

Now, we prove that if a given subproblem $(D, R, C, Y, k^*, t)$ is a "yes"-instance, then the probability that the algorithm returns "no" is $p_t \leq 1/4$. We prove this by induction on $t$, and if $t < \log k$ the algorithm solves the problem by brute force and correctness follows. If $t \geq \log k$, consider an out-forest $T$ that satisfies the conditions of DROF. Consider the run of the algorithm where the separator $S$ is guessed correctly.

Now, there are two possible reasons why the algorithm fails to answer "yes". Reason (a) is that the random partition $(N_L, N_R)$ of $N \cup Z'$ could be done in the wrong way, that is $N_L^T \not\subseteq N_L$ or $N_R^T \not\subseteq N_R$. Reason (b) is that even though $N_L$ and $N_R$ are guessed correctly, in the iteration of the algorithm where the guesses for $C_L = \zeta_{Z'}(T_L)$, $C_R = \zeta_{Z'}(T_R)$, $Y_L$, $Y_R$, $k_L^*$, $k_R^*$, $t_L$ and $t_R$ are correct, the algorithm could fail to recognize either $\mathcal{P}_L$ or $\mathcal{P}_R$ as "yes" instances.

The probability of the first event is at most $1 - 2^{-t}$. Recall that $t_L, t_R \leq t/2$, since the algorithm uses a perfectly balanced separator to split $N^T \setminus Z'$. Hence, by the union bound, the probability of event (b) is at most $2^{-t} 2p_{t/2}$. Altogether $p_t$ satisfies

$$p_t \leq \left(1 - 2^{-t} + 2^{-t+1} p_{t/2}\right)^{3 \cdot 2^t}.$$

Therefore, by the inductive hypothesis,

$$p_t \leq \left(1 - 2^{-t} + 2^{-t+1}/4\right)^{3 \cdot 2^t} = \left(\left(1 - 1/2^{t+1}\right)^{2^{t+1}}\right)^{1.5} \leq e^{-1.5} \leq \frac{1}{4}.$$

Consider now the running time of the algorithm. Observe that in the beginning $t = 2k$ and that $t$ always drops by a factor of one half in the recursive steps. Furthermore the algorithm stops when $t$ drops below $\log k$. Hence the recursion depth is at most $\log(2k)$. For each new level of the recursion the size of $Z'$

increases by $\log 2k$. Hence $|Z'|$ never grows over $\log^2(2k)$. In the base case we try all possible subsets of $A$ of size $|Z|' + t$. Since $D$ has at most $O(k^2)$ vertices it has at most $O(k^4)$ arcs and hence in the base we need to try at most $O(\binom{k^4}{\log^2 2k}) = O(2^{o(k)})$ different possibilities, each of which can be checked in $O(k^{O(1)})$ time.

Consider now the recursive step. There are $\binom{O(k^2)}{\log 2k}$ choices for the separator. For each choice of the separator the number of random partitions tried is $3 \cdot 2^k$. For each random partition, $|Z'|^{O(|Z'|)} = O(2^{\log^3 k})$ pairs $(\mathcal{P}_L, \mathcal{P}_R)$ of instances are generated. Let $T(n,t)$ be the running time of the DROF algorithm on an instance where $D$ has $n$ nodes and the number of nodes in the tree searched for that are not in $Z'$ is $t$. Then the following recurrence holds.

$$T(n,t) \leq n^{O(\log^3 2k)} \cdot 3 \cdot 2^t \cdot (2T(n, t/2) + n^{O(1)})$$
$$\leq n^{O(t \log^3 2k)} 2^k \cdot T(n, t/2)$$
$$= O((n^{O(t \log^3 2k)})^{\log k} \cdot 2^{(\sum_{i=0}^{\log t} \frac{t}{2^i})}) = O(4^t \cdot n^{O(\log^4 k)}).$$

Since we first run the kernelization algorithm from [13], the $k$-RDIOT instance we solve recursively has $O(k^2)$ nodes. Since $t = 2k$ in the instance of DROF we construct from this $k$-RDIOT instance, the total running time for the algorithm is bounded from above by $O(4^{2k} \cdot (k^2)^{O(\log^4 k)} + n^{O(1)}) = O(16^{k+o(k)} + n^{O(1)})$. □

Our algorithm for $k$-INTERNAL OUT-BRANCHING can be derandomized using the method presented by Chen et al. [5], which is based on the construction of $(n, k)$-*universal sets* [20]. The main idea is to replace the random partitioning of the host graph $H$ by a partitioning that uses universal sets. Lemmas 4 and 5 together imply Theorem 2.

**Lemma 5.** [⋆] *There is a deterministic algorithm for* $k$-INTERNAL OUT-BRANCHING *running in time* $O(16^{k+o(k)} + n^{O(1)})$ *and requiring* $O(4^k k^{O(\log k)})$ *space.*

## 4    Degree Constrained Spanning Tree

In this section we present our $O(2^{n+o(n)})$-time algorithm for the DEGREE CON-STRAINED SPANNING TREE problem (DCSS). We recall that in this problem we are given an undirected graph $G = (V, E)$, and a list of *desirable* degrees $\mathcal{D}(v)$ for each vertex $v$. The aim is finding a spanning tree $T$ of $G$ which maximizes the number of *hits*, i.e. the number of vertices $v$ with $d_T(v) \in \mathcal{D}(v)$.

Our algorithm is based on the divide-and-conquer approach, and has several similarities with the algorithm for $k$-Internal Spanning Tree. The main differences are that the random partitioning and kernelization parts are no longer required, and that the Sharp Separation theorem is used to divide the problem into *very unbalanced* subproblems. Consider a subproblem on the graph $H = (V, E)$. In the divide step we guess a proper (logarithmic-size) separator $S$ of the optimum solution, and the corresponding two sides $V_L$ and $V_R$ of the partition. Set $S$ is chosen such that $V_L$ is sufficiently small to make the guessing of $S$, $V_L$ and $V_R$ cheap enough. The existence of $S$ is guaranteed by our Sharp Separation Theorem. The two sub-problems induced by $V_L \cup S$ and $V_R \cup S$ are then solved recursively.

Just as for the case of $k$-Internal Spanning Tree there are two aspects of sub-problems which do not show up in the original problem. First of all, the solution to a subproblem is not necessarily a spanning tree: it is a spanning forest in general. Still, the union of such forests must induce a tree. In order to take this fact into account, we introduce a *constraint forest* $C = (Z, Q)$, defined over a proper subset of nodes $Z \subseteq V$. The set $Z$ includes all the separators guessed from the original problem down to the current subproblem. The components of $C$ describe which pairs of nodes of $Z$ must and must not be connected in the desired forest.

Second, in order to obtain two independent maximization sub-problems, we need to guess the degree of the separator nodes in the optimum solution, and force the solution to have that degree on those nodes. This is modeled via an auxiliary function $\mathcal{A} : V \to 2^{\{1,\dots,n\}}$. For $z \in Z$, $\mathcal{A}(z)$ is a singleton set containing the mentioned guessed degree, while $\mathcal{A}$ coincides with $\mathcal{D}$ on the remaining nodes. We remark that it might be that $\mathcal{A}(z) \not\subseteq \mathcal{D}(z)$ for some $z \in Z$, since not all the nodes of $Z$ need to be hits in the optimum solution. Altogether, a subproblem $(H, C, \mathcal{A})$ can be defined as follows.

DEGREE-CONSTRAINED CUT & CONNECT (DCCC). Given a graph $H = (V, E)$, a forest $C = (Z, Q)$, $Z \subseteq V$, and a function $\mathcal{A} : V \to 2^{\{1,\dots,n\}}$, $|\mathcal{A}(z)| = 1$ for $z \in Z$. Find a spanning forest $F$ of $H$ maximizing the number of hits, i.e. $|\{v \in V : d_F(v) \in \mathcal{A}(v)\}|$, such that: (i) every connected component of $F$ contains at least one vertex of $Z$; (ii) for any $u, v \in Z$, $u$ and $v$ are connected in $C$ if and only if they are connected in $F$; (iii) $d_F(z) \in \mathcal{A}(z)$ for all $z \in Z$.

Observe that the original DEGREE CONSTRAINED SPANNING TREE instance $(G, \mathcal{D})$ is equivalent to a DEGREE-CONSTRAINED CUT & CONNECT instance where $H = G$, $C = (\{z\}, \emptyset)$ for an arbitrary vertex $z$ of $G$, $\mathcal{A}(z) = \{d_{OPT}(z)\}$ where $d_{OPT}(z)$ is the degree of $z$ in an optimum solution $OPT$, and $\mathcal{A}(v) = \mathcal{D}(v)$ for any vertex $v \neq z$. We remark that we can guess $d_{OPT}(z)$ by trying all the possibilities.

We give a memoization based algorithm for DCST. Initially the algorithm encodes the input problem into a DCCC problem as described above. The latter problem is then solved recursively. The solution to each subproblem generated is stored in a memoization table, which is used to avoid to solve the same subproblem twice.

Let us describe the recursive algorithm for DCCC. Consider a given subproblem $\mathcal{P} = (H, C, \mathcal{A})$, with $H = (V, E)$ and $C = (Z, Q)$. If $|V| \leq n/\log^2 n$, the problem is solved in a brute force manner by enumerating all the spanning forests $F$ of $H$. Otherwise, the algorithm splits the problem in two smaller independent sub-problems $\mathcal{P}_L = (H_L, C_L, \mathcal{A}_L)$ and $\mathcal{P}_R = (H_R, C_R, \mathcal{A}_R)$, which are solved recursively. The desired solution $F$ is obtained by merging the two solutions obtained for the two sub-problems.

We next describe how $\mathcal{P}_L$ and $\mathcal{P}_R$ are obtained. Consider the optimum solution $OPT = OPT(H, C, \mathcal{A})$ to $(H, C, \mathcal{A})$. For $x = n/\log^2 n$, by the Sharp Separation Theorem there is a separator $S$ of $OPT$, $|S| \leq t \log n = \log n$, which splits $V \setminus S$ in two subsets $V_L$ and $V_R$, with $|V_L| \leq x$ and $|V_R| \leq |V| - x$. Let $Z' = S \cup Z$. The algorithm guesses $S$, $V_L$ and $V_R$, and sets $H_L = H[V_L \cup Z']$ and $H_R = H[V_R \cup Z']$.

Consider the forest $C'$ obtained from $OPT$ by iteratively contracting the edges with one endpoint not in $Z'$. Note that, if we further contract $C'$ on vertices $S \setminus Z$, we must obtain the forest $C$. Each edge of $C'$ corresponds to a path in $H$ whose vertices belong entirely either to $V_L \cup Z'$ or to $V_R \cup Z'$. (In order to simplify the algorithm, we assume that edges between adjacent nodes of $Z'$ belong to the first class). Let $Q_L$ and $Q_R$ be the edges of the first and second type, respectively. The algorithm guesses $C'$, $Q_L$ and $Q_R$, and sets $C_L = (Z', Q_L)$ and $C_R = (Z', Q_R)$.

It remains to specify $\mathcal{A}_L$ and $\mathcal{A}_R$. Consider the two forests $OPT_L$ and $OPT_R$, on vertex set $V_L \cup Z'$ and $V_R \cup Z'$, respectively, obtained from $OPT$ by inserting every edge of $OPT$ with both endpoints in $V_L \cup Z'$ in $OPT_L$, and all the remaining edges in $OPT_R$. Note that $d_{OPT}(z') = d_{OPT_L}(z') + d_{OPT_R}(z')$ for all $z' \in Z'$. The algorithm guesses $d_{OPT_L}(z')$ (resp., $d_{OPT_R}(z')$) for all $z' \in Z'$, and sets $\mathcal{A}_L(z') = \{d_{OPT_L}(z')\}$ (resp., $\mathcal{A}_R(z') = \{d_{OPT_R}(z')\}$). Moreover, it sets $\mathcal{A}_L(v) = \mathcal{A}(v)$ (resp, $\mathcal{A}_R(v) = \mathcal{A}(v)$) for all the remaining nodes $v$.

Summarizing the discussion above, the following recurrence holds, where the maximum, computed with respect to the number of hits, is taken over all the possible choices of $(H_L, C_L, \mathcal{A}_L)$ and $(H_R, C_R, \mathcal{A}_R)$ such that the pair of feasible solutions to the smaller instances can be combined to a feasible solution for the original instance $(H, C, \mathcal{A})$.

$$OPT(H, C, \mathcal{A}) = \arg\max\{OPT(H_L, C_L, \mathcal{A}_L) \cup OPT(H_R, C_R, \mathcal{A}_R)\}. \quad (1)$$

In particular, the maximum considers all the possible choices of the separator $S$ and of the partition $(V_L, V_R)$, of the forest $C'$ and of the partition $(Q_L, Q_R)$ of its edges, and of the degrees $d_{OPT_L}(z')$ and $d_{OPT_R}(z')$.

Due to correctness of Recurrence (1) the algorithm described above solves DEGREE CONSTRAINED SPANNING TREE. What remains for a complete proof of Theorem 3 is a running time analysis, which has been omitted due to space restrictions.

*Remark:* The algorothm for DEGREE CONSTRAINED SPANNING TREE can be applied to find spanning subgraphs of treewidth $t$ satisfying degree constraints in time $O(2^{n+o(n)})$ for every fixed constant $t$. In addition to the degree constraints one could require the spanning subgraph to belong to a minor-closed graph family. Our approach is also easily generalizable to handle super-polynomial edge weights.

# References

1. Arnborg, S., Corneil, D.G., Proskurowski, A.: Complexity of finding embeddings in a $k$-tree. SIAM J. Algebraic Discrete Methods 8, 277–284 (1987)
2. Björklund, A., Husfeldt, T.: Inclusion–Exclusion Algorithms for Counting Set Partitions. In: FOCS, pp. 575–582 (2006)

3. Björklund, A., Husfeldt, T., Kaski, P., Koivisto, M.: Fourier meets Möbius: Fast Subset Convolution. In: STOC, pp. 67–74 (2007)
4. Bodlaender, H.L.: A partial k-arboretum of graphs with bounded treewidth. Theor. Comp. Sc. 209, 1–45 (1998)
5. Chen, J., Lu, S., Sze, S.-H., Zhang, F.: Improved Algorithms for Path, Matching, and Packing Problems. In: SODA, pp. 298–307 (2007)
6. Cohen, N., Fomin, F.V., Gutin, G., Kim, E.J., Saurabh, S., Yeo, A.: Algorithm for Finding k-Vertex Out-trees and its Application to k-Internal Out-branching Problem. In: Ngo, H.Q. (ed.) COCOON 2009. LNCS, vol. 5609, pp. 37–46. Springer, Heidelberg (2009)
7. Diestel, R.: Graph Theory. Springer, Heidelberg (2005)
8. Fernau, H., Raible, D., Gaspers, S., Stepanov, A.A.: Exact Exponential Time Algorithms for Max Internal Spanning Tree. In: Paul, C. (ed.) WG 2009. LNCS, vol. 5911, pp. 100–111. Springer, Heidelberg (2009)
9. Fomin, F.V., Grandoni, F., Kratsch, D.: A Measure & Conquer Approach for the Analysis of Exact Algorithms. Journal of ACM 56(5) (2009)
10. Fomin, F.V., Gaspers, S., Saurabh, S., Thomassé, S.: A Linear Vertex Kernel for Maximum Internal Spanning Tree. In: Dong, Y., Du, D.-Z., Ibarra, O. (eds.) ISAAC 2009. LNCS, vol. 5878, pp. 275–282. Springer, Heidelberg (2009)
11. Gaspers, S., Saurabh, S., Stepanov, A.A.: A Moderately Exponential Time Algorithm for Full Degree Spanning Tree. In: Agrawal, M., Du, D.-Z., Duan, Z., Li, A. (eds.) TAMC 2008. LNCS, vol. 4978, pp. 479–489. Springer, Heidelberg (2008)
12. Goemans, M.X.: Minimum bounded degree spanning trees. In: FOCS, pp. 273–282 (2006)
13. Gutin, G., Razgon, I., Kim, E.J.: Minimum Leaf Out-Branching Problems. In: Fleischer, R., Xu, J. (eds.) AAIM 2008. LNCS, vol. 5034, pp. 235–246. Springer, Heidelberg (2008)
14. Held, M., Karp, R.M.: A dynamic programming approach to sequencing problems. Journal of SIAM 10, 196–210 (1962)
15. Kneis, J., Molle, D., Richter, S., Rossmanith, P.: Divide-and-color. In: Fomin, F.V. (ed.) WG 2006. LNCS, vol. 4271, pp. 58–67. Springer, Heidelberg (2006)
16. Khuller, S., Bhatia, R., Pless, R.: On local search and placement of meters in networks. SIAM J. Comput. 32, 470–487 (2003)
17. Koivisto, M.: An $O(2^n)$ Algorithm for Graph Colouring and Other Partitioning Problems via Inclusion-Exclusion. In: FOCS, pp. 583–590 (2006)
18. Lawler, E.L.: A Note on the Complexity of the Chromatic Number Problem. Inform. Proc. Letters 5(3), 66–67 (1976)
19. Lawler, E.L.: Applications of a planar separator theorem. SIAM J. Comput. 9, 615–627 (1980)
20. Naor, M., Schulman, L.J., Srinivasan, A.: Splitters and Near-Optimal Derandomization. In: FOCS, pp. 182–193 (1995)
21. Nederlof, J.: Fast polynomial-space algorithms using Mobius inversion: Improving on Steiner Tree and related problems. In: Albers, S., Marchetti-Spaccamela, A., Matias, Y., Nikoletseas, S., Thomas, W. (eds.) ICALP 2009, Part I. LNCS, vol. 5555, pp. 713–725. Springer, Heidelberg (2009)
22. Prieto, E., Sloper, C.: Reducing to Independent Set Structure – the Case of k-Internal Spanning Tree. Nord. J. Comput. 12(3), 308–318 (2005)
23. Niedermeier, R.: Invitation to fixed-parameter algorithms. Oxford Lecture Series in Mathematics and its Applications, vol. 31. Oxford University Press, Oxford (2006)
24. Reed, B., Vetta, A., Smith, K.: Finding Odd Cycle Transversals. Operations Research Letters 32, 229–301 (2004)

# Finding the Minimum-Distance Schedule
# for a Boundary Searcher with a Flashlight

Tsunehiko Kameda[1,*], Ichiro Suzuki[2,**], and John Z. Zhang[3]

[1] School of Computing Science, Simon Fraser University, Canada
[2] Dept. of Computer Science, University of Wisconsin at Milwaukee, USA
[3] Dept. of Mathematics and Computer Science, University of Lethbridge, Canada

**Abstract.** Consider a dark polygonal region in which intruders move freely, trying to avoid detection. A robot, which is equipped with a flashlight, moves along the polygon boundary to illuminate all intruders. We want to minimize the total distance traveled by the robot until all intruders are detected in the worst case. We present an $O(n \log n)$ time and $O(n)$ space algorithm for optimizing this metric, where $n$ is the number of vertices of the given polygon. This improves upon the best known time and space complexities of $O(n^2)$ and $O(n^2)$, respectively. The *distance graph* plays a critical role in our analysis and algorithm design.

## 1   Introduction

In this paper we study path planning for a special kind of mobile robot. In the *polygon search* problem, first formulated by Suzuki and Yamashita [11], mobile *searchers* and *intruders* are represented by moving points in a polygonal area. We consider a special case, where a single searcher can move around on the polygon boundary with bounded speed, carrying a flashlight which he can direct in any direction with bounded rotational speed. In other words, it is a *boundary 1-searcher* in the terminology defined in [11]. The intruders try to evade detection by dodging the light beam, and can move faster than the angular movement of the flashlight beam. Tan [12] characterized the class of polygons searchable by a boundary searcher, by identifying a set of "forbidden patterns."

Our objective in this paper is to minimize the distance that needs to be traveled by the searcher to illuminate all intruders in the worst case. Our motivation is that the energy expended by the searching robot is an increasing function of the distance it travels. It is known that the minimum range of travel is less than twice the perimeter of the polygon [8]. Recently, Fukami *et al.* designed an algorithm for finding the minimum distance schedule, which runs in $O(n^2)$ time and uses $O(n^2)$ space [2]. Since the data structure they use to represent relevant information is of size $O(n^2)$, it is not possible to improve the time complexity beyond $O(n^2)$, using their approach. Our algorithm is based on a totally different approach, and runs in $O(n \log n)$ time and uses $O(n)$ space. The concepts of

* The first and last authors are supported in part by the NSERC of Canada.
** Supported in part by UWM Research Growth Initiative.

A. López-Ortiz (Ed.): LATIN 2010, LNCS 6034, pp. 84–95, 2010.

the *landmark, skeleton visibility diagram* and *distance graph* play central roles in our analysis and algorithm design. Due to lack of space, we omit the proofs of some lemmas and theorems. The interested reader is referred to [7].

## 2    Preliminaries

### 2.1    Notation and Definitions

A (simple) polygon $P$ is defined by a sequence of vertices and the edges that connect adjacent vertices. The edge between vertices $u$ and $v$ is denoted by $(u, v)$. The *boundary* of $P$, denoted by $\partial P$, consists of all its edges and vertices. The vertices immediately preceding (resp. succeeding) vertex $v$ in the clockwise[1] direction are denoted by $Pred(v)$ (resp. $Succ(v)$). For any two points $a, b \in \partial P$, the closed (resp. open) cw section of $\partial P$ from $a$ to $b$ is denoted by $\partial P[a, b]$ (resp. $\partial P(a, b)$). For three points $a, b, c \in \partial P$ we sometimes write $a \prec b \prec c$ if $b \in \partial P(a, c)$.

A point $p$ in $P$ is *visible* from another point $q$ if the line segment $\overline{pq}$ is totally contained in $P$ [8].[2] The role of the light beam is to separate the *clear* part of polygon that is free of intruders from the *contaminated* part that may contain intruders. The contiguous section of polygon boundary that is in contact with the cleared part of the polygon starts at the searcher position, $s$, extends cw, and ends at a point $p \in \partial P$, where the beam touches the polygon boundary. This point may not coincide with the *beam head*, where the beam leaves the polygon for the first time. The pair $(p, s)$ is called a *configuration*, and we sometimes denote it by $\overrightarrow{sp}$.

A vertex whose interior angle between its two incident edges is more than $180°$ is called a *reflex vertex*. For a reflex vertex $r$, extend the edge $(Succ(r), r)$ (resp. $(Pred(r), r)$) in the direction $Succ(r) \rightarrow r$ (resp. $Pred(r) \rightarrow r$), and let $B(r) \in \partial P$ (resp. $F(r) \in \partial P$) denote the point where it hits $\partial P$ for the first time. We call $B(r)$ (resp. $F(r)$) the *backward* (resp. *forward*) *extension point* associated with $r$. We call the closed polygonal area enclosed by $\partial P[r, B(r)]$ (resp. $\partial P[F(r), r]$) and line segment $\overline{B(r)r}$ (resp. $\overline{rF(r)}$) vertex $r$'s *cw* (resp. *ccw*) *component*, denoted by $C_{cw}(r)$ (resp. $C_{ccw}(r)$).

**Lemma 1.** [1] *Given a polygon with $n$ vertices, all its extension points can be computed in $O(n \log n)$ time and $O(n)$ space.*                                          □

### 2.2    Searchability and Performance Metric

Let a continuous function $I : [0, \infty) \rightarrow P$ represent the unpredictable moves of an intruder. The value $I(t)$ at time $t \geq 0$ is not known to the searcher. Let $\gamma : [0, T] \rightarrow \partial P$ be a continuous function such that $\gamma(t)$ is the position of the

---

[1] From now on, we use cw (resp. ccw) as an abbreviation for clockwise (resp. counter-clockwise).

[2] Another definition of visibility is adopted in [4], which requires that $\overline{pq} \in P - \partial P$.

searcher at time $t \geq 0$, where $T$ is a positive constant. Finally, let $\theta(t) \in \partial P$ denote a point on $\partial P$ that the beam touches at time $t$, so that $\overline{\gamma(t)\theta(t)} \subseteq P$. Thus $\theta : [0,T] \to \partial P$ is a piecewise continuous function. The function pair $(\gamma, \theta)$ is a *search schedule*, or just *schedule*, if for every continuous function $I : [0,\infty) \to P$, there exists a time $t \in [0,T]$ such that $I(t) \in \overline{\gamma(t)\theta(t)}$ [10]. The polygon $P$ is *searchable* if there exists a schedule $(\gamma, \theta)$ for it.

The *distance* of a schedule is the cumulative distance along the polygon boundary traversed by the searcher. A search schedule that minimizes the distance is called an *optimal schedule*.

## 2.3   Visibility Diagram

Let $x, y \in \mathbf{R}$, where $\mathbf{R}$ is the set of all real numbers. The *visibility space*, denoted by $\mathcal{V}$, consists of the infinite area between and including the lines $Y = X$ (*start line* $\mathcal{S}$) and $Y = X - |\partial P|$ (*goal line* $\mathcal{G}$), as shown in Fig. 1, where $|\partial P|$ denotes the perimeter of $P$ [8]. We have $(x, y) \in \mathcal{V}$ if and only if $x - |\partial P| \leq y \leq x$. We assume that an arbitrary point on $\partial P$ has been chosen as the origin. Let $0 \leq x' < |\partial P|$. Then $x'$ and $x = x' + k|\partial P|$ represent the same point on $\partial P$ that is at distance $x'$ cw from the origin for any integer $k$. The *visibility diagram*, *V-diagram* or just *VD* for short, for a given polygon is drawn in $\mathcal{V}$ by shading some areas in it gray as follows: point $(x, y) \in \mathcal{V}$ is gray if points $x, y \in \partial P$ are not mutually visible [8]. Thus, the VD has a period $|\partial P|$, and each reflex vertex $r$ gives rise to two shaded areas in each period, as shown in an idealized form in Fig. 1. The V-diagram is a generalization of the "visibility obstruction diagram" [4], which essentially represents one period of the VD.

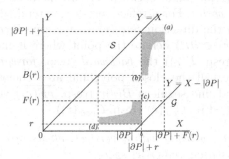

**Fig. 1.** SE and NW barriers due to reflex vertex $r$

For convenience, we refer to the directions in the VD, up, down, left and east as *north*, *south*, *west* and *east*, respectively. We call each gray area in Fig. 1 a *barrier* and a barrier whose corner is touching line $\mathcal{S}$ (resp. $\mathcal{G}$) is called a *southeast* (*SE*) (resp. *northwest* (*NW*)) *barrier* [8]. The SE (resp. NW) barrier due to reflex vertex $r$ is denoted by $SE(r)$ (resp. $NW(r)$). There are infinitely many copies of an SE (resp. NW) barrier touching line $\mathcal{S}$ (resp. $\mathcal{G}$) at $(k|\partial P|+|r|, k|\partial P|+|r|)$ for any integer $k$, where $|r|$ denotes the distance of vertex $r$ from the origin. We call some corner points of barriers *landmarks*. See points (a), (b), (c) and (d) in Fig. 1, which represent configurations $\overrightarrow{rF(r)}$, $\overrightarrow{B(r)r}$, $\overrightarrow{F(r)r}$ and $\overrightarrow{rB(r)}$, respectively. A maximal white area in the VD is called a *cell*. Note that a finite cell is a closed area, since the barriers are open. If $\overrightarrow{sp}$ is a configuration, then point $(p, s)$ in the VD is called a *configuration point*. A landmark to (resp. from) which there is a straight horizontal path from line $\mathcal{S}$ (resp. to line $\mathcal{G}$) within a cell in the VD is called an *initial landmark* (resp. *terminal landmark*). Figure 2(b) shows the VD for the polygon in Fig. 2(a). Configuration points $C$ and $C'$ represent the corresponding configurations in Fig. 2(a).

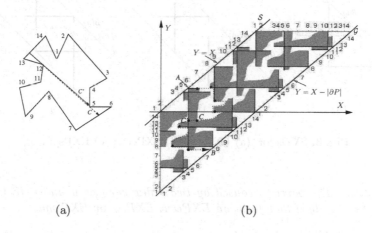

(a)                              (b)

**Fig. 2.** (a) A polygon; (b) Its VD and SVD

## 2.4   Skeleton Visibility Diagram (SVD)

We replace the SE (resp. NW) barrier caused by reflex vertex $r$ by two line segments, the horizontal line segment from landmark $(a)$ to line $\mathcal{S}$ (resp. $(d)$ to line $\mathcal{G}$) in Fig. 1 and the vertical line segment from landmark $(b)$ to $\mathcal{S}$ (resp. $(c)$ to line $\mathcal{G}$). These line segments are called the *backbones* of the barrier. A backbone does not include the landmark, i.e., it is open at that end. We assume that the two backbones of a barrier touch each other. The *skeleton visibility diagram*, or *SVD* for short, is obtained from a VD by replacing the barriers by their backbones. For example, the black line segments on the barrier faces in Fig. 2(b) constitute an SVD.

## 2.5   Other Relevant Facts

Reflex vertex $v$ *cw-excludes* (resp. *ccw-excludes*) point $p \in \partial P$ if $p \notin C_{cw}(v)$ (resp. $p \notin C_{ccw}(v)$). Two reflex vertices $u$ and $v$ form

1. a *cw exclusion pair*, *EXPcw* for short, if $u$ and $v$ cw-exclude each other.
2. a *ccw exclusion pair*, *EXPccw* for short, if $u$ and $v$ ccw-exclude each other.
3. a *symmetric exclusion pair*, *EXPsym* for short, if one of them cw-excludes the other and is ccw-excluded by the other. If $u$ cw-excludes $v$ and $v$ ccw-excludes $u$, then $\partial P(v, u)$ is called a *trap*, and any point $p \in \partial P(v, u)$ is said to be *trapped*.          □

The SVDs for the three types of EXPs are shown in Fig. 3.

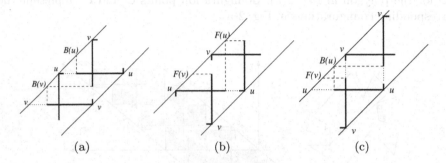

(a)                 (b)                 (c)

**Fig. 3.** SVDs for: (a) EXPcw; (b) EXPccw; (c) EXPsym

**Lemma 2.** [9] *The barriers caused by two reflex vertices $u$ and $v$ in the VD intersect if and only if they form an EXPccw, EXPcw or EXPsym.*          □

**Lemma 3.** [9] *The backbones due to two reflex vertices $u$ and $v$ in the SVD intersect if and only if $u$ and $v$ form an EXPccw, EXPcw or EXPsym.*          □

## 3   Canonical Search Path

In what follows, a *path* shall mean a simple (i.e., non-self-intersecting), directed path. A *legal path* is a path in the VD (or SVD) that stays within cells, except that it may cross a gray area (or a vertical backbone) horizontally from east to west any number of times [8]. A *search path* is a legal path from $S$ to $G$. For example, the dotted path in Fig. 2(b) from $A$ to $B$ is a search path. It crosses a barrier from point $C$ to $C'$.

**Theorem 1.** [8,10] *A given polygon is searchable by a boundary 1-searcher if and only if there is a search path in its VD.*          □

From Theorem 1 and Lemmas 2 and 3, we get

**Theorem 2.** *A given polygon is searchable by a boundary 1-searcher if and only if there is a search path in its SVD.*          □

A path in the VD is said to be *canonical* if it satisfies the following conditions:

1. It moves along barrier faces (i.e., cell boundaries) with two exceptions: It may contain horizontal eastward line segments through the interiors of cells, and it may contain horizontal westward line segments across gray areas.
2. It does not move along the curved face or east face of any SE barrier. When it moves along some other (straight) face of an SE barrier, it moves away from $S$.
3. It does not move along the east face any NW barrier, and when it moves along the north face of an NW barrier it moves towards $G$. □

The search path from $A$ to $B$ in Fig. 2(b) is canonical. All landmarks on it, as well as points $C$ and $C'$, are indicated by black dots. Whenever a canonical path reverses its vertical direction, it goes through a landmark. We thus have

**Proposition 1.** *The $Y$ coordinate of a canonical path between two successive landmarks is either non-increasing or non-decreasing.* □

**Lemma 4.** *If there is a canonical path from a landmark $L$ to another landmark $L'$ such that there is no other landmark on it, then the path is unique.*

*Proof.* This lemma is trivially true if the eastward line from $L$ hits $L'$. If this line hits the curved face or the west face of a barrier, then by Proposition 1, the canonical path may make a southward or northward[3] turn there, but the location of $L'$ uniquely determines which turn should be made. The rest of the path is clearly unique. □

**Theorem 3.** [7] *There is a canonical search path representing an optimal schedule.* □

## 4  Distance Graph

A canonical search path may go through a number of landmarks. We will first construct canonical paths between pairs of landmarks, and then synthesize a canonical search path by connecting some of them. *Distance graph $G(N, A)$* for polygon $P$ is a directed graph whose node set $N$ consists of the landmarks. For a pair of nodes, $L_1, L_2 \in N$ we have arc $(L_1, L_2) \in A$ if there is a canonical path from $L_1$ to $L_2$ in the VD of $P$ such that there is no other landmark on this path. This arc has a weight that equals the difference in the $Y$ coordinates of $L_1$ and $L_2$. By Proposition 1, the arc weight is the distance that the searcher must travel to move from the position given by the $Y$ coordinate of $L_1$ to that given by the $Y$ coordinate of $L_2$. Clearly, any path from an initial landmark to a final landmark in graph $G(N, A)$ unique specifies a canonical search path.

To find the arcs from each non-terminal landmark $L \in N$, we look for at most two landmarks to which there are canonical paths from $L$. One is the first landmark $L'$, if any, encountered when we trace a canonical path in a cell from

---

[3] From now on southward (resp. northward) will be abbreviated as sw (resp. nw).

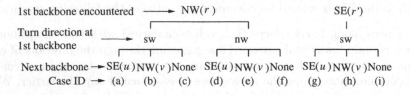

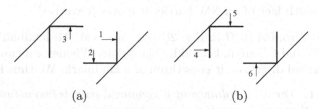

**Fig. 4.** Canonical paths from a landmark to next

**Fig. 5.** (a) Approaching backbones from the curved side of a barrier; (b) Approaching backbones from the straight side of a barrier

$L$ such that when we first hit a barrier, we turn southward (sw) there. The other landmark $L''$, if any, is the first landmark encountered when we turn northward (nw) instead. We will show below how to find $L'$ and $L''$ using the SVD, instead of the VD. All the possible cases are illustrated by the two trees in Fig. 4. The left (resp. right) tree shows the case where the first vertical backbone encountered in the SVD belongs to $NW(r)$ (resp. $SE(r')$). This is shown in Fig. 5 by arrows 1 and 4, respectively. The second level of the trees indicates the direction of the canonical path once it hits $NW(r)$ or $SE(r')$. If the first backbone encountered is due to $SE(r')$ (right tree), then there is no nw branch under $SE(r')$, since no canonical search path goes in the nw direction from there. If the first level is $NW(r)$ (left tree) and the second level is nw, as a special case of the arrow 1 in Fig. 5, the nw path moves nw immediately from $L$ for a landmark $L$ of type (d) in Fig. 1. In the third level, we show the horizontal backbones that are encountered next. If it is "None", then the next landmark is at a tip of $NW(r)$ or $SE(r')$. If it is $SE(u)$ (i.e., cases (a), (d), and (g)), there is no next landmark. Finally, if it is $NW(v)$, then the next landmark is at the southwestern (in cases (b), (h)) or northeastern (in case (e)) tip of $NW(v)$. Cases (a), (b), (d), (e), (g), and (h) at the bottom level of Fig. 4 are illustrated by arrows 5, 2, 3, 6, 5, and 2 in Fig. 5, respectively. The conditions leading to different cases, (a), (b), etc., will be given in Lemma 5 below (and Lemma 6 in [7]). Let $L = (l_x, l_y)$, and let us find the first landmark $L'$ to which there is a canonical eastward (ew) then sw path from $L$. (Direction = sw in the second level of Fig. 4.) We discuss two cases, depending on which of reflex vertices $r, r' \in \partial P(l_x, l_y)$ defined below is first encountered as we move cw from $l_x$ on $\partial P$.

$$l_y \notin C_{ccw}(r) \quad (r \text{ ccw-excludes } l_y) \tag{1}$$
$$l_y \notin C_{cw}(r') \quad (r' \text{ cw-excludes } l_y) \tag{2}$$

Note that (1) (resp. (2)) holds if, as we move eastward horizontally from $L$ in the SVD, we first hit the vertical backbone of $NW(r)$ (resp. $SE(r')$), as shown by the left (resp. right) tree in Fig. 4.

**Lemma 5.** *Suppose that* $l_x \prec r \prec r' \prec l_y$ *holds, where* $r$ *and* $r'$ *are defined by (1) and (2), respectively. We assume that* $r$ *exists, but* $r'$ *may not. (See the left tree in Fig. 4.)*

- *Let* $u$ *be the first reflex vertex encountered, if any, moving ccw from* $l_y$ *on* $\partial P(r, l_y)$, *such that* $r \prec F(u) \prec u$ *($r$ and $u$ form an EXPccw as in Fig. 3(b)).*
- *Let* $v$ *be the first reflex vertex encountered, if any, moving ccw from* $l_y$ *on* $\partial P(r, l_y)$, *such that* $v \prec B(v) \prec r$ *($r$ and $v$ form an EXPsym as in Fig. 3(c)).*

(a) *If neither* $u$ *nor* $v$ *exists, then* $L' = (B(r), r)$. *(Case (c) in Fig. 4.)*

(b) *If we have* $l_x \prec u \prec v \prec l_y$, *where* $u$ *may not exist, then we have* $L' = (B(v), v)$. *(Case (b) in Fig. 4.)*

(c) *If we have* $l_x \prec v \prec u \prec l_y$, *where* $v$ *may not exist, then* $L'$ *does not exist. (Case (a) in Fig. 4.)*

*Proof.* Vertex $r$ is the first vertex such that the vertical backbone of $NW(r)$ is hit by a horizontal line extending east from $L$ in the SVD. (a) In this case, there is no SE barrier or NW barrier that intersects $NW(r)$ to the south of line $Y = l_y$, and the canonical sw path from $L$ eventually reaches the southwest landmark of $NW(r)$. (b) Vertex $v$ is the first vertex on $\partial P$ ccw from $l_y$ such that $NW(r)$ and $NW(v)$ intersect. As we move sw on reaching an NW barrier, we may encounter $NW(v)$ such that $r \prec v \prec u \prec l_y$, but the southwest landmark of $NW(v)$ is eventually reached, i.e., $L' = (B(v), v)$. Note that the sw search path along the west face of $NW(v)$ can simply move across any SE barrier from east to west. (c) $u$ is the first vertex ccw from $l_y$ that forms an EXPccw with $r$. The vertical backbone of $NW(r)$ and the horizontal backbone of $SE(u)$ meet. Any potential canonical path is blocked by $SE(u)$. □

For example, in Fig. 6, we have case (b) with $L = (F(10), 10)$, $r = 3$, $v = 7$, and $L' = (B(7), 7)$. Note that the canonical path from $L$ to $L'$ may not first hit $NW(r)$ in the VD. In fact, in this example, it first hits $NW(5)$ at $A$ and moves along $L \to A \to B \to L'$, reaching landmark $L'$ given by case (b) in Lemma 5.[4]

## 5  Finding Arcs of $G(N, A)$ Efficiently

For $i = 1, 2, \ldots$, we give two aliases, $q_i$ and $p_i$, to reflex vertex $r_i$. For convenience, we consider that $q_i$ and $B(q_i)$ (resp. $p_i$ and $F(p_i)$) are the end points of $C_{cw}(r_i)$ (resp. $C_{ccw}(r_i)$). We say that $q_i$ (resp. $p_i$) is an item of *type* $q$ (resp. $p$).

Let $\mathcal{L}$ denote the set of all landmarks, except for the terminal landmarks. As we discussed in the previous section, for each $L \in \mathcal{L}$ we want to find the barrier $NW(r)$ or $SE(r')$ whose vertical backbone is first encountered by the eastward

---

[4] Lemma 6 in [7] deals with the cases represented by the right tree in Fig. 4.

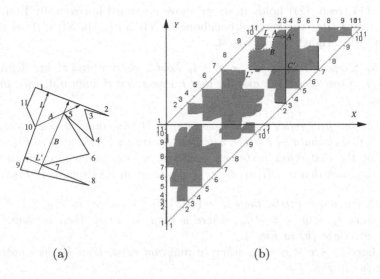

(a)                                             (b)

**Fig. 6.** Case (b): (a) A polygon; (b) Its VD showing the configuration points corresponding to configurations $L, A$, etc., in part (a)

line from $L$ in the SVD. For a particular $L \in \mathcal{L}$, we need to identify the first backbone of an NW barrier and that of an SE barrier, and then select the one that is the nearer from $L$. We want to do this collectively for all landmarks in $\mathcal{L}$ in an efficient manner. Let us first consider the backbones of NW barriers encountered by the eastward lines from them. In this case, only the ccw components caused by items of type $p$ are relevant. We construct a circular linked list $\mathcal{K}_p$ that initially lists all items of type $p$, $\{p_j \mid j = 1, 2, \ldots\}$, and their forward extension points, $\{F(p_j) \mid j = 1, 2, \ldots\}$, in the cw order. List $\mathcal{K}_p$ also contains $x_l$'s and $y_l$'s of a set of configurations of interest, $\mathcal{C} = \{\overline{y_l x_l} \mid l = 1, 2, \ldots\}$, where $x_l, y_l \in \partial P$ for each $l$. In the current setting, we have $\mathcal{C} = \mathcal{L}$.

Each item in $\mathcal{K}_p$ has links to its predecessor and successor so that if any item is deleted, the affected links can be updated in $O(1)$ time. Each item of type $p$ on $\mathcal{K}_p$ is considered to be the root of a tree, which initially consists of just the root. We start scanning $\mathcal{K}_p$ at an arbitrary item of type $p$, and as we examine each item, we *mark* it. We maintain a pointer $I$ to the first marked item of type $p$, which is updated when the item it points to is deleted. We also maintain set $S$, which is initially empty. When point $x_l$ is scanned, we detach it from $\mathcal{K}_p$, mark it, and put it in $S$. When an item $p_j$ is then scanned, we mark it, and make each point $x_l$ in $S$ a child node of the tree root at $p_j$. With this arrangement, it is easy to find the next marked item of type $p$ clockwise from $x_l$ in the future, as we will explain shortly.

We will need a family of algorithms, $NxtTouch(\mathcal{C}, t, d)$, where $t \in \{p, q\}$ denotes the type and $d \in \{ew(\text{eastward}), nw(\text{northward}), sw(\text{southward})\}$ denotes the direction. (See the arrows in Fig. 5.) For concreteness we present $NxtTouch(\mathcal{C}, p, ew)$.

Algorithm $NxtTouch(\mathcal{C}, p, ew)$

   *Construct $\mathcal{K}_p$ in the cw order, using $\mathcal{C}$ and $\{p_j, F(p_j) \mid j = 1, 2, \ldots\}$. Start at an arbitrary item of type $p$ in $\mathcal{K}_p$, and set pointer $I$ to it. Scan $\mathcal{K}_p$, carrying out the following operations, depending on the item scanned, until $\mathcal{K}_p$ becomes empty:*

1. *Unmarked $x_l$: Delete it from $\mathcal{K}_p$, mark it, and put it in set $S$.*
2. *Unmarked $p_j$: Mark it, and attach the items in $S$ to it as its child nodes and empty $S$.*
3. *$F(p_j)$: If $p_j$ is marked, then delete both $F(p_j)$ and $p_j$ from $\mathcal{K}_p$, and attach the tree at $p_j$ to the root of the tree at the next item of type $p$ from $p_j$, if it exists.*
4. *$y_l$: If $x_l$ is marked, identify the root, $p_k$, if any, of the tree that $x_l$ belongs to, delete both $y_l$ and $x_l$. If $x_l$ is not marked, let $p_k$ be the item pointed to by $I$. If $p_k$ exists, output it as the vertex causing the nearest backbone from $(x_l, y_l)$ such that $y_l \notin C_{ccw}(p_k)$ holds. (See Eq. (1).)*     □

In performing operation 4 above, if $x_l$ is marked and belongs to a tree, then $p_k$ can be found by a FIND operation of the UNION-FIND algorithm of Hopcroft and Ullman [5]. An example for the above algorithm can be found in [7].

**Lemma 6.** *Given the initialized $\mathcal{K}_p$ as input, Algorithm $NxtTouch(\mathcal{C}, p, ew)$ outputs, for each pair $(x_l, y_l) \in \mathcal{C}$, the reflex vertex $p_k$, if any, closest cw from $x_l$ such that $y_l \notin C_{ccw}(p_k)$. It runs in $O(n \log^* n)$ time.*

*Proof.* First suppose that $p_k$ is output by operation 4 of $NxtTouch(\mathcal{C}, p, ew)$ when $x_l$ is already marked. Then all $(p_j, F(p_j))$ pairs such that $\partial P(p_j, F(p_j)) \subset \partial P(x_l, y_l)$ have already been deleted. Therefore, we have $y_l \notin C_{ccw}(p_k)$. Moreover, $p_k$ is the first vertex satisfying this condition clockwise from $x_l$. Next suppose that $p_k$ was reported when $x_l$ was not marked, but no vertex is reported when $x_l$ is marked.[5] This implies that there is no $p_j \in \partial P(x_l, p_k)$ that ccw-excludes $y_l$. If no $p_k$ is output, it implies that there is no $p_j \in \partial P(x_l, y_l)$ that ccw-excludes $y_l$. Clearly, the algorithm stops before the entire $\mathcal{K}_p$ is scanned twice, and the number of UNIONs and FINDs is $O(n)$. Therefore, it runs in $O(n \log^* n)$ total time [5].     □

We had $\mathcal{C} = \mathcal{L}$ above. However, in case (b) in Fig. 4, for example, we need to use a different $\mathcal{C}$ to identify vertex $v$ defined in Lemma 5. Algorithm $NxtTouch(\mathcal{C}, p, ew)$ was tailer-made for the case shown by arrow 1 in Fig. 5(a). With minor modifications, it can also deal with the cases shown by arrows 2 and 3. To find vertex $v$ in case (b), for example, let $\mathcal{C}$ be the set of configurations on the backbones of $NW(r)$ reached from the landmarks in $\mathcal{L}$. Then for each member of $\mathcal{C}$, $v$ can be computed by $NxtTouch(\mathcal{C}, q, sw)$ with the following changes: (i) $\mathcal{K}_p \to \mathcal{K}_q$ ($\mathcal{K}_q$ contains items of type $q$ in the ccw order), and (ii) $x_l \leftrightarrow y_l$. To see this intuitively, note that if we flip the diagram in Fig. 5(a) relative to line $Y = X$, arrow 1 becomes arrow 3. This implies that we are interchanging $x_l$ and $y_l$, and

---

[5] For example, this would be the case with $x_2$ in Fig. 11 in [7] if $p_5$ were absent.

this can be reflected in operations 1 and 4 of $NxtTouch(\mathcal{C}, p, ew)$. If we flip the diagram relative to line $Y = 0$ and then relative to line $X = 0$, arrow 3 turns into arrow 2. This implies that we are changing the direction of movement from cw to ccw, as well as changing the type from $p$ to $q$. If we reflect these flipping operations in $NxtTouch$, we obtain algorithms that can find the next backbones in all the cases shown in Fig. 5(a).

However, $NxtTouch$ cannot handle the cases shown in Fig. 5(b). In [7], we present Algorithm $NxtReach$ to deal with them. We need to invoke both Algorithm $NxtTouch(\mathcal{C}, t, d)$ and $NxtReach(\mathcal{C}, \bar{t}, d)$ and for each member of $\mathcal{C}$, to identify the vertex of type $p$ or $q$, whichever is encountered first, where if $t = p$ (resp. $q$) then $\bar{t} = q$ (resp. $p$). Minor changes are needed to stop looking for $NW(v)$ or $SE(u)$ in cases (d) and (e) in Fig. 4, for example, when the north tip of $NW(r)$ is reached.

## 6   Optimal Search Path

This section presents our main result, namely an $O(n \log n)$ time and $O(n)$ space algorithm for finding an optimal search path by a 1-searcher that minimizes the distance traveled by the searcher. Now that we have constructed distance graph $G(N, A)$, the remaining task is straightforward and can be carried out as follows:

1. Modify $G(N, A)$ by adding the source node and the sink node. Introduce an arc with weight 0 to (resp. from) each node representing an initial (resp. terminal) landmark from (resp. to) the source (resp. sink) node.
2. Apply Dijkstra's algorithm to find the shortest path from the source node to the sink node.                                                                 □

We now show that the distance graph is planar, so that the shortest path in it can be found in linear time. To see this let us look at subdiagram $D$ of the V-diagram, consisting of one period between the lines $Y = c$ and $Y = c + |\partial P|$ for some $c$. The distance graph is "almost planar", except that some arcs may wrap vertically around $D$. Now map line $\mathcal{S}$ in $D$ onto a circle, and map line $\mathcal{G}$ in $D$ onto a larger concentric circle. Then the top line of $D$ and the bottom line of $D$ naturally meet, and the resulting diagram, hence the distance graph, is planar.

By Lemma 1 we can compute all extension points in $O(n \log n)$ time. This implies that we can construct graph $G(N, A)$ in $O(n \log n)$ time. Graph $G(N, A)$ contains $O(n)$ arcs, since there are at most two arcs out of each node. Since it is a planar graph, the shortest path can be computed in $O(n)$ time.

**Theorem 4.** *Given a polygon of $n$ vertices, an optimal search path can be constructed in $O(n \log n)$ time and $O(n)$ space, where $n$ is the number of vertices in the given polygon.*                                                                 □

Since our output schedule is given at a higher level than "instructions" used in [6], their bound of $\Omega(n^2)$ on the schedule length does not apply. Each step of our schedule is of the form "Move from landmark $L$ to landmark $L'$."

# 7    Conclusion and Discussion

The main contribution of this paper is to present an $O(n \log n)$ time and $O(n)$ space algorithm for minimizing the distance traveled by a searcher moving on the boundary. The time is dominated by $O(n \log n)$ that is required for finding all extension points. Once they are known, the remaining task can be carried out in $O(n \log^* n)$ time. If the more sophisticated UNITE-FIND algorithm of Gabor and Tarjan [3] is used, then this can be improved to $O(n)$. We made an extensive use of the VD and SVD throughout the paper, demonstrating their usefulness.

# References

1. Chazelle, B., Guibas, L.: Visibility and intersection problems in plane geometry. In: Proc. 3rd Annual Symp. on Computational Geometry, pp. 135–146 (1985)
2. Fukami, K., Ono, H., Sadakane, K., Yamashita, M.: Optimal polygon search by a boundary 1-searcher. In: Proc. Workshop on Foundations of Theoretical Computer Science: For New Computational View, RIMS, Kyoto University, January 2008, pp. 182–188 (2008)
3. Gabor, H., Tarjan, R.: A linear-time algorithm for a special case of disjoint set union. Journal of Computer and System Sciences 30, 209–221 (1985)
4. Guibas, L., Latombe, J., LaValle, S., Lin, D., Motwani, R.: A visibility-based pursuit-evasion problem. Int'l. J. of Computational Geometry and Applications 9(5), 471–494 (1999)
5. Hopcroft, J., Ullman, J.: Set merging algorithm. SIAM J. Computing 2, 294–303 (1973)
6. Icking, C., Klein, R.: The two guards problem. Int'l. J. of Computational Geometry and Applications 2(3), 257–285 (1992)
7. Kameda, T., Suzuki, I., Zhang, Z.: Finding the minimum-distance schedule for a boundary searcher with a flashlight. Tech. Rep. TR 2009-24, School of CS, Simon Fraser Univ. (December 2009)
8. Kameda, T., Yamashita, M., Suzuki, I.: On-line polygon search by a seven-state boundary 1-searcher. IEEE Trans. on Robotics 22, 446–460 (2006)
9. Kameda, T., Zhang, J.Z., Yamashita, M.: Simple characterization of polygons searchable by 1-searcher. In: Proc. 18th Canadian Conf. on Computational Geometry, August 2006, pp. 113–116 (2006)
10. Suzuki, I., Tazoe, Y., Yamashita, M., Kameda, T.: Searching a polygonal region from the boundary. Int'l. J. of Computational Geometry and Applications 11(5), 529–553 (2001)
11. Suzuki, I., Yamashita, M.: Searching for a mobile intruder in a polygonal region. SIAM J. on Computing 21(5), 863–888 (1992)
12. Tan, X.: A characterization of polygonal regions searchable from the boundary. In: Akiyama, J., Baskoro, E.T., Kano, M. (eds.) IJCCGGT 2003. LNCS, vol. 3330, pp. 200–215. Springer, Heidelberg (2005)

# The Language Theory of Bounded Context-Switching

S. La Torre[1], P. Madhusudan[2], and G. Parlato[1,2]

[1] Università degli Studi di Salerno, Italy
[2] University of Illinois at Urbana-Champaign, USA

**Abstract.** Concurrent compositions of recursive programs with finite data are a natural abstraction model for concurrent programs. Since reachability is undecidable for this class, a restricted form of reachability has become popular in the formal verification literature, where the set of states reached within $k$ context-switches, for a fixed small constant $k$, is explored. In this paper, we consider the language theory of these models: concurrent recursive programs with finite data domains that communicate using shared memory and work within $k$ round-robin rounds of context-switches, and where further the stack operations are made visible (as in visibly pushdown automata). We show that the corresponding class of languages, for any fixed $k$, forms a robust subclass of context-sensitive languages, closed under all the Boolean operations. Our main technical contribution is to show that these automata are *determinizable* as well. This is the first class we are aware of that includes non-context-free languages, and yet has the above properties.

## 1 Introduction

Concurrent threads with recursive procedures that communicate using shared memory is a natural and attractive model, as it models concurrent imperative programs naturally. While message-passing is more common in the distributed computing world where processes run on different machines, the advent of multi-core computing has led to an increased interest in shared-memory programs.

In the methodology of model-checking for program verification, a common paradigm is to analyze a program by verifying a model of it, where the model is obtained by abstracting or simplifying the *data-domain* used by a program, but preserving control flows accurately. Many program analysis frameworks, such as data-flow analysis or predicate abstraction, fall under this category. Hence concurrent recursive programs where variables range over a finite data-domain are an attractive model of study.

The automata-theoretic model of a concurrent program with recursion and shared-memory communication is simply an automaton with multiple stacks. (Note that in such a model, the shared memory, and hence the communication between processes, resides in the control state of the automaton.) Since Turing machines can be simulated by an automaton with two stacks, the *emptiness* problem for these automata, and hence the model-checking problem, is undecidable.

A. López-Ortiz (Ed.): LATIN 2010, LNCS 6034, pp. 96–107, 2010.

In order to overcome this barrier, a recent proposal is to search only the space reached by these automata using a bounded number of context-switches. In other words, we view the computation as occurring in $k$ consecutive stages (for a fixed constant $k$), where in each stage only one of the concurrent threads is active. This restriction in the automaton model translates to restricting the computation to $k$ stages, where in each stage, only one of the stacks is manipulated. It turns out that in this model the reachability problem is decidable (for any fixed $k$) [16].

The idea of checking and testing concurrent programs under a context-switching bound has gained attention in recent years for several reasons. First, there is an intuitive appeal that one expects most concurrency errors to manifest themselves within a few context-switches. This has been argued fairly effectively in recent experimental studies (see [14]). Second, the model-checking problem for bounded context-switching is decidable [16], thus yielding exact algorithmic methods to solve the reachability problem. And third, checking concurrent programs under a context-bound can be done *compositionally*— we can search the state space by avoiding to build explicitly the product of local states of all automata, and instead work with only the space defined by a single thread and $k$ copies of shared variables. The last aspect is in fact a very appealing (and the least articulated) aspect of bounded context-switching that has been exploited in recent work: model-checking tools for concurrent Boolean programs have been developed [8,12,17], and translations of concurrent programs to sequential programs have been developed that reduce bounded context-switching reachability to sequential reachability, even for general C-programs [9,12]. In recent work, the above translation has even been used to verify concurrent programs under a context-bound using deductive verification tools for sequential software [11].

In this paper, we undertake a language and automata-theoretic study of the concurrent programs with recursion under a context-switching bound. While research has so far concentrated on the computation of reachable states, we instead look at the *class of languages* accepted by these automata. In doing so, we make the calls and returns to procedures in the concurrent program *visible*— for sequential programs this yields the class of *visibly pushdown automata* [1], which has been shown to define a robust class of context-free languages, and has led to a flurry of research (see [19] for a list of papers in this area).

We consider automata with $n$ stacks, where an execution goes through $k$ *round-robin schedules* of computation, i.e., a round is a fixed sequence of exactly $n$ contexts, one for each stack. In a *context* for a stack $i$, for $i = 1, \ldots, n$, the automaton can only read letters pertaining to stack $i$ and manipulate stack $i$.

The visibility of actions on the stack immediately implies that the class of languages is closed under union and intersection. Surprisingly, we show that the nondeterministic and deterministic versions of these automata are equivalent. The determinization construction is the key technical theorem in this paper, and crucially uses the compositional reasoning of the automata using interfaces of tuples of global states that we alluded to earlier. Determinizability gives us closure under complement as well, and hence shows that the class of automata with $k$-rounds of round-robin scheduling is a robust class closed under all Boolean operations.

The class of languages accepted by visibly multi-stack pushdown automata with a bounded number of context-switching rounds is the only class we are aware of that includes non-context-free languages, has a decidable emptiness (and membership) problem, is closed under all Boolean operations, and is further determinizable. (It is easy to see that this class is a subclass of context-sensitive languages, i.e. languages accepted by nondeterministic Turing machines with linear space). Note that standard *complexity* classes defined using resource constraints seldom have a decidable emptiness problem (one can prove undecidability by padding input).

We make several other observations regarding these automata. First, when the automata are generalized to have no bound on the number of round-robins of schedule, they are *not* determinizable. Second, it is well-known now that the emptiness problem restricted to only the words that can be accepted up to a bounded number of round-robins is decidable for these automata and in fact is NP-complete [13,16]. Thus, from the closure under boolean operations, it follows that universality and inclusion are decidable. Third, since these automata define a subclass of bounded-phase multi-stack pushdown automata [10], it follows that the Parikh theorem holds for the bounded context-switching class as well. Further, we show that the monadic second-order logic on $n$-nested words with $k$ rounds, where the logic has $n$ binary relations corresponding to the $n$ nesting relations on the word, corresponds exactly to the class of languages introduced in this paper.

Notice that our results show that the $k$-round-robin executions of multi-stack automata, which define a subclass of *context-sensitive languages* is determinizable. In [10], we show that even if there is *one phase* where a 2-stack automaton can push onto both stacks, followed by two phases where the automaton can pop from one stack only in each phase, is non-determinizable. This one example of non-determinizability rules out most natural extensions of multi-stack automata (where restrictions are based only on the patterns of pushes and pops) from being determinizable, and leads us to conjecture that considerably larger and determinizable sub-classes of context-sensitive languages defined using multi-stack nondeterministic automata are unlikely to exist.

In conclusion, the contribution of this paper is to exhibit a class of languages (those accepted by multi-stack automata with bounded rounds of context-switches), the first that we are aware of, that includes certain non-context-free languages, and has all the desirable properties that regular languages possess: closure under Boolean operations, decidable problems of membership, emptiness, and inclusion, determinizability and an MSO characterization.

**Related Work:** The classes of multi-stack automata studied in this paper are proper subclasses of the multi-stack automata which work in a bounded number of phases, where in each phase, symbols can be pushed into all stacks but popped only from one stack [10]. Though the class of languages accepted by bounded phase multi-stack automata is closed under all Boolean operations as well, such automata are *not determinizable*, while the class of automata we consider here are. To the best of our knowledge, the class of automata we have introduced in this paper is the first extension of visibly pushdown automata with multiple stacks which is determinizable. [4] gives a wrong determinization construction for 2-stack visibly pushdown automata, which are indeed not determinizable (even if the stacks usage is constrained such that pop operations on the second stack are allowed only if the first stack is empty) as shown in [6]. Also, the class of 2-stack visibly pushdown automata is in general not closed under complement [2], while the answer is not known when such automata are constrained with an ordering on the usage of stacks (the proof of such closure property given in [4] relied on the determinizability of the model). Our determinization construction uses tuples of global states to capture the points at which context-switching occurs, similar to earlier papers [12,9] that reduce reachability in concurrent programs to reachability in sequential programs. However, the determinizability result does not follow from such conversions as the latter only preserve reachability, and not language equivalence (after all, sequential programs with finite data domains define only context-free languages). Besides the papers we have already cited, bounded context-switching has also been exploited for systems with heaps [3], systems communicating using queues [7], and weighted pushdown systems [13].

## 2   Multi-Stack Pushdown Automata

In this section we give the notation and definitions to introduce the model of automata we will use in the rest of the paper.

Given two positive integers $i$ and $j$, $i \leq j$, we denote with $[i,j]$ the set of integers $k$ with $i \leq k \leq j$, and with $[j]$ the set $[1,j]$.

An $n$-stack call-return alphabet is a tuple $\widetilde{\Sigma}_n = \langle \Sigma_c^i, \Sigma_r^i, \Sigma_{int}^i \rangle_{i=1}^n$ of pairwise disjoint finite alphabets. For any $i \in [n]$, $\Sigma_c^i$ is a finite set of *calls of stack* $i$, $\Sigma_r^i$ is a finite set of *returns of stack* $i$, and $\Sigma_{int}^i$ is a finite set of *internal actions of stack* $i$. For any such $\widetilde{\Sigma}_n$, let

- $\Sigma^i = \Sigma_c^i \cup \Sigma_r^i \cup \Sigma_{int}^i$, for every $i \in [n]$;
- $\Sigma_c = \bigcup_{i=1}^n \Sigma_c^i$, $\Sigma_r = \bigcup_{i=1}^n \Sigma_r^i$, and $\Sigma_{int} = \bigcup_{i=1}^n \Sigma_{int}^i$;
- $\Sigma = \Sigma_c \cup \Sigma_r \cup \Sigma_{int}$.

A multi-stack visibly pushdown automaton over such an alphabet must push on stack $i$ exactly one symbol when it reads a call of the $i$-th alphabet, and pop exactly one symbol from stack $i$ when it reads a return of the $i$-th alphabet. Also, it cannot touch any stack when reading an internal symbol.

**Definition 1.** (MULTI-STACK VISIBLY PUSHDOWN AUTOMATON) *A multi-stack visibly pushdown automaton (MVPA) over the n-stack call-return alphabet* $\widetilde{\Sigma}_n = \langle \Sigma_c^i, \Sigma_r^i, \Sigma_{int}^i \rangle_{i=1}^n$, *is a tuple* $M = (Q, Q_I, \Gamma, \delta, Q_F)$ *where $Q$ is a finite set of states, $Q_I \subseteq Q$ is the set of initial states, $\Gamma$ is a finite stack alphabet that contains a special bottom-of-stack symbol* $\bot$, $\delta \subseteq (Q \times \Sigma_c \times Q \times (\Gamma \setminus \{\bot\})) \cup (Q \times \Sigma_r \times \Gamma \times Q) \cup (Q \times \Sigma_{int} \times Q)$, *and $Q_F \subseteq Q$ is the set of final states.*

*Moreover, M is* deterministic *if* $|Q_I| = 1$, *and* $|\{(q, a, q') \in \delta\} \cup \{(q, a, q', \gamma) \in \delta\} \cup \{(q, a, \gamma', q') \in \delta\}| \leq 1$, *for every given* $q \in Q$, $a \in \Sigma$ *and* $\gamma' \in \Gamma$.

Let us fix an $n$-stack alphabet $\widetilde{\Sigma}_n$ for the rest of the paper.

A transition $(q, a, q', \gamma)$, for $a \in \Sigma_c^i$ and $\gamma \neq \bot$, is a push-transition where on input $a$, $\gamma$ is pushed onto stack $i$ and the control changes from $q$ to $q'$. Similarly, $(q, a, \gamma, q')$ for $a \in \Sigma_r^i$ is a pop-transition where on input $a$, $\gamma$ is read from the top of stack $i$ and popped (except for $\gamma = \bot$, which is never popped), and the control changes from $q$ to $q'$. A transition $(q, a, q')$, for $a \in \Sigma_{int}$, is an internal transition where on input $a$ the control changes from $q$ to $q'$.

A *stack content* $\sigma$ is a nonempty finite sequence over $\Gamma$ where the bottom-of-stack symbol $\bot$ appears always in the end, i.e., $\sigma \in (\Gamma \setminus \{\bot\})^*.\{\bot\}$. A *configuration* of an MVPA $M$ over $\widetilde{\Sigma}_n$ is a tuple $C = \langle q, \sigma_1, \ldots, \sigma_n \rangle$, where $q \in Q$ and each $\sigma_i$ is a stack content. Moreover, $C$ is *initial* if $q \in Q_I$ and $\sigma_i = \bot$ for every $i \in [n]$, and *accepting* if $q \in Q_F$. *Transitions* between configurations are defined as follows: $\langle q, \sigma_1, \ldots, \sigma_n \rangle \xrightarrow{a}_M \langle q', \sigma_1', \ldots, \sigma_n' \rangle$ if one of the following holds ($M$ is omitted whenever it is clear from the context):

**[Push]** If $a \in \Sigma_c^i$ (i.e., $a$ is a call of stack $i$), then $\exists \gamma \in \Gamma \setminus \{\bot\}$ such that $(q, a, q', \gamma) \in \delta$, $\sigma_i' = \gamma \cdot \sigma_i$, and $\sigma_h' = \sigma_h$ for every $h \in ([n] \setminus \{i\})$.

**[Pop]** If $a \in \Sigma_r^i$ (i.e., $a$ is a return of stack $i$), then $\exists \gamma \in \Gamma$ such that $(q, a, \gamma, q') \in \delta$, $\sigma_h' = \sigma_h$ for every $h \in ([n] \setminus \{i\})$, and either $\gamma \neq \bot$ and $\sigma_i = \gamma \cdot \sigma_i'$, or $\gamma = \sigma_i = \sigma_i' = \bot$.

**[Internal]** If $a \in \Sigma_{int}$, then $(q, a, q') \in \delta$, and $\sigma_h' = \sigma_h$ for every $h \in [n]$.

For a word $w = a_1 \ldots a_m$ in $\Sigma^*$, a *run* of $M$ on $w$ from $C_0$ to $C_m$, denoted $C_0 \xrightarrow{w}_M C_m$, is a sequence of transitions $C_{i-1} \xrightarrow{a_i} C_i$ for $i \in [m]$ where each $C_i$ is a configuration. A word $w \in \Sigma^*$ is accepted by an MVPA $M$ if there is an initial configuration $C$ and an accepting configuration $C'$ such that $C \xrightarrow{w}_M C'$. The language accepted by $M$ is denoted with $L(M)$.

A visibly pushdown automaton [1] is an MVPA with just one stack.

**Definition 2.** (VISIBLY PUSHDOWN AUTOMATON) *A visibly pushdown automaton, denoted* VPA, *is an* MVPA *over* $\widetilde{\Sigma}_n$ *with* $n = 1$. *A language over* $\Sigma$ *accepted by a* VPA *is a* visibly pushdown language. *With* VPL *we denote the class of visibly pushdown languages.*

## 2.1 Restricting to a Bounded Number of Rounds

A *context* over $\Sigma^i$, with $i \in [n]$, is a word in $(\Sigma^i)^*$. A *round* over $\widetilde{\Sigma}_n$ is a word $w$ of $\Sigma^*$ of the form $w_1 w_2 \ldots w_n$ where for each $i \in [n]$, $w_i$ is a context over $\Sigma^i$. A *k-round word* over $\widetilde{\Sigma}_n$ is a word of $\Sigma^*$ that can be obtained as the concatenation of $k$ rounds over $\widetilde{\Sigma}_n$. Let $Round(\widetilde{\Sigma}_n, k)$ denote the set of all the $k$-round words over $\widetilde{\Sigma}_n$.

$\Sigma_c^1 = \{a\}, \ \Sigma_r^1 = \{b\}, \ \Sigma_{int}^1 = \emptyset, \ \Sigma_c^2 = \{x\}, \ \Sigma_r^2 = \{y\}, \ \Sigma_{int}^2 = \emptyset$

$A = (\, 2, \{q_i | i \in [0,5]\}, \{q_0\}, \{\#, \$\}, \delta, \{q_5\}\, )$

$\delta = \{\, (q_0, a, q_1, \$), \ (q_1, a, q_1, \#), \ (q_1, x, q_2, \$), \ (q_2, x, q_2, \#), \ (q_2, b, \#, q_3),$
$\quad (q_3, b, \#, q_3), \ (q_3, b, \$, q_4), \ (q_2, b, \$, q_4), \ (q_4, y, \#, q_4), \ (q_4, y, \$, q_5)\,\}$

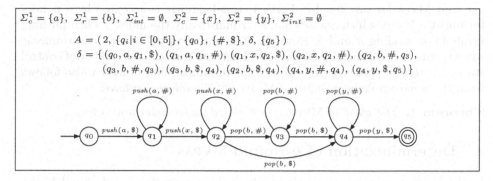

**Fig. 1.** A 2–round MVPA recognizing the language $\{\, a^t x^s b^t y^s \mid t, s \geq 1 \}$

**Definition 3.** (MULTI-STACK VISIBLY PUSHDOWN LANGUAGES WITH $k$-ROUNDS)
*For any $k$, a $k$-round multi-stack visibly pushdown automaton (k-round MVPA)
over $\widetilde{\Sigma}_n$ is a tuple $A = (k, Q, Q_I, \Gamma, \delta, Q_F)$ where $M = (Q, Q_I, \Gamma, \delta, Q_F)$ is an MVPA
over $\widetilde{\Sigma}_n$. Moreover, $A$ is deterministic iff $M$ is deterministic. The language accepted
by $A$ is $L(A) = L(M) \cap Round(\widetilde{\Sigma}_n, k)$ and is called a $k$-round multi-stack visibly
pushdown language. The class of $k$-round multi-stack visibly pushdown languages
is denoted with $k$-RVPL. The set $\bigcup_{k \geq 1} k$-RVPL is denoted with RVPL (the class of
multi-stack visibly pushdown languages with a bounded number of rounds).*

*Example 1.* Figure 1 gives a formal definition of a 2–round MVPA $A$ over $\widetilde{\Sigma}_2$
that accepts the language $\{a^t x^s b^t y^s | t, s \geq 1\}$. (Note that this language is not
context-free and $A$ is deterministic.) The automaton $A$ checks whether the input
word has the form $a^+ x^+ b^+ y^+$ using its control states. $A$ starts in the control
state $q_0$. When it reads the first call symbol $a$ it pushes the symbol $\$$ onto the
stack $S_1$; for all the remaining $a$'s $A$ pushes the symbol $\#$ onto $S_1$. Stack $S_1$ will
contain as many symbols as the number of read $a$'s. When the first call symbol
$x$ of stack 2 is read a $\$$ symbol is pushed onto stack $S_2$, for the remaining $x$'s
the symbol $\#$ is pushed onto stack $S_2$. As in the previous case, stack $S_2$ will
contain as many symbols as the $x$'s which are read. Stack $S_1$ is then popped for
each return symbol $b$ until $S_1$ is empty (read the symbol $\$$). Then only return
symbols $y$ can be read. Stack $S_2$ is popped for each read $y$ until it gets empty
(popping the symbol $\$$). After that $A$ moves into the accepting state $q_5$.     □

The main result on MVPAs with a bounded number of rounds, that we show in
this paper, is that the class of languages accepted by the deterministic and the
nondeterministic models coincide. Notice that the boundedness of the number
of rounds is crucial in our proof. In fact, determinizability does not hold in gen-
eral for MVPAs. To see this consider the language $L = \{(ab)^i c^j d^{i-j} x^j y^{i-j} | i \in
\mathbb{N}, j \in [i]\}$ over $\widetilde{\Sigma}_2 = (\Sigma_c^1, \Sigma_r^1, \emptyset, \Sigma_c^2, \Sigma_r^2, \emptyset)$, with $\Sigma_c^1 = \{a\}, \ \Sigma_r^1 = \{c, d\}, \ \Sigma_c^2 =
\{b\}, \ \Sigma_r^2 = \{x, y\}$. First, observe that since the number of occurrences of $ab$ is
unbounded in $L$ and $a$ and $b$ are from different stacks, this language contains
words with an unbounded number of rounds and thus cannot be accepted by a

$k$-round MVPA for any fixed $k$. Further, this language is accepted by a nondeterministic MVPA which guesses nondeterministically the index $j$ when pushing symbols on reading $a$ and $b$. However, it is not accepted by any deterministic MVPA, since a deterministic MVPA would need an unbounded number of control states to store the index $j$ (see [10]). Nondeterminizability of MVPA also follows from the non-complementability of MVPA [2]. Therefore, we have:

**Theorem 1.** *The class of* MVPA *is not closed under determinization.*

## 3   Determinization of $k$-Round MVPAs

In this section we prove the main result of this paper: if $A$ is a $k$-round MVPA over $\widetilde{\Sigma}_n$, then there exists a deterministic $k$-round MVPA $A^D$ over $\widetilde{\Sigma}_n$ such that $L(A) = L(A^D)$.

Fix a $k$-round MVPA $A = (k, Q, Q_I, \Gamma, \delta, Q_F)$ over $\widetilde{\Sigma}_n = \langle \Sigma_c^j, \Sigma_r^j, \Sigma_{int}^j \rangle_{j=1}^n$ and a word $w \in Round(\widetilde{\Sigma}_n, k)$.

For ease of presentation, in the rest of this section we assume that each context of $w$ is not empty. Also, we denote with $w[i,j]$ the $j$-th context of the $i$-th round in $w$, and with $A_j$, $j \in [n]$, the VPA $(Q, Q_I, \Gamma, \delta', Q_F)$ over $\langle \Sigma_c^j, \Sigma_r^j, \Sigma_{int}^j \rangle$ where $\delta' \subseteq \delta$ is the set of all moves of $\delta$ on symbols of $\Sigma^j$ (i.e., the VPA which equals $A$ on the $j$-th stack).

The main idea behind our construction of $A^D$ is to look at the executions of $A$ on $w$ as shown in Fig. 2. The automaton $A$ is seen as a composition of the $A_j$'s. Initially $A$ is in control state $q_{\langle 1,1 \rangle}$. Then, it starts the computation by passing $q_{\langle 1,1 \rangle}$ to $A_1$. $A_1$ reads $w[1,1]$ and reaches a control state $q'_{\langle 1,1 \rangle}$ with stack content $\sigma_{\langle 1,1 \rangle}$. At this point, $A$ stops $A_1$ and passes $q_{\langle 1,2 \rangle} = q'_{\langle 1,1 \rangle}$ to $A_2$. $A_2$ reads $w[1,2]$ and reaches a state $q'_{\langle 1,2 \rangle}$ with stack content $\sigma_{\langle 1,2 \rangle}$. And so on, from $A_3$ through $A_n$, until $q'_{\langle 1,n \rangle}$ is reached. Now, $A$ passes $q_{\langle 2,1 \rangle} = q'_{\langle 1,n \rangle}$ to $A_1$. Since this is the first time $A_1$ is re-activated after reading $w[1,1]$, its stack (i.e., the first one) has not changed in the meantime. Thus $A_1$ starts now from control state $q_{\langle 2,1 \rangle}$ and stack content $\sigma_{\langle 1,1 \rangle}$. Then, again by reading $w[2,1]$, $A_1$ reaches a control state $q'_{\langle 2,1 \rangle}$ and $A_2$ is started from control state $q_{\langle 2,2 \rangle} = q'_{\langle 2,1 \rangle}$ and stack content $\sigma_{\langle 1,2 \rangle}$, and so on, until completion of the whole run.

The salient aspect in the above description is that each run of $A$ on $w$ can be computed by running each $A_j$ individually on $w[1,j], \ldots, w[k,j]$, provided that $q_{\langle 1,j \rangle}, \ldots, q_{\langle k,j \rangle}$ are fed. Also, note that $A_j$ computes a relation of state pairs $\langle q_{\langle i,j \rangle}, q'_{\langle i,j \rangle} \rangle$ which are connected by a run of $A_j$ over words $w[i,j]$ for $i \in [k]$, and thus, a relation of tuples $\langle q_{\langle i,j \rangle}, q'_{\langle i,j \rangle} \rangle_{i=1}^k$ corresponding to words $\langle w[i,j] \rangle_{i=1}^k$. We call such tuples *switching vectors*. Note that the switching vectors store all the information we need to stitch together the local runs of all $A_j$'s in order to build a global run of $A$.

This suggests the following scheme to construct $A^D$. First, for each $A_j$, construct a VPA $A'_j$ that computes the switching vectors corresponding to $\langle w[i,j] \rangle_{i=1}^k$ when reading $w[1,j]\# \ldots \#w[k,j]\#$, where $\#$ is a fresh internal symbol (computed switching vectors are stored in the final states). Construct for each $A'_j$ an

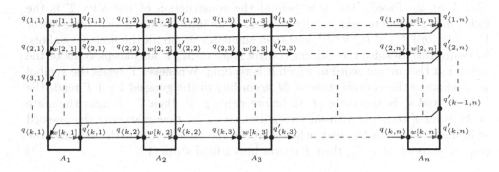

**Fig. 2.** Decomposition of a $k$-round MVPA

equivalent deterministic VPA $A_j^D$, and then, $A^D$ by composing them such that:
(1) the states of $A^D$ are the cross product of the states of the $A_j^D$'s; (2) in each
context over $\Sigma^j$, except for the first symbol, only $A_j^D$ is executed, and on reading
the first symbol $a$ of a context $j+1$ of a round $i$, $A_j^D$ is executed on input $\#$ and
$A_{j+1}^D$ is executed on input $a$; (3) a word $w$ is accepted if the computed switching
vectors for each $A_j^D$ can be composed according to a scheme such as in Fig. 2,
i.e., they form a *sequence of compatible tuples*.

The above sketched construction is formally addressed in the rest of this
section (more details are available in the Appendix). We start by defining the
switching vectors, and then construct the VPA computing the switching vectors
for a given VPA. We then define the concept of compatible tuples and prove that
acceptance of $A$ can be checked by verifying the existence of a sequence of compatible switching vectors of $A_1, \ldots, A_n$. Finally, we construct $A^D$ by composing
the VPAs computing the switching vectors of $A_1, \ldots, A_n$ and argue its soundness
and completeness.

### 3.1   Visibly Pushdown Automata Computing Switching-Vectors

**Definition 4.** (SWITCHING VECTORS) *Let $M$ be a VPA over $\widetilde{\Sigma}_1$ with set of
control states $Q$, and $u = \langle u_i \rangle_{i=1}^k$ be a tuple of $k$ words in $\Sigma^*$. The tuple $V =
\langle (q_i, q_i') \rangle_{i=1}^k \in (Q \times Q)^k$, is a switching-vector of $M$ with respect to $u$ if there
exist $k$ pairs of $M$ configurations $\langle (C_i, C_i') \rangle_{i=1}^k$, with $C_i = \langle q_i, \sigma_i \rangle$ and $C_i' =
\langle q_i', \sigma_i' \rangle$, such that (1) $\sigma_1 = \perp$, (2) $\sigma_i' = \sigma_{i+1}$, for every $i \in [k-1]$ (3) $C_i \xrightarrow{u_i}_M C_i'$,
for every $i \in [k]$.*

In the next lemma we prove the existence of a VPA $T$, called *switching automaton*,
that computes switching-vectors of a given VPA.

**Lemma 1.** (SWITCHING AUTOMATA) *Let $M$ be a VPA over $\langle \Sigma_c, \Sigma_r, \Sigma_{int} \rangle$, and
let $\#$ be a fresh symbol not in $\Sigma$. Then, there exists a nondeterministic VPA $T$
over $\langle \Sigma_c, \Sigma_r, \Sigma_{int} \cup \{\#\} \rangle$ such that $V$ is a switching-vector of $M$ with respect
to $\langle u_i \rangle_{i=1}^k \in \Sigma^k$ iff while reading the word $u_1 \# u_2 \# \ldots \# u_k \#$, $T$ enters a state
which contains $V$ (denoted as $\langle V \rangle$).*

*Sketch of the Proof.* The idea behind the construction of the VPA $T$ is the following. $T$ nondeterministically guesses, in its initial state, a switching-vector $V = \langle (q_i, q_i') \rangle_{i=1}^k \in (Q \times Q)^k$ and then simulates $M$ on all the non-# symbols. In doing this, besides the current control state of $M$, $T$ also keeps track of the index $i$ of the current word $u_i$ which it is reading. Whenever $T$ reads the symbol #, it changes the control state of $M$ according to the guessed $V$: if $T$ reads the $i$-th #, and $q_i'$ is the state of $M$ before reading #, then $T$ changes the state of $M$ from $q_i'$ to $q_{i+1}$, with an internal move on #, thus matching the guessed switching-vector. In the end, when $T$ reads the last # (the $k$-th one) and the control state of $M$ is $q_k'$, then $T$ moves into a final state $\langle V \rangle$.    □

## 3.2   Compatible Tuples

**Definition 5.** (COMPATIBLE TUPLES) *Let* $V_j = \langle (q_{\langle i,j \rangle}, q_{\langle i,j \rangle}') \rangle_{i=1}^k$ *for* $j \in [n]$, *a sequence of* compatible *tuples* $V_1, V_2, \ldots, V_n$ *is such that*

- $q_{\langle i,j \rangle}' = q_{\langle i,j+1 \rangle}$, *for every* $i \in [k], j \in [n-1]$, *and*
- $q_{\langle i,n \rangle}' = q_{\langle i+1,1 \rangle}$, *for every* $i \in [k-1]$.

The following lemma is used in the next section to argue soundness and completeness of the determinization construction. It relates the acceptance of a word by a $k$-round MVPA to the existence of a sequence of compatible switching-vectors.

**Lemma 2.** *Let* $w \in Round(\widetilde{\Sigma}_n, k)$, $A = (k, Q, Q_I, \Gamma, \delta, Q_F)$ *be a $k$-round* MVPA *over* $\widetilde{\Sigma}_n$, *and* $w_j = \langle w[i,j] \rangle_{i=1}^k$, *for* $j \in [n]$. *The word* $w \in L(A)$ *iff for each* $j \in [n]$, *there exists a switching-vector* $V_j = \langle (q_{\langle i,j \rangle}, q_{\langle i,j \rangle}') \rangle_{i=1}^k$ *of the* VPA $A_j$ *with respect to* $w_j$ *such that* $V_1, V_2, \ldots, V_n$ *is a sequence of compatible tuples*, $q_{\langle 1,1 \rangle} \in Q_I$, *and* $q_{\langle k,n \rangle} \in Q_F$.

## 3.3   Determinization of $k$-Round MVPAs

**Theorem 2.** *(*DETERMINIZABILITY*) If $A$ is a $k$-round* MVPA *over* $\widetilde{\Sigma}_n$, *then there exists a deterministic $k$-round* MVPA $A^D$ *over* $\widetilde{\Sigma}_n$ *such that* $L(A) = L(A^D)$. *Moreover, the size of* $A^D$ *is doubly exponential in the number of rounds, and singly exponential in the number of stacks and the number of states of $A$.*

*Proof.* For $j \in [n]$, let $T_j$ be a switching automaton which accepts the same language as $A_j$ and is constructed according to Lemma 1, and $S_j$ be the deterministic VPA such that $L(S_j) = L(T_j)$ which is obtained via the construction given in [1]. Thus, a state of $S_j$ contains the set of reachable control states of $T_j$. From the definition of $T_j$, it is easy to see that the $S_j$ state reached on input $u_1 \# u_2 \# \ldots \# u_k \#$ is the set of all the switching-vectors of $A_j$ with respect to $\langle u_i \rangle_{i=1}^k \in \Sigma^k$.

The idea behind the construction of $A^D$ is the following. $A^D$ simulates each $S_j$, by keeping track of the control state of every $S_j$. The entire simulation mimics the schema shown in Fig. 2. After the input word $w$ is completely read, $A^D$

reaches a state storing the set of all switching-vectors of each $A_j$. The states of $A^D$ which contain a sequence of compatible tuples are defined final. Thus, from Lemma 2, $A^D$ accepts the input word $w$ if and only if $A$ also does.

An issue that has to be addressed in the simulation of all $S_j$ is the following. Let $w$ be the input word of $A$. $S_j$ needs to read a symbol $\#$ when a context-switch happens, i.e., at the end of each context $j$ of each round. Thus, the idea is to simulate the move on $\#$ meanwhile $A^D$ processes the first symbol of the next context. This solves the problem for all the occurrences of $\#$ but the last one (no other context follows). Thus, for the simulation of $S_n$ in the last round, we keep a pair $(q, q_\#)$ where $q$ is the current state of $S_n$ in the simulation, and $q_\#$ is the state computed from $q$ by applying the transition on $\#$. Thus, $q$ is used to simulate the moves of $S_n$, and $q_\#$ is considered only for acceptance.     □

## 4   Discussion

The class of languages studied in this paper is closely related to the class of *bounded-phase* MVPAs studied in [10]; using this relationship, we can derive many properties for RVPLs. Intuitively, a phase is a stage of computation of a multi-stack automaton where push actions are allowed on all stacks while pop actions are allowed only on one (hence phases generalize contexts).

MVPA *with a bounded number of phases.* Given a word $w \in \Sigma^*$, we denote with $Ret(w)$ the set of all returns in $w$. A word $w$ is a *phase* if $Ret(w) \subseteq \Sigma_r^i$, for some $i \in [n]$. For any $k$, a $k$-phase word is a word $w \in \Sigma^*$ such that $w$ can be factorized as $w = w_1 w_2 \ldots w_{k'}$ where $k' \le k$ and $w_h$ is a phase, for every $h \in [k']$. With $Phases(\widetilde{\Sigma}_n, k)$ we denote the set of all $k$-phase words over $\widetilde{\Sigma}_n$.

For any $k$, a $k$-*phase multi-stack visibly pushdown automaton* ($k$-phase MVPA) over $\widetilde{\Sigma}_n$ is a tuple $A = (k, Q, Q_I, \Gamma, \delta, Q_F)$ where $M = (Q, Q_I, \Gamma, \delta, Q_F)$ is an MVPA over $\widetilde{\Sigma}_n$. The language accepted by $A$ is $L(A) = L(M) \cap Phases(\widetilde{\Sigma}_n, k)$. The class of languages accepted by $k$-phase MVPAs is denoted with $k$-PVPL, and the set $\bigcup_{k>0} k$-PVPL is denoted with PVPL (the class of all languages accepted by a $k$-phase MVPA for some $k$).

**Theorem 3 ([10]).** *Let $k$ be any positive integer. $k$-PVPLs are closed under union, intersection, and complement. The membership, emptiness, inclusion, equivalence, and universality problems are decidable for $k$-PVPLs. $k$-PVPLs are not determinizable.*

The notion of phase is less restrictive than the notion of context, i.e., a context is a phase, and hence a round of context-switching can be simulated using a bounded number of phases. Hence:

**Lemma 3.** *Let the number of stacks be $n$. Then $k$-RVPL $\subset (k \cdot n)$-PVPL and RVPL $\subset$ PVPL.*

*Closure properties and decision problems.* Closure under union and intersection of $k$-RVPL can be shown with standard constructions, and decidability

| | Closure properties | | | | Decision Problems | |
|---|---|---|---|---|---|---|
| | ∪ | ∩ | Compl. | Determ. | Emptiness | Univ./ Equiv./Incl. |
| Reg. | Yes | Yes | Yes | Yes | NLOG-C | PSPACE-C |
| VPL | Yes | Yes | Yes | Yes | PTIME-C | EXPTIME-C |
| CFL | Yes | No | No | No | PTIME-C | Undecidable |
| **Rvpl** | **Yes** | **Yes** | **Yes** | **Yes** | **NP-c** | **2Exptime** |
| PVPL | Yes | Yes | Yes | No | 2ETIME-C | 3EXPTIME 2EXPTIME-HARD |
| CSL | Yes | Yes | Yes | Unknown | Undecidable | Undecidable |

**Fig. 3.** Summary of main closure properties and decision problems

of decision problems such as membership and emptiness can be inherited from $k$-PVPL. Notice that since complementation of $k$-RVPL is defined with respect to words with bounded rounds of context-switching, closure under complement does not immediately follow from closure under complement for $k$-PVPL. However, closure under complement for $k$-RVPL follows from determinizability of the corresponding class of automata. Therefore, we get the following results:

**Theorem 4.** *Let $k$ be any positive integer. $k$-RVPLs are closed under union, intersection, and complement. The membership, emptiness, inclusion, equivalence, and universality problems are decidable for $k$-RVPLs.*

The table in Figure 3 summarizes the closure properties and decision problems for CSLs, CFLs, VPLs, PVPLs, RVPLs, and regular languages (see [1] for VPLs, and [5] for CSLs,CFLs and regular languages). In the table, NLOG-C stands for NLOG-complete, and so on.

***Parikh Theorem.*** The Parikh mapping $\Phi(w)$, introduced by Parikh [15], associates a word with the vector of natural numbers that reflect the number of occurrences of the symbols in the word. This mapping extends to languages in the natural way. Since a Parikh theorem holds for $k$-phase MVPAS [10], from Lemma 3 we get:

**Corollary 1.** *For every RVPL $L$ over $\widetilde{\Sigma}_n$, there exists a regular language $L'$ over $\Sigma$ such that $\Phi(L') = \Phi(L)$. Moreover, $L'$ can be effectively computed.*

***A Logical Characterization.*** Consider the *monadic second-order logic* ($\mathrm{MSO}_\mu$) over $\widetilde{\Sigma}_n$ defined by:

$$\varphi := P_a(x)|x \in X|x \leq y|\mu_j(x,y)|\neg\varphi|\varphi \vee \varphi|\exists x\varphi|\exists X\varphi$$

where $j \in [n]$, $a \in \Sigma$, $x, y$ are a first-order variables and $X$ is a set variable [10].

The models are words over $\Sigma$. Each of the $n$ binary relations $\mu_j$ ($j \in [n]$) is interpreted as the nested matching relation of calls and returns of $\Sigma^j$. We denote with $R_k(\varphi)$ the set of all words of $Round(\widetilde{\Sigma}_n, k)$ that satisfy a sentence $\varphi$. By standard techniques that convert MSO to automata (given that the automata are closed under boolean operations and projection), we get (see [18]):

**Theorem 5.** *A language $L$ is a $k$-RVPL over $\widetilde{\Sigma}_n$ iff there is an $\mathrm{MSO}_\mu$ sentence $\varphi$ over $\widetilde{\Sigma}_n$ with $R_k(\varphi) = L$.*

# References

1. Alur, R., Madhusudan, P.: Visibly pushdown languages. In: STOC, pp. 202–211. ACM, New York (2004)
2. Bollig, B.: On the expressive power of 2-stack visibly pushdown automata. LMCS 4(4:16), 1–35 (2008)
3. Bouajjani, A., Fratani, S., Qadeer, S.: Context-bounded analysis of multithreaded programs with dynamic linked structures. In: Damm, W., Hermanns, H. (eds.) CAV 2007. LNCS, vol. 4590, pp. 207–220. Springer, Heidelberg (2007)
4. Carotenuto, D., Murano, A., Peron, A.: 2-visibly pushdown automata. In: Harju, T., Karhumäki, J., Lepistö, A. (eds.) DLT 2007. LNCS, vol. 4588, pp. 132–144. Springer, Heidelberg (2007)
5. Hopcroft, J.E., Ullman, J.D.: Introduction to Automata Theory, Languages, and Computation. Addison-Wesley, Reading (1979)
6. La Torre, S., Madhusudan, P., Parlato, G.: 2-VPAs are not determinizable (2007), http://www.cs.uiuc.edu/homes/~madhu/vpa/wrong-proof-CMP07.html
7. La Torre, S., Madhusudan, P., Parlato, G.: Context-bounded analysis of concurrent queue systems. In: Ramakrishnan, C.R., Rehof, J. (eds.) TACAS 2008. LNCS, vol. 4963, pp. 299–314. Springer, Heidelberg (2008)
8. La Torre, S., Madhusudan, P., Parlato, G.: Analyzing recursive programs using fixed-point calculus. In: PLDI, pp. 211–222. ACM, New York (2009)
9. La Torre, S., Madhusudan, P., Parlato, G.: Reducing context-bounded concurrent reachability to sequential reachability. In: Bouajjani, A., Maler, O. (eds.) CAV 2009. LNCS, vol. 5643, pp. 477–492. Springer, Heidelberg (2009)
10. La Torre, S., Madhusudan, P., Parlato, G.: A robust class of context-sensitive languages. In: LICS, pp. 161–170. IEEE Computer Society, Los Alamitos (2007)
11. Lahiri, S., Qadeer, S., Rakamaric, Z.: Static and precise detection of concurrency errors in systems code using SMT solvers. In: Bouajjani, A., Maler, O. (eds.) CAV 2009. LNCS, vol. 5643, pp. 509–524. Springer, Heidelberg (2009)
12. Lal, A., Reps, T.W.: Reducing concurrent analysis under a context bound to sequential analysis. In: Gupta, A., Malik, S. (eds.) CAV 2008. LNCS, vol. 5123, pp. 37–51. Springer, Heidelberg (2008)
13. Lal, A., Touili, T., Kidd, N., Reps, T.W.: Interprocedural analysis of concurrent programs under a context bound. In: Ramakrishnan, C.R., Rehof, J. (eds.) TACAS 2008. LNCS, vol. 4963, pp. 282–298. Springer, Heidelberg (2008) (See also Tech Report TR-1598, Comp. Sc. Dept., Univ. of Wisconsin, Madison)
14. Musuvathi, M., Qadeer, S.: Iterative context bounding for systematic testing of multithreaded programs. In: PLDI, pp. 446–455. ACM, New York (2007)
15. Parikh, R.: On context-free languages. J. ACM 13(4), 570–581 (1966)
16. Qadeer, S., Rehof, J.: Context-bounded model checking of concurrent software. In: Halbwachs, N., Zuck, L.D. (eds.) TACAS 2005. LNCS, vol. 3440, pp. 93–107. Springer, Heidelberg (2005)
17. Suwimonteerabuth, D., Esparza, J., Schwoon, S.: Symbolic context-bounded analysis of multithreaded Java programs. In: Havelund, K., Majumdar, R., Palsberg, J. (eds.) SPIN 2008. LNCS, vol. 5156, pp. 270–287. Springer, Heidelberg (2008)
18. Thomas, W.: Languages, automata, and logic. In: Handbook of Formal Languages, vol. 3, pp. 389–455 (1997)
19. VPL, http://www.cs.uiuc.edu/homes/~madhu/vpa/

# Local Search Performance Guarantees for Restricted Related Parallel Machine Scheduling

Diego Recalde[1,*], Cyriel Rutten[2], Petra Schuurman[3], and Tjark Vredeveld[2]

[1] Escuela Politécnica Nacional, Department of Mathematics,
Ladrón de Guevara E11-253, Quito, Ecuador
drecalde@math.epn.edu.ec
[2] Maastricht University, Department of Quantitative Economics,
P.O. Box 616, 6200 MD Maastricht, The Netherlands
c.rutten@maastrichtuniversity.nl, t.vredeveld@maastrichtuniversity.nl
[3] Sterk spel, Eindhoven, The Netherlands
petraschaakt@xs4all.nl

**Abstract.** We consider the problem of minimizing the makespan on restricted related parallel machines. In restricted machine scheduling each job is only allowed to be scheduled on a subset of machines. We study the worst-case behavior of local search algorithms. In particular, we analyze the quality of local optima with respect to the jump, swap, push and lexicographical jump neighborhood.

**Keywords:** Local Search, Performance Guarantee, Restricted Machines, Eligibility Constraints.

## 1 Introduction

We consider the problem of minimizing the makespan in restricted related parallel machine scheduling. In this setting, each job is only allowed to be scheduled on a subset of machines. The problem is also known as the related parallel machine scheduling problem with eligibility constraints [15,16,24] or as the restricted assignment model for related parallel links [1,7,8]. It has applications in, among others, operating systems, communication networks [13], semiconductor manufacturing [3], and the throughput management of hospital operating rooms [25].

The problem is defined as follows. Given is a set $J = \{1, ..., n\}$ of $n$ jobs and a set $M = \{1, ..., m\}$ of $m$ machines. Each job $j$ needs to be scheduled on one of its *eligible machines* $\mathcal{M}_j \subseteq M$. We will refer to the family $\{\mathcal{M}_j\}$ as *eligibility sets*. We also say that job $j$ is *allowable* on machine $i$ if $i \in \mathcal{M}_j$. A machine $i \in M$ can process at most one job at a time, and all jobs and machines are available at time 0. Each machine $i \in M$ is characterized by a processing speed $s_i > 0$. Similarly, each job has a processing requirement $p_j > 0$. If a job $j$ is allowable on a machine $i$, then the processing time of job $j$ on machine $i$, $p_{ij}$, equals $p_j/s_i$. If job $j$ is

---

* Partially supported by a grant of the Ecuadorian Organization for Science and Technology (SENACYT).

A. López-Ortiz (Ed.): LATIN 2010, LNCS 6034, pp. 108–119, 2010.

not allowable on machine $i$, then $p_{ij}$ is set to infinity. We refer to the setting as stated above as *restricted related parallel machines*. The term "restricted" refers to jobs being restricted in the sense that they are only allowed to be processed on a subset of machines. The objective is to schedule the jobs in such a way that the *makespan* is minimized, i.e., we seek the last job to be completed as early as possible. The *latency* of a machine is the ratio of the processing requirements of all jobs assigned to the machine over its speed. Then, the *makespan* equals the maximum latency over machines. In absence of the eligibility constraints, that is $\mathcal{M}_j = M$ for all jobs $j$, the model is known as the *(uniform) related parallel machine scheduling model*. In the special case of *restricted identical parallel machines* the processing speed of each machine equals one. Furthermore, we refer to the special case wherein the processing requirements of all jobs equal one as the *restricted related parallel machines with identical (unit-length) jobs*. Adapting the standard notation introduced by [11], the problems of minimizing the makespan on restricted identical parallel machines, restricted related parallel machines with identical jobs and restricted related parallel machines are denoted by $P|\mathcal{M}_j|C_{\max}$, $Q|\mathcal{M}_j, p_j = 1|C_{max}$ and $Q|\mathcal{M}_j|C_{\max}$, respectively, see e.g., [14,21]. Standard scheduling problems which minimize the makespan on identical or related parallel machines are both known to be strongly NP-hard [9]. Hence, the problems with eligibility constraints are strongly NP-hard as well. One way to deal with NP-hard problems is to find approximative solutions. If an algorithm is guaranteed to deliver a solution that has a value at most $\rho$ times the optimal solution value, we call it a *$\rho$-approximation* algorithm. $\rho$ is called the *performance guarantee*.

A way to find approximate solutions is through *local search*. Local search methods iteratively search through the set of feasible solutions. Starting from an initial solution, a local search procedure moves from a feasible solution to a neighboring solution until some stopping criteria are met. A *neighborhood function* defines for each feasible solution a set of solutions which are in some sense close to it. This set is called a *neighborhood*. The choice of a suitable neighborhood function has an important influence on the performance of local search. The simplest form of local search is *iterative improvement*, also called local improvement algorithms. This method iteratively chooses a better solution in the neighborhood of the current solution and it terminates when no better solution is found; we say that the final solution is a *local optimum*.

**Neighborhoods.** In this paper, we investigate the performance guarantees of four different neighborhoods for various restricted related parallel machines settings, namely the *jump, swap, push* and *lexicographical jump (lexjump) neighborhood*.

Before discussing the neighborhoods, we first describe our representation of a schedule. Since the order in which the jobs are processed on a machine does not influence the latency of the corresponding machine, we will represent a schedule by an assignment. An *assignment* $A$ is uniquely determined by a partition of the set of jobs $J$ into $m$ disjoint subsets $J_1^A, J_2^A, \ldots, J_m^A$ where $J_i^A$ denotes the set of jobs assigned to machine $i \in M$ in assignment $A$. Let $A(j) \in \mathcal{M}_j$ denote the machine to which job $j$ is assigned in assignment $A$, that is, $A(j) = i$ implies $j \in J_i^A$ and vice versa. The *load of a machine* is the total processing requirement assigned to the machine for some assignment $A$, i. e., $L_i^A = \sum_{j \in J_i^A} p_j$, for all $i \in M$. The *latency of a machine* is the total processing time needed by a machine to process all jobs which are assigned to it, i. e., $\Lambda_i^A = \sum_{j \in J_i^A} p_{ij} = L_i^A / s_i$. Obviously, for identical parallel machines, $\Lambda_i^A = L_i^A$ for all machines $i \in M$. A *critical machine* is a machine with maximum latency. The makespan of some given assignment $A$, $C_{\max}^A$, i. e., the latest completion time of a job, equals the latency of the critical machine(s). Thus, $C_{\max}^A = \max_{i \in M} \Lambda_i^A$.

The first neighborhood we consider is the *jump neighborhood*, also known as the *move neighborhood*. A *jump* is defined as jumping or moving a job from the machine to which it is currently assigned to another machine (on which it is allowed). In the jump neighborhood, jobs are iteratively jumped from a critical machine to a non-critical machine. We say that an assignment is a *jump optimal assignment* if no jump decreases the makespan or the number of critical machines without increasing the makespan.

The second neighborhood we consider is the *swap neighborhood*. Select two jobs, $j$ and $k$, assigned to different machines, i. e., $j \in J_i^A$, $k \in J_h^A$ and $i \neq h$, such that $h \in \mathcal{M}_j$ and $i \in \mathcal{M}_k$. A *swap* is performed by interchanging the machine allocations of the jobs. If all jobs are assigned on the same machine, then no swap neighbor exists. Therefore, we define the swap neighborhood as one that consists of all possible jumps which jump a job from a critical machine to a non-critical machine and all possible swaps which swap a job from a critical machine with another job from a non-critical machine. We say that an assignment is a *swap optimal assignment* if no jump or swap decreases the makespan or the number of critical machines without increasing the makespan.

Next, we consider the *push neighborhood* introduced by Schuurman and Vredeveld [23]. A *push* consists of a sequence of jumps. Starting with an assignment $A$ with makespan $C_{\max}^A$, a push is initiated by selecting a job $k$ on a critical machine and a machine $i \in \mathcal{M}_k$ to move it. We say that $k$ *fits* on $i$ if $p_{ik} + \sum_{j \in J_i^A : p_{ij} \geq p_{ik}} p_{ij} < C_{\max}^A$. If a job $k$ fits on some machine $i$, then we move $j$ to $i$ and iteratively remove the smallest job from $i$ until the latency of $i$ is less than $C_{\max}^A$. The removed jobs are gathered in a queue. We now have a queue of pending jobs and a partial assignment that has lower makespan or fewer critical machines. If the queue is non-empty, then the largest job in the queue is removed and moved to one of its eligible machine on which it fits, in the same way as the first job was pushed. Thus, if necessary, we allow some smaller jobs to be removed. If the largest job in the queue does not fit on any eligible machine, then we say that the push is *unsuccessful*. We repeat the procedure of moving the largest job in the queue to a machine until the queue is empty or until we have determined that the push is unsuccessful. If none of the jobs on any of the critical machines can succesfully be pushed, then we are in a *push optimal assignment*. The push neighborhood is explained in more detail in [23].

The last neighborhood we consider is the *lexicographical jump (lexjump) neighborhood*. Define the *experienced latency of a job* as the latency of the machine to which the job is currently assigned. An assignment $A$ is lexjump optimal if no jump exists which decreases the latency of a machine $i$ without increasing the latency of another machine $h \neq i$ to a value exceeding the original latency of machine $i$. In other words, no job can decrease its experienced latency by jumping to another machine. That is, $A$ is lexjump optimal if $p_{hj} + \Lambda_h^A \geq \Lambda_i^A$ for all $i \in M$, $j \in J_i^A$, $h \in \mathcal{M}_j$. Notice that the notion of a lexjump optimal assignment corresponds to the notion of pure Nash equilibrium in the context of load balancing games, see e. g., [26].

**Related Work.** Worst case analysis of local search has become increasingly popular in the last decade. A summary of the best known upperbounds on the performance guarantees for the jump, swap, push and the lexjump neighborhood for (unrestricted) identical parallel machines and (unrestricted) related parallel machines is provided in Table 1. Schuurman and Vredeveld [23] provided examples showing that the bounds of the jump and the swap neighborhood are tight for identical and related parallel machines. In addition to the bounds provided in Table 1, Brueggemann, Hurink, Vredeveld, and Woeginger [2] introduced the so-called split neighborhood and considered the corresponding performance guarantees. They showed that this exponentially sized neighborhood has a performance guarantee of at most $2 - 2/(m+1)$. Moreover, combining this neighborhood with the jump neighborhood improves the guarantee only to $2 - 4/(m+3)$. However, if the split neighborhood is combined with the lexjump neighborhood, the performance guarantee drops to $3/2$.

A lexjump optimal assignment corresponds to a pure Nash equilibrium for the appropriate defined game. Therefore, the results on the price of anarchy carry over to performance guarantees for lexjump optimal assignments. Czumaj and Vöcking [5] showed the performance guarantee of a lexjump optimal assignment on related parallel machines is in $O(\min\{\log m/\log\log m), \log(s^+/s^-)\}$, where $s^+ := \max_i s_i$ and $s^- := \min_i s_i$, and that this bound is aymptotically tight.

**Table 1.** Local search performance guarantees for unrestricted parallel machines which are shown to be tight. "LB" and "UB" denote a lowerbound and an upperbound on the performance guarantee respectively if the performance guarantee is not shown to be tight.

| Setting | Jump | Swap | Push | Lexjump |
|---|---|---|---|---|
| 2 identical machines | $\frac{4}{3}$[6] | $\frac{4}{3}$[6] | $\frac{8}{7}$[23] | $\frac{4}{3}$[26] |
| Identical machines $(m > 2)$ | $2 - \frac{2}{m+1}$[6] | $2 - \frac{2}{m+1}$[6] | UB $= \frac{4}{3} - \frac{1}{3m}$[23]  LB $= \frac{4m}{3m+1}$[23] | $2 - \frac{2}{m+1}$[26] |
| Related machines | $\frac{1+\sqrt{4m-3}}{2}$[4] | $\frac{1+\sqrt{4m-3}}{2}$[4] | UB $= 2 - \frac{2}{m+1}$[23]  LB $= \frac{3}{2} - \epsilon$[23] | $O\left(\frac{\log m}{\log\log m}\right)$[26] |

For unrelated parallel machines, Awerbuch, Azar, Richter and Tsur [1] showed that the performance guarantee is in $\Theta(p^+ + (\log m/\log(2 + (\log m)/p^+)))$ where $p^+ := \max_{j,i,h:p_{ij}<\infty} p_{ij}/p_{hj}$.

Recently, results for lexjump optimal assignments on restricted parallel machines have been developed. Awerbuch et al. [1] proved that the performance guarantee for identical machines is bounded by $\Theta(\log m/(r \cdot \log(2 + (\log m)/r)))$, where $r$ denotes the ratio between the makespan of the optimal schedule and the largest task (note $r \geq 1$). Note that the general bound for identical machines of $\Theta(\log m/\log\log m)$ are obtained by setting $r = 1$, i.e., when making no assumptions on the largest job in the system. Hoefer and Souza [12] provided an alternative upper bound for the performance guarantee: $1 + m^2/\sum_{j\in J} p_j$. Gairing, Lücking, Mavronicolas and Monien [7] showed that the performance guarantee for restricted related parallel machines and identical jobs is in $\Omega(\log n/\log\log n)$. For restricted related parallel machines, they show that the performance guarantee is bounded from below by $m - 1$ and bounded from above by $m$. Since the counterexample of Gairing et al. [7], which shows that the performance guarantee for restricted related machines can be as bad as $m - 1$, is somewhat artificial, Lu and Yu [18] introduced the concept of $\lambda$-*goodness instances* to develop an alternative performance guarantee. An instance is $\lambda$-good if and only if every job can use at least one machine which has a speed of no less than $s^+/\lambda$. Lu and Yu show that for $\lambda$-good instances, the performance guarantee is in $\Theta\left(\min\left\{\frac{\log \lambda m}{\log\log \lambda m}, m\right\}\right)$.

For a more elaborate overview of worst case analysis and other theoretical aspects of local search, we refer to the book of Michiels, Aarts, and Korst [19].

In [21,17,15] polynomial time algorithms to solve several special cases for scheduling unit-length jobs on restricted related parallel machines to optimality are given. Glass and Kellerer [10] gave several polynomial time approximation algorithms for special cases of the problem of restricted parallel machines with performance guarantees of $2 - 1/m$ or better. A PTAS for the identical parallel machines cases with a special type of eligibility sets is given by Ou, Leung, and

**Table 2.** Local search performance guarantees for restricted parallel machines. Let $s^- := \min_i s_i$, $s^+ := \max_i s_i$, $\tilde{s}_i := s_i/s^-$ and $\tilde{s} := s^+/s^-$.

| Setting | Jump/Swap/Push | Lexjump |
|---|---|---|
| Identical Machines | $1/2 + \sqrt{m - 3/4}$ | $O\left(\frac{\log m}{\log\log m}\right)$ [1] |
| Identical Jobs | $\sqrt{\left(1 + \frac{m-1}{n}\right)\sum_{i\in M} \tilde{s}_i}$ | $O\left(\frac{\log n}{\log\log n}\right)$ [7] |
| Related machines | $1/2 + \sqrt{1/4 + (m - 1)\tilde{s}}$ | $O\left(\frac{\log \sum_{i\in M} \tilde{s}_i}{\log\log \sum_{i\in M} \tilde{s}_i}\right)$ |

Li [20]. We refer to Leung and Li [14] for a survey on results on polynomial time algorithms, complexity issues and approximation schemes concerning scheduling problems with restricted machines.

**Our Contribution.** In this paper we consider the following neighborhoods: the jump, swap, push and lexicographical jump neighborhood. We analyze the quality of each neighborhood by establishing worst-case performance guarantees for the restricted identical parallel machines, restricted related parallel machines with identical jobs and restricted related parallel machines problems. The new performance guarantees are summarized in Table 2, see the unreferenced bounds. Furthermore, we provide examples to show that these performance guarantees are tight or almost tight.

## 2  Performance Guarantees for Restricted Identical Parallel Machines

In this section, we provide performance guarantees for the scheduling problem of minimizing makespan on restricted identical parallel machines. For the jump neighborhood we obtain the following result.

**Theorem 1.** *A jump-optimal assignment for restricted identical parallel machines has makespan at most $1/2 + \sqrt{m - 3/4}$ times the optimal makespan.*

Theorem 1 follows straightforward from Theorem 5 since for the case of identical machines $s_i = 1$ for all machines $i \in M$. The following example shows that there exist instances for which the performance guarantee is tight.

*Example 1.* Let $k$ be an arbitrary positive integer and consider an instance with $n = k(k-1)+1$ jobs and $m = n$ machines. All jobs have processing time $p_j = 1$. Jobs $1, \ldots, k$ can only be processed on the first $k$ machines. The remaining jobs are allowable on all machines. Consider the following assignment which is depicted in Figure 1. Jobs $1, \ldots, k$ are assigned to machine 1. Machines $2, \ldots, k$ process each $k - 1$ of the remaining jobs. This assignment is jump optimal and

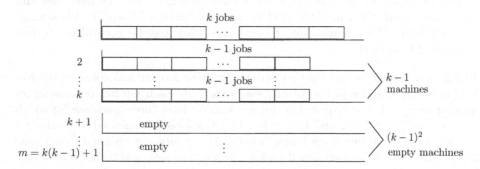

**Fig. 1.** Jump Optimal Assignment

has a makespan of $C^A_{\max} = k$, whereas in an optimal assignment, each machine processes only one job and $C^{OPT}_{\max} = 1$. Hence, $C^A_{\max}/C^{OPT}_{\max} = k = 1/2 + \sqrt{m - 3/4}$.

The following theorem has been established independently by Awerbuch, Azar, Richter and Tsur [1] and Gairing, Lücking, Mavronicolas and Monien [7].

**Theorem 2 (Awerbuch et al. [1], Gairing et al. [7]).** *The performance guarantee of lexjump optimal assignments for the problem of minimizing the makespan on restricted identical parallel machines is $O\left(\log m/\log\log m\right)$.*

Gairing et al. [7] also provide an example showing that the bound of Theorem 2 is tight up to a constant factor.

## 3    Performance Guarantees for Restricted Related Parallel Machines with Identical (Unit-Length) Jobs

In this section, we discuss performance guarantees on restricted related parallel machines for the special case of identical (unit-length) jobs. The general case for arbitrary jobs will be discussed in the next section. For now we will assume that $p_j = 1$ for all jobs $j \in J$. Denote by $s^-$ the minimum speed among all machines, i. e., $s^- := \min_{i \in M} s_i$. For the jump neighborhood we have the following result:

**Theorem 3.** *A jump optimal assignment for restricted related parallel machines with identical (unit-length) jobs has a makespan of at most*

$$\sqrt{\left(1 + \frac{m - 1}{n}\right) \sum_{i \in M} \widetilde{s}_i}$$

*where $\widetilde{s}_i$ denotes the relative speed, i. e., $\widetilde{s}_i := s_i/s^-$.*

The proof of Theorem 3 is not provided here due to space limitations. Instead we refer to [22]. The proof uses similar arguments as the proof of Theorem 5.

The example below shows that there exist instances of three machines and particular speeds for which the performance guarantee of Theorem 3 is asymptotically tight. We remark that the example below can be generalized to any number of machines.

*Example 2.* Let $k$ be an arbitrary strictly positive integer and consider the following instance and a jump-optimal assignment $A$. Each job has processing requirement $p_j = 1$ as is required in this section. We have three machines for which $s_1 = s^- = 1$, $s_2 = k - 1$ and $s_3 = k(k - 1) - 1$. $k$ jobs are assigned to machine 1 but are allowed on machines 1 and 2. $k(k - 1) - 1$ jobs are assigned to machine 2 but are allowed on machines 2 and 3. No jobs are assigned to machine 3. We have $\Lambda^A_1 = C^A_{\max} = k$, $\Lambda^A_2 = k - 1/(k - 1)$ and $\Lambda^A_3 = 0$. An optimal assignment is obtained by assigning $k - 1$ jobs, which are in $A$ assigned to machine 1, to machine 2 and by assigning all jobs which are in $A$ assigned to machine 2, to machine 3. Consequently, $C^A_{\max}/C^{OPT}_{\max} = k/1 = k$. Theorem 3 yields the following upper bound on the performance guarantee

$$\sqrt{\left(1 + \frac{m-1}{n}\right) \sum_{i \in M} \widetilde{s}_i} = \sqrt{\left(1 + \frac{3-1}{k^2-1}\right)(k^2-1)} = \sqrt{k^2+1} \stackrel{k \to \infty}{\longrightarrow} k = \frac{C^A_{\max}}{C^{OPT}_{\max}}. \quad (1)$$

The results from Gairing, Lücking, Mavronicolas and Monien [7] yield the following result.

**Theorem 4 (Gairing et al. [7], Theorem 3.1 and Theorem 4.2).** *The performance guarantee of a lexjump optimal assignment for the problem of minimizing the makespan on restricted related parallel machines and identical jobs is $\Theta(\log n/\log\log n)$.*

## 4 Performance Guarantees for Restricted Related Parallel Machines

In this section, we establish performance guarantees for the scheduling problem of minimizing the makespan on restricted related parallel machines. Recall that in this machine environment $p_{ij} = p_j/s_i$ for $i \in \mathcal{M}_j$ and $p_{ij} = \infty$ otherwise. Let $s^+ := \max_{i \in M} s_i$ and let $s^- := \min_{i \in M} s_i$. For the jump neighborhood we obtain the following result.

**Theorem 5.** *A jump optimal assignment for restricted related parallel machines has makespan at most $1/2 + \sqrt{1/4 + (m-1)\tilde{s}}$ times the optimal solution value; where $\tilde{s} := s^+/s^- = \max_{i,h \in M} s_i/s_h$.*

*Proof:* Consider a jump optimal assignment $A$ having makespan $C^A_{\max}$. Assume w.l.o.g. that machine 1 is a critical machine, i.e., $\Lambda^A_1 = C^A_{\max}$. Let $\mathcal{M}^A_1$ be the set of machines to which a job, currently assigned to machine 1 for assignment $A$, can be moved, i.e., $\mathcal{M}^A_1 = \bigcup_{j \in J^A_1} \mathcal{M}_j$. Let $x := |\mathcal{M}^A_1|$ and $p^+ := \max_{j \in J} p_j$. Consider a machine $i \in \mathcal{M}^A_1$ such that $i \neq 1$. Then, there exists at least one job $j \in J^A_1$ such that $i \in \mathcal{M}_j$. By jump optimality of $A$ we have, $\Lambda^A_i + p_j/s_i \geq C^A_{\max}$. Consequently, $\Lambda^A_i \geq C^A_{\max} - p^+/s_i$ for all $i \in \mathcal{M}^A_1 \setminus \{1\}$. Multiplying the last inequality by $s_i$ and accumulating over all machines $i \in \mathcal{M}^A_1$ we obtain,

$$\sum_{i \in \mathcal{M}^A_1} L^A_i = \sum_{i \in \mathcal{M}^A_1} s_i \Lambda^A_i \geq s_1 C^A_{\max} + \sum_{i \in \mathcal{M}^A_1 \setminus \{1\}} s_i \left(C^A_{\max} - \frac{p^+}{s_i}\right). \quad (2)$$

To convert assignment $A$ to an optimal assignment with makespan $C^{OPT}_{\max}$, we need to move at least a load of

$$s_1(C^A_{\max} - C^{OPT}_{\max}) + \sum_{i \in \mathcal{M}^A_1 : i \neq 1} s_i \left(C^A_{\max} - C^{OPT}_{\max} - \frac{p^+}{s_i}\right) \quad (3)$$

from the machines in $\mathcal{M}_1^A$ to the machines in $M \backslash \mathcal{M}_1^A$. Therefore,

$$(m-x)s^+ C_{\max}^{OPT} \geq \sum_{i \in M \backslash \mathcal{M}_1^A} s_i \Lambda_i^{OPT} \tag{4}$$

$$\overset{(3)}{\geq} s_1 (C_{\max}^A - C_{\max}^{OPT}) \tag{5}$$

$$+ \sum_{i \in \mathcal{M}_1^A : i \neq 1} s_i \left( C_{\max}^A - C_{\max}^{OPT} \right) - \sum_{i \in \mathcal{M}_1^A : i \neq 1} p^+ \tag{6}$$

$$= \sum_{i \in \mathcal{M}_1^A} s_i \left( C_{\max}^A - C_{\max}^{OPT} \right) - (x-1)p^+ \tag{7}$$

$$\geq \sum_{i \in \mathcal{M}_1^A} s_i \left( C_{\max}^A - C_{\max}^{OPT} \right) - (x-1)s^+ C_{\max}^{OPT}, \tag{8}$$

since $p^+/s^+ \leq C_{\max}^{OPT}$. Then,

$$\frac{C_{\max}^A}{C_{\max}^{OPT}} \leq \frac{(m-x)s^+ + \sum_{i \in \mathcal{M}_1^A} s_i + (x-1)s^+}{\sum_{i \in \mathcal{M}_1^A} s_i} = \frac{(m-1)s^+}{\sum_{i \in \mathcal{M}_1^A} s_i} + 1. \tag{9}$$

As in an optimal assignment the jobs in $J_1^A$ must be assigned to the machines in $\mathcal{M}_1^A$, we have $s_1 C_{\max}^A \leq \sum_{i \in \mathcal{M}_1^A} s_i C_{\max}^{OPT}$ and consequently

$$\frac{C_{\max}^A}{C_{\max}^{OPT}} \leq \sum_{i \in \mathcal{M}_1^A} \frac{s_i}{s_1} \leq \sum_{i \in \mathcal{M}_1^A} \frac{s_i}{s^-}. \tag{10}$$

Combining (9) and (10) yields

$$\frac{C_{\max}^A}{C_{\max}^{OPT}} \left( \frac{C_{\max}^A}{C_{\max}^{OPT}} - 1 \right) \overset{(10)}{\leq} \sum_{i \in \mathcal{M}_1^A} \frac{s_i}{s^-} \left( \frac{C_{\max}^A}{C_{\max}^{OPT}} - 1 \right) \overset{(9)}{\leq} (m-1)\frac{s^+}{s^-}. \tag{11}$$

From this it follows that,

$$\frac{C_{\max}^A}{C_{\max}^{OPT}} \leq \frac{1}{2} + \sqrt{\frac{1}{4} + (m-1)\frac{s^+}{s^-}}. \tag{12}$$

$\square$

Note that the bound given in Theorem 5 corresponds to the bound given in Theorem 1 by setting $s_i = 1$ for all machines $i \in M$. Therefore, Example 1 shows that the bound of Theorem 5 is tight for $\tilde{s} = 1$. The example below shows that there exist instances with non-identical speeds for which a jump optimal assignment has a makespan of at least $\sqrt{\tilde{s}(m-1) + 1/4} \cdot C_{\max}^{OPT}$, leaving a gap of less than $1/2$ between the upper and the lower bound on the performance guarantee.

*Example 3.* Let $k > 1$ be an arbitrary strictly positive integer and consider the following instance. Let there be $m = k + 3$ machines having speeds $s_1 = 1$ and $s_2 = \ldots = s_m = k$. Let there be $k + 1$ jobs of size $p_j = 1$ for which $\mathcal{M}_j = \{1, 2\}$ and $k + 1$ jobs of size $p_j = k$ for which $\mathcal{M}_j = M$. Additionally, there is one job of size $\epsilon > 0$ which is only allowed on machine 1. In an optimal assignment, one job of size 1 and the one job of size $\epsilon$ are assigned to machine 1, $k$ jobs of size 1 are assigned to machine 2 and 1 job of size $k$ is assigned to each of the remaining machines. Then, $C_{\max}^{OPT} = 1 + \epsilon$. Consider the following jump optimal assignment $A$: $k + 1$ jobs of size 1 and the one job of size $\epsilon$ are assigned to machine 1, $k + 1$ jobs of size $k$ are assigned to machine 2 and all the other machines remain empty. Then $C_{\max}^A = k + 1 + \epsilon$. Hence, when $\epsilon$ tends to zero, $C_{\max}^A / C_{\max}^{OPT}$ tends to $k + 1$. Since $k + 1 > \sqrt{k(k + 2) + 1/4} = \sqrt{\tilde{s}(m - 1) + 1/4}$, we have established a lower bound of $\sqrt{\tilde{s}(m - 1) + 1/4}$ on the performance guarantee of the jump neighborhood for related parallel machines by taking $\epsilon$ small enough.

For the lexjump neighborhood for restricted related parallel machines we have the following result.

**Theorem 6.** *The performance guarantee of lexjump optimal assignments for the problem of minimizing the makespan on restricted related parallel machines is*

$$O\left(\frac{\log \sum_i \tilde{s}_i}{\log \log \sum_i \tilde{s}_i}\right)$$

*where $\tilde{s}_i$ denotes the relative speed, i. e., $\tilde{s}_i = s_i / s^-$.*

Due to space limitations the proof of Theorem 6 is not provided here. Instead we refer to [22]. Note that the bound given in the above theorem confirms to the bound given in Theorem 2 by setting $s_i = 1$ for all $i \in M$. The following example shows that there exist instances for which the bound of Theorem 6 is tight up to a constant factor.

*Example 4.* Let $k > 1$ be an arbitrary strictly positive integer and let $s > 1$. Consider the following instance and assignment. Each job has a processing requirement $p_j = s$. The machines are partitioned into $sk + 1$ groups, $S_0, S_1, \ldots, S_{sk}$. Group $S_0$ consists of only one machine which has a speed of one. For $l = 1, \ldots, sk$, group $S_l$ contains $k \Pi_{i=1}^{l-1}(sk - i)$ machines each having a processing speed of $s$. In assignment $A$, each machine in group $S_l$, for $l \geq 1$, processes $sk - l$ jobs. $k$ jobs are assigned to the machine in $S_0$. Each job $j \in J_i^A$ with $i \in S_l$ has $\mathcal{M}_j = S_l \cup S_{l+1}$. $A$ is lexjump optimal with makespan $sk$, whereas $C_{\max}^{OPT} = 1$. The optimal solution is attained by assigning one job to each machine $i \in S_l$ for $l \geq 1$ and leaving the machine in $S_0$ empty. Moreover,

$$\sum_i \tilde{s}_i \leq 1 + sk \sum_{i=0}^{sk-1} \frac{(sk - 1)!}{i!} \leq 1 + (sk)! \sum_{i=0}^{+\infty} \frac{1}{i!} \tag{13}$$

$$\leq 1 + e\,(sk)! \leq (sk + 2)! = \Gamma\,(sk + 3). \tag{14}$$

where $\Gamma(n)$ denotes the gamma function, i.e., $\Gamma(n) = (n-1)!$ for some integer $n$. Hence, using the fact that the inverse of the Gamma function is monotonically increasing, $C_{\max}^{A}/C_{\max}^{OPT} = sk \geq \Gamma^{-1}\left(\sum_i \widetilde{s}_i\right) - 3$. Thus, we are able to provide instances for which $C_{\max}^{A}/C_{\max}^{OPT}$ is in $\Omega(\log \sum_i \widetilde{s}_i / \log \log \sum_i \widetilde{s}_i)$.

Gairing, Lücking, Mavronicolas and Monien [7] established the following result.

**Theorem 7 (Gairing et al. [7]).** *A lexjump optimal assignment for restricted related parallel machines has a performance guarantee of at most $m$.*

Furthermore, they provide an example which establishes a lower bound of $m-1$ on the performance guarantee. Theorems 6 and 7 both establish an upper bound on the performance guarantee of a lexjump optimal solution. It can be shown that neither result implies the other result, i.e., we can construct examples for which the bound of Theorem 6 is tight and the bound of Theorem 7 is not and vice versa.

## 5    Concluding Remarks

Since swap optimal assignments as well as push optimal assignments are both jump optimal, we have that Theorems 1, 3 and 5 directly carry over to the swap and the push neighborhood. Moreover, the Examples and 1, 2 and 3 are swap and push optimal. Hence, the tightness results for the jump neighborhood apply to the swap and the push neighborhood as well.

## References

1. Awerbuch, B., Azar, Y., Richter, Y., Tsur, D.: Tradeoffs in worst-case equilibria. Theoretical Computer Science 361, 200–209 (2006)
2. Brueggemann, T., Hurink, J.L., Vredeveld, T., Woeginger, G.J.: Very large-scale neighborhoods with performance guarantees for minimizing makespan on parallel machines. In: Kaklamanis, C., Skutella, M. (eds.) WAOA 2007. LNCS, vol. 4927, pp. 41–55. Springer, Heidelberg (2008)
3. Centeno, G., Armacost, R.L.: Minimizing makespan on parallel machines with release time and machine eligibility restrictions. International Journal of Production Research 42(6), 1243–1256 (2004)
4. Cho, Y., Sahni, S.: Bounds for list schedules on uniform processors. SIAM Journal on Computing 9, 91–103 (1980)
5. Czumaj, A., Vocking, B.: Tight bounds for worst-case equilibria. ACM Transactions on Algorithms 3(1), 4 (2007)
6. Finn, G., Horowitz, E.: A linear time approximation algorithm for multiprocessor scheduling. BIT 19, 312–320 (1979)
7. Gairing, M., Lücking, T., Mavronicolas, M., Monien, B.: The price of anarchy for restricted parallel links. Parallel Processing Letters 16(1), 117–131 (2006)
8. Gairing, M., Lücking, T., Mavronicolas, M., Monien, B.: Computing nash equilibria for scheduling on restricted parallel links. Theory of Computing Systems (2009), http://www.springerlink.com/content/q7n8267q76716v55/fulltext.pdf (February 12, 2009)

9. Garey, M.R., Johnson, D.S.: Computers and Intractability: A Guide to the Theory of NP-Completeness. Freeman, New York (1979)
10. Glass, C.A., Kellerer, H.: Parallel machine scheduling with job assignment restrictions. Naval Research Logistics 54(3), 250–257 (2007)
11. Graham, R.L., Lawler, E.L., Lenstra, J.K., Rinnooy Kan, A.H.G.: Optimization and approximation in deterministic sequencing and scheduling: A survey. Annals of Discrete Mathematics 5, 287–326 (1979)
12. Hoefer, M., Souza, A.: The influence of link restriction on (random) selfish routing. In: Monien, B., Schroeder, U.-P. (eds.) SAGT 2008. LNCS, vol. 4997, pp. 22–32. Springer, Heidelberg (2008)
13. Immorlica, N., Li, L., Mirrokni, V., Schulz, A.: Coordination mechanism for selfish scheduling. Theoretical Computer Science 410, 1589–1598 (2009)
14. Leung, J.Y.T., Li, C.L.: Scheduling with processing set restrictions: A survey. International Journal of Production Economics 116, 251–262 (2008)
15. Li, C.L.: Scheduling unit-length jobs with machine eligibility restrictions. European Journal of Operational Research 174, 1325–1328 (2006)
16. Liao, L.W., Sheen, G.J.: Parallel machine scheduling with machine availability and eligibility constraints. European Journal of Operational Research 184, 458–467 (2008)
17. Lin, Y., Li, W.: Parallel machine scheduling of machine-dependent jobs with unit-length. European Journal of Operational Research 156, 261–266 (2004)
18. Lu, P., Yu, C.: Worst-case nash equilibria in restricted routing. In: Papadimitriou, C., Zhang, S. (eds.) WINE 2008. LNCS, vol. 5385, pp. 231–238. Springer, Heidelberg (2008)
19. Michiels, W., Aarts, E., Korst, J.: Theoretical Aspects of Local Search. Springer, Berlin (2007)
20. Ou, J., Leung, J.Y.T., Li, C.L.: Scheduling parallel machines with inclusive set restrictions. Naval Research Logistics 55(4), 328–338 (2008)
21. Pinedo, M.: Scheduling: Theory, Algorithms and Systems, 3rd edn. Springer, New York (2008); Original edition published by Prentice Hall, Englewood Cliffs, NJ, (1995)
22. Recalde, D., Rutten, C., Vredeveld, T., Schuurman, P.: Local search performance guarantees for restricted related parallel machine scheduling. Technical report, Maastricht University, METEOR Research Memoranda RM/09/061, Maastricht, The Netherlands (2009)
23. Schuurman, P., Vredeveld, T.: Performance guarantees of local search for multiprocessor scheduling. INFORMS Journal of Computing 19(1), 52–63 (2007)
24. Shchepin, E.V., Vakhania, N.: An optimal rounding gives a better approximation for scheduling unrelated machines. Operations Research Letters 33(8), 2266–2278 (2005)
25. Vairaktarakis, G.L., Cai, X.: The value of processing flexibility in multipurpose machines. IIE Transactions 35(8), 763–774 (2003)
26. Vöcking, B.: Selfish load balancing. In: Nisan, N., Roughgarden, T., Tardos, E., Vazirani, V. (eds.) Algorithmic Game Theory, ch. 20. Cambridge University Press, New York (2007)

# Packet Routing on the Grid*

Britta Peis, Martin Skutella, and Andreas Wiese

Technische Universität Berlin, Straße des 17. Juni 136, 10623 Berlin, Germany
{peis,skutella,wiese}@math.tu-berlin.de

**Abstract.** The packet routing problem, i.e., the problem to send a
given set of unit-size packets through a network on time, belongs to
one of the most fundamental routing problems with important practical applications, e.g., in traffic routing, parallel computing, and the design of communication protocols. The problem involves critical routing
and scheduling decisions. One has to determine a suitable (short) origin-
destination path for each packet and resolve occurring conflicts between
packets whose paths have an edge in common. The overall aim is to find
a path for each packet and a routing schedule with minimum makespan.
  A significant topology for practical applications are grid graphs. In
this paper, we therefore investigate the packet routing problem under
the restriction that the underlying graph is a grid. We establish approximation algorithms and complexity results for the general problem on
grids, and under various constraints on the start and destination vertices or on the paths of the packets.

## 1 Introduction

In this paper, we study the packet routing problem on grid graphs. In an instance
of this problem we are given a set of unit-size packets with specified source and
destination vertices. First, we need to find a path for each packet along which
we want to route it. Then we need to find a routing schedule to transfer the
packets through the network such that each link in the network can be used
by at most one packet at a time. The overall goal is to find a path assignment
and routing schedule that minimizes the makespan, i.e., the time when the last
packet has reached its destination. We study also the special case that the paths
of the packets are given and we only need to find the routing schedule.

The packet routing problem has several applications in practice, e.g., in parallel computing or in cell structured networks. In those settings, packets of information need to be transferred through the network. In order for the network to
operate efficiently, it is required that the packets reach their respective destina-
tions as quickly as possible. Therefore, the paths of the packets and the routing
schedule need to be computed such that the packets encounter as few delay as
possible. One of the most common natural topologies of the routing problem

---

* This work was partially supported by Berlin Mathematical School, by DFG research
  center MATHEON, and by the DFG Focus Program 1307 within the project "Algorithm Engineering for Real-time Scheduling and Routing".

A. López-Ortiz (Ed.): LATIN 2010, LNCS 6034, pp. 120–130, 2010.

in practical applications are grid graphs, e.g. in parallel computing. Therefore, we take this special structure into account when looking for efficient solution methods.

## 1.1 Packet Routing Problem

The packet routing problem is defined as follows: Let $G = (V, E)$ be an undirected graph (in our case this will usually be a grid graph). A packet $M_i = (s_i, t_i)$ is a tuple consisting of a start vertex $s_i \in V$ and a destination vertex $t_i \in V$. Let $\mathcal{M} = \{M_1, M_2, M_3, ..., M_{|\mathcal{M}|}\}$ be a set of packets. Then $(G, \mathcal{M})$ is an instance of the *packet routing problem with variable paths*. The problem has two parts: First, for each packet $M_i$ we need to find a path $P_i = (s_i = v_0, v_1, ...v_{\ell-1}, v_\ell = t_i)$ from $s_i$ to $t_i$ such that $\{v_i, v_{i+1}\} \in E$ for all $i$ with $0 \leq i \leq \ell - 1$. Assuming that it takes one timestep to send a packet along an edge, we need to find a routing schedule for the packets such that each message $M_i$ follows its path $P_i$ from $s_i$ to $t_i$ and each edge is used by at most one packet at a time. We assume that time is discrete and that all packets take their steps simultaneously. The objective is to minimize the makespan, i.e., the time when the last packet has reached its destination vertex. For each packet $M_i$ we define $\bar{D}_i$ to be the length of the shortest path from $s_i$ to $t_i$, assuming that all edges have unit length. Moreover, the *minimal dilation* $\bar{D}$ is defined by $\bar{D} := \max_i \bar{D}_i$. It holds that $\bar{D}$ is a lower bound for the length of an optimal schedule.

Since there are algorithms known to determine paths for routing the packets (see [4,13,24] or simply take shortest paths) we will also consider the *packet routing problem with fixed paths*. An instance of this problem is a tuple $(G, \mathcal{M}, \mathcal{P})$ such that $G$ is a (grid) graph, $\mathcal{M}$ is a set of packets and $\mathcal{P}$ is a set of predefined paths, one for each packet. Since the paths of the packets are given in advance they do not need to be computed here. The aim is to find a schedule with the properties described above such that the makespan is minimized. For each packet $M_i$ we define $D_i$ to be the length of the path $P_i$, again assuming that all edges have unit length. Like above we define the *dilation* $D$ by $D := \max_i D_i$. For each edge $e$ we define $C_e$ to be the number of paths that use $e$. Then we define the *congestion* $C$ by $C := \max_e C_e$. It holds that $C$ and $D$ are lower bounds for the length of an optimal schedule.

We distinguish between grid graphs in which two packets are allowed to use an edge in opposite directions at the same time, or not. The infinite grid graph $G_\# = (V_\#, E_\#)$ is the undirected graph consisting of the vertices $V_\# = \{v_{i,j} | i, j \in \mathbb{Z}\}$ and the edges

$$E_\# = \{\{v_{i,j}, v_{i',j'}\} \mid |i - i'| + |j - j'| = 1\}$$

The directed graph $\overleftrightarrow{G}_\# = \left(V_\#, \overleftrightarrow{E}_\#\right)$ is the bidirected infinite grid graph with

$\overleftrightarrow{E}_\# = \{(u, v), (v, u) \mid \{u, v\} \in E_\#\}$. We will consider infinite grid graphs rather than finite grids because we want the borders of a finite grid not to have any impact on the problem. However, for our algorithms it would be sufficient to assume that each start and each destination vertex has a certain minimum distance to the boundary.

Throughout the paper we will use the notation $|S|$ for the length of a schedule $S$. For a packet routing instance $I$ with fixed or variable paths, let $OPT(I)$ denote a schedule with minimum makespan. For an algorithm $\mathcal{A}$ for the packet routing problem denote by $\mathcal{A}(I)$ the schedule computed by $\mathcal{A}$ for the instance $I$. The algorithm $\mathcal{A}$ is an $\alpha$-approximation algorithm if it runs in polynomial time and for all instances $I$ it holds that $|\mathcal{A}(I)| \leq \alpha \cdot |OPT(I)|$. We call $\alpha$ the approximation ratio or performance ratio of $\mathcal{A}$. We denote by $n$ the length of the overall input where we assume that the coordinates of the start and destination vertices are given in binary representation. We assume that the grid itself is not given explicitly in the input. Some of our algorithms are dealing with the special case that all packets have the same start vertex. In this case we assume that this start vertex has the coordinates $s = (0,0)$.

## 1.2   Related Work

Packet routing and related problems are widely studied in the literature. Di Ianni shows that the delay routing problem [12] is $NP$-hard. The proof implies that the packet routing problem on general graphs is $NP$-hard as well. Leung et al. [17, chapter 37] study packet routing on different graph classes. In [3] Busch et al. study the direct routing problem, that is the problem of finding a routing schedule such that a packet is never delayed once it has left its start vertex. They give complexity results and algorithms for finding direct schedules. Peis et al. [20] present non-approximability results for the packet routing problem on several graph classes and algorithms for the problem on trees.

In [14] Leighton et al. show that there is always a routing schedule that finishes in $O(C + D)$ steps. In [15] Leighton et al. present an algorithm that finds such a schedule in polynomial time. However, this algorithm is not suitable for practical applications since the hidden constants in the schedule length are very large. Using the algorithm by Leighton et al. as a subroutine, Srinivasan and Teo [24] present an algorithm that solves the packet routing problem with variable paths with a constant approximation factor. Koch et al. [13] improve and generalize this algorithm for the more general message routing problem (where each message consists of several packets).

Leighton, Makedon and Tollis [16] show that the permutation routing problem on an $n \times n$ grid can be solved in $2n - 2$ steps using constant size queues. Rajasekaran [21] presents several randomized algorithms for packet routing on grids. They also give their bounds in terms of the grid size. For the case that each vertex of the grid is the start vertex of at most one packet Mansour and Patt-Shamir [18] present an algorithm with constant approximation factor which uses the algorithm by Leighton et al. [15] as a subroutine.

The packet routing problem is related to the multi-commodity flow over time problem [5,6,9,10,11]. In particular, Hall et al. [9] show that the latter problem is $NP$-hard, even in the very restricted case of series-parallel networks. It is equivalent to the packet routing problem if we additionally require unit edge capacities, unit transit times, and integral flow values. If there is only one start and one destination vertex then the packet routing problem can be solved optimally in polynomial time, e.g., using the Ford-Fulkerson algorithm for the maximum flow over time problem [7,8,22] together with a binary search framework. For the quickest transshipment problem with multiple sources and multiple sinks Hoppe and Tardos [11] present a polynomial time algorithm. As a consequence, the packet routing problem with a single start vertex or a single destination vertex can be solved optimally.

Finally, Adler et al. [1,2] study the problem of scheduling as many packets as possible through a given network in a certain time frame. They give approximation algorithms and $NP$-hardness results.

### 1.3   Our Contributions

For the case of the bidirectional grid $\overleftrightarrow{G}_\#$ with the start- and the destination vertices of all packets being pairwise different, we present an optimal algorithm which always computes a schedule of length $D$. For the undirected grid $G_\#$ it can be adapted to a 2-approximation algorithm (we will show later that in $G_\#$ this setting is $NP$-hard). We also show that on the grid a bad choice for the paths of the packets can yield an arbitrary high approximation factor for the entire problem. This holds even we use an optimal scheduling algorithm once the paths of the packets are fixed.

We investigate the case that there is only one start vertex but arbitrarily many destination vertices. Note here that the algorithm by Hoppe et al. [11] does not necessarily run in polynomial time since in our case the graph is not part of the input but it is given implicitly. We present an algorithm that finds a schedule with length at most $OPT + 8$. Then we improve this algorithm to a $1 + \epsilon$ approximation while still guaranteeing an absolute error of eight. Denote by $k$ the number of sink vertices and by $n$ the length of the overall input. The runtime of our algorithm is bounded by $O(n \log n + f(1/\epsilon))$ (for an exponential function $f$). For the same setting we give an optimal algorithm with running time $O(k^6 n)$. We achieve the polynomial bound on the runtime by not considering the full grid but only certain subgrids.

Finally, we study the complexity of the packet routing problem on grids. We prove that if the paths are fixed, it is $NP$-hard on the bidirected grid $\overleftrightarrow{G}_\#$, even if there is only one start vertex and all predefined paths are shortest paths. Allowing the paths to be variable, we show that the problem is still $NP$-hard on the undirected grid $G_\#$ even if no two packets share their start or destination vertex.

Due to space constraints the descriptions of the algorithms and the proofs had to be shortened significantly. For full details we refer to our technical report [19].

## 2    Unique Start and Destination Vertices

In this section we assume that in the given instance of the packet routing problem no two packets have the same start or destination vertex. We present an optimal algorithm for the bidirectional grid $\overleftrightarrow{G}_{\#}$. The algorithm is very simple and can therefore be implemented very efficiently. A slight modification yields a factor 2 approximation algorithm on $G_{\#}$. Note that the problem is $NP$-hard on $G_{\#}$, as we will show in Section 4. Finally, we show that a bad choice of the paths can result in arbitrarily bad schedules, even if an optimal schedule is used once the paths of the packets are fixed.

### 2.1    Optimal Algorithm for $\overleftrightarrow{G}_{\#}$

Let $I = \left( \overleftrightarrow{G}_{\#}, \mathcal{M} \right)$ be an instance of the packet routing problem with variable paths. Assume that for each pair of packets $M = (s, t)$ and $M' = (s', t')$ it holds that $s \neq s'$ and $t \neq t'$. We first need to specify the paths of the packets and then the routing schedule.

Let $M = ((s_r, s_c), (t_r, t_c))$ be a packet ($s_r$ denotes the row and $s_c$ the column coordinate of $s$, $t_r$ and $t_c$ are to be understood respectively). The path of each packet $M = ((s_r, s_c), (t_r, t_c))$ is defined as follows: First, $M$ moves vertically on the (unique) shortest path from $(s_r, s_c)$ to $(t_r, s_c)$ (call this the vertical part) and then horizontally on the (unique) shortest path from $(t_r, s_c)$ to $(t_r, t_c)$ (call this the horizontal part).

Since each vertex is the start vertex of at most one packet, there can be no delay in the vertical part of a packet's path. For scheduling the horizontal part we use the farthest-destination-first rule: if there are two packets competing for an edge we give priority to the packet which has still the longer way to go to its destination. Denote by $GRID(I)$ the obtained schedule for $I$ and by $OPT(I)$ a schedule with minimum makespan.

**Theorem 1.** *Let $I = \left( \overleftrightarrow{G}_{\#}, \mathcal{M} \right)$ be an instance of the packet routing problem such that no two packets share a start- or destination vertex. Then it holds that $|GRID(I)| = D = |OPT(I)|$.*

*Proof.* Due to space constraints we refer to our technical report [19].    □

On $G_{\#}$ this schedule can be simulated as follows: In all even timesteps we move packets whose next edge goes up or to the right. In all odd timesteps we move packets whose next edge goes down or to the left. This stretches the length of the schedule by at most a factor of two and thus we obtain a 2-approximation algorithm for $G_{\#}$.

### 2.2    Choice of the Paths

The choice of the paths in Theorem 1 might seem pretty simple. However, in the above setting a bad choice of (shortest) paths can result in an arbitrarily high approximation factor.

**Theorem 2.** *For every* $k \geq 1$ *there is an instance* $\Pi_k = \left( \overleftrightarrow{G}_{\#}, \mathcal{M}_k \right)$ *of the packet routing problem with variable paths with the following properties:*

- *The start and destination vertices of all packets are pairwise different.*
- *Let* $OPT_k$ *denote an optimal solution for* $\Pi_k$. *There is a set of shortest paths for the packets* $\mathcal{M}_k$ *such that for the best possible makespan* $OPT'_k$ *using these paths it holds that* $\frac{OPT'_k}{OPT_k} \in \Omega(k)$.

# 3   Single Start Multiple Destination Vertices

If we allow only one start vertex the packet routing problem with variable paths on arbitrary graphs can be solved with the algorithm for the quickest transshipment problem presented by Hoppe et al. [11]. However, if we are dealing with grid graphs the graph itself is not part of the input but it is given implicitly. Thus, this algorithm does not necessarily run in polynomial time.

First, we present an approximation algorithm for this setting with an approximation factor of $1 + \epsilon$ and an absolute error of 8. Then, we present an optimal algorithm with a higher bound on the runtime than our PTAS.

## 3.1   Absolute Approximation

Let $I = (G_{\#}, \mathcal{M})$ be an instance of the packet routing problem with variable paths such that there is only one start and arbitrarily many destination vertices. Our algorithm has two phases which we sketch in the sequel.

First, we order the packets decreasingly by the length of the shortest paths from their start to their destination vertices. W.l.o.g. let $M_0, M_1, \ldots \ldots, M_{|\mathcal{M}|-1}$ be this order. Now we schedule the packets exactly in this order. Since $s$ has four adjacent edges, we schedule four packets at each timestep. To be precise, at time $t$ we schedule the packets $M_{4t}, M_{4t+1}, M_{4t+2}$, and $M_{4t+3}$ (if packets with the respective indices exist). We assign the four packets arbitrarily to the four outgoing edges of $s$: these edges form the first edges on the paths of the packets. Then we define a path for each packet such that all *collisions* (i.e., occasions where two packets need to use an edge at the same time) are caused by two packets which use an edge $e$ in opposite directions at the same time. Our choice of the paths ensures that each packet takes a detour of at most 8 edges.

In the second phase we adjust the schedule $S_0(I)$ in order to eliminate all collisions. By doing this, we do not change the makespan of the overall schedule. Denote by $S(I)$ the resulting schedule.

**Theorem 3.** *For the schedule* $S(I)$ *it holds that*

$$|S(I)| \leq |OPT(I)| + 8$$

*Moreover, the length of* $S(I)$ *can be computed in* $O(n \log n)$.

Furthermore, if there are at most $k$ different destination vertices, in time $O(n + k \log k)$ we can compute a bound $T$ on the length of the optimal schedule such that $|OPT(I)| \leq T \leq |OPT(I)| + 8$. The $(k \log k)$-term in the runtime results from sorting the destination vertices by their distance to the source.

We can obtain an approximation scheme for this problem by solving instances with small optimal values by enumeration and using the algorithm above for instances with large optimal values. This results in a runtime of $O(n \log n + f(1/\epsilon))$ for an approximation factor of $1 + \epsilon$.

## 3.2    Optimal Algorithm

In this section we present an optimal algorithm for solving the packet routing problem with one start and many destination vertices on the grid. It is based on the algorithm by Hoppe et al. [11] which solves the quickest transshipment problem with multiple sources and multiple sinks on arbitrary graphs in polynomial time. Our main technique is that we do not consider the whole grid but only certain subgrids of polynomial size.

The algorithm by Hoppe et al. computes a dynamic flow, i.e. a flow over time. For an introduction to dynamic flows we refer to [11] and [22]. Now we introduce the quickest transshipment problem.

**Definition 1.** *An instance of the* quickest transshipment problem *consists of a dynamic network* $\mathcal{N} = (G, u, \tau, S)$ *and a vector* $v$ *where* $G = (V, E)$ *is a graph, the maps* $u$ *and* $\tau$ *denote the capacities and the transit times on the edges and the vector* $v$ *denotes the demand/supply of each vertex in* $S \subset V$ *(the set* $S$ *contains all source and sink vertices). We ask for the smallest time horizon* $T^*$ *such that there is a dynamic flow with time horizon* $T^*$ *which satisfies all demands and supplies.*

First we give an outline of the algorithm by Hoppe and Tardos for the quickest transshipment problem. Then we show how it can be adjusted in order to obtain a strongly polynomial time algorithm for the packet routing problem on the grid with a single start vertex. The algorithm by Hoppe and Tardos works as follows:

1. The algorithm checks for several values of $T$ whether the instance of the quickest transshipment problem has a solution with makespan at most $T$. The optimal makespan $T^*$ is then determined by binary search.
2. In order to check whether there is a solution with makespan at most $T$, the algorithm needs to compute lexicographically maximum dynamic flows with time horizon $T$ according to certain orderings of the terminals.
3. The computation of the lexicographically maximum dynamic flows can be reduced to several minimum-cost-flow computations on $G$. The source and sink vertices of these instances correspond to the vertices in $G$ which have non-zero supply/demand.
4. It can be shown that if the input network is integral (i.e., all capacities and all transit times are integral) then the resulting dynamic flow is integral as well.

Our adjustment for $G_\#$ addresses the third point: Computing the necessary minimum cost flows on the whole grid or even in the smallest subgrid which contains all source/sink vertices would lead to a runtime exponential in the input size. Instead, we perform them on subgrids which we call *thinned out grids*. In the sequel, we will first define the type of minimum cost flow instances which need to be solved. Then we will show that if we restrict our grid graph to thinned out grids we can obtain flows with the same cost as in the whole grid.

For the min-cost-flow computations in the third step we study the following problem:

**Definition 2.** *The* One-Source-MinCostFlow-Problem (OSMCF) *is defined as follows: Let $G_\# = (V_\#, E_\#)$ be the grid graph with a source vertex $s' \in V_\#$ and sink vertices $S \subseteq V_\#$. Let $T$ be a given integer. We introduce a super sink $\psi$ and for each $\bar{s} \in S$ we introduce an edge $e_{\bar{s}} := (\bar{s}, \psi)$ with transit time $c(e_{\bar{s}}) := -(T+1)$ and infinite capacity. The problem is to find a minimum cost flow from the source $s'$ to the sink $\psi$ in the network. The cost for each edge equals its transit time.*

**Lemma 1.** *All mincost-flow computations which are needed in the algorithm by Hoppe and Tardos applied to $G_\#$ with a single source vertex can be reduced to the OSMCF-problem.*

Now we describe how to solve the OSMCF-problem in polynomial time. Instead of solving the OSMCF-problems arising in the packet routing problem with the full grid we resort to *thinned out grids*. The intuition is the following: Given a set of vertices $V'$ from $G_\#$, we create a grid graph $G_\#(V')$ whose rows and columns are exactly the rows and columns of the vertices in $V'$. The lengths of the edges are set according to the distance of their respective end-vertices in $G_\#$. Formally, let $R(V') = \{r_1, r_2, ..., r_R\}$ and $C(V') = \{c_1, c_2, ..., c_C\}$ be the sets of the row and column indices of the vertices in $V'$. We define $V_\#(V') := \{(x, y) \,|\, x \in C(V') \wedge y \in R(V')\}$ and $E_\#(V') := \bigcup_{i \in R} E_{H,i} \cup \bigcup_{j \in C} E_{V,j}$ with $E_{H,i} := \{\{(i, c_j), (i, c_{j+1})\} \,|\, 1 \leq j \leq C - 1\}$ and $E_{V,j} := \{\{(r_i, j), (r_{i+1}, j)\} \,|\, 1 \leq i \leq R - 1\}$. We define the cost of an edge $e = \{(i, c_j), (i, c_{j+1})\}$ by $c(e) = (c_{j+1} - c_j)$ and analogously for an edge $e' = \{(r_i, j), (r_{i+1}, j)\}$ by $c(e') = (r_{i+1} - r_i)$. We define $G_\#(V') := (V_\#(V'), E_\#(V'))$. We call a graph $\bar{G}$ a *thinned out grid graph* if there is a set $V'$ such that $\bar{G} = G_\#(V')$.

Now let $\bar{G} = (\bar{V}, \bar{E})$ be a thinned out grid graph. We define an operation $dense(\bar{G})$ which, intuitively, makes the thinned out grid a little "denser" by adding new rows and columns next to already existing ones. Let $\bar{V}' := \{(r-1, c-1), (r, c), (r+1, c+1) \,|\, (r, c) \in \bar{V}\}$. We define $dense(\bar{V}) := G_\#(\bar{V}')$. Note that the resulting graph is again a thinned out grid graph.

Recursively, we define $dense^{(i)}(\bar{V}) := dense(dense^{(i-1)}(\bar{V}))$.

**Lemma 2.** *Let $S \subseteq V_\#$ be a set of sink vertices. Then the OSMCF-problems on $G_\#$ and $dense^{(4)}(S)$ have the same objective value.*

*Proof (sketch).* The proof relies on the fact that in the thinned out grid one can always find the necessary augmenting paths for computing the mincost-flow.    □

**Theorem 4.** *Let $I = (G_\#, \mathcal{M})$ be an instance of the packet routing problem on $G_\#$ with at most $k$ destination vertices and exactly one start vertex. The optimal makespan for $I$ can be computed in $O\left(k^6 n\right)$.*

*Proof (sketch).* The proof relies on the framework by Hoppe and Tardos, Lemma 2, and a runtime analysis which uses the fact that for solving the OSMCF-problems we need at most four augmenting paths. Moreover, employing the algorithm presented in Section 3.1 we reduce the number of possible values for the optimal makespan to 8. Thus, we can omit the binary search.    □

Note that the above implies that there is an optimal solution for the instance which uses only vertices and edges in $dense^{(4)}(S)$. Thus, we can compute an optimal routing schedule (and not only its length) in polynomial time using the algorithm presented in [11] for computing an optimal dynamic flow for the quickest transshipment problem.

# 4    Complexity Results

Due to space constraints we give only brief sketches of the proof of our complexity results for the packet routing problem on the grid.

**Theorem 5.** *The packet routing problem with fixed paths is $NP$-hard on grid graphs, even if there is only one start vertex. This holds even if all predefined paths are shortest paths.*

*Proof.* In this proof we employ a technique which was used in [3,23]. We reduce the 3-COLORING problem to the packet routing problem. We introduce a packet for each vertex of the coloring instance. The paths of two packets share an edge if and only if their respective vertices in $G$ are connected by an edge. All paths have the same length. The number of colors needed in the coloring instance equals the maximum number of delays which are necessary for each packet.    □

**Theorem 6.** *The packed routing problem with variable paths is $NP$-hard on the undirected grid $G_\#$. This holds even if no two packets have the same start and destination vertices.*

*Proof.* We reduce to this problem from MONOTONE-NOT-ALL-EQUAL-3-SAT [25]. The idea is that for each clause there is one packet which ensures that there is a true literal and one packet which ensures that there is a false literal in the clause. The reduction heavily uses the fact that two packets delay each other if they need to use an edge in opposite directions.    □

# References

1. Adler, M., Khanna, S., Rajaraman, R., Rosén, A.: Time-constrained scheduling of weighted packets on trees and meshes. Algorithmica 36, 123–152 (2003)
2. Adler, M., Sitaraman, R., Rosenberg, A., Unger, W.: Scheduling time-constrained communication in linear networks. In: Proceedings of the 10th annual ACM symposium on Parallel algorithms and architectures, pp. 269–278 (1998)
3. Busch, C., Magdon-Ismail, M., Mavronicolas, M., Spirakis, P.: Direct routing: Algorithms and complexity. Algorithmica 45, 45–68 (2006)
4. Busch, C., Magdon-Ismail, M., Xi, J.: Optimal oblivious path selection on the mesh. IEEE Transactions on Computers 57, 660–671 (2008)
5. Fleischer, L., Skutella, M.: Minimum cost flows over time without intermediate storage. In: Proceedings of the 14th Annual Symposium on Discrete Algorithms (2003)
6. Fleischer, L., Skutella, M.: Quickest flows over time. SIAM Journal on Computing 36, 1600–1630 (2007)
7. Ford, L.R., Fulkerson, D.R.: Constructing maximal dynamic flows from static flows. Operations Research 6, 419–433 (1958)
8. Ford, L.R., Fulkerson, D.R.: Flows in Networks. Princeton University Press, Princeton (1962)
9. Hall, A., Hippler, S., Skutella, M.: Multicommodity flows over time: Efficient algorithms and complexity. Theoretical Computer Science 2719, 397–409 (2003)
10. Hall, A., Langkau, K., Skutella, M.: An FPTAS for quickest multicommodity flows with inflow-dependent transit times. Algorithmica 47, 299–321 (2007)
11. Hoppe, B., Tardos, É.: The quickest transshipment problem. Mathematics of Operations Research 25, 36–62 (2000)
12. Di Ianni, M.: Efficient delay routing. Theoretical Computer Science 196, 131–151 (1998)
13. Koch, R., Peis, B., Skutella, M., Wiese, A.: Real-time message routing and scheduling. In: Dinur, I., Jansen, K., Naor, J., Rolim, J. (eds.) RANDOM 2009 and APPROX 2009. LNCS, vol. 5687, pp. 217–230. Springer, Heidelberg (2009)
14. Leighton, F.T., Maggs, B.M., Rao, S.B.: Packet routing and job-scheduling in $O(congestion + dilation)$ steps. Combinatorica 14, 167–186 (1994)
15. Leighton, F.T., Maggs, B.M., Richa, A.W.: Fast algorithms for finding $O(congestion + dilation)$ packet routing schedules. Combinatorica 19, 375–401 (1999)
16. Leighton, F.T., Makedon, F., Tollis, I.G.: A $2n - 2$ step algorithm for routing in an $n \times n$ array with constant size queues. In: Proceedings of the 1st Annual Symposium on Parallel Algorithms and Architectures, pp. 328–335 (1989)
17. Leung, J.Y.-T.: Handbook of Scheduling: Algorithms, Models and Performance Analysis (2004)
18. Mansour, Y., Patt-Shamir, B.: Many-to-one packet routing on grids. In: Proceedings of the 27th Annual Symposium on Theory of Computing, pp. 258–267 (1995)
19. Peis, B., Skutella, M., Wiese, A.: Packet routing on the grid. Technical Report 012-2009, Technische Universität Berlin (March 2009)
20. Peis, B., Skutella, M., Wiese, A.: Packet routing: Complexity and algorithms. In: Proceedings of the 7th Workshop on Approximation and Online Algorithms. LNCS. Springer, Heidelberg (to appear, 2010)

21. Rajasekaran, S.: Randomized algorithms for packet routing on the mesh. Technical Report MS-CIS-91-92, Dept. of Computer and Information Sciences, Univ. of Pennsylvania, Philadelphia, PA (1991)
22. Skutella, M.: An introduction to network flows over time. In: Cook, W., Lovász, L., Vygen, J. (eds.) Research Trends in Combinatorial Optimization, pp. 451–482. Springer, Berlin (2009)
23. Spenke, I.: Complexity and approximation of static k-splittable flows and dynamic grid flows. PhD thesis, Technische Universität Berlin (2006)
24. Srinivasan, A., Teo, C.-P.: A constant-factor approximation algorithm for packet routing and balancing local vs. global criteria. SIAM Journal on Computing 30 (2001)
25. Williamson, D.P., Hall, L.A., Hoogeveen, J.A., Hurkens, C.A.J., Lenstra, J.K., Sevast'janov, S.V., Shmoys, D.B.: Short shop schedules. Operations Research 45, 288–294 (1997)

# Faithful Representations of Graphs by Islands in the Extended Grid

Michael D. Coury[1], Pavol Hell[1], Jan Kratochvíl[2], and Tomáš Vyskočil[2]

[1] School of Computing Science, Simon Fraser University,
Burnaby, B.C., Canada V5A 1S6*
mdcoury@gmail.com, pavol@cs.sfu.ca
[2] Department of Applied Mathematics
and Institute for Theoretical Computer Science, Charles University,
Malostranské nám. 25, 118 00 Praha 1, Czech Republic**
{honza,whisky}@kam.mff.cuni.cz

**Abstract.** We investigate embeddings of graphs into the infinite extended grid graph. The problem was motivated by computation on adiabatic quantum computers, but it is related to a number of other well studied grid embedding problems. Such problems typically deal with representing vertices by grid points, and edges by grid paths, while minimizing some objective function such as the area or the maximum length of the grid paths representing the edges. Our particular model, while expressible in this language, is more naturally viewed as one where the vertices are represented by subtrees of the grid (called islands), and the edges are represented by the usual grid edges joining the islands. Somewhat unexpectedly, these graphs turn out to unify such seemingly unrelated graph classes as the string graphs and the induced subgraphs of the extended grid. The connection is established by limiting the size (number of vertices) $k$ of the representing islands. We study the problem of representability of an input graph $G$ by islands of size at most $k$. We conjecture that this problem is NP-complete for any positive integer $k$, and prove the conjecture for $k < 3$ and $k > 5$; the cases $k = 3, 4, 5$ remain open.

## 1 Introduction

The *extended grid* is the set of all points $(X, Y) \in \mathbb{Z}^2$ in which two distinct points $A = (X_A, Y_A)$ and $B = (X_B, Y_B)$ are *adjacent*, or *neighbors*, if $|X_A - X_B| \leq 1$ and $|Y_A - Y_B| \leq 1$. Moreover, we say that two points $A, B$ are *horizontal* (respectively *vertical*, respectively *diagonal*) neighbors if $X_A = X_B$ and $|Y_A - Y_B| = 1$ (respectively $Y_A = Y_B$ and $|X_A - X_B| = 1$, respectively $|X_A - X_B| = 1$ and $|Y_A - Y_B| = 1$). The extended grid differs from the usual *plane grid* only

* Parts of this research were conducted with financial and logistic support from IRMACS, MITACS, and NSERC; all are gratefully acknowledged.
** Supported by Research projects MSM0021620838 and 1M0545 of the Ministry of Education of the Czech Republic.

A. López-Ortiz (Ed.): LATIN 2010, LNCS 6034, pp. 131–142, 2010.

in the diagonal adjacency; in other words, the plane grid is obtained from the extended grid by deleting the edges joining diagonal neighbors.

An *island* in the extended grid is a connected set of points in the grid, i.e., a set of points in which the relation of adjacency defines a connected graph. We say that two islands $i, i'$ are *adjacent* if there is a pair of points $P \in i, P' \in i'$, that are adjacent in the extended grid. Given a set of islands $\mathcal{I}$ in the extended grid, the *graph of $\mathcal{I}$* is the graph in which vertices are the islands from $\mathcal{I}$, and adjacency is defined as above. Equivalently, the graph of $\mathcal{I}$ is the intersection graph of the sets $i^+, i \in \mathcal{I}$, where $i^+$ is obtained from $i$ by expanding each point of $i$ to a ball of suitable radius $r$. (Any $r$ strictly between $1/\sqrt{2}$ and 1 will do.)

A *faithful representation* of a graph $G$ by islands is a set $\mathcal{I}$ of vertex disjoint islands in the extended grid, such that the graph of $\mathcal{I}$ is isomorphic to $G$. The set of graphs which have faithful representations by islands will be denoted ISLAND. We also denote by ISLAND the decision problem of recognizing ISLAND graphs.

An island with at most $k$ vertices will be called a *$k$-island*, and a *$k$-path* if it is a path. The sets and problems $k$-ISLAND and $k$-PATH are defined in the obvious way as the sets of graphs that can be represented by $k$-islands or $k$-paths, and the problems of their recognition.

Note that a faithful representation of $G$ is essentially an embedding of $G$ in which vertices are mapped to islands and edges to edges in the extended grid. A number of embedding problems have been investigated, mostly in the case of the plane grid, e.g., [3,15,16,18]. In these problems, for a given graph $G$, one seeks an embedding (or representation, or layout) where each vertex of $G$ is represented by a point of the plane grid, and each edge of $G$ by a path in the plane grid. One such problem is the VLSI Layout problem (see e.g. [5,6,10,17]), where one typically tries to minimize the area or the width of the smallest rectangle in which there exists a layout of $G$. Alternately one may want to minimize the length or number of bends of the paths representing the edges of $G$; see the book [15]. We note that the idea of representing vertices by islands (in this case in the plane grid) also occurs in [11], where islands are called fragments.

It is interesting to note that our representation problem is related to the embedding problem in the extended grid, in which one is minimizing the length of the representing paths. Indeed, we can replace each island by a vertex at the cost of replacing the edges by paths; if the islands are small, then the paths will be short. (Conversely, one can construct the islands from the representing points and paths.)

Most of the embedding problems mentioned above are NP-complete [15]. Our problem ISLAND will also turn out to be NP-complete.

When $k = 1$ the set 1-ISLAND (or 1-PATH) consists of graphs which are induced subgraphs of the extended grid. We shall prove that 1-ISLAND (or 1-PATH) is an NP-complete problem. We conjecture that recognizing $k$-ISLAND graphs is NP-complete for every $k$. We show that it is indeed so for $k < 3$ and $k > 5$. The unusual aspect of our proof is the fact that the techniques for small values of $k$ (that is $k = 1, 2$) are different from the techniques for large values of $k$ ($k > 5$). Although we think both these techniques can possibly be stretched a little bit more, we don't see how either of them can cover the other case. In any event, we leave the three missing cases $k = 3, 4, 5$ open.

In the plane grid, the problem corresponding to 1-ISLAND is also NP-complete (this is mentioned, e.g., in [8]). We believe that each of the problems $k$-ISLAND with a finite $k$ also behaves similarly from the computational complexity point of view in the plane and the extended grid, namely, we think both versions are NP-complete for all finite $k$.

Somewhat surprisingly, these versions behave quite differently when $k = \infty$, i.e., when the sizes of the islands are not bounded. In the planar grid, ISLAND consists exactly of planar graphs, and hence these graphs can be recognized in polynomial (even linear) time. On the other hand we will show that in the extended grid the set ISLAND coincides with the set of string graphs. A *string graph* is a graph $G$ whose vertices can be represented by *strings*, i.e., curves in the plane, such that two vertices are adjacent in $G$ if and only if the strings intersect. String graphs were introduced in [4]; Sinden [21] showed that not all graphs are string graphs. Finally, it was shown in [12] that recognizing string graphs is NP-hard, while Schaefer et al. [20] proved the problem is in NP. (This was quite unexpected and is rather nontrivial). Hence recognizing whether or not a given graph $G$ has a faithful representation in the extended grid is also NP-complete. We shall show that each problem $k$-ISLAND with $k > 5$ is NP-complete using the connection to string graphs.

The paper is organized as follows. In Section 2 we introduce our motivation which stems from quantum computation. Then in Section 3 the connection to string graphs is revealed. Graphs representable by large islands are dealt with in Section 4. The cases of 1-ISLAND and 2-ISLAND are presented in Section 5. Most of the technical proofs are left for the journal version of the paper.

## 2   Motivation from Adiabatic Quantum Computing

Our questions are motivated by problems arising in the proposed adiabatic quantum computer AQC [9]. An AQC relies upon a process known as quantum annealing, analogous to classical simulated annealing, to find a global minimum to an optimization problem. (Rather than relying on thermal fluctuations to escape local minima, quantum annealing relies on quantum fluctuations which escape local minima through a process known as tunneling.) Known theoretical results indicate the rate at which annealing can occur, and for some classes of problems exponential speedups over classical optimization seem possible. (It was shown in [1] that the standard and adiabatic quantum models are computationally equivalent.)

A realization of an AQC might consist of qubits and couplers, which can be naturally seen to correspond to vertices and edges of a graph. This abstraction away from the underlying physics allows us to translate computational problems into a graph theoretic language. An AQC depends upon its hardware to find a binary vector that optimizes some objective function. In one model, the goal is to maximize an unconstrained quadratic objective function

$$f(x) = \sum_{i=1}^{n} h_i x_i + \lambda \sum_{i=1}^{n} \sum_{j=i+1}^{n} J_{ij} x_i x_j,$$

where $\lambda$ is a (Lagrange) multiplier, $h$ is a vector, and $J$ is an upper-triangular matrix, describing respectively the linear and quadratic objective functions. For instance, by taking for $h$ the all-one vector and for $J$ the (upper half of the) adjacency matrix of $G$, and setting $\lambda$ to be smaller than $-2$, we can formulate the maximum independent set problem in $G$.

One specific AQC is layed out like a finite portion of the extended grid. Each qubit in the hardware is a vertex in the extended grid and each coupler an edge [2,19]. It may seem that only a limited set of problems can be posed in this specific graph (the extended grid). However, it is possible to constrain certain qubits to take on the same spin, and in this way to create *islands*, i.e., connected subgraphs (or equivalently subtrees) of the extended grid, which can be treated as 'supervertices'. To represent an arbitrary graph $G$ then amounts to assigning to each vertex of $G$ an island of the extended grid, and to each edge of $G$ an edge connecting the corresponding islands. Note that adjacent vertices of $G$ must be assigned to islands that are sufficiently near to be connected by an edge of the extended grid. If we additionally want to ensure that no noise or information in the quantum system transfer between islands through inactive couplers, we further require that nonadjacent vertices of $G$ be assigned islands that are not sufficiently near to be connected by an edge of the extended grid. In such a case, we seek a faithful representation of $G$ as defined earlier.

## 3   Unbounded Islands

It is well known that every string graph has a representation by curves in which no three curves intersect in the same point, and in every intersection point the two curves sharing this point cross each other in this point (i.e., they do not touch). It can be easily seen that such a representation is homeomorphic to a system of curves that are piece-wise linear and their segments follow the vertical and horizontal lines of a planar grid. Given such a representation of a graph $G$, one can rotate it by 45 degrees, and blow it up so that the segments of the representation follow the diagonal lines of the (now) extended grid in such a way that the segments intersect inside the grid squares but not in the grid points. Then one may represent each vertex of $G$ by the island consisting of the grid points lying on the corresponding curve. If we further blow up so that the segments of the curves use only every second diagonal in each direction, any two islands are adjacent if and only if their corresponding curves cross. (See an illustration in Fig. 1.)

On the other hand, it is also well known (cf. e.g. [12]) that string graphs are exactly the intersection graphs of arc-connected sets in the plane. Since for every island $i$ the expanded set $i^+$ is arc-connected, it follows that every ISLAND graph is a string graph. Therefore we have the following claim.

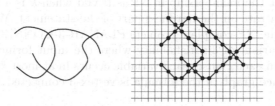

**Fig. 1.** A string representation and an ISLAND representation of $K_3$

**Proposition 1.** *The ISLAND graphs are exactly the string graphs. Hence the problem ISLAND is NP-complete.*

There is an even more straightforward way to transform a faithful representation of a graph in the extended grid into a string representation. For every island $i$ (which represents a vertex $u_i \in V(G)$), consider a spanning tree $T_i$ of the subgraph of the extended grid induced by the points of $i$. If two islands $i, j$ are adjacent via a grid edge $e = pq$ (vertical, horizontal, or diagonal), with vertices $p \in T_i$ and $q \in T_j$, one can create a new vertex $P_e$ in the middle (of the drawing) of $e$ and add the edge $pP_e$ to $T_i$ and the edge $qP_e$ to $T_j$. The newly added point $P_e$ may belong to more than just these two trees (if $e$ was a diagonal edge), but in such a case all the (at most) four islands involved in the corners of the square whose diagonal is $e$ are adjacent in the extended grid and so are the corresponding vertices in $G$. Hence $G$ is the intersection graph of a family of trees in the plane. Replacing every tree by a curve running around the drawing of the tree at a very small distance, we obtain a representation of $G$ by intersecting curves, where every common point $P_e$ will give rise to 2 crossing points of the curves, and every crossing of diagonal edges of two different trees gives rise to 4 crossing points. A $k$-island would thus result in a tree with a linear number of nodes (linear in $k$), and hence the total number of crossing points in the string representation obtained in this way would be $O(nk)$. It is known that there are string graphs on $n$ vertices requiring at least $\Omega(2^{cn})$ crossing points in any string representation [13]. Hence one obtains the following result.

**Proposition 2.** *There are ISLAND graphs which require exponentially many grid points in any faithful representation.*

We conclude that the natural guess-and-verify approach to obtaining a faithful representation of a graph, by guessing the representing islands point by point, fails. This observation highlights the fact that the membership of ISLAND in NP is due to deeper reasons [20].

## 4  Large Islands

In this section we sketch the proof of the following main theorem.

**Theorem 1.** *For every $k > 5$, the problem $k$-ISLAND is NP-complete.*

*Proof.* Note that the membership in NP is trivial when $k$ is a fixed parameter. Hence we concentrate on the hardness part of the statement. We will describe a reduction from the NP-complete problem PLANAR-3-CONNECTED (3,4)-SAT [14]. This is a variant of 3-satisfiability where the input formula has exactly 3 distinct literals in each clause, each variable occurs in at most 4 clauses, and the bipartite incidence graph of the formula is vertex-3-connected and planar.

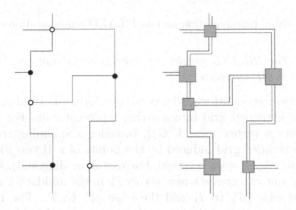

**Fig. 2.** An illustration for the proof of Theorem 1. A rectilinear drawing of the bipartite incidence graph of the formula (left) is turned into a 'city of squares and streets' in the grid (right).

Given such a formula $\Phi$, we construct a graph $G_\Phi$ such that $G_\Phi \in$ 6-ISLAND when $\Phi$ is satisfiable while $G_\Phi \notin$ ISLAND when $\Phi$ is not. Then the NP-hardness of the problem $k$-ISLAND for every $k > 5$ immediately follows.

The way we construct $G_\Phi$ is similar to the reduction introduced in [12]. We first fix a rectilinear drawing of the incidence graph of $\Phi$ such that both the variable and the clause vertices are located in points of the planar grid, and the edges connecting clauses to their variables are piece-wise linear and following the grid lines. Then we consider a refinement of the grid so that the variable and clause vertices are replaced by disjoint squares. Based on this drawing we construct $G_\Phi$ by replacing every variable vertex by a copy of a variable gadget, every clause vertex by a copy of a clause gadget, and the edges of the drawing by pairs of parallel paths of length derived from the drawing. We succeeded in constructing these gadgets using islands of size at most six. If $\Phi$ is satisfiable, a 6-ISLAND representation is built by representing the gadgets inside the corresponding squares and the connecting paths by chains of 6-paths. The details of the reduction will be given in the journal version of the paper.

## 5 Small Islands

In this section we focus on representations by 1-islands and 2-islands (i.e., by 1-paths and 2-paths). As pointed out above, 1-ISLAND graphs are exactly the induced subgraphs of the extended grid. Already representations by islands of size at most 2 provide much more flexibility than those by 1-islands. A 2-island may be positioned along an edge of the planar grid (i.e., horizontally or vertically) or along a diagonal; this results in a different number of grid points adjacent to the 2-island. Moreover a vertex may be represented by a 1-island or a 2-island, thus occupying a different area. As a consequence of this flexibility, it becomes more difficult to construct rigid structures, or at least to prove their rigidity. Thus 2-ISLAND graphs are bridging the features of subgraphs and intersection graphs and they best capture the concept of bounded island representations. We prove that both problems are computationally hard.

**Theorem 2.** *The problem 1-ISLAND is NP-complete, i. e., it is NP-complete to decide whether a given graph is an induced subgraph of the extended grid.*

**Theorem 3.** *The problem 2-ISLAND is NP-complete.*

### 5.1   The Logic Engine

Our proof is based on the so-called *logic engine* construction pioneered by Eades and Whitesides [7]. The logic engine offers an intuitive model of a standard reduction from the problem not-all-equal 3-SAT. Every Boolean formula can be assigned a logic engine. The physical model of the engine consists of a U-shaped outer *frame* that holds a *shaft* to which are attached rotating U-shaped *armatures*, nested into each other (cf. Fig. 3). The armatures can rotate independently of each other and the shaft bisects each of them. The tips of each armature are connected by a *chain* of fixed length, whose *links* can flip independently of each other. Some links have *flags* (also referred to as *strikers*) attached to them. By flipping the strikers while simultaneously rotating the armatures one can encode a NAE-satisfying assignment to a given formula. The tricky part is to make the engine rigid and flexible at the same time. Given a formula and a problem, the

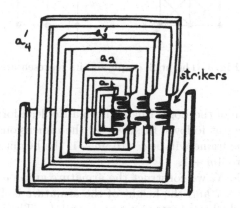

**Fig. 3.** The logic engine. Reprinted with permission of the authors from [7].

point is to model the Logic Engine by building an instance of the problem from building blocks that on one hand are rigid enough (fixing the shape of the outer frame and the armatures, the lengths of the chains and the shape of the flags) while allowing the desired flexibility (rotating the armatures and flipping the flags). In such a case the original formula will be NAE-satisfiable if and only if the corresponding logic engine can be laid out flat so that the strikers do not overlap. For more details and first examples see [7].

In the rest of the section we describe the constructions of the frame, shafts, armatures, and strikers. When representing a graph $G$, we usually name the vertices of $G$ using lower case letters $(a, b, c, \ldots)$, and name the corresponding islands in a representation by the respective upper case letters $(A, B, C, \ldots)$.

## 5.2   1-ISLAND Graphs: The Proof of Theorem 2

It is easy to see that the only 1-island representation of $K_4$ is formed by four points positioned in the four corners of a grid square. It follows that the location of the points representing any three vertices of $K_4$ uniquely determines the location of the fourth one. Based on this simple observation we build several rigid blocks. The basic one is the *brick graph* depicted in Fig. 4, which has only one representation (up to obvious symmetries). Overlapping bricks yield rigid structures which are used for the construction of the frame and the armatures. They are combined by *joints* formed by cut-vertices allowing rotations and flips along the shaft and on the rods. A close-up of the *rod* and *striker* graphs is shown in Fig. 5 and an overview of the entire construction in Fig. 6. A formal description of this construction and the corresponding proof will be given in the journal vestion of the paper.

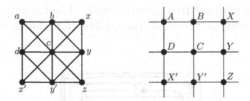

**Fig. 4.** The brick graph and its representation

The construction of the Logic Engine from a Boolean formula with $m$ clauses and $n$ variables goes as follows. First we set the dimensions of the frame graph $f$. The width of the frame will be set to $9n - 1$, the height to $4m + 4n + 3$, and the striker height to $4m + 1$.

For each variable $X_i$ we construct the armature graph $a_i$ where the width of the armature is $8i - 7$ and the height of the armature is $4m + 4i - 1$. We also construct a chain which connects vertices $u_{a_i}$ and $d_{a_i}$. The chain has three parts: the first part contain $m$ rods $g_1, g_2, \ldots, g_m$ ($g$ as gation); the next part is a cross

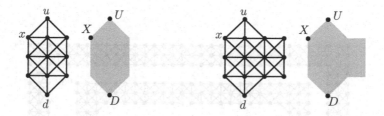

**Fig. 5.** The rod graph, the striker graph, and the shapes of their representations

graph $c_i$ (if $i$ is $n$ then the short cross graph $c_n$); and the last part contains $m$ rods $n_m, \ldots, n_1$ ($n$ as negation). We connect these structures sequentially into a chain, in such way that if we have two consecutive objects $a$ and $b$, then we connect $d_a$ with $u_b$. Thus the order of the structures in the chain will be $g_1, \ldots, g_m, c_i, n_m, \ldots, n_1$. This part of the construction is general for all instances with $n$ variables and $m$ clauses.

We now focus on the parts of the construction that are specific to the given instance. For a clause $c_j$, all variables $X_i$ not appearing in $c_j$ have their rod $g_j$ (for $X_i$) replaced by the striker graph; all variables $X_i$ for which $\neg X_i$ is not appearing in $c_j$ have their rod $n_j$ replaced by the striker. (Both can happen.) It follows from the construction that the original formula is NAE-satisfiable if and only if the resulting graph has a faithful representation by 1-islands.

## 5.3   2-ISLAND Graphs: The Proof of Theorem 3

As in the previous section, we will describe a reduction using the logic engine. Again, the crucial step is to construct a rigid gadget, i.e., a graph which has only one 2-island representation (up to obvious symmetries).

This role will be played the *2-brick graph* depicted in Fig. 7. It can be shown that in any 2-island representation of this graph, the four vertices of the outer four-cycle are represented by diagonal 2-islands as depicted in the figure, while the islands that represent the inner $K_4$ contain the four corners of the inner grid square of the representation (these islands may or may not extend to diagonal 2-islands as illustrated by the shaded lines). Connecting two bricks together by edges of a complete bipartite graph as depicted in Fig. 8 we obtain the *2-brickjoint* graph which provides both rigidity and flexibility that we need. Every 2-island representation looks the same from the geometrical point of view, but the pairs of islands $A, B$ and $A', B'$ can be swapped independently. Brickjoints are then combined into *2-bricklines* and *2-bricksquares* which are also rigid in the geometric sense and are used for the construction of the frame, shaft and armatures. Finally, 2-bricks connected to the rods by 2-brickjoints are also used as strikers. In this way we construct a graph which has a 2-island representation if and only if the original formula was NAE-satisfiable. The details will be included in the journal version of the paper.

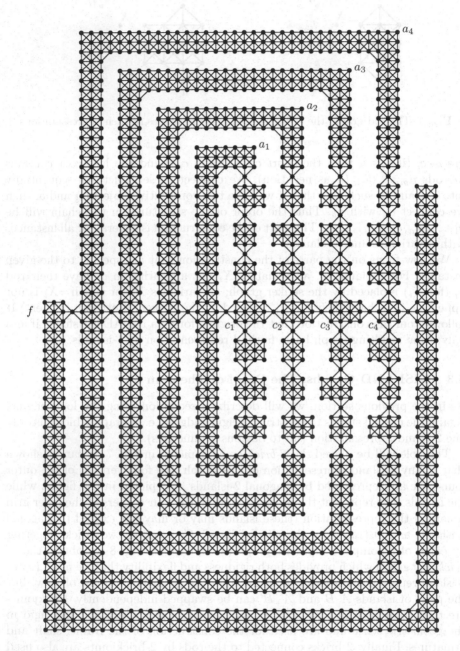

**Fig. 6.** The construction for the formula $(X_1 \lor X_2 \lor \neg X_3) \land (\neg X_1 \lor X_2 \neg X_4) \lor (\neg X_2 \lor \neg X_3 \lor X_4)$

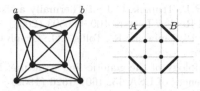

**Fig. 7.** The 2-brick graph, with a representation

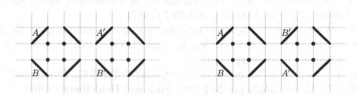

**Fig. 8.** The 2-brickjoint graph and its two possible representations

## 6 Conclusions

We have introduced the sets $k$-ISLAND, for positive integers $k$. The set 1-ISLAND consists of induced subgraphs of the extended grid. The set ISLAND = ∪$k$-ISLAND consists of string graphs. Both have NP-complete recognition problems. However, the nature of the problems, and the proofs of their NP-completeness, are very different. For ISLAND, and similarly for $k$-ISLAND with $k > 5$, we employed the techniques of string graphs [12]; for 1-ISLAND, and similarly for 2-ISLAND, we gave a proof based on the logic engine of [7]. We suspect that all the $k$-ISLAND recognition problems are NP-complete, but do not see a unified reduction using just one of these techniques. Perhaps a new approach can cover both ends of the spectrum. In any event, we leave the cases $k = 3, 4, 5$ open.

The problems discussed make sense in higher-dimensional grids as well. We observe that in three-dimensional grids, ISLAND consists of all graphs, and hence is trivially polynomial; on the other hand recognizing 1-ISLAND graphs can be shown NP-complete using similar logic engine techniques as in Section 3. As we remarked earlier, the situation is similar in the usual plane grid: there ISLAND consists of all planar graphs and so is polynomial, while 1-ISLAND is NP-complete.

**Acknowledgments.** We thank Tomás Feder for suggesting the connection to string graphs, and William G. Macready for improving our explanation of the adiabatic quantum computer.

## References

1. Aharonov, D., van Dam, W., Kempe, J., Landau, Z., Lloyd, S., Regev, O.: Adiabatic quantum computation is equivalent to standard quantum computation. In: Proceedings of the $45^{th}$ FOCS, pp. 42–51 (2004)

2. Amin, M.H.S., Love, P.J., Truncik, C.J.S.: Thermally assisted adiabatic quantum computation. Physical Review Letters 100, 060503 (2008)
3. Di Battista, G., Eades, P., Tamassia, R., Tollis, I.G.: Graph Drawing. Prentice-Hall, Englewood Cliffs (1999)
4. Benzer, S.: On the topology of the genetic fine structure. Proceedings of the National Academy of Sciences of USA 45, 1607–1620 (1959)
5. Bhatt, S.N., Leighton, F.T.: A framework for solving VLSI and graph layout problems. J. Comput. System Sciences 28, 300–343 (1984)
6. Bhatt, S.N., Cosmadakis, S.S.: The complexity of minimizing wire length in VLSI layouts. Inform. Process. Lett. 25, 263–267 (1987)
7. Eades, P., Whitesides, S.: Nearest neighbour graph realizability is NP-hard. In: Baeza-Yates, R., Poblete, P.V., Goles, E. (eds.) LATIN 1995. LNCS, vol. 911, pp. 245–256. Springer, Heidelberg (1995)
8. Eppstein, D.: Isometric Diamond Subgraphs. In: Tollis, I.G., Patrignani, M. (eds.) GD 2008. LNCS, vol. 5417, pp. 384–389. Springer, Heidelberg (2009)
9. Farhi, E., Goldstone, J., Gutmann, S., Sipser, M.: Quantum computation by adiabatic evolution (2000), http://arxiv.org/abs/quant-ph/0001106
10. Formann, M., Wagner, F.: The VLSI layout problem in various embedding models. In: Möhring, R.H. (ed.) WG 1990. LNCS, vol. 484, pp. 130–139. Springer, Heidelberg (1991)
11. Garey, M.R., Johnson, D.S., So, H.C.: An application of graph coloring to printed circuit testing. IEEE Trans. on Circuits and Systems 23, 591–599 (1976)
12. Kratochvíl, J.: String graphs II. Recognizing string graphs is NP-hard. J. Comb. Theory, Ser. B 52, 67–78 (1991)
13. Kratochvíl, J.: String graphs requiring exponential representations. J. Comb. Theory, Ser. B 53, 1–4 (1991)
14. Kratochvíl, J.: A Special Planar Satisfiability Problem and a Consequence of its NP-completeness. Discr. Appl. Math. 52, 233–252 (1994)
15. Leighton, F.T.: Introduction to parallel algorithms and architectures. Morgan Kaufmann, San Mateo (1993)
16. Lin, Y.B., Miller, Z., Perkel, M., Pritikin, D., Sudborough, I.H.: Expansion of layouts of complete binary trees into grids. Discr. Appl. Math. 131, 611–642 (2003)
17. Matsubayashi, A.: VLSI layout of tree into grids of minimum width. In: Proc. 15th Annual ACM Symposium on Parallel Algorithms and Architectures, pp. 75–84 (2003)
18. Opatrny, J., Sotteau, D.: Embeddings of complete binary trees into grids and extended grids with total vertex-congestion 1. Discr. Appl. Math. 98, 237–254 (2000)
19. Rose, G., Bunyk, P., Coury, M.D., Macready, W., Choi, V.: Systems, Devices, and Methods for Interconnected Processor Topology, US Patent Application No. 20080176750 (January 2008)
20. Schaefer, M., Sedgwick, E., Stefankovic, D.: Recognizing string graphs in NP. J. Comput. Syst. Sci. 67, 365–380 (2003)
21. Sinden, A.: Topology of thin film RC-circuits. Bell System Technical Journal, 1639–1662 (1966)

# The I/O Complexity of Sparse Matrix Dense Matrix Multiplication

Gero Greiner and Riko Jacob

Technische Universität München

**Abstract.** We consider the multiplication of a sparse $N \times N$ matrix $\mathbf{A}$ with a dense $N \times N$ matrix $\mathbf{B}$ in the I/O model. We determine the worst-case non-uniform complexity of this task up to a constant factor for all meaningful choices of the parameters $N$ (dimension of the matrices), $k$ (average number of non-zero entries per column or row in $\mathbf{A}$, i.e., there are in total $kN$ non-zero entries), $M$ (main memory size), and $B$ (block size), as long as $M \geq B^2$ (tall cache assumption).

For large and small $k$, the structure of the algorithm does not need to depend on the structure of the sparse matrix $\mathbf{A}$, whereas for intermediate densities it is possible and necessary to find submatrices that fit in memory and are slightly denser than on average.

The focus of this work is asymptotic worst-case complexity, i.e., the existence of matrices that require a certain number of I/Os and the existence of algorithms (sometimes depending on the shape of the sparse matrix) that use only a constant factor more I/Os.

## 1   Introduction

Traditionally, the aim of a good algorithmic design is to achieve a task with as few CPU-operations as possible. In a setting where the main calculation is phrased as matrix and vector operations, this usually means that the matrices are kept sparse and it is exploited that many entries are zero. This reduction of CPU-operation sometimes comes at the price of irregular access patterns induced by the sparse matrix operations, which can lead to a situation where memory access is the real bottleneck of the computation.

One successful way of modeling this bottleneck is the so called I/O-model (also known as Disk-Access-Model DAM) introduced by Aggarwal and Vitter [1]. It assumes that the CPU can only operate on a main memory of size $M$ whereas further intermediate results (just like input and final result) must be stored on an infinite disk, that is organized in blocks of size $B$. The resulting performance measure counts the number of read/write operations of the disk, the so called I/O-operations (or I/Os for short). By now this model is accepted and the I/O-complexity of many tasks is well understood.

We consider the multiplication of a sparse $N \times N$ matrix $\mathbf{A}$ containing $kN$ non-zero entries ($\mathbf{A}$ is called $k$-sparse) with a dense $N \times N$ matrix $\mathbf{B}$, computing $\mathbf{C} = \mathbf{A} \cdot \mathbf{B}$. Throughout the paper, we abbreviate this task as $\mathrm{SDM}_k$. We study the worst case complexity of this task in the I/O-model, i.e., we determine up

A. López-Ortiz (Ed.): LATIN 2010, LNCS 6034, pp. 143–156, 2010.

to some constant factor the number of I/Os that are necessary and sufficient to compute **C**. Here, we use the so-called semiring I/O-machine, where algorithms can only use addition and multiplication, but cannot rely upon subtraction or division (as detailed in Section 2 "Model of Computation"). We consider worst-case complexities, where the worst case is taken over the shape (or conformation) of the matrix **A** as given by the positions of the non-zero entries. Our notion of an algorithm is non-uniform in the sense that we ask for a program that, depending on the conformation of **A**, computes **C** with few I/Os irrespective of the complexity to create this program.

This model of computation and notion of sparseness has been successfully used in [3] to study multiplying a sparse matrix with a dense vector. It also coincides with the notion of "independent evaluation" of Hong and Kung [5] to study the multiplication of two dense matrices. In this case (in our notation $k = N$), we do know that this is a restriction as it disallows algorithms like Strassen's. However, the algebraic complexity of dense matrix multiplication is still unknown [7]. In this computational model, our notion of $k$-sparseness of **A** has the effect that the matrix multiplication requires precisely $kN^2$ multiplications of numbers.

In this paper, we determine the complexity to multiply two $N \times N$ matrices, one of which is $k$-sparse in the semiring I/O-model with tall cache $M \geq B^2$ to be

$$\Theta\left(\max\left\{\frac{kN^2}{B\Delta}, \frac{kN^2}{B\sqrt{M}}, \frac{N^2}{B}, 1\right\}\right)$$

where

$$\Delta = \max\left\{\frac{\ln\frac{N}{M}}{\ln\frac{N\ln^2\frac{N}{M}}{Mk}}, \sqrt{\frac{kM}{N}}\right\},$$

and $\ln$ is defined by $\ln x = \max\{1, \ln x\}$.

This expression yields three interesting ranges of the parameter $k$, but for all $k$ the I/O-complexity boils down to the question how many of the $kN^2$ elementary products can be performed on $M$ elements that are simultaneously in memory (i.e. in one so called Hong-Kung round). For large $k$ the situation is basically that of two dense matrices, in particular, for $k = N$ it coincides with the classical result of Hong and Kung [5] that multiplying two dense square matrices has complexity $\Theta\left(\frac{N^3}{B\sqrt{M}}\right)$, i.e, that at most $M^{\frac{3}{2}}$ multiplications per round are possible and can be achieved by using $\sqrt{M} \times \sqrt{M}$ tiles. For small $k$ it resembles that of **A** being a permutation matrix where $M$ multiplications per round are best possible (i.e. loaded elements cannot be reused).

Additionally, there is a density from which point on the complexity (reuse of loaded operands) can be described by above average dense submatrices consisting of $M$ entries and having on average $\min\{\Delta, \sqrt{M}\}$ entries per row and column. Our complexity analysis proceeds by showing that there exist matrices that have essentially no denser submatrices. We get a matching upper bound by showing that every matrix that has sufficiently many entries must have such dense submatrices. The resulting algorithm hence depends upon the conformation (shape) of the sparse input matrix in a complicated manner (which does not influence the theoretical statement). How difficult it is to actually compute a good program is not completely understood. We only have preliminary results [8] showing that finding such a program is NP-complete, and that determining the maximum possible density cannot be approximated within an arbitrarily small constant factor. This limited structural insight is already more than what is known for the multiplication of a sparse matrix with a dense vector [3], where it is only clear that difficult matrices exist (by a counting argument), but there is no characterization of which (permutation) matrices are difficult to multiply with. One key difference in the consideration of this paper and [3] is that here the block size $B$ is basically irrelevant (it is only the scaling factor to translate I/O-volume to number of I/Os), whereas matrix vector multiplication becomes trivial if $B = 1$.

For the sake of clarity, throughout the paper we stick to the case where all matrices are square. The results naturally generalize to non-square situations as long as the smallest dimension (side-length) is at least $\sqrt{M} \geq B$.

Clearly, the results presented here are theoretical in nature, and the presented algorithms are stated more to complement the lower bounds than to be implemented. The real goal is to devise and evaluate practical algorithms, and the results presented here give important limits on what to expect from them. Further, such practical algorithms exist and are implemented in many sparse matrix libraries, which indicates that the considered problem is of a certain practical relevance. Also there it has been recognized that memory access patterns are an important factor in the overall execution time, and memory aware algorithms have been proposed [2,9,4].

Here, it is worth noting that our notion of $k$-sparseness fits to the established experimental performance measure "number of floating point operations performed per second". Because the number of floating point operations is precisely $kN^2$ this immediately translates to running time, and hence if memory is the bottleneck, to number of I/Os.

In contrast to the mentioned practical considerations, our work is theoretical in nature, with all the well known consequences: The presented results are mathematical theorems, stating all assumptions and having a clear conclusion. This generality of course comes at a price, in our context mainly the abstraction of the model of computation (neglecting everything but the memory access patterns of a program, and disallowing Strassen-like algorithms) and the focus on the worst-case input (where practitioners can and have to exploit the special structure in the input at hand). Nevertheless, we believe that our theoretical findings and understanding give an important reference point also for the practical work: What are the difficult instances? In what respect are practical inputs easy? What is a good worst-case behavior of an algorithm?

*Outline of the paper.* We introduce the precise model of computation in Section 2. This is followed by Section 3 where some easy to derive inequalities, so called Observations are given. In Section 4 we describe the different algorithms, depending on the density parameter $k$, whereas Section 5 shows that there exist matrices that require the stated number of I/Os.

## 2    Model of Computation

We consider the number of I/Os induced by a program as a measurement of costs. Therefore, we use the model described in [3] which consists of two memory layers, as is standard. We assume a single processing unit with a fast memory of limited capacity $M$ assigned to it. Calculations can only be performed on the elements residing in this *internal memory*, whereas the programs input and any (intermediate) results are stored on an *external memory* of infinite size which is organized in blocks of size $B$. Elements are moved between memory layers in blocks, where the movement of a block incurs costs 1.

Memory elements are to belong to a *commutative semiring* $S = (\mathbb{R}, +, \cdot)$, i.e., a set $\mathbb{R}$ with operations addition $(+)$ and multiplication $(\cdot)$ that are associative, distributive and commutative. Further, there is a neutral element 0 for addition, 1 for multiplication and 0 is annihilating with respect to multiplication. In contrast to rings and fields, inverse elements are neither guaranteed for addition nor for multiplication, i.e., the program is not allowed to use subtraction and division.

*Definition [3]:* The *semiring I/O machine* consists of an internal memory which can hold up to $M$ elements of a commutative semiring $S$, and an external memory of infinite size which is organized in blocks of $B$ consecutive elements. The current *configuration* of a machine is described by the content $\mathcal{M} = (m_1, \ldots, m_M)$, $m_i \in \mathbb{R}$ of internal memory, and an infinite sequence of blocks $t_i \in \mathbb{R}^B$, $i \in \mathbb{N}$ of external memory. An *operation* is a transformation of one configuration into another, which can be one of the following types

- *Computation*, performs any operation of $S$ on elements in $\mathcal{M}$.
- *Input*, replaces some chosen elements $m_{i_1}, \ldots, m_{i_B}$ in $\mathcal{M}$ by a block $t_i$.
- *Output*, replaces a block $t_i$ of external memory by some chosen elements $m_{i_1}, \ldots, m_{i_B}$ of $\mathcal{M}$.

Using this, we define a *program* $P$ as a finite sequence of operations. The number of input and output operations describes the I/O costs of $P$. An *algorithm* is a family of programs where the program can be chosen according to the parameters $N$, $k$, and the *conformation* of $\mathbf{A}$, i.e., the position of the non-zero entries in $\mathbf{A}$. By $L(k, N)$, we denote the required number of I/Os induced by an algorithm for $\mathrm{SDM}_k$ with $N \times N$ matrices, and $kN$ non-zero entries in $\mathbf{A}$ for all kinds of conformations.

The value of an element $c_{ij}$ of the result matrix $\mathbf{C} := \mathbf{A} \cdot \mathbf{B}$ is given by $\sum_{l \in A_i} a_{il} \cdot b_{lj}$ where $A_i \subseteq \{1, \ldots, N\}$ describes the positions of non-zero elements in the $i$-th row of $\mathbf{A}$. Since our model is based on a semiring, the computation of $\mathbf{C}$ includes the calculation of exactly $kN^2$ *elementary products* $a_{il} \cdot b_{lj}$. Further, any intermediate result can be seen as a sum $\sum_{l \in S} a_{il} \cdot b_{lj}$ for a subset $S \subseteq A_i$. If $|S| < |A_i|$, we refer to this as a *partial result* of $c_{ij}$.

Since we can assume that every program requires at least one I/O, when writing complexity using $\mathcal{O}$, $\Theta$, or $\Omega$ at least 1 is meant.

# 3    Math

For the proofs provided in Section 4 and 5, the following Observations are necessary.

**Observation 1.** *For $0 \leq x \leq 1/2$, it holds that $\ln(1 - x) \geq -2x$.*

**Observation 2.** *For $0 < a \leq e$, for any $x > 0$ it holds $x \geq a \ln x$.*

**Observation 3.** *For $x \geq y \geq 1$ it holds*

$$\left(\frac{x}{y}\right)^y \overset{(a)}{\leq} \binom{x}{y} \overset{(b)}{\leq} \left(\frac{ex}{y}\right)^y.$$

**Observation 4.** *For $n \geq k \geq a \geq 1$ it holds*

$$\binom{n}{k} \geq \left(\frac{n-k}{k}\right)^a \binom{n}{k-a}.$$

*Proof.* By definition of binomial coefficients

$$\binom{n}{k} \cdot \binom{n}{k-a}^{-1} = \frac{n!(n-k+a)!(k-a)!}{(n-k)!k!n!} = \prod_{i=1}^{a} \frac{n-k+i}{k-a+i}.$$

By showing that for each $1 \leq i \leq a$, $\frac{n-k+i}{k-a+i} \geq \frac{n-k}{k}$ the statement follows. This is equivalent to $(a-2i)k \leq (a-i)n$ which holds for any $1 \leq k \leq n$ and $1 \leq i \leq a$. □

**Observation 5.** *For $D, f, x > 0$, the inequality $D \ln fD \leq x$ is satisfied for $D \leq \frac{x}{\overline{\ln} xf}$ where $\overline{\ln} x = \max\{1, \ln x\}$.*

*Proof.* By substitution, we obtain

$$D \ln fD \leq \frac{x}{\overline{\ln} xf} \ln\left(f\frac{x}{\overline{\ln} xf}\right) \leq \frac{x}{\overline{\ln} xf} \ln\left(f\frac{x}{1}\right) \leq x. \qquad \square$$

**Observation 6.** *For $D, f, x \geq 0$, the inequality $D \ln fD > x$ is fulfilled if $D > \frac{2x}{\ln 2xf}$.*

*Proof.* Substituting $D$ yields

$$D \ln fD > \frac{2x}{\ln 2xf} \ln\left(f\frac{2x}{\ln 2xf}\right) \geq \frac{2x}{\ln 2xf} \ln \sqrt{2xf} = x$$

where we use $\sqrt{2xf} \geq 2\ln\sqrt{2xf}$ given by Observation 2. □

# 4    Algorithms

**Theorem 1.** *For $1 \leq k \leq N$, $SDM_k$ is possible with*

$$\mathcal{O}\left(\max\left\{\frac{kN^2}{B\Delta}, \frac{kN^2}{B\sqrt{M}}, \frac{N^2}{B}\right\}\right)$$

*I/Os for*

$$\Delta = \max\left\{\frac{\ln\frac{N}{M}}{\ln\frac{N\ln^2\frac{N}{M}}{Mk}}, \sqrt{\frac{kM}{N}}\right\}$$

*if $M = \Omega\left(B^2\right)$. Note that $\Delta$ is lower bounded by a constant.*

For the sake of clarity, we omit the use of ceiling functions from now on. Since all the fractions are greater or equal 1, this only increases the bounds by constant factors.

## 4.1    Layouts

For the algorithms presented, we either need the dense matrices **B** and **C** in column major layout where elements are saved column wise, or row major layout. As we prove by lower bounds, a different layout chosen by the algorithm does not lead to an asymptotic speed up. For the first and the third algorithm presented in this section, a row major layout is required, i.e., elements are saved row wise in external memory. If **B** is saved in column major layout, the matrix has to be transposed first. In [1], Aggarwal and Vitter showed that this is possible with $\mathcal{O}\left(N^2/B\right)$ I/Os assuming a tall cache, i.e., $M \geq B^2$. Similarly, if required, **C** has to be transposed in the end. For the tile-based approach, the matrices are read in tiles. However, assuming a tall cache, the algorithm is applicable if **B** is in column major layout. This also applies for small instances where $M \geq kN + N$.

For all algorithms presented, we assume that **A** is a list of the non-zero entries in arbitrary order. For the direct algorithm, the ordering is indeed unimportant, whereas the desired layout for the other algorithms can be obtained by sorting. Sorting the elements of **A** is possible with $\mathcal{O}\left(\frac{kN}{B}\log_{M/B}\frac{kN}{M}\right)$ I/Os [1]. Since $k \leq N$, this is at most $2c\frac{kN}{B}\log_{M/B}\frac{N}{\sqrt{M}} \leq c\frac{kN^2}{B\sqrt{M}}$ since $M \geq 2B$. Note that this is less than the number of I/Os required for the computation itself.

## 4.2    Direct Algorithm

The direct algorithm for permuting can simply be extended to any $1 \leq k \leq N$. Let $b_i$ be the $i$-th row of **B**, and $c_i$ the $i$-th row of **C**. By one scan of **A** while adding for each non-zero entry $a_{ij}$ the product $a_{ij} \cdot b_j$ to $c_i$, the result matrix **C** is computed with $\mathcal{O}\left(\frac{kN^2}{B}\right)$ I/Os. As we will see, for $k \leq (N/M)^{1-\epsilon}$ and any constant $\epsilon > 0$ this algorithm is asymptotically optimal since $\Delta$ in Theorem 1 becomes a constant.

### 4.3 Tile-Based Algorithm

For denser cases of $\mathbf{A}$, there is a modification of the tile-based algorithm in [6] that clearly outperforms the direct algorithm. More precisely, this algorithm works for any $k \geq \frac{N}{M}$ and $M \leq kN$. For the ease of notation, let $3M$ be the size of internal memory. For this approach, matrices $\mathbf{B}$ and $\mathbf{C}$ have to be partitioned into tiles of size $a \times M/a$ for $a = \sqrt{MN/k}$, while $\mathbf{A}$ is partitioned into tiles of size $a \times a$ (cf. Fig. 1). Let $\mathbf{A}_{ij}, \mathbf{B}_{ij}$, and $\mathbf{C}_{ij}$ denote the $j$-th tile within the $i$-th tile row of the corresponding matrix. Clearly, it holds $\mathbf{C}_{ij} = \sum_{l=1}^{n} \mathbf{A}_{il}\mathbf{B}_{lj}$ for $n = N/a$. Throughout the calculation of a certain $\mathbf{C}_{ij}$, partial results can be kept in internal memory while $\mathbf{A}_{il}$ and $\mathbf{B}_{lj}$ are loaded consecutively for each $l$. Since each $\mathbf{B}$-tile contains exactly $M$ elements, each such tile can be loaded in a whole and only once per calculation of a $\mathbf{C}_{ij}$. The same holds for tiles of $\mathbf{A}$ containing at most $M$ entries. For tiles with more non-zero entries, elements are loaded in bunches of $M$ elements while the corresponding $\mathbf{B}$-tile is kept in memory.

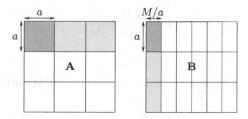

**Fig. 1.** An illustration of the tiles in $\mathbf{A}$ and $\mathbf{B}$. $\mathbf{C}$ is partitioned similarly to $\mathbf{B}$.

Hence, for saving all $\mathbf{C}$-tiles no more than $\frac{N^2}{B}$ I/Os are required. Loading $\mathbf{B}$-tiles costs at most $\frac{N^2}{M} \cdot n\frac{M}{B}$ I/Os. Each tile, and thus, each element in $\mathbf{A}$ has to be loaded for the calculation of no more than $N/\frac{M}{a}$ tiles of $\mathbf{C}$. Therefore, loading the non-zero elements of $\mathbf{A}$ requires at most $\frac{kN}{B}\frac{N\sqrt{N}}{\sqrt{kM}}$ I/Os. Altogether, this sums up to

$$\mathcal{O}\left(\sqrt{\frac{kN}{M}}\frac{N^2}{B}\right)$$

I/Os for the computation of $\mathbf{C}$.

### 4.4 Using Dense Parts of A

In this section, we show that by loading $M$ elements from each matrix, even for $k < \frac{N}{M}$ a number of $\omega(M)$ elementary products can be obtained, i.e., more than the direct algorithm achieves. This is done by loading dense parts from $\mathbf{A}$.

For the sake of illustration, we consider the matrix $\mathbf{A}$ as an adjacency matrix of a bipartite graph $G = (U \cup V, E)$, where $a_{ij} \neq 0$ constitutes a connection between the $i$-th node of $U$ and the $j$-th node of $V$. If there are sufficiently many subgraphs containing $\mathcal{O}(M)$ edges, with average degree $\Omega(D)$, $\text{SDM}_k$ is possible in time $\mathcal{O}\left(\frac{kN^2}{BD}\right)$.

**Lemma 1.** *Given a bipartite graph $G = (U \cup V, E)$, $|U| = |V| = N$, $|E| = kN$, i.e., $k$ is the average degree.*
*Then, for $2 \le k \le \frac{N}{32M} \ln^2 \frac{N}{M}$ and $M \le N/4$ there exist two subsets $X \subseteq U$, $Y \subseteq V$, such that the subgraph induced by $X$ and $Y$ has average degree at least*

$$D = \min\left\{ \frac{\ln \frac{N}{M}}{2\ln \frac{N \ln^2 \frac{N}{M}}{4Mk}}, \sqrt{\frac{M}{2}} \right\}$$

*and it holds that $|X|, |Y| \le M/D$. This $D$ satisfies $D \le k/2$.*

Before showing this, we need the following preliminary lemma.

**Lemma 2.** *For $k \le \frac{N}{32M} \ln^2 \frac{N}{M}$, $M \le N$, and $D$ according to Lemma 1 the inequality*

$$k \le \frac{ND}{4M} \ln \frac{N}{M} \tag{1}$$

*is satisfied.*

*Proof.* For $D = \frac{\ln \frac{N}{M}}{2\ln \frac{N}{4Mk} \ln^2 \frac{N}{M}}$, (1) holds as follows. Substituting $D$ yields

$$k \ln \frac{N \ln^2 \frac{N}{M}}{4Mk} \le \frac{N}{8M} \ln^2 \frac{N}{M}.$$

Observe, that for $\frac{x}{k} \ge e$, the term $k \ln \frac{x}{k}$ for $x > 0$ is monotonously increasing in $k$. Its derivative is $\ln \frac{x}{k} - 1$, and hence, positive for $\frac{x}{k} \ge e$. Since by assumption $\frac{N \ln^2 \frac{N}{M}}{4Mk} \ge e$, we can substitute $k$ resulting in

$$\frac{N \ln 8}{32M} \ln^2 \frac{N}{M} \le \frac{N}{6M} \ln^2 \frac{N}{M}.$$

which is obviously true.

If the second term of the minimum in $D$ applies, i.e. $D = \sqrt{M/2}$, we have to distinguish the cases $\frac{N}{32M} \ln^2 \frac{N}{M} \le N$, i.e., $M \ge \frac{\ln^2 \frac{N}{M}}{32}$ and vice versa, i.e. only $k \le N$ has to hold. By substituting $D$ and $k$ in (1) both cases hold within the desired range. $\square$

*Proof (Lemma 1).* In order to make a statement about the minimal degree of a node, we transform $G$ such that the maximal degree in $V$ is restricted to at most $k$. Therefore, split each node $v_i \in V$ with degree $d_i > k$ into $v_{i,1}, \ldots, v_{i,\lceil d_i/k \rceil}$ such that each node has degree no more than $k$. Let $V'$ denote this transformation of $V$, $E'$ the transformed set of edges and $G' = (U \cup V', E')$ the created graph. By construction, the size of $V$ will increase by no more than $N$, i.e., $|V'| \le 2N$. From this, we can conclude that there are at least $N/2$ nodes with degree no less than $k/2$: Suppose that $c$ nodes in the original set $V$ have degree less than $k/2$. Hence, the degrees of the remaining $N - c$ nodes sum up to at least $(N - c/2)k$. By construction of $V'$, for each node with degree $d_i > k$, there will be at most one new node with degree less than $k/2$. This leads to no less than $(N - c/2)k - (N - c)k/2 = kN/2$ edges that have to belong to nodes with degree at least $k/2$. Since all nodes have degree at most $k$, there have to be at least $N/2$ nodes of degree at least $k/2$. We call the subset of these nodes $V'_{k/2}$.

Observe that any subgraph $G'_S$ in $G'$ with average degree $D$ consisting of nodes $X \subseteq U$ and $Y \subseteq V'$ can be transformed into a subgraph $G_S$ of $G$ with average degree at least $D$ by simply replacing any $v_{i,j} \in Y$ by the corresponding node $v_i$ of the original graph. The subgraph $G_S$ contains at least the amount of edges of $G'_S$, but no more nodes than $G'_S$. Hence, it suffices to show the existence of the desired $X$ and $Y$ for $G'$. To this end, we will prove that in $G'$ for a random $X \subseteq U, |X| = M/D$, the expected number of nodes in $V'$ that have degree $\geq D$ into $X$ is at least $M/D$.

Now, choose $X \subseteq U$ uniformly at random and consider a vertex $v_i \in V'_{k/2}$. The number of vertices chosen for $X$ in the neighborhood of $v_i$ is given by a hypergeometric distribution, resembling drawing at least $k/2$ times without replacement from an urn with $N$ marbles, $M/D$ of which are black. The event we are interested in is that at least $D$ of the drawn marbles are black.

We lower bound this probability by considering only the case of drawing precisely $D$ black marbles. The probability can be expressed in the following way: Consider drawing the $k/2$ marbles one after another, and choose precisely $D$ positions where black marbles are drawn. The probability of such a drawing can then be calculated as the product of the fractions of black (white) marbles that are left in the urn before each drawing. For black marbles the fraction is at least $p = (\frac{M}{D} - D)/N$, for white it is at least $q = 1 - \frac{M}{D}/(N - \frac{k}{2})$. In the following, we use $D \leq \sqrt{M/2}$, i.e., $D \leq \frac{M}{2D}$, and $k \leq N$ to simplify these expressions.

The overall probability of drawing $D$ black marbles can then be bounded by summing the probabilities of all possible choices to position the $D$ black marbles in the consecutive drawing. Let $X_i$ be the number of black marbles drawn, i.e. the number of edges from $v_i$ into $X$. Thus, we can lower bound the probability similar to a binomial distribution:

$$P(X_i \geq D) \geq \binom{k/2}{D} p^D q^{k/2 - D} > \left( \frac{k}{2D} \frac{M}{2DN} \right)^D \left( 1 - \frac{2M}{DN} \right)^{k/2}.$$

Taking logarithm we get

$$\ln P(X_i \geq D) \geq D \ln \frac{Mk}{4ND^2} + \frac{k}{2} \ln \left( 1 - \frac{2M}{ND} \right) \geq D \ln \frac{Mk}{4ND^2} - k \frac{2M}{ND}$$

where the last inequality is justified by Observation 1 and $4M \leq N$.

Since we consider at least $N/2$ nodes, the goal is now to choose the biggest $D$ satisfying

$$D \ln \frac{4ND^2}{Mk} + k \frac{2M}{ND} \leq \ln \frac{ND}{2M}.$$

By Lemma 2, $k \le \frac{ND}{4M} \ln \frac{N}{M}$, i.e., $k \frac{2M}{ND} \le \frac{1}{2} \ln \frac{N}{M}$, holds. Hence, we are interested in

$$D \ln \frac{4ND^2}{Mk} \le \frac{1}{2} \ln \frac{N}{M} + \ln \frac{D}{2}$$

which is implied by

$$D \ln \frac{2\sqrt{ND}}{\sqrt{Mk}} \le \frac{1}{4} \ln \frac{N}{M} \tag{2}$$

for $D \ge 2$. Otherwise, since $k \ge 2$ one can obtain the desired subgraph by choosing $M$ adjacent edges in $G$. This yields a subgraph consisting of at most $M + 1$ vertices, i.e. with average degree at least $2/(1 + 1/M)$.

Now, we can use Observation 5 with $f = \sqrt{\frac{4N}{Mk}}$ and $x = \frac{1}{4} \ln \frac{N}{M}$, and get the approximation

$$D \le \frac{x}{\ln xf} = \frac{\ln \frac{N}{M}}{2 \ln \frac{N \ln^2 \frac{N}{M}}{4Mk}}$$

for which inequality (2) holds.

Finally, we check $D \le k/2$. To this end, we plug $k/2$ as $D$ into (2) and get

$$\frac{k}{2} \ln \frac{4N \cdot k^2}{Mk \cdot 4} \le \frac{1}{2} \ln \frac{N}{M}$$

$\frac{k}{2} \ln \frac{Nk}{M} \le \frac{1}{2} \ln \frac{N}{M}$, assuming $k \le 1$, contradicting one of the assumptions.     □

In the following, we assume an internal memory of size $2M$ to ease notation. Now consider a subgraph $G_S = (U_S \cup V_S, E_S)$ with $M$ edges and average degree $D$. By construction of $G = (U \cup V, E)$, we considered a non-zero entry $a_{ij}$ as an edge between $u_i$ and $v_j$. Let $I_U$, $I_V$ be the set of indices of vertices in $U_S$, $V_S$ respectively. In order to create elementary products corresponding to $E_S$, the $M$ corresponding non-zero entries $a_{ij}$ with $i \in I_U$, $j \in I_V$ have to be loaded. Then, for each column $1 \le k \le N$, by loading all elements $b_{jk}$ with row indices $j \in I_V$ together, $M$ elementary products for rows with indices $I_U$ in $\mathbf{C}$ can be obtained.

To efficiently load certain elements of a column in $\mathbf{B}$, we extract these rows into a separate $|I_V| \times N$ matrix and transpose it to column major layout. This is possible with $2\frac{NM}{DB}$ I/Os, since we assume $\mathbf{B}$ to be in row major layout. Then, elements corresponding to a certain column can be loaded with at most $\frac{M}{DB}$ I/Os. Similarly, partial products can be stored into a $|I_U| \times N$ matrix in column major layout. Transposing this, and adding the rows to the corresponding rows in $\mathbf{C}$ requires no more than $3\frac{NM}{DB}$ I/Os. Hence, given a subgraph with $M$ edges and average degree $D$, $NM$ elementary products can be created with at most $6\frac{NM}{DB} + \frac{M}{B} = \mathcal{O}\left(\frac{NM}{DB}\right)$ I/Os.

Lemma 1 only states the existence of at least one dense subgraph. However, after creating all the elementary products corresponding to the edges of a dense subgraph, one can think of removing these edges. This will decrease the number of edges by no more than $M$, and we can use Lemma 1 for graphs with $kN - M$ edges again. Clearly, half of the elementary products can be obtained by subgraphs with average degree $D(k/2)$. This is possible with $\mathcal{O}\left(\frac{kN^2}{2D(k/2)B}\right)$ I/Os.

Let $D_1(k) = \ln\frac{N}{M}/(2\ln\frac{N\ln^2\frac{N}{M}}{4Mk})$, i.e., the first argument of the minimum of $D$ in Lemma 1. Altogether, the number of I/Os necessary to create all elementary products for $\mathbf{C}$ is bounded from above by

$$L(k,N) \leq \frac{kN}{B} + \sum_{i=1}^{\infty} 6\max\left\{\frac{2kN^2}{2^i B\sqrt{M}}, \frac{kN^2}{2^i D_1(k/2^i)B}\right\}$$

$$\leq \frac{kN}{B} + \frac{12kN^2}{B\sqrt{M}} + 6kN^2 \sum_{i=0}^{\infty} \frac{\ln\frac{N\ln^2\frac{N}{M}}{4Mk} + \ln 2^i}{2^i B\ln\frac{N}{M}} = \mathcal{O}\left(\frac{kN^2}{DB}\right).$$

Observe that for $k \geq \frac{N}{32M}\ln^2\frac{N}{M}$, $\sqrt{\frac{kN}{M}} = \Omega\left(\ln\frac{N}{M}\right)$ and thus, the tile-based algorithm is asymptotically better.

### 4.5  Small Instances

For smaller instances where $M \geq kN$, neither the tile-based algorithm is applicable, nor does the proof of dense subgraphs hold. Once $M \geq kN + N$, a degenerated version of the tile-based approach becomes the one of choice: Initially, all non-zero entries of $\mathbf{A}$ are loaded into internal memory. Afterwards, $\mathbf{B}$ is loaded in whole columns. For each column, the corresponding column in $\mathbf{C}$ can be calculated directly and written to external memory. This requires only $\mathcal{O}\left(\frac{N^2}{B}\right)$ I/Os.

## 5   Lower Bounds

**Theorem 2.** *For $1 \leq k \leq N$ any program for SDM$_k$ needs*

$$\Omega\left(\max\left\{\frac{kN^2}{B\Delta}, \frac{kN^2}{B\sqrt{M}}, \frac{N^2}{B}\right\}\right)$$

*I/Os, with $\Delta$ according to Theorem 1.*

Theorem 2 will be proven throughout this section. Therefore, we make use of the following technique introduced by Hong and Kung in [5].

**Lemma 3.** *A round-based program consists of $q$ rounds where each round consists of $M/B$ input operations, followed by $M/B$ output operations such that after the round internal memory is empty. A lower bound on the number of rounds $q_{min}$ of any round-based program with internal memory of size $2M$ can be transformed into a lower bound on the number of I/Os $l$ of any (normal) program with internal memory of size $M$ by $l \geq \frac{M}{B} \cdot (q_{min} - 1)$.*

Recall that the overall number of elementary products that have to be produced for SDM$_k$ is $kN^2$. Thus, given an upper bound on the number of elementary products that can be made during one round, a lower bound on the number of necessary rounds is obtained. We will do this by showing that there are matrices with only few dense parts. In the following, we consider the matrix $\mathbf{A}$ again as an adjacency matrix.

**Lemma 4.** *Let $\mathcal{G}$ be the family of bipartite graphs $G = (U \cup V, E)$ with $|U| = |V| = N$ and $|E| = kN$ for $k \leq N/2$.*

*For any $M \leq kN$ there is a graph $G \in \mathcal{G}$ such that $G$ contains no subgraph $G_S = (U_S \cup V_S, E_S)$ with $|E_S| = M$ and average degree*

$$D_M' > \max \left\{ \frac{8 \ln \frac{N}{M}}{\ln \frac{16N \ln^2 \frac{N}{M}}{Mk}}, e^4 \cdot \sqrt{\frac{kM}{N}} \right\}. \tag{3}$$

*Proof.* We will show this by upper bounding the number of graphs containing at least one such dense subgraph and compare this to the cardinality of $\mathcal{G}$. The upper bound is given by the number of possibilities to choose $2M/D_M'$ vertices from $U \cup V$ and the number of possibilities to insert $M$ edges between the selected vertices. Furthermore, the remaining $kN - M$ edges are chosen uniformly within the graph. The former presumes $M/D_M' \leq N$. However, since $M \leq kN$ and $D_M' > \sqrt{\frac{kM}{N}}$ this is implied. Further, we can assume $D_M' \leq \sqrt{M}$ since this is the maximum degree of a subgraph consisting of $M$ edges. Hence, if the inequality

$$\binom{2N}{2M/D_M'} \binom{(M/D_M')^2}{M} \binom{N^2}{kN - M} < \binom{N^2}{kN}$$

holds for the parameters given, Lemma 4 is proven. Observation 4 yields

$$\binom{2N}{2M/D_M'} \binom{(M/D_M')^2}{M} < \left( \frac{N^2 - kN}{kN} \right)^M.$$

Estimating binomial coefficients according to Observation 3, taking logarithms and multiplying by $D_M'/M$, we obtain

$$2 \ln \frac{eD_M'N}{M} + D_M' \ln \frac{eM}{D_M'^2} < D_M' \ln \frac{N^2 - kN}{kN} = D_M' \ln \frac{N}{k} + D_M' \ln \left( 1 - \frac{k}{N} \right).$$

The last term can be estimated for $k \leq N/2$ by Observation 1 resulting in

$$2 \ln \frac{eD_M'N}{M} + D_M' \ln \frac{eM}{D_M'^2} < D_M' \ln \frac{N}{k} - D_M' \frac{2k}{N}.$$

And by simple transformations, we obtain

$$D_M' \ln \frac{D_M'^2 N}{kM} > \underbrace{2 \ln \frac{N}{M}}_{\text{Term 1}} + \underbrace{2 \ln eD_M' + D_M' \left( 1 + 2\frac{k}{N} \right)}_{\text{Term 2}}. \tag{4}$$

Equation 4 is implied if Terms 1 and 2 are both bounded by $\frac{1}{2} D_M' \ln \frac{D_M'^2 N}{kM}$. We first check this for Term 2 only. By Observe 2 $\ln eD_M' \leq D_M'$. Thus,

$$\frac{1}{2} D_M' \ln \frac{D_M'^2 N}{kM} > 2 \ln eD_M' + 2D_M'$$

is implied by $D'_M > e^4 \cdot \sqrt{\frac{kM}{N}}$. For any such $D'_M$ Inequality 4 holds if

$$D'_M \ln \frac{D'_M \sqrt{N}}{\sqrt{kM}} > 2 \ln \frac{N}{M}. \tag{5}$$

By substitution of $D'_M$ by $e^4 \cdot \sqrt{\frac{kM}{N}}$, (5) already holds for $\sqrt{k} > \frac{1}{2e^4} \sqrt{\frac{N}{M}} \ln \frac{N}{M}$, i.e. especially for $M \geq N$. For $\sqrt{k} \leq \frac{1}{2e^4} \sqrt{\frac{N}{M}} \ln \frac{N}{M}$, we use Observation 6. Altogether, for

$$D'_M > \max \left\{ \frac{8 \ln \frac{N}{M}}{\overline{\ln} \frac{16N \ln^2 \frac{N}{M}}{Mk}}, e^4 \cdot \sqrt{\frac{kM}{N}} \right\}$$

not all possible graphs in $\mathcal{G}$ are covered and therefore, Lemma 4 holds. Since the second term is a sufficient bound for any $\sqrt{k} > \frac{1}{2e^4} \sqrt{\frac{N}{M}} \ln \frac{N}{M}$, we use $\overline{\ln}$ instead of $\ln$ to derive a closed formula by bounding the first term. Finally, note that $D'_M > 4$ holds for $k > 1$. $\qquad\square$

**Lemma 5.** *Let $\mathcal{G}$ be the family of bipartite graphs $G = (U \cup V, E)$ with $|U| = |V| = N$ and $|E| = kN$ for $k \leq N/2$.*
*For any $M \leq kN$, there is a graph $G \in \mathcal{G}$ such that $G$ contains at most $M - 1$ edges in subgraphs $G_S = (U_S \cup V_S, E_S)$ with $|E_S| \leq M$ and average degree $D' \geq 2e^4 \Delta$ where $\Delta$ is defined according to Theorem 2.*

*Proof.* By Lemma 4, this holds already for subgraphs consisting of exactly $M$ edges. For smaller subgraphs, we prove the statement by contradiction.

Suppose that there are at least $M$ edges in subgraphs with average degree at least $D'$ consisting of less than $M$ edges. Let $S$ be the set of such subgraphs. Since each subgraph in $S$ has less than $M$ edges, there exists a subset $S'$ of subgraphs in $S$ with a total number of $cM$ edges for $1 \leq c < 2$. The subgraph $G_{S'} = (U_{S'} \cup V_{S'}, E_{S'})$ induced by $S'$ has obviously still average degree at least $D'$.

Wlog let $|U_{S'}| \geq |V_{S'}|$ and consider the vertices $U_{S'}$ in $G_{S'}$. Now choose the $\lceil \frac{M}{D'} \rceil$ vertices in $U_{S'}$ with highest degree, and let $U'_{S'}$ denote the set of these. Since the vertices $U_{S'}$ have average degree $D'$ in $G_{S'}$, the subset $U'_{S'}$ cannot have a lower average degree. Hence, the subgraph $G'_{S'}$ induced by $U'_{S'}$ and $V_{S'}$ contains at least $M$ edges, but consists of no more than $\frac{M}{D'} + \frac{cM}{D'} + 1$ vertices. Therefore, any subgraph induced by exactly $M$ edges of $G'_{S'}$ has average degree at least $\frac{2MD'}{M + cM + D'}$. Since $D' \leq \sqrt{M}$, the average degree is at least $\frac{2D'}{2+c} \geq \frac{1}{2} D'$. This contradicts Lemma 4 for any $D' \geq 2D'_M$. $\qquad\square$

Using this, we can finally prove Theorem 2. Recall that Lemma 4, and thus, 5 fails for $D'_M > \sqrt{M}$. However, the maximum degree of a subgraph with $M$ edges is $\sqrt{M}$. The total number of elementary products, necessary for SDM$_k$ is $kN^2$. By Lemma 5, there are at most $N(M - 1)$ elementary products which might be calculated faster than the rest. For the remaining $kN^2 - NM + N$ elementary products, the following holds. Consider any round-based program for SDM$_k$.

Within each round, there are at most $2M$ elements of $\mathbf{B}$ and $\mathbf{C}$ loaded. Let $s_{ij}$, $t_{ij}$ be the number of elements from the $j$-th column of $\mathbf{B}$, $\mathbf{C}$ respectively, loaded in round $i$. By Lemma 5 and the observation that any subgraph has degree at most $\sqrt{M}$, there can be made no more than $\sum_{j=1}^{N} \min\{D', \sqrt{M}\} \cdot s_{ij} t_{ij} = 2M \cdot \min\{D', \sqrt{M}\}$ elementary products during each round. Hence, there have to be at least

$$\frac{kN^2 - MN + N}{2M \cdot \min\{D', \sqrt{M}\}}$$

rounds. This yields a lower bound of

$$\frac{M}{B}\left(\frac{kN^2 - MN + N}{2M \cdot \min\{D', \sqrt{M}\}} - 1\right) = \Omega\left(max\left\{\frac{kN^2}{B\Delta}, \frac{kN^2}{B\sqrt{M}}\right\}\right)$$

I/Os for $SDM_k$.

### 5.1   Closing the Parameter Range

Recall that Lemma 5 only holds for $k \leq N/2$. However, $\Omega\left(max\left\{\frac{kN^2}{B\Delta}, \frac{kN^2}{B\sqrt{M}}\right\}\right)$ is a lower bound for $N/2 \leq k \leq N$ as well since increasing the number of non-zero entries in $A$ cannot decrease the number of I/Os. For $M \geq kN$ a scanning bound of $\Omega\left(\frac{N^2}{B}\right)$ holds for the output of $\mathbf{C}$.

## References

1. Aggarwal, A., Vitter, J.S.: The input/output complexity of sorting and related problems. Communications of the ACM 31(9), 1116–1127 (1988)
2. Bader, M., Heinecke, A.: Cache oblivious dense and sparse matrix multiplication based on peano curves. In: PARA 2008. LNCS. Springer, Heidelberg (accepted for publication)
3. Bender, M.A., Brodal, G.S., Fagerberg, R., Jacob, R., Vicari, E.: Optimal sparse matrix dense vector multiplication in the I/O-model. In: Proceedings of SPAA 2007, pp. 61–70. ACM, New York (2007)
4. Bulucc, A., Gilbert, J.R.: Challenges and advances in parallel sparse matrix-matrix multiplication. In: Proceedings of ICPP 2008, Portland, Oregon, USA, September 2008, pp. 503–510 (2008)
5. Hong, J.-W., Kung, H.T.: I/O complexity: The red-blue pebble game. In: Proceedings of STOC 1981, pp. 326–333. ACM, New York (1981)
6. Kowarschik, M., Weiß, C.: An overview of cache optimization techniques and cache-aware numerical algorithms. Algorithms for Memory Hierarchies, 213–232 (2003)
7. Landsberg, J.M.: Geometry and the complexity of matrix multiplication. Bulletin of the American Mathematical Society 45, 247–284 (2008)
8. Lieber, T.: Combinatorial approaches to optimizing sparse matrix dense vector multiplication in the I/O-model. Master's thesis, Informatik Technische Universität München (2009)
9. Navarro, J.J., García-Diego, E., Larriba-Pey, J.-L., Juan, T.: Block algorithms for sparse matrix computations on high performance workstations. In: Proceedings of ICS 1996, pp. 301–308. ACM, New York (1996)

# Sparse Recovery Using Sparse Random Matrices

Piotr Indyk

MIT Computer Science and Artificial Intelligence Laboratory
32 Vassar Street, Cambridge, MA 02139, USA
indyk@theory.lcs.mit.edu

**Abstract.** Over the recent years, a new linear method for compressing high-dimensional data (e.g., images) has been discovered. For any high-dimensional vector $x$, its sketch is equal to $Ax$, where $A$ is an $m \times n$ matrix (possibly chosen at random). Although typically the sketch length $m$ is much smaller than the number of dimensions $n$, the sketch contains enough information to recover an approximation to $x$. At the same time, the linearity of the sketching method is very convenient for many applications, such as data stream computing and compressed sensing.

The major sketching approaches can be classified as either combinatorial (using sparse sketching matrices) or geometric (using dense sketching matrices). They achieve different trade-offs, notably between the compression rate and the running time. Thus, it is desirable to understand the connections between them, with the goal of obtaining the "best of both worlds" solution. Several recent results established such connections, indicating that the two approaches are just different manifestations of the same underlying phenomenon. This enabled the development of novel algorithms, including the first algorithms that provably achieve the (asymptotically) optimal compression rate and near-linear recovery time simultaneously.

In this talk we give an overview of the results in the area, as well as look at some of them in more detail. In particular, we will describe a new algorithm, called "Sequential Sparse Matching Pursuit (SSMP)". In addition to having the aforementioned theoretical guarantees, the algorithm works well on real data, with the recovery quality often outperforming that of more complex algorithms, such as $l_1$ minimization.

Joint work with: Radu Berinde, Anna Gilbert, Howard Karloff, Milan Ruzic and Martin Strauss.

A. López-Ortiz (Ed.): LATIN 2010, LNCS 6034, p. 157, 2010.
© Springer-Verlag Berlin Heidelberg 2010

# Optimal Succinctness for
# Range Minimum Queries

Johannes Fischer

Universität Tübingen, Sand 14, 72076 Tübingen, Germany
fischer@informatik.uni-tuebingen.de

**Abstract.** For a static array $A$ of $n$ totally ordered objects, a *range minimum query* asks for the position of the minimum between two specified array indices. We show how to preprocess $A$ into a scheme of size $2n + o(n)$ bits that allows to answer range minimum queries on $A$ in constant time. This space is asymptotically optimal in the important setting where access to $A$ is not permitted after the preprocessing step. Our scheme can be computed in linear time, using only $n + o(n)$ additional bits for construction. We also improve on LCA-computation in BPS- or DFUDS-encoded trees.

## 1 Introduction

For an array $A[1, n]$ of $n$ natural numbers or other objects from a totally ordered universe, a *range minimum query* $\mathrm{RMQ}_A(i, j)$ for $i \leq j$ returns the *position* of a minimum element in the sub-array $A[i, j]$; i.e., $\mathrm{RMQ}_A(i, j) = \mathrm{argmin}_{i \leq k \leq j}\{A[k]\}$. This fundamental algorithmic problem has numerous applications, e.g., in text indexing [1, 2], document retrieval [3], and position-restricted pattern matching [4], just to mention a few.

In all of these applications, the array $A$ in which the range minimum queries (RMQs) are performed is static and known in advance, which is also the scenario considered in this article. In this case it makes sense to preprocess $A$ into a (preprocessing-) *scheme* such that future RMQs can be answered quickly. We can hence formulate the following problem.

*Problem 1 (RMQ-Problem).*

**Given:** a static array $A[1, n]$ of $n$ totally ordered objects.
**Compute:** an (ideally small) data structure, called *scheme*, that allows to answer RMQs on $A$ in constant time.

The problem of most previous schemes for $O(1)$-RMQs [5, 6, 7] is their huge space consumption of $O(n \log n)$ bits. A *succinct data structure*, on the other hand, uses space that is close to the information-theoretic lower bound, in the sense that objects from a universe of cardinality $L$ are stored in $(1 + o(1)) \lg L$ bits.[1] Research on succinct data structures is very active, and we just mention

---

[1] Throughout this article, space is measured in bits, and lg denotes the binary logarithm.

A. López-Ortiz (Ed.): LATIN 2010, LNCS 6034, pp. 158–169, 2010.
© Springer-Verlag Berlin Heidelberg 2010

**Table 1.** Preprocessing schemes for $O(1)$-RMQs, where $|A|$ denotes the space for the (read-only) input array. All schemes can be constructed in $O(n)$ time.

| reference | final space | construction space | comments |
|---|---|---|---|
| [5] | $O(n \lg n) + |A|$ | $O(n \lg n) + |A|$ | via LCAs in Cartesian Tree |
| [6] | $O(n \lg n) + |A|$ | $O(n \lg n) + |A|$ | simpler than previous schemes |
| [7] | $O(n \lg n) + |A|$ | $O(n \lg n) + |A|$ | not based on Cartesian Trees |
| [15] | $2n + o(n) + |A|$ | $2n + o(n) + |A|$ | generalizes to $\frac{2}{c}n + o(n) + |A|$ bits |
| [16] | $O(nH_k) + o(n)$ | $O(nH_k) + o(n)$ | $H_k$: empirical entropy of $A$ |
| [1] | $n + o(n)$ | $n + o(n)$ | only for $\pm 1$RMQ; |
| [3] | $4n + o(n)$ | $O(n \lg n) + |A|$ | only non-systematic data structure |
| **this article** | **$2n + o(n)$** | **$3n + o(n) + |A|$** | **final space optimal** |

some examples from the realm of trees [8,9,10,11,12] and strings [13,14], being well aware of the fact that this list is far from complete. This article presents the first succinct data structure for $O(1)$-RMQs in the standard word-RAM model of computation (which is also the model used in all LCA- and RMQ-schemes cited in this article).

Before detailing our contribution, we first classify and summarize existing solutions for $O(1)$-RMQs.

## 1.1 Previous Solutions for RMQ

In accordance with common nomenclature, preprocessing schemes for $O(1)$-RMQs can be classified into two different types: *systematic* and *non-systematic*. Systematic schemes must store the input array $A$ verbatim along with the additional information for answering the queries. In such a case the query algorithm can consult $A$ when answering the queries; this is indeed what all systematic schemes make heavy use of. On the contrary, non-systematic schemes must be able to obtain their final answer without consulting the array. This second type is important for at least two reasons:

1. In some applications, e.g., in algorithms for document retrieval [3] or position restricted substring matching [4], only the *position* of the minimum matters, but *not* the value of this minimum. In such cases it would be a waste of space (both in theory and in practice) to keep the input array in memory, just for obtaining the final answer to the RMQs, as in the case of systematic schemes.
2. If the time to access the elements in $A$ is $\omega(1)$, this slowed-down access time propagates to the time for answering RMQs if the query algorithm consults the input array. As a prominent example, in string processing RMQ is often used in conjunction with the array of *longest common prefixes* of lexicographically consecutive suffixes, the so-called *LCP-array* [17]. However, storing the LCP-array efficiently in $2n + o(n)$ bits [1] increases the access-time to the time needed to retrieve an entry from the corresponding *suffix array* [17], which is $\Omega(\lg^\epsilon n)$ (constant $\epsilon > 0$) at the very best if the suffix

array is also stored in compressed form [14]. Hence, with a systematic scheme the time needed for answering RMQs on LCP could never be $O(1)$ in this case. But exactly this would be needed for constant-time navigation in RMQ-based compressed suffix trees [2] (where for different reasons the LCP-array is still needed, so this is not the same as the above point).

In the following, we briefly sketch previous solutions for RMQ schemes. For a summary, see Table 1, where, besides the final space consumption, in the third column we list the peak space consumption at construction time of each scheme, which sometimes differs from the former term.

**Systematic Schemes.** Most schemes are based on the Cartesian Tree [18], the only exception being the scheme due to Alstrup et al. [7]. All direct schemes [1, 6, 7, 15] are based on the idea of splitting the query range into several sub-queries, all of which have been precomputed, and then returning the overall minimum as the final result. The schemes from the first three rows of Table 1 have the same theoretical guarantees, with Bender et al.'s scheme [6] being less complex than the previous ones, and Alstrup et al.'s [7] being even simpler (and most practical). The only $O(n)$-bit scheme is due to Fischer and Heun [15] and achieves $2n + o(n)$ bits of space in addition to the space for the input array $A$. It is based on an "implicit" enumeration of Cartesian Trees only for very small blocks (instead of the whole array $A$). Its further advantage is that it can be adapted to achieve entropy-bounds for compressible inputs [16]. For systematic schemes, no lower bound on space is known.[2]

An important special case is Sadakane's $n + o(n)$-bit solution [1] for $\pm 1$RMQ, which we will describe it in greater detail in Sect. 2.2.

**Non-Systematic Schemes.** The only existing scheme is due to Sadakane [3] and uses $4n + o(n)$ bits. It is based on the balanced-parentheses-encoding (BPS) [8] of the Cartesian Tree $T$ of the input array $A$ and a $o(n)$-LCA-computation therein [1]. The difficulty that Sadakane overcomes is that in the "original" Cartesian Tree, there is no natural mapping between array-indices in $A$ and positions of parentheses (basically because there is no way to distinguish between left and right nodes in the BPS of $T$); therefore, Sadakane introduces $n$ "fake" leaves to get such a mapping. There are two main drawbacks of this solution.

1. Due to the introduction of the "fake" leaves, it does not achieve the *information-theoretic lower bound* (for non-systematic schemes) of $2n - \Theta(\lg n)$ bits. This lower bound is easy to see because any scheme for RMQs allows to reconstruct the Cartesian Tree by iteratively querying the scheme for the

---

[2] The claimed lower bound of $2n + o(n) + |A|$ bits under the "min-probe-model" [15] turned out to be wrong, as was kindly pointed out to the authors by S. Srinivasa Rao (personal communication, November 2007). In fact, it is easy to lower the space consumption of [15] to $\frac{2}{c}n + o(n) + |A|$ bits (constant integer $c > 0$) by grouping $c$ adjacent elements in $A$'s blocks together, and "building" the Cartesian Trees only on the minima of these groups.

minimum (in analogy to the definition of the Cartesian Tree); and because the Cartesian Tree is binary and each binary tree is a Cartesian Tree for some input array, any scheme must use at least $\lg(\binom{2n-1}{n-1}/(2n-1)) = 2n - \Theta(\lg n)$ bits [19].

2. For getting an $O(n)$-time construction algorithm, the (modified) Cartesian Tree needs to be first constructed in a pointer-based implementation, and then converted to the space-saving BPS. This leads to a *construction space requirement* of $O(n \lg n)$ bits, as each node occupies $O(\lg n)$ bits in memory. The problem why the BPS cannot be constructed directly in $O(n)$ time (at least we are not aware of such an algorithm) is that a "local" change in $A$ (be it only appending a new element at the end) does not necessarily lead to a "local" change in the tree; this is also the intuitive reason why maintaining dynamic Cartesian Trees is difficult [20].

## 1.2   Our Results

We address the two aforementioned problems of Sadakane's solution [3] and resolve them in the following way:

1. We introduce a new preprocessing scheme for $O(1)$-RMQs that occupies only $2n + o(n)$ bits in memory, thus being the first that asymptotically achieves the information-theoretic lower bound for non-systematic schemes. The critical reader might call this "lowering the constants" or "micro-optimization," but we believe that data structures using the smallest possible space are of high importance, both in theory and in practice. And indeed, there are many examples of this in literature: for instance, Munro and Raman [8] give a $2n + o(n)$-bit-solution for representing ordered trees, while supporting most navigational operations in constant time, although a $O(n)$-bit-solution (roughly $10n$ bits [8]) had already been known for some 10 years before [19].
2. We give a *direct* construction algorithm for the above scheme that needs only $n + o(n)$ bits of space in addition to the space for the final scheme, thus lowering the construction space for non-systematic schemes from $O(n \lg n)$ to $O(n)$ bits (on top of $A$). This is a significant improvement, as the space for storing $A$ is not necessarily $\Theta(n \lg n)$; for example, if the numbers in $A$ are integers in the range $[1, \lg^{O(1)} n]$, $A$ can be stored as an array of packed words using only $O(n \lg \lg n)$ bits of space. See Sect. 6 for a different example. The construction space is an important issue and often limits the practicality of a data structure, especially for large inputs (as they arise nowadays in web-page-analysis or computational biology).

The intuitive explanation why our scheme works better than Sadakane's scheme [3] is that ours is based on a new tree in which the preorder-numbers of the nodes correspond to the array-indices in $A$, thereby rendering the introduction of "fake" leaves (as described earlier) unnecessary. In summary, this article is devoted to proving

**Theorem 1.** *For an array $A$ of $n$ objects from a totally ordered universe, there is a preprocessing scheme for $O(1)$-RMQs on $A$ that occupies only $2n + O(\frac{n \lg \lg n}{\lg n})$ bits of memory, while not needing access to $A$ after its construction, thus meeting the information-theoretic lower bound. This scheme can be constructed in $O(n)$ time, using only $n + o(n)$ bits of space in addition to the space for the input and the final scheme.*

This result is not only appealing in theory, but also important in practice. For example, when RMQs are used in conjunction with sequences of DNA (genomic data), where the alphabet size $\sigma$ is 4, storing the DNA even in *uncompressed* form takes only $2n$ bits, already less than the $4n$ bits of Sadakane's solution [3]. Hence, halving the space for RMQs leads to a significant reduction of total space. Further, because $n$ is typically very large ($n \approx 2^{32}$ for the human genome), a construction space of $O(n \lg n)$ bits is much higher than the $O(n \lg \sigma)$ bits for the DNA itself. An implementation in C++ of our new scheme is available at http://www-ab.informatik.uni-tuebingen.de/people/fischer/ optimalRMQ.tgz.

## 2    Preliminaries

We use the standard *word-RAM* model of computation, where fundamental arithmetic operations on words consisting of $\Theta(\lg n)$ consecutive bits can be computed in $O(1)$ time.

### 2.1    Rank and Select on Binary Strings

Consider a *bit-string* $S[1,n]$ of length $n$. We define the fundamental *rank*- and *select*-operations on $S$ as follows: $rank_1(S,i)$ gives the number of 1's in the prefix $S[1,i]$, and $select_1(S,i)$ gives the position of the $i$'th 1 in $S$, reading $S$ from left to right ($1 \le i \le n$). Operations $rank_0(S,i)$ and $select_0(S,i)$ are defined similarly for 0-bits. There are data structures of size $O(\frac{n \lg \lg n}{\lg n})$ bits in addition to $S$ that support rank- and select-operations in $O(1)$ time [21].

### 2.2    Data Structures for ±1RMQ

Consider an array $E[1,n]$ of natural numbers, where the difference between consecutive elements in $E$ is either $+1$ or $-1$ (i.e. $E[i] - E[i-1] = \pm 1$ for all $1 < i \le n$). Such an array $E$ can be encoded as a bit-vector $S[1,n]$, where $S[1] = 0$, and for $i > 1$, $S[i] = 1$ iff $E[i] - E[i-1] = +1$. Then $E[i]$ can be obtained by $E[1] + rank_1(S,i) - rank_0(S,i) + 1 = E[1] + i - 2rank_0(S,i) + 1$. Under this setting, Sadakane [1] shows how to support RMQs on $E$ in $O(1)$ time, using $S$ and additional structures of size $O(\frac{n \lg^2 \lg n}{\lg n})$ bits. We will improve this space to $O(\frac{n \lg \lg n}{\lg n})$ in Sect. 5. A technical detail is that $\pm 1\text{RMQ}(i,j)$ yields the position of the *leftmost* minimum in $E[i,j]$ if there are multiple occurrences of this minimum.

## 2.3 Sequences of Balanced Parentheses

A string $B[1, 2n]$ of $n$ opening parentheses '(' and $n$ closing parentheses ')' is called *balanced* if in each prefix $B[1, i]$, $1 \leq i \leq 2n$, the number of ')''s is no more than the number of '('s. Operation *findopen*$(B, i)$ returns the position $j$ of the "matching" opening parenthesis for the closing parenthesis at position $i$ in $B$. This position $j$ is defined as the largest $j < i$ for which $rank_((B, i) - rank_)(B, i) = rank_((B, j) - rank_)(B, j)$. The *findopen*-operation can be computed in constant time [8]; the most space-efficient data structure for this needs $O(\frac{n \lg \lg n}{\lg n})$ bits [10].

## 2.4 Depth-First Unary Degree Encoding of Ordered Trees

The Depth-First Unary Degree Sequence (DFUDS) $U$ of an ordered tree $T$ is defined as follows [9]. If $T$ is a leaf, $U$ is given by '()'. Otherwise, if the root of $T$ has $w$ subtrees $T_1, \ldots, T_w$ in this order, $U$ is given by the juxtaposition of $w + 1$ '('s, a ')', and the DFUDS's of $T_1, \ldots, T_w$ in this order, with the first '(' of each $T_i$ being omitted. It is easy to see that the resulting sequence is balanced, and that it can be interpreted as a preorder-listing of $T$'s nodes, where, ignoring the very first '(', a node with $w$ children is encoded in *unary* as '$(^w)$' (hence the name DFUDS).

# 3 The New Preprocessing Scheme

We are now ready to dive into the technical details of our new preprocessing scheme. The basis will be a new tree, the *2d-Min-Heap*, defined as follows. Recall that $A[1, n]$ is the array to be preprocessed for RMQs. For technical reasons, we define $A[0] = -\infty$ as the "artificial" overall minimum.

**Definition 1.** *The* 2d-Min-Heap $\mathcal{M}_A$ *of* $A$ *is a labeled and ordered tree with vertices* $v_0, \ldots, v_n$, *where* $v_i$ *is labeled with* $i$ *for all* $0 \leq i \leq n$. *For* $1 \leq i \leq n$, *the parent node of* $v_i$ *is* $v_j$ *iff* $j < i$, $A[j] < A[i]$, *and* $A[k] \geq A[i]$ *for all* $j < k \leq i$. *The order of the children is chosen such that their labels are increasing from left to right.*

Observe that this is a well-defined tree with the root being always labeled as 0, and that a node $v_i$ can be uniquely identified by its label $i$, which we will do henceforth. See Fig. 1 for an example.

We note the following useful properties of $\mathcal{M}_A$.

**Lemma 1.** *Let* $\mathcal{M}_A$ *be the 2d-Min-Heap of* $A$.

1. *The node labels correspond to the preorder-numbers of* $\mathcal{M}_A$ *(starting at 0).*
2. *Let* $i$ *be a node in* $\mathcal{M}_A$ *with children* $x_1, \ldots, x_k$. *Then* $A[i] < A[x_j]$ *for all* $1 \leq j \leq k$.
3. *Again, let* $i$ *be a node in* $\mathcal{M}_A$ *with children* $x_1, \ldots, x_k$. *Then* $A[x_j] \leq A[x_{j-1}]$ *for all* $1 < j \leq k$.

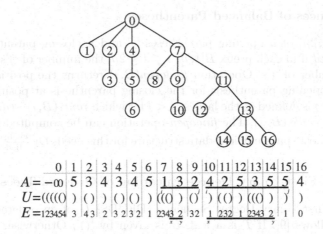

**Fig. 1.** Top: The 2d-Min-Heap $\mathcal{M}_A$ of the input array $A$. Bottom: $\mathcal{M}_A$'s DFUDS $U$ and $U$'s excess sequence $E$. Two example queries $\text{RMQ}_A(i,j)$ are underlined, including their corresponding queries $\pm 1\text{RMQ}_E(x,y)$.

*Proof.* Because the root of $\mathcal{M}_A$ is always labeled with 0 and the order of the children is induced by their labels, property 1 holds. Property 2 follows immediately from Def. 1. For property 3, assume for the sake of contradiction that $A[x_j] > A[x_{j-1}]$ for two children $x_j$ and $x_{j-1}$ of $i$. From property 1, we know that $i < x_{j-1} < x_j$, contradicting the definition of the parent-child-relationship in $\mathcal{M}_A$, which says that $A[k] \geq A[x_j]$ for all $i < k \leq x_j$.                    □

Properties 2 and 3 of the above lemma explain the choice of the name "2d-Min-Heap," because $\mathcal{M}_A$ exhibits a minimum-property on both the parent-child- and the sibling-sibling-relationship, i.e., in two dimensions.

   The following lemma will be central for our scheme, as it gives the desired connection of 2d-Min-Heaps and RMQs.

**Lemma 2.** *Let $\mathcal{M}_A$ be the 2d-Min-Heap of $A$. For arbitrary nodes $i$ and $j$, $1 \leq i < j \leq n$, let $\ell$ denote the LCA of $i$ and $j$ in $\mathcal{M}_A$ (recall that we identify nodes with their labels). Then if $\ell = i$, $\text{RMQ}_A(i,j)$ is given by $i$, and otherwise, $\text{RMQ}_A(i,j)$ is given by the child of $\ell$ that is on the path from $\ell$ to $j$.*

*Proof.* For an arbitrary node $x$ in $\mathcal{M}_A$, let $T_x$ denote the subtree of $\mathcal{M}_A$ that is rooted at $x$. There are two cases to prove.

$\ell = i$. This means that $j$ is a descendant of $i$. Due to property 1 of Lemma 1, this implies that all nodes $i, i+1, \ldots, j$ are in $T_i$, and the recursive application of property 2 implies that $A[i]$ is the minimum in the query range $[i,j]$.

$\ell \neq i$. Let $x_1, \ldots, x_k$ be the children of $\ell$. Further, let $\alpha$ and $\beta$ ($1 \leq \alpha \leq \beta \leq k$) be defined such that $T_{x_\alpha}$ contains $i$, and $T_{x_\beta}$ contains $j$. Because $\ell \neq i$ and property 1 of Lemma 1, we must have $\ell < i$; in other words, the LCA is not in the query range. But also due to property 1, every node in $[i,j]$ is in $T_{x_\gamma}$

for some $\alpha \leq \gamma \leq \beta$, and in particular $x_\gamma \in [i,j]$ for all $\alpha < \gamma \leq \beta$. Taking this together with property 2, we see that $\{x_\gamma : \alpha < \gamma \leq \beta\}$ are the only candidate positions for the minimum in $A[i,j]$. Due to property 3, we see that $x_\beta$ (the child of $\ell$ on the path to $j$) is the position where the overall minimum in $A[i,j]$ occurs.                                                    □

Note that (unlike for $\pm 1$RMQ) this algorithm yields the *rightmost* minimum in the query range if this is not unique. However, it can be easily arranged to return the leftmost minimum by adapting the definition of the 2d-Min-Heap, if this is desired.

To achieve the optimal $2n + o(n)$ bits for our scheme, we represent the 2d-Min-Heap $\mathcal{M}_A$ by its DFUDS $U$ and $o(n)$ structures for *rank$_\rangle$*-, *select$_\rangle$*-, and *findopen*-operations on $U$ (see Sect. 2). We further need structures for $\pm 1$RMQ on the *excess-sequence* $E[1,2n]$ of $U$, defined as $E[i] = rank_((U,i) - rank_\rangle(U,i)$. This sequence clearly satisfies the property that subsequent elements differ by exactly 1, and is already encoded in the right form (by means of the DFUDS $U$) for applying the $\pm 1$RMQ-scheme from Sect. 2.2.

The reasons for preferring the DFUDS over the BPS-representation [8] of $\mathcal{M}_A$ are (1) the operations needed to perform on $\mathcal{M}_A$ are particularly easy on DFUDS (see the next corollary), and (2) we have found a fast and space-efficient algorithm for constructing the DFUDS directly (see the next section).

**Corollary 1.** *Given the DFUDS $U$ of $\mathcal{M}_A$, $\mathrm{RMQ}_A(i,j)$ can be answered in $O(1)$ time by the following sequence of operations $(1 \leq i < j \leq n)$.*

1. $x \leftarrow select_\rangle(U, i+1)$
2. $y \leftarrow select_\rangle(U, j)$
3. $w \leftarrow \pm 1\mathrm{RMQ}_E(x,y)$
4. *if $rank_\rangle(U, findopen(U,w)) = i$ then return $i$*
5. *else return $rank_\rangle(U,w)$*

*Proof.* Let $\ell$ be the true LCA of $i$ and $j$ in $\mathcal{M}_A$. Inspecting the details of how LCA-computation in DFUDS is done [11, Lemma 3.2], we see that after the $\pm 1$RMQ-call in line 3 of the above algorithm, $w+1$ contains the starting position in $U$ of the encoding of $\ell$'s child that is on the path to $j$.[3] Line 4 checks if $\ell = i$ by comparing their preorder-numbers and returns $i$ in that case (case 1 of Lemma 2) — it follows from the description of the parent-operation in the original article on DFUDS [9] that this is correct. Finally, in line 5, the preorder-number of $\ell$'s child that is on the path to $j$ is computed correctly (case 2 of Lemma 2).        □

We have shown these operations so explicitly in order to emphasize the simplicity of our approach. Note in particular that not all operations on DFUDS have to be "implemented" for our RMQ-scheme, and that we find the correct child of the LCA $\ell$ directly, without finding $\ell$ explicitly. We encourage the reader to work on the examples in Fig. 1, where the respective RMQs in both $A$ and $E$ are underlined and labeled with the variables from Cor. 1.

---

[3] In line 1, we correct a minor error in the original article [11] by computing the starting position $x$ slightly differently, which is necessary in the case that $i = \mathrm{LCA}(i,j)$ (confirmed by K. Sadakane, personal communication, May 2008).

# 4   Construction of 2d-Min-Heaps

We now show how to construct the DFUDS $U$ of $\mathcal{M}_A$ in linear time and $n + o(n)$ bits of extra space. We first give a general $O(n)$-time algorithm that uses $O(n \lg n)$ bits (Sect. 4.1), and then show how to reduce its space to $n + o(n)$ bits, while still having linear running time (Sect. 4.2).

## 4.1   The General Linear-Time Algorithm

We show how to construct $U$ (the DFUDS of $\mathcal{M}_A$) in linear time. The idea is to scan $A$ from *right to left* and build $U$ from right to left, too. Suppose we are currently in step $i$ ($n \geq i \geq 0$), and $A[i+1, n]$ have already been scanned. We keep a stack $S[1, h]$ (where $S[h]$ is the top) with the properties that $A[S[h]] \geq \cdots \geq A[S[1]]$, and $i < S[h] < \cdots < S[1] \leq n$. $S$ contains exactly those indices $j \in [i+1, n]$ for which $A[k] \geq A[j]$ for all $i < k < j$. Initially, both $S$ and $U$ are empty. When in step $i$, we first write a ')' to the current beginning of $U$, and then pop all $w$ indices from $S$ for which the corresponding entry in $A$ is strictly greater than $A[i]$. To reflect this change in $U$, we write $w$ opening parentheses '(' to the current beginning of $U$. Finally, we push $i$ on $S$ and move to the next (i.e. preceding) position $i-1$. It is easy to see that these changes on $S$ maintain the properties of the stack. If $i = 0$, we write an initial '(' to $U$ and stop the algorithm.

The correctness of this algorithm follows from the fact that due to the definition of $\mathcal{M}_A$, the degree of node $i$ is given by the number $w$ of array-indices to the right of $i$ which have $A[i]$ as their closest smaller value (properties 2 and 3 of Lemma 1). Thus, in $U$ node $i$ is encoded as '$(^w)$', which is exactly what we do. Because each index is pushed and popped exactly once on/from $S$, the linear running time follows.

## 4.2   $O(n)$-Bit Solution

The only drawback of the above algorithm is that stack $S$ requires $O(n \lg n)$ bits in the worst case. We solve this problem by representing $S$ as a *bit-vector* $S'[1, n]$. $S'[i]$ is 1 if $i$ is on $S$, and 0 otherwise. In order to maintain constant time access to $S$, we use a standard blocking-technique as follows. We logically group $s = \lceil \frac{\lg n}{2} \rceil$ consecutive elements of $S'$ into *blocks* $B_0, \ldots, B_{\lfloor \frac{n-1}{s} \rfloor}$. Further, $s' = s^2$ elements are grouped into *super-blocks* $B'_0, \ldots, B'_{\lfloor \frac{n-1}{s'} \rfloor}$.

For each such (super-)block $B$ that contains at least one 1, in a new table $M$ (or $M'$, respectively) at position $x$ we store the block number of the leftmost (super-)block to the right of $B$ that contains a 1, in $M$ only relative to the beginning of the super-block. These tables need $O(\frac{n}{s} \lg(s'/s)) = O(\frac{n \lg \lg n}{\lg n})$ and $O(\frac{n}{s'} \lg(n/s)) = O(\frac{n}{\lg n})$ bits of space, respectively. Further, for all possible bit-vectors of length $s$ we maintain a table $P$ that stores the position of the leftmost 1 in that vector. This table needs $O(2^s \cdot \lg s) = O(\sqrt{n} \lg \lg n) = o(n)$ bits. Next, we show how to use these tables for constant-time access to $S$, and how to keep $M$ and $M'$ up to date.

When entering step $i$ of the algorithm, we known that $S'[i+1] = 1$, because position $i + 1$ has been pushed on $S$ as the last operation of the previous step. Thus, the top of $S$ is given by $i + 1$. For finding the leftmost 1 in $S'$ to the right of $j > i$ (position $j$ has just been popped from $S$), we first check if $j$'s block $B_x$, $x = \lfloor \frac{i-1}{s} \rfloor$, contains a 1, and if so, find this leftmost 1 by consulting $P$. If $B_x$ does not contain a 1, we jump to the next block $B_y$ containing a 1 by first jumping to $y = x + M[x]$, and if this block does not contain a 1, by further jumping to $y = M'[\lfloor \frac{i-1}{s'} \rfloor]$. In block $y$, we can again use $P$ to find the leftmost 1. Thus, we can find the new top of $S$ in constant time.

In order to keep $M$ up to date, we need to handle the operations where (1) elements are pushed on $S$ (i.e., a 0 is changed to a 1 in $S'$), and (2) elements are popped from $S$ (a 1 changed to a 0). Because in step $i$ only $i$ is pushed on $S$, for operation (1) we just need to store the block number $y$ of the former top in $M[x]$ ($x = \lfloor \frac{i-1}{s} \rfloor$), if this is in a different block (i.e., if $x \neq y$). Changes to $M'$ are similar. For operation (2), nothing has to be done at all, because even if the popped index was the last 1 in its (super-)block, we know that all (super-)blocks to the left of it do not contain a 1, so no values in $M$ and $M'$ have to be changed. Note that this only works because elements to the right of $i$ will never be pushed again onto $S$. This completes the description of the $n + o(n)$-bit construction algorithm.

## 5   Lowering the Second-Order-Term

Until now, the second-order-term is dominated by the $O(\frac{n \lg^2 \lg n}{\lg n})$ bits from Sadakane's preprocessing scheme for $\pm 1$RMQ (Sect. 2.2), while all other terms (for rank, select and findopen) are $O(\frac{n \lg \lg n}{\lg n})$. We show in this section a simple way to lower the space for $\pm 1$RMQ to $O(\frac{n \lg \lg n}{\lg n})$, thereby completing the proof of Theorem 1.

As in the original algorithm [1], we divide the input array $E$ into $n' = \lfloor \frac{n-1}{s} \rfloor$ blocks of size $s = \lceil \frac{\lg n}{2} \rceil$. Queries are decomposed into at most three non-overlapping sub-queries, where the first and the last sub-queries are inside of the blocks of size $s$, and the middle one exactly spans over blocks. The two queries inside of the blocks are answered by table lookups using $O(\sqrt{n} \lg^2 n)$ bits, as in the original algorithm.

For the queries spanning exactly over blocks of size $s$, we proceed as follows. Define a new array $E'[0, n']$ such that $E'[i]$ holds the minimum of $E$'s $i$'th block. $E'$ is represented only *implicitly* by an array $E''[0, n']$, where $E''[i]$ holds the position of the minimum in the $i$'th block, relative to the beginning of that block. Then $E'[i] = E[is + E''[i]]$. Because $E''$ stores $n/\lg n$ numbers from the range $[1, s]$, the size for storing $E'$ is thus $O(\frac{n \lg \lg n}{\lg n})$ bits. Note that unlike $E$, $E'$ does not necessarily fulfill the $\pm 1$-property. $E'$ is now preprocessed for constant-time RMQs with the systematic scheme of Fischer and Heun [15], using $2n' + o(n') = O(\frac{n}{\lg n})$ bits of space. Thus, by querying $\text{RMQ}_{E'}(i, j)$ for $1 \leq i \leq j \leq n'$, we can also find the minima for the sub-queries spanning exactly over the blocks in $E$.

Two comments are in order at this place. First, the used RMQ-scheme [15] does allow the input array to be represented implicitly, as in our case. And second, it does not use Sadakane's solution for $\pm 1$RMQ, so there are no circular dependencies.

As a corollary, this approach also lowers the space for LCA-computation in BPS [1] and DFUDS [11] from $O(\frac{n \lg^2 \lg n}{\lg n})$ to $O(\frac{n \lg \lg n}{\lg n})$, as these are based on $\pm 1$RMQ:

**Corollary 2.** *Given the BPS or DFUDS of an ordered tree $T$, there is a data structure of size $O(\frac{n \lg \lg n}{\lg n})$ bits that allows to answer LCA-queries in $T$ in constant time.*

## 6    Application in Document Retrieval Systems

We now sketch a concrete example of where Theorem 1 lowers the construction space of a different data structure. We consider the following problem:

*Problem 2 (Document Listing Problem [22]).*

**Given:** a collection of $k$ text documents $\mathcal{D} = \{D_1, \dots, D_k\}$ of total length $n$.
**Compute:** an index that, given a search pattern $P$ of length $m$, returns all $d$ documents from $\mathcal{D}$ that contain $P$, in time proportional to $m$ and $d$ (in contrast to *all* occurrences of $P$ in $\mathcal{D}$).

Due to a lack of space, we only state the final result; for a more complete exposition, see the full version of this paper [23].

**Theorem 2.** *The construction space for Sadakane's Index for Document Listing [3] is lowered from $O(n \lg n)$ bits to $O(n + k \lg n)$ bits (constant alphabet) or $O(n \lg |\Sigma| + k \lg n)$ bits (arbitrary alphabet $\Sigma$) with our scheme for RMQs from Theorem 1, while not increasing the construction time.*

## 7    Concluding Remarks

We have given the first optimal preprocessing scheme for $O(1)$-RMQs under the important assumption that the input array is not available after preprocessing. To the expert, it might come as a surprise that our algorithm is *not* based on the Cartesian Tree, a concept that has proved to be very successful in former schemes. Instead, we have introduced a new tree, the 2d-Min-Heap, which seems to be better suited for our task. We hope to have thereby introduced a new versatile data structure to the algorithms community. And indeed, we are already aware of the fact that the 2d-Min-Heap, made public via a preprint of this article [23], is pivotal to a new data structure for succinct trees [12].

We leave it as an open research problem whether the $3n + o(n)$-bit construction space be lowered to an optimal $2n + o(n)$-bit "in-place" construction algorithm. (A simple example shows that it is *not* possible to use the leading $n$ bits of the DFUDS for the stack.)

# References

1. Sadakane, K.: Compressed suffix trees with full functionality. Theory of Computing Systems 41(4), 589–607 (2007)
2. Fischer, J., Mäkinen, V., Navarro, G.: Faster entropy-bounded compressed suffix trees. Theor. Comput. Sci. 410(51), 5354–5364 (2009)
3. Sadakane, K.: Succinct data structures for flexible text retrieval systems. J. Discrete Algorithms 5(1), 12–22 (2007)
4. Crochemore, M., Iliopoulos, C.S., Kubica, M., Rahman, M.S., Walen, T.: Improved algorithms for the range next value problem and applications. In: Proc. STACS, IBFI Schloss Dagstuhl, pp. 205–216 (2008)
5. Berkman, O., Vishkin, U.: Recursive star-tree parallel data structure. SIAM J. Comput. 22(2), 221–242 (1993)
6. Bender, M.A., Farach-Colton, M., Pemmasani, G., Skiena, S., Sumazin, P.: Lowest common ancestors in trees and directed acyclic graphs. J. Algorithms 57(2), 75–94 (2005)
7. Alstrup, S., Gavoille, C., Kaplan, H., Rauhe, T.: Nearest common ancestors: A survey and a new distributed algorithm. In: Proc. SPAA, pp. 258–264 (2002)
8. Munro, J.I., Raman, V.: Succinct representation of balanced parentheses and static trees. SIAM J. Comput. 31(3), 762–776 (2001)
9. Benoit, D., Demaine, E.D., Munro, J.I., Raman, R., Raman, V., Rao, S.S.: Representing trees of higher degree. Algorithmica 43(4), 275–292 (2005)
10. Geary, R.F., Rahman, N., Raman, R., Raman, V.: A simple optimal representation for balanced parentheses. Theor. Comput. Sci. 368(3), 231–246 (2006)
11. Jansson, J., Sadakane, K., Sung, W.K.: Ultra-succinct representation of ordered trees. In: Proc. SODA, pp. 575–584. ACM/SIAM (2007)
12. Sadakane, K., Navarro, G.: Fully-functional succinct trees. Accepted for SODA 2010. See also CoRR, abs/0905.0768v1 (2009)
13. Ferragina, P., Manzini, G., Mäkinen, V., Navarro, G.: Compressed representations of sequences and full-text indexes. ACM TALG 3(2), Article No. 20 (2007)
14. Sadakane, K.: New text indexing functionalities of the compressed suffix arrays. J. Algorithms 48(2), 294–313 (2003)
15. Fischer, J., Heun, V.: A new succinct representation of RMQ-information and improvements in the enhanced suffix array. In: Chen, B., Paterson, M., Zhang, G. (eds.) ESCAPE 2007. LNCS, vol. 4614, pp. 459–470. Springer, Heidelberg (2007)
16. Fischer, J., Heun, V., Stühler, H.M.: Practical entropy bounded schemes for O(1)-range minimum queries. In: Proc. DCC, pp. 272–281. IEEE Press, Los Alamitos (2008)
17. Manber, U., Myers, E.W.: Suffix arrays: A new method for on-line string searches. SIAM J. Comput. 22(5), 935–948 (1993)
18. Vuillemin, J.: A unifying look at data structures. Comm. ACM 23(4), 229–239 (1980)
19. Jacobson, G.: Space-efficient static trees and graphs. In: Proc. FOCS, pp. 549–554. IEEE Computer Society Press, Los Alamitos (1989)
20. Bialynicka-Birula, I., Grossi, R.: Amortized rigidness in dynamic Cartesian Trees. In: Durand, B., Thomas, W. (eds.) STACS 2006. LNCS, vol. 3884, pp. 80–91. Springer, Heidelberg (2006)
21. Golynski, A.: Optimal lower bounds for rank and select indexes. Theor. Comput. Sci. 387(3), 348–359 (2007)
22. Muthukrishnan, S.: Efficient algorithms for document retrieval problems. In: Proc. SODA, pp. 657–666. ACM/SIAM (2002)
23. Fischer, J.: Optimal succinctness for range minimum queries. CoRR, abs/0812.2775 (2008)

# Compact Rich-Functional Binary Relation Representations

Jérémy Barbay[1], Francisco Claude[2,*], and Gonzalo Navarro[1,**]

[1] Department of Computer Science, University of Chile
{jbarbay,gnavarro}@dcc.uchile.cl
[2] David R. Cheriton School of Computer Science, University of Waterloo
fclaude@cs.uwaterloo.ca

**Abstract.** Binary relations are an important abstraction arising in a number of data representation problems. Each existing data structure specializes in the few basic operations required by one single application, and takes only limited advantage of the inherent redundancy of binary relations. We show how to support more general operations efficiently, while taking better advantage of some forms of redundancy in practical instances. As a basis for a more general discussion on binary relation data structures, we list the operations of potential interest for practical applications, and give reductions between operations. We identify a set of operations that yield the support of all others. As a first contribution to the discussion, we present two data structures for binary relations, each of which achieves a distinct tradeoff between the space used to store and index the relation, the set of operations supported in sublinear time, and the time in which those operations are supported. The experimental performance of our data structures shows that they not only offer good time complexities to carry out many operations, but also take advantage of regularities that arise in practical instances to reduce space usage.

## 1 Introduction

Binary relations appear everywhere in Computer Science. Graphs, trees, inverted indexes, strings and permutations are just some examples. They have also been used as a tool to complement existing data structures (such as trees [3] or graphs [2]) with additional information, such as weights or labels on the nodes or edges, that can be indexed and searched. Interestingly, the data structure support for binary relations has not undergone a systematic study, but rather one triggered by particular applications: we aim to remedy this fact.

Let us say that a binary relation $\mathcal{B}$ relates *objects* in $[1, n]$ with *labels* in $[1, \sigma]$, containing $t$ pairs out of the $n\sigma$ possible ones. A simple entropy measure using these parameters and ignoring any other possible regularity is $H(\mathcal{B}) = \log \binom{n\sigma}{t} = t \log \frac{n\sigma}{t} + O(t)$ bits ($\log = \log_2$ in this paper). Fig. 1 (top left) illustrates a binary relation (identifying labels with rows and objects with columns henceforth).

---

* Funded by NSERC of Canada and Go-Bell Scholarships Program.
** Funded in part by Fondecyt Grant 1-080019, Chile.

A. López-Ortiz (Ed.): LATIN 2010, LNCS 6034, pp. 170–183, 2010.

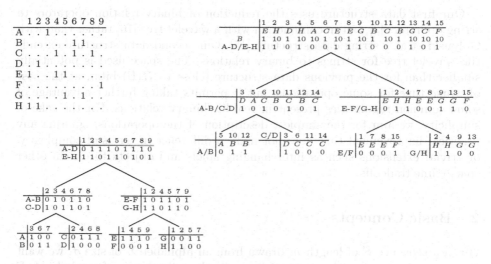

**Fig. 1.** An example of binary relation (top left), its representation according to Sec. 4 (right) and according to Sec. 5 (bottom). Note that the labels and object numbers are included in each node solely for ease of reading; in the encoding they are implicit.

Previous work focused on relatively basic primitives for binary relations: extract the list of all labels associated to an object or of all objects associated to a label (an operation called access), or extracting the $r$-th such element (an operation called select), or counting how many of these are there up to some object/label value (called operation rank).

The first representation specifically designed for binary relations [3] supports rank, select and access on the rows (labels) of the relation, for the purpose of supporting faster joins on labels. It was later extended to index text [13], and to separate the content from the index [4], which in turn allows supporting labeled operations on planar and quasi-planar labeled graphs [2].

Ad-hoc compressed representations for inverted lists [22] and Web graphs [12] can also be considered as supporting binary relations. The idea here is to write the objects of the pairs, in label-major order, and support extracting substrings of the resulting string, that is, little more than access on labels. The string can be compressed by different means depending on the application.

In this paper we aim at describing the foundations of efficient compact data structures for binary relations. We list operations of potential interest for practical applications; we give various reductions between operators, thus identifying a core set of operations which support all others; we present two data structures for binary relations, each of which achieves a distinct tradeoff between the space used to store and index the relation and the time in which the operations are supported; and we compare the practical performances of our data structures with the state of the art, showing that our data structures not only offer good time complexities to carry out many operators, but also reduce the space used by taking advantage of the redundancy of practical instances.

Our first data structure uses the reduction of binary relation operators to string operators [3], but in conjunction with a wavelet tree [16] rather than with Golynski et al.'s string data structure [15]. Our second data structure extends the wavelet tree for strings to binary relations. The space used is potentially smaller than for the previous data structure (close to $H(\mathcal{B})$ bits), at the cost of worse time for some operations, but it permits taking further advantage of some common regularities present in real-life binary relations. For the sake of simplicity, we aim for the simplest description of the operations, ignoring any practical improvement that does not make a difference in terms of complexity, or trivial extensions such as interchanging labels and objects to obtain other space/time tradeoffs.

## 2   Basic Concepts

Given a sequence $S$ of length $n$, drawn from an alphabet $\Sigma$ of size $\sigma$, we want to answer the queries: (1) $\mathrm{rank}_a(S, i)$ counts the occurrences of symbol $a \in \Sigma$ in $S[1, i]$; (2) $\mathrm{select}_a(S, i)$ finds the $i$-th occurrence of symbol $a \in \Sigma$ in $S$; and (3) $\mathrm{access}(S, i) = S[i]$. We omit $S$ if clear from context.

For the special case $\Sigma = \{0, 1\}$, the problem has been solved using $n + o(n)$ bits of space while answering the three queries in constant time [10]. This was later improved to use $nH_0(S) + o(n)$ bits [21]. Here $H_0(S)$ is the *zero-order entropy* of sequence $S$, defined as $H_0(S) = \sum_{a \in \Sigma} \#_a/n \log(n/\#_a)$, where $\#_a$ is the number of occurrences of symbol $a$ in $S$.

The wavelet tree [16] reduces the three operations on general alphabets to those on binary sequences. It is a perfectly balanced tree that stores a bitmap of length $n$ at the root; every position in the bitmap is either 0 or 1 depending on whether the symbol at this position belongs to the first half of the alphabet or to the second. The left child of the root will handle the subsequence of $S$ marked with a 0 at the root, and the right child will handle the 1s. This decomposition into alphabet subranges continues recursively until reaching level $\lceil \log \sigma \rceil$, where the leaves correspond to individual symbols. We call $B_v$ the bitmap at node $v$.

The $\mathrm{access}$ query $S[i]$ can be answered by following the path described for position $i$. At the root $v$, if $B_v[i] = 0/1$, we descend to the left/right child, switching to the bitmap position $\mathrm{rank}_{0/1}(B_v, i)$ in the left/right child, which then becomes the new $v$. This continues recursively until reaching the last level, when we arrive at the leaf corresponding to the answer symbol. Query $\mathrm{rank}_a(S, i)$ can be answered similarly to $\mathrm{access}$, except that we descend according to $a$ and not to the bit of $B_v$. We update position $i$ for the child node just as before. At the leaves, the final bitmap position $i$ is the answer. Query $\mathrm{select}_a(S, i)$ proceeds as $\mathrm{rank}$, but upwards. We start at the leaf representing $a$ and update $i$ to $\mathrm{select}_{0/1}(B_v, i)$ where $v$ is the parent node, depending on whether the current node is its left/right child. At the root, position $i$ is the result.

Wavelet trees require $n \log \sigma + o(n) \log \sigma$ bits of space, while answering all the queries in $O(\log \sigma)$ time. If the bitmaps $B_v$ are represented using the technique of Raman et al. [21], the wavelet tree uses $nH_0(S) + o(n) \log \sigma$ bits. Fig. 1 (right) illustrates the structure. Wavelet trees are not only used to represent strings [14], but also grids [7], permutations [5], and many other structures.

# 3   Operations

## 3.1   Definition of Operations

Data structures for binary relations which support efficiently the `rank` and `select` operations on the row (label) yield faster searches in relational databases and text search engines [3] and, in combination with data structures for ordinal trees, yield faster searches in multi-labeled trees, such as those featured by semi-structured documents [3] (e.g. XML). A similar technique [2] combining various data structures for graphs with binary relations yields a family of data structures for edge-labeled and vertex-labeled graphs that support labeled operations on the neighborhood of each vertex. The extension of those operations to the union of labels in a given range allows them to handle more complex queries, such as conjunctions of disjunctions.

As a simple example, an inverted index [22] can be seen as a relation between vocabulary words (the labels) and the documents where they appear (the objects). Apart from the basic operation of extracting the documents where a word appears (`access` on the row), we want to intersect rows (implemented on top of row `rank` and `select`) for phrase and conjunctive queries (popular in Google-like search engines). Extending these operations to a range of words allows for stemmed and/or prefix searches (by properly ordering the words). Extracting a column gives important *summarization* information on a document: the list of its different words. Intersecting columns allows for analysis of content between documents (e.g. plagiarism or common authorship detection). Handling ranges of documents allows for considering hierarchical document structures such as XML or filesystems (search within a subtree or subdirectory).

As another example, a directed graph is just a binary relation between vertices. Extracting rows or columns supports direct and reverse navigation from a node. In Web graphs, where the nodes (Web pages) are usually sorted by URL, ranges of nodes correspond to domains and subdirectories. For example, counting the number of connections between two ranges of nodes allows estimating the connectivity between two domains. In general, considering domain ranges permits the analysis and navigation of the Web graph at a coarser granularity (e.g. as a graph of hosts, or institutions).

Several text indexing data structures [8,13,17,18,20] resort to a grid, which relates for example text suffixes (in lexicographical order) with their text positions, or phrase prefixes and suffixes in grammar compression, or two labels that form a rule in straight-line programs, etc. The most common operation needed is counting and returning all the points in a range.

Obviously, the case where the relation represents a geometric grid, where objects and labels are simply coordinates, and where pairs of the relation are points at those coordinates, is useful for GIS and other geometric applications.

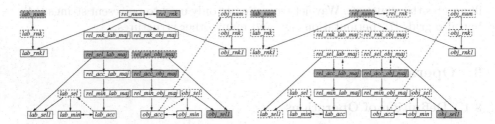

**Fig. 2.** Results achieved by reducing to strings (left) and by the binary-relation wavelet tree (right). Grayed boxes are the operations we adressed directly; all the others are supported via reductions given in Theorem 1: constant-time ones are represented by solid arrows, and non-constant-time ones by dashed arrows. Operations supported in time $O(\log \sigma)$ are in solid squares. Dashed squares represent operations supported in higher time. We draw only the dashed arrows needed to follow the source of the operations supported via non-constant-time reductions.

The generalization of the basic operations to ranges allows for counting the number of points in a rectangular area, and retrieving them in different orders.

These examples illustrate several useful ways to extend the definition of the **rank** and **select** operations from single rows (labels) or columns (objects) to ranges over both rows and columns. Consider for instance the extension of **select** to ranges of labels: $select(\alpha, r)$ yields the position of the $r$-th 1 in the row $\alpha$ of the matrix (see Fig. 1 (top left)), corresponding to the $r$-th object associated to label $\alpha$. On the range of rows $[\alpha, \beta]$, the expression "the $r$-th 1" requires a total order on the two-dimensional area defined by the range (e.g. label-major or object-major), which yields two distinct extensions of the operation. Other applications require instead a **select** operation that retrieves the $r$-th object associated to any label from a given range, regardless of how many pairs the object participates in.

We generalize the **access** operation to ranges of labels and objects by supporting the search for the minimal (resp. maximal) label or object that participates in a given rectangular area of the relation, and the search for the first related pair (in label-major or object-major order) in this area. Among other applications, this supports the search for the highest (resp. lowest) neighbor of a point, when the binary relation encodes the levels of points in a planar graph representing a topography map [2].

Those examples require many distinct extensions for each of the **rank**, **select** and **access** operations. Table 1 lists their formal definitions.

## 3.2  Reductions between Operations

The solid arrows in Fig. 2 (left) show the constant-time reductions that we identified among the operations; disregard the rest for now. A solid arrow $op \to op'$ means that solving $op$ we also solve $op'$. First, **rel_rnk** is a particular case of **rel_num**, whereas the latter can be supported by adding/subtracting four **rel_rnk** queries at the corners of the rectangle. Hence they are equivalent.

**Table 1.** Operations of interest for binary relations on $[1,\sigma] \times [1,n]$ (labels × objects). $x, y, z$ are objects (usually $x \leq z \leq y$); $\alpha, \beta, \gamma$ are labels (usually $\alpha \leq \gamma \leq \beta$); $r$ is an integer (typically an index, parameter of `select`) and '#' is short for 'number of'. The solutions for maxima are similar to those for minima. The last two columns are the complexities we achieve in Section 4 and 5, respectively, per delivered datum.

| Operation | Meaning | String | BRWT |
|---|---|---|---|
| $\texttt{rel\_num}(\alpha,\beta,x,y)$ | # pairs in $[\alpha,\beta] \times [x,y]$ | $O(\log \sigma)$ | $O(\beta-\alpha+\log \sigma)$ (*) |
| $\texttt{rel\_rnk}(\alpha,x)$ | # pairs in $[1,\alpha] \times [1,x]$ | $O(\log \sigma)$ | $O(\alpha + \log \sigma)$ |
| $\texttt{rel\_rnk\_lab\_maj}(x,y,\alpha,z)$ | # pairs in $[x,y]$, up to $(\alpha,z)$ † | $O(\log \sigma)$ | $O(\alpha + \log \sigma)$ (*) |
| $\texttt{rel\_sel\_lab\_maj}(\alpha,r,x,y)$ | $r$-th pair in $[x,y] \times [\alpha,\sigma]$ † | $O(\log \sigma)$ | $O(r \log \sigma)$ (*) |
| $\texttt{rel\_acc\_lab\_maj}(\alpha,x,y)$ | consecutive pairs in $[\alpha,\sigma] \times [x,y]$ † | $O(\log \sigma)$ | $O(\log \sigma)$ |
| $\texttt{rel\_min\_lab\_maj}(\alpha,x,y)$ | minimum pair in $[\alpha,\sigma] \times [x,y]$ † | $O(\log \sigma)$ | $O(\log \sigma)$ |
| $\texttt{rel\_rnk\_obj\_maj}(\alpha,\beta,\gamma,x)$ | # pairs in $[\alpha,\beta]$, up to $(\gamma,x)$ ‡ | $O(\log \sigma)$ | $O(\beta - \alpha + \log \sigma)$ |
| $\texttt{rel\_sel\_obj\_maj}(\alpha,\beta,x,r)$ | $r$-th pair in $[\alpha,\beta] \times [x,n]$ ‡ | (+) | $O(r \log \sigma)$ |
| $\texttt{rel\_acc\_obj\_maj}(\alpha,\beta,x)$ | consecutive pairs in $[\alpha,\beta] \times [x,n]$ ‡ | $O(\log \sigma)$ | $O(\log \sigma)$ |
| $\texttt{rel\_min\_obj\_maj}(\alpha,\beta,x)$ | minimum pair in $[\alpha,\beta] \times [x,n]$ ‡ | $O(\log \sigma)$ | $O(\log \sigma)$ |
| $\texttt{lab\_num}(\alpha,\beta,x,y)$ | # distinct labels in $[\alpha,\beta] \times [x,y]$ | $O(\beta-\alpha+\log \sigma)$ | $O(\beta-\alpha+\log \sigma)$ |
| $\texttt{lab\_rnk}(\alpha,x,y)$ | # distinct labels in $[1,\alpha] \times [x,y]$ | $O(\alpha + \log \sigma)$ | $O(\alpha + \log \sigma)$ |
| $\texttt{lab\_sel}(\alpha,r,x,y)$ | $r$-th distinct label in $[\alpha,\sigma] \times [x,y]$ | $O(r \log \sigma)$ | $O(r \log \sigma)$ |
| $\texttt{lab\_acc}(\alpha,x,y)$ | consecutive labels in $[\alpha,\sigma] \times [x,y]$ | $O(\log \sigma)$ | $O(\log \sigma)$ |
| $\texttt{lab\_min}(\alpha,x,y)$ | minimum label in $[\alpha,\sigma] \times [x,y]$ | $O(\log \sigma)$ | $O(\log \sigma)$ |
| $\texttt{obj\_num}(\alpha,\beta,x,y)$ | # distinct objects in $[\alpha,\beta] \times [x,y]$ | $O(r \log \sigma)$ | $O(r \log \sigma)$ |
| $\texttt{obj\_rnk}(\alpha,\beta,x)$ | # distinct objects in $[\alpha,\beta] \times [1,x]$ | $O(r \log \sigma)$ | $O(r \log \sigma)$ |
| $\texttt{obj\_sel}(\alpha,\beta,x,r)$ | $r$-th distinct object in $[\alpha,\beta] \times [x,n]$ | $O(r \log \sigma)$ | $O(r \log \sigma)$ |
| $\texttt{obj\_acc}(\alpha,\beta,x)$ | consecutive objects in $[\alpha,\beta] \times [x,n]$ | $O(\log \sigma)$ | $O(\log \sigma)$ |
| $\texttt{obj\_min}(\alpha,\beta,x)$ | minimum object in $[\alpha,\beta] \times [x,n]$ | $O(\log \sigma)$ | $O(\log \sigma)$ |
| $\texttt{lab\_rnk1}(\alpha,x)$ | # distinct labels in $[1,\alpha] \times x$ | $O(\log \sigma)$ | $O(r \log \sigma)$ |
| $\texttt{lab\_sel1}(\alpha,r,x)$ | $r$-th distinct label in $[\alpha,\sigma] \times x$ | $O(\log \sigma)$ | $O(r \log \sigma)$ |
| $\texttt{obj\_rnk1}(\alpha,x)$ | # distinct objects in $\alpha \times [1,x]$ | $O(\log \sigma)$ | $O(\log \sigma)$ |
| $\texttt{obj\_sel1}(\alpha,x,r)$ | $r$-th distinct object in $\alpha \times [x,n]$ | $O(\log \sigma)$ | $O(\log \sigma)$ |

(+) $O(\min(r, \log n, \log r \log(\beta - \alpha + 1)) \log \sigma)$     † in label-major order
(*) $O(\log \sigma)$ if $[x,y] = [1,n]$     ‡ in object-major order

With a constant number of any of these we also cover the areas described by `rel_rnk_obj_maj` and `rel_rnk_lab_maj`, and vice versa, thus these are equivalent too. Obviously, `obj_rnk1` and `lab_rnk1` are particular cases of `rel_num`. Also, `lab_rnk1` is a particular case of `lab_rnk`, itself a particular case of `lab_num`. Note that `lab_num` does not reduce to `lab_rnk` because a label could be related with objects inside and outside the range $[x,y]$. Similar reductions hold for objects.

Obviously the support for the select operation `rel_sel_lab_maj` implies the support for the access operation `rel_acc_lab_maj`, and accessing the first result of the latter gives the solution for the minimum operation `rel_min_lab_maj`. In turn this gives the minimum label in a range $[\alpha,\sigma] \times [x,y]$, thus if this label is $\beta$, we get the next label by rerunning the query on $[\beta + 1, \sigma] \times [x,y]$, this way supporting `lab_acc`. The latter, in turn, gives the solution to `lab_min` in its first iteration, whereas successive invocations to `lab_min` (in a fashion similar to `rel_min_lab_maj`) solves `lab_acc`. Also analogously as before, `lab_sel` allows supporting `lab_min` by asking the first occurrence, and `lab_sel1` is a particular case of `lab_sel`. Note also that `rel_sel_lab_maj` allows supporting `lab_sel1`, by requiring the pairs starting at the desired rows, and extracting the resulting objects. By symmetry, analogous reductions hold for objects instead of labels.

The rest of the following theorem stems from inverse-function relations between `rank` and `select` queries, as well as one-by-one solutions to counting and direct-access problems.

**Theorem 1.** *All the arrows in Figure 2 (left) represent constant-time reductions that hold for the operations. In addition, the pairs* (`lab_num, lab_sel`) *support each other with an* $O(\log \sigma)$ *penalty factor,* (`obj_num, obj_sel`) *with an* $O(\log n)$ *penalty factor, and* (`rel_rnk_lab_maj, rel_sel_lab_maj`) *and* (`rel_rnk_obj_maj, rel_sel_obj_maj`) *with an* $O(\log(\sigma n))$ *penalty factor. Finally, in pairs* (`lab_acc, lab_sel`), (`rel_acc_lab_maj, rel_sel_lab_maj`), (`obj_acc, obj_sel`), *and* (`rel_acc_obj_maj, rel_sel_obj_maj`), *the first operation supports the second with an* $O(r)$ *penalty factor, where r is the parameter of the* `select` *operation. Finally, the* `access` *operations support the corresponding* `rank` *(and counting) operations in time proportional to the answer of the latter.*

# 4   Reduction to Strings

A simple representation for binary relations [3,13] consists in a bitmap $B[1, n+t]$ and a string $S[1, t]$ over the alphabet $[1, \sigma]$. The bitmap $B[n+t]$ concatenates the consecutive cardinalities of the $n$ columns of the relation in unary. The string $S$ contains the rows (labels) of the pairs of the relation in column (object)-major order (see Fig. 1 (right)). Barbay *et al.* showed [3] that an easy way to support the `rank` and `select` operations on the rows of the binary relation is to support the `rank` and `select` operations on $B$ and $S$, using any of the several data structures known for bitmaps and strings. We show that representing $S$ using a wavelet tree yields the support for more complex operations. For this purpose, we define the mapping from a column number $x$ to its last element in $S$ as $\text{map}(x) = \text{rank}_1(B, \text{select}_0(B, x))$. The inverse, from a position in $S$ to its column number, is $\text{unmap}(m) = \text{rank}_0(B, \text{select}_1(B, m)) + 1$. Both mappings take constant time. Finally, let us also define for shortness $\text{rank}_c(B, x, y) = \text{rank}_c(B, y) - \text{rank}_c(B, x-1)$. The following operations are supported efficiently, and many others are derived with Theorem 1.

- `rel_rnk`$(\alpha, x)$ **in** $O(\log \sigma)$ **time.** This is $\text{rank}_{\leq \alpha}(S, \text{map}(x))$, where operation $\text{rank}_{\leq \alpha}(S, m)$ counts the number of symbols $\leq \alpha$ in $S[1, m]$. It can be supported in time $O(\log \sigma)$ in a string wavelet tree by following the root-to-leaf branch corresponding to $\alpha$, while counting at each node the number of objects preceding position $m$ that are related with a label preceding $\alpha$, as follows. Start at the root $v$ with counter $c \leftarrow 0$. If $\alpha$ corresponds to the left subtree, then enter the left subtree with $m \leftarrow rank_0(B_v, m)$. Else enter the right subtree with $c \leftarrow c + rank_0(B_v, m)$ and $m \leftarrow \text{rank}_1(B_v, m)$. When a leaf is reached (indeed, that of $\alpha$), the answer is $c + m$.

- `rel_sel_lab_maj`$(\alpha, r, x, y)$ **in** $O(\log \sigma)$ **time.** We first get rid of $\alpha$ by setting $r \leftarrow r + \text{rel_num}(1, \alpha - 1, x, y)$ and thus reduce to the case $\alpha = 1$. Furthermore we map $x$ and $y$ to the domain of $S$ by $x \leftarrow \text{map}(x - 1) + 1$ and $y \leftarrow \text{map}(y)$.

We first find which is the symbol $\beta$ whose row contains the $r$-th element. For this sake we first find the $\beta$ such that $\mathtt{rank}_{\leq\beta-1}(S,x,y) < r \leq \mathtt{rank}_{\leq\beta}(S,x,y)$. This is achieved in time $O(\log\sigma)$ as follows. Start at the root $v$ and set $r' \leftarrow r$. If $\mathtt{rank}_0(B_v,x,y) \geq r$, then continue to the left subtree with $x \leftarrow \mathtt{rank}_0(B_v,x-1)+1$ and $y \leftarrow \mathtt{rank}_0(B_v,y)$. Else continue to the right subtree with $r' \leftarrow r' - \mathtt{rank}_0(B_v,x,y)$, $x \leftarrow \mathtt{rank}_1(B_v,x-1)+1$, and $y \leftarrow \mathtt{rank}_1(B_v,y)$. The leaf arrived at is $\beta$. Finally, we set $r \leftarrow r - r'$, and answer $(\beta, \mathtt{unmap}(\mathtt{select}_\beta(S, r + \mathtt{rank}_\beta(S, x-1))))$.

- $\mathtt{rel\_sel\_obj\_maj}(\alpha,\beta,x,r)$ in $O(\min(\log n, \log r \log(\beta - \alpha + 1))\log\sigma)$ **time.** Remember that the elements are written in $S$ in object major order. First, we note that the particular case where $[\alpha,\beta] = [1,\sigma]$ is easily solved in $O(\log\sigma)$ time, by doing $r' \leftarrow r + \mathtt{rel\_num}(1,\sigma,1,x-1)$ and returning $(S[r'], \mathtt{unmap}(r'))$. In the general case, one can obtain time $O(\log n \log\sigma)$ by binary searching the column $y$ such that $\mathtt{rel\_num}(\alpha,\beta,x,y) < r \leq \mathtt{rel\_num}(\alpha,\beta,x,y+1)$. Then the answer is $(\mathtt{lab\_sel1}(\alpha, r - \mathtt{rel\_num}(\alpha,\beta,x,y),y), y)$. Finally, to obtain the other complexity, we find the $O(\log(\beta - \alpha + 1))$ wavelet tree nodes that cover the interval $[\alpha,\beta]$; let these be $v_1, v_2, \ldots, v_k$. We map position $x$ from the root towards those $v_i$s, obtaining all the mapped positions $x_i$ in $O(k + \log\sigma)$ time. Now the answer is within the positions $[x_i, x_i + r - 1]$ of some $i$. We cyclically take each $v_i$, choose the middle element of its interval, and map it towards the root, obtaining position $y$, corresponding to pair $(S[y], \mathtt{unmap}(y))$. If $\mathtt{rel\_rnk\_obj\_maj}(\alpha,\beta,S[y],\mathtt{unmap}(y)) - \mathtt{rel\_rnk}(\alpha,\beta,1,x-1) = r$, the answer is $(S[y], \mathtt{unmap}(y))$. Otherwise we know whether $y$ is before or after the answer. So we discard the left or right interval in $v_i$. After $O(k \log r)$ such iterations we have reduced all the intervals of length $r$ of all the nodes $v_i$, finding the answer. Each iteration costs $O(\log\sigma)$ time.

- $\mathtt{rel\_acc\_obj\_maj}(\alpha,\beta,x)$ in $O(\log\sigma)$ **time per pair output.** Just as for the last solution of the previous operator, we obtain the positions $x_i$ at the nodes $v_i$ that cover $[\alpha,\beta]$. The first element to deliver is precisely one of those $x_i$. We have to merge the results, choosing always the smaller, as we return from the recursion that identifies the $v_i$ nodes. If we are in $v_i$, we return $y = x_i$. Else, if the left child of $v$ returned $y$, we map it to $y' \leftarrow \mathtt{rank}_0(B_v,y)$. Similarly, if the right child of $v$ returned $y$, we map it to $y'' \leftarrow \mathtt{rank}_1(B_v,y)$. If we have only $y'$ ($y''$), we return $y = y'$ ($y = y''$); if we have both we return $y = \min(y',y'')$. The process takes $O(\log\sigma)$ time. When we arrive at the root we have the next position $y$ where a label in $[\alpha,\beta]$ occurs in $S$. We can then report all the pairs $(S[y+j], \mathtt{unmap}(y))$, for $j = 0, 1, \ldots$, as long as $\mathtt{unmap}(y+j) = \mathtt{unmap}(y)$ and $S[y+j] \leq \beta$. Once we have reported all the pairs corresponding to object $\mathtt{unmap}(y)$, we can obtain those of the next objects by repeating the procedure from $\mathtt{rel\_acc\_obj\_maj}(\alpha,\beta,\mathtt{unmap}(y)+1)$.

- $\mathtt{lab\_num}(\alpha,\beta,x,y)$ in $O(\beta - \alpha + \log\sigma)$ **time.** After mapping $x$ and $y$ to positions in $S$, we descend in the wavelet tree to find all the leaves in $[\alpha,\beta]$ while remapping $[x,y]$ appropriately. We count one more label each time we arrive at a leaf, and we stop descending from an internal node if its range $[x,y]$ is empty.

- obj_sel1($\alpha, x, r$) in $O(\log \sigma)$ time: This is a matter of selecting the $r$-th occurence of the label $\alpha$ in $S$, after the position of the pair $(\alpha, x)$. The formula is unmap(select$_\alpha(S, r + $ obj_rnk1($\alpha, x - 1$))).

The overall result is stated in the next theorem and illustrated in Fig. 2 (left).

**Theorem 2.** *There is a representation for a binary relation $\mathcal{B}$, of $t$ pairs over $[1, \sigma] \times [1, n]$, using $t \log \sigma + o(t) \log \sigma + O(\min(n, t) \log(\frac{n+t}{\min(n,t)}))$ bits of space. The structure supports operations* rel_rnk($\alpha, x$), rel_sel_lab_maj($\alpha, r, x, y$), rel_sel_obj_maj($1, \sigma, x, r$) *(note the limitation)*, rel_acc_obj_maj($\alpha, \beta, x$), *and* obj_sel1($\alpha, x, r$), *in time $O(\log \sigma)$, plus* rel_sel_obj_maj($\alpha, \beta, x, r$) *in time $O(\min(\log n, \log r \log(\beta - \alpha + 1)) \log \sigma)$, and* lab_num($\alpha, \beta, x, y$) *in time $O(\beta - \alpha + \log \sigma)$. The other operations are supported via the reductions from Theorem 1.*

*Proof.* The operations have been obtained throughout the section. For the space, $B$ contains $n$ 0s out of $n + t$, so a compressed representation [21] requires $O(\min(t, n) \log \frac{n+t}{\min(t,n)})$ bits. The wavelet tree for $S[1, t]$ requires $t \log \sigma + o(t) \log \sigma$ bits. □

Note that the particular case rel_num($1, \sigma, x, y$) can be answered in $O(1)$ time using $B$'s succinct encoding. In general the space result is incomparable with $tH(\mathcal{B})$: if all the $n\sigma$ pairs are related, then $tH_0(S) = n\sigma \log \sigma$ and $H(\mathcal{B}) = 0$; but if all the pairs are within a row, then $tH_0(S) = 0$ and $H(\mathcal{B}) > 0$. In the particular case where $t \leq n$, $t \log \sigma \leq tH(\mathcal{B}) + O(t)$, while the wavelet tree for $S$ requires $tH_0(S) \leq t \log \sigma$ bits: this difference can be relevant depending on the distribution of pairs across the rows.

## 5   Binary Relation Wavelet Trees (BRWT)

We propose now a special wavelet tree structure to represent binary relations. This wavelet tree contains two bitmaps per level at each node $v$, $B_v^l$ and $B_v^r$. At the root, $B_v^l[1, n]$ has the $x$-th bit set to 1 iff there exists a pair of the form $(\alpha, x)$ for $\alpha \in [1, \lfloor \sigma/2 \rfloor]$, and $B_v^r$ has the $x$-th bit set to 1 iff there exists a pair of the form $(\alpha, x)$ for $\alpha \in [\lfloor \sigma/2 \rfloor + 1, \sigma]$. Left and right subtrees are recursively built on the positions set to 1 in $B_v^l$ and $B_v^r$, respectively. The leaves (where no bitmap is stored) correspond to individual rows of the relation. We store a bitmap $B[1, \sigma + t]$ recording in unary the number of elements in each row. See Fig. 1 (bottom) for an example. For ease of notation, we define the following functions on $B$, trivially supported in constant-time: lab($r$) = $1 + $ rank$_0(B, $ select$_1(B, r))$ gives the label of the $r$-th pair in a label-major traversal of $R$; while its inverse poslab($\alpha$) = rank$_1(B, $ select$_0(B, \alpha))$ gives the position in the traversal where the pairs for label $\alpha$ start.

Note that, because an object $x$ may propagate both left and right, the sizes of the second-level bitmaps may add up to more than $n$ bits. Indeed, the last level contains $t$ bits and represents all the pairs sorted in row-major order.

The following operations can be carried out efficiently on this structure.

• $\texttt{rel\_num}(\alpha, \beta, x, y)$ in $O(\beta - \alpha + \log \sigma)$ **time.** We project the interval $[x, y]$ from the root to each leaf in $[\alpha, \beta]$, adding up the resulting interval sizes at leaves. Of course we can stop earlier if the interval becomes empty. Note that we can only count pairs at the leaves. In the case $[x, y] = [1, n]$ we can achieve $O(1)$ time, as the answer is simply $\texttt{poslab}(\beta) - \texttt{poslab}(\alpha - 1)$. Note this allows solving the restricted case $\texttt{rel\_rnk\_lab\_maj}(1, n, \alpha, z)$ in $O(\log \sigma)$ time.

• $\texttt{rel\_sel\_lab\_maj}(\alpha, r, 1, n)$ in $O(\log \sigma)$ **time.** Let $r' \leftarrow r + \texttt{poslab}(\alpha - 1)$ and $\beta \leftarrow \texttt{lab}(r')$, thus $\beta$ is the row where the answer is. Now we start at position $y = r' - \texttt{poslab}(\beta - 1)$ in leaf $\beta$ and walk the wavelet tree upwards while mapping $y \leftarrow \texttt{select}_1(B_v^l, y)$ or $y \leftarrow \texttt{select}_1(B_v^r, y)$, depending on whether we are left or right child of our parent $v$, respectively. When we reach the root, the answer is $(\beta, y)$. Note we are only solving the particular case $[x, y] = [1, n]$.

• $\texttt{rel\_acc\_lab\_maj}(\alpha, x, y)$ in $O(\log \sigma)$ **time per pair output.** Map $[x, y]$ from the root to each leaf in $[\alpha, \sigma]$, abandoning a path when $[x, y]$ becomes empty. (Because left and right child cannot become simultaneously empty, the total amount of work is proportional to the number of leaves that contain pairs to report.) Now, for each leaf $\gamma$ arrived at with interval $[x', y']$, map each $z' \in [x', y']$ up to the root, to discover the associated object $z$, and return $(\gamma, z)$.

• $\texttt{rel\_acc\_obj\_maj}(\alpha, \beta, x)$ in $O(\log \sigma)$ **time per pair output.** Just as in Section 4, we cover $[\alpha, \beta]$ with $O(\log \sigma)$ wavelet tree nodes $v_1, v_2, \ldots$, and map $x$ to $x_i$ at each such $v_i$, all in $O(\log \sigma)$ time. Now, on the way back of this recursion, we obtain the next $y \geq x$ in the root associated to some label in $[\alpha, \beta]$, by following a process analogous to that for $\texttt{rel\_acc\_obj\_maj}$ in Section 4. Finally, we start from position $y' - y$ at the root $v$ and report all the pairs related to $y$: Recursively, we descend left if $B_v^l[y'] = 1$, and then right if $B_v^r[y'] = 1$, remapping $y'$ appropriately at each step, and keeping within the interval $[\alpha, \beta]$. Upon reaching each leaf $\gamma$ we report $(\gamma, y)$. Then we continue from $\texttt{rel\_acc\_obj\_maj}(\alpha, \beta, y+1)$.

• $\texttt{lab\_num}(\alpha, \beta, x, y)$ in $O(\beta - \alpha + \log \sigma)$ **time.** Map $[x, y]$ from the root to each leaf in $[\alpha, \beta]$, adding one per leaf where the interval is nonempty. Recursion can stop when $[x, y]$ becomes empty.

• $\texttt{obj\_sel1}(\alpha, x, r)$ in $O(\log \sigma)$ **time.** Map $x - 1$ from the root to $x'$ in leaf $\alpha$, then walk upwards the path from $x' + r$ to the root and report the position obtained.

We have obtained the following theorem, illustrated in Fig. 2 (right; we ignore the particular cases).

**Theorem 3.** *There is a representation for a binary relation $\mathcal{B}$, of $t$ pairs over $[1, \sigma] \times [1, n]$, using $\log(1 + \sqrt{2})tH(\mathcal{B}) + o(tH(\mathcal{B})) + O(t + n + \sigma)$ bits of space. The structure supports operations $\texttt{rel\_num}(\alpha, \beta, 1, n)$, $\texttt{rel\_rnk\_lab\_maj}(1, n, \alpha, z)$, $\texttt{rel\_sel\_lab\_maj}(\alpha, r, 1, n)$ (note the limitations of these three), $\texttt{rel\_acc\_lab\_maj}(\alpha, x, y)$, $\texttt{rel\_acc\_obj\_maj}(\alpha, \beta, x)$, and $\texttt{obj\_sel1}(\alpha, x, y)$, in time $O(\log \sigma)$, plus $\texttt{rel\_num}(\alpha, \beta, x, y)$ and $\texttt{lab\_num}(\alpha, \beta, x, y)$ in time $O(\beta - \alpha + \log \sigma)$. This yields the support for other operations via the reductions from Theorem 1.*

*Proof.* The operations have been obtained throughout the section. For the space, $B$ contains $\sigma$ 0s out of $\sigma + t$, so a compressed representation [21] requires $O(\min(\sigma,t) \log \frac{\sigma+t}{\min(\sigma,t)}) = O(\max(\sigma,t))$ bits. The space of the wavelet tree can be counted as follows. Except for the $2n$ bits in the root, each other bit is induced by the presence of a pair. Each pair has a unique representative bit in a leaf, and also induces the presence of bits up to the root. Yet those leaf-to-root paths get merged, so that not all those bits are different. Consider an element $x$ related to $t_x$ labels. It induces $t_x$ bits at $t_x$ leaves, and their paths of bits towards the single $x$ at the root. At worst, all the $O(t_x)$ bits up to level $\log t_x$ are created for these elements, and from there on all the $t_x$ paths are different, adding up a total of $O(t_x) + t_x \log \frac{\sigma}{t_x}$. Adding over all $x$ we get $O(t) + \sum_x t_x \log \frac{\sigma}{t_x}$. This is maximized when $t_x = t/n$ for all $x$, yielding $O(t) + t \log \frac{\sigma n}{t} = tH(\mathcal{B}) + O(t)$ bits.

Instead of representing two bitmaps (which would multiply the above value by 2), we can represent a single sequence $B_v$ with the possible values of the two bits at each position, 00, 01, 10, 11. Only at the root 00 is possible. Except for those $2n$ bits, we can represent the sequence over an alphabet of size 3 following Ferragina et al. [14], to achieve at worst $(\log 3)tH(\mathcal{B}) + o(tH(\mathcal{B}))$ bits for this part while retaining constant-time `rank` and `select` over each $B_v^l$ and $B_v^r$. (To achieve this, we maintain the directories for the original bitmaps, of sublinear-size.)

To improve the constant $\log 3$ to $\log(1+\sqrt{2})$, we consider that the representation by Ferragina et al. actually achieves $|B_v|H_0(B_v)$ bits. We call $\ell_x = |B_v| \leq t_x$ and $H_x = |B_v|H_0(B_v)$. After level $\log t_x$, there is space to put all the $t_x$ bits separately, thus using only 01 and 10 symbols we achieve $\ell_x = t_x$ and $H_x = t_x$ bits. Yet, this is not the worst that can happen. $H_x$ can be increased by collapsing some 01's and 10's into 11's (thus reducing $\ell_x$). Note that collapsing further 01's or 10's or 11's with 11's effectively removes one symbol from $B_v$, which cannot increase $H_x$, thus we do not consider these. Assume the $t_x$ bits are partitioned into $t_{01}$ 01's, $t_{10}$ 10's, and $t_{11}$ 11's, so that $t_x = t_{01}+t_{10}+2t_{11}$, $\ell_x = t_{01}+t_{10}+t_{11}$, and $H_x = t_{01} \log \frac{\ell_x}{t_{01}} + t_{10} \log \frac{\ell_x}{t_{10}} + t_{11} \log \frac{\ell_x}{t_{11}}$. As $t_{11} = (t_x - t_{01} - t_{10})/2$, the maximum of $H_x$ as a function of $t_{01}$ and $t_{10}$ yields the worst case at $t_{01} = t_{10} = \frac{\sqrt{2}}{4}t_x$, so $t_{11} = (\frac{1}{2} - \frac{\sqrt{2}}{4})t_x$ and $\ell_x = (\frac{1}{2} + \frac{\sqrt{2}}{4})t_x$, where $H_x = \log(1+\sqrt{2})t_x$ bits. This can be achieved separately at each level. Using the same distribution of 01's, 10's, and 11's for all $x$ we add up to $(1 + \sqrt{2})t \log \frac{\sigma n}{t} + O(t)$ bits.  $\square$

Note that this is a factor of $\log(1 + \sqrt{2}) \approx 1.272$ away of the entropy of $\mathcal{B}$. On the other hand, it is actually better if the $t_x$ do not distribute uniformly.

## 6    Exploiting Regularities

Real-life binary relations exhibit regularities that permit compressing them far more than to $tH(\mathcal{B})$ bits. For example, social networks, Web graphs, and inverted indexes follow well-known properties such as clustering of the matrix, uneven distribution of 1s, similarity across rows and/or columns, etc. [6,1,9].

The space $tH_0(S)$ achieved in Theorem 2 can indeed be improved upon certain regularities. The wavelet tree of $S$, when bitmaps are compressed with local encoding methods [21], achieves locality in the entropy [19]. That is, if $S = S_1 S_2 \ldots S_n$

then the space achieved is $\sum_x |S_x| H_0(S_x) + O(n \log t)$. In particular, if $S_x$ corresponds to the labels related to object $x$, then the space will benefit from *clustering* in the binary relation: If each object is related only to a small subset of labels, then its $S_x$ will have a small alphabet and thus a small entropy. Alternatively, similar columns (albeit not rows) induce copies in string $S$. This is not captured by the zero-order entropy, but it is by grammar compression methods. Some have been exploited for graph compression [12].

The space formula in Theorem 3 can also be refined: If some objects are related to many labels and others to few, then $\sum_x t_x \log \frac{\sigma}{t_x}$ can be smaller than $tH(\mathcal{B})$. This second approach can be easily modified to exploit several other regularities. Imagine we represent bitmaps $B_v^l$ and $B_v^r$ separately, but instead of $B_v^r$ we store $B_v' = B_v^l$ xor $B_v^r$, while keeping the original sublinear structures for `rank` and `select`. Any access to $O(\log n)$ contiguous bits in $B_v^r$ is achieved in constant time under the RAM model by xor-ing $B_v^l$ and $B_v'$.

The following regularities turn into a highly compressible $B_v'$, that is, one with few or many 0's: (1) Row-wise similarities between nearby rows, extremely common on Web graphs [6], yield an almost-all-zero $B_v'$; (2) (sub)relations that are actually permutations or strings, that is, with exactly one 1 per column, yield an almost-all-one $B_v'$. This second kind of (sub)relations are common in relational databases, e.g., when objects or labels are primary keys in the table.

As there exists no widely agreed-upon notion of entropy for binary relations finer than $\log \binom{n\sigma}{t}$, we show now some experiments on the performance of these representations on some real-life relations. We choose instances of three types of binary relations: (1) Web graphs, (2) social networks, (3) inverted indexes.

For (1), we downloaded two crawls from the WebGraph project [6], `http://law.dsi.unimi.it`. Crawl EU (2005) contains $n = \sigma = 862,664$ nodes and $t = 19,235,140$ edges. Crawl Indochina (2004) contains $n = \sigma = 7,414,866$ nodes and $t = 194,109,311$ edges. For (2), we downloaded a coauthorship graph from DBLP (`http://dblp.uni-trier.de/xml`), which is a symmetric relation, and kept the upper triangle of the symmetric matrix. The result contains $n = \sigma = 452,477$ authors and $t = 1,481,877$ coauthorships. For (3), we consider the relation FT, the inverted index for all of the Financial Times collections from TREC-4 (`http://trec.nist.gov`), converting the terms to lowercase. It relates $\sigma = 502,259$ terms with $n = 210,139$ documents, using $t = 51,290,320$ pairs.

Table 2 shows, for these relations $\mathcal{B}$, their entropy $H(\mathcal{B})$, their *gap complexity* (defined below), the space of the string representation of Section 4, the space of the BRWT representation of Section 5, and that using the *xor*-improvement described above. All spaces are measured in bits per pair of the relation.

The *gap complexity* is the sum of the logarithms of the consecutive differences of objects associated to each label. It is upper bounded by the entropy and gives a more refined measure that accounts for clustering in the matrix. The string representation of Section 4 already improves upon the entropy, but not much. Although it has more functionality, this representation requires significantly more space than the BRWT, which takes better advantage of regularities. Note, however, that for example Web graphs are much more amenable than the

**Table 2.** Entropy and space consumption, in bits per pair, of different binary relation representations over relations from different applications. Ad-hoc representations have limited functionality.

| $\mathcal{B}$ | $H(\mathcal{B})$ | Gap | String | BRWT | $+xor$ | Best Ad-Hoc |
|---|---|---|---|---|---|---|
| EU | 16.68 | 5.52 | 12.57 | 7.72 | 6.87 | 4.38 (WebGraph) |
| Indochina | 19.55 | 3.12 | 12.81 | 4.07 | 3.93 | 1.47 (WebGraph) |
| DBLP | 18.52 | 6.18 | 15.97 | 13.54 | 11.67 | 21.9 (WebGraph) |
| FT | 12.45 | 3.54 | 13.91 | 9.32 | 7.85 | 6.20 (Rice) |

social network to exploiting such regularities, while the inverted index is in between. The *xor* improvement has a noticeable additional effect on the BRWT space, reducing it by about 5%–15%. Particularly on the Web graphs, this latter variant becomes close to the gap complexity.

The last column of the table shows the compression achieved by the best ad-hoc alternatives, which support a very restricted set of operations (namely, extracting all the labels associated to an object). The results for crawls **Indochina** and **EU** are the best reported in the WebGraph Project page, and they even break the gap complexity. For **FT** we measured the space required by Rice encoding of the differential inverted lists, plus pointers from the vocabulary to the sequence. This state-of-the-art in inverted indexes [22]. Finally, in absence of available software specifically targeted at compressing social networks, we tried WebGraph v. 1.7 (default parameters) on **DBLP**. As this is an undirected graph, we duplicate each edge $\{i, j\}$ as $(i, j)$ and $(j, i)$. This is not necessary on our representations, as we can extract direct and reverse neighbors. Our representations are by far the best in this case where no specific compressors exist.

# 7   Future Work

The times we have achieved for most operations is $O(\log \sigma)$, where $\sigma$ is the number of labels. These can probably be improved to $O(\frac{\log \sigma}{\log \log t})$ by using recent techniques on multiary wavelet trees [7], which would reach the best results achieved with wavelet trees for much simpler problems [14]. Our representations allow dynamic variants, where new pairs and/or objects can be inserted in/deleted from the waveleet trees [19,11]. Adding/removing labels, instead, is an open challenge, as it alters the wavelet tree shape. The space of our structures is close but does not reach the entropy of the binary relation, $H(\mathcal{B})$, in the worst theoretical case. An ambitious goal is to support all the operations we have defined in logarithmic time and within $H(\mathcal{B})(1+o(1))$ bits of space. A related issue is to define a finer notion of binary relation entropy that captures regularities that arise in real life, and express the space we achieve in those terms.

Finally, there is no reason why our list of operations should be exclusive. For example, determining whether a pair is related in the *transitive closure* of $\mathcal{B}$ is relevant for many applications (e.g. ancestorship in trees, paths in graphs). Alternatively one could enrich the data itself, e.g., associating a *tag* to each object/label pair, so that one can not only ask for the tag of a pair but also find pairs with some tag range within a range of the relation, and so on. This extension has already found applications, e.g. [13]. Another extension is $n$-ary relations, which would more naturally capture joins in the relational model.

## References

1. Baeza-Yates, R., Navarro, G.: Modeling text databases. In: Recent Advances in Applied Probability, pp. 1–25. Springer, Heidelberg (2004)
2. Barbay, J., Aleardi, L.C., He, M., Munro, J.I.: Succinct representation of labeled graphs. In: Tokuyama, T. (ed.) ISAAC 2007. LNCS, vol. 4835, pp. 316–328. Springer, Heidelberg (2007)
3. Barbay, J., Golynski, A., Munro, I., Rao, S.S.: Adaptive searching in succinctly encoded binary relations and tree-structured documents. TCS 387(3), 284–297 (2007)
4. Barbay, J., He, M., Munro, I., Rao, S.S.: Succinct indexes for strings, binary relations and multi-labeled trees. In: SODA, pp. 680–689 (2007)
5. Barbay, J., Navarro, G.: Compressed representations of permutations, and applications. In: STACS. pp. 111–122 (2009)
6. Boldi, P., Vigna, S.: The WebGraph framework I: compression techniques. In: WWW, pp. 595–602 (2004)
7. Bose, P., He, M., Maheshwari, A., Morin, P.: Succinct orthogonal range search structures on a grid with applications to text indexing. In: Dehne, F., Gavrilova, M.L., Sack, J.-R., Tóth, C.D. (eds.) WADS 2009. LNCS, vol. 5664, pp. 98–109. Springer, Heidelberg (2009)
8. Chien, Y.F., Hon, W.K., Shah, R., Vitter, J.: Geometric Burrows-Wheeler transform: Linking range searching and text indexing. In: DCC. pp. 252–261 (2008)
9. Chierichetti, F., Kumar, R., Lattanzi, S., Mitzenmacher, M., Panconesi, A., Raghavan, P.: On compressing social networks. In: KDD, pp. 219–228 (2009)
10. Clark, D.: Compact Pat Trees. Ph.D. thesis, Univ. of Waterloo, Canada (1996)
11. Claude, F.: Compressed Data Structures for Web Graphs. MSc. thesis, U. Chile (2008)
12. Claude, F., Navarro, G.: A fast and compact Web graph representation. In: Ziviani, N., Baeza-Yates, R. (eds.) SPIRE 2007. LNCS, vol. 4726, pp. 105–116. Springer, Heidelberg (2007)
13. Claude, F., Navarro, G.: Self-indexed text compression using straight-line programs. In: Královič, R., Niwiński, D. (eds.) MFCS 2009. LNCS, vol. 5734, pp. 235–246. Springer, Heidelberg (2009)
14. Ferragina, P., Manzini, G., Mäkinen, V., Navarro, G.: Compressed representations of sequences and full-text indexes. ACM TALG 3(2), article 20 (2007)
15. Golynski, A., Munro, J.I., Rao, S.S.: Rank/select operations on large alphabets: a tool for text indexing. In: SODA, pp. 368–373 (2006)
16. Grossi, R., Gupta, A., Vitter, J.: High-order entropy-compressed text indexes. In: SODA. pp. 841–850 (2003)
17. Kärkkäinen, J.: Repetition-Based Text Indexing. Ph.D. thesis, U. Helsinki, Finland (1999)
18. Mäkinen, V., Navarro, G.: Rank and select revisited and extended. TCS 387(3), 332–347 (2007)
19. Mäkinen, V., Navarro, G.: Dynamic entropy-compressed sequences and full-text indexes. ACM TALG 4(3), article 32 (2008)
20. Navarro, G.: Indexing text using the Ziv-Lempel trie. JDA 2(1), 87–114 (2004)
21. Raman, R., Raman, V., Rao, S.S.: Succinct indexable dictionaries with applications to encoding $k$-ary trees and multisets. In: SODA, pp. 233–242 (2002)
22. Witten, I., Moffat, A., Bell, T.: Managing Gigabytes, 2nd edn. Morgan Kaufmann Publishers, San Francisco (1999)

# Radix Cross-Sections for Length Morphisms

Sylvain Lombardy[1] and Jacques Sakarovitch[2]

[1] LIGM, Université Paris-Est
[2] LTCI, CNRS / ENST, Paris

**Abstract.** We prove that the radix cross-section of a rational set for a length morphism, and more generally for a rational function from a free monoid into $\mathbb{N}$, is rational. This property no longer holds if the image of the function is a subset of a free monoid with two or more generators.

The proof is based on several results on finite automata, such as the lexicographic selection of synchronous relations and the iterative decomposition of unary rational series with coefficients in the tropical semiring. It also makes use of a structural construction on weighted transducers that we call the length difference unfolding.

The purpose of this paper is to clarify the deep structure of relations realised by finite transducers. Its starting point is a positive answer to a problem left open in an old paper [12] and we prove the following property, a refinement of the Cross-Section Theorem [4]:

**Proposition 1.** *The radix cross-section of a rational set for a length morphism is rational.*

By 'rational set' we mean *rational set of a free monoid $A^*$* and by 'length morphism', a morphism from $A^*$ into $\mathbb{N}$, or, which is the same, into $\{x\}^*$, the one-generator free monoid. Let us take for instance the alphabet $A = \{a, b\}$, the morphism $\theta \colon A^* \to \{x\}^*$ defined by $a\theta = x^2$ and $b\theta = x^3$ and the rational set $R = (ba^*)^*$. The *lexicographic cross-section* of $R$ for $\theta$ (assuming that $a < b$) is

$$1 + ba^*(1 + b),$$

which is the set of words in $R$ such that each one is the smallest in the lexicographic order in its class modulo the map equivalence of $\theta$ (this smallest element exists in this case, even if the lexicographic order is not a well-ordering, since every class is finite). The *radix cross-section* of $R$ for $\theta$ is

$$1 + b(1 + a + a^2)b^*,$$

which is the set of words in $R$ obtained if we replace lexicographic order by *radix order* in the definition above. (The radix order is sometimes called *shortlex* or *length-lexicographic* order — definition of both lexicographic and radix orders will be recalled below.) That the lexicographic cross-section of a rational set is rational follows from general results that are recalled in the next section. That

A. López-Ortiz (Ed.): LATIN 2010, LNCS 6034, pp. 184–195, 2010.
© Springer-Verlag Berlin Heidelberg 2010

the radix cross-section of a rational set is rational is not true in general (*cf.* Example 2 below), but holds for length morphisms.

As for the Cross-Section Theorem, Proposition 1 has a dual, and equivalent formulation, the following generalisation of which is established in this paper.

**Theorem 1.** *The radix uniformisation of a rational relation from $\{x\}^*$ into $A^*$ is a rational function.*

Indeed, the radix cross-section of a rational set $R$ of $A^*$ for the morphism $\alpha \colon A^* \to \{x\}^*$ is the image of the radix uniformisation of the composition of the inverse of $\alpha$ with the intersection by $R$ and Proposition 1 directly follows from Theorem 1. The proof of the latter heavily relies on results and constructions presented by the first author in his thesis [10] and hardly publicised yet [9].

Before going to this proof we want to sketch here, we first recall a series of statements developed from the Cross-Section Theorem. On the one hand, they shed some light on the meaning of Theorem 1 by describing similarities and differences with this statement. On the other hand, they lead rather naturally to the notion of (rational) relation 'with bounded length-discrepancy with respect to a given ratio', rbld-relation for short.

The proof is then split in two main steps. We first remark that every transducer $T$ from $A^*$ into $B^*$ can be mapped onto an automaton $A$ over $A^*$ with multiplicity in the tropical (or *min-plus*) semiring $\mathcal{N}$ by replacing the output by their length. We call $A$ the *min-plus projection* of $T$. The core of that part of the proof amounts to establishing that the *minimal length selection* of a relation realised by a transducer $T$ is realised by an *immersion* in the product of $T$ with a min-plus automaton $A$ which realises its *min-plus projection*, under the condition that $A$ be unambiguous. The second step boils down to the effective decomposition of *unary* min-plus automata into domain-disjoint deterministic automata, which, in the case of unary relations, insures both that the min-plus projection is unambiguous and the minimal length selection is a rbld-relation.

# 1   The Cross-Section and Uniformisation Theorems

In his treatise [4], Eilenberg defines the 'Rational Cross-Section Property' (RCSP): a map $\alpha \colon A^* \to E$ has the RCSP if for every rational set $R$ of $A^*$ there exists a rational set $T$ which is a set of representatives for the trace on $R$ of the map equivalence of $\alpha$, that is, $T \subseteq R$, $T\alpha = R\alpha$, and $\alpha$ is injective on $T$ — the property being that such a $T$ can be chosen *rational*.

**Theorem 2 (Eilenberg [4]).** *Every morphism $\alpha \colon A^* \to B^*$ has the RCSP.*

An immediate corollary is that *every rational function*[1] $\alpha \colon A^* \to B^*$ *has the RCSP.* As already quoted above, this last statement can be given a *dual*, and immediately equivalent, formulation. A *uniformisation* of a relation $\theta \colon A^* \to B^*$ is a function $\tau \colon A^* \to B^*$ which has the same domain as $\theta$ and whose graph is contained in the graph of $\theta$. The dual statement then reads:

---

[1] *i.e.,* functional rational relation.

**Theorem 3.** *Every rational relation* $\theta \colon A^* \to B^*$ *has a uniformisation which is an unambiguous rational function.*

From now on and for sake of simplicity, we speak of uniformisation only (but in some examples).

## 1.1  Lexicographic and Radix Uniformisations

Building a uniformisation amounts to a *choice function*: the choice of a representative in the image of every element of the domain. The foregoing statements insure that the corresponding choice can be made in such a way that the whole function is rational, but tell nothing on how this choice is made. Even worse, the constructive proofs given to these statements are likely to produce uniformisation functions which will depend upon the representation chosen for the relation and the arbitrary choices made in the corresponding algorithms. The next step in the theory is then to characterise choice functions which yield rational uniformisation or conversely to describe rational relations for which natural choice functions are rational.

For sake of completeness, let us first recall the definition of the lexicographic and radix orders. Let $A = \{a_1, a_2, \ldots, a_n\}$ be a *totally ordered* alphabet with $a_1 < a_2 < \cdots < a_n$. The relation $\preccurlyeq$ defined on $A^*$ by:

$$f \preccurlyeq g \quad \Longleftrightarrow \quad \begin{cases} g = fh & \text{with} \quad h \in A^* \quad \text{or} \\ f = ua_iv, g = ua_jw & \text{with} \quad i < j \end{cases}$$

is a total order called *lexicographic order*, but is not a well ordering. The relation $\sqsubseteq$ defined on $A^*$ by:

$$f \sqsubseteq g \quad \Longleftrightarrow \quad \begin{cases} |f| < |g| & \text{or} \\ |f| = |g| & \text{and} \quad f \preccurlyeq g \, . \end{cases}$$

is a well order called *radix order*.

**Definition 1.** *Let* $\theta \colon A^* \to B^*$ *be a relation from* $A^*$ *to* $B^*$. *We call* radix uniformisation *of* $\theta$, *denoted* $\theta_{\mathsf{rad}}$, *the function from* $A^*$ *to* $B^*$ *obtained by choosing for every* $f$ *in* $A^*$ *the smallest element of* $f\theta$ *in the radix order. As the radix order is a well-order, such a smallest element always exists and* $\mathsf{Dom}\,\theta_{\mathsf{rad}} = \mathsf{Dom}\,\theta$.

*We call* lexicographic selection *of* $\theta$, *denoted* $\theta_{\mathsf{lex}}$, *the function from* $A^*$ *to* $B^*$ *obtained by choosing for every* $f$ *in* $A^*$ *the smallest element of* $f\theta$ *in the lexicographic order. As the lexicographic order is not a well-order, such a smallest element may not exist,* $\mathsf{Dom}\,\theta_{\mathsf{lex}} \subseteq \mathsf{Dom}\,\theta$ *and* $\theta_{\mathsf{lex}}$ *is not necessarily a uniformisation.*

*We call* minimal-length selection *of* $\theta$, *denoted* $\theta_{\mathsf{ml}}$, *the relation from* $A^*$ *to* $B^*$ *obtained by choosing for every* $f$ *in* $A^*$ *the elements of* $f\theta$ *of minimal length. Obviously* $\mathsf{Dom}\,\theta_{\mathsf{ml}} = \mathsf{Dom}\,\theta$ *and the following holds:*

$$\theta_{\mathsf{rad}} = (\theta_{\mathsf{ml}})_{\mathsf{lex}} \, . \tag{1}$$

There exist rational relations such that neither the radix uniformisation nor the lexicographic uniformisation is rational.

*Example 1.* Let $A = \{a < b < c\}$ be an ordered alphabet and $\theta \colon A^* \to A^*$ defined by $(a^n b^m)\theta = \{a^n b, a^m c\}$. It is immediate that

$$(a^n b^m)\theta_{\mathsf{rad}} = \begin{cases} a^n b & \text{if } n \leqslant m, \\ a^m c & \text{otherwise,} \end{cases} \quad \text{and} \quad (a^n b^m)\theta_{\mathsf{lex}} = \begin{cases} a^n b & \text{if } n \geqslant m, \\ a^m c & \text{otherwise.} \end{cases}$$

And neither $\theta_{\mathsf{rad}}$ nor $\theta_{\mathsf{lex}}$ is a rational function.

In [12], and by means of tedious calculations, we proved that the lexicographic cross-section of a *morphism* $\alpha \colon A^* \to B^*$ is rational, a result that has then been generalised, and given a more readable proof, by H. Johnson.

**Theorem 4 ([8]).** *The lexicographic selection of a deterministic[2] rational relation is a deterministic rational function.*

The same is not true of the radix uniformisation of deterministic rational relations, even of inverse morphisms, as shown by the following example (taken from [12] and that can be found in [13, Exer. V.3.7] as well).

*Example 2.* Let $\alpha \colon \{a, b, c, d, e, f\}^* \to \{x, y, z\}^*$ be defined by $a\alpha = x$, $b\alpha = yxyx$, $c\alpha = xy$, $d\alpha = yz$, $e\alpha = zyzy$, $f\alpha = z$. The radix cross-section $T$ of $\{a, b, c, d, e, f\}^*$ for $\alpha$ is such that $T \cap ab^*(d^2)^*d = \{ab^n d^{2m+1} \mid m \leqslant n\}$, and is not rational.

It is thus natural to turn to an even more restricted family of rational relations.

## 1.2 Uniformisation of Synchronous Relations

Synchronous relations are those rational relations which can be realised by transducers whose transitions are labelled by *pairs* of letters. *Stricto sensu*, this definition yields *length preserving* relations only. This last constraint is relaxed by allowing the replacement of a letter by a padding symbol, in either component (but under the 'padding condition', that is, no letter can appear after the padding symbol on the same component). Even if this is more than a technical trick since in particular it is not decidable whether a given relation is synchronous or not (see [6]), synchronous relations are a very natural subfamily of rational relations. They have been given a logical characterisation in [5] and are considered in many instances (automatic structures, operations on numbers, *etc.*). They form a Boolean algebra, the largest one considered so far inside the rational relations. As the lexicographic order and radix order are synchronous relations, it then follows:

---

[2] N.B. Deterministic rational relations are those relations realised by *deterministic transducers*, that is, deterministic 2-tape automata, which are distinct from *sequential transducers*, that is, transducers with deterministic underlying input automata (see [8,13]). Morphisms, and then inverse of morphisms, are deterministic rational relations.

**Proposition 2 (see [13]).** *The lexicographic selection and radix uniformisation of a synchronous relation are synchronous functions.*

Proposition 2 and its variants are commonly used to give simple proofs for statements involving rational sets and lexicographic or radix order, *e.g.* '*if K is a rational set, then the set of words of K which are maximal for every length is rational*' (see [14]), or '*the radix enumeration of a rational set is a rational function*' (see [3,1]).

## 1.3   Rational ρ-bld-Relations

A synchronous relation is, by definition, realised by a transducer whose transitions are labelled by pairs of letters, that is, such that the ratio between the length of the 'input' and the the length of the 'output' is fixed and equal to 1. For every rational number $\rho = \frac{p}{q}$ ($p$ and $q$ are co-prime integers), we define the *ρ-synchronous relations* as those rational relations that are realised by transducers whose transitions are labelled by pairs of words $(f, g)$ where $|f| = q$ and $|g| = p$ (as above, we allow the use of a padding symbol, under the padding condition). It is straightforward to verify that the ρ-synchronous relations form a Boolean algebra. Moreover, standard constructions show that the *composition* of a ρ-synchronous relation with a (1-)synchronous relation is a ρ-synchronous relation, hence the same proof as above yield the following.

**Proposition 3.** *The lexicographic selection and radix uniformisation of a ρ-synchronous relation are ρ-synchronous functions.*

Let $\theta \colon A^* \to B^*$ be a relation with the property that there exists a rational number $\rho$ and an integer $k$ such that, for every $f$ in $A^*$ and every $g$ in $f\theta$, then $\big| \rho|f| - |g| \big| \leqslant k$. If $\rho = 1$ and $k = 0$, $\theta$ is a *length preserving relation*. If $\rho = 1$ and $k$ is arbitrary, $\theta$ has been called a *bounded length difference relation* ([6]) or *bounded length discrepancy relation* ([13]), *bld-relation* for short in any case; for arbitrary $\rho$ (and $k$), let us call $\theta$ a *ρ-bld-relation*. And let us say that $\theta$ is an *rbld-relation* if there exists a $\rho$ such that $\theta$ is a *ρ-bld-relation*.

It is not difficult to verify that a rational relation is ρ-bld if, and only if, any transducer $\mathcal{T}$ (without padding symbol!) which realises $\theta$ has the property that the label of every circuit in $\mathcal{T}$ is such that the ratio between the length of the 'input' and the the length of the 'output' is fixed and equal to $\rho$, a property which is thus decidable. The following result is essentially due to Eilenberg who proves it for length-preserving relations ([4]); it has been extended to bld-relations in [6], the generalisation to ρ-bld-relations can be found in [13, IV.6].

**Theorem 5.** *A rational ρ-bld-relation is a ρ-synchronous relation.*

Theorem 1 will thus be proved once we have established the following.

**Theorem 6.** *The minimal length selection of a rational relation from $\{x\}^*$ into $A^*$ is an effectively computable finite union of domain-disjoint rational rbld-relations.*

Indeed, if $\theta_{\mathsf{ml}} = \bigcup \theta_i$ where every $\theta_i$ is $\rho_i$-bld, then $\theta_{\mathsf{rad}} = \bigcup (\theta_i)_{\mathsf{lex}}$ by (1) and since the $\theta_i$ are domain disjoint: $\theta_{\mathsf{rad}}$ is then rational by Theorem 5 and Proposition 3 and since the union is finite.

## 2  Minimal-Length Selection of Relations

The construction underlying the proof of Theorem 6 is reminiscent of that of the *Schützenberger covering* that allowed us to give a new proof of the rational uniformisation theorem ([13]). There, the product of a transducer with the determinisation of its underlying input automaton was used to choose among computations with same inputs. Here, we consider a transducer $\mathcal{T}$ and a min-plus automaton $\mathcal{A}$ that produces for every input the minimal length of its images in $\mathcal{T}$. Then, a *length-difference covering* of the product $\mathcal{T} \times \mathcal{A}$ allows to select the computations in $\mathcal{T}$ that yield the outputs of minimal length.

### 2.1  The Min-Plus Projection

All automata or transducers that we consider in this paper are 'real-time', that is, every transition is labelled with a letter (in automata) or by a pair of words whose first component is a letter (in transducers). As a consequence, they can equally be described as 'tuples' or as 'representations'. A $\mathbb{K}$-automaton $\mathcal{A}$ over $A^*$ is described either as $\mathcal{A} = \langle Q, A, \mathbb{K}, E, I, T \rangle$ where $E$ is the set of transitions of the form $(p, a, k, q)$ and $I$ and $T$ are functions from $Q$ into $\mathbb{K}$ or as a triple $\langle I, \mu, T \rangle$, where $\mu$ is a morphism from $A^*$ into the monoid of matrices $\mathbb{K}^{Q \times Q}$ and $I$ (*resp.* $T$) is a row (*resp.* column) vector of $\mathbb{K}^Q$. The automaton $\mathcal{A}$ *realises* the $\mathbb{K}$-rational series $|\mathcal{A}| = \sum_{f \in A^*} (I \cdot f \mu \cdot T) f$, that is, an element of $\mathbb{K}\mathrm{Rat}\, A^*$.

A transducer $\mathcal{T}$ over $A^*$ with outputs in $B^*$ and set of states $Q$ is described either as $\mathcal{T} = \langle Q, A, B^*, E, I, T \rangle$ where $E$ is the set of transitions of the form $(p, a, L, q)$ with $L$ in $\mathrm{Rat}\, B^*$ and $I$ and $T$ are functions from $Q$ into $\mathrm{Rat}\, B^*$, or as a triple $\langle I, \mu, T \rangle$, where $\mu$ is a morphism from $A^*$ into the monoid of matrices $(\mathrm{Rat}\, B^*)^{Q \times Q}$ and $I$ (*resp.* $T$) is a row (*resp.* column) vector of $(\mathrm{Rat}\, B^*)^Q$. The image of a word $f$ in $A^*$ by $\mathcal{T}$ is $f|\mathcal{T}| = I \cdot f \mu \cdot T$ and $|\mathcal{T}|$ the relation realised by $\mathcal{T}$ is rather seen as a series over $A^*$ with coefficients in $\mathfrak{P}(B^*)$ (in $\mathrm{Rat}\, B^*$ in fact) than as a (rational) subset of $A^* \times B^*$.

The *tropical semiring* is denoted with $\mathcal{N} = \langle \mathbb{N} \cup \{+\infty\}, \min, +, +\infty, 0 \rangle$. For any alphabet $B$, the set $\mathfrak{P}(B^*)$ equipped with union and product is also a semiring and the map $\psi : \mathfrak{P}(B^*) \to \mathcal{N}$ defined by $R\psi = \min\{|f| \mid f \in R\}$ for every $R \subseteq B^*$ is a semiring morphism, which we call the *min-plus projection*.

Let $\mathcal{T} = \langle I, \mu, T \rangle$ be a transducer, $\tau = |\mathcal{T}|$ the relation it realises and $\sigma = \tau \psi$. Then, by definition, for every $f$ in $A^*$,

$$f\sigma = \min\{|u| \mid u \in f\tau\}\ .$$

Moreover, if $\mathcal{A} = \mathcal{T}\psi$ is the image of $\mathcal{T}$ by $\psi$, that is, the weight of every transition of $\mathcal{A}$ is obtained by taking the image by $\psi$ of the output of the corresponding transition in $\mathcal{T}$, then $\sigma$ is realised by $\mathcal{A}$ (as $\psi$ is a semiring morphism). A transducer and its min-plus projection are shown below, at Figure 1.

## 2.2  Product of a Transducer by a Min-Plus Automaton

Let us first define *min-plus transducers*, that is, transducers where every word of the output of a transition is weighted by an element of $\mathcal{N}$, or, which is equivalent, transducers which are realised by representations in $\mathcal{N}\mathrm{Rat}\,B^*$. Let us then note that both $\mathrm{Rat}\,B^*$ and $\mathcal{N}$ are subsemirings of $\mathcal{N}\mathrm{Rat}\,B^*$, by assigning a weight $0 = 1_\mathcal{N}$ to every element of any subset of $B^*$ on one hand, and by considering every element $k$ in $\mathcal{N}$ as the monomial $k\varepsilon$ on the other hand (in this context where there are 0's and 1's from $\mathcal{N}$ going around, we rather denote the empty word as $\varepsilon$). Moreover, we remark that after these embeddings, every element of $\mathrm{Rat}\,B^*$ commutes with every element of $\mathcal{N}$ within $\mathcal{N}\mathrm{Rat}\,B^*$ (which is not itself a commutative semiring).

It follows then that every transducer $\mathcal{T}$ from $A^*$ to $B^*$ and every $\mathcal{N}$-automaton $\mathcal{A}$ may be seen as min-plus transducers, that their *product* $\mathcal{T} \otimes \mathcal{A}$ is well-defined and is a min-plus transducer described by the *tensor product* of their representations.

By Schützenberger's Theorem, the product of $\mathcal{T}$ by $\mathcal{A}$ realises the Hadamard product of their respective behaviours (see [2,13]). Hence the following.

**Proposition 4.** *Let* $\mathcal{T}$ *be a transducer and* $\mathcal{A}$ *a* $\mathcal{N}$-*automaton which realises* $|\mathcal{T}|\psi$. *Then* $\mathcal{U} = \mathcal{T} \otimes \mathcal{A}$ *is the* $\mathcal{N}$-*transducer which associates to every word* $f$ *of* $A^*$ *the following series:*

$$f|\mathcal{U}| = \sum \{k\,u \mid u \in f|\mathcal{T}|,\quad k = \min\{|v| \mid v \in f|\mathcal{T}|\}\}\ .$$

*Example 3.* Figure 1 shows the product of a transducer $\mathcal{T}_1$ (vertical, left) with its min-plus projection (horizontal, above).

## 2.3  Length-Difference Unfolding of a Min-Plus Transducer

Let $\mathcal{U} = \langle\, S, A, B^*, \mathcal{N}, E, I, T\,\rangle$ be a min-plus transducer. We call *length-difference unfolding* of $\mathcal{U}$ the (possibly infinite) transducer[3]

$$\mathcal{V} = \langle\, S \times \mathbb{Z}, A, B^*, E', I', T'\,\rangle \qquad\qquad \text{defined by}$$

$$E' = \big\{\,\big((p, z), (a, u), (q, z + |u| - k)\big) \mid (p, (a, k\,u), q) \in E, z \in \mathbb{Z}\,\big\}\ ,$$

$$I' = \{(i, -k) \mid i \in S,\, (i)I = k\} \quad \text{and} \quad T' = \{(t, k) \mid t \in S,\, (t)T = k\}\ .$$

The transducer $\mathcal{V}$ is an *immersion* in $\mathcal{U}$ and not a *covering* of $\mathcal{U}$: it would be one if every state $(t, z)$ for $t$ such that $(t)T \neq +\infty$ were final. Along a computation of $\mathcal{V}$, and at every step, the second component of the state gives the difference between the length of the word which is output and its coefficient; as in our case, the coefficient will represent the length of the shortest output for the input read so far, the name 'length-difference' is justified. This construction is associated with the product of transducer with min-plus automata in the following statement.

---

[3] For the sake of simplicity of notations, we suppose that the functions $I$ and $T$ have values in $\mathcal{N}$ and not in $\mathcal{N}\langle\!\langle B^*\rangle\!\rangle$. It would be straightforward to generalise both the definition and the proof.

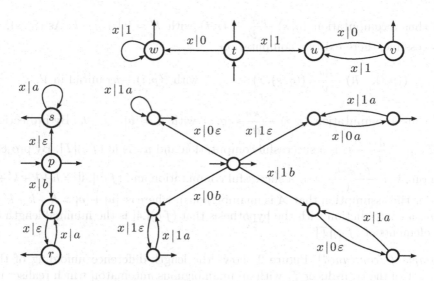

**Fig. 1.** Product of a unary transducer with its min-plus projection

**Proposition 5.** *Let $T$ be a transducer, $A$ an* unambiguous *$N$-automaton which realises $|T|\psi$ and $U = T \otimes A$. Let $V$ be the length-difference unfolding of $U$. Then:*

(i)  *$V$ realises $|T|_{\mathsf{ml}}$;*
(ii) *the trim part of $V$ is finite, and effectively computable.*

*Proof.* For every $f$ in $\mathrm{Dom}\,T$, there exists a *unique* computation $i \xrightarrow[A]{f|k} t$ labelled by $f$ in $A = \langle J, k, U \rangle$. If $(i)J = h$ and $(t)U = l$, then $f|A| = h + k + l$ is the minimal length of the outputs of $T$ for the input $f$. For every successful computation $p \xrightarrow[T]{f|u} q$, there exists then a *unique* computation

$$(p, i) \xrightarrow[U]{f|k\,u} (q, t)$$

and for every $x$ in $\mathbb{Z}$, a *unique* computation

$$((p, i), x) \xrightarrow[V]{f|u} ((q, t), y) \qquad \text{where } y = x + |u| - k.$$

This computation is successful if, and only if, $x = -h$ and $y = l$, and thus $|u| = h + k + l$ that is, if, and only if, $u$ is an output of $f$ by $T$ of minimal length.

Let $((q, s), x)$ and $((q, s), y)$ be two accessible states in $V$, with $x < y = x + n$. We prove that $((q, s), y)$ is not co-accessible in $V$. Suppose, by way of contradiction, that it is: there exists a computation

$$((q, s), y) \xrightarrow[V]{g|v} ((r, t), l) \qquad \text{with } ((r, t), l) \text{ final in } V,$$

and thus a computation $(q,s) \xrightarrow[u]{g|k'\,v} (r,t)$ with $k' = |v| + y - l$. As $((q,s),x)$ is accessible, there exists a computation

$$((p,i),-h) \xrightarrow[\mathcal{V}]{f|u} ((q,s),x) \qquad \text{with } ((p,i),-h) \text{ initial in } \mathcal{V},$$

and thus a computation $(p,i) \xrightarrow[\mathcal{U}]{f|k\,u} (q,s)$ with $k = |u| - x - h$. By projection on $\mathcal{T}$, $p \xrightarrow[\mathcal{T}]{f\,g|u\,v} r$ is a successful computation and $uv$ is in $(f\,g)|\mathcal{T}|$. By projection on $\mathcal{A}$, $i \xrightarrow[\mathcal{A}]{f\,g|k+k'} t$ is a successful computation and $(f\,g)|\mathcal{A}| = h+k+k'+l$ (under the assumption that $\mathcal{A}$ is unambiguous), whereas $|u| + |v| = h+k+k'+l-n$, a contradiction with the hypothesis that $(f\,g)|\mathcal{A}|$ is the minimal length of the elements of $(f\,g)|\mathcal{T}|$.

*Example 4 (continued).* Figure 2 shows the length-difference unfolding of the product of the transducer $\mathcal{T}_1$ with an unambiguous automaton which realises its min-plus projection.

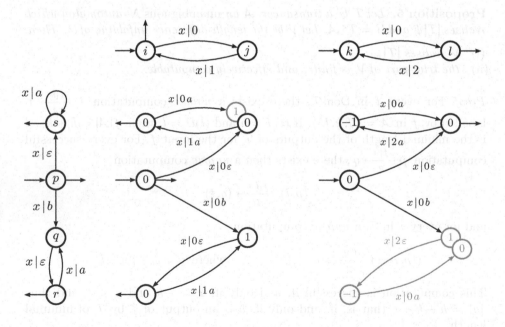

**Fig. 2.** A length-difference unfolding for $\mathcal{T}_1$

With stronger hypotheses, we get an interesting corollary of Proposition 5.

**Definition 2.** *We call* yield *of a circuit in a min-plus automaton the ratio between the weight of the circuit and its length.*

**Proposition 6.** *With the hypotheses and notation of Proposition 5, and if more-over all circuits in $\mathcal{A}$ have the same yield $\rho$, then $\mathcal{V}$ realises a $\rho$-bld-relation.*

*Proof.* A circuit in $\mathcal{V}$ comes from a circuit in $\mathcal{U}$ which itself projects onto a circuit in $\mathcal{A}$ with yield $\rho$. A circuit in $\mathcal{U}$ gives rise to a circuit in $\mathcal{V}$ if, and only if, the length of its output is equal to its weight and thus the ratio between the lengths of the output and input in a circuit of $\mathcal{V}$ is fixed and equal to $\rho$.

Our last step in the proof of Theorem 6 will then be to establishing that any rational relation from $\{x\}^*$ into $A^*$ has a min-plus projection which can be realised by finite union of pairwise domain disjoint min-plus automata, each of which with the property that all circuits have the same yield.

# 3   Sequential Decomposition of Min-Plus Unary Automata

The result we have in mind has first been stated in an existential way (for any commutative dioid); we specialise it for the semiring $\mathcal{N}$. For sake of simplicity, let us call a series over $\{x\}^*$ a *unary series*, an automaton over $\{x\}^*$ a *unary automaton*. Let us say that a unary min-plus series $s$ is *ultimately arithmetic* if it is finite or if there exist non negative integers $N$ and $r$, and an integer $k$ such that, for every $n$ larger than $N$, $\langle s, x^n \rangle = nr + k$. (In $\mathcal{N}$, these series are actually *geometric* series, since the sum is the multiplication law of the semiring, the product is an exponentiation in $\mathcal{N}$.) A series $s$ is the *merge* of $p$ series $s_0$, $s_1, \ldots, s_{p-1}$ if for every pair $(n, i)$ in $\mathbb{N} \times [0; p-1]$, it holds $\langle s, x^{np+i} \rangle = \langle s_i, x^n \rangle$.

**Proposition 7 ([7]).** *A unary min-plus rational series is a merge of ultimately arithmetic series.*

The original proof of Proposition 7 consists in proving that the family of merges of ultimately arithmetic series is closed under rational operations. In terms of automata, it means that every unary min-plus rational series can be realised by a finite union of domain disjoint deterministic min-plus automata.

**Proposition 8 ([10]).** *A min-plus unary automaton can be effectively decomposed into a finite union of deterministic min-plus (unary) automata whose supports are pairwise disjoint.*

## 3.1   Sequentialisation of Unary Tight Automata

In [11], we have described a folklore 'generalised sequentialisation procedure' which applies to every min-plus automata, but which does not always yield a finite (sequential) automaton, even for automata that realise a sequential series.

**Definition 3.** *A circuit in a min-plus automaton $\mathcal{A}$ is* critical *if its yield is minimal (among the yields of all circuits in $\mathcal{A}$). The* critical part *of $\mathcal{A}$ is the union of its critical circuits. A computation in $\mathcal{A}$ is* critical *if it meets the critical part of $\mathcal{A}$. An automaton is* tight *if almost all words in its domain label a successful critical computation.*

*Remark 1.* The critical part of an automaton is easily computable, since critical circuits are either primitive or composed of smaller critical circuits. It is therefore sufficient to detect primitive critical circuits.

**Theorem 7** ([10]). *A unary min-plus automaton realises a sequential min-plus series if, and only if, it is tight.*

## 3.2    Iterative Decomposition of Unary Min-Plus Automata

The following proposition is the core of the iterative method. The first point describes a step of the iteration, the second one guarantees that the iteration ends.

**Proposition 9.** *Let $\mathcal{A}$ be a min-plus (unary) automaton, $\rho$ its minimal yield, and $L$ the set of words which label a successful critical computation. Then:*

(i)  *the restriction of $s = |\mathcal{A}|$ to $L$ is sequential, realised by a deterministic automaton whose loop has a yield which is equal to $\rho$;*
(ii)  *the restriction of $s$ to the complement of $L$ is realised by an automaton $\mathcal{B}$ which is an immersion in $\mathcal{A}$ and whose critical yield is strictly larger than $\rho$.*

*Proof.* i) Let $\mathsf{supp}(\mathcal{A})$ be the support of $\mathcal{A}$ in which the critical part of $\mathcal{A}$ is tagged. Let $\mathcal{B}$ be an automaton formed by two copies of $\mathsf{supp}(\mathcal{A})$; the first one does not contain any final state and the second one does not contain any initial state; it is possible to jump from a state of the first copy to the corresponding state in the second copy if and only if this state is tagged. Therefore $\mathcal{B}$ recognises $L$. Thus the product $\mathcal{A} \times \mathcal{B}$ realises the restriction of $s$ to these words.

ii) Let $\mathcal{C}$ be a deterministic automaton that recognises the complement of $L$. The product $\mathcal{A} \times \mathcal{C}$ realises the restriction of $s$ to the complement of $L$. This product is an *immersion*, each of its circuits corresponds to a circuit of $\mathcal{A}$ with the same yield.

The space constraint does not allow to develop every step of the decomposition of the min-plus projection $\tau_1 = |\mathcal{T}_1|\psi$ of the transducer $\mathcal{T}_1$. Anyway, Figure 2 above already showed a length unfolding of $\mathcal{T}_1$ that is obtained by the product of $\mathcal{T}_1$ by two unary deterministic tight automata with disjoint domains.

## 3.3    Proof of Theorem 6

Let $\mathcal{T}$ be a transducer that realises a rational relation $\theta \colon \{x\}^* \to B^*$ and $\mathcal{A}$ its min-plus projection. The iterative application of Proposition 9 effectively gives $k$ unary deterministic min-plus automata $\mathcal{A}_1, \mathcal{A}_2, \ldots, \mathcal{A}_k$, with disjoint domains $K_1, K_2, \ldots, K_k$, and with yield $\rho_1, \rho_2, \ldots, \rho_k$. The length unfolding of every $\mathcal{T} \times \mathcal{A}_i$ realises the length selection of the restriction of $\theta$ to $K_i$ which is, by Proposition 9, a $\rho_i$-bld relation.

# References

1. Angrand, P.-Y., Sakarovitch, J.: Radix enumeration of rational languages. RAIRO Theor. Informatics and Appl. (to appear)
2. Berstel, J., Reutenauer, C.: Les séries rationnelles et leurs langages, Masson (1984); English translation: Rational Series and Their Languages. Springer, Heidelberg (1988)
3. Berthé, V., Frougny, C., Rigo, M., Sakarovitch, J.: On the cost and complexity of the successor function. In: Arnoux, P., Bédaride, N., Cassaigne, J. (eds.) Proc. of WORDS 2007, CIRM, Luminy, Marseille (2007)
4. Eilenberg, S.: Automata, Languages and Machines, vol. A. Academic Press, London (1974)
5. Eilenberg, S., Elgot, C.C., Shepherdson, J.C.: Sets recognized by n-tape automata. J. Algebra 13, 447–464 (1969)
6. Frougny, C., Sakarovitch, J.: Synchronized relations of finite and infinite words. Theoret. Comp. Sci. 108, 45–82 (1993)
7. Gaubert, S.: Rational series over dioids and discrete event systems. In: Proc. of the 11th Conf. on Anal. and Opt. of Systems: Discrete Event Systems. LNCIS, vol. 199. Springer, Heidelberg (1994)
8. Johnson, H.J.: Rational equivalence relations. Theoret. Computer Sci. 47, 39–60 (1986)
9. Lombardy, S.: Sequentialization and unambiguity of (max,+) rational series over one letter. In: Gaubert, S., Loiseau, J.-J. (eds.) Proc. of a Workshop on Max-Plus Algebras. IFAC. Elsevier, Amsterdam (2001)
10. Lombardy, S.: Approche structurelle de quelques problèmes de la théorie des automates. Thèse Doctorat Informatique ENST, Paris (2001)
11. Lombardy, S., Sakarovitch, J.: Sequential? Theoret. Computer Sci. 356, 224–244 (2006)
12. Sakarovitch, J.: Deux remarques sur un théorème de S. Eilenberg. RAIRO Theor. Informatics and Appl. 17, 23–48 (1983)
13. Sakarovitch, J.: Éléments de théorie des automates. Vuibert (2003); Corrected English edition: Elements of Automata Theory. Cambridge University Press, Cambridge (2009)
14. Shallit, J.: Numeration systems, linear recurrences, and regular sets. Inform. and Comput. 113, 331–347 (1994)

# Pairs of Complementary Unary Languages with "Balanced" Nondeterministic Automata

Viliam Geffert[1,*] and Giovanni Pighizzini[2,**]

[1] Department of Computer Science, P. J. Šafárik University
Jesenná 5, SK-04001 Košice, Slovakia
viliam.geffert@upjs.sk
[2] Dipartimento di Informatica e Comunicazione, Università degli Studi di Milano
via Comelico 39, I-20135 Milano, Italy
pighizzini@dico.unimi.it

**Abstract.** For each sufficiently large $N$, there exists a unary regular language $L$ such that both $L$ and its complement $L^c$ are accepted by unambiguous nondeterministic automata with at most $N$ states while the smallest deterministic automata for these two languages require a superpolynomial number of states, at least $e^{\Omega(\sqrt[3]{N \cdot \ln^2 N})}$. Actually, $L$ and $L^c$ are accepted by nondeterministic machines sharing the same transition graph, differing only in the distribution of their final states. As a consequence, the gap between the sizes of unary unambiguous self-verifying automata and deterministic automata is also superpolynomial.

**Keywords:** finite state automata, state complexity, unary regular languages.

## 1 Introduction

Finite state automata with a one-letter input alphabet have been studied extensively, mainly in the last decade (see, e.g., [1,3,6,11,12]). This research revealed many important differences between the regular languages over the unary input alphabet and the general case, using at least two different input symbols.

For example, it is well known that each $n$-state nondeterministic automaton can be simulated by an equivalent deterministic automaton with $2^n$ states [17], and this cost was shown to be tight for binary or any larger input alphabet [10,13,15]. However, in the unary case, the optimal cost of this simulation is strictly smaller, as shown by Chrobak [1] (see also [3,20]): it essentially reduces to the *Landau Function* $F(n)$ [8,9], with the subexponential but superpolynomial growth rate $e^{\Theta(\sqrt{n \cdot \ln n})}$ [19].

* Supported by the Slovak Grant Agency for Science (VEGA) under contract 1/0035/09 "Combinatorial Structures and Complexity of Algorithms".
** Partially supported by the Italian MIUR under project "Aspetti matematici e applicazioni emergenti degli automi e dei linguaggi formali: metodi probabilistici e combinatori in ambito di linguaggi formali".

A. López-Ortiz (Ed.): LATIN 2010, LNCS 6034, pp. 196–207, 2010.

We also know that, for each $n$, there exists a language $L$ over the binary (or any larger) input alphabet recognizable by an $n$-state nondeterministic automaton, such that even $L^c$, the complement of $L$, can be accepted by a nondeterministic machine using no more than $n+1$ states, but the smallest deterministic automaton for $L$ must still use $2^n$ states [11]. That is, even the gap between the total size of the two nondeterministic automata for the pair $L, L^c$ and the size of the corresponding deterministic machines is exponential.

On the other hand, this phenomenon does not happen in the unary case. In fact, Mera and Pighizzini [11] proved that if a unary language $L$ is a witness of the maximal state gap, i.e., if $L$ is accepted by an $n$-state nondeterministic automaton, while the smallest deterministic automaton for $L$ requires $F(n) = e^{\Theta(\sqrt{n \cdot \ln n})}$ states, then the smallest nondeterministic automaton for $L^c$ must also use $F(n)$ states. (Therefore, it coincides with the smallest deterministic automaton for $L^c$.) Thus, taking into account the total number of states in machines for $L$ and $L^c$, the superpolynomial gap disappears.

In this paper, we show that if the state gap is a little bit smaller than $F(n)$, it can be achieved by both $L$ and $L^c$ even in the unary case. More precisely, for any sufficiently large $N$, there exists a unary language $L$ such that both $L$ and $L^c$ are recognized by nondeterministic automata with at most $N$ states, while the smallest deterministic automaton for $L$ or $L^c$ requires $e^{\Omega(\sqrt[3]{N \cdot \ln^2 N})}$ states. Moreover, $L$ and $L^c$ are accepted by *unambiguous* nondeterministic machines with isomorphic transition graphs (therefore, with equal number of states), differing only in the distribution of their final states.

This result has an important consequence concerning *self-verifying automata*, a special kind of finite automata with a symmetric form of nondeterminism, introduced in [2]. Recently, Jirásková and Pighizzini [7] proved that an $n$-state self-verifying automaton could be converted into an equivalent deterministic machine with $O(3^{n/3}) \approx O(1.45^n)$ states. This cost was also shown to be tight, *if the input alphabet consists of at least of two letters*, leaving open the unary case. As observed in [7], the cost of the corresponding conversion for unary automata must be smaller than $F(n)$.

In this paper, we show that the cost of the conversion of unary self-verifying $n$-state automata into deterministic ones is still superpolynomial, namely, at least $e^{\Omega(\sqrt[3]{N \cdot \ln^2 N})}$ states, even if the self-verifying automata are unambiguous. This is not too far from $F(n) = e^{\Theta(\sqrt{n \cdot \ln n})}$.

## 2   Preliminaries

Let $\ln x$ denote the natural logarithm of $x$. The $i$-th prime number will be denoted by $p_i$, starting with $p_1 = 2$, $p_2 = 3$, .... The *greatest common divisor* of positive integers $a_1, \ldots, a_s$ is denoted by $\gcd(a_1, \ldots, a_s)$, their *least common multiple* by $\operatorname{lcm}(a_1, \ldots, a_s)$.

A *nondeterministic finite automaton* (NFA) is a quintuple $A = (Q, \Sigma, \delta, q_1, F)$, where $Q$ is a finite set of states, $\Sigma$ a finite set of input symbols, $\delta : Q \times \Sigma \to 2^Q$ a transition function, $q_1 \in Q$ an initial state, and $F \subseteq Q$ a set of final (accepting) states. The automaton $A$ is *deterministic* (DFA), if the cardinality of $\delta(q, s)$ is equal to 1, for each $q \in Q$ and $s \in \Sigma$, i.e., if $A$ has exactly one computation path for each input $w$, not halting before reading the entire input. An NFA $A$ is called *unambiguous*, if it has exactly one accepting computation path for each accepted input $w$. All automata in this paper are unary, so we can fix the input alphabet to $\Sigma = \{a\}$.

Clearly, the transition graph of a unary DFA consists of an initial path starting in an initial state, of length $\varphi \geq 0$, followed by some loop, of length $\lambda \geq 1$.

A unary language $L$ is $\lambda$-*cyclic*, if it is accepted by a unary DFA whose transition graph is a loop of $\lambda$ states without any initial path, i.e., with $\varphi = 0$. Equivalently, $L$ is $\lambda$-cyclic if, for each $K \geq 0$, $a^K \in L$ if and only if $a^{K+\lambda} \in L$. $L$ is *properly* $\lambda$-*cyclic*, if it is $\lambda$-cyclic but not $\lambda'$-cyclic for any $\lambda' < \lambda$. (In this case, $\lambda$ is equal to the number of the states in the smallest DFA for $L$.)

An NFA is in *Chrobak Normal Form*, if its graph consists of an initial deterministic path and some $s$ disjoint deterministic loops. The last state of the initial path is connected via $s$ leaving edges to each of the $s$ loops: this is the only nondeterministic choice ever made. (If $s = 1$, we get an ordinary deterministic machine.)

Even though this form is very simple and the use of nondeterminism is restricted to a single computation step, each $n$-state unary NFA $A$ can be transformed into an NFA $A'$ in Chrobak Normal Form with at most $n^2 - 2$ states along the initial path and at most $n - 1$ states in the loops [1,3].[1] Moreover, by [6], if the language accepted by $A$ is $\lambda$-cyclic, then $A$ can be converted into $A'$ without increasing the number of states, either with the initial path consisting of a single state, or the entire $A'$ becomes a single deterministic loop.

## 3   Witness Languages and Their Automata

In this section, we introduce the pairs of complementary unary languages used to state our main result and study their state complexity, both for DFAs and NFAs.

Let $\lambda_0, \ldots, \lambda_{s-1}$ be some given powers of different prime numbers, for some $s \geq 1$. That is, $\lambda_i = p_{h_i}^{\alpha_i}$, for some positive integer $\alpha_i$ and some prime $p_{h_i}$. We also suppose that $\lambda_0, \ldots, \lambda_{s-1}$ are sorted with respect to their underlying prime numbers, i.e., $p_{h_{i'}} < p_{h_{i''}}$ for $i' < i''$. Since the underlying primes are all different, $\lambda_i = p_{h_i}^{\alpha_i} \geq p_{h_i} \geq p_i > i$. The product of these prime powers will be denoted by $\lambda = \lambda_0 \cdots \lambda_{s-1}$.

Now, fix $\hat{s}$ as the smallest integer satisfying $\hat{s} \geq s$ and dividing some $\lambda_\ell$ from among $\lambda_0, \ldots, \lambda_{s-1}$. That is, $\hat{s} = p_{h_\ell}^{\hat{\alpha}} \geq s$, for some $\ell \in \{0, \ldots, s-1\}$ and $\hat{\alpha} \leq \alpha_\ell$. Since $\lambda_{s-1} > s-1$, such $\hat{s}$ and $\lambda_\ell$ must exist.

Finally, for the given $\lambda_0, \ldots, \lambda_{s-1}$, define the language $L$ as follows:

$$L = \{a^K \mid \exists i \in \{0, \ldots, s-1\} : K \bmod \lambda_{K \bmod \hat{s}} = K \bmod \hat{s} = i\}.$$

---

[1] A subtle error in [1] has been fixed in [20]. The upper bounds $n^2 - 2$ for the number of states on the initial path and $n-1$ for the total number of states in the loops are presented, with a different proof of the Chrobak Normal Form, in [3].

It is not hard to see that the language $L$ can be expressed in the following form, which will turn out to be useful to prove our results:

$$L = \{a^K \mid P(K)\}, \quad \text{where}$$
$$P(K) \equiv \bigvee_{i=0}^{s-1} [\, K \bmod \hat{s} = i \,\wedge\, K \bmod \lambda_i = i \,]. \tag{1}$$

Since $\hat{s} \geq s$, the (conjunctive) clauses in the above predicate $P(K)$ are pairwise disjoint for each input string $a^K$.

**Lemma 1.** Let $a^K$ be a string with $K \bmod \hat{s} = j$, for some $j \in \{0, \ldots, s-1\}$. Then $a^K \in L$ if and only if $K \bmod \lambda_j = j$.

*Proof.* Assume that $K \bmod \hat{s} = j$, for some $j \in \{0, \ldots, s-1\}$. Then

$$P(K) \equiv \bigvee_{i=0}^{s-1} [\, K \bmod \hat{s} = i \,\wedge\, K \bmod \lambda_i = i \,]$$
$$\equiv [\, K \bmod \hat{s} = j \,\wedge\, K \bmod \lambda_j = j \,] \equiv [\, K \bmod \lambda_j = j \,],$$

using $[K \bmod \hat{s} = i] \equiv \mathsf{false}$ for $i \neq j$, together with $[K \bmod \hat{s} = j] \equiv \mathsf{true}$. □

Note that even for the special case of $j = \ell$, in which $\hat{s}$ divides $\lambda_j = \lambda_\ell$, the above lemma says that if $K \bmod \hat{s} = \ell$, then $a^K \in L$ if and only if $K \bmod \lambda_\ell = \ell$.

On the other hand, the lemma does not cover the case of $K \bmod \hat{s} = j \in \{s, \ldots, \hat{s}-1\}$, in which $a^K \notin L$.

**Theorem 1.** *The language $L$, defined by (1) above, is properly $\lambda$-cyclic, i.e., the transition graph of its optimal DFA is a loop consisting of $\lambda = \lambda_0 \cdots \lambda_{s-1}$ states.*

*Proof.* From the definition, it is easy to see that $L$ is $\lambda$-cyclic. To show it is properly $\lambda$-cyclic, it is enough to prove that, for each $\lambda' \geq 1$, if $L$ is $\lambda'$-cyclic then $\lambda'$ must be a multiple of $\lambda_i$, for $i = 0, \ldots, s-1$.

For such given $\lambda'$, let $\nu$ be the smallest integer such that $\hat{s}$ divides $\nu\lambda'$. Hence, $\nu = \frac{\hat{s}}{\gcd(\hat{s},\lambda')}$. Since $\nu$ divides $\hat{s}$ which in turn divides $\lambda_\ell$, $\gcd(\nu, \lambda_j) = 1$ for each $j \in \{0, \ldots, s-1\} \setminus \{\ell\}$. It is also easy to see that $a^j \in L$ for $j \in \{0, \ldots, s-1\}$, since $j < s \leq \hat{s}$ and $j < \lambda_j$, which gives $j \bmod \hat{s} = j$ and $j \bmod \lambda_j = j$. But then also $a^{j+\nu\lambda'} \in L$. Now, since $\hat{s}$ divides $\nu\lambda'$, we get $(j+\nu\lambda') \bmod \hat{s} = j \bmod \hat{s} = j$. Combining this with $a^{j+\nu\lambda'} \in L$, we thus get $(j+\nu\lambda') \bmod \lambda_j = j$. But then $(\nu\lambda') \bmod \lambda_j = ((j+\nu\lambda')-j) \bmod \lambda_j = j - j = 0$, and hence $\lambda_j$ divides $\nu\lambda'$. In the case of $j \neq \ell$, with $\gcd(\nu, \lambda_j) = 1$, this implies that $\lambda_j$ divides $\lambda'$.

It only remains to show that $\lambda_\ell$ also divides $\lambda'$. Let us fix some $K \geq 0$ so that $K \bmod \lambda_j = (j-1) \bmod \lambda_j$, for each $j \in \{0, \ldots, s-1\} \setminus \{\ell\}$, but $K \bmod \lambda_\ell = \ell$. By the Chinese Remainder Theorem, such $K$ must exist. Since $\lambda_\ell$ is an integer multiple of $\hat{s}$ and $\ell < \hat{s}$, we have also $K \bmod \hat{s} = \ell$. This gives that $a^K \in L$. But then $a^{K+\lambda/\lambda_\ell \cdot \lambda'}$ must also be in $L$. However, for $j \in \{0, \ldots, s-1\} \setminus \{\ell\}$, $\lambda_j$ divides $\lambda/\lambda_\ell$, and hence $(K + \lambda/\lambda_\ell \cdot \lambda') \bmod \lambda_j = K \bmod \lambda_j = (j-1) \bmod \lambda_j \neq j$. Summing up, $a^{K+\lambda/\lambda_\ell \cdot \lambda'} \in L$ but $(K + \lambda/\lambda_\ell \cdot \lambda') \bmod \lambda_j \neq j$ for $j \neq \ell$, which leaves us a single satisfiable possibility, namely, that $(K + \lambda/\lambda_\ell \cdot \lambda') \bmod \lambda_\ell = \ell$. Combining this with $K \bmod \lambda_\ell = \ell$, we get that $(\lambda/\lambda_\ell \cdot \lambda') \bmod \lambda_\ell = 0$. However, $\gcd(\lambda_\ell, \lambda/\lambda_\ell) = 1$, and hence $\lambda_\ell$ must divide $\lambda'$. □

**Theorem 2.** *The language $L$, defined by (1) above, can be accepted by an unambiguous* NFA *in Chrobak Normal Form consisting of at most $1 + \lambda_\ell + \hat{s} \cdot \sum_{i=0, i \neq \ell}^{s-1} \lambda_i$ states.*

*Proof.* From the definition of $L$ by (1), it turns out that $L$ can be accepted by an NFA in Chrobak Normal Form with the initial path consisting just of a single initial state, in which it nondeterministically chooses one of $s$ loops. The $i$-th loop, for $i \in \{0, \ldots, s-1\}$, verifies the truth of the $i$-th clause in (1), i.e., whether $[K \bmod \hat{s} = i \wedge K \bmod \lambda_i = i]$, for the given input $a^K$. This condition can be verified by a loop counting modulo $\hat{s}$ and, at the same time, modulo $\lambda_i$, thus consisting of $\mathrm{lcm}(\hat{s}, \lambda_i)$ states. Clearly, $\mathrm{lcm}(\hat{s}, \lambda_i) = \hat{s} \cdot \lambda_i$ for $i \neq \ell$, but $\mathrm{lcm}(\hat{s}, \lambda_\ell) = \lambda_\ell$. Hence, the total number of the states in our NFA is $1 + \lambda_\ell + \hat{s} \cdot \sum_{i=0, i \neq \ell}^{s-1} \lambda_i$.

In addition, it is easy to see that the clauses in (1) are pairwise disjoint, and hence the above machine does not have more than one accepting computation path, for any accepted input $a^K$. Therefore, our NFA is unambiguous. □

Now we focus our attention on the complement of $L$.

**Theorem 3.** *The complement of the language $L$, defined by (1) above, can be accepted by an unambiguous* NFA *in Chrobak Normal Form with the same transition graph as the* NFA *of Theorem 2, and hence with at most $1 + \lambda_\ell + \hat{s} \cdot \sum_{i=0, i \neq \ell}^{s-1} \lambda_i$ states.*

*Proof.* By the definition of $L$, $a^K \in L$ if and only if $P(K) \equiv \mathsf{true}$, where $P(K)$ denotes the predicate introduced in (1). In order to cover all possible cases, including the case of $K \bmod \hat{s} \in \{s, \ldots, \hat{s}-1\}$, let us rewrite $P(K)$ as follows.

$$P(K) \equiv \bigvee_{i=0}^{s-1} [K \bmod \hat{s} = i \wedge K \bmod \lambda_i = i]$$
$$\vee \quad [K \bmod \hat{s} \in \{s, \ldots, \hat{s}-1\} \wedge \mathsf{false}].$$

Now $P(K)$ is expressed in the form $P(K) \equiv \bigvee_{i=0}^{s} [\alpha_i(K) \wedge \beta_i(K)]$, where

- $\alpha_i(K) \equiv [K \bmod \hat{s} = i]$ for $i = 0, \ldots, s-1$, but $\alpha_s(K) \equiv [K \bmod \hat{s} \in \{s, \ldots, \hat{s}-1\}]$,
- $\beta_i(K) \equiv [K \bmod \lambda_i = i]$ for $i = 0, \ldots, s-1$, but $\beta_s(K) \equiv [\mathsf{false}]$.

Note that exactly one of the predicates $\alpha_0(K), \ldots, \alpha_s(K)$ evaluates to true. More precisely, for each $K$, $\bigvee_{i=0}^{s} \alpha_i(K) \equiv \mathsf{true}$, but $\alpha_i(K) \wedge \alpha_j(K) \equiv \mathsf{false}$, for $i \neq j$. For such exhaustive pairwise disjoint enumeration of all possibilities, it is easy to see that the negated predicate is $\neg P(K) \equiv \bigvee_{i=0}^{s} [\alpha_i(K) \wedge \neg \beta_i(K)]$. This gives

$$\neg P(K) \equiv \bigvee_{i=0}^{s-1} [K \bmod \hat{s} = i \wedge K \bmod \lambda_i \neq i]$$
$$\vee \quad [K \bmod \hat{s} \in \{s, \ldots, \hat{s}-1\} \wedge \mathsf{true}]$$
$$\equiv \bigvee_{i=0}^{s-2} [K \bmod \hat{s} = i \wedge K \bmod \lambda_i \neq i]$$
$$\vee \quad [[K \bmod \hat{s} = s-1 \wedge K \bmod \lambda_{s-1} \neq s-1]$$
$$\vee \ K \bmod \hat{s} \in \{s, \ldots, \hat{s}-1\}].$$

But then the complementary language $L^c = \{a^K \mid \neg P(K)\}$ can also be accepted by an NFA in Chrobak Normal Form choosing nondeterministically one of $s$ loops in its initial state. The $i$-th loop, for $i \in \{0, \ldots, s-2\}$, verifies whether $[\, K \bmod \hat{s} = i \wedge K \bmod \lambda_i \neq i\,]$, while the $(s-1)$-st loop verifies the truth of the clause $[\,[\, K \bmod \hat{s} = s-1 \wedge K \bmod \lambda_{s-1} \neq s-1\,] \vee K \bmod \hat{s} \in \{s, \ldots, \hat{s}-1\}\,]$. Clearly, these $s$ clauses are pairwise disjoint, and hence our NFA is unambiguous.

For each $i \in \{0, \ldots, s-1\}$, including $i = s-1$, the $i$-th clause can be verified by a loop counting modulo $\hat{s}$ and $\lambda_i$, with $\mathrm{lcm}(\hat{s}, \lambda_i)$ states. Thus, the number of loops and their lengths are the same as in the NFA of Theorem 2, which results in the same transition graph and in the same number of states, namely, $1 + \lambda_\ell + \hat{s} \cdot \sum_{i=0, i\neq\ell}^{s-1} \lambda_i$. However, the NFAs for $L$ and $L^c$ differ in the distribution of their final states.                                                                                    □

It should be easily seen that the number of loops (hence, the total number of states as well) can be reduced, if $\hat{s} = \lambda_\ell$ and $s > 1$. Then $\lambda_\ell$ divides the length of some other loop, namely, $\mathrm{lcm}(\hat{s}, \lambda_i)$, for some $i \neq \ell$, and hence the loop of length $\lambda_\ell$ can be removed from the list $\lambda_0, \ldots, \lambda_{s-1}$. This holds for NFAs presented both in Theorem 2 and in Theorem 3.

We also point out that, if $s > 1$ then the general lower bound on the number of states for any NFA accepting $L$ or $L^c$ is $1 + \sum_{i=0}^{s-1} \lambda_i$. This follows from [6, Cor. 2.1], because of the fact that the *smallest deterministic machine* for $L$, presented in Theorem 1, uses a loop of length $\lambda = \lambda_0 \cdots \lambda_{s-1}$, where the list $\lambda_0, \ldots, \lambda_{s-1}$ represents the factorization of $\lambda$.

## 4    The Superpolynomial Gap

Here we shall prove our main result, namely, the superpolynomial gap in the number of the states, between two NFAs accepting some unary language $L$ and its complement $L^c$ on one side, and the minimal DFAs for these two languages on the other one. The gap holds even if we restrict ourselves to cyclic unary languages and, moreover, both $L$ and $L^c$ require the corresponding NFAs to be unambiguous.

To achieve this result, we shall make use of the *Landau Function* $F(n)$ [8,9]. This function, the investigation of which was initially motivated by the group theory, plays an important role in analysis of simulations among various models of unary automata [1,3,12]. We now recall the definition of $F(n)$ and present some of its properties, required later. Given a positive integer $n$, let

$$F(n) = \max\{\mathrm{lcm}(\lambda_0, \ldots, \lambda_{s-1}) \mid \lambda_0 + \cdots + \lambda_{s-1} = n\},$$

where $\lambda_0, \ldots, \lambda_{s-1}$ denote, for the time being, arbitrary positive integers. Szalay [19] gave a sharp estimation of $F(n)$ that, after some simplification, can be formulated as follows:[2]

---

[2] The asymptotic notation in the formula (and, throughout the paper) is interpreted as follows. The exact value of $F(n)$ can be expressed in the form $F(n) = e^{(1+r(n))\cdot\sqrt{n\cdot\ln n}}$, for some real function $r(n)$ satisfying $\lim_{n\to\infty} r(n) = 0$, and $r(n) \geq 0$ for each $n$. If we do not claim $r(n) \geq 0$, we use $e^{(1\pm o(1))\cdot\sqrt{n\cdot\ln n}}$ instead. The meaning of $e^{(1-o(1))\cdot\sqrt{n\cdot\ln n}}$ should be obvious.

$$F(n) = e^{(1+o(1))\cdot\sqrt{n\cdot\ln n}}.\tag{2}$$

Before proceeding further, we need to express $F(n)$ in a slightly different form. First, since $\mathrm{lcm}(\lambda_0,\ldots,\lambda_{s-1}) = \mathrm{lcm}(\lambda_0,\ldots,\lambda_{s-1},1,\ldots,1)$, the condition $\lambda_0 + \cdots + \lambda_{s-1} = n$ can be replaced by $\lambda_0 + \cdots + \lambda_{s-1} \leq n$ in the definition of $F(n)$. Second, if a number $\lambda_i$ is a product of some two integers $a > 1$ and $b > 1$, with $\gcd(a,b) = 1$, it can be replaced in the list $\lambda_0,\ldots,\lambda_{s-1}$ by these two numbers. This does not change the least common multiple of $\lambda_0,\ldots,\lambda_{s-1}$ nor, since $a+b \leq a\cdot b$, does it increase their sum. Third, if a number $\lambda_i$ divides $\lambda_j$, for some $i \neq j$, then the number $\lambda_i$ can be removed from the list $\lambda_0,\ldots,\lambda_{s-1}$ for the same reasons. In this way, we can restrict our attention to the lists of prime powers, with the underlying prime numbers all different. Finally, we can also assume that such lists are sorted with respect to their underlying primes. This gives:

$$F(n) = \max\left\{\prod_{i=0}^{s-1} p_{h_i}^{\alpha_i} \;\middle|\; \sum_{i=0}^{s-1} p_{h_i}^{\alpha_i} \leq n,\; p_{h_0}<\ldots<p_{h_{s-1}},\; \alpha_0,\ldots,\alpha_{s-1}>0\right\}.$$

The above formula clearly shows that $F(n) \leq F(n+1)$ for each $n$, i.e., $F(n)$ is monotone nondecreasing. (For more details, see [8,9,14,16].)

From this point forward, for a given integer $n$, let $\lambda_0,\ldots,\lambda_{s-1}$ denotes the list of prime powers, ordered with respect to their underlying primes, for which the function $F(n)$ reaches its maximum.[3] Clearly, $\lambda_i = p_{h_i}^{\alpha_i}$, for some positive integer $\alpha_i$ and some prime $p_{h_i}$. Now we recall some facts about these numbers:

**Lemma 2.**

  i. $p_{h_{s-1}} \sim \sqrt{n\cdot\ln n}$, that is,[4] $p_{h_{s-1}} = (1 \pm o(1)) \cdot \sqrt{n\cdot\ln n}$.
 ii. $s \sim 2\cdot\sqrt{n/\ln n}$, that is, $s = (2 \pm o(1)) \cdot \sqrt{n/\ln n}$.
iii. There can exist at most one prime smaller than $p_{h_{s-1}}/2$ that does not divide $F(n)$.

*Proof.* The items (i) and (ii) have been proved by Nicolas [16]. In the same paper, he proved also that, given three distinct primes $p, p', q$, with $p+p' \leq q$, such that $q$ divides $F(n)$, at least one of $p, p'$ must also divide $F(n)$. As an immediate corollary, Grantham [4] stated (iii).    □

Given an integer $n$, hence, given also a list of prime powers $\lambda_0,\ldots,\lambda_{s-1}$ for which the function $F(n)$ reaches its maximum, fix $\hat{s}$ and $\lambda_\ell$ in the same way as done in Section 3. Recall that $\hat{s}$ is the smallest integer satisfying $\hat{s} \geq s$ and, moreover,

---

[3] By the Fundamental Theorem of Arithmetic, this factorization of $F(n)$ into $\lambda_0,\ldots,\lambda_{s-1}$ is unique.

[4] The statement gives only the asymptotic behavior of $p_{h_{s-1}}$. For the sake of completeness, we mention that Grantham [4] proved that the largest prime $p_{h_{s-1}}$ in the factorization of $F(n)$ never exceeds the value $1.328\cdot\sqrt{n\cdot\ln n}$, for any integer $n$. It is also interesting to mention that the exponent for this largest prime is always $\alpha_{s-1} = 1$, with the only exception of $n = 4$ [16].

dividing some $\lambda_\ell$ from among $\lambda_0, \ldots, \lambda_{s-1}$. That is, we have $\lambda_\ell = p_{h_\ell}^{\alpha_\ell}$, for some $\ell \in \{0, \ldots, s-1\}$, and also $\hat{s} = p_{h_\ell}^{\hat{\alpha}} \geq s$, for some $\hat{\alpha} \leq \alpha_\ell$.

Now, by the use of the construction presented in Section 3, we can define a language $L$ such that:

- the language $L$ and its complement $L^c$ can be accepted by two NFAs using the same number of states, namely, $N \leq 1 + \lambda_\ell + \hat{s} \cdot \sum_{i=0, i \neq \ell}^{s-1} \lambda_i$, as shown by Theorems 2 and 3, respectively,
- the smallest DFA accepting $L$ (or $L^c$) must use $F(n) = \lambda = \lambda_0 \cdots \lambda_{s-1}$ states, by Theorem 1.

In order to evaluate the gap, let us now express $F(n)$, the number of states in the DFAs for $L$ and $L^c$, as a function of $N$, the number of states in our NFAs for these two languages. To this aim, we prove the following bounds for $\hat{s}$:

**Lemma 3.** *For each sufficiently large $n$, $2 < s \leq \hat{s} \leq 2s$.*

*Proof.* First, by (ii) and (i) in Lemma 2, we have $s \leq (2 + o(1)) \cdot \sqrt{n / \ln n}$, but $p_{h_{s-1}} \geq (1 - o(1)) \cdot \sqrt{n \cdot \ln n}$. From this we can conclude, for each sufficiently large $n$, that $2s < p_{h_{s-1}}/2$. Similarly, since $s \geq (2 - o(1)) \cdot \sqrt{n / \ln n}$, we get also $2 < \frac{11}{2} \leq s$.

But, by [18], for each $s \geq \frac{11}{2}$, there must exist at least two different primes $p', p''$ between $s$ and $2s$. We thus get two primes satisfying $s \leq p' < p'' \leq 2s < p_{h_{s-1}}/2$.

Now, by statement (iii) in Lemma 2, we can conclude that at least one of $p', p''$ must be a factor of $F(n)$, thus dividing some $\lambda_\ell$ from among $\lambda_0, \ldots, \lambda_{s-1}$ and, moreover, this divisor is positioned in between $s$ and $2s$.

But then $\hat{s}$, the smallest divisor of $F(n)$ above $s$, must also satisfy $s \leq \hat{s} \leq 2s$. $\qquad \square$

Using Lemmas 3 and 2, we are now able to derive the following inequalities:

$$N \leq 1 + \lambda_\ell + \hat{s} \cdot \sum_{i=0, i \neq \ell}^{s-1} \lambda_i \leq \hat{s} \cdot \sum_{i=0}^{s-1} \lambda_i \leq \hat{s} \cdot n \leq 2s \cdot n$$
$$\leq (4 + o(1)) \cdot \sqrt{n / \ln n} \cdot n \leq (4 + o(1)) \cdot n^{3/2} / \ln^{1/2} n \qquad (3)$$

Note that, for sufficiently large $n$, we must get $N \leq n^{3/2}$. But then:

$$\ln^2 N \leq (\ln n^{3/2})^2 \leq \tfrac{9}{4} \cdot \ln^2 n \qquad (4)$$

Now, by combining (3) with (4), we obtain:

$$N \cdot \ln^2 N \leq (4 + o(1)) \cdot n^{3/2} / \ln^{1/2} n \cdot \tfrac{9}{4} \cdot \ln^2 n \leq (9 + o(1)) \cdot (n \cdot \ln n)^{3/2}$$

This gives:

$$(n \cdot \ln n)^{1/2} \geq \left( \tfrac{1}{9 + o(1)} \cdot N \cdot \ln^2 N \right)^{1/3} \geq \left( (\tfrac{1}{9} - o(1)) \cdot N \cdot \ln^2 N \right)^{1/3}$$
$$\geq \left( \tfrac{1}{9^{1/3}} - o(1) \right) \cdot N^{1/3} \cdot \ln^{2/3} N$$

Now, using the approximation for the Landau Function, presented by (2), we obtain:

$$F(n) = e^{(1+o(1))\cdot\sqrt{n\cdot\ln n}} \geq e^{\sqrt{n\cdot\ln n}} \geq e^{(1/\sqrt[3]{9}-o(1))\cdot\sqrt[3]{N\cdot\ln^2 N}}$$

Summing up, for each integer $n$, we have found an integer $N = N_n$ and a cyclic unary language $L_{N_n}$ such that:

- the languages $L_{N_n}$ and its complement $L_{N_n}^c$ can be accepted by two unambiguous NFAs using at most $N_n$ states,
- the smallest DFA accepting the language $L_{N_n}$ (or $L_{N_n}^c$) must use at least $F(n) \geq e^{(1/\sqrt[3]{9}-o(1))\cdot\sqrt[3]{N_n\cdot\ln^2 N_n}}$ states.

It is not difficult to prove that the sequence $N_1, N_2, N_3, \ldots$ must contain infinitely many different integers. Actually, we can extend the gap to *all sufficiently large* integers:

**Theorem 4.** *For each sufficiently large $N$, there exists a cyclic unary language $L_N$ such that:*

- *the languages $L_N$ and its complement $L_N^c$ can be accepted by two unambiguous NFAs using at most $N$ states,*
- *the smallest DFA accepting the language $L_N$ (or $L_N^c$) must use at least $\frac{1}{2}\cdot e^{(1/\sqrt[3]{9}-o(1))\cdot\sqrt[3]{N\cdot\ln^2 N}} \geq e^{\Omega(\sqrt[3]{N\cdot\ln^2 N})}$ states.*

*Proof.* First, the sequence $F(1), F(2), \ldots$ is unbounded, i.e., $\lim_{n\to\infty} F(n) = \infty$, by (2). Second, the smallest DFA accepting the language $L_{N_n}$ must use at least $F(n)$ states but, at the same time, $L_{N_n}$ can be accepted by an NFA using no more than $N_n$ states. But then the sequence $N_1, N_2, \ldots$ must also be unbounded. Hence, for each sufficiently large $N$, satisfying among others $N \geq N_1$, we can find an integer $n$ such that $N_n \leq N < N_{n+1}$.

Now we choose $L_N = L_{N_n}$. Clearly, $L_N$ and $L_N^c$ are accepted by two unambiguous NFAs using at most $N$ states. Furthermore, by [16, Cor. p. 319], $F(n) \geq \frac{1}{2}\cdot F(n+1)$. Hence, the number of the states in the smallest DFA accepting $L_N$ (or $L_N^c$) is

$$F(n) \geq \tfrac{1}{2}\cdot F(n+1) \geq \tfrac{1}{2}\cdot e^{(1/\sqrt[3]{9}-o(1))\cdot\sqrt[3]{N_{n+1}\cdot\ln^2 N_{n+1}}} \geq \tfrac{1}{2}\cdot e^{(1/\sqrt[3]{9}-o(1))\cdot\sqrt[3]{N\cdot\ln^2 N}}.$$
□

## 5    Self-verifying Automata

Let us now consider a special kind of finite automata with a symmetric form of nondeterminism, introduced in [2]. A *self-verifying automaton* (SVFA) is a nondeterministic automaton equipped with two kinds of final states, namely, the set of *accepting* states $F^+$ and the set of *rejecting* states $F^-$. All remaining states in $Q$ are *neutral*. By definition, for each $w \in L$, there must exist at least one accepting computation path halting in some $q \in F^+$, but no path may halt in a state $q \in F^-$. Symmetrically, for each $w \in L^c$, at least one rejecting path halts in $q \in F^-$, but no path may halt in $q \in F^+$.

We shall consider even a more restricted form of nondeterminism: the given SVFA is called *unambiguous*, if it has exactly one computation path ending in $F^+$ or $F^-$, for any input $w$.

As a special case of NFAs, SVFAs can be converted into equivalent DFAs by the usual techniques. However, the cost of the conversion can be reduced [7]: each $n$-state SVFA can be converted into a DFA with $O(3^{n/3}) \approx O(1.45^n)$ states. Moreover, this cost was shown to be tight, *if the input alphabet consists of at least of two letters*, leaving open the unary case. We remind that the tight upper bound for the conversion of unary NFAs into DFAs is given by the Landau Function [1], with the growth rate only $F(n) = e^{(1+o(1))\cdot\sqrt{n\cdot\ln n}}$. However, as observed in [7], the cost for the unary SVFAs must be strictly smaller than $F(n)$. Here we show that this cost is still superpolynomial, with a lower bound not too far from $F(n)$.

Before doing this, let us begin with a general construction, converting two given NFAs $A^+$ and $A^-$, accepting a language $L$ and its complement $L^c$, respectively, into equivalent SVFA $A$ for $L$. As pointed out in [5], $A$ can be obtained by introducing a new initial state, connected to copies of $A^+$ and $A^-$ with suitable transitions. The total number of the states in $A$ is thus the sum of the corresponding numbers in the two original automata, plus 1. In the next lemma (which holds for any input alphabet), we show that if $A^+$ and $A^-$ share the same transition graph, we do not need to duplicate this graph.

**Lemma 4.** *Let $A^+$ and $A^-$ be two NFAs accepting, respectively, a language $L$ and its complement $L^c$. If $A^+$ and $A^-$ have the same transition graph, then $L$ can be accepted by an SVFA $A$ that also uses the same transition graph.*

*Moreover, if both $A^+$ and $A^-$ are unambiguous, then so is $A$.*

*Proof.* First, without loss of generality, we assume that neither $A^+$ nor $A^-$ have unreachable states. Now we define $A$ by taking the same states and the same transitions as in $A^+$ and $A^-$, choosing as a set of accepting states $F^+$ the set of final states in $A^+$, and as a set of rejecting states $F^-$ the set of final states in $A^-$. All remaining states become neutral. Using the fact that there are no unreachable states and that each input string $w$ must be accepted either by $A^+$ or by $A^-$, but not by both, the sets of accepting and rejecting states must be disjoint, and the resulting automaton must be an SVFA for $L$. (See Figure 1 for an example.)

Moreover, if both $A^+$ and $A^-$ are unambiguous, i.e., if $A^+$ has exactly one accepting computation path for any $w \in L$ and, at the same time, $A^-$ has exactly one such path for any $w \in L^c$, then $A$ must have exactly one path ending in $F^+$ or $F^-$ for any input.                                                                                      □

The two unambiguous NFAs used in Theorem 4 have the same transition graph. Thus, by Lemma 4, we get the following:

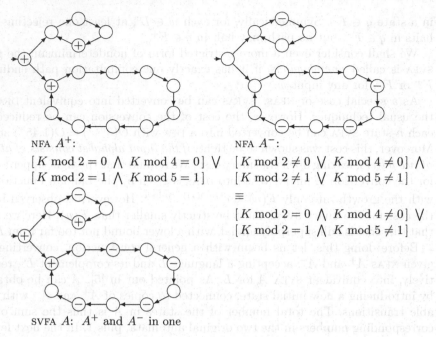

NFA $A^+$:

$[K \bmod 2 = 0 \wedge K \bmod 4 = 0] \bigvee$
$[K \bmod 2 = 1 \wedge K \bmod 5 = 1]$

NFA $A^-$:

$[K \bmod 2 \neq 0 \bigvee K \bmod 4 \neq 0] \wedge$
$[K \bmod 2 \neq 1 \bigvee K \bmod 5 \neq 1]$

$\equiv$

$[K \bmod 2 = 0 \wedge K \bmod 4 \neq 0] \bigvee$
$[K \bmod 2 = 1 \wedge K \bmod 5 \neq 1]$

SVFA $A$: $A^+$ and $A^-$ in one

**Fig. 1.** An example with $n = 9$ and $F(n) = 20 = 2^2 \cdot 5$. This gives $\lambda_0 = 4$ and $\lambda_1 = 5$, and hence $\hat{s} = 2$, $\mathrm{lcm}(\hat{s}, \lambda_0) = 4$, $\mathrm{lcm}(\hat{s}, \lambda_1) = 10$. The NFA $A^+$ for $L$, top left, has its final states marked by "+", while the NFA $A^-$ for $L^c$, top right, by "−". The resulting SVFA $A$ is depicted on bottom. The smallest DFA uses $\lambda_0 \cdot \lambda_1 = 20$ states.

**Corollary 1.** *For each sufficiently large $N$, there exists a cyclic unary language $L_N$ such that:*

- *the languages $L_N$ and its complement $L_N^c$ can be accepted by two unambiguous NFAs using the same transition graph, of at most $N$ states,*
- *$L_N$ (and $L_N^c$) can also be accepted by an unambiguous SVFA with the same transition graph, and hence with the same number of states, at most $N$,*
- *the smallest DFA accepting the language $L_N$ (or $L_N^c$) must use at least $\frac{1}{2} \cdot e^{(1/\sqrt[3]{9} - o(1)) \cdot \sqrt[3]{N \cdot \ln^2 N}} \geq e^{\Omega(\sqrt[3]{N \cdot \ln^2 N})}$ states.*

The above corollary permits us to conclude that the following gaps are super-polynomial, even for unary cyclic languages:

- between the total size of two NFAs accepting a language and its complement on one side, and the size of the corresponding DFA on the other one,
- between the sizes of unambiguous NFAs and DFAs,
- between the sizes of SVFAs and DFAs,
- between the sizes of unambiguous SVFAs and DFAs.

**Acknowledgment.** The authors would like to thank the anonymous referee for helping to simplify the proof of Theorem 1.

# References

1. Chrobak, M.: Finite automata and unary languages. Theoretical Computer Science 47, 149–158 (1986); Corrigendum: *IBID.* 302, 497–498 (2003)
2. Ďuriš, P., Hromkovič, J., Rolim, J., Schnitger, G.: Las Vegas versus determinism for one-way communication complexity, finite automata, and polynomial-time computations. In: Reischuk, R., Morvan, M. (eds.) STACS 1997. LNCS, vol. 1200, pp. 117–128. Springer, Heidelberg (1997)
3. Geffert, V.: Magic numbers in the state hierarchy of finite automata. Information and Computation 205, 1652–1670 (2007)
4. Grantham, J.: The largest prime dividing the maximal order of an element of $S_n$. Math. Comp. 64, 407–410 (1995)
5. Hromkovič, J., Schnitger, G.: On the power of Las Vegas for one-way communication complexity, OBDDs, and finite automata. Information and Computation 2, 284–296 (2001)
6. Jiang, T., McDowell, E., Ravikumar, B.: The structure and complexity of minimal nfa's over a unary alphabet. International Journal of Foundations of Computer Science 2, 163–182 (1991)
7. Jirásková, G., Pighizzini, G.: Converting self-verifying automata into deterministic automata. In: Dediu, A.H., Ionescu, A.M., Martín-Vide, C. (eds.) LATA 2009. LNCS, vol. 5457, pp. 458–468. Springer, Heidelberg (2009)
8. Landau, E.: Über die Maximalordnung der Permutation gegebenen Grades. Archiv der Math. und Phys. 3, 92–103 (1903)
9. Landau, E.: Handbuch der Lehre von der Verteilung der Primzahlen I. Teubner, Leipzig/Berlin (1909)
10. Lupanov, O.B.: A comparison of two types of finite automata. Problemy Kibernetiki 9, 321–326 (1963) (in Russian); German translation: Über den Vergleich zweier Typen endlicher Quellen, Probleme der Kybernetik 6, 329–335 (1966)
11. Mera, F., Pighizzini, G.: Complementing unary nondeterministic automata. Theoretical Computer Science 330, 349–360 (2005)
12. Mereghetti, C., Pighizzini, G.: Optimal simulations between unary automata. SIAM J. Computing 30, 1976–1992 (2001)
13. Meyer, A.R., Fischer, M.J.: Economy of description by automata, grammars, and formal systems. In: Proc. 12th Ann. IEEE Symp. on Switching and Automata Theory, pp. 188–191 (1971)
14. Miller, W.: The maximum order of an element of a finite symmetric group. American Mathematical Monthly 94, 497–506 (1987)
15. Moore, F.: On the bounds for state-set size in the proofs of equivalence between deterministic, nondeterministic, and two-way finite automata. IEEE Transactions on Computers C-20(10), 1211–1214 (1971)
16. Nicolas, J.-L.: Sur l'ordre maximum d'un élément dans le groupe $S_n$ des permutations. Acta Aritmetica 14, 315–332 (1968)
17. Rabin, M., Scott, D.: Finite automata and their decision problems. IBM J. Res. Develop. 3, 114–125 (1959)
18. Ramanujan, S.: A proof of Bertrand's postulate. Journal of the Indian Mathematical Society 11, 181–182 (1919)
19. Szalay, M.: On the maximal order in $S_n$ and $S_n^*$. Acta Aritmetica 37, 321–331 (1980)
20. To, A.W.: Unary finite automata vs. arithmetic progressions. Information Processing Letters 109, 1010–1014 (2009)

# Quotient Complexity of Ideal Languages[*]

Janusz Brzozowski[1], Galina Jirásková[2], and Baiyu Li[1]

[1] David R. Cheriton School of Computer Science, University of Waterloo,
Waterloo, ON, Canada N2L 3G1
{brzozo@,b5li@student.cs.}uwaterloo.ca
[2] Mathematical Institute, Slovak Academy of Sciences,
Grešákova 6, 040 01 Košice, Slovakia
jiraskov@saske.sk

**Abstract.** We study the state complexity of regular operations in the class of ideal languages. A language $L \subseteq \Sigma^*$ is a right (left) ideal if it satisfies $L = L\Sigma^*$ ($L = \Sigma^* L$). It is a two-sided ideal if $L = \Sigma^* L\Sigma^*$, and an all-sided ideal if $L = \Sigma^* \shuffle L$, the shuffle of $\Sigma^*$ with $L$. We prefer "quotient complexity" to "state complexity", and we use quotient formulas to calculate upper bounds on quotient complexity whenever it is convenient. We find tight upper bounds on the quotient complexity of each type of ideal language in terms of the complexity of an arbitrary generator and of its minimal generator, the complexity of the minimal generator, and also on the operations union, intersection, set difference, symmetric difference, concatenation, star and reversal of ideal languages.

**Keywords:** automaton, complexity, ideal, language, quotient, state complexity, regular expression, regular operation, upper bound.

## 1 Introduction

The *state complexity of a regular language* $L$ is the number of states in the minimal deterministic finite automaton (dfa) recognizing $L$. The *state complexity of an operation* $f(K, L)$ or $g(L)$ in a subclass $\mathcal{C}$ of regular languages is the maximal state complexity of the language $f(K, L)$ or $g(L)$, respectively, when $K$ and $L$ have state complexities $m$ and $n$, respectively, and range over all languages in $\mathcal{C}$. For a detailed discussion of general issues of state complexity see [5,27] and the reference lists in those papers. Here we briefly mention the previous work on the complexity of operations in the class of regular languages and its subclasses.

The bound $mn$ on the complexity of intersection was noted in 1959 by Rabin and Scott [22]. The state complexity of union, product and star was first studied in 1970 by Maslov [17]. He stated the upper bounds on these operations and gave examples of languages meeting these bounds, but provided no proofs. In 1981 Leiss [14] showed that the $2^n$ bound for reversal can be met. In 1991

---

[*] This work was supported by the Natural Sciences and Engineering Research Council of Canada grant OGP0000871 and by VEGA grant 2/0111/09.

A. López-Ortiz (Ed.): LATIN 2010, LNCS 6034, pp. 208–221, 2010.

Birget [2] introduced the term "state complexity" and studied the complexity of the intersection of $n$ languages. In 1994 Yu, Zhuang and K. Salomaa [28] examined the complexity of concatenation, star, left and right quotients, reversal, intersection, and union in the class or regular languages.

The complexity of operations was also considered in several subclasses of regular languages: unary [21,28], finite [9,27], prefix-free [13], suffix-free [12], and closed [7]. These studies of subclasses show that the complexity can be significantly lower in a subclass of regular languages than in the general case.

Here we continue the study of the state complexity of operations in subclasses. Specifically, we examine four related classes of regular languages: right, left, two-sided, and all-sided ideals, defined below. Ideals are chosen for several reasons. They are special subsets of semigroups, well-known in semigroup theory [16,24]. They appear in the theoretical computer science literature as early as 1965 [19] and continue to be of interest [1,3]. Ideal languages are complements of prefix-, suffix-, factor-, and subword-closed languages—another interesting class [1,7], and are closed with respect to the "has a word as a prefix" (respectively, suffix, factor, subword) relation [1]. They are special cases of convex languages [1,25], a much larger class. The fact that the four classes of ideals are related to each other permits us to obtain many complexity results using similar methods.

Our interest is in regular ideal languages. Left and right ideals were studied by Paz and Peleg [19] in 1965 under the names "ultimate definite" and "reverse ultimate definite events". The results in [19] include closure properties, decision procedures, and canonical representations for these languages. All-sided ideals were used by Haines [11] (not under that name) in 1969 in connection with subword-free and subword-closed languages, and by Thierrin [25] in 1973 in connection with subword-convex languages. De Luca and Varricchio [15] showed in 1990 that a language is factor-closed (also called "factorial") if and only if it is the complement of a two-sided ideal. In [28] there are two results about left and right ideals. In 2001 Shyr [24] studied right, left, and two-sided ideals and their generators in connection with codes. In 2007 Okhotin [18] presented a result concerning all-sided ideals. In 2008 all four types of ideals were considered by Ang and Brzozowski [1] in the framework of languages convex with respect to arbitrary binary relations. Decision problems for various classes of convex languages, including ideals, were addressed in [8]. Complexity issues of nfa to dfa conversion in right, left, and two-sided ideals were studied in 2008 by Bordihn, Holzer, and Kutrib [3], under the names "ultimate definite", "reverse ultimate definite", and "central definite" languages, respectively. The closure properties of ideals were analyzed in [1].

Our approach to state complexity follows closely that of [5]. Since state complexity is a property of a language, it is more appropriately defined in language-theoretic terms. Hence we use the equivalent concept of *quotient complexity* as explained below. Quotient complexity was introduced in 2009 [5], but quotient methods were used there mainly to derive known results in a new way. In the present paper, we derive many new complexity bounds using quotients.

## 2   Ideal Languages and Quotient Complexity

We assume that the reader is familiar with basic concepts of regular languages and finite automata, as described in [20,26], for example, or in many textbooks. For general properties of ideal languages we refer the reader to [16,24].

If $\Sigma$ is a non-empty finite alphabet, then $\Sigma^*$ is the free monoid generated by $\Sigma$. A *word* is any element of $\Sigma^*$, and the empty word is $\varepsilon$. The length of a word $w \in \Sigma^*$ is $|w|$. A *language* over $\Sigma$ is any subset of $\Sigma^*$.

If $u, v, w \in \Sigma^*$ and $w = uxv$, then $u$ is a *prefix* of $w$, $v$ is a *suffix* of $w$, and $x$ is a *factor* of $w$. If $w = w_0 a_1 w_1 \cdots a_n w_n$, where $a_1, \ldots, a_n \in \Sigma$, and $w_0, \ldots, w_n \in \Sigma^*$, then $v = a_1 \cdots a_n$ is a *subword* of $w$.

A language $L$ is *prefix-free* (*prefix-closed*) if $w \in L$ implies that no proper prefix of $w$ is in $L$ (that every prefix of $w$ is in $L$). In the same way, we define *suffix-free*, *factor-free*, and *subword-free*, and the corresponding closed versions.

A language $L \subseteq \Sigma^*$ is a *right ideal* (*left ideal, two-sided ideal, all-sided ideal*) if it is non-empty and satisfies $L = L\Sigma^*$ ($L = \Sigma^* L$, $L = \Sigma^* L \Sigma^*$, $L = \Sigma^* \shuffle L$, respectively), where $\shuffle$ is the shuffle operator. We refer to all four types as *ideal languages* or simply *ideals*.

The following set operations are defined on languages: *complement* ($\overline{L} = \Sigma^* \setminus L$), *union* ($K \cup L$), *intersection* ($K \cap L$), *difference* ($K \setminus L$), and *symmetric difference* ($K \oplus L$). A general *boolean operation* with two arguments is denoted by $K \circ L$. We also define the *product*, usually called *concatenation* or *catenation*, ($KL = \{w \in \Sigma^* \mid w = uv, u \in K, v \in L\}$), *positive closure* ($L^+ = \bigcup_{i \geq 1} L^i$), and *star* ($L^* = \bigcup_{i \geq 0} L^i$). The reverse $w^R$ of a word $w \in \Sigma^*$ is defined as follows: $\varepsilon^R = \varepsilon$, and $(wa)^R = aw^R$. The *reverse* of a language $L$ is denoted by $L^R$ and defined as $L^R = \{w^R \mid w \in L\}$.

*Regular languages* over an alphabet $\Sigma$ are languages that can be obtained from the *basic languages* $\emptyset$, $\{\varepsilon\}$, and $\{a\}$, $a \in \Sigma$, using a finite number of operations of union, product and star. Such languages are usually denoted by regular expressions. If $E$ is a regular expression, then $\mathcal{L}(E)$ is the language denoted by that expression. For example, $E = (\varepsilon \cup a)^* b$ denotes $L = (\{\varepsilon\} \cup \{a\})^* \{b\}$. We use the regular expression notation for both expressions and languages.

The *left quotient*, or simply *quotient*, of a language $L$ by a word $w$ is the language $L_w = \{x \in \Sigma^* \mid wx \in L\}$. The *quotient complexity* of $L$ is the number of distinct quotients of $L$, and is denoted by $\kappa(L)$.

We now describe the computation of quotients of a regular language. First, the $\varepsilon$-*function* $L^\varepsilon$ of a regular language $L$ is equal to $\varepsilon$ if $\varepsilon \in L$, and to $\emptyset$ otherwise. The quotient by a letter $a$ in $\Sigma$ is computed by structural induction: $b_a = \emptyset$ if $b \in \{\emptyset, \varepsilon\}$ or $b \in \Sigma$ and $b \neq a$, and $b_a = \varepsilon$ if $b = a$; $(\overline{L})_a = \overline{L_a}$; $(K \cup L)_a = K_a \cup L_a$; $(KL)_a = K_a L \cup K^\varepsilon L_a$; $(K^*)_a = K_a K^*$. The quotient by a word $w \in \Sigma^*$ is computed by induction on the length of $w$: $L_\varepsilon = L$; $L_w = L_a$ if $w = a \in \Sigma$; $L_{wa} = (L_w)_a$. Quotients computed this way are indeed the left quotients of a regular language [4,5].

A quotient $L_w$ is *accepting* if $\varepsilon \in L_w$; otherwise it is *rejecting*.

A *deterministic finite automaton* (dfa) is a quintuple $\mathcal{D} = (Q, \Sigma, \delta, q_0, F)$, where $Q$ is a finite, non-empty set of *states*, $\Sigma$ is a finite, non-empty *alphabet*,

$\delta : Q \times \Sigma \to Q$ is the *transition function*, $q_0 \in Q$ is the *initial state*, and $F \subseteq Q$ is the set of *final states*.

A *nondeterministic finite automaton* (nfa) is a quintuple $\mathcal{N} = (Q, \Sigma, \eta, S, F)$, where $Q$, $\Sigma$, and $F$ are as in a dfa, $\eta : Q \times \Sigma \to 2^Q$ is the *transition function*, and $S \subseteq Q$ is the *set of initial states*.

The *quotient automaton* of a regular language $L$ is $\mathcal{D} = (Q, \Sigma, \delta, q_0, F)$, where $Q = \{L_w \mid w \in \Sigma^*\}$, $\delta(L_w, a) = L_{wa}$, $q_0 = L_\varepsilon = L$, $F = \{L_w \mid L_w^\varepsilon = \varepsilon\}$, and $L_w^\varepsilon = (L_w)^\varepsilon$. So the number of states in the quotient automaton of $L$ is the quotient complexity of $L$. A quotient automaton can be conveniently represented by *quotient equations* [4]:

$$L_w = \bigcup_{a \in \Sigma} a L_{wa} \cup L_w^\varepsilon, \tag{1}$$

where there is one equation for each distinct quotient $L_w$.

We use the following formulas [4,5] for quotients of regular languages to establish upper bounds on quotient complexity:

**Proposition 1.** *If $K$ and $L$ are regular languages and $w \in \Sigma^*$, then*

$$(\overline{L})_w = \overline{L_w}, \quad (K \circ L)_w = K_w \circ L_w, \tag{2}$$

$$(KL)_w = K_w L \cup K^\varepsilon L_w \cup (\bigcup_{\substack{w=uv \\ u,v \in \Sigma^+}} K_u^\varepsilon L_v), \tag{3}$$

$$(L^*)_\varepsilon = \varepsilon \cup LL^*, \quad (L^*)_w = (L_w \cup \bigcup_{\substack{w=uv \\ u,v \in \Sigma^+}} (L^*)_u^\varepsilon L_v) L^* \text{ for } w \in \Sigma^+. \tag{4}$$

## 3   Ideals, Generators, and Minimal Generators

If $L$ is a right (respectively, left, two-sided, all-sided) ideal, any language $G \subseteq \Sigma^*$ such that $L = G\Sigma^*$ (respectively, $L = \Sigma^* G$, $L = \Sigma^* G \Sigma^*$, $L = \Sigma^* \sqcup G$) is a *generator* of $L$. The quotients of generated ideals $G\Sigma^*$, $\Sigma^* G$, and $\Sigma^* G \Sigma^*$ are given below, where in $w = uv$ and $v = xy$ we assume that $u, v, x, y \in \Sigma^+$:

$$(G\Sigma^*)_w = (G_w \cup G^\varepsilon \cup \bigcup_{w=uv} G_u^\varepsilon)\Sigma^*, \tag{5}$$

$$(\Sigma^* G)_w = \Sigma^* G \cup G_w \cup \bigcup_{w=uv} G_v. \tag{6}$$

$$(\Sigma^* G \Sigma^*)_w = \Sigma^*(G\Sigma^*) \cup (G\Sigma^*)_w \cup \bigcup_{w=uv} (G\Sigma^*)_v = \tag{7}$$

$$[\Sigma^* G \cup (G_w \cup \bigcup_{w=uv} G_v) \cup \bigcup_{w=uv} [G_u^\varepsilon \cup (\bigcup_{v=xy} G_x^\varepsilon)]]\Sigma^*. \tag{8}$$

We use these formulas to establish upper bounds on the complexity of the ideals $G\Sigma^*$, $\Sigma^* G$, and $\Sigma^* G \Sigma^*$ generated by $G$.

**Theorem 1 (Complexity of Ideals in Terms of Generators).** *Let $G$ be any generator of the right ideal $G\Sigma^*$ (left ideal $\Sigma^*G$, two-sided ideal $\Sigma^*G\Sigma^*$, or all-sided ideal $\Sigma^* \sqcup\!\sqcup G$) with $\kappa(G) = n \geq 2$, and assume that $\varepsilon \notin G$. Then*

1. *$\kappa(G\Sigma^*) \leq n$, and the bound is tight if $|\Sigma| \geq 1$.*
2. *$\kappa(\Sigma^*G) \leq 2^{n-1}$, and the bound is tight if $|\Sigma| \geq 2$.*
3. *$\kappa(\Sigma^*G\Sigma^*) \leq 2^{n-2} + 1$, and the bound is tight if $|\Sigma| \geq 3$.*
4. *$\kappa(\Sigma^* \sqcup\!\sqcup G) \leq 2^{n-2} + 1$, and the bound is tight if $|\Sigma| \geq 2n$.*

*Proof.* The first two items follow from [28]. We give short proofs using quotients.

1. From (5), if $w$ has no prefix in $G$, then $(G\Sigma^*)_w = G_w\Sigma^*$. As $G$ is non-empty and $\kappa(G) = n$, there can be at most $n - 1$ such quotients, for there must be at least one quotient $G_w$ with $w \in G$. However, for every word $w$ with a prefix $x$ in $G$, we have $(G\Sigma^*)_w = (G\Sigma^*)_x = \Sigma^*$. Hence there are at most $n$ different quotients of $G\Sigma^*$. The unary language $G = a^{n-1}a^*$ meets the bound.

2. One of the $n$ quotients of $G$, namely $G_\varepsilon = G$, always appears in (6). Thus there are at most $2^{n-1}$ subsets of quotients of $G$ to be added to $\Sigma^*G$. For $n \geq 2$ consider the language $G = (b \cup a(a \cup b)^{n-1})^*a(a \cup b)^{n-2}$. Then $L = \Sigma^*G = \{w \mid w$ has an $a$ in position $(n-1)$ from the end$\}$. Let $x$ and $y$ be two different words of length $n - 1$, and let $u$ be their longest common prefix. Then, for some $v, w \in \Sigma^*$, we have $x = uav$ and $y = ubw$, and $a^{|u|} \in L_x \setminus L_y$. Hence all the quotients of $L$ by words of length $n - 1$ are distinct, and $L$ has at least $2^{n-1}$ distinct quotients. In view of the bound, $L$ has exactly $2^{n-1}$ quotients.

3. Let $L = \Sigma^*G\Sigma^*$. Since $G$ is always present in the expression $L_w$ in (8), there are $2^{n-1}$ unions of quotients of $G$ possible. Since $G$ is non-empty, it has at least one accepting quotient, and that quotient is not $G$, since $\varepsilon \notin G$. Hence at least $2^{n-2}$ unions contain an accepting quotient of $G$ and the corresponding quotients of $L$ are $\Sigma^*$. Thus $2^{n-2} + 1$ is an upper bound. If $n = 2$ and $\Sigma = \{a\}$, then $G = aa^* = a^*aa^*$ meets the bound. If $n = 3$, use $\Sigma = \{a\}$ and $G = aaa^* = a^*aaa^*$. For $n \geq 4$, consider the language $G$ defined by the quotient automaton in Fig. 1. Let $x = uav$ and $y = ubw$ be two different words of length $n - 2$, and let $z = a^{|u|}c$; then $z \in L_x \setminus L_y$. This gives $2^{n-2}$ distinct quotients. Adding $L_{a^{n-2}c} = \Sigma^*$, which is the only quotient of $L$ containing $\varepsilon$, we have the bound.

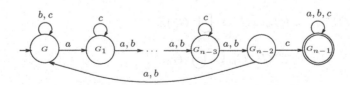

**Fig. 1.** Quotient automaton of $G$ with $\kappa(G) = n$ satisfying $\kappa(\Sigma^*G\Sigma^*) = 2^{n-2} + 1$

4. This result for all-sided ideals was proved by Okhotin [18]. For $n \geq 3$, he used $\Sigma = \{a_1, \ldots, a_t\}$, where $t = n - 2$, and $G = \Sigma^*(a_1a_1 \cup \cdots \cup a_ta_t)\Sigma^*$. He also showed that the bound cannot be reached if $n - 3$ letters are used.    $\square$

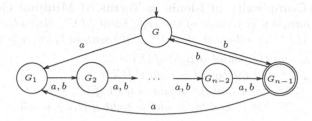

**Fig. 2.** Quotient automaton of $G$ with $\kappa(\Sigma^*G) = 2^{n-1} - 2^{n-3} + 1$

**Theorem 2 (Generators of Left Ideals with Special Properties).** *Let $G$ be any generator of the left ideal $\Sigma^*G$ with $\kappa(G) = n \geq 3$.*

1. *If $G_a \neq G_b$, $G_a \neq G$ and $G_b \neq G$, then $\kappa(\Sigma^*G) \leq 2^{n-1} - 2^{n-3} + 1$.*
2. *If $G_a = G_b$ and $G_a \neq G$, then $\kappa(\Sigma^*G) \leq 2^{n-2} + 1$.*

*Both bounds are tight if $|\Sigma| \geq 2$.*

*Proof.* 1. Since $G$ always appears in (6), we have at most $2^{n-1}$ subsets of quotients of $G$. Moreover, $(\Sigma^*G)_{wa}$ contains $G_a$ and $(\Sigma^*G)_{wb}$ contains $G_b$. Therefore the quotient of $G$ by any word of length greater than zero contains either $G_a$ or $G_b$. Let $S = \{G_1, \ldots, G_{n-1}\}$ be the set of quotients of $G$ other than $G$ itself. There are $2^{n-3} - 1$ non-empty subsets of $S$ containing neither $G_a$ nor $G_b$. These subsets can never appear in the union in (6); hence we have the upper bound. Now consider the language $G$ defined by the quotient automaton in Fig. 2. Here $L = \Sigma^*G = \{w \mid w$ ends in $b$ or has an $a$ in position $(n-1)$ from the end$\}$. The quotients $L_\varepsilon = L$, $L_{aw}$, where $|w| = n-2$, and $L_{bva}$, where $|v| = n-3$, are all distinct: First, we have $\varepsilon \in L_{aw} \setminus L$, $a^{n-2} \in L_{bva} \setminus L$, and $\varepsilon \in L_{aw} \setminus L_{bva}$. Second, consider two words $x = auaz$ and $y = aubz'$ of the form $aw$; then $a^{|au|} \in L_x \setminus L_y$. Third, consider two words $x = buaza$ and $y = bubz'a$ of the form $bva$; then $a^{|bu|} \in L_x \setminus L_y$. Thus the $2^{n-1} - 2^{n-3} + 1$ quotients are distinct.

2. For all $u \in \Sigma^*$, the quotient $(\Sigma^*G)_{ua}$ always contains $G_a$, by Equation (6). Since $G_a = G_b$, we know that $G_w$ contains $G_a$ for any $w \in \Sigma^+$. Hence the number of possibilities is reduced from $2^{n-1}$ to $2^{n-2} + 1$, where 1 is added for $(\Sigma^*G)_\varepsilon$. Now let $\Sigma = \{a, b\}$ and $G' = \Sigma G$, where $G = (b \cup a(a \cup b)^{n-1})^*a(a \cup b)^{n-2}$. Then $L = \Sigma^*G' = \{w \mid w$ has an $a$ in position $(n-2)$ from the end$\}$. The $2^{n-2} + 1$ quotients $L_\varepsilon$ and $L_{aw}$ with $|w| = n-2$ are all distinct. $\square$

We now consider the complexities of ideals in terms of their minimal generators. The following are well-known properties of ideals [16]: If $L$ is a right ideal, the *minimal generator* of $L$ is $M = L \setminus (L\Sigma^+)$, and $M$ is prefix-free. If $L$ is a left ideal, the *minimal generator* of $L$ is $M = L \setminus (\Sigma^+L)$, and $M$ is suffix-free. If $L$ is a two-sided ideal, the *minimal generator* of $L$ is $M = L \setminus (\Sigma^+L\Sigma^* \cup \Sigma^*L\Sigma^+)$, and $M$ is factor-free. If $L$ is an all-sided ideal, the *minimal generator* of $L$ is the set $M$ of all words of $L$ that have no proper subwords in $L$, and thus $M$ is subword-free.

**Theorem 3 (Complexity of Ideals in Terms of Minimal Generators).**
Let $M$ be the minimal generator of the right ideal $M\Sigma^*$, (left ideal $\Sigma^* M$, two-sided ideal $\Sigma^* M\Sigma^*$, or all-sided ideal $\Sigma^* \sqcup\!\sqcup M$) with $\kappa(M) = n \geq 3$. Then

1. $\kappa(M\Sigma^*) \leq n$ and the bound is tight if $|\Sigma| \geq 2$.
2. $\kappa(\Sigma^* M) \leq 2^{n-2}$ and the bound is tight if $|\Sigma| \geq 2$.
3. $\kappa(\Sigma^* M\Sigma^*) \leq 2^{n-3} + 1$ and the bound is tight if $|\Sigma| \geq 2$.
4. $k(\Sigma^* \sqcup\!\sqcup M) \leq 2^{n-3} + 1$, and the bound is tight if $|\Sigma| \geq n - 3$.

*Proof.* 1. The upper bound $n$ follows from Theorem 1. Let $\Sigma = \{a, b\}$, and let $M = a\Sigma^{n-3}$. Then $M$ has $n$ quotients and generates the right ideal $L = a\Sigma^{n-3}\Sigma^*$, which also has $n$ quotients. The minimal generator of $L$ is $L \setminus L\Sigma^+ = M\Sigma^* \setminus M\Sigma^*\Sigma^+ = (M \cup M\Sigma^+) \setminus M\Sigma^+ = M \setminus M\Sigma^+ = M$, since $M$ is prefix-free. Hence $M$ is indeed the minimal generator of $M\Sigma^*$, and the bound is tight.

2. Let $G$ be $M$ in (6). One of the $n$ quotients of $M$, namely $M_\varepsilon = M$, always appears in the union. Thus there are at most $2^{n-1}$ subsets of quotients of $M$ to be added to $\Sigma^* M$. Moreover, since $M$ is suffix-free, $M$ has $\emptyset$ as a quotient [12]. Since each union of the $n-1$ quotients other than $M$ that contains $\emptyset$ is equivalent to a union without $\emptyset$, there are at most $2^{n-2}$ quotients of $\Sigma^* M$. Let $\Sigma = \{a, b\}$ and $M = a\Sigma^{n-3}$; then $M$ has $n$ quotients, and generates the ideal $L = \Sigma^* M = \{w \mid w \text{ has an } a \text{ in position } (n-2) \text{ from the end}\}$. Thus the $2^{n-2}$ quotients of $L$ by words of length $n-2$ are distinct, and $\kappa(L) = 2^{n-2}$. The minimal generator of $L$ is $L \setminus \Sigma^+ L = \Sigma^* M \setminus \Sigma^+\Sigma^* M = (M \cup \Sigma^+ M) \setminus \Sigma^+ M = M \setminus \Sigma^+ M = M$, since $M$ is suffix-free. Hence $M$ is indeed the minimal generator of $\Sigma^* M$.

3. Let $G$ be $M$ in (8). Since $M_\varepsilon = M$ is always present, there are at most $2^{n-1}$ subsets of quotients of $M$ to add to $M_\varepsilon$. Since $M$ is the minimal generator of $L$, it is factor-free, and hence prefix-free. Thus it has only one accepting quotient, $\varepsilon$, and also has $\emptyset$ as a quotient. So we have at most $2^{n-2}$ subsets, because each subset containing $\emptyset$ is equivalent to another subset without $\emptyset$. Finally, half of those $2^{n-2}$ subsets contain $\Sigma^*$, and hence are equivalent to $\Sigma^*$. This leaves $2^{n-3} + 1$ subsets, and $\kappa(L) \leq 2^{n-3} + 1$. For $n = 3$, let $\Sigma = \{a\}$ and $M = a$; then $M$ is the minimal generator of $a^* a a^*$ and meets the bound. For $n \geq 4$, let $\Sigma = \{a, b\}$, $M = a\Sigma^{n-4}a$, and $L = \Sigma^* M\Sigma^*$. Then $M$ has $n$ quotients. We now show that the quotients $L_w$, where $|w| = n - 3$, and $L_{a^{n-2}}$ are all distinct. The only quotient containing $\varepsilon$ is $L_{a^{n-2}}$. If $x = uav$ and $y = ubw$ are two different words of length $n - 3$ and $z = a^{|u|}a$, then $z \in L_x \setminus L_y$. The minimal generator of $L$ is $L \setminus (\Sigma^+ L\Sigma^* \cup \Sigma^* L\Sigma^+) = \Sigma^* M\Sigma^* \setminus (\Sigma^+ M\Sigma^* \cup \Sigma^* M\Sigma^+) = M$, since $M$ is factor-free. Hence $M$ is indeed the minimal generator of $\Sigma^* M\Sigma^*$.

4. Since an all-sided ideal is a two-sided ideal, the bound of $2^{n-3} + 1$ applies. Let $\Sigma = \{a_1, \ldots, a_t\}$, where $t = n - 3$, and let $M = a_1 a_1 \cup \cdots \cup a_t a_t$. Now let $L = \Sigma^* \sqcup\!\sqcup M = \bigcup_{i=1}^{t} \Sigma^* a_i \Sigma^* a_i \Sigma^*$, let $k \geq 0$, let $S = \{a_{i_1}, \ldots, a_{i_k}\} \subseteq \Sigma$, where $i_1 < i_2 < \cdots < i_k$, let $w_S = a_{i_1} a_{i_2} \cdots a_{i_k}$, and add the word $a_1 a_1$. The quotients of $L$ by these $2^t + 1$ words are all distinct. Thus $\kappa(L) = 2^{n-3} + 1$. Now $x \in L$ if and only if $x = ua_i v a_i w$ for some $a_i \in \Sigma$, $u, v, w \in \Sigma^*$, and all words of this form are generated by $a_i a_i$. Hence $M$ is indeed the minimal generator of $\Sigma^* \sqcup\!\sqcup M$. □

We now consider the converse problem: Given an ideal $L$ of quotient complexity $n$, what is the quotient complexity of its minimal generator?

**Theorem 4 (Complexity of Minimal Generators).** *Let $L$ be an ideal language with $\kappa(L) = n \geq 1$. Let $M$ be its minimal generator.*

1. *If $L$ is a right ideal, then $\kappa(M) \leq n + 1$.*
2. *If $L$ is a left ideal, then $\kappa(M) \leq n(n - 1)/2 + 2$.*
3. *If $L$ is a two-sided or an all-sided ideal, then $\kappa(M) \leq n + 1$.*

*The bound for left ideals is tight if $|\Sigma| \geq 2$; the other bounds are tight if $|\Sigma| \geq 1$.*

*Proof.* 1. If $n = 1$, then $L = \Sigma^*$, $M = \varepsilon$, $\kappa(L) = 1$, $\kappa(M) = 2$, and the bound is satisfied. Assume now that $\varepsilon \notin L$. Since $L = L\Sigma^*$, we have $L\Sigma = L\Sigma^+$. Now $M_\varepsilon = L \setminus L\Sigma$ and, for $a \in \Sigma$, $x \in \Sigma^*$, $M_{xa} = L_{xa} \setminus (L\Sigma)_{xa}$. By (3), since $\varepsilon \notin L$, we have $L^\varepsilon = \emptyset$ and (recalling that $u, v \in \Sigma^+$),

$$(L\Sigma)_{xa} = L_{xa}\Sigma \cup L^\varepsilon \Sigma_{xa} \cup \Big( \bigcup_{xa=uv} L_u^\varepsilon \Sigma_v \Big) = L_{xa}\Sigma \cup L_x^\varepsilon \varepsilon,$$

because the only non-empty quotient of $\Sigma$ by a non-empty word occurs when $v = a$. Thus $M_{xa} = L_{xa} \setminus (L_{xa}\Sigma \cup L_x^\varepsilon \varepsilon)$. We know that $L$ has only one accepting quotient, namely $\Sigma^*$. If $L_{xa} \neq \Sigma^*$, then $\varepsilon \notin L_{xa}$ and $L_x \neq \Sigma^*$, which implies that $L_x^\varepsilon = \emptyset$; thus $M_{xa} = L_{xa} \setminus L_{xa}\Sigma$, and there are $n - 1$ such quotients of $M$. If $L_{xa} = \Sigma^*$, then there are two cases: $(i)$ $L_x = \Sigma^*$: we have $\varepsilon \in L_x$ and $M_{xa} = \Sigma^* \setminus (\Sigma^+ \cup \varepsilon) = \emptyset$; $(ii)$ $L_x \neq \Sigma^*$: we have $\varepsilon \notin L_x$ and $M_{xa} = \Sigma^* \setminus \Sigma^+ = \varepsilon$. In this case $M_{xa}$ has the form $L_{xa} \setminus L_{xa}\Sigma$, and this has already been counted. Altogether we have $M_\varepsilon$, $\emptyset$, and $n - 1$ other quotients. Hence $\kappa(M) \leq n + 1$. Let $\Sigma = \{a\}$, and let $L = a^{n-1}a^*$, for $n \geq 1$. Then $L$ is a right ideal, $\kappa(L) = n$, and the minimal generator is $M = a^{n-1}$ with $\kappa(M) = n + 1$.

2. If $L$ is a left ideal and $u, v \in \Sigma^*$, then $L_v \subseteq L_{uv}$. Since $L = \Sigma^* L$, we have $\Sigma L = \Sigma^+ L$, showing that $M = L \setminus \Sigma L$. Let $L$ have quotients $L_1, L_2, \ldots, L_n$. If $w = av$ is a nonempty word, then $M_w = L_{av} \setminus L_v$, which is a difference of two quotients of $L$. Next, we have $L_v \subseteq L_{av}$. This means, that if $i \neq j$, then at most one of $L_i \setminus L_j$ and $L_j \setminus L_i$ may be a non-empty quotient of $M$. Also, $L_i \setminus L_i = \emptyset$ for all $i$. Hence there are at most $n(n-1)/2 + 2$ quotients of $M$: $M_\varepsilon$, at most one quotient for each $i \neq j$, and $\emptyset$. Let $n \geq 3$ and let $L = (b \cup ab)^* a(ab^*)^{n-3} a(a \cup b)^*$. Note that $w \in L$ if and only if $w = xa(ab^*)^{n-3}ay$ for some $x, y \in \Sigma^*$ because every quotient of $L$ contains $a(ab^*)^{n-3}a$. Thus $L$ is a left ideal with $\kappa(L) = n$. Now let $M = L \setminus \Sigma L$. Consider the quotients of $M$ by the following $n(n-1)/2 + 2$ words: $\varepsilon$, $b$, $a(ab)^i a^j$, where $i = 0, 1, \ldots, n - 2$ and $j = 0, 1, \ldots, n - 2 - i$. Let $x = a(ab)^i a^j$ and $y = a(ab)^k a^l$, where $(i, j) \neq (k, l)$. If $i + j < k + \ell$, then $a^{n-2-(k+\ell)}$ is in $M_y \setminus M_x$. If $i + j = k + \ell$ and $j < \ell$, then $a^{n-1-\ell}$ is in $M_x \setminus M_y$. Next, $a^{n-1}$ is in $M_\varepsilon \setminus M_b$, and $a^{n-2-(i+j)}$ is in $M_x \setminus M_\varepsilon$ as well as in $M_x \setminus M_b$. Thus the quotients of the minimal generator $M$ by these words are all distinct, and the bound is tight.

3. Since every two-sided ideal is a right ideal, the bound of $n + 1$ applies. Let $\Sigma = \{a\}$ and let $L = \Sigma^* \sqcup a^{n-1}$ for $n \geq 1$. Then $L$ is an all-sided ideal, $\kappa(L) = n$, and the minimal generator is $M = a^{n-1}$ with $\kappa(M) = n + 1$. $\qquad \square$

## 4    Basic Operations on Ideals

We now examine the complexity of operations on ideal languages. If $K$ and $L$ are regular languages with quotient complexities $m$ and $n$, respectively, then the following bounds, tight in the binary case, are known [5,14,17,28]: $\kappa(K \cup L)$, $\kappa(K \cap L), \kappa(K \setminus L), \kappa(K \oplus L) \leq mn$, $\kappa(KL) \leq m2^n - 2^{n-1}$, $\kappa(K^*) \leq 3/4 \cdot 2^n$, and $\kappa(L^R) \leq 2^n$. In this section, we show that the bounds for ideals are generally lower, and tight for languages over small fixed alphabets, except for reversal of all-sided ideals, which requires a growing alphabet.

**Theorem 5 (Boolean Operations).** *Let $K$ and $L$ be right ideals (respectively, two-sided ideals, or all-sided ideals) with $\kappa(K) = m$, $\kappa(L) = n$, and $\varepsilon \notin K \cup L$.*

*1. $\kappa(K \cap L) \leq mn$,    $\kappa(K \oplus L) \leq mn$.*
*2. $\kappa(K \cup L) \leq mn - (m + n - 2)$.*
*3. $\kappa(K \setminus L) \leq mn - (m - 1)$.*

*If $K$ and $L$ are left ideals, then $\kappa(K \circ L) \leq mn$. The languages $K = (b^*a)^{m-1}\Sigma^*$ and $L = (a^*b)^{n-1}\Sigma^*$ meet all the bounds for right ideals, and also for intersection and symmetric difference of left ideals. For union and difference of left ideals, the following languages $K$ and $L$ over the alphabet $\{a, b, c, d\}$ meet the bound:*

$$K = (b \cup c \cup d)K \cup aK_1,$$
$$K_i = (b \cup d)K_i \cup aK_{i+1} \cup cK \text{ for } i = 1, \ldots, m - 2,$$
$$K_{m-1} = (a \cup b \cup d)K_{m-1} \cup cK \cup \varepsilon.$$

$$L = (a \cup c \cup d)L \cup bL_1,$$
$$L_i = (a \cup c)L_i \cup bL_{i+1} \cup dL \text{ for } i = 1, \ldots, n - 2,$$
$$L_{n-1} = (a \cup b \cup c)L_{n-1} \cup dL \cup \varepsilon.$$

See [6] for the proof of this result.

**Theorem 6 (Product).** *Let $K$ and $L$ be ideals of the same type with $\kappa(K) = m$ and $\kappa(L) = n$. Then the following are tight bounds:*

*1. If $K$ and $L$ are right ideals, then $\kappa(KL) \leq m + 2^{n-2}$.*
*2. If $K$ and $L$ are left, two-sided, or all-sided ideals, then $\kappa(KL) \leq m + n - 1$.*

*Proof.* The following claim can be proved: If $N = \Sigma^*L$ is a left ideal with $\kappa(N) = r$ and $K$ is any non-empty language with $\kappa(K) = m$, then $\kappa(KN) \leq m + r - 1$, and the bound is tight; see [6].

1. Suppose that $K = K\Sigma^*$ and $L = L\Sigma^*$ are right ideals. Then $KL = K\Sigma^*L\Sigma^* = KN$, where $N = \Sigma^*L\Sigma^*$. Let $\kappa(N) = r$. By our claim, $\kappa(KN) \leq m + r - 1$, and our problem reduces to that of finding $r = \kappa(N)$ as a function of $n = \kappa(L)$. By Equation (6) with $G = L$ and $N = \Sigma^*L$, $N_w$ is the union of $\Sigma^*L$ and some quotients of $L$. Since $L$ is always present in the union, we have at most $2^{n-1}$ different unions. Since one of the quotients of $L$ is $\Sigma^*$, and $\Sigma^* \cup L_v = \Sigma^*$, we have at most $r = 2^{n-2} + 1$ distinct quotients of $N$. Thus $\kappa(KL) \leq m + 2^{n-2}$. Now let $\Sigma = \{a, b, c\}$, $K = \Sigma^{m-1}\Sigma^*$, and let $L$ be the right ideal in the proof of Theorem 1 (Fig. 1). Then $\kappa(K) = m$, $\kappa(L) = n$, $\kappa(\Sigma^*L\Sigma^*) = 2^{n-2} + 1$, and $\kappa(KL) = \kappa(\Sigma^{m-1}\Sigma^*L\Sigma^*) = m - 1 + 2^{n-2} + 1 = m + 2^{n-2}$.

2. Suppose $K = \Sigma^* K$ and $L = \Sigma^* L$ are left ideals. If $\varepsilon \in K$, then $K = \Sigma^*$, $m = 1$, $KL = L$, and $\kappa(KL) = n = m + n - 1$. Otherwise, by our claim, $\kappa(KL) \leq m + n - 1$, and this bound is tight. Since every all-sided or two-sided ideal is also a left ideal, the upper bound applies in these cases as well. The bound is met for $\Sigma = \{a\}$ and all-sided ideals $a^* \sqcup a^{m-1}$ and $a^* \sqcup a^{n-1}$.    □

**Theorem 7 (Star).** *Let $L$ be an ideal with $\kappa(L) = n$. If $\varepsilon \in L$ then $\kappa(L^*) = 1$. Otherwise, $\kappa(L^*) \leq n + 1$, and this bound is tight if $|\Sigma| \geq 2$.*

*Proof.* If $L$ is an ideal with $\varepsilon \in L$, then $L = \Sigma^*$. Consider right ideals first, and suppose $\varepsilon \notin L$. If $L = L\Sigma^*$, then $L^* = \varepsilon \cup L\Sigma^*$. We have $(L^*)_\varepsilon = L^* = \varepsilon \cup L\Sigma^*$. For every word $w$ in $\Sigma^+$, from Equation (5)

$$(L^*)_w = (L_w \cup \bigcup_{w=uv} L_u^\varepsilon)\Sigma^*,$$

where $u, v \in \Sigma^+$. Assume that a non-empty word $w$ has no prefix in $L$. In that case, $(L^*)_w = L_w \Sigma^*$. If $w = uv$ and $u \in L$, then $L_u = \Sigma^*$, hence also $L_w = \Sigma^*$. Then $(L^*)_w = L_w \Sigma^*$ as well. There are at most $n$ such quotients of $L^*$. So $\kappa(L^*) \leq n + 1$. Since every all-sided ideal and every two-sided ideal is a right ideal, we have an upper bound for these three classes of ideals.

If $L$ is a left ideal, and $\varepsilon \notin L$, then $(L^*)_\varepsilon = \varepsilon \cup \Sigma^* L$, and if $w \in \Sigma^+$, then $(L^*)_w$ is given by Equation (6). We have

$$(L^*)_w = (\Sigma^* L)_w = \Sigma^* L \cup L_w \cup \bigcup_{w=uv} L_v = L \cup L_w \cup \bigcup_{w=uv} L_v = L_w,$$

with $u, v \in \Sigma^+$, since $L_v \subseteq L_w$ for a left ideal. Hence there are at most $n + 1$ quotients of $L^*$.

The bound is met for $\Sigma = \{a, b\}$ and all sided ideal $L = \Sigma^* \sqcup a^{n-1}$ since $\kappa(L) = n$, $L^* = L \cup \varepsilon$, and $\kappa(L^*) = n + 1$.    □

To deal with reversal, we start with the dfa of $L$, reverse it, and use the subset construction to obtain a dfa for $L^R$ with at most $2^n$ states.

**Theorem 8 (Reversal).** *Let $L$ be an ideal language with $\kappa(L) = n \geq 3$.*

1. *If $L$ is a right ideal, then $\kappa(L^R) \leq 2^{n-1}$.*
2. *If $L$ is a left ideal, then $\kappa(L^R) \leq 2^{n-1} + 1$.*
3. *If $L$ is a two-sided ideal, then $\kappa(L^R) \leq 2^{n-2} + 1$.*
4. *If $L$ is an all-sided ideal, then $\kappa(L^R) \leq 2^{n-2} + 1$.*

*The bound is tight for right ideals if $|\Sigma| \geq 2$, for left and two-sided ideals if $|\Sigma| \geq 3$, and for all-sided ideals if $|\Sigma| \geq 2n - 4$.*

*Proof.* 1. Since $L$ is a right ideal, it has only one accepting quotient, $f = \Sigma^*$. This quotient becomes the initial state of the nfa for $L^R$. Since $L^R$ is a left ideal, we can add a loop for every letter of $\Sigma$ from $f$ to $f$ in the nfa. Therefore $f$ appears in every subset of states of the nfa reachable from $q$. Hence there are at most

$2^{n-1}$ states of the equivalent subset automaton. Let $\Sigma = \{a, b\}$, and consider the right ideal $L = (\Sigma^{n-2}b)^*\Sigma^{n-2}a\Sigma^*$. Then $\kappa(L) = n$. In the nfa obtained by reversing the dfa for $L$, a word $w$ of length at most $2n-3$ is accepted if and only if $w$ has an $a$ in position $n-1$ from the end. Now, if $x, y \in \Sigma^{n-1}$ and $x = uav$, $y = ubw$, then $|uava^{|u|}| \le 2n-3$ since $|u| \le n-2$. Similarly, $|ubva^{|u|}| \le 2n-3$. Hence $a^{|u|} \in (L^R)_x \setminus (L^R)_y$ and all the quotients of $L^R$ by the $2^{n-1}$ words of length $n-1$ are distinct.

2. The quotient $L_\varepsilon = L$, which is the initial state of the quotient automaton of $L$, is the only accepting state in the nfa for $L^R$. In the corresponding subset automaton, $L$ appears in $2^{n-1}$ subsets. All these subsets are accepting states of the subset automaton and all accept $\Sigma^*$, since $L^R$ is a right ideal. Hence $\kappa(L^R) \le 1 + 2^{n-1}$. If $n = 1$, respectively, $n = 2$, $n = 3$, then $L = a^*$, respectively, $L = aa^*$, $L = (a \cup b)^*c(c \cup (a \cup b)b^*(a \cup c))^*$, meets the bound. If $n = 4$, then the bound is met for $\Sigma = \{a, b, c\}$ and $L$ defined by $L = (a \cup b)L \cup cL_1$, $L_1 = (a \cup b)L_2 \cup cL_1 \cup \varepsilon$, $L_2 = (a \cup b)L_3 \cup cL_1$, $L_3 = (b \cup c)L_1 \cup aL_3$. For $n \ge 5$, let $\mathcal{D} = (\{0, 1, \ldots, n-1\}, \{a, b, c\}, \delta, 0, \{n-1\})$ be the dfa shown in Fig. 3, where the unspecified transitions under $c$ are all to state 1. It was proved in [23] that the reverse of $\mathcal{D}' = (\{1, \ldots, n-1\}, \{a, b\}, \delta', 1, \{n-1\})$ with $\delta'$ being $\delta$ restricted to $\{a, b\}$, has $2^{n-1}$ states. It follows that the reverse of dfa $\mathcal{D}$ has $2^{n-1}+1$ states. Since $L(\mathcal{D})$ is a left ideal, the theorem holds.

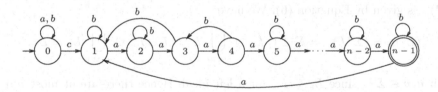

**Fig. 3.** The dfa $\mathcal{D}$ for Theorem 8. States $1, 2, \ldots, n-1$ go to state 1 under $c$.

3. Since $L$ is a right ideal, its quotient automaton has exactly one accepting state $f$, and this state is not initial because $\varepsilon \notin L$. Now $f$ is the only initial state of the nfa for $L^R$. Since $L^R$ is a left ideal, we can add a loop for every letter of $\Sigma$ from $f$ to $f$ in the nfa. Therefore $f$ appears in every subset of states of the nfa reachable from $f$. Hence there are at most $2^{n-1}$ subsets of states of the nfa to consider when using the subset construction. Since $L$ is a left ideal, the initial state of the quotient automaton of $L$ is the only accepting state of the nfa for $L^R$, and it appears in $2^{n-2}$ of the subsets of states of the nfa. All these subsets are accepting states of the corresponding subset automaton and all accept $\Sigma^*$, since $L^R$ is a right ideal. Hence $\kappa(L^R) \le 2^{n-2}+1$. If $n = 1$, then $L = \Sigma^*$ and $\kappa(L^R) = 1$. If $n = 2$, the $2^{n-2}+1$ bound is met for $\Sigma = \{a\}$ and $L = a^*aa^*$. If $n = 3$, the bound is met for $\Sigma = \{a\}$ and $L = a^*aa^*aa^*$. For $n \ge 4$, and $\Sigma = \{a, b, c\}$, consider the language $L$ accepted by the quotient automaton in Fig. 4. The language $L$ is a two-sided ideal. Now construct the nfa for $L^R$. Note that a word $w$ in $(a \cup b)^*c$ of length at most $2n-4$ is accepted by this nfa if and only if $w$ has an $a$ in position $n-1$ from the end and $w$ ends with $c$. We claim

that $\{w \in \{a,b\}^* \mid |w| = n - 2\} \cup \{a^{n-2}c\}$ all define distinct quotients: For let $x = uav$ and $y = ubw$ with $|u| \leq n - 3$, be two different words of length $n - 2$ and let $z = a^{|u|}c$. Then $|xz| = |yz| \leq 2n - 4$ and $z \in (L^R)_x \setminus (L^R)_y$. Also, the quotient of $L^R$ by $a^{n-2}c$ is accepting, while the other are rejecting.

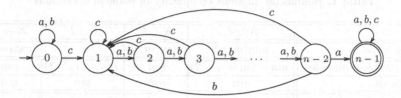

**Fig. 4.** Quotient automaton of two-sided ideal $L$ with $\kappa(L^R) = 2^{n-2} + 1$

4. Since an all-sided ideal is a two-sided ideal, the $2^{n-2} + 1$ bound applies. If $n = 2$, let $\Sigma = \{a\}$ and $L = a^*aa^*$. For $n \geq 3$, let $t = n - 2$ and $\Sigma = \{a_1, \ldots, a_t, b_1, \ldots, b_t\}$. Also, let $A = (a_1 \cup \cdots \cup a_t)$, $B = (b_1 \cup \cdots \cup b_t)$, and $B \setminus b_i = (b_1 \cup \cdots \cup b_{i-1} \cup b_{i+1} \cup \cdots \cup b_t)$. Let $L$ be the language defined by the equations below and shown as an automaton in the figure on the right for $n = 5$.

$$L = BL \cup \bigcup_{i=1}^{t} a_i L_i,$$

$$L_i = (B \setminus b_i)L_i \cup (A \cup b_i)L_{n-1}$$

$$\text{for } i = 1, 2, \ldots, t,$$

$$L_{n-1} = (A \cup B)L_{n-1} \cup \varepsilon.$$

We claim that $L$ is an all-sided ideal; for this, it suffices to show that if $w = uv \in L$ for $u, v \in \Sigma^*$, then $uav \in L$ for every $a \in \Sigma$. If $u = \varepsilon$ and $a \in B$, then $L_a = L$, and if $a = a_i$, then $L_{a_i} = L_i$. However $L_i \supseteq L$; hence all words of the form $\varepsilon av$ are in $L$. If $L_u = \Sigma^*$, then $uav$ is in $L$. Finally, suppose that $L_u = L_i$ for some $i$. Since $(L_i)_a$ is either $L_i$ or $\Sigma^*$, it follows that $uav \in L$. Thus $L$ is an all-sided ideal.

Now $L^R = \bigcup_{i=1}^{t}(\Sigma^*(A \cup b_i)(B \setminus b_i)^* a_i B^*)$. Consider the set of $2^{n-2} + 1$ words $\{b_{i_1} b_{i_2} \cdots b_{i_k} \mid 0 \leq k \leq n - 2, \ 1 \leq i_1 < i_2 < \cdots < i_k \leq n - 2\} \cup \{a_1 a_1\}$. If $x = b_{i_1} \cdots b_{i_l} b_i u$ and $y = b_{i_1} \cdots b_{i_l} b_j v$ with $i < j$, then $a_i \in (L^R)_x \setminus (L^R)_y$. Also, $(L^R)_{a_1 a_1}$ is the only accepting quotient. Hence $L^R$ has $2^{n-2} + 1$ quotients. $\square$

## 5   Conclusions

Tables 1 and 2 summarize our complexity results. The complexities for regular languages are from [28]; difference and symmetric difference are considered in [5].

In Table 2, $k$ is the number of accepting quotients of $K$ in column $KL$, and the number of accepting quotients of $L$ other than $L$ in column $L^*$. The results for unary languages can be found in [6].

**Table 1.** Bounds on quotient complexity of boolean operations

|  | $K \cup L$ | $K \cap L$ | $K \setminus L$ | $K \oplus L$ |
|---|---|---|---|---|
| unary ideals | $\min(m,n)$ | $\max(m,n)$ | $n$ | $\max(m,n)$ |
| right, 2-sided, all-sided | $mn - (m+n-2)$ | $mn$ | $mn - (m-1)$ | $mn$ |
| left ideals | $mn$ | $mn$ | $mn$ | $mn$ |
| regular languages | $mn$ | $mn$ | $mn$ | $mn$ |

**Table 2.** Bounds on quotient complexity of generation, product, star and reversal

|  | $f(G)$ | $f(M)$ | $\kappa(M)$ | $KL$ | $L^*$ | $L^R$ |
|---|---|---|---|---|---|---|
| unary | $n$ | $n-1$ | $n+1$ | $m+n-1$ | $n-1$ | $n$ |
| right | $n$ | $n$ | $n+1$ | $m+2^{n-2}$ | $n+1$ | $2^{n-1}$ |
| 2-sided | $2^{n-2}+1$ | $2^{n-3}+1$ | $n+1$ | $m+n-1$ | $n+1$ | $2^{n-2}+1$ |
| all-sided | $2^{n-2}+1$ | $2^{n-3}+1$ | $n+1$ | $m+n-1$ | $n+1$ | $2^{n-2}+1$ |
| left | $2^{n-1}$ | $2^{n-2}$ | $\frac{n(n-1)}{2}+2$ | $m+n-1$ | $n+1$ | $2^{n-1}+1$ |
| regular | $-$ | $-$ | $-$ | $m2^n - k2^{n-1}$ | $2^{n-1}+2^{n-k-1}$ | $2^n$ |

# References

1. Ang, T., Brzozowski, J.: Languages convex with respect to binary relations, and their closure properties. Acta Cybernet. 19, 445–464 (2009)
2. Birget, J.C.: Intersection of regular languages and state complexity. ACM SIGACT News 22(2), 49 (1980)
3. Bordihn, H., Holzer, M., Kutrib, M.: Determination of finite automata accepting subregular languages. Theoret. Comput. Sci. 410, 3209–3249 (2009)
4. Brzozowski, J.: Derivatives of regular expressions. J. ACM 11, 481–494 (1964)
5. Brzozowski, J.: Quotient complexity of regular languages. In: Dassow, J., Pighizzini, G., Truthe, B. (eds.) 11th International Workshop on Descriptional Complexity of Formal Systems. pp. 25–42. Otto-von-Guericke-Universität, Magdeburg (2009), http://arxiv.org/abs/0907.4547
6. Brzozowski, J., Jirásková, G., Li, B.: Quotient complexity of ideal languages, http://arxiv.org/abs/0908.2083
7. Brzozowski, J., Jirásková, G., Zou, C.: Quotient complexity of closed languages, http://arxiv.org/abs/0912.1034
8. Brzozowski, J., Shallit, J., Xu, Z.: Decision procedures for convex languages. In: Dediu, A.H., Ionescu, A.M., Martín-Vide, C. (eds.) LATA 2009. LNCS, vol. 5457, pp. 247–258. Springer, Heidelberg (2009)
9. Câmpeanu, C., Salomaa, K., Culik II, K., Yu, S.: State complexity of basic operations on finite languages. In: Boldt, O., Jürgensen, H. (eds.) WIA 1999. LNCS, vol. 2214, pp. 60–70. Springer, Heidelberg (2001)

10. Crochemore, M., Hancart, C.: Automata for matching patterns. In: Rozenberg, G., Salomaa, A. (eds.) Handbook of Formal Languages, vol. 2, pp. 399–462. Springer, Heidelberg (1997)
11. Haines, L.H.: On free monoids partially ordered by embedding. J. Combin. Theory 6, 94–98 (1969)
12. Han, Y.-S., Salomaa, K.: State complexity of basic operations on suffix-free regular languages. Theoret. Comput. Sci. 410, 2537–2548 (2009)
13. Han, Y.-S., Salomaa, K., Wood, D.: Operational state complexity of prefix-free regular languages. In: Automata, Formal Languages, and Related Topics, pp. 99–115. University of Szeged, Hungary (2009)
14. Leiss, E.: Succinct representation of regular languages by boolean automata. Theoret. Comput. Sci. 13, 323–330 (1981)
15. de Luca, A., Varricchio, S.: Some combinatorial properties of factorial languages. In: Capocelli, R. (ed.) Sequences, pp. 258–266. Springer, Heidelberg (1990)
16. de Luca, A., Varricchio, S.: Finiteness and Regularity in Semigroups and Formal Languages. Springer, Heidelberg (1999)
17. Maslov, A.N.: Estimates of the number of states of finite automata. Dokl. Akad. Nauk SSSR 194, 1266–1268 (1970) (in Russian); English translation: Soviet Math. Dokl. 11, 1373–1375 (1970)
18. Okhotin. A.: On the state complexity of scattered substrings and superstrings. Turku Centre for Computer Science Technical Report No. 849 (2007)
19. Paz, A., Peleg, B.: Ultimate-definite and symmetric-definite events and automata. J. ACM 12(3), 399–410 (1965)
20. Perrin, D.: Finite automata. In: van Leewen (ed.) Handbook of Theoretical Computer Science, vol. B, pp. 1–57. Elsevier, Amsterdam (1990)
21. Pighizzini, G., Shallit, J.: Unary language operations, state complexity and Jacobsthal's function. Internat. J. Found. Comput. Sci. 13, 145–159 (2002)
22. Rabin, M., Scott, D.: Finite automata and their decision problems. IBM J. Res. Develop. 3, 114–129 (1959)
23. Salomaa, A., Wood, D., Yu, S.: On the state complexity of reversals of regular languages. Theoret. Comput. Sci. 320, 315–329 (2004)
24. Shyr, H.J.: Free Monoids and Languages. Hon Min Book Co., Taiwan (2001)
25. Thierrin, G.: Convex languages. In: Nivat, M. (ed.) Automata, Languages and Programming, pp. 481–492. North-Holland, Amsterdam (1973)
26. Yu, S.: Regular languages. In: Rozenberg, G., Salomaa, A. (eds.) Handbook of Formal Languages, pp. 41–110. Springer, Heidelberg (1997)
27. Yu, S.: State complexity of regular languages. J. Autom. Lang. Comb. 6, 221–234 (2001)
28. Yu, S., Zhuang, Q., Salomaa, K.: The state complexities of some basic operations on regular languages. Theoret. Comput. Sci. 125, 315–328 (1994)

# Complexity of Operations on Cofinite Languages

Frédérique Bassino[1], Laura Giambruno[2], and Cyril Nicaud[3]

[1] LIPN UMR 7030, Université Paris 13 - CNRS, 93430 Villetaneuse, France
[2] Dipartimento di Matematica e Applicazioni, Università di Palermo, Italy
[3] IGM, UMR CNRS 8049, Université Paris-Est, 77454 Marne-la-Vallée, France
bassino@lipn.univ-paris13.fr, lgiambr@math.unipa.it, nicaud@univ-mlv.fr

**Abstract.** We study the worst case complexity of regular operations on cofinite languages (*i.e.*, languages whose complement is finite) and provide algorithms to compute efficiently the resulting minimal automata.

## 1 Introduction

Regular languages are possibly infinite sets of words that can be encoded in different ways by finite objects. One such encoding is based on finite automata that are very convenient to efficiently answer to most natural algorithmic questions, such as membership, emptiness, regular constructions, *etc.* It is why softwares handling regular languages given by another kind of representation, such as regular expressions, often start with the computation of an automaton recognizing the same language.

In this framework all questions about the size of automata in usual algorithms are important and directly related to the space complexity needed for the computations. One such issue could be: "Given a regular language $L$ of size $n$, what is the number of states required to encode the language $L^*$?" Here the notion of size of a language must be specified, since the results depend upon the representation (regular expression, nondeterministic automata, deterministic automata, two way automata, *etc.*) used.

The starting point is often a deterministic automaton. Although deterministic automata may require more space than nondeterministic ones, they have good algorithmic and algebraic properties that make them very useful. In particular to every regular language is associated its minimal automaton, which is the smallest deterministic automaton recognizing it and which is unique. The *state complexity* of a regular language is then defined as the number of states of its minimal automaton. The previous question can be reformulated in the following way: "Given a regular language of state complexity $n$, what is the state complexity of its star?" This topic is intensively studied since almost the beginning of automata theory (see [1,2,3,4] for recent results). Researchers are mainly interested in automata that represent natural subclasses of regular languages such as finite languages [5] or prefix-free regular languages [6]. Most articles focus on the worst case state complexity of basic regular operations such as set constructions, concatenation, Kleene star and reversal.

A. López-Ortiz (Ed.): LATIN 2010, LNCS 6034, pp. 222–233, 2010.

In this paper we analyze the worst case complexity of regular operations on cofinite languages (*i.e.*, languages whose complement is finite) and provide algorithms to compute efficiently the resulting minimal automata. We now present the measure chosen for the size of the input. As finite languages can be described by the finite list of the words they contain, cofinite languages can be described by the finite list of words they do not contain. Though elementary, this representation, is natural and often corresponds to the way the input (a finite or cofinite language) is given for further algorithmic treatments. Consequently given a finite language its *size* is defined as the sum of the length of its words. In other words it is the number of letters needed to write all the words of the language. The size of a cofinite language is then defined as the size of its complement. In the framework of finite languages, our question is changed into: "Given a finite language of size $n$, what is the state complexity of its star?". Several articles study that kind of question. For instance, in [7] it is proved that the state complexity of the star can be exponential and in [8] the authors study the average state complexity of the star operation.

In the following we focus on the set of cofinite languages. It is stable for the regular operations, namely union, concatenation and star and has interesting absorbing properties since combinations of regular and cofinite languages often produce a cofinite language. To roughly summarize our results the regular operations applied to cofinite languages tend to produce cofinite languages of small state complexities, and their minimal automata can be computed quickly. Therefore in constructions involving cofinite languages together with other regular languages one can use dedicated algorithms instead of the general ones, improving significantly the complexity of the computations. This can be seen as a heuristic method consisting in identifying simpler cases to use specific algorithms on them.

The paper is organized as follows. In Section 2, we present cofinite languages and their associated automata, along with algorithms to handle them. In Section 3 we study the state complexities of basic operations on cofinite languages, and provide algorithms to compute the resulting minimal automaton. In Section 4, we briefly consider operations involving both cofinite languages and regular languages.

## 2   Cofinite Languages and Automata

### 2.1   Automata and State Complexity

In this section we introduce objects and notations used in the sequel. For a general presentation of automata and regular languages we refer the reader to [9].

We denote *automata* by tuples $(A, Q, T, I, F)$ where $Q$ is a finite set of *states*, $A$ is a finite set of *letters* called *alphabet*, the *transition relation* $T$ is a subset of $Q \times A \times Q$, $I \subset Q$ is the set of *initial states* and $F \subset Q$ is the set of final states. An automaton is *complete* when for every $(a, q) \in A \times Q$, there exist $p \in Q$ quch that $(q, a, p) \in T$. An automaton is *deterministic* if $|I| = 1$ and for every $(a, q) \in A \times Q$, if both $(q, a, p)$ and $(q, a, p')$ belong to $T$, then $p = p'$. We denote deterministic automata by tuples $(A, Q, \cdot, q_0, F)$, where $q_0$ is the unique initial state and $(q, a, p) \in T$ is denoted by $q \cdot a$. For any word $u \in A^*$ and any state $q \in Q$ in a deterministic automaton, $q \cdot u$ is recursively defined by $q \cdot \varepsilon = q$ ($\varepsilon$ is the empty word) and $q \cdot ua = (q \cdot u) \cdot a$. The regular language recognized by an automaton $\mathcal{A}$ is denoted by $L(\mathcal{A})$.

Let $L$ be a regular language in $A^*$. Let $\mathcal{M}_L$ be the deterministic automaton having the nonempty sets, called the *left quotients* $u^{-1}L = \{w \in A^* \mid uw \in L\}$ for $u \in A^*$ as states, $L$ as initial state, and the states containing the empty word as final states. Define the transition function, for a state $Y = u^{-1}L$ and a letter $a \in A$, by $Y \cdot a = a^{-1}Y = (ua)^{-1}L$. The automaton $\mathcal{M}_L$ is the unique smallest deterministic and complete automaton recognizing $L$, it is called the *minimal automaton* of $L$. If $\mathcal{A}$ is an automaton, we denote by $\mathcal{M}_{\mathcal{A}}$ the minimal automaton of $L(\mathcal{A})$. There is another way to define the minimal automaton of a regular language $L$. Let $\mathcal{A} = (A, Q, \cdot, i, F)$ be a deterministic automaton recognizing $L$. For each $q \in Q$, let $L_q = \{w \in A^* \mid q \cdot w \in F\}$. Two states $p, q \in Q$ are called *inseparable* if $L_p = L_q$, and *separable* otherwise. The Myhill-Nerode equivalence on Q is the relation defined by $p \sim q \iff p$ and $q$ are inseparable. Let $q \in Q$ and let $u \in A^*$ such that $i \cdot u = q$. Then $L_q = u^{-1}X$. The automaton obtained by merging the states belonging to the same equivalence class is isomorphic to the minimal automaton of $L(\mathcal{A})$ (it is in fact a quotient of $\mathcal{A}$ by the Myhill-Nerode equivalence). The *state complexity* of a regular language is the number of states of its minimal automaton.

Given a deterministic and complete automaton $\mathcal{A} = (A, Q, \cdot, q_0, F)$, the *complement* of $\mathcal{A}$ is the automaton $\overline{\mathcal{A}} = (A, Q, \cdot, q_0, Q \setminus F)$. One can check that $L(\overline{\mathcal{A}}) = \overline{L(\mathcal{A})}$, where $\overline{L(\mathcal{A})} = A^* \setminus L(\mathcal{A})$. If $\mathcal{A}$ is a deterministic but not complete automaton, the complement of $\mathcal{A}$ is the complement of the automaton obtained by completing $\mathcal{A}$ with a sink state.

If $u = u_1 \cdots u_n$ is a word, the *reversal* (or the *mirror image*) of $u$ is the word $\tilde{u} = u_n \cdots u_1$. For any language $X$ the set of all prefixes of words in $X$ is denoted by $\mathrm{Pr}(X)$.

## 2.2  Cofinite Languages

A language $X$ in $A^*$ is said to be *cofinite* if $\overline{X} = A^* \setminus X$ is a finite language. For a finite language $X = \{u_1, \cdots, u_m\}$, denote by $\|X\| = \sum_{i=1}^{m} |u_i|$ the sum of lengths of its elements.

The following properties will be useful throughout this article:

- If $X$ and $Y$ are two languages such that $X \subset Y$ and $X$ is cofinite, then $Y$ is cofinite and $\|\overline{Y}\| \leq \|\overline{X}\|$.
- A language $X$ is cofinite if and only if there exists an integer $p$ such that every word of length at least $p$ is in $X$.

Cofinite languages have nice stability properties that are described below.

**Lemma 1.** *The set of cofinite languages is stable for the union, the concatenation and the star operation. It is also stable by mirror image, left quotient and right quotient. The union of a cofinite language and an arbitrary language is cofinite.*

## 2.3   Automata Recognizing Cofinite Languages

Given a cofinite language $X$, let $\mathcal{T}_X$ be the deterministic automaton defined by
$\mathcal{T}_X = (A, Q_X, \cdot, \{\varepsilon\}, F)$, where $Q_X = \mathrm{Pr}(\overline{X}) \cup \{p_X\}$ (where $p_X$ is an accepting
sink state), the transitions are defined, for all $a \in A$ and for all $u \in \mathrm{Pr}(\overline{X})$, by

$$\begin{cases} u \cdot a = ua, & \text{if } ua \in \mathrm{Pr}(\overline{X}) \\ u \cdot a = p_X, & \text{if } ua \notin \mathrm{Pr}(\overline{X}) \\ p_X \cdot a = p_X \end{cases}$$

and the set of final states by $F = (Q_X \cap X) \cup \{p_X\}$. As $\mathcal{T}_X$ is the complement
of the tree automaton of $\overline{X}$, one obtains the following lemma:

**Lemma 2.** *For any cofinite language $X$, the automaton $\mathcal{T}_X$ is a deterministic
and complete automaton with at most $\|\overline{X}\| + 2$ states that recognizes $X$.*

The bound is tight. When $X = \overline{\{u\}}$, $u$ being a nonempty word, $\|\mathrm{Pr}(\{u\})\| = |u| + 1$ and one must add the accepting sink state.

A state of an automaton is *useless* if it is either not reachable from an initial
state or no final state can be reached from it.

**Lemma 3.** *Let $\mathcal{A}$ be a deterministic automaton. The language $L(\mathcal{A})$ is cofinite if
and only if the automaton obtained by removing useless states from $\overline{\mathcal{A}}$ is acyclic.*

**Lemma 4.** *Let $\mathcal{A}$ be a complete minimal automaton. The language $L(\mathcal{A})$ is
cofinite if and only if $\mathcal{A}$ contains an accepting sink state and the graph obtained
after removing it is acyclic.*

Using a result on acyclic automata due to Revuz [10], one gets the following
complexities:

**Proposition 1.** *The following computations can be done in linear time:*

- *Checking if $L(\mathcal{A})$ is cofinite, where $\mathcal{A}$ is a deterministic automaton.*
- *Computing the minimal automaton and therefore the state complexity of a
  cofinite language $L(\mathcal{A})$ given by a deterministic automaton $\mathcal{A}$.*
- *Computing the minimal automaton and therefore the state complexity of a
  cofinite language $X$ given by a set of words.*

*Proof.* First one can compute the automaton $\mathcal{B}$ obtained by removing useless
states from $\overline{\mathcal{A}}$, the complement of $\mathcal{A}$. Then check whether $\mathcal{B}$ is acyclic, which
can be done in linear time. Second compute $\mathcal{B}$ as before, minimize it in $\mathcal{M}_{\mathcal{B}}$
in linear time using Revuz' algorithm [10]. Then compute $\overline{\mathcal{M}_{\mathcal{B}}}$ which is the
minimal automaton of $L(\mathcal{A})$. Indeed, if it is not minimal, computing its complement produce a deterministic and complete automaton recognizing $\overline{L(\mathcal{A})}$ which
is strictly smaller than $\mathcal{M}_{\mathcal{B}}$, which is not possible. Finally starting from a cofinite

language $X$, compute $\mathcal{T}_X$ whose number of states is linear in $\|\overline{X}\|$, then proceed as in the previous construction.

**Proposition 2.** *Let $X$ be a language such that $\overline{X} \subset \{x_1, \cdots, x_m\}$, then $X$ is cofinite and the state complexity of $X$ is at most equal to $\|\{x_1, \cdots, x_m\}\| + 2$.*

*Proof.* Let $Z = \{x_1, \ldots, x_m\}$ with $\|Z\| = n$ and let $\mathcal{T}_{\overline{Z}}$ be the automaton associated to $\overline{Z}$, with $|\mathcal{T}_{\overline{Z}}| \leq n + 2$. Consider the automaton $\mathcal{B}_X$ obtained from $\mathcal{T}_{\overline{Z}}$ by adding the elements of $Z \setminus \overline{X}$ to the set of final states of $\mathcal{T}_{\overline{Z}}$. As every state in $\mathcal{T}_{\overline{Z}}$ that is not the accepting sink state can be reached by reading only one word, no other words than the ones of $Z \setminus \overline{X}$ have been added to the recognized language. Hence $\mathcal{B}_X$ recognizes $X$ and $|\mathcal{B}_X| = |\mathcal{T}_{\overline{Z}}| \leq n + 2$.

Note that, under this only hypothesis, it is not true in general that $X \subset Y$ implies $|\mathcal{M}_Y| \leq |\mathcal{M}_X|$. For example, if $\overline{X} = \{a^2b, bab, a^2, ba\}$ and $\overline{Y} = \{a^2b, bab, ba\}$, the minimal automaton $\mathcal{M}_Y$ of $Y$ of size 7 is bigger than the one of $X$ that is of size 5.

# 3   Operations on Cofinite Languages

In this section we investigate the state complexities of regular operations between cofinite languages and provide *ad hoc* algorithms to make the computations.

## 3.1   Union, Reversal and Quotient

These operations on cofinite languages given by the list of the words they do not contain are easy to realize.

**Proposition 3 (Union of two cofinite languages).** *Let $X_1$ and $X_2$ be two cofinite languages, with $\|\overline{X_1}\| = n_1$ and with $\|\overline{X_2}\| = n_2$. The union $X_1 \cup X_2$ is a cofinite language of state complexity at most $\min(n_1, n_2) + 2$. The minimal automaton of $X_1 \cup X_2$ can be computed in time $\mathcal{O}(n_1 + n_2)$.*

*Proof.* Assume by symmetry that $n_1 \leq n_2$. The first part follows from $X_1 \subset X_1 \cup X_2$ and Proposition 2. Using a classical lexicographic sort [11], one can compute $\overline{X_1} \cap \overline{X_2}$ in time $\mathcal{O}(n_1 + n_2)$: Form a list made of the element of $\overline{X_1}$ then those of $\overline{X_2}$, sort it, and extract words that appear twice. Then build the associated tree and minimize it in linear time using Proposition 1.

As the the reversal of the complement of a language is equal to the complement of the reversal the language, we obtain:

**Lemma 5 (Reversal of a cofinite language).** *Let $X$ be a cofinite language with $\|\overline{X}\| = n$, the reversal of $X$ is of state complexity at most $n + 2$ and its minimal automaton can be computed in linear time.*

**Proposition 4 (Quotient of a cofinite language).** *Let $X$ be a cofinite language with $\|\overline{X}\| = n$ and let $u \in A^*$. The state complexities of $u^{-1}X$ and $Xu^{-1}$ are at most $n + 2$. Their minimal automata can be computed in time $\mathcal{O}(n)$.*

*Proof.* For the state complexity, let $q_u$ be the state $\varepsilon \cdot u$ in $\mathcal{T}_X$. Consider the automaton $\mathcal{A}$ obtained by taking $q_u$ as initial state and by keeping only the accessible part of the automaton. Then $\mathcal{A}$ recognizes $u^{-1}X$, its number of states is at most $|\mathcal{T}_X| \leq n + 2$. This construction is linear in $n$ as one can stop reading $u$ in $\mathcal{T}_X$ as soon as $p_X$ is reached or if $|u| > n$.

## 3.2    Star

Note that the result given in Theorem 1 below shows that the behaviors of finite and cofinite languages are very different. Recall that Ellul, Krawetz, Shallit and Wang gives in [7] a finite language $X_h$, with $\|X_h\| = \Theta(h^2)$, such that the state complexity of $X_h^*$ is in $\Theta(h2^h)$.

**Theorem 1 (Star of a cofinite language).** *For any cofinite language $X$ with $\|\overline{X}\| = n$, the state complexity of $X^*$ is at most $n + 2$ in the worst case. There exists an algorithm that build the minimal automaton of $X^*$ in quadratic time.*

*Proof.* Since $X \subset X^*$, the state complexity of $X^*$ is at most $n + 2$ by Proposition 2. The time complexity is given in the following.

We propose two algorithms, one based on usual automata constructions and another one, easier to implement, related to dynamic programming. Both algorithms produce in time $\mathcal{O}(n^2)$ a deterministic automaton with at most $n + 2$ states that recognizes $X^*$. Once computed, the automaton can been minimized in linear time from Proposition 1.

**First Algorithm:** First associate to the language $X$ the automaton $\mathcal{T}_X = (A, Q_X, T, \{\varepsilon\}, F)$, as defined in Section 2.3. Build from $\mathcal{T}_X$ the nondeterministic automaton $\mathcal{A}(\mathcal{T}_X) = (A, Q_X, T', \{\varepsilon\}, F \cup \{\varepsilon\})$, where $T'$ is defined by:

$$T' = T \cup \{(u, a, a) \mid u \in F, \ a \in A \cap Q_X\} \cup \{(u, a, p_X) \mid u \subset F, a \in A \setminus Q_X\}$$

The automaton $\mathcal{A}(\mathcal{T}_X)$ is obtained from $\mathcal{T}_X$ by adding, for every $a \in A$, the transitions labelled by $a$ from every final state to the state $a$, when it exists.

**Lemma 6.** *For any cofinite language $X$, the nondeterministic automaton $\mathcal{A}(\mathcal{T}_X)$ recognizes the language $X^*$.*

To obtain a deterministic automaton recognizing $X^*$ from $\mathcal{A}(\mathcal{T}_X)$, we apply a tuned version of the accessible subset construction: Since all reachable subsets $Q$ of $Q_X$ containing $p_X$ are inseparable (one always has $L_Q = A^*$), we first create a state $P_X$ and assimilate every built subset containing $p_X$ to $P_X$. Hence the automaton is partially minimized on the fly, while doing the subset construction. Let $\mathcal{D}_X$ denote the resulting deterministic automaton, that recognizes $X^*$.

**Lemma 7.** *Let $X$ be a cofinite language and let $Q \neq P_X$ be a state of $\mathcal{D}_X$. The longest word in $Q$ is the label $u$ of the unique path $\pi$ from $\varepsilon$ to $Q$, and $u$ belongs to $Pr(\overline{X})$.*

*Proof.* By construction if $Q = \{u_1, \cdots, u_i\}$, with $i \in \mathbb{N}$ and each $u_i \in \mathrm{Pr}(\overline{X})$, is a state of $\mathcal{D}_X$ and $a$ is a letter in $A$ such that neither $Q$ nor $Q \cdot a$ are equal to $P_X$ then $Q \cdot a = \{a, u_1a, \cdots, u_ia\}$ if $Q \cap F \neq \emptyset$, and $Q \cdot a = \{u_1a, \cdots, u_ia\}$ if

---

**Algorithm 1.** inStar($X$,$u$)

---

1  **if** $u \in X$ *or* $u \in S$ **then return**
    *True*
2  **if** $u \in N$ **then return** *False*

3  **forall** $i \in \{1, \cdots, |u| - 1\}$ **do**
4  $\quad$ $v$ = prefix of length $i$ of $u$
5  $\quad$ $w$ = word such that $u = vw$
6  $\quad$ **if** *inStar(X,v) and $w \in X$*
      **then**
7  $\quad\quad$ Add $u$ in $S$
8  $\quad\quad$ **return** *True*
9  $\quad$ **end**
10 **end**
11 Add $u$ in $N$
12 **return** *False*

---

The algorithm inStar($X$,$u$) is called for every word $u \in \overline{X}$. To check whether a word is in $X^*$, first test if it is in $X$ or in $S$. $S$ is initially empty, and we store in $S$ every already tested word that belongs to $X^*$. If it is not in $X \cup S$, check whether $u \in N$ or not, *i.e.*, that is whether from previous computations it known that $u$ is not in $X^*$ or not. Initially $N = A \cap \overline{X}$ is made of letters that do not belong to $X$. If the algorithm continues to Step 3, $u$ is split in all possible ways in $u = vw$, with $v$ and $w$ nonempty, and we recursively check if $v \in X^*$ and $w \in X$. If it is true for one prefix $v$, $u$ is known to be in $X^*$ and added to $S$. If for all prefixes $v \notin X^*$ or $w \notin X$, then $u$ is known not to be in $X^*$ and added to $N$.

**Fig. 1.** Second algorithm: computation of $X^*$ with a dynamic programming approach

$Q \cap F = \emptyset$. As the states of $\mathcal{D}_X$ are the state $P_X$ and the states reachable from $\{\varepsilon\}$, the result is obtained by induction on the size of $u$.

If one labels the states with integers to avoid handling words, the implementation of this method can be done in $\mathcal{O}(n^2)$ since from Lemma 7 there are at most $n + 1$ states of the form $Q \neq P_X$ and each state is a set containing at most $n$ elements.

**Second Algorithm:** (see Fig.1)
Remark that as $X \subset X^*$, it is sufficient to determine which words of $\overline{X}$ are in $X^*$, that is, which words $u \in \overline{X}$ can be written $u = vw$, with $v$ and $w$ different from $\varepsilon$, such that $v \in X^*$ and $w \in X$. As such a $w$ is strictly smaller in size than $u$, this can be checked inductively as described in Fig. 1.

In practice we do not use two sets $S$ and $N$, but flags on the states of $\mathcal{T}_X$, to mark whether they correspond to words that are in $S$ or $N$. In this way, Step 1 and Step 2 are checked in time $\mathcal{O}(|u|)$ by reading the path labelled by $u$ in $\mathcal{T}_X$. So it is linear in the length of $u$ when the result is already known. When it is not, for each $i$, Step 6 is done in time $\mathcal{O}(i) + \mathcal{O}(|u| - i) = \mathcal{O}(|u|)$ if $v$ is in $X \cup S \cup N$. Counting separately the first calls for every word, the overall complexity is therefore $\mathcal{O}(\sum_u |u|)$, where the summation is done on all $u$ such that inStar($X$,$u$) is called. Since it can be called only for prefixes of elements in $\overline{X}$, the complexity is upper bounded by $\mathcal{O}(\sum_{u \in \Pr(\overline{X})} |u|) = \mathcal{O}(n^2)$.

### 3.3  Concatenation

In this section, we shall show that the state complexity of the concatenation of two cofinite languages is linear, and propose an algorithm to build the minimal automaton in linear time.

Let $Set_{n,m}$ be the set of sets of $m$ nonempty words whose sum of lengths is $n$: $Set_{n,m} = \{X = \{u_1, \cdots, u_m\} \mid \|X\| = n, \forall i \in \{1, \cdots, m\}, u_i \in A^*\}$.

**Theorem 2 (Concatenation of two cofinite languages).** *Let $X_1$ and $X_2$ be two cofinite languages such that $\overline{X_1} \in Set_{n_1,m_1}$ and $\overline{X_2} \in Set_{n_2,m_2}$. The state complexity of $X_1 \cdot X_2$ is at most $n_1 + 1 + \min(2^{m_2}, n_2 + 2)$. Moreover, in the particular case where $\varepsilon \in X_2$ (resp. $\varepsilon \in X_1$), the state complexity of $X_1 \cdot X_2$ is at most $n_1 + 2$ (resp. $n_2 + 2$). The minimal automaton of $X_1 \cdot X_2$ can be computed in time $\mathcal{O}(n_1 + n_2)$.*

*Proof.* First if $\varepsilon \in X_2$, then $X_1 \subset X_1 \cdot X_2$, and the state complexity of $X_1 \cdot X_2$ is at most $n_1 + 2$ by Proposition 2. Similarly, if $\varepsilon \in X_1$ then the state complexity of $X_1 \cdot X_2$ is at most $n_2 + 2$. The general upper bound for the state complexity of $X_1 \cdot X_2$ is proved in Lemmas 9 and 12.

Associate to the cofinite languages $X_1$ and $X_2$ the automata $\mathcal{T}_1 = \mathcal{T}_{X_1} = (A, Q_1, \cdot, \{\varepsilon\}, F_1)$ and $\mathcal{T}_2 = \mathcal{T}_{X_2} = (A, Q_2, *, \{\varepsilon\}, F_2)$, as defined in Section 2.3. Then use the classical construction of the concatenation of two automata: From each final state $q$ of the automaton $\mathcal{T}_1$ and for each letter $a \in A$, add a transition from $q$ to the state $\varepsilon * a$ in $\mathcal{T}_2$. Formally consider the nondeterministic automaton $\mathcal{A}_{X_1 X_2} = (A, (Q_1 \times \{\emptyset\}) \cup (\{\emptyset\} \times Q_2), T_1 \cup T_2 \cup T, \{(\varepsilon, \emptyset)\}, F)$, with:

- $T_1 = \{((u, \emptyset), a, (u \cdot a, \emptyset)) \mid u \in \mathrm{Pr}(\overline{X_1}) \cup \{p_{X_1}\}, a \in A\}$ to form a copy of $\mathcal{T}_1$ on the first coordinate.
- $T_2 = \{((\emptyset, u), a, (\emptyset, u * a)) \mid u \in \mathrm{Pr}(\overline{X_2}) \cup \{p_{X_2}\}, a \in A\}$ to form a copy of $\mathcal{T}_2$ on the second coordinate.
- $T = \{((u, \emptyset), a, (\emptyset, \varepsilon * a)) \mid u \in F_1, a \in A\}$.
- $F = F_1 \times \{\emptyset\} \cup \{\emptyset\} \times F_2$ if $\varepsilon \in F_2$ and $F = \{\emptyset\} \times F_2$ if $\varepsilon \notin F_2$.

Then $\mathcal{A}_{X_1 X_2}$ recognizes $X_1 \cdot X_2$. Consider the automaton $\mathcal{D}_{X_1 X_2}$ obtained from $\mathcal{A}_{X_1 X_2}$ using the accessible subset construction. The states of $\mathcal{D}_{X_1 X_2}$ are sets of pairs. But since only transitions from the copy of $\mathcal{T}_1$ to the copy of $\mathcal{T}_2$ have been added, the automaton is deterministic on its first coordinate. Hence each state $Q$ of $\mathcal{D}_{X_1 X_2}$ can be rewritten as $(u, Q)$, where $u$ is the unique value of the first coordinate in $Q_1$ and $Q \subset Q_2$ is the set of values of the second coordinate.

**Lemma 8.** *Let $(u, Q)$ be a state of $\mathcal{D}_{X_1 X_2}$, with $u \in Pr(\overline{X_1})$. Every word in $Q \setminus \{p_{X_2}\}$ is a suffix of $u$. In particular, $Q \setminus \{p_{X_2}\}$ is a suffix chain: for every $v, w \in Q \setminus \{p_{X_2}\}$ either $v$ is suffix of $w$ or $w$ is suffix of $v$.*

*Proof.* In $\mathcal{T}_1$, for every state $u$ that is not $p_{X_1}$, there is only one path from the initial state that reaches it, which is the path of label $u$. Hence, the path labelled by $u$ is only one path in $\mathcal{D}_{X_1 X_2}$ that reaches the state $(u, Q)$. Every state $v$ in $Q$ that is not $p_{X_2}$ is the label of a path from a final state of $\mathcal{T}_1$, reading the first letter $a$ of $v$ while going from the copy of $\mathcal{T}_1$ to the copy of $\mathcal{T}_2$, then following the path labelled by $a^{-1}v$ in $\mathcal{T}_2$. Hence, as it has been done while following the path labelled by $u$ in $\mathcal{T}_1$, $v$ is a suffix of $u$.

**Lemma 9.** *Let $X_1$ and $X_2$ be two cofinite languages, with $\overline{X_1}$ in $Set_{n_1,m_1}$ and $\overline{X_2}$ in $Set_{n_2,m_2}$. If $(p_{X_1}, Q)$ is a state of $\mathcal{D}_{X_1 X_2}$, the language $L_{(p_{X_1},Q)}$ contains $A^* X_2$. The state complexity of $X_1 \cdot X_2$ is at most $n_1 + 2^{m_2} + 1$.*

*Proof.* As stated above, for every $u \in \mathrm{Pr}(\overline{X_1})$ there is a unique state $(u, Q)$ in $\mathcal{D}_{X_1 X_2}$. Hence there are at most $n_1 + 1$ states in $\mathcal{D}_{X_1 X_2}$ such that the first coordinate is not $p_{X_1}$.

Let $(p_{X_1}, Q)$ be a state of $\mathcal{D}_{X_1 X_2}$ and let $L_{(p_{X_1},Q)}$ be the language recognized by taking $(p_{X_1}, Q)$ as initial state. Every word $u$ of $A^* X_2$ is in $L_{(p_{X_1},Q)}$: Let $v \in A^*$ and $w \in X_2$ be such that $u = vw$, one can loop on $p_{X_1}$ on the first coordinate while reading $v$, then since in $\mathcal{A}_{X_1 X_2}$ there is a transition from $(p_{X_1}, \emptyset)$ labelled by the first letter $a$ of $w$ to $(\emptyset, \varepsilon * a)$, that is the starting point of a path labelled by $a^{-1} w$ in the copy of $T_2$, $u$ is recognized by $\mathcal{D}_{X_1 X_2}$. Hence $X_2 \subset A^* X_2 \subset L_{(p_{X_1},Q)}$. Since $\overline{X_2}$ contains $m_2$ elements, there are at most $2^{m_2}$ distinct languages of the form $L_{(p_{X_1},Q)}$. Hence the states of the form $(p_{X_1}, Q)$ are in at most $2^{m_2}$ equivalence classes of Myhill-Nerode equivalence, concluding the proof.

The property $A^* X_2 \subset L_{(p_{X_1},Q)}$ in Lemma 9 is the key of the next results. The following lemma characterizes languages of the form $\overline{A^* X_2}$.

**Lemma 10.** *A word $u$ belongs to $\overline{A^* X_2}$ if and only if all its suffixes belong to $\overline{X_2}$. Consequently $\overline{A^* X_2}$ is the greatest suffix-closed subset of $\overline{X_2}$ and if $\varepsilon \in X_2$, then $A^* X_2 = A^*$.*

**Lemma 11.** *Each language $L_{(p_{X_1},Q)}$ is the set of labels of the successful paths in the minimal automaton of $A^* X_2$ taking as initial state the initial state of the automaton if $Q$ is empty, the state corresponding to the equivalence class of $p_{X_2}$ if $p_{X_2} \in Q$ or to the equivalence class of the longest word of $Q$ otherwise.*

*Proof.* Setting $S = \overline{A^* X_2}$, let $\mathcal{M}_{\overline{S}}$ be the minimal automaton of $\overline{S}$ that is $A^* X_2$. The state corresponding to the class of the state $p_{X_2}$ in $\mathcal{M}_{\overline{S}}$ is still denoted by $p_{X_2}$. Note that if $Q$ is the empty set, $L_{(p_{X_1},Q)} = A^* X_2$ and if $p_{X_2} \in Q$ then $L_{(p_{X_1},Q)} = A^*$; otherwise $L_{(p_{X_1},Q)} = \cup_{u \in Q} u^{-1} \overline{S}$ or $\overline{L_{(p_{X_1},Q)}} = \cap_{u \in Q} u^{-1} S$. As $S$ is suffix-closed, for any $u \in Q$, $u^{-1} S \subset S$. Moreover If $u$ is a suffix of $v$, then when $vw \in S$, then $uw \in S$ and $v^{-1} S \subseteq u^{-1} S$. So $\overline{L_{(p_{X_1},Q)}} = w^{-1} S$ where $w$ is the longest word of $Q$.

Therefore each language $L_{(p_{X_1},Q)}$ is the set of labels of paths in $\mathcal{M}_{\overline{S}}$ from one of the state, says $q$, to the final states. If $Q$ is the empty set, $q = \varepsilon$, if $p_{X_2} \in Q$ then $q = p_{X_2}$, otherwise $q$ is the state reached in $\mathcal{M}_{\overline{S}}$ reading the longest word $w$ of $Q$ from the initial state.

Note that as $S$ is suffix-closed the size of the minimal automaton of $\overline{S}$ is smaller or equal to $2^{|S|}$.

When $Q$ is not empty all the information we need about the state $(u, Q)$ is given by $(u, p_{X_2})$ if $p_{X_2}$ belongs to $Q$ and by $(u, w)$ otherwise, $w$ being the longest word of $Q$.

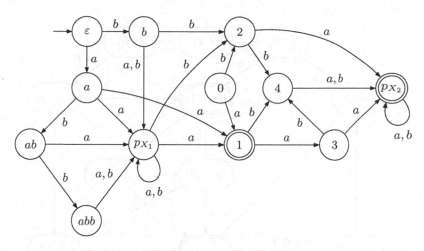

**Fig. 2.** The automaton $\mathcal{A}'_{X_1X_2}$ for $\overline{X_1} = \{\varepsilon, b, ab, abb\}$ and $\overline{X_2} = \{c, b, aa, ab, bb, aab\}$

**Lemma 12.** *Let $X_1$ and $X_2$ be two cofinite languages, with $\overline{X_1}$ in $Set_{n_1,m_1}$ and $\overline{X_2}$ in $Set_{n_2,m_2}$. The minimal automaton of $X_1 \cdot X_2$ has at most $n_1 + n_2 + 3$ states and can be computed in time $\mathcal{O}(n_1 + n_2)$.*

*Proof.* We will construct a deterministic automaton recognizing $X_1 \cdot X_2$ equivalent to $\mathcal{D}_{X_1X_2}$.

First build the automaton $\mathcal{M}_{\overline{S}}$ is time $\mathcal{O}(n_2)$ from the automaton $\mathcal{A}_{X_2}$: the reversal $\widetilde{X_2}$ of $X_2$ is cofinite and a depth-first search in $\mathcal{T}_{\widetilde{X_2}}$ is enough to obtain the greatest prefix-closed subset of the complement of $\widetilde{X_2}$, which is reversal of the greatest suffix-closed subset of $\overline{X_2}$. Denote by $X$ the cofinite set such that $\overline{X}$ is the greatest suffix-closed subset of $\overline{X_2}$. The next step is to build $\mathcal{T}_X$ and to minimize it in time $\mathcal{O}(n_2)$, using Proposition 1.

Then construct the automaton $\mathcal{A}'_{X_1 \cdot X_2}$ as the automaton $\mathcal{A}_{X_1 \cdot X_2}$ but from the automata $\mathcal{A}_{X_1}$ and $\mathcal{M}_{\overline{S}}$ instead of $\mathcal{A}_{X_2}$. This is done in time $\mathcal{O}(n_1 + n_2)$.

Apply the subset construction to $\mathcal{A}'_{X_1 \cdot X_2}$ until finding states with $p_{X_1}$ as first component. The intermediate states of this automaton do not have $p_{X_1}$ as first component, so their number is at most $n_1 + 1$. Moreover the second component can be reduced to $\emptyset$, $p_{X_2}$ and its longest element otherwise. For each state $(p_{X_1}, Q)$ continue to apply the subset construction to $\mathcal{A}'_{X_1 \cdot X_2}$ until finding state $(p_{X_1}, p_{X_2})$. This part of the algorithm is just the traversal of acyclic paths in $\mathcal{M}_{\overline{S}}$ and can be done in $\mathcal{O}(n_2)$. Finally add a loop from $(p_{X_1}, p_{X_2})$ to itself for every letter of the alphabet. The final states are the one whose second component is final in $\mathcal{M}_{\overline{S}}$. The automaton obtained recognizes $X_1 \cdot X_2$ and the total complexity of the construction is $\mathcal{O}(n_1 + n_2)$, and at most $(n_1+1)+(n_2+2) = n_1 + n_2 + 3$ states have been built.

*Example 1.* Fig. 2 and Fig. 3 depict an example of the constructions used in the proofs of this section.

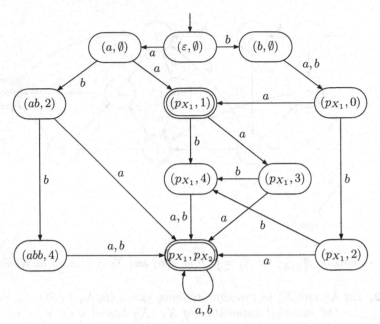

**Fig. 3.** The automaton $\mathcal{D}_{X_1 X_2}$, for $\overline{X_1} = \{\varepsilon, b, ab, abb\}$ and $\overline{X_2} = \{\varepsilon, b, aa, ab, bb, aab\}$

## 4    Remarks

As shown in Lemma 1 the union of a cofinite language $X$, given by the list of the words in $\overline{X}$, with $\|\overline{X}\| = n$, and a regular language $L$ given by a deterministic automaton $\mathcal{A}$ is a cofinite language. Its minimal automaton can be computed in time $\mathcal{O}(n)$: Compute the standard product automaton of $\mathcal{T}_X$ and $\mathcal{A}$, identifying on the fly all states whose first component is the accepting sink state of $\mathcal{T}_X$.

For the concatenation, we have not establish any interesting state complexity result, leaving it as an open problem, but the following lemma characterizes the conditions in which the concatenation of a cofinite language and a regular language is cofinite.

**Lemma 13.** *Let $Y$ be a cofinite language and $X$ be a regular language. The language $X \cdot Y$ (resp. $Y \cdot X$) is cofinite if and only if there exists a positive integer $p$ such that, for every word $w$ of length at least $p$, there exists a prefix (resp. suffix) of $w$ that is in $X$.*

*Proof.* From Section 2.2 if $X \cdot Y$ is cofinite then there exists a positive number $p$ such that for each word $w$ of length greater than $p$, $w$ is in $X \cdot Y$. Then there exists a prefix of $w$ in $X$.

Conversely as $Y$ is cofinite there exists a positive number $p_Y$ such that every word of length at least $p_Y$ is in $Y$. Let $w$ be a word of length greater than or equal to $p + p_Y$. Let $u$ be the prefix of length $p$ of $w$. By hypothesis, there exists a prefix $u'$ of $u$ that belongs to $X$. $u'$ is also a prefix of $w$, hence there exists a word $v$, with

$|v| \geq p_Y$, such that $w = u'v$. Therefore, $v \in Y$ and $w \in X \cdot Y$. Consequently, $X \cdot Y$ is cofinite since it contains every word of length at least $p + n_Y$.

Note that the condition of Lemma 13 is equivalent to say that $X$ contains a maximal prefix (resp. suffix) code [12]. Testing whether $X$ contains a maximal prefix code can be done in time $\mathcal{O}(|\mathcal{A}|)$, where $\mathcal{A}$ is a given deterministic automaton recognizing $X$, by removing final states in $\mathcal{A}$ and then checking if there is no cycle accessible from the initial state in the remaining graph.

# References

1. Yu, S., Zhuang, Q., Salomaa, K.: The state complexities of some basic operations on regular languages. Theor. Comput. Sci. 125(2), 315–328 (1994)
2. Holzerand, M., Kutrib, M.: State complexity of basic operations on nondeterministic finite automata. In: Champarnaud, J.-M., Maurel, D. (eds.) CIAA 2002. LNCS, vol. 2608, pp. 148–157. Springer, Heidelberg (2001)
3. Gruber, H., Holzer, M.: On the average state and transition complexity of finite languages. Theor. Comput. Sci. 387(2), 155–166 (2007)
4. Jirásková, G., Okhotin, A.: On the state complexity of operations on two-way finite automata. In: [13], pp. 443–454
5. Campeanu, C., Culik II, K., Salomaa, K., Yu, S.: State complexity of basic operations on finite languages. In: Boldt, O., Jürgensen, H. (eds.) WIA 1999. LNCS, vol. 2214, pp. 60–70. Springer, Heidelberg (2001)
6. Han, Y.S., Salomaa, K., Wood, D.: Nondeterministic state complexity of basic operations for prefix-free regular languages. Fundam. Inf. 90(1-2), 93–106 (2009)
7. Ellul, K., Krawetz, B., Shallit, J., wei Wang, M.: Regular expressions: New results and open problems. Journal of Automata, Languages and Combinatorics 10(4), 407–437 (2005)
8. Bassino, F., Giambruno, L., Nicaud, C.: The average state complexity of the star of a finite set of words is linear. In: [13], pp. 134–145
9. Hopcroft, J.E., Ullman, J.D.: Introduction to Automata Theory, Languages, and Computation. Addison-Wesley, Reading (1979)
10. Revuz, D.: Minimisation of acyclic deterministic automata in linear time. Theor. Comput. Sci. 92(1), 181–189 (1992)
11. Crochemore, M., Hancart, C., Lecroq, T.: Algorithms on strings. Cambridge University Press, Cambridge (2007)
12. Berstel, J., Perrin, D.: Theory of Codes. Academic Press, London (1985)
13. Ito, M., Toyama, M. (eds.): DLT 2008. LNCS, vol. 5257. Springer, Heidelberg (2008)

# Fast Set Intersection and Two-Patterns Matching

Hagai Cohen and Ely Porat*

Department of Computer Science, Bar-Ilan University, 52900 Ramat-Gan, Israel
{cohenh5,porately}@cs.biu.ac.il

**Abstract.** In this paper we present a new problem, the *fast set intersection* problem, which is to preprocess a collection of sets in order to efficiently report the intersection of any two sets in the collection. In addition we suggest new solutions for the *two-dimensional substring indexing* problem and the *document listing* problem for two patterns by reduction to the *fast set intersection* problem.

## 1 Introduction and Related Work

The intersection of large sets is a common problem in the context of retrieval algorithms, search engines, evaluation of relational queries and more. Relational databases use indices to decrease query time, but when a query involves two different indices, each one returning a different set of results, we have to intersect these two sets to get the final answer. The running time of this task depends on the size of each set, which can be large and make the query evaluation take longer even if the number of results is small. In information retrieval there is a great use of inverted index as a major indexing structure for mapping a word to the set of documents that contain that word. Given a word, it is easy to get from the inverted index the set of all the documents that contain that word. Nevertheless, if we would like to search for two words to get all documents that contain both, the inverted index doesn't help us that much. We have to calculate the occurrences set for each word and intersect these two sets. The problem of intersecting sets finds its motivation also in web search engines where the dataset is very large.

Various algorithms to improve the problem of intersecting sets have been introduced in the literature. Demaine et al. [1] proposed a method for computing the intersection of $k$ sorted sets using an adaptive algorithm. Baeza-Yates [2] proposed an algorithm to improve the multiple searching problem which is related directly to computing the intersection of two sets. Barbay et al. [3] showed that using interpolation search improves the performance of adaptive intersection algorithms. They introduced an intersection algorithm for two sorted sequences that is fast on average. In addition Philip et al. [4] presented a solution for

---

* This work was supported by BSF and ISF.

A. López-Ortiz (Ed.): LATIN 2010, LNCS 6034, pp. 234–242, 2010.

computing expressions on given sets involving unions and intersections. A special case of their result is the intersectin of $m$ sets containing $N$ elements in total, which they solve in expected time $O(N(\log \omega)^2/\omega + m \cdot output)$ for word size $\omega$ where *output* is the number of elements in the intersection.

In this paper we present a new problem, the *fast set intersection* problem. This problem is to preprocess a databases of size $N$ consisting of a collection of $m$ sets to answer queries in which we are given two set indices $i, j \leq m$, and wish to find their intersection. This problem has lots of applications where there is a need to intersect two sets in a lot of different fields like Information Retrieval, Web Searching, Document Indexing, Databases etc. An optimal solution for this problem will bring better solutions to various applications.

We solve this problem using minimal space and still decrease the query time by using a preprocessing part. Our solution is the first non-trivial algorithm for this problem. We give a solution that requires linear space with worst case query time bounded by $O(\sqrt{N\,output} + output)$ where *output* is the intersection size.

In addition, we present a solution for the *two-dimensional substring indexing* problem, introduced by Muthukrishnan et al. [5]. In this problem we preprocess a database $D$ of size $N$. So when given a string pair $(\sigma_1, \sigma_2)$, we wish to return all the database string pairs $\alpha_i \in D$ such that $\sigma_1$ is a substring of $\alpha_{i,1}$ and $\sigma_2$ is a substring of $\alpha_{i,2}$. Muthukrishnan et al. suggested a tunable solution for this problem which uses $O(N^{2-y})$ space for a positive fraction $y$ and query time of $O(N^y + output)$ where *output* is the number of such string pairs. We present a solution for this problem, based on solving the *fast set intersection* problem, that uses $O(N \log N)$ space with $O((\sqrt{N \log N\,output} + output) \log^2 N)$ query time.

In the *document listing* problem which was presented by Muthukrishnan [6], we are given a collection of size $N$ of text documents which may be preprocessed so when given a pattern $p$ we want to return the set of all the documents that contain that pattern. Muthukrishnan suggested an optimal solution for this problem which requires $O(N)$ space with $O(|p| + output)$ query time where *output* is the number of documents that contain the pattern. However, there is no optimal solution when given a query consists of two patterns $p, q$ to return the set of all the documents that contain them both. The only known solution for this problem is of Muthukrishnan [6] which suggested a solution that uses $O(N\sqrt{N})$ space which supports queries in time $O(|p| + |q| + \sqrt{N} + output)$. We present a solution for the *document listing* problem when the query consists of two patterns. Our solution uses $O(N \log N)$ space with $O(|p| + |q| + (\sqrt{N \log N\,output} + output) \log^2 N)$ query time.

The paper is structured as follows: In Sect. 2 we describe the *fast set intersection* problem. In Sect. 3 we describe our solution for this problem. In Sect. 4 we present similar problems with their solutions. In Sect. 5 we present our solution for the *two-dimensional substring indexing* problem and the *document listing* problem for two patterns. In Sect. 6 we present some concluding remarks.

## 2    Fast Set Intersection Problem

We formally define the fast set intersection (FSI) problem.

**Definition 1.** *Let $D$ be a database of size $N$ consisting of a collection of $m$ sets. Each set has elements drawn from $1 \ldots c$. We want to preprocess $D$ so that given a query of two indices $i, j \leq m$, we will be able to calculate the intersection between sets $i, j$ efficiently.*

A naive solution for this problem is to store the sets sorted. Given a query of two sets $i, j$, go over the smaller set and check for each element if it exists in the second set. This costs $O(min(|i|, |j|) \log max(|i|, |j|))$. This solution can be further improved using hash tables. A static hash table [7] can store $n$ elements using $O(n)$ space and build time, with $O(1)$ query time. For each set we can build a hash table to check in $O(1)$ time if an element is in the set or not. This way the query time is reduced to $O(min(|i|, |j|))$ using linear space. The disadvantage of using this solution is that on the worst case we go over a lot of elements even if the intersection is small. A better query time can be gained by using more space for saving the intersection between every two sets. Using $O(m^2 c)$ space we get an optimal query time of $O(output)$ where $output$ is the size of the intersection. Nevertheless, this solution uses extremely more space. In the next section we present our solution for the *fast set intersection* problem which bounds the query time on the worst case.

## 3    Fast Set Intersection Solution

Here we present our algorithm for solving the FSI problem. We call *result set* to the output of the algorithm, i.e., the intersection of the two sets. By *output* we denote the size of the result set.

### 3.1    Preprocessing

For each set in $D$ we store a hash table to know in $O(1)$ time if an element is in that set or not. In addition, we store the inverse structure, i.e., for each element we store a hash table to know in $O(1)$ time if it belongs to a given set or not.

Our main data structure consists of an unbalanced binary tree. Starting from the root node at level 0, each node in that tree handles number of subsets of the original sets from $D$. The cost of a node in that tree is the sum of the sizes of all the subsets it handles. The root node handles all the $m$ sets in $D$, therefore, it costs $N$.

**Definition 2.** *Let $d$ be a node which costs $n$. A large set in $d$ is a set which has more than $\sqrt{n}$ elements.*

**Lemma 1.** *By definition, a node $d$ which costs $n$, can handle at most $\sqrt{n}$ large sets.*

A *set intersection matrix* is a matrix that stores for each set if it has an intersection with any other set. For $\hat{m}$ sets this matrix costs $O(\hat{m}^2)$ bits space with $O(1)$ query time for answering if set $i$ and set $j$ have a non-empty intersection.

For each node we construct a set intersection matrix for the large sets in that node. By lemma 1, saving the set intersection matrix only for the large sets in a node that costs $n$ space will cost only another $n$ space.

Now we describe how we divide sets between the children of a node. Only large sets in a node will be propagated down to its two children, we call them the *propagated group*. Let $d$ be a node which costs $n$ and let $G$ be its propagated group. Then, $G$ costs at most $n$ as well. Let $E$ be the set of all elements in the sets of $G$. We partition $E$ into two disjoint sets $E_1, E_2$. For a given set $S \in G$ we partition it between the two children as following: The left child will handle $S \cap E_1$ and the right child will handle $S \cap E_2$. We want each child of $d$ to cost at most $\frac{n}{2}$. Nevertheless, finding such a partition of $E$ is a hard problem, if even possible at all. To overcome this difficulty we shall add elements to $E_1$ until adding another element will make the left child cost more than $\frac{n}{2}$. The next element, which we denote by $e$, will be remarked in $d$ for checking, during query time, whether it lies in the intersection. We now take $E_2 = E - E_1 - \{e\}$, i.e., the remaining elements. This way each child costs at most $\frac{n}{2}$.

A leaf in this binary tree is a node which is in constant size. Because each node in the tree costs half the space of its parent then this tree has $\log N$ levels.

**Theorem 1.** *The space needed for this data structure is $O(N)$ space.*

*Proof.* The hash tables for all the sets cost $O(N)$ space. As well the inverse hash tables for all the elements cost $O(N)$ space.

The binary tree structure space cost is as follows: The root costs $O(N)$ bits for saving the set intersection matrix. In each level we store only another $O(N)$ bits because every two children don't cost more than their parent. Hence, the total cost of this tree structure is $O(N \log N)$ bits which is $O(N)$ space in term of words. □

### 3.2  Query Answering

Given sets $i, j$ (without loss of generality we assume $|i| \leq |j|$), we start traversing the tree from the root node. If $i$ is not a large set in the root we check each element from it in the hash table of $j$. As there can be at most $\sqrt{N}$ elements in $i$ because it is not a large set, this will cost $O(\sqrt{N})$. If both $i, j$ are large sets we do as follows: We check in the set intersection matrix of the root wether there is a non-empty intersection between $i$ and $j$. If there is not there is nothing to add to the result set so we stop traversing down. If there is an intersection we check the hash table of the element which is remarked in that node if it belongs to $i$ and $j$ and add that element to the intersection if it belongs to both. Next we go down to the children of the root and continue the traversing recursively.

Elements are added to the result set when we get to a node which in that node $i$ is not a large set. In this case, we stop traversing down the tree from that node. Instead we step over all the elements of $i$ in that node checking for each one of them if it belongs to $j$. We call such a node a *stopper node*.

**Theorem 2.** *The query time is bounded by* $O(\sqrt{N\,output} + output)$.

*Proof.* The query computation consists of two parts. The tree traversal part and the time we spend on stopper nodes.

There are *output* elements in the result set, therefore, there can be at most $O(output)$ stopper nodes. Because the tree height is $\log N$, for each stopper node we visit at most $\log N$ nodes for the tree traversal until we get to it. Therefore, the tree traversal part adds at most $O(output \log N)$ to the query time. But this is more than what we actually pay for the tree traversal because some stopper nodes share their path from the root. This can be bounded better. Because the tree is a binary tree if we fully traverse the tree till $\log output$ height it will cost $O(output)$ time. Now, from this height if we continue traverse the tree we visit for each stopper node at most $\log N - \log output$ nodes because we are already at $\log output$ height. Thus, the tree traversal part is bounded by $O(output + output(\log N - \log output))$. By log rules this equals to $O(output + output \log \frac{N}{output})$.

Now, we calculate how much time we spent on all the stopper nodes. A stopper node is a node which during the tree traversal we have to go over all elements of a non-large set in that node. The size of a non-large set in a stopper at level $l$ is $\sqrt{\frac{N}{2^l}}$. Consider there are $x$ stopper nodes. We denote by $l_i$ the level for stopper node $i$. For all stopper nodes we pay at most:

$$\sum_{i=1}^{x} \sqrt{\frac{N}{2^{l_i}}} = \sqrt{N} \sum_{i=1}^{x} 2^{-\frac{1}{2}l_i} = \sqrt{N} \sum_{i=1}^{x} 1 \cdot 2^{-\frac{1}{2}l_i}$$

The Cauchy-Schwarz inequality is that $(\sum_{i=1}^{n} x_i y_i)^2 \le (\sum_{i=1}^{n} x_i^2)(\sum_{i=1}^{n} y_i^2)$. We use it in our case to get:

$$\le \sqrt{N} \sqrt{\sum_{i=1}^{x} 1^2} \sqrt{\sum_{i=1}^{x} (2^{-\frac{1}{2}l_i})^2}$$

$$= \sqrt{N} \sqrt{x} \sqrt{\sum_{i=1}^{x} 2^{-l_i}}$$

Kraft inequality from Information Theory states that for any binary tree:

$$\sum_{l \in leaves} 2^{-depth(l)} \le 1$$

Because we never visit a subtree rooted by a stopper node, then in our case each stopper node can be viewed as a leaf in the binary tree. Therefore, we can transform Kraft inequality for all the stopper nodes instead of all tree leaves to get that $\sum_{i=1}^{x} 2^{-l_i} \leq 1$. Using this inequality gives us that:

$$\leq \sqrt{N}\sqrt{x} = \sqrt{Nx} \leq \sqrt{Noutput} = output\sqrt{\frac{N}{output}}$$

Thus, we pay $O(output\sqrt{\frac{N}{output}})$, for the time we spend in the stopper nodes.

Therefore, the tree traversal part and the time we spend on all stopper nodes is $O(output + output \log \frac{N}{output} + output\sqrt{\frac{N}{output}})$. Hence, the final query time is bounded by $O(\sqrt{Noutput} + output)$. □

**Corollary 1.** *The fast set intersection problem can be solved in linear space with worst case query time of $O(\sqrt{Noutput} + output)$.*

## 4 Intersection-Empty Query and Intersection-Size Query

In the FSI problem given a query we want to return the result set, i.e., the intersection between two sets. What if we only want to know if there is any intersection between two sets? We call that the *intersection-empty query* problem. Moreover, sometimes we would like only to know the size of the intersection without calculating the actual result set. We define these problems as follows:

**Definition 3.** *Let D be a database of size N consisting of a collection of m sets. Each set has elements drawn from 1...c. The* intersection-empty query *problem is to preprocess D so that given a query of two indices $i, j \leq m$, we want to calculate if sets $i, j$ have any intersection. In the* intersection-size query *problem when given a query we want to calculate the size of the result set.*

A naive solution for the intersection-empty query problem is to build a matrix saving if there is any intersection between every two sets. This solution uses $O(m^2)$ bits space with query time of $O(1)$. For the intersection-size query problem we store the intersection size for every two sets by using slightly more space, $O(m^2)$ space, with query time of $O(1)$.

We can use part of our FSI solution method to solve the intersection-empty query problem using $O(N)$ space with $O(\sqrt{N})$ query time. Instead of the whole tree structure we store only the root node with its set intersection matrix using $O(N)$ space. Given sets $i, j$ (without loss of generality let's assume $|i| \leq |j|$), if $i$ is not large set in the root we check each element from it in the hash table of $j$. Because $i$ is not large set, this will cost at most $O(\sqrt{N})$ time. If $i$ is a large set then we check in the set intersection matrix of the root to see if there is any intersection in $O(1)$ time. Hence, we can solve the intersection-empty query problem in $O(\sqrt{N})$ time using $O(N)$ space.

With the same method we can solve the intersection-size query problem by saving the size of the intersection instead of saving if there is any intersection in the set intersection matrix. This way we can solve the intersection-size query problem in $O(\sqrt{N})$ time using $O(N)$ space.

# 5    Two-Dimensional Substring Indexing Solution

In this section, we show how to solve the *two-dimensional substring indexing* problem and the *document listing* problem for two patterns using our FSI solution. The *two-dimensional substring indexing* problem was showed by Muthukrishnan et al. [5]. It is defined as follows:

**Definition 4.** *Let $D$ be a database consisting of a collection of string pairs $\alpha_i = (\alpha_{i,1}, \alpha_{i,2}), 1 \leq i \leq c$, which may be preprocessed. Given a query string pair $(\sigma_1, \sigma_2)$, the 2-d substring indexing problem is to identify all string pairs $\alpha_i \in D$, such that $\sigma_i$ is a substring of $\alpha_{i,1}$ and $\sigma_2$ is a substring of $\alpha_{i,2}$.*

Muthukrishnan et al. [5] reduced the *two-dimensional substring indexing* problem to the *common colors query* problem which is defined as follows:

**Definition 5.** *We are given an array $A[1 \ldots N]$ of colors drawn from $1 \ldots C$. We want to preprocess this array so that the following query can be answered efficiently: Given two non-overlapping intervals $I_1, I_2$ in $[1, N]$, list the distinct colors that occur in both intervals $I_1$ and $I_2$.*

The common colors query (CCQ) problem is another intersection problem where we have to intersect two intervals on the same array. We now show how to solve the CCQ problem by solving the FSI problem. By that we solve the *two-dimensional substring indexing* problem as well.

Given array $A$ of size $N$, we build a data structure consisting of $\log N$ levels over this array. In the top level we partition $A$ into two sets of size at most $\frac{N}{2}$, the first set containing colors, i.e., elements, of $A$ in range $A[1 \ldots \frac{N}{2}]$ and the second set containing colors in range $A[\frac{N}{2} + 1 \ldots N]$. As well, each level $i$ is partitioned into $2^i$ sets, each respectively, containing a successive set of $\frac{N}{i}$ colors from $A$. The bottom level, in similar fashion, is therefore partitioned into $N$ sets each containing one different color from array $A$. The size of all the sets in each level is $O(N)$. Therefore, the size needed for all the sets in all levels is $O(N \log N)$.

**Lemma 2.** *An interval $I$ on $A$ can be covered by at most $2 \log N$ sets.*

*Proof.* Assume, by contradiction, that there exists an interval for which at least $m > 2 \log N$ sets are needed. This implies that there is some level that at least 3 (consecutive) sets are selected. However, for every 2 consecutive sets there have to be a set in the upper level that contains them both, so we can take it instead, and cover the same interval with only $m - 1$ sets, in contradiction to the assumption that at least $m$ sets are required for the cover.    □

**Theorem 3.** *The CCQ problem can be solved using $O(N \log N)$ space with $O((\sqrt{N \log N output} + output) \log^2 N)$ query time where output is the number of distinct colors that occur in both $I_1$ and $I_2$.*

*Proof.* Given two intervals $I_1, I_2$ we want to calculate their intersection, By lemma 2, $I_1, I_2$ are each covered by a group of $2 \log n$ sets at the most. To get the intersection of $I_1, I_2$ we will take each set from the first group and intersect it with each set from the second group using our FSI solution. Hence, we have to solve the FSI problem $O(\log^2 N)$ times. Our FSI solution takes $O(\sqrt{N\,output} + output)$ time and $O(N)$ space for dataset which costs $O(N)$ space. Here the dataset costs $O(N \log N)$ space, therefore, we can solve the common colors query problem in $O((\sqrt{N \log N\,output} + output) \log^2 N)$ time using $O(N \log N)$ space. $\qquad\square$

As showed in [5] to solve the two-dimensional substring problem we can solve a CCQ problem. As a result, the two-dimensional substring problem can be solved in $O((\sqrt{N \log N\,output} + output) \log^2 N)$ time using $O(N \log N)$ space.

### 5.1  Document Listing Solution for Two Patterns

The document listing problem was presented by Muthukrishnan [6]. In this problem we are given a collection $D$ of text documents $d_1, \ldots, d_c$, with $\sum_i |d_i| = N$, which may be preprocessed, so when given a query comprising of a pattern $p$ our goal is to return the set of all documents that contain one or more copies of $p$. Muthukrishnan presented an optimal solution for this problem by building a suffix tree for $D$, searching the suffix tree for $p$ and getting an interval $I$ on an array with all the occurrences of $p$ in $D$. Then they solve the colored range query problem on $I$ to get each document only once. This solution requires $O(N)$ space with optimal query time of $O(|p| + output)$ where $output$ is the number of documents that contain $p$.

We are interested in solving this problem for a two patterns query. Given two patterns $p, q$, our goal is to return the set of all documents that contain both $p$ and $q$. In [6] there is a solution that uses $O(N\sqrt{N})$ space with $O(|p| + |q| + \sqrt{N} + output)$ query time. Their solution is based on searching a suffix tree of all the documents for the two patterns $p, q$ in $O(|p| + |q|)$ time. From this they get two intervals: $I_1$ with $p$ occurrences and $I_2$ with $q$ occurrences.. On these intervals they solve a CCQ problem to get the intersection between $I_1$ and $I_2$ for all the documents that contain both $p$ and $q$.

We suggest a new solution based on solving the FSI problem. We use the same method as Muthukrishnan [6] until we get the two intervals: $I_1$ with $p$ occurrences and $I_2$ with $q$ occurrences. Now, we have to solve a CCQ problem which can be solved as shown above in theorem 3. Therefore, the document listing problem for two patterns can be solved in $O(|p| + |q| + (\sqrt{N \log N\,output} + output) \log^2 N)$ time using $O(N \log N)$ space where $output$ is the number of documents that contain both $p$ and $q$.

## 6  Conclusions

In this paper we developed a method to improve algorithms which intersects sets as a common task. We solved the fast set intersection problem using $O(N)$ space with query time bounded by $O(\sqrt{N\,output} + output)$. We showed how to improve some other problems, the two-dimensional substring indexing problem and the document listing problem for two patterns, using the fast set intersection problem.

There is still a lot of research to be done in regards to the fast set intersection problem. It is open if the query time can be bounded better. Moreover, we showed only two applications for the fast set intersection problem. We are sure that the fast set intersection problem can be useful in other fields as well.

# References

1. Demaine, E.D., López-Ortiz, A., Munro, J.I.: Adaptive set intersections, unions, and differences. In: SODA 2000: Proceedings of the eleventh annual ACM-SIAM symposium on Discrete algorithms, pp. 743–752. Society for Industrial and Applied Mathematics, Philadelphia (2000)
2. Baeza-Yates, R.A.: A fast set intersection algorithm for sorted sequences. In: Sahinalp, S.C., Muthukrishnan, S., Dogrusöz, U. (eds.) CPM 2004. LNCS, vol. 3109, pp. 400–408. Springer, Heidelberg (2004)
3. Barbay, J., López-Ortiz, R., Lu, T.: Faster adaptive set intersections for text searching. In: Àlvarez, C., Serna, M. (eds.) WEA 2006. LNCS, vol. 4007, pp. 146–157. Springer, Heidelberg (2006)
4. Bille, P., Pagh, A., Pagh, R.: Fast evaluation of union-intersection expressions. CoRR abs/0708.3259 (2007)
5. Ferragina, P., Koudas, N., Srivastava, D., Muthukrishnan, S.: Two-dimensional substring indexing. In: PODS 2001: Proceedings of the twentieth ACM SIGMOD-SIGACT-SIGART symposium on Principles of database systems, pp. 282–288. ACM Press, New York (2001)
6. Muthukrishnan, S.: Efficient algorithms for document retrieval problems. In: SODA 2002: Proceedings of the thirteenth annual ACM-SIAM symposium on Discrete algorithms, pp. 657–666. Society for Industrial and Applied Mathematics, Philadelphia (2002)
7. Fredman, M.L., Komlós, J., Szemerédi, E.: Storing a sparse table with 0(1) worst case access time. J. ACM 31(3), 538–544 (1984)

# Counting Reducible, Powerful, and Relatively Irreducible Multivariate Polynomials over Finite Fields

## (Extended Abstract)

Joachim von zur Gathen[1], Alfredo Viola[2], and Konstantin Ziegler[1]

[1] B-IT, Universität Bonn
D-53113 Bonn, Germany
{gathen,zieglerk}@bit.uni-bonn.de
[2] Instituto de Computación, Universidad de la República
Montevideo, Uruguay
Associate Member of LIPN - CNRS UMR 7030,
Université de Paris-Nord, F-93430 Villetaneuse, France
viola@fing.edu.uy

**Abstract.** We present counting methods for some special classes of multivariate polynomials over a finite field, namely the reducible ones, the $s$-powerful ones (divisible by the $s$th power of a nonconstant polynomial), and the relatively irreducible ones (irreducible but reducible over an extension field). One approach employs generating functions, another one a combinatorial method. They yield approximations with relative errors that essentially decrease exponentially in the input size.

## 1 Introduction

Classical results describe the distribution of prime numbers and of irreducible univariate polynomials over a finite field. Randomly chosen integers up to $x$ or polynomials of degree up to $n$ are prime or irreducible with probability about $1/\ln x$ or $1/n$, respectively.

In two or more variables, the situation changes dramatically. Most multivariate polynomials are irreducible. This question was first studied by Leonard Carlitz, later by Stephen Cohen and others. In the bivariate case, von zur Gathen (2008) gave precise approximations with an exponentially decreasing relative error.

This paper makes the following contributions.

- Precise approximations to the number of reducible, $s$-powerful, and relatively irreducible polynomials, with rapidly decaying relative errors. Only reducible polynomials have been treated in the literature, usually with much larger error terms.
- Two orthogonal methodologies to obtain such bounds: generating functions and combinatorial counting.

A. López-Ortiz (Ed.): LATIN 2010, LNCS 6034, pp. 243–254, 2010.

- The classical approach of analytic combinatorics with complex coefficients leads to series that diverge everywhere (except at 0). We use symbolic coefficients, namely rational functions in a variable representing the field size, and manage to extract substantial information by coefficient comparisons.
- Our combinatorial counting follows the aforementioned approach for bivariate polynomials. It is technically slightly involved, but yields explicit constants in our error estimates.

For perspective, we also give an explicit formula which is easy to obtain but hard to use.

## 2   Notation and an Exact Formula

We work in the polynomial ring $F[x_1, \ldots, x_r]$ in $r \geq 2$ variables over a field $F$ and consider polynomials with total degree equal to some nonnegative integer $n$:

$$P_{r,n}^{\text{all}}(F) = \{f \in F[x_1, \ldots, x_r] : \deg f = n\}.$$

The polynomials of degree at most $n$ form an $F$-vector space of dimension

$$b_{r,n} = \binom{r+n}{r}.$$

Over a finite field $\mathbb{F}_q$ with $q$ elements, we have

$$\#P_{r,n}^{\text{all}}(\mathbb{F}_q) = q^{b_{r,n}} - q^{b_{r,n-1}} = q^{b_{r,n}}(1 - q^{-b_{r-1,n}}).$$

The property of a certain polynomial to be reducible, squareful or relatively irreducible is shared with all polynomials associated to the given one. For counting them, it is sufficient to take one representative. We choose an arbitrary monomial order, so that the monic polynomials are those with leading coefficient 1, and write

$$P_{r,n}(F) = \{f \in P_{r,n}^{\text{all}}(F) : f \text{ is monic}\}.$$

Then

$$\#P_{r,n}(\mathbb{F}_q) = \frac{\#P_{r,n}^{\text{all}}(\mathbb{F}_q)}{q-1} = q^{b_{r,n}-1}\frac{1 - q^{-b_{r-1,n}}}{1 - q^{-1}}. \tag{1}$$

The product of two monic polynomials is again monic.

To study reducible polynomials, we consider the following subsets of $P_{r,n}(F)$:

$$I_{r,n}(F) = \{f \in P_{r,n}(F) : f \text{ irreducible}\},$$
$$R_{r,n}(F) = P_{r,n}(F) \setminus I_{r,n}(F).$$

Carlitz (1963) provided the first count of irreducible multivariate polynomials. In Carlitz (1965), he went on to study the fraction of irreducibles when bounds on the degrees in each variable are prescribed; see also Cohen (1968). We opt for the total degree because it has the charm of being invariant under invertible linear transformations. Gao & Lauder (2002) considered our problem in yet another model, namely where one variable occurs with maximal degree. The natural generating function (or zeta function) for the irreducible polynomials in two or more variables does not converge anywhere outside of the origin. Wan (1992) notes that this explains the lack of a simple combinatorial formula for the number of irreducible polynomials. But he gives a $p$-adic formula, and also a (somewhat complicated) combinatorial formula.

In the remainder of this section we will focus on $R_{r,n}(F)$ in the special case of a finite field $F = \mathbb{F}_q$ and omit it from the notation. We want to count exactly how many polynomials there are with a given factorization pattern. Our approach is to look at inclusion-exclusion from the bottom up, that is, considering each factorization pattern. A general polynomial of degree $n$ is described by a partition

$$
\begin{aligned}
(\mathbf{m} \colon \mathbf{e}, \#\mathbf{d}) = (m_1 \colon e_{11}, \#d_{11}, \dots, e_{1s_1}, \#d_{1s_1}; \dots; \\
m_t \colon e_{t1}, \#d_{t1}, \dots, e_{ts_t}, \#d_{ts_t})
\end{aligned}
\tag{2}
$$

of $n$ with

$$
n = \sum_{1 \le i \le t} m_i \sum_{1 \le j \le s_i} e_{ij} d_{ij}.
\tag{3}
$$

All $s_i$, $m_i$, $e_{ij}$, $d_{ij}$, and $t$ are positive integers, $m_1, \dots, m_t$ are pairwise distinct, and for each $i$, $e_{i1}, \dots, e_{is_i}$ are pairwise distinct.

We write $P_{r,n}(\mathbf{m} \colon \mathbf{e}, \#\mathbf{d})$ for the set of polynomials in $P_{r,n}$ with exactly $d_{ij}$ distinct irreducible factors of degree $m_i$ and multiplicity $e_{ij}$. Then

$$
\#P_{r,n}(\mathbf{m} \colon \mathbf{e}, \#\mathbf{d}) = \prod_{1 \le i \le t} \binom{\#I_{r,m_i}}{d_{i1}, \dots, d_{is_i}}.
\tag{4}
$$

**Theorem 1**

$$
R_{r,n} = \bigcup_{(\mathbf{m} \colon \mathbf{e}, \#\mathbf{d})} P_{r,n}(\mathbf{m} \colon \mathbf{e}, \#\mathbf{d}),
$$

$$
\#R_{r,n} = \sum_{(\mathbf{m} \colon \mathbf{e}, \#\mathbf{d})} \#P_{r,n}(\mathbf{m} \colon \mathbf{e}, \#\mathbf{d}),
$$

*where the union is disjoint and over all partitions as in (2), except $(\mathbf{m} \colon \mathbf{e}, \#\mathbf{d}) = (n \colon 1, \#1)$ which corresponds to the irreducible polynomials.*

These formulas provide an explicit way of calculating the number of irreducible polynomials inductively. Cohen (1968) notes that, compared to the univariate case, "the situation is different and much more difficult. In this case, no explicit formula [...] is available".

Bodin (2008) gives a recursive formula for $\#I_{2,n}$ and remarks on a generalization for more than 2 variables.

The formula of Theorem 1 is easily derived but quite impractical (and error-prone) for hand calculations already at small sizes, and also cumbersome to program. The next section presents a more elegant approach.

## 3    Generating Functions for Reducible Polynomials

We now present a generating series that is easy to implement in any computer algebra system and gives exact values in lightning speed. This is modeled on the analytic combinatorial approach that is presented in Flajolet & Sedgewick (2009) by two experts who created large parts of this theory. We first recall a few general primitives from this theory that enable one to set up symbolic equations for generating functions starting from combinatorial specifications. Given a combinatorial object $\omega$, the formal identity

$$\frac{1}{1-\omega} = 1 + \omega + \omega^2 + \dots$$

generates symbolically arbitrary sequences composed of $\omega$. Let $\mathcal{I}$ be a family of primitive combinatorial objects. Then the product

$$\mathcal{P} = \prod_{\omega \in \mathcal{I}} (1 - \omega)^{-1}$$

generates the class of all finite multisets of elements taken from $\mathcal{I}$.

We take $\mathcal{I}$ to be the collection of all monic irreducible polynomials. Then, by the unique factorization property of polynomials, $\mathcal{P}$ generates the collection of all monic polynomials. Let $z$ be a variable, and $|\omega|$ be the total degree of the polynomial $\omega$. The substitution $\omega \mapsto z^{|\omega|}$ in formal sums and products of objects gives rise to counting generating functions. Since $\mathcal{I}$ is identified with the formal sum $\sum_{\omega \in \mathcal{I}} \omega$, the corresponding power series is

$$\mathsf{I} = \sum_{\omega \in \mathcal{I}} z^{|\omega|} = \sum_{n \geq 1} \mathcal{I}_n z^n, \tag{5}$$

where $\mathcal{I}_n = \#I_{r,n}$ is the number of polynomials in $\mathcal{I}$ of total degree $n$. In a similar way, the generating function for all monic polynomials is

$$\mathsf{P} = \prod_{\omega \in \mathcal{I}} \left(1 - z^{|\omega|}\right)^{-1} = \prod_{n \geq 1} (1 - z^n)^{-\mathcal{I}_n} = \sum_{n \geq 0} \mathcal{P}_n z^n. \tag{6}$$

We now apply the ideas of this theory in a different setting. For multivariate polynomials over a finite field $\mathbb{F}_q$, the natural generating functions diverge in all points except 0. We replace the usual complex analytic scenario, where, e.g., $\mathsf{P}$ is a power series with complex coefficients, by a symbolic setting. Namely, for any $r$ and $n$, we consider $\mathcal{P}_n$ to be a polynomial in a variable $\mathsf{q}$ with integer coefficients. Substituting a prime power $q$ for $\mathsf{q}$ yields the integer $\#P_{r,n}(\mathbb{F}_q)$. Allowing, more generally, rational functions in $\mathsf{q}$ leads to power series in $\mathbb{Q}(\mathsf{q})[\![z]\!]$. Thus now

$$\mathcal{P}_n = \mathsf{q}^{b_{r,n}-1} \frac{1 - \mathsf{q}^{-b_{r-1,n}}}{1 - \mathsf{q}^{-1}} \in \mathbb{Q}(\mathsf{q}), \tag{7}$$

in accordance with (1), and similarly for $\mathcal{I}_n$. Interpreting the second product in (6) as $\exp\left(-\sum_{n \geq 1} \mathcal{I}_n \log(1 - z^n)\right)$, taking logarithms, and applying Möbius inversion, with the Möbius function $\mu$, yields

$$\log \mathsf{P} = \sum_{k \geq 1} \frac{\mathsf{l}(z^k)}{k} \quad \text{and} \quad \mathsf{l} = \sum_{k \geq 1} \frac{\mu(k)}{k} \log \mathsf{P}(z^k). \tag{8}$$

P is known from (7), and we take the second equation in (8) as the definition of l. Similarly, R = P − l is the series of reducible polynomials. The coefficients of l can be calculated by expanding the logarithm and equating the appropriate powers of $z$ on both sides.

An 8-line Maple implementation of the resulting algorithm is described in Figure 1. It is easy to program and execute and was used to calculate the number of bivariate reducible polynomials in von zur Gathen (2008, Table 2.1).

```
allp := proc(z,n,r) local i: options remember:
   sum('simplify((q^binomial(i+r,r)-q^binomial(i+r-1,r))/
                              (q-1))*z^i',i=0..n):
end:

reducible := proc(n,r) local k: options remember:
   convert(taylor(allp(z,n,r)-sum('mobius(k)/k*log(allp(z^k,n,r))',
                              k=1..n),z,n+1),polynom):
end:
```

**Fig. 1.** Maple program to compute the generating function of the number of reducible polynomials in $r$ variables up to degree $n$

For $f \in \mathbb{Q}(\mathsf{q})$, $\deg f$ is the degree of $f$, that is, the numerator degree minus the denominator degree. The appearance of $O(\mathsf{q}^{-1})$ in an equation means the existence of some $f$ with negative degree that makes the equation valid. The charm of our approach is that we obtain results for any ("fixed") $r$ and $n$. If a term $O(\mathsf{q}^{-1})$ appears, then we have an asymptotic result for growing prime powers $q$. We denote by $[z^n]\mathsf{F} \in \mathbb{Q}(\mathsf{q})$ the coefficient of $z^n$ in $\mathsf{F} \in \mathbb{Q}(\mathsf{q})[\![z]\!]$.

**Lemma 1.** *(i) For $i, j \geq 0$, we have $\deg_q(\mathcal{P}_i \cdot \mathcal{P}_j) \leq \deg_q \mathcal{P}_{i+j}$.*
*(ii) For $1 \leq k \leq n/2$, the sequence of integers $\deg_q(\mathcal{P}_k \cdot \mathcal{P}_{n-k})$ is strictly decreasing in $k$.*

*Proof.* (i) The claim is equivalent to the binomial inequality

$$\binom{r+i}{r} + \binom{r+j}{r} - 1 \leq \binom{r+i+j}{r}, \tag{9}$$

which is easily seen by considering the choices of $r$-element subsets from a set with $r + i + j$ elements.
(ii) Using (7), we define a function $u$ as

$$u_{r,n}(k) = \deg_q(\mathcal{P}_k \cdot \mathcal{P}_{n-k}) = \binom{r+k}{r} + \binom{r+n-k}{r} - 2. \tag{10}$$

We extend the domain of $u_{r,n}(k)$ from positive integers $k$ between 1 and $n/2$ to real numbers $k$ by means of falling factorials $x^{\underline{r}} = x \cdot (x-1) \cdots (x-r+1)$:

$$u_{r,n}(k) = \frac{(k+r)^{\underline{r}}}{r!} + \frac{(n-k+r)^{\underline{r}}}{r!} - 2.$$

It is sufficient to show that the affine transformation $\bar{u}$ with

$$\bar{u}(k) = r! \cdot (u_{r,n}(k) + 2) = (k+r)^{\underline{r}} + (n-k+r)^{\underline{r}}$$

is strictly decreasing. The first derivative with respect to $k$ is

$$\bar{u}'(k) = \sum_{1 \le i \le r} \Big( (k+1) \cdots \widehat{(k+i)} \cdots (k+r)$$

$$- (n-k+1) \cdots \widehat{(n-k+i)} \cdots (n-k+r) \Big).$$

Since $(k+j) < (n-k+j)$ for $1 \le j \le r$, each difference is negative, and so is $\bar{u}'(k)$. $\qquad\square$

**Theorem 2.** *Let $r, n \ge 2$, and*

$$\rho_{r,n}(q) = q^{\binom{n+r-1}{r}+r-1}\frac{1-q^{-r}}{(1-q^{-1})^2} \in \mathbb{Q}(q), \tag{11}$$

$$\alpha_n = \begin{cases} 1/2 & \text{if } n = 4, \\ 1 & \text{otherwise.} \end{cases}$$

*Then for $n \ge 4$*

$$R_n = \rho_{r,n}(q) \cdot \Big( 1 + \alpha_n q^{-\binom{n+r-2}{r-1}+r(r+1)/2}(1 + O(q^{-1})) \Big),$$

$$R_2 = \frac{\rho_{r,2}(q)}{2}\left( 1 - q^{-r-1} \right),$$

$$R_3 = \rho_{r,3}(q)\left( 1 - q^{-r} + q^{-r(r-1)/2}\frac{1 - 2q^{-r} + 2q^{-2r-1} - q^{-2r-2}}{3(1-q^{-1})} \right).$$

*Proof.* Let $F = P - 1$. The Taylor expansion of $\log(1 + F(z^k))$ in (8) yields

$$I = \sum_{k \ge 1} \frac{\mu(k)}{k} \sum_{i \ge 1} (-1)^{i+1}\frac{F(z^k)^i}{i}$$

$$= F + \sum_{i \ge 2} (-1)^{i+1}\frac{F^i}{i} + \sum_{k \ge 2} \frac{\mu(k)}{k} \sum_{i \ge 1} (-1)^{i+1}\frac{F(z^k)^i}{i},$$

$$R = P - I = 1 + \sum_{i \ge 2} (-1)^i\frac{F^i}{i} + \sum_{k \ge 2} \frac{\mu(k)}{k} \sum_{i \ge 1} (-1)^i\frac{F(z^k)^i}{i}.$$

Since $\mathcal{R}_2 = [z^2]R$, we directly find $\mathcal{R}_2 = (\mathcal{P}_1{}^2 + \mathcal{P}_1)/2$. Similarly, $\mathcal{R}_3 = \mathcal{P}_2\mathcal{P}_1 - (\mathcal{P}_1{}^3 - \mathcal{P}_1)/3$, which implies the claim.

When $n \geq 4$, then the main contribution of $\mathcal{R}_n = [z^n]R$ comes from the term $i = 2$ in the first sum, according to Lemma 1. For $n = 4$, the main contribution is $\mathcal{P}_3\mathcal{P}_1 + \mathcal{P}_2{}^2/2$, giving the asymptotic expansion stated in the theorem using the bounds of Lemma 1. For $n \geq 5$ it leads to $\mathcal{R}_n \sim (2\mathcal{P}_{n-1}\mathcal{P}_1 + 2\mathcal{P}_{n-2}\mathcal{P}_2)/2 = \mathcal{P}_{n-1}\mathcal{P}_1(1 + \mathcal{P}_{n-2}\mathcal{P}_2/(\mathcal{P}_{n-1}\mathcal{P}_1))$. The error estimate follows by taking the main term in $P_{n-2}\mathcal{P}_2/(\mathcal{P}_{n-1}\mathcal{P}_1)$, again using Lemma 1. $\qquad\square$

Bodin (2008, Theorem 7) shows (in our notation)

$$1 - \frac{\#I_{r,n}}{\#P_{r,n}} \sim q^{-b_{r-1,n}-r}\frac{1 - q^{-r}}{1 - q^{-1}}.$$

Hou & Mullen (2009) provide results for $\#I_{r,n}(\mathbb{F}_q)$. These do not yield error bounds for the approximation of $\#R_{r,n}(\mathbb{F}_q)$. Bodin (2009) claims a result similar to Theorem 2, for values of $n$ that tend to infinity and with an unspecified multiplicative factor of the error term, but without proving the required bounds on the various terms, as in Lemma 1.

## 4    Explicit Bounds for Reducible Polynomials

We now describe a third approach to counting the reducible polynomials. The derivation is somewhat more involved. The payoff of this additional effort are explicit relative error bounds in Theorem 3, replacing the asymptotic $O(\mathsf{q}^{-1})$ by bounds like $3q^{-m}$ for some explicit positive integer $m$.

We consider, for integers $1 \leq k < n$, the multiplication map

$$\mu_{r,n,k}\colon P_{r,k} \times P_{r,n-k} \to P_{r,n}$$
$$(g, h) \mapsto g \cdot h.$$

Without loss of generality, we assume $k \leq n/2$. Then

$$\# \operatorname{im} \mu_{r,n,k} \leq \#P_{r,k} \cdot \#P_{r,n-k}$$
$$= q^{u_{r,n}(k)} \frac{(1 - q^{-b_{r-1,k}})(1 - q^{-b_{r-1,n-k}})}{(1 - q^{-1})^2}$$
$$< q^{u_{r,n}(k)} \frac{1 - q^{-b_{r-1,k}}}{(1 - q^{-1})^2}, \tag{12}$$

with $u_{r,n}(k) = b_{r,k} + b_{r,n-k} - 2$ as in (10). The asymptotic behavior of this upper bound is dominated by the behavior of $u_{r,n}(k)$. From Lemma 1(ii) we know that for any $r, n \geq 2$, $u_{r,n}(k)$ is strictly decreasing for $1 \leq k \leq n/2$. As $u_{r,n}(k)$ takes only integral values for integers $k$ we conclude that

$$\sum_{2 \leq k \leq n/2} q^{u_{r,n}(k)} < q^{u_{r,n}(2)} \sum_{k \geq 0} q^{-k} = \frac{q^{u_{r,n}(2)}}{1 - q^{-1}}. \tag{13}$$

**Proposition 1.** *In the notation of Theorem 2, we have for $n \geq 2$:*

*(i)*
$$\#R_{r,n}(\mathbb{F}_q) \leq \rho_{r,n}(q)\left(1 + 3q^{-b_{r-1,n-1}+b_{r-1,2}}\right),$$

*(ii)*
$$\#R_{r,n}(\mathbb{F}_q) \geq \rho_{r,n}(q)\left(1 - 7q^{-b_{r-1,n-1}+r}\right),$$

*(iii)*
$$\#I_{r,n}(\mathbb{F}_q) \geq \#P_{r,n}(\mathbb{F}_q)\left(1 - 6q^{-b_{r-1,n}+r}\right).$$

*Proof.* We start with the proof of (i). Observing

$$R_{r,n} = \bigcup_{1 \leq k \leq n/2} \operatorname{im} \mu_{r,n,k}, \tag{14}$$

and using (12), we find

$$\#R_{r,n} \leq \sum_{1 \leq k \leq n/2} \#\operatorname{im}\mu_{r,n,k}$$

$$\leq \frac{1}{(1-q^{-1})^2} \sum_{1 \leq k \leq n/2} q^{u_{r,n}(k)}(1 - q^{-b_{r-1,k}}). \tag{15}$$

For this sum we have from (13)

$$\sum_{1 \leq k \leq n/2} q^{u_{r,n}(k)}(1 - q^{-b_{r-1,k}}) = q^{u_{r,n}(1)}(1 - q^{-r}) + \sum_{2 \leq k \leq n/2} q^{u_{r,n}(k)}(1 - q^{-b_{r-1,k}})$$

$$\tag{16}$$

$$\leq q^{u_{r,n}(1)}(1 - q^{-r}) + \frac{q^{u_{r,n}(2)}}{1 - q^{-1}}$$

$$= q^{u_{r,n}(1)}(1 - q^{-r})\left(1 + \frac{q^{-u_{r,n}(1)+u_{r,n}(2)}}{(1 - q^{-1})(1 - q^{-r})}\right).$$

Since $u_{r,n}(1) = b_{r,n-1} + r - 1$ and $-u_{r,n}(1) + u_{r,n}(2) = -b_{r-1,n-1} + b_{r-1,2}$, we conclude

$$\#R_{r,n} \leq \frac{q^{b_{r,n-1}+r-1}(1 - q^{-r})}{(1 - q^{-1})^2}\left(1 + \frac{q^{-b_{r-1,n-1}+b_{r-1,2}}}{(1 - q^{-1})(1 - q^{-r})}\right)$$

$$= \rho_{r,n}(q)\left(1 + \frac{q^{-b_{r-1,n-1}+b_{r-1,2}}}{(1 - q^{-1})(1 - q^{-r})}\right).$$

We note that the sum on the right-hand side of (16) is empty for $n \leq 3$ and furthermore that

$$\frac{1}{(1 - q^{-1})(1 - q^{-r})} \leq \frac{8}{3} < 3$$

for all $q, r \geq 2$. This proves (i).

We proceed with (iii). For linear polynomials we have $I_{r,1} = P_{r,1}$, hence (iii) holds for $n = 1$. For $n \geq 2$, we find from (i) that

$$\#I_{r,n} = \#P_{r,n} - \#R_{r,n}$$

$$\geq \#P_{r,n}\left(1 - \rho_{r,n}(q)\frac{1 + 3q^{-b_{r-1,n-1}+b_{r-1,2}}}{\#P_{r,n}}\right)$$

$$= \#P_{r,n}\left(1 - q^{-b_{r-1,n}+r}\frac{(1 + 3q^{-b_{r-1,n-1}+b_{r-1,2}})(1 - q^{-r})}{(1 - q^{-b_{r-1,n}})(1 - q^{-1})}\right)$$

$$> \#P_{r,n}\left(1 - q^{-b_{r-1,n}+r}\frac{(1 + 3q^{-b_{r-1,n-1}+b_{r-1,2}})}{(1 - q^{-1})}\right),$$

since $(1 - q^{-r})/(1 - q^{-b_{r-1,n}}) < 1$. It is now sufficient to note that for $n \geq 2$

$$\frac{1 + 3q^{-b_{r-1,n-1}+b_{r-1,2}}}{1 - q^{-1}} \leq 6.$$

We conclude with the proof of (ii). Using (11), (14), and the injectivity of $\mu_{r,n,1}$ on $P_{r,1} \times I_{r,n-1}$ for $n \geq 3$, it follows that

$$\#R_{r,n} \geq \#\operatorname{im}\mu_{r,n,1} \tag{17}$$

$$\geq \#P_{r,1} \cdot \#I_{r,n-1}$$

$$\geq q^{b_{r,1}-1}\frac{1 - q^{-r}}{1 - q^{-1}} \cdot \#P_{r,n-1}\left(1 - 6q^{-b_{r-1,n-1}+r}\right)$$

$$\geq \rho_{r,n}(q)(1 - q^{-b_{r-1,n-1}})\left(1 - 6q^{-b_{r-1,n-1}+r}\right)$$

$$\geq \rho_{r,n}(q)(1 - 7q^{-b_{r-1,n-1}+r}).$$

For $n = 2$ the claimed lower bound is negative. $\qquad\square$

We combine the upper and lower bounds of Proposition 1 for $n \geq 5$ and argue separately for smaller $n$ to obtain the following result.

**Theorem 3.** *In the notation of Theorem 2 we have for $n \geq 4$,*

$$|\#R_{r,n}(\mathbb{F}_q) - \rho_{r,n}(q)| \leq \rho_{r,n}(q) \cdot 3q^{-\binom{n+r-2}{r-1}+r(r+1)/2},$$

$$|\#R_{r,3}(\mathbb{F}_q) - \rho_{r,3}(q)| \leq \rho_{r,3}(q) \cdot 2q^{-r(r-1)/2},$$

$$\#R_{r,2}(\mathbb{F}_q) = \frac{\rho_{r,2}(q)}{2} \cdot (1 - q^{-r-1}).$$

This result supplements the bound of Theorem 2, by replacing $O(q^{-1})$ with 2 (for $n \geq 5$).

For $r = 2$, these results agree with those in von zur Gathen (2008), but the added generality leads to a slightly larger relative error term $3q^{-n+3}$, compared to the older bound $2q^{-n}$. Our relative error bound is exponentially decaying in the same sense as there.

We conclude this section with a bound on the irreducible polynomials.

**Corollary 1.** *Let $q, r \geq 2$ and $\rho_{r,n}(q)$ as in (11). We have*

$$\#I_{r,n}(\mathbb{F}_q) \geq \#P_{r,n}(\mathbb{F}_q) - 2\rho_{r,n}(q).$$

## 5  Powerful Polynomials

For a positive integer $s$, a polynomial is called $s$-*powerful* if it is divisible by the $s$th power of some nonconstant polynomial, and $s$-*powerfree* otherwise; it is *squarefree* if $s = 2$. Let

$$Q_{r,n,s}(F) = \{f \in P_{r,n}(F): f \text{ is } s\text{-powerful}\}.$$

Similar to Section 3, let $\mathcal{Q}_{n,s}$ denote the corresponding polynomial in $\mathbb{Q}(q)$. The exact count of Section 2 can be immediately applied to $s$-powerful polynomials, by taking $d_{11} = s$ in 2. After proving a statement similar to, but more involved than, Lemma 1, the generating function approach of Section 3 yields the following result.

**Theorem 4.** *Let* $r \geq 2$, $n \geq s \geq 2$, *and let*

$$\eta_{r,n,s}(q) = q^{\binom{n-s+r}{r}+r-1}\frac{(1-q^{-r})(1-q^{-\binom{n+r-s-1}{r-1}})}{(1-q^{-1})^2} \in \mathbb{Q}(q),$$

$$\delta = \binom{n-s+r}{r} - \binom{n-2s+r}{r} - \frac{r(r+1)}{2}.$$

*Then* $\delta > 0$, *and we have*

$$\mathcal{Q}_{n,s} = \begin{cases} \eta_{r,n,s}(q)(1-q^{-\delta}(1+O(q^{-1}))) & \text{if } (n,s) \neq (6,2), \\ \eta_{r,n,s}(q)(1-q^{-\binom{r+3}{4}-r}(1+O(q^{-1}))) & \text{if } (n,s) = (6,2). \end{cases}$$

The combinatorial approach yields the following explicit error bounds.

**Theorem 5.** *With the notation of Theorem 4, we have*

$$|\#Q_{r,n,s}(\mathbb{F}_q) - \eta_{r,n,s}(q)| \leq \begin{cases} \eta_{r,n,s}(q) \cdot 6q^{-\delta} & \text{if } (n,s) \neq (6,2), \\ \eta_{r,6,2}(q) \cdot 3q^{-\binom{r+3}{4}-r+1} & \text{if } (n,s) = (6,2). \end{cases}$$

## 6  Relatively Irreducible Polynomials

A polynomial over $F$ is called *relatively irreducible* if it is irreducible but factors over some extension field of $F$. We define

$$E_{r,n}(F) = \{f \in P_{r,n}(F): f \text{ is relatively irreducible}\}.$$

Similar to Section 3, let $\mathcal{E}_n$ denote the corresponding polynomial in $\mathbb{Q}(q)$. With the appropriate preparation à la Lemma 1, the approach by generating functions gives the following result.

**Theorem 6.** *Let* $r, n \geq 2$, *let* $\ell$ *be the smallest prime divisor of* $n$, *and*

$$\epsilon_{r,n}(q) = q^{\ell(\binom{r+n/\ell}{r}-1)}\frac{1-q^{-\ell\binom{r-1+n/\ell}{r-1}}}{\ell(1-q^{-\ell})} \in \mathbb{Q}(q). \tag{18}$$

$$\gamma_{r,,n}(\mathsf{q}) = \begin{cases} 1/2 & \text{if } n = 4, r \leq 4, \\ 3/2 & \text{if } n = 4, r = 5, \\ \mathsf{q}^{r(r-5)/2} & \text{if } n = 4, r \geq 6, \\ 1 & \text{otherwise.} \end{cases}$$

*Then the following hold.*

*(i) If n is prime, then*

$$\mathcal{E}_n = \epsilon_{r,n}(\mathsf{q}) \left( 1 - \mathsf{q}^{-r(n-1)} \frac{(1 - \mathsf{q}^{-r})(1 - \mathsf{q}^{-n})}{(1 - \mathsf{q}^{-1})(1 - \mathsf{q}^{-nr})} \right).$$

*(ii) If n is composite, then*

$$\mathcal{E}_n = \epsilon_{r,n}(\mathsf{q}) \left( 1 - \gamma_{r,n}(\mathsf{q}) \cdot \mathsf{q}^{-\ell((\binom{r-1+n/\ell}{r-1})-r)} (1 + O(\mathsf{q}^{-1})) \right).$$

The combinatorial approach yields the following explicit error bounds.

**Theorem 7.** *In the notation of Theorem 6, the following holds.*

*(i) If $n \neq 4$ is composite, then*

$$|\#E_{r,n}(\mathbb{F}_q) - \epsilon_{r,n}(q)| \leq \epsilon_{r,n}(q) \cdot 2q^{-\ell(b_{r-1,n/\ell}-r)},$$

*where the 2 can be omitted unless $n = 6$.*
*(ii) For $n = 4$, we have*

$$|\#E_{r,4}(\mathbb{F}_q) - \epsilon_{r,4}(q)| \leq \epsilon_{r,4}(q) \cdot q^{-(r^2+3r-6)/2}.$$

# Acknowledgments

Joachim von zur Gathen and Alfredo Viola thank Philippe Flajolet for useful discussions about Bender's method in analytic combinatorics of divergent series in April 2008. The work of Joachim von zur Gathen and Konstantin Ziegler was supported by the B-IT Foundation and the Land Nordrhein-Westfalen.

A complete version of this Extended Abstract is available as von zur Gathen, Viola & Ziegler (2009).

# References

Bodin, A.: Number of irreducible polynomials in several variables over finite fields. American Mathematical Monthly 115(7), 653–660 (2008)

Bodin, A.: Generating series for irreducible polynomials over finite fields (2009), http://arxiv.org/abs/0910.1680v2 (Accessed December 10, 2009)

Carlitz, L.: The distribution of irreducible polynomials in several indeterminates. Illinois Journal of Mathematics 7, 371–375 (1963)

Carlitz, L.: The distribution of irreducible polynomials in several indeterminates II. Canadian Journal of Mathematics 17, 261–266 (1965)

Cohen, S.: The distribution of irreducible polynomials in several indeterminates over a finite field. Proceedings of the Edinburgh Mathematical Society 16, 1–17 (1968)

Flajolet, P., Sedgewick, R.: Analytic Combinatorics. Cambridge University Press, Cambridge (2009)

Gao, S., Lauder, A.G.B.: Hensel Lifting and Bivariate Polynomial Factorisation over Finite Fields. Mathematics of Computation 71(240), 1663–1676 (2002)

von zur Gathen, J.: Counting reducible and singular bivariate polynomials. Finite Fields and Their Applications 14(4), 944–978 (2008)
http://dx.doi.org/10.1016/j.ffa.2008.05.005; Extended abstract in Proceedings of the International Symposium on Symbolic and Algebraic Computation ISSAC 2007, Waterloo, Ontario, Canada, pp. 369-376 (2007)

von zur Gathen, J., Viola, A., Ziegler, K.: Counting reducible, powerful, and relatively irreducible multivariate polynomials over finite fields (2009),
http://arxiv.org/abs/0912.3312

Hou, X.-D., Mullen, G.L.: Number of Irreducible Polynomials and Pairs of Relatively Prime Polynomials in Several Variables over Finite Fields. Finite Fields and Their Applications 15(3), 304–331 (2009),
http://dx.doi.org/10.1016/j.ffa.2008.12.004

Wan, D.: Zeta Functions of Algebraic Cycles over Finite Fields. Manuscripta Mathematica 74, 413–444 (1992)

# A Larger Lower Bound on the OBDD Complexity of the Most Significant Bit of Multiplication

Beate Bollig

LS2 Informatik, TU Dortmund,
44221 Dortmund, Germany

**Abstract.** Ordered binary decision diagrams (OBDDs) are one of the most common dynamic data structures for Boolean functions. The reachability problem, i.e., computing the set of nodes reachable from a predefined vertex $s \in V$ in a digraph $G = (V, E)$, is an important problem in computer-aided design, hardware verification, and model checking. Sawitzki (2006) has presented exponential lower bounds on the space complexity of a restricted class of symbolic OBDD-based algorithms for the reachability problem. Here, these lower bounds are improved by presenting a larger lower bound on the OBDD complexity of the most significant bit of integer multiplication.

## 1 Introduction

### 1.1 Motivation

Some modern application require huge graphs such that explicit representations by adjacency matrices or adjacency lists are not any longer possible. Since time and space do not suffice to consider individual nodes, one way out seems to be to deal with sets of nodes and edges represented by their characteristic functions. Ordered binary decision diagrams, denoted OBDDs, introduced by Bryant [4], are one of the most often used data structures supporting all fundamental operations on Boolean functions, therefore, in the last years a research branch has emerged which is concerned with the theoretical design and analysis of so-called symbolic algorithms for classical graph problems on OBDD-represented graph instances (see, e.g., [10,11], [16], and [21]). Symbolic algorithms have to solve problems on a given graph instance by efficient functional operations offered by the OBDD data structure. At the beginning the OBDD-based algorithms have been justified by analyzing the number of executed OBDD operations (see, e.g., [10,11]). Since the number of OBDD operations is not directly proportional to the running time of an algorithm, as the running time for one OBDD operation depends on the sizes of the OBDDs on which the operations are performed, newer research tries to analyze the over-all running time of symbolic methods including the analysis of all OBDD sizes occurring during such an algorithm (see, e.g., [21]).

A. López-Ortiz (Ed.): LATIN 2010, LNCS 6034, pp. 255–266, 2010.

In reachability analysis the task is to compute the set of states of a state-transition system that are reachable from a set of initial states. Besides explicit methods for traversing states one by one and SAT-based techniques for deciding distance-bounded reachability between pairs of state sets, symbolic methods are one of the most commonly used approaches to this problem (see, e.g., [6,7]). In the OBDD-based setting our aim is to compute the characteristic function $\mathcal{X}_R$ of the solution set $R \subseteq V$. To be more precise, the input consists of an OBDD representing the characteristic function of the edge set of a graph $G = (V, E)$ and a predefined vertex $s \in V$ and the output is an OBDD representing the characteristic function of the node set $R$ which contains all nodes reachable from the source $s$ via a directed paths. BFS-like approaches using $O(|V|)$ OBDD operations [12] and iterative squaring methods using $O(\log^2 |V|)$ operations [16] are known. In [17] Sawitzki has proved that algorithms solving the reachability problem by computing intermediate sets of nodes reachable from $s$ via directed paths of length at most $2^p$, $p \in \{1, \ldots, \lfloor \log |V| \rfloor\}$, need exponential space if the variable order is not changed during the algorithms. Here, the challenge has been to find graph instances whose OBDD size is small, but for which during the computation of the reachability algorithms exponentially large OBDDs have to be represented. First, Sawitzki has proved the first exponential lower bound on the size of OBDDs representing the most significant bit of integer multiplication for one predefined variable order. Afterwards, he has defined pathological graph instances for the reachability problem, such that during the computation of the investigated restricted class of algorithms, representations for the negation of the most significant bit of integer multiplication are necessary. Since the negation of a Boolean function cannot be represented by smaller OBDDs than the function itself, the proof has been done. Hence, an enlargement of the lower bound on the OBDD size of the most significant bit of integer multiplication leads to larger lower bounds on the space and time complexity of the considered reachability algorithms.

In [3] Sawitzki's result has already been improved by presenting a larger lower bound on the OBDD size of the most significant bit for the variable order chosen in [17]. Lower bounds on the size of OBDDs for a predefined variable order do not rule out the possibility that there are other variable orders leading to OBDDs of small size. Since Sawitzki's assumption that the variable order is not changed during the computation is not realistic because in application reordering heuristics are used in order to minimize the OBDD size for intermediate OBDD results, in [1,2] the result has been improved by presenting general exponential lower bounds on the OBDD size of the most significant bit of integer multiplication. Here, we improve Sawitzki's result once more by presenting larger lower bounds on the OBDD size of the most significant bit.

## 1.2   Integer Multiplication and OBDDs

Integer multiplication is certainly one of the most important functions in computer science and a lot of effort has been spent in designing good algorithms and small circuits and in determining its complexity. For some computation models integer multiplication is a quite simple function. It is contained in $NC^1$ and even in $TC^{0,3}$ (polynomial-size threshold circuits of depth 3) but neither in $AC^0$ (polynomial-size $\{\lor, \land, \neg\}$-circuits of unbounded fan-in and constant depth) nor in $TC^{0,2}$ [13]. For more than 35 years the algorithm of Schönhage-Strassen [18] has been the fastest method for integer multiplication running in time $O(n \log n \log \log n)$. Recently algorithms running in time $n \log n \cdot 2^{O(\log^* n)}$ have been presented [8,9]. Until now it is open whether integer multiplication is possible in time $O(n \log n)$.

**Definition 1.** *Let $B_n$ denote the set of all Boolean functions $f : \{0,1\}^n \to \{0,1\}$. The Boolean function $\mathrm{MUL}_{i,n} \in B_{2n}$ maps two n-bit integers $x = x_{n-1} \ldots x_0$ and $y = y_{n-1} \ldots y_0$ to the ith bit of their product, i.e., $\mathrm{MUL}_{i,n}(x,y) = z_i$, where $x \cdot y = z_{2n-1} \ldots z_0$ and $x_0, y_0, z_0$ denote the least significant bits.*

Boolean circuits, formulae, and binary decision diagrams (BDDs), sometimes called branching programs, are standard representations for Boolean functions. (For a history of results on binary decision diagrams see, e.g., the monograph of Wegener [19]). Besides the complexity theoretical viewpoint people have used restricted binary decision diagrams in applications and OBDDs have become one of the most popular data structures for Boolean functions. Among the many areas of application are verification, model checking, computer-aided design, relational algebra, and symbolic graph algorithms.

**Definition 2.** *Let $X_n = \{x_1, \ldots, x_n\}$ be a set of Boolean variables. A variable order $\pi$ on $X_n$ is a permutation on $\{1, \ldots, n\}$ leading to the ordered list $x_{\pi(1)}, \ldots, x_{\pi(n)}$ of the variables.*

In the following a variable order $\pi$ is sometimes identified with the corresponding order $x_{\pi(1)}, \ldots, x_{\pi(n)}$ of the variables if the meaning in clear from the context.

**Definition 3.** *A $\pi$-OBDD on $X_n$ is a directed acyclic graph $G = (V, E)$ whose sinks are labeled by Boolean constants and whose non sink (or inner) nodes are labeled by Boolean variables from $X_n$. Each inner node has two outgoing edges one labeled by 0 and the other by 1. The edges between inner nodes have to respect the variable order $\pi$, i.e., if an edge leads from an $x_i$-node to an $x_j$-node, $\pi^{-1}(i) \le \pi^{-1}(j)$ ($x_i$ precedes $x_j$ in $x_{\pi(1)}, \ldots, x_{\pi(n)}$). Each node $v$ represents a Boolean function $f_v : \{0,1\}^n \to \{0,1\}$ defined in the following way. In order to evaluate $f_v(b)$, $b \in \{0,1\}^n$, start at $v$. After reaching an $x_i$-node choose the outgoing edge with label $b_i$ until a sink is reached. The label of this sink defines $f_v(b)$. The size of the $\pi$-OBDD $G$ is equal to the number of its nodes and $\pi$-OBDD$(f)$ denotes the size of the minimal $\pi$-OBDD representing $f$.*

It is well known that the size of an OBDD representing a function $f$ depends on the chosen variable order and may vary between linear and exponential size. Since in applications the variable order is not given in advance we have the freedom (and the problem) to choose a good or even an optimal order for the representation of $f$.

**Definition 4.** *The OBDD size or OBDD complexity of $f$ (denoted by $OBDD(f)$) is the minimum of all $\pi$-OBDD$(f)$.*

Lower bounds for integer multiplication are motivated by the general interest in the complexity of important arithmetic functions. Moreover, lower bounds on the OBDD complexity of the most significant bit of integer multiplication are interesting because of the following observation. If $z_{2n-1}$ cannot be computed with size $s(n)$, then any other output bit of integer multiplication $z_i$, $2n-1 > i \geq 0$, cannot be computed with size $s(i/4)$.

Although many exponential lower bounds on the OBDD size of Boolean functions are known and the lower bound methods are simple, it is often a more difficult task to prove large lower bounds for some predefined and interesting functions. The most significant bit of integer multiplication is a good example. Despite the well-known lower bounds on the OBDD size of the middle bit of multiplication $MUL_{n-1,n}$ ([5], [20]), only recently it has been shown that the OBDD complexity of the most significant bit of multiplication $MUL_{2n-1,n}$ is also exponential [1] answering an open question posed by Wegener [19].

### 1.3   Results and Organization of the Paper

Using techniques from analytical number theory Sawitzki [17] has presented a lower bound of $\Omega(2^{n/6})$ on the size of $\pi$-OBDDs representing the most significant bit of integer multiplication for the variable order $\pi$ where the variables are tested according to increasing significance, i.e., $\pi = (x_0, y_0, x_1, y_1, \ldots, x_{n-1}, y_{n-1})$. In [3] this lower bound has been improved up to $\Omega(2^{n/4})$ without analytical number theory. Here, in Section 3 we enlarge this lower bound using a simple proof.

In [2] the lower bound proof on the OBDD complexity of $MUL_{2n-1,n}$ has been improved up to $\Omega(2^{n/72})$. In Section 4 we present a larger general lower bound. As a result we gain more insight into the structure of the most significant bit of integer multiplication.

Our results can be summarized as follows.

**Theorem 1.** *Let $\pi = (x_0, y_0, x_1, y_1, \ldots, x_{n-1}, y_{n-1})$. The $\pi$-OBDD size for the representation of $MUL_{2n-1,n}$ is $\Omega(2^{n/3})$.*

**Theorem 2.** $OBDD(MUL_{2n-1,n}) = \Omega(2^{n/45})$.

## 2   Preliminaries

### 2.1   Notation

In the rest of the paper we use the following notation.

Let $[x]_r^l$, $n-1 \geq l \geq r \geq 0$, denote the bits $x_l \ldots x_r$ of a binary number $x = (x_{n-1}, \ldots, x_0)$. For the ease of description we use the notation $[x]_r^l = z$ if $(x_l, \ldots, x_r)$ is the binary representation of the integer $z \in \{0, \ldots, 2^{l-r+1} - 1\}$. Sometimes, we identify $[x]_r^l$ with $z$ if the meaning is clear from the context. We use the notation $(z)_r^l$ for an integer $z$ to identify the bits at position $l, \ldots, r$ in the binary representation of $z$.

Let $\ell \in \{0, \ldots, 2^m - 1\}$, then $\bar{\ell}$ denotes the number $(2^m - 1) - \ell$. For a binary number $x = (x_{n-1}, \ldots, x_0)$ we use the notation $\bar{x}$ for the binary number $(\bar{x}_{n-1}, \ldots, \bar{x}_0)$.

Let $a_S$ be an assignment to variables in a set $S$ and $a_S(x_k) \in \{0, 1\}$ be the assignment to $x_k \in S$, then we define $\|a_S\| := \sum_{x_k \in S} a_S(x_k) \cdot 2^k$.

## 2.2  One-Way Communication Complexity and the Size of OBDDs

In order to obtain lower bounds on the size of OBDDs one-way communication complexity has become a standard technique (see, e.g., [14,15] for the theory of communication complexity.)

One central notion of communication complexity are fooling sets.

**Definition 5.** *Let* $f : \{0,1\}^{|X_A|} \times \{0,1\}^{|X_B|} \to \{0,1\}$. *A set* $S \subseteq \{0,1\}^{|X_a|} \times \{0,1\}^{|X_B|}$ *is called fooling set for* $f$ *if* $f(a,b) = c$ *for all* $(a,b) \in S$ *and some* $c \in \{0,1\}$ *and if for different pairs* $(a_1, b_1), (a_2, b_2) \subset S$ *at least one of* $f(a_1, b_2)$ *and* $f(a_2, b_1)$ *is unequal to* $c$.

The following theorem is well known.

**Theorem 3.** *If* $f : \{0,1\}^{|X_A|} \times \{0,1\}^{|X_B|} \to \{0,1\}$ *has a fooling set of size* $t$ *and* $\pi$ *is a variable order where the* $X_A$*-variables are before the* $X_B$*-variables or vice versa the size of a* $\pi$*-OBDD for* $f$ *is at least* $t$.

Now, we present a function $f_n$ with a large fooling set which is a main ingredient in our general lower bound proof on the OBDD size of the most significant bit of integer multiplication.

The function $f_n \in B_{3n}$ is defined on the variables $a = (a_1, \ldots, a_n)$, $b = (b_1, \ldots, b_n)$, and $c = (c_1, \ldots, c_n)$:

$$f_n(a, b, c) := (\mathrm{EQ}_n(a, \bar{c}) \wedge \overline{\mathrm{GT}_n}(a, b)) \vee \mathrm{GT}_n(a, \bar{c}),$$

where $\mathrm{EQ}_n(a, b) = 1$ iff the vectors $a = (a_1, \ldots, a_n)$ and $b = (b_1, \ldots, b_n)$ are equal, $\overline{\mathrm{GT}_n}(a, b) = 1$ iff $[a]_1^n \leq [b]_1^n$, and $\mathrm{GT}_n(a, b) = 1$ iff $[a]_1^n > [b]_1^n$.

Using case inspection on the distribution of the $c$-variables in [2] it has been shown that for a partition, where for all $i \in \{1, \ldots, n\}$ the set $X_A$ contains exactly one of the variables $a_i$ and $b_i$, there exists a fooling set of size $2^n$ for $f_n$.

## 3  A Larger Lower Bound on the $\pi$-OBDD Size of MUL$_{2n-1,n}$ for Some Predefined Variable Order

In this section we prove Theorem 1. We start with the following useful observations.

**Fact 1.** *For a number* $2^{n-1} + \ell 2^{n/3}$ *the corresponding smallest integer such that the product of the two numbers is at least* $2^{2n-1}$ *is* $2^n - \ell 2^{n/3+1} + \left\lceil \frac{\ell^2 2^{(2/3)n+1}}{2^{n-1} + \ell 2^{n/3}} \right\rceil$.

**Fact 2.** *For $\ell \in \mathbb{N}$ and $\ell \le 2^{n/3-2}$:*

$$\left\lceil \frac{\ell^2 2^{(2/3)n+1}}{2^{n-1} + \ell 2^{n/3}} \right\rceil \le 2^{n/3-1}.$$

**Fact 3.** *For $\ell_1, \ell_2 \in \{2^{n/3-3}, \ldots, 2^{n/3-2}\}$ and $\ell_2 - \ell_1 = c \ge 4$:*

$$\frac{\ell_2^2 2^{(2/3)n+1}}{2^{n-1} + \ell_2 2^{n/3}} - \frac{\ell_1^2 2^{(2/3)n+1}}{2^{n-1} + \ell_1 2^{n/3}} > c/4 \ge 1.$$

Now, it is not difficult to construct a fooling set of size $2^{n/3-3}/4 = \Omega(2^{n/3})$:

Let $X_U := \{x_{n-1}, x_{n-2}, \ldots, x_{n/3}\}$, $Y_U := \{y_{n-1}, y_{n-2}, \ldots, y_{n/3}\}$, $X_L := \{x_{n/3-1}, x_{n/2-2}, \ldots, x_0\}$, and $Y_L := \{y_{n/3-1}, y_{n/2-2}, \ldots, y_0\}$. The Set $S$ contains all pairs $(a, b)$ for $\ell \in \{2^{n/3-3}, \ldots, 2^{n/3-2}\}$ and $(\ell \bmod 4) = 0$ with the following properties:

1. $a$ is an assignment that consists of a partial assignment $a_{X_U}$ to the variables in $X_U$ and a partial assignment $a_{Y_U}$ to the $Y_U$-variables, where $\|a_{X_U}\| = 2^{n-1} + \ell 2^{n/3}$ and $\|a_{Y_U}\| = 2^n - \ell 2^{n/3+1}$ and
2. $b$ is an assignment that consists of a partial assignment $b_{X_L}$ to the variables in $X_L$ and a partial assignment $b_{Y_L}$ to the $Y_L$-variables, where $\|b_{X_L}\| = 0$ and $\|b_{Y_L}\| = \left\lceil \frac{\ell^2 2^{(2/3)n+1}}{2^{n-1} + \ell 2^{n/3}} \right\rceil.$

Because of Fact 2 we know that the partial assignment $b_{Y_L}$ exists. Using Fact 1 we know that the function value of $\mathrm{MUL}_{2n-1,n}$ is 1 for all pairs in $S$. Let $(a_1, b_1)$ and $(a_2, b_2)$ be two different pairs in $S$. If the value of the partial assignment of the $X_U$-variables according to $a_1$ is $2^{n-1} + \ell_1 2^{n/3}$ and the value of the partial assignment of the $X_U$-variables according to $a_2$ is $2^{n-1} + \ell_2 2^{n/3}$, where w.l.o.g. $\ell_1 < \ell_2$, the function value of $\mathrm{MUL}_{2n-1,n}(a_2, b_1)$ is 0 since

$$\left\lceil \frac{\ell_2^2 2^{(2/3)n+1}}{2^{n-1} + \ell_2 2^{n/3}} \right\rceil > \left\lceil \frac{\ell_1^2 2^{(2/3)n+1}}{2^{n-1} + \ell_1 2^{n/3}} \right\rceil$$

according to Fact 3. Therefore, $S$ is a fooling set of size $2^{n/3-5}$.

Because of the symmetric definition of fooling sets we also obtain a lower bound of $2^{n/3-5}$ on the size of $\pi'$-OBDDs for the most significant bit, where $\pi' = (x_{n-1}, y_{n-1}, x_{n-2}, y_{n-2}, \ldots, x_0, y_0)$.

## 4 A Larger Lower Bound on the OBDD Complexity of the Most Significant Bit of Integer Multiplication

In this section we prove Theorem 2 and determine the lower bound of $\Omega(2^{n/45})$ on the size of OBDDs for the representation of the most significant bit of integer multiplication. We use some of the ideas presented in [2] but we have to apply them in a more clever way to obtain a larger lower bound. Our aim is to show for an arbitrary variable order $\pi$ that a $\pi$-OBDD for $\text{MUL}_{2n-1,n}$ contains a $\pi$-OBDD for the Boolean function $f_{n'}$ defined in Section 2.2:

$$f_{n'}(a,b,c) = (\text{EQ}_{n'}(a,\bar{c}) \wedge \overline{\text{GT}_{n'}}(a,b)) \vee \text{GT}_{n'}(a,\bar{c}),$$

where for each position $i$ the variables $a_i$ and $b_i$ are suitably separated in $\pi$ and $n' = \Theta(n)$. Therefore, the size of the $\pi$-OBDD for $\text{MUL}_{2n-1,n}$ has to be large. The vector $a$ is a subvector of one of the inputs $x$ and $y$ for $\text{MUL}_{2n-1,n}$, the vectors $b$ and $c$ of the other input.

In the sequel for the sake of simplicity we do not apply floor or ceiling functions to numbers even when they need to be integers whenever this is clear from the context and has no bearing on the essence of the proof.

In the following let $\ell$ be an integer in $\{1, \ldots, 2^{n/3-1} - 1\}$. A key observation is the following one. $\text{MUL}_{2n-1,n}$ answers the question whether the product of two integers is at least $2^{2n-1}$. As already mentioned in Section 3, for a number $2^{n-1} + \ell 2^{n/3}$, the corresponding smallest number such that the product of the two numbers is at least $2^{2n-1}$ is

$$2^n - \ell 2^{n/3+1} + \left\lceil \frac{\ell^2 2^{(2/3)n+1}}{2^{n-1} + \ell 2^{n/3}} \right\rceil = 2^n - \ell 2^{n/3+1} + \left\lceil \ell^2 2^{-n/3+2} - \frac{4\ell^3}{2^{n-1} + \ell 2^{n/3}} \right\rceil.$$

We notice that the number $\frac{4\ell^3}{2^{n-1}+\ell 2^{n/3}}$ is smaller than 1 if $\ell \leq 2^{n/3-1}$.

Next, we investigate requirements that have to be fulfilled for inputs $x$ and $y$, where $\text{MUL}_{2n-1,n}(x,y) = 1$.

If $x$ represents an integer $2^{n-1} + \ell 2^{n/3}$, $1 \leq \ell < 2^{n/3-1}$, the upper part of $y$ has to represent an integer of at least $2^{(2/3)n} - 2\ell$, i.e., $[y]_{n/3}^{n-1} \geq 2^{(2/3)n} - 2\ell$.

- If $[y]_{n/3}^{n-1} > 2^{(2/3)n} - 2\ell$, the function value $\text{MUL}_{2n-1,n}(x,y)$ is 1.
- Let $j$ be the minimum integer of the set

$$\{i \mid n/3 \leq i < (2/3)n - 1 \text{ and } x_i = 1\}.$$

   i) If $[\bar{y}]_{j|2}^{n-1} > [x]_{j+1}^{n-2}$, the function value $\text{MUL}_{2n-1,n}$ is 0.
   ii) If $[\bar{y}]_{j+2}^{n-1} < [x]_{j+1}^{n-2}$, the function value $\text{MUL}_{2n-1,n}$ is 1.
   iii) If $y_{j+1} = 1$, $[\bar{y}]_{j+2}^{n-1} = [x]_{j+1}^{n-2}$, and $[y]_{n/3}^j = 0$, $[y]_0^{n/3-1}$ has to represent an integer of at least $\left\lceil \ell^2 2^{-n/3+2} - \frac{4\ell^3}{2^{n-1}+\ell 2^{n/3}} \right\rceil$. In our lower bound proof the following observation will be helpful. Since $\frac{4\ell^3}{2^{n-1}+\ell 2^{n/3}}$ is smaller than 1 for $\ell \leq 2^{n/3-1}$, its influence can be limited by choosing $\ell$ carefully. As a result $\ell^2 2^{-n/3+2}$ is the term which is the most important one to decide whether the function value is 1.

Now, we investigate some square numbers more closely. The reason is the following one. As we have seen, if some of the $x$- and $y$-variables fulfill certain properties, then the function value of $\text{MUL}_{2n-1,n}(x,y)$ is 1 if $[y]_0^{n/3-1}$ is at least $\left\lceil \ell^2 2^{-n/3+2} - \frac{4\ell^3}{2^{n-1}+\ell 2^{n/3}} \right\rceil$. Next, we restrict the assignments carefully such that

$\ell^2$ satisfies certain requirements. Among others the $n/3-2$ least significant bits in the binary representation of the considered square numbers $\ell^2$ are less important for us.

In the rest of the proof let $\ell = u2^m + w$, for integers $w < 2^m$ and $u < 2^{(7/8)m}$ and $m := (8/45)n - 8/15$. Then $\ell^2 = u^2 2^{2m} + uw2^{m+1} + w^2$. In our lower bound proof the integer $u$ will be fixed and $w$ will be chosen in such a way that no carry is generated by the addition of $w^2$, $uw2^{m+1}$, and $u^2 2^{2m}$. Since the $n/3 - 2$ least significant bits in the binary representation of $\ell^2$ are more or less unimportant in our lower bound proof, the square number $w^2$ is too small to be of much influence. In our lower bound proof the most decisive part is $uw2^{m+1}$.

Furthermore, we choose the assignments for $\ell$ in such a way that for different integers $\ell_1$ and $\ell_2$, where $\ell_1 = u2^m + w_1$ and $\ell_2 = u2^m + w_2$, $w_1 < w_2$,

$$(u^2 2^{2m} + uw_1 2^{m+1}) \text{ div } 2^{2m+1} < (u^2 2^{2m} + uw_2^{2m+1}) \text{ div } 2^{2m+1}.$$

Moreover, $(u^2 2^{2m} + uw_1 2^{m+1} + w_1^2) \text{ mod } 2^{2m+1}$ and $(u^2 2^{2m} + uw_2 2^{m+1} + w_2^2) \text{ mod } 2^{2m+1}$ are less than $2^{2m}$. Therefore,

$$(u^2 2^{2m} + uw_1 2^{m+1} + 2^{2m}) \text{ div } 2^{2m} > (u^2 2^{2m} + uw_1 2^{m+1} + w_1^2) \text{ div } 2^{2m} \text{ but}$$

$$(u^2 2^{2m} + uw_1 2^{m+1} + 2^{2m}) \text{ div } 2^{2m} < (u^2 2^{2m} + uw_2 2^{m+1} + w_2^2) \text{ div } 2^{2m}.$$

We can conclude that

$$(u^2 2^{2m} + uw_1 2^{m+1} + 2^{2m}) \text{ div } 2^{n/3-2} > \left\lceil \ell_1^2 2^{-n/3+2} - \frac{4\ell_1^3}{2^{n-1} + \ell_1 2^{n/3}} \right\rceil \text{ and}$$

$$(u^2 2^{2m} + uw_1 2^{m+1} + 2^{2m}) \text{ div } 2^{n/3-2} < \left\lceil \ell_2^2 2^{-n/3+2} - \frac{4\ell_2^3}{2^{n-1} + \ell_2 2^{n/3}} \right\rceil$$

since $2m > n/3 - 2$.

Altogether, we have seen that we can get rid of the influence of $w^2$ and $\frac{4\ell^3}{2^{n-1}+\ell 2^{n/3}}$ provided that it is possible to choose $\ell$ as discussed above.

Next, we make our proof ideas more precise. We rename $[x]_{n/3}^{n/3+m-1}$ by $[w]_0^{m-1}$ and $[x]_{n/3+m}^{(2/3)n-1}$ by $[u]_0^{(7/8)m} - 1$. (Note, that $n/3 + m - 1 = n/3 + (8/45)n - 2$ and $n/3 + m + (7/8)m - 1 = (2/3)n - 1$.) (See Figure 1 for the composition of the number $x$.) The main idea is to replace some of the $x$-variables and the corresponding $y$-variables by constants, where $y_{i+1}$ is the corresponding $y$-variable to $x_i$, such that a certain part of the upper half of the binary representation of $u \cdot w$ is equal to a certain part of $2^d \cdot w$ for $d$ suitably chosen.

Now, the crucial part is to choose an appropriate subset of the input variables in order to show that there exists a large fooling set. In other words we have to choose the variables for the $a$, $b$-, and $c$-variables in the reduction from $f_{n'}$ to $\text{MUL}_{2n-1,n}$ carefully.

Let $S := \{w_{m/2}, \ldots, w_{m-1}, y_{(5/2)m-n/3+2}, \ldots, y_{2m+1-n/3+2}\}$ and $T$ be the set of the first $|T|$ variables according to $\pi$ where there are $m/2$ variables from $S$ and $B$ be the set of the remaining variables. (Note, that $(5/2)m - n/3 + 2 = n/9 + 1/2$

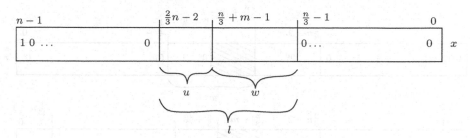

**Fig. 1.** The composition of the input $x$

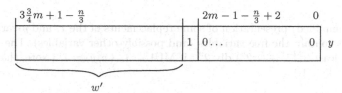

**Fig. 2.** The effect of the replacements of some of the $y$-variables, where $u = [u]_0^{(7/8)m-1}$ and $[w]_0^{m-1}$ ($w'$ has to be at least $u^2 2^{-1} + (uw)$ div $2^m = 2^{2d-1} + (w2^d)$ div $2^m$ )

and $2m+1-n/3+2 = n/45+1$.) Let $W_{S,T}$ be the $w$-variables in $S \cap T$, $W_{S,B}$ the $w$-variables in $S \cap B$. Similar the sets $Y_{S,T}$ and $Y_{S,B}$ are defined. Using simple counting arguments we can prove that there exists a distance parameter $d$ such that there are at least $m/8$ pairs $(w_i, y_{m+1+i+d-n/3+2})$ in $W_{S,T} \times Y_{S,B} \cup W_{S,B} \times Y_{S,T}$ (for a similar proof see, e.g., [5]). (Note, that $m + 1 + i + d - n/3 + 2 = i + d - (7/45)n + 2$.) Let $I$ be the set of indices, where $w_i$ belongs to such a pair. We replace the $u$-variables such that $[u]_0^{(7/8)m-1} = 2^d$.

The variables $x_{n/3+i}$, $i \in I$, are called free $x$-variables, the variables $y_{n/3+i+1}$ and $y_{i+d-(7/45)n+2}$, $i \in I$, free $y$-variables. The free $x$-variables will play the role of the $a$-variables, the free variables $y_{n/3+i+1}$, $i \in I$, the role of the $c$-, and the remaining free $y$-variables the role of the $b$-variables in the reduction from the function $f_{n'}$ mentioned above to $\text{MUL}_{2n-1,n}$. Finally, we are ready to present the reduction. (Figure 2 and 3 show some of the replacements to the inputs $x$ and $y$ of $\text{MUL}_{2n-1,n}$.)

- The variables $y_{n-1}$ and $x_{n-1}$ are set to 1,
- the variables $x_{n/3}$ and $y_{n/3+1}$ are set to 1,
- $x_{n/3+m-d-1}$ (which corresponds to $w_{m-d-1}$) is set to 0 and $y_{n/3+m-d}$ and $y_{n/45}$ are set to 1 (note that $2m + d - n/3 + 2 = n/45$),
- $x_{n/3+m+d}$ (which corresponds to $u_d$) is set to 1, the corresponding variable $y_{n/3+m+d+1}$ is set to 0, and $y_{n/45+2d-1}$ to 1 (note, that $2m+2d-1-n/3+2 = n/45 + 2d - 1$).
- The variables $y_{n/2}, \ldots, y_{n/2+m-d-1}$ are set to 0.
- Besides the free $y$-variables in $\{y_0, \ldots, y_{n/3}\}$ the remaining $y$-variables in $\{y_0, \ldots, y_{n/3}\}$ are replaced by 0.

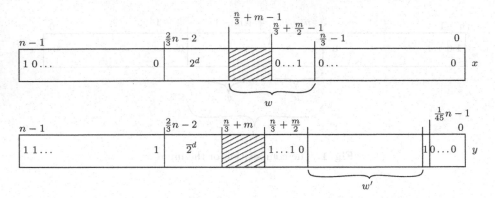

**Fig. 3.** A (simplified) presentation of some replacements of the $x$- and $y$-variables. The shaded areas contain the free variables (and possibly other variables). The number $w'$ has to be at least $2^{2d-1} + (w2^d) \operatorname{div} 2^m$, if $\operatorname{MUL}_{2n-1,n}(x, y) = 1$. (Note, that $2m - 1 - n/3 + 2 = n/45 - 1$.)

- Besides the free $x$-variables the remaining $x$-variables are replaced by 0.
- Besides the free $y$-variables the remaining $y$-variables are replaced by 1.

In the following we describe the effect of these replacements.

- The inputs $x$ and $y$ represent integers that are at least $2^{n-1}$, since otherwise the function value $\operatorname{MUL}_{2n-1,n}(x, y)$ is 0.
- Since $y_{n/3} = 0$, $[u]_0^{(7/8)m-1} = 2^d$, $w_{m-d-1} = 0$ but $y_{2m+d-n/3+2} = 1$, $[y]_0^{n/3-1}$ has to be at least $(u^2 2^{2m} + uw2^{m+1} + 2^{2m}) \operatorname{div} 2^{n/3-2} = (2^{2m+2d} + w2^{m+d+1} + 2^{2m}) \operatorname{div} 2^{n/3-2}$ to represent an integer of at least $\left\lceil \ell^2 2^{-n/3+2} - \frac{4\ell^3}{2^{n-1}+\ell 2^{n/3}} \right\rceil$, where $\ell = u2^m + w = 2^{m+d} + w$.
- Since $x_{n/3} = 1$ and $y_{n/3+1} = 1$, $[x]_{n/3+m}^{n-2} = [\bar{y}]_{n/3+m+1}^{n-1}$, $[x]_{n/3+1}^{n/3+m-1}$ has to be at least $[\bar{y}]_{n/3+2}^{n/3+m}$ for inputs $x$ and $y$, where $\operatorname{MUL}_{2n-1,n}(x, y) = 1$. If $[x]_{n/3+1}^{n/3+m-1} > [\bar{y}]_{n/3+2}^{n/3+m}$, $\operatorname{MUL}_{2n-1,n}(x, y) = 1$.
- Since $y_{n/45+2d-1} = 1$ (and because of some of the other replacements), the product of $x$ and $y$ is at least $2^{2n-1}$ and therefore $\operatorname{MUL}_{2n-1,n}(x, y) = 1$, where $[y]_{n/3}^{n-1} = 2^{(2/3)n} - 2\ell$ and $[x]_{n/3}^{n-1} = 2^{(2/3)n-1} + \ell$, if $[y]_{n/45+1}^{n/45+d}$ is at least $(w2^d) \operatorname{div} 2^m$. (Note, that $2m + 1 - n/3 + 2 = n/45 + 1$.)

Therefore, the correctness of our reduction follows from our considerations above. Summarizing, we have shown that for an arbitrary variable order $\pi$ a $\pi$-OBDD for $\operatorname{MUL}_{2n-1,n}$ contains a $\pi$-OBDD for the functions $f_{n'}$, where $n'$ is at least $m/8$, and for each $i \in \{1, \dots, n'\}$ exactly one of the variables $a_i$ and $b_i$ is in $T$, in other words $\pi$ is a bad variable order for $f_{n'}$. Considering the fact that $m = (8/45)n - 1$, we get the result that the OBDD complexity of $\operatorname{MUL}_{2n-1,n}$ is at least $\Omega(2^{n/45})$.

## 5   Concluding Remarks

We have already learned in primary school how to multiply integers, nevertheless, the complexity of integer multiplication is a fascinating subject. The next challenge is to improve the lower bound on the OBDD complexity of $MUL_{2n-1,n}$ significantly. The method presented in this paper seems to be not strong enough. Moreover, the complexity of $MUL_{2n-1,n}$ for more general (non-oblivious) models than OBDDs is unknown.

## Acknowledgment

The author would like to thank the anonymous referees for their helpful comments.

## References

1. Bollig, B.: On the OBDD complexity of the most significant bit of integer multiplication. In: Agrawal, M., Du, D.-Z., Duan, Z., Li, A. (eds.) TAMC 2008. LNCS, vol. 4978, pp. 306–317. Springer, Heidelberg (2008)
2. Bollig, B.: Larger lower bounds on the OBDD complexity of integer multiplication. In: Dediu, A.H., Ionescu, A.M., Martín-Vide, C. (eds.) LATA 2009. LNCS, vol. 5457, pp. 212–223. Springer, Heidelberg (2009)
3. Bollig, B., Klump, J.: New results on the most significant bit of integer multiplication. In: Hong, S.-H., Nagamochi, H., Fukunaga, T. (eds.) ISAAC 2008. LNCS, vol. 5369, pp. 129–140. Springer, Heidelberg (2008)
4. Bryant, R.E.: Graph-based algorithms for Boolean function manipulation. IEEE Trans. on Computers 35, 677–691 (1986)
5. Bryant, R.E.: On the complexity of VLSI implementations and graph representations of Boolean functions with application to integer multiplication. IEEE Trans. on Computers 40, 205–213 (1991)
6. Burch, J.R., Clarke, E.M., Long, D.E., Mc Millan, K.L., Dill, D.L., Hwang, L.J.: Symbolic model checking: $10^{20}$ states and beyond. In: Proc. of Symposium on Logic in Computer Science, pp. 428–439 (1990)
7. Coudert, O., Berthet, C., Madre, J.C.: Verification of synchronous sequential machines based on symbolic execution. In: Sifakis, J. (ed.) CAV 1989. LNCS, vol. 407, pp. 365–373. Springer, Heidelberg (1989)
8. De, A., Kurur, P., Saha, C., Sapthariski, R.: Fast integer multiplication using modular arithmetic. In: Proc. of 40th STOC, pp. 499–506 (2008)
9. Fürer, M.: Faster integer multiplication. In: Proc. of 39th STOC, pp. 57–66 (2007)
10. Gentilini, R., Piazza, C., Policriti, A.: Computing strongly connected components in a linear number of symbolic steps. In: Proc. of SODA, pp. 573–582. ACM Press, New York (2003)
11. Gentilini, R., Piazza, C., Policriti, A.: Symbolic graphs: linear solutions to connectivity related problems. Algorithmica 50, 120–158 (2008)
12. Hachtel, G.D., Somenzi, F.: A symbolic algorithm for maximum flow in $0-1$ networks. Formal Methods in System Design 10, 207–219 (1997)
13. Hajnal, A., Maass, W., Pudlák, P., Szegedy, M., Turán, G.: Threshold circuits of bounded depth. In: Proc. 28th FOCS, pp. 99–110 (1987)

14. Hromkovič, J.: Communication Complexity and Parallel Computing. Springer, Heidelberg (1997)
15. Kushilevitz, E., Nisan, N.: Communication Complexity. Cambridge University Press, Cambridge (1997)
16. Sawitzki, D.: Implicit flow maximization by iterative squaring. In: Van Emde Boas, P., Pokorný, J., Bieliková, M., Štuller, J. (eds.) SOFSEM 2004. LNCS, vol. 2932, pp. 301–313. Springer, Heidelberg (2004)
17. Sawitzki, D.: Exponential lower bounds on the space complexity of OBDD-based graph algorithms. In: Correa, J.R., Hevia, A., Kiwi, M. (eds.) LATIN 2006. LNCS, vol. 3887, pp. 781–792. Springer, Heidelberg (2006)
18. Schönhage, A., Strassen, V.: Schnelle Multiplikation großer Zahlen. Computing 7, 281–292 (1971)
19. Wegener, I.: Branching Programs and Binary Decision Diagrams - Theory and Applications. SIAM Monographs on Discrete Mathematics and Applications (2000)
20. Woelfel, P.: New bounds on the OBDD-size of integer multiplication via universal hashing. Journal of Computer and System Science 71(4), 520–534 (2005)
21. Woelfel, P.: Symbolic topological sorting with OBDDs. Journal of Discrete Algorithms 4(1), 51–71 (2006)

# Modelling the LLL Algorithm by Sandpiles

Manfred Madritsch and Brigitte Vallée

GREYC, CNRS and University of Caen, 14032 Caen Cedex (France)

**Abstract.** The LLL algorithm aims at finding a "reduced" basis of a Euclidean lattice and plays a primary role in many areas of mathematics and computer science. However, its general behaviour is far from being well understood. There are already many experimental observations about the number of iterations or the geometry of the output, that raise challenging questions which remain unanswered and lead to natural conjectures which are yet to be proved. However, until now, there exist few experimental observations about the precise execution of the algorithm. Here, we provide experimental results which precisely describe an essential parameter of the execution, namely the "logarithm of the decreasing ratio". These experiments give arguments towards a "regularity" hypothesis $(R)$. Then, we propose a simplified model for the LLL algorithm based on the hypothesis $(R)$, which leads us to discrete dynamical systems, namely sandpiles models. It is then possible to obtain a precise quantification of the main parameters of the LLL algorithm. These results fit the experimental results performed on general input bases, which indirectly substantiates the validity of such a regularity hypothesis and underlines the usefulness of such a simplified model.

## Introduction

Lenstra, Lenstra, and Lovász designed the LLL algorithm [10] in 1982 for solving integer programming problems and factoring polynomials. This algorithm belongs to the general framework of lattice basis reduction algorithms and solves a general problem: Given a basis for a lattice, how to find a basis for the same lattice, which enjoys good euclidean properties? Nowadays, this algorithm has a wide area of applications and plays a central algorithmic role in many areas of mathematics and computer science, like cryptology, computer algebra, integer linear programming, and number theory. However, even if its overall structure is simple (see Figure 1), its general probabilistic behaviour is far from being well understood. A precise quantification of the main parameters which are characteristic of the algorithms —principally, the number of iterations and the geometry of reduced bases— is yet unknown. The works of Gama, Nguyen and Stehlé [6,11] provide interesting experiments, which indicate that the geometry of the output seems to be largely independent of the input distribution, whereas the number of iterations is highly dependent on it. The article of Daudé and Vallée [5] provides a precise description of the probabilistic behaviour of these parameters (number of iterations, geometry of the output), but only in the particular case in which the vectors of the input basis are independently chosen in the unit ball. This

A. López-Ortiz (Ed.): LATIN 2010, LNCS 6034, pp. 267–281, 2010.

input distribution does not arise naturally in applications. In summary, the first works [6,11] study general inputs, but do not provide proofs, whereas the second one [5] provides proofs, but for non realistic inputs. Furthermore, none of these studies is dedicated to the fine understanding of the internal structure of the algorithm.

The LLL algorithm is a multidimensional extension, in dimension $n$, of the Euclid algorithm (obtained for $n = 1$) or the Gauss algorithm (obtained for $n = 2$). In these small dimensions, the dynamics of the algorithms is now well understood and there exist precise results on the probabilistic behaviour of these algorithms [12,13,14] which are obtained by using the dynamical systems theory, as well as its related tools. However, even in these small dimensions, the dynamics is rather complex and it does not seem possible to directly describe the fine probabilistic properties of the internal structure of the LLL algorithm in an exact way.

This is why we introduce here a simplified model of the LLL algorithm, which is based on a regularity hypothesis: Whereas the classical version deals with a decreasing factor which may vary during the algorithm, the simplified version assumes *this decreasing factor to be constant*. Of course, this appears to be a strong assumption, but we provide arguments towards this simplification. This assumption leads us to a classical model, the *sandpile model*, and this provides another argument for such a simplification.

Sandpile models are instances of dynamical systems which originate from observations in Nature [9]. They were first introduced by Bak, Tang and Wiesenfeld [3] for modelling sandpile formations, snow avalanches, river flows, etc.. By contrast, the sandpiles that arise in a natural way from the LLL algorithm are not of the same type as the usual instances, and the application of sandpiles to the LLL algorithm thus needs an extension of classical results.

**Plan of the paper.** Section 1 presents the LLL algorithm, describes a natural class of probabilistic models, and introduces the simplified models, based on the regularity assumption. Section 2 provides arguments for the regularity assumption. Then, Section 3 studies the main parameters of interest inside the simplified models, namely the number of iterations, the geometry of reduced bases, and the independence between blocks. Section 4 then returns to the actual LLL algorithm, within the probabilistic models of Section 1, and exhibits an excellent fitting between two classes of results : the proven results in the simplified model, and the experimental results that hold for the actual LLL algorithm. This explains why these "regularized" results can be viewed as a first step for a probabilistic analysis of the LLL algorithm.

# 1   The LLL Algorithm and Its Simplified Version

## 1.1   Description of the Algorithm

The LLL algorithm considers a Euclidean lattice $\mathcal{L}$ given by a system $B$ of $n$ linearly independent vectors in the ambient space $\mathbb{R}^p$ $(n \leq p)$. It aims at finding a reduced basis $\widehat{B}$ formed with vectors that are *almost orthogonal and short enough*. The algorithm operates with the matrix $\mathcal{P}$ which expresses the system $B$ as a function of the Gram–Schmidt orthogonalized system $B^*$; the generic coefficient of the matrix $\mathcal{P}$ is denoted by $m_{i,j}$. The lengths $\ell_i$ of the vectors of the basis $B^*$ and the ratios $r_i$ between successive $\ell_i$, *i.e.*

$$r_i := \frac{\ell_{i+1}}{\ell_i}, \qquad \text{with} \quad \ell_i := \|b_i^*\|. \tag{1}$$

play a fundamental role in the algorithm. The algorithm aims at obtaining lower bounds on these ratios, by computing a $s$–*Siegel reduced* basis $\widehat{B}$ that fulfills, for any $i \in [1..n-1]$, the Siegel condition $\mathcal{S}_s(i)$,

$$|\widehat{m}_{i+1,i}| \leq \frac{1}{2}, \quad \widehat{r}_i := \frac{\widehat{\ell}_{i+1}}{\widehat{\ell}_i} \geq \frac{1}{s}, \qquad \text{with} \quad s > s_0 = \frac{2}{\sqrt{3}}. \tag{2}$$

In the classical LLL algorithm, a stronger condition, the Lovasz condition $\mathcal{L}_t(i)$,

$$|\widehat{m}_{i+1,i}| \leq \frac{1}{2}, \quad \widehat{\ell}_{i+1}^2 + \widehat{m}_{i+1,i}^2\,\widehat{\ell}_i^2 \geq \frac{1}{t^2}\widehat{\ell}_i^2 \qquad (\text{with} \quad t > 1), \tag{3}$$

must be fulfilled for all $i \in [1..n-1]$. When $s$ and $t$ are related by the equality $1/t^2 = (1/4) + 1/s^2$, Condition $\mathcal{L}_t(i)$ implies Condition $\mathcal{S}_s(i)$.

The version of the LLL algorithm studied here directly operates with the Siegel conditions (2). However, the behaviours of the two algorithms are similar, as it is shown in [2], and they perform the same two main types of operations:

(i) *Translation* $(i,j)$ *(for $j < i$).*[1] The vector $b_i$ is translated with respect to the vector $b_j$ by : $b_i := b_i - \lfloor m_{i,j} \rceil b_j$, with $\lfloor x \rceil :=$ the integer closest to $x$. This translation does not change $\ell_i$, and entails the inequality $|m_{i,j}| \leq (1/2)$.

(ii) *Exchange* $(i, i+1)$. When the condition $\mathcal{S}_s(i)$ is not satisfied, there is an exchange between $b_i$ and $b_{i+1}$, which modifies the lengths $\ell_i, \ell_{i+1}$. The new value $\check{\ell}_i$ is multiplied by a factor $\rho$ and satisfies

$$\check{\ell}_i^2 := \ell_{i+1}^2 + m_{i+1,i}^2\,\ell_i^2, \qquad \text{so that} \quad \check{\ell}_i = \rho\,\ell_i \quad \text{with} \quad \rho^2 = \frac{\ell_{i+1}^2}{\ell_i^2} + m_{i+1,i}^2, \tag{4}$$

while the determinant invariance implies the relation $\check{\ell}_i\,\check{\ell}_{i+1} = \ell_i\,\ell_{i+1}$, hence the equality $\check{\ell}_{i+1} = (1/\rho)\,\ell_{i+1}$. This entails that $\rho$ defined in (4) satisfies

$$\rho \leq \rho_0(s) \qquad \text{with} \quad \rho_0^2(s) = \frac{1}{s^2} + \frac{1}{4} < 1; \tag{5}$$

The "decreasing factor" $\rho$ plays a crucial rôle in the following.

Figure 1 describes the standard strategy for the LLL algorithm, where the index $i$ is incremented or decremented at each step. However, there exist other strategies which perform other choices for the next position of reduction, which can be any index $i$ for which Condition $\mathcal{S}_s(i)$ does not hold (See Section 2). Each execution conducted by a given strategy leads to a random walk. See Figure 9 for some instances of random walks in the standard strategy.

---

[1] In the usual LLL algorithm, all the translations $(i+1, j)$ are performed at each step when the condition $\mathcal{S}_s(i)$ is satisfied. These translations do not change the length $\ell_{i+1}$, but are useful to keep the length of $b_{i+1}$ small. Here, we look at the trace of the algorithm only on the $\ell_i$, and the translations $(i+1, j)$, with $j < i$, are not performed.

**RLLL** $(\rho, s)$
with $s > 2/\sqrt{3}$, $\rho \le \rho_0(s) < 1$

**Input.** A sequence $(\ell_1, \ell_2, \ldots \ell_n)$
**Output.** A sequence $(\widehat{\ell}_1, \widehat{\ell}_2, \ldots \widehat{\ell}_n)$
with $\widehat{\ell}_{i+1} \ge (1/s)\widehat{\ell}_i$.

```
i := 1;
While i < n do
    If ℓ_{i+1} ≥ (1/s)ℓ_i, then i := i+1
    else ℓ_i := ρℓ_i;
         ℓ_{i+1} := (1/ρ) ℓ_{i+1};
         i := max(i − 1, 1);
```

**ARLLL** $(\alpha)$    with $\alpha > \alpha_0(s)$.

**Input.** A sequence $(q_1, q_2, \ldots q_n)$
**Output.** A sequence $(\widehat{q}_1, \widehat{q}_2, \ldots \widehat{q}_n)$
with $\widehat{q}_i - \widehat{q}_{i+1} \le 1$.

```
i := 1;
While i < n do
    If q̂_i − q̂_{i+1} ≤ 1, then i := i+1
    else q_i := q_i − α;
         q_{i+1} := q_{i+1} + α;
         i := max(i − 1, 1);
```

**Fig. 1.** Two versions of the LLL algorithm. On the left, the classical version, which depends on parameters $s, \rho$, with $\rho_0(s)$ defined in (5). On the right, the additive version, which depends on the parameter $\alpha := -\log_s \rho$, with $\alpha_0 := -\log_s \rho_0(s)$.

### 1.2    What Is Known about the Analysis of the LLL Algorithm?

The main parameters of interest are the number of iterations and the quality of the output basis. These parameters depend a priori on the strategy. There are classical bounds, which are valid for any strategy, and involve the potential $D(B)$ and the determinant $\det B$ defined as

$$D(B) = \prod_{i=1}^{n} \ell_i^i, \qquad \det(B) = \prod_{i=1}^{n} \ell_i.$$

*Number of iterations.* This is the number of steps $K$ where the test in step 2 is negative. There is a classical upper bound for $K$ which involves the potential values, the initial one $D(B)$ and the final one $D(\widehat{B})$, together with the constant $\rho_0(s)$ defined in (5). We observe that $K$ *can be exactly expressed* with the potential values and the mean $\overline{\alpha}$ of the values $\alpha := -\log_s \rho$ used at each iteration

$$K(B) = \frac{1}{\overline{\alpha}(B)} \log_s \frac{D(B)}{D(\widehat{B})}, \qquad \text{so that} \quad K(B) \le \frac{1}{\alpha_0} \log_s \frac{D(B)}{D(\widehat{B})}, \qquad (6)$$

where $\alpha_0 := -\log_s \rho_0(s)$ is the minimal value of $\alpha$.

*Quality of the output.* The first vector $\widehat{b}_1$ of a $s$-Siegel reduced basis $\widehat{B}$ is short enough; there is an upper bound for the ratio $\gamma(B)$ between its length and the $n$-th root of the determinant,

$$\gamma(B) := \frac{\|\widehat{b}_1\|}{\det(B)^{1/n}} \le s^{(n-1)/2}. \qquad (7)$$

The two main bounds previously described in (6) and (7) are worst–case bounds, and we are interested here in the "average" behaviour of the algorithm: What are the mean values of the number $K$ of steps and of the output parameter $\gamma$?

## 1.3 Our Probabilistic Model

We first define a probabilistic model for input bases, which describe realistic instances, of variable difficulty. We directly choose a distribution on the actual input instance, which is formed with the coefficients $m_{i,j}$ of the matrix $\mathcal{P}$, together with the ratios $r_i$. As Ajtai in [1], we consider lattice bases of full–rank (i.e, $n = p$) whose matrix $B$ is triangular: in this case, the matrix $\mathcal{P}$ and the ratios $r_i$ are easy to compute as a function of $b_i := (b_{i,j})$,

$$r_i = \frac{b_{i+1,i+1}}{b_{i,i}}, \qquad m_{i,j} = \frac{b_{i,j}}{b_{j,j}}.$$

Furthermore, it is clear that the main parameters are the ratios $r_i$, whereas the coefficients $m_{i,j}$ only play an auxilliary rôle. As Ajtai suggests it, we choose them (for $j < i$) independently and uniformly distributed in the interval $[-1/2, +1/2]$. Since Ajtai is interested in worst-case bounds, he chooses very difficult instances where the input ratios $r_i$ are *fixed* and very small, of the form $r_i \sim 2^{-(a+1)(2n-i)^a}$ with $a > 0$. Here, we design a model where each ratio $r_i$ is now a random variable which follows a power law:

$$\forall i \in [1..n-1], \; \exists \theta_i > 0 \text{ for which } \quad \mathbb{P}\left[r_i \leq x\right] = x^{1/\theta_i} \quad \text{ for } x \in [0, 1]. \quad (8)$$

This model produces instances with variable difficulty, which increases when the parameters $\theta_i$ become large. This distribution arises in a natural way in various frameworks, in the two dimensional case [13] or when the initial basis is uniformly chosen in the unit ball. See [14] for a discussion about this probabilistic model.

## 1.4 An Additive Version

First, we adopt an additive point of view, and thus consider the logarithms of the main variables ($\log_s$ is the logarithm to base $s$),

$$q_i := \log_s \ell_i, \qquad c_i := -\log_s r_i = q_i - q_{i+1} \qquad \alpha := -\log_s \rho, \qquad (9)$$

Then, the Siegel condition becomes $q_i \leq q_{i+1} + 1$ or $c_i \leq 1$, and the exchange in the LLL algorithm is rewritten as (see Figure 1. right)

$$\textbf{If} \quad q_i > q_{i+1} + 1, \quad \textbf{then} \quad [\tilde{q}_i = q_i - \alpha, \quad \tilde{q}_{i+1} = q_{i+1} + \alpha]. \quad (10)$$

In our probabilistic model, each $c_i$ follows an exponential law of the form

$$\mathbb{P}[c_i \geq y] = s^{-y/\theta_i} \quad \text{ for } \quad y \in [0, +\infty[ \quad \text{ with } \quad \mathbb{E}[c_i] = \frac{\theta_i}{\log s}. \quad (11)$$

This model is then called the Exp-Ajtai$(\theta)$ model. Remark that, if we restrict ourselves to non-reduced bases, we deal with the Mod-Exp-Ajtai$(\theta)$ distribution,

$$\mathbb{P}[c_i \geq y + 1] = s^{-y/\theta_i} \quad \text{ for } \quad y \in [0, +\infty[, \quad \text{ with } \quad \mathbb{E}[c_i] = 1 + \frac{\theta_i}{\log s}. \quad (12)$$

## 1.5 The Regularized Version of the LLL Algorithm

The main difficulty of the analysis of the LLL algorithm is due to the fact that the decreasing factor $\rho$ defined in (4) can vary throughout the interval $[0, \rho_0(s)]$. For simplifying the behaviour of the LLL algorithm, we assume that the following Regularity Hypothesis holds $(R)$:

($R$). *The decreasing factor $\rho$ (and thus its logarithm $\alpha := -\log_s \rho$) are constant.*

Then, Equation (10) defines a sandpile model which is studied in Section 3.

There are now three main questions:
– Is Hypothesis ($R$) reasonable? This is discussed in Section 2.
– What are the main features of the regularized versions of the LLL algorithm, namely sandpiles? This is the aim of Section 3.
– What consequences can be deduced for the probabilistic behaviour of the LLL algorithm? This is done in Section 5 that transfers results of Section 4 to the framework described in Section 2 with the arguments discussed in Section 4.

## 2    Is the LLL Algorithm Regular?

### 2.1    General Bounds for $\alpha$

Since the evolution of the coefficients $m_{i+1,i}$ seems less "directed" by the algorithm, we may suppose them to be uniformly distributed inside the $[-1/2, +1/2]$ interval, and independent of the Siegel ratios. The average of the square $m^2$ is then equal to $1/12$, and if we assume $m^2$ to be constant and equal to $1/12$, then the value of $\alpha$ satisfies (with $s$ near $s_0 = 2/\sqrt{3}$),

$$-\frac{1}{2}\log_{s_0}\left(\frac{3}{4}+\frac{1}{12}\right) \leq \alpha := -\frac{1}{2}\log_{s_0}\left(r^2+\frac{1}{12}\right) \leq -\frac{1}{2}\log_{s_0}\left(\frac{1}{12}\right).$$

Then $\alpha \in [0.5, 8.5]$ most of the time. This fits with our experiments.

### 2.2    General Study of Parameter $\alpha$

We must make precise the regularity assumption. Of course, we cannot assume that there is a universal value for $\alpha := -\log_s \rho$, and we describe the possible influence of four variables on the parameter $\alpha$, when the dimension $n$ becomes large:

(a) The input distribution of Exp-Ajtai type is described by $\Theta = (\theta_1, \ldots, \theta_{n-1})$.
(b) The position $i \in [1..n(B) - 1]$ is the index where the reduction occurs.
(c) The discrete time $j \in [1..K(B)]$ is the index when the reduction occurs,
(d) The strategy defines the choice of the position $i$ at the $j$-th iteration, inside the set $\mathcal{N}(j)$ which gathers the indices for which Condition $\mathcal{S}(i)$ is not satisfied. We consider three main strategies $\Sigma$ : – The standard strategy $\Sigma_s$ chooses $i := \text{Min}\,\mathcal{N}(j)$ – The random strategy $\Sigma_r$ chooses $i$ at random in $\mathcal{N}(j)$ – The greedy strategy $\Sigma_g$ chooses the index $i \in \mathcal{N}(j)$ for which the ratio $r_i$ is minimum.

The study of $\alpha$ decomposes into two parts. First, we study the variations of $\alpha$ during one execution, due to the position $i$ or the time $j$. Second, we consider the variable $\overline{\alpha}$, defined as the mean value of $\alpha$ during one execution, and study the influence of the input distribution, the strategy, and the dimension on $\overline{\alpha}$.

We consider a set $\mathcal{B}$ of input bases, and we determine a maximal value $M$ of $\alpha$ for this set of inputs. In order to deal with fixed intervals for positions, times, and values, we choose three integers $X, Y, Z$, and we divide

- the interval $[1..n]$ of positions into $X$ equal intervals of type $I_x$ with $x \in [1..X]$,
- the interval $[1..K]$ of times into $Y$ equal intervals of type $J_y$ with $y \in [1..Y]$,
- the interval $[0, M]$ of values into $Z$ equal intervals. of type $L_z$ with $z \in [1..Z]$

Then the parameters $\alpha_{\langle x \rangle}, \alpha^{\langle y \rangle}$ are respectively defined as the restriction of $\alpha$ for $i \in I_x$, (resp. for $j \in J_y$).

### 2.3  Distribution of the Variable $\alpha$

Here, the parameter $\Theta$ of the input distribution and the strategy $\Sigma \in \{\Sigma_s, \Sigma_r, \Sigma_g\}$ are fixed, . and we consider a set $\mathcal{N}$ of dimensions. We first consider the global variable $\alpha$, study its distribution, and its mean, for each $n \in \mathcal{N}$ [See Figure 2(1)]. We observe that the distribution of $\alpha$ gets more and more concentrated when the dimension grows, around a value which appears to tend to 2.5.

### 2.4  Variations of $\alpha$ During an Execution

Figure 2(2) describes the two functions $x \mapsto \overline{\alpha}_{\langle x \rangle}$ and $y \mapsto \overline{\alpha}^{\langle y \rangle}$, for each dimension $n \in \mathcal{N}$. Figure 2(3) provides (for $n = 20$) a description of the distribution of parameters $\alpha_{\langle x \rangle}, \alpha^{\langle y \rangle}$ for various values of $(x, y)$. We observe that the variations of the functions $y \mapsto \overline{\alpha}^{\langle y \rangle}$ and $x \mapsto \overline{\alpha}_{\langle x \rangle}$ are small, and become smaller when the dimension $n$ increases. The distributions of $\alpha_{\langle x \rangle}$ and $\alpha^{\langle y \rangle}$ are also concentrated, at least for $y$'s not too small and for central values of $x$.

### 2.5  Influence of the Strategy

Here, for $n = 20$, we investigate the influence of the strategy on the functions $x \mapsto A_{\langle x \rangle}, \quad y \mapsto \overline{\alpha}^{\langle y \rangle}, \quad z \mapsto \mathbb{P}[\alpha \in L_z]$.

The experimental results, reported in Figure 2(4), show the important influence of the strategy on the parameter $\alpha$. They are of independent interest, since, to the best of our knowledge, the strategy is not often studied. There are two groups: On the one hand, the standard strategy[2] is the least efficient: it performs a larger number of steps, and deals with a parameter $\alpha$ whose value is concentrated below $\alpha = 5$. On the other hand, the other two ones, (random and greedy) are much more efficient, with a much smaller number of steps; they deal with values of $\alpha$ which vary in the whole interval $[5, 20]$ and decrease with the discrete time. These two strategies (random and greedy) must be used if we wish more efficient algorithms. If we wish simulate with sandpiles the LLL algorithm under these two strategies, we have to consider different values of $\alpha$, for instance, at the beginning, in the middle and at the end of the execution.

---

[2] We have not reported the results relative to the anti-standard strategy which chooses $i := \text{Max}\,\mathcal{N}(j)$, but they are of the same type as the standard one.

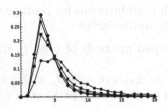 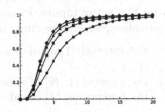

(1) The distribution of the parameter $\alpha$ for $n = 5\,(\bullet); n = 10\,(\blacksquare); n = 15\,(\blacktriangle); n = 20\,(\blacklozenge)$

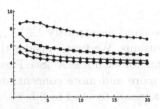 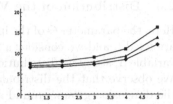

(2) The two functions $y \mapsto \overline{\alpha}^{\langle y \rangle}$ (left, with $Y = 20$) and $x \mapsto \overline{\alpha}_{\langle x \rangle}$ (right, with $X = 5$), for $n = 5\,(\bullet); n = 10\,(\blacksquare); n = 15\,(\blacktriangle); n = 20\,(\blacklozenge)$

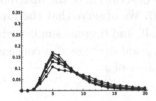 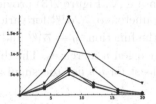

(3) On the left, the distribution of $\alpha^{\langle y \rangle}$ with $Y = 20$ and $y = 2\,(\bullet); 5\,(\blacksquare); 10\,(\blacktriangle);$ $15\,(\blacklozenge); 20\,(\blacktriangledown)$
On the right, the distribution of $\alpha_{\langle x \rangle}$ with $X = 5$ and $x = 1\,(\bullet); 2\,(\blacksquare); 3\,(\blacktriangle); 4\,(\blacklozenge); 5\,(\blacktriangledown)$

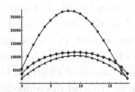 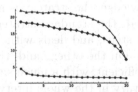 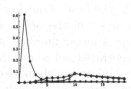

(4) The curves are associated to $\bullet$ for $\Sigma_s$ (standard), $\blacktriangle$ for $\Sigma_g$ (greedy), and $\blacklozenge$ for $\Sigma_r$ (random). On the right, the functions $x \mapsto A_{\langle x \rangle}$. In the middle, the functions $y \mapsto \overline{\alpha}^{\langle y \rangle}$. On the left, the distribution of $\alpha$.

**Fig. 2.** Experiments about the Regularity Hypothesis: Study of the global parameter $\alpha$. Influence of position and time. Influence of the strategy for $n = 20$.

## 2.6   Influence of the Input Distribution

We study the influence of the parameter $\Theta$ of the `Exp-Ajtai` distribution on $\overline{\alpha}$. We first recall what happens in two dimensions, where the LLL algorithm coincides with the Gauss algorithm. The paper [13] studies this algorithm when the input $c := -\log_s r$ follows an exponential law with mean $\theta$ and proves that the number of steps $K$ of the Gauss algorithm follows a geometric law of ratio $\lambda(1 + 1/\theta)$, where $\lambda(s)$ is the dominant eigenvalue of the transfer operator associated to the Gauss algorithm.

$$\text{The relations} \qquad -\log_s \mathbb{P}[K \geq k] \sim -\log_s \mathbb{P}[c \geq k\overline{\alpha}] \sim \frac{\mathbb{E}_\theta[\overline{\alpha}]}{\theta} k$$

entail that the mean $\mathbb{E}_\theta[\overline{\alpha}]$ depends on $\theta$ as $\quad \mathbb{E}_\theta[\overline{\alpha}] \sim -\theta \log_s \lambda \left(1 + \dfrac{1}{\theta}\right).$

Then, properties of the pressure[3] imply that the function $\mathbb{E}_\theta[\overline{\alpha}]$ satisfies

$$\mathbb{E}_\theta[\overline{\alpha}] \sim \frac{|\lambda'(1)|}{\log s} \quad \text{for } \theta \to \infty, \qquad \text{and} \quad \mathbb{E}_\theta[\overline{\alpha}] \sim \frac{2}{\log s} \log(1 + \sqrt{2}) \quad \text{for } \theta \to 0,$$

where $|\lambda'(1)| \sim 3.41$ equals the entropy of the Euclid centered algorithm. This entails that, in two dimensions, the mean value $\mathbb{E}[\overline{\alpha}]$ varies in the interval $[14, 23]$. Led by the dynamical point of view, we set a conjecture which extends the previous two–dimensional property to any dimensions.

**Entropy Conjecture.** *Consider the probabilistic* `Exp-Ajtai`*$(\theta)$ model in $n$ dimensions. Then, for $\theta \to \infty$, the mean of the variable $\overline{\alpha}$ tends to the entropy $\mathcal{E}_n$ of the dynamical system underlying the LLL algorithm.*

$$\lim_{\theta \to \infty} \mathbb{E}_{(\theta, n)}[\overline{\alpha}] = \frac{\mathcal{E}_n}{\log s}$$

# 3   Study of the Sandpile Model

There are three main questions about the RLLL algorithm:

(Q1) Does the RLLL algorithm depend on the strategy?

(Q2) How does the behaviour of the RLLL algorithm depend on the value of parameter $\alpha$? What about the number of iterations? the output configuration? Are there lower bounds on average in relations (6) and (7)?

(Q3) Does there exist a characterisation for two blocks to be independent in the RLLL algorithm? We can run the execution of the LLL algorithm, both on the block $B_-$ formed with the first vectors and on the block $B_+$ formed on the last vectors The two blocks $B_-$ and $B_+$ are said to be independent if the total basis formed by concatenating the two reduced bases $\widehat{B}_-$ and $\widehat{B}_+$ is reduced.

Here, we answer these three main questions. As we already said previously, the additive version of the regularized algorithm (see Figure 1.right) deals with the sandpile model. Even if this model is very well known, the modelling of the RLLL algorithm gives rise to non classical instances of sandpile models.

---

[3] In dynamical systems theory, the pressure is the logarithm of the dominant eigenvalue.

## 3.1 Main Objects for Sandpiles

Here, $H, h$ denote strictly positive real numbers.

A sandpile model $\mathbf{Q}_n(\mathbf{q}, H, h)$ describes all the possible evolutions of the configuration $\mathbf{q} = (q_1, \ldots, q_n)$ under the action of functions $f_i$

$$f_i(\mathbf{q}) = \begin{cases} q_j - h & \text{if } j = i \text{ and } q_i - q_{i+1} > H, \\ q_j + h & \text{if } j = i+1 \text{ and } q_i - q_{i+1} > H, \\ q_j & \text{else.} \end{cases}$$

We associate to $\mathbf{q} := (q_1, \ldots, q_n)$ the configuration $\mathbf{c} := \Delta(\mathbf{q})$ formed with the differences between the components, $c_i = q_i - q_{i+1}$ for $i \in [1..n-1]$.

The strategy graph, denoted by $\mathcal{G}(\mathbf{q}, H, h)$, is a directed graph whose vertices are all the configurations that are reachable from $\mathbf{q}$; there is an edge from $\mathbf{u}$ to $\mathbf{v}$ (with $\mathbf{u} \neq \mathbf{v}$) if there exists an index $i \in [1..n-1]$ for which $\mathbf{v} = f_i(\mathbf{u})$.

The energy $E$ and the total mass $M$ of the configuration $\mathbf{q}$ are defined by

$$E(\mathbf{q}) = \sum_{i=1}^{n} i \cdot q_i, \quad \text{and} \quad M(\mathbf{q}) = \sum_{i=1}^{n} q_i, \tag{13}$$

and satisfy $\quad M(f_i(\mathbf{q})) = M(\mathbf{q}), \qquad E(f_i(\mathbf{q})) = E(\mathbf{q}) + h.$

## 3.2 Various Kinds of Sandpiles

The usual sandpiles are basic and decreasing:

**Definition 1.** (*i*) A sandpile $\mathbf{q}$ is basic if the configuration $\Delta(\mathbf{q})$ is integral and parameters $(H, h)$ equal $(1, 1)$

(*ii*) A sandpile is $(H, h)$–integral if the components $c_i$ of $\mathbf{c} := \Delta(\mathbf{q})$ belong to the same discrete line $H + \mathbb{Z}h$

(*iii*) A basic sandpile $\mathbf{q}$ is decreasing if the components of $\mathbf{c} := \Delta(\mathbf{q})$ are positive ($c_i \geq 0$). It is strictly decreasing if $\mathbf{c}$ is strictly positive. On the contrary, it is increasing if all the components of $\mathbf{c}$ are negative.

The sandpiles used in the RLLL algorithm are not basic. However, the following result shows that any general sandpile is isomorphic to a basic sandpile.

**Proposition 1.** The mapping $\psi : \mathbf{q} \mapsto \mathbf{q}'$ defined by

$$c_i' := 1 - \left\lfloor \frac{H - c_i}{h} \right\rfloor, \qquad q_n' = 0 \tag{14}$$

transforms a general sandpile into a basic sandpile. Moreover, the two graphs $\mathcal{G}(\mathbf{q}, H, h)$ and $\mathcal{G}(\psi(\mathbf{q}), 1, 1)$ are isomorphic.

A general sandpile $\mathbf{q}$ is decreasing (resp. strictly decreasing, increasing) if $\psi(\mathbf{q})$ is decreasing (resp. strictly decreasing, increasing). A general sandpile decomposes into strictly decreasing configurations, separated by increasing configurations.

**Definition 2.** Two adjacent strictly decreasing sandpiles $\mathbf{q}^-, \mathbf{q}^+$ are independent if the configuration obtained by concatenating the two final configurations $\widehat{\mathbf{q}}^-$ and $\widehat{\mathbf{q}}^+$ is a final configuration.

### 3.3  Study of a General Sandpile

Here, we obtain (easy) extensions of results of Goles and Kiwi who considered only in [8] basic decreasing sandpiles.

**Theorem 1.** *The following holds for any sandpile $\mathcal{Q}(\mathbf{q}, H, h)$:*

*(i) The graph $\mathcal{G}(\mathbf{q}, H, h)$ is finite, with a unique final configuration $\widehat{\mathbf{q}}$. The length of a path $\mathbf{q} \to \widehat{\mathbf{q}}$ is the same for any path. This is the number of steps $T(\mathbf{q})$,*

$$T(\mathbf{q}) = \frac{1}{h}\left[E(\widehat{\mathbf{q}}) - E(\mathbf{q})\right] = \frac{1}{2h}\sum_{i=1}^{n-1} i(n-i)\,(c_i - \widehat{c}_i)$$

*(ii) If $\mathcal{Q}_n(\mathbf{q}, H, h)$ is decreasing, then the components of the output configuration $\widehat{\mathbf{c}}$ satisfy $H - 2h < \widehat{c}_i \leq H$, and the number of iterations satisfy*

$$0 \leq T(\mathbf{q}) - \frac{1}{2h}\sum_{i=1}^{n-1} i(n-i)\,(c_i - H) \leq 2A(n) \qquad \text{with} \quad A(n) := n\frac{n^2 - 1}{12}$$

*(iii) If $\mathcal{Q}_n(\mathbf{q}, H; h)$ is strictly decreasing, then there exists $j \in [1..n-1]$ for which*

$$\forall i \neq j, \ H - h < \widehat{c}_i \leq H, \qquad \text{and} \quad H - 2h < \widehat{c}_j \leq H - h,$$

*and the number of steps $T(\mathbf{q})$ satisfies*

$$0 \leq T(\mathbf{q}) - \left[A(n) + \frac{1}{2h}\sum_{i=1}^{n-1} i(n-i)\,(c_i - H)\right] \leq \frac{1}{8}n^2$$

*(iv) For a general sandpile $\mathcal{Q}_n(\mathbf{q}, H, h)$, the output configuration satisfies*

$$H - 2h < \widehat{c}_i \leq H \ \ \text{if} \ \ c_i > H - h, \qquad \widehat{c}_i \geq c_i \ \ \text{if} \ \ c_i \leq H - h$$

*and the number of steps $T(\mathbf{q})$ satisfies*

$$\frac{1}{2h}\sum_{i=1}^{n-1} i(n-i)(c_i - H + h) \leq T(\mathbf{q}) \leq \frac{1}{2h}\sum_{i=1}^{n-1} i(n-i)\max(c_i - H + h, 0)$$

*(v) A sufficient condition for two adjacent sandpiles $\mathcal{Q}_p(\mathbf{q}_-, H, h), \mathcal{Q}_n(\mathbf{q}_+, H, h)$ to satisfy the independence condition of Definition 2 is*

$$\frac{1}{p}M(\mathbf{q}_-) - \frac{1}{n}M(\mathbf{q}_+) < \left(\frac{n \mid p}{2}\right)(H - h) - h$$

*and for a sandpile $(H, h)$–integral:* $\quad \dfrac{1}{p}M(\mathbf{q}_-) - \dfrac{1}{n}M(\mathbf{q}_+) \leq \left(\dfrac{n+p}{2}\right)H - 2h.$

## 4   Returning to Lattices

We now return to the LLL algorithm, with the framework of Section 1, and apply the results of Section 3 to the so–called $\rho$–regular executions of the LLL algorithm, for which the decreasing factor is constant and equal to $\rho$. We recall that, in this case, the execution of the algorithm in dimension $n$ can be viewed as a sandpile model $\mathbf{Q}_n(\mathbf{q}, 1, \alpha)$ associated to a parameter $\alpha := -\log_s \rho$, and an initial configuration $\mathbf{q}$ related to the lengths $\ell_i$ of the orthogonalised basis $B^\star$ of the input basis $B$ via the equalities $q_i := \log_s \ell_i$. The main objects associated to the basis $B$, namely the potential $D(B)$ or the determinant $\det(B)$ are then related to the energy $E(\mathbf{q})$ or the total mass $M(\mathbf{q})$,

$$E(\mathbf{q}) = \log_s D(B), \qquad M(\mathbf{q}) = \log_s \det(B).$$

We are interested in two kinds of input bases:

($i$) We first study totally non-reduced bases, for which Condition $\mathcal{S}_s(i)$ is never satisfied on the input. In this case, the sandpile is strictly decreasing. [Sections 4.1 and 4.2]

($ii$) We then study a general input basis, which is a sequence of blocks, some of them are totally non-reduced, and other ones are totally reduced [Section 4.3]

We compare here the results that are proven for regular executions of the LLL algorithm, (by an easy transfer of results of Section 3) and the experimental results that are performed on general executions of the algorithm. We will see that there is a good fitting between these two kinds of results. This good fitting has two main consequences:

- This provides an indirect validation of the property : "The executions of the LLL algorithm are very often regular enough".
- This shows that long experiments on the LLL algorithm can be simulated by fast computations in the sand pile model (with a good choice of parameter $\alpha$).

As in Section 3, we study the final configurations, the number of steps, and the independence of blocks.

## 4.1   Final Configurations

When the initial basis is totally non reduced, the sandpile is strictly decreasing. Then, with Theorem 1 ($ii$), each output Siegel ratio $\widehat{r}_i$ and the first vector of the output basis satisfy

$$\rho s \le \frac{1}{\widehat{r}_i} = \frac{\widehat{\ell}_i}{\widehat{\ell}_{i+1}} \le s, \qquad \rho(s \cdot \rho)^{(n-1)/2} \le \gamma(\widehat{B}) = \frac{\|\widehat{b}_1\|}{(\det L)^{1/n}} \le s^{(n-1)/2}. \tag{15}$$

Then, we have proven:

**Theorem 2.** *Consider a totally non reduced basis $B$ on which the execution of the LLL algorithm is $\rho$–regular. Then, the output parameter $\gamma(\widehat{B})$ defined in (7) satisfies*

$$\frac{2}{n-1} \log_s \gamma(\widehat{B}) \in [1 - \alpha, 1], \qquad \text{with } \alpha := -\log_s \rho.$$

This is compatible with experiments done on general executions by Nguyen and Stehlé [11], which show that there is a mean value $\beta \sim 1.04$ , such that, for most of the output bases $\widehat{B}$, the ratio $\gamma(\widehat{B})$ is close to $\beta^{(n-1)/2}$. The relation $\beta \sim s\sqrt{\rho}$ is then plausible, so that the "usual" $\rho$ would be close to 0.81.

## 4.2   Number of Iterations

Suppose that the (totally non reduced) input basis follows the Mod-Exp-Ajtai($\theta$) distribution. Then, the configuration $\mathbf{c}'$ associated to $\mathbf{c}$ via Theorem 1 follows a geometric law,

$$\mathbb{P}[c_i' \geq 1 + k] = \rho^{k/\theta}, \qquad \mathbb{E}[c_i' - 1] = \frac{\rho^{1/\theta}}{1 - \rho^{1/\theta}},$$

and Theorem 1 (*iii*) entails:

**Theorem 3.** *Consider an input basis $B$, which follows the* Mod-Exp-Ajtai *distribution of parameter $\theta$. If the execution of the LLL algorithm in dimension $n$ is $\rho$–regular on the basis $B$, the number of iterations satisfies*

$$K_n(\rho, \theta) \sim \frac{n^3}{12\alpha} \left( \frac{\rho^{1/\theta}}{1 - \rho^{1/\theta}} \right) \qquad (n \to \infty).$$

*If the Entropy Conjecture of Section 3.6 is true, then*

$$\lim_{\theta \to \infty} K_n(\theta) \sim \left( \frac{\theta \log s}{12} \right) \frac{n^3}{\mathcal{E}_n^2} \qquad \text{where $\mathcal{E}_n$ is the entropy of the LLL algorithm.}$$

This results fits with the experiments done for general executions of the LLL algorithm by Nguyen and Stehlé [11]. In particular, for the choice of Ajtai, namely $\theta = n^a$, the experiments show a number of iterations of order $\Theta(n^{3+a})$.

### 4.3 An Instance of the Independence Property

The question of the independence between blocks is important. We now describe such an instance of this phenomenon in the framework of Coppersmith's method. In the paper [4], Boneh and Durfee present a method for breaking the RSA cryptosystem based on Coppersmith's method. Coppersmith's method uses the LLL algorithm in order to find a small root of a polynomial modulo an integer $E$. For the cryptanalysis of RSA, one deals with the public exponent $E$. We let $L := \log_s E$.

The initial configuration is formed with $m + 1$ blocks, indexed from $k = 0$ to $m$. The $k$-th block has length $k + 1$, is $(1, \alpha)$-integral, and the components $c_i$ of the configuration $\mathbf{c}$ are equal to $L/2$. However, the total configuration is not totally decreasing, but the (second) sufficient condition of Theorem 1 (*v*) is always true. Then, Theorem 1(*v*) entails:

**Theorem 4.** *Suppose that the execution of the LLL algorithm is $\rho$-regular on the Coppersmith lattice described in [4]. Then, the blocks of the lattice are always independent, and the reduction can be done in parallel on each block. The number of iterations $K_p$ performed in this parallel strategy is then*

$$K_p = \frac{m^3}{12\alpha} \left( \frac{L}{2} - 1 \right) \qquad \text{to be compared to} \quad K_s = \sum_{i=1}^{m} K_i \sim \frac{m^4}{48\alpha} \left( \frac{L}{2} - 1 \right),$$

*which is the number of steps in the sequential strategy.*

Of course, the execution of the LLL algorithm on the Boneh-Durfee lattice cannot be totally regular : the first vector of the reduced lattice basis would be the first vector of the initial basis, and the method would fail! However, it is possible to compare (see [7]) the result of Theorem 4 to an execution of the actual LLL

 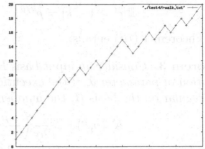

**Fig. 3.** On the left, the random walk of the actual LLL algorithm on a Coppersmith lattice of dimension 21 (related to $m = 5$). On the right, the random walk of the execution of the LLL algorithm on the basis formed by the concatenation of the reduced blocks.

algorithm on a Boneh–Durfee lattice (see Figure 3 left). We first see that, on each block, the number of iterations is quite large (the blocks are totally non reduced) and this fits with the order $\Theta(k^3)$ which is proven for a $\rho$–regular execution. We also remark that the blocks are not independent but almost independent: the basis obtained by concatening the reduced bases of each block is not totally reduced, but few reduction steps are needed for reducing it, as Figure 3 (right) shows it. Such a strategy, whose first step can be performed in a parallel way, is very efficient in this case.

## 5    Conclusion

This paper presents a simplified model of the LLL algorithm, under a "regularity" hypothesis which assumes that the decreasing factor $\rho$ is constant. Of course, this hypothesis does not exactly hold in the reality, and we have provided experimental results about its validity. We have also explained why this simplified model is very useful for understanding the LLL algorithm in an intuitive way, and for testing (at least qualitative) conjectures on the algorithm. The excellent fitting of this model on a class of Coppersmith lattices is also striking. In fact, the sandpile model represents a good compromise between simplicity and adequation to the reality.

**Acknowledgements.** This research was supported by the LAREDA Project (LAttice REDuction Algorithms: Dynamics, Probability, Experiments) of the ANR (French National Research Agency). The authors thank Ali Akhavi, Julien Clément, Mariya Georgieva, Fabien Laguillaumie, Loïck Lhote, Damien Stehlé, Antonio Vera, and the whole group LAREDA for interesting discussions on the subject.

# References

1. Ajtai, M.: Optimal lower bounds for the Korkine-Zolotareff parameters of a lattice and for Schnorr's algorithm for the shortest vector problem. Theory of Computing 4(1), 21–51 (2008)
2. Akhavi, A.: Random lattices, threshold phenomena and efficient reduction algorithms. Theoret. Comput. Sci. 287(2), 359–385 (2002)
3. Bak, P., Tang, C., Wiesenfeld, K.: Self-organized criticality: An explanation of the 1/f noise. Phys. Rev. Lett. 59(4), 381–384 (1987)
4. Boneh, D., Durfee, G.: Cryptanalysis of RSA with private key d less than N ≤ 0.292. IEEE Trans. Inform. Theory 46(4), 1339–1349 (2000)
5. Daudé, H., Vallée, B.: An upper bound on the average number of iterations of the LLL algorithm. Theoretical Computer Science 123(1), 95–115 (1994)
6. Gama, N., Nguyen, P.: Predicting Lattice Reduction. In: Smart, N.P. (ed.) EUROCRYPT 2008. LNCS, vol. 4965, pp. 31–51. Springer, Heidelberg (2008)
7. Georgieva, M.: Étude expérimentale de l'algorithme LLL sur certaines bases de Coppersmith, Master Thesis, University of Caen (2009)
8. Goles, E., Kiwi, M.A.: Games on line graphs and sandpiles. Theoret. Comput. Sci. 115(2), 321–349 (1993)
9. Jensen, H.J.: Self-organized criticality. In: Emergent complex behavior in physical and biological systems. Cambridge Lecture Notes in Physics, vol. 10. Cambridge University Press, Cambridge (1998)
10. Lenstra, A.K., Lenstra Jr., H.W., Lovász, L.: Factoring polynomials with rational coefficients. Math. Ann. 261(4), 515–534 (1982)
11. Nguyen, P., Stehlé, D.: LLL on the average. In: Hess, F., Pauli, S., Pohst, M. (eds.) ANTS 2006. LNCS, vol. 4076, pp. 238–256. Springer, Heidelberg (2006)
12. Vallée, B.: Euclidean Dynamics. Discrete and Continuous Dynamical Systems 15(1), 281–352 (2006)
13. Vallée, B., Vera, A.: Probabilistic analyses of lattice reduction algorithms. In: ch.3. The LLL Algorithm. Collection Information Security and Cryptography Series. Springer, Heidelberg (2009)
14. Vera, A.: Analyses de l'algorithme de Gauss. Applications à l'analyse de l'algorithme LLL, PhD Thesis, Universiy of Caen (2009)

# Communication-Efficient Construction of the Plane Localized Delaunay Graph[*]

Prosenjit Bose[1], Paz Carmi[2], Michiel Smid[1], and Daming Xu[1]

[1] School of Computer Science, Carleton University, Ottawa, Canada
[2] Department of Computer Science, Ben-Gurion University, Beer-Sheva, Israel

**Abstract.** Let $V$ be a finite set of points in the plane. We present a 2-local algorithm that constructs a plane $\frac{4\pi\sqrt{3}}{9}$-spanner of the unit-disk graph $UDG(V)$. Each node can only communicate with nodes that are within unit-distance from it. This algorithm only makes one round of communication and each point of $V$ broadcasts at most 5 messages. This improves on all previously known message-bounds for this problem.

## 1   Introduction

A *wireless ad hoc network* consists of a finite set $V$ of wireless nodes. Each node $u$ in $V$ is a point in the plane that can communicate directly with all points of $V$ within $u$'s communication range. If this range is one unit for each point, then the network is modeled by the *unit-disk graph* $UDG(V)$ of $V$. This (undirected) graph has $V$ as its vertex set and any two distinct vertices $u$ and $v$ are connected by an edge if and only if the Euclidean distance $|uv|$ between $u$ and $v$ is at most one unit. In order for two points that are more than one unit apart to be able to communicate, the points of $V$ use a so-called *local algorithm* to construct a subgraph $G$ of $UDG(V)$. This subgraph should support efficient routing of messages, i.e., there should be a simple and efficient protocol that allows any point of $V$ to send a message to any other point of $V$.

In this paper, we present a local algorithm that constructs a subgraph $G$ of $UDG(V)$ that satisfies the following properties:

1. Each point $u$ of $V$ stores a set $E(u)$ of edges that are incident on $u$. The edge set of $G$ is equal to $\cup_{u\in V} E(u)$.
2. The edge sets $E(u)$ with $u \in V$ are *consistent*: For any two points $u$ and $v$ in $V$, $(u,v)$ is an edge in $E(u)$ if and only if $(u,v)$ is an edge in $E(v)$.
3. The graph $G$ is *plane*. This property is useful, because several algorithms are known for routing messages in a plane subgraph of $UDG(V)$; see, e.g., [4,6].
4. The graph $G$ is a *t-spanner* of $UDG(V)$, for some constant $t > 1$: For each edge $(u,v)$ of $UDG(V)$, the graph $G$ contains a path between $u$ and $v$ whose Euclidean length is at most $t|uv|$. Observe that this implies that shortest-path distances in $UDG(V)$ are approximated, within a factor of $t$, by shortest-path distances in $G$. Our construction shows that $t \leq \frac{4\pi\sqrt{3}}{9}$.

---

[*] This work was supported by the Natural Sciences and Engineering Research Council of Canada.

A. López-Ortiz (Ed.): LATIN 2010, LNCS 6034, pp. 282–293, 2010.

As mentioned above, we model a wireless ad hoc network by the unit-disk graph $UDG(V)$, where $V$ is a finite set of points in the plane. The points of $V$ want to construct a communication graph $G$ (which is a subgraph of $UDG(V)$) using a distributed and local algorithm. The points of $V$ can communicate with each other by broadcasting messages. If a point $u$ of $V$ broadcasts a message, then each point of $V$ within Euclidean distance one from $u$ receives the message. Each point of $V$ can perform computations based on its location and all information received from other points. Let $\delta_{UDG}(u, v)$ denote the minimum Euclidean length of any path between the points $u$ and $v$ in the graph $UDG(V)$. For any integer $k \geq 1$, let

$$N_k(u) = \{v \in V : \delta_{UDG}(u, v) \leq k\}.$$

(Observe that $u \in N_k(u)$.) An algorithm is called $k$-*local*, if the computation performed at each point $u$ of $V$ is based only on the points in the set $N_k(u)$. Thus, in a local algorithm, information cannot "travel" over a "large" distance.

A $k$-local algorithm runs in parallel on all points of $V$, where each point $u$ performs an alternating sequence of computation steps and broadcasting steps in a synchronized manner. In a *computation step*, point $u$ performs some computation based on the subset of $N_k(u)$ that is known to $u$ at that moment. For example, $u$ may compute the Delaunay triangulation of this subset; we consider this to be one computation step. We assume that each such computation step works in the algebraic computation model (see, e.g., [9] for a description of this model). In a *broadcasting step*, point $u$ broadcasts a (possibly empty) sequence of messages, which is received by all points in $N_1(u)$. A *message* is defined to be the location of a point in the plane. A message broadcast by $u$ need not be an element of $V$, but it must have been computed, based on the subset of $N_k(u)$ that is known to $u$ at that moment, in the algebraic computation model. (Thus, bit-manipulation cannot be used to encode several points, or any other information, in one message.)

The efficiency of a local algorithm will be expressed in terms of the following measures: (i) The value of $k$. The smaller the value of $k$, the "more local" the algorithm is. (ii) The maximum number of messages that are broadcast by any point of $V$. The goal is to minimize this number. (iii) The number of *communication rounds*, which is defined to be the maximum number of broadcasting steps performed by any point in $V$. This number measures the (parallel) time for the entire algorithm to complete its computation. Again, the goal is to minimize this number.

Above, we have defined the notion of a $t$-spanner of the unit-disk graph $UDG(V)$. For a real number $t > 1$, a graph $G$ is called a $t$-spanner of the *point set* $V$ if for any two elements $u$ and $v$ of $V$, there exists a path in $G$ between $u$ and $v$ whose length is at most $t|uv|$. The problem of constructing $t$-spanners for point sets has been studied intensively in computational geometry; see the book by Narasimhan and Smid [9] for a survey.

Since we are concerned with plane spanners of the unit-disk graph, our algorithm will be based on the *Delaunay Triangulation* $DT(V)$ of $V$. Keil and Gutwin [7] have shown that $DT(V)$ is a $\frac{4\pi\sqrt{3}}{9}$-spanner of $V$. To extend this result to unit-disk graphs, it is natural to consider subgraphs of $UDel(V)$, which is defined to be the intersection of the Delaunay triangulation and the unit-disk graph of $V$. It has been shown by Bose *et al.* [3] that $UDel(V)$ is a $\frac{4\pi\sqrt{3}}{9}$-spanner of $UDG(V)$. Unfortunately, constructing $UDel(V)$ using a $k$-local algorithm, for any constant value of $k$, is not possible: Consider an edge $(u, v)$ in $UDel(V)$ whose empty disk $D$ is very large. In order for a $k$-local algorithm to verify that no point of $V$ is in the interior of $D$, information about the points of $V$ must travel over a large distance to $u$ or $v$. Clearly, this is possible only if the value of $k$ is very large. Because of this, researchers have considered the problem of designing local algorithms that construct a plane subgraph of $UDG(V)$ which is a *supergraph* of $UDel(V)$. Obviously, by the result of [3], such a graph is also a $\frac{4\pi\sqrt{3}}{9}$-spanner of $UDG(V)$.

As is common in this field, we assume that, at the start of the algorithm, each point $u$ of $V$ knows the locations (i.e., the $x$- and $y$-coordinates) of all points in $N_1(u)$. Gao *et al.* [5] proposed a 2-local algorithm that constructs a plane subgraph of $UDG(V)$ which is a supergraph of $UDel(V)$. However, the number of messages broadcast by a single point of $V$ can be as large as $\Theta(n)$, where $n$ is the number of elements of $V$. This result was improved by Li *et al.* [8]: They presented a 2-local algorithm that constructs such a graph in four communication rounds and in which each point broadcasts at most 49 messages.

Currently, the best result for computing a plane $t$-spanner (for some constant $t$) of the unit-disk graph $UDG(V)$ is by Araújo *et al.* [1]. They presented a 2-local algorithm which computes such a spanner in one communication round and in which each point broadcasts at most 11 messages.

In this paper, we improve the upper bound of Araújo *et al.* [1]:

**Theorem 1.** *Let $V$ be a finite set of points in the plane. There exists a 2-local algorithm that computes a plane and consistent $\frac{4\pi\sqrt{3}}{9}$-spanner of the unit-disk graph of $V$. This algorithm makes one communication round and each point of $V$ broadcasts at most 5 messages.*

Because of lack of space, we can only give an outline of the proof of this result. Complete proofs can be found in [2].

Throughout the rest of this paper, we assume that the points in the set $V$ are in general position (meaning that no three points of $V$ are collinear and no four points of $V$ are cocircular). We also assume that the unit-disk graph $UDG(V)$ is connected. We will use the following notation:

- $D(a, b, c)$ denotes the disk having the three points $a$, $b$, and $c$ on its boundary.
- $D(c; r)$ denotes the disk centered at the point $c$ and having radius $r$.
- $\Delta(a, b, c)$ denotes the triangle with points $a$, $b$, and $c$ as its vertices.
- $\partial D$ denotes the boundary of the disk $D$.
- $int(D)$ denotes the interior of the disk $D$.
- Let $v$, $x$, and $y$ be points of $V$, where $v \neq y$. Assume there exists a disk $D$ such that $N_1(x) \cap \partial D = \{v, y\}$ and $N_1(x) \cap int(D) = \emptyset$. We denote such a disk $D$ by $Del_x(v, y)$. Observe that $Del_x(v, y)$ is a certificate for the fact that $(v, y)$ is an edge in the Delaunay triangulation of the point set $N_1(x)$.

# 2    A Preliminary Algorithm

In this section, we present a 2-local algorithm that constructs a graph, called the *plane localized Delaunay graph* $PLDG(V)$, whose vertex set is a finite set $V$ of points in the plane. The algorithm computes $PLDG(V)$ in one communication round and each point of $V$ broadcasts at most 6 messages. We will prove that $PLDG(V)$ is a plane and consistent supergraph of $UDel(V)$.

In the construction, each point $v$ of $V$ runs algorithm PLDG($v$) in parallel. Let $N_v = N_1(v)$, i.e., $N_v = \{u \in V : |uv| \leq 1\}$. Recall that we assume that, at the start of the algorithm, point $v$ knows the locations of all points in $N_v$. Algorithm PLDG($v$) first computes the Delaunay triangulation $LDT(v)$ of the set $N_v$. Then, for each triangular face $\Delta(u, v, w)$ in $LDT(v)$ for which $\angle uvw > \frac{\pi}{3}$, algorithm PLDG($v$) broadcasts the location $v$ together with the center of the disk $D(u, v, w)$ containing $u$, $v$, and $w$ on its boundary.

In the final step, algorithm PLDG($v$) checks the validity of all edges that are incident on $v$ in $LDT(v)$ and removes those edges which cause a crossing. To be more precise, let $x$ be a point in $N_v$, and assume that $v$ receives a center $c_i'$ from $x$. Algorithm PLDG($v$) considers the unit-disk $D(v; 1)$ centered at $v$ and the disk $D(c_i'; |c_i'x|)$ centered at $c_i'$ that contains $x$ on its boundary. The algorithm knows that $\partial D(c_i'; |c_i'x|)$ contains exactly three points which define a triangular face in the Delaunay triangulation $LDT(x)$ of $N_x$. Point $x$ is one of these three points; let $p$ and $q$ be the other two points. Assume that the set $N_v$ contains exactly two points of $\{x, p, q\}$, say $x$ and $p$. Thus, algorithm PLDG($v$) knows the points $x$ and $p$, but it does not know $q$. The algorithm computes $arc_i$, which is defined to be the (open) portion of $\partial D(c_i'; |c_i'x|)$ which is not contained in $D(v; 1)$. Even though the algorithm does not know the exact location of the third point $q$, it does know that $q$ is on $arc_i$. The algorithm chooses an arbitrary point $z'$ on $arc_i$ such that $|xz'| \leq 1$ or $|pz'| \leq 1$ and *acts as if* $\Delta(x, p, z')$ is a triangular face in $LDT(x)$. (Observe that, since $q \in arc_i$ and $|xq| \leq 1$, the algorithm can choose such a point $z'$. Also, $z'$ is not necessarily a point of $V$.) The algorithm now considers each edge $(v, y)$ in $LDT(v)$ (where, possibly, $v = p$, $y = p$, or $y = x$) and uses the triangle $\Delta(x, p, z')$ to decide whether or not to remove $(v, y)$: Since $(v, y)$ is an edge in $LDT(v)$, algorithm PLDG($v$) can compute a disk $D = Del_v(v, y)$ such that (i) $v$ and $y$ are the only points of $N_v$ that are on the boundary of $D$ and (ii) the interior of $D$ does not contain any point of $N_v$. If $arc_i$ is fully contained in the interior of $Del_v(v, y)$, then the algorithm knows that $q$ is contained in the interior of $Del_v(v, y)$ (even though it does not know the exact location of $q$) and, therefore, $Del_v(v, y)$ is not a certificate that $(v, y)$ is an edge in the Delaunay triangulation of the entire set $V$. Therefore, the algorithm checks if (i) $arc_i$ is fully contained in the interior of $Del_v(v, y)$ and (ii) the line segment $vy$ crosses any of the two line segments $xz'$ and $pz'$. If both (i) and (ii) hold, the algorithm removes the edge $(v, y)$. Observe that if $(v, y)$ is not an edge of the Delaunay triangulation $DT(V)$, the algorithm still keeps it as long as it does not cross any other edge. The formal algorithm is given below. An illustration, for the case when $y \neq x$, $y \neq p$, and $v \neq p$, is given in Figure 1.

**Algorithm.** PLDG($v$)

1.   let $N_v = \{u \in V : |uv| \le 1\}$;
2.   compute the Delaunay triangulation $LDT(v)$ of $N_v$;
3.   let $E(v)$ be the set of all edges in $LDT(v)$ that are incident on $v$;
4.   let $\Delta_v$ be the set of all triangular faces $\Delta(u, v, w)$ in $LDT(v)$ for which $\angle uvw > \frac{\pi}{3}$;
5.   let $k$ be the number of elements in $\Delta_v$;
6.   **if** $k \ge 1$
7.       **then** let $c_1, \ldots, c_k$ be the centers of the circumcircles of all triangles in $\Delta_v$;
8.           broadcast the sequence $(v, c_1, \ldots, c_k)$;
9.   **for** each sequence $(x, c'_1, \ldots, c'_m)$ received
10.    **do for** $i = 1$ **to** $m$
11.      **do** let $D(c'_i; |c'_i x|)$ be the disk with center $c'_i$ that contains $x$ on its boundary;
12.        **if** $\partial D(c'_i; |c'_i x|)$ contains exactly two points of $N_v$
13.          **then** let $p$ be the point in $(N_v \setminus \{x\}) \cap \partial D(c'_i; |c'_i x|)$;
14.          let $arc_i$ be the (open) arc on $\partial D(c'_i; |c'_i x|)$ that is not contained in the unit-disk $D(v; 1)$ centered at $v$;
15.          let $z'$ be an arbitrary point on $arc_i$ with $|xz'| \le 1$ or $|pz'| \le 1$;
16.          **for** each edge $(v, y)$ in $E(v)$
17.            **do** let $Del_v(v, y)$ be a disk $D$ such that $N_v \cap \partial D = \{v, y\}$ and $N_v \cap int(D) = \emptyset$;
18.            **if** $arc_i$ is contained in the interior of $Del_v(v, y)$ and the line segment $vy$ crosses at least one of the line segments $xz'$ and $pz'$
19.            **then** remove $(v, y)$ from $E(v)$

Running algorithm PLDG($v$) for all points $v$ of $V$ in parallel will be referred to as running algorithm PLDG($V$). We denote by $E(v)$ the edge set that is computed by algorithm PLDG($v$). Observe that each edge in $E(v)$ is incident on the point $v$. Let $E = \cup_{v \in V} E(v)$ and let $PLDG(V)$ denote the graph with vertex set $V$ and edge set $E$.

**Lemma 1.** *Let $S = \{u, v, w, z\}$ be a set of four points in the plane in general position, such that $|uv| \le 1$, $|wz| \le 1$, and the line segments $uv$ and $wz$ cross. Then there exists a point $x$ in $S$ such that $|xy| \le 1$ for all $y$ in $S$.*

**Lemma 2.** *Let $p$ and $q$ be two points with $|pq| \le 1$, let $D$ be a disk containing $p$ and $q$ on its boundary, and let $D_{cap}$ be the part of $D$ that is bounded by the line segment $pq$ and the minor arc $\widehat{pq}$ on $\partial D$ between $p$ and $q$. Then $|xy| \le 1$ for all $x$ and $y$ in $D_{cap}$.*

The following lemma implies that for every edge $(v, y)$ in $E(v)$, the edge $(v, y)$ is in the Delaunay triangulation $LDT(y)$ of the set $N_y$.

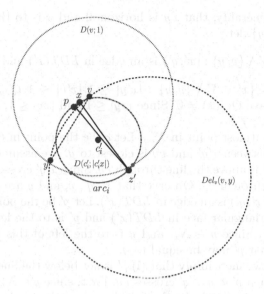

**Fig. 1.** Illustrating algorithm $PLDG(v)$. Edge $(v, y)$ is removed, where $y \neq x$, $y \neq p$, and $v \neq p$

**Lemma 3.** *Let $v$ and $y$ be two distinct points of $V$ and assume that $(v, y)$ is not an edge in $LDT(y)$. Then, after algorithm $PLDG(V)$ has terminated, $(v, y)$ is not an edge in $E(v)$.*

*Proof.* If $(v, y)$ is not an edge in $LDT(v)$, then, since $E(v)$ is a subset of the edge set of $LDT(v)$, $(v, y)$ is not an edge in $E(v)$. Assume that $(v, y)$ is an edge in $LDT(v)$. Observe that $|vy| \leq 1$. Since $(v, y)$ is not an edge in $LDT(y)$, there exist two points $p$ and $q$ in $V$ such that the triangle $\Delta(y, p, q)$ is a triangular face in $LDT(y)$ and $vy$ crosses $pq$. Observe that the points $p$, $q$, $v$, and $y$ are pairwise distinct. The lemma follows by proving the following two claims: First, algorithm $PLDG(y)$ broadcasts the center of the circumcircle of $\Delta(y, p, q)$. Since $|vy| \leq 1$, $v$ will receive this center. Second, when algorithm $PLDG(v)$ considers the center of $\Delta(y, p, q)$, it deletes the edge $(v, y)$. As a result, the edge $(v, y)$ is not in $E(v)$. □

**Lemma 4.** *Let $x$, $q$, $v$, and $y$ be four pairwise distinct points of $V$. Assume that $|xq| \leq 1$, $|xv| \leq 1$, $|xy| \leq 1$, $|vy| \leq 1$, $xq$ crosses $vy$, $(x, q)$ is an edge in $LDT(x)$, and $(v, y)$ is an edge in $LDT(y)$. Then, after algorithm $PLDG(V)$ has terminated, $(v, y)$ is not an edge in $E(y)$.*

*Proof.* If $(v, y)$ is not an edge in $LDT(v)$, then the claim follows from Lemma 3. Assume that $(v, y)$ is an edge in $LDT(v)$. We have to show that algorithm $PLDG(y)$ removes the edge $(v, y)$ from $E(y)$. Thus, we have to show that there exists a point $x'$ in $N_y$ which broadcasts the center of the circumcircle of some triangular face in $LDT(x')$ and, based on this information, $PLDG(y)$ removes $(v, y)$. We will use the edge $(x, q)$ to prove that such a point $x'$ exists. We assume, without loss of generality, that $vy$ is horizontal and $v$ is to the right of $y$. For each $x' \in V \setminus \{v, y\}$, let

$$Q_{vy}(x') = \{q' \in V \setminus \{v, y\} : (x', q') \text{ is an edge in } LDT(x') \text{ and } x'q' \text{ crosses } vy\}.$$

We define $X_{vy} = \{x' \in V \setminus \{v, y\} : |x'y| \leq 1, |x'v| \leq 1, Q_{vy}(x') \neq \emptyset\}$. Since $q \in Q_{vy}(x)$, we have $Q_{vy}(x) \neq \emptyset$. Since $|xy| \leq 1$ and $|xv| \leq 1$, we have $x \in X_{vy}$ and, therefore, $X_{vy} \neq \emptyset$.

Let $x'$ be the leftmost point in $X_{vy}$. Let $q'$ be the point in $Q_{vy}(x')$ such that the intersection between $x'q'$ and $vy$ is closest to $y$. We assume, without loss of generality, that $x'$ is above the line through $vy$. Since $x'q'$ crosses $vy$, the point $q'$ is below the line through $vy$. Observe that $x'$, $q'$, $v$, and $y$ are pairwise distinct.

By definition, $(x', q')$ is an edge in $LDT(x')$. Let $p'$ be the point of $V$ such that $\Delta(x', p', q')$ is a triangular face in $LDT(x')$ and $p'$ is to the left of the directed line from $q'$ to $x'$. Since $y \in N_{x'}$ and $y$ is to the left of this line, the point $p'$ exists. Observe that $p'$ may be equal to $y$.

The following two facts imply that (i) $p'$ is not below the line through $vy$, and (ii) in the case when $p' \neq y$, $p'q'$ crosses $vy$: First, since $y \in N_{x'}$ and $\Delta(x', p', q')$ is a triangular face in $LDT(x')$, the point $y$ cannot be in $\Delta(x', p', q')$. Second, by our choice of $q'$, the line segments $x'p'$ and $vy$ do not cross.

The lemma follows by proving the following two claims. First, algorithm $PLDG(x')$ broadcasts the center of the circumcircle of $\Delta(x', p', q')$. Since $|x'y| \leq 1$, $y$ will receive this center. Second, when algorithm $PLDG(y)$ considers the center of the circumcircle of $\Delta(x', p', q')$, it deletes the edge $(v, y)$. As a result, the edge $(v, y)$ is not in $E(y)$.                                                          □

**Lemma 5.** *$PLDG(V)$ is a plane graph.*

*Proof.* The proof is by contradiction. Assume that $PLDG(V)$ contains two crossing edges $(v, y)$ and $(x, q)$. By Lemma 1, we may assume without loss of generality that $|xq| \leq 1$, $|xv| \leq 1$, and $|xy| \leq 1$. By Lemma 3, $(v, y)$ is an edge in $LDT(v)$ and in $LDT(y)$, and $(x, q)$ is an edge in $LDT(x)$.

Since all conditions in Lemma 4 are satisfied, $(v, y)$ is not an edge in $E(y)$. Also, the conditions in Lemma 4, with $v$ and $y$ interchanged, are satisfied. Therefore, $(v, y)$ is not an edge in $E(v)$. Thus, $(v, y)$ is not an edge in $PLDG(V)$, which is a contradiction.                                                          □

The following lemma summarizes the different scenarios when algorithm $PLDG(v)$ removes an edge $(v, y)$ from the edge set $E(v)$. The proof is omitted; the main difficulty is in proving the third claim.

**Lemma 6.** *Let $v$ and $y$ be two distinct points of $V$ such that $(v, y)$ is an edge in $LDT(v)$. Assume that algorithm $PLDG(v)$ removes $(v, y)$ from $E(v)$. Then, there exist three pairwise distinct points $x$, $p$, and $q$ in $V$ such that*

1. *$\Delta(x, p, q)$ is a triangular face in $LDT(x)$,*
2. *$v \neq x$, $|vx| \leq 1$, $|vp| \leq 1$, $|vq| > 1$,*
3. *neither $v$ nor $y$ is in the interior of the disk $D(x, p, q)$, and*

4. (a) if $y \neq x$, $v \neq p$, and $y \neq p$, the line segment $vy$ crosses both the line segments $xq$ and $pq$,

   (b) if $y = x$, the line segment $vy$ crosses the line segment $pq$,

   (c) if $v = p$, the line segment $vy$ crosses the line segment $xq$,

   (d) if $y = p$, the line segment $vy$ crosses the line segment $xq$.

**Lemma 7.** *The graph $PLDG(V)$ is consistent: For any two distinct points $v$ and $y$ of $V$, $(v, y)$ is an edge in $E(v)$ if and only if $(v, y)$ is an edge in $E(y)$.*

*Proof.* The proof is by contradiction. Assume there is a pair $(v, y)$ which is an edge in $E(y)$ but not in $E(v)$. Then $(v, y)$ is an edge in $LDT(y)$ and, by Lemma 3, $(v, y)$ is an edge in $LDT(v)$. Since $(v, y)$ is not an edge in $E(v)$, it has been removed by algorithm $PLDG(v)$. Thus, by Lemma 6, there exist three pairwise distinct points $x$, $p$, and $q$ in $V$ such that (i) $\Delta(x, p, q)$ is a triangular face in $LDT(x)$, (ii) $v \neq x$, $|vx| \leq 1$, $|vq| > 1$, and (iii) the line segment $vy$ crosses at least one of the line segments $pq$ and $xq$.

Assume that $vy$ does not cross $xq$. Then $vy$ crosses $pq$ and, by the fourth claim in Lemma 6, $y = x$. Thus, since $(v, y)$ is an edge in $LDT(y) = LDT(x)$ and using (i), it follows that $LDT(x)$ is not plane, which is a contradiction.

Thus, $vy$ crosses $xq$. This implies that the points $x$, $q$, $v$, and $y$ are pairwise distinct. It follows from (i) that $(x, q)$ is an edge in $LDT(x)$ and $|xq| \leq 1$. Since $|vy| \leq 1$, $|xq| \leq 1$, $|vq| > 1$, and since $vy$ crosses $xq$, it follows from Lemma 1 that $|xy| \leq 1$. Thus, all conditions in Lemma 4 are satisfied. As a result, algorithm $PLDG(y)$ deletes the edge $(v, y)$ from $E(y)$. This is a contradiction.    □

Recall that $UDel(V)$ denotes the intersection of the Delaunay triangulation and the unit-disk graph of $V$. We next show that $PLDG(V)$ contains $UDel(V)$.

**Lemma 8.** *The graph $UDel(V)$ is a subgraph of $PLDG(V)$.*

*Proof.* Let $(v, y)$ be an edge of $UDel(V)$. We will show that $(v, y)$ is an edge in $E(v)$. By definition, $|vy| \leq 1$ and $(v, y)$ is an edge in the Delaunay triangulation of $V$. Therefore, $(v, y)$ is also an edge in the Delaunay triangulation $LDT(v)$ of $N_v$ and, thus, $(v, y)$ is added to the edge set $E(v)$ in line 3 of algorithm $PLDG(v)$. We have to show that algorithm $PLDG(v)$ does not remove $(v, y)$ in line 19.

Assume that $(v, y)$ is removed in line 19 of algorithm $PLDG(v)$. By Lemma 6, there exist three pairwise distinct points $x$, $p$, and $q$ in $V$ such that (i) neither $v$ nor $y$ is in the interior of the disk $D(x, p, q)$ and (ii) the line segment $vy$ crosses at least one of the line segments $xq$ and $pq$.

Assume that $vy$ crosses $xq$. Then, the points $v$, $y$, $x$, and $q$ are pairwise distinct. Observe that $p$ may be equal to $v$ or $y$. Let $D$ be an arbitrary disk having $v$ and $y$ on its boundary, and assume that neither $x$ nor $q$ is contained in $D$. Then it follows from (i) and (ii) that the boundaries of $D$ and $D(x, p, q)$ intersect more than twice, which is a contradiction. Thus, $D$ contains at least one of $x$ and $q$. Since $D$ was arbitrary, this contradicts the fact that $(v, y)$ is an edge in the Delaunay triangulation of $V$.

By a symmetric argument, the case when $vy$ crosses $pq$ also leads to a contradiction to the fact that $(v, y)$ is an edge in the Delaunay triangulation of $V$.  □

In the next lemma, we summarize the results obtained in this section.

**Lemma 9.** *Let $V$ be a finite set of points in the plane. The distributed algorithm $\mathrm{PLDG}(v)$, where $v$ ranges over all points in $V$, is a 2-local algorithm that computes a plane and consistent $\frac{4\pi\sqrt{3}}{9}$-spanner $PLDG(V)$ of the unit-disk graph of $V$. This algorithm makes one communication round and each point of $V$ broadcasts at most 6 messages.*

## 3   The Final Algorithm

We have seen that in algorithm PLDG, each point of $V$ broadcasts at most 6 messages. In this section, we improve this upper bound to 5. We obtain this improvement, by making the following modification to the algorithm: The sequence $(v, c_1, \ldots, c_k)$ that is broadcast in line 8 of algorithm $\mathrm{PLDG}(v)$ contains the location of the sender $v$. In our new algorithm, point $v$ sends only the sequence $(c_1, \ldots, c_k)$ of centers. Thus, any point that receives this sequence does not know that the sequence was broadcast by $v$. Assume that $v$ receives a center $c_i'$ from some node $x$ in $N_v$. Since $v$ does not know that $c_i'$ was broadcast by $x$, line 11 in algorithm $\mathrm{PLDG}(v)$ has to be modified. In the new algorithm, $v$ computes a point $x'$ in $N_v \setminus \{v\}$ that is closest to $c_i'$ and uses the disk $D(c_i'; |c_i'x'|)$ to decide whether or not to remove an edge $(v, y)$. The new algorithm, which we denote by PLDG', is given in Figure 2.

We denote by $E'(v)$ the edge set that is computed by algorithm $\mathrm{PLDG}'(v)$. Let $E' = \cup_{v \in V} E'(v)$ and let $PLDG'(V)$ denote the graph with vertex set $V$ and edge set $E'$.

Recall that $E(v)$ denotes the edge set that is computed by algorithm $\mathrm{PLDG}(v)$ and $PLDG(V)$ denotes the graph with vertex set $V$ and edge set $\cup_{v \in V} E(v)$. We claim that $PLDG(V) = PLDG'(V)$; thus, the new algorithm PLDG' computes the same graph as algorithm PLDG. In order to prove this claim, it suffices to show that algorithm $\mathrm{PLDG}(v)$ removes an edge $(v, y)$ from $E(v)$ if and only if algorithm $\mathrm{PLDG}'(v)$ removes the edge $(v, y)$ from $E'(v)$. We will show this in the following two lemmas.

**Lemma 10.** *Let $v$ be an element of $V$ and let $(v, y)$ be an edge of the Delaunay triangulation $LDT(v)$ of the set $N_v$. If algorithm $\mathrm{PLDG}(v)$ removes $(v, y)$ from $E(v)$, then algorithm $\mathrm{PLDG}'(v)$ removes $(v, y)$ from $E'(v)$.*

*Proof.* By Lemma 6, there exist three pairwise distinct points $x$, $p$, and $q$ in $V$ such that (i) $\Delta(x, p, q)$ is a triangular face in $LDT(x)$, (ii) $v \neq x$, $|vx| \leq 1$, $|vp| \leq 1$, $|vq| > 1$, (iii) neither $v$ nor $y$ is in the interior of the disk $D(x, p, q)$.

In fact, in algorithm $\mathrm{PLDG}(v)$, $v$ receives from $x$ the center $c_i'$ of the disk $D(c_i'; |c_i'x|) = D(x, p, q)$. Since $|vx| \leq 1$, in algorithm $\mathrm{PLDG}'(v)$, $v$ receives the center $c_i'$, but does not know that it was broadcast by $x$. Consider the point

**Algorithm** PLDG$'(v)$
1.   let $N_v = \{u \in V : |uv| \le 1\}$;
2.   compute the Delaunay triangulation $LDT(v)$ of $N_v$;
3.   let $E'(v)$ be the set of all edges in $LDT(v)$ that are incident on $v$;
4.   let $\Delta_v$ be the set of all triangular faces $\Delta(u, v, w)$ in $LDT(v)$ for which $\angle uvw > \frac{\pi}{3}$;
5.   let $k$ be the number of elements in $\Delta_v$;
6.   **if** $k \ge 1$
7.   **then** let $c_1, \ldots, c_k$ be the centers of the circumcircles of all triangles in $\Delta_v$;
8.               broadcast the sequence $(c_1, \ldots, c_k)$;
9.   **for** each sequence $(c'_1, \ldots, c'_m)$ received
10.     **do for** $i = 1$ **to** $m$
11.         **do** compute a point $x'$ in $N_v \setminus \{v\}$ that is closest to $c'_i$;
12.             let $D(c'_i; |c'_i x'|)$ be the disk with center $c'_i$ that contains $x'$ on its boundary;
13.             **if** $\partial D(c'_i; |c'_i x'|)$ contains exactly two points of $N_v$
14.                 **then** let $p'$ be the point in $(N_v \setminus \{x'\}) \cap \partial D(c'_i; |c'_i x'|)$;
15.                     let $arc_i = (\partial D(c'_i; |c'_i x'|)) \setminus D(v; 1)$;
16.                     let $Z = \{z' \in arc_i : |x'z'| \le 1 \text{ or } |p'z'| \le 1\}$;
17.                     **if** $arc_i \ne \emptyset$ and $Z \ne \emptyset$
18.                         **then** let $z'$ be an arbitrary element of $Z$;
19.                             **for** each edge $(v, y)$ in $E'(v)$
20.                                 **do** let $Del_v(v, y)$ be a disk $D$ such that $N_v \cap \partial D = \{v, y\}$ and $N_v \cap int(D) = \emptyset$;
21.                                     **if** $arc_i$ is contained in the interior of $Del_v(v, y)$ and the line segment $vy$ crosses at least one of the line segments $x'z'$ and $p'z'$
22.                                         **then** remove $(v, y)$ from $E'(v)$

**Fig. 2.** The final algorithm PLDG$'(v)$

$x'$ that is computed in line 11 of algorithm PLDG$'(v)$. Thus, $x'$ is a point in $N_v \setminus \{v\}$ that is closest to $c'_i$. Since $x \in N_v \setminus \{v\}$, we have $|c'_i x'| \le |c'_i x|$. We claim that $|c'_i x'| = |c'_i x|$.

To prove this claim, assume, by contradiction, that $|c'_i x'| < |c'_i x|$. Then $x'$ is in the interior of the disk $D(x, p, q)$. Since $\Delta(x, p, q)$ is a triangular face in $LDT(x)$, we have $|xx'| > 1$.

Consider the disk $Del_v(v, y)$ that is computed in line 17 of algorithm PLDG$(v)$. Recall that $N_v \cap \partial Del_v(v, y) = \{v, y\}$ and $N_v \cap int(Del_v(v, y)) = \emptyset$. Since both $x$ and $x'$ are in $N_v$, neither of these two points is in the interior of $Del_v(v, y)$. It follows from line 18 of algorithm PLDG$(v)$ that $q$ is in the interior of $Del_v(v, y)$ (because $q \in arc_i$).

Assume, without loss of generality that $vy$ is horizontal, $v$ is to the right of $y$, and $q$ is below the line through $v$ and $y$; refer to Figure 3.

Since $|vq| > 1$ and $|vy| \le 1$, we have $\angle yqv < \pi/2$. Let $\widehat{yv}$ be the arc on $\partial Del_v(v, y)$ with endpoints $y$ and $v$ that contains the north pole of $\partial Del_v(v, y)$. Then $\widehat{yv}$ is a minor arc.

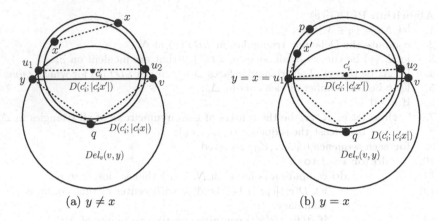

(a) $y \neq x$            (b) $y = x$

**Fig. 3.** Illustrating the proof of Lemma 10

Let $u_1$ and $u_2$ be the intersections between $\partial Del_v(v, y)$ and $\partial D(x, p, q)$, where $u_1$ is to the left of $u_2$. Then both $u_1$ and $u_2$ are contained in $\widehat{yv}$ and, therefore, by Lemma 2, $|u_1 u_2| \leq 1$.

Let $\widehat{u_1 u_2}$ be the arc on $\partial D(x, p, q)$ with endpoints $u_1$ and $u_2$ that contains the north pole of $\partial D(x, p, q)$. Since $\angle u_1 q u_2 \leq \angle y q v < \pi/2$, $\widehat{u_1 u_2}$ is a minor arc.

Recall that $x \notin int(Del_v(v, y))$. Also, if $x \in \partial Del_v(v, y)$, then $x = y$. It follows that $x$ is not below the line through $u_1$ and $u_2$. By a similar argument, $x'$ is not below this line. Since both $x$ and $x'$ are in $D(x, p, q)$ and since $\widehat{u_1 u_2}$ is a minor arc, it follows from Lemma 2 that $|xx'| \leq 1$, which is a contradiction.

Thus, we have shown that $|c_i' x'| = |c_i' x|$. Recall that $p$ is the point that is computed in line 13 of algorithm PLDG$(v)$. Consider the point $p'$ that is computed in line 14 of algorithm PLDG$'(v)$. Since $D(c_i'; |c_i' x|) = D(c_i'; |c_i' x'|)$, we have $\{x, p\} = \{x', p'\}$. In other words, algorithm PLDG$'(v)$ knows the points $x$ and $p$, but does not know which of them is $x$ and which of them is $p$.

Since lines 15–19 of algorithm PLDG$(v)$ are symmetric in $x$ and $p$, and since lines 16–22 of algorithm PLDG$'(v)$ are symmetric in $x'$ and $p'$, it follows that the behaviors of PLDG$(v)$ and PLDG$'(v)$ with respect to the edge $(v, y)$ are identical. Therefore, algorithm PLDG$'(v)$ removes the edge $(v, y)$ from $E'(v)$. □

**Lemma 11.** *Let $v$ be an element of $V$ and let $(v, y)$ be an edge of the Delaunay triangulation $LDT(v)$ of the set $N_v$. If algorithm PLDG$'(v)$ removes $(v, y)$ from $E'(v)$, then algorithm PLDG$(v)$ removes $(v, y)$ from $E(v)$.*

*Proof.* Since algorithm PLDG$'(v)$ removes $(v, y)$ from $E'(v)$, there exist three pairwise distinct points $x$, $p$, and $q$ in $V$ such that (i) $\Delta(x, p, q)$ is a triangular face in $LDT(x)$, (ii) algorithm PLDG$'(x)$ broadcasts the center $c_i'$ of the disk $D(x, p, q) = D(c_i'; |c_i' x|)$, (iii) $v \neq x$, $|vx| \leq 1$, (iv) $v$ receives the center $c_i'$ (but does not know that it was broadcast by $x$).

Consider the point $x'$ that is computed in line 11 of algorithm PLDG$'(v)$. Thus, $x'$ is a point in $N_v \setminus \{v\}$ that is closest to $c_i'$. Since $x \in N_v \setminus \{v\}$, we have

$|c_i'x'| \leq |c_i'x|$. We claim that $|c_i'x'| = |c_i'x|$. As in the proof of Lemma 10, this implies that algorithm PLDG($v$) removes the edge $(v, y)$ from $E(v)$.    □

By Lemmas 10 and 11, algorithms PLDG($v$) and PLDG$'$($v$) compute the same graph. Therefore, the proof of Theorem 1 can be completed as in the proof of Lemma 9 and by observing that the sequence that is broadcast in line 8 of algorithm PLDG$'$($v$) contains at most 5 points.

# References

1. Araújo, F., Rodrigues, L.: Fast localized Delaunay triangulation. In: Higashino, T. (ed.) OPODIS 2004. LNCS, vol. 3544, pp. 81–93. Springer, Heidelberg (2005)
2. Bose, P., Carmi, P., Smid, M., Xu, D.: Communication-efficient construction of the plane localized Delaunay graph (2008), http://arxiv.org/abs/0809.2956
3. Bose, P., Maheshwari, A., Narasimhan, G., Smid, M., Zeh, N.: Approximating geometric bottleneck shortest paths. Computational Geometry: Theory and Applications 29, 233–249 (2004)
4. Bose, P., Morin, P., Stojmenović, I., Urrutia, J.: Routing with guaranteed delivery in ad hoc wireless networks. Wireless Networks 7, 609–616 (2001)
5. Gao, J., Guibas, L.J., Hershberger, J., Zhang, L., Zhu, A.: Geometric spanners for routing in mobile networks. IEEE Journal on Selected Areas in Communications 23, 174–185 (2005)
6. Karp, B., Kung, H.T.: GPSR: greedy perimeter stateless routing for wireless networks. In: Proceedings of the 6th Annual International Conference on Mobile Computing and Networking, pp. 243–254 (2000)
7. Keil, J.M., Gutwin, C.A.: Classes of graphs which approximate the complete Euclidean graph. Discrete & Computational Geometry 7, 13–28 (1992)
8. Li, X.-Y., Calinescu, G., Wan, P.-J., Wang, Y.: Localized Delaunay triangulation with applications in wireless ad hoc networks. IEEE Transactions on Parallel and Distributed Systems 14, 1035–1047 (2003)
9. Narasimhan, G., Smid, M.: Geometric Spanner Networks. Cambridge University Press, Cambridge (2007)

# Time Complexity of Distributed Topological Self-stabilization: The Case of Graph Linearization*

Dominik Gall[1], Riko Jacob[1], Andrea Richa[2], Christian Scheideler[3], Stefan Schmid[4], and Hanjo Täubig[1]

[1] Institut für Informatik, TU München, Garching, Germany
[2] Dept. Computer Science and Engineering, Arizona State University, Tempe, USA
[3] Dept. Computer Science, University of Paderborn, Paderborn, Germany
[4] TU Berlin / Deutsche Telekom Laboratories, Berlin, Germany

**Abstract.** Topological self-stabilization is an important concept to build robust open distributed systems (such as peer-to-peer systems) where nodes can organize themselves into meaningful network topologies. The goal is to devise distributed algorithms that converge quickly to such a desirable topology, independently of the initial network state. This paper proposes a new model to study the parallel convergence time. Our model sheds light on the achievable parallelism by avoiding bottlenecks of existing models that can yield a distorted picture. As a case study, we consider local graph linearization—i.e., how to build a sorted list of the nodes of a connected graph in a distributed and self-stabilizing manner. We propose two variants of a simple algorithm, and provide an extensive formal analysis of their worst-case and best-case parallel time complexities, as well as their performance under a greedy selection of the actions to be executed.

## 1 Introduction

Open distributed systems such as peer-to-peer systems are often highly transient in the sense that nodes join and leave at a fast pace. In addition to this natural churn, parts of the network can be under attack, causing nodes to leave involuntarily. Thus, when designing such a system, a prime concern is *robustness*. Over the last years, researchers have proposed many interesting approaches to build robust overlay networks. A particularly powerful concept in this context is (distributed) *topological self-stabilization*: a self-stabilizing system guarantees that from *any* connected topology, eventually an overlay with desirable properties will result.

In this paper, we address one of the first and foremost questions in distributed topological self-stabilization: *How to measure the parallel time complexity?* We consider a very strong adversary who presents our algorithm with an arbitrary

---

* For a complete technical report, we refer the reader to [8]. Research supported by the DFG project SCHE 1592/1-1, and NSF Award number CCF-0830704.

A. López-Ortiz (Ed.): LATIN 2010, LNCS 6034, pp. 294–305, 2010.

connected network. We want to investigate how long it takes until the topology reaches a (to be specified) desirable state. While several solutions have been proposed in the literature over the last years, these known models are inappropriate to adequately model parallel efficiency: either they are overly pessimistic in the sense that they can force the algorithm to work serially, or they are too optimistic in the sense that contention or congestion issues are neglected.

Our model is aware of bottlenecks in the sense that nodes cannot perform too much work per time unit. Thus, we consider our new model as a further step to explore the right level of abstraction to measure parallel execution times. As a case study, we employ our tools to the problem of *graph linearization* where nodes—initially in an arbitrary connected graph—are required to *sort* themselves with respect to their identifiers.

As the most simple form of topological self-stabilization, linearization allows to study the main properties of our model. As we will see in our extensive analysis, graph linearization under our model is already non-trivial and reveals an interesting structure. This paper focuses on two natural linearization algorithms, such that the influence of the modeling becomes clear. For our analysis, we will assume the existence of some hypothetical schedulers. In particular, we consider a scheduler that always makes the worst possible, one that always makes the best possible, one that makes a random, and one that makes a "greedy" selection of actions to execute at any time step. Since the schedulers are only used for the *complexity analysis* of the protocols proposed, for ease of explanation, we treat the schedulers as global entities and we make no attempt to devise distributed, local mechanisms to implement them.[1]

## 1.1  Related Work

The construction and maintenance of a given network structure is of prime importance in many distributed systems, for example in peer-to-peer computing [6,7,11,13,14]; e.g., Kuhn et al. [11] showed how to maintain a hypercube under dynamic worst-case joins and leaves, and Scheideler et al. [13] presented the SHELL network which allows peers to join and leave efficiently. In the technical report of the distributed hash table Chord [15], stabilization protocols are described which allow the topology to recover from certain degenerate situations. Unfortunately, however, in these papers, no algorithms are given to recover from *arbitrary* states.

Also skip graphs [1] can be repaired from certain states, namely states which resulted from node faults and inconsistencies due to churn. In a recent paper, we have shown [9] that a modified and locally checkable version of the skip graph can be built from any connected network. Interestingly, the algorithm introduced in [9] self-stabilizes in polylogarithmic time. Unfortunately, however, while the resulting structure is indeed scalable, the number of edges can become quadratic

---

[1] In fact, most likely no such local mechanism exists for implementing the worst-case and best-case schedulers, while we believe that local distributed implementations that closely approximate—within a constant factor of the parallel complexity—the randomized and greedy schedulers presented here would not be hard to devise.

during the execution of that algorithm. Moreover, the execution model does not scale either.

In this paper, in order to be able to focus on the main phenomena of our model, we chose to restrict ourselves to topological *linearization*. Linearization and the related problem of organizing nodes in a ring is also subject to active research. The *Iterative Successor Pointer Rewiring Protocol* [5] and the *Ring Network* [14] organize the nodes in a sorted ring. Unfortunately, both protocols have a large runtime. We have recently started to work on 2-dimensional linearization problems for a different (classic) time-complexity model [10].

The papers closest to ours are by Onus et al. [12] and by Clouser et al. [4]. In [12], a local-control strategy called *linearization* is presented for converting an arbitrary connected graph into a sorted list. However, it is only studied in a synchronous environment, and the strategy does not scale since in one time round it allows a node to communicate with an arbitrary number of its neighbors (which can be as high as $\Theta(n)$ for $n$ nodes). Clouser et al. [4] formulated a variant of the linearization technique for arbitrary asynchronous systems in which edges are represented as Boolean shared variables. Any node may establish an undirected edge to one of its neighbors by setting the corresponding shared variable to true, and in each time unit, a node can manipulate at most one shared variable. If these manipulations never happen concurrently, it would be possible to emulate the shared variable concept in a message passing system in an efficient way. However, concurrent manipulations of shared variables can cause scalability problems because even if every node only modifies one shared variable at a time, the fact that the other endpoint of that shared variable has to get involved when emulating that action in a message passing system implies that a single node may get involved in up to $\Theta(n)$ many of these variables in a time unit.

## 1.2   Our Contributions

The contribution of this paper is two-fold. First, we present an alternative approach to modeling scalability of distributed, self-stabilizing algorithms that does not require synchronous executions like in [12] and also gets rid of the scalability problems in [4,12] therefore allowing us to study the parallel time complexity of proposed linearization approaches. Second, we propose two variants of a simple, local linearization algorithm. For each of these variants, we present extensive formal analyses of their worst-case and best-case parallel time complexities, and also study their performance under a random and a greedy selection of the actions to be executed.

## 2   Model

We are given a system consisting of a fixed set $V$ of $n$ nodes. Every node has a unique (but otherwise arbitrary) integer *identifier*. In the following, if we compare two nodes $u$ and $v$ using the notation $u < v$ or $u > v$, we mean that the identifier of $u$ is smaller than $v$ or vice versa. For any node $v$, $\text{pred}(v)$ denotes the predecessor of $v$ (i.e., the node $u \in V$ of largest identifier with $u < v$) and $\text{succ}(v)$ denotes the successor of $v$ according to "$<$". Two nodes $u$ and $v$ are called *consecutive* if and only if $u = \text{succ}(v)$ or $v = \text{succ}(u)$.

Connections between nodes are modeled as shared variables. Each pair $(u, v)$ of nodes shares a Boolean variable $e(u, v)$ which specifies an undirected adjacency relation: $u$ and $v$ are called *neighbors* if and only if this shared variable is true. The set of neighbor relations defines an undirected graph $G = (V, E)$ among the nodes. A variable $e(u, v)$ can only be changed by $u$ and $v$, and both $u$ and $v$ have to be involved in order to change $e(u, v)$. (E.g., node $u$ sends a change request message to $u$. More details on this will be given below.) For any node $u \in V$, let $u.L$ denote the set of left neighbors of $u$—the neighbors which have smaller identifiers than $u$—and $u.R$ the set of right neighbors (with larger IDs) of $u$.

In this paper, $\deg(u)$ will denote the degree of a node $u$ and is defined as $\deg(u) = |u.L \cup u.R|$. Moreover, the distance between two nodes $\text{dist}(u, v)$ is defined as $\text{dist}(u, v) = |\{w : u < w \le v\}|$ if $u < v$ and $\text{dist}(u, v) = |\{w : v < w \le u\}|$ otherwise. The length of an edge $e = \{u, v\} \in E$ is defined as $\text{len}(e) = \text{dist}(u, v)$.

We consider *distributed algorithms* which are run by each node in the network. The algorithm or program executed by each node consists of a set of *variables* and *actions*. An action has the form

$$< \text{name} > \quad : \quad < \text{guard} > \quad \rightarrow \quad < \text{commands} >$$

where $<$ name $>$ is an *action label*, $<$ guard $>$ is a Boolean predicate over the (local and shared) variables of the executing node and $<$ commands $>$ is a sequence of commands that may involve any local or shared variables of the node itself or its neighbors. Given an action $A$, the set of all nodes involved in the commands is denoted by $V(A)$. Every node that either owns a local variable or is part of a shared variable $e(u, v)$ accessed by one of the commands in $A$ is part of $V(A)$. Two actions $A$ and $B$ are said to be *independent* if $V(A) \cap V(B) = \emptyset$. For an action execution to be scalable we require that the number of interactions a node is involved in (and therefore $|V(A)|$) is independent of $n$. An action is called *enabled* if and only if its guard is true. Every enabled action is passed to some underlying scheduling layer (to be specified below). The scheduling layer decides whether to accept or reject an enabled action. If it is accepted, then the action is executed by the nodes involved in its commands.

We model distributed computation as follows. The assignments of all local and shared variables defines a *system state*. Time proceeds in *rounds*. In each round, the scheduling layer may select any set of independent actions to be executed by the nodes. The *work* performed in a round is equal to the number of actions selected by the scheduling layer in that round. A *computation* is a sequence of states such that for each state $s_i$ at the beginning of round $i$, the next state $s_{i+1}$ is obtained after executing all actions that were selected by the scheduling layer in round $i$. A distributed algorithm is called *self-stabilizing* w.r.t. a set of system states $S$ and a set of *legal* states $L \subseteq S$ if for any initial state $s_1 \in S$ and any fair scheduling layer, the algorithm eventually arrives (and stays) at a state $s \in L$.

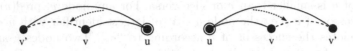

**Fig. 1.** Left and right linearization step

Notice that this model can cover arbitrary asynchronous systems in which the actions are implemented so that the sequential consistency model applies (i.e., the outcome of the executions of the actions is equivalent to a sequential execution of them) as well as parallel executions in synchronous systems. In a round, the set of enabled actions selected by the scheduler must be independent as otherwise a state transition from one round to another would, in general, not be unique, and further rules would be necessary to handle dependent actions that we want to abstract from in this paper.

### 2.1 Linearization

In this paper we are interested in designing distributed algorithms that can transform any initial graph into a sorted list (according to the node identifiers) using only local interactions between the nodes. A distributed algorithm is called *self-stabilizing* in this context if for any initial state that forms a connected graph, it eventually arrives at a state in which for all node pairs $(u, v)$,

$$e(u, v) = 1 \quad \Leftrightarrow \quad u = \mathrm{succ}(v) \ \lor \ v = \mathrm{succ}(u)$$

i.e., the nodes indeed form a sorted list. Once it arrives at this state, it should stay there, i.e., the state is the (only) *fixpoint* of the algorithm. In the distributed algorithms studied in this paper, each node $u \in V$ repeatedly performs simple linearization steps in order to arrive at that fixpoint.

A linearization step involves three nodes $u$, $v$, and $v'$ with the property that $u$ is connected to $v$ and $v'$ and either $u < v < v'$ or $v' < v < u$. In both cases, $u$ may command the nodes to move the edge $\{u, v'\}$ to $\{v, v'\}$. If $u < v < v'$, this is called a *right* linearization and otherwise a *left* linearization (see also Figure 1). Since only three nodes are involved in such a linearization, this can be formulated by a scalable action. In the following, we will also call $u$, $v$, and $v'$ a *linearization triple* or simply a *triple*.

### 2.2 Schedulers

Our goal is to find linearization algorithms that spend as little time and work as possible in order to arrive at a sorted list. In order to investigate their worst, average, and best performance under concurrent executions of actions, we consider different schedulers.

1. Worst-case scheduler $\mathcal{S}_{\mathrm{wc}}$: This scheduler must select a maximal independent set of enabled actions in each round, but it may do so to enforce a runtime (or work) that is as large as possible.

2. Randomized scheduler $\mathcal{S}_{\mathrm{rand}}$: This scheduler considers the set of enabled actions in a random order and selects, in one round, every action that is independent of the previously selected actions in that order.

3. Greedy scheduler $\mathcal{S}_{\mathrm{greedy}}$: This scheduler orders the nodes according to their degrees, from maximum to minimum. For each node that still has enabled actions left that are independent of previously selected actions, the scheduler picks one of them in a way specified in more detail later in this paper when our self-stabilizing algorithm has been introduced. (Note, that 'greedy' refers to a greedy behavior w.r.t. the degree of the nodes; large degrees are preferred. Another meaningful 'greedy' scheduler could favor triples with largest gain w.r.t. the potential function that sums up all link lengths.)

4. Best-case scheduler $\mathcal{S}_{\mathrm{opt}}$: The enabled actions are selected in order to minimize the runtime (or work) of the algorithm. (Note, that 'best' in this case requires maximal independent sets although there might be a better solution without this restriction.)

The worst-case and best-case schedulers are of theoretical interest to explore the parallel time complexity of the linearization approach. The greedy scheduler is a concrete algorithmic selection rule that we mainly use in the analysis as an upper bound on the best-case scheduler. The randomized scheduler allows us to investigate the average case performance when a local-control randomized symmetry breaking approach is pursued in order to ensure sequential consistency while selecting and executing enabled actions.

As noted in the introduction, for ease of explanation, we treat the schedulers as global entities and we make no attempt to formally devise distributed, local mechanisms to implement them (that would in fact be an interesting, orthogonal line for future work). The schedulers are used simply to explore the parallel time complexity limitations (e.g., worst-case, average-case, best-case behavior) of the linearization algorithms proposed. In practice the algorithms $\mathrm{LIN_{all}}$ and $\mathrm{LIN_{max}}$ to be presented below may rely on any local-control rule (scheduler) to decide on a set of locally independent actions—which trivially leads to global independence—to perform at any given time.

## 3   Algorithms and Analysis

We now introduce our distributed and self-stabilizing linearization algorithms $\mathrm{LIN_{all}}$ and $\mathrm{LIN_{max}}$. Section 3.1 specifies our algorithms formally and gives correctness proofs. Subsequently, we study the algorithms' runtime.

### 3.1   $\mathrm{LIN_{all}}$ and $\mathrm{LIN_{max}}$

We first describe $\mathrm{LIN_{all}}$. Algorithm $\mathrm{LIN_{all}}$ is very simple. Each node constantly tries to linearize its neighbors according to the *linearize left* and *linearize right* rules in Figure 1. In doing so, *all* possible triples on both sides are proposed to the scheduler. More formally, in $LIN_{all}$ every node $u$ checks the following actions for every pair of neighbors $v$ and $w$:

**linearize left**$(v, w)$:
$(v, w \in u.L \land w < v < u) \rightarrow e(u, w) := 0, e(v, w) := 1$

**linearize right**$(v, w)$:
$(v, w \in u.R \land u < v < w) \rightarrow e(u, w) := 0, e(v, w) := 1$

$LIN_{max}$ is similar to $LIN_{all}$: instead of proposing all possible triples on each side, $LIN_{max}$ only proposes the triple which is the furthest (w.r.t. IDs) on the corresponding side. Concretely, every node $u \in V$ checks the following actions for every pair of neighbors $v$ and $w$:

**linearize left**$(v, w)$:
$(v, w \in u.L) \land w < v < u \land \nexists x \in u.L \setminus \{w\} : x < v) \rightarrow e(u, w) := 0, e(v, w) := 1$

**linearize right**$(v, w)$:
$(v, w \in u.R) \land u < v < w \land \nexists x \in u.R \setminus \{w\} : x > v) \rightarrow e(u, w) := 0, e(v, w) := 1$

We first show that these algorithms are correct in the sense that eventually, a linearized graph will be output.

**Theorem 1.** *$LIN_{all}$ and $LIN_{max}$ are self-stabilizing and converge to the sorted list.*

### 3.2   Runtime

We first study the worst case scheduler $\mathcal{S}_{wc}$ for both $LIN_{all}$ and $LIN_{max}$.

**Theorem 2.** *Under a worst-case scheduler $\mathcal{S}_{wc}$, $LIN_{max}$ terminates after $O(n^2)$ work (single linearization steps), where $n$ is the total number of nodes in the system. This is tight in the sense that there are situations where under a worst-case scheduler $\mathcal{S}_{wc}$, $LIN_{max}$ requires $\Omega(n^2)$ rounds.*

*Proof.* Due to space constraints, we only give a proof sketch for the upper bound. Let $\zeta_l(v)$ denote the length of the longest edge out of node $v \in V$ to the left and let $\zeta_r(v)$ denote the length of the longest edge out of node $v$ to the right. If node $v$ does not have any edge to the left, we set $\zeta_l(v) = 1/2$, and similarly for the right. We consider the potential function $\Phi$ which is defined as

$$\Phi = \sum_{v \in V} [(2\zeta_l(v) - 1) + (2\zeta_r(v) - 1)] = \sum_{v \in V} 2(\zeta_l(v) + \zeta_r(v) - 1).$$

Observe that initially, $\Phi_0 < 2n^2$, as $\zeta_l(v) + \zeta_r(v) < n$ for each node $v$. We show that after round $i$, the potential is at most $\Phi_i < 2n^2 - i$. Since $LIN_{max}$ terminates (cf. also Theorem 1) with a potential $\Phi_j > 0$ for some $j$ (the term of each node is positive, otherwise the node would be isolated), the claim follows. In order to see why the potential is reduced by at least one in every round, consider a triple $u, v, w$ which is right-linearized and where $u < v < w$, $\{u, v\} \in E$, and $\{u, w\} \in E$. (Left-linearizations are similar and not discussed further here.) During the linearization step, $\{u, w\}$ is removed from $E$ and the edge $\{v, w\}$ is added if it did not already exist.

We distinguish two cases.

*Case 1:* Assume that $\{u, w\}$ was also the longest edge of $w$ to the left. This implies that during linearization of the triple, we remove two longest edges (of nodes $u$ and $w$) of length len($\{u, w\}$) from the potential function. On the other hand, we may now have the following increase in the potential: $u$ has a new longest edge $\{u, v\}$ to the right, $v$ has a new longest edge $\{v, w\}$ to the right, and $w$ has a new longest edge of length up to len($\{u, w\}) - 1$ to the left. Summarizing, we get

$$\Delta \Phi \leq (2 \cdot \text{len}(\{u, v\}) - 1) + (2 \cdot \text{len}(\{v, w\}) - 1) +$$
$$(2(\text{len}(\{u, w\}) - 1) - 1) - (4 \cdot \text{len}(\{u, w\}) - 2)$$
$$\leq -3$$

since len($\{u, w\}$) = len($\{u, v\}$) + len($\{v, w\}$).

*Case 2:* Assume that $\{u, w\}$ was not the longest edge of $w$ to the left. Then, by this linearization step, we remove edge $\{u, w\}$ from the potential function but may add edges $\{u, v\}$ (counted from node $u$ to the right) and $\{v, w\}$ (counted from node $v$ to the right). We have

$$\Delta \Phi \leq (2 \cdot \text{len}(\{u, v\}) - 1) + (2 \cdot \text{len}(\{v, w\}) - 1) - (2 \cdot \text{len}(\{u, w\}) - 1) \leq -1$$

since len($\{u, w\}$) = len($\{u, v\}$) + len($\{v, w\}$). Since in every round, at least one triple can be linearized, this concludes the proof.

For the LIN$_{\text{all}}$ algorithm, we obtain a slightly higher upper bound. In the analysis, we need the following helper lemma.

**Lemma 1.** *Let $\Xi$ be any positive potential function, where $\Xi_0$ is the initial potential value and $\Xi_i$ is the potential after the $i^{th}$ round of a given algorithm ALG. Assume that $\Xi_i < \Xi_{i-1} \cdot (1 - 1/f)$ and that ALG terminates if $\Xi_j \leq \Xi_{stop}$ for some $j \in \mathbb{N}$. Then, the runtime of ALG is at most $O(f \cdot \log(\Xi_0/\Xi_{stop}))$ rounds.*

**Theorem 3.** *LIN$_{all}$ terminates after $O(n^2 \log n)$ many rounds under a worst-case scheduler $\mathcal{S}_{wc}$, where $n$ is the network size.*

*Proof.* We consider the potential function $\Psi = \sum_{e \in E} \text{len}(e)$, for which it holds that $\Psi_0 < n^3$. We show that in each round, this potential is multiplied by a factor of at most $1 - \Omega(1/n^2)$.

Consider an arbitrary triple $u, v, w \in V$ with $u < v < w$ which is right-linearized by node $u$. (Left-linearizations are similar and not discussed further here.) During a linearization step, the sum of the edge lengths is reduced by at least one. Similarly to the proof of Theorem 4, we want to calculate the amount of blocked potential in a round due to the linearization of the triple $(u, v, w)$. Nodes $u, v$, and $w$ have at most $\widehat{\deg}(u) + \widehat{\deg}(v) + \widehat{\deg}(w) < n$ many independent neighbors. In the worst case, when the triple's incident edges are removed (blocked potential at most $O(n^2)$), these neighbors fall into different disconnected components which cannot be linearized further in this round; in other words, the remaining components form sorted lines. The blocked potential amounts to at most $\Theta(n^2)$. Thus, together with Lemma 1, the claim follows. $\square$

Besides $\mathcal{S}_{wc}$, we are interested in the following type of greedy scheduler. In each round, both for $LIN_{all}$ and $LIN_{max}$, $\mathcal{S}_{greedy}$ orders the nodes with respect to their *remaining degrees*: after a triple has been fired, the three nodes' incident edges are removed. For each node $v \in V$ selected by the scheduler according to this order (which still has enabled actions left which are independent of previously selected actions), the scheduler greedily picks the enabled action of $v$ which involves the two most distant neighbors on the side with the larger remaining degree (if the number of remaining left neighbors equals the number of remaining neighbors on the right side, then an arbitrary side can be chosen.) The intuition behind $\mathcal{S}_{greedy}$ is that neighborhood sizes are reduced quickly in the linearization process.

Under this greedy scheduler, we get the following improved bound on the time complexity of $LIN_{all}$.

**Theorem 4.** *Under a greedy scheduler $\mathcal{S}_{greedy}$, $LIN_{all}$ terminates in $O(n \log n)$ rounds, where $n$ is the total number of nodes in the system.*

Finally, we have also investigated an optimal scheduler $\mathcal{S}_{opt}$.

**Theorem 5.** *Even under an optimal scheduler $\mathcal{S}_{opt}$, both $LIN_{all}$ and $LIN_{max}$ require at least $\Omega(n)$ rounds in certain situations.*

## 3.3  Degree Cap

It is desirable that the nodes' neighborhoods or degrees do not increase much during the sorting process. We investigate the performance of $LIN_{all}$ and $LIN_{max}$ under the following *degree cap model*. Observe that during a linearization step, only the degree of the node in the middle of the triple can increase (see Figure 1). We do not schedule triples if the middle node's degree would increase, with one exception: during left-linearizations, we allow a degree increase if the middle node has only one left neighbor, and during right-linearizations we allow a degree increase to the right if the middle node has degree one. In other words, we study a *degree cap of two*.

We find that both our algorithms $LIN_{all}$ and $LIN_{max}$ still terminate with a correct solution under this restrictive model.

**Theorem 6.** *With degree cap, $LIN_{max}$ terminates in at most $O(n^2)$ many rounds under a worst-case scheduler $\mathcal{S}_{wc}$, where $n$ is the total number of nodes in the system. Under the same conditions, $LIN_{all}$ requires at most $O(n^3)$ rounds.*

# 4    Experiments

In order to improve our understanding of the parallel complexity and the behavior of our algorithms, we have implemented a simulation framework which allows us to study and compare different algorithms, topologies and schedulers. In this section, some of our findings will be described in more detail.

We will consider the following graphs.

1. *Random graph:* Any pair of nodes is connected with probability $p$ (Erdös-Rényi graph), i.e., if $V = \{v_1, \ldots, v_n\}$, then $\mathbb{P}[\{v_i, v_j\} \in E] = p$ for all $i, j \in \{1, \ldots, n\}$. If necessary, edges are added to ensure connectivity.
2. *Bipartite backbone graph (k-BBG):* This seems to be a "hard" graph which also allows to compare different models. For $n = 3k$ for some positive integer $k$ define the following $k$-bipartite backbone graph on the node set $V = \{v_1, \ldots, v_n\}$. It has $n$ nodes that are all connected to their respective successors and predecessors (except for the first and the last node). This structure is called the graph's *backbone*. Additionally, there are all $(n/3)^2$ edges from nodes in $\{v_1, \ldots, v_k\}$ to nodes in $\{v_{2k+1}, \ldots, v_n\}$.
3. *Spiral graph:* The spiral graph $G = (V, E)$ is a sparse graph forming a spiral, i.e., $V = \{v_1, \ldots, v_n\}$ where $v_1 < v_2 < \ldots < v_n$ and

$$E = \{\{v_1, v_n\}, \{v_n, v_2\}, \{v_2, v_{n-1}\}, \{v_{n-1}, v_3\}, \ldots, \{v_{\lceil n/2 \rceil}, v_{\lceil n/2 \rceil + 1}\}\}.$$

4. *k-local graph:* This graph avoids long-range links. Let $V = \{v_1, \ldots, v_n\}$ where $v_i = i$ for $i \in \{1, \ldots, n\}$. Then, $\{v_i, v_j\} \in E$ if and only if $|i - j| \le k$.

We will constrain ourselves to two schedulers here: the greedy scheduler $S_{\mathrm{greedy}}$ which we have already considered in the previous sections, and a randomized scheduler $S_{\mathrm{rand}}$ which among all possible enabled actions chooses one *uniformly at random*.

Many experiments have been conducted to shed light onto the parallel runtime of $\mathrm{LIN}_{\mathrm{all}}$ and $\mathrm{LIN}_{\mathrm{max}}$ in different networks. Figure 2 depicts some of our results for $\mathrm{LIN}_{\mathrm{all}}$. As expected, in the $k$-local graphs, the execution is highly parallel and yields a constant runtime—independent of $n$. The sparse spiral graphs appear to entail an almost linear time complexity, and also the random graphs perform better than our analytical upper bounds suggest. Among the graphs we tested, the *BBG* network yielded the highest execution times. Figure 2 gives the corresponding results for $\mathrm{LIN}_{\mathrm{max}}$.

A natural yardstick to measure the quality of a linearization algorithm—besides the parallel runtime—is the node degree. For instance, it is desirable that an initially sparse graph will remain sparse during the entire linearization process. It turns out that $\mathrm{LIN}_{\mathrm{all}}$ and $\mathrm{LIN}_{\mathrm{max}}$ indeed maintain a low degree. Figure 3 (top) shows how the maximal and average degrees evolve over time both for $\mathrm{LIN}_{\mathrm{all}}$ and $\mathrm{LIN}_{\mathrm{max}}$ on two different random graphs. Note that the average degree cannot increase because the rules only move or remove edges. The random graphs studied in Figure 3 have a high initial degree, and it is interesting to analyze what happens in case of sparse initial graphs. Figure 3

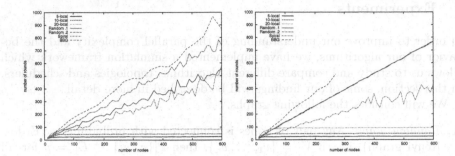

**Fig. 2.** *Left:* Parallel runtime of LIN$_{all}$ for different graphs under $S_{rand}$: two $k$-local graphs with $k = 5$, $k = 10$ and $k = 20$, two random graphs with $p = .1$ and $p = .2$, a spiral graph and a $n/3$-BBG. *Right:* Same experiments with LIN$_{max}$.

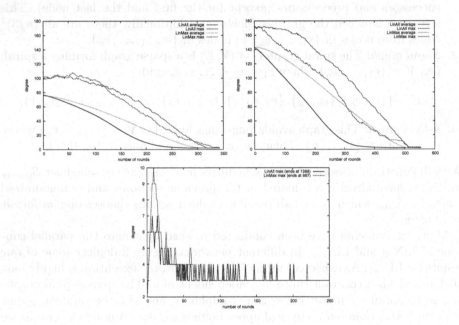

**Fig. 3.** *Top left:* Maximum and average degree during a run of LIN$_{all}$ and LIN$_{max}$ on a random graph with edge probability $p = .1$. *Top right:* The same experiment on a random graph with $p = .2$. *Bottom:* Evolution of maximal degree on spiral graphs under a randomized scheduler $S_{rand}$.

(bottom) plots the maximal node degree over time for the spiral graph. While there is an increase in the beginning, the degree is moderate at any time and declines again quickly.

# References

1. Aspnes, J., Shah, G.: Skip graphs. In: Proc. 14th Annual ACM-SIAM Symposium on Discrete Algorithms (SODA), pp. 384–393 (2003)
2. Blumofe, R.D., Leiserson, C.E.: Space-e cient scheduling of multithreaded computations. SIAM Journal on Computing 27(1), 202–229 (1998)
3. Blumofe, R.D., Leiserson, C.E.: Scheduling multithreaded computations by work stealing. Journal of the ACM 46(5), 720–748 (1999)
4. Clouser, T., Nesterenko, M., Scheideler, C.: Tiara: A self-stabilizing deterministic skip list. In: Kulkarni, S., Schiper, A. (eds.) SSS 2008. LNCS, vol. 5340, pp. 124–140. Springer, Heidelberg (2008)
5. Cramer, C., Fuhrmann, T.: Self-stabilizing ring networks on connected graphs. Technical Report 2005-5, System Architecture Group, University of Karlsruhe (2005)
6. Dolev, D., Hoch, E.N., van Renesse, R.: Self-stabilizing and byzantine-tolerant overlay network. In: Tovar, E., Tsigas, P., Fouchal, H. (eds.) OPODIS 2007. LNCS, vol. 4878, pp. 343–357. Springer, Heidelberg (2007)
7. Dolev, S., Kat, R.I.: Hypertree for self-stabilizing peer-to-peer systems. Distributed Computing 20(5), 375–388 (2008)
8. Gall, D., Jacob, R., Richa, A., Scheideler, C., Schmid, S., Täubig, H.: Modeling scalability in distributed self-stabilization: The case of graph linearization. Technical Report TUM-I0835, Technische Universität München, Computer Science Dept. (November 2008)
9. Jacob, R., Richa, A., Scheideler, C., Schmid, S., Täubig, H.: A distributed polylogarithmic time algorithm for self-stabilizing skip graphs. In: Proc. ACM Symp. on Principles of Distributed Computing, PODC (2009)
10. Jacob, R., Ritscher, S., Scheideler, C., Schmid, S.: A self-stabilizing and local delaunay graph construction. In: Dong, Y., Du, D.-Z., Ibarra, O. (eds.) ISAAC 2009. LNCS, vol. 5878. Springer, Heidelberg (2009)
11. Kuhn, F., Schmid, S., Wattenhofer, R.: A self-repairing peer-to-peer system resilient to dynamic adversarial churn. In: Castro, M., van Renesse, R. (eds.) IPTPS 2005. LNCS, vol. 3640, pp. 13–23. Springer, Heidelberg (2005)
12. Onus, M., Richa, A., Scheideler, C.: Linearization: Locally self-stabilizing sorting in graphs. In: Proc. 9th Workshop on Algorithm Engineering and Experiments (ALENEX). SIAM, Philadelphia (2007)
13. Scheideler, C., Schmid, S.: A distributed and oblivious heap. In: Albers, S., Marchetti-Spaccamela, A., Matias, Y., Nikoletseas, S., Thomas, W. (eds.) ICALP 2009. LNCS, vol. 5556, pp. 571–582. Springer, Heidelberg (2009)
14. Shaker, A., Reeves, D.S.: Self-stabilizing structured ring topology P2P systems. In: Proc. 5th IEEE International Conference on Peer-to-Peer Computing, pp. 39–46 (2005)
15. Stoica, I., Morris, R., Karger, D., Kaashoek, M.F., Balakrishnan, H.: Chord: A scalable peer-to-peer lookup service for internet applications. Technical Report MIT-LCS-TR-819. MIT (2001)

# Randomised Broadcasting: Memory vs. Randomness

Petra Berenbrink[1], Robert Elsässer[2,*], and Thomas Sauerwald[3]

[1] School of Computing Science, Simon Fraser University,
Burnaby B.C. V5A 1S6, Canada
petra@cs.sfu.ca
[2] Institute for Computer Science,
University of Paderborn, 33102 Paderborn, Germany
elsa@upb.de
[3] International Computer Science Institute, Berkeley, CA 94704, USA
sauerwal@icsi.berkeley.edu

**Abstract.** In this paper we analyse broadcasting in $d$-regular networks with good expansion properties. For the underlying communication, we consider modifications of the so called random phone call model. In the standard version of this model, each node is allowed in every step to open a channel to a randomly chosen neighbour, and the channels can be used for bi-directional communication. Then, broadcasting on the graphs mentioned above can be performed in time $O(\log n)$, where $n$ is the size of the network. However, every broadcast algorithm with runtime $O(\log n)$ needs in average $\Omega(\log n/\log d)$ message transmissions per node. This lower bound even holds for random $d$-regular graphs, and implies a high amount of message transmissions especially if $d$ is relatively small.

In this paper we show that it is possible to save significantly on communications if the standard model is modified such that nodes can avoid opening channels to the same neighbours in consecutive steps. We consider the so called RR model where we assume that every node has a cyclic list of all of its neighbours, ordered in a random way. Then, in step $i$ the node communicates with the $i$-th neighbour from that list. We provide an $O(\log n)$ time algorithm which produces in average $O(\sqrt{\log n})$ transmissions per node in networks with good expansion properties. Furthermore, we present a related lower bound of $\Omega(\sqrt{\log n/\log\log n})$ for the average number of message transmissions. These results show that by using memory it is possible to reduce the number of transmissions per node by almost a quadratic factor.

## 1 Introduction

We consider randomised broadcasting in (almost) regular graphs[1] with good expansion properties. In the broadcasting problem, the goal is to spread a message

---

[*] Partly supported by the German Research Foundation under contract EL 399/2-1.
[1] Almost regular means that the smallest and largest degree in the graph differ by at most a constant factor.

A. López-Ortiz (Ed.): LATIN 2010, LNCS 6034, pp. 306–319, 2010.

of a certain node among all vertices of a network by using local communication only. Our interest in the graphs mentioned before is motivated by overlay topologies in peer to peer (P2P) systems. Important topological properties of these networks include good connectivity, high expansion, and small diameter; all these properties are perfectly fulfilled by the graphs considered here. Our aim is to develop time-efficient broadcasting algorithms which produce a minimal number of message transmissions in the graphs described above. Since P2P systems are significant decentralised platforms for sharing data and computing resources, it is very important to provide efficient, simple, and robust broadcasting algorithms for P2P overlays. Minimising the number of transmissions is important in applications such as the maintenance of replicated databases in which broadcasts are necessary to deal with frequent updates in the system.

In this paper we assume the so called *phone call model* (see [17]). In this model, each node $v$ may perform the following actions in every step. 1) create a new message to be broadcasted. 2) establish a communication channel between $v$ itself and one of its neighbours. 3) transmit messages over incident channels opened by himself or by its neighbours. At the end of each step, the nodes close all open channels. Note that open channels can be used for bi-directional (push&pull) communications. In the case of push transmissions calling nodes (i.e., the nodes that opened the channels) send their messages to their neighbours. In the case of pull transmissions messages are transmitted from called nodes to the calling ones. Note that nodes can combine several broadcast messages to larger ones which can be sent over a channel in one time step.

In the standard phone call model it is assumed that nodes open a channel to a randomly chosen neighbour, and the nodes have to decide whether to transmit a specific message over a channel, without knowing if they opened a channel to the corresponding node in earlier steps. In this paper we assume that every node has a cyclic list of all of its neighbours, ordered in a random way. In step $i$ the node opens a channel to the $i$ (modulo $d$)-th neighbour from that list. This model is called RR *model* in the following. The RR rule prevents a node to open a channel to the same neighbour again and again. The question we address in this paper is whether remembering the communication partners from earlier rounds helps or not. We give a positive answer to this question and provide further evidence for the power of memory in randomised broadcasting (see [10]). More precisely, we present an algorithm, and show that w.r.t. the average number of transmissions per node this algorithm performs significantly better than any algorithm in the so called RANDOM[$c$]-model introduced in [10] (i.e. we achieve an almost quadratic improvement). The later model is similar to the standard random phone call model, however, every node may open channels to $c$ different randomly chosen neighbours in each step. Our algorithm is *address oblivious*, i.e., the send decisions of the nodes do not depend on the IDs of the nodes to which they open channels in the actual step. However, the nodes are allowed to remember with which nodes they communicated in the steps before [17].

## 1.1 Related Work

There is a huge amount of work considering epidemic type (broadcasting) algorithms on proper graph models for P2P overlays. Most of these papers deal with the empirical analysis of these algorithms e.g. [18,20]. Due to space constraints, we can only describe here the results which focus on the analytical study of push&pull algorithms.

*Runtime.* Most randomised broadcasting results analyse the runtime of the push algorithm. For complete graphs of size $n$, Frieze and Grimmett [14] present an algorithm that broadcasts a message in time $\log_2(n) + \ln(n) + o(\log n)$ with a probability of $1 - o(1)$. Later, Pittel [21] shows that (with probability $1 - o(1)$) it is possible to broadcast a message in time $\log_2(n) + \ln(n) + f(n)$ [21], where $f(n)$ can be any slow growing function. In [13], Feige et al. determine asymptotically optimal upper bounds for the runtime of the push algorithm on $G(n, p)$ graphs (i.e., traditional Erdös-Rényi random graphs [11,12]), bounded degree graphs, and Hypercubes. In [9] Elsässer and Sauerwald consider arbitrary graphs and show that the expected broadcasting time is bounded, up to a logarithmic factor, by the mixing time of a corresponding Markov chain. Additionally, a new class of Cayley graphs is introduced on which the push algorithm has optimal performance. Boyd et al. consider the combined push&pull model in arbitrary graphs of size $n$, and show that the running time is asymptotically bounded by the mixing time of a corresponding Markov chain plus an $O(\log n)$ value [2]. Sauerwald shows that the same result also holds if only the push algorithm is considered [22]. In [6] Doerr et al. analyse the so called quasi-random rumor spreading in an adversarial version of the RR model where the order of the lists is assumed to be given by an adversary. However, the nodes choose a random position in their lists to start with communication. They show for hypercubes and $G(n, p)$ graphs that $O(\log n)$ push steps suffice to inform every node, w.h.p.[2]. These bounds are similar to the ones in traditional randomised broadcasting (push model). Recently, these results have been extended to further graph classes with good expansion properties [7].

*Number of transmissions.* Karp et al. [17] note that in complete graphs the pull approach is inferior to the push approach, until roughly $n/2$ nodes receive the message, and then the pull approach becomes superior. They present a push&pull algorithm, together with a termination mechanism, which reduces the number of total transmissions to $O(n \log \log n)$ (w.h.p.), and show that this result is asymptotically optimal. They also consider communication failures and analyse the performance of their method in cases where the connections are established using arbitrary probability distributions.

For sparser graphs it is not possible to get $O(n \log \log n)$ message transmissions together with a broadcast time of $O(\log n)$ in the standard phone call model. In [8] Elsässer considers random $G(n, p)$ graphs, and shows a lower bound of $\Omega(n \log n / \log(pn))$ message transmissions for broadcast algorithms with a runtime of $O(\log n)$. On the positive side, for $p > \log^2 n/n$ he develops an algorithm

---

[2] W.h.p. or "with high probability" means with probability $1 - o(n^{-1})$.

that broadcasts in time $O(\log n)$ using $O(n \cdot (\log \log n + \log n / \log(pn)))$ transmissions, w.h.p.

In [10] the authors consider a simple modification of the standard phone call model called RANDOM[$c$] above. For $G(n,p)$ graphs with $p > \log^2 n / n$, they show that this modification results in a reduction of the number of message transmissions down to $O(n \log \log n)$. In [1] the authors show similar results for random $d$-regular graphs with $d = O(\log n)$.

## 1.2 Models and Results

We assume that every node has an estimation of $n$ which is accurate to within a constant factor. We also assume that all nodes have access to a global clock, and that they work synchronously. In each step every node can create an arbitrary amount of messages to be broadcasted. The number of message transmissions for a certain message is defined as the the number of open channels traversed by the message during the execution of the algorithm. As in [17], we assume here that new pieces of information are generated frequently in the network, and then the cost of establishing communication channels amortises over all message transmissions. However, we only concentrate on the distribution and lifetime of a single message, and consider broadcasting in the following graphs.

*Edge-Node Expanders.* Let $G = (V, E)$ be a $d$-regular graph of size $n$. For $A \subset V$, let $E(A, \overline{A})$ denote the set of edges between $A$ and $\overline{A} = V \setminus A$, and let $N(A) = \{v \in \overline{A} \mid \exists u \in A \text{ with } (u, v) \in E\}$. For a constant $\alpha$, $G$ is called $\alpha$-*Edge-Node expander* (or simply Edge-Node expander) if the following holds:

1. For any set $A \subset V$ with $|A| \leq n/2$ we have $|E(A, \overline{A})| \geq \alpha d \cdot |A|$.
2. For any set $A \subset V$ with $|A| \leq \min\{\phi n / d, n/2\}$ it holds that $|N(A)| \geq \alpha d \cdot |A|$, where $\phi$ is large but constant.

In [10,1] we showed that in random graphs one can save on the number of message transmissions if the nodes avoid communication with the neighbours chosen already in some recent steps. In the analysis we used the randomised construction of these graphs, and integrated the dynamical behaviour of the RANDOM[$c$] model (i.e., the parallelised version of the model described above) into the random structure of the underlying topology. However, the methods derived in [10,1] cannot be generalised to non-random graphs with similar expansion and connectivity properties, not even to pseudo-random graphs [19]. Therefore, the main question is whether the same result also holds in graphs with random graph like properties. To answer this question, we show that in certain Edge-Node expanders the RANDOM[$c$] model requires $\Omega(n \log n / \log \log n)$ message transmissions for constant $c$, if the broadcast time is $O(\log n)$ and $d = \log^{O(1)} n$ (Theorem 1). Then, we present an algorithm for the RR model, and introduce a new combinatorial technique which only uses the structural properties of Edge-Node expanders to show that this algorithm completes broadcasting in time $O(\log n)$ and generates $O(n\sqrt{\log n})$ message transmissions (see Sections 2.1-2.2). Finally we establish a lower bound of $\Omega(n\sqrt{\log n / \log \log n})$ on the number of message transmissions in the RR model (if the broadcast time is $O(\log n)$ and $d = \log^{O(1)} n$), showing that our analysis is tight up to a $\sqrt{\log \log n}$ factor.

The upper bound on the number of message transmissions in the RR model is significantly smaller than the lower bound in the RANDOM[c] model, which substantiates the importance of memory in randomised broadcasting. Notice that all (regular) graphs $G$ for which $\lambda_2 = d - O(\sqrt{d})$ ($\lambda_2$ is the second smallest eigenvalue of the Laplacian of $G$) obey the properties described above (cf. [4,16,23]). Examples for such graphs are so called Ramanujan graphs. By using the techniques of this paper, our results can be extended to graphs in which the ratio between the maximum and minimum degree is bounded by a constant.

## 2    Broadcasting on Edge-Node Expanders

For simplicity, we prove our results for the case $d \in \{\omega(\log^{3/2} n), 2^{o(\sqrt{\log n})}\}$. The more general case with no restriction on $d$ can be shown using similar techniques, however, this would require an elaborate case analysis which is omitted in this extended abstract. In this section we first consider the following lower bound w.r.t. the performance of the RANDOM[c] model in Edge-Node Expanders. The proof of this theorem is omitted due to space limitations.

**Theorem 1.** *Assume A is a broadcasting algorithm with runtime $O(\log n)$ in the* RANDOM[c] *communication model, where c is a constant. There exists a family of Edge-Node Expanders for which A needs $\Omega(n \log n / \log \log n)$ message transmissions, w.h.p.*

### 2.1    The Algorithm

We assume that every node $v$ stores a cyclic list $\ell_v$ with a random permutation of all its neighbours. Let $\ell_v(t)$ be the $t$-th entry in the list. Then we assume that $v$ communicates with $\ell_v(t)$ in step $t$ (we omit the division by $d$ for $t > d$). We also define $L_v[t, t'] = \{\ell_v(t), \ell_v(t+1), \dots \ell_v(t')\}$ for $t' > t$. For a fixed node $v$, push sends the message over the outgoing channel to $\ell_v(t)$, and pull sends the message over all incoming channels of $v$. We should note that a node has to decide whether to transmit a message (by push or pull) without knowing if this message have already been received by the node at the other end of the channel. If several messages are to be broadcasted, then the algorithm will be run for every message on every node. Each node combines all messages it has to transmit via push (pull) to a single message and forwards it over all open outgoing (incoming) channels.

The following algorithm describes the behavior of the nodes w.r.t. one specific message. The age of a message (i.e., the difference between the current time step and the time step in which the message has been generated) is used by the nodes to decide which of the following phases applies for the message. A node is called *informed* if it got a copy of that message. We assume that $\rho > 40/\alpha^2$ is a constant. We also assume that the message is generated at time step 0 (i.e., at step $t$ the age of the message equals $t$). Recall that $\alpha$ is the expansion value of the graph.

*Phase 0: [age $\leq \lceil \rho \log n \rceil$]* The node which generates the message performs push in each step of this phase. No other node transmits the message in this phase.

*Phase 1: [$\lceil \rho \log n \rceil + 1 \leq age \leq 2 \cdot \lceil \rho \log n \rceil + \lceil 80/\alpha^2 \rceil$]* Nodes that received the message in Phase 0 use the first $\lceil 80/\alpha^2 \rceil$ steps of this phase to perform push in each of these steps. If a node receives a message for the *first* time at time step $t \in \{\lceil \rho \log n \rceil + 1, \ldots, 2 \cdot \lceil \rho \log n \rceil\}$, then the node will use the next $\lceil 80/\alpha^2 \rceil$ steps to perform push in each of these steps.

*Phase 2: [$2 \cdot \lceil \rho \log n \rceil + \lceil 80/\alpha^2 \rceil + 1 \leq age \leq 2 \cdot \lceil \rho \log n \rceil + \lceil \rho \log d \rceil$]* Every *informed* node performs push in each step of this phase.

*Phase 3: [$2 \cdot \lceil \rho \log n \rceil + \lceil \rho \log d \rceil + 1 \leq age \leq 2 \cdot \lceil \rho \log n \rceil + 2 \cdot \lceil \rho \log d \rceil$]* Every *informed* node performs pull in each step of this phase.

*Phase 4: [$2 \cdot \lceil \rho \log n \rceil + 2 \cdot \lceil \rho \log d \rceil + 1 \leq age \leq 2 \cdot \lceil \rho \log n \rceil + 2 \cdot \lceil \rho \log d \rceil + \lceil \rho \sqrt{\log n} \rceil$]* Every *informed* node performs pull in each step of this phase.

*Phase 5: [$2 \cdot \lceil \rho \log n \rceil + 2 \cdot \lceil \rho \log d \rceil + \lceil \rho \sqrt{\log n} \rceil + 1 \leq age \leq 3 \cdot \lceil \rho \log n \rceil$]* Every node that receives the message *in Phase 4 or 5* performs pull in each step of this phase. The other informed nodes flip a coin and perform pull with probability $1/\sqrt{\log n}$.

*Phase 6: [$3 \cdot \lceil \rho \log n \rceil + 1 \leq age \leq 3 \cdot \lceil \rho \log n \rceil + \lceil \rho \sqrt{\log n} \rceil$]* Every *informed* node performs pull in each step of this phase.

We assume that there is only one message in the network. However, the algorithm will work exactly the same as if there were several broadcast messages. We say a node is *active* in a phase if it performs transmissions in that phase. The algorithm consists of 7 phases. In the first 3 Phases the algorithm performs push transmissions, in the remaining 4 phases it performs pull transmissions.

push *Phases.* In Phase 0 the node which generated the message performs a push transmission in every step. At the end of the phase $O(\log n)$ nodes are informed, w.h.p. In Phase 1 every informed node performs *a constant number* of push transmissions. After that we have w.h.p. $n/d$ informed nodes. The restriction to a constant number of transmissions per node helps to reduce the transmission number. The purpose of the next phase is to inform $n/2$ nodes. In this phase every informed node performs a push transmission in every step of the phase. Note that this phase consists only of $O(\log d)$ many steps.

pull *Phases.* In every step of Phase 3 and Phase 4 every informed node performs a pull transmission. At the end of the Phase 3 we have $n - n/d^3$ informed nodes, w.h.p. Note that here is no algorithmic difference between Phase 3 and Phase 4. We introduce these two phases since they will be analysed separately. In Phase 5 the nodes that were informed during phases 4 and 5 become active. All remaining nodes will become active with a probability of $1/\sqrt{\log n}$. This again helps to keep the number of transmissions down. In Phase 5 every active node performs a pull transmission in every step. Phase 4 and Phase 5 inform w.h.p. all uninformed nodes that have, in turn, many uninformed neighbours at the beginning of Phase 4. The remaining nodes are informed in Phase 6 where every informed node performs a pull transmission in every step.

## 2.2    Analysis of the Algorithm

The analysis of the algorithm is more or less divided into the same phases as the algorithm. First we show (Lemma 1 and Lemma 2) that the algorithm informs w.h.p. at least $n/2$ nodes during the first $O(\log n)$ steps, using $O(n)$ message transmissions. More precisely, we show that (w.h.p.) in a constant number of steps the number of informed nodes increases by a constant factor, as long as the number of informed nodes is less than $n/2$. As soon as the number of informed nodes is larger than $n/2$ the analysis becomes much more complicated. If we were only interested in the running time of our algorithm, then we could apply a backward analysis as in e.g. [10,22] to show that the algorithm completes broadcasting in $O(\log n)$ steps, w.h.p. However, this would result in a bound of $\Theta(n \log n)$ on the number of message transmissions. Since our goal is to significantly reduce the number of message transmission per node we need new analytical techniques for this case. Thus, we first analyse the distribution of edges in the set of uninformed nodes as well as the distribution of the so called cut edges separating informed and uninformed nodes from each other. To obtain the desired result, we design a new combinatorial technique that combines the information flow from informed to uninformed vertices with the distribution of the cut edges.

In our proofs we assume for simplicity that $\rho \log n$, $\rho \log d$, and $\rho \sqrt{\log n}$ are all integers. We also assume that $d = \Omega(f(n) \log^{3/2} n)$ with $f : \mathbb{N} \to \mathbb{R}$ being a function such that $\lim_{n \to \infty} f(n) = \infty$. Whenever we analyse a phase of our algorithm we assume that all earlier phases were successfull in the sense that they informed the right number of nodes. We can do that since all results hold with high probability. To get the failure probability of Algorithm RR one has to add up the failure probability of all 6 phases. We will use the following definitions.

- $I(t)$ is the set of informed nodes at the beginning of step $t$.
- Let $I^+(t) = I(t+1) \setminus I(t)$, that is, the nodes that get informed in step $t$.
- Let $\tau = (t, t+1, \ldots t')$ be some consecutive steps of our algorithm. Then $I^+(\tau)$ is the set of nodes that get informed in steps $t, t+1, \ldots t'$ from one of the nodes of $I(t)$.
- $H(t)$ is the set of uninformed nodes $V \setminus I(t)$ at time $t$.
- $E(S, \overline{S})$ is the set of edges between $S$ and $\overline{S}$.
- $N(S, S')$ is the set of neighbours of $S$ that are in $S'$. Accordingly, $N(u, S')$ is the set of neighbours of $u \in V$ in $S'$.

It is easy to see that in Phase 0 the node, on which the message is generated, informs $\rho \log n$ different neighbours, which results in the following observation.

**Observation 1.** *At the end of Phase 0 there are $\rho \log n$ informed nodes.*

**Lemma 1.** *With a probability of $1 - n^{-2}$ at least $n/d$ nodes are informed at the end of Phase 1.*

*Proof.* Let $t$ be the beginning of Phase 1 and define $\ell = 40/\alpha^2$. To show the result we will prove that $|I^+(t + \rho \log n)| \geq n/d$ with a probability of $1 - n^{-2}$.

We divide Phase 1 into $k = (\rho \log n)/\ell + 2$ time intervals $\tau_1, \ldots, \tau_k$ of length $\ell$ each. We assume that $t_i$ is the first step of time interval $\tau_i$. Let $|I^+(\tau_i)|$ be the random variable that counts the number of nodes in $I^+(\tau_i)$. Note that all nodes in $I^+(\tau_i)$ will perform push transmissions in every step of interval $\tau_{i+1}$ (for $i < k-1$). Since $\rho > 4$ we can assume that we have already $4 \log n$ informed nodes at the beginning of Phase 1. None of these nodes has transmitted the message yet. The proof of the lemma is based on Claim A 1 which can be found in the appendix. The claim shows that with probability $1 - n^{-3}$ it holds that $|I^+(\tau_i)| \geq (8/\alpha) \cdot |I(t_i)|$, which implies $|I(t_{i+1})| \geq (8/\alpha) \cdot |I(t_i)|$. By repeatedly applying Claim A 1 we conclude that if $\rho$ is large enough, then after $k$ many time intervals at least $n/d$ nodes are informed with a probability of $1 - n^{-2}$. □

Next we consider Phase 2 where informed nodes perform **push** for roughly $\rho \log d$ many steps. Note that the statement of this lemma holds conditioned on the event that Phase 1 was successfull.

**Lemma 2.** *With a probability of $1 - n^{-2}$ at least $(n/2)$ nodes are informed at the end of Phase 2.*

The proof of this lemma is omitted due to space limitations.

Now we consider Phase 3 in the next lemma. To proof of this lemma is omitted due to space limitations.

**Lemma 3.** *With a probability of $1 - n^{-2}$ at least $n - n/d^3$ nodes are informed at the end of Phase 3.*

Now we focus on Phases 4 and 5. Assume that $t$ is the end of Phase 3 and that there are at least $n - n/d^3$ informed nodes at that time. Recall that, in Phase 4 (age $t$ to $t + \rho\sqrt{\log n}$)[3] all informed nodes perform pull transmissions. In Phase 5 (age $t + \rho\sqrt{(\log n)} + 1$ to $3\rho \log n$) every node that was informed in Phase 1-3 performs pull transmissions with probability $1/\sqrt{\log n}$, and every node that was informed in Phase 4 or Phase 5 performs pull transmissions with probability 1.

**Lemma 4.** *Let $t$ be the beginning of Phase 4. A node $v \in H(t)$ with $|N(v, I(t))| \leq d/2$ receives the message with probability $1 - n^{-2}$ by the end of Phase 5.*

*Proof.* To prove the lemma we need some definitions first. The node $\ell_v(\tau)$ is called the $\tau$-*active* neighbour of $v$, and the nodes $L_v[\tau_1, \tau_2]$ are called $(\tau_1, \tau_2)$-*active* neighbours of $v$ (notice that $\tau$ denotes a single time step here rather than a time period as in Lemma 1). A node $w$ is called $\tau$-*predecessor* of $v$ if there exists $k \leq \tau$, some nodes $w_1, \ldots, w_k$ and time steps $t_0 < t_1 < \cdots < t_k \leq \tau$ such that $w$ is the $t_0$-active neighbour of $w_1$, node $w_i$ is the $t_i$-active neighbour of $w_{i+1}$ $(1 \leq i < k)$, and $w_k$ the $t_k$-active neighbour of $v$. This means that $v$ is connected to $w$ by a path consisting of edges that were active in the time

---

[3] Although Phase 4 begins at time $t + 1$, we assume for simplicity in our analysis that $t$ also belongs to Phase 4.

interval $[t_0, \tau]$. Note that, if $w$ is a $\tau$-predecessor of $v$ and $w$ is informed at step $t_0$, then $v$ becomes informed at time $\tau$ at the latest. For different choices of $k$ and $t_0, t_1, \ldots, t_k$ one might regard $v$ as being a node of a tree consisting of $\tau$-predecessors. If one of the nodes is informed at the right time, $v$ will get the message via the corresponding path in the tree. We define

$$T_0 = t + \rho\sqrt{\log n}, \ T_1 = t + (\rho \cdot \log n)/4, \ T_2 = t + (\rho \cdot \log n)/2, \text{ and } T_3 = t + \rho \log n.$$

Although there is a small overlap between the intervals $[T_0, T_1]$, $[T_1, T_2]$, and $[T_2, T_3]$ (i.e., $T_1 = [T_0, T_1] \cap [T_1, T_2]$ and $T_2 = [T_1, T_2] \cap [T_2, T_3]$), we can ignore these side effects in our further analysis. We show the following claim stating that every uninformed node with many uninformed neighbours has at least $(\rho \cdot \log n)/8$ many $(T_2, T_3)$-active neighbours in $H(t)$.

**Claim 1.** Let $v$ be a node in $H(t)$ with $|N(v, I(t))| \leq d/2$. Then $H(t) \cap L_v[T_2, T_3] \geq \frac{\rho \log n}{8}$, with probability $1 - O(n^{-4})$.

The proof of this claim is omitted due to space limitations. Applying the claim, we can assume for the rest of the proof that $v$ has at least $(\rho \log n)/8$ many $(T_2, T_3)$-active neighbours in $H(t)$. We say a node $v \in H(t)$ is $I(t)$-*good* if it has at least $d/2$ of its neighbours in $I(t)$ (meaning $|N(v, I(t))| \geq d/2$). Otherwise the node is called $I(t)$-*bad*. In the following we show that every node that is $I(t)$-bad ($|N(v, I(t))| \leq d/2$) will receive the message from a node in $I(t)$ either directly via one of its $I(t)$-good neighbours, or via a longer path to a node in $I(t)$ consisting of nodes which are in $H(t)$. Note that this shows the lemma since it only states that nodes $v \in H(t)$ with $|N(v, I(t))| \leq d/2$ will be informed by the end of Phase 5. We consider two cases.

*Case 1:* In $L_v[T_2, T_3] \cap H(t)$ are at least $\rho \cdot \sqrt{\log n}$ $I(t)$-*good* nodes. Let $U$ be the set of $I(t)$-good neighbours of $L_v[T_2, T_3]$. Note that $v$ receives the message in $[T_2, T_3]$ if there exists a node $u \in U$ that received the message in Phase 4. This holds since all nodes which receive the message for the first time in Phase 4 or Phase 5 respond to every pull request. The probability that a node $w \in U$ is still uninformed at the end of Phase 4 (step $t + \rho\sqrt{\log n}$) is at most $(3/8)^{-\rho\sqrt{\log n}}$. This holds since for every $\tau \in [t, t + \rho\sqrt{\log n}]$ the probability that $\ell_w(\tau) \in I(t)$ is at least $(d/2 - \rho \log d - \rho\sqrt{\log n})/d \geq 3/8$ ($\rho \log d$ edges in Phase 3 and at most $\rho\sqrt{\log n}$ edges in Phase 4 might have already been used for pull requests). For $\rho > 4$, all nodes of $U$ are still uninformed at time $t + \rho\sqrt{\log n}$ with a probability of at most $(3/8)^{-\rho^2(\sqrt{\log n})^2} = o(n^{-4})$.

*Case 2:* In $L_v[T_2, T_3] \cap H(t)$ are fewer than $\rho \cdot \sqrt{\log n}$ $I(t)$-*good* nodes. Due to our assumption that $v$ has at least $\rho \log n/8$ many $(T_2, T_3)$-active neighbours in $H(t)$, at least $(\rho/8) \cdot \log n - \rho\sqrt{\log n}$ of $v$'s $(T_2, T_3)$-active neighbours (in $H(t)$) are $I(t)$-bad. Let $U = w_1, \ldots w_k$ be an arbitrary subset of size $k = \sqrt{f(n)} \log n$ of the $I(t)$-bad neighbours of $v$. Now we step back in time and consider the time interval $[T_1, T_2]$. Let $U' = \bigcup_{w \in U} L_w(T_1, T_2)$. We show that $|U' \cap H(t)| = \Omega(\sqrt{f(n)}(\log n)^{3/2})$ with a probability of $1 - o(n^{-3})$. To bound the size of $U' \cap$

$H(t)$ we consider one node of $U$ after the other. We define $U'_0 = \emptyset$ and for $1 \le i \le k$

$$U'_i = \begin{cases} U'_{i-1} & \text{if } |U'_{i-1}| \ge \sqrt{f(n)} \cdot (\log n)^{3/2} \\ \bigcup_{j=1}^i L_{w_j}[T_1, T_2] \cap H(t) & \text{otherwise.} \end{cases}$$

Then we calculate the probability that the construction ends with $|U'_k| < \sqrt{f(n)} \cdot (\log n)^{3/2}$.

Assume $|U'_{i-1}| < \sqrt{f(n)} \cdot (\log n)^{3/2}$ for some $i$. Then, for $w_i \in U$ and $\ell_j \in L_{w_i}[T_1, T_2]$ we have

$$\mathbf{Pr}\left[\ell_j \in U'_{i-1}\right] \le \frac{\sqrt{f(n)} \cdot (\log n)^{3/2} + \cdot \rho \cdot \log n/2}{d - \rho \log n}$$

$$\le \frac{\sqrt{f(n)} \cdot (\log n)^{3/2} + (\rho \cdot \log n)/2}{f(n) \cdot (\log n)^{3/2} - \rho \log n} \le \frac{2}{\sqrt{f(n)}}.$$

Here, the term $\sqrt{f(n)} \cdot (\log n)^{3/2}$ stands for the maximum size of $U'_{i-1}$, and the additional term $\rho \cdot \log n/2$ for $|L_{w_i}[T_0, T_2]|$. The term $\rho \log n$ in the denominator represents an upper bound on the number of neighbours chosen already in Phases 3, 4, and 5.

By time step $\tau \in [T_1, T_2]$, $w_i$ used at most $\rho \log d + (\rho \log n)/2 < \rho \log n$ many edges for pull transmissions. Since we assumed that $d = \omega(\log^{3/2} n)$, we can argue that

$$\mathbf{Pr}\left[\ell_j \in I(t)\right] \le \frac{d/2}{d - (\rho \log n)} \le \frac{5}{8}.$$

Hence,

$$\mathbf{Pr}\left[\ell_j \in I(t) \cup U'_{i-1}\right] \le \frac{2}{\sqrt{f(n)}} + \frac{5}{8} \le \frac{3}{4},$$

regardless of the sets $U_{i-1}$ and $I(t)$. The expected number of nodes in $L_{w_i}[T_1, T_2] \cap (I(t) \cup U'_{i-1})$ is at most $(3\rho \log n)/16$ and we can apply Chernoff bounds [3] to conclude that with probability $1 - O(n^{-4}/\sqrt{f(n) \log n})$ we have $L_{w_i}[T_1, T_2] \cap (I(t) \cup U'_{i-1}) \le 7\rho \log n/32$, and thus, $|U'_i| \ge |U'_{i-1}| + (\rho \log n)/4 - 7\rho \log n/32$. Since $k = \sqrt{f(n)} \cdot \log n$, with a probability of $1 - o(n^{-3})$ we have $|U' \cap H(t)| \ge (\rho\sqrt{f(n)} \cdot (\log n)^{3/2})/32$.

Next we show that the set of predecessors of any node $w \in U' \cap H(t)$ grows the further we go backwards from $T_1$ to $T_0$. We define $\ell = 40/\alpha$ and divide the time interval $[T_0, T_1]$ into $k' = (T_1 - T_0)/\ell$ rounds of equal length (although the interval $[T_0, T_1]$ consists of $T_1 - T_0 + 1$ time steps we ignore the $+1$ in our analysis). For $0 \le i \le k' - 1$ we define $\tilde{T}_i = [T_1 - i \cdot \ell, T_1 - (i+1) \cdot \ell]$. Let

$$U_0^H = \bigcup_{w \in U' \cap H(t)} L_w[\tilde{T}_0] \cap H(t) \quad \text{and} \quad U_0^I = \bigcup_{w \in U' \cap H(t)} L_w[\tilde{T}_0] \cap I(t)$$

be the corresponding ($\tilde{T}_0$-active) set of uninformed and informed neighbours of $w \in U' \cap H(t)$, respectively. For $1 \leq i \leq k' - 1$ we define

$$U_i^H = \bigcup_{w \in U_{i-1}^H} L_w[\tilde{T}_i] \cap H(t) \text{ and } U_i^I = \bigcup_{w \in U_{i-1}^H} L_w[\tilde{T}_i] \cap I(t).$$

For every node $\overline{w}_i \in U_i^I$ there is a path $P = (\overline{w}_i, \ldots, \overline{w}_0, w', w)$ to a node $w \in U$, where $\overline{w}_{i-1}, \ldots, \overline{w}_0, w', w \in H(t)$, and $\overline{w}_{j+1}$ is an active neighbour of $\overline{w}_j$ in $\tilde{T}_{j+1}$. Hence, together $U_i^H$ and $U_i^I$ can be regarded as the $i$-th level of a tree routed in $w$ and, consequently (since $w$ is an active neighbour of $v$), also in $v$. Nodes in $H(t)$ are inner nodes of the tree, and the leaves are nodes in $I(t)$ or nodes in $H(t)$ on level $k' - 1$. Note that nodes can occur several times on several different levels of the tree.

In the following we argue that the amount of different leaves (informed nodes) in the tree is w.h.p. at least $\rho \cdot (\log n)^{3/2}$. Then we will show that the informed leaves are sufficient to inform node $v$. Let

$$U_{0 \to i}^H = \left[ \bigcup_{j=0}^{i} U_j^H \right] \text{ and } U_{0 \to i}^I = \left[ \bigcup_{j=0}^{i} U_j^I \right].$$

According to Claim A 1, $U_0^H \cup U_0^I \geq \frac{8}{\alpha} |U' \cap H(t)|$. Then, the following claim shows that w.h.p. there exists a time interval $i$ with $|U_{0 \to i}^I| \geq \rho \cdot (\log n)^{3/2}$.

**Claim 2.** *With a probability of $1 - O(n^{-3})$ there exists $i \in \{1, \ldots, k' - 1\}$ with $U_{0 \to i}^I > \rho \cdot (\log n)^{3/2}$.*

The proof of this claim is omitted due to space limitations. Now let $i \leq k' - 1$ be the value with $U_{0 \to i}^I > \rho \cdot (\log n)^{3/2}$. Using the claim we can easily argue that $v$ will get the message over a path starting at one of the nodes in $U_{0 \to i}^I$. Recall that every node $u \in U_{0 \to i}^I$ answers pull requests with a probability of $1/\sqrt{\log n}$. Since for such a $u$ there is a path $P = (u, \overline{w}_s, \ldots, \overline{w}_0, w', w, v)$ consisting of nodes in $H(t)$, all the nodes on this path answer every pull request. Hence, if there exists a node $u \in U_{0 \to i}^I$ which answers the pull request at the time $u$ was the active neighbour of $\overline{w}_s$, $v$ will be informed. This happens with a probability of

$$1 - \left( 1 - \frac{1}{\sqrt{\log n}} \right)^{\rho \cdot (\log n)^{3/2}} \leq 1 - e^{-\rho \log n} \leq 1 - o(n^{-3}),$$

and the lemma follows.                                                                    □

**Lemma 5.** *At the end of Phase 6 every node is informed with probability $1 - n^{-2}$.*

The proof of this lemma is omitted due to space limitations. By summarizing the results of the lemmas of this section we obtain the following theorem.

**Theorem 2.** *Assume $G$ is a Edge-Node Expander. Algorithm RR broadcasts a message in $G$ in time $O(\log n)$ by using $O(n\sqrt{\log n})$ transmissions, w.h.p.*

In the next theorem we show that the result of Theorem 2 is tight up to a $\sqrt{\log\log n}$ factor. For the following bound we also assume the *oblivious communication model*. In this model, a node's decision whether to transmit in a fixed step can depend on the age of the message and on any information the node might have aquired before the current step. However, the node's decision is not influenced by the ID's of the nodes at the other end of a currently open channel. We can assume that the nodes decide if they want to transmit in a fixed step or not *before* the channels become opened. The proof of the theorem is omitted due to space limitations.

**Theorem 3.** *Assume A is a broadcast algorithm with runtime $O(\log n)$ in the* RR *communication model. There exists a family of Edge-Node Expanders for which A needs $\Omega(n\sqrt{(\log n)/\log\log n})$ message transmissions, w.h.p.*

## 3   Conclusion

In this paper we presented upper and lower bounds for broadcasting in Edge-Node Expanders. The results of this paper (together with the results of [10]) show that choosing *different* neighbours is very important to save on broadcast communication. In this sense, model RR can be regarded as more advantageous than RANDOM[c], which provides further evidence for the power of memory in randomised broadcasting.

## References

1. Berenbrink, P., Elsässer, R., Friedetzky, T.: Efficient Randomised Broadcasting in Random Regular Networks with Applications in Peer-to-Peer Systems. In: Proc. of PODC 2008, pp. 155–164 (2008)
2. Boyd, S., Ghosh, A., Prabhakar, B., Shah, D.: Randomized Gossip Algorithms. IEEE Transactions on Information Theory and IEEE/ACM Transactions on Networking 52, 2508–2530 (2006)
3. Chernoff, H.: A measure of asymptotic efficiency for tests of a hypothesis based on the sum of observations. The Annals of Mathematical Statistics 23, 493–507 (1952)
4. Chung, F.R.K.: Spectral Graph Theory. American Mathematical Society, Providence (1985)
5. Demers, A., Greene, D., Hauser, C., Irish, W., Larson, J., Shenker, S., Sturgis, H., Swinehart, D., Terry, D.: Epidemic algorithms for replicated database maintenance. In: Proc. of PODC 1987, pp. 1–12 (1987)
6. Doerr, B., Friedrich, T., Sauerwald, T.: Quasirandom Rumor Spreading. In: Proc. of SODA 2008, pp. 773–781 (2008)
7. Doerr, B., Friedrich, T., Sauerwald, T.: Quasirandom rumor spreading: expanders, push vs. pull, and robustness. In: Proc. of ICALP 2009, track A, pp. 366–377 (2009)
8. Elsässer, R.: On the communication complexity of randomized broadcasting in random-like graphs. In: Proc. of SPAA 2006, pp. 148–157 (2006)
9. Elsässer, R., Sauerwald, T.: Broadcasting vs. mixing and information dissemination on Cayley graphs. In: Thomas, W., Weil, P. (eds.) STACS 2007. LNCS, vol. 4393, pp. 163–174. Springer, Heidelberg (2007)

10. Elsässer, R., Sauerwald, T.: The power of memory in randomized broadcasting. In: Proc. of SODA 2008, pp. 218–227 (2008)
11. Erdős, P., Rényi, A.: On random graphs I. Publ. Math. Debrecen 6, 290–297 (1959)
12. Erdős, P., Rényi, A.: On the evolution of random graphs. Publ. Math. Inst. Hungar. Acad. Sci. 5, 17–61 (1960)
13. Feige, U., Peleg, D., Raghavan, P., Upfal, E.: Randomized broadcast in networks. Random Structures and Algorithms 1(4), 447–460 (1990)
14. Frieze, A.M., Grimmett, G.R.: The shortest-path problem for graphs with random arc-lengths. Discrete Applied Mathematics 10, 57–77 (1985)
15. Hagerup, T., Rüb, C.: A guided tour of Chernoff bounds. Information Processing Letters 36(6), 305–308 (1990)
16. Kahale, N.: Eigenvalues and expansion of regular graphs. Journal of the ACM 42, 1091–1106 (1995)
17. Karp, R., Schindelhauer, C., Shenker, S., Vöcking, B.: Randomized rumor spreading. In: Proc. of FOCS 2000, pp. 565–574 (2000)
18. Kermarrec, A.-M., Massouli, L., Ganesh, A.J.: Probabilistic reliable dissemination in large-scale systems. IEEE Transactions on Parallel and Distributed Systems 14(3), 248–258 (2003)
19. Krivelevich, M., Sudakov, B.: Pseudo-random graphs. More Sets, Graphs and Numbers. Bolyai Society Mathematical Studies 15, 199–262 (2006)
20. Melamed, R., Keidar, I.: Araneola: A scalable reliable multicast system for dynamic environments. In: Proc. of NCA 2004, pp. 5–14 (2004)
21. Pittel, B.: On spreading a rumor. SIAM Journal on Applied Mathematics 47(1), 213–223 (1987)
22. Sauerwald, T.: On mixing and edge-expansion properties in randomized broadcasting. In: Tokuyama, T. (ed.) ISAAC 2007. LNCS, vol. 4835, pp. 196–207. Springer, Heidelberg (2007)
23. Tanner, M.: Explicit concentrators from generalized n-gons. SIAM J. Algebraic and Discrete Methods 5, 287–293 (1984)

# A    Appendix

Before we show Claim A 1 below, we first define the so called ADVRR model. This model is similar to RR, but the lists of the nodes are ordered by an adversary. Then, each node chooses at the beginning a random place in the list to start the communication. That is, if $1 \leq l \leq d$ is the random choice of node $v$, then $v$ communicates in step $i$ with the neighbour $(i + l) \mod d + 1$ from the list. To create the list, the adversary may have total knowledge about the topology of the network, but she cannot foresee any node's random choice w.r.t. the position selected at the beginning. Obviously, the result of the following claim also holds in the RR model.

**Claim A 1.** *Let $\tau_1, \ldots, \tau_i, \ldots$ be the time intervals of Phase 1 and let $t_i$ be the beginning of $\tau_i$. Furthermore, we assume that the underlying communication model is the ADVRR model, and there may be $o(|I(t_i)|)$ nodes in $I^+(\tau_{i-1})$ which do not send the message in the whole time interval $\tau_i$. Assume further that*

$$|I(t_i)| \leq \frac{n}{d} \quad and \quad |I(t_i)| \geq \frac{8}{\alpha} \cdot |I(t_{i-1})|.$$

*Then with a probability of $1 - n^{-3}$ we have*

$$|I^+(\tau_i)| \geq \frac{8}{\alpha} \cdot |I(t_i)|.$$

*Proof.* We assumed that there are at most $o(I(t_i)) = o(|I^+(\tau_{i-1})|)$ nodes in $I^+(\tau_{i-1})$ which do not transmit in time interval $\tau_i$. We denote this subset of nodes by $I^-(\tau_{i-1})$. Using the expansion properties of the graph and $H(t_i) \subset H(t_{i-1})$ we obtain that $|N(I^+(\tau_{i-1}) \setminus I^-(\tau_{i-1}), H(t_i))|$ is at least

$$
\begin{aligned}
&\quad |N(I(t_i), H(t_i))| - |N(I(t_{i-1}), H(t_i))| - |N(I^-(\tau_{i-1}), H(t_i))| \\
&\geq |N(I(t_i), H(t_i))| - |N(I(t_{i-1}), H(t_i))| - |N(I^-(\tau_{i-1}), H(t_i))| \\
&\geq \alpha \cdot d \cdot |I(t_i)| - d \cdot |I(t_{i-1})| - d \cdot o(|I(t_i)|) \\
&\geq \alpha \cdot d \cdot |I(t_i)| - d \cdot \frac{\alpha}{8} \cdot |I(t_i)| - o(d|I(t_i)|) \\
&= \frac{7}{8} \cdot \alpha \cdot d \cdot |I(t_i)|(1 - o(1)).
\end{aligned}
$$

The last inequality holds due to the second precondition of the claim. Every node $v \in I^+(\tau_{i-1}) \setminus I^-(\tau_{i-1})$ performs $\ell = 40/\alpha^2$ push transmissions in time interval $\tau_i$. Now fix node $u \in N(I^+(\tau_{i-1}) \setminus I^-(\tau_{i-1}))$ and assume the node has $r$ neighbours in $I^+(\tau_{i-1}) \setminus I^-(\tau_{i-1})$. Then

$$\mathbf{Pr}\left[u \in I^+(\tau_i)\right] = 1 - \left(1 - \frac{40}{\alpha^2 d}\right)^r \geq 1 - \left(1 - \frac{40}{\alpha^2 d}\right) = \frac{40}{\alpha^2 d}.$$

By linearity of expectations we get

$$\mathbf{E}\left[|I^+(\tau_i)|\right] \geq \left(\frac{7}{8} \cdot \alpha \cdot d \cdot |I(t_i)|(1 - o(1))\right) \cdot \left(\frac{40}{\alpha^2 d}\right) \geq \frac{35}{\alpha} \cdot |I(t_i)|(1 - o(1)).$$

For $v \in I^+(\tau_{i-1})$, let $S_v = \{s_v^1, \ldots s_v^\ell\}$ be the random variables determining the choices of $v$, i.e. determining the nodes to which $v$ opens a channel in the interval $\tau_i$, and let $S = \bigcup_{v \in I^+(\tau_{i-1}) \setminus I^-(\tau_{i-1})} S_v$. Note that the choices in $S_v$ and $S_w$ are independent from each other for $v \neq w$. Since every $v \in I^+(\tau_{i-1})$ can only inform at most $40/\alpha^2$ in time interval $\tau_i$, $I^+(\tau_i)$ satisfies the $40/\alpha^2$-Lipschitz condition and the method of independent bounded differences [17] gives

$$\mathbf{Pr}\left[I^+(\tau_i) \leq \mathbf{E}\left[I^+(\tau_i)\right] - \lambda\right] \leq \exp\left(-\frac{\lambda^2}{2|I^+(\tau_i)|(40/\alpha^2)^2}\right).$$

With $\lambda = 27|I(t_i)|/\alpha$ we can conclude that

$$\mathbf{Pr}\left[I^+(\tau_i) \leq \frac{8}{\alpha} \cdot |I(t_i)|\right] \leq \exp\left(-O(\log n)\right) \leq n^{-3},$$

since $|I(t_i)| \geq \rho \cdot \log n$ with a sufficiently large $\rho$. $\qquad \square$

# Limit Theorems for Random MAX-2-XORSAT

Vonjy Rasendrahasina[1] and Vlady Ravelomanana[2,*]

[1] LIPN - UMR CNRS 7030, Université de Paris Nord. 93430 Villetaneuse, France
vonjy-at-lipn.univ-paris13.fr
[2] LIAFA - UMR CNRS 7089, Université Denis Diderot. 75205 Paris Cedex 13, France
vlad-at-liafa.jussieu.fr

**Abstract.** We consider random instances of the MAX-2-XORSAT optimization problem. A 2-XOR formula is a conjunction of Boolean equations (or clauses) of the form $x \oplus y = 0$ or $x \oplus y = 1$. The MAX-2-XORSAT problem asks for the maximum number of clauses which can be satisfied by any assignment of the variables in a 2-XOR formula. In this work, formula of size $m$ on $n$ Boolean variables are chosen uniformly at random from among all $\binom{n(n-1)}{m}$ possible choices. Denote by $X_{n,m}$ the *minimum* number of clauses that can not be satisfied in a formula with $n$ variables and $m$ clauses. We give precise characterizations of the r.v. $X_{n,m}$ around the critical density $\frac{m}{n} \sim \frac{1}{2}$ of random 2-XOR formula. We prove that for random formulas with $m$ clauses $X_{n,m}$ converges to a Poisson r.v. with mean $-\frac{1}{4} \log(1 - 2c) - \frac{c}{2}$ when $m = cn$, $c \in ]0, 1/2[$ constant. If $m = \frac{n}{2} - \frac{\mu}{2} n^{2/3}$, $\mu$ and $n$ are both large but $\mu = o(n^{1/3})$, $\frac{X_{n,m} - \lambda}{\sqrt{\lambda}}$ with $\lambda = \frac{\log n}{12} - \frac{\log \mu}{4}$ is normal. If $m = \frac{n}{2} + O(1)n^{2/3}$, $\frac{X_{n,m} - \frac{\log n}{12}}{\sqrt{\frac{\log n}{12}}}$ is normal. If $m = \frac{n}{2} + \frac{\mu}{2} n^{2/3}$ with $1 \ll \mu = o(n^{1/3})$ then $\frac{12 X_{n,m}}{2\mu^3 + \log n - 3\log(\mu)} \xrightarrow{\text{dist.}} 1$. For any absolute constant $\varepsilon > 0$, if $\mu = \varepsilon n^{1/3}$ then $\frac{8(1+\varepsilon)}{n(\varepsilon^2 - \sigma^2)} X_{n,m} \xrightarrow{\text{dist.}} 1$ where $\sigma \in (0, 1)$ is the solution of $(1 + \varepsilon)e^{-\varepsilon} = (1 - \sigma)e^{\sigma}$. Thus, our findings describe phase transitions in the optimization context similar to those encountered in decision problems.

**Keywords:** MAX XORSAT, Constraint Satisfaction Problem, Phase transition, Random graph, Analytic Combinatorics.

## 1 Introduction

### 1.1 Context and Previous Works

The last decade has seen a growth of interest in phase transition for Boolean satisfiability (SAT) and more generally for Constraint Satisfaction Problems (CSP). For $k \geq 2$, the random version of $k$-SAT is known to exhibit sharp phase transition [19] : for a given number of variables and as the number of clauses in the

---

* This research was partially done while the second author was at LIPN – University of Paris-Nord. The research of the second author is supported by ANR Boole (ANR-09-BLAN6011-04).

A. López-Ortiz (Ed.): LATIN 2010, LNCS 6034, pp. 320–331, 2010.

formula is increased, it shows a phase transition from almost-sure satisfiability to almost-sure unsatisfiability around a critical threshold point. For general CSP, determining the nature of the phase transition (sharp/coarse [8,9]), locating it (see e.g. [1,14,20]), determining a precise scaling window (cf. [5,10]) and understanding the structure of the space of solutions (cf. [2,6]) represent the main tasks. These challenging problems have involved different communities including those from computer science, mathematics and physics (see e.g. [15,18]). As far as we know, the results of Coppersmith, Hajiaghayi, Gamarnik and Sorkin [6] are the closest to ours in the litterature. Given a conjunctive normal form (CNF) formula $F$ on $n$ variables and with $m$ clauses, denote by $\max F$ the maximum number of clauses satisfiable by a single assignment of the variables. By means of method of moments, algorithmic analysis and martingale arguments, the authors of [6] were able to quantify the expectation of $\max F$. In [10], Daudé and Ravelomanana were able to give the precise description of the SAT/UNSAT transition of 2-XORSAT as the density of clauses is increased. If [10] quantifies the SAT/UNSAT probabilities corresponding to the polynomial-time decision problem 2-XORSAT, in the current paper, we deal with the corresponding hard optimization problem MAX-2-XORSAT.

## 1.2   Our Contribution and Organization of the Paper

As mentioned by Coppersmith *et al.* in their paper (see [6, Section 9. Conclusion and open problems]) which studied MAX CUT, MAX $k$-XORSAT is an obvious candidate of these families to study. The XORSAT problem is a variant of SAT and has been introduced in [7]. Since then, it has aroused a lot of interest both as an alternative to the SAT problem [10,14] and as a generalization of MAX CUT.

In this paper, we consider random instances of MAX 2-XORSAT. More precisely, formula of size $m$ on $n$ Boolean variables are chosen uniformly at random from among all $\binom{n(n-1)}{m}$ possible choices. In order to cope with the maximum number of satisfiable clauses, denote by $X_{n,m}$ the *minimum* number of clauses that can not be satisfied in a formula with $n$ variables and $m$ clauses. We then provide precise results about the random variable $X_{n,m}$ as $m \equiv m(n)$ increases. Since the maximum number of satisfiable clauses is $m - X_{n,m}$, the characterization of the r.v. $X_{n,m}$ is obviously crucial. We will consider formula in different regions according to the density of their underlying graph, the *sub-critical* region whenever $n - 2m \gg n^{2/3}$, the *critical* region when $n - 2m = \pm O(n^{2/3})$ and the *super-critical* region when $2m - n \gg n^{2/3}$.

Using enumerative and analytic combinatorics [17], we obtain our results condensed in the following:

**Theorem 1.** *Let* $X_{n,m}$ *be the minimum number of unsatisfiable clauses in a random 2-XOR formula with $n$ variables and $m$ clauses. Then, $X_{n,m}$ satisfies:*

- *For* $m = cn$ *s.t. $c$ constant and* $0 < c < \frac{1}{2}$,

$$X_{n,m} \xrightarrow{\text{dist.}} \text{Poisson}\left(-\frac{1}{4}\log(1-2c) - \frac{c}{2}\right). \tag{1}$$

• *For $m = \frac{n}{2}(1 - \mu n^{-1/3})$ where $1 \ll \mu = o(n^{1/3})$,*

$$\frac{X_{n,m} - \frac{\log n}{12} + \frac{\log \mu}{4}}{\sqrt{\frac{\log n}{12} - \frac{\log \mu}{4}}} \xrightarrow{\text{dist.}} \mathcal{N}(0,1). \tag{2}$$

• *For $m = \frac{n}{2} + O(1)n^{2/3}$,*

$$\frac{X_{n,m} - \frac{\log n}{12}}{\sqrt{\frac{\log n}{12}}} \xrightarrow{\text{dist.}} \mathcal{N}(0,1). \tag{3}$$

• *For $m = \frac{n}{2}(1 + \mu n^{-1/3})$ where $1 \ll \mu = o(n^{1/3})$,*

$$\frac{12\,X_{n,m}}{2\mu^3 + \log n - 3\log(\mu)} \xrightarrow{\text{dist.}} 1. \tag{4}$$

• *For $m = \frac{n}{2}(1 + \varepsilon)$,*

$$\frac{8(1 + \varepsilon)}{n(\varepsilon^2 - \sigma^2)} X_{n,m} \xrightarrow{\text{dist.}} 1, \tag{5}$$

*where $\varepsilon > 0$ is any positive absolute constant and $\sigma$ is the unique solution of $(1 + \varepsilon)e^{-\varepsilon} = (1 - \sigma)e^{\sigma}$ with $\sigma \in (0,1)$.*

Theorem 1 shows that just as random instances of the decision problem 2-XORSAT exhibit a phase transition from almost-sure satisfiability to almost-sure unsatisfiability (around the critical density $\frac{m}{n} = \frac{1}{2}$), the corresponding maximization problem undergoes a transition at the same point. The expectation of the minimum number of unsatisfiable clauses rises quickly from $\Theta(1)$ to $\Theta(n)$. Though our results draw on the structure of the components of the classical random graph $G(n, m)$ (e.g. [4,25]), we note that the MAX-2-XORSAT problem is an NP-complete problem. Thus, obtaining any combinatorial description of the quantity of interest is extremely difficult. Consequently, the results described by the theorem above are not trivial, especially in the super-critical region. In fact, if (1), (2) and (3) can be viewed as 'subsequent' results from random graph theory, to obtain (4) and (5) many combinatorial arguments are needed as shown in Section 3. We should point out that by (1), (2), (3), (4) and (5), $X_{n,m}$ evolves in a 'continuous' fashion. Note also that in a previous work and using different methods, Coppersmith, Hajiaghayi, Gamarnik and Sorkin [6] obtained the *expectation* of the *maximum* number of satisfiable clauses in a CNF formula. In our paper, we restrict our attention on the MAX-2-XORSAT problem, but we give more precise results since Theorem 1 offers the *limit distributions* of the *minimum* number of unsatisfiable clauses in 2-XORSAT formula with various densities. Note that due the lack of precise enumerative results, we are enable to obtain non-degenerate limit laws for the cases (4) and (5).

This paper is organized as follows. Section 2 gives some enumerative background for our investigation. It also offers the proofs of (1) and of (2). The proof of (3) is omitted due to lack of space. In Section 3, we will enumerate connected cactus graphs or Husimi trees (see e.g. [22,23]) according to their number of vertices and edges. Then, using these enumerative results together with combinatorial arguments and analytical tools, we will prove (4) and (5).

## 2 The Sub-critical and Critical Regions

In order to study random MAX-2-XORSAT, we will use enumerative/analytic methods by means of generating functions [22,17]. The correspondence between satisfiable formula and edge-weighted 0/1 graphs (or shortly weighted graphs) has been shown in [10]. In the following, we will recall briefly that a formula is satisfiable if and only if its associated weighted graphs has no cycle of odd weight.

### 2.1  2-XORSAT, Weighted Graphs and MAX-2-XORSAT

A 2-*XOR-clause* or shortly a *clause* is a linear equation over the finite field $GF(2)$ using exactly 2 variables, $C = ((x_1 \oplus x_2) = \varepsilon)$ where $\varepsilon = 0$ or 1. A 2-*XOR-formula* or simply a *formula* is a conjunction of distinct 2-XOR-clauses.

A *truth assignment* $I$ is a mapping that assigns 0 or 1 to each variable in its domain. It satisfies a 2-XOR-clause $C = ((x_1 \oplus x_2) = \varepsilon)$ iff $(I(x_1) + I(x_2))$ mod $2 = \varepsilon$. A truth assignment satisfies a formula $s$ iff it satisfies every clause in $s$. The set of 2-XOR-formula that are satisfiable (resp., unsatisfiable) will be denoted shortly SAT (resp. UNSAT). Observe that every formula that contains an unsatisfiable sub-formula is itself unsatisfiable. In the terminology of random graph theory, SAT (resp. UNSAT) is a decreasing (resp. increasing) property.

Throughout this paper we reserve $n$ for the number of variables ($\{x_1, \ldots, x_n\}$ denotes the set of variables). There are $N = 2\binom{n}{2} = n(n-1)$ different 2-XOR-clauses over $n$ variables. We consider random formula obtained by choosing uniformly, independently and without replacement $m$ clauses from the $N$ possible 2-clauses. This defines a discrete probability space $(\Omega(n,m), P_{n,m})$ whose associated probability is the uniform law : $\forall s \in \Omega(n,m)$   $P_{n,m}(s) = \binom{n(n-1)}{m}^{-1}$. In [10], the authors gave very precise estimates of the probability that a formula drawn at random from $\Omega(n,m)$ is SAT, that is in estimating

$$p(n,m) := P_{n,m}\Big(2XORSAT\Big).$$

Each formula $s$ in $\Omega(n,m)$ can be represented by a weighted graph $G(s)$ with $n$ vertices (one for each variable) and $m$ weighted (0 or 1) edges. For each equation $x_i \oplus x_j = \varepsilon$, we add the edge $\{x_i, x_j\}$ weighted by $\varepsilon$. The *weight* of a graph is the sum of the weights of its edges. Observe that our weighted graphs are without self-loops and that the underlying graph of a satisfiable formula is always simple (without multiple edges). We have, as noticed in [8,10]:

$$p(n,m) = P_{n,m}\Big(G(s) \text{ has no cycle of odd weight}\Big).$$

We are interested in the *maximum* number of satisfiable clauses or equivalently the *minimum* number of unsatisfiable clauses. According to the correspondence between random 2-XORSAT formula and its underlying weighted graphs, the random variable $X_{n,m}$ defined in Theorem 1 can be (simply) described as follows:

**Proposition 1.** *Given a formula s and its graph $G(s)$, $X_{n,m}$ is the minimum number of edges to suppress in $G(s)$ in order to obtain a weighted graph without cycle of odd weight.*

## 2.2   Exponential Generating Functions

As shown in [16] and [24], exponential generating functions (EGFs) can lead to stringent results about the main characteristics of random graphs when they apply. In this paragraph, let us recall briefly the main EGFs involved in our proofs. In particular, we refer the reader to Harary and Palmer [22] for EGFs related to graphical enumeration.

The number of (unweighted) trees on $n$ labelled vertices is given by Cayley's formula $n^{n-2}$ and its EGF is given by

$$W_{-1}(z) = t(z) - \frac{t(z)^2}{2} \text{ where } t(z) = ze^{t(z)} = \sum_{n=1}^{\infty} n^{n-1}\frac{z^n}{n!}. \tag{6}$$

The EGF of unicyclic components (connected components with $n$ vertices and $n$ edges) is given by

$$W_0(z) = -\frac{1}{2}\log{(1 - t(z))} - \frac{t(z)}{2} - \frac{t(z)^2}{4}. \tag{7}$$

As in [24], the *excess* of a connected component is the number of its edges minus the number of its vertices. Thus, the excess of the trees enumerated by (6) is $-1$, the excess of unicyclic components (7) is $0$ and so on. For any $\ell \geq 1$, Wright [29] gave a method to compute all the EGFs of components of excess $\ell$. As in Janson, Knuth, Łuczak and Pittel [24], connected components of excess $> 0$ are called *complex*. An $\ell$-component is a connected component with $\ell$ edges more than vertices, i.e. it has excess $\ell$.

For our purpose, let us decompose the random variable $X_{n,m}$ into two parts: Given a random 2-XOR-formula $s$ and its underlying weighted graph $G(s)$, let $Y_{n,m}$ (resp. $Z_{n,m}$) be the minimum number of edges to be removed from $G(s)$ in order to suppress all odd weighted unicyclic (resp. complex) components. By definition, we have $X_{n,m} = Y_{n,m} + Z_{n,m}$.

## 2.3   Proof of (1) and (2)

In [10, Theorem 3.2], it is shown that the probability that a random unweighted graph built with $n$ vertices and $m$ edges has no complex components is

$$1 - O\left(\frac{n^2}{(n-2m)^3}\right), \tag{8}$$

as $n$ and $m$ tend to infinity but $n - 2m \gg n^{2/3}$. Consequently, in these ranges[1] $Z_{n,m} = O_p\left(\frac{n^2}{(n-2m)^3}\right)$. Thus, (2) is proved by means of the Theorem 2 below and the dominated convergence theorem.

**Theorem 2.** *Let $m = \frac{n}{2}(1 - \mu n^{-1/3})$. If $n$ and $\mu$ tend to infinity but $\mu = o(n^{1/3})$ then the following holds:*
*(a) Define*

$$\beta(n) = \frac{1}{12}\log(n) - \frac{1}{4}\log(\mu) - \frac{1}{4} + \frac{1}{4}\mu n^{-1/3} = -\frac{1}{4}\log\left(1 - \frac{2m}{n}\right) - \frac{m}{2n}. \quad (9)$$

*For all $R = O(\beta(n))$, we have*

$$\mathbb{P}\left(Y_{n,m} = R\right) = e^{-\beta(n)}\frac{\beta(n)^R}{R!}\left(1 + O\left(\frac{1}{\mu^3}\right)\right). \quad (10)$$

*(b) There exists $R_0$ and constants $C, \varepsilon > 0$, such that $\forall R > R_0$*

$$\mathbb{P}\left(Y_{m,n} = R\right) \leq Ce^{-\varepsilon R}. \quad (11)$$

*Proof.* Let $C_{s,\ell}$ be the generating functions of connected weighted graphs of excess $\ell$ such that exactly $s$ edges have to be suppressed to obtain graphs without odd weight cycles. Using symbolic methods of EGFs [17], we easily obtain

$$C_{0,-1}(z) = T(z) - T(z)^2, \quad C_{1,0}(z) = \frac{1}{4}\log\left(\frac{1}{1 - 2T(z)}\right) - \frac{T(z)}{2} \quad \text{and}$$

$$C_{0,0}(z) = \frac{1}{4}\log\left(\frac{1}{1 - 2T(z)}\right) - \frac{T(z)}{2} - \frac{T(z)^2}{2},$$

where $T(z) = z\exp(2T(z)) = \sum_{n \geq 1}(2n)^{n-1}z^n/n!$ is the EGF for weighted rooted trees. Therefore, the probability that a random weighted graph contains only trees, unicyclic components of even weight and $R$ unicyclic components of odd weight is[2]

$$p_R(n,m) = \frac{n!}{\binom{n(n-1)}{m}}[z^n]\frac{\left(T(z) - T(z)^2\right)^{n-m}}{(n-m)!}\exp(C_{0,0}(z))\frac{C_{1,0}(z)^R}{R!}. \quad (12)$$

Substituting $u = 2T(z)$, we get

$$p_R(n,m) = \frac{1}{\binom{n(n-1)}{m}}\frac{n!}{(n-m)!}\frac{4^m}{2^n}\frac{1}{R!}\frac{1}{2\pi i}\oint g_1(u)g_{2,R}(u)e^{nh(u)}\frac{du}{u}, \quad (13)$$

---

[1] As in Janson, Łuczak and Ruciński [25, p. 10], if $X_n$ are r. v. and $a_n$ are real numbers we write $X_n = O_p(a_n)$ as $n \to \infty$ if for every $\delta > 0$ there exists constants $C_\delta$ and $n_0$ s.t. $\mathbb{P}(|X_n| \leq C_\delta a_n) > 1 - \delta$ for every $n \geq n_0$.
[2] If $A(z) = \sum_n a_n z^n$ is a power series, $[z^n]A(z) = a_n$.

with

$$g_1(u) = (1 - u)^{3/4} \exp(-u/4 - u^2/8) ,$$
$$g_{2,R}(u) = (-\log(1 - u) - u)^R /4^R ,$$
$$h(u) = u - \log(u) + (1 - m/n) \log(2u - u^2) .$$

We have $h'(1) = 0$ , $h'(2m/n) = 0$ and $h''(2m/n) = \frac{1}{4} \frac{(-n+2\,m)n}{m(-n+m)} > 0$. Applying the saddle-point method for (13) by choosing a path of integration $\{u/|u| = 2m/n\}$, using the probability that the random graph has no complex components, viz. (8), after some algebra we get (10). The part **(b)** of the theorem is proved by bounding (13).

**Corollary 1.** *If* $n$, $m = \frac{n}{2}(1 - \mu n^{-1/3}) \to \infty$ *such that* $1 \ll \mu = o(n^{1/3})$ *then for* $x \in \mathbb{R}$ *we have*

$$\mathbb{P}\left(\frac{Y_{n,m} - \beta(n)}{\sqrt{\beta(n)}} \leq y\right) \to \frac{1}{\sqrt{2\pi}} \int_{-\infty}^{y} e^{-x^2/2} dx , \tag{14}$$

*where* $\beta(n)$ *is given by (9).*

Using similar methods, we obtain the following result for values of $m$ s.t. $\lim \frac{m}{n} < \frac{1}{2}$:

**Theorem 3.** *If* $m = cn$ *with* $\limsup_{n \to \infty} c < 1/2$ *then for* $R \in \mathbb{N}$

$$\mathbb{P}(Y_{n,m} = R) = e^{-\gamma(c)} \frac{\gamma(c)^R}{R!} \left(1 + O\left(\frac{1}{n}\right)\right) , \tag{15}$$

*where*

$$\gamma(c) = -\frac{1}{4} \log(1 - 2c) - \frac{c}{2} . \tag{16}$$

**Remark.** Due to the page limitation, the proof of (3) is omitted in this extended abstract.

## 3    The Super-Critical MAX-2-XORSAT

As the number of clauses of $s$ reaches $m = \frac{n}{2} + \mu n^{2/3}$ with $\mu \equiv \mu(n) \gg 1$, almost surely there is a unique giant component in $G(s)$ of excess $O(\mu^3)$ [24,27]. Hence, in contrary to the previous sections, we need to take into account the odd weighted cycles in this unique complex component. Recall that $C_{s,\ell}$ is the EGF of connected weighted graphs of excess $\ell$ such that exactly $s$ edges have to be suppressed to obtain graphs without odd weight cycles. The power series $C_{0,\ell}$, $\ell \geq -1$, have been computed in [10] and we have:

$$C_{0,\ell}(z) = \frac{W_\ell(2z)}{2} , \quad \forall \ell \geq -1 , \tag{17}$$

where $W_\ell$ is the Wright's EGFs of connected unweighted graphs of excess $\ell$ (see [29]). We find $C_{\ell+1,\ell}$ (see paragraph 3.2 below) but it turns out that finding the EGF $C_{s,\ell}$ for general $\ell$ and $s \geq 1$ is an extremely difficult enumeration problem (probably the difficulty is intrinsic to the NP-completeness of MAX-2-XORSAT). Nevertheless, we succeed to find (4) and (5) of Theorem 1 and the main ideas of their proofs relies on the following:

- First, we will find an upper-bound of $C_{s,\ell}$. This bound implies that with high probability, the minimum number of edges to delete in $C_{s,\ell}$ (in order to suppress all cycles of odd weight) is *at least* $\frac{\ell}{4} - o(\ell)$.
- Next, we count connected weighted *cactus graphs*. Cactus graphs are defined as graphs where every edge belongs to at most one cycle[3]. By using the EGF of connected cactus graphs, we will show that in connected components of excess $\ell$, with high probability the number of edges to suppress is *at most* $\frac{\ell}{4} + o(\ell)$.

## 3.1   Upper-Bound of $C_{s,\ell}$ and Lower-Bound of the Number of Edges to Delete

In this paragraph, we will prove that the EGFs $C_{s,\ell}$ satisfy the following:

**Lemma 1**

$$\forall s \geq 0, \qquad C_{s,\ell}(z) \preceq \sum_{i=s}^{2s} \binom{\ell+1}{i} C_{0,\ell}(z) + B_{s,\ell}(z). \qquad (18)$$

*with $B_{s,\ell}(z)$ is the EGF of all $\ell$-components with multiedges.*

*Proof.* Given a connected graph $G$ of excess $\ell$, it can be associated to its $\ell + 1$ fundamental cycles. In fact, let $T$ be a spanning tree of $G$, and $a$ be any of the $(\ell + 1)$ edges in $G \setminus T$. The graph $T \cup a$ has a unique cycle $C_a$ and in graph terminology (see for instance [13]) the $(\ell + 1)$ cycles $C_a$ form a basis of the cycle space associated to $G$. An edge can be shared by many fundamental cycles. Suppose that initially $G$ has no cycles of odd weight. We can modify this graph by flipping the weight of an edge $e$ in a path. The fact of choosing $e$ and changing its weight is similar to the one of changing the weights of all the cycles containing this path. This is equivalent to the choice of at most 2 of the fundamental cycles in order to modify their weights. Recursively, the choice of $s$ edges in $s$ different paths in order to change their weights is similar to the choice of at least $s$ and at most $2s$ cycles among the $\ell + 1$ fundamental cycles. The components containing multiple edges are not taken into account by this construction. Hence, we have the additional term $B_{s,\ell}(z)$ denoting the EGF of these components.

As a consequence, using the concentration of the binomial coefficients and since $[z^n] B_{s,\ell} \ll [z^n] C_{0,\ell}$, it yields:

**Lemma 2.** *Let $Z_\ell$ be the number of edges to remove from an $\ell$-component in order to obtain a component without odd cycles. Define $c(\ell)$ as a function of*

---

[3] Also known as Husimi trees or cactus trees.

$\ell$ such that $\sqrt{\log(\ell+1)} \ll c(\ell) \ll (\ell+1)^{1/6}$ for large values of $\ell$. For any $s \le \frac{\ell+1}{4} - c(\ell)\sqrt{\ell+1}$ and large $\ell$, we have

$$\mathbb{P}\left(Z_\ell = s\right) \le e^{-4c(\ell)^2 + \frac{1}{2}\log(\ell+1)}. \tag{19}$$

## 3.2   Enumeration of Spanning Cactus Graphs and Probabilistic Upper-Bound of the Excess of Complex Components

Let $\Xi_s$ be the EGFs of *unweighted* connected cactus with $s$ cycles. Clearly, $C_{s,s-1}(z) = \Xi_s(z)$ and $\Xi_s(z)$ can be obtained via standard enumerative tools. Thus, using analytic combinatorics it can be proved that the EGF $\Xi_s(z)$ 'behaves' as $\frac{\xi_s}{(1-t(z))^{3s-3}}$ with

$$\xi_s = \frac{1}{6}\left(\frac{3}{2}\right)^{s-1} \frac{3^{s/2}}{\sqrt{2\pi s^3}(s-1)}\left(1 + O\left(\frac{1}{s}\right)\right). \tag{20}$$

Now, we are ready to show that almost surely a cactus graph $\Lambda$ of excess $\frac{\ell}{2} + o(\ell)$ can not span[4] a connected component of excess $\ell$. In fact, we have the following lemmas:

**Lemma 3.** *Let $\Upsilon_{s,\ell}$ be the EGFs of connected graphs of excess $\ell$ obtained by adding $\ell - s$ edges to cacti of $\Xi_{s+1}$ (of excess $s$), then as both $\ell$ and $s$ are large*

$$\Upsilon_{s,\ell}(z) \preceq \xi_{s+1}\left(\frac{9}{2}\right)^{\ell-s} \frac{\Gamma(\ell)\,\Gamma\left(\ell+\frac{2}{3}\right)}{\Gamma(s)\,\Gamma\left(s+\frac{2}{3}\right)} \frac{1}{(1-t(z))^{3\ell}}. \tag{21}$$

*Proof.* The proof makes use of combinatorial operators corresponding to the marking of an edge and then unselecting it (see e.g. [21,3,17]). Recall that $\Upsilon_{s,\ell}$ is the EGF of connected graphs of excess $\ell$ obtained by adding $\ell - s$ edges to cacti of $\Xi_{s+1}$. We repeatedly proceed $\ell - s$ times the two following combinatorial operations:

(OP$_1$). We add an edge to $\Xi_{s+1}$. To do so, we have to choose 2 distinguished and not connected vertices. For a graph of size $n$, the number of such choices is less than $n!\binom{n}{2}[z^n]\Xi_{s+1}(z)$. Thus, the combinatorial operator is $\frac{z^2}{2!}\frac{\partial^2 z}{\partial z^2}$ applied on $\Xi_{s+1}(z)$.

(OP$_2$). The edge added in (OP$_1$) is marked. We have to unmark this edge. We know that the last added edge belongs to a smooth graph (obtained by the pruning, i.e., reducing recursively all vertices of degree 1 from the original graph of $\Xi_{s+1}$). Hence, the operators (OP$_1$) and (OP$_2$) applied on graphs of $\Xi_s$ leads to an upper-bound of the form : $(1 - t(z))\frac{z^2}{2!}\frac{\partial^2 z}{\partial z^2}\Xi_{s+1}(z)$.

**Lemma 4.** *For any fixed real number $x$, the probability that a cactus from $\Xi_s = \frac{\ell}{2} + \left(\frac{3}{8}\log 3 + \frac{1}{2}\log 2 + x\right)\frac{\ell}{\log \ell}$ spans an $\ell$-component is at most $e^{-2x\ell}$.*

---

[4] A spanning cactus of a graph $G$ is subgraph of $G$ that is a cactus and connects all of the vertices of $G$.

*Proof.* (Sketch.) The proof involves the enumerative results from paragraph 3.2 and Lemma 3.

**Corollary 2.** *Let* $s_{\max} \equiv s_{\max}(\ell)$ *be the r.v. denoting the maximum excess of a cactus spanning an $\ell$-component, then for large values of $\ell$*

$$\forall x > 0, \quad \mathbb{P}\left[s_{\max} \leq \frac{\ell}{2} + \left(\frac{3}{8}\log 3 + \frac{1}{2}\log 2 + x\right)\frac{\ell}{\log \ell}\right] \geq 1 - e^{-2x\ell}. \quad (22)$$

### 3.3  Analytic Results for the Super-Critical Region

**Lemma 5.** *If $s$ edges have to be removed in an $\ell$-component in order to suppress all cycles of odd weight then the component has at most $s$ fundamental odd weighted and pairwise disjoint[5] cycles.*

*Proof.* Recall that an $\ell$-component contains no odd weighted cycle if and only if it has no fundamental cycle of odd weight [10]. If we need to remove $s$ edges, it has obviously at least $s$ fundamental cycles of odd weight. Only $s$ of these cycles are pairwise disjoint. If at least $s + 1$ of these cycles are pairwise disjoint, they cannot be eliminated by removing at most $s$ edges.

**Lemma 6.** *Let $Z_\ell$ the r.v. given in Lemma 2. For any real positive $x$, if $\ell \to \infty$ we have*

$$\mathbb{P}\left(\frac{\ell}{4} - c(\ell)\sqrt{\ell} \leq Z_\ell \leq \frac{\ell}{4} + (c_0 + x)\frac{\ell}{2\log(\ell)}\right) \geq 1 - e^{-2x\ell} - e^{-2c(\ell)^2 + \frac{1}{4}\log(\ell)},$$

*where $c(\ell)$ is any function satisfying $\log \ell \ll c(\ell)^2 \ll \ell^{1/3}$ and $c_0 = \frac{3}{8}\log 3 + \frac{1}{2}\log 2$.*

*Proof.* By means of lemma 2, corollary 2 and lemma 5.

Finally as $\mu$ is large but $m = O(n)$, $Y_{n,m}$ is at most $O_p(\log n)$ and for the r.v $Z_{n,m}$, we have the following result:

**Theorem 4.** *Let $m = \frac{n}{2} + \mu\frac{n^{2/3}}{2}$. If $\mu = O(n^{1/3})$ and tend to infinity, we have*

$$\mathbb{P}\left(Z_{n,m} \sim \frac{\ell}{4}\right) \geq 1 - e^{-O(\ell)} - e^{-2c(\ell)^2 + \frac{1}{4}\log(\ell)}, \quad (23)$$

*where $\ell \sim \frac{2}{3}\frac{(2m-n)^3}{n^2}$ if $\mu = o(n^{1/3})$ and $\ell \sim \frac{\varepsilon^2 - \sigma^2}{2(1+\varepsilon)}n$ if $\mu = \varepsilon n^{1/3}$. $\sigma \in (0,1)$ is the unique solution of $(1+\varepsilon)e^{-\varepsilon} = (1-\sigma)e^\sigma$ and $c(\ell)$ is any function satisfying $\log \ell \ll c(\ell)^2 \ll \ell^{1/3}$.*

*Proof.* Equation (23) follows from Lemma 6. We know from the theory of random graphs [27] that the excess of the (almost surely) unique giant component of $\mathbb{G}(n, m = \frac{n}{2} + \mu\frac{n^{2/3}}{2})$ is Gaussian (throughout the whole supercritical phase, viz. from $\mu = o(n^{1/3})$ to $\mu = \varepsilon n^{1/3}$). For $\mu = o(n^{1/3})$, the expected excess of the giant component is $\sim \frac{2}{3}\mu^3$ whereas if $\mu = \varepsilon n^{1/3}$ the excess of the giant component is centered at $\frac{\varepsilon^2 - \sigma^2}{2(1+\varepsilon)}n$.

---

[5] We say two cycles are disjoints if and only if they have no common edge.

# 4   Conclusion

We have presented a work on random MAX-2-XORSAT based on enumerative/analytic approaches. Our results establish that as the random formula are enriched with more clauses, the number of satisfiable clauses diminishes dramatically around the critical density $\frac{1}{2}$. In particular, we have shown how one can quantify the *minimum* number of unsatisfiable clauses for random instances of the MAX-2-XORSAT problem when the ratio number of clauses over number of variables is about $\frac{1}{2}$. For the supercritical cases, our results are obtained by counting cactus spanning subgraphs. These techniques that allow us to compute the most probable excess of these specific subgraphs are of independent interest and can be generalized in order to study other instances of MAX-2-CSP problems.

# References

1. Achlioptas, D., Moore, C.: Random $k$-SAT: Two moments suffice to cross a sharp threshold. SIAM Journal of Computing 36(3), 740–762 (2006)
2. Achlioptas, D., Naor, A., Peres, Y.: On the maximum satisfiability of random formulas. Journal of the ACM 54(2) (2007)
3. Bergeron, F., Labelle, G., Leroux, P.: Combinatorial Species and Tree-like Structures. Cambridge University Press, Cambridge (1997)
4. Bollobás, B.: Random Graphs. Cambridge Studies in Advanced Mathematics (1985)
5. Bollobás, B., Borgs, C., Chayes, J.T., Kim, J.H., Wilson, D.B.: The scaling window of the 2-SAT transition. Random Structures and Algorithms 18, 201–256 (2001)
6. Coppersmith, D., Hajiaghayi, M.T., Gamarnik, D., Sorkin, G.B.: Random MAX-SAT, random MAX-CUT, and their phase transitions. Random Structures and Algorithms 24(4), 502–545 (2004)
7. Creignou, N., Daudé, H.: Satisfiability threshold for random XOR-CNF formula. Discrete Applied Mathematics 96-97(1-3), 41–53 (1999)
8. Creignou, N., Daudé, H.: Coarse and sharp thresholds for random k-XOR-CNF satisfiability. Theoretical Informatics and Applications 37(2), 127–147 (2003)
9. Creignou, N., Daudé, H.: Coarse and sharp transitions for random generalized satisfiability problems. In: Proc. of the third Colloquium on Mathematics and Computer Science, pp. 507–516. Birkhäuser, Basel (2004)
10. Daudé, H., Ravelomanana, V.: Random 2-XORSAT phase transition. Algorithmica (2009); In: Laber, E.S., Bornstein, C., Nogueira, L.T., Faria, L. (eds.) LATIN 2008. LNCS, vol. 4957, pp. 12–23. Springer, Heidelberg (2008)
11. de Bruin, N.G.: Asymptotic Methods in Analysis. Dover, New York (1981)
12. DeLaurentis, J.: Appearance of complex components in a random bigraph. Random Structures and Algorithms 7(4), 311–335 (1995)
13. Diestel, R.: Graph Theory. Springer, Heidelberg (2000)
14. Dubois, O., Mandler, J.: The 3-XOR-SAT threshold. In: Proceedings of the 43th Annual IEEE Symposium on Foundations of Computer Science, pp. 769–778 (2002)
15. Dubois, O., Monasson, R., Selman, B., Zecchina, R.: Phase transitions in combinatorial problems. Theoretical Computer Science 265(1-2) (2001)

16. Flajolet, P., Knuth, D.E., Pittel, B.: The first cycles in an evolving graph. Discrete Mathematics 75(1-3), 167–215 (1989)
17. Flajolet, P., Sedgewick, R.: Analytic Combinatorics. Cambridge University Press, Cambridge (2009)
18. Franz, S., Leone, M.: Replica bounds for optimization problems and diluted spin systems. Journal of Statistical Physics 111, 535–564 (2003)
19. Friedgut, E.: Sharp thresholds of graph properties, and the k-SAT problem. appendix by J. Bourgain. Journal of the A.M.S. 12(4), 1017–1054 (1999)
20. Goerdt, A.: A sharp threshold for unsatisfiability. Journal of Computer and System Sciences 53(3), 469–486 (1996)
21. Goulden, I.P., Jackson, D.M.: Combinatorial enumeration. John Wiley and Sons, Chichester (1983)
22. Harary, F., Palmer, E.: Graphical enumeration. Academic Press, New-York (1973)
23. Harary, F., Uhlenbeck, G.: On the number of Husimi trees, I. Proceedings of the National Academy of Sciences 39, 315–322 (1953)
24. Janson, S., Knuth, D.E., Łuczak, T., Pittel, B.: The birth of the giant component. Random Structures and Algorithms 4(3), 233–358 (1993)
25. Janson, S., Luczak, T., Ruciński, A.: Random Graphs. Wiley-Interscience, Hoboken (2000)
26. Kolchin, V.F.: Random graphs. Cambridge University Press, Cambridge (1999)
27. Pittel, B., Wormald, N.C.: Counting connected graphs inside-out. Journal of Comb. Theory, Ser. B 93(2), 127–172 (2005)
28. Ravelomanana, V.: Another proof of Wright's inequalities. Information Processing Letters 104(1), 36–39 (2007)
29. Wright, E.M.: The number of connected sparsely edged graphs. Journal of Graph Theory 1, 317–330 (1977)
30. Wright, E.M.: The number of connected sparsely edged graphs III: Asymptotic results. Journal of Graph Theory 4(4), 393–407 (1980)

# On Quadratic Threshold CSPs

Per Austrin[1], Siavosh Benabbas[2], and Avner Magen[2]

[1] Courant Institute of Mathematical Sciences
New York University
austrin@cims.nyu.edu
[2] Department of Computer Science
University of Toronto
{siavosh,avner}@cs.toronto.edu

**Abstract.** A predicate $P : \{-1,1\}^k \to \{0,1\}$ can be associated with a constraint satisfaction problem MAX CSP($P$). $P$ is called "approximation resistant" if MAX CSP($P$) cannot be approximated better than the approximation obtained by choosing a random assignment, and "approximable" otherwise. This classification of predicates has proved to be an important and challenging open problem. Motivated by a recent result of Austrin and Mossel (Computational Complexity, 2009), we consider a natural subclass of predicates defined by signs of quadratic polynomials, including the special case of predicates defined by signs of linear forms, and supply algorithms to approximate them as follows.

In the quadratic case we prove that every *symmetric* predicate is approximable. We introduce a new rounding algorithm for the standard semidefinite programming relaxation of MAX CSP($P$) for any predicate $P : \{-1,1\}^k \to \{0,1\}$ and analyze its approximation ratio. Our rounding scheme operates by first manipulating the optimal SDP solution so that all the vectors are nearly perpendicular and then applying a form of hyperplane rounding to obtain an integral solution. The advantage of this method is that we are able to analyze the behaviour of a set of $k$ rounded variables together as opposed to just a pair of rounded variables in most previous methods.

In the linear case we prove that a predicate called "Monarchy" is approximable. This predicate is not amenable to our algorithm for the quadratic case, nor to other LP/SDP-based approaches we are aware of.

## 1   Introduction

This paper studies the approximability of *constraint satisfaction problems* (CSPs). Given a predicate $P : \{-1,1\}^k \to \{0,1\}$, the MAX CSP($P$) problem is defined as follows. An instance is given by a list of $k$-tuples (clauses) of literals over some set of variables $x_1, \ldots, x_n$, where a literal is either a variable or its negation. A clause is satisfied by an assignment to the variables if $P$ is satisfied when applied to the $k$ literals of the clause. The goal is then to find an assignment to the variables that maximizes the number of satisfied clauses. Our specific interest is predicates of the form $P(x) = \frac{1+\text{sign}(Q(x))}{2}$ where $Q : \mathbb{R}^k \to \mathbb{R}$ is a quadratic

A. López-Ortiz (Ed.): LATIN 2010, LNCS 6034, pp. 332–343, 2010.

polynomial with no constant term, i.e., $Q(x) = \sum_{i=1}^{k} a_i x_i + \sum_{i \neq j} b_{ij} x_i x_j$ for some set of coefficients $a_1, \ldots, a_n$ and $b_{11}, \ldots, b_{nn}$. While this special case is arguably very rich and interesting in its own right, we give some further motivations below. But first, we give some background to the study of MAX CSP($P$) problems in general.

A canonical example of a MAX CSP($P$) problem is when $P(x_1, x_2, x_3) = x_1 \vee x_2 \vee x_3$ is a disjunction of three variables, in which case MAX CSP($P$) is the classic MAX 3-SAT problem. Another well-known example is the MAX 2-LIN(2) problem in which $P(x_1, x_2) = x_1 \oplus x_2$. As MAX CSP($P$) is NP-hard for almost all choices of $P$ (the only case for which it is not NP-hard is when $P$ depends on at most 1 variable), much effort has been put into understanding the best possible approximation ratio achievable in polynomial time. A (randomized) algorithm is said to have approximation ratio $\alpha \leq 1$ if, given an instance with optimal value Opt, it produces an assignment with (expected) value at least $\alpha \cdot$ Opt.

The arguably simplest approximation algorithm is to pick a uniformly random assignment. As this algorithm satisfies each constraint with probability $\frac{|P^{-1}(1)|}{2^k}$ it follows that it gives an approximation ratio of $\frac{|P^{-1}(1)|}{2^k}$. In their classic paper [5], Goemans and Williamson used semidefinite programming to obtain improved approximation algorithms for predicates on two variables. For instance, for MAX 2-LIN(2) they gave an algorithm with approximation ratio $\alpha_{GW} \approx 0.878$. Following [5], many new approximation algorithms were found for various specific predicates, improving upon the random assignment algorithm. However, for some cases, perhaps most prominently the MAX 3-SAT problem, no such progress was made. Then, in another classic paper [7], Håstad proved that MAX 3-SAT is in fact NP-hard to approximate within $7/8 + \epsilon$, showing that a random assignment in fact gives the best possible worst-case approximation that can be obtained in polynomial time.

Predicates which exhibit this behavior are called *approximation resistant*. One of the main open questions along this line of research is to characterize which predicates admit a non-trivial approximation algorithm, and which predicates are approximation resistant. For predicates on three variables, the work of Håstad together with work of Zwick [13] shows that a predicate is resistant iff it is implied by an XOR of the three variables, or the negation thereof, where a predicate $P$ is said to imply a predicate $P'$ if $P(x) = 1 \Rightarrow P'(x) = 1$. For four variables, Hast [6] made an extensive classification leaving open the status of 46 different predicates.

There have been several papers [11,3,12], mainly motivated by the soundness-query trade off for PCPs, giving increasingly general conditions under which predicates are approximation resistant. In a recent paper [2], the first author and Mossel proved that, if there exists an unbiased pairwise independent distribution on $\{-1, 1\}^k$ whose support is contained in $P^{-1}(1)$, then $P$ is approximation resistant under the Unique Games Conjecture [10]. This condition is very general and turned out to give many new cases of resistant predicates [1]. A related result by Georgiou et al [4] that is independent of complexity assumptions, shows that under the same condition on $P$, the so-called Sherali-Admas hierarchy—which is in some sense the strongest version of the Linear Programming approach—does not beat a random assignment. Indeed, when it comes to algorithms, there are very few systematic results that give algorithms for large classes of predicates. One such result can be found in [6]. Given the result of [2], such systematic results can only work for predicates that do not support pairwise independence. A very natural subclass of these predicates are those of the form $\frac{1+\text{sign}(Q)}{2}$ for $Q$ a quadratic polynomial as described above. To be more precise, the following fact from [1] is our main motivation for studying this type of predicates.

**Fact 1.** *A predicate $P$ does not support pairwise independence if and only if there exists a quadratic polynomial $Q : \{-1,1\}^k \to \mathbb{R}$ with no constant term that is positive on all of $P^{-1}(1)$ (in other words, $P$ implies a predicate of such form).*

Given that the main tool for approximation algorithms—semidefinite programming—works by optimizing quadratic forms, it seemed natural and intuitive to hope that predicates of this form are always approximable. This however turns out to be false—Håstad [8] constructs a predicate that is the sign of a quadratic polynomial and still approximation resistant. Loosely speaking, the main crux is that semidefinite programming is good for optimizing the degree-2 part of the Fourier expansion of a predicate, which unfortunately can behave very differently from $P$ itself or the quadratic polynomial used to define $P$ (we elaborate on this below.) However, it turns out that when we restrict our attention to the special case of symmetric predicates, this can not happen, and we can obtain an approximation algorithm, which is our first result.

**Theorem 1.** *Let $P : \{-1,1\}^k \to \{0,1\}$ be a predicate that is of the form $P(x) = \frac{1+\text{sign}(Q(x))}{2}$ where $Q$ is a symmetric quadratic polynomial with no constant term. Then $P$ is not approximation resistant.*

A very natural special case of the signs of quadratic polynomials is the case when $P(x) = \frac{1+\text{sign}(\sum a_i x_i)}{2}$ is simply the sign of a linear form, i.e., a linear threshold function. While we cannot prove that linear threshold predicates are approximable in general, we do believe this is the case, and make the following conjecture.

*Conjecture 1. Let $P : \{-1,1\}^k \to \{0,1\}$ be a predicate that is a sign of a linear form with no constant term. Then $P$ is not approximation resistant.*

We view the resolution of this conjecture as a very natural and interesting open problem. As in the quadratic case, the difficulty stems from the fact that the low-degree part of $P$ can be unrelated to the linear form used to define $P$. Specifically, it can be the case that the low-degree part of the arithmetization of $P$ vanishes or becomes negative for some inputs where the linear/quadratic polynomial is positive (i.e. accepting inputs), and unfortunately this seems to make the standard SDP approach fail. The perhaps most extreme case of this phenomenon is exhibited by the predicate Monarchy : $\{-1,1\}^k \to \{0,1\}$ suggested by Håstad [8], in which the first variable (the "monarch") decides the outcome, unless all the other variables unite against it. In other words,

$$\text{Monarchy}(x) = \frac{1 + \text{sign}((k-2)x_1 + \sum_{i=2}^{k} x_i)}{2}.$$

Now, for the input $x_1 = -1$, $x_2 = \ldots = x_k = 1$, the linear part of the Fourier expansion of Monarchy takes value $-1 + o_k(1)$, whereas the linear form used to define monarchy is positive on this input, hence the value of the predicate is 1. Again, we stress that this means that known algorithms and techniques do not apply. However, in this case we are still able to achieve an approximation algorithm, which is our second result.

**Theorem 2.** *The predicate* Monarchy *is not approximation resistant.*

This shows that there is some hope in overcoming the apparent barriers to proving Conjecture 1.

**Techniques:** Our starting point in both our algorithms is the standard SDP relaxation of MAX CSP($P$). The main difficulty in rounding the solutions of these SDPs is that current rounding algorithms offer no analysis of the joint distribution of the outcome of the rounding for $k$ variables, when $k > 2$. (Interestingly, when some modest value $k = 3$ is used, often some numerical methods are employed to complete the analysis [9,13].) Unfortunately, such analysis seems essential to understanding the performance of the algorithm for MAX CSP($P$) as each constraint depends on $k$ variables. Indeed, even a local argument would have to argue about the outcome of the rounding algorithm for $k$ variables together.

For Theorem 1, we give a new, simpler proof of a theorem by Hast [6], giving a general condition on the low-degree part of the Fourier Expansion which guarantees a predicate is approximable (Theorem 4). We then show that this condition holds for predicates which are defined by symmetric quadratic polynomials. The basic idea behind our new algorithm is as follows. First, observe that the SDP solution in which all vectors are perpendicular is easy to analyze when the usual hyperplane rounding is employed, as in this case the obtained integral values are distributed uniformly. This motivates the following approach: start with the perpendicular configuration and then perturb the vectors in the direction of the optimal SDP solution. This perturbation acts as a differentiation operator, and as such allows for a "linear snapshot" of what is typically a complicated system. For each clause we analyze the probability that hyperplane rounding outputs a satisfying assignment, as a function of the inner products of vectors involved. Now, the object of interest is the gradient of this function at "zero". The hope is that since the optimal SDP solution (almost) satisfies this clause, it has a positive inner product with the gradient, and so can act as a global recipe that works for all clauses. It is important to stress that since we are only concerned with getting an approximation algorithm that works slightly better than random we can get away with this linear simplification. We show that this condition on the gradient translates into a condition on the low-degree part of the Fourier expansion of the predicate.

As it turns out, the predicate Monarchy which we tackle in Theorem 2 does not exhibit the aforementioned desirable property. In other words, the gradient above does not generally have a positive inner product with an optimal SDP solution. Instead, we show that when all vectors are sufficiently far from $\pm \mathbf{v}_0$ it is possible to get a similar guarantee on the gradient using high (but not-too-high) moments of the vectors. We can then handle vectors which are very close to $\pm \mathbf{v}_0$ separately by rounding them deterministically to $\pm 1$.

**Organization:**  The rest of the paper is organized as follows. First, we introduce some definitions and preliminaries including the standard SDP relaxation of MAX CSP$(P)$ in Section 2. Then, in Section 3 we give our new algorithm for this SDP relaxation and characterize the predicates for which it gives an approximation ratio better than a random assignment. We then take a closer look at signs of symmetric quadratic forms in Section 4 and show that these satisfy the condition of the previous section, proving Theorem 1. In Section 5 we give the approximation algorithm for the Monarchy predicate and the ideas behind its somewhat tedious analysis. Finally, we give a discussion and some directions for future work in Section 6.

## 2    Preliminaries

In what follows $\mathbb{E}$ stands for expectation. For any positive integer $n$ we use the notation $[n]$ for the set $\{1, \ldots, n\}$. For a finite set $S$ (often a subset of $[n]$) we use the notation $\{-1, 1\}^S$ for the set of all $-1, 1$ vectors indexed by elements of $S$. For example, $|\{-1, 1\}^S| = 2^{|S|}$.

We use $\varphi$ and $\Phi$ for the probability density function and the cumulative distribution function of a standard normal random variable, respectively. We use the notation $\mathbb{S}^{n-1}$ for the $n - 1$ dimensional sphere, i.e. , the set of unit vectors in $\mathbb{R}^n$.

Throughout the paper, we take the convention that $\text{sign}(0) = -1$.

### 2.1    Fourier Representation

Consider the set of real functions with domain $\{-1, 1\}^k$ as a vector space. It is well known that the following set of functions called the characters form a complete basis for this space, $\chi_S(x) \stackrel{\text{def}}{=} \prod_{i \in S} x_i$. In fact if we define inner products of functions as $f \cdot g \stackrel{\text{def}}{=} \mathbb{E}_x [f(x)g(x)]$ this basis will be orthonormal and every function will have a unique *Fourier expansion* when written in this basis,

$$f = \sum_{S \subseteq [k]} \widehat{f}(S)\chi_S, \qquad\qquad \widehat{f}(S) \stackrel{\text{def}}{=} f \cdot \chi_S.$$

$\widehat{f}(S)$'s are often called the *Fourier coefficients* of $f$. We write $f^{=d}$ for the part of the function that is of degree $d$, i.e.,

$$f^{=d}(x) = \sum_{|S|=d} \widehat{f}(\{S\})\chi_S(x).$$

## 2.2  Semidefinite Relaxation

For any fixed $P$, MAX CSP($P$) has a natural SDP relaxation that can be seen in Figure 1. The essence of this relaxation is that each $I_{S,*}$ is a distribution, often called a *local distribution*, over all possible assignments to the variables in set $S$ as enforced by (1). Whenever, $S_1$ and $S_2$ intersect (2) guarantees that their marginal distributions on the intersection agree. Also, (3) and (4) ensure that $\mathbf{v}_0 \cdot \mathbf{v}_i$ and $\mathbf{v}_i \cdot \mathbf{v}_j$ are equal to the bias of variable $x_i$ and the correlation of the variables $x_i$ and $x_j$ in the local distributions respectively. The clauses of the instance are $C_1, \ldots, C_m$, with $C_i$ being an application of $P$ (possibly with some variables negated) on the set of variables $T_i$. The objective function is the fraction of the clauses that are satisfied.

Observe that the reason this SDP is not an exact formulation but a relaxation is that these distributions are defined only on sets of size up to $k$. It is worth mentioning that this program is weaker than the $k$th round of the Lasserre hierarchy for this problem while stronger than the $k$th round of the Sheralli-Adams hierarchy. From here on the only things we use in the rounding algorithms are the vectors $\mathbf{v}_0, \ldots, \mathbf{v}_n$ and the existence of the local distributions.

$$
\begin{array}{lll}
\text{Maximize} & & \dfrac{1}{m} \displaystyle\sum_{i=1}^{m} \sum_{\omega \in \{-1,1\}^{T_i}} C_i(\omega)\, I_{T_i,\omega} \\[2ex]
\text{Where,} & \forall S \subset [n], |S| \le k, \omega \in \{-1,1\}^S & I_{S,\omega} \in [0,1] \\[1ex]
& \forall i \subset [n] & \mathbf{v}_i \in \mathbb{R}^{n+1}, \|\mathbf{v}_i\|_2^2 - 1 \\[1ex]
& & \mathbf{v}_0 \in \mathbb{R}^{n+1}, \|\mathbf{v}_0\|_2^2 = 1 \\[1ex]
\text{Subject to} & \forall S \subset [n], |S| \le k & \displaystyle\sum_{\omega \in \{-1,1\}^S} I_{S,\omega} = 1 \qquad (1) \\[2ex]
& \forall S \subset S' \subset [n], |S'| \le k, \omega \in \{-1,1\}^S & \displaystyle\sum_{\omega' \in \{-1,1\}^{S'}} I_{S',\omega'} = I_{S,\omega} \quad (2) \\
& & \omega' \text{ is an extension of } \omega \\[1ex]
& \forall i \in [n] & I_{\{i\},(1)} - I_{\{i\},(-1)} = \mathbf{v}_0 \cdot \mathbf{v}_i \qquad (3) \\[1ex]
& \forall i, j \in [n] & I_{\{i,j\},(1,1)} - I_{\{i,j\},(-1,1)} \\[1ex]
& & -I_{\{i,j\},(1,-1)} + I_{\{i,j\},(-1,-1)} = \mathbf{v}_i \cdot \mathbf{v}_j (4)
\end{array}
$$

**Fig. 1.** Standard SDP relaxation of MAX CSP($P$)

# 3  $(\epsilon, \eta)$-Hyperplane Rounding

In this section we define a rounding scheme for the semidefinite program of MAX CSP($P$) and proceed to analyze its performance. The rounding scheme is based on the usual hyperplane rounding but is more flexible in that it uses two parameters $\epsilon$ and $\eta$ where $\epsilon$ is a sufficiently small constant and $\eta$ is an arbitrary real number. We will then formalize a (sufficient) condition involving $P$ and $\eta$ under which our approximation algorithm has approximation factor better than that of a random assignment. In the next section we show that this condition is satisfied (for some $\eta$) by signs of symmetric quadratic polynomials.

Given an instance of MAX CSP($P$), our algorithm first solves the standard SDP relaxation of the problem (Figure 1.) Then, it employs the rounding scheme outlined in Figure 2 to get an integral solution.

---

INPUT: $\mathbf{v}_0, \mathbf{v}_1, \ldots, \mathbf{v}_n \in \mathbb{S}^n$.
OUTPUT: $x_1, \ldots, x_n \in \{-1, 1\}$.
1. Define unit vectors $\mathbf{w}_0, \mathbf{w}_1, \ldots, \mathbf{w}_n \in \mathbb{S}^n$ such that for all $0 \leq i \neq j$,

$$\mathbf{w}_i \cdot \mathbf{w}_j = \epsilon(\mathbf{v}_i \cdot \mathbf{v}_j),$$

2. Let $g \in \mathbb{R}^{n+1}$ be a random $(n+1)$-dimensional Gaussian.
3. Assign each $x_i$ as,

$$x_i = \begin{cases} 1 & \text{if } \mathbf{w}_i \cdot g > -\eta(\mathbf{w}_0 \cdot \mathbf{w}_i), \\ -1 & \text{otherwise.} \end{cases}$$

---

**Fig. 2.** $(\epsilon, \eta)$-Hyperplane Rounding

Note that when $\epsilon = 0$ the rounding scheme above simplifies to assigning all $x_i$'s uniformly and independently at random which satisfies $\frac{|P^{-1}(1)|}{2^k}$ fraction of all clauses in expectation. For non-zero $\epsilon$, $\eta$ will determine how much weight is given to the position of $\mathbf{v}_0$ compared to the correlation of the variables.

Notice that in the pursuit of a rounding algorithm that has approximation ratio better than $\frac{|P^{-1}(1)|}{2^k}$ it is possible to assume that the optimal integral solution is arbitrary close to 1 (in terms of $k$) as otherwise random assignment already delivers an approximation factor better than $\frac{|P^{-1}(1)|}{2^k}$. In particular, the optimal vector solution can be assumed to be that good. This observation is in fact essential to our analysis. But, for the sake of simplicity first consider the case where the value of the vector solution is precisely 1. Fix a clause, say $P(x_1, x_2, \ldots, x_k)$. (In general, without loss of generality we can assume that the current clause is on $k$ *variables* as opposed to $k$ *literals*. This is simply because one can assume that $\neg x_i$ is a separate variable from $x_i$ with SDP vector $-\mathbf{v}_i$.) Since the SDP value is 1, every clause (and this clause in particular) is completely satisfied by the SDP, hence the local distribution $I_{[k],*}$ is supported on the set of satisfying assignments of $P$. The hope now is that when $\epsilon$ increases from zero to some small positive value this distribution helps to boost the probability of satisfying the clause (a little) beyond $\frac{|P^{-1}(1)|}{2^k}$. This becomes a question of differentiation. Specifically, consider the probability of satisfying the clause at hand as a function of $\epsilon$. We want to show that for some $\epsilon > 0$ the value of this function is bigger than its value at zero which is closely related to the derivative of the function at zero. We can show,

**Theorem 3.** *For any fixed $\eta$, the probability that $P(x_1, \ldots, x_k)$ is satisfied by the assignment behaves as follows at $\epsilon = 0$:*

$$\Pr\left[(x_1, \ldots, x_k) \in P^{-1}(1)\right] = \frac{|P^{-1}(1)|}{2^k}$$

$$\frac{d}{d\epsilon} \Pr\left[(x_1, \ldots, x_k) \in P^{-1}(1)\right] = \frac{2\eta}{\sqrt{2\pi}} \sum_{i=1}^{k} \widehat{P}(\{i\}) \mathbf{v}_0 \cdot \mathbf{v}_i + \frac{2}{\pi} \sum_{i<j} \widehat{P}(\{i,j\}) \mathbf{v}_i \cdot \mathbf{v}_j.$$

$$(5)$$

Now, the inner products $\mathbf{v}_i \cdot \mathbf{v}_j$ are equal to the moments of the local distributions $I_{\{i,j\},*}$, which in turn agree with those of the local distribution $I_{[k],*}$. It follows that,

$$\frac{2\eta}{\sqrt{2\pi}} \sum_{i=1}^{k} \widehat{P}(\{i\}) \mathbf{v}_0 \cdot \mathbf{v}_i + \frac{2}{\pi} \sum_{i<j} \widehat{P}(\{i,j\}) \mathbf{v}_i \cdot \mathbf{v}_j = \mathop{\mathbb{E}}_{\omega \sim I_{[k],*}} \left[\frac{2\eta}{\sqrt{2\pi}} P^{=1}(\omega) + \frac{2}{\pi} P^{=2}(\omega)\right].$$

$$(6)$$

Thus, in order for the derivative in (5) to be positive for all possible values of the $\mathbf{v}_i$'s, it is necessary and sufficient that $\frac{2\eta}{\sqrt{2\pi}} P^{=1}(\omega) + \frac{2}{\pi} P^{=2}(\omega)$ is positive for every $\omega \in P^{-1}(1)$. We can then show the following Theorem.

**Theorem 4.** *Suppose that there exists an $\eta \in \mathbb{R}$ such that*

$$\frac{2\eta}{\sqrt{2\pi}} P^{=1}(\omega) + \frac{2}{\pi} P^{=2}(\omega) > 0 \qquad (7)$$

*for every $\omega \in P^{-1}(1)$. Then $P$ is approximable.*

As mentioned in the Techniques section, this theorem is not new. It was previously found by Hast [6]. However, his algorithm and analysis is completely different from ours (using different algorithms to optimize the linear and quadratic parts of the predicate, and case analysis depending on the behaviour of the integral solution). Our algorithm is considerably more direct, and its analysis is simpler.

The general strategy for the proof, which is deferred to the full version is as follows. We will concentrate on a clause that is almost satisfied by the SDP solution. By the condition and Theorem 3 the first derivative of the probability that this clause is satisfied by the rounded solution is at least some positive global constant (say $\delta$) at $\epsilon = 0$. We will then show that provided that $\epsilon$ is small enough the second derivative of this probability is bounded in absolute value by, say, $\Gamma$ at any point in $[0, \epsilon]$. Now we can apply Taylor's theorem to show that if $\epsilon$ is small enough the probability of success is at least $\frac{|P^{-1}(1)|}{2^k} + \delta\epsilon - \Gamma\epsilon^2/2$ which for $\epsilon = \delta/\Gamma$ is at least $\frac{|P^{-1}(1)|}{2^k} + \delta^2/2\Gamma$.

## 4   Signs of Symmetric Quadratic Polynomials

In this section we study signs of symmetric quadratic polynomials, and give a proof of Theorem 1. Consider a predicate $P : \{-1, 1\}^k \to \{0, 1\}$ that is the sign of a symmetric quadratic polynomial with no constant term, i.e.,

$$P(x) = \frac{1 + \text{sign} \left( \alpha \sum x_i + \beta \sum x_i x_j \right)}{2}$$

for some constants $\alpha$ and $\beta$. We would like to apply the $(\epsilon, \eta)$-rounding scheme to MAX CSP($P$), which in turn requires us to understand the low-degree Fourier coefficients of $P$. Note that because of symmetry, the value of a Fourier coefficient $\widehat{P}(S)$ depends only on $|S|$.

We will prove that "morally" the degree-2 Fourier coefficient of $P$ and $\beta$ have the same sign and that if one of them is 0 then so is the other. This statement is not quite true (consider for instance the predicate $P(x_1, x_2) = \frac{1+\text{sign}(x_1+x_2)}{2} = \frac{1+x_1+x_2+x_1x_2}{4}$), however it is always true that by slightly adjusting $\beta$ (without changing $P$), we can assure that this is the case.

**Theorem 5.** *For any $P$ of the above form, there exists $\beta'$ with the property that $\beta' \cdot \widehat{P}(\{1,2\}) \geq 0$ and $\beta' = 0$ iff $\widehat{P}(\{1,2\}) = 0$, satisfying*

$$P(x) = \frac{1 + \text{sign} \left( \alpha \sum x_i + \beta' \sum x_i x_j \right)}{2}.$$

Due to space considerations the proof of this theorem is deferred to the full version. We are now ready to prove Theorem 1.

**Theorem 1 (restated).** *Let $P : \{-1, 1\}^k \to \{0, 1\}$ be a predicate that is of the form $P(x) = \frac{1+\text{sign}(Q(x))}{2}$ where $Q$ is a symmetric quadratic polynomial with no constant term. Then $P$ is not approximation resistant.*

*Proof.* Without loss of generality, we can take $Q(x) = \alpha \sum x_i + \beta \sum x_i x_j$ where $\beta$ satisfies the property of $\beta'$ in Theorem 5.

If $\widehat{P}(\{1,2\}) = \beta = 0$, we set $\eta = \alpha/\widehat{P}(\{1\})$ (note that in this case we can assume that $\alpha$, hence $\widehat{P}(\{1\})$ is non-zero as otherwise $P$ is the trivial predicate that is always false). We then have, for every $x \in P^{-1}(1)$,

$$\frac{2\eta}{\sqrt{2\pi}} P^{=1}(x) + \frac{2}{\pi} P^{=2}(x) = \frac{2\alpha}{\sqrt{2\pi}} \sum x_i,$$

which is positive by the definition of $P$. If $\widehat{P}(\{1,2\}) \neq 0$, we set $\eta = \sqrt{\frac{2}{\pi}} \frac{\alpha}{\widehat{P}(\{1\})} \cdot \frac{\widehat{P}(\{1,2\})}{\beta}$. In this case for every $x \in P^{-1}(1)$,

$$\frac{2\eta}{\sqrt{2\pi}} P^{=1}(x) + \frac{2}{\pi} P^{=2}(x) = \frac{2\widehat{P}(\{1,2\})}{\pi\beta} \left( \alpha \sum x_i + \beta \sum x_i x_j \right) > 0,$$

since $\beta$ agrees with $\widehat{P}(\{1,2\})$ in sign and $Q(x) > 0$. In either cases, using Theorem 4 and the respective choices of $\eta$ we conclude that $P$ is approximable.

## 5   Monarchy

In this section we prove that for $k > 4$ the Monarchy predicate is not approximation resistant. Notice that Monarchy is defined only for $k > 2$, and that the case $k = 3$ coincides with the predicate majority that is known not to be approximation resistant. Further, the case $k = 4$ is handled by [6].[1]

Just like the algorithm for symmetric predicates we first solve the natural semidefinite program of Monarchy, and then use a rounding algorithm to construct an integral solution out of the vectors. The rounding algorithm, which is given in Figure 3, has two parameters $\epsilon > 0$ and an odd positive integer $\ell$, both depending on $k$. These will be fixed in the proof.

---

INPUT: "biases" $b_1 = \mathbf{v}_0 \cdot \mathbf{v}_1, \ldots, b_n = \mathbf{v}_0 \cdot \mathbf{v}_n$.
OUTPUT: $x_1, \ldots, x_n \in \{-1, 1\}$.
1. Choose a parameter $\tau \in [1/2(k-2)^2, 1/(k-2)^2]$ uniformly at random.
2. For all $i$,
   (a) If $b_i > 1 - \tau$ or $b_i < -1 + \tau$, set $x_i$ to 1 or $-1$ respectively.
   (b) Otherwise, set $x_i$ (independent of all other $x_j$'s), randomly to $-1$ or 1 such that $\mathbb{E}[x_i] = \epsilon b_i^\ell$. In particular, set $x_i = 1$ with probability $(1 + \epsilon b_i^\ell)/2$ and $x_i = -1$ with probability $(1 - \epsilon b_i^\ell)/2$.

---

**Fig. 3.** Rounding SDP solutions for Monarchy

*Remark 1.* As the reader may have observed, the "geometric" power of SDP is not used in the above rounding scheme, and indeed a linear programming relaxation of the problem would suffice for the algorithm we propose. However, in the interest of consistency and being able to describe the the techniques in a language comparable to Theorem 1 we elected to use the SDP framework.

Here is the outline of the analysis. We first ignore the greedy ingredient (2a above). Notice that for $\epsilon = 0$ the rounding gives a uniform assignment to the variables, hence the expected value of the obtained solution is $1/2$. As long as $\epsilon > 0$ is small enough, the probability of success for a clause is essentially only affected by the degree-one Fourier coefficients of Monarchy. Now, fix a clause and assume that the SDP solution completely satisfies it. Specifically, consider the clause Monarchy$(x_1, \ldots, x_k)$, and define $b_1, \ldots, b_k$ as the corresponding biases. Notice that all Fourier coefficients of Monarchy are positive. This implies that the rounding scheme above will succeed with probability that is essentially $1/2$ plus some *positive linear combination* of the $\epsilon b_i^\ell$. Our objective is then to fix $\ell$ that would make the value of this combination positive (and independent from $n$). It turns out that the maximal $b_i$ in magnitude (call it $b_j$) is always positive in this case. Oversimplifying, imagine that $|b_j| \geq |b_i| + \xi$ for all $i$ different than $j$ where $\xi$ is some positive constant. Clearly in this setting it is easy to take $\ell$ (a function of $k$) that makes the effect of all $b_i$ other than $b_j$ vanish, ensuring a positive addition to the probability as desired so that overall the expected fraction of satisfied clauses is more than $1/2$.

---

[1] In the notation of [6], Monarchy on 4 variables is the predicate 0000000101111111, which is listed as approximable in Table 6.6. We remark that this is not surprising since Monarchy in this case is simply a majority in which $x_1$ serves as a tie-breaker variable.

More realistically, the above slack $\xi$ does not generally exist. However, we can show that a similar condition holds provided that the $|b_i|$ are bounded away from 1. This condition suffices to prove that the rounding algorithm works for clauses that do not have any variables with bias very close to $\pm 1$. The case where there are $b_i$ that are very close to 1 in magnitude is where the greedy ingredient of the algorithm (2a) is used, and it can be shown that when $\tau$ is roughly $1/k^2$, this ingredient works.In particular, we can show that for each clause, if rule (2a) is used to round one of the variables, it is used to round essentially every variable in the clause. Also, if this happens, the clause is going to be satisfied with high probability by the rounded solution.

The last complication stems from the fact that the clauses are generally not completely satisfied by the SDP solution. However, using an averaging argument, it is enough to only deal with clauses that are *almost* satisfied by the SDP solution. For any such clause the SDP induces a probability distribution on the variables that is mostly supported on satisfying assignments, compared to *only* on satisfying assignments in the above ideal setting. As such, the corresponding $b_i$'s can be thought of as a perturbed version of the biases in that ideal setting. Unfortunately, the greedy ingredient of the algorithm is very sensitive to such small perturbations. In particular, if the biases are very close to the set threshold, $\tau$, a small perturbation can break the method. To avoid this, we choose the actual threshold randomly, and we manage to argue that only a small fraction of the clauses end up in such unfortunate configurations.

This completes the high level description of the proof of the following theorem. Due to space considerations we leave the complete proof for the full version.

**Theorem 2 (restatement of Theorem 2).** *The predicate* Monarchy *is not approximation resistant.*

# 6  Discussion

We have given algorithms for two cases of MAX CSP($P$) problems not previously known to be approximable. The first case, signs of symmetric quadratic forms, follows from the condition that the low-degree part of the Fourier expansion behaves "roughly" like the predicate in the sense of Theorem 4. The second case, Monarchy, is interesting since it does not satisfy the condition of Theorem 4. As far as we are aware, this is the first example of a predicate which does not satisfy this property but is still approximable. Monarchy is of course only a very special case of Conjecture 1, and we leave the general form open.

A further interesting special case of the conjecture is a generalization of Monarchy called "republic", defined as $\mathrm{sign}(\frac{k}{2}x_1 + \sum_{i=2}^{k} x_i)$. In this case the $x_1$ variable needs to get a $1/4$ fraction of the other variables on its side. We do not even know how to handle this seemingly innocuous example.

It is interesting that the condition on $P$ for our $(\epsilon, \eta)$-rounding to succeed turned out to be precisely the same as the condition previously found by Hast [6], with a completely different algorithm. It would be interesting to know whether this is a coincidence or whether there is a larger picture there that we can not yet see.

As we mentioned in the introduction, there are very few results which give approximation algorithms for large classes of predicates, and it would be very interesting if new such algorithms could be devised.

# References

1. Austrin, P., Håstad, J.: Randomly Supported Independence and Resistance. In: ACM Symposium on Theory of Computing (STOC), pp. 483–492 (2009)
2. Austrin, P., Mossel, E.: Approximation Resistant Predicates from Pairwise Independence. Computational Complexity 18(2), 249–271 (2009)
3. Engebretsen, L., Holmerin, J.: More Efficient Queries in PCPs for NP and Improved Approximation Hardness of Maximum CSP. In: Diekert, V., Durand, B. (eds.) STACS 2005. LNCS, vol. 3404, pp. 194–205. Springer, Heidelberg (2005)
4. Georgiou, K., Magen, A., Tulsiani, M.: Optimal sherali-adams gaps from pairwise independence. In: Dinur, I., Jansen, K., Naor, J., Rolim, J. (eds.) APPROX 2009 and RANDOM 2009. LNCS, vol. 5687, pp. 125–139. Springer, Heidelberg (2009)
5. Goemans, M.X., Williamson, D.P.: Improved Approximation Algorithms for Maximum Cut and Satisfiability Problems Using Semidefinite Programming. Journal of the ACM 42, 1115–1145 (1995)
6. Hast, G.: Beating a Random Assignment – Approximating Constraint Satisfaction Problems. Ph.D. thesis, KTH – Royal Institute of Technology (2005)
7. Håstad, J.: Some Optimal Inapproximability Results. Journal of the ACM 48(4), 798–859 (2001)
8. Håstad, J.: Personal communication (2009)
9. Karloff, H., Zwick, U.: A 7/8-Approximation Algorithm for MAX 3SAT? In: IEEE Symposium on Foundations of Computer Science (FOCS), p. 406 (1997)
10. Khot, S.: On the Power of Unique 2-prover 1-round Games. In: ACM Symposium on Theory of Computing (STOC), pp. 767–775 (2002)
11. Samorodnitsky, A., Trevisan, L.: A PCP characterization of NP with optimal amortized query complexity. In: ACM Symposium on Theory of Computing (STOC), pp. 191–199 (2000)
12. Samorodnitsky, A., Trevisan, L.: Gowers Uniformity, Influence of Variables, and PCPs. In: ACM Symposium on Theory of Computing (STOC), pp. 11–20 (2006)
13. Zwick, U.: Approximation Algorithms for Constraint Satisfaction Problems Involving at Most Three Variables Per Constraint. In: ACM-SIAM Symposium on Discrete Algorithms, SODA (1998)

# Finding Lower Bounds on the Complexity of Secret Sharing Schemes by Linear Programming

Carles Padró and Leonor Vázquez

Universitat Politècnica de Catalunya
{cpadro,leonor}@ma4.upc.edu

**Abstract.** Determining the optimal complexity of secret sharing schemes for every given access structure is a difficult and long-standing open problem in cryptology. Lower bounds have been found by a combinatorial method that uses polymatroids. In this paper, we point out that the best lower bound that can be obtained by this method can be determined by using linear programming, and this can be effectively done for access structures on a small number of participants. By applying this linear programming approach, we present better lower bounds on the optimal complexity and the optimal average complexity of several access structures. Finally, by adding the Ingleton inequality to the previous linear programming approach, we find a few examples of access structures for which there is a gap between the optimal complexity of linear secret sharing schemes and the combinatorial lower bound.

**Keywords:** Secret sharing, Optimization of secret sharing schemes for general access structures.

## 1 Introduction

A *secret sharing scheme* is a method to distribute a *secret value* among a set $P$ of participants, in such a way that only some *qualified subsets* can recover the secret value. Secret sharing schemes were introduced in an independent way by Blakley [9] and Shamir [33]. In this work we consider only *unconditionally secure perfect secret sharing schemes*, in which the shares of the participants in a non-qualified set do not provide any information about the secret. This paper is self-contained. Nevertheless, the reader that is unfamiliar with secret sharing will find more information about the topic in the survey by Stinson [34]. In addition, some of the concepts appearing in this paper are described in more detail in [29].

The collection of qualified subsets $\Gamma$ is called the *access structure* of the secret sharing scheme. In general, access structures are considered to be *monotone*, that is, every superset of a qualified set of participants is also qualified. Then an access structure is completely determined by the family $\min \Gamma$ of its *minimal qualified subsets*.

The length of the shares, when compared to the length of the secret value, is usually considered as a measure of the efficiency of a secret sharing scheme. Specifically, the *complexity* $\sigma(\Sigma)$ of a secret sharing scheme is defined as the

A. López-Ortiz (Ed.): LATIN 2010, LNCS 6034, pp. 344–355, 2010.

ratio between the maximum length of the shares and the length of the secret. The *average complexity* $\widetilde{\sigma}(\Sigma)$ is defined analogously, but considering the average length of the shares instead of the maximum one.

In every secret sharing scheme, the length of every share is at least the length of the secret [28]. Therefore, $1 \leq \widetilde{\sigma}(\Sigma) \leq \sigma(\Sigma)$. A scheme $\Sigma$ is said to be *ideal* if $\sigma(\Sigma) = 1$, that is, if all shares have the same length as the secret. The access structures of ideal secret sharing schemes are called *ideal* as well.

Ito, Saito and Nishizeki [25] constructively proved that there exists a secret sharing scheme for every access structure. Nevertheless, the complexity of the obtained schemes is exponential in the number of participants. A natural problem arises at this point: to determine the most efficient secret sharing scheme for every given access structure. The *optimal complexity* $\sigma(\Gamma)$ of an access structure $\Gamma$ is defined as the infimum of the complexities of all secret sharing schemes for $\Gamma$. The *optimal average complexity* $\widetilde{\sigma}(\Gamma)$ is defined analogously. Determining the values of these parameters is one of the main open problems in secret sharing. Very few is known about it and there is a huge gap between the best known upper and lower bounds.

Because of the difficulty of finding general results, this problem has been considered for several particular families of access structures in [10,14,15,16,17,20,27] among other works. But the value of this parameter is not known even for very simple access structures, as some structures on five [27] and six [17,20] participants. On the other hand, a great achievement has been obtained recently by Csirmaz and Tardos [16] by determining the optimal complexity of all access structures defined by trees.

The homomorphic properties of *linear secret sharing schemes* are very important for some of the main applications of secret sharing as, for instance, secure multiparty computation. On the other hand, linear secret sharing schemes are obtained when applying the best known techniques to construct efficient schemes, as the decomposition method in [35]. Because of that, it is also interesting to consider the parameters $\lambda(\Gamma)$ and $\widetilde{\lambda}(\Gamma)$, the infimum of the (average) complexities of all *linear* secret sharing schemes for $\Gamma$. Obviously, $\sigma(\Gamma) \leq \lambda(\Gamma)$ and $\widetilde{\sigma}(\Gamma) \leq \widetilde{\lambda}(\Gamma)$.

Most of the known lower bounds on the optimal complexity have been found by implicitly or explicitly using a combinatorial method based on polymatroids. The parameter $\kappa(\Gamma)$ was introduced in [29] to denote the best lower bound on $\sigma(\Gamma)$ that can be obtained by this method. We introduce here the corresponding parameter $\widetilde{\kappa}(\Gamma)$ for the combinatorial lower bounds on $\widetilde{\sigma}(\Gamma)$.

Since $\kappa(\Gamma) \leq \sigma(\Gamma) \leq \lambda(\Gamma)$, the parameters $\kappa$ and $\lambda$ are used to find, respectively, lower and upper bounds on the optimal complexity. The same applies to the corresponding parameters for the average optimal complexity. Nevertheless, these bounds are not tight in general because of the separation results among those parameters that have been obtained in [3,5,7,15]. See [29] for more information about these results.

In this paper, we discuss how linear programming can be used to determine the values of parameters $\kappa$ and $\widetilde{\kappa}$ for access structures on small sets of participants, and also to find better lower bounds for the other parameters. In particular, we improve the known bounds for these parameters for several particular access structures over five participants, and also for some graph access structures over six and eight participants. We think that some refinements of this method could be applied in the future to determine the values of those parameters for broader families of access structures.

## 2  Secret Sharing Schemes and Polymatroids

Let $Q$ be a finite set with a distinguished element $p_0 \in Q$ called *dealer*, and let $P = Q - \{p_0\}$ be the set of *participants*. Consider a finite set $E$ with a probability distribution on it and, For every $i \in Q$, consider a finite set $E_i$ and a surjective mapping $\pi_i \colon E \to E_i$. We notate $E_0 = E_{p_0}$ and $\pi_0 = \pi_{p_0}$. Those mappings induce random variables on the sets $E_i$. Let $H(E_i)$ be the Shannon entropy of one of these random variables. For a subset $A = \{i_1, \dots, i_r\} \subseteq Q$, we write $H(E_A)$ for the joint entropy $H(E_{i_1} \dots E_{i_r})$, and a similar convention is used for conditional entropies as, for instance, in $H(E_i|E_A) = H(E_i|E_{i_1} \dots E_{i_r})$.

The mappings $\pi_i$ define a *secret sharing scheme* $\Sigma$ with *access structure* $\Gamma$ on the set of participants $P$ if, for every $A \subseteq P$,

- $H(E_0|E_A) = 0$ if $A \in \Gamma$,
- while $H(E_0|E_A) = H(E_0)$ if $A \notin \Gamma$.

In this situation, every random choice of an element $\mathbf{x} \in E$ according to the given probability distribution results in a *distribution of shares* $((s_i)_{i \in P}, s_0)$, where $s_i = \pi_p(\mathbf{x}) \in E_i$ is the *share* of the participant $i \in P$ and $s_0 = \pi_0(\mathbf{x}) \in E_0$ is the *shared secret value*.

Since the security of a system decreases with the amount of information that must be kept secret, the length of the shares is an important parameter in secret sharing. We define the *complexity* of a secret sharing scheme $\Sigma$ as $\sigma(\Sigma) = \max_{i \in P} H(E_i)/H(E_0)$, that is, the maximum length of the shares in relation to the length of the secret. The *average complexity* is defined by $\widetilde{\sigma}(\Sigma) = \sum_{i \in P} H(E_i)/(nH(E_0))$, where $n = |P|$ is the number of participants. The *optimal complexity* $\sigma(\Gamma)$ of an access structure $\Gamma$ is defined as the infimum of the complexities $\sigma(\Sigma)$ of the secret sharing schemes for $\Gamma$. The *optimal average complexity* $\widetilde{\sigma}(\Gamma)$ is defined analogously. It is not difficult to check that $H(E_i) \geq H(E_0)$ for every participant $i \in P$, and hence $\sigma(\Sigma) \geq \widetilde{\sigma}(\Sigma) \geq 1$. Secret sharing schemes with $\sigma(\Sigma) = 1$ are said to be *ideal* and their access structures are called *ideal* as well.

A secret sharing scheme $\Sigma$ is said to be *linear* if the sets $E$ and $E_i$ are vector spaces over some finite field $\mathbb{K}$, the mappings $\pi_i$ are linear mappings, and the uniform probability distribution is considered on $E$. As a consequence of the general construction in [25], every access structure admits a linear secret sharing scheme. For an access structure $\Gamma$, we notate $\lambda(\Gamma)$ for the infimum of the complexities of the linear secret sharing schemes for $\Gamma$. We consider as well the corresponding parameter $\widetilde{\lambda}(\Gamma)$ for the average complexity.

Fujishige [23] proved that the joint Shannon entropies of a set of random variables define a polymatroid. Csirmaz [13] used that result to provide a unified description for the methods previously used to find most of the known lower bounds on the optimal complexity. Namely, they can be obtained by using the fact that the access structure of a secret sharing scheme implies certain restrictions on the polymatroid derived from the random variables involved in the scheme. Some new results about this connection between secret sharing and polymatroids have been given in [29]. Some basic facts about this topic are presented in the following.

Let $\mathcal{P}(Q)$ denote the power set of $Q$. A *polymatroid* is a pair $\mathcal{S} = (Q, h)$, where $Q$ is a finite set, the *ground set*, and the $h$, the *rank function*, is a mapping $h \colon \mathcal{P}(Q) \to \mathbb{R}$ satisfying:

1. $h(\emptyset) = 0$;
2. $h$ is *monotone increasing*: if $A \subseteq B \subseteq Q$, then $h(A) \leq h(B)$;
3. $h$ is *submodular*: $h(A \cup B) + h(A \cap B) \leq h(A) + h(B)$ for every $A, B \subseteq Q$.

An element $p_0 \in Q$ is said to be an *atomic point* of the polymatroid $\mathcal{S} = (Q, h)$ if, for every $X \subseteq Q$, either $h(X \cup \{p_0\}) = h(X)$ or $h(X \cup \{p_0\}) = h(X) + 1$.

For a polymatroid $\mathcal{S} = (Q, h)$ with an atomic point $p_0 \in Q$, we define on the set $P = Q - \{p_0\}$ the access structure $\Gamma_{p_0}(\mathcal{S}) = \{A \subseteq P : h(A \cup \{p_0\}) = h(A)\}$, which is clearly a monotone increasing family of subsets of $P$. For an access structure $\Gamma$ on $P$, a polymatroid $\mathcal{S} = (Q, h)$ with $Q = P \cup \{p_0\}$ is said to be a $\Gamma$-*polymatroid* if $p_0$ is an atomic point of $\mathcal{S}$ and $\Gamma = \Gamma_{p_0}(\mathcal{S})$.

A secret sharing scheme $\Sigma$ defines a polymatroid $\mathcal{S}(\Sigma) = (Q, h)$ by considering the mapping $h \colon \mathcal{P}(Q) \to \mathbb{R}$ defined by $h(A) = H(E_A)/H(E_0)$. Clearly, $\mathcal{S}(\Sigma)$ determines the access structure $\Gamma$ of $\Sigma$ because $\mathcal{S}(\Sigma)$ is a $\Gamma$-polymatroid. For a polymatroid $\mathcal{S} = (Q, h)$ and $p_0 \in Q$, we define

$$\sigma_{p_0}(\mathcal{S}) = \max\{h(\{i\}) : i \in Q - \{p_0\}\} \quad \text{and} \quad \widetilde{\sigma}_{p_0}(\mathcal{S}) = \frac{1}{n} \sum_{i \in Q - \{p_0\}} h(\{i\}),$$

where $n = |Q| - 1$. Clearly, $\sigma(\Sigma) = \sigma_{p_0}(\mathcal{S}(\Sigma))$ and $\widetilde{\sigma}(\Sigma) = \widetilde{\sigma}_{p_0}(\mathcal{S}(\Sigma))$ for every secret sharing scheme $\Sigma$. Finally, given an access structure $\Gamma$, we define

$$\kappa(\Gamma) = \inf\{\sigma_{p_0}(\mathcal{S}) : \mathcal{S} \text{ is a } \Gamma\text{-polymatroid}\},$$

while $\widetilde{\kappa}(\Gamma)$ is defined analogously, that is, the infimum of the values $\widetilde{\sigma}_{p_0}(\mathcal{S})$ over all $\Gamma$-polymatroids. It is obvious that $\kappa(\Gamma) \leq \sigma(\Gamma)$ and $\widetilde{\kappa}(\Gamma) \leq \widetilde{\sigma}(\Gamma)$. These inequalities are not equalities in general because there can exist $\Gamma$-polymatroids that are not defined from any secret sharing scheme for $\Gamma$. The first examples of access structures with $\kappa(\Gamma) < \sigma(\Gamma)$ were presented in [5].

## 3   Linear Programming Approach

We discuss here how the values $\kappa(\Gamma)$ and $\widetilde{\kappa}(\Gamma)$ can be obtained by solving linear programming problems. Nevertheless, the number of variables and of constraints is exponential in the number of participants, and hence, this only can be done if the set of participants is not too large.

Observe that, by ordering in some way the elements in $\mathcal{P}(Q)$, a polymatroid $\mathcal{S} = (Q, h)$ can be represented as a vector $(h(A))_{A \subseteq Q} \in \mathbb{R}^k$, where $k = |\mathcal{P}(Q)| = 2^{n+1}$. The polymatroid axioms imply a number of linear constraints on this vector. If, in addition, we assume that $\mathcal{S}$ is a $\Gamma$-polymatroid for some access structure $\Gamma$ on $P = Q - \{p_0\}$, other linear constraints appear. Since $\tilde{\sigma}_{p_0}(\mathcal{S})$ is also a linear function on the vector $h$, one can find $\tilde{\sigma}(\Gamma)$ by solving the linear programming problem

$$\text{Minimize}\ \ \tilde{\sigma}_{p_0}(\mathcal{S}) = (1/n) \sum_{i \in P} h(\{i\})$$
$$\text{subject to}\ \ \ \ \ \ \ \mathcal{S} \text{ is a } \Gamma\text{-polymatroid.}$$

Observe that $\sigma_{p_0}(\mathcal{S})$ is not linear. Because of that, we introduce a new real variable $v$ in such a way that $\kappa(\Gamma)$ is the solution of the next linear programming problem

$$\text{Minimize}\ \ \ \ \ \ \ \ \ \ \ \ \ \ v$$
$$\text{subject to } \mathcal{S} \text{ is a } \Gamma\text{-polymatroid and}$$
$$v \geq h(\{i\}) \text{ for every } i \in Q$$

The feasible region for the first linear programming problem is

$$\Omega = \Omega(\Gamma) = \{h \in \mathbb{R}^k : h \text{ is the rank function of a } \Gamma\text{-polymatroid}\}.$$

Since there exist $\Gamma$-polymatroids for every access structure, $\Omega \neq \emptyset$. For the other linear programming problem, the feasible region is

$$\Omega' = \{(h, v) \in \mathbb{R}^{k+1} : h \in \Omega \text{ and } v \geq h(\{i\}) \text{ for every } i \in Q\},$$

which is obviously nonempty as well.

The number of constraints to define these feasible regions can be reduced by using the following characterization of polymatroids given by Matúš [31]. Namely, $h : \mathcal{P}(Q) \to \mathbb{R}$ is the rank function of a polymatroid $\mathcal{S} = (Q, h)$ if and only if

1. $h(\emptyset) = 0$,
2. $h(Q - \{i\}) \leq h(Q)$ for every $i \in Q$, and
3. $h(A \cup \{i\}) + h(A \cup \{j\}) \geq h(A \cup \{i, j\}) + h(A)$ for every $i, j \in Q$ with $i \neq j$ and for every $A \subseteq Q - \{i, j\}$.

Moreover, we can further reduce the number of constraints by taking into account that a polymatroid $\mathcal{S} = (Q, h)$ is a $\Gamma$-polymatroid if and only if

4. $h(\{p_0\}) = 1$,
5. $h(A \cup \{p_0\}) = h(A)$ if $A \subseteq P$ is a minimal qualified subset of $\Gamma$, and
6. $h(B \cup \{p_0\}) = h(B) + 1$ if $B \subseteq P$ is a maximal unqualified subset of $\Gamma$.

For every $A \subseteq Q$, we consider the vector $\mathbf{e}_A \in \mathbb{R}^k$ with $\mathbf{e}_A(A) = 1$ and $\mathbf{e}_A(B) = 0$ for every $B \in \mathcal{P}(Q) - \{A\}$. At this point, we can present a set of linear constraints defining the feasible region $\Omega$ (vectors are considered as columns).

- $\mathbf{e}_{\emptyset}^{T} h = 0$.
- $(\mathbf{e}_{Q-\{i\}} - \mathbf{e}_Q)^T h \le 0$ for every $i \in Q$.
- $(\mathbf{e}_{A\cup\{i,j\}} + \mathbf{e}_A - \mathbf{e}_{A\cup\{i\}} - \mathbf{e}_{A\cup\{j\}})^T h \le 0$ for every $i, j \in Q$ with $i \ne j$ and for every $A \subseteq Q - \{i,j\}$.
- $\mathbf{e}_{\{p_0\}}^{T} h = 1$.
- $(\mathbf{e}_{A\cup\{p_0\}} - \mathbf{e}_A)^T h = 0$ for every $A \in \min \Gamma$.
- $(\mathbf{e}_{B\cup\{p_0\}} - \mathbf{e}_B)^T h = 1$ for every maximal unqualified subset $B$.

Both the number of variables and the number of constraints grow exponentially on the number $n$ of participants. The number of variables is $k = 2^n$. If $m = |\min \Gamma|$ and $m'$ is the number of all maximal unqualified subsets, then the number $N_c$ of constraints is $N_c = \binom{n+1}{2} \cdot 2^{n-1} + n + 2(m + m') + 5$, where $m, m' \le \binom{n}{\lfloor n/2 \rfloor}$ by the Sperner's Theorem [1].

## 4   New Bounds

In this section we present some improvements on previous bounds on the optimal complexity of some access structures that have been obtained by using a computer to solve the linear programming problems described in the previous section.

Jackson and Martin [27] determined the optimal (average) complexities of all access structures on five participants except a few ones, for which upper and lower bounds were given. Specifically, there are 180 non-isomorphic access structures with five participants. The optimal complexities of 170 of them and the optimal average complexities of 165 of them are determined in [27].

Duality plays an important role in the problems that are considered here. For an access structure $\Gamma$ on a set $P$, the *dual access structure* $\Gamma^*$ is defined by $\Gamma^* = \{A \subseteq P : P - A \notin \Gamma\}$. From every linear secret sharing scheme for an access structure $\Gamma$, a linear scheme with the same (average) complexity can be obtained for the dual access structure $\Gamma^*$ [19,26], and hence $\lambda(\Gamma^*) = \lambda(\Gamma)$ and $\widetilde{\lambda}(\Gamma^*) = \widetilde{\lambda}(\Gamma)$. In addition, it was proved in [29] that $\kappa(\Gamma^*) = \kappa(\Gamma)$ and, by using the same arguments, it is not difficult to check that $\widetilde{\kappa}(\Gamma^*) = \widetilde{\kappa}(\Gamma)$. Therefore, the results we obtained for those access structures on five participants apply as well to their dual access structures. Nevertheless the behavior of the parameters $\sigma, \widetilde{\sigma}$ with respect the dual access structures is unknown.

The results in [27] are obtained by finding lower bounds on $\kappa(\Gamma)$ and $\widetilde{\kappa}(\Gamma)$ and upper bounds on $\lambda(\Gamma)$ and $\widetilde{\kappa}(\Gamma)$. The values of $\sigma(\Gamma)$ that are determined in [27] correspond to exactly to the cases in which the lower bound on $\kappa(\Gamma)$ is equal to the upper bound on $\lambda(\Gamma)$. The same applies to the average complexity. Because of that, the value of $\sigma(\Gamma)$ (or $\widetilde{\kappa}(\Gamma)$) is determined in [27] if and only if $\sigma(\Gamma^*)$ (respectively, $\widetilde{\kappa}(\Gamma^*)$) is determined.

By using our linear programming approach, we are able to improve the results in [27] for the access structures on five participants given in the following. We enumerate them as in [27]. Since our results deal with the values of $\kappa$ and $\widetilde{\kappa}$, they apply as well to the corresponding dual access structures.

- $\min \Gamma_{73} = \{\{1,2\},\{1,3\},\{2,4\},\{3,5\},\{1,4,5\}\}$.
- $\min \Gamma_{80} = \{\{1,2\},\{1,3\},\{2,3,4\},\{2,3,5\},\{4,5\}\}$.
- $\min \Gamma_{82} = \{\{1,2\},\{1,3\},\{2,3,4\},\{2,3,5\},\{1,4,5\},\{2,4,5\}\}$.
- $\min \Gamma_{83} = \{\{1,2\},\{1,3\},\{2,3,4\},\{2,3,5\},\{1,4,5\},\{2,4,5\},\{3,4,5\}\}$.
- $\min \Gamma_{86} = \{\{1,2\},\{1,3\},\{2,3,4\},\{4,5\}\}$.
- $\min \Gamma_{88} = \{\{1,2\},\{1,3\},\{2,3,4\},\{1,4,5\},\{2,4,5\}\}$.
- $\min \Gamma_{89} = \{\{1,2\},\{1,3\},\{2,3,4\},\{1,4,5\},\{1,4,5\},\{3,4,5\}\}$.
- $\min \Gamma_{150} = \{\{1,2,3\},\{1,2,4\},\{1,3,4\},\{1,2,5\},\{2,3,5\}$.
- $\min \Gamma_{152} = \{\{1,2,3\},\{1,2,4\},\{1,3,4\},\{1,2,5\},\{3,4,5\}$.
- $\min \Gamma_{153} = \{\{1,2,3\},\{1,2,4\},\{1,3,4\},\{1,2,5\},\{2,3,4,5\}\}$.

We determined the values of $\kappa(\Gamma)$ and $\widetilde{\kappa}(\Gamma)$ for all these access structures (and hence also for the dual structures) by using linear programming. The obtained results are given in Table 1. The entries with an interval correspond to a lower and an upper bound. Observe that we improved some of the bounds on $\widetilde{\sigma}(\Gamma)$ but could not improve the bounds on $\sigma(\Gamma)$ for any of these access structures. Nevertheless, the exact values of $\kappa(\Gamma)$ and $\widetilde{\kappa}(\Gamma)$ have been determined. Therefore, we know now that no better lower bounds can be obtained by using this combinatorial technique. That is, whether better constructions of secret sharing schemes are obtained for those structures, or better lower bounds have to be searched by considering information inequalities other than the basic Shannon inequalities, as it is discussed in Section 5. We also included in the table some information about the running time of our implementation (in mili-seconds) and the number of constraints that define the feasible region.

A graph defines an access structure on the set of vertices by considering the edges as the minimal qualified subsets. Secret sharing for graph access structures have been studied by many works, as for instance [10,14,12,16,17,20,34,35]. The 112 non-isomorphic graph access structures on six vertices were listed by van Dijk [17], and the optimal complexities of all of them, except 18, were determined.

**Table 1.** Our results for access structures on five participants

| Access structure | $\sigma$ from [27] | $\widetilde{\sigma}$ from [27] | $\kappa$ with LP (time in ms) | $\widetilde{\kappa}$ with LP | Current $\widetilde{\sigma}$ | Number of constraints |
|---|---|---|---|---|---|---|
| $\Gamma_{73}$ | $[3/2, 5/3]$ | $[7/5, 8/5]$ | $3/2$ (121) | $3/2$ | $[3/2, 8/5]$ | 272 |
| $\Gamma_{80}$ | $3/2$ | $[6/5, 7/5]$ | $3/2$ (122) | $13/10$ | $[13/10, 7/5]$ | 274 |
| $\Gamma_{82}$ | $[3/2, 5/3]$ | $[6/5, 7/5]$ | $3/2$ (145) | $13/10$ | $[13/10, 7/5]$ | 274 |
| $\Gamma_{83}$ | $3/2$ | $[6/5, 7/5]$ | $3/2$ (139) | $13/10$ | $[13/10, 7/5]$ | 280 |
| $\Gamma_{86}$ | $3/2$ | $[6/5, 7/5]$ | $3/2$ (121) | $13/10$ | $[13/10, 7/5]$ | 268 |
| $\Gamma_{88}$ | $3/2$ | $[6/5, 7/5]$ | $3/2$ (122) | $7/5$ | $7/5$ | 270 |
| $\Gamma_{89}$ | $3/2$ | $[6/5, 7/5]$ | $3/2$ (99) | $13/10$ | $[13/10, 7/5]$ | 274 |
| $\Gamma_{150}$ | $[3/2, 12/7]$ | $7/5$ | $3/2$ (138) | $7/5$ | $7/5$ | 272 |
| $\Gamma_{152}$ | $[3/2, 5/3]$ | $[7/5, 8/5]$ | $3/2$ (156) | $3/2$ | $[3/2, 8/5]$ | 272 |
| $\Gamma_{153}$ | $[3/2, 5/3]$ | $7/5$ | $3/2$ (122) | $7/5$ | $7/5$ | 274 |

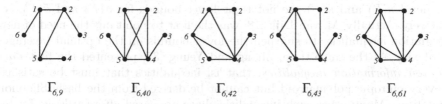

$$\Gamma_{6,9} \qquad \Gamma_{6,40} \qquad \Gamma_{6,42} \qquad \Gamma_{6,43} \qquad \Gamma_{6,61}$$

**Fig. 1.** Graph access structures with six vertices

Better upper bounds for some of these 18 structures were obtained in [12,20]. By using linear programming, we have been able to improve the lower bounds for five of them, the ones depicted in Figure 1. The results are shown in Table 2.

**Table 2.** Our results for graph access structures on six vertices

| Access structure | $\sigma$ from [17] | $\sigma$ from [12,20] | $\kappa$ with LP (time in sec) | Current $\sigma$ | Number of constraints |
|---|---|---|---|---|---|
| $\Gamma_{6,9}$ | $[5/3, 2]$ | $[5/3, 7/4]$ (from [12] and [20]) | $7/4$ (1.43) | $7/4$ | 703 |
| $\Gamma_{6,40}$ | $[5/3, 9/5]$ | $[5/3, 7/4]$ (from [12]) | $7/4$ (1.53) | $7/4$ | 707 |
| $\Gamma_{6,42}$ | $[5/3, 7/4]$ | no improvement | $7/4$ (1.41) | $7/4$ | 707 |
| $\Gamma_{6,43}$ | $[5/3, 7/4]$ | no improvement | $7/4$ (1.51) | $7/4$ | 707 |
| $\Gamma_{6,61}$ | $[5/3, 2]$ | $[5/3, 16/9]$ (from [12]) | $7/4$ (1.52) | $[7/4, 16/9]$ | 707 |

## 5   Sharpening the Feasible Region

In the previous sections, we applied linear programming to determine the values of the parameters $\kappa$ and $\widetilde{\kappa}$, which provide lower bounds on the parameters $\sigma$ and $\widetilde{\sigma}$, respectively. Nevertheless, these lower bounds are not tight in general and, in order to determine the values of $\sigma$ and $\widetilde{\sigma}$, the so called non-Shannon information inequalities must be considered in some situations.

A $\Gamma$-polymatroid $\mathcal{S} = (Q, h)$ is said to be *entropic* if there exists a secret sharing scheme $\Sigma$ with access structure $\Gamma$ such that $\mathcal{S} = \mathcal{S}(\Sigma)$, that is, if there exists a set of random variables such that their normalized joint entropies coincide with the values of the rank function $h$. There exists a secret sharing scheme for every access structure, and hence, for every structure $\Gamma$, there exist an entropic $\Gamma$-polymatroid. Nevertheless, an access structure $\Gamma$ admits many different $\Gamma$-polymatroids, and some of them may not be entropic. Because of that, $\kappa(\Gamma)$ or $\widetilde{\kappa}(\Gamma)$ may be smaller than $\sigma(\Gamma)$ or $\widetilde{\kappa}(\Gamma)$, respectively.

Consider the region

$$\Omega^* = \Omega^*(\Gamma) = \{h \in \mathbb{R}^k : \mathcal{S} = (Q, h) \text{ is an entropic polymatroid}\}.$$

Then $\sigma(\Gamma) = \inf\{\sigma_{po}(\mathcal{S}) : \mathcal{S} = (Q, h) \text{ with } h \in \Omega^*(\Gamma)\}$. Clearly, $\Omega^*(\Gamma) \subseteq \Omega(\Gamma)$, but these two regions are not equal in general [36]. In addition, the region $\Omega^*$ cannot be described by linear constraints [30]. Therefore, it seems that in general $\kappa(\Gamma)$ and $\widetilde{\kappa}(\Gamma)$ are not tight lower bounds for $\sigma(\Gamma)$ and $\widetilde{\sigma}(\Gamma)$, respectively. Actually, M. van Dijk [18] was the first to point out the need of new information inequalities to sharpen the lower bounds on the optimal (average) complexity. At the same time, Zhang and Yeung [37] presented the first *non-Shannon information inequalities*, that is, inequalities that must be satisfied by every entropic polymatroid but cannot be derived from the basic Shannon inequalities. Many other such inequalities have appeared afterwards as, for instance, the ones in [21,32]. These inequalities actually prove that $\Omega^* \neq \Omega$. Moreover, they were used in [5] to present the first examples of access structures with $\kappa(\Gamma) < \sigma(\Gamma)$. Those non-Shannon information inequalities are linear, and hence, they can be used to sharpen the feasible region to find better lower bounds on $\sigma(\Gamma)$ and $\widetilde{\sigma}(\Gamma)$ by using linear programming. Even though Beimel and Orlov [6] proved that all known non-Shannon information inequalities cannot improve our knowledge on the asymptotic values of the optimal complexity, the use of these inequalities in combination with the linear programming approach can provide better bounds for some particular families of access structures.

If one is interested only in linear secret sharing schemes, then the feasible region defined by the *linearly entropic* polymatroids, that is the polymatroids that are defined from linear secret sharing schemes, should be considered:

$$\Omega_L^* = \Omega_L^*(\Gamma) = \{h \in \mathbb{R}^k : \mathcal{S} = (Q, h) \text{ is an linearly entropic } \Gamma\text{-polymatroid}\}.$$

The parameters $\lambda$ and $\widetilde{\lambda}$ are obtained by minimizing the corresponding functions over $\Omega_L^*$. Clearly, $\Omega_L^*(\Gamma) \subseteq \Omega^*(\Gamma)$, and these regions are different because of the separation results between linear and non-linear schemes in [3,7]. A non-Shannon information inequality that must be satisfied by every linearly entropic polymatroid was given by Ingleton [24]. Specifically, if $\mathcal{S} = (Q, h)$ is a linearly entropic polymatroid, then $J(h; A, B, C, D) \leq 0$ for every $A, B, C, D \subseteq Q$, where

$$J(h; A, B, C, D) = h(A) + h(B) + h(C \cup D) + h(A \cup B \cup C) + h(A \cup B \cup D)$$
$$-h(A \cup B) - h(A \cup C) - h(A \cup D) - h(B \cup C) - h(B \cup D)$$

As a consequence of results in [11], a minimal set of Ingleton inequalities can be found by adding to the axioms of the rank function the next property:

$$J^{In}(h; A, B, C, D, X) = J(h; AX, BX, CX, DX) \leq 0,$$

where $A, B, C, D, X$ are disjoint nonempty sets.

If $H_{IN}$ is the region in $\mathbb{R}^k$ determined by a well defined set of Ingleton inequalities, then the region $\Omega_{IN}(\Gamma) = \Omega(\Gamma) \cap H_N$ is such that $\Omega_{LIN}^*(\Gamma) \subseteq \Omega_{IN}(\Gamma)$. As a consequence, the solution $\lambda_{IN}(\Gamma)$ of the next linear programming problem is such that $\lambda_{IN}(\Gamma) \leq \lambda(\Gamma)$. As before, we added here a new variable $v$.

$$\begin{array}{ll} \text{Minimize} & v \\ \text{subject to} & h \in \Omega_{IN}(\Gamma) \text{ and} \\ & v \geq h(\{i\}) \text{ for every } i \in Q, \end{array}$$

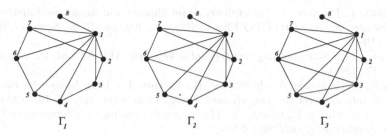

**Fig. 2.** Graph access structures with 8 vertices: $\Gamma_1, \Gamma_2$ and $\Gamma_3$

This idea can be applied to any access structure in order to find separations between the parameters $\kappa$ and $\lambda$. By considering the Ingleton inequalities of the form $J^{In}(h; \{a\}, \{b\}, \{c\}, \{d\}, X) \leq 0$, where $a, b, c, d \in P$ are distinct participants and either $X = \emptyset$ or $X = \{x\}$ with $x \in Q$, we solved this linear program problem for the access structures in Tables 1 and 2. We obtained that $\lambda_{IN}(\Gamma) = \kappa(\Gamma) < \lambda(\Gamma)$ in all those cases, and hence no new information about the value of $\lambda(\Gamma)$ is obtained for those structures. Nevertheless, by using this method, we are able to present three examples of graph access structures such that $\kappa(\Gamma) < \lambda(\Gamma)$. Specifically, we consider the access defined by the graphs in Figure 2 and we apply the above method by considering the Ingleton inequalities of the form $J^{In}(h; \{a\}, \{b\}, \{c\}, \{d\}, X) \leq 0$, where $a, b, c, d \in P$ are distinct participants and either $X = \emptyset$ or $X = \{x\}$ with $x \in Q$. The solutions to the corresponding linear programming problems are $\lambda_{IN}(\Gamma_1) = 19/10$, and $\lambda_{IN}(\Gamma_2) = \lambda_{IN}(\Gamma_3) = 13/7$. On the other hand, we used linear programming as well to compute $\kappa(\Gamma_1) = 11/6$, and $\kappa(\Gamma_2) = \kappa(\Gamma_3) = 9/5$.

An infinite family of graphs such that $\kappa(\Gamma) < \lambda(\Gamma)$ has been found recently by Csirmaz [15] by using other techniques.

# References

1. Anderson, I.: Combinatorics of Finite Sets. Oxford University Press, Oxford (1987)
2. Bazaraa, M., Jarvis, J., Sherali, H.: Linear Programming and Network Flows, 2nd edn. John Wiley& Sons, Chichester (1990)
3. Beimel, A., Ishai, Y.: On the power of nonlinear secret sharing schemes. SIAM J. Discrete Math. 19, 258–280 (2005)
4. Beimel, A., Livne, N.: On matroids and non-ideal secret sharing. In: Halevi, S., Rabin, T. (eds.) TCC 2006. LNCS, vol. 3876, pp. 482–501. Springer, Heidelberg (2006)
5. Beimel, A., Livne, N., Padró, C.: Matroids can be far from ideal secret sharing. In: Canetti, R. (ed.) TCC 2008. LNCS, vol. 4948, pp. 194–212. Springer, Heidelberg (2008)
6. Beimel, A., Orlov, I.: Secret Sharing and Non-Shannon Information Inequalities. In: Reingold, O. (ed.) TCC 2009. LNCS, vol. 5444, pp. 539–557. Springer, Heidelberg (2009)
7. Beimel, A., Weinreb, E.: Separating the power of monotone span programs over different fields. SIAM J. Comput. 34, 1196–1215 (2005)

8. Benaloh, J., Leichter, J.: Generalized secret sharing and monotone functions. In: Goldwasser, S. (ed.) CRYPTO 1988. LNCS, vol. 403, pp. 27–35. Springer, Heidelberg (1990)
9. Blakley, G.R.: Safeguarding cryptographic keys. In: AFIPS Conf. P., vol. 48, pp. 313–317 (1979)
10. Blundo, C., de Santis, A., de Simone, R., Vaccaro, U.: Tight bounds on the information rate of secret sharing schemes. Design Code Cryptogr. 11, 107–122 (1997)
11. Chan, T.H., Guillé, L., Grant, A.: The minimal set of Ingleton inequalities (2008), http://arXiv.org/abs/0802.2574
12. Chen, B.-L., Sun, H.-M.: Weighted Decomposition Construction for Perfect Secret Sharing Schemes. Comput. Math. Appl. 43(6-7), 877–887 (2002)
13. Csirmaz, L.: The size of a share must be large. J. Cryptology 10, 223–231 (1997)
14. Csirmaz, L.: Secret sharing on the d-dimensional cube. Cryptology ePrint Archive. Report 2005/177 (2005), http://eprint.iacr.org
15. Csirmaz, L.: An impossibility result on graph secret sharing. Designs, Codes and Cryptography 53, 195–209 (2009)
16. Csirmaz, L., Tardos, G.: Secret sharing on trees: problem solved. Cryptology ePrint Archive (preprint, 2009), http://eprint.iacr.org/2009/071
17. van Dijk, M.: On the information rate of perfect secret sharing schemes. Des. Codes Cryptogr. 6, 143–169 (1995)
18. van Dijk, M.: More information theoretical inequalities to be used in secret sharing? Inform. Process. Lett. 63, 41–44 (1997)
19. van Dijk, M., Jackson, W.-A., Martin, K.M.: A note on duality in linear secret sharing schemes. Bull. Inst. Combin. Appl. 19, 93–101 (1997)
20. van Dijk, M., Kevenaar, T., Schrijen, G., Tuyls, P.: Improved constructions of secret sharing schemes by applying-($\lambda$, $\omega$)-decompositions. Inform. Process. Lett. 99 (4), 154–157 (2006)
21. Dougherty, R., Freiling, C., Zeger, K.: Six new non-Shannon information in- equalities. In: ISIT 2006, pp. 233–236 (2006)
22. Fehr, S.: Linear VSS and Distributes Commitments Based on Secret Sharing and Pairwise Checks. In: Yung, M. (ed.) CRYPTO 2002. LNCS, vol. 2442, pp. 565–580. Springer, Heidelberg (2002)
23. Fujishige, S.: Polymatroidal Dependence Structure of a Set of Random Variables. Inform. and Control. 39, 55–72 (1978)
24. Ingleton, A.W.: Conditions for representability and transversability of matroids. In: Proc. Fr. Br. Conf. 1970, pp. 62–27 (1971)
25. Ito, M., Saito, A., Nishizeki, T.: Secret sharing scheme realizing any access structure. In: Proc. IEEE Globecom 1987, pp. 99–102 (1987)
26. Jackson, W.-A., Martin, K.M.: Geometric secret sharing schemes and their duals. Des. Codes Cryptogr. 4, 83–95 (1994)
27. Jackson, W.-A., Martin, K.M.: Perfect secret sharing schemes on five participants. Des. Codes Cryptogr. 9, 267–286 (1996)
28. Karnin, E.D., Greene, J.W., Hellman, M.E.: On secret sharing systems. IEEE Trans. Inform. Theory 29, 35–41 (1983)
29. Martí-Farré, J., Padró, C.: On Secret Sharing Schemes, Matroids and Polymatroids. In: Vadhan, S.P. (ed.) TCC 2007. LNCS, vol. 4392, pp. 273–290. Springer, Heidelberg (2007), Cryptology ePrint Archive, http://eprint.iacr.org/2006/077
30. Matúš, F.: Piecewise linear conditional information inequality. On IEEE Trans. Inform. Theory 44, 236–238 (2006)
31. Matúš, F.: Adhesivity of polymatroids. Discrete Math. 307, 2464–2477 (2007)

32. Matúš, F.: Infinitely many information inequalities. In: IEEE International Symposium on Information Theory 2007, pp. 41–44 (2007)
33. Shamir, A.: How to share a secret. Commun. of the ACM 22, 612–613 (1979)
34. Stinson, D.R.: An explication of secret sharing schemes. Des. Codes Cryptogr. 2, 357–390 (1992)
35. Stinson, D.R.: Decomposition constructions for secret-sharing schemes. IEEE Trans. Inform. Theory 40, 118–125 (1994)
36. Yeung, R.: A First Course in Information Theory. Kluwer Academic/Plenum Publishers (2002)
37. Zhang, Z., Yeung, R.W.: On characterization of entropy function via information inequalities. IEEE Trans. Inform. Theory 44, 1440–1452 (1998)

# Finding the Best CAFE Is NP-Hard*

Elizabeth Maltais** and Lucia Moura***

University of Ottawa,
Ottawa, Ontario, Canada, K1N 6N5

**Abstract.** In this paper, we look at covering arrays with forbidden
edges (CAFEs), which are used in testing applications (software, networks, circuits, drug interaction, material mixtures, etc.) where certain
combinations of parameter values are forbidden. Covering arrays are classical objects used in these applications, but the situation of dealing with
forbidden configurations is much less studied. Danziger et. al. [8] have
recently studied this problem and shown some computational complexity
results, but left some important open questions. Around the same time,
Martinez et al. [18] defined and studied error-locating arrays (ELAs),
which are very related to CAFEs, leaving similar computational complexity questions. In particular, these papers showed polynomial-time
solvability of the existence of CAFEs and ELAs for binary alphabets
($g = 2$), and the NP-hardness of these problems for $g \geq 5$. This not only
left open the complexity of determining optimum CAFEs and ELAs for
$g = 2, 3, 4$, but some suspicion that the binary case might be solved by
a polynomial-time algorithm. In this paper, we prove that optimizing
CAFEs and ELAs is indeed NP-hard even when restricted to the case of
binary alphabets. We also provide a hardness of approximation result.
The proof strategy uses a reduction from edge clique covers of graphs
(ECCs) and covers all cases of $g$. We also explore important relationships
between ECCs and CAFEs and give some new bounds for uniform ECCs
and CAFEs.

## 1   Introduction

Before a company releases a new product, much testing needs to occur in order to ensure high quality standards. Whether the product is a software-based
electronic device or a new prescription drug, there are often various components
or factors involved, each having several options, which should be tested in some
sensible way. To model the general situation, we use the following definition.

A *testing problem* is a system with $k$ components called *factors*, which we
label by the indices $1, ..., k$. Each factor $i \in [1, k]$ has $g_i$ possible options, called

---

  * The results presented in this paper are part of the Master's thesis of E. Maltais [17]
    under the supervision of L. Moura.
 ** E. Maltais was supported by the Department of Mathematics and Statistics, Uni-
    versity of Ottawa.
*** L. Moura was supported by NSERC.

A. López-Ortiz (Ed.): LATIN 2010, LNCS 6034, pp. 356–371, 2010.

**Table 1.** Mobile phone product line

| Factors | 1 = display | 2 = email viewer | 3 = camera | 4 = video camera | 5 = video ringtones |
|---------|-------------|------------------|------------|------------------|---------------------|
| Values  | 0 = 16 million colours<br>1 = 8 million colours<br>2 = black and white | 0 = graphical<br>1 = text<br>2 = none | 0 = 2 megapixels<br>1 = 1 megapixel | 0 = yes<br>1 = no | 0 = yes<br>1 = no |

values[1]. Typically, we use the alphabet $[0, g_i - 1]$ to denote the values of factor $i$. For convenience, we denote such a testing problem as $TP(k, (g_1, ..., g_k))$. If the alphabet size is constant, that is, if $g_1 = g_2 = \cdots = g_k = g$ for some $g \in \mathbb{Z}$, then we shorten the notation to $TP(k, g)$. We represent a *test* by a $k$-tuple $T = (a_1, ..., a_k) \in [0, g_1 - 1] \times \cdots \times [0, g_k - 1]$, to mean that value $a_i$ has been selected for factor $i$ for each $i \in [1, k]$. For example, Table 1 shows a $TP(5, (3, 3, 3, 2, 2))$ for possible options on a mobile phone taken from Cohen et. al. [4]. We assume that the nature of the system is such that the *outcome* of each test is either *pass* or *fail*. If a test fails, we conclude that a fault is present in the system and that this fault is responsible for the test's failure. Our goal is thus to design a suite of tests which can reveal the faults of the system.

In practice, exhaustively testing a $TP(k, (g_1, ..., g_k))$ is too costly. Even for a moderately small testing problem, say a $TP(5, 4)$, exhaustive testing would require $4^5 = 1024$ tests, which could be infeasible depending on budget and time. So we must look for more reasonably sized test suites, but at the same time, we want the tests to cover a wide range of possibilities. Since the purpose of testing products is to eliminate problems, we have to consider the causes of problems. It may be that one specific value of one of the factors causes a fault. However, with systems involving several components, faults are often due to unexpected interactions that occur between a specific combination of the options [5,21]. Therefore, one alternative to exhaustive testing would be to design a suite of tests in which every $t$-way interaction between any $t$ of the factors is covered. To be precise we use the following definition.

**Definition 1.** *Let $TP(k, (g_1, ..., g_k))$ be a testing problem, and let $t$ be a positive integer such that $1 \leq t \leq k$. A $t$-way interaction is a set of values assigned to $t$ distinct factors. We denote a $t$-way interaction as $I = \{(f_1, a_{f_1}), ..., (f_t, a_{f_t})\}$ where $f_i \in [1, k]$, $f_i \neq f_j$ for $i \neq j$, and $a_{f_i} \in [0, g_{f_i} - 1]$ for $1 \leq i \leq t$. If $t = 2$, we refer to a 2-way interaction as a pairwise interaction. If $t = 1$, we refer to a 1-way interaction as a pointwise interaction. We say that a test $T = (T_1, ..., T_k) \in [0, g_1 - 1] \times \cdots \times [0, g_k - 1]$ covers interaction $I = \{(f_1, a_{f_1}), ..., (f_t, a_{f_t})\}$ if $T_{f_i} = a_{f_i}$ for each $i \in [1, t]$.*

As an alternative to exhaustive testing, test suites designed to cover all $t$-way interactions for some small value of $t$ can be applied. Indeed, research has shown that testing all pairwise interactions in a testing problem finds a large percentage of existing faults, thus offers a good compromise to exhaustive testing [2,7,13,14].

---

[1] Throughout, this paper, for integers $i, j$, we denote $\{i, i+1, ..., j\}$ by $[i, j]$.

We focus on (mixed) covering arrays which have the desired properties of corresponding to test suites that guarantee the coverage of all $t$-way interactions.

**Definition 2.** *A* mixed covering array (MCA) *is an $N \times k$ array $A$, with each column $i \in [1, k]$ having symbols from the alphabet $[0, g_i - 1]$, that satisfies the following requirement. For each $\{i_1, ..., i_t\} \subseteq \{1, ..., k\}$, consider the $N \times t$ subarray of $A$ obtained by selecting columns $i_1, ..., i_t$; there are $\prod_{j=1}^{t} g_{i_j}$ distinct $t$-tuples that could appear as a row, and we require that each appear at least once. We denote such an array as an $MCA(N; t, k, (g_1, g_2, ..., g_k))$. The minimum integer $N$ for which an $MCA(N; t, k, (g_1, ..., g_k))$ exists is called the* MCA number *and we denote it by $MCAN(t, k, (g_1, ..., g_k))$. When $g_i = g$ for all $i \in [1, k]$ then we call it a* covering array, *denoted by $CA(N; t, k, g)$, with* CA number $CAN(t, k, g)$.

For a survey of constructions for CAs and MCAs and their applications to testing, see [5]. The number of tests in a test suite built from an MCA is much smaller than in exhaustive testing, since for fixed $k$ and $g \geq \max\{g_i | i \in [1, k]\}$, $MCAN(t, k, (g_1, ..., g_k)) \leq CAN(t, k, g) \in O(\log k)$ (see [5]), while exhaustive testing would use $\prod_{i=1}^{k} g_i \in O(g^k)$ tests.

Unfortunately, in practice, testing problems are even more complicated, and they frequently come with extra constraints. For many reasons, some particular $t$-way interactions of a given testing problem may need to be forbidden from all tests. For example, some combinations of components in a highly-configurable software system can be invalid. Consider the example in Table 1 This system contains some inherent constraints, as given in Table 2. For example, video ringtones cannot be used without the presence of a video camera. In this case, the system has seven forbidden pairwise interactions and one forbidden 3-way interaction. An $MCA(N; 2, 5, (3, 3, 3, 2, 2))$ would provide a suite of tests which guarantees the coverage of all pairwise interactions, but would ignore these constraints. Consequently, some of the tests generated by the MCA simply could not take place, resulting in wasted tests. Thus it is desirable to design a minimal suite of tests which cover all permitted interactions, but which avoid the forbidden interactions. Experiments involving material mixtures provide an example of a testing problem where ignoring forbidden interactions could be deadly. These types of experiments may combine materials in order to produce mixtures with improved properties such as strength and flexibility, but absolutely must avoid creating known explosive or toxic combinations. Cawse [3] supports the use of covering arrays for the design of such experiments, but the ability to avoid the dangerous combinations is essential.

Hartman and Raskin [11] address the need for forbidden configurations in testing applications, although their proposed solution requires an exhaustive list of all invalid tests (not simply a list of the forbidden interactions themselves). Cohen et. al. [4] define constrained covering arrays and present a general technique for representing constraints so that existing algorithms can now handle constraints. Danziger et. al. [8] use graphs to represent forbidden pairwise interactions of a testing problem and define covering arrays avoiding forbidden edges (CAFEs); we follow this approach in this paper.

**Table 2.** Constraints on the mobile phone product line

| Constraints | Forbidden Interactions |
|---|---|
| (C1) graphical email viewer **requires** a colour display | $\{(1,2),(2,0)\}$ |
| (C2) 2 megapixel camera **requires** a colour display | $\{(1,2),(3,0)\}$ |
| (C3) graphical email viewer is **not supported** with 2 megapixel camera | $\{(2,0),(3,0)\}$ |
| (C4) 8 million colour display **does not support** a 2 megapixel camera | $\{(1,1),(3,0)\}$ |
| (C5) video camera **requires** a camera and a colour display | $\{(3,2),(4,0)\}$ $\{(1,2),(4,0)\}$ |
| (C6) video ringtones **cannot occur** without a video camera | $\{(4,1),(5,0)\}$ |
| (C7) the combination of 16 million colours, text, and 2 megapixel camera will **not be supported** | $\{(1,0),(2,1),(3,0)\}$ |

In general, the constraints imposed on a testing problem can result in forbidden interactions of any size (forbidden $t$-way interactions for any $t \in [1,k]$). For example, the constraint (C7) of Table 2 yields a forbidden 3-way interaction. These situations can be modeled using hypergraphs to represent the forbidden interactions; however, we concentrate solely on the simpler case, where all forbidden interactions are pairwise interactions. From now on, we only consider the problem of covering all pairwise interactions of a given testing problem, that is, we fix $t = 2$. Moreover, if a testing problem has an associated forbidden interaction set, we assume all forbidden interactions are pairwise interactions. Unless otherwise stated, we refer to pairwise interactions simply as interactions.

In this paper, we study CAFEs and prove several hardness results. In Section 2, we give definitions of CAFEs and related objects, including uniform edge clique covers and error-locating arrays. In Section 3 and 4, we provide some new results and bounds on uniform ECCs and on CAFEs, respectively. In Section 5, we show that the problem of finding a CAFE of minimum size is NP-hard, even for the binary alphabet case (where all $g_i = 2$, for all $i \in [1,k]$). We use this result to show that the problem of finding a minimum error-locating array is also NP-hard, even for the binary alphabet case. We also provide a hardness of approximation result for CAFEs.

## 2   Definitions and Preliminaries

Given a testing problem $\mathrm{TP}(k, (g_1, ..., g_k))$ and an associated *forbidden (pairwise) interaction set*, say $\mathfrak{F} = \{I | I$ is a forbidden interaction of $\mathrm{TP}(k, (g_1, ..., g_k))\}$, we represent the forbidden interactions using a $k$-partite graph $G$ that is a member of the following family of graphs. The *family of forbidden edges graphs*, denoted by $\mathcal{G}_{(g_1,...,g_k)}$, is the family of $k$-partite graphs having parts of sizes $g_1, ..., g_k$. Furthermore the vertices of any $G \in \mathcal{G}_{(g_1,...,g_k)}$ are labeled $v_{i,a_i}$ where $i \in [1,k]$

and $a_i \in [0, g_i - 1]$, so that the respective parts are of the form $P_i = \{v_{i,a_i} | a_i \in [0, g_i - 1]\}$ for each $i \in [1, k]$. The edge set of any $G \in \mathcal{G}_{(g_1,...,g_k)}$ is a subset of the edge set of the complete $k$-partite graph $K_{(g_1,...,g_k)}$ with vertices labeled likewise. In the particular case when $g_1 = g_2 = \cdots = g_k = g$, we denote the family of forbidden edges graphs with uniform alphabet size $g$ as $\mathcal{G}_{k,g}$, and the complete $k$-partite graph in $\mathcal{G}_{k,g}$ as $K_{k,g}$.

For each $G \in \mathcal{G}_{(g_1,...,g_k)}$ there is a corresponding testing problem TP $(k, (g_1, ..., g_k))$ with forbidden interaction set $\mathfrak{F}$ such that $G$ contains one vertex for each value of each factor in the corresponding testing problem, and interaction $I = \{(i, a_i), (j, a_j)\} \in \mathfrak{F}$ if and only if $\{v_{i,a_i}, v_{j,a_j}\} \in E(G)$. In this case, we call $G$ the *forbidden edges graph* for the testing problem TP$(k, (g_1, ..., g_k))$ with forbidden interaction set $\mathfrak{F}$. It is sometimes convenient for us to refer to an interaction $I = \{(i, a_i), (j, a_j)\}$ simply as the pair of vertices $\{v_{i,a_i}, v_{j,a_j}\}$. Then, if $\{v_{i,a_i}, v_{j,a_j}\} \notin E(G)$, we have a *non-forbidden interaction*, and if $\{v_{i,a_i}, v_{j,a_j}\} \in E(G)$, we have a *forbidden interaction*.

Although we never allow a test to assign two distinct values of a given factor simultaneously, we do not add these "implicitly forbidden" interactions to the forbidden edges graph. However, it is sometimes convenient to consider these edges. We denote by $G^|$ the graph obtained from a forbidden edges graph $G \in \mathcal{G}_{(g_1,...,g_k)}$ by adding to $G$ all edges of the form $\{v_{i,a_i}, v_{i,b_i}\}$ for each factor $i \in [1, k]$ and for every two distinct values $a_i \neq b_i$ such that $a_i, b_i \in [0, g_i - 1]$.

Given a testing problem TP$(k, (g_1, ..., g_k))$, we say that a $k$-tuple $T = (T_1, ..., T_k)$ *avoids* interaction $I = \{(f_1, a_{f_1}), ..., (f_t, a_{f_t})\}$ if $T$ does not cover $I$. A $k$-tuple $T = (T_1, ..., T_k) \in [0, g_1 - 1] \times \cdots \times [0, g_k - 1]$ is said to *avoid* the forbidden edges graph $G \in \mathcal{G}_{(g_1,...,g_k)}$ if for all $i, j \in [1, k]$, we have $\{v_{i,T_i}, v_{j,T_j}\} \notin E(G)$. Note that if a $k$-tuple $T = (T_1, ..., T_k)$ avoids a graph $G$, then the set of vertices $\{v_{1,T_1}, ..., v_{k,T_k}\}$ is an independent set of $G$ and of $G^|$.

We now define CAFEs, a generalization of MCAs that considers forbidden interactions.

**Definition 3.** *[8] A covering array avoiding forbidden edges (CAFE) of a graph $G \in \mathcal{G}_{(g_1,...,g_k)}$, is an $N \times k$ array $A$, with each column $i$ having symbols from the alphabet $[0, g_i - 1]$, and denoted CAFE$(N, G)$, such that:*

1. *each row of $A$ forms a $k$-tuple avoiding $G$, and*
2. *for all $v_{i,a_i}, v_{j,a_j} \in V(G)$ with $i \neq j$, if $\{v_{i,a_i}, v_{j,a_j}\} \notin E(G)$, then there exists a row $R_l = (R_l(1), ..., R_l(k))$ of $A$ such that $R_l(i) = a_i$ and $R_l(j) = a_j$.*

*The* CAFE *number of a forbidden edges graph $G$, denoted by CAFEN$(G)$, is the minimum integer $N$ for which a CAFE$(N, G)$ exists, if a CAFE of $G$ exists, or $+\infty$ otherwise.*

Not every graph $G \in \mathcal{G}_{(g_1,...,g_k)}$ admits a CAFE. An interaction $I = \{(i, a_i), (j, a_j)\}$ is said to be *consistent* with $G$ if there exists a $k$-tuple $T$ with $T_i = a_i$ and $T_j = a_j$ that avoids $G$. The graph $G$ is *consistent* if all interactions $\{(i, a_i), (j, a_j)\}$ such that $i \neq j$ and $\{v_{i,a_i}, v_{j,a_j}\} \notin E(G)$ are consistent with $G$. Indeed, there exists a CAFE$(n, G)$ for a forbidden edges graph $G$ if and only if $G$ is consistent.

For the particular case of *binary* forbidden edges graphs, that is, for graphs $G \in \mathcal{G}_{k,2}$ corresponding to CAFEs with binary alphabets, we have the following result by Danziger et. al. [8] which characterizes their consistency.

**Proposition 1.** *(Danziger et. al. [8]) Let $G \in \mathcal{G}_{k,2}$ be a forbidden edges graph with vertex set $V(G) = \{v_{i,a} | 1 \leq i \leq k, a \in \{0, 1\}\}$. Then $G$ is consistent if and only if*

1. *$\{v_{i,a}, v_{j,b}\} \in E(G)$ whenever $i \neq j$ and there exist vertices in the same factor, say $v_{l,c}$ and $v_{l,1-c}$, such that $l \neq i, l \neq j$ and $\{v_{i,a}, v_{l,c}\} \in E(G)$ and $\{v_{j,b}, v_{l,1-c}\} \in E(G)$, and*
2. *$\{v_{i,a}, v_{j,b}\} \in E(G)$ for all $j \in [1, k] \setminus \{i\}$ whenever there exist vertices in the same factor, say $v_{l,0}$ and $v_{l,1}$ such that $l \neq i$, $l \neq j$, and $\{v_{i,a}, v_{l,0}\} \in E(G)$ and $\{v_{i,a}, v_{l,1}\} \in E(G)$.*

**Proposition 2.** *(Danziger et. al. [8]) Let $G \in \mathcal{G}_{(g_1,\ldots,g_k)}$ be a consistent forbidden edges graph. Let $E^{i,j}(G)$ denote the set of edges with one end in factor $i$ and the other end in factor $j$. Then*

$$\max_{1 \leq i < j \leq k} \{g_i g_j - |E^{i,j}(G)|\} \leq CAFEN(G) \leq \sum_{1 \leq i < j \leq k} (g_i g_j - |E^{i,j}(G)|).$$

The lower and upper bounds of Proposition 2 are attained for all forbidden edges graphs $G \in \mathcal{G}_{(g_1,g_2)}$ with only $k = 2$ factors, since in this case lower and upper bounds match.

In Section 4, we prove that the upper bound is never attained by any consistent forbidden edges graph with $k \geq 3$ factors. The lower bound, however, can be attained for all $k \geq 3$ by a specific consistent graph $G \in \mathcal{G}_{(g_1,g_2,1,\ldots,1)}$ such that all of its forbidden interactions lie between factors 1 and 2.

Other closely related objects are error-locating arrays (ELAs), which are generalization of covering arrays that allows the determination of the pairwise failing interactions [18].

**Definition 4.** *Let $G \in \mathcal{G}_{(g_1,\ldots,g_k)}$. A $k$-tuple $T = (T_1, ..., T_k) \in [0, g_i - 1] \times \cdots \times [0, g_k - 1]$ is said to* locate *interaction $I = \{(i, a_i), (j, a_j)\}$ if $T_i = a_i$ and $T_j = a_j$, and for every other interaction $\{(p, a_p), (q, a_q)\} \neq I$ that $T$ covers we have $\{v_{p,a_p}, v_{q,a_q}\} \notin E(G)$. In this case we say that interaction $I$ is located by $T$. An* error-locating array *(ELA) for a graph $G \in \mathcal{G}_{(g_1,\ldots,g_k)}$, denoted by $ELA(N, G)$, is an $N \times k$ array $A$, with each column $i$ having symbols from the alphabet $[0, g_i - 1]$, such that every interaction $\{(i, a_i), (j, a_j)\}$ corresponding to a pair of vertices $v_{i,a_i}, v_{j,a_j} \in V(G)$ with $i \neq j$ (corresponding to an edge, or a non-edge of $G$) is located by a $k$-tuple corresponding to some row of $A$. If for some $N \in \mathbb{Z}$ there exists an $ELA(N, G)$, then we say that $G$ is* locatable. *The ELA number of $G$, denoted by $ELAN(G)$, is the smallest $N$ such that an $ELA(N, G)$ exists, if there exists an ELA for $G$, or $+\infty$ otherwise.*

The next result shows a strong relationship between ELAs and CAFEs.

**Theorem 1.** *(Danziger et. al. [8]) If $G \in \mathcal{G}_{(g_1,...,g_k)}$ is locatable, then*

$$ELAN(G) = CAFEN(G) + |E(G)|.$$

We now look at the relationship of CAFEs and edge clique covers of graphs.

**Definition 5.** *Let $G$ be a simple graph. If $C$ is a clique of $G$ and $e$ is an edge of $G$, we say that the clique $C$ covers $e$ if the ends of $e$ belong to $C$. That is, if $e = \{u,v\}$ and $u,v \in C$, then $C$ covers $e$. If $\mathfrak{C} = \{C_1,...,C_N\}$ is a collection of $N$ cliques of $G$ such that for every edge $e \in E(G)$ there is at least one clique $C_i \in \mathfrak{C}$ that covers $e$, then we say that $\mathfrak{C}$ is an edge clique cover (ECC) of $G$. We say that an ECC of $G$, $\mathfrak{C}$, is optimal if there is no ECC of $G$, say $\mathfrak{C}'$, such that $|\mathfrak{C}'| < |\mathfrak{C}|$. The number of cliques in an optimal ECC of $G$ is called the ECC number of $G$, and is denoted by $\theta'(G)$.*

For results on edge clique covers and $\theta'(G)$ see [1,9,10,12,15,20]. A variation on the ECC problem is the restriction on the size of all the cliques.

**Definition 6.** *Let $G$ be a simple graph and $k$ be an integer. An ECC of $G$, $\mathfrak{C}$, is said to be $k$-uniform if every clique in $\mathfrak{C}$ has cardinality $k$. We call $\mathfrak{C}$ a $k$-ECC of $G$ for short. We define the $k$-uniform ECC number of $G$ to be the size of a minimum $k$-ECC of $G$ if it exists, or $+\infty$ if one does not exist, and denote it by $\theta'_k(G)$. An ECC of $G$ is said to be uniform if it is $k$-uniform for some integer $k$.*

Comparing the $k$-uniform ECC number of a graph $G$ to its ECC number, we always have $\theta'(G) \le \theta'_k(G)$, since $\theta'_k(G) = +\infty$ if a $k$-ECC of $G$ does not exist, and if $\theta'_k(G) \ne +\infty$, then every $k$-uniform ECC of $G$ is in particular an ECC of $G$. Strict inequality holds, for example, in the case where $G = K_n$ for some $n > k$. Then $\theta'(K_n) = 1 < \theta'_k(K_n)$. Now, suppose for a graph $G$ we know that $G$ admits a $k$-uniform ECC, then when do we have equality? That is, when does $\theta'(G) = \theta'_k(G)$ hold? Proposition 3 and Theorem 2 give classes of graphs for which this holds.

**Proposition 3.** *Let $k$ be a positive integer. Then $\theta'(K_{(g_1,...,g_k)}) = \theta'_k(K_{(g_1,...,g_k)})$.*

Since an $MCA(N; 2, k, (g_1,...,g_k))$ can be shown to be equivalent to a $k$-uniform ECC, containing $N$ cliques, of the complete $k$-partite graph $K_{(g_1,...,g_k)}$, we obtain the following corollary.

**Corollary 1.** *$MCAN(2, k, (g_1,...,g_k)) = \theta'_k(K_{(g_1,...,g_k)}) = \theta'(K_{(g_1,...,g_k)})$. In particular, we have $CAN(2, k, g) = \theta'_k(K_{k,g}) = \theta'(K_{k,g})$.*

Indeed, we can now improve on Orlin's bound for $\theta'(K_{k,2})$ [20], which is linear in $k$, using the exact value for binary strength 2 covering arrays which is known to be in $O(\log k)$ (see [5]).

**Corollary 2.** *Let $k$ be a positive integer. Then*

$$\theta'(K_{k,2}) = \theta'_k(K_{k,2}) = CAN(2,k,2) = \min\left\{ N \in \mathbb{Z} \Big| \binom{N-1}{\lfloor\frac{N}{2}\rfloor - 1} \ge k \right\} = O(\log k).$$

The next proposition gives us the equivalence between a $CAFE(N, G)$, where $G \in \mathcal{G}_{(g_1,...,g_k)}$, and a $k$-uniform ECC of the complement of $G^{|}$.

**Proposition 4.** *(Danziger et. al.[8]) Let $k$ be a positive integer and let $G \in \mathcal{G}_{(g_1,...,g_k)}$ be a forbidden edges graph. Then there exists a $CAFE(N, G)$ if and only if there exists a $k$-uniform ECC, containing $N$ cliques, of the graph $\overline{G^{|}}$.*

**Corollary 3.** *(Danziger et. al. [8]) Let $G \in \mathcal{G}_{(g_1,...,g_k)}$. Then $CAFEN(G) = \theta'_k(\overline{G^{|}}) \geq \theta'(\overline{G^{|}})$.*

In fact, in the case of binary alphabets the above inequality is an equality.

**Theorem 2.** *(Danziger et. al. [8]) Let $G \in \mathcal{G}_{k,2}$ be a binary forbidden edges graph. If $G$ is consistent, then $CAFEN(G) = \theta'_k(\overline{G^{|}}) = \theta'(\overline{G^{|}})$.*

# 3   New Results for Uniform ECCs

In this section, we provide a number of results for uniform ECCs, but the proofs are omitted here. Using the hand-shaking Lemma for graphs, we obtain the following result.

**Proposition 5.** *Let $G$ be a simple graph and let $k \geq 2$ be an integer. If $G$ admits a $k$-uniform edge clique cover then $|E(G)| \geq \frac{n(k-1)}{2}$, where $n$ is the number of non-isolated vertices of $G$.*

The next propositions shows classes of graphs in which $k$-ECC and ECC numbers are the same.

**Proposition 6.** *If a simple graph $G$ satisfies $\omega(G) = 2$ (triangle-free graphs), then $\theta'(G) = \theta'_2(G)$.*

**Proposition 7.** *Let $G$ be a simple graph satisfying $\omega(G) = 3$. Furthermore, assume that $G$ admits a 3-ECC, that is, assume $\theta'_3(G) \neq +\infty$. Then $\theta'(G) = \theta'_3(G)$.*

The statement that a graph $G$ satisfies $\theta'(G) = \theta'_k(G)$ whenever $\omega(G) = k$ and $\theta'_k(G) \neq +\infty$, however, does not hold in general, as we show next.

**Theorem 3.** *Let $k \geq 4$ and $m \geq 1$ be integers. Then there exists a graph $G$ such that $\omega(G) = k$ and $\theta'_k(G) = \theta'(G) + m$.*

In the following results, we give upper bounds on the $k$-uniform ECC number.

**Proposition 8.** *Let $k \geq 4$ be a positive integer and let $G$ be a simple graph such that $\omega(G) = k$ and $\theta'_k(G) \neq +\infty$. Then*

$$\theta'_k(G) \leq \binom{k-1}{2}\theta'(G).$$

**Proposition 9.** *Let $G$ be a simple graph and let $k \geq 2$ be a positive integer. If $G$ admits a $k$-uniform ECC then*

$$\theta'_k(G) \leq |E(G)| - \binom{k}{2} + 1.$$

## 4   New Results for CAFEs

Next, we give a necessary condition for a CAFE to exist, as well as some new upper bounds on the CAFE number, based on its relationship with the $k$-ECC number. We omit their proofs.

**Proposition 10.** *Let $G \in \mathcal{G}_{(g_1,\ldots,g_k)}$ and assume that for every vertex $v_{i,a_i} \in V(G)$ there exists at least one vertex $v_{j,a_j} \in V(G)$ such that $i \neq j$ and $\{v_{i,a_i}, v_{j,a_j}\} \notin E(G)$ (i.e., there are no "dummy" vertices). Then $CAFEN(G) \neq +\infty$ implies*

$$|E(G)| \leq \sum_{1 \leq i < j \leq k} g_i g_j - \left(\frac{k-1}{2}\right) \sum_{i=1}^{k} g_i.$$

The following result gives an upper bound on the CAFE number, and for $k \geq 3$, it is a strict improvement on the upper bound given in Proposition 2.

**Proposition 11.** *Let $G \in \mathcal{G}_{(g_1,\ldots,g_k)}$. If $CAFEN(G) \neq +\infty$, then*

$$CAFEN(G) \leq \sum_{1 \leq i < j \leq k} g_i g_j - |E(G)| - \binom{k}{2} + 1.$$

Here we translate Proposition 8 and Proposition 7 into results for CAFEs, respectively.

**Proposition 12.** *Let $k \geq 4$ and let $G \in \mathcal{G}_{(g_1,\ldots,g_k)}$. If $CAFEN(G) \neq +\infty$ then,*

$$CAFEN(G) \leq \binom{k-1}{2} \theta'(\overline{G^{\mathsf{T}}}).$$

**Proposition 13.** *Let $G \in \mathcal{G}_{(g_1,g_2,g_3)}$ be a consistent forbidden edges graph with $k = 3$ factors. Then $CAFEN(G) = \theta'_3(\overline{G^{\mathsf{T}}}) = \theta'(\overline{G^{\mathsf{T}}})$.*

## 5   Complexity of Problems Related to CAFEs

### 5.1   Decision Problems Related to CAFEs and Previous Results

We consider the following decision problems corresponding to the existence of a $CAFE(n, G)$ and the determination of $CAFEN(G)$, for a graph $G \in \mathcal{G}_{(g_1,\ldots,g_k)}$:

COVER&AVOID$=\{G \in \mathcal{G}_{(g_1,\ldots,g_k)} \mid$ for some $N$ there exists a $CAFE(N, G)\}$,

CAFEN $= \{(G, N) \in \mathcal{G}_{(g_1,\ldots,g_k)} \times \mathbb{Z} \mid$ there exists a $CAFE(N, G)\}$.

Furthermore, for each language $L$ defined above and below, we use the notation $g$-$L$ to describe the language where the graph input $G$ is of the particular form $G \in \mathcal{G}_{k,g}$. For example, 2-CAFEN$=\{(G, N) \in \mathcal{G}_{k,2} \times \mathbb{Z} \mid$ there exists a $CAFE(N, G)\}$.

Danziger et. al. [8] has shown that $g$-COVER&AVOID and $g$-CAFEN are NP-complete for all $g \geq 5$. On the other hand, they show that 2-COVER&AVOID $\in$ $P$, leaving the suspicion that 2-CAFEN be polynomial-time solvable. Indeed in the next section we show that $g$-CAFEN is NP-complete for all $g \geq 2$, providing an answer to the open cases $g = 2, 3, 4$.

We also consider the languages related to the ECC problems:

$$\text{ECCN} = \{(G, N) \in \mathbb{G} \times \mathbb{Z} \mid \theta'(G) \leq N\},$$

$$\text{UNIFORM-ECCN} = \{(G, N, k) \in \mathbb{G} \times \mathbb{Z} \times \mathbb{Z} \mid \theta'_k(G) \leq N\}.$$

Kou, Stockmeyer and Wong [12] and independently Orlin [20] have shown that ECCN is NP-complete. In the next section, we obtain that UNIFORM-ECCN is also NP-complete.

We now look at the related problem of error-locating arrays. We consider the following decision problems corresponding to the existence of a $\text{ELA}(N, G)$ and the determination of $\text{ELAN}(G)$, for a graph $G \in \mathcal{G}_{(g_1,...,g_k)}$:

$$\text{LOCATE} = \{G \in \mathcal{G}_{(g_1,...,g_k)} \mid \text{there exists an } \text{ELA}(N, G) \text{ for some } N \in \mathbb{Z} \text{ ($G$ is locatable)}\},$$

$$\text{ELAN} = \{(G, N) \in \mathcal{G}_{(g_1,...,g_k)} \times \mathbb{Z} \mid \text{there exists an } \text{ELA}(N, G)\}.$$

Martinez et. al. [18] have shown that $g$-LOCATE is NP-complete for $g \geq 5$, which implies $g$-ELAN is NP-complete for $g \geq 5$. However, since 2-LOCATE $\in P$, we could conceive that 2-ELAN might be in P. In Section 5.3, we show that $g$-ELAN is NP-complete for all $g \geq 2$, providing an answer to the open cases $g = 2, 3, 4$.

## 5.2 NP-Completeness and Hardness of Approximation of g-CAFEN, for $g \geq 2$

Throughout this section we assume without loss of generality that the original graph $G$ is nonempty. Let $N_G(v)$ denote the neighbours of $v$ in $G$, that is $N_G(v) = \{u \in V(G) : \{u, v\} \in E(G)\}$. The reduction algorithm we use to prove our main result is given next.

**Algorithm 1.** *Let $\nu \geq 2$ and let $G$ be a simple nonempty graph on $\nu$ vertices. We construct another simple graph, $G_{UV}$, on $2(k + 2)$ vertices, where $k$ is the number of non-isolated vertices in $G$, such that $\theta'(G) + 2 = CAFEN(G_{UV})$.*

1. *Remove all isolated vertices from $G$ to obtain a new graph $G_k$ on $k$ non-isolated vertices, which we denote by $\{v_1, v_2, ..., v_k\}$. Since $G$ is nonempty, $k \geq 2$.*
2. *Take the complement, $\overline{G_k}$, of $G_k$.*
3. *Add two extra vertices, $v_{k+1}$ and $v_{k+2}$, and add edges joining $v_{k+1}$ to each $v_i$ for $1 \leq i \leq k$. Moreover, join the vertex $v_{k+2}$ to all vertices $v_i$ for $1 \leq i \leq k + 1$. Refer to the resulting graph as $G_V$ and denote its vertex set as $V = \{v_1, v_2, ..., v_k, v_{k+1}, v_{k+2}\}$.*
4. *Construct graph $G_{UV}$ from $G_V$ by adding the vertex set $U = \{u_1, u_2, ..., u_{k+2}\}$ to $G_V$, and adding edges according to the **Across Edge Rule**: we add edge $\{v_i, u_j\}$ to $E(G_{UV})$ if and only if $i \neq j$ and $N_{G_V}(v_j) \subseteq N_{G_V}(v_i)$. Any edge joining a vertex in $U$ to a vertex in $V$, we refer to as an across edge. Any pair of vertices $u_i \in U$ and $v_j \in V$ which are not joined to each other by an edge we refer to as an across non-edge.*

Now, let us observe a few properties of $G_{UV}$.

**Proposition 14.** *Let $G$ be a simple nonempty graph, and let $G_{UV}$ be the graph obtained from $G$ using Algorithm 1. Let $i \in [1, k+2]$. Then,*

1. $\{v_i, u_{k+1}\} \notin E(G_{UV})$, *and* $\{v_i, u_{k+2}\} \notin E(G_{UV})$,
2. $\{u_i, v_{k+1}\} \notin E(G_{UV})$ *and* $\{u_i, v_{k+2}\} \notin E(G_{UV})$,
3. *if* $i \notin \{k+1, k+2\}$, *then* $N_{G_V}(v_i) \neq V \setminus \{v_i\}$.

The following result gives an equivalent statement for the Across Edge Rule.

**Corollary 4.** *For two distinct vertices $v_i, v_j \in V$, $\{v_i, u_j\} \in E(G_{UV})$ if and only if $\{v_i, v_j\} \in E(G_k)$ and $N_{G_k}(v_i) \setminus \{v_j\} \subseteq N_{G_k}(v_j) \setminus \{v_i\}$.*

Now, by Proposition 14, we get the following lemma.

**Lemma 1.** *Let $G$ be a simple graph. Then $I = \{u_1, \ldots, u_k, u_{k+1}, v_{k+2}\}$ is the only independent set of size $k+2$ of $G_{UV}$ that contains both vertices $u_{k+1}$ and $v_{k+2}$, and $I' = \{u_1, \ldots, u_k, v_{k+1}, u_{k+2}\}$ is the only independent set of size $k+2$ of $G_{UV}$ that contains both vertices $v_{k+1}$ and $u_{k+2}$.*

*Remark 1.* The graph $G_{UV}$ is a graph in $\mathcal{G}_{k+2,2}$, with each vertex $v_i$ of $G_{UV}$ representing the vertex $v_{i,1}$ and each vertex $u_i$ representing $v_{i,0}$, where $1 \leq i \leq k+2$. Given an interaction $I = \{(i, a_i), (j, a_j)\}$, we refer to $I$ as a **zero-zero, zero-one,** or **one-one** interaction, if $a_i = a_j = 0$, $\{a_i, a_j\} = \{0, 1\}$, or $a_i = a_j = 1$, respectively. Moreover, if $I$ is not forbidden, that is, if $\{v_{i,a_i}, v_{j,a_j}\} \notin E(G_{UV})$, then we call $I$ a **required** interaction.

**Lemma 2.** *Let $G$ be a simple graph and let $A$ be a $CAFE(N, G_{UV})$, for some $N$. Then each row of $A$, $R_i$, corresponds to an independent set of $G_{UV}$, namely $I_i = \{v_j | R_i(j) = 1\} \cup \{u_j | R_i(j) = 0\}$ and $|I_i| = k+2$.*

**Lemma 3.** *Let $G$ be a simple graph. A required one-one interaction of the graph $G_{UV}$ corresponds to an edge of the original graph $G$.*

**Proposition 15.** *Let $G$ be a nonempty simple graph. Then the graph $G_{UV}$, constructed by Algorithm 1, is consistent.*

*Proof.* Since $G_{UV}$ is a loopless binary CAFE graph and also has the property that none of its zero vertices $u_i$ are joined by any edge, there are only three possibilities for an inconsistency to occur according to Proposition 1. First, by condition 2. of Proposition 1, we would have an inconsistency if for some $i \neq j$ we have $\{v_i, u_j\} \in E(G_{UV})$ and $\{v_i, v_j\} \in E(G_{UV})$, but $\{v_i, x\} \notin E(G_{UV})$ for some $x \in V(G_{UV})$. However, this would never occur because the across edge $\{v_i, u_j\}$ would not be added when computing $G_{UV}$ since $N_{G_V}(v_j) \not\subseteq N_{G_V}(v_i)$. By condition 1. of Proposition 1, $G_{UV}$ would not be consistent, if for three distinct indices $i, j, l \in [1, k+2]$ we have $\{v_i, u_l\} \in E(G_{UV})$, $\{v_j, v_l\} \in E(G_{UV})$, and $\{v_i, v_j\} \notin E(G_{UV})$. However, since the across edge $\{v_i, u_l\}$ is an edge of $G_{UV}$, we know that $N_{G_V}(v_l) \subseteq N_{G_V}(v_i)$. This is a contradiction since $v_j \in N_{G_V}(v_l)$ but $v_j \notin N_{G_V}(v_i)$. Thus, $G_{UV}$ cannot contain such an inconsistency. By condition 1.

of Proposition 1, we could also have an inconsistency if for three distinct indices $i, j, l \in [1, k+2]$ we have $\{v_i, u_l\} \in E(G_{UV})$, $\{v_l, u_j\} \in E(G_{UV})$, but $\{v_i, u_j\} \notin E(G_{UV})$. By the Across Edge Rule, we have $N_{G_V}(v_l) \subseteq N_{G_V}(v_i)$ and $N_{G_V}(v_j) \subseteq N_{G_V}(v_l)$. Therefore, we have $N_{G_V}(v_j) \subseteq N_{G_V}(v_i)$, and so $\{v_i, u_j\}$ must be an edge of $G_{UV}$. Therefore, $G_{UV}$ is consistent.                                  $\square$

**Proposition 16.** *Let $G$ be a nonempty simple graph and let $G_{UV}$ be graph constructed by Algorithm 1. Let $C$ be a clique of $G$ that is maximal with respect to set inclusion such that $|C| > 1$. Then, $I = \{v_i \in V | v_i \in C\} \cup \{u_i \in U | v_i \notin C\}$ forms an independent set of size $k+2$ of $G_{UV}$.*

**Theorem 4.** *Let $G$ be a nonempty simple graph and let $G_{UV}$ be the graph obtained from $G$ by applying Algorithm 1. Then $\theta'(G) + 2 = CAFEN(G_{UV})$.*

*Proof.* First we show that $CAFEN(G_{UV}) \leq \theta'(G) + 2$. Let $N = \theta'(G)$ and let $\mathfrak{C} = \{C_1, ..., C_N\}$ be an optimal ECC of $G$. Assume w.l.o.g. that each $C_i$ is a maximal clique with respect to set inclusion. By Proposition 16, for $1 \leq i \leq N$, we can build a row, $R_i = (R_i(1), R_i(2), ..., R_i(k+2))$, corresponding to each clique $C_i \in \mathfrak{C}$, by taking $R_i(j) = 1$ whenever $v_j \in C_i$ and $R_i(j) = 0$ whenever $v_j \notin C_i$. Since $G_{UV}$ is constructed so that $v_{k+1}$ and $v_{k+2}$ are both joined by an edge to every other vertex in $V$, we see that covering all interactions of the form $\{v_i, v_j\}$ where $i, j \in [1, k]$ is sufficient to cover all the required one-one interactions of $G_{UV}$. Since the required one-one interactions of $G_{UV}$ correspond exactly to the edges of the original graph $G$, we see that the $N$ rows, $R_1, ..., R_N$ do indeed cover the required one-one interactions of $G_{UV}$. Note that every row $R_i$ corresponding to the clique $C_i \in \mathfrak{C}$ must also cover the interaction $\{u_{k+1}, u_{k+2}\}$ since $v_{k+1} \notin C_i$ and $v_{k+2} \notin C_i$ for each $C_i \in \mathfrak{C}$.

Now, we build row $R_{N+1}$ corresponding to the independent set $I_{N+1} = \{u_1, ..., u_{k+1}, v_{k+2}\}$, possible by Lemma 1. The row $R_{N+1}$ is sufficient to cover all the required zero-zero interactions between the vertices $u_1, ..., u_{k+1}$, as well as all the required interactions of the form $\{u_i, v_{k+2}\}$ where $1 \leq i \leq k+1$. Finally, we build row $R_{N+2}$ corresponding to the independent set $I_{N+2} = \{u_1, ..., u_k, v_{k+1}, u_{k+2}\}$, possible by Lemma 1. The row $R_{N+2}$ is sufficient to cover all the required zero-zero interactions of the form $\{u_i, u_{k+2}\}$, as well as all the required interactions of the form $\{u_i, v_{k+1}\}$ where $1 \leq i \leq k$ or $i = k+2$.

We claim that $R_1, ..., R_{N+2}$ are sufficient to cover all the across non-edges of $G_{UV}$, i.e., all required zero-one interactions. Let $v_i \in V$ and $u_j \in U$ such that $\{v_i, u_j\} \notin E(G_{UV})$. We have two possible cases.

*Case 1:* $\{v_i, v_j\} \notin E(G_{UV})$. By the Across Edge Rule, we know that $N_{G_V}(v_j) \not\subseteq N_{G_V}(v_i)$, otherwise $\{v_i, u_j\}$ would be an edge of $G_{UV}$. Hence, there exists a vertex $v_l \in N_{G_V}(v_j)$ such that $v_l \notin N_{G_V}(v_i)$. This means that $\{v_i, v_l\}$ is a non-edge corresponding to a required one-one interaction and therefore was already covered by $R_p$, for some $p \in [1, N]$. Row $R_p$ must also cover $\{v_i, u_j\}$.

*Case 2:* $\{v_i, v_j\} \in E(G_{UV})$. If $i = k+1$, we are done since row $R_{N+2}$ covers all required interactions of the form $\{v_{k+1}, u_j\}$ where $1 \leq j \leq k$ or $j = k+2$. Similarly, if $i = k+2$ we are done since the row $R_{N+1}$ covers all required interactions of the form $\{v_{k+2}, u_j\}$ where $1 \leq j \leq k+1$. If $1 \leq i \leq k$ then by Proposition 14 we know that $N_{G_V}(v_i) \neq V \setminus \{v_i\}$. So there is a vertex $v_l \in V$ such that $v_l \notin N_{G_V}(v_i)$. Thus $\{v_i, v_l\} \notin E(G_{UV})$ is a required one-one interaction and it is covered by row $R_p$, for some $p \in [1, N]$, which forces the across non-edge $\{v_i, u_j\}$ to be covered.

Hence, we have $\mathrm{CAFEN}(G_{UV}) \leq N + 2$ whenever $\theta'(G) = N$.

Next we show that $\theta'(G) + 2 \leq \mathrm{CAFEN}(G_{UV})$. Suppose we have an optimal $\mathrm{CAFE}(N, G_{UV})$ with $N$ rows. By Lemma 1, the only row that can cover the interaction $\{u_{k+1}, v_{k+2}\}$ is the one corresponding to the independent set $I_1 = \{u_1, ..., u_{k+1}, v_{k+2}\}$. Call this row $R_1$. In addition, the only row that can cover the interaction $\{v_{k+1}, u_{k+2}\}$ is the one corresponding to the independent set $I_2 = \{u_1, ..., u_k, v_{k+1}, u_{k+2}\}$. Call this row $R_2$. Since neither $R_1$ nor $R_2$ cover any one-one interactions, we observe that the remaining $N - 2$ rows of the optimal $\mathrm{CAFE}(n, G_{UV})$ must be sufficient to cover all the one-one interactions of $G_{UV}$. Call these remaining $N - 2$ rows $R_3, ..., R_N$, and name the corresponding independent sets of $G_{UV}$ of size $k+2$, $I_3, ..., I_N$, respectively. Then for $3 \leq i \leq N$, we have an independent set $C_i \subseteq I_i$ where $C_i = \{v_j | v_j \in I_i \text{ and } j \in [1, k]\}$. Thus, each $C_i$ is an independent set of $G_{UV}$ containing only vertices from the set $\{v_1, ..., v_k\}$. In other words, each $C_i$ corresponds to a clique of the original graph $G$, and $C_3, ..., C_N$ cover all the edges of $G$. Therefore, $N - 2 \geq \theta'(G)$. Equivalently, $\theta'(G) + 2 \leq \mathrm{CAFEN}(G_{UV})$.     □

**Corollary 5.** *2-CAFEN is NP-complete.*

*Proof.* Algorithm 1 is a polynomial-time reduction, and Theorem 4 shows ECCN $\leq_P$ 2-CAFEN. Since ECCN is NP-complete, the result follows.     □

**Corollary 6.** *UNIFORM-ECCN is NP-complete.*

*Proof.* An instance $(G, N)$ for 2-CAFEN with $G \in \mathcal{G}_{k,2}$ is a particular instance for UNIFORM-ECCN, namely $(\overline{G^|}, N, k)$, so the result follows from Corollary 5.     □

To obtain a hardness of approximation for 2-CAFEN, we use the following result.

**Theorem 5.** *(Lund and Yannakakis [16]) There exists a $\delta > 0$ such that there does not exist a polynomial-time approximation algorithm $A$ that satisfies $A(G) \leq \nu^\delta \theta'(G)$ for all simple graphs $G$ on $\nu$ vertices, unless $P = NP$.*

The next result follows from Theorem 4 and Theorem 5, taking $\delta' = \delta/3$.

**Theorem 6.** *There exists a $\delta' > 0$ such that there does not exist a polynomial-time approximation algorithm $A'$ that satisfies $A'(G) \leq k^{\delta'} \mathrm{CAFEN}(G)$ for all $G \in \mathcal{G}_{k,2}$, unless $P = NP$.*

Now we look at $g$-CAFEN, for $g \geq 3$.

**Proposition 17.** *For $g \geq 2$ we have $g$-CAFEN $\leq_P$ $(g+1)$-CAFEN.*

*Proof.* Let $G \in \mathcal{G}_{k,g}$ be a graph that is an instance for $g$-CAFEN. Without loss of generality, assume the vertices of $G$ are labeled as $v_{i,a}$ where $i \in [1,k]$ and $a \in [0, g-1]$. Construct a new graph $G'$ from $G$ as follows. Add a new vertex, $v_{i,g}$, to each factor $i \in [1,k]$. Add edges of the form $\{v_{i,g}, v_{j,a}\}$ for all $i \neq j$, $i,j \in [1,k]$ and for all $a \in [0, g-1]$. Moreover, add edges of the form $\{v_{i,g}, v_{j,g}\}$ for all $i \neq j$, $i,j \in [1,k]$. So $G' \in \mathcal{G}_{k,g+1}$. Clearly the non-forbidden interactions of $G'$ correspond exactly to the non-forbidden interactions of $G$, thus CAFEN$(G)$ = CAFEN$(G')$. Therefore $(G, N) \in g$-CAFEN if and only if $(G', N) \in (g+1)$-CAFEN. It is easy to see that $G'$ can be computed from $G$ in polynomial time with respect to the size of the graph $G$, and thus, $g$-CAFEN $\leq_P$ $(g+1)$-CAFEN. $\qquad \square$

**Corollary 7.** *For $g \geq 2$, $g$-CAFEN is NP-complete.*

## 5.3 NP-Completeness of g-ELAN, for $g \geq 2$

We now give a proof for the NP-completeness of $g$-ELAN, for $g \geq 2$, by reducing from ECCN using the following reduction algorithm. The proof is based on Theorem 1.

**Algorithm 2.** *Let $G$ be a simple graph on $\nu$ vertices and let $n$ be the number of non-isolated vertices of $G$. We construct from $G$ the graph $Y_G$ on $g(2n+2)$ vertices as follows.*

1. *Remove all isolated vertices of $G$ to obtain a new graph $G_n$ on $n$ vertices. Denote the vertices of $G_n$ as $V(G_n) = \{v_1, v_2, ..., v_n\}$.*
2. *Add a new set of vertices $V' = \{v_{n+1}, v_{n+2}, ..., v_{2n}\}$ and join by an edge each $v_i$ to the corresponding vertex $v_{n+i} \in V'$ for $1 \leq i \leq n$. In addition, form an $n$-clique between all the vertices of $V'$ by adding the edges $\{v_{n+i}, v_{n+j}\}$ to the graph for $1 \leq i < j \leq n$. Denote this graph by $H_G$.*
3. *Apply Algorithm 1 to the graph $H_G$ obtaining $(H_G)_{UV}$.*
4. *Create $(g-1)$ copies of $(H_G)_{UV}$, and refer to the vertices in copy $i$ as $V^i = \{v_1^i, ..., v_{2n+2}^i\}$ and $U^i = \{u_1^i, ..., u_{2n+2}^i\}$. For $1 \leq j \leq 2n+2$ identify vertices $u_j^i$, for all $1 \leq i \leq g-1$. For every $i_1 \neq i_2$, $1 \leq i_1, i_2 \leq g-1$, and for every $1 \leq j_1, j_2 \leq 2n+2$, add the edge $\{v_{j_1}^{i_1}, v_{j_2}^{i_2}\}$. Refer to this as graph $Y_G$.*

**Theorem 7.** *$g$-ELAN is NP-complete, for $g \geq 2$.*

*Sketch of the Proof.* Let $G$ be any simple graph and let $g \geq 2$. Use Algorithm 2 to create $Y_G$. We first prove that $(H_G)_{UV}$ built in Step 3 has no across edges, and so every $u_i \in U$ is isolated, which implies that $Y_G$ is locatable. Using Theorem 1, we get that ELAN$(Y_G)$ = CAFEN$(Y_G) + |E(Y_G)|$. Moreover, we can show CAFEN$(Y_G) = (g-1)$[CAFEN$((H_G)_{UV})$]. We also use that CAFE$((H_G)_{UV})$ = $\theta'(H_G) + 2$, by Theorem 4, and show that $\theta'(H_G) = \theta'(G) + n + 1$. It is easy to see that

$$|E(Y_G)| = (g-1)|E((H_G)_{UV})| + \binom{g-1}{2}(2n+1)(2n+2).$$

Therefore,

$$\text{ELAN}(Y_G) = (g-1)[\theta'(G) - |E(G)|] + (g-1)\left[\frac{3n^2}{2} + \frac{7n}{2} + 4\right] + \binom{g-1}{2}(2n+1)(2n+2),$$

and so $(G, N) \in$ ECCN if and only if $(Y_G, N') \in g$-ELAN, where

$$N' = (g-1)[N - |E(G)|] + (g-1)\left[\frac{3n^2}{2} + \frac{7n}{2} + 4\right] + \binom{g-1}{2}(2n+1)(2n+2).$$

Since Algorithm 2 runs in polynomial time with respect to the size of $G$, and ECCN is NP-complete, we get that $g$-ELAN is NP-complete. $\qquad\square$

# References

1. Brigham, R., Dutton, R.: On clique covers and independence numbers of graphs. Discrete Math. 44, 139–144 (1983)
2. Burr, K., Young, W.: Combinatorial test techniques: table-based automation, test generation and code coverage. In: Proc. of the Intl. Conf. on Software Testing Analysis and Review, pp. 503–513 (1998)
3. Cawse, J.: Experimental design for combinatorial and high throughput materials development. GE Global Research Technical Report 29, pp. 769–781 (2002)
4. Cohen, M., Dwyer, M., Shi, J.: Interaction testing of highly-configurable systems in the presence of constraints. In: International Symposium on Software Testing and Analysis (ISSTA), London, pp. 129–139 (2007)
5. Colbourn, C.: Combinatorial aspects of covering arrays. Le Matematiche, Catania 58, 121–167 (2004)
6. Colbourn, C., Dinitz, J.: Handbook of Combinatorial Designs, Second Edition. Chapman and Hall/CRC (2007)
7. Dalal, S., Jain, A., Karunanithi, N., Leaton, J., Lott, C., Patton, G., Horowitz, B.: Model-based testing in practice. In: Proc. of the Intl. Conf. on Software Engineering (ICSE 1999), New York, pp. 285–294 (1999)
8. Danziger, P., Mendelsohn, E., Moura, L., Stevens, B.: Covering arrays avoiding forbidden edges. Theor. Comput. Sci. 410, 5403–5414 (2009)
9. Erdös, P., Goodman, A., Pósa, L.: The representation of a graph by set intersections. Can. J. Math. 18, 106–112 (1966)
10. Gyárfás, A.: A simple lower bound on edge coverings by cliques. Discrete Math. 85, 103–104 (1990)
11. Hartman, A., Raskin, L.: Problems and algorithms for covering arrays. Discrete Math. 284, 149–156 (2004)
12. Kou, L., Stockmeyer, L., Wong, C.: Covering edges by cliques with regard to keyword conflicts and intersection graphs. Commun. ACM 21(2), 135–139 (1978)
13. Kuhn, D., Reilly, M.: An investigation into the applicability of design of experiments to software testing. In: Proc. 27th Annual NASA/IEEE Software Engineering Workshop, NASA Goddard Space Flight Center, pp. 91–95 (2002)

14. Kuhn, R., Wallace, D., Gallo, A.: Software fault interactions and implications for software testing. IEEE T. Software Eng. 30(6), 418–421 (2004)
15. Lovász, L.: On covering of graphs. In: Theory of Graphs (Proc. Colloq., Tihany, 1966), pp. 231–236. Academic Press, New York (1968)
16. Lund, C., Yannakakis, M.: On the hardness of approximating minimization problems. J. Assoc. Comput. Mach. 41(5), 960–981 (1994)
17. Maltais, E.: Covering arrays avoiding forbidden edges and edge clique covers. MSc thesis, University of Ottawa (2009)
18. Martinez, C., Moura, L., Panario, D., Stevens, B.: Locating errors using ELAs, covering arrays, and adaptive testing algorithms. SIAM J. Discrete Math. 23, 1776–1799 (2009)
19. Moura, L., Stardom, J., Stevens, B., Williams, A.: Covering arrays with mixed alphabet sizes. J. Comb. Des. 11, 413–432 (2003)
20. Orlin, J.: Contentment in graph theory: covering graphs with cliques. Nederl. Akad. Wetensch. Proc. Ser. A 80(5), 406–424 (1977)
21. Williams, A., Probert, R.: A measure for component interaction test coverage. In: Proc. ACS/IEEE International Conference on Computer Systems and Applications, pp. 301–311 (2001)

# The Size and Depth of Layered Boolean Circuits

Anna Gál* and Jing-Tang Jang**

Dept. of Computer Science, University of Texas at Austin,
Austin, TX 78712-1188, USA
{panni,keith}@cs.utexas.edu

**Abstract.** We consider the relationship between size and depth for layered Boolean circuits, synchronous circuits and planar circuits as well as classes of circuits with small separators. In particular, we show that every layered Boolean circuit of size $s$ can be simulated by a layered Boolean circuit of depth $O(\sqrt{s \log s})$. For planar circuits and synchronous circuits of size $s$, we obtain simulations of depth $O(\sqrt{s})$. The best known result so far was by Paterson and Valiant [16], and Dymond and Tompa [6], which holds for general Boolean circuits and states that $D(f) = O(C(f)/\log C(f))$, where $C(f)$ and $D(f)$ are the minimum size and depth, respectively, of Boolean circuits computing $f$. The proof of our main result uses an adaptive strategy based on the two-person pebble game introduced by Dymond and Tompa [6]. Improving any of our results by polylog factors would immediately improve the bounds for general circuits.

**Keywords:** Boolean circuits, circuit size, circuit depth, pebble games.

## 1   Introduction

In this paper, we study the relationship between the size and depth of fan-in 2 Boolean circuits over the basis $\{\vee, \wedge, \neg\}$. Given a Boolean circuit $C$, the size of $C$ is the number of gates in $C$, and the depth of $C$ is the length of the longest path from any input to the output. We will use the following notation for complexity classes. $DTIME(t(n))$ and $SPACE(s(n))$ are the classes of languages decidable by deterministic multi-tape Turing machines in time $O(t(n))$ and space $O(s(n))$, respectively. Given a Boolean function $f : \{0,1\}^n \to \{0,1\}$, define $C(f)$ to be the smallest size of any circuit over $\{\vee, \wedge, \neg\}$ computing $f$, and define $D(f)$ to be the smallest depth of any circuit over $\{\vee, \wedge, \neg\}$ computing $f$. Note that $C(f)$ and $D(f)$ are not necessarily achieved by the same circuit. We also need the following conventions to compare computation times in uniform models (Turing machines) and in non-uniform models (Boolean circuits and PRAMs). Given $L \subseteq \{0,1\}^*$, let $L_n = L \cap \{0,1\}^n$. We will also view $L_n$ as the following Boolean function: $L_n : \{0,1\}^n \to \{0,1\}$ such that $L_n(x_1, \ldots, x_n) = 1$

---

* Supported in part by NSF Grant CCF-0830756.
** Supported in part by MCD fellowship from Dept. of Computer Science, University of Texas at Austin, and NSF Grant CCF-0830756.

A. López-Ortiz (Ed.): LATIN 2010, LNCS 6034, pp. 372–383, 2010.
© Springer-Verlag Berlin Heidelberg 2010

iff $x_1 \ldots x_n \in L_n$. In other words, we will use the notation $L_n$ to denote both the set $L_n$ and its characteristic function. Now define $SIZE(t(n)) = \{L : C(L_n) = O(t(n))\}$ and $DEPTH(s(n)) = \{L : D(L_n) = O(s(n))\}$. We will also consider uniform versions of these classes, i.e. $logspace\text{-}uniform\text{-}SIZE(t(n))$ and $logspace\text{-}uniform\text{-}DEPTH(s(n))$.

Pippenger and Fischer [17] showed that for $t(n) \geq n$, $DTIME(t(n)) \subseteq logspace\text{-}uniform\text{-}SIZE(t(n) \log t(n))$. Thus, circuit size is related to sequential computation time. Furthermore, Borodin [4] showed that for $s(n) \geq \log n$, $logspace\text{-}uniform\text{-}DEPTH(s(n))$ is a subset of $SPACE(s(n))$, and $SPACE(s(n))$ is a subset of $logspace\text{-}uniform\text{-}DEPTH(s^2(n))$. Thus, circuit depth is closely related to sequential computation space.

For the PRAM model, define $P_{unit}(f)$ and $P_{log}(f)$ to be the minimum computation time to compute $f$ in the unit-cost and log-cost PRAM models, respectively. For our purposes, it is sufficient to consider CRCW PRAMs. Define $PRAM_{unit}(t(n)) = \{L : P_{unit}(L_n) = O(t(n))\}$ and $PRAM_{log}(t(n)) = \{L : P_{log}(L_n) = O(t(n))\}$. Similarly to circuit classes, we will also consider uniform versions of PRAM classes.

There is a tight connection between circuit depth and PRAM time. Stockmeyer and Vishkin [21] showed that $PRAM_{log}(t(n)) \subseteq DEPTH(t(n) \log m(n))$ and $DEPTH(s(n)) \subseteq PRAM_{log}(s(n))$, where $m(n)$ is the maximum of $t(n)$, the number of processors, and the input length $n$. These results hold even for the unit-cost PRAM model as long as multiplication is not counted as a unit-cost instruction.

The above results show that the study of circuit size versus depth helps to investigate the relationship between sequential and parallel computation time, as well as time versus space in sequential computation. However, very little is known about the size versus depth question for general Boolean circuits. The best known result so far is the following theorem, which was first proved by Paterson and Valiant [16], and later proved by Dymond and Tompa [6] using another method.

**Theorem A.** *[16,6] Given a Boolean function* $f : \{0,1\}^n \rightarrow \{0,1\}$, *we have* $D(f) = O(C(f)/\log C(f))$, *or* $SIZE(t(n)) \subseteq DEPTH(t(n)/\log t(n))$.

On the other hand, it can be easily shown that $D(f) = \Omega(\log C(f))$. Theorem A leaves a huge gap ($\log C(f)$ versus $C(f)/\log C(f)$) for circuits of any size. McColl and Paterson [13] showed that every Boolean function depending on $n$ variables has circuit depth at most $n + 1$. There is an even stronger result by Gaskov [7] showing that circuit depth is at most $n - \log \log n + 2 + o(1)$. This gives a much stronger bound on depth than Theorem A for functions that require circuits of large size. In particular, for $f : \{0,1\}^n \rightarrow \{0,1\}$ such that $C(f)$ is exponential in $n$, [13] and [7] give essentially tight bounds on depth. However, for functions that can be computed by subexponential-size circuits, there is still a large gap. Note that Theorem A gives a stronger result than [13] and [7] only when $C(f) = o(n \log n)$. Improving Theorem A would yield improvements over [13] and [7] for larger $C(f)$ as well.

Because of the connections mentioned above, there are other important consequences if Theorem A can be improved. Hopcroft, Paul, and Valiant [10] proved the following analogous theorem about sequential time and space, and Adleman and Loui [1] later gave an alternative proof.

**Theorem B.** *[10,1]* $DTIME(t(n)) \subseteq SPACE(t(n)/\log t(n))$.

By the results of Pippenger and Fischer, and Borodin mentioned above, improving Theorem A by at least a polylog factor immediately improves Theorem B.

Dymond and Tompa [6] showed that $DTIME(t(n)) \subseteq PRAM_{unit}(\sqrt{t(n)})$ for the unit-cost PRAM model. (This also holds for logspace uniform unit-cost PRAM.) However, no such result is known for the log-cost PRAM model. Since $DEPTH(s(n)) \subseteq PRAM_{log}(s(n))$, improving Theorem A by at least a polylog factor will also imply non-trivial relationship between $DTIME$ and log-cost $PRAM$ computation time.

For general Boolean circuits, the simulating depth $O(t(n)/\log t(n))$ in Theorem A is very close to the circuit size. On the other extreme, consider tree-like circuits, where every gate has fan-out at most 1. Spira [20] showed that given any tree-like Boolean circuit $C$ of size $t(n)$, we can always simulate $C$ by another tree-like Boolean circuit of depth $O(\log t(n))$. Note that tree-like circuits are commonly referred to as *formulas* in circuit complexity. We will use the term tree-like circuits to avoid any ambiguity. It is unlikely that Spira's result holds for general Boolean circuits, since that would imply $P = NC_1$. Still, it is possible that Theorem A can be improved. We indeed achieve improved simulations for special classes of Boolean circuits.

## 1.1    Our Results

We consider the size versus depth problem for special classes of Boolean circuits. As far as we know, previously no better bounds were known for these classes than what follows from the bounds for general circuits [16,6]. We obtain significant improvements over these general bounds for layered circuits, synchronous circuits, and planar circuits as well as classes of circuits with small separators. Informally, a circuit is layered if its set of gates can be partitioned into subsets called layers, such that every wire in the circuit is between adjacent layers. A circuit is synchronous if for any gate $g$, every path from the inputs to $g$ has the same length. Synchronous and planar circuits have been extensively studied before. Synchronous circuits were introduced by Harper [9]. Planar circuits were introduced by Lipton and Tarjan [12]. Layered circuits are a natural generalization of synchronous circuits, but as far as we know they have not been explicitly studied. Layered graphs have been studied by Paul, Tarjan, and Celoni [14] (they call these "level graphs" in their paper). Belaga [3] defined locally synchronous circuits, which is a subclass of layered circuits, with the extra condition that each input variable can appear at most once. The synchronous circuits form a proper subset of layered circuits. (See next section for more details.) Furthermore, Turán [22] showed that there exists a function $f_n$ such that any synchronous circuit for $f_n$ has size $\Omega(n\log n)$, but there exists a layered circuit for $f_n$ with size $O(n)$.

See Belaga [3] for the same gap for functions with multiple outputs. This distinguishes synchronous circuits and layered circuits with respect to their computational powers. Notice that every Boolean function can be computed by circuits from each of the classes we consider. Furthermore, these classes of circuits are quite frequently used in various situations.

Our main result is for layered circuits.

**Theorem 1.** *Every layered Boolean circuit of size s can be simulated by a layered Boolean circuit of depth $O(\sqrt{s}\log s)$ computing the same function.*

We obtain slightly better bounds for synchronous circuits and planar circuits.

**Theorem 2.** *Every synchronous Boolean circuit of size s can be simulated by a synchronous Boolean circuit of depth $O(\sqrt{s})$ computing the same function.*

A circuit is planar if its underlying graph can be embedded in the plane without crossings of the wires.

**Theorem 3.** *Every planar Boolean circuit of size s can be simulated by a planar Boolean circuit of depth $O(\sqrt{s})$ computing the same function.*

For planar circuits, we use the fact that every planar circuit of size $s$ has a separator of size $O(\sqrt{s})$ [11]. Informally, the separator of a graph is a subset of the nodes whose removal yields two subgraphs of comparable sizes. This allows us to use a divide-and-conquer strategy. Graphs with small separators include trees, planar graphs [11], graphs with bounded genus [8], and graphs with excluded minors [2]. In fact, we can get similar results for arbitrary classes of circuits with small separators.

On the other hand, not all synchronous circuits and layered circuits have small separators. See [19] for many examples. So we need strategies other than the divide-and-conquer approach. Our idea is to consider *cuts*, which separate the graph into two subgraphs that are not necessarily comparable in size. For synchronous circuits, our technique is to find a relatively small cut such that the function can be computed by the composition of two circuits of small depths. This gives a simple proof for synchronous circuits, but the same method cannot be applied to the more general layered circuits. For layered circuits, we develop an adaptive strategy in the two-person pebble game, such that the sizes of the cuts are taken into account during the game. Note that both [16] and [6] use the notion of separators in their proofs. Our results for synchronous circuits and layered circuits show that the minimum circuit depth does not necessarily grow with the separator size of the minimum-size circuit.

Finally we note that an arbitrary circuit of size $s$ can be converted to either a planar or a synchronous circuit of size $O(s^2)$ [24]. Thus improving our results by polylog factors for any of the classes we considered would also yield improvements over the best known bounds for general circuits.

# 2 Definitions and Backgrounds

## 2.1 The Circuit Model

A Boolean circuit is a labeled directed acyclic graph (DAG), where every node is labeled by either a variable from $\{x_1, \ldots, x_n\}$, or an operation from $\{\wedge, \vee, \neg\}$. The inputs of a Boolean circuit are the nodes with in-degree (fan-in) zero, and the outputs of a Boolean circuit are the nodes with out-degree (fan-out) zero. The size of a Boolean circuit is the number of its gates. We will consider Boolean circuits with gates of fan-in at most 2 from the basis $\{\wedge, \vee, \neg\}$. We refer interested readers to [24] for more background on Boolean circuits.

A circuit is *planar* if we can find an embedding in the plane for the circuit such that no two edges cross [12].

**Definition 1.** *[9] Synchronous circuits: A circuit is* synchronous *if for any gate $g$, all paths from the inputs to $g$ have the same length.*

**Definition 2.** *Layered circuits: A circuit is* layered, *if its set of gates can be partitioned into subsets called* layers, *such that every wire in the circuit is between adjacent layers. For circuits with one output, the following is an equivalent definition: A circuit with one output is layered if for any gate $g$ all paths from $g$ to the output have the same length.*

**Definition 3.** *Depth and height: Let $C$ be a circuit, and let $g$ be any gate in $C$. The* depth *of $g$ is the length of the longest path from any input to $g$. The* depth *of $C$ is the depth of the output gate.*

*For circuits with one output, the* height *of $g$ is the length of the longest path from $g$ to the output.*

**Definition 4.** *Levels and layers: The $i$th* level *of a circuit consists of all gates with depth equal to $i$. For circuits with one output, the $i$th* layer *of the circuit consists of all gates with height equal to $i$.*

Note that the 0th layer in a circuit with one output consists of the output gate, and the 0th level in any circuit consists of the inputs. Also note that "levels" and "layers" are usually used interchangeably in the literature, and distinguishing them this way is just our terminology. The following lemma is straightforward from the definitions, and it shows that every synchronous circuit is layered. A simple example shows that the converse is not true: consider the circuit with inputs $x_1, x_2, x_3$, gates $g_1 = x_1 \wedge x_2$, $g_2 = g_1 \wedge x_3$, and $g_2$ being the output.

**Lemma 1.** *A circuit $C$ is synchronous if and only if every wire in $C$ is between adjacent levels. Thus, every synchronous circuit is also a layered circuit.*

## 2.2 The Two-Person Pebble Game

Several variants of pebble games have been invented to study questions related to the space requirements of computation, e.g. [15,5]. See [18] for a survey. Here we focus on the two-person pebble game, which was defined by Dymond and Tompa [6]. The game is played on a DAG $G$. There are two players, the challenger and the pebbler. The challenger starts the game by challenging any single node of $G$, then the pebbler puts some pebbles on a subset of the nodes. From this point on, the challenger can only challenge a node that was either challenged or pebbled in the previous round. The game continues until at the beginning of the pebbler's move, all the predecessors of the currently challenged node $w$ are already pebbled. Then we say the challenger *loses* $G$ *at* $w$. If, under the best defense of the challenger, the pebbler can win with $t$ number of pebble placements, then we say that $G$ can be *two-person pebbled in time* $t$. Notice that the pebbler does not remove pebbles once a node is pebbled.

The next two theorems give an alternative proof of Theorem A.

**Theorem C.** *[6] Let $G$ be a DAG with node set $V$. Then the pebbler can win the game in time $O(|V|/\log|V|)$.*

**Theorem D.** *(Theorem 3 in [6]) Let $C$ be a Boolean circuit computing a function $f$. If $C$ can be two-person pebbled with $t$ pebbles, then there exists a tree-like circuit of depth $2t + 1$ that also computes $f$.*

We will use Theorem D to obtain our results for layered circuits and circuits with small separators. Note that Paul, Tarjan, and Celoni [14] gave a pebbling strategy for layered graphs but used the rules of a different pebble game, which does not imply bounds on the depth.

## 3   Size versus Depth for Layered Circuits

The following lemma gives an adaptive strategy in the two-person pebble game for layered circuits.

**Lemma 2.** *Let $C$ be a layered circuit of size $s$. Then $C$ can be two-person pebbled in time $O(\sqrt{s\log s})$. That is, the pebbler can win by using $O(\sqrt{s\log s})$ pebbles.*

*Proof.* First note that at any point in the game, we only need to consider the subcircuit whose single output is the currently challenged node. Thus, in the proof we can assume without loss of generality that the circuit $C$ has only one output, and that the first move of the challenger is to challenge the output gate.

Let $L(0), \ldots, L(d)$ be all the layers, where $d$ is the depth of $C$, and $L(i)$ is the set of gates with height $i$. (See previous section for definitions.) Note that $L(0)$ consists of the output gate. We say that a layer $L(i)$ is large if $|L(i)| > y$ and small otherwise. We shall determine the value of $y$ later.

The strategy of the pebbler has two phases. During the first phase, the pebbler forces the challenger to go into a subcircuit between two small layers such that every layer between the two small layers is large, or into a subcircuit such that all nodes of the subcircuit belong to large layers. In the second phase, the pebbler will win the game within that subcircuit.

During the first phase, the pebbler always pebbles a small layer $S_\alpha$ with $\alpha > \beta$, where $S_\beta$ is the layer where the challenged node resides in that round. The pebbler continues this phase until the small layer $S_\alpha$ with $\alpha > \beta$ closest to the challenged node is pebbled, or until there are no more such small layers. Note that the pebbler pebbles the small layers $S_0, S_1, \ldots, S_m$ in a divide-and-conquer way depending on the location of the challenged node in each round. Since there are at most $s$ small layers, the number of pebbles used in the first phase is at most $y\lceil \log s \rceil$.

*Phase I.* Let $S_0, S_1, \ldots, S_m$ be the small layers numbered starting from the output. Note that $S_0 = L(0)$ since $L(0)$ contains only one gate. Define $h(j)$ to be the height of the gates in the $j$th small layer $S_j$. We shall define the strategy inductively.

At the beginning of the game (round 1), the challenger challenges the output node, which belongs to $S_0 = L(0)$.

Suppose that for $r \geq 1$ at the beginning of round $r$ the challenger challenges a node $w \in S_j$. Note that since during phase I pebbles are only placed on nodes in small layers, the challenged node $w$ belongs to a small layer in every round within phase I. We have three cases.

1. No small layer $L(h(b)) = S_b$ with $b > j$ exists. That is, every layer $L(k)$ with $k > h(j)$ is a large layer. Then the pebbler continues with the second phase.
2. None of the small layers $L(h(b)) = S_b$ with $b > j$ is pebbled. The pebbler then puts a pebble on each node of $S_{\lceil \frac{m+j}{2} \rceil}$.
3. There exists a small layer $L(h(b)) = S_b$ with $b > j$, such that all nodes in $S_b$ are pebbled, and none of the small layers between $S_j$ and $S_b$ is pebbled. The pebbler then puts a pebble on each node of $S_{\lfloor \frac{b+j}{2} \rfloor}$ if $\lfloor \frac{b+j}{2} \rfloor \neq b$. If $\lfloor \frac{b+j}{2} \rfloor = b$, then there are no small layers between $S_j$ and $S_b$, and the pebbler continues with the second phase.

*Phase II.* The pebbler's strategy in the second phase is as follows: Suppose that the challenger challenges node $w$ in the beginning of the $k$th round for some $k$. Then the pebbler puts pebbles on the two inputs of $w$, say $u$ and $v$. In the $(k+1)$st round, if the challenger stays on $w$, then the pebbler wins the game. On the other hand, if the challenger challenges one of the inputs of $w$, WLOG $u$, then the pebbler puts pebbles on the two inputs of $u$ in the $(k+1)$st round. The game continues inductively this way until at the beginning of pebbler's move either the currently challenged node $w$ is an input of $C$, or the two immediate predecessors of $w$ are already pebbled. Thus the pebbler wins in the second phase.

Note that in this phase, the pebbler only spends at most two pebbles in each round, and the two pebbles are put on nodes in large layers of $C$. Moreover, during $k$ rounds of the second phase, the pebbler pebbles nodes from $k$ different large layers. Since the number of large layers in $C$ is at most $\frac{s}{y}$, the second phase must terminate in at most $\frac{s}{y}$ rounds. Thus, the number of pebbles used in this phase is at most $\frac{2s}{y}$.

The total number of pebbles used throughout the game is at most $p = y\lceil \log s \rceil + 2s/y$. The minimum of this expression is $p = 2\sqrt{2s\lceil \log s \rceil}$, achieved when $y = \sqrt{\frac{2s}{\lceil \log s \rceil}}$. This proves the lemma.    □

*Proof of Theorem 1.* Follows immediately from Theorem D and Lemma 2.

# 4    Size versus Depth for Synchronous Circuits

The following simple lemma was given in [24]. The results by McColl and Paterson [13], and Gaskov [7] mentioned in the introduction give stronger results. But for our purposes, this slightly weaker bound is sufficient, and we include a simple proof for completeness.

**Lemma 3.** *[24] For every function* $f : \{0,1\}^n \to \{0,1\}$, *there exists a synchronous circuit of depth at most* $n + \log n + 1$ *computing* $f$.

*Proof.* The proof is based on considering any DNF of $f$. The terms can be computed in parallel with depth at most $\log n + 1$, and the number of terms is at most $2^n$. This gives the desired depth. Note that any circuit can be made synchronous without increasing its depth.    □

Next we prove Theorem 2. Note that in the proof, we use the property of synchronous circuits that given any level $LV(i)$, $f$ is a function of exactly those functions computed at the gates in $LV(i)$. This property allows us to do function composition in terms of two circuits. However, for layered circuits, inputs could be in the $j$th layer for $j > i$. Thus the property no longer holds for layered circuits that are not synchronous.

Note that the notation $LV$ stands for "levels", while in the previous section we used $L$ for "layers".

*Proof of Theorem 2.* Let $f$ be the function computed by $C$. (If $C$ has more than one output, the proof can be applied by considering each output function separately, and combine the resulting small depth circuits.) Since $C$ is synchronous, every level in $C$ forms a cut. Furthermore, given any level $LV(i)$, $f$ is a function of exactly those functions computed at the gates in $LV(i)$. We shall use this special property of synchronous circuits to compute $f$ by the composition of two circuits.

Let $LV(0), LV(1), \ldots, LV(d)$ be the levels in $C$, where $d$ is the depth of $C$, and $LV(0)$ contains the inputs. Let $y$ be an integer whose value will be determined later. We say that a level $LV(i)$ is *small* if $|LV(i)| \leq y$ and *large* otherwise.

If $C$ has many outputs, then it is possible that all the levels are large, but then the depth of $C$ is at most $\frac{s}{y}$. Assume that $C$ has at least one small level. Let $LV(k_0)$ be the small level farthest from the output of $C$.

Now let $g_1, \ldots, g_m$ be the gates in $LV(k_0)$, and let $\gamma_i$ be the function computed at $g_i$. As noted above, $f$ is a function of $\gamma_1, \ldots, \gamma_m$. Let $f = f'(\gamma_1, \ldots, \gamma_m)$. Then by Lemma 3, given $\gamma_1, \ldots, \gamma_m$ as inputs, $f'$ can be computed by a synchronous circuit $F$ of depth $O(|LV(k_0)|) = O(y)$.

Let $C'$ be the multiple-output subcircuit of $C$ with outputs $g_1, \ldots, g_m$. That is, $C'$ consists of the levels $LV(0), \ldots, LV(k_0)$ in $C$, where $LV(k_0)$ contains the outputs of $C'$. (If $LV(k_0) = LV(0)$, then $C'$ consists of only one level, formed by the inputs of $C$.) We use the outputs of $C'$ as inputs for the circuit $F$. The resulting combined circuit $F'$ is a synchronous circuit computing $f$. Note that all the levels $LV(0), \ldots, LV(k_0 - 1)$ are large. Since there are at most $\frac{s}{y}$ large levels in $C$, the depth of $F'$ is at most $O(y + \frac{s}{y})$.

Thus, we obtain a synchronous circuit of depth at most $O(y + \frac{s}{y})$. Letting $y = \sqrt{s}$, we can simulate $C$ by a synchronous circuit of depth $O(\sqrt{s})$.    □

## 5    Size versus Depth for Planar Circuits and Circuits with Small Separators

Informally, a node separator of a graph $G$ is a set of nodes whose removal yields two disjoint subgraphs of $G$ that are comparable in size. The following gives a formal definition of a node separator in the fashion of Ullman [23].

**Definition 5.** *A DAG $G = (V, E)$ has an $h(t)$-separator, and we say $G$ is $h(t)$-separable, if $G$ has only one node, or it satisfies the following two properties.*

1. *There exists an $S \subseteq V$ of size at most $h(|V|)$ whose removal disconnects $G$ into two subDAGs, $G_1 = (V_1, E_1)$ and $G_2 = (V_2, E_2)$, that satisfy $|V_i| \geq \frac{1}{3}|V|$ for $i = 1, 2$.*
2. *The two subDAGs $G_1$ and $G_2$ have $h(t)$-separators.*

Note that the two subDAGs could be disconnected within themselves.

Although the separator serves as a natural tool to define a divide-and-conquer strategy, we need something stronger than the separator for the two-person pebble game. This is because the challenged nodes in the game could be arbitrary nodes in the graph. Thus we need to consider every possible subDAG of $G$.

**Definition 6.** *A DAG $G$ is everywhere-$h(t)$-separable, if any subDAG $H$ of $G$ with one output such that the underlying undirected graph of $H$ is connected, also has an $h(t)$-separator. A circuit $C$ is everywhere-$h(t)$-separable if its underlying DAG is everywhere-$h(t)$-separable.*

**Theorem 4.** *Let $C = (V, E)$ be an everywhere-$h(t)$-separable Boolean circuit, where $h(t) = o(t)$. Then $C$ can be two-person pebbled in time*

$$O\left( \sum_{i=0}^{\lceil \log_{3/2} |V| \rceil} h\left( (2/3)^i |V| \right) \right).$$

*Proof.* Let $p(|V|)$ be the number of pebbles required by the pebbler to win in the two-person pebble game on $C$. In each round, the pebbler will separate $C$ into two subDAGs to see which subDAG the currently challenged node belongs to, and then recurse on that subDAG. Notice that in general, both components may contain several disjoint subcomponents.

We now define the strategy recursively. If $C$ has only one node, then the pebbler wins immediately after the challenger's initial move. Now suppose that the challenger puts a challenge on a node $g \in C$. Then the pebbler places pebbles on the nodes of some appropriately chosen separator $S$ of $C_1$, where $C_1$ is the unique maximal subcircuit of $C$ with $g$ as its output. Let $i \geq 1$. There are two cases in the $(i + 1)$th round.

1. The challenger re-challenges $g$. Let $S_i$ be the separator of $C_i$ chosen in the $i$th round. Let $C_{i+1}$ be the unique maximal subcircuit of $C_i$ with $g$ as its output, such that the inputs of $C_{i+1}$ are some inputs of $C_i$ upon which $g$ depends, or some nodes in the separator $S_i$, and the underlying undirected graph of $C_{i+1}$ is connected. Furthermore, we require that every path from the inputs of $C_{i+1}$ to $g$ does not contain any node in $S_i$. Then the pebbler applies this strategy recursively, and in the next round, looks for an appropriate separator of $C_{i+1}$.
2. The challenger puts a new challenge on a node $w \in S$. Let $C_{i+1}$ be defined as above, but with $w$ as its output. Then, as above, the pebbler applies this strategy recursively.

Let $C_{i1}$ and $C_{i2}$ be the two subDAGs of $C_i$ defined by the separator $S_i$. Note that $C_{i1}$ and $C_{i2}$ might be disconnected, but we defined $C_{i+1}$ to be a connected subcircuit of either $C_{i1}$ or $C_{i2}$. We claim that the pebbler will win after at most $O(\log |V|)$ rounds. To see this, notice that after the first round, the challenger can only challenge a node that has just been challenged, or a node in the separator. So the challenged node is restricted to either $C_{i1} \cup S_i$ or $C_{i2} \cup S_i$ that have sizes at most $(\frac{2}{3} + o(1))|C_i|$ because by assumption $C$ is everywhere-$h(t)$-separable and $h(t) = o(t)$. This also implies that the size of $C_{i+1}$ is at most $(\frac{2}{3} + o(1))|C_i|$, and the game must terminate in at most $O(\log |V|)$ rounds.

For the number of pebbles used, we have the following recursion:

$$p(1) = 0, p(|C_i|) \leq h(|C_i|) + p\left(\left(\frac{2}{3} + o(1)\right)|C_i|\right).$$

Solving the above recursion yields $p(|V|) = O\left(\sum_{i=0}^{\lceil \log_{3/2}|V| \rceil} h\left((2/3)^i |V|\right)\right).$ □

For planar graphs, the following theorem is due to Lipton and Tarjan [11].

**Theorem E.** *[11] Given a planar graph $G$ with node set $V$, $G$ has a separator of size $O(\sqrt{|V|})$.*

Since any subgraph of a planar graph is still planar, every planar graph is everywhere-$O(\sqrt{t})$-separable.

*Proof of Theorem 3.* The claim follows by Theorem 4, Theorem D, and the observation that any tree-like circuit is also planar.    □

For graphs with bounded genus, we have the following theorem due to Gilbert, Hutchinson, and Tarjan [8].

**Theorem F.** *[8] Given a graph $G$ with node set $V$ and genus $g$, $G$ has a separator of size $O(\sqrt{g|V|})$.*

Since any subgraph of a graph $G$ with genus $g$ cannot have genus more than $g$, $G$ is everywhere-$O(\sqrt{gt})$-separable. By Theorem 4 and Theorem D we obtain:

**Theorem 5.** *Let $C$ be a Boolean circuit of size $s$ such that its underlying graph of $C$ has genus $g$. Then there exists a tree-like circuit $F$ of depth $O(\sqrt{gs})$ that computes the same function.*

Let $K_k$ denote the complete graph on $k$ nodes. For graphs having no $K_k$ as a minor, we have the following theorem due to Alon, Seymour, and Thomas [2].

**Theorem G.** *[2] Given a graph $G$ with node set $V$ and having no $K_k$ as a minor, $G$ has a separator of size $O(k^{\frac{3}{2}}|V|^{\frac{1}{2}})$.*

If a graph $G$ does not have $K_k$ as its minor, neither can any subgraph of $G$. So $G$ is everywhere-$O(k^{3/2}t^{1/2})$-separable. As above, this gives:

**Theorem 6.** *Let $C$ be a Boolean circuit of size $s$ such that its underlying graph of $C$ does not contain $K_k$ as a minor. Then there exists a tree-like circuit $F$ of depth $O(k^{\frac{3}{2}}s^{\frac{1}{2}})$ that computes the same function.*

## Acknowledgements

We thank the anonymous referees for helpful comments.

## References

1. Adleman, L.M., Loui, M.C.: Space-bounded simulation of multitape turing machines. Theory of Computing Systems 14(1), 215–222 (1981)
2. Alon, N., Seymour, P., Thomas, R.: A Separator Theorem for Graphs with an Excluded Minor and its Applications. In: Proceedings of the ACM Symposium on Theory of Computing, pp. 293–299 (1990)
3. Belaga, E.G.: Locally Synchronous Complexity in the Light of the Trans-Box Method. In: Fontet, M., Mehlhorn, K. (eds.) STACS 1984. LNCS, vol. 166, pp. 129–139. Springer, Heidelberg (1984)
4. Borodin, A.: On Relating Time and Space to Size and Depth. SIAM Journal on Computing 6(4), 733–744 (1977)
5. Cook, S.A.: An Observation on Time-Storage Trade Off. Journal of Computer and System Sciences 9(3), 308–316 (1974)
6. Dymond, P., Tompa, M.: Speedups of Deterministic Machines by Synchronous Parallel Machines. J. Comp. and Sys. Sci. 30(2), 149–161 (1985)
7. Gaskov, S.B.: The depth of Boolean functions. Probl. Kibernet. 34, 265–268 (1978)

8. Gilbert, J.R., Hutchinson, J.P., Tarjan, R.E.: A Separator Theorem for Graphs of Bounded Genus. Journal of Algorithms 5(3), 391–407 (1984)
9. Harper, L.H.: An $n \log n$ Lower Bound on Synchronous Combinational Complexity. Proc. AMS 64(2), 300–306 (1977)
10. Hopcroft, J., Paul, W., Valiant, L.: On Time Versus Space. Theory of Computation 24(2), 332–337 (1977)
11. Lipton, R., Tarjan, R.E.: A Separator Theorem for Planar Graphs. SIAM J. Appl. Math. 36, 177–189 (1979)
12. Lipton, R., Tarjan, R.E.: Applications of a Planar Separator Theorem. SIAM Journal on Computing 9(3), 615–627 (1980)
13. McColl, W.F., Paterson, M.S.: The depth of all Boolean functions. SIAM J. on Comp. 6, 373–380 (1977)
14. Paul, W., Tarjan, R.E., Celoni, J.: Space Bounds for a Game on Graphs. Mathematical Systems Theory 10, 239–251 (1977)
15. Paterson, M.S., Hewitt, C.: Comparative Schematology. MIT AI Memo 464 (1978)
16. Paterson, M.S., Valiant, L.G.: Circuit Size is Nonlinear in Depth. Theoretical Computer Science 2(3), 397–400 (1976)
17. Pippenger, N., Fischer, M.: Relations Among Complexity Measures. Journal of the ACM 26, 361–381 (1979)
18. Pippenger, N.: Pebbling. Mathematical Foundations of Computer Science (1980)
19. Rosenberg, A., Heath, L.: Graph Separators with Applications (2001)
20. Spira, P.M.: On time-hardware complexity tradeoffs for Boolean functions. In: Proc. 4th Hawaii Symp. on System Sciences, pp. 525–527 (1971)
21. Stockmeyer, L., Vishkin, U.: Simulation of Parallel Random Access Machines by Circuits. SIAM Journal on Computing 13(2), 409–422 (1984)
22. Turán, G.: On restricted Boolean circuits. Fund. of Comp. Theory, 460–469 (1989)
23. Ullman, J.D.: Computational Aspects of VLSI (1984)
24. Wegener, I.: The Complexity of Boolean Functions (1987)

# Lipschitz Unimodal and Isotonic Regression on Paths and Trees

Pankaj K. Agarwal[1], Jeff M. Phillips[2], and Bardia Sadri[3]

[1] Duke University
[2] University of Utah
[3] University of Toronto

**Abstract.** We describe algorithms for finding the regression of $t$, a sequence of values, to the closest sequence $s$ by mean squared error, so that $s$ is always increasing (isotonicity) and so the values of two consecutive points do not increase by too much (Lipschitz). The isotonicity constraint can be replaced with a unimodular constraint, for exactly one local maximum in $s$. These algorithm are generalized from sequences of values to trees of values. For each we describe near-linear time algorithms.

## 1 Introduction

Let $\mathbb{M}$ be a triangulation of a polygonal region $P \subseteq \mathbb{R}^2$ in which each vertex is associated with a real valued height (or elevation)[1]. Linear interpolation of vertex heights in the interior of each triangle of $\mathbb{M}$ defines a piecewise-linear function $t : P \to \mathbb{R}$, called a *height function*. A height function (or its graph) is widely used to model a two-dimensional surface in numerous applications (e.g. modeling the terrain of a geographical area). With recent advances in sensing and mapping technologies, such models are being generated at an unprecedentedly large scale. These models are then used to analyze the surface and to compute various geometric and topological properties of the surface. For example, researchers in GIS are interested in extracting river networks or computing visibility or shortest-path maps on terrains modeled as height functions. These structures depend heavily on the topology of the level sets of the surface and in particular on the topological relationship between its critical points (maxima, minima, and saddle points). Because of various factors such as measurement or sampling errors or the nature of the sampled surface, there is often plenty of noise in these surface models which introduces spurious critical points. This in turn leads to misleading or undesirable artifacts in the computed structures, e.g., artificial breaks in river networks. These difficulties have motivated extensive work on topological simplification and noise removal through modification of the height function $t$ into another one $s : \mathbb{M} \to \mathbb{R}$ that has the desired set of critical points and that is as close to $t$ as possible [8,20,23]. A popular approach is to decompose the

---

[1] Supported by grants NSF: CNS-05-40347, CFF-06-35000, DEB-04-25465, 0937060-CIF32 to CRA; ARO: W911NF-04-1-0278, W911NF-07-1-0376; NIH: 1P50-GM-08183-01; DOE: OEG-P200A070505; and U.S.–Israel Binational Science Foundation.

A. López-Ortiz (Ed.): LATIN 2010, LNCS 6034, pp. 384–396, 2010.

surface into pieces and modify each piece so that it has a unique minimum or maximum [23]. In some applications, it is also desirable to impose the additional constraint that the function $s$ is Lipschitz; see below for further discussion.

**Problem statement.** Let $\mathbb{M} = (V, A)$ be a planar graph with vertex set $V$ and arc (edge) set $A \subseteq V \times V$. We may treat $\mathbb{M}$ as undirected in which case we take the pairs $(u, v)$ and $(v, u)$ as both representing the same undirected edge connecting $u$ and $v$. Let $\gamma \geq 0$ be a real parameter. A function $s : V \to \mathbb{R}$ is called

(L)  $\gamma$-*Lipschitz* if $(u, v) \in A$ implies $s(v) - s(u) \leq \gamma$.

Note that if $\mathbb{M}$ is undirected, then Lipschitz constraint on an edge $(u, v) \in A$ implies $|s(u) - s(v)| \leq \gamma$. For an undirected planar graph $\mathbb{M} = (V, A)$, a function $s : V \to \mathbb{R}$ is called

(U)  *unimodal* if $s$ has a unique *local maximum*, i.e. only one vertex $v \in V$ such that $s(v) > s(u)$ for all $(u, v) \in A$.

For a directed planar graph $\mathbb{M} = (V, A)$, a function $s : V \to \mathbb{R}$ is called

(I)  *isotonic* if $(u, v) \in A$ implies $s(u) \leq s(v)$.[2]

For an arbitrary function $t : V \to \mathbb{R}$ and a parameter $\gamma$, the $\gamma$-*Lipschitz unimodal regression ($\gamma$-LUR)* of $t$ is a function $s : V \to \mathbb{R}$ that is $\gamma$-Lipschitz and unimodal on $\mathbb{M}$ and minimizes $\|s - t\|^2 = \sum_{v \in V} (s(v) - t(v))^2$. Similarly, if $\mathbb{M}$ is a directed planar graph, then $s$ is the $\gamma$-*Lipschitz isotonic regression ($\gamma$-LIR)* of $t$ if $s$ satisfies (L) and (I) and minimizes $\|s - t\|^2$. The more commonly studied *isotonic regression (IR)* and *unimodal regression* [3,5,7,22] are the special cases of LIR and LUR, respectively, in which $\gamma = \infty$, and therefore only the condition (I) or (U) is enforced.

Given a planar graph $\mathbb{M}$, a parameter $\gamma$, and $t : V \to \mathbb{R}$, the LIR (resp. LUR) problem is to compute the $\gamma$-LIR (resp. $\gamma$-LUR) of $t$. In this paper we propose near-linear-time algorithms for the LIR and LUR problems for two special cases: when $\mathbb{M}$ is a path or a tree. We study the special case where $\mathbb{M}$ is a path prior to the more general case where it is tree because of the difference in running time and because doing so simplifies the exposition to the more general case.

**Related work.** As mentioned above, there is extensive work on simplifying the topology of a height function while preserving the geometry as much as possible. Two widely used approaches in GIS are the so-called *flooding* and *carving* techniques [1,11,23]. The former technique raises the height of the vertices in "shallow pits" to simulate the effect of flooding, while the latter lowers the value of the height function along a path connecting two pits so that the values along

---

[2]  A function $s$ satisfying the isotonicity constraint (I) must assign the same value to all the vertices of a directed cycle of $\mathbb{M}$ (and indeed to all vertices in the same strongly connected component). Therefore, without loss of generality, we assume $\mathbb{M}$ to be a directed *acyclic* graph (DAG).

the path vary monotonically. As a result, one pit drains to the other and thus one of the minima ceases to exist. Various methods based on Laplacian smoothing have been proposed in the geometric modeling community to remove unnecessary critical points; see [6,17,20] and references therein.

A prominent line of research on topological simplification was initiated by Edelsbrunner et. al. [13] who introduced the notion of *persistence*; see also [12,28]. Roughly speaking, each homology class of the contours in sublevel sets of a height function is characterized by two critical points at one of whom the class is born and at the other it is destroyed. The persistence of this class is then the height difference between these two critical points. Efficient algorithms [13,14,4] simplify topology based on persistence, optimally eliminating all critical points of persistence below a threshold measured as $\|s - t\|_\infty = \max_{v \in V} |s(v) - t(v)|$. No efficient algorithm is known to minimize $\|s - t\|_2$.

The isotonic-regression (IR) problem has been studied in statistics [5,7,22] since the 1950s. It has many applications ranging from statistics [25] to bioinformatics [7], and from operations research [19] to differential optimization [16]. The *pool adjacent violator algorithm* (PAVA) [5] solves the IR problem on paths in $O(n)$ time, by merging consecutive level sets of vertex values that are out of order. Brunk [10] and Thompson [27] initiated the study of the IR problem on general DAGs and trees, respectively. Jewel [18] introduced the problem of Lipschitz isotonic regression on DAGs and showed connections between this problem and network flow with quadratic cost functions. Stout [26] solves the UR problem on paths in $O(n)$ time. Pardalos and Xue [21] give an $O(n \log n)$ algorithm for the IR problem on trees. For the special case when the tree is a star they give an $O(n)$ algorithm. Spouge et. al.[24] give an $O(n^4)$ time algorithm for the IR problem on DAGs. The problems can be solved under the $L_1$ and $L_\infty$ norms on paths [26] and DAGs [3] as well, with an additional $\log n$ factor for $L_1$. To our knowledge there is no prior work on efficient algorithms for Lipschitz isotonic/unimodal regressions in the literature.

**Our results.** We present efficient exact algorithms for LIR and LUR problems on two special cases of planar graphs: paths and trees. In particular, we present an $O(n \log n)$ algorithm for computing the LIR on a path of length $n$ (Section 4), and an $O(n \log n)$ algorithm on a tree with $n$ nodes (Section 6). We present an $O(n \log^2 n)$ algorithm for computing the LUR problem on a path of length $n$ (Section 5). Our algorithm can be extended to solve the LUR problem on an unrooted tree in $O(n \log^3 n)$ time (Section 7). The LUR algorithm for a tree is particularly interesting because of its application in the aforementioned carving technique [11,23]. The carving technique modifies the height function along a number of trees embedded on the terrain where the heights of the vertices of each tree are to be changed to vary monotonically towards a chosen "root" for that tree. In other words, to perform the carving, we need to solve the IR problem on each tree. The downside of doing so is that the optimal IR solution happens to be a step function along each path toward the root of the tree with potentially large jumps. Enforcing the Lipschitz condition prevents sharp jumps in function value and thus provides a more natural solution to the problem.

Section 3 presents a data structure, called *affine composition tree* (ACT), for maintaining a $xy$-monotone polygonal chain, which can be regarded as the graph of a monotone piecewise-linear function $F : \mathbb{R} \to \mathbb{R}$. Besides being crucial for our algorithms, ACT is interesting in its own right. A special kind of binary search tree, an ACT supports a number of operations to query and update the chain, each taking $O(\log n)$ time. Besides the classical insertion, deletion, and query (computing $F(x)$ or $F^{-1}(x)$ for a given $x \in \mathbb{R}$), one can apply an INTERVAL operation that modifies a contiguous subchain provided that the chain remains $x$-monotone after the transformation, i.e., it remains the graph of a monotone function. For space, several proofs are missing; see the full version [2].

## 2   Energy Functions

On a discrete set $U$, a real valued function $s : U \to \mathbb{R}$ can be viewed as a point in the $|U|$-dimensional Euclidean space in which coordinates are indexed by the elements of $U$ and the component $s_u$ of $s$ associated to an element $u \in U$ is $s(u)$. We use the notation $\mathbb{R}^U$ to represent the set of all real-valued functions defined on $U$.

Let $\mathbb{M} = (V, A)$ be a directed acyclic graph on which we wish to compute $\gamma$-Lipschitz isotonic regression of an *input function* $t \in \mathbb{R}^V$. For any set of vertices $U \subseteq V$, let $\mathbb{M}[U]$ denote the subgraph of $\mathbb{M}$ induced by $U$, i.e. the graph $(U, A[U])$, where $A[U] = \{(u, v) \in A : u, v \in U\}$. The set of $\gamma$-Lipschitz isotonic functions on the subgraph $\mathbb{M}[U]$ of $\mathbb{M}$ constitutes a convex subset of $\mathbb{R}^U$, denoted by $\Gamma(\mathbb{M}, U)$. It is the common intersection of all half-spaces determined by the isotonicity and Lipschitz constraints associated with the edges in $A[U]$, i.e., $s_u \leq s_v$ and $s_v \leq s_u + \gamma$ for all $(u, v) \in A[U]$.

For $U \subseteq V$ we define $E_U : \mathbb{R}^V \to \mathbb{R}$ as $E_U(s) = \sum_{v \in U} (s(v) - t(v))^2$. The $\gamma$-Lipschitz isotonic regression of the input function $t$ is $\sigma = \arg\min_{s \in \Gamma(\mathbb{M}, V)} E_V(s)$. For a subset $U \subseteq V$ and $v \in U$ define the function $E^v_{\mathbb{M}[U]} : \mathbb{R} \to \mathbb{R}$ as

$$E^v_{\mathbb{M}[U]}(x) = \min_{s \in \Gamma(\mathbb{M}, U); s(v) = x} E_U(s).$$

**Lemma 1.** *For any $U \subseteq V$ and $v \in U$, the function $E^v_{\mathbb{M}[U]}$ is continuous and strictly convex.*

## 3   Affine Composition Tree

In this section we introduce a data structure, called *affine composition tree* (ACT), for representing an $xy$-monotone polygonal chain in $\mathbb{R}^2$, which is being deformed dynamically. Such a chain can be regarded as the graph of a piecewise-linear monotone function $F : \mathbb{R} \to \mathbb{R}$, and thus is bijective. A *breakpoint* of $F$ is the $x$-coordinate of a *vertex* of the graph of $F$ (a vertex of $F$ for short), i.e., a $b \in \mathbb{R}$ at which the left and right derivatives of $F$ disagree. The number of breakpoints of $F$ will be denoted by $|F|$. A continuous piecewise-linear function $F$ with breakpoints $b_1 < \cdots < b_n$ can be characterized by its vertices $q_i = (b_i, F(b_i)), i = 1, \ldots, n$ together with the *slopes* $\mu_-$ and $\mu_+$ of its *left* and *right unbounded pieces*, respectively, extending to $-\infty$ and $+\infty$. An *affine transformation* of $\mathbb{R}^2$ is a map $\phi : \mathbb{R}^2 \to \mathbb{R}^2$, $q \mapsto M \cdot q + c$ where $M$ is a nonsingular $2 \times 2$ matrix, (a *linear transformation*) and $c \in \mathbb{R}^2$ is a *translation vector* — in our notation we treat $q \in \mathbb{R}^2$ as a column vector.

An ACT supports the following operations on a monotone piecewise-linear function $F$ with vertices $q_i = (b_i, F(b_i)), i = 1, \ldots, n$:

1. EVALUATE$(a)$ and EVALUATE$^{-1}(a)$: Given any $a \in \mathbb{R}$, return $F(a)$ or $F^{-1}(a)$.
2. INSERT$(q)$: Given a point $q = (x, y) \in \mathbb{R}^2$ insert $q$ as a new vertex of $F$. If $x \in (b_i, b_{i+1})$, this operation removes the segment $q_i q_{i+1}$ from the graph of $F$ and replaces it with two segments $q_i q$ and $q q_{i+1}$, thus making $x$ a new breakpoint of $F$ with $F(x) = y$. If $x < b_1$ or $x > b_k$, then the affected unbounded piece of $F$ is replaced with one parallel to it but ending at $q$ and a segment connecting $q$ and the appropriate vertex of $F$ ($\mu_+$ and $\mu_-$ remain intact). We assume that $F(b_i) \le y \le F(b_{i+1})$.
3. DELETE$(b)$: Given a breakpoint $b$ of $F$, removes the vertex $(b, F(b))$ from $F$; a delete operation modifies $F$ in a manner similar to insert.
4. AFFINE$(\psi)$: Given an affine transformation $\psi : \mathbb{R}^2 \to \mathbb{R}^2$, modify the function $F$ to one whose graph is the result of application of $\psi$ to the graph of $F$. See Figure 1(Left).
5. INTERVAL$_p(\psi, \tau^-, \tau^+)$: Given an affine transformation $\psi : \mathbb{R}^2 \to \mathbb{R}^2$ and $\tau^-, \tau^+ \in \mathbb{R}$ and $p \in \{x, y\}$, this operation applies $\psi$ to all vertices $v$ of $F$ whose $p$-coordinate is in the range $[\tau^-, \tau^+]$. Note that AFFINE$(\psi)$ is equivalent to INTERVAL$_p(\psi, -\infty, +\infty)$ for $p = \{x, y\}$. See Figure 1(Right).

AFFINE and INTERVAL are applied with appropriate choice of transformation parameters so that the resulting chain remains $xy$-monotone.

An ACT $\mathcal{T} = \mathcal{T}(F)$ is a red-black tree that stores the vertices of $F$ in the sorted order, i.e., the $i$th node of $\mathcal{T}$ is associated with the $i$th vertex of $F$. However, instead of storing the actual coordinates of a vertex, each node $z$ of $\mathcal{T}$ stores an

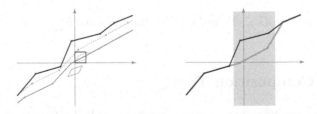

**Fig. 1.** Left: the graph of a monotone piecewise linear function $F$ (in solid black). The dashed curve is the result of applying the linear transform $\psi_0(q) = Mq$ where $M = \left(\begin{smallmatrix} 1 & 1/3 \\ 1/3 & 2/3 \end{smallmatrix}\right)$. The gray curve, a translation of the dashed curve under vector $c = \left(\begin{smallmatrix} -1 \\ -2 \end{smallmatrix}\right)$, is the result of applying AFFINE$(\psi)$ to $F$ where $\psi(q) = Mq + c$. Right: INTERVAL$(\psi, \tau^-, \tau^+)$, only the vertices of curve $F$ whose $x$-coordinates are in the marked interval $[\tau^-, \tau^+]$ are transformed. The resulting curve is the thick gray curve.

affine transformation $\phi_z : q \mapsto M_z \cdot q + c_z$. If $z_0, \ldots, z_k = z$ is the path in $\mathcal{T}$ from the root $z_0$ to $z$, then let $\Phi_z(q) = \phi_{z_0}(\phi_{z_1}(\ldots \phi_{z_k}(q) \ldots))$. Notice that $\Phi_z$ is also an affine transformation. The actual coordinates of the vertex associated with $z$ are $(q_x, q_y) = \Phi_z(\overline{0})$ where $\overline{0} = (0, 0)$.

Given a value $b \in \mathbb{R}$ and $p \in \{x, y\}$, let $\mathrm{PRED}_p(b)$ (resp. $\mathrm{SUCC}_p(b)$) denote the rightmost (resp. leftmost) vertex $q$ of $F$ such that the $p$-coordinate of $q$ is at most (resp. least) $b$. Using ACT $\mathcal{T}$, $\mathrm{PRED}_p(b)$ and $\mathrm{SUCC}_p(b)$ can be computed in $O(\log n)$ time by following a path in $\mathcal{T}$, composing the affine transformations along the path, evaluating the result at $\overline{0}$, and comparing its $p$-coordinate with $b$. $\mathrm{EVALUATE}(a)$ determines the vertices $q^- = \mathrm{PRED}_x(a)$ and $q^+ = \mathrm{SUCC}_x(a)$ of $F$ immediately preceding and succeeding $a$ and interpolates $F$ linearly between $q^-$ and $q^+$; if $a < b_1$ (resp. $a > b_k$), then $F(a)$ is calculated using $F(b_1)$ and $\mu_-$ (resp. $F(b_k)$ and $\mu_+$). Since $b^-$ and $b^+$ can be computed in $O(\log n)$ time and the interpolation takes constant time, $\mathrm{EVALUATE}(a)$ is answered in time $O(\log n)$. Similarly, $\mathrm{EVALUATE}^{-1}(a)$ is answered using $\mathrm{PRED}_y(a)$ and $\mathrm{SUCC}_y(a)$.

A key observation of ACT is that a standard rotation on any edge of $\mathcal{T}$ can be performed in $O(1)$ time by modifying the stored affine transformations in a constant number of nodes (see Figure 2(left)) based on the fact that an affine transformation $\phi : q \mapsto M \cdot q + c$ has an inverse affine transformation $\phi^{-1} : q \mapsto M^{-1} \cdot (q - c)$; provided that the matrix $M$ is invertible. A point $q \in \mathbb{R}^2$ is inserted into $\mathcal{T}$ by first computing the affine transformation $\Phi_u$ for the node $u$ that will be the parent of the leaf $z$ storing $q$. To determine $\phi_z$ we solve, in constant time, the system of (two) linear equations $\Phi_u(\phi_z(\overline{0})) = q$. The result is the translation vector $c_z$. The linear transformation $M_z$ can be chosen to be an arbitrary invertible linear transformation, but for simplicity, we set $M_z$ to the identity matrix. Deletion of a node is handled similarly.

To perform an $\mathrm{INTERVAL}_p(\psi, \tau^-, \tau^+)$ query, we first find the nodes $z^-$ and $z^+$ storing the vertices $\mathrm{PRED}_p(\tau^-)$ and $\mathrm{SUCC}_p(\tau^+)$, respectively. We then successively rotate $z^-$ with its parent until it becomes the root of the tree. Next, we do the same with $z^+$ but stop when it becomes the right child of $z^-$. At this stage, the subtree rooted at the left child of $z^+$ contains exactly all the vertices $q$ for which $q_p \subset [\tau^-, \tau^+]$. Thus we compose $\psi$ with the affine transformation

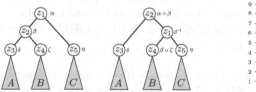 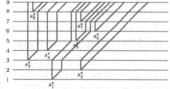

**Fig. 2.** Left: Rotation in affine composition trees, with each node's affine function in greek. When rotating $(z_1, z_2)$, changing the functions as shown the leaves remain unchanged. Right: The breakpoints of the function $F_i = d\tilde{E}_i/dx$. For each $i$, $s_{i-1}^*$ and $s_{i-1}^* + \gamma$ are the "new" breakpoints of $F_i$. All other breakpoints of $F_i$ come from $F_{i-1}$ where those smaller than $s_i^*$ remain unchanged and those larger are increased by $\gamma$.

at that node and issue the performed rotations in the reverse order to put the tree back in its original (balanced) position. Since $z^-$ and $z^+$ were both within $O(\log n)$ steps from the root of the tree, and since performing each rotation on the tree can only increase the depth of a node by one, $z^-$ is taken to the root in $O(\log n)$ steps and this increases the depth of $z^+$ by at most $O(\log n)$. Thus the whole operation takes $O(\log n)$ time.

We can augment $\mathcal{T}(F)$ with additional information so that for any $a \in \mathbb{R}$ the function $E(a) = E(b_1) + \int_{b_1}^a F(x)\,dx$, where $b_1$ is the leftmost breakpoint and $E(b_1)$ is value associated with $b_1$, can be computed in $O(\log n)$ time; we refer to this operation as INTEGRATE$(a)$. We provide the details in the full version [2].

**Theorem 1.** *A continuous piecewise-linear monotonically increasing function $F$ with $n$ breakpoints can be maintained using a data structure $\mathcal{T}(F)$ such that*

1. EVALUATE *and* EVALUATE$^{-1}$ *queries can be answered in $O(\log n)$ time,*
2. *an* INSERT *or a* DELETE *operation can be performed in $O(\log n)$ time,*
3. AFFINE *and* INTERVAL *operations can be performed in $O(1)$ and $O(\log n)$ time, respectively.*
4. INTEGRATE *operation can be performed in $O(\log n)$ time.*

One can use the above operations to compute the sum of two increasing continuous piecewise-linear functions $F$ and $G$ as follows: we first compute $F(b_i)$ for every breakpoint $b_i$ of $G$ and insert the pair $(b_i, F(b_i))$ into $\mathcal{T}(F)$. At this point the tree still represents $\mathcal{T}(F)$ but includes all the breakpoints of $G$ as *degenerate* breakpoints (at which the left and right derivates of $F$ are the same). Finally, for every consecutive pair of breakpoints $b_i$ and $b_{i+1}$ of $G$ we apply an INTERVAL$_x(\psi_i, b_i, b_{i+1})$ operation on $\mathcal{T}(F)$ where $\psi_i$ is the affine transformation $q \mapsto Mq + c$ where $M = \left(\begin{smallmatrix} 1 & 0 \\ \alpha & 1 \end{smallmatrix}\right)$ and $c = \left(\begin{smallmatrix} 0 \\ \beta \end{smallmatrix}\right)$, in which $G_i(x) = \alpha x + \beta$ is the linear function that interpolates between $G(b_i)$ at $b_i$ and $G(b_{i+1})$ at $b_{i+1}$ (similar operation using $\mu_-$ and $\mu_+$ of $G$ for the unbounded pieces of $G$ must can be applied in constant time). It is easy to verify that after performing this series of INTERVAL's, $\mathcal{T}(F)$ turns into $\mathcal{T}(F + G)$. The total running time of this operation is $O(|G| \log |F|)$. Note that this runtime can be reduced to $O(|G|(1 + \log |F| / \log |G|))$, for $|G| < |F|$, by using an algorithm of Brown and Tarjan [9] to insert all breakpoints and then applying all INTERVAL operations in a bottom up manner. Furthermore, this process can be reversed (i.e. creating $\mathcal{T}(F - G)$ without the breakpoints of $G$, given $\mathcal{T}(F)$ and $\mathcal{T}(G)$) in the same runtime. We therefore have shown:

**Lemma 2.** *Given $\mathcal{T}(F)$ and $\mathcal{T}(G)$ for a piecewise-linear isotonic functions $F$ and $G$ where $|G| < |F|$, $\mathcal{T}(F + G)$ or $\mathcal{T}(F - G)$ can computed in $O(|G|(1 + \log |F| / \log |G|))$ time.*

**Tree sets.** We will be repeatedly computing the sum of two functions $F$ and $G$. It will be too expensive to compute $\mathcal{T}(F + G)$ explicitly using Lemma 2, therefore we represent it implicitly. More precisely, we use a *tree set* $\mathcal{S}(F) =$

$\{\mathcal{T}(F_1), \ldots, \mathcal{T}(F_k)\}$ consisting of affine composition trees of monotone piecewise-linear functions $F_1, \ldots, F_k$ to represent the function $F = \sum_{j=1}^{k} F_j$. We perform several operations on $F$ or $\mathcal{S}$ similar to those of a single affine composition tree.

EVALUATE$(x)$ on $F$ takes $O(k \log n)$ time, by evaluating $\sum_{j=1}^{k} F_j(x)$. And EVALUATE$^{-1}(y)$ on $F$ takes $O(k \log^2 n)$ time using Frederickson and Johnson [15].

Given the ACT $\mathcal{T}(F_0)$, we can convert $\mathcal{S}(F)$ to $\mathcal{S}(F + F_0)$ in two ways: an INCLUDE$(\mathcal{S}, F_0)$ operation sets $\mathcal{S} = \{\mathcal{T}(F_1), \ldots, \mathcal{T}(F_k), \mathcal{T}(F_0)\}$ in $O(1)$ time. A MERGE$(\mathcal{S}, F_0)$ operations sets $\mathcal{S} = \{\mathcal{T}(F_1 + F_0), \mathcal{T}(F_2), \ldots, \mathcal{T}(F_k)\}$ in time $O(|F_0| \log |F_1|)$. We can also perform UNINCLUDE$(\mathcal{S}, F_0)$ and UNMERGE$(\mathcal{S}, F_0)$, operations that reverse the respective above operations in the same runtimes.

We can perform an AFFINE$(\mathcal{S}, \psi)$ where $\psi$ describes a linear transform $M$ and a translation vector $c$. To update $F$ by $\psi$ we update $F_1$ by $\psi$ and for $j \in [2, k]$ update $F_j$ by just $M$. This takes $O(k)$ time. It follows that we can perform INTERVAL$(\mathcal{S}, \psi, \tau^-, \tau^+)$ in $O(k \log n)$ time, where $n = |F_1| + \ldots + |F_k|$. Here we assume that the transformation $\psi$ is such that each $F_i$ remains monotone after the transformation.

## 4    LIR on Paths

In this section we describe an algorithm for solving the LIR problem on a path, represented as a directed graph $P = (V, A)$ where $V = \{v_1, \ldots, v_n\}$ and $A = \{(v_i, v_{i+1}) : 1 \leq i < n\}$. A function $s : V \to \mathbb{R}$ is isotonic (on $P$) if $s(v_i) \leq s(v_{i+1})$, and $\gamma$-Lipschitz for some real constant $\gamma$ if $s(v_{i+1}) \leq s(v_i) + \gamma$ for each $i = 1, \ldots, n-1$. For the rest of this section let $t : V \to \mathbb{R}$ be an input function on $V$. For each $i = 1, \ldots, n$, let $V_i = \{v_1, \ldots, v_i\}$, let $P_i$ be the subpath $v_1, \ldots, v_i$, and let $E_i$ and $\tilde{E}_i$, respectively, be shorthands for $E_{V_i}$ and $E_{P_i}^{v_i}$. By definition, if we let $\tilde{E}_0 = 0$, then for each $i \geq 1$:

$$\tilde{E}_i(x) = (x - t(v_i))^2 + \min_{x - \gamma \leq y \leq x} \tilde{E}_{i-1}(y). \tag{1}$$

By Lemma 1, $\tilde{E}_i$ is convex and continuous and thus has a unique minimizer $s_i^*$.

**Lemma 3.** *For $i \geq 1$, the function $\tilde{E}_i$ is given by the recurrence relation:*

$$\tilde{E}_i(x) = (x - t(v_i))^2 + \begin{cases} \tilde{E}_{i-1}(x - \gamma) & x > s_{i-1}^* + \gamma \\ \tilde{E}_{i-1}(s_{i-1}^*) & x \in [s_{i-1}^*, s_{i-1}^* + \gamma] \\ \tilde{E}_{i-1}(x) & x < s_{i-1}^*. \end{cases} \tag{2}$$

Thus by Lemmas 1 and 3, $\tilde{E}_i$ is strictly convex and piecewise quadratic. We call a value $x$ that determines the boundary of two neighboring quadratic pieces of $\tilde{E}_i$ a *breakpoint* of $\tilde{E}_i$. Since $\tilde{E}_1$ is a simple (one-piece) quadratic function, it has no breakpoints. For $i > 1$, the breakpoints of the function $\tilde{E}_i$ consist of $s_{i-1}^*$ and $s_{i-1}^* + \gamma$, as determined by recurrence (2), together with breakpoints that arise from recursive applications of $\tilde{E}_{i-1}$. Examining equation (2) reveals that all breakpoints of $\tilde{E}_{i-1}$ that are smaller than $s_{i-1}^*$ remain breakpoints in $\tilde{E}_i$ and all those larger than $s_{i-1}^*$ are increased by $\gamma$ and these form all of the breakpoints of $\tilde{E}_i$ (see Figure 2(right)). Thus $\tilde{E}_i$ has precisely $2i - 2$ breakpoints. To compute the point $s_i^*$ at which $\tilde{E}_i(x)$ is minimized, it is enough to scan over these $O(i)$ quadratic pieces and find the unique piece whose minimum lies between its two ending breakpoints.

**Lemma 4.** *Given the sequence* $s_1^*, \ldots, s_n^*$, *one can compute the $\gamma$-LIR $s$ of input function $t$ in $O(n)$ time.*

One can compute the values of $s_1^*, \ldots, s_n^*$ in $n$ iterations. The $i$th iteration computes the value $s_i^*$ at which $\tilde{E}_i$ is minimized and then uses it to compute $\tilde{E}_{i+1}$ via (2) in $O(i)$ time. Hence, the $\gamma$-LIR of $t \in \mathbb{R}^V$ can be computed in linear time. However, this gives an $O(n^2)$ algorithm for computing the $\gamma$-LIR of $t$. We now show how this running time can be reduced to $O(n \log n)$.

For simplicity we assume each $s_i^*$ is none of the breakpoints of $\tilde{E}_i$. Hence $s_i^*$ belongs to the interior of some interval on which $\tilde{E}_i$ is quadratic, and its derivative is zero at $s_i^*$. If we know to which quadratic piece of $\tilde{E}_i$ the point $s_i^*$ belongs, we can determine $s_i^*$ by setting the derivative of that piece to zero.

**Lemma 5.** *The derivative of $\tilde{E}_i$ is continuous isotonic and piecewise-linear.*

Let $F_i$ denote the derivative of $\tilde{E}_i$ with the recurrence (via (2)):

$$F_{i+1} = 2(x - t(v_{i+1})) + \hat{F}_i(x); \tag{3}$$

$$\hat{F}_i(x) = \begin{cases} F_i(x - \gamma) & x > s_i^* + \gamma, \\ 0 & x \in [s_i^*, s_i^* + \gamma], \\ F_i(x) & x < s_i^*. \end{cases} \tag{4}$$

if we set $F_0 = 0$. As mentioned above, $s_i^*$ is simply the solution of $F_i(x) = 0$, which, by Lemma 5 always exists and is unique. Intuitively, $\hat{F}_i$ is obtained from $F_i$ by splitting it at $s_i^*$, shifting the right part by $\gamma$, and connecting the two pieces by a horizontal edge from $s_i^*$ to $s_i^* + \gamma$ (lying on the $x$-axis).

In order to find $s_i^*$ efficiently, we use an ACT $\mathcal{T}(F_i)$ to represent $F_i$. It takes $O(\log |F_i|) = O(\log i)$ time to compute $s_i^* = \text{EVALUATE}^{-1}(0)$ on $\mathcal{T}(F_i)$. Once $s_i^*$ is computed, we store it in a separate array for back-solving through Lemma 4. We turn $\mathcal{T}(F_i)$ into $\mathcal{T}(\hat{F}_i)$ by performing a sequence of $\text{INSERT}((s_i^*, 0))$, $\text{INTERVAL}_x(\psi, s_i^*, \infty)$, and $\text{INSERT}((s_i^*, 0))$ operations on $\mathcal{T}(F_i)$ where $\psi(q) = q + \binom{\gamma}{0}$; the two insert operations add the breakpoints at $s_i^*$ and $s_i^* + \gamma$ and the interval operation shifts the portion of $F_i$ to the right of $s_i^*$ by $\gamma$. We then turn $\mathcal{T}(\hat{F}_i)$ into $\mathcal{T}(F_{i+1})$ by performing $\text{AFFINE}(\phi_{i+1})$ operation on $\mathcal{T}(\hat{F}_i)$ where $\phi_{i+1}(q) = Mq + c$ where $M = \left(\begin{smallmatrix} 1 & 0 \\ 2 & 1 \end{smallmatrix}\right)$ and $c = \left(\begin{smallmatrix} 0 \\ -2t(v_{i+1}) \end{smallmatrix}\right)$.

Given ACT $\mathcal{T}(F_i)$, $s_i^*$ and $\mathcal{T}(F_{i+1})$ can be computed in $O(\log i)$ time. Hence, we can compute $s_1^*, \ldots, s_n^*$ in $O(n \log n)$ time. By Lemma 4, we can conclude:

**Theorem 2.** *Given a path $P = (V, A)$, a function $t \in \mathbb{R}^V$, and a constant $\gamma$, the $\gamma$-Lipschitz isotonic regression of $t$ on $P$ can be found in $O(n \log n)$ time.*

**Update operation.** We define a procedure UPDATE($\mathcal{T}(\hat{F}_i), t(v_{i+1}), \gamma$) that encapsulates the process of turning $\mathcal{T}(\hat{F}_i)$ into $\mathcal{T}(\hat{F}_{i+1})$ and returning $s^*_{i+1}$. Specifically, it performs AFFINE($\phi_{i+1}$) of $\mathcal{T}(\hat{F}_i)$ to produce $\mathcal{T}(F_{i+1})$, then it outputs $s^*_{i+1} =$ EVALUATE$^{-1}(0)$ on $\mathcal{T}(F_{i+1})$, and finally a sequence of INSERT($(s^*_{i+1}, 0)$), INTERVAL($\psi, s^*_{i+1}, \infty$), and INSERT($(s^*_{i+1}, 0)$) operations on $\mathcal{T}(F_{i+1})$ for $\mathcal{T}(\hat{F}_{i+1})$. Performed on $\mathcal{T}(F)$ where $F$ has $n$ breakpoints, an UPDATE takes $O(\log n)$ time. An unUPDATE($\mathcal{T}(\hat{F}_{i+1}), t(v_{i+1}), \gamma$) reverts affects of UPDATE($\mathcal{T}(\hat{F}_i), t(v_{i+1}), \gamma$). This requires that $s^*_{i+1}$ is stored for the reverted version. Similarly, we can perform UPDATE($\mathcal{S}, t(v_i), \gamma$) and unUPDATE($\mathcal{S}, t(v_i), \gamma$) on a tree set $\mathcal{S}$, in $O(k \log^2 n)$ time, the bottleneck coming from EVALUATE$^{-1}(0)$.

**Lemma 6.** *For $\mathcal{T}(\hat{F}_i)$,* UPDATE($\mathcal{T}(\hat{F}_i), t(v_i), \gamma$) *and* unUPDATE($\mathcal{T}(\hat{F}_i), t(v_i), \gamma$) *take $O(\log n)$ time.*

## 5   LUR on Paths

Let $P = (V, A)$ be an *undirected* path, $V = \{v_1, \ldots, v_n\}$, $A = \{\{v_i, v_{i+1}\}, 1 \leq i < n\}$, and $t \in \mathbb{R}^V$. For $v_i \in V$ let $P_i = (V, A_i)$ be a directed graph in which all edges are directed towards $v_i$; that is, for $j < i$, $(v_j, v_{j+1}) \in A_i$ and for $j > i$ $(v_j, v_{j-1}) \in A_i$. For each $i = 1, \ldots, n$, let $\Gamma_i = \Gamma(P_i, V) \subseteq \mathbb{R}^V$ and let $\sigma_i = \arg\min_{s \in \Gamma_i} E_V(s)$. If $\kappa = \arg\min_i E_V(\sigma_i)$, then $\sigma_\kappa$ is the $\gamma$-LUR of $t$ on $P$.

We find $\sigma_\kappa$ in $O(n \log^2 n)$ time by solving the LIR problem, then traversing the path while maintaining the optimal solution using UPDATE and unUPDATE. Specifically, for $i = 1, \ldots, n$, let $V_i^- = \{v_1, \ldots, v_{i-1}\}$ and $V_i^+ = \{v_{i+1}, \ldots, v_n\}$. For $P_i$, let $\hat{F}_{i-1}^-$, $\hat{F}_{i+1}^+$ be the functions on directed paths $P[V_i^-]$ and $P[V_i^+]$, respectively, as defined in (3). Set $\bar{F}_i(x) = 2(x - t(v_i))$. Then the function $F_i(x) = dE_{v_i}(x)/dx$ can be written as $F_i(x) = \hat{F}_{i-1}^-(x) + \hat{F}_{i+1}^+(x) + \bar{F}_i(x)$. We store $F_i$ as the tree set $S_i = \{\mathcal{T}(\hat{F}_{i-1}^-), \mathcal{T}(\hat{F}_{i+1}^+), \mathcal{T}(\bar{F}_i)\}$. By performing EVALUATE$^{-1}(0)$ we can compute $s^*_i$ in $O(\log^2 n)$ time (the rate limiting step), and then we can compute $E_V(s^*_i)$ in $O(\log n)$ time using INTEGRATE($s^*_i$). Assuming we have $\mathcal{T}(\hat{F}_{i-1}^-)$ and $\mathcal{T}(\hat{F}_{i+1}^+)$, we can construct $\mathcal{T}(\hat{F}_i^-)$ and $\mathcal{T}(\hat{F}_{i+2}^+)$ in $O(\log n)$ time be performing UPDATE($\mathcal{T}(F_{i-1}^-), t(v_i), \gamma$) and unUPDATE($\mathcal{T}(F_{i+1}^+), t(v_{i+1}), \gamma$). Since $S_1 = \{\mathcal{T}(F_0^-), \mathcal{T}(F_2^+), \mathcal{T}(\bar{F}_1)\}$ is constructed in $O(n \log n)$ time, finding $\kappa$ by searching all $n$ tree sets takes $O(n \log^2 n)$ time.

**Theorem 3.** *Given an undirected path $P = (V, A)$ and a $t \in \mathbb{R}^V$ together with a real $\gamma \geq 0$, the $\gamma$-LUR of $t$ on $P$ can be found in $O(n \log^2 n)$ time.*

## 6   LIR on Rooted Trees

Let $T = (V, A)$ be a rooted tree with root $r$ and let for each vertex $v$, $T_v = (V_v, A_v)$ denote the subtree of $T$ rooted at $v$. Similar to the case of path LIR, for each vertex $v \in V$ we use the shorthands $E_v = E_{V_v}$ and $\tilde{E}_v = E_{T_v}^v$. Since the subtrees rooted at distinct children of a node $v$ are disjoint, one can write an equation corresponding to (1) in the case of paths, for any vertex $v$ of $T$:

$$\tilde{E}_v(x) = (x - t(v))^2 + \sum_{u \in \delta^-(v)} \min_{x - \gamma \leq y \leq x} \tilde{E}_u(y), \qquad (5)$$

where $\delta^-(v) = \{u \in V \mid (u, v) \in A\}$. An argument similar to that of Lemma 5 together with Lemma 1 implies that for every $v \in V$, the function $\tilde{E}_v$ is convex and piecewise quadratic, and its derivative $F_v$ is continuous, monotonically increasing, and piecewise linear. We can prove that $F_v$ satisfies the following recurrence where $\hat{F}_u$ is defined analogously to $\hat{F}_i$ in (4):

$$F_v(x) = 2(x - t(v)) + \sum_{u \in \delta^-(v)} \hat{F}_u(x). \qquad (6)$$

Thus to solve the LIR problem on a tree, we post-order traverse the tree (from the leaves toward the root) and when processing a node $v$, we compute and sum the linear functions $\hat{F}_u$ for all children $u$ of $v$ and use (6) to compute the function $F_v$. We then solve $F_v(x) = 0$ to find $s_v^*$. As in the case of path LIR, $\hat{F}_v$ can be represented by an ACT $\mathcal{T}(\hat{F}_v)$. For simplicity, we assume each non-leaf vertex $v$ has two children $h(v)$ and $l(v)$, where $|\hat{F}_{h(v)}| \geq |\hat{F}_{l(v)}|$. Given $\mathcal{T}(\hat{F}_{h(v)})$ and $\mathcal{T}(\hat{F}_{l(v)})$, we can compute $\mathcal{T}(\hat{F}_v)$ and $s_v^*$ with the operation $\text{UPDATE}(\mathcal{T}(\hat{F}_{h(v)}) + \hat{F}_{l(v)}), t(v), \gamma)$ in $O(|\hat{F}_{l(v)}|(1 + \log |\hat{F}_{h(v)}| / \log |\hat{F}_{l(v)}|))$ time. The merging of two functions dominates the time for the update. By careful, global analysis of all merge costs we can achieve the following result:

**Theorem 4.** *Given rooted tree $T = (V, A)$, function $t \in \mathbb{R}^V$, and Lipschitz constant $\gamma$, we can find in $O(n \log n)$ time, the $\gamma$-LIR of $t$ on $T$.*

## 7   LUR on Trees

We extend this framework to solve the LUR on trees problem. Given an undirected input tree, we direct the tree towards an arbitrary vertex chosen as the root and invoke Theorem 4. Similar to LUR on paths, we traverse the tree, letting each vertex be the root, and maintaining enough information to recompute the LIR solution for the new root in $O(\log^3 n)$ time. We return the solution for the root which minimizes the error. The details are left for the full version [2].

**Theorem 5.** *Given unrooted tree $T = (V, A)$, function $t \in \mathbb{R}^V$, and Lipschitz constraint $\gamma$, we can find in $O(n \log^3 n)$ time the $\gamma$-LUR of $t$ on $T$.*

## References

1. Agarwal, P.K., Arge, L., Yi, K.: I/O-efficient batched union-find and its applications to terrain analysis. In: Proc. 22 ACM Symp. on Comp. Geometry (2006)
2. Agarwal, P.K., Phillips, J.M., Sadri, B.: Lipschitz unimodal and isotonic regression on paths and trees. Technical report, arXiv:0912.5182 (2009)

3. Angelov, S., Harb, B., Kannan, S., Wang, L.-S.: Weighted isotonic regression under the $l_1$ norm. In: Proc. 17th ACM-SIAM Symp. on Discrete Algorithms (2006)
4. Attali, D., Glisse, M., Hornus, S., Lazarus, F., Morozov, D.: Persistence-sensitive simplification of functions on surfaces in linear time. INRIA (2008) (manuscript)
5. Ayer, M., Brunk, H.D., Ewing, G.M., Reid, W.T., Silverman, E.: An empirical distribution function for sampling with incomplete information. Annals of Mathematical Statistics 26, 641–647 (1955)
6. Bajaj, C., Pascucci, V., Schikore, D.: Visualization of scalar topology for structural enhancement. In: IEEE Visualization, pp. 51–58 (1998)
7. Barlow, R.E., Bartholomew, D.J., Bremmer, J.M., Brunk, H.D.: Statistical Inference Under Order Restrictions: The Theory and Application of Isotonic Regression. John Wiley and Sons, Chichester (1972)
8. Bremer, P., Edelsbrunner, H., Hamann, B., Pascucci, V.: A multi-resolution data structure for two-dimensional morse functions. In: Proceedings 14th IEEE Visualization Conference, pp. 139–146 (2003)
9. Brown, M.R., Tarjan, R.E.: A fast merging algorithm. J. Alg. 15, 416–446 (1979)
10. Brunk, H.D.: Maximum likelihood estimates of monotone parameters. Annals of Mathematical Statistics 26, 607–616 (1955)
11. Danner, A., Mølhave, T., Yi, K., Agarwal, P.K., Arge, L., Mitasova, H.: Terrastream: from elevation data to watershed hierarchies. In: 15th ACM International Symposium on Advances in Geographic Information Systems (2007)
12. Edelsbrunner, H., Harer, J.: Persistent Homology: A Survey. Contemporary Mathematics, vol. 453. American Mathematical Society, Providence (2008)
13. Edelsbrunner, H., Letscher, D., Zomorodian, A.: Topological persistence and simplification. In: Proc. 41 Symp. on Foundatons of Computer Science (2000)
14. Edelsbrunner, H., Morozov, D., Pascucci, V.: Persistence-sensitive simplification functions on 2-manifolds. In: Proc. 22 ACM Symp. on Comp. Geometry (2006)
15. Frederickson, G.N., Johnson, D.B.: The complexity of selection and ranking in x + y and matrices with sorted columns. J. Comput. Syst. Sci. 24, 192–208 (1982)
16. Grotzinger, S.J., Witzgall, C.: Projection onto order simplexes. Applications of Mathematics and Optimization 12, 247–270 (1984)
17. Guskov, I., Wood, Z.J.: Topological noise removal. In: Graphics Interface (2001)
18. Jewel, W.S.: Isotonic optimization in tariff construction. ASTIN 8, 175–203 (1975)
19. Maxwell, W.L., Muckstadt, J.A.: Establishing consistent and realistic reorder intervals in production-distribution systems. Operations Res. 33, 1316–1341 (1985)
20. Ni, X., Garland, M., Hart, J.C.: Fair Morse functions for extracting the topological structure of a surface mesh. ACM Transact. Graphics 23, 613–622 (2004)
21. Pardalos, P.M., Xue, G.: Algorithms for a class of isotonic regression problems. Algorithmica 23, 211–222 (1999)
22. Preparata, F.P., Shamos, M.I.: Computational geometry: an introduction. Springer, New York (1985)
23. Soille, P., Vogt, J., Cololmbo, R.: Carbing and adaptive drainage enforcement of grid digital elevation models. Water Resources Research 39(12), 1366–1375 (2003)
24. Spouge, J., Wan, H., Wilbur, W.J.: Least squares isotonic regression in two dimensions. Journal of Optimization Theory and Applications 117, 585–605 (2003)
25. Stout, Q.F.: Optimal algorithms for unimodal regression. Computing Science and Statistics 32, 348–355 (2000)

26. Stout, Q.F.: Unimodal regression via prefix isotonic regression. Computational Statistics and Data Analysis 53, 289–297 (2008)
27. Thompson Jr., W.A.: The problem of negative estimates of variance components. Annals of Mathematical Statistics 33, 273–289 (1962)
28. Zomorodian, A.: Computational topology. In: Atallah, M.J., Blanton, M. (eds.) Algorithms and Theory of Computation Handbook. Chapman & Hall/CRC Press (2009)

# Ambiguity and Deficiency in Costas Arrays and APN Permutations*

Daniel Panario, Brett Stevens, and Qiang Wang

School of Mathematics and Statistics, Carleton University
Ottawa, Canada, K1S 5B6
{daniel,brett,wang}@math.carleton.ca

**Abstract.** We introduce the concepts of weighted ambiguity and deficiency for a mapping between two finite Abelian groups of the same size. Then we study the optimum lower bounds of these measures for a permutation of $\mathbb{Z}_n$ and give a construction of permutations meeting the lower bound by modifying some permutation polynomials over finite fields. These permutations are also APN permutations.

## 1 Introduction

The injectivity and surjectivity of maps between groups and rings can be of crucial importance to various applications where these maps appear. One simple example of this is polynomials over finite rings. Since the domain and codomain are identical and finite the concepts of injectivity and surjectivity are equivalent. A polynomial is a permutation polynomial over a finite ring $R$ if it induces a bijective map from $R$ to $R$. We are interested in the finite field $\mathbb{F}_q$ or the integer ring $\mathbb{Z}_n$. In recent years, there has been a lot of interest in studying permutation polynomials, partly due to their applications in coding theory, combinatorics and cryptography. For more background material on permutation polynomials over finite fields we refer to Chapter 7 of [7]. For a detailed survey of open questions and recent results see [5,6,9]. For permutation polynomials over $\mathbb{Z}_n$, we refer the readers to [10,12,14].

Two other instances of the importance of injectivity and surjectivity have received a lot of attention in recent years. Let $f : G_1 \to G_2$ be a map, or partial map, between two Abelian groups. For $a \in G_1, a \neq 0$ we can define a difference map

$$g_{f,a}(x) = f(x + a) - f(x)$$

which can measure the degree of "linearity" of $f$. A *Costas array* is a permutation $f : [1, n] \subset \mathbb{Z} \to [1, n] \subset \mathbb{Z}$ such that $g_{f,a} : \mathbb{Z} \to \mathbb{Z}$ is injective.

Costas arrays were first considered by Costas [2] as $n \times n$ permutation matrices with ambiguity functions taking only the values 0 and (possibly) 1, which were applied to the processing of radar and sonar signals. The injectivity of $g_{f,a}$ reduces the ambiguity of locating a time and frequency shifted echo of the original signal.

---

* The authors are partially supported by NSERC.

A. López-Ortiz (Ed.): LATIN 2010, LNCS 6034, pp. 397–406, 2010.
© Springer-Verlag Berlin Heidelberg 2010

Similarly for maps between Abelian groups of the same cardinality, a function $f$ is called *perfectly non-linear (PN)* if $g_{f,a}$ is injective and *almost perfectly non-linear (APN)* if $g_{f,a}$ is at worst 2 to 1. These functions have received significant attention because they are resistant to linear cryptanalysis and differential cryptanalysis (see [11]). In particular, we note that APN functions of $\mathbb{Z}_{256}$ was used in SAFER family of cryptosystem proposed by Massey [8].

In these examples and applications injectivity and surjectivity of $g_{f,a}$ are no longer equivalent as they are for the function $f$ itself (due to either the sizes of domain and codomain being finite but different or infinite) but they are still strongly correlated. In this paper we attempt to understand the injectivity and surjectivity of $g_{f,a}$. In Section 2 we define two generalized measures of injectivity and surjectivity of $g_{f,a}$ which we call the *ambiguity* and the *deficiency* of $f$, respectively. These help to make strong connections between permutations, Costas arrays and almost perfectly non-linear functions. In Section 3 we prove bounds on these measures which then allow us to define notions of optimality with respect to them. We finish Section 3 by constructing an infinite family of permutations which achieve these lower bounds for functions $f : \mathbb{Z}_n \to \mathbb{Z}_n$. These are provably better and more frequently existing than other recent constructions for this Abelian group [3].

## 2   Definitions and Connections

### 2.1   Definitions

Let $G_1$ and $G_2$ be finite Abelian groups of the same cardinality $n$ and $f : G_1 \to G_2$. Let $G_1^* = G_1 \setminus \{0\}$ and $G_2^* = G_2 \setminus \{0\}$. For any $a \in G_1^*$ and $b \in G_2$, we denote $g_{f,a}(x) = f(x+a) - f(x)$ and $\lambda_{a,b}(f) = \# g_{f,a}^{-1}(b)$. Let $n_i(f) = \#\{(a,b) \in G_1^* \times G_2 \mid \lambda_{a,b}(f) = i\}$ for $0 \le i \le n$. We call $n_0(f)$ the *deficiency* of $f$, denoted by $D(f)$. Hence $D(f) = n_0(f)$ measures the number of pairs $(a,b)$ such that $g_{f,a}(x) = b$ has no solutions. This is a measure of the surjectivity of $g_{f,a}$; the lower the deficiency the closer the $g_{f,a}$ are to surjective.

Moreover, we define the *ambiguity* of $f$ as

$$A(f) = \sum_{0 \le i \le n} n_i(f) \binom{i}{2}.$$

From this definition, we can see that the weighted ambiguity of $f$ measures the total replication of pairs of $x$ and $x'$ such that $g_{f,a}(x) = g_{f,a}(x')$ for some $a \in G_1^*$. This is a measure of the injectivity of the functions $g_{f,a}$; the lower the ambiguity of $f$ the closer the $g_{f,a}$ are to injective.

For a fixed $a \neq 0$ the values of $g_{f,a}(x)$ are the entries in the $a$th row of what is often referred to as the *difference triangle* of $f$ (when the domain of $f$ is $\mathbb{Z}$ [3,13]) or what we might call the *difference array* (when the domain of $f$ is $\mathbb{Z}_n$). Thus for a fixed $a \neq 0$, we define the *row-a-ambiguity of $f$* as

$$A_{r=a}(f) = \sum_b \binom{\lambda_{a,b}(f)}{2}.$$

This measures the injectivity of the individual $g_{f,a}$. Similarly, we define the *row-a-deficiency* as $D_{r=a}(f) = \#\{b \mid \lambda_{a,b}(f) = 0, b \in G_2\}$, which measures the number of $b$'s such that $g_{f,a}(x) = b$ has no solutions for a fixed $a$. We have that $D_{r=a}(f) = n - \#\{g_{f,a}(x) \mid x \in G_1\}$ (and $D_{r=a}(f) = n - 1 - \#\{g_{f,a}(x) \mid x \in G_1\}$ when $f$ is a bijection). Likewise, we define *column-b-deficiency* as $D_{c=b}(f) = \#\{a \mid \lambda_{a,b}(f) = 0, a \in G_1^*\}$, which measures the number of $a$'s such that $g_{f,a}(x) = b$ has no solutions for a fixed $b$.

In this paper we restrict our attention to $f : \mathbb{Z}_n \rightarrow \mathbb{Z}_n$ that are bijections. This has the implication that $g_{f,a}(x) = b$ can never have solutions for $b = 0$ thus we use the corresponding form of all our definitions that restrict $b \in G_2^* = \mathbb{Z}_n^*$ and this includes summations and universal quantifiers. Another effect of this is that the domain and co-domain of $g_{f,a}$ are now sizes $n$ and $n - 1$, respectively; this is particularly important to remember when reading the proofs otherwise our references to "$n - 1$" will seem odd.

It is clear that the ambiguity and deficiency are positively correlated although they are not exactly related. In the context of $b \in G_2^*$ we can give bounds on their relationship.

**Lemma 1.** *Let* $f : G_1 \rightarrow G_2$ *be a bijection. If a row-a-deficiency of* $f$ *is* $D_{r=a}(f) = d$, *then row-a-ambiguity of* $f$ *satisfies*

$$d + 1 \leq A_{r=a}(f) \leq \binom{d+2}{2}.$$

*Proof.* Because $D_{r=a}(f) = n - 1 - \#\{g_{f,a}(x) \mid x \in G_1\}$, the size of the value set $\{g_{f,a}(x) \mid x \in G_1\}$ is $n - 1 - d$ for a given row-a-deficiency $d$. The maximum row-a-ambiguity, $A_{r=a}(f) = \binom{d+2}{2}$, occurs when the $n$ images, $g_{f,a}(x)$, are distributed with $n - 2 - d$ values of $x$ giving distinct images and the remaining $d + 2$ values all agreeing. The minimum value, $A_{r=a}(f) = d + 1$, occurs when the $n$ images are distributed with $d + 1$ pairs of $\{x, x'\}$ having $g_{f,a}(x) = g_{f,a}(x')$ and the remaining $n - 2(d + 1)$ images distinct. □

We note that two conditions are necessary when $A_{r=a}(f)$ achieves its minimum:

- $d \leq n/2 - 1$;
- the sets $g_{f,a}^{-1}(b)$ having cardinality 0, 1 and 2.

## 2.2 Connections

Costas arrays were first considered by Costas [2] for applications to radar and sonar signal processing. His definition is equivalent to an $n \times n$ permutation matrix with ambiguity function taking only the values 0 and (possibly) 1. We remark that the domain and co-domain were both $\mathbb{Z}$ for Costas.

**Definition 1.** *A Costas array is a permutation matrix (that is, a square matrix with precisely one 1 in each row and column and all other entries 0) for which all the vectors joining the pairs of 1's are distinct.*

It is clear that a permutation $f$, from the columns to the rows (i.e. to each column $x$ we assign one and only one row $f(x)$), gives a Costas array if and only if for $x \neq y$ and $k \neq 0$, $f(x + k) - f(x) \neq f(y + k) - f(y)$. We note that in the standard definition of Costas array, the arithmetic takes place inside $\mathbb{Z}$ and the vectors are in $\mathbb{Z} \times \mathbb{Z}$. The Costas array definition is precisely the property of $A(f) = 0$ when $f : [1, n] \subset \mathbb{Z} \to [1, n] \subset \mathbb{Z}$.

A special class of Costas arrays is the so called singly periodic Costas arrays, which is an $n \times \infty$ matrix built by infinitely and repeatedly horizontally concatenating an $n \times n$ Costas array with the property that any $n \times n$ window is a Costas array. This is equivalent to considering the injection $f : \mathbb{Z}_n \to \mathbb{Z}$ and asking again that $f$ have zero ambiguity.

If we consider $f : \mathbb{Z}_n \to \mathbb{Z}_n$, the bounds from our Theorem 1 show that zero ambiguity is impossible and thus "doubly periodic Costas arrays" cannot exist. However the bounds from Theorem 1 also tell us precisely what it means to be as close as possible to a "doubly periodic Costas array": we require the ambiguity (and correspondingly the deficiency) to be as small as possible. In Theorem 2 we build a family of permutations $f$ for an infinite number of orders, $n$, which are optimum with respect to both the ambiguity and deficiency.

Perfect and almost perfect non-linear functions can also be defined within the terminology of ambiguity and deficiency.

**Definition 2.** *[4] Let $G_1$ and $G_2$ be finite Abelian groups of the same cardinality and $f : G_1 \to G_2$. We say $f$ is a perfect non-linear function if*

$$f(x + a) - f(x) = b$$

*has exactly one solution for all $a \neq 0 \in G_1$ and all $b \in G_2$.*

This corresponds again to zero ambiguity. This property is often too strong to require and particularly in the case of bijections, it can never be satisfied. Thus a relaxed definition is frequently useful.

**Definition 3.** *[4] Let $G_1$ and $G_2$ be finite Abelian groups of the same cardinality and $f : G_1 \to G_2$. We say $f$ is an almost perfect non-linear function if*

$$f(x + a) - f(x) = b$$

*has at most two solutions for all $a \neq 0 \in G_1$ and all $b \in G_2$.*

When $f$ is a bijection we only consider $b \in G_2^*$ and APN functions are clearly functions with small ambiguity. Since a function can be APN and still have an ambiguity anywhere between 0 and $(n-1)\lfloor n/2 \rfloor$, our definition of ambiguity has a higher resolution power than just the definition of APN and thus can usefully be regarded as a refinement of the concept. Indeed, we show in Section 3 that ambiguity meeting our bounds implies the APN property for permutations of $\mathbb{Z}_n$ and thus provides a general construction of such APN permutations.

The two subjects of Costas arrays and APN functions have been connected before by Drakakis, Gow and McGuire in [4] where they use the Welch construction of singly periodic Costas arrays to build APN permutations, $f : \mathbb{Z}_{p-1} \to \mathbb{Z}_{p-1}$

for $p$ a prime. We note that our constructions have optimum and therefore lower ambiguity than those coming from the Welch construction and thus are closer to being PN functions. Additionally our constructions are defined on the larger set of $n = q - 1$ where $q$ is a prime power. Our methods in Section 3 modify known families of permutation polynomials of finite fields. Our permutations are optimum in both ambiguity and deficiency.

## 3    Results

In this section we determine a lower bound on the ambiguity and the deficiency and then construct permutations of $\mathbb{Z}_n$ achieving these bounds for an infinite number of values of $n$.

**Theorem 1.** *Let $n \in \mathbb{N}$ and $f : \mathbb{Z}_n \to \mathbb{Z}_n$ be a bijection. The ambiguity of $f$ is at least $2(n - 1)$ when $n$ is odd and $2(n - 2)$ when $n$ is even. The deficiency of $f$ is at least $n - 1$ if $n$ is odd and at least $n - 3$ when $n$ is even.*

*Proof.* Since $f$ is a permutation of $\mathbb{Z}_n$, $f(x + a) \neq f(x)$ and so the $n$ elements of the multiset $\{f(x + a) - f(x)\}$ are elements of $\mathbb{Z}_n^*$. The pigeon hole principle says there must be at least one pair, $x$ and $x'$ such that $f(x + a) - f(x) = f(x' + a) - f(x')$.

The lower bound on deficiency when $n$ is odd is straightforward. Suppose for a given $a$ that the set $\{f(x + a) - f(x)\} = \mathbb{Z}_n^*$, that is, the contribution by this $a$ to the deficiency is zero. Since there are $n$ elements $f(x + a) - f(x)$ which span $n - 1$ values there can only be a single repeated value, say $r \neq 0$. Thus since $f$ is a permutation and $n$ is odd

$$0 = \sum_{x \in \mathbb{Z}_n} f(x + a) - \sum_{x \in \mathbb{Z}_n} f(x) = \sum_{x \in \mathbb{Z}_n} (f(x + a) - f(x)) = r + \sum_{x \in \mathbb{Z}_n^*} x = r,$$

which is a contradiction. Thus among the values of $g_{f,a}(x) = f(x + a) - f(x)$, at least one value from $\mathbb{Z}_n^*$ is missed. Hence $D_{r=a}(f) \geq 1$. Summing these over all non-zero $a$ gives the required lower bound $D(f) \geq n - 1$. By Lemma 1, $A_{r=a}(f) \geq 2$. Summing these over all non-zero $a$ gives the required lower bound for ambiguity of $f$, that is, $A(f) \geq 2(n - 1)$.

When $n$ even, and for a fixed $a \neq 0$, we suppose that the set $\{g_{f,a}(x) = f(x + a) - f(x)\} = \mathbb{Z}_n^*$ and the single repeated element is $r$. We get

$$0 = \sum_{x \in \mathbb{Z}_n} f(x + a) - \sum_{x \in \mathbb{Z}_n} f(x) = \sum_{x \in \mathbb{Z}_n} (f(x + a) - f(x))$$

$$= r + \sum_{x \in \mathbb{Z}_n^*} x = r + n/2,$$

and thus it is possible for $\{g_{f,a}(x) = f(x + a) - f(x)\} = \mathbb{Z}_n^*$ if the repeated value is $n/2$. We put an upper bound on the number of times this can happen.

For a given $f$ let $R$ be the set of $a \in \mathbb{Z}_n^*$ such that $\{f(x + a) - f(x)\} = \mathbb{Z}_n^*$ and let $C$ be the set of $b$ such that $\{f^{-1}(x + b) - f^{-1}(x)\} = \mathbb{Z}_n^*$.

As we have seen earlier, the row-$a$-deficiency is the cardinality of

$$\mathbb{Z}_n^* \setminus \{f(x+a) - f(x) \mid x \in \mathbb{Z}_n\}.$$

Similarly the column-$b$-deficiency is the cardinality

$$\mathbb{Z}_n^* \setminus \{f^{-1}(y+b) - f^{-1}(y) \mid y \in \mathbb{Z}_n\}.$$

It is clear that the deficiency can be computed from either the row or column deficiencies

$$\sum_{a \neq 0} D_{r=a}(f) = \sum_{b \neq 0} D_{c=b}(f).$$

For $a \in \mathbb{Z}_n$ if $D_{r=a}(f) = 0$ then for all $n/2 \neq b \in \mathbb{Z}_n^*$ there exists $x_b$ such that $f(x_b + a) - f(x) = b$ and there exist $x, x'$ such that $f(x + a) - f(x) = f(x' + a) - f(x') = n/2$. Let $y = f(x)$ and $y' = f(x')$. Thus

$$f^{-1}(y + n/2) - f^{-1}(y) = f^{-1}(y' + n/2) - f^{-1}(y') = a$$

Therefore, for every row-$a$-deficiency value of 0, there is a pair of repeated elements, $a$, in $\{f^{-1}(y + n/2) - f^{-1}(y)\}$. Also $D_{r=a}(f) = D_{r=-a}(f)$, and hence

$$D_{c=n/2}(f) \geq n - 1 - (|R \setminus \{n/2\}| + (n - 2|R \setminus \{n/2\}|)) = |R \setminus \{n/2\}| - 1,$$

and similarly

$$D_{r=n/2}(f) \geq |C \setminus \{n/2\}| - 1.$$

So the deficiency, $D(f)$ is calculated

$$D(f) = \frac{1}{2} \left( \sum_{a \neq 0} D_{r=a}(f) + \sum_{b \neq 0} D_{c=b}(f) \right)$$

$$= \frac{1}{2} \left( \sum_{a \neq 0, n/2} D_{r=a}(f) + \sum_{b \neq 0, n/2} D_{c=b}(f) + D_{c=n/2}(f) + D_{r=n/2}(f) \right)$$

$$\geq \frac{1}{2} \left( (n - 2 - |R \setminus \{n/2\}|) + (n - 2 - |C \setminus \{n/2\}|) + |R \setminus \{n/2\}| - 1 + |C \setminus \{n/2\}| - 1 \right)$$

$$= n - 3.$$

Again, by Lemma 1, a row-$a$-deficiency value of $d$ contributes at least $d + 1$ to the ambiguity, so we get that the total ambiguity for $f$ is at least $n - 1 + n - 3 = 2(n - 2)$. $\qquad\square$

If, for a fixed $n$, a permutation $f : \mathbb{Z}_n \to \mathbb{Z}_n$ has the smallest possible ambiguity we say it has *optimum ambiguity* and similarly we define *optimum deficiency* for a permutation if it achieves the smallest possible deficiency over all permutations of $\mathbb{Z}_n$. Clearly any permutations meeting the bounds from Theorem 1 will be *optimum*.

For the optimum ambiguity, all the sets $g_{f,a}^{-1}(b)$ have cardinality 2. It is these observations that allow us to connect our notions of ambiguity to APN functions.

**Corollary 1.** *If a permutation $f : \mathbb{Z}_n \to \mathbb{Z}_n$ achieves the bound on ambiguity from Theorem 1, then $f$ is Almost Perfect Non-linear.*

*Proof.* Consideration of the forced equalities throughout the proof of Theorem 1 give that the number of pairs of $(a, b)$ such that $|g_{f,a}^{-1}(b)| \geq 2$ is exactly the ambiguity and each inverse image size $0, 1$ or $2$. Thus $f$ is APN [4].    □

However a permutation which is APN could have ambiguity as large as $(n - 1)\lfloor n/2 \rfloor$ and correspondingly a deficiency as large as $(n-1)(\lfloor n/2 \rfloor - 1)$.

*Example 1.* One APN permutation constructed in $\mathbb{Z}_{10}$ from the Welch Costas array constructions is $f(x) = (2^x \bmod 11) - 1$ or $f = (0)(1)(23768)(4)(59)$ and has ambiguity $19 > 2(10-2) = 16$ and deficiency $12 > (10-3) = 7$ although this construction does not attain the worst possible values for APN permutations.    □

Next we provide our main construction which produces permutations that achieve the minimum ambiguity and deficiency whenever $n$ is precisely one less than a prime power. In order to do so, we first introduce a way to obtain a permutation polynomial of fixed point $0$ over a finite field $\mathbb{F}_q$ from another permutation polynomial of $\mathbb{F}_q$ which does not fix $0$. Namely, let $h(x)$ be a permutation polynomial of $\mathbb{F}_q$ such that $h(0) = a \neq 0$ and $h(b) = 0$. Then we define $g(x)$ as

$$g(x) = \begin{cases} h(b) = 0, & x = 0; \\ h(0) = a, & x = b; \\ h(x), & x \neq 0, b. \end{cases}$$

It is obvious that $g(x)$ is again a permutation polynomial of $\mathbb{F}_q$ which fixes $0$.

*Example 2.* For any positive integer $e$ such that $\gcd(e, q - 1) = 1$ and $m, a \neq 0 \in \mathbb{F}_q$, the polynomial $h(x) = mx^e + a$ is a permutation polynomial of $\mathbb{F}_q$ which does not fix $0$. Let $b$ be the unique (non-zero) field element such that $h(b) = 0$. Using the above construction, we let

$$g(x) = \begin{cases} h(b) = 0, & x = 0; \\ h(0) = a, & x = b; \\ h(x) = mx^e + a, & x \neq 0, b. \end{cases}$$

Then $g(x)$ is a permutation polynomial of $\mathbb{F}_q$ which fixes $0$.    □

This twist of permutation polynomials can be very useful in constructing permutations of $\mathbb{Z}_{q-1}$ with optimum deficiency and optimum ambiguity.

**Theorem 2.** *Let $q$ be a prime power, $n = q-1$ and $\alpha$ a primitive element in $\mathbb{F}_q$. For $\gcd(e, n) = 1$ and $m, a \neq 0 \in \mathbb{F}_q$, let $h : \mathbb{F}_q \to \mathbb{F}_q$ be defined by $h(x) = mx^e + a$ and let $b$ be the unique (non-zero) field element such that $h(b) = 0$. Let*

$$g(x) = \begin{cases} h(b) = 0, & x = 0; \\ h(0) = a, & x = b; \\ h(x) = mx^e + a, & x \neq 0, b. \end{cases}$$

*Then, $f : \mathbb{Z}_n \to \mathbb{Z}_n$ defined by $f(i) = \log_\alpha(g(\alpha^i))$ meets the lower bounds from Theorem 1 and therefore has optimum ambiguity and optimum deficiency.*

*Proof.* We have $f(i+a) - f(i) = \log_\alpha(g(\alpha^{i+a})) - \log_\alpha(g(\alpha^i)) = \log_\alpha(\frac{g(\alpha^{i+a})}{g(\alpha^i)})$. Let $d = \alpha^a$. We need to study the size $v_d$ of the value set of $g(dx)/g(x)$ for $x \neq 0$. From the definition of $g(x)$, we have

$$\frac{g(dx)}{g(x)} = \begin{cases} \frac{m(db)^e + a}{a}, & x = b; \\ \frac{a}{m(b/d)^e + a}, & x = b/d; \\ \frac{m(dx)^e + a}{mx^e + a}, & x \neq b, b/d. \end{cases}$$

First we show that $v_d \geq q - 3$ for any $d \neq 0, 1$. Let $x, y$ be both different from $b, b/d$. Assume that

$$\frac{m(dx)^e + a}{mx^e + a} = \frac{m(dy)^e + a}{my^e + a}.$$

Then

$$m^2 d^e x^e y^e + amy^e + amd^e x^e + a^2 = m^2 d^e x^e y^e + amd^e y^e + amx^e + a^2.$$

Since $m, a \neq 0$, we obtain $(d^e - 1)y^e = (d^e - 1)x^e$. Because $\gcd(e, q - 1) = 1$, we have $d^e \neq 1$ if $d \neq 1$. Hence $x^e = y^e$. Again, by $\gcd(e, q - 1) = 1$, we obtain $x = y$. Hence $v_d \geq q - 3$ for any $d \neq 0, 1$.

Moreover, if

$$\frac{m(db)^e + a}{a} = \frac{m(dx)^e + a}{mx^e + a},$$

then

$$m^2 d^e b^e x^e + amx^e + amd^e b^e + a^2 = amd^e x^e + a^2.$$

Hence

$$(m^2 d^e b^e + am - amd^e)x^e = -amd^e b^e.$$

Since $mb^e = -a$, we obtain

$$(am - 2amd^e)x^e = -amd^e b^e.$$

Again, $m, a \neq 0$. This implies that $(2d^e - 1)x^e = d^e b^e$. If $q$ is odd, we can find a solution for $x$ as long as $2d^e - 1 \neq 0$. On the other hand, there exists a unique $d$ such that $d^e = 1/2$ and

$$\frac{m(db)^e + a}{a} \neq \frac{m(dx)^e + a}{mx^e + a}.$$

Similarly, there exists a unique $d$ such that $d^e = 2$ and

$$\frac{a}{m(b/d)^e + a} \neq \frac{m(dx)^e + a}{mx^e + a}.$$

Hence $v_d = q - 3 = n - 2$ if $d^e \neq 2$ or $1/2$, and $v_d = q - 2 = n - 1$ if $d^e = 2$ or $1/2$. (We observe that if $\text{char}(\mathbb{F}_q) = 3$ then $\frac{a}{m(b/d)^e + a} = \frac{m(db)^e + a}{a}$). Hence there are two rows with row deficiency 0 and the rest rows have deficiency 1. Thus $D(f) = n - 3$ for even $n$. It is obvious that $A(f) = n - 1 + n - 3 = 2(n - 2)$ in this case.

If $q$ is even, we always find $x$ such that

$$\frac{m(db)^e + a}{a} = \frac{m(dx)^e + a}{mx^e + a},$$

and

$$\frac{a}{m(b/d)^e + a} = \frac{m(dx)^e + a}{mx^e + a}.$$

Hence $v_d = q - 3$, and $D(f) = \sum_{a \in \mathbb{Z}_n^*} D_{r=a}(f) = (n-1)(n-1-(q-3)) = n-1$ when $n$ is odd. It is obvious that $A(f) = 2(n-1)$ in this case.     □

## 4   Conclusions

Our construction can produce a permutation $f : \mathbb{Z}_n \to \mathbb{Z}_n$ that has optimum ambiguity and deficiency when $n$ is one less than a prime power. We currently know of a permutation for only one other $n$ which is not one less than a prime power.

*Example 3.* Let $f : \mathbb{Z}_5 \to \mathbb{Z}_5$ be the permutation $f = (0)(1)(2)(34)$. The multiset of differences that are covered by the map is

$$\{(1,1),(1,1),(1,2),(1,4),(1,2),(2,2),(2,3),(2,1),(2,1),(2,3)\}$$

and their inverses. The differences that are missing, including inverses, are

$$\{(1,3),(2,4),(3,1),(4,2)\}.$$

Thus the ambiguity is $4 + 4 = 8$ and the deficiency is 4 both optimum with respect to the bounds on Theorem 1.     □

We have done exhaustive searches for optimum permutations for all $n \leq 16$ and $n = 5$ is the only case where they exist and $n$ is not one less than a prime power. This could either be a case of optimality in general only for $n = q - 1$ for $q$ a prime power and $n = 5$ excusable with the "law of small numbers" or there may be other exceptions.

We also hoped that the permutations constructed in Theorem 2 might give Costas arrays (vectors in $\mathbb{Z} \times \mathbb{Z}$ rather than in $\mathbb{Z}_n \times \mathbb{Z}_n$). The first prime power for which this construction does not yield Costas arrays is $q = 23$ and we found none for the more interesting range $29 \leq q \leq 101$.

There are some other interesting open questions:

- Can general lower bounds be proven for bijections $f : G_1 \to G_2$?
- Can constructions similar to Theorem 2 be applied to other known families of permutation polynomials?

# References

1. Colbourn, C.J., Dinitz, J.H. (eds.): Handbook of Combinatorial Designs, 2nd edn. Discrete Mathematics and its Applications. Chapman & Hall/CRC, Boca Raton (2007)
2. Costas, J.P.: A study of a class of detection waveforms having nearly ideal range-doppler ambiguity properties. Proceedings of IEEE 72, 996–1009 (1984)
3. Drakakis, K.: A review of Costas arrays. J. Appl. Math., Art. ID 26385, 32 (2006)
4. Drakakis, K., Gow, R., McGuire, G.: APN permutations on $\mathbb{Z}_n$ and Costas arrays. Discrete Applied Mathematics 157(15), 3320–3326 (2009)
5. Lidl, R., Mullen, G.L.: Unsolved problems: when does a polynomial over a finite field permute the elements of the field? Amer. Math. Monthly 95(3), 243–246 (1988)
6. Lidl, R., Mullen, G.L.: Unsolved problems: when does a polynomial over a finite field permute the elements of the field? II. Amer. Math. Monthly 100(1), 71–74 (1993)
7. Lidl, R., Niederreiter, H.: Finite Fields, 2nd edn. Encyclopedia of Mathematics and its Applications, vol. 20. Cambridge University Press, Cambridge (1997); With a foreword by P. M. Cohn
8. Massey, J.: SAFER $K_{64}$: A byte-oriented block-ciphering algorithm. Fast Software Encryption, 1–17 (1993)
9. Mullen, G.L.: Permutation polynomials over finite fields. In: Finite fields, coding theory, and advances in communications and computing. LNPAM, vol. 141, pp. 131–151. Dekker, New York (1993)
10. Mullen, G.L., Stevens, H.: Polynomial functions (mod m). Acta Mathematica, Hungarica 44(3-4), 237–241 (1984)
11. Nyberg, K.: Differentially uniform mappings for cryptography. In: Helleseth, T. (ed.) EUROCRYPT 1993. LNCS, vol. 765, pp. 55–64. Springer, Heidelberg (1994)
12. Rivest, R.L.: Permutation polynomials modulo $2^w$. Finite Fields and their Applications 7, 287–292 (2001)
13. Shearer, J.B.: Difference triangle sets. In: Colbourn, C.J., Dinitz, J. (eds.) [1], ch. VI.19, pp. 436–440
14. Sun, J., Takeshita, O.Y.: Interleavers for Turbo codes using permutation polynomials over integer rings. IEEE Trans. Inform. Theory 51(1), 101–119 (2005)

# Iterated Shared Memory Models*

## (Invited Talk)

Sergio Rajsbaum

Instituto de Matemáticas
Universidad Nacional Autónoma de México
D.F. 04510, Mexico
rajsbaum@math.unam.mx

**Abstract.** In centralized computing we can compute a function composing a sequence of elementary functions, where the output of the $i$-th function in the sequence is the input to the $i + 1$-st function in the sequence. This computation is done without persistent registers that could store information of the outcomes of these function invocations. In distributed computing, a *task* is the analogue of a function. An *iterated model* is defined by some base set of tasks. Processes invoke a sequence of tasks from this set. Each process invokes the $i + 1$-st task with its output from the $i$-th task. Processes access the sequence of tasks, one-by-one, in the same order, and asynchronously. Any number of processes can crash. In the most basic iterated model the base tasks are read/write registers. Previous papers have studied this and other iterated models with more powerful base tasks or enriched with failure detectors, which have been useful to prove impossibility results and to design algorithms, due to the elegant recursive structure of the runs. This talk surveys results in this area, contributed mainly by Borowsky, Gafni, Herlihy, Raynal, Travers and the author.

## 1 Introduction

A distributed model of computation consists of a set of $n$ processes communicating through some medium, satisfying specific timing and failure assumptions. The communication medium can be message passing or some form of shared memory. The processes can run synchronously or run at arbitrarily varying speeds. The failure assumptions describe how many processes may fail, and it what way. Along each one of these three dimensions, there are many variants, which when combined, give rise to a wide variety of distributed computing models, e.g. [6,29,37]. And for each model, we would like to know which distributed tasks can be solved, and at what cost, in terms of time and communication. Thus, work for developing a theory of distributed computing has been concerned with finding ways of unifying results, impossibility techniques, and algorithm design paradigms of different models.

---

* Partially supported by PAPIIT and PAPIME UNAM projects.

A. López-Ortiz (Ed.): LATIN 2010, LNCS 6034, pp. 407–416, 2010.

## 1.1    In Search of a Fundamental Model

In early stages of distributed computing theory similar result needed different proofs for different models. Consider for example the *consensus* task, where processes need to agree on one of their input values. In an asynchronous system, the problem is impossible to solve even if only one process may crash. This was proved in [18] for the case where the processes communicate by sending messages to one another. Read/write shared memory is in principle a more powerful communication media than message passing, so proving that consensus is also impossible here, is a stronger result. Indeed, the same impossibility result holds, if processes communicate through read/write memory, as proved in [28]. A first approach towards the goal of unifying distributed computing theory, was to derive direct simulations from one model to another, e.g., [2,5,8]. In particular, [2] shows how to transform a protocol running in an asynchronous message passing model to one for a shared memory model. This implies that it is sufficient to prove the consensus impossibility result in the shared memory model, to get the impossibility in the message passing model.

Later on, the approach of devising models of a higher level of abstraction, where results about various more specific models can be derived, e.g., [19,24,30] was explored. For instance, [30] described a generic layered model of computation where a consensus impossibility result is proved, and as specific cases, consensus impossibilities can be derived for both a message passing and a shared memory model, and even for several synchronous models. Recently, the approach proved useful also for randomized algorithms, e.g. [3].

## 1.2    From Graph Connectivity to Topology

The 1 failure asynchronous consensus impossibility results [18,28] mentioned above, lead to a characterization of the tasks that can be solved in an asynchronous system where at most 1 process can crash [11]. Thus, the case of 1 failure seemed to be the base case from which one should generalize to any number of failures. After all, dealing with many failures is more complicated than dealing with few failures, isn't it? A major step in the development of the theory was taken in 1993, by three works presented in the ACM STOC conference [8,25,36], that uncovered a deep relationship between distributed computing and topology. This lead to the realization that, instead of the case of 1 failure, the read/write *wait-free* case is fundamental. In a system where any number of processes can crash, each process must complete the protocol in a finite number of its own steps, and "wait statements" to hear from another process are not useful. Roughly speaking, when we want to study the complexity and solvability of a task, we should first study it in the wait-free model, and then generalize the results to stronger models (either with more powerful communication primitives or stronger synchrony assumptions). For example, reductions from the case where at most $t$ processes can crash, to the wait-free model have been presented in [8,10,20].

In the topology approach one considers the *simplicial complex of global states* of the system after a finite number of steps, and then proves topological invariants

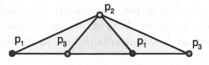

**Fig. 1.** Three simplexes

about the structure of such a complex, to derive impossibility results. The notion of *indistinguishability*, which has played a fundamental role in nearly every lower bound in distributed computing, is hence generalized from graph connectivity, to higher dimensions. Two global states are indistinguishable to a set of processes if they have the same local states in both. In the next figure there is a complex with three triangles, each one is a *simplex* representing a global state; the corners of a simplex represent local states of processes in the global state. The center simplex and the rightmost simplex represent global states that are indistinguishable to processes $p_1$ and $p_2$, which is why the two triangles share an edge. Only process $p_3$ can distinguish between the two global states.

### 1.3   The Importance of the Wait-Free Snapshot Model

We have seen that the read/write wait-free model is fundamental. However, there are several ways of defining read/write registers, like single-writer/multi-reader, multi-writer/multi-reader, and others. A research branch in distributed computing theory has been concerned with finding the simplest read/write communication abstraction. Several variants of read/write registers were studied early on [6,27] and proved to be equivalent. We now use *snapshots* [1], or even *immediate snapshots* [7], as such abstractions are wait-free equivalent to read/write registers (although at a complexity cost), but give rise to cleaner and more structured models. In a snapshot object each process has a component where it can write, and the process can read all components with a single atomic read operation that returns an instantaneous snapshot of its contents. An immediate snapshot object provides a single write-snapshot operation, guaranteeing that the snapshot is executed immediately after the write. The complex corresponding to a snapshot object for three processes is in the first part of Figure 2 and will be explained below.

## 2   The Basic Iterated Model

We have explained above that the snapshot wait-free model is fundamental. However, in this paper we argue that a specific variant of this model is especially suitable for a central role in distributed computing theory. Most attempts at unifying models of various degrees of asynchrony restrict attention to a subset of well-behaved, *round-based* executions. The approach in [9] goes beyond that and defines an *iterated* model, where each communication object can be accessed only once by each process. In the paper only the basic case of snapshot objects is considered. The sequence of snapshot objects are accessed asynchronously, and one after the other by each process. It is shown in [9] that this iterated model is equivalent (for bounded wait-free task solvability) to the usual read/write shared memory model.

## 2.1   Recursive Structure

The iterated model has an elegant recursive structure. In each iteration, the only information transmitted can be the local state of processes after each snapshot– the snapshot objects are not persistent, we may think they exist only during an iteration. The result in [9] can be thought of as a variant of the result in [2] that shows that shared-memory can be emulated over message passing. In message passing too, there are no persistent objects. The recursive structure is clearly expressed in the complex of global states of a protocol. The complex of global states after $i + 1$ rounds is obtained by replacing each simplex by a one round complex, see Figure 2. Indeed, this iterated model was the basis for the proof in [9] of the main characterization theorem of [25].

In more detail, the properties of an immediate snapshot are represented in the first image of Figure 2, for the case of three processes. The image represents a *simplicial complex*, i.e. a family of sets closed under containment; each set is called a *simplex*, and it represents the views of the processes after accessing the *IS* object. The *vertices* are the 0-simplexes, of size one; edges are 1-simplexes, of size two; triangles are of size three (and so on). Each vertex is associated with a process $p_i$, and is labeled with $sm_i$ (the *view* $p_i$ obtains from the object).

In the first complex of Figure 2, the highlighted 2-dimensional simplex, represents a run where $p_1$ and $p_3$ access the object concurrently, both get the same views seeing each other, but not seeing $p_2$, which accesses the object later, and gets back a view with the 3 values written to the object. But $p_2$ can't tell the order in which $p_1$ and $p_3$ access the object; the other two runs are indistinguishable to $p_2$, where $p_1$ accesses the object before $p_3$ and hence gets back only its own value or the opposite. These two runs are represented by the corner 2-simplexes.

Recall that in the iterated immediate snapshot model the objects are accessed sequentially and asynchronously by each process. In Figure 2 one can see that the complex after one round is constructed recursively by replacing each simplex by the one round complex. Thus, the highlighted 2-simplex in the second complex of the figure, represents global states after two snapshots, given that in the first snapshot, $p_1$ and $p_3$ saw each other, but they did not see $p_2$.

## 2.2   On the Meaning of Failures

Notice that the runs of the iterated model are not a subset of the runs of a standard (non-iterated) model. Consider a run where processes, $p_1, p_2, p_3$, execute an infinite number of rounds, but $p_1$ is scheduled before $p_2, p_3$ in every round. The triangles at the left-bottom corners of the complexes in Figure 2 represent such a situation; $p_1$, at the corner, never hears from the two other processes. Of course, in the usual (non-iterated read/write shared memory) asynchronous model, two

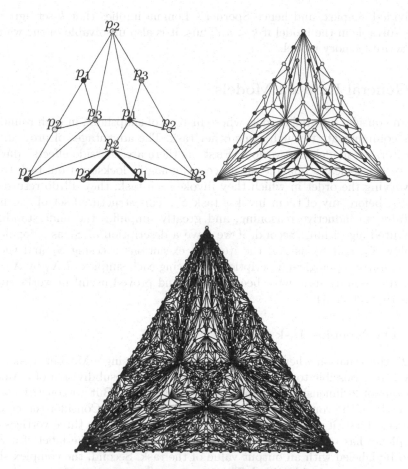

**Fig. 2.** One, two and three rounds in the IIS model

correct processes can always eventually communicate with each other. Thus, in an iterated model, the set of *correct processes* of a run, may be defined as the set of processes that observe each other directly or indirectly infinitely often (a formal definition is given in [33]).

## 2.3 Equivalence with the Standard Model

Recall that in the $k$-set agreement task each of the $n$ processes in the system starts with an input value of some domain of at least $n$ values, and must decide on at most $k$ of their input values. It was proved in [9] that if a task (with a finite number of inputs) is solvable wait-free in the read/write memory model then it is solvable in the snapshot iterated model (the other direction is trivial), using an algorithm that simulates the read/write model in the iterated model. Recently another simulation was described in [21], somewhat simpler. As can be seen in Figure 2, the complex of global states at any round of this model is a subdivided simplex, and hence Sperner's Lemma implies that $k$-set agreement is not solvable in the model if $k < n$. Thus, it is also unsolvable in the wait-free read/write memory model.

## 3    General Iterated Models

We can consider iterated models where instead of snapshots, in each round processes communicate through some other task. The advantages of programming in an iterated model are two fold: First, as there are no "side effects" during a run, we can logically imagine all processes going in lockstep from task to task just varying the order in which they invoke each task; they all do return from task $S_1$, before any of them invokes task $S_2$. This structured set of executions facilitates an inductive reasoning, and greatly simplifies the understanding of distributed algorithms. Second, if we have a description of $S_1$ as a topological complex $X_1$, and $S_2$ as $X_2$, the iterated executions accesing $S_1$ and then $S_2$ have a simple topological description: replacing each simplex of $X_1$ by $X_2$. This iterated reasoning style have been studied and proved useful in works such as [15,16,19,24,30,35,34].

### 3.1    The Moebius Task

In [22], the situation where processes communicate using a Moebius task is considered. It is possible to construct a complex that is a subdivision of a Moebius band out of 2-dimensional simplexes, as in Figure 3. But to construct a task specification, the complex should satisfy two properties. Consider the case of 3 processes. First, it must be chromatic. That is, each of the three vertices of its 2-simplexes has to be labeled with a different process id. The label of a vertex is also be labeled with an output value of the task. Second, the complex should have a *span* structure identified. That is, a specification stating what is the output of the task when each of the processes runs solo, what is the output when pairs of processes run solo, and what are the possible outputs when all three run concurrently. Figure 4 from [22], is the task specification for three processes, that corresponds to a Moebius band. Notice that its boundary is identical to the boundary of a chromatic subdivided simplex.

The Moebius task was introduced in [22] because it is a manifold: if we consider any of the edges of its complex, either it belongs to one or to two triangles. The one-round Moebius task is a manifold task, so composing the Moebius task with itself in an iterated model, with read/write rounds, or with any other manifold task yields a manifold task. Thus, any protocol in such an iterated model yields a protocol complex that is also a manifold. And as mentioned above, we can apply Sperner's lemma to a manifold to prove that $k$-set agreement is not solvable in the model if $k < n$. Furthermore, [22] shows that the Moebius task can be used to prove that set agreement is strictly more difficult than renaming [4], in the iterated model.

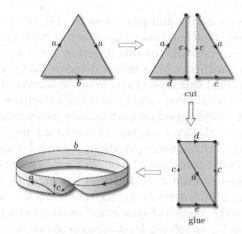

cut

glue

**Fig. 3.** Construction of a Mobius band

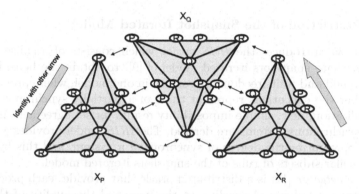

$X_Q$

Identify with other arrow

$X_P$    $X_R$

**Fig. 4.** One-round Moebius task protocol complex for 3 processes

## 3.2 Equivalence of More General Models with the Standard Model

Recall that the standard read/write memory model is equivalent to the iterated snapshot model (for bounded task solvability) [9]. Recently another simulation was described in [21], that shows that both models are also equivalent, when enriched with tasks $T$ more powerful than read/write registers– for any task $T$ solvable by set agreement, the power of the standard and the iterated model coincide. This implies that set agreement is strictly more difficult than renaming also in the standard non-iterated model.

## 4 Iterated Models and Failure Detectors

In the construction of a distributed computing theory, a central question has been understanding how the degree of synchrony of a system affects its power to solve distributed tasks. The degree of synchrony has been expressed in various ways, typically either by specifying a bound $t$ on the number of processes that can crash, as bounds on delays and process steps [17], or by a failure detector [12]. It has been shown multiple times that systems with more synchrony can solve more tasks. Previous works in this direction have mainly considered an asynchronous system enriched with a failure detector that can solve consensus. Some works have identified this type of synchrony in terms of fairness properties [38]. Other works have considered round-based models with no failure detectors [19]. Some other works [26] focused on performance issues mainly about consensus. Also, in some cases, the least amount of synchrony required to solve some task has been identified, within some paradigm. A notable example is the weakest failure detector to solve consensus [13] or $k$-set agreement [40]. Set agreement [14] represents a desired coordination degree to be achieved in the system, and hence is natural to use it as a measure for the *synchrony degree* in the system. A clear view of what exactly "degree of synchrony" means is still lacking. For example, the same power as far as solving $k$-set agreement can be achieved in various ways, such as via different failure detectors [31] or $t$-resilience assumptions.

## 4.1   A Restriction of the Snapshot Iterated Model

The paper [35] introduces the *IRIS model,* which consists of a subset of runs of the immediate snapshots iterated model of [9], to obtain the benefits of the round by round and wait-freedom approaches in one model, where processes run wait-free but the executions represent those of a partially synchronous model. As an application, new, simple impossibility results for set agreement in several partially synchronous systems are derived. The *IRIS* model provides a mean of precisely representing the degree of synchrony of a system, and this by considering particular subsets of runs of the snapshots iterated model.

A *failure detector* [12] is a distributed oracle that provides each process with hints on process failures. According to the type and the quality of the hints, several classes of failure detectors have been defined (e.g., [31,40]). Introducing a failure detector directly into an iterated model is not useful [34]. Instead, the the *IRIS* model of [35] represents a failure detector as a restriction on the set of possible runs of the iterated system. As an example, the paper considers the family of *limited scope* accuracy failure detectors, denoted $\Diamond \mathcal{S}_x$ [23,39]. They are a generalization of the class denoted $\Diamond \mathcal{S}$ that has been introduced in [12].

Consider the read/write computation model enriched with a failure detector $C$ of the class $\Diamond \mathcal{S}_x$. An *IRIS* model that precisely captures the synchrony provided by the asynchronous system equipped with $C$ is described in [35]. To show that the synchrony is indeed captured, the paper presents two simulations. The first is a simulation from the shared memory model with $C$ to the *IRIS* model. The second shows how to extract $C$ from the *IRIS* model, and then simulate the read/write model with $C$. For this, a generalization of the wait-free simulation of [9] is described, that preserves consistency with the simulated failure detector.

## 4.2 Equivalence of Failure Detector Enriched Models with Iterated Models

As a consequence of these simulations, we get: an agreement task is wait-free solvable in the read/write model enriched with $C$ if and only if it is wait-free solvable in the corresponding $IRIS$ model. Then, using a simple topological observation, it is easy to derive the lower bound of [23] for solving $k$-set agreement in a system enriched with $C$. In the approach presented in this paper, the technically difficult proofs are encapsulated in algorithmic reductions between the shared memory model and the $IRIS$ model, while in the proof of [23] combinatorial topology techniques introduced in [24] are used to derive the topological properties of the runs of the system enriched with $C$ directly. A companion technical report [32] extends the equivalence presented in [35] to other failure detector classes.

# References

1. Afek, Y., Attiya, H., Dolev, D., Gafni, E., Merritt, M., Shavit, N.: Atomic Snapshots of Shared Memory. J. ACM 40(4), 873–890 (1993)
2. Attiya, H., Bar-Noy, A., Dolev, D.: Sharing Memory Robustly in Message Passing Systems. J. ACM 42(1), 124–142 (1995)
3. Attiya, H., Censor, K.: Tight bounds for asynchronous randomized consensus. J. ACM 55(5) (2008)
4. Attiya, H., Bar-Noy, A., Dolev, D., Peleg, D., Reischuk, R.: Renaming in an Asynchronous Environment. Journal of the ACM 37(3), 524–548 (1990)
5. Awerbuch, B.: Complexity of network synchronization. J. ACM 32, 804–823 (1985)
6. Attiya, H., Welch, J.: Distributed Computing: Fundamentals, Simulations, and Advanced Topics. Wiley, Chichester (2004)
7. Borowsky, E., Gafni, E.: Immediate Atomic Snapshots and Fast Renaming. In: Proc. 12th ACM Symp. on Principles of Distributed Computing (PODC 1993), pp. 41–51 (1993)
8. Borowsky, E., Gafni, E.: Generalized FLP Impossibility Results for $t$-Resilient Asynchronous Computations. In: Proc. 25th ACM STOC, pp. 91–100 (1993)
9. Borowsky, E., Gafni, E.: A Simple Algorithmically Reasoned Characterization of Wait-free Computations. In: Proc. 16th ACM PODC, pp. 189–198 (1997)
10. Borowsky, E., Gafni, E., Lynch, N., Rajsbaum, S.: The BG distributed simulation algorithm. Distributed Computing 14(3), 127–146 (2001)
11. Biran, O., Moran, S., Zaks, S.: A Combinatorial Characterization of the Distributed 1-solvable Tasks. J. Algorithms 11, 420–440 (1990)
12. Chandra, T., Toueg, S.: Unreliable Failure Detectors for Reliable Distributed Systems. J. ACM 43(2), 225–267 (1996)
13. Chandra, T., Hadzilacos, V., Toueg, S.: The Weakest Failure Detector for Solving Consensus. J. ACM 43(4), 685–722 (1996)
14. Chaudhuri, S.: More Choices Allow More Faults: Set Consensus Problems in Totally Asynchronous Systems. Information and Computation 105, 132–158 (1993)
15. Elrad, T., Francez, N.: Decomposition of Distributed Programs into Communication-Closed Layers. Sci. Comput. Program. 2(3), 155–173 (1982)
16. Chou, C.-T., Gafni, E.: Understanding and Verifying Distributed Algorithms Using Stratified Decomposition. In: PODC 1988, pp. 44–65 (1988)
17. Dwork, C., Lynch, N., Stockmeyer, L.: Consensus in the Presence of Partial Synchrony. J. ACM 35(2), 288–323 (1988)

18. Fischer, M., Lynch, N., Paterson, M.: Impossibility of Distributed Consensus with One Faulty Process. J. ACM 32(2), 374–382 (1985)
19. Gafni, E.: Round-by-round Fault Detectors: Unifying Synchrony and Asynchrony (Extended Abstract). In: Proc. 17th ACM Symp. on Principles of Distributed Computing (PODC), pp. 143–152 (1998)
20. Gafni, E.: The extended BG-simulation and the characterization of t-resiliency. In: Proc. of the 41st ACM Symp. on Theory of Computing (STOC), pp. 85–92 (2009)
21. Gafni E., Rajsbaum S., Loopless Programming with Tasks (Manuscript November 5, 2009) (Submitted for publication)
22. Gafni, E., Rajsbaum, S., Herlihy, M.: Subconsensus Tasks: Renaming is Weaker than Set Agreement. In: Dolev, S. (ed.) DISC 2006. LNCS, vol. 4167, pp. 329–338. Springer, Heidelberg (2006)
23. Herlihy, M., Penso, L.D.: Tight Bounds for $k$-Set Agreement with Limited Scope Accuracy Failure Detectors. Distributed Computing 18(2), 157–166 (2005)
24. Herlihy, M.P., Rajsbaum, S., Tuttle, M.: Unifying Synchronous and Asynchronous Message-Passing Models. In: Proc. 17th ACM PODC, pp. 133–142 (1998)
25. Herlihy, M., Shavit, N.: The Topological Structure of Asynchronous Computability. J. ACM 46(6), 858–923 (1999)
26. Keidar, I., Shraer, A.: Timeliness, Failure-detectors, and Consensus Performance. In: Proc. 25th ACM PODC, pp. 169–178 (2006)
27. Lamport, L.: On Interprocess Communication. Distributed Computing 1(2), 77–101 (1986)
28. Loui, M.C., Abu-Amara, H.H.: Memory Requirements for Agreement among Unreliable Asynchronous Processes. Advances in Computing Research 4, 163–183 (1987)
29. Lynch, N.A.: Distributed Algorithms, 872 pages. Morgan Kaufmann, San Francisco (1997)
30. Moses, Y., Rajsbaum, S.: A Layered Analysis of Consensus. SICOMP 31(4), 989–1021 (2002)
31. Mostefaoui, A., Rajsbaum, S., Raynal, M., Travers, C.: Irreducibility and Additivity of Set Agreement-oriented Failure Detector Classes. In: Proc. PODC 2006, pp. 153–162. ACM Press, New York (2006)
32. Rajsbaum, S., Raynal, M., Travers, C.: Failure Detectors as Schedulers. Tech. Report # 1838, IRISA, Université de Rennes, France (2007)
33. Rajsbaum, S., Raynal, M., Travers, C.: The Iterated Restricted Immediate Snapshot Model. Tech. Report # 1874, IRISA, Université de Rennes, France (2007)
34. Rajsbaum, S., Raynal, M., Travers, C.: An impossibility about failure detectors in the iterated immediate snapshot model. Inf. Process. Lett. 108(3), 160–164 (2008)
35. Rajsbaum, S., Raynal, M., Travers, C.: The Iterated Restricted Immediate Snapshot Model. In: Hu, X., Wang, J. (eds.) COCOON 2008. LNCS, vol. 5092, pp. 487–497. Springer, Heidelberg (2008)
36. Saks, M., Zaharoglou, F.: Wait-Free $k$-Set Agreement is Impossible: The Topology of Public Knowledge. SIAM Journal on Computing 29(5), 1449–1483 (2000)
37. Santoro, N.: Design and Analysis of Distributed Algorithms. Wiley Interscience, Hoboken (2006)
38. Völzer, H.: On Conspiracies and Hyperfairness in Distributed Computing. In: Fraigniaud, P. (ed.) DISC 2005. LNCS, vol. 3724, pp. 33–47. Springer, Heidelberg (2005)
39. Yang, J., Neiger, G., Gafni, E.: Structured Derivations of Consensus Algorithms for Failure Detectors. In: Proc. 17th ACM PODC, pp. 297–308 (1998)
40. Zieliński, P.: Anti-Omega: the Weakest Failure Detector for Set Agreement. Tech. Rep. # 694, University of Cambridge

# Optimal Polygonal Representation of Planar Graphs

E.R. Gansner[1], Y.F. Hu[1], M. Kaufmann[2], and S.G. Kobourov[3]

[1] AT&T Research Labs, Florham Park, NJ
{erg,yifanhu}@research.att.com
[2] Wilhelm-Schickhard-Institut for Computer Science, Tübingen University
mk@informatik.uni-tuebingen.de
[3] Dept. of Computer Science, University of Arizona
kobourov@cs.arizona.edu

**Abstract.** In this paper, we consider the problem of representing graphs by polygons whose sides touch. We show that at least six sides per polygon are necessary by constructing a class of planar graphs that cannot be represented by pentagons. We also show that the lower bound of six sides is matched by an upper bound of six sides with a linear time algorithm for representing any planar graph by touching hexagons. Moreover, our algorithm produces convex polygons with edges with slopes 0, 1, -1.

## 1 Introduction

For both theoretical and practical reasons, there is a large body of work considering how to represent planar graphs as *contact graphs*, i.e., graphs whose vertices are represented by geometrical objects with edges corresponding to two objects touching in some specified fashion. Typical classes of objects might be curves, line segments or isothetic rectangles, and an early result is Koebe's theorem [20], which shows that all planar graphs can be represented by touching disks.

In this paper, we consider contact graphs whose objects are simple polygons, with an edge occurring whenever two polygons have non-trivially overlapping sides. As with treemaps [3], such representations are preferred in some contexts [4] over the standard node-link representations for displaying relational information. Using adjacency to represent a connection can be much more compelling, and cleaner, than drawing a line segment between two nodes. For ordinary users, this representation suggests the familiar metaphor of a geographical map.

It is clear that any graph represented this way must be planar. As noted by de Fraysseix *et al.* [7], it is also easy to see that all planar graphs have such representations for sufficiently general polygons. Starting with a straight-line planar drawing of a graph, we can create a polygon for each vertex by taking the midpoints of all adjacent edges and the centers of all neighboring faces. Note that the number of sides in each such polygon is proportional to the degree of its vertex. Moreover, these polygons are not necessarily convex; see Figure 1.

It is desirable, for aesthetic, practical and cognitive reasons, to limit the complexity of the polygons involved, where "complexity" here means the number of sides in the polygon. Fewer sides, as well as wider angles in the polygons, make for simpler

A. López-Ortiz (Ed.): LATIN 2010, LNCS 6034, pp. 417–432, 2010.
© Springer-Verlag Berlin Heidelberg 2010

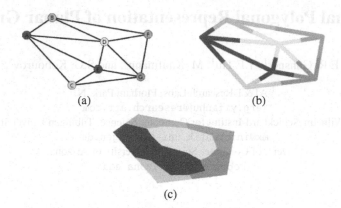

**Fig. 1.** Given a drawing of a planar graph(a), we apportion the edges to the endpoints by cutting each edge in half (b), and then apportion the faces to form polygons (c).

and cleaner drawings. In related applications such as floor-planning [24], physical constraints make undesirable polygons with very small angles or many sides. One is then led to consider how simple can such representations be. How many sides do we really need? Can we insist that the polygons be convex, perhaps with a lower bound on the size of the angles or the edges? If limiting some of these parameters prevents the drawings of all planar graphs, which ones can be drawn?

## 1.1  Our Contribution

This paper provides answers to some of these questions. Previously, it was known [12,24] that triangulated planar graphs can be represented using non-convex octagons. On the other hand, it is not hard to see that one cannot use triangles (e.g., $K_5$ minus one edge cannot be represented with triangles).

Our main result is showing that hexagons are necessary and sufficient for representing all planar graphs. For necessity we construct a class of graphs that cannot be represented using five or fewer sides. For sufficiency, we describe a linear-time algorithm that produces a representation using convex hexagons all of whose sides have slopes 1, 0, or -1. Finally, we describe an alternative algorithm for generating convex hexagonal representations for general planar graphs that leads to $O(n) \times O(n)$ drawing area. Note that if the input graph is triangulated, our output corresponds to a tiling of the plane with convex heagons; otherwise, there might be convex holes present.

## 1.2  Related Work

As remarked above, there is a rich literature related to various types of contact graphs. There are many results considering curves and line segments as objects (cf. [13,14]). For closed shapes such as polygons, results are rarer, except for axis-aligned (or *isothetic*) rectangles. In a sense, results on representing planar graphs as "contact systems" can be dated back to Koebe's 1936 theorem [20] which states that any planar graph can be represented as a contact graph of disks in the plane.

The focus of this paper is side-to-side contact of polygons. The algorithms of He [12] and Liao *et al.* [24] produce contact graphs of this type for triangulated graphs, with nodes represented by the union of at most two isothetic rectangles, thus giving a polygonal representation by non-convex octagons.

We now turn to contact graphs using isothetic rectangles, which are often referred to as *rectangular layouts*. This is the most extensively studied class of contact graphs, due in part to its relation to application areas such as VLSI floor-planning [22,31], architectural design [28] and geographic information systems [10], but also due to the mathematical ramifications and connections to other areas such as rectangle-of-influence drawings [25] and proximity drawings [1,16].

Graphs allowing rectangular layouts have been fully characterized [26,30] with linear algorithms for deciding if a rectangular layout is possible and, if so, constructing one. The simplest formulation [4] notes that a graph has a rectangular layout if and only if it has a planar embedding with no filled triangles. Thus, $K_4$ has no rectangular layout. Buchsbaum *et al.* [4] also show, using results of Biedl *et al.* [2], that graphs that admit rectangular layouts are precisely those that admit a weaker variation of planar rectangle-of-influence drawings.

Rectangular layouts required to form a partition of a rectangle are known as *rectangular duals*. In a sense, these are "maximal" rectangular layouts; many of the results concerning rectangular layouts are built on results concerning rectangular duals. Graphs admitting rectangular duals have been characterized [11,21,23] and there are linear-time algorithms [11,19] for constructing them.

Another view of rectangular layouts arises in VLSI floorplanning, where a rectangle is partitioned into rectilinear regions so that region adjacencies correspond to a given planar graph. It is natural to try to minimize the complexities of the resulting regions. The best known results are due to He [12] and Liao *et al.* [24] who show that regions need not have more than 8 sides. Both of these algorithms run in $O(n)$ time and produce layouts on an integer grid of size $O(n) \times O(n)$, where $n$ is the number of vertices.

Rectilinear cartograms can be defined as rectilinear contact graphs for vertex-weighted planar graphs, where the area of a rectilinear region must be proportional to the weight of its corresponding node. Even with this extra condition, de Berg *et al.* [6] show that rectilinear cartograms can always be constructed in $O(n \log n)$ time, using regions having at most 40 sides. The resulting regions, however, are highly non-convex and can have poor aspect ratio.

Although not considered by the authors, an upper bound of six for the minimum number of sides in a touching polygon representation of planar graphs might be obtained from the vertex-to-side triangle contact graphs of de Fraysseix *et al.* [7]. The top edge of each triangle can be converted into a raised 3-segment polyline, clipping the tips of the triangles touching it from above, thereby turning the triangles into side-touching hexagons. It is likely to be difficult to use this approach for generating hexagonal representations as it involves computing the amounts by which each triangle may be raised so as to become a hexagon without changing any of the adjacencies. Moreover, by the nature of such an algorithm, there would be many "holes," potentially making such drawings less appealing, or requiring further modifications to remove them.

### 1.3 Preliminaries

**Touching Hexagons Graph Representation:** Throughout this paper, we assume we are dealing with a connected planar graph $G = (V, E)$. We would like to construct a set of closed simple polygons $R$ whose interiors are pairwise disjoint, along with an isomorphism $\mathcal{R} : V \to R$, such that for any two vertices $u, v \in V$, the boundaries of $\mathcal{R}(u)$ and $\mathcal{R}(v)$ overlap non-trivially if and only if $\{u, v\} \in E$. For simplicity, we adopt a convention of the cartogram community and define the *complexity* of a polygonal region as the number of sides it has. We call the set of all graphs having such a representation where each polygon in $R$ has complexity 6 *touching hexagons graphs*.

**Canonical Labeling:** Our algorithms begin by first computing a *planar embedding* of the input graph $G = (V, E)$ and using that to obtain a *canonical labeling* of the vertices. A planar embedding of a graph is simply a clockwise order of the neighbors of each vertex in the graph. Obtaining a planar embedding can be done in linear time using the algorithm by Hopcroft and Tarjan [15]. The canonical labeling or order of the vertices of a planar graph was defined by de Fraysseix *et al.* [9] in the context of straight-line drawings of planar graphs on an integer grid of size $O(n) \times O(n)$. While the first algorithm for computing canonical orders required $O(n \log n)$ time [8], Chrobak and Payne [5] have shown that this can be done in $O(n)$ time.

In this section we review the canonical labeling of a planar graph as defined by de Fraysseix *et al.* [8]. Let $G = (V, E)$ be a fully triangulated planar graph embedded in the plane with exterior face $u, v, w$. A canonical labeling of the vertices $v_0 = u, v_1 = v, v_2, \ldots, v_{n-1} = w$ is one that meets the following criteria for every $2 < i < n$:

1. The subgraph $G_{i-1} \subseteq G$ induced by $v_0, v_1, \ldots, v_{i-1}$ is 2-connected, and the boundary of its outer face is a cycle $C_{i-1}$ containing the edge $(u, v)$;
2. The vertex $v_i$ is in the exterior face of $G_{i-1}$, and its neighbors in $G_{i-1}$ form an (at least 2-element) subinterval of the path $C_{i-1} - (u, v)$.

The canonical labeling of a planar graph $G$ allows for the incremental placement of the vertices of $G$ on a grid of size $O(n) \times O(n)$ so that when the edges are drawn as straight-line segments there are no crossings in the drawing. The two criteria that define a canonical labeling are crucial for the region creation step of our algorithm.

Kant generalized the definition for triconnected graphs. In this case, the vertices are partitioned into sets $V_1$ to $V_K$ which can be either singleton vertices or chains of vertices [18].

## 2    Lower Bound of Six Sides

Here we show that at least six sides per polygon are needed in touching polygon representations of planar graphs. We begin by constructing a class of planar graphs that cannot be represented by four-sided polygons and then extend the argument to show that there exists a class of planar graphs that cannot be represented by five-sided regions.

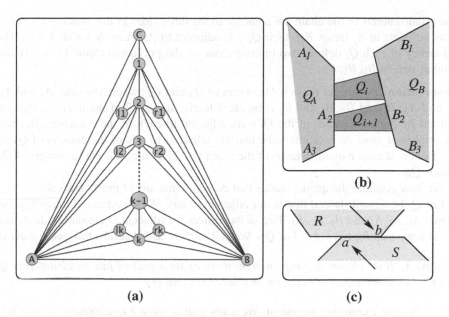

**Fig. 2.** (a) The graph that provides the counterexample. (b) A pair of subsequent fair quadrilaterals adjacent to the same sides of $Q_A$ and $Q_B$. (c) Illustration for Lemma 2 shows one of three possible cases for two touching regions.

## 2.1  Four Sides Are Not Enough

Consider the fully triangulated graph $G$ in Figure 2(a). $G$ has three nodes on the outer face $A$, $B$ and $C$, and contains a chain of nodes $1, ..., k$ which are all adjacent to $A$ and $B$. Consecutive nodes in the chain, $i$ and $i + 1$, are also adjacent. The remaining nodes of $G$ are degree-3 nodes $l_i$ and $r_i$ inside the triangles $\Delta(A, i, i + 1)$ and $\Delta(B, i, i + 1)$.

**Theorem 1.** *For k sufficiently large, there does not exist a touching polygon represen-tation for G in which all regions have complexity 4 or less.*

*Proof:* Assume, for the sake of contradiction, that we are given a touching polygon drawing for $G$ in which all regions have complexity 4 or less. Without loss of generality, we assume that the drawing has an embedding that corresponds to the one shown in Figure 2(a). Let $Q_A$ and $Q_B$ denote the quadrilaterals representing nodes $A$ and $B$, and $Q_i$ denotes the quadrilateral representing node $i$. Once again, without loss of generality, let $Q_A$ lie in the left corner, $Q_B$ in the right corner and $Q_C$ at the top of the drawing.

We start with a couple of observations:

**Observation 1:** For simplicity, assume that the three quadrilaterals $Q_A, Q_B, Q_C$ that are adjacent to the outer face are convex. Then a complete side of each quadrilateral must be adjacent to the outer face.

From this observation, we conclude that at most three sides of each of the outer quadrilaterals are inside of the drawing. We consider the three sides $A_1, A_2, A_3$ and $B_1, B_2, B_3$ of $Q_A$ and $Q_B$, respectively, numbered from top to bottom; see Figure 2(b).

The quadrilaterals of the chain are adjacent to the three sides in this order, such that if $Q_i$ is adjacent to $A_j$ (resp. $B_j$), then $Q_{i+1}$ is adjacent to $A_k$ (resp. $B_k$) with $k \geq j$. The adjacency of each $Q_i$ defines two intervals, one on the polygonal chain $A_1, A_2, A_3$ and another one on $B_1, B_2, B_3$.

**Observation 2:** Consider the $c(= 4)$ corners of $Q_A$ and $Q_B$, where the sides $A_1$ and $A_2$, $A_2$ and $A_3$, $B_1$ and $B_2$, $B_2$ and $B_3$ coincide. Clearly, at most 2 of the intervals that are defined by the adjacencies of the $Q_i$'s are adjacent to each of the $c$ corners. In total, this makes at most $2c = 8$ intervals, that are adjacent to any of the corners of $Q_A$ or $Q_B$. Hence, at most 8 quadrilaterals of the chain $Q_1, ..., Q_k$ are adjacent to corners of $Q_A$ and/or $Q_B$.

We now consider the quadrilaterals that do *not* define any of those intervals.

Let $Q_i$ be a quadrilateral that is not adjacent to any of the corners of the polygonal chains $A_1, A_2, A_3$ and $B_1, B_2, B_3$. Two of its corners are adjacent to the same side $A_k$ and to the same side $B_l$, $1 \leq k, l \leq 3$ of $Q_B$. We call such a quadrilateral a *fair quadrilateral*.

**Lemma 1.** *If we choose k large enough, there exists a pair of fair quadrilaterals $Q_i$ and $Q_{i+1}$ that are adjacent to the same sides of $Q_A$ and $Q_B$.*

*Proof:* We use a counting argument. We know that at most 8 quadrangles are not fair. Hence, for $k \geq 2 \cdot 2c + 2 = 18$, there must be a pair of subsequent fair quadrilaterals. The worst case happens for $k = 17$ if $Q_2, Q_4, Q_6, ... Q_{16}$ are not fair. We can state even more precisely that there are at least $k - 17$ pairs of subsequent fair quadrilaterals. Note that the pair $(Q_i, Q_{i+1})$ of fair quadrilaterals where $Q_i$ is adjacent to the sides $A_1$ and $B_1$, but $Q_{i+1}$ is not adjacent to $A_1$ and $B_1$ does not have the property claimed in the lemma. We call such a pair transition pair.

We can partition the set of fair quadrilaterals into at most 5 equivalence classes $C_1, ..., C_5$ that denote the sets of fair quadrilaterals, which are adjacent to the same sides of $Q_A$ and $Q_B$. When we sweep through the chain of middle quadrilaterals, we simultaneously proceed through the equivalence classes. Hence there exist at most $t = 4$ transition pairs, namely pairs of subsequent fair quadrilaterals that are in different equivalence classes.

These equivalence classes denote the pairs of sides $(A_i, B_j)$ that are used, beginning from the top with, say, $(A_1, B_1)$, then $(A_1, B_2), (A_2, B_2), (A_3, B_2)$ and finally $(A_3, B_3)$. Note that this is not the only possible set of equivalence classes, but by planarity, it is not possible to have $(A_2, B_3)$ and $(A_3, B_1)$ simultaneously. Hence, there are at most 5 classes.

We repeat our counting argument from above and argue that for $k \geq 23$ there are at least 5 or more pairs of subsequent fair quadrilaterals, so at least one has the property claimed in the lemma.                                                                      □

Before we continue with the proof of the theorem, we include the following Lemma, partially illustrated in Figure 2(c):

**Lemma 2.** *If there are two regions $R, S$ touching in some nontrivial interval $I = (a, b)$ then at a, there is a corner of $R$ or $S$. The same holds for corner b.*

Now, let $(Q_i, Q_{i+1})$ be a pair of fair same-sided quadrilaterals, touching sides $A_p$ and $B_q$. Since $Q_i$ and $Q_{i+1}$ have to be adjacent, the two sides next to each other touch. We can use the above Lemma 2 to show that each interval that is shared by two polygons ends at two of the corners of the two polygons. Since there exist the polygonal regions representing $r_i$ and $l_i$, it is clear that the interval where $Q_i$ and $Q_{i+1}$ touch is disjoint from the regions $Q_A$ and $Q_B$. Hence the corners derived from Lemma 2 are not the corners of $Q_i$ or $Q_{i+1}$ that are incident to sides $A_p$ and $B_q$. This is a contradiction, since then both $Q_i$ and $Q_{i+1}$ must have at least 5 corners, or one of them has even 6 corners.    □

## 2.2    Five Sides Are Not Enough

If we allow the regions to be pentagons, we have to sharpen the argument a little more.

**Lemma 3.** *If we choose k large enough, there exists a triple of fair pentagons $P_i$, $P_{i+1}$, $P_{i+2}$ that is adjacent to the same sides of $P_A$ and $P_B$.*

*Proof:* We prove this along the same lines as before. Now we have four sides with $c = 6$ inner corners of the pentagons $P_A$ and $P_B$. As before, we can see that at most 12 pentagons of the inner chain are not fair. Since we aim now for triples and not just for pairs, we get a worst case where every third pentagon is not fair. Hence for $k \geq 3 \cdot 2c + 3$, we get at least $k - 38$ fair subsequent pentagons. Next, we estimate the number of transition triples. The number of equivalence classes of pentagons with sides solely on the same side of $P_A$ and $P_B$ is seven. As we deal with triples, this makes a bound of at most 14 transition triples, since we can differentiate transition points between the first two and the last two pentagons of the triple.

Hence, we have to grow $k$ to $38 + 14 = 52$ to ensure that a triple of fair same-sided pentagons exists.    □

**Theorem 2.** *For k sufficiently large, there does not exist a touching polygon representation for G in which all regions have complexity five or less.*

*Proof:* We choose $k$ to be at least 52. Now, let $(P_i, P_{i+1}, P_{i+2})$ be a triple of fair same-sided pentagons, touching sides $A_p$ and $B_q$. Since $P_i$ and $P_{i+1}$ have to be adjacent, the two sides next to each other touch. We can use Lemma 2 that each interval that is shared by two polygons ends at two of the corners of the two polygons. Since there exist the polygonal regions representing $r_i$ and $l_i$, it is clear that the interval where $Q_i$ and $Q_{i+1}$ touch is disjoint from the regions $P_A$ and $P_B$. Hence the corners derived from Lemma 2 are not the corners of $P_i$ or $P_{i+1}$ that are incident to sides $A_p$ and $B_q$. This is a contradiction, since both $P_i$ and $P_{i+1}$ have at least 5 corners, or one of them has even 6 corners. In the case, that $P_i$ and $P_{i+1}$ have exactly 5 corners, we repeat the same argument for $P_{i+1}$ and $P_{i+2}$. From the second application, we prove the existence of a second additional corner at $P_{i+1}$ or that $P_{i+2}$ has two additional corners at the side opposite to $P_{i+1}$. In both cases, we get a contradiction. There exists a region with at least 6 corners.    □

Note that six-sided polygons are indeed sufficient to represent the graph in Figure 2(a). In particular, for subsequent fair polygons $P_i$ and $P_{i+1}$, we can use three segments on the lower side of $P_i$, while the upper side of $P_{i+1}$ consists of only one segment which completely overlaps the middle of the three segments from the lower side of $P_i$.

## 3 Touching Hexagons Representation

In this section, we present a linear time algorithm that takes as input a planar graph $G = (V, E)$ and which produces a representation of $G$ in which all regions are convex hexagons, thus proving that planar graphs belong to the class of touching hexagons graphs.

### 3.1 Algorithm Overview

We assume that the input graph $G = (V, E)$ is a fully triangulated planar graph with $|V| = n$ vertices. If the graph is planar but not fully triangulated, we can augment it to a fully triangulated graph with the help of dummy vertices and edges, run the algorithm below and remove the polygons that correspond to dummy vertices. Traditionally, planar graphs are augmented to fully triangulated graphs by adding edges to each non-triangular face. Were we to take this approach, however, when we remove the dummy edges we have to perturb the resulting space partition to remove polygonal adjacencies. As this is difficult to do, we convert our input graph to a fully triangulated one by adding one additional vertex to each face and connecting it to all vertices in that face. The above approach works if the input graph is biconnected. Singly-connected graphs must first be augmented to biconnected graphs as follows. Consider any articulation vertex $v$, and let $u$ and $w$ be consecutive neighbors of $v$ in separate biconnected components. Add new vertex $z$ and edges $(z, u)$ and $(z, w)$. Iterating for every articulation point biconnects $G$ and results in an embedding in which each face is bounded by a simple cycle.

The algorithm has two main phases. The first phase computes the canonical labeling. In the second phase we create regions with slopes 0, 1, -1 out of an initial isosceles right-angle triangle, by processing vertices in the canonical order. Each time a new vertex is processed, a new region is carved out of one or more already existing regions. At the end of the second phase of the algorithm we have a right-angle isosceles triangle which has been partitioned into exactly $n = |V|$ convex regions, each with at most 6 sides. We will show that creating and maintaining the regions requires linear time in the size of the input graph. We illustrate the algorithm with an example; see Figure 3.

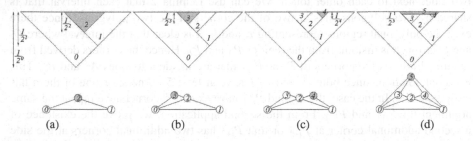

(a)                    (b)                    (c)                    (d)

**Fig. 3.** Incremental construction of the touching hexagons representation of a graph. Shaded vertices on the bottom row and shaded regions on the top row are processed at this step. In general, the region defined at step $i$ is carved at distance $1/2^i$ from the active front on the top. Note that the top row forms a horizontal line at all times.

## 3.2  Region Creation

In this section we describe the $n$-step incremental process of inserting new regions in the order given by the canonical labeling, where $n = |V|$. The regions will be carved out of an initial triangle with coordinates $(0,0), (-1,1), (1,1)$. The process begins by the creation of $R_0, R_1$, and $R_2$, which correspond to the first three vertices, $v_0, v_1, v_2$; see Figure 3(a). Note that the first three vertices in the canonical order form a triangular face in $G$ and hence must be represented as mutually touching regions.

At step $i$ of this process, where $2 < i < n$, region $R_i$ will be carved out from the current set of regions. Define a region as "active" at step $i$ if it corresponds to a vertex that has not yet been connected to all its neighbors. An invariant of the algorithm is that all active regions are non-trivially tangent to the top side of the initial triangle, which we refer to as the "active front."

New vertices are created in one of two ways, depending on the degree of the current node, $v_i$, in the graph induced by the first $i$ vertices, $G_i$. By the property of the canonical ordering and the active regions invariant, $v_i$ is connected to 2 or more consecutive vertices on the outer face of $G_{i-1}$:

1. If $d_{G_i}(v_i) > 2$ then $R_i$, the region corresponding to $v_i$, is a quadrilateral carved out of all but the leftmost and rightmost regions, by a horizontal line segment that is at distance $1/2^i$ from the active front; see Figure 3(d). Note that all but the leftmost and rightmost neighbors of $v_i$ are removed from the set of active regions as their corresponding vertices have been connected to all their neighbors. Region $R_i$ is added to the new set of active regions. Call this a "type 1 carving."
2. If $d_{G_i}(v_i) = 2$, let $R_a$ and $R_b$ be its neighbors on the frontier. Region $R_i$ is then carved out as a triangle from either $R_a$ or $R_b$.

**Lemma 4.** *The regions produced by the above algorithm are convex and have at most 6 sides.*

*Proof:* First note that the above algorithm leads to the creation of at most fifteen different types of regions; see Figure 4. Each region has a horizontal top segment, a horizontal bottom segment (possibly of length 0), and sides with slopes -1 or 1. Moreover, each region can be characterized as either opening (the first two), static (the next six), or closing (the last 7), depending on the angles of the two sides connecting it to the top horizontal segment. Opening and static regions give rise to new regions via type 1 carvings (dashed arrows) and type 2 carvings (solid arrows). Closing regions only give rise to type 1 carvings.

We show that the regions produced as a result of type 1 and type 2 carvings from the initial triangle are convex polygons with at most 6 sides with slopes 0, 1, -1 by induction on the number of steps. Assume that the claim is true until right before step $i$; we will show that the claim is true after step $i$.

If $d_{G_i}(v_i) > 2$ then the new region $R_i$ is created by a type 1 carving. Recall that $R_i$ is created by the addition of the horizontal line segment at distance $1/2^i$ from the top of the triangle, cutting through all but the leftmost and rightmost neighbors of $v_i$. It remains to show that the resulting region $R_i$ has exactly four sides and that the complexity of the all other regions is unchanged. By construction, $R_i$ has a top and bottom horizontal

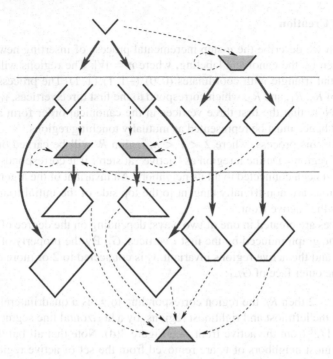

**Fig. 4.** There are a fifteen possible region shapes, falling into three categories: 2 opening, 6 static, and 7 closing. Solid arrows indicate type 2 (triangular) carving and dashed arrows indicate type 1 carving (a horizontal strip from the top of the current region). The four filled quadrilateral regions are the only types created due to type 1 carving.

segments and exactly one line segment on the left and one line segment on the right. The construction of $R_i$ resulted in modifications in the regions representing all but the leftmost and rightmost neighbors of $v_i$ in $G_i$, and there is at least one such neighbor. The changes in these regions are the same: each such region had its top carved off by the bottom horizontal side of the new region $R_i$. These changes do not affect the number of sides defining the regions. Regions corresponding to nodes that are not adjacent to $v_i$ in $G_i$ are unchanged.

Otherwise, if $d(v_i) = 2$ we must create a new region $R_i$ between two adjacent regions $R_a$ and $R_b$. By construction, the complexity of the new region $R_i$ is 3, as we carve off a new triangle between regions $R_a$ and $R_b$ with a horizontal top side and apex at distance $1/2^i$ from the active front. As a result of this operation either the $R_a$ or $R_b$ was modified and all other regions remain unchanged. Specifically, the complexity of either $R_a$ or $R_b$ must increase by exactly one. Without loss of generality, let $R_a$ be the region from which $R_i$ will be carved; see Figure 5. It is easy to see that if $R_a$ had complexity 6 then it must have been a "closing" region (one of the rightmost two in the last row on Fig. 4. Then the new region $R_i$ would have been carved out of $R_b$ which must have complexity 5 or less as it is impossible to have $R_a$ and $R_b$ both "closing" and adjacent. Therefore, at the end of step $i$ the complexity of $R_a$ has increased by one but is still no greater than 6. □

## 3.3   Running Time

The above algorithm can be implemented in linear time. The linear time algorithm for computing a canonical labeling of a planar graph [5] requires a planar embedding as an input. Recall that planar embedding of a graph is simply a clockwise order of the neighbors of each vertex in the graph. Obtaining a planar embedding can be done in linear time using the algorithm by Hopcroft and Tarjan [15].

Creating and maintaining the regions in the second phase of our algorithm can also be done in linear time. We next prove this by showing that each region requires $O(1)$ time to create and requires $O(1)$ number of modifications.

Consider the creation of new regions. By the properties of canonical labeling, when we process the current vertex $v_i$, it is adjacent to at least two consecutive vertices on the outer face of $G_{i-1}$. By construction of our algorithm the vertices in the outer face of $G_{i-1}$ correspond to active regions and so have a common horizontal tangent. If $d_{G_i}(v_i) = 2$, then a new region $R_i$ is carved out of one of the neighboring regions $R_a$ or $R_b$. Determining the coordinates of $R_i$ takes constant time, given the coordinates of $R_a$ and $R_b$ and the fact that $R_i$ will have height $1/2^i$ and will be tangent to the active frontier. If $d_{G_i}(v_i) > 2$, then all but the leftmost and rightmost neighbors of $v_i$ have their corresponding regions carved, in order to create the new region $R_i$. In this case the coordinates of the $R_i$ can also be determined in constant time given the coordinates of the leftmost and rightmost neighbors and the fact that $R_i$ will have height $1/2^i$ and will be tangent to the active frontier. Note that the updates of the regions between the leftmost and rightmost are considered in the modification step.

Consider the modifications of existing regions. As can be seen from the hierarchy of regions on Figure 4, there are exactly 15 different kinds of regions and each region begins as a triangle and undergoes at most 4 modifications (e.g., from triangle, to quadrilateral, to pentagon, to hexagon, to quadrilateral). Moreover, once a region goes from one type to the next, it can never change back to the same type (i.e., all the arrows point downward). Finally note that the total number of region modifications is proportional to $|E|$ and since $G$ is planar, $|E| = O(|V|)$ Thus, each region needs at most a constant number of modifications from the time it is created to the end of the algorithm.

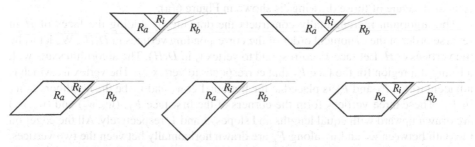

**Fig. 5.** Introducing region $R_i$ between $R_a$ and $R_b$, assuming $R_i$ is carved out of $R_a$. All the possible cases are shown, assuming that $R_a$ and $R_b$ were convex, at most 6-sided regions with slopes 0, 1, -1. (There are five more symmetric cases when $R_i$ is carved out of $R_b$.) Note that these five regions correspond to the non-filled regions from the region-creating hierarchy in Fig. 4 with two static regions in the first row and the three closing regions in the second row.

The algorithm described in this section, yields the following theorem:

**Theorem 3.** *A planar graph can be converted into a set of touching convex polygons with complexity at most six, in linear time in the number of vertices of the graph.*

As defined, the above algorithm requires exponential area, if polygonal endpoints are to be placed at integer grid points. We show in the Appendix how to compact the initial exponential area drawings. However, the compaction approach is not guaranteed to always find a small area drawing. Therefore, we next show with a different algorithm that, in fact, $O(n) \times O(n)$ area suffices.

## 4  Hexagonal Representation of Planar Graphs Using $O(n) \times O(n)$ Area

One drawback to the algorithms described in Sections 3 is it is not easy to obtain a good bound on the drawing area. Using a different approach, we can show that any general $n$-vertex planar graph can be represented by touching convex hexagons, drawn on the $O(n) \times O(n)$ grid. This approach is based on Kant's algorithm for hexagonal grid drawing of 3-connected, 3-planar graphs [17]. In Kant's algorithm the drawing is obtained by looking at the dual graph, and processing its vertices in the canonical order. In the final drawing, however, there are two non-convex faces, separated by an edge which is not drawn as a straight-line segment. These problems can be addressed by adding several extra vertices in a pre-processing step. When the dual of this augmented graph is embedded, the faces corresponding to the extra vertices can be removed to yield the desired grid drawing on area $O(n) \times O(n)$.

Let $H = (V, E)$ be a 3-connected, 3-planar graph. Note that the dual $D(H)$ is fully triangulated, as each face in the dual corresponds to exactly one vertex in $H$. So, for $f$ faces in $H$, we have $f$ vertices in $D(H)$. We first compute a canonical ordering on the vertices of $D(H)$ as defined by de Fraysseix et al. [7]. Let $v_1, ..., v_f$ be the vertices in $D(H)$ in this canonical order.

Kant's algorithm now constructs a drawing for $H$ such that all edges but one have slopes $0°$, $60°$, or $-60°$, with the one edge with bends lying on the outer face. The typical structure of those drawings is shown in Figure 6(a).

The algorithm incrementally constructs the drawing by adding the faces of $H$ in reverse order of the canonical order of the corresponding vertices in $D(H)$. We let $w_i$ be the vertices of $H$. Let face $F_i$ correspond to vertex $v_i$ in $D(H)$. The algorithm starts with a triangular region for the face $F_f$ that corresponds to vertex $v_f$. The vertex $w_x$ which is adjacent to $F_f$, $F_1$ and $F_2$ is placed at the bottom. Let $w_y$ and $w_z$ be the neighbors of $w_x$ in $F_f$. These three vertices form the corners of the first face $F_f$. $(w_x, w_z)$ and $(w_x, w_y)$ are drawn upward with equal lengths and slopes -1 and 1, respectively. All the edges on the path between $w_y$ and $w_z$ along $F_f$ are drawn horizontally between the two vertices. From this first triangle, all other faces are added in reverse canonical order to the upper boundary of the drawing region. If a face is completed by only one vertex $w_i$, this vertex is placed appropriately above the upper boundary such that it can be connected by two edges with slopes -1 and 1, respectively. If the face is completed by a path, then the two end segments of the path have slopes -1 and 1, while the other edges are horizontal.

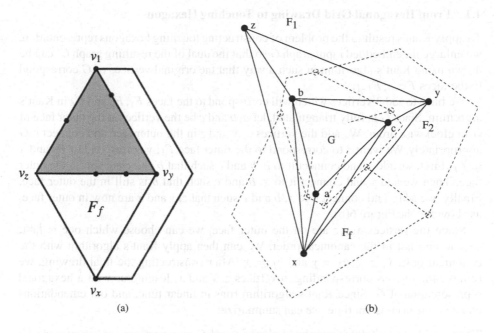

**Fig. 6.** (a) Polygonal structure obtain from Kant's algorithm. (b) Graph $G$ augmented by vertices $z, y$ and $x$ together with its dual which serves as input graph for Kant's algorithm.

The construction ends when $w_1$ is inserted, corresponding to the outer face $F_1$. Note that there is an edge between $w_1$ and $w_x$, which is drawn using some bends. This edge is adjacent to the faces $F_1$ (the outer face) and $F_2$.

From this construction, we can observe that the angles at faces $F_f, ..., F_3$ have size $\leq \pi$ as the first two edges do not enter the vertex from above, and the last edge leaves the vertex upwards. Hence, we have the following result.

**Lemma 5.** *The faces $F_f, ..., F_3$ are convex, and as the slopes of the edges are -1,0 or 1, they are drawn with at most 6 sides.*

This property is exactly what we are aiming for, as the vertices of our input graph $G$ should be represented by convex regions of at most 6 sides. Unfortunately, Kant's algorithm creates two non-convex faces $F_1$ and $F_2$ separated by an edge which is not drawn as a line segment. Furthermore, the face $F_f$ is drawn as large as all the remaining faces $F_3, ...F_{f-1}$ together.

Kant also gave an area estimate for the result of his algorithm. A corollary of Kant's algorithm is the following.

**Corollary 1.** *For a given 3-connected, 3-planar graph $H$ of $n$ vertices, $H - w_x$ can be drawn within an area of $n/2 - 1 \times n/2 - 1$.*

## 4.1    From Hexagonal Grid Drawing to Touching Hexagons

To apply Kant's result to the problem of constructing touching hexagons representation, we enlarge the embedded input graph $G$ so that the dual of the resulting graph $G'$ can be drawn using Kant's algorithm in such a way that the original vertices of $G$ correspond to the faces $F_3, ..., F_{f-1}$.

We have to add 3 vertices which will correspond to the faces $F_1, F_2$ and $F_f$ in Kant's algorithm. Since $G$ is fully triangulated, let $a, b$ and $c$ be the vertices at the outer face of $G$ in clockwise order. We add the vertices $x$, $y$ and $z$ in the outer face and connect to $G$ appropriately. We want $z$ to correspond to the outer face $F_1$, $y$ correspond to $F_2$ and $x$ to $F_f$. First, we add $x$ and connect it to $a$, $b$ and $c$ such that $b$ and $c$ are still in the outer face. Then we add $y$ and connect it to $x$, $b$ and $c$ such that $b$ is still in the outer face. Finally, we add $z$ and connect it to $x$, $b$ and $c$ such that $z$, $y$ and $x$ are now in outer face, as shown in the Figure 6(b).

Since the vertices $x, y, z$ are on the outer face, we can choose which one is first, second and last in the canonical order. We can then apply Kant's algorithm with the canonical order $v_1 = z, v_2 = y$ and $v_f = x$. After constructing the final drawing, we remove the regions corresponding to vertices $z$, $y$ and $x$, leaving us with a hexagonal representation of $G$. Since Kant's algorithm runs in linear time, and our emendations can be done in constant time, we can summarize:

**Theorem 4.** *For a fully triangulated planar graph $G$ on $n$ vertices, we can construct a contact graph of convex hexagons in time $O(n)$. The sides of the hexagons have slope 1, 0, or -1.*

Given any planar graph $G$, if it is not biconnected, we can make it biconnected using a procedure attributed to Read [27], adding a vertex and two edges at each articulation point. Once biconnected, we can fully triangulate the graph by adding a vertex inside each non-triangular face and connecting that vertex to each vertex on the face. We can then apply Theorem 4, to get a hexagonal representation of the extended graph. Finally, removing the added vertices and their edges, we obtain a hexagonal representation of $G$. This gives us:

**Theorem 5.** *For any planar graph $G$ on $n$ vertices, we can construct a contact graph of convex hexagons in time $O(n)$. The sides of the hexagons have slope 1, 0, or -1.*

## 4.2    Area Estimation

For a triangulated input graph $G = (V, E)$, we have $n$ vertices and, by Euler's formula, $2n - 4$ faces. Since we enhanced our graph to $n + 3$ vertices, we have $f = 2n + 2$ faces. Those faces are the vertices in the dual $D(G)$ which is the input to Kant's algorithm. His area estimation gives an area of $n/2 - 1 \times n/2 - 1$ for $f = n$ vertices when we coalesce the faces $F_1, F_2$ and $F_f$ into a single outer face by removing the corresponding vertices and edges. Thus, we get an area bound of $n \times n$ using exactly the same argument as he did.

**Theorem 6.** *For a fully triangulated planar graph $G$ of $n$ vertices, we can achieve a contact representation of convex hexagons with area $n \times n$.*

# 5   Conclusion and Future Work

Thomassen [29] had shown that not all planar graphs can be represented by touching pentagons, where the external boundary of the figure is also a pentagon and there are no holes. Our results in this paper are more general, as we do not insist on the external boundary being a pentagon or on there being no holes between pentagons. Finally, it is possible to derive algorithms for convex hexagonal representations for general planar graphs from several earlier papers, e.g., de Fraysseix *et al.* [7], Thomassen [29], and Kant [17]. However, these do not immediately lead to algorithmic solutions to the problem of graph representation with convex low-complexity touching polygons. To the best of our knowledge, this problem has never been formally considered.

In this paper we presented several results about touching n-sided graphs. We showed that for general planar graph six sides are necessary. Then we presented an algorithm for representing general planar graphs with convex hexagons. Finally, we discussed a different algorithm for general planar graphs which also yields an $O(n) \times O(n)$ drawing area.

Several interesting related problems are open. What is the complexity of the deciding whether a given planar graph can be represented by touching triangles, quadrilaterals, or pentagons? In the context of rectilinear catrograms the vertex-weighted problem has been carefully studied. However, the same problem without the rectilinear constraint has received less attention. Finally, it would be interesting to characterize the subclasses of planar graphs that allow for touching triangles, touching quadrilaterals, and touching pentagons representations.

## Acknowledgments

We would like to thank Therese Biedl for pointing out the very relevant work by Kant and Thomassen and Christian Duncan for discussions about this problem.

# References

1. Battista, G.D., Lenhart, W., Liotta, G.: Proximity drawability: A survey. In: Tamassia, R., Tollis, I.G. (eds.) GD 1994. LNCS, vol. 894, pp. 328–339. Springer, Heidelberg (1995)
2. Biedl, T., Bretscher, A., Meijer, H.: Rectangle of influence drawings of graphs without filled 3-cycles. In: Kratochvíl, J. (ed.) GD 1999. LNCS, vol. 1731, pp. 359–368. Springer, Heidelberg (1999)
3. Bruls, M., Huizing, K., van Wijk, J.J.: Squarified treemaps. In: Proc. Joint Eurographics/IEEE TVCG Symp. Visualization, VisSym., pp. 33–42 (2000)
4. Buchsbaum, A.L., Gansner, E.R., Procopiuc, C.M., Venkatasubramanian, S.: Rectangular layouts and contact graphs. ACM Transactions on Algorithms 4(1) (2008)
5. Chrobak, M., Payne, T.: A linear-time algorithm for drawing planar graphs. Inform. Process. Lett. 54, 241–246 (1995)
6. de Berg, M., Mumford, E., Speckmann, B.: On rectilinear duals for vertex-weighted plane graphs. Discrete Mathematics 309(7), 1794–1812 (2009)
7. de Fraysseix, H., de Mendez, P.O., Rosenstiehl, P.: On triangle contact graphs. Combinatorics, Probability and Computing 3, 233–246 (1994)

8. de Fraysseix, H., Pach, J., Pollack, R.: Small sets supporting Fary embeddings of planar graphs. In: Procs. 20th Symposium on Theory of Computing (STOC), pp. 426–433 (1988)
9. de Fraysseix, H., Pach, J., Pollack, R.: How to draw a planar graph on a grid. Combinatorica 10(1), 41–51 (1990)
10. Gabriel, K.R., Sokal, R.R.: A new statistical approach to geographical analysis. Systematic Zoology 18, 54–64 (1969)
11. He, X.: On finding the rectangular duals of planar triangular graphs. SIAM Journal of Computing 22(6), 1218–1226 (1993)
12. He, X.: On floor-plan of plane graphs. SIAM Journal of Computing 28(6), 2150–2167 (1999)
13. Hliněný, P.: Classes and recognition of curve contact graphs. Journal of Comb. Theory (B) 74(1), 87–103 (1998)
14. Hliněný, P., Kratochvíl, J.: Representing graphs by disks and balls (a survey of recognition-complexity results). Discrete Mathematics 229(1-3), 101–124 (2001)
15. Hopcroft, J., Tarjan, R.E.: Efficient planarity testing. Journal of the ACM 21(4), 549–568 (1974)
16. Jaromczyk, J.W., Toussaint, G.T.: Relative neighborhood graphs and their relatives. Proceedings of the IEEE 80, 1502–1517 (1992)
17. Kant, G.: Hexagonal grid drawings. In: Mayr, E.W. (ed.) WG 1992. LNCS, vol. 657, pp. 263–276. Springer, Heidelberg (1993)
18. Kant, G.: Drawing planar graphs using the canonical ordering. Algorithmica 16, 4–32 (1996) (special issue on Graph Drawing, edited by G. Di Battista and R. Tamassia)
19. Kant, G., He, X.: Regular edge labeling of 4-connected plane graphs and its applications in graph drawing problems. Theoretical Computer Science 172, 175–193 (1997)
20. Koebe, P.: Kontaktprobleme der konformen Abbildung. Berichte über die Verhandlungen der Sächsischen Akademie der Wissenschaften zu Leipzig. Math.-Phys. Klasse 88,141–164 (1936)
21. Koźmiński, K., Kinnen, W.: Rectangular dualization and rectangular dissections. IEEE Transactions on Circuits and Systems 35(11), 1401–1416 (1988)
22. Lai, Y.-T., Leinwand, S.M.: Algorithms for floorplan design via rectangular dualization. IEEE Transactions on Computer-Aided Design 7, 1278–1289 (1988)
23. Lai, Y.-T., Leinwand, S.M.: A theory of rectangular dual graphs. Algorithmica 5, 467–483 (1990)
24. Liao, C.-C., Lu, H.-I., Yen, H.-C.: Compact floor-planning via orderly spanning trees. Journal of Algorithms 48, 441–451 (2003)
25. Liotta, G., Lubiw, A., Meijer, H., Whitesides, S.H.: The rectangle of influence drawability problem. Computational Geometry: Theory and Applications 10, 1–22 (1998)
26. Rahman, M., Nishizeki, T., Ghosh, S.: Rectangular drawings of planar graphs. Journal of Algorithms 50(1), 62–78 (2004)
27. Read, R.C.: A new method for drawing a graph given the cyclic order of the edges at each vertex. Congressus Numerantium 56, 31–44 (1987)
28. Steadman, P.: Graph-theoretic representation of architectural arrangement. In: March, L. (ed.) The Architecture of Form, pp. 94–115. Cambridge University Press, Cambridge (1976)
29. Thomassen, C.: Plane representations of graphs. In: Bondy, J.A., Murty, U.S.R. (eds.) Progress in Graph Theory, pp. 43–69 (1982)
30. Thomassen, C.: Interval representations of planar graphs. Journal of Comb. Theory (B) 40, 9–20 (1988)
31. Yeap, G.K., Sarrafzadeh, M.: Sliceable floorplanning by graph dualization. SIAM Journal on Discrete Mathematics 8(2), 258–280 (1995)

# Minimum-Perimeter Intersecting Polygons

Adrian Dumitrescu[1],* and Minghui Jiang[2],**

[1] Department of Computer Science, University of Wisconsin-Milwaukee
Milwaukee, WI 53201-0784, USA
ad@cs.uwm.edu
[2] Department of Computer Science, Utah State University
Logan, UT 84322-4205, USA
mjiang@cc.usu.edu

**Abstract.** Given a set $S$ of segments in the plane, a polygon $P$ is an intersecting polygon of $S$ if every segment in $S$ intersects the interior or the boundary of $P$. The problem MPIP of computing a minimum-perimeter intersecting polygon of a given set of $n$ segments in the plane was first considered by Rappaport in 1995. This problem is not known to be polynomial, nor it is known to be NP-hard. Rappaport (1995) gave an exponential-time exact algorithm for MPIP. Hassanzadeh and Rappaport (2009) gave a polynomial-time approximation algorithm with ratio $\frac{\pi}{2} \approx 1.58$. In this paper, we present two improved approximation algorithms for MPIP: a 1.28-approximation algorithm by linear programming, and a polynomial-time approximation scheme by discretization and enumeration. Our algorithms can be generalized for computing an approximate minimum-perimeter intersecting polygon of a set of convex polygons in the plane. From the other direction, we show that computing a minimum-perimeter intersecting polygon of a set of (not necessarily convex) simple polygons is NP-hard.

## 1 Introduction

A polygon is an *intersecting polygon* of a set of segments in the plane if every segment in the set intersects the interior or the boundary of the polygon. In 1995, Rappaport [11] proposed the following geometric optimization problem:

**MPIP:** Given a set $S$ of $n$ (possibly intersecting) segments in the plane, compute a minimum-perimeter intersecting polygon $P^*$ of $S$.

The problem MPIP was originally motivated by the theory of geometric transversals; see [13] for a recent survey on related topics. As of now, MPIP is not known to be polynomial, nor it is known to be NP-hard. Rappaport [11] gave an exact algorithm for MPIP that runs in $O(n \log n)$ time when the input segments are constrained to a constant number of orientations, but the running time becomes exponential in the general case. Recently, Hassanzadeh and Rappaport [8]

---

* Supported in part by NSF CAREER grant CCF-0444188.
** Supported in part by NSF grant DBI-0743670.

A. López-Ortiz (Ed.): LATIN 2010, LNCS 6034, pp. 433–445, 2010.

presented the first polynomial-time constant-factor approximation algorithm for MPIP with ratio $\frac{\pi}{2} \approx 1.58$.

In this paper, we present two improved approximation algorithms for MPIP. Our first result (in Section 2), is a 1.28-approximation algorithm for MPIP which is based on linear programming:

**Theorem 1.** *For any $\varepsilon > 0$, a $\frac{4}{\pi}(1 + \varepsilon)$-approximation for minimum-perimeter intersecting polygon of $n$ segments in the plane can be computed by solving $O(1/\varepsilon)$ linear programs, each with $O(n)$ variables and $O(n)$ constraints. In particular, a 1.28 approximation can be computed by solving a constant number of such linear programs.*

Our second result (in Section 3) is a polynomial-time approximation scheme (PTAS) for MPIP which is based on discretization and enumeration:

**Theorem 2.** *For any $\varepsilon > 0$, a $(1+\varepsilon)$-approximation for minimum-perimeter intersecting polygon of $n$ segments in the plane can be computed in $O(1/\varepsilon) \operatorname{poly}(n) + 2^{O((1/\varepsilon)^{2/3})}n$ time.*

Both algorithms can be generalized for computing an approximate minimum-perimeter intersecting polygon of a set of (possibly intersecting) convex polygons in the plane. Details appear in Section 4. From the other direction, we show that computing a minimum-perimeter intersecting polygon of a set of (possibly intersecting but not necessarily convex) simple polygons is NP-hard. Details appear in Section 5.

**Theorem 3.** *Computing a minimum-perimeter intersecting polygon of a set of simple polygons, or that of a set of simple polygonal chains, is NP-hard.*

While the problem MPIP has been initially formulated for segments, it can be formulated for any finite collection of connected (say, polygonal) regions in the plane, as input, for which the problem is to find a minimum-perimeter intersecting polygon. Natural subproblems to consider are the cases when the input is a set of line segments, a set of convex or non-convex polygons, and a set of polygonal chains (as we do in the hardness reduction).

*Preliminaries.* Denote by $\operatorname{conv}(A)$ the convex hull of a planar set $A$. For a polygon $P$, let $\operatorname{perim}(P)$ denote its perimeter. Let $P^*$ denote a minimum-perimeter intersecting polygon of $\mathcal{S}$. We can assume without loss of generality that not all segments in $\mathcal{S}$ are concurrent at a common point (this can be easily checked in linear time), thus $\operatorname{perim}(P^*) > 0$. The following two facts are easy to prove; see also [8,11].

**Proposition 1.** *$P^*$ is a convex polygon with at most $n$ vertices.*

**Proposition 2.** *If $P_1$ is an intersecting polygon of $\mathcal{S}$, and $P_1$ is contained in another polygon $P_2$, then $P_2$ is also an intersecting polygon of $\mathcal{S}$.*

## 2    A $\frac{4}{\pi}(1 + \varepsilon)$-Approximation Algorithm

In this section we prove Theorem 1. We present a $\frac{4}{\pi}(1 + \varepsilon)$-approximation algorithm for computing a minimum-perimeter intersecting polygon of a set $S$ of line segments. The idea is to first prove that every convex polygon $P$ is contained in some rectangle $R = R(P)$ that satisfies $\mathrm{perim}(R) \leq \frac{4}{\pi} \mathrm{perim}(P)$, then use linear programming to compute a $(1 + \varepsilon)$-approximation for the minimum-perimeter intersecting rectangle of $S$.

**Algorithm A1.**

Let $m = \lceil \frac{\pi}{4\varepsilon} \rceil$. For each direction $\alpha_i = i \cdot 2\varepsilon$, $i = 0, 1, \ldots, m - 1$, compute a minimum-perimeter intersecting rectangle $R_i$ of $S$ with orientation $\alpha_i$. Return the rectangle with the minimum perimeter over all $m$ directions.

We now show how to compute the rectangle $R_i$ by linear programming. By a suitable rotation of the set $S$ of segments in each iteration $i \geq 1$, we can assume for convenience that the rectangle $R_i$ is axis-parallel. For $i = 1, \ldots, n$, let $p_i = (a_i, b_i)$ and $q_i = (c_i, d_i)$ be the two endpoints of the $i$th segment in $S$. Then a point $p$ in the plane belongs to the $i$th segment if and only if $p$ is a convex combination of the two endpoints of the segment, that is, $p = (1 - t_i)p_i + t_i q_i$ for some parameter $t_i \in [0, 1]$. To satisfy the intersecting requirement, each segment in $S$ must have a point contained the rectangle $R_i$. The objective of minimum perimeter is naturally expressed as a linear function. The resulting linear program has $n$ variables $t_i$, $i = 1, \ldots, n$, for the $n$ segments in $S$, and 4 variables $x_1, x_2, y_1, y_2$ for the rectangle $R_i = [x_1, x_2] \times [y_1, y_2]$:

$$\text{minimize} \quad 2(x_2 - x_1) + 2(y_2 - y_1) \tag{LP1}$$

$$\text{subject to} \quad \begin{cases} x_1 \leq (1 - t_i)a_i + t_i c_i \leq x_2, \ 1 \leq i \leq n \\ y_1 \leq (1 - t_i)b_i + t_i d_i \leq y_2, \ 1 \leq i \leq n \\ 0 \leq t_i \leq 1, \qquad\qquad\qquad 1 \leq i \leq n \end{cases}$$

A key fact in the analysis of the algorithm is the following lemma. This inequality is also implicit in [12], where a slightly different proof is given. Nevertheless we present our own proof for completeness.

**Lemma 1.** *Let $P$ be a convex polygon. Then the minimum-perimeter rectangle $R$ containing $P$ satisfies $\mathrm{perim}(R) \leq \frac{4}{\pi} \mathrm{perim}(P)$.*

*Proof.* Let $R(\alpha)$ denote a minimum-perimeter rectangle with orientation $\alpha$ that contains the polygon $P$. For any direction $\alpha \in [0, \pi)$ in the plane, let $w(\alpha)$ denote the *width* of $P$ in the direction $\alpha$, that is, the width of the smallest parallel strip in the direction $\alpha$ that contains $P$. We have

$$\mathrm{perim}(R(\alpha)) = 2 \left( w(\alpha) + w \left( \alpha + \frac{\pi}{2} \right) \right). \tag{1}$$

According to Cauchy's surface area formula [14], we have

$$\int_0^\pi w(\alpha) \, d\alpha = \mathrm{perim}(P). \tag{2}$$

We obviously have

$$\int_0^\pi w(\alpha)\, d\alpha = \int_0^\pi w\left(\alpha + \frac{\pi}{2}\right) d\alpha. \tag{3}$$

By substituting (3) into (1) and by integrating (1) over the interval $[0, \pi]$ we get that

$$\pi \cdot \operatorname{perim}(R) \le \int_0^\pi \operatorname{perim}(R(\alpha)) = 4 \int_0^\pi w(\alpha)\, d\alpha = 4 \cdot \operatorname{perim}(P). \tag{4}$$

From this the claimed inequality follows:

$$\operatorname{perim}(R) \le \frac{4}{\pi} \cdot \operatorname{perim}(P). \qquad \square$$

Let $R^*$ be a minimum-perimeter intersecting rectangle of $\mathcal{S}$. To account for the error made by discretization, we need the following:

**Lemma 2.** *For all* $i = 0, 1, \ldots, m - 1$, $\operatorname{perim}(R_i) \le \sqrt{2}\operatorname{perim}(R^*)$. *Moreover, there exists an* $i \in \{0, 1, \ldots, m - 1\}$ *such that* $\operatorname{perim}(R_i) \le (1 + \varepsilon)\operatorname{perim}(R^*)$.

*Proof.* Refer to Figure 1. Consider any rectangle $R_i$, $i \in \{0, 1, \ldots, m - 1\}$. Let $\beta$ be the minimum angle difference between the orientations of $R_i$ and $R^*$, $0 \le \beta \le \pi/4$. Let $R_i'$ be the minimum-perimeter rectangle with the same orientation as $R_i$ such that $R_i'$ contains $R^*$. An easy trigonometric calculation shows that

$$\operatorname{perim}(R_i') = (\cos\beta + \sin\beta)\operatorname{perim}(R^*).$$

It follows that

$$\operatorname{perim}(R_i) \le \operatorname{perim}(R_i') = (\cos\beta + \sin\beta)\operatorname{perim}(R^*) \le \sqrt{2}\operatorname{perim}(R^*).$$

Since the directions $\alpha_i = i \cdot 2\varepsilon$ are discretized with consecutive difference at most $2\varepsilon$, there exists an $i \in \{0, 1, \ldots, m\}$ such that the difference $\beta$ between the orientations of $R_i$ and $R^*$ is at most $\varepsilon$. For this $i$, we have

$$\operatorname{perim}(R_i) \le (\cos\beta + \sin\beta)\operatorname{perim}(R^*) \le (1 + \beta)\operatorname{perim}(R^*) \le (1 + \varepsilon)\operatorname{perim}(R^*),$$

as required. $\qquad \square$

**Fig. 1.** The discretization error

Let $R_i$ be the rectangle returned by Algorithm A1. Let $P^*$ be a minimum-perimeter intersecting polygon of $S$. Then

$$\text{perim}(R_i) \leq (1 + \varepsilon) \, \text{perim}(R^*) \leq \frac{4}{\pi}(1 + \varepsilon) \, \text{perim}(P^*),$$

where the two inequalities follow by Lemma 2 and Lemma 1, respectively. This completes the proof of Theorem 1.

## 3   A Polynomial-Time Approximation Scheme

In this section we prove Theorem 2. We present a $(1 + \varepsilon)$-approximation algorithm for computing a minimum-perimeter intersecting polygon of a set $S$ of line segments. The idea is to first locate a region $Q$ that contains either an optimal polygon $P^*$ or a good approximation of it, then enumerate a suitable set of convex grid polygons in this region $Q$ to approximate $P^*$.

**Algorithm A2.**

STEP 1. Let $\varepsilon_1 = \frac{\varepsilon}{2+\varepsilon}$. Run Algorithm A1 to compute a rectangle $R$ that is a $(1 + \varepsilon_1)$-approximation for minimum-perimeter intersecting rectangle of $S$. Let $Q$ be a square of side length $3 \, \text{perim}(R)$ that is concentric with $R$ and parallel to $R$.

STEP 2. Let $k = \lceil 48/\varepsilon \rceil$. Divide the square $Q$ into a $k \times k$ grid $Q_\delta$ of cell length $\delta = 3 \, \text{perim}(R)/k$. Enumerate all convex grid polygons with grid vertices from $Q_\delta$. Find an intersecting polygon $P_\delta$ of the minimum perimeter among these grid polygons. If $\text{perim}(P_\delta) < \text{perim}(R)$, return $P_\delta$. Otherwise return $R$.

Let the *distance* between two compact sets $A$ and $B$ in the plane be the minimum distance between two points $a \in A$ and $b \in B$ ($A$ and $B$ intersect if and only if their distance is zero). Let $P^*$ be a minimum-perimeter intersecting polygon of $S$ that has the smallest distance to $R$. The choice of the square region $Q$ in STEP 1 is justified by the following lemma:

**Lemma 3.** *Suppose that* $\text{perim}(R) \geq (1 + \varepsilon) \, \text{perim}(P^*)$. *Then* $P^* \subseteq Q$.

*Proof.* The length of any side of the rectangle $R$ is at most $\frac{1}{2} \, \text{perim}(R)$. Consider the smallest rectangle $R'$ that is parallel to $R$ and contains the polygon $P^*$. Then the length of any side of $R'$ is at most $\frac{1}{2} \, \text{perim}(P^*)$. Suppose first that $R$ intersects $P^*$. Then $R$ must also intersect $R'$. It follows that $R'$ is contained in a square of side length $\frac{1}{2} \, \text{perim}(R) + \frac{1}{2} \, \text{perim}(P^*) + \frac{1}{2} \, \text{perim}(P^*)$ that is concentric with $R$ and parallel to $R$. Since $\text{perim}(P^*) \leq \text{perim}(R)$, we have

$$\frac{1}{2} \, \text{perim}(R) + \frac{1}{2} \, \text{perim}(P^*) + \frac{1}{2} \, \text{perim}(P^*) \leq \frac{3}{2} \, \text{perim}(R) < 3 \, \text{perim}(R).$$

Thus $P^*$ is contained in the square $Q$ of side length $3 \, \text{perim}(R)$ that is concentric with $R$ and parallel to $R$. In the following we assume that $R$ and $P^*$ are disjoint.

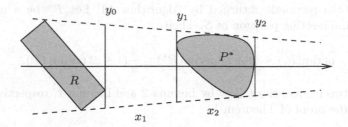

**Fig. 2.** The two dashed lines are tangent to both the rectangle $R$ and the polygon $P^*$, and are symmetric about the $x$ axis. The three vertical segments between the two dashed lines have lengths $y_0$, $y_1$, and $y_2$, respectively, and have distances $x_1$ and $x_2$ between consecutive segments. The left vertical segment is tangent to the rectangle $R$. The middle and right vertical segments are tangent to the polygon $P^*$.

Refer to Figure 2. Consider the two common supporting lines of $R$ and $P^*$ such that each line is tangent to both $R$ and $P^*$ on one side. Assume without loss of generality that the two lines are symmetric about the $x$ axis. Denote by $x_{\min}(A)$ and $x_{\max}(A)$, respectively, the minimum and the maximum $x$-coordinates of a point in a planar set $A$. Without loss of generality, further assume that $x_{\max}(R) \leq x_{\max}(P^*)$. We next consider two cases: (1) $x_{\max}(R) < x_{\min}(P^*)$, and (2) $x_{\max}(R) \geq x_{\min}(P^*)$. Figure 2 illustrates the first case.

*Case 1:* $x_{\max}(R) < x_{\min}(P^*)$. Refer to Figure 2. The crucial observation in this case is the following:

($\star$) Since both $R$ and $P^*$ are intersecting polygons of $\mathcal{S}$, each segment in $\mathcal{S}$ has at least one point in $R$ and one point in $P^*$.

Therefore, by the convexity of each segment in $\mathcal{S}$, any vertical segment between the two supporting lines with $x$-coordinate at least $x_{\max}(R)$ and at most $x_{\min}(P^*)$ is a (degenerate) intersecting polygon of $\mathcal{S}$. For example, the left and middle vertical segments of lengths $y_0$ and $y_1$ in Figure 2 correspond to two intersecting polygons of perimeters $2y_0$ and $2y_1$, respectively. We must have $y_0 > y_1$ because otherwise there would be an intersecting polygon of perimeter $2y_0 \leq 2y_1 \leq \operatorname{perim}(P^*)$ that is closer to $R$ than $P^*$ is, which contradicts our choice of $P^*$.

Recall that $R$ is a $(1 + \varepsilon_1)$-approximation for minimum-perimeter intersecting polygon of $\mathcal{S}$, and observe that the middle vertical segment of length $y_1$ in Figure 2 corresponds to a (degenerate) intersecting rectangle of $\mathcal{S}$. Thus $\operatorname{perim}(R) \leq (1 + \varepsilon_1)2y_1$. Since $\operatorname{perim}(R) \geq 2y_0$, it follows that $y_0 \leq (1 + \varepsilon_1)y_1$. Also note that $2y_2 \leq \operatorname{perim}(P^*)$. If $y_2 > (1 - \varepsilon_1)y_1$, then we would have

$$\operatorname{perim}(R) \leq (1 + \varepsilon_1)2y_1 < \frac{1 + \varepsilon_1}{1 - \varepsilon_1}2y_2 = (1 + \varepsilon)2y_2 \leq (1 + \varepsilon)\operatorname{perim}(P^*),$$

which contradicts the assumption of the lemma. Therefore we must have $y_2 \leq (1 - \varepsilon_1)y_1$.

Let $x_1 = x_{\min}(P^*) - x_{\max}(R)$ and $x_2 = x_{\max}(P^*) - x_{\min}(P^*)$. By triangle similarity, we have

$$\frac{x_1}{x_2} = \frac{y_0 - y_1}{y_1 - y_2} \le \frac{(1+\varepsilon_1)y_1 - y_1}{y_1 - (1-\varepsilon_1)y_1} = 1.$$

Thus $x_1 + x_2 \le 2x_2 \le \operatorname{perim}(P^*) \le \operatorname{perim}(R)$. Note that the distance from the center of $R$ to the left vertical segment of length $y_0$ is at most half the diagonal of $R$, which is at most $\frac{1}{4}\operatorname{perim}(R)$. Thus $P^*$ is contained in an axis-parallel rectangle concentric with $R$, of width $2(\frac{1}{4}\operatorname{perim}(R) + \operatorname{perim}(R)) = \frac{5}{2}\operatorname{perim}(R)$ and of height $\operatorname{perim}(R)$. Since $(\frac{5}{2})^2 + 1^2 < 3^2$, this axis-parallel rectangle is contained in the square $Q$ of side length $3\operatorname{perim}(R)$ that is concentric with $R$ and parallel to $R$ (recall that $R$ is not necessarily axis-parallel). Thus $P^* \subseteq Q$, as required.

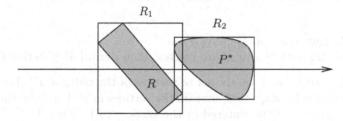

**Fig. 3.** The case that $x_{\max}(R) \ge x_{\min}(P^*)$

*Case 2:* $x_{\max}(R) \ge x_{\min}(P^*)$. Refer to Figure 3. Consider the smallest axis-parallel rectangle $R_1$ that contains $R$ and the smallest axis-parallel rectangle $R_2$ that contains $P^*$. Then the two rectangles $R_1$ and $R_2$ intersect. The length of any side of $R_1$ is at most $\frac{1}{2}\operatorname{perim}(R)$; the length of any side of $R_2$ is at most $\frac{1}{2}\operatorname{perim}(P^*)$. Thus $R_2$ is contained in an axis-parallel square of side length $\frac{1}{2}\operatorname{perim}(R) + \frac{1}{2}\operatorname{perim}(P^*) + \frac{1}{2}\operatorname{perim}(P^*) \le \frac{3}{2}\operatorname{perim}(R)$ that is concentric with $R_1$. Note that $R_1$ is concentric with $R$. Since $\sqrt{2} \cdot \frac{3}{2}\operatorname{perim}(R) < 3\operatorname{perim}(R)$, this axis-parallel square is contained in the square $Q$ of side length $3\operatorname{perim}(R)$ that is concentric with $R$ and parallel to $R$ (recall that $R$ is not necessarily axis-parallel). Thus again $P^* \subseteq Q$, as required.  □

The following lemma justifies the enumeration of convex grid polygons in $Q$:

**Lemma 4.** *Suppose that $P^* \subseteq Q$. Then there exists a convex grid polygon $P_\delta$ with grid vertices from $Q_\delta$ such that $P^* \subseteq P_\delta$ and $\operatorname{perim}(P_\delta) \le (1+\varepsilon)\operatorname{perim}(P^*)$.*

*Proof.* We will use the following two well-known facts[1] on any two compact sets $A$ and $B$ in the plane:

---

[1] Fact 1 is trivial. For convex polygons $A$ and $B$, fact 2 can be easily proved by repeatedly "shaving" $B$ into a smaller convex polygon by a supporting line of $A$ and applying the triangle inequality. Since convex compact sets can be approximated arbitrarily well by convex polygons, a limiting argument completes the proof of fact 2 for convex compact sets.

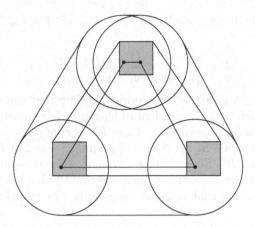

**Fig. 4.** $V$ (dots), $C$ (shaded squares), and $D$ (disks)

1. If $A \subseteq B$, then $\mathrm{conv}(A) \subseteq \mathrm{conv}(B)$.
2. If $A \subseteq B$ and both $A$ and $B$ are convex, then $\mathrm{perim}(A) \leq \mathrm{perim}(B)$.

Refer to Figure 4. Let $V$ be the set of vertices of the polygon $P^*$. Let $C$ be the union of the grid cells of $Q_\delta$ that contain the vertices in $V$. Let $D$ be the union of the disks of radii $r = \sqrt{2}\delta$ centered at the vertices in $V$. Then $V \subseteq C \subseteq D$. It follows by Fact 1 that $\mathrm{conv}(V) \subseteq \mathrm{conv}(C) \subseteq \mathrm{conv}(D)$. Note that $P^* = \mathrm{conv}(V)$. Let $P_\delta = \mathrm{conv}(C)$. Then $P^* = \mathrm{conv}(V) \subseteq \mathrm{conv}(C) = P_\delta$. By Fact 2, we also have

$$\mathrm{perim}(P_\delta) = \mathrm{perim}(\mathrm{conv}(C)) \leq \mathrm{perim}(\mathrm{conv}(D))$$
$$= \mathrm{perim}(\mathrm{conv}(V)) + 2\pi r = \mathrm{perim}(P^*) + 2\pi\sqrt{2}\delta.$$

Recall that $\delta = 3\,\mathrm{perim}(R)/k = 3\,\mathrm{perim}(R)/\lceil 48/\varepsilon \rceil \leq \frac{\varepsilon}{16}\,\mathrm{perim}(R)$. Let $R^*$ the a minimum-perimeter intersecting rectangle of $\mathcal{S}$. By Lemma 2 and Lemma 1 in the previous section, we have $\mathrm{perim}(R) \leq \sqrt{2}\,\mathrm{perim}(R^*)$ and $\mathrm{perim}(R^*) \leq \frac{4}{\pi}\,\mathrm{perim}(P^*)$. Thus

$$2\pi\sqrt{2}\delta \leq 2\pi \cdot \sqrt{2} \cdot \frac{\varepsilon}{16} \cdot \mathrm{perim}(R) \leq 2\pi \cdot \sqrt{2} \cdot \frac{\varepsilon}{16} \cdot \sqrt{2} \cdot \frac{4}{\pi} \cdot \mathrm{perim}(P^*) = \varepsilon \cdot \mathrm{perim}(P^*).$$

So we have $\mathrm{perim}(P_\delta) \leq (1 + \varepsilon)\,\mathrm{perim}(P^*)$. □

By Lemma 3 and Lemma 4, Algorithm A2 indeed computes an $(1 + \varepsilon)$-approximation for the minimum-perimeter intersecting polygon of $\mathcal{S}$. We now analyze its running time. STEP 1 runs in $O(1/\varepsilon_1)\,\mathrm{poly}(n) = O(1/\varepsilon)\,\mathrm{poly}(n)$ time. In STEP 2, each convex grid polygon in a $k \times k$ grid has $O(k^{2/3})$ grid vertices [1,2]. It follows from a result of Bárány and Pach [4] that there are $2^{O(k^{2/3})}$ such polygons in a $k \times k$ grid. Moreover, all these polygons can be enumerated in $2^{O(k^{2/3})}$ time because the proof in [4] is constructive. For each convex grid polygon, computing its perimeter takes $O(1/\varepsilon)$ time, and checking whether it is an intersecting polygon of $\mathcal{S}$ takes $O(n/\varepsilon)$ time, by simply checking each segment for intersection in

$O(1/\varepsilon)$ time. Thus STEP 2 runs in $2^{O((1/\varepsilon)^{2/3})}O(n/\varepsilon) = 2^{O((1/\varepsilon)^{2/3})}n$ time, and the total running time of Algorithm A2 is $O(1/\varepsilon)\operatorname{poly}(n) + 2^{O((1/\varepsilon)^{2/3})}n$.

## 4   Generalization for Convex Polygons

Both Algorithm A1 and Algorithm A2 can be generalized for computing an approximate minimum-perimeter intersecting polygon of a set $\mathcal{C}$ of $n$ (possibly intersecting) convex polygons in the plane.

To generalize Algorithm A1, for each direction $\alpha_i = i \cdot 2\varepsilon$, where $i = 0, 1, \ldots, m - 1$, we simply replace the linear program LP1 by another linear program LP2, that computes a minimum-perimeter intersecting rectangle $R_i$ of $\mathcal{C}$ with orientation $\alpha_i$. As earlier, by a suitable rotation of the set $\mathcal{C}$ of polygons in each iteration $i \geq 1$, we can assume that the rectangle $R_i$ is axis-parallel. For $1 \leq j \leq n$, let $C_j$ be the $j$th convex polygon in $\mathcal{C}$, and let $n_j$ be the number of vertices of $C_j$. Each convex polygon $C_j$ can be represented as intersection of $n_j$ linear constraints (halfplanes). The linear program requires the existence of $n$ points, $p_j = (s_j, t_j)$, $1 \leq j \leq n$, such that $p_j$ is contained in $R_i \cap C_j$, for each $j = 1, \ldots, n$. It can be written symbolically as follows:

$$\text{minimize}\quad 2(x_2 - x_1) + 2(y_2 - y_1) \qquad \text{(LP2)}$$
$$\text{subject to}\quad \begin{cases} p_j \in C_j, 1 \leq j \leq n \\ p_j \in R_i, 1 \leq j \leq n \end{cases}$$

There are $2n$ variables for the point coordinates, $s_j, t_j$, $1 \leq j \leq n$, and 4 variables $x_1, x_2, y_1, y_2$ for the rectangle $R_i = [x_1, x_2] \times [y_1, y_2]$. There are $n_j$ linear constraints corresponding to $p_j \in C_j$, and 4 linear constraints corresponding to $p_j \in R_i$. So the resulting linear program has $2n + 4$ variables and $4n + \sum_j n_j$ constraints.

To generalize Algorithm A2, use the generalized Algorithm A1 in STEP 1, then in STEP 2 replace the checking for intersection between a convex polygon and a segment by the checking for intersection between two convex polygons. Lemma 3 is still true because the crucial observation $(\star)$ remains valid when segments are generalized to convex polygons; since convexity is a property shared by segments and convex polygons, the argument based on this observation continues to hold. Lemma 4 is utterly unaffected by the generalization.

## 5   NP-Hardness

In this section we prove Theorem 3, namely that computing a minimum-perimeter intersecting polygon of a set of simple polygons is NP-hard. For simplicity, we will prove the stronger result that computing a minimum-perimeter intersecting polygon of a set of simple polygonal chains is NP-hard. Note that a simple polygonal chain is a degenerate simple polygon with area zero; by slightly "fattening" the polygonal chains, our reduction also works for simple polygons. The reduction is from the NP-hard problem *Vertex Cover* [7]:

INSTANCE: A graph $G = (V, E)$ with a set $V$ of $n$ vertices and a set $E$ of $m$ edges, and a positive integer $k \leq n$.

QUESTION: Is there a subset $S \subseteq V$ of $k$ vertices such that $S$ contains at least one vertex from each edge in $E$?

Assume that $n \geq 5$. We will construct a set $\mathcal{Z}$ of $n+m$ polygonal chains. Refer to Figure 5. Let $V = \{0, 1, \ldots, n-1\}$. Let $V_n$ be a regular $n$-gon centered at the origin, with vertices $v_i = (\cos \frac{i \cdot 2\pi}{n}, \sin \frac{i \cdot 2\pi}{n})$, $i = 0, 1, \ldots, n-1$. Let $W_n$ be another regular $n$-gon centered at the origin, with vertices $w_i = (4\cos \frac{i \cdot 2\pi}{n}, 4\sin \frac{i \cdot 2\pi}{n})$, $i = 0, 1, \ldots, n-1$. Let $U_n = u_0 u_1 \ldots u_{n-1}$ be a regular $n$-gon inscribed in $V_n$ such that the vertices of $U_n$ are midpoints of the edges of $V_n$, that is, $u_i = \frac{1}{2}(v_i + v_{i+1 \bmod n})$, $i = 0, 1, \ldots, n-1$. The set $\mathcal{Z}$ includes $n$ polygonal chains that degenerate into the $n$ points $u_i$, $i = 0, 1, \ldots, n-1$, and $m$ polygon chains that represent the $m$ edges in $E$, where each edge $\{i, j\}$ in $E$, $0 \leq i < j \leq n-1$, is represented by the polygonal chain $v_i w_i \ldots w_j v_j$ in $\mathcal{Z}$. The following lemma establishes the reduction:

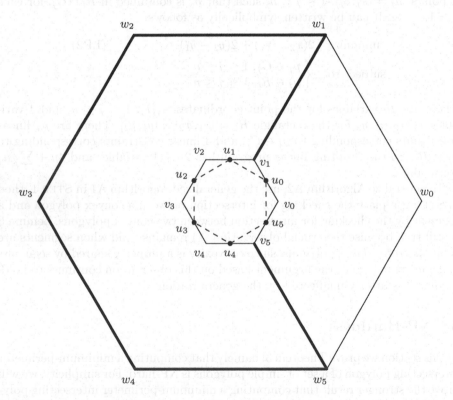

**Fig. 5.** Reduction from vertex cover. For $n = 6$, $V_6 = v_0 v_1 v_2 v_3 v_4 v_5$ and $W_6 = w_0 w_1 w_2 w_3 w_4 w_5$ are two regular hexagons centered at the origin. The vertices of the dashed hexagon $U_6 = u_0 u_1 u_2 u_3 u_4 u_5$ are midpoints of the edges of $V_6$. The edge $\{1, 5\}$ is represented by the polygonal chain $v_1 w_1 w_2 w_3 w_4 w_5 v_5$.

**Lemma 5.** *There is a vertex cover $S$ of $k$ vertices for $G$ if and only if there is an intersecting polygon $P$ of perimeter at most $2n \sin \frac{\pi}{n} \cos \frac{\pi}{n} + 2k \sin \frac{\pi}{n}(1 - \cos \frac{\pi}{n})$ for $\mathcal{Z}$.*

*Proof.* We first prove the direction implication. Suppose there is a vertex cover $S$ of $k$ vertices for $G$. We will find an intersecting polygon $P$ of perimeter at most $2n \sin \frac{\pi}{n} \cos \frac{\pi}{n} + 2k \sin \frac{\pi}{n}(1 - \cos \frac{\pi}{n})$ for $\mathcal{Z}$. For each vertex $i$ in $S$, select the corresponding vertex $v_i$ of $V_n$. Let $P$ be the convex hull of the $k$ selected vertices of $V_n$ and the $n$ vertices of $U_n$. Then $P$ contains at least one point from each polygonal chain in $\mathcal{Z}$. An easy trigonometric calculation shows that the perimeter of $P$ is exactly

$$(n - k) \cdot 2 \cos \frac{\pi}{n} \sin \frac{\pi}{n} + k \cdot 2 \sin \frac{\pi}{n} = 2n \sin \frac{\pi}{n} \cos \frac{\pi}{n} + 2k \sin \frac{\pi}{n} \left(1 - \cos \frac{\pi}{n}\right). \quad (5)$$

We next prove the reverse implication. Suppose there is an intersecting polygon $P$ of perimeter at most $2n \sin \frac{\pi}{n} \cos \frac{\pi}{n} + 2k \sin \frac{\pi}{n}(1 - \cos \frac{\pi}{n})$ for $\mathcal{Z}$. We will find a vertex cover $S$ of $k$ vertices for $G$. Assume without loss generality that $P$ is a minimum-perimeter intersecting polygon of $\mathcal{Z}$. Then $P$ must be convex because otherwise the convex hull of $P$ would be an intersecting polygon of even smaller perimeter. Since the $n$ vertices of $U_n$ are included in $\mathcal{Z}$ as $n$ degenerate polygonal chains, the convex polygon $P$ must contain the regular $n$-gon $U_n$. It follows that $P$ contains the origin. For $n \geq 3$, we have

$$\mathrm{perim}(P) \leq 2n \sin \frac{\pi}{n} \cos \frac{\pi}{n} + 2k \sin \frac{\pi}{n} \left(1 - \cos \frac{\pi}{n}\right)$$
$$\leq 2n \sin \frac{\pi}{n} \cos \frac{\pi}{n} + 2n \sin \frac{\pi}{n} \left(1 - \cos \frac{\pi}{n}\right)$$
$$= 2n \sin \frac{\pi}{n}$$
$$\leq 2\pi.$$

The distance from the origin to each edge of the regular $n$-gon $W_n$ is $4 \cos \frac{\pi}{n}$. For $n \geq 5$, this distance is greater than $\pi$:

$$4 \cos \frac{\pi}{n} \geq 4 \cos \frac{\pi}{5} = 3.236 \ldots > \pi.$$

Thus $P$ cannot intersect the boundary of $W_n$, although it may intersect some segments $v_i w_i$. For each segment $v_i w_i$ that $P$ intersects, the vertex $v_i$ must be contained in $P$. This is because $P$ is convex and contains the origin, while each segment $v_i w_i$ is on a line through the origin. Since $P$ is a minimum-perimeter intersecting polygon, it must be the convex hull of some vertices of $V_n$ and the $n$ vertices of $U_n$. Then the same calculation as in (5) shows that $P$ contains at most $k$ vertices of $V_n$, which, by construction, correspond to a vertex cover of at most $k$ vertices of $V$. By adding more vertices as necessary we obtain a vertex cover $S$ of exactly $k$ vertices for $G$. □

The polygonal chains in $\mathcal{Z}$ can be slightly fattened into simple polygons with rational coordinates. Then Lemma 5 still holds with some slight change in the threshold value $2n \sin \frac{\pi}{n} \cos \frac{\pi}{n} + 2k \sin \frac{\pi}{n}(1 - \cos \frac{\pi}{n})$. Note that the multiplicative coefficient of $k$ in this threshold value is

$$2 \sin \frac{\pi}{n}\left(1 - \cos \frac{\pi}{n}\right) = 2 \sin \frac{\pi}{n} \cdot 2 \sin^2 \frac{\pi}{2n} = \Theta\left(\frac{1}{n^3}\right).$$

The reduction can clearly be made polynomial. This completes the proof of Theorem 3.

## 6    Concluding Remarks

The problem MPIP is related to two other geometric optimization problems called largest and smallest convex hulls for imprecise points [10]. Given a set $\mathcal{R}$ of $n$ regions that model $n$ imprecise points in the plane, the problem largest (resp. smallest) convex hull is that of selecting one point from each region such that the convex hull of the resulting set $P$ of $n$ points is the largest (resp. smallest) with respect to area or perimeter. Note that MPIP is equivalent to smallest-perimeter convex hull for imprecise points as segments. The dual problem of largest convex hull for imprecise points as segments has been recently shown to be NP-hard [10] for both area and perimeter measures, and to admit a PTAS [9] for the area measure. We note that the core-set technique used in obtaining the PTAS for largest-area convex hull [9] cannot be used for MPIP because, for minimization, there could be many optimal or near-optimal solutions that are far from each other. For example, consider two long parallel segments that are very close to each other. Our Algorithm A2 overcomes this difficulty by Lemma 3.

The problem of computing a minimum-perimeter intersecting polygon of a set of simple polygons is also related to the traveling salesman problem with neighborhoods (TSPN), see [3,6,5]. Given a set $\mathcal{R}$ of $n$ regions (neighborhoods) in the plane, the problem TSPN is that of finding the shortest tour that visits at least one point from each neighborhood. Note that the shortest tour is the boundary of a (possibly degenerate) simple polygon but this polygon need not be convex. Although established independently, we noticed afterwards a certain similarity between our NP-hardness reduction and the APX-hardness reduction for TSPN in [5]. However, the convexity of an optimal solution, in our case, seemed to limit this similarity and prohibit us in obtaining a similar inapproximability result for MPIP with arbitrary polygons (or chains).

We conclude with two open questions:

1. Is the problem of finding a minimum-perimeter intersecting polygon of a set of segments NP-hard?
2. Does the problem of finding a minimum-perimeter intersecting polygon of a set of simple polygons admit a PTAS or a constant factor approximation algorithm?

*Acknowledgment.* The authors would like to thank the anonymous reviewers for pertinent comments.

# References

1. Andrews, G.E.: An asymptotic expression for the number of solutions of a general class of Diophantine equations. Transactions of the American Mathematical Society 99, 272–277 (1961)
2. Andrews, G.E.: A lower bound for the volume of strictly convex bodies with many boundary lattice points. Transactions of the American Mathematical Society 106, 270–279 (1963)
3. Arkin, E.M., Hassin, R.: Approximation algorithms for the geometric covering salesman problem. Discrete Applied Mathematics 55, 197–218 (1994)
4. Bárány, I., Pach, J.: On the number of convex lattice polygons. Combinatorics, Probability & Computing 1, 295–302 (1992)
5. de Berg, M., Gudmundsson, J., Katz, M.J., Levcopoulos, C., Overmars, M.H., van der Stappen, A.F.: TSP with neighborhoods of varying size. Journal of Algorithms 57, 22–36 (2005)
6. Dumitrescu, A., Mitchell, J.S.B.: Approximation algorithms for TSP with neighborhoods in the plane. Journal of Algorithms 48, 135–159 (2003)
7. Garey, M.R., Johnson, D.S.: Computers and Intractability: A Guide to the Theory of NP-Completeness. W.H. Freeman and Company, New York (1979)
8. Hassanzadeh, F., Rappaport, D.: Approximation algorithms for finding a minimum perimeter polygon intersecting a set of line segments. In: Dehne, F., Gavrilova, M.L., Sack, J.-R., Tóth, C.D. (eds.) WADS 2009. LNCS, vol. 5664, pp. 363–374. Springer, Heidelberg (2009)
9. van Kreveld, M., Löffler, M.: Approximating largest convex hulls for imprecise points. Journal of Discrete Algorithms 6, 583–594 (2008)
10. Löffler, M., van Kreveld, M.: Largest and smallest convex hulls for imprecise points. Algorithmica (2008), doi:10.1007/s00453-008-9174-2
11. Rappaport, D.: Minimum polygon transversals of line segments. International Journal of Computational Geometry and Applications 5(3), 243–265 (1995)
12. Welzl, E.: The smallest rectangle enclosing a closed curve of length $\pi$ (1993) (manuscript), http://www.inf.ethz.ch/personal/emo/SmallPieces.html
13. Wenger, R.: Helly-type theorems and geometric transversals. In: Handbook of Discrete and Computational Geometry, 2nd edn., pp. 73–96. CRC Press, Boca Raton (2004)
14. Yaglom, I.M., Boltyanski, V.G.: Convex Figures. Holt, Rinehart and Winston, New York (1961)

# Finding the Smallest Gap between Sums of Square Roots*

Qi Cheng and Yu-Hsin Li

School of Computer Science
The University of Oklahoma
Norman, OK 73019, USA
{qcheng,yli}@cs.ou.edu

**Abstract.** Let $k$ and $n$ be positive integers, $n > k$. Define $r(n, k)$ to be the minimum positive value of

$$|\sqrt{a_1} + \cdots + \sqrt{a_k} - \sqrt{b_1} - \cdots - \sqrt{b_k}|$$

where $a_1, a_2, \cdots, a_k, b_1, b_2, \cdots, b_k$ are positive integers no larger than $n$. It is important to find a tight bound for $r(n, k)$, in connection to the sum-of-square-roots problem, a famous open problem in computational geometry. The current best lower bound and upper bound are far apart. In this paper, we present an algorithm to find $r(n, k)$ *exactly* in $n^{k+o(k)}$ time and in $n^{\lceil k/2 \rceil + o(k)}$ space. As an example, we are able to compute $r(100, 7)$ exactly in a few hours on one PC. The numerical data indicate that the known upper bound seems closer to the truth value of $r(n, k)$.

## 1   Introduction

In computational geometry, one often needs to compare lengths of two polygonal paths, whose nodes are on an integral lattice, and whose edges are measured according to the Euclidean norm. The geometrical question can be reduced to a numerical problem of comparing two sums of square roots of integers. In computational geometry one sometimes assumes a model of real-number machines, where one memory cell can hold one real number. It is then assumed that an algebraic operation, taking a square root as well as a comparison between real numbers can be done in one operation. There is a straight-forward way to compare sums of square roots in real-number machines. But this model is not realistic, as shown in [11,9].

If we consider the problem in the model of the Turing machine, then we need to design an algorithm to compare two sums of square roots of integers with low bit complexity. One approach would be approximating the sums by decimal numbers up to a certain precision, and then hopefully we can learn which one is larger. Formally define $r(n, k)$ to be the minimum positive value of

$$|\sqrt{a_1} + \cdots + \sqrt{a_k} - \sqrt{b_1} - \cdots - \sqrt{b_k}|$$

* This research is partially supported by NSF grant CCF-0830522 and CCF-0830524.

A. López-Ortiz (Ed.): LATIN 2010, LNCS 6034, pp. 446–455, 2010.

where $a_1, a_2, \cdots, a_k, b_1, b_2, \cdots, b_k$ are positive integers no larger than $n$. The time complexity of the approximation approach is polynomial on $-\log r(n, k)$, since an approximation of a sum of square roots of integers can be computed in time polynomial in the number of precisions. One would like to know if $-\log r(n, k)$ is bounded from above by a polynomial function in $k$ and $\log n$. If so, the approximation approach to compare two sums of square roots of integers runs in polynomial time. Note that even if the lower bound of $-\log r(n, k)$ is exponential, it does not necessarily rule out a polynomial time algorithm.

Although this problem was put forward during the 1980s [4], not many results have been reported. In [3], it is proved that

$$-\log r(n, k) = O(2^{2k} \log n)$$

using the root separation method. Qian and Wang [8] presented a constructive upper bound for $r(n, k)$ at $O(n^{-2k+\frac{3}{2}})$. The constant hidden in the big-O can be derived from their paper and it depends on $k$. Taking it into account, one can show that

$$-\log r(n, k) \geq 2k \log n - 8k^2 + O(\log n + k \log k).$$

See Section 2 for details. Hence the bound is nontrivial only when $n \geq 2^{4k}$. There is another upper bound for $r(n, k)$ using solutions for the Prouhet–Tarry–Escott problem [8]. However, the Prouhet–Tarry–Escott problem is hard to solve by itself.

There is a wide gap between the known upper bound and lower bound of $r(n, k)$. For example by the root separation method, one has

$$r(100, 7) \geq (14 * \sqrt{100})^{-2^{13}} \approx 10^{-17581}.$$

One can not derive a nontrivial upper bound for $r(100, 7)$ either from Qian and Wang's method, or from the Prouhet–Tarry–Escott method.

## 1.1  Our Contribution

The lack of strong bounds for $r(n, k)$ after many years of study indicates that finding a tight bound is likely to be very hard. We feel that the situation calls for an extensive numerical study of $r(n, k)$. So far only a few toy examples have been reported and they can be found easily using an exhaustive search:

$$r(20, 2) \approx .0002 = \sqrt{10} + \sqrt{11} - \sqrt{5} - \sqrt{18}.$$

$$r(20, 3) \approx .000005 = \sqrt{5} + \sqrt{6} + \sqrt{18} - \sqrt{4} - \sqrt{12} - \sqrt{12}.$$

Computing power has gradually increased which makes it feasible for us to go beyond toy examples. In addition, there are other motivations for a numerical study of the sum-of-square-roots problem:

1. The numerical data shed light on the type of integers whose square roots summations are extremely close.
2. In many practical situations, especially in the exact geometric computation, $n$ and $k$ are small. Explicit bounds like one we produce here help to speed up the computation, as they are better than the bounds predicted by the root separation method.
3. Since the upper bound is so far away from the lower bound, the numerical data may provide us some hints on which bound is closer to the truth and may inspire us to formulate a reasonable conjecture on a tight bound of $r(n, k)$.

How can we find the exact value of $r(n, k)$? The naive exhaustive search uses little space but requires $n^{2k}$ time. If $n = 100$ and $k = 7$, the algorithm needs about $100^{14} \approx 2^{93}$ operations, which is prohibitive. A better approach would be first sorting all the summations of $\sqrt{a_1} + \cdots + \sqrt{a_k}$ ( $1 \leq a_i \leq n$ for all $1 \leq i \leq k$) and then going through the sorted list to find the smallest gap between two consecutive elements. It runs in time at least $n^k$ and in space at least $n^k$. If $n = 100$ and $k = 7$, then the approach would use at least $100^7 = 10^{14} \approx 10000$ Gbytes of space, under an overly optimistic assumption that we use only one byte to hold one value of the summation. The space complexity makes the computation of $r(100, 7)$ very expensive, to say the least.

We present an algorithm to compute $r(n, k)$ exactly based on the idea of enumerating summations using heap. Our algorithm uses much less space than the sorting approach while preserving the time complexity, which makes computing $r(100, 7)$ feasible. Indeed it has the space complexity at most of $n^{\lceil k/2 \rceil + o(k)}$. Our search reveals that

$$r(100, 7) = 1.88 \times 10^{-19},$$

which is reached by

$$\sqrt{7} + \sqrt{14} + \sqrt{39} + \sqrt{70} + \sqrt{72} + \sqrt{76} + \sqrt{85} = 47.42163068019049036900034846$$

and

$$\sqrt{13} + \sqrt{16} + \sqrt{46} + \sqrt{55} + \sqrt{67} + \sqrt{73} + \sqrt{79} = 47.42163068019049036881196876.$$

We also prove a simple lower bound for $-\log r(n, k)$ based on a pigeonhole argument:

$$-\log r(n, k) \geq k \log n - k \log k + O(k + \log n).$$

In comparison to Qian–Wang's bound, it is weaker when $n$ is very large, but it is better when $n$ is polynomial on $k$, hence it has wider applicability. For example, when $n = 100$ and $r = 7$, it can give us a meaningful upper bound:

$$r(100, 7) \leq 7.2 \times 10^{-8}.$$

## 1.2   Related Work

The use of heaps to enumerate sums in a sorted order appeared quite early [6, Section 5.2.3]. Let $P$ be a sorted list of $p$ real numbers whose $i$-th element is denoted by $P[i]$. Let $Q$ be another sorted list of $q$ real numbers whose $i$-th element is denoted by $Q[i]$. Consider the following way of enumerating elements of form $P[i] + Q[j]$ in a sorted order:

## Algorithm 1

*Build a heap for $P[i] + Q[1]$, $1 \leq i \leq p$;*
*while the heap is not empty do*
    *Remove the element $P[i] + Q[j]$ at the root from the heap*
    *if $j < q$*
        *then put $P[i] + Q[j + 1]$ at the root of the heap*
    *endif*
    *reheapify.*
*endwhile*

Note that for the program to work, one needs to keep track of the indexes $i$ and $j$ for the summation $P[i] + Q[j]$. The algorithm uses space to store $p + q$ elements but produces a stream of $pq$ elements in a sorted order. Schroeppel and Shamir [10] applied this idea to attack cryptosystems based on knapsack. Number theorists have been using this idea as a space-saving mechanism to test difficult conjectures on computers. For example, consider the following Diophantine equation:

$$a^4 + b^4 + c^4 = d^4.$$

Euler conjectured that the equation had no positive integer solutions. It was falsified with a explicit counterexample by Elkies [5] using the theory of elliptic curves with help from a computer search. Bernstein [1] was able to find all the solutions with $d \leq 2.1 \times 10^7$. His idea was to build two streams of sorted integers, one for $a^4 + b^4$ and another one for $d^4 \quad c^4$, and then look for collisions. To find solutions with $d \leq H$, the algorithm needs only $H^{1+o(1)}$ space and runs in time $H^{2+o(1)}$. A similar idea can be used to find integers which can be written in many ways as summations of certain powers. Our approach is inspired by this work. Essentially we use heap to enumerate all the summations of form $\sum_{i=1}^{k} \sqrt{a_i}$ ( $1 \leq a_i \leq n$ for all $1 \leq i \leq k$ ) and try to find the smallest gap between two consecutive elements. In our case, equality (i.e. gap $= 0$) is not interesting in the view of Proposition 1, while in the power summation applications, only equality (collision) is desired. There are other important differences:

- In the power sum case it will deal with only integers, while in our case, we have to deal with float-point numbers. The precision of real numbers plays an important role. Sometimes two equal sums of square roots can result in different float point numbers. For example, using `double double` type to represent real numbers, the evaluation of

$$(\sqrt{1} + \sqrt{8} + \sqrt{8}) + (\sqrt{24} + \sqrt{83} + \sqrt{83} + \sqrt{89})$$

differs from the evaluation of

$$(\sqrt{1} + \sqrt{6} + \sqrt{6}) + (\sqrt{32} + \sqrt{83} + \sqrt{83} + \sqrt{89})$$

by about $8 \times 10^{-28}$, even though they are clearly equal to each other.
- In the power sum case, $p$-adic restriction can often be applied to speed up the search, while unfortunately we do not have it here.

## 2 An Upper Bound from the Pigeonhole Principle

Qian-Wang's upper bound was derived from the inequality:

$$0 < \left| \sum_{i=0}^{2k-1} \binom{2k-1}{i} (-1)^i \sqrt{t+i} \right| \leq \frac{1 * 3 * 5 * \cdots * (4k-5)}{2^{2k-1} t^{2k - \frac{3}{2}}}.$$

Let $a_i = \binom{2k-1}{2i-2}^2 (t + 2i - 2)$ for $1 \leq i \leq k$ and $b_i = \binom{2k-1}{2i-1}^2 (t + 2i - 1)$ for $1 \leq i \leq k$, we have

$$0 < \left| \sum_{i=1}^{k} \sqrt{a_i} - \sum_{i=1}^{k} \sqrt{b_i} \right| \leq \frac{1 * 3 * 5 * \cdots * (4k-5)}{2^{2k-1} t^{2k - \frac{3}{2}}}.$$

Note that $\binom{2k-1}{i}$ can be as large as $\binom{2k-1}{k} \geq 2^{2k-1}/(2k)$. To get an upper bound for $r(n,k)$, assign

$$n = \binom{2k-1}{k}^2 (t + k), \tag{1}$$

thus we have

$$- \log r(n,k) \geq 2k \log n - 8k^2 + O(\log n + k \log k).$$

Hence Qian and Wang's result only applies when $n$ is much greater than $2^{4k}$. In particular it does not give a meaningful bound for $r(100, 7)$.

Another interesting upper bound depends on the Prouhet–Tarry–Escott problem, which is to find a solution for a system of equations:

$$\sum_{i=1}^{k} a_i^t = \sum_{i=1}^{k} b_i^t, 1 \leq t \leq k - 1$$

under the condition that $a_1 \leq a_2 \cdots \leq a_k$ and $b_1 \leq b_2 \cdots \leq b_k$ are distinct lists of integers. However no such solutions have been found for $k = 11$ and $k > 13$ [2]. Therefore the approach based on the Prouhet–Tarry–Escott problem is not scalable.

Here we present an upper bound based on the pigeonhole argument.

**Definition 1.** *We call an integer $n$ square-free if there is no integer $a > 1$ such that $a^2 | n$. We use $s(n)$ to denote the number of positive square free integers less than $n$, e.g. $s(100) = 61$.*

**Proposition 1.** *Suppose that $s_1, s_2, \cdots, \cdots$, and $s_k$ are distinct positive square-free integers. Then $\sqrt{s_1}, \sqrt{s_2}, \cdots$, and $\sqrt{s_k}$ are linear independent over $\mathbf{Q}$.*

**Theorem 1.** *We have*

$$r(n,k) \leq \frac{k\sqrt{n} - k}{\binom{s(n)+k-1}{k} - 1}.$$

*Proof.* Consider the set

$$\{(a_1, a_2, \cdots, a_k) | a_i \text{ is squarefree }, 1 \leq a_1 \leq a_2 \leq \cdots \leq a_k \leq n.$$

The set has cardinality $\binom{s(n)+k-1}{k}$. For each element $(a_1, a_2, \cdots, a_k)$ in the set, the sum $\sum_{i=1}^{k} \sqrt{a_i}$ is distinct by Proposition 1. Hence there are $\binom{s(n)+k-1}{k}$ many distinct sums in the range $[k, k\sqrt{n}]$. There must be two points within the distance $\frac{k\sqrt{n}-k}{\binom{s(n)+k-1}{k}-1}$ from each other. The theorem follows. ∎

Plugging in $n = 100$ and $k = 7$, we have

$$r(100, 7) \leq \frac{(70 - 7)}{\binom{67}{61} - 1} = 7.2 \times 10^{-8}.$$

It is well known that $s(n) = \frac{6n}{\pi^2} + O(\sqrt{n})$ [7]. From this one can derive

**Corollary 1.**

$$-\log r(n, k) \geq k \log n - k \log k + O(\log(nk))$$

Note that when $n$ is much larger than $k$, then this bound is not as good as Qian and Wang's bound.

## 3  Algorithm for Finding $r(n, k)$

We first sketch the algorithm. It takes two positive integers $n$ and $k$ as input. Assume that $k < n$.

**Algorithm 2.** *Input: Two positive integers $n, k$ $(n > k)$.*

*Store all the lists $(a_1, a_2, \ldots, a_A)$, where $1 \leq a_1 \leq a_2 \leq \cdots \leq a_A \leq n$, into an array $P$, and then sort the array $P$ according to the sum $\sum_{i=1}^{A} \sqrt{a_i}$. Assume that there are $p$ elements in the list;*

*Store all the lists $(a_1, a_2, \ldots, a_{k-A})$, where $1 \le a_1 \le a_2 \le \cdots \le a_{k-A} \le n$,*
*into an array $Q$, and then sort the array $Q$ according to the sum*
$\sum_{i=1}^{k-A} \sqrt{a_i}$. *Assume that there are $q$ many elements in $Q$;*
*current_small_gap $= \infty$;*
*previous_smallest_element $= k$;*
*Build a heap for $(P[i], Q[1])$, $1 \le i \le p$, where two lists are compared according*
*to the sum of square roots of the integers in the lists;*
*While the heap is not empty do*
*Let $(P[i], Q[j])$ be the element at the root of the heap;*
*current_top_element $= \sum_{l=1}^{A} \sqrt{P[i][l]} + \sum_{l=1}^{k-A} \sqrt{Q[j][l]}$*
*if $0 <$ current_top_element $-$ previous_top_element $<$ current_small_gap*
*then current_small_gap $=$ current_top_element$-$ previous_top_element;*
*endif*
*remove $(P[i], Q[j])$ from the heap;*
*previous_top_element $= (P[i], Q[j])$;*
*if there exist integers $j'$ such that $j < j' \le q$ and $P[i][A] \le Q[j'][1]$*
*let $j'$ be the smallest one and put $(P[i], Q[j'])$ at the root*
*endif*
*reheapify*
*endwhile*
*Ouput $r(n, k) =$ current_small_gap*

Note that in the above algorithm, unlike in Algorithm 1, we replace $(P[i], Q[j])$
at the root by $(P[i], Q[j'])$, which is not necessarily $(P[i], Q[j+1])$. In many
cases, $j'$ is much bigger than $j+1$. This greatly improves the efficiency of the
algorithm. Now we prove the correctness of the algorithm.

**Theorem 2.** *When the algorithm halts, it outputs $r(n, k)$;*

*Proof.* For any $1 \le a_1 \le a_2 \cdots \le a_A \le n$, define

$$S_{a_1, a_2, \cdots, a_A} = \{(a_1, a_2, \cdots, a_k) | a_A \le a_{A+1} \le a_{A+2} \le \cdots \le a_k \le n\}$$

Partition the set

$$S = \{(a_1, a_2, \cdots, a_k) | 1 \le a_1 \le a_2 \le \cdots \le a_k \le n\}$$

into subsets according to the first $A$ elements, namely,

$$S = \bigcup_{1 \le a_1 \le a_2 \le \cdots a_A \le n} S_{a_1, a_2, \cdots, a_A}.$$

As usual, we order two lists of integers by their sums of square roots. Consider
the following procedure: select the smallest element among all the the minimum
elements in all the subsets, and remove it from the subset. If we repeat the
procedure, we generate a stream of elements in $S$ in a sorted order.

It can be verified that in our algorithm, the heap consists of exactly all the
minimum elements from all the subsets. The root of the heap contains the minimum element of the heap. After we remove the element at the root, we put

the next element from its subset into the heap. Hence the algorithm produces a stream of elements from $S$ in a sorted order. The minimum gap between two consecutive elements in the stream is $r(n, k)$ by definition.

**Theorem 3.** *The algorithm runs in time at most* $n^{k+o(k)}$ *and space at most* $n^{\max(A, k-A)+o(k)}$.

*Proof.* Using the root separation bound, we need at most $O(2^{2k} \log n)$ bit to represent a sum of square roots for comparison purposes. So comparing two elements takes time $(2^{2k} \log n)^{O(1)}$. Since every element in $S$ appears at the root of the heap at most once and $|S| \leq n^k$, the main loop has at most $n^k$ iterations. For each iteration, the time complexity is

$$(2^{2k} \log n)^{O(1)} \log(n^A).$$

The complexity of other steps are much smaller comparing to the loop. Hence the time complexity is $n^{k+o(k)}$. The space complexity is clearly $n^{\max(A, k-A)+o(k)}$.

## 4   Numerical Data and Observations

To implement our algorithm, the main issue is to decide the precision when computing the square roots and their summations. We need to pay attention to two possibilities:

- First, two summations may be different, but if the precision is set too small, then they appear to be equal numerically. Keep in mind that we have not ruled out that $r(n, k)$ can be as small as $n^{-2^k}$.
- Secondly two expressions may represent the same real number, but after the numerical calculation, they are different. This is the issue of numerical stability.

In either case, we may get a wrong $r(n, k)$. Our strategy is to set the precision at about $2k \log n$ decimal digits. For example, to compute $r(100, 7)$, we use the data type which has precision about 32 decimal digits. Whenever the difference of two summations is smaller than $k^2 n^{-2k}$, we call a procedure based on Proposition 1 to decide whether the two numbers are equal or not.

We produce some statistics data about the sums of square roots and the gaps between two consecutive sums. The computation takes about 18 hours on a high-end PC. There are 17940390852 real numbers in $[7, 100]$ which can be written as summations of 7 square roots of positive integers less than 100. Hence there are 17940390851 gaps between two consecutive numbers after we sort all the sums.

In Table 1, we list an integer $7 \leq a \leq 70$ with the number of reals in $[a, a+1)$ which can be represented as $\sqrt{a_1} + \sqrt{a_2} + \cdots + \sqrt{a_7}$ ( $1 \leq a_1 \leq a_2 \cdots \leq a_7 \leq 100$ ). Note that if two summations have the same value, they are counted only once. From the table, we see that there are 1163570911 sums in the $[48, 49)$, which gives us a more precise pigeonhole upper bound for $r(n, k)$ at $1/1163570911 = 8.6 \times 10^{-10}$, which is still several magnitudes away from $r(n, k)$.

In Table 2, for each range, we list the number of gaps between consecutive numbers in the range. From the table, we see that there are 7 gaps which have magnitude at $10^{-19}$.

**Table 1.** Statistics on the summations of square roots

| 7 | 8 | 9 | 10 | 11 | 12 |
|---|---|---|---|---|---|
| 4 | 17 | 57 | 161 | 418 | 1003 |
| **13** | **14** | **15** | **16** | **17** | **18** |
| 2259 | 4865 | 10044 | 20061 | 38742 | 72903 |
| **19** | **20** | **21** | **22** | **23** | **24** |
| 133706 | 239593 | 420279 | 722739 | 1218852 | 2017818 |
| **25** | **26** | **27** | **28** | **29** | **30** |
| 3280805 | 5239096 | 8218857 | 12664315 | 19165803 | 28482325 |
| **31** | **32** | **33** | **34** | **35** | **36** |
| 41554376 | 59503519 | 83607939 | 115241837 | 155784865 | 206478894 |
| **37** | **38** | **39** | **40** | **41** | **42** |
| 268254403 | 341520055 | 425961992 | 520334126 | 622307266 | 728445926 |
| **43** | **44** | **45** | **46** | **47** | **48** |
| 834229563 | 934295227 | 1022797808 | 1093860379 | 1142175328 | 1163570911 |
| **49** | **50** | **51** | **52** | **53** | **54** |
| 1155526520 | 1117588507 | 1051539385 | 961294902 | 852549403 | 732208073 |
| **55** | **56** | **57** | **58** | **59** | **60** |
| 607649679 | 486014737 | 373475729 | 274666260 | 192383944 | 127511613 |
| **61** | **62** | **63** | **64** | **65** | **66** |
| 79264404 | 45637971 | 23914891 | 11119037 | 4410314 | 1398655 |
| **67** | **68** | **69** | **70** | | |
| 316043 | 40172 | 1476 | 1 | | |

**Table 2.** Statistics about the gaps

| $10^{-19} \sim 10^{-18}$ | $10^{-18} \sim 10^{-17}$ | $10^{-17} \sim 10^{-16}$ | $10^{-16} \sim 10^{-15}$ | $10^{-15} \sim 10^{-14}$ |
|---|---|---|---|---|
| 7 | 47 | 1245 | 14139 | 129248 |
| $10^{-14} \sim 10^{-13}$ | $10^{-13} \sim 10^{-12}$ | $10^{-12} \sim 10^{-11}$ | $10^{-11} \sim 10^{-10}$ | $10^{-10} \sim 10^{-9}$ |
| 1459473 | 13100265 | 132767395 | 1272832428 | 8256755966 |
| $10^{-9} \sim 10^{-8}$ | $10^{-8} \sim 10^{-7}$ | $10^{-7} \sim 10^{-6}$ | $10^{-6} \sim 10^{-5}$ | $10^{-5} \sim 10^{-4}$ |
| 7766837445 | 463570895 | 30415764 | 2314151 | 176109 |
| $10^{-4} \sim 10^{-3}$ | $10^{-3} \sim 10^{-2}$ | $10^{-2} \sim 10^{-1}$ | $10^{-1} \sim 1$ | |
| 14890 | 1300 | 80 | 5 | |

## 5   Conclusion Remarks

In this paper we have proposed a space-efficient algorithm to compute $r(n, k)$ exactly. Our numerical data seem to suggest that the upper bound is closer to the truth than the root separation bounds. Further investigations, both experimental and theoretical, are needed.

# References

1. Bernstein, D.: Enumerating solutions to $p(a) + q(b) = r(c) + s(d)$. Math. of Comp. 70, 389–394 (2001)
2. Borwein, P.: Computational Excursions in Analysis and Number Theory. Springer, Heidelberg (2002)
3. Burnikel, C., Fleischer, R., Mehlhorn, K., Schirra, S.: A strong and easily computable separation bound for arithmetic expressions involving radicals. Algorithmica 27(1), 87–99 (2000)
4. Demaine, E.D., Mitchell, J.S.B., O'Rourke, J.: The open problems project: Problem 33, http://maven.smith.edu/~orourke/TOPP/
5. Elkies, N.: On $a^4 + b^4 + c^4 = d^4$. Math. of Comp. 51, 825–835 (1988)
6. Knuth, D.: The Art of Computer Programming, vol. 3. Addison-Wesley, Reading (1973)
7. Pappalardi, F.: A survey on k-power freeness. In: Proceeding of the Conference in Analytic Number Theory in Honor of Prof. Subbarao. Ramanujan Math. Soc. Lect. Notes Ser., vol. 1, pp. 71–88 (2002)
8. Qian, J., Wang, C.A.: How much precision is needed to compare two sums of square roots of integers? Inf. Process. Lett. 100(5), 194–198 (2006)
9. Schönhage, A.: On the power of random access machines. In: Maurer, H.A. (ed.) ICALP 1979. LNCS, vol. 71, pp. 520–529. Springer, Heidelberg (1979)
10. Schroeppel, R., Shamir, A.: A T = $o(2^{n/2})$, S = $o(2^{n/4})$ algorithm for certain NP-complete problems. SIAM journal on Computing 10(3), 456–464 (1981)
11. Shamir, A.: Factoring numbers in O(log n) arithmetic steps. Information Processing Letters 1, 28–31 (1979)

# Matching Points with Things

Greg Aloupis[1], Jean Cardinal[1], Sébastien Collette[1,*], Erik D. Demaine[2],
Martin L. Demaine[2], Muriel Dulieu[3], Ruy Fabila-Monroy[4], Vi Hart[5],
Ferran Hurtado[6], Stefan Langerman[1,**], Maria Saumell[6], Carlos Seara[6],
and Perouz Taslakian[1]

[1] Université Libre de Bruxelles, CP212, Bld. du Triomphe, 1050 Brussels, Belgium
{galoupis,jcardin,secollet,slanger,ptaslaki}@ulb.ac.be
Supported by the Communauté française de Belgique - ARC.
[2] MIT Computer Science and Artificial Intelligence Laboratory, 32 Vassar St.,
Cambridge, MA 02139, USA
{edemaine,mdemaine}@mit.edu
[3] Polytechnic Institute of NYU, USA
mdulieu@gmail.com
[4] Instituto de Matemáticas, Universidad Nacional Autónoma de México
ruy@ciencias.unam.mx
[5] Stony Brook University, Stony Brook, NY 11794, USA
vi@vihart.com
[6] Universitat Politècnica de Catalunya, Jordi Girona 1–3, E-08034 Barcelona, Spain
{ferran.hurtado,maria.saumell,carlos.seara}@upc.edu
Partially supported by projects MTM2009-07242 and Gen. Cat. DGR 2009SGR1040.

**Abstract.** Given an ordered set of points and an ordered set of geometric objects in the plane, we are interested in finding a non-crossing
matching between point-object pairs. We show that when the objects we
match the points to are finite point sets, the problem is NP-complete in
general, and polynomial when the objects are on a line or when their
number is at most 2. When the objects are line segments, we show that
the problem is NP-complete in general, and polynomial when the segments form a convex polygon or are all on a line. Finally, for objects
that are straight lines, we show that the problem of finding a min-max
non-crossing matching is NP-complete.

## 1 Introduction

Finding a matching between pairs of planar objects, that is connecting them by
a set of non-crossing line segments, is a natural problem that has been frequently
studied in computational geometry. It is well known, for instance, that given two
sets of $n$ points in the plane, say $n$ red points and $n$ blue points, there always
exists a non-crossing matching between red and blue points. In particular, it
is not difficult to show that the minimum Euclidean length matching is non-
crossing. Kaneko and Kano [22] survey a number of related results. Algorithms
for finding minimum sum and minimum bottleneck distance red-blue matchings
are given in [15,27].

---

* Chargé de Recherches du FRS-FNRS.
** Maître de Recherches du FRS-FNRS.

A. López-Ortiz (Ed.): LATIN 2010, LNCS 6034, pp. 456–467, 2010.
© Springer-Verlag Berlin Heidelberg 2010

In this paper, we investigate related questions for general planar objects instead of points. Again, matchings are represented by line segments, but here the endpoints can be placed anywhere inside the corresponding matched objects. Note that as a consequence of the aforementioned result on points, there always exists a non-crossing matching between two sets of objects. Here we consider the problem where we are given object *pairs* (i.e. a point and the geometric object it must be matched to) and need to find a set of non-crossing matching edges, if one exists. This can be seen as a 1-regular graph drawing problem with constraints on the location of vertices.

**Related work.** Problems on matchings have an important role in combinatorial graph theory, both for theoretical and applied aspects; hence a lot of research is devoted to the study of these problems (for example, see [24]).

Suppose we are given an embedding of a graph in the Euclidean plane, where the vertices are points in the plane, edges are rectilinear line segments, and weights on these edges represent the Euclidean distance between the vertices they connect. Elementary geometry tells us that the sum of any pair of opposite sides of a convex quadrilateral is strictly smaller than the sum of the diagonals. Remarkably, this implies that the minimum weight matching in any realization of the complete graphs $K_{2n}$ and $K_{n,n}$ consists of pairwise non-crossing segments. These geometric graph problems can be solved using generic algorithms for weighted graphs. However, in the planar case just mentioned, Vaidya [27] proved that it is possible to obtain specialized algorithms with better running times (the title of his paper is especially suggestive: *Geometry helps in matching*). In particular, in [27] the running time of the generic algorithm for the bipartite case was reduced from $O(n^3)$ to $O(n^{2.5} \log n)$. This was later improved to $O(n^{2+\varepsilon})$ by Agarwal et al. [1]. Similar results have been obtained for other matching variations, such as *bottleneck matching* or *uniform matching*, in the work of Efrat, Itai and Katz [15]. The authors consider matchings as an approach for the problem of matching a point set $A$ with a point set $B$, where $A$ must be moved in some way to coincide as much as possible with $B$ or one of its subsets. This is a fundamental problem in pattern recognition [5,7,8,10,11,12,19,20,21].

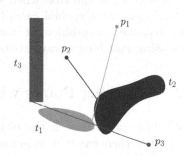

**Fig. 1.** A non-crossing matching for a set $P=\{p_1,p_2,p_3\}$ of points and a set $T=\{t_1,t_2,t_3\}$ of planar objects

The non-crossing requirement in our problems is quite natural in geometric scenarios (see for example [25,2,3]), and the family of geometric problems that we consider has several applications; these applications include geometric shape matching [4,13,17,18], colour-based image retrieval [13], music score matching [26], and computational biology [14,16].

**Our results.** Throughout the paper, we let $P := \{p_1, p_2, \ldots, p_n\}$ be a set of points in the plane and $T := \{t_1, t_2, \ldots, t_n\}$ be a set of planar objects. A *matching*

for a pair $(P, T)$ consists of a set of line segments, called *edges*, of the form $\{p_1m_1, p_2m_2, \ldots, p_nm_n\}$, where $m_i \in t_i$. A matching is said to be *non-crossing* if no pair of matching edges properly cross. This is illustrated in Figure 1.

We consider the problem of deciding whether a non-crossing matching exists for a given pair $(P, T)$. In cases where a non-crossing matching always exists, we consider the problem of finding the matching that minimizes either the length of the longest edge, or the sum of the lengths of all the edges.

In Section 2, we study the case where the objects $t_i$ are finite point sets. We prove that the decision problem is NP-complete in general, but becomes polynomial when every $t_i$ has size at most two, or when all the $t_i$s are on a line. In Section 3 we consider $T$ to be a set of line segments and prove that the $(P, T)$ matching problem is NP-complete. We also consider special cases, such as the case when the line segments form a convex polygon surrounding all points in $P$ (Section 4), or the case when segments belong to a single line (Section 5). We show that these special cases have polynomial solutions. Finally, in Section 6, we consider the problem of matching points with lines. In this variation, a non-crossing matching always exists, but the optimization problems are NP-hard.

## 2 Matching Points with Finite Point Sets

We first prove that if the objects $t_i$ are pairs of points, then we can decide whether there exists a non-crossing matching in polynomial time. On the other hand, if the sets $t_i$ may contain three points or more, the problem becomes NP-complete. This situation is similar to that of the $k$-satisfiability problem ($k$-SAT). In $k$-SAT we are given a boolean formula $f$ of the form $C_1 \wedge C_2 \wedge \cdots \wedge C_m$ (where each $C_i$ is an OR clause of $k$ variables), and we are required to find a truth assignment of its variables that satisfy the formula. It is well-known that 2-SAT has a polynomial-time solution whereas $k$-SAT is NP-complete for $k \geq 3$. The 2-SAT problem can be solved in polynomial time by exploiting the fact that, if in a clause a variable is set to false, it forces the other variable to be set to true. This dependency between the variables can be represented by an *implication graph*.

An implication graph for the formula $f$ is a directed graph having two vertices for each variable $x_i$ of $f$, one of these vertices is labeled $x_i$ while the other is labeled $\neg x_i$. The vertex $x_i$ represents setting $x_i$ to true while $\neg x_i$ represents setting $x_i$ to false. Dependencies between literals in $f$ are represented by directed edges. Thus if $(x_i \vee \neg x_j)$ is a clause in $f$, in the implication graph there would be a directed edge from $\neg x_i$ to $\neg x_j$ and a directed edge from $x_j$ to $x_i$. These edges represent the fact that if $x_i$ is set to false then $x_j$ must also be set to false in order for the formula to be satisfied. Likewise if $x_j$ is set to true then $x_i$ must be set to true.

There exists a truth assignment satisfying $f$ if and only if no strong component of the implication graph contains both a vertex and its negation. The implication graph can be constructed in $O(m)$ time, where $m$ is the number of clauses. The previous condition can be verified in $O(m)$ time, and in general the strong components of a directed graph of $v$ vertices and $e$ edges can be computed in $O(e+v)$ time. A similar implication graph can be constructed for our problem when $t_i$ is a pair of points. Using this graph, it is possible to decide in $O(n^2)$ time whether $(P, T)$ has a non-crossing perfect matching.

**Theorem 1.** *Given an ordered set $P$ of points and an ordered set $T$ of pairs of points, there is an algorithm that decides in $O(n^2)$ time whether $(P, T)$ has a non-crossing matching.*

*Proof.* Assume that the elements of each $t_i$ are labeled arbitrarily "$T_i$" and "$F_i$" (thus $t_i = \{T_i, F_i\}$). We think of each $p_i$ as a boolean variable, so that if we match $p_i$ with $T_i$ then $p_i$ is set to "true", and if $p_i$ is matched with $F_i$, it is set to "false". We construct a directed implication graph $G$ as follows: For each $p_i$ we have vertices $p_{i,T_i}$ and $p_{i,F_i}$ in $G$. For every $i, j = 1, 2, \ldots, n$, we add the edge $(p_{i,X}, p_{j,Y})$ ($X$ equal to $T_i$ or $F_i$, and $Y$ equal to $T_j$ or $F_j$) to $G$ if and only if the line segments $\overline{p_i, X}$ and $\overline{p_j, \neg Y}$ intersect. For example if $\overline{p_i, T_i}$ intersects $\overline{p_j, F_j}$, we add the edge $(p_{i,T_i}, p_{j,T_j})$ to $G$ (since if $p_i$ is matched to $T_i$, $p_j$ must be matched to $T_j$ as well). So $(P, T)$ has a non-crossing complete matching if and only if for every $p_i$, $p_{i,T_i}$ and $p_{j,F_j}$ lie in different strong connected components of $G$. Since $G$ is constructed in $O(n^2)$ time and has $O(n^2)$ edges, the overall complexity of the algorithm is $O(n^2)$. □

### 2.1 Matching Points with Triples

By a reduction from Planar 3-SAT, we can prove that the problem of matching points with triples is NP-complete, even when the points within each triple are horizontally collinear. Details are omitted due to space limitations.

**Theorem 2.** *Given an ordered set $P$ of points and an ordered set $T$ of triples of points, it is NP-complete to decide whether $(P, T)$ has a non-crossing matching. The problem remains NP-complete even if each triple is horizontally collinear.*

### 2.2 Matching Points with $k$-Tuples on a Line

**Theorem 3.** *Given an ordered set $P$ of points and an ordered set $T$ of $k$-tuples of points on a line, we can decide in $O(k^3 n^2)$ time whether $(P, T)$ has a non-crossing matching.*

*Proof.* Without loss of generality, assume all the tuples are on a horizontal line $L$. Assume also that all points are on one side of $L$; otherwise we may consider each problem separately as the matching edges on each side of $L$ do not interact. We now show how to build a dynamic programming table that solves the problem.

In any solution to the problem, if a matching edge $e$ is part of the solution, then there is no matching edge that intersects $e$. Therefore, we can consider the regions on each side of $e$ (sub-problems) separately and determine whether they in turn have a valid solution. Thus, a sub-problem $(P', T')$ is defined as follows (see Figure 2): given a simple quadrilateral $A$ with one face adjacent to $L$, we want to decide if it is possible to find a non-crossing matching completely contained in the region $A$ for all the points contained in $A$, i.e., we want to solve

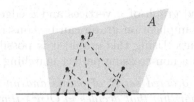

**Fig. 2.** Definition of a sub-problem

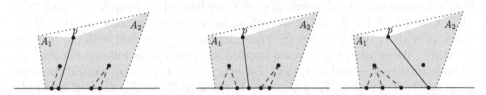

**Fig. 3.** In this example, there are three pairs of sub-problems to consider to decide if $p$ can be matched

the problem with $P' = P \cap A$ and $T'$ containing the subsets of the tuples of $T$ contained in $A$. If $A$ does not contain at least one point of $P$ (sub-problem of size 0), it is trivially true that there is a non-crossing matching. Otherwise, to solve the sub-problem we consider the topmost point $p$ in $A$. It has at most $k$ possible matching edges. If it has no possible matching edge, i.e., if all points that $p$ could be matched to in $T$ are out of $A$, then there is no valid matching.

Each of the possible matching edges defines two new independent sub-problems (see Figure 3) in the quadrilaterals $A_1$ and $A_2$, whose sizes are strictly smaller than that of the original problem, as there is one less point to match. To decide whether a matching exists for the original sets $P$ and $T$, we solve the sub-problem defined by the bounding box of both $P$ and $T$. Notice that all the sub-problems correspond to quadrilaterals defined by a pair of possible matching edges (or by the edges of the bounding box). Moreover, since in every sub-problem $A$ the $y$-coordinates of the corners of the bounding box are at least as large as that of every point in $A$, then the union of the regions of the sub-problems of $A$ will contain all the points in $A$. Thus no point will be left out.

The dynamic programming table has $kn + 2$ rows and $kn + 2$ columns, each of which corresponds to a possible matching edge or one of the left and right edges of the bounding box; the cells correspond to sub-problems (a pair of non-adjacent edges defines a quadrilateral), and we fill them with true or false depending on whether or not a matching exists for the considered sub-problem. Filling a cell of the table corresponds to solving at most $k$ pairs of sub-problems, which implies at most $2k$ lookups in the table for each of the $O(k^2 n^2)$ cells. Therefore, the total time and space required to solve the problem is $O(k^3 n^2)$.    □

The additional restriction of having points on a line greatly simplifies the problem, because the problem is NP-hard in the general case, but is polynomial for points on a line.

# 3   Matching Points with Line Segments: General Case

In this section we show that deciding the existence of a non-crossing matching between a set of points and a set of line segments is NP-complete, even if the segments are all horizontal or all have equal length. The proof uses appropriate gadgets to show that this problem reduces from the problem of matching points to triples (Theorem 2). It is omitted due to space limitations.

**Theorem 4.** *Given an ordered set $P$ of points and an ordered set $T$ of line segments, it is NP-complete to decide whether $(P, T)$ has a non-crossing matching. The problem remains NP-complete even if all line segments in $T$ are horizontal or all have equal length.*

# 4   Matching Points to an Enclosing Convex Polygon

In this special case of matching points with line segments, we assume the segments are the edges of a convex polygon and the points to be matched are inside the polygon, in general position.

We first describe some geometric properties of the input of this problem. We then describe an algorithm that runs in $O(n \log^2 n)$ time, and finds a non-crossing matching (if one exists) between a given set of point-segment pairs where the line segments form a convex polygon enclosing the points. This algorithm allows a minimum-length matching to be extracted easily.

Let $D^o = \{\Delta_1^o, \Delta_2^o, \ldots, \Delta_n^o\}$ be a set of triangles where each $\Delta_i^o$ is the triangle with apex $p_i$ and base $t_i$. Any valid matching edge $e_i$ must lie completely inside $\Delta_i^o$. Depending on the positions of other triangles in $D^o$, some candidate positions for $e_i$ can be identified as invalid because they would always cross other matching edges. By identifying such cases, triangle $\Delta_i^o$ can be reduced to a smaller triangle $\Delta_i$. At any time, the *reduced triangle* $\Delta_i$ has apex $p_i$ but its opposite base is a subsegment of $t_i$. Initially, $\Delta_i = \Delta_i^o$.

There are four ways in which two triangles $\Delta_i$ and $\Delta_j$ interact. The second case leads to a *reduction rule*. We describe the four cases below (see Figure 4):

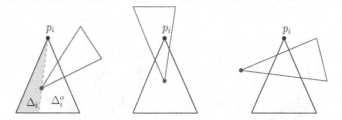

**Fig. 4.** Left: $\Delta_i^o$ is reduced to $\Delta_i$ (case 2). Middle: one of the two combinatorially distinct solutions (case 3). Right: no solution exists (case 4).

1. $\Delta_i, \Delta_j$ are disjoint. In this case there will never be a direct interaction between the two.
2. $p_j$ is in $\Delta_i$, but $p_i$ is not in $\Delta_j$. In this case $\Delta_i$ should be reduced so that the two triangles become tangent (so that $p_j$ is no longer in $\Delta_i$).
3. $p_i$ is in $\Delta_i$ and $p_j$ is in $\Delta_j$. We call $\Delta_i$ and $\Delta_j$ *inverted triangles*, and cannot immediately make a reduction.
4. Both edges incident to each of $p_i$ and $p_j$ pairwise intersect. Then no non-crossing matching exists.

Note that in case (2) there is no choice but to reduce. The matching edge $e_j$ that is finally chosen will block any candidate $e_i$ that is outside the newly reduced $\Delta_i$. In case (3) there are two combinatorially valid placements for $e_i, e_j$, with respect to the positions of $p_i, p_j$. There is no reason to choose arbitrarily before verifying that neither triangle will be reduced further.

If we exhaustively apply our reduction rule to the triangles based on the cases described above, we would end up with a set having certain properties. Due to lack of space, we omit a detailed discussion of these properties.

Let two (three) mutually inverted triangles be called an *inverted pair (triple)*. Let a *unit* be a (possibly reduced) triangle, an inverted pair, or an inverted triple.

**Theorem 5.** *Given an ordered set $P$ of points inside a convex polygon having an ordered set $T$ of line segments as edges, deciding whether $(P, T)$ admits a non-crossing matching can be done in $O(n \log^2 n)$ time.*

*Proof.* We provide an algorithm where we employ a divide-and-conquer technique. Suppose that we have solved the problem separately on two consecutive convex chains (we can transform a chain into a polygon by adding 3 fake edges and points; thus, solving the problem on a chain is equivalent to solving the polygonal version).

We claim that we can merge the two solutions in $O(n \log n)$ time. Each solution is a set of disjoint triangles and inverted pairs or triples. Refer to Figure 5.

Let $A$ and $B$ be two solved sub-problems of size $k$. We construct a standard point-location data structure on each in $O(k)$ time [23] by first triangulating $A$ and $B$ using Chazelle's linear-time triangulation algorithm [9]. Now, for every

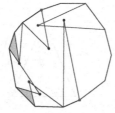 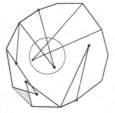

**Fig. 5.** Merging two solved sub-problems. In the left diagram, the grey regions in the left (black) sub-problem cannot contain points from the right (blue) sub-problem if there is a valid solution. In the right diagram, we see the type of event that we must check for after some initial reductions.

point $p_i$ in $B$, we locate $p_i$ in $A$ to determine if it is inside a unit in $A$. Note that $p_i$ can be in at most one unit. If it is, we determine if $\Delta_i$ reduces this unit by case (2). Likewise, for every point $p_j$ in $A$, we locate $p_j$ in $B$ to determine if it is inside a unit in $B$ and apply the appropriate reductions. Note that if at some moment $\Delta_i$ (belonging to $B$) gets reduced, this will not affect its corresponding unit in $A$; the same holds for all $\Delta_j$ in $A$ that get reduced.

Of course, it is possible that $\Delta_i$ will be inverted with a triangle in $A$. In this case we simply determine if there are reductions and, if applicable, we merge the two units. Therefore a constant number of reductions are applied per point, which means we spend $O(\log k)$ time per point for the point-location step.

The only unresolved issue is to detect if case (4) will occur between triangles of $A$ and $B$ (see Figure 5-right). Given that all triangles have been reduced and merged into units, essentially we are verifying that no segments intersect. For this we can use the Bentley-Ottmann line segment intersection algorithm and stop as soon as a bad intersection is found [6]. For $k$ segments, such queries take $O(\log k + h \log k)$ time, where $h$ is the number of intersections reported. As we stop as soon as we report one intersection, $h = 1$ and hence the total time is $O(\log k)$ per point. Thus our merge procedure takes $O(k \log k)$ time. By a simple recurrence analysis, we determine that the entire algorithm takes $O(n \log^2 n)$ time.                                                                    □

The algorithm described in the proof of Theorem 5 either decides that no solution exists, or otherwise produces a final set of reduced triangles that represents all valid solutions to the problem. In the latter case, every resulting unit is disjoint and thus independent of all others. So in each triangle we can easily pick the shortest joining segment, and in each inverted pair/triple, we try out the two possible choices and take the best matching. Therefore, after the algorithm finds a solution, the min-max and min-sum optimization problems can be solved in linear time.

## 5    Matching Points with Segments on a Line

As another special case of matching points to line segments, we now consider the case when the input line segments belong to one single line $L$. Throughout this section we will assume, without loss of generality, that $L$ is horizontal. As no matching edge will cross over $L$, our problem is split into two disjoint sub-problems, and we focus on points above $L$. We consider two cases, depending on whether the segments are disjoint or not.

### 5.1    Matching Points with Disjoint Segments on a Line

**Theorem 6.** *Given an ordered set $P$ of points above a horizontal line $L$ and an ordered set $T$ of disjoint line segments belonging to $L$, deciding whether $(P, T)$ admits a non-crossing matching can be done in linear time. In the affirmative, the matching that minimizes the sum of the lengths of the edges can be found within the same time bound.*

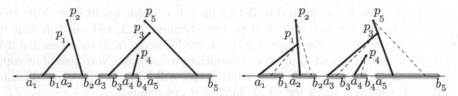

**Fig. 6.** Leftmost non-crossing matching (right) obtained from an initial non-crossing matching (left)

*Proof.* We denote by $[a_i, b_i]$ the interval corresponding to segment $t_i$, for $i = 1, ..., n$. Since the intervals are given in sorted order, we have $a_1 \leq b_1 < a_2 \leq b_2 < ... < a_n \leq b_n$.

If $(P, T)$ admits some non-crossing matching $\{p_1 m_1, p_2 m_2, ..., p_n m_n\}$, where $a_i \leq m_i \leq b_i$ for all $i = 1, 2, ..., n$, we can always *slide* the point $m_i$ inside $t_i$ to a position $m_i^L$ as far to the left as possible (see Figure 6). This gives the unique *leftmost non-crossing matching* for $(P, T)$, $\{p_1 m_1^L, p_2 m_2^L, ..., p_n m_n^L\}$. Notice that either $m_i^L = a_i$, or $p_i$ and $m_i^L$ are collinear with some $p_j$ with $j < i$.

Next we describe an algorithm for finding the leftmost non-crossing matching, if it exists. The algorithm considers points in a sequential greedy fashion, in the left-to-right order of the corresponding segments.

For $p_1$, the leftmost matching is simply given by the segment $p_1 a_1$. We then consider the rays from the endpoints of this segment in the direction of the negative semiaxis of abscissae; their points at infinity can be symbolically described as $q_0 = (-\infty, 0)$ and $q_1 = (-\infty, y(p_1))$.

The *forbidden region* is the (unbounded) region enclosed by an alternating sequence of horizontal line segments and subsegments of matched edges (See Figure 7). This region is updated at every step of the algorithm. Initially, it is described clockwise by its vertices, namely $q_1 p_1 a_1 q_0$. Observe that if $p_2$ is inside the forbidden region, then a non-crossing matching $(P, T)$ would be impossible. If $p_2$ is outside the forbidden region, a matching is possible if and only if there is some point $m_2$ in the interval $a_2 b_2$ such that the segment $p_2 m_2$ does not cross the forbidden region. In the affirmative, we slide $m_2$ to its leftmost possible position, and shoot a ray from $p_2$ in the direction of the negative semiaxis of abscissae, which may go to infinity, or stop by hitting the segment $p_1 a_1$. The forbidden

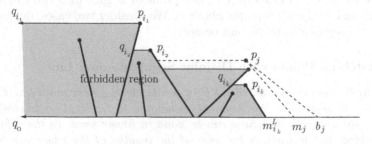

**Fig. 7.** Forbidden region and incremental step

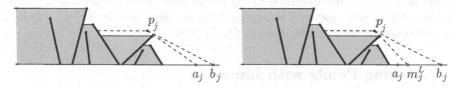

**Fig. 8.** Moving the new edge to the leftmost position

region is updated in each case, and is always defined by alternating horizontal edges with portions of segments from the matching.

Assume that, in a generic step, we have obtained the leftmost matching $\{p_1 m_1^L, p_2 m_2^L, \ldots, p_{j-1} m_{j-1}^L\}$ and we are processing $p_j$. Let $q_{i_1} p_{i_1} q_{i_2} p_{i_2} \ldots q_{i_k} p_{i_k} m_{i_k}^L q_0$ be the current forbidden region (refer to Figure 7). Observe that if there is some $m_j \in [a_j, b_j]$ such that the segment $p_j m_j$ can be added to the edges found so far, getting a non-crossing matching, the segment $p_j b_j$ is also valid. We show next how to check the validity of $p_j b_j$.

We first check the $y$ coordinates of the points $m_{i_k}, p_{i_k}, p_{i_{k-1}} \ldots$, which form an increasing sequence, until we find that $y(p_{i_t}) \geq y(p_j) \geq y(p_{i_{t+1}})$ (the case in which $y(p_j)$ is a maximum is completely analogous). Then, we check whether the segment $p_j b_j$ crosses the segments $m_{i_k} p_{i_k}, q_{i_{k-1}} p_{i_{k-1}}, \ldots, q_{i_{t-1}} p_{i_{t-1}}$. In the affirmative, the algorithm is over, as no crossing-free matching is possible. Otherwise, the segment $p_j b_j$ is valid. We slide the point matched with $p_j$ as much to the left as possible (Figure 8), which can be done by finding the angularly closest point among $p_{i_{t+1}}, p_{i_{t+2}}, \ldots, p_{i_k}, a_j$.

If we shoot a ray from $p_j$ in the direction of the negative semiaxis of abscissae, we hit the boundary of the forbidden region in a point $q_j$, possibly at infinity, and the forbidden region is updated to be $q_{i_1} p_{i_1} q_{i_2} p_{i_2} \ldots p_{i_t} q_j p_j m_j^L q_0$.

The cost of the step for $p_j$ is proportional to the size of the forbidden polygonal region that disappears, and that will never be processed again. Therefore, the amortized cost of one step is constant and the global cost of the algorithm is $O(n)$. At the end we obtain the leftmost matching $\{p_1 m_1^L, p_2 m_2^L, \ldots, p_n m_n^L\}$, unless no matching is possible.

If $(P, T)$ admits a non-crossing matching, with a symmetrical algorithm we can obtain the rightmost matching $\{p_1 m_1^R, p_2 m_2^R, \ldots, p_n m_n^R\}$. Then any points $m_i$ in the intervals $[m_i^L, m_i^R]$ provide a non-crossing matching $\{p_1 m_1, p_2 m_2, \ldots, p_n m_n\}$. In particular, in each interval $[m_i^L, m_i^R]$ we can pick the matching point $m_i$ which is closest to $p_i$, and hence obtain the matching that minimizes the sum of the lengths of the edges in additional $O(n)$ time. $\qquad\square$

## 5.2   Matching Points with Arbitrary Segments on a Line

In this section, we show that when the given segments are confined to a line and possibly intersect, we can determine the existence of a non-crossing matching in polynomial time. The proof first discretizes the problem, and then uses the same approach as in the proof of Theorem 3 for $k$-tuples with $k = O(n^2)$. Details of the proof are omitted from this version of the paper.

**Theorem 7.** *Given an ordered set $P$ of points above a horizontal line $L$ and an ordered set $T$ of line segments belonging to $L$, deciding whether $(P, T)$ admits a non-crossing matching can be done in $O(n^8)$ time.*

# 6    Matching Points with Lines

In the case where points are matched with lines, it is easy to see that a non-crossing matching always exists: choose an arbitrary direction, not parallel to any line, and project each point on its corresponding line in that direction.

Here we consider the optimization problem of minimizing the maximum length over all matching edges. The proof, omitted here due to space limitations, uses a reduction from the problem of deciding the existence of a non-crossing matching between a set of points and a set of segments.

**Theorem 8.** *Given an ordered set $P$ of points and an ordered set $T$ of lines, finding a min-max non-crossing matching of $(P, T)$ is NP-complete.*

# References

1. Agarwal, P., Aronov, B., Sharir, M., Suri, S.: Selecting distances in the plane. Algorithmica 9(5), 495–514 (1993)
2. Aichholzer, O., Bereg, S., Dumitrescu, A., García, A., Huemer, C., Hurtado, F., Kano, M., Márquez, A., Rappaport, D., Smorodinsky, S., Souvaine, D., Urrutia, J., Wood, D.R.: Compatible geometric matchings. Computational Geometry: Theory and Applications 42, 617–626 (2009)
3. Aichholzer, O., Cabello, S., Fabila-Monroy, R., Flores-Penaloza, D., Hackl, T., Huemer, C., Hurtado, F., Wood, D.R.: Edge-Removal and Non-Crossing Configurations in Geometric Graphs. In: Proceedings of 24th European Conference on Computational Geometry, pp. 119–122 (2008)
4. Alt, H., Guibas, L.: Discrete geometric shapes: Matching, interpolation, and approximation. In: Handbook of computational geometry, pp. 121–154 (1999)
5. Arkin, E., Kedem, K., Mitchell, J., Sprinzak, J., Werman, M.: Matching points into noise regions: combinatorial bounds and algorithms. In: Proceedings of the second annual ACM-SIAM symposium on Discrete algorithms, pp. 42–51 (1991)
6. Bentley, J.L., Ottmann, T.A.: Algorithms for reporting and counting geometric intersections. IEEE Transactions on Computers 28(9), 643–647 (1979)
7. Cabello, S., Giannopoulos, P., Knauer, C., Rote, G.: Matching point sets with respect to the Earth Mover's Distance. Computational Geometry: Theory and Applications 39(2), 118–133 (2008)
8. Cardoze, D., Schulman, L.: Pattern matching for spatial point sets. In: Proceedings. 39th Annual Symposium on Foundations of Computer Science (FOCS), 1998, pp. 156–165 (1998)
9. Chazelle, B.: Triangulating a simple polygon in linear time. Discrete and Computational Geometry 6(1), 485–542 (1991)
10. Chew, L., Dor, D., Efrat, A., Kedem, K.: Geometric pattern matching in d-dimensional space. Discrete and Computational Geometry 21(2), 257–274 (1999)

11. Chew, L., Goodrich, M., Huttenlocher, D., Kedem, K., Kleinberg, J., Kravets, D.: Geometric pattern matching under Euclidean motion. Computational Geometry: Theory and Applications 7(1-2), 113–124 (1997)
12. Chew, L., Kedem, K.: Improvements on geometric pattern matching problems. In: Nurmi, O., Ukkonen, E. (eds.) SWAT 1992. LNCS, vol. 621, pp. 318–325. Springer, Heidelberg (1992)
13. Cohen, S.: Finding color and shape patterns in images. PhD thesis, Stanford University, Department of Computer Science (1999)
14. Colannino, J., Damian, M., Hurtado, F., Iacono, J., Meijer, H., Ramaswami, S., Toussaint, G.: An O(n logn)-time algorithm for the restriction scaffold assignment problem. Journal of Computational Biology 13(4), 979–989 (2006)
15. Efrat, A., Itai, A., Katz, M.: Geometry helps in bottleneck matching and related problems. Algorithmica 31(1), 1–28 (2001)
16. Formella, A.: Approximate point set match for partial protein structure alignment. In: Proceedings of Bioinformatics: Knowledge Discovery in Biology (BKDB 2005), Facultade Ciencias Lisboa da Universidade de Lisboa, pp. 53–57 (2005)
17. Giannopoulos, P., Veltkamp, R.: A pseudo-metric for weighted point sets. In: Heyden, A., Sparr, G., Nielsen, M., Johansen, P. (eds.) ECCV 2002. LNCS, vol. 2352, pp. 715–730. Springer, Heidelberg (2002)
18. Grauman, K., Darrell, T.: Fast contour matching using approximate earth mover's distance. In: Proceedings of the 2004 IEEE Computer Society Conference on Computer Vision and Pattern Recognition, pp. 220–227 (2004)
19. Heffernan, P.: Generalized approximate algorithms for point set congruence. In: Dehne, F., Sack, J.-R., Santoro, N. (eds.) WADS 1993. LNCS, vol. 709, pp. 373–373. Springer, Heidelberg (1993)
20. Heffernan, P., Schirra, S.: Approximate decision algorithms for point set congruence. In: Proceedings of the eighth annual Symposium on Computational geometry, pp. 93–101 (1992)
21. Huttenlocher, D., Kedem, K.: Efficiently computing the Hausdorff distance for point sets under translation. In: Proceedings of the Sixth ACM Symposium on Computational Geometry, pp. 340–349 (1990)
22. Kaneko, A., Kano, M.: Discrete geometry on red and blue points in the plane—a survey. Discrete & Computational Geometry 25, 551–570 (2003)
23. Kirkpatrick, D.: Optimal search in planar subdivisions. SIAM Journal on Computing 12(1), 28–35 (1983)
24. Lovász, L., Plummer, M.D.: Matching theory. Elsevier Science Ltd., Amsterdam (1986)
25. Rappaport, D.: Tight bounds for visibility matching of f-equal width objects. In: Akiyama, J., Kano, M. (eds.) JCDCG 2002. LNCS, vol. 2866, pp. 246–250. Springer, Heidelberg (2002)
26. Typke, R., Giannopoulos, P., Veltkamp, R., Wiering, F., Van Oostrum, R.: Using transportation distances for measuring melodic similarity. In: Proceedings of the 4th International Conference on Music Information Retrieval (ISMIR 2003), pp. 107–114 (2003)
27. Vaidya, P.: Geometry helps in matching. In: STOC 1988: Proceedings of the twentieth annual ACM symposium on Theory of computing, pp. 422–425. ACM, New York (1988)

# Homotopic Rectilinear Routing
# with Few Links and Thick Edges*

Bettina Speckmann and Kevin Verbeek

Dept. of Mathematics and Computer Science, TU Eindhoven, The Netherlands
speckman@win.tue.nl, k.a.b.verbeek@tue.nl

**Abstract.** We study the NP-hard problem of finding non-crossing thick minimum-link rectilinear paths which are homotopic to a set of input paths in an environment with rectangular obstacles. We present a 2-approximation that runs in $O(n^3 + k_{in} \log n + k_{out})$ time, where $n$ is the total number of input paths and obstacles and $k_{in}$ and $k_{out}$ are the total complexities of the input and output paths. Our algorithm not only approximates the minimum number of links, but also simultaneously minimizes the total length of the paths. We also show that an approximation factor of 2 is optimal when using *smallest paths* as lower bound.

## 1 Introduction

**Motivation.** Schematic maps are a well-known cartographic tool; they visualize a set of nodes and edges (for example, highway or metro networks) in simplified form to communicate connectivity information as effectively as possible. Many schematic maps deviate substantially from the underlying geography since edges and vertices of the original network are moved in the simplification process. This can be a problem if we want to integrate the schematized network with a geographic map, for example, when creating a thematic map depicting traffic flow on highways. In this scenario the schematized network has to be drawn with thick edges, using few orientations and links, while critical features (cities, lakes, etc.) of the base map are not obscured and retain their correct topological position with respect to the network. There has been little algorithmic work on network schematization under geometric embedding restrictions; in this paper we address one of the fundamental underlying problems for the first time.

Our input is a set of $n_o$ rectangular obstacles and $n_p$ pairs of points $(a_i, b_i)$ together with non-crossing paths $\pi_i$ that connect $a_i$ to $b_i$. We want to find a collection of $n_p$ non-crossing rectilinear thick $a_i$-$b_i$ paths which are homotopic to the paths $\pi_i$. Our goal is to minimize the total number of horizontal and vertical segments (*links*) of the paths. We refer to this problem as the *thick routing problem* (see Fig. 1). The obstacles model the (bounding boxes of) critical map features and the paths $\pi_i$ correspond to the actual geographic location of the highways we want to schematize. Note that the endpoints of the paths—the cities connected by the highways—necessarily also constitute obstacles.

---

* B. Speckmann and K. Verbeek are supported by the Netherlands Organisation for Scientific Research (NWO) under project no. 639.022.707.

A. López-Ortiz (Ed.): LATIN 2010, LNCS 6034, pp. 468–479, 2010.

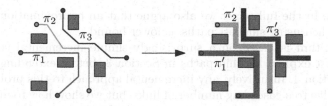

**Fig. 1.** The thick routing problem, input and output

**Related Work.** The thick routing problem can be seen as a variation on the *thick non-crossing paths problem* studied by Mitchell and Polishchuk [13]. They find shortest non-crossing thick paths in a polygonal domain, that is, the points $a_i$ and $b_i$ lie on the boundary of a simple polygon. We consider general input paths, albeit with fixed homotopy classes, and study minimum-link rectilinear instead of shortest paths. There are several papers [2,7] that find shortest paths homotopic to a given collection of input paths. However, while a set of shortest paths homotopic to a set of non-crossing input paths is necessarily non-crossing, the same does not hold for minimum-link (rectilinear) paths. Hence the problem of finding non-crossing minimum-link paths differs substantially from the problem of finding shortest paths. Our problem is also related to *drawing graphs with fat edges* [6] and to wire routing in VLSI design [5,8,10,12], although none of these papers strives to minimize the number of links.

Many variants of the thick routing problem, even without obstacles and with thin paths, are proven NP-hard by Bastert and Fekete [1] if the homotopy classes of the paths are not specified. Yang *et al.* [15] find a pair of non-crossing minimum-link rectilinear paths inside a rectilinear polygon in linear time. Their result does not generalize to more than two paths. Cabello *et al.* [3] schematize a given network using 2 or 3 links per paths, if that is possible. Nöllenburg and Wolff [14] use a method based on mixed-integer programming to generate metro maps using one edge per path. These methods do not incorporate obstacles and are restricted to a small constant number of links per path. Our problem allows an arbitrary number of paths with an arbitrary number of links per path. Gupta and Wenger [9] do allow many paths with many links and present an approximation algorithm (with a factor $\geq 120$) for finding non-crossing minimum-link paths inside a simple polygon; here all endpoints lie on the boundary of the polygon. The endpoints of our paths can lie anywhere in the plane.

**Results.** The thin (or thick) routing problem is NP-hard (the proof can be found in the full version of the paper). In the following we present a 2-approximation algorithm which runs in $O(n^3 + k_{in} \log n + k_{out})$ time, where $n = n_o + n_p$ is the total number of obstacles and input paths and $k_{in}$ and $k_{out}$ are the total complexities of the input and output paths. As a lower bound for the minimum number of links any solution must have, we use the total number of links of *smallest paths* (rectilinear paths that are both shortest and minimum-link) that are homotopic to the input paths. These smallest paths always exist. Our algorithm not only approximates the minimum number of links, but also minimizes the total length of the paths. In the full paper we also argue that an approximation factor of 2 is optimal when using smallest paths as lower bound.

Our algorithm is based on a surprisingly simple incremental construction, which we first explain for thin paths in Section 3, before extending it to thick paths in Section 4. Intuitively, any incremental approach to this problem should be doomed due to a cascading number of links, but we show how to move already inserted paths without increasing the number of links or creating crossings to make room for each new path. We compute smallest paths using variants of the algorithms described in [6,7]. Our main contribution is the algorithm that "untangles" these smallest paths while keeping the number of links low. Background material and supporting lemmas concerning homotopy and rectilinear paths are given in the next section; all omitted proofs and many additional details can be found in the full version of the paper.

## 2   Preliminaries

Our input is a set of $n_o$ rectangular obstacles and $n_p$ pairs of points $(a_i, b_i)$ together with non-crossing paths $\pi_i$ that connect $a_i$ to $b_i$ (note that non-crossing paths can overlap). The endpoints of the paths together with the rectangular obstacles form a set $\mathcal{B}$ of obstacles. No path can contain an element of $\mathcal{B}$, except its own endpoints. Two paths $\pi_1, \pi_2$ with the same endpoints are *homotopic* (denoted by $\pi_1 \sim_h \pi_2$) with respect to $\mathcal{B}$ if there exists a continuous function $\Gamma : [0,1] \times [0,1] \to \mathbb{R}^2$ with the following properties: (i) $\Gamma(0,t) = \pi_1(t)$ and $\Gamma(1,t) = \pi_2(t)$ for $0 \le t \le 1$, (ii) $\Gamma(s,0) = \pi_1(0) = \pi_2(0)$ and $\Gamma(s,1) = \pi_1(1) = \pi_2(1)$ for $0 \le s \le 1$, and (iii) $\Gamma(s,t) \notin \mathcal{B}$ for $0 \le s \le 1$ and $0 < t < 1$. Since the homotopic relation is an equivalence relation, we can speak of the equivalence class of a path, its *homotopy class*. A homotopy class $\mathcal{C}$ is said to be *non-crossing* if there is a path $\pi \in \mathcal{C}$ which is non-crossing. Similarly, two homotopy classes $\mathcal{C}_1$ and $\mathcal{C}_2$ are called *pairwise non-crossing* if there are two paths $\pi_1 \in \mathcal{C}_1$ and $\pi_2 \in \mathcal{C}_2$ such that $\pi_1$ and $\pi_2$ are pairwise non-crossing.

**Observation 1.** *Let $\mathcal{C}_1$ and $\mathcal{C}_2$ be two non-crossing homotopy classes which are pairwise non-crossing. Then the shortest paths of these homotopy classes $\sigma_1 \in \mathcal{C}_1$ and $\sigma_2 \in \mathcal{C}_2$ are non-crossing.*

A collection of $n$ non-crossing homotopy classes is non-crossing if there is a path in each homotopy class such that these paths are non-crossing. Observation 1 implies that a collection of $n$ non-crossing homotopy classes is non-crossing if the homotopy classes are pairwise non-crossing. Since the input paths are non-crossing, the homotopy classes of our input are always non-crossing.

The following lemma is about y-monotone paths. Note that a path is y-monotone if every horizontal line crosses the path only once.

**Lemma 1.** *Given a collection of y-monotone paths with non-crossing homotopy classes and a collection of obstacles $\omega_j$, the homotopy classes of all paths can be characterized by a total order $\prec$ on the paths and the obstacles.*

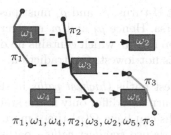

$$\pi_1, \omega_1, \omega_4, \pi_2, \omega_3, \omega_2, \omega_5, \pi_3$$

**Fig. 2.** Order on paths and obstacles

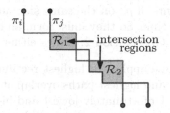

**Fig. 3.** Intersection regions

This order is illustrated in Figure 2. It can be seen as the rotated analogue of the aboveness order defined in [4].

If $\pi_i \prec \pi_j$, then an *intersection region* of $\pi_i$ and $\pi_j$ is a region enclosed by $\pi_i$ and $\pi_j$ at $y$-coordinates where the paths are out of order, i.e. where $\pi_j$ is to the left of $\pi_i$. Intersection regions are $y$-monotone polygons; two paths can have more than one intersection region (Fig. 3). The following observation follows directly from Lemma 1.

**Observation 2.** *Intersection regions do not contain obstacles.*

Every rectilinear path $\pi$ consists of a sequence of horizontal and vertical links. If a horizontal link $\ell_i$ is above or below both its neighboring (vertical) links $\ell_{i-1}$ and $\ell_{i+1}$, then we call $\ell_{i-1}\ell_i\ell_{i+1}$ a *horizontal U-turn*. Vertical U-turns are defined correspondingly. If there is an obstacle on the inside of a U-turn touching the middle link, then we call this U-turn a *tight U-turn*. The *support* of a tight U-turn is the part of the U-turn supported by the obstacles. We say that two paths share a tight U-turn, if the supports of the U-turns are equal. If we split a rectilinear path at the supports of its U-turns, then we get a collection of monotone *staircase chains*. As every staircase chain between two points has the same length, the length of a rectilinear path depends only on its U-turns. Hence a rectilinear path of a given homotopy class is shortest iff it has only tight U-turns. Rectilinear shortest path are not unique. Further, not every collection of rectilinear shortest paths of non-crossing homotopy classes is non-crossing. So we consider rectilinear shortest paths with "lowest" staircase chains. We say a staircase chain $\rho$ is *lowest* if for every staircase chain $\rho'$ ($\rho' \sim_h \rho$), $\rho$ is completely below or on $\rho'$. A rectilinear shortest path with only lowest staircase chains—a *lowest path* in short—always exists and is unique.

**Lemma 2.** *Lowest paths* $(\pi_i)_{i=1}^n$ *of non-crossing homotopy classes are non-crossing.*

*Proof (sketch).* Since the homotopy classes are non-crossing, there exist rectilinear shortest paths $\sigma_i$ ($\sigma_i \sim_h \pi_i$) that are non-crossing. Because the paths $\pi_i$ are shortest, they share the (tight) U-turns with $\sigma_i$. Now consider two crossing staircase chains $\rho_j$ and $\rho_k$ and let $\rho'_j$ and $\rho'_k$ be the corresponding non-crossing chains in $\sigma_i$. Note that $\rho_j$ and $\rho'_j$ must have the same endpoints. Furthermore, since $\rho_j \sim_h \rho_j'$ and there are obstacles at tight U-turns, $\rho_j$ and $\rho_j'$ must pass the endpoints of $\rho_k$ on the same side and vice versa. Hence $\rho_j$ and $\rho_k$ must cross a second time. So they must form an intersection region, which contains no obstacles by Observation 2. Hence either $\rho_j$ or $\rho_k$ is not lowest. Contradiction.    □

The same applies to highest rectilinear shortest paths (*highest paths* in short). Lowest and highest paths overlap at every U-turn and differ only in the staircase chains. Unfortunately lowest and highest paths can have $\Omega(n)$ times more links than a homotopic minimum-link path. Hence we use *smallest paths*—paths that are both minimum-link and rectilinear shortest. We use lowest and highest paths to compute smallest paths, as every smallest path must lie in between the lowest and highest paths.

We distinguish two types of staircase chains: *positive staircase chains* which go right and down (or left and up) and *negative staircase chains* which go left and down (or right and up). Positive staircase chains and negative staircase chains can cross only once due to monotonicity. Following the proof (sketch) of Lemma 2, this immediately implies the following.

**Lemma 3.** *A positive staircase chain and a negative staircase chain of rectilinear shortest paths with non-crossing homotopy classes cannot cross.*

Our paths are rectilinear shortest, so we need to consider only crossings between two positive or two negative staircase chains; this is crucial for our algorithm.

**Thick paths.** We assume that we are given a required thickness $\Delta_i$ for each thick path. For $\Delta > 0$ let $\mathcal{S}_\Delta$ be the square of size $\Delta$ centered at the origin. A *thick path* is defined as the Minkowski sum of a *spine* $\pi_i$ and $\mathcal{S}_{\Delta_i}$: $(\pi_i)^{\Delta_i} = \pi_i \oplus \mathcal{S}_{\Delta_i}$. For convenience we refer also to a thick path by $\pi_i$ instead of $(\pi_i)^{\Delta_i}$. When working with thick paths, fixing one path might make it impossible for another path to be routed (Fig. 4). We say that a

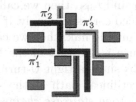

**Fig. 4.** Path $\pi_2'$ is not feasible

path $\pi_i' \sim_h \pi_i$ is *feasible* if there exist paths $\pi_j' \sim_h \pi_j$ for $1 \le j \le n_p$ and $i \ne j$ such that all thick paths are interior disjoint. In a valid solution of the thick routing problem every path is feasible. Hence we use feasible smallest paths as a starting point for the thick routing problem.

Lemma 3 holds only for thin paths, but can easily be extended to feasible thick paths. Therefore positive and negative staircase chains of feasible rectilinear shortest paths are interior disjoint.

## 3    Thin Paths

To illustrate our techniques we first present a solution for the *thin routing problem* where all paths have thickness zero. Algorithm THINROUTING consists of two steps: (i) computing smallest paths that are homotopic to the input paths, and (ii) untangling smallest paths in such a way that the output paths have no more than twice the number of links of the input paths. The preprocessing Step (i) relies mostly on previous work, Step (ii)—the crucial part of the algorithm—is the main contribution of this paper.

## 3.1 Computing Smallest Paths

To compute smallest paths efficiently, we first compute the lowest and highest paths homotopic to the input paths. We use a variant of the algorithm by Efrat *et al.* [7] which computes shortest (non-rectilinear) paths that are homotopic to a collection of non-crossing input paths. This algorithm can easily be adapted to compute lowest and highest paths. It also allows us to *bundle* the homotopic $y$-monotone chains of the lowest and highest paths into $O(n)$ $y$-monotone chains with $O(n)$ links each. This bundled representation can be computed in $O(n^2 + k_{in} \log n)$ time.

We now compute a $y$-monotone smallest path for each bundle using a plane sweep. This is possible, since every bundle of lowest chains corresponds to a bundle of highest chains representing the same $y$-monotone chains of the input paths (chains are only bundled when they are homotopic). So we compute a smallest path for each corresponding pair of bundles and obtain a bundled representation of all smallest paths in $O(n^2 + k_{in} \log n)$ time.

Although the lowest and highest paths are non-crossing, smallest paths are generally not. Actually, finding non-crossing smallest paths is NP-complete (the proof can be found in the full version of the paper). For the approximation algorithm we use a set of canonical smallest paths where all horizontal links are pushed down as much as possible and all vertical links are pushed right (left) as much as possible for positive (negative) staircase chains. These canonical smallest paths have the nice property that all intersection regions are rectangular[1].

## 3.2 Untangling Smallest Paths

The second step untangles the (bundled) $y$-monotone chains. By Lemma 3 we can treat positive and negative staircase chains independently, so we restrict the description of algorithm UNTANGLE to positive staircase chains. By Lemma 1 we can assume the input paths to be ordered positive staircase chains $\pi_i$ ($1 \le i \le n$).

We untangle the paths incrementally, adding paths in order from left to right. The first two paths $\pi_1$ and $\pi_2$ can have only rectangular intersection regions. We can add two links to $\pi_2$ (for each intersection region) to remove the crossings. Unfortunately this might make intersection regions with $\pi_k$ ($k > 2$) non-rectangular such that rerouting requires more links. To avoid these cascading effects we keep track of where paths are rerouted using *reroute boxes* (Fig. 5). Reroute boxes are rectangular and contain no obstacles.

---

[1] This is easy to prove using Obs. 2. Note that there can be two different smallest paths with the given requirements: one starting with a horizontal link and one starting with a vertical link. If both are possible, we refer to the latter as the canonical path.

We maintain three invariants after adding path $\pi_k$ (rerouted paths are denoted by $\pi_i'$): (i) $\pi_1' \ldots \pi_k'$ are non-crossing, (ii) $\pi_1 \ldots \pi_k$ have at most one reroute box per lower-left bend, and (iii) all paths $\pi_i'$ pass an upper-right bend of a path $\pi_j$ ($i < j$) on the left side. All invariants hold initially.

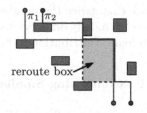

reroute box

**Fig. 5.** Untangling two paths

Path $\pi_{k+1}$ is added one L-segment $\mathcal{L}$ at a time. $\mathcal{L}$ can cross multiple other paths. The original paths $\pi_i$ have only rectangular intersection regions. $\mathcal{L}$ is unchanged and we have added only reroute boxes to $\pi_1' \ldots \pi_k'$, hence there are only a few shapes possible for the intersection regions with $\mathcal{L}$ (Fig. 6): one, two, or three upper-right bends.

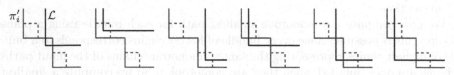

**Fig. 6.** The possible shapes of intersection regions. Dashed lines denote reroute boxes.

Now assume that $\mathcal{L}$ is crossed by the paths $\pi_a' \ldots \pi_b'$ ($a \leq b$). We want to make the intersection region of $\mathcal{L}$ with $\pi_b'$ rectangular such that we can add a reroute box to $\mathcal{L}$ to remove all crossings. We do this incrementally for all intersection regions, starting with $\pi_a'$ and ending with $\pi_b'$. Assume the intersection region $\mathcal{R}_i$ of $\mathcal{L}$ and $\pi_i' \in [\pi_a' \ldots \pi_b']$ is rectangular. To make the intersection region $\mathcal{R}_{i+1}$ rectangular, we move the links of $\pi_{i+1}'$ through $\mathcal{R}_{i+1}$ (intersection regions contain no obstacles), but not through $\mathcal{R}_i$ (this would introduce crossings with $\pi_i'$). By Invariant (i) $\pi_{i+1}'$ does not cross $\mathcal{R}_i$ initially. If $\mathcal{R}_{i+1}$ contains one upper-right bend, we do nothing as $\mathcal{R}_{i+1}$ is already rectangular. If $\mathcal{R}_{i+1}$ contains two upper-right bends, then we either move the first vertical link of $\pi_{i+1}'$ to the left (onto $\mathcal{L}$) or the last horizontal link down. Due to the rectangular shape of $\mathcal{R}_i$, we can always perform one of the two moves without crossing $\mathcal{R}_i$. If $\mathcal{R}_{i+1}$ contains three upper-right bends, then either $\mathcal{R}_i$ is in the middle corner and we move the first vertical link of $\pi_{i+1}'$ to the left and the last horizontal link down, or $\mathcal{R}_i$ is not in the middle corner, in which case we can simplify the middle corner and handle this as a case with two upper-right bends (see Fig. 7 for an example).

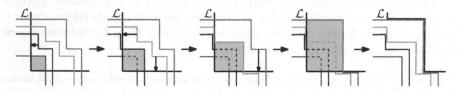

**Fig. 7.** Making intersection regions rectangular and rerouting $\mathcal{L}$

The algorithm maintains all invariants and hence the output paths are non-crossing. They are also homotopic to the input paths, because we change paths only by adding reroute boxes or by moving links through intersection regions. We add only two links per reroute box and there is at most one reroute box per lower-left bend. Because a positive staircase chain with $L$ links has at most $L/2$ lower-left bends, the algorithm at most doubles the number of links.

We can extend this algorithm to work on the y-monotone chains of the smallest paths. Since we can add an L-segment in $O(n)$ time and we have only $O(n)$ chains with at most $O(n)$ links each, we can untangle the chains in $O(n^3)$ time. Finally we can unbundle the paths in $O(k_{out})$ time. So in total we can compute a 2-approximation for the thin routing problem in $O(n^3 + k_{in} \log n + k_{out})$ time.

## 4  Thick Paths

We now extend THINROUTING to thick paths. For thin paths, non-crossing homotopy classes imply that a solution exists. This is not the case for thick paths. If there is an input path for which there is no homotopic feasible path, then there is no valid solution for that problem instance and our algorithm will detect this fact. Otherwise a solution always exists and our algorithm will find it.

Algorithm THICKROUTING also consists of two steps. First we compute *feasible* canonical smallest paths that are homotopic to the input paths. Then we untangle the paths such that they are interior disjoint, while making sure that the number of links is no more than doubled. To simplify the discussion, we describe the algorithm for the spines instead of the thick paths. Clearly the thick paths are interior disjoint if their spines are sufficiently separated.

### 4.1  Feasible Smallest Paths

We first compute feasible lowest and highest paths by adapting the growing approach by Duncan *et al.* [6][2]. We start with the thin lowest and highest paths and "grow" the paths until they have the required thickness. During the growing process we keep the paths disjoint, shortest, and lowest (or highest). The resulting paths are feasible (if the problem instance has a solution). The algorithm as described in [6] works for shortest paths and point obstacles, we extend it to rectilinear paths and rectangular obstacles. The details are many and somewhat intricate, although not difficult; they can be found in the full version of the paper. This algorithm takes $O(n^3 + k_{in} \log n)$ time and returns an ordered bundled representation of the y-monotone chains of the feasible lowest and highest paths.

Using the bundled representation of feasible lowest and highest paths, we can easily compute a bundled representation of the feasible canonical smallest paths, because every feasible smallest path lies between the feasible lowest and highest paths. Since the chains in a bundle are homotopic, we can compute a feasible canonical smallest path for a whole bundle to represent the feasible canonical smallest paths of all chains in the bundle.

---

[2] We cannot use the algorithm from [7], as the feasibility of a path depends on the homotopy classes and thickness of the other paths.

## 4.2   Untangling Thick Paths

The second step of the algorithm untangles the feasible smallest paths while ensuring that the output paths are interior disjoint. We cannot directly use the approach for thin paths, but we can do something similar.

By extending Lemma 3 to thick paths, we can again restrict the discussion to positive staircase paths. To keep the thick paths feasible we use *fences*. A *fence* $\gamma_i(j)$ of a path $\pi_i$ with respect to another path $\pi_j$ is a path representing the closest position at which $\pi_j$ can be with respect to $\pi_i$ (see Fig. 8(a)). A fence ignores all obstacles. The fence $\gamma_i(j)$ defines the region where $\pi_j$ is feasible, if we fix $\pi_i$ and ignore the obstacles. So a fence $\gamma_i(j)$ is a shifted version of $\pi_i$, where the distance depends on the thickness of the paths between $\pi_i$ and $\pi_j$, which can differ along $\pi_i$. We do not compute fences explicitly, but only the parts we need.

**Lemma 4.** *If $\mathcal{R}$ is an intersection region between $\gamma_i(j)$ and $\pi_j$, then any routing of $\pi_j$ inside $\mathcal{R}$ is homotopic to $\pi_j$ and feasible.*

*Proof.* To prove that any routing is homotopic, we prove that $\mathcal{R}$ does not contain any obstacles. We consider two parts of $\mathcal{R}$: the intersection region of $\pi_i$ and $\pi_j$ and the remaining part (Fig. 8(b)). The first part does not necessarily exist. If it exists, then by Observation 2, it is free of obstacles. Now consider the remaining part and assume there is an obstacle $\omega$ in it. Both $\pi_i$ and $\pi_j$ must be routed on the same side of $\omega$. But $\gamma_i(j)$ shows that $\pi_i$ forces $\pi_j$ to be on the other side of $\omega$ (because $\omega$ lies inside $\mathcal{R}$). That implies that $\pi_i$ is not feasible. Contradiction.

Because $\pi_i$ is feasible, following $\gamma_i(j)$ is also feasible for $\pi_j$ by definition of $\gamma_i(j)$. As $\pi_j$ itself is also feasible, every routing of $\pi_j$ inside $\mathcal{R}$ must be as well, because $\mathcal{R}$ does not contain any obstacles.                                    □

An intersection region of a fence $\gamma_i(j)$ and a path $\pi_j$ is also called a *forbidden region*. Note that a forbidden region $\mathcal{R}_{ij}$ between $\gamma_i(j)$ and $\pi_j$ has exactly the same shape as the forbidden region $\mathcal{R}_{ji}$ between $\gamma_j(i)$ and $\pi_i$, because both fences depend on the thickness of the paths in between $\pi_i$ and $\pi_j$.

**Lemma 5.** *If $\pi_i$ and $\pi_j$ are canonical smallest paths, then any forbidden region between $\gamma_i(j)$ and $\pi_j$ is rectangular.*

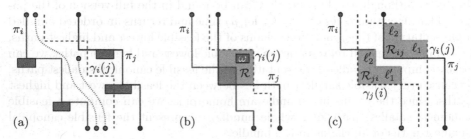

**Fig. 8.** An example of a fence $\gamma_i(j)$ (a), the dotted paths lie between $\pi_i$ and $\pi_j$. Forbidden regions are free of obstacles (b) and rectangular (c).

*Proof.* Without loss of generality we have $i < j$. Define $\mathcal{R}_{ij}$ and $\mathcal{R}_{ji}$ as above (see Fig. 8(c)). Assume that $\mathcal{R}_{ij}$ is not rectangular. Then there must be a horizontal link $\ell_1$ and a vertical link $\ell_2$ that are completely part of the boundary of $\mathcal{R}_{ij}$. If $\ell_1$ and $\ell_2$ are part of $\pi_j$, then by Lemma 4 $\ell_2$ is not rightmost. If $\ell_1$ and $\ell_2$ are part of $\gamma_i(j)$, then there are also corresponding links $\ell_1'$ and $\ell_2'$ which are part of $\pi_i$ that are completely part of the boundary of $\mathcal{R}_{ji}$. Again by Lemma 4 $\ell_1'$ is not downmost. Contradiction.    □

To untangle thick paths, we use an adapted version of UNTANGLE. The rerouted paths are denoted by $\pi_i'$. We maintain three invariants: (i) $\pi_1' \ldots \pi_k'$ are interior disjoint, (ii) $\pi_1 \ldots \pi_k$ have at most one reroute box per lower-left bend, and (iii) all paths $\pi_i'$ pass an upper-right bend of a fence $\gamma_j(i)$ $(i < j)$ on the left side. We change the paths in only two ways: we move links of paths $\pi_1' \ldots \pi_k'$ (only left and down) and we add reroute boxes to the current path (and hence add links). We add paths one L-segment $\mathcal{L}$ at a time, always turning the forbidden regions involving $\mathcal{L}$ into a rectangle. Then we add a reroute box to $\mathcal{L}$ to make the paths interior disjoint. The challenge here is to efficiently turn the forbidden regions into rectangles. For this we first need to compute a certain subset of the fences for $\mathcal{L}$. Recall the situation for thin paths. To create a rectangular intersection region between $\mathcal{L}$ and a previously inserted path $\pi_i'$, we moved segments of $\pi_i'$ to coincide with $\mathcal{L}$. Now we move these segments to coincide with the fence $\gamma_{\mathcal{L}}(i)$. In the following we sketch our approach to compute the relevant fences.

When inserting path $\pi_{k+1}$ we go through the paths $\pi_i'$ $(1 \leq i \leq k)$ from right to left and accumulate their thickness in $\Delta$ to compute the fences $\gamma_{\mathcal{L}}(i)$. Recall that a fence $\gamma_{\mathcal{L}}(i)$ is a shifted version of $\mathcal{L}$ where the distance depends on the total thickness $\Delta$ of the paths between $\mathcal{L}$ and $\pi_i'$. Hence we can compute the fences by maintaining $\Delta$ going from right to left and for every path we encounter, add its thickness to $\Delta$. But although all paths $\pi_j'$ $(i < j \leq k)$ are between $\mathcal{L}$ and $\pi_i'$, a path $\pi_j'$ might have a $y$-range such that it does not influence $\mathcal{L}$ or its fences (see Fig. 9(b)). In those cases, the thickness of $\pi_j'$ should not be added to $\Delta$. To determine whether the thickness $\Delta_j$ of a path $\pi_j'$ should be added to $\Delta$, we use the following simple rule: if the forbidden region of $\gamma_{\mathcal{L}}(j)$ with $\pi_j'$ does not exist, then $\Delta_j$ is not added to $\Delta$, otherwise it is.

**Lemma 6.** *The above rule correctly computes the forbidden regions with $\mathcal{L}$.*

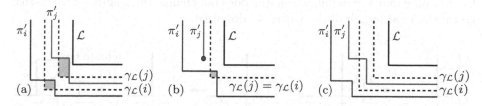

**Fig. 9.** Computing fences for $\mathcal{L}$. If a forbidden region exists for $\pi_j'$, $\Delta_j$ is added to $\Delta$ (a), otherwise it is not (b). If the forbidden region for $\pi_j'$ does not exist, the forbidden region for $\pi_i'$ will also not exist (c).

*Proof.* First assume the forbidden region of $\gamma_{\mathcal{L}}(j)$ with $\pi'_j$ does not exist. If $\pi'_j$ does not influence $\mathcal{L}$ or its fences, then the rule is correct (Fig. 9(b)). Otherwise, because all paths $\pi'_i$ ($1 \leq i \leq j$) are interior disjoint (Invariant (i)), the forbidden regions for all $\pi'_i$ do not exist either (even if we added $\Delta_j$ to $\Delta$) and hence the rule is correct in this case as well (Fig. 9(c)). If the forbidden region with $\pi'_j$ does exist (Fig. 9(a)), then we should be able to add $\Delta_j$ to $\Delta$. Recall that a fence $\gamma_{\mathcal{L}}(i)$ ($i < j$) defines the region where $\pi'_i$ is feasible, if we fix $\mathcal{L}$ and ignore the obstacles. Now assume we fix $\mathcal{L}$, then $\pi'_j$ must follow the lower-left bend of $\gamma_{\mathcal{L}}(j)$. Consider a line between the lower-left bend of $\mathcal{L}$ and the lower-left bend of a fence $\gamma_{\mathcal{L}}(i)$ ($i < j$). For the fence to be correct, it should include all paths crossed by this line, like $\pi'_j$. Hence it is correct to add $\Delta_j$ to $\Delta$.     □

After the fences and forbidden regions have been computed, we work from left to right, considering only the paths $\pi'_a \ldots \pi'_b$ for which a forbidden region exists. The different possible forbidden regions are handled as is described in Section 3. The cases for the forbidden regions are exactly the same as those shown in Fig. 6 (the cases are implied by Invariant (ii) and (iii)). Finally we reroute $\mathcal{L}$ along the fence $\gamma_b(\mathcal{L})$ which can trivially be computed. In fact this is essentially the same as adding a reroute box (see Fig. 10).

**Lemma 7.** *The resulting paths are interior disjoint and homotopic to the input paths.*

We can extend this algorithm to untangle complete $y$-monotone chains. After the $y$-monotone bundles of the feasible smallest paths have been untangled, we can easily extract the resulting paths from the bundles.

**Theorem 1.** *The algorithm* THICKROUTING *computes a 2-approximation for the thick routing problem that also minimizes the lengths of the paths in* $O(n^3 + k_{in} \log n + k_{out})$ *time.*

*Proof.* The correctness of the algorithm follows from Lemma 7. The approximation ratio of the algorithm can be argued exactly as for thin paths. The feasible smallest paths can be found in $O(n^3 + k_{in} \log n)$ time. The feasible smallest paths are given as $O(n)$ $y$-monotone bundles with at most $O(n)$ links each. Untangling thick paths is just as fast asymptotically as untangling thin paths, hence it takes $O(n^3)$ time. Finally, we need to unbundle the paths and reconnect chains, which takes $O(k_{out})$ time. Hence the total running time is $O(n^3 + k_{in} \log n + k_{out})$. Finally note that untangling the paths does not change the lengths of the paths, hence the total length of the paths is minimized.     □

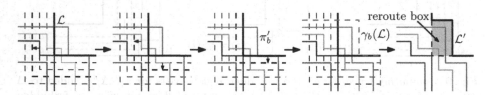

**Fig. 10.** Growing a rectangle for thick paths

## 5    Conclusions and Open Problems

We presented an algorithm that computes a 2-approximation for the thick routing problem. Our algorithm not only approximates the minimum number of links, but also simultaneously minimizes the total length of the paths. This approximation factor is optimal when smallest paths are used as a lower bound.

Our motivation to study these problems relates to the construction of schematic maps with geometric embedding restrictions. In this context we would like to extend our results to $c$-oriented paths, implement our algorithm, and evaluate the cartographic quality of the resulting maps experimentally.

## References

1. Bastert, O., Fekete, S.P.: Geometrische Verdrahtungsprobleme. Technical Report 247, Mathematisches Institut, Universität zu Köln (1996)
2. Bespamyatnikh, S.: Computing homotopic shortest paths in the plane. Journal of Algorithms 49(2), 284–303 (2003)
3. Cabello, S., de Berg, M., van Kreveld, M.: Schematization of networks. Computational Geometry: Theory and Applications 30(3), 223–238 (2005)
4. Cabello, S., Liu, Y., Mantler, A., Snoeyink, J.: Testing homotopy for paths in the plane. Discrete & Computational Geometry 31, 61–81 (2004)
5. Cole, R., Siegel, A.: River routing every which way, but loose. In: Proc. 25th Symp. Foundations of Computer Science, pp. 65–73 (1984)
6. Duncan, C.A., Efrat, A., Kobourov, S.G., Wenk, C.: Drawing with fat edges. Intern. Journal of Foundations of Computer Science 17(5), 1143–1163 (2006)
7. Efrat, A., Kobourov, S.G., Lubiw, A.: Computing homotopic shortest paths efficiently. Computational Geometry: Theory and Applications 35(3), 162–172 (2006)
8. Gao, S., Jerrum, M., Kaufmann, M., Mehlhorn, K., Rülling, W.: On continuous homotopic one layer routing. In: Proc. 4th Symp. Comp. Geom., pp. 392–402 (1988)
9. Gupta, H., Wenger, R.: Constructing pairwise disjoint paths with few links. ACM Transactions on Algorithms 3(3), 26 (2007)
10. Leiserson, C.E., Maley, F.M.: Algorithms for routing and testing routability of planar VLSI layouts. In: Proc. 17th Symp. Theory of Comp., pp. 69–78 (1985)
11. Lichtenstein, D.: Planar Formulae and Their Uses. SIAM Journal on Computing 11(2), 329–343 (1982)
12. Maley, M.: Single-layer wire routing and compaction. MIT Press, Cambridge (1990)
13. Mitchell, J.S., Polishchuk, V.: Thick non-crossing paths and minimum-cost flows in polygonal domains. In: Proc. 23rd Symp. Comp. Geom., pp. 56–65 (2007)
14. Nöllenburg, M., Wolff, A.: A mixed-integer program for drawing high-quality metro maps. In: Healy, P., Nikolov, N.S. (eds.) GD 2005. LNCS, vol. 3843, pp. 321–333. Springer, Heidelberg (2005)
15. Yang, C.D., Lee, D.T., Wong, C.K.: The smallest pair of noncrossing paths in a rectilinear polygon. IEEE Transactions on Computers 46(8), 930–941 (1997)

# Tilings Robust to Errors

Alexis Ballier, Bruno Durand, and Emmanuel Jeandel

Laboratoire d'informatique fondamentale de Marseille (LIF)
Aix-Marseille Université, CNRS
39 rue Joliot-Curie, 13 453 Marseille Cedex 13, France

**Abstract.** We study the error robustness of tilings of the plane. The fundamental question is the following: given a tileset, what happens if we allow a small probability of errors? Are the objects we obtain close to an error-free tiling of the plane?

We prove that tilesets that produce only periodic tilings are robust to errors. For this proof, we use a hierarchical construction of islands of errors (see [6,7]). We also show that another class of tilesets, those that admit countably many tilings is not robust and that there is no computable way to distinguish between these two classes.

## 1 Introduction

Tilings are a basic and intuitive way to express geometrical constraints. They have been used as static geometrical models of computation since Berger proved the undecidability of the so-called domino problem [2] by capturing geometric aspects of computation [19,2,11,14,5,10,3]. The model assumes the reliability of local color-constraints of tilings, hence several research tracks were aimed at constructing tilesets that are reliable to errors, be it as a computing model [7,8] or as a model for DNA self-assembly [22,18].

In this paper we are interested in error robustness of tilesets. We can see an erroneous tiling as an usual tiling by Wang tiles [21] where we allow a small proportion of colors on adjacent edges of tiles not to match. We give a more formal definition of this notion in Section 2. Some constructions using fixed point methods construct tilesets that are robust to a small proportion of errors [7,8]. Our goal here is to focus on general properties that imply error-robustness or non-error-robustness of some classes of tilesets.

In Section 3 we give a construction of "error-cleaning functions" for tilesets that allow only periodic tilings. We prove that for this class of tilesets it is possible to apply well known hierarchical constructions [9,6] so that we can repair every erroneous tiling with probability one granted the probability of errors is sufficiently small.

On the other hand, we prove that the family of tilesets that produce only a countable number of tilings are not locally robust: the correction of a finite number of errors in those tilings may always require a modification of an infinite number of tiles. This is incompatible with local error correction and thus robustness. Locally robust tilesets can be expressed as properties of their ground state configurations as used to model crystals [16,17].

A. López-Ortiz (Ed.): LATIN 2010, LNCS 6034, pp. 480–491, 2010.

Finally, by using classical constructions [3,19,15] for encoding Turing machines into tilesets, we prove in Section 5 that these two classes of tilesets are recursively inseparable, which shows that it is not possible to obtain a simple (recursive) characterization of tilesets robust to errors.

## 2   Definitions

We present notations and definitions for tilings; we focus our study on tilings of the plane but most of the results naturally extend to higher dimensions. We define tilings by *local constraints* as they give the most straightforward definitions for tiling errors. Different models are used in literature, such as Wang tiles [20] or subshifts of finite type [12], but one can easily transform one formalism into another (see [4] for more details and proofs).

In our definition of tilings, we first associate a color to each cell of the plane. Then we impose a local constraint on them, that is we describe which colorings are allowed and which are not. More formally, $Q$ is a finite set, called the *set of colors*. A *configuration* $c$ consists of cells of the plane with colors, thus $c$ is an element of $Q^{\mathbb{Z}^2}$. We denote by $c(i)$ the color of $c$ at the cell $i \in \mathbb{Z}^2$. For an element $x$ of $\mathbb{Z}^2$ we denote by $c + x$ the configuration $i \to c(i + x)$.

**Definition 1 (Patterns).** *A pattern $P$ is a finite restriction of a configuration i.e., an element of $Q^V$ for some finite domain $V$ of $\mathbb{Z}^2$. A pattern appears in a configuration $c$ if it can be found somewhere in $c$; i.e., if there exists a vector $t \in \mathbb{Z}^2$ such that $c(x + t) = P(x)$ on the domain of $P$.*

**Definition 2 (Local constraints).** *A local constraint is a pair $\tau = (V, \delta)$. $V$ is a finite domain of $\mathbb{Z}^2$ and is called the neighborhood. $\delta$ is the constraint function from $Q^V$ to $\{0, 1\}$.*

The idea behind this formalism of local constraints is to define which patterns are allowed and which are not. A pattern is allowed if and only if it maps to 0 by the constraint function. The local constraint $\tau = (V, \delta)$ naturally extends to a global constraint function $\Delta_\tau : Q^{\mathbb{Z}^2} \to \{0, 1\}^{\mathbb{Z}^2}$ by applying it uniformly on every cell of the plane: $\Delta_\tau(c)(x) = \delta((c + x)_{|V})$. This corresponds to a sliding block code [13, Chapter 1, § 5].

**Definition 3 (Tilings).** *A configuration $c \in Q^{\mathbb{Z}^2}$ is said to be valid for $\tau = (V, \delta)$ (or a tiling by $\tau$, or allowed by $\tau$) when it satisfies the local constraint everywhere, that is for every cell $x$, $\Delta_\tau(c)(x) = 0$.*

The set of tilings by $\tau$ is denoted by $\mathcal{T}_\tau$. In this paper we only consider tilesets that can tile the plane, thus $\mathcal{T}_\tau \neq \emptyset$ is an implicit condition of all the results.

As we want to study tilings with some errors, this definition of classical tilings naturally extends to tilings with errors by considering an error repartition where some cells are not correctly tiled:

**Definition 4 (Tiling with errors).** *Let $e$ be an element of $\{0,1\}^{\mathbb{Z}^2}$ and $c$ a configuration, we say that $c$ is a tiling by $\tau$ with error repartition $e$ if $\Delta_\tau(c) = e$.*

One may remark that with this definition, repairing an error is different from replacing the erroneous tile with a correct one as errors may have consequences that require replacing other tiles, even if those tiles are locally correct.

In this paper we first prove that the consequences of such a correction are not problematic in the case of tilesets that allow only periodic tilings (Section 3) but may have consequences on infinitely many cells in the case of tilesets that allow countably many tilings (Section 4).

Before entering the core of the problem of tilings with errors we need to recall a couple of structural results on regular tilings of the plane. We embed $Q$ with the discrete topology, $Q^{\mathbb{Z}^2}$ with the induced product topology. A classical result on the set of configurations is its compactness, as a direct application of Tychonoff's theorem:

**Proposition 1.** *$Q^{\mathbb{Z}^2}$ is a compact perfect metric space (a Cantor space).*

The metric we consider, as induced by the product topology, is defined by $d(x,y) = 2^{-|i|}$ where $|(a,b)|$ denotes $|a| + |b|$ and $i$ is the point closest to $(0,0)$ (i.e., with minimal norm) such that $x(i) \neq y(i)$. The ball of center $x$ and radius $k$ is the set of points $y$ such that $|x - y| \leq k$. In this paper we reformulate the compactness of sets of tilings in a way that suits our needs:

**Lemma 1.** *For each finite subset $D$ of $\mathbb{Z}^2$, there exists $C$ such that if a pattern defined on $D$ can be extended to a pattern on $C$ while respecting the local constraints then $P_{|D}$ appears in a tiling of $\mathbb{Z}^2$.*

This means that if we can tile sufficiently large but finite patterns we are sure that a small part of it will appear in a valid tiling of the whole plane. The function that is given a tileset $\tau$ and outputs the set $C$, even with a fixed $D = \{(0,0)\}$, is uncomputable since this would allow us to decide the domino problem which is a well known non-recursive problem [2,15,19].

*Proof.* We prove that for every pattern $P$ there exists a finite domain $C_P$ of $\mathbb{Z}^2$ such that there exists a valid pattern defined on $C_P$ that contains $P$ if and only if $P$ appears in a tiling of the plane. This proves the lemma by taking $C$ to be the finite union of all $C_P$ where $P$ is a pattern of domain $D$.

If $P$ appears in a tiling of the plane take $C_P$ to be the domain of $P$. Suppose that there is a pattern $P$ that does not appear in a tiling of the plane but such that there exists arbitrary large valid extensions of it. By compactness, this sequence of patterns can be extracted to converge towards a configuration. This configuration contains $P$ and is a valid tiling of the plane. Therefore, if $P$ does not appear in a tiling of the plane, there exists a finite domain such that any correct tiling of this domain does not contain $P$; we take $C_P$ to be this domain.    □

# 3   Robustness

In this section we show how, in the case of tilesets that allow only periodic tilings, it is possible to reconstruct a valid tiling from one with a small proportion of errors by a local modification. The method is the same that was used to prove the error robustness of strongly aperiodic tilesets in [7,8]; since sets of periodic tilings have a very simple structure we are able to easily apply the same methods. We consider error distributions $e_\varepsilon$ such that each cell of $\mathbb{Z}^2$ has probability $\varepsilon$ to be equal to 1 independently of other points, i.e., a Bernoulli distribution.

We first describe a generic process [6,7] of sorting errors into "islands of errors" that given a repartition taken with a Bernoulli distribution transforms it into an error free repartition with probability one (Section 3.1). Then we present some structural results on tilesets that will allow us to apply this generic process to these tilesets (Section 3.2).

## 3.1   Iterative Cleaning Process

When $\varepsilon$ is small the intuition is that cells with value 1 will be sparse; we will not have big clusters of 1's; however big clusters have a non-null probability to appear thus they almost surely appear on the infinite plane $\mathbb{Z}^2$. Even if there is such an unavoidable problem, we will see that we can decompose error repartitions into different layers then repair each layer incrementally so that eventually everything gets repaired. First of all, let us define what a layer is:

**Definition 5.** *For an error repartition $e \in \{0,1\}^{\mathbb{Z}^2}$ and a point $x \in \mathbb{Z}^2$ such that $e(x) = 1$, $x$ is said to be in an $(i,j)$–island of $e$ if there exists no point at distance between $i$ and $j$ in $e$ that has value 1, that is:*

$$\forall y \in \mathbb{Z}^2, e(y) = 1 \Rightarrow |x - y| \notin [i;j]$$

We denote by $\mathcal{I}_{i,j}(e)$ the set of points of $\mathbb{Z}^2$ that are in an $(i,j)$–island of $e$. Figure 1 depicts some islands. These islands can be seen as isolated clusters of errors. The idea now is to remove the islands, hence our definition of an erasing function:

**Definition 6 (Erasing function).** *The function $\mathcal{E}_{i,j}$ from $\{0,1\}^{\mathbb{Z}^2}$ into itself erases the $(i,j)$–islands of an error repartition:*

$$\{0,1\}^{\mathbb{Z}^2} \to \qquad \{0,1\}^{\mathbb{Z}^2}$$
$$e \quad \to c : c(x) = \begin{cases} 0 & \text{if } x \in \mathcal{I}_{i,j}(e) \\ e(x) & \text{otherwise} \end{cases}$$

Now if we consider integer sequences $(\alpha_n)_{n \in \mathbb{N}}$ and $(\beta_n)_{n \in \mathbb{N}}$, we can think about applying successively the functions $\mathcal{E}_{\alpha_n, \beta_n}$ to an error repartition so that we obtain an iterative cleaning process [6,7] that erases small islands of errors, then bigger ones, then even bigger ones, etc.

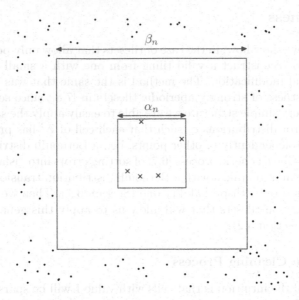

**Fig. 1.** Islands of rank $n$

**Definition 7 (Iterative cleaning process).** *Let $e$ be an error repartition and consider two sequences $\alpha_n$ and $\beta_n$. We define the iterative cleaning of $e$ by $(\alpha, \beta)$ denoted $(e_n^{\alpha,\beta})_{n \in \mathbb{N}}$ by:*

- $e_0^{\alpha,\beta} = e$
- $e_{n+1}^{\alpha,\beta} = \mathcal{E}_{\alpha_n, \beta_n}(e_n^{\alpha,\beta})$.

We call *"islands of rank $n$"* the points that we corrected at the $n^{th}$ iteration. This operation is pictured on Figure 1 where the black points are the points with value 1, those that will be "cleaned" at this step are marked by a cross, and we can see that we have a kind of security belt between $\alpha_n$ and $\beta_n$. The important part of this process is that cleaning the islands of rank $n$ at the $n^{th}$ iteration creates more islands that we will catch at the $n+1^{th}$ step. The following theorem catches this phenomenon:

**Theorem 1 ([6,7]).** *If the sequences $\alpha_n$ and $\beta_n$ match the conditions:*

- $\forall i, 8(\beta_1 + ... + \beta_i) < \alpha_{i+1} \leq \beta_{i+1}$
- $\sum_i \frac{\log(\beta_i)}{2^i} < \infty$

*Then there exists $\varepsilon > 0$ (sufficiently small) such that, almost surely, for any element of $\{0,1\}^{\mathbb{Z}^2}$ taken with the Bernoulli distribution of probability $\varepsilon$, this iterative cleaning process removes all the ones.*

We do not give a proof of this result and refer the reader to the original [6,7]. It is easy to check that the following sequences match the conditions of Theorem 1 for $n$ sufficiently large:

$$\alpha_n = 2^{n^2}$$

$$\beta_n = n\alpha_n$$

Moreover, what interests us is the fact that $\lim_{n \to \infty} \beta_n - \alpha_n = \infty$, which means that we can have arbitrary large security belts.

## 3.2   Error-Robustness of Periodic Tilesets

In this section we consider tilesets such that their only valid tilings are periodic. We describe how to use Theorem 1 for tilings.

With the process described in Section 3.1, when $n$ is sufficiently large, $\alpha_n - \beta_n$ is also large. In our application to tilings with errors, this means that the islands of rank $n$ are surrounded by a belt of width $\alpha_n - \beta_n$ that is correctly tiled. However, this zone is only locally tiled correctly. Lemma 1 ensures that when we have a sufficiently large correctly tiled pattern then a smaller part of it appears in a tiling of the whole plane. Hence, if $n$ is sufficiently large, the islands of rank $n$ are surrounded by a belt of large width, say $k$, where each pattern of size $k \times k$ appears in a tiling of the whole plane. One may remark that $k$ may be much smaller than $\alpha_n - \beta_n$ but $k$ can still be as big as we would like it to be if we take $n$ sufficiently big. This is depicted on Figure 2.

Now we want to replace the erroneous zone by other tiles so that the whole zone is tiled correctly but we are facing a problem: even if we can ensure that the islands of rank $n$ are surrounded by a belt of patterns that appear in a valid tiling of the whole plane, like on Figure 2, how can we be certain that the zone surrounded by the belt can be properly filled by tiles so that no error remains?

A pattern $P$ defined on $D$ is said to be $k$−extensible if there exists a decomposition of $D$ in $(\mathcal{D}_i)_{1 \le i \le n}$ such that:

 − $\cup_{1 \le i \le n} \mathcal{D}_i = D$
 − For every $i$, $P_{|\mathcal{D}_i}$ appears in a valid tiling of the plane
 − For every $i < n$, $\mathcal{D}_i \cap \mathcal{D}_{i+1}$ contains a ball of radius $k$.

With the help of the previous remarks, it is clear that for any $k$ there exists a sufficiently large $n$ such that the islands of rank $n$ are surrounded by a belt that is $k$−extensible.

It is already known that a tileset allows only a finite number of tiling if and only if it allows only periodic tilings [1, Theorem 3.8], hence we can focus on this simpler case:

**Lemma 2.** *If a tileset $\tau$ admits only a finite number of tilings then there exists $k$ such that any $k$−extensible pattern $P$ can be found in a tiling of the whole plane.*

*Proof.* There exists $\varepsilon > 0$ such that the distance between two different tilings is greater than $\varepsilon$. Let $k$ be an integer such that any two configurations that are equal at their center on the pattern defined on $[-k; k] \times [-k; k]$ are at distance strictly smaller than $\varepsilon$. We now prove that any $k$−extensible pattern can be found in a tiling of the whole plane.

Let $n$ and $(\mathcal{D}_i)_{1 \leq i \leq n}$ be such that $P$ is $k$−extensible over $(\mathcal{D}_i)_{1 \leq i \leq n}$. We may assume that $n = 2$, the result for any $n$ being obtained by an easy induction of this simpler case.

Let $a$ be a tiling containing $P_{|\mathcal{D}_1}$ and $b$ one containing $P_{|\mathcal{D}_2}$. There exists $a'$ and $b'$ that are shifted forms of, respectively, $a$ and $b$ such that $a'$ and $b'$ are equal on $P_{|\mathcal{D}_1} \cap P_{|\mathcal{D}_2}$ at their center. $a'$ and $b'$ are both tilings by $\tau$. Since $P_{|\mathcal{D}_1} \cap P_{|\mathcal{D}_2}$ contains a ball of radius $k$, by our choice of $k$ we obtain that $a' = b'$ because the distance between them is strictly lower than $\varepsilon$. Therefore $P$ appears in $a'$ which is a tiling of the whole plane.                                    □

Now, with $n$ sufficiently large, Lemma 2 tells us a bit more on the belt surrounding islands of rank $n$: there exists such a surrounding belt that appears in a valid tiling of the plane. On Figure 2 this means that the outer (blue) belt is in fact part of a tiling of the plane. This allows us to state our main theorem of robustness:

**Theorem 2.** *If a tileset allows only periodic tilings then it is robust to a small probability of errors.*

*Proof.* By Theorem 1 from [1], tilesets that allow only periodic tilings are exactly the tilesets that allow only a finite number of tilings. Hence Lemma 2 applies to these tilesets. By Lemmas 1 and 2, there exists $N$ such that any belt of width $N$ contains a smaller belt that appears in a tiling of the plane.

If we take $\alpha_n = 2^{(n+N)^2}$ and $\beta_n = (n + N)\alpha_n$, for every $n$, the belt has width at least $N$, hence the (finite) set of points surrounded by this belt can be filled

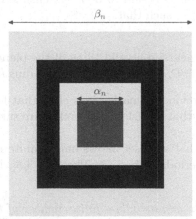

◼ : Locally correct.

■ : Possibly incorrect.

■ : Locally correct and appears in a tiling of the plane.

**Fig. 2.** Example of what happens with a sufficiently large security belt

by tiles from the tiling where the belt appears. Moreover these sequences match
the conditions of Theorem 1 thus we can repair every tiling taken with a small
probability of errors.                                                          □

We remark that this method of obtaining a valid tiling by surrounding the errors
with a security belt also works obviously for a tileset that allows everything
($\mathcal{T}_\tau = Q^{\mathbb{Z}^2}$). We can complicate it a little bit by considering $Q = \{0, 1, 2\}$ and
the tilings may contain only 1's and 2's or only 0's. Moreover it is possible to
use this method to obtain aperiodic tilings robust to errors [7,8].

## 4  Non Robustness

In the previous section we proved that there exist ways of correcting errors for
some tilesets. It would not make much sense to define exactly what we call a
"tileset robust to errors" since there may exist other methods for correcting
errors but it seems natural that every way of correcting errors has to be local,
hence our definition of local robustness:

**Definition 8 (Locally robust).** *We say that a tileset is locally robust if for any
finite repartition of errors (that contains only finitely many 1's), any tiling with
this error repartition can be transformed in a tiling without error by modifying
only finitely many cells.*

This definition can be related to the ground state configurations used by C.
Radin to model crystalline order [16,17]. Recall that a configuration is said to
be ground state if whenever we modify a finite part of it we do not decrease the
number of tiling errors in it. In that sense, locally robust tilesets are the tilesets
for which ground state configurations are either tilings without any error or have
an infinite number of tiling errors.

The structural results from Section 3 ensure that tilesets which allow only periodic tilings are locally robust: find $n$ such that all the errors are in the same island
of rank $n$ and replace the finite area surrounded by the belt by a valid one.

**Theorem 3.** *Tilesets that allow countably many tilings are not locally robust.*

First recall a structural result about such tilesets:

**Theorem 4 ([1]).** *If a tileset allows a countable number of tilings then it allows
one with exactly one direction of periodicity.*

**Corollary 1.** *If a tileset $\tau$ allows a countable number of tilings then it allows
one with exactly one direction of periodicity that can be seen as a bi-infinite word
$^{\omega}xyz^{\omega}$ with $|x| = |y| = |z|$ and $y \neq z$.*

*Proof.* Take a configuration $c$ with exactly one direction of periodicity $v$ from
Theorem 4. We can represent $c$ as a bi-infinite word $w$ on a finite alphabet
$\Sigma$. By representation we mean that we can decode letters of $\Sigma$ in blocks of tiles

such that when the blocks from $w$ are repeated along $v$ we obtain $c$. It is easy to prove that we can obtain from $\tau$ a tileset $\tau'$ over $\mathbb{Z}$ such that any tiling by $\tau'$ can be decoded in such a way in a tiling by $\tau$. Without loss of generality we may also assume that the neighborhood of $\tau'$ is $\{-1, 0, 1\}$: it suffices to code the tileset $\tau'$ into Wang tiles [4]. Even after all these codings, $\Sigma$ is still finite.

Since $\Sigma$ is finite, there exists $j > i > 0$ such that $w(i) = w(j)$. Therefore, the word $w$ is of the form $xayaz$ where $a$ is a letter of $\Sigma$ and $x$ and $z$ are infinite words.

If $x$ is equal to $^{\omega}(ay)$, $w$ is equal to $^{\omega}(ay)z'$. If $x$ is not equal to $^{\omega}(ay)$, $x(ay)^{\omega}$ is also allowed by $\tau$. In both cases, we obtain a bi-infinite word $w = x'(z')^{\omega}$ that is not periodic and $x'$ is an infinite word (the case where $x$ is equal to $^{\omega}(ay)$ being symmetric). Since $w$ is not periodic it can be written as $w = xy(z)^{\omega}$ where $|y| = |z|$ and $y \neq z$. We repeat the same argument on the infinite $x$ part of $w$ to write it in the desired form.                                                              □

This corollary gives us a configuration that we prove to be incompatible with local robustness. Such a configuration is depicted on Figure 3. Imagine now that we shift half a plane horizontally by one cell. We obtain a configuration like on Figure 4(a). This configuration has only a finite number of tiling errors: around the cell where we broke the vertical line of $y$'s. Now, what can we do in order to correct it? The only solution seems to shift back half a plane in order to restore the vertical line of $y$'s. We now prove it is indeed the only solution by obtaining a contradiction with our hypothesis that the set of all possible tilings is countable.

| $x$ | $x$ | $x$ | $x$ | $x$ | $y$ | $z$ | $z$ | $z$ | $z$ |
|-----|-----|-----|-----|-----|-----|-----|-----|-----|-----|
| $x$ | $x$ | $x$ | $x$ | $x$ | $y$ | $z$ | $z$ | $z$ | $z$ |
| $x$ | $x$ | $x$ | $x$ | $x$ | $y$ | $z$ | $z$ | $z$ | $z$ |
| $x$ | $x$ | $x$ | $x$ | $x$ | $y$ | $z$ | $z$ | $z$ | $z$ |
| $x$ | $x$ | $x$ | $x$ | $x$ | $y$ | $z$ | $z$ | $z$ | $z$ |
| $x$ | $x$ | $x$ | $x$ | $x$ | $y$ | $z$ | $z$ | $z$ | $z$ |

**Fig. 3.** Example of a tiling from Corollary 1

*Proof (of Theorem 3).* Let $c$ be a tiling as in Corollary 1: $c$ is of the form $^{\omega}xyz^{\omega}$ with $|x| = |y| = |z| = p$ and $y \neq z$. We modify $c$ by shifting half a plane by $p$ to obtain a configuration like depicted on Figure 4(a). Since this transformation breaks only finitely many tiling rules, suppose that we can correct this by modifying only a finite number of cells. We obtain a tiling like on Figure 4(b) where we have a semi-infinite line of $y$'s in one direction and a shifted semi-infinite line of $y$'s in the other direction, separated by a pattern that repaired the error.

We obtained a transformation of the tiling and can repeat it on every sufficiently long vertical line of $y$'s. This transformation gives us a different tiling each time we apply it since $y \neq z$. Such vertical lines of $y$'s appear infinitely many times, therefore we obtain $2^{\aleph_0}$ valid tilings for $\tau$, a contradiction.     □

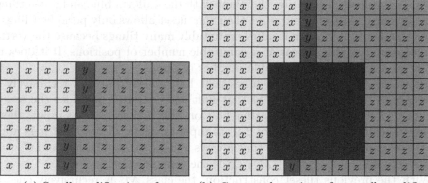

(a) Small modification of $c$     (b) Corrected version of a small modification of $c$

**Fig. 4.** Transformations of $c$

# 5 Recursive Inseparability of Robust and Non-locally Robust Tilesets

Tilesets that allow countably many tilings are not locally robust (Theorem 3) while tilesets that allow only periodic tilings are robust to errors (Theorem 2). In this section we prove that those classes of tilesets are not recursively separable, hence neither are robust and non locally robust tilesets.

**Theorem 5.** *Robust tilesets and non-locally robust tilesets are recursively inseparable.*

*Proof.* We will assume the reader familiar with the encoding of Turing machines into tilesets à la Berger [2], if not please refer to the detailed constructions in [3,15]. Every tiling by such a tileset contains arbitrary large squares on which we can force the bottom left corner. This corner is where we put the start of a Turing machine computation. We can see the rows of these squares as the Turing machine's tape, time is going bottom-up. Then we do the following:

– If the machine halts and outputs 1 then we force a periodic tiling: when this halting state reaches the border of the square we force a new border such that the only way to tile the plane is to repeat periodically this square. This is exactly what is done in [3, Appendix A].
– If the machine does not halt, the tileset tiles aperiodically with an infinite computation of the Turing machine inside.
– If the machine halts and outputs 2 then force the periodicity vertically but allow only a new color, blue, that forces a monochromatic half plane at its left and another color at its right, green, that also forces a monochromatic half plane.

This new tileset always tiles the plane with the uniform blue and green tilings. If the machine halts and outputs 1 then the tileset allows only periodic tilings. If it halts and outputs 2 then it allows countably many tilings because the vertical computation line can appear at a countable number of positions. If it does not halt then it allows an uncountable number of tilings.

The class $\mathcal{M}_1$ of Turing machines that halt and output 1 is recursively inseparable of the class $\mathcal{M}_2$ of TM that halt and output 2: consider the sets $C_i$ ($i \in \{1,2\}$) of Turing machines that halt on $i$ with their code as input, then a Turing machine $M$ that outputs 2 if its input is not in $C_2$ and 1 if it is not in $C_1$; there exists such a machine that always halts if we suppose $C_1$ and $C_2$ to be recursively separable and $M(\langle M \rangle)$ gives the contradiction.

With the previous tileset construction, the class $\mathcal{M}_1$ is recursively encoded into tilesets that are robust to errors and the class $\mathcal{M}_2$ into tilesets that are not locally robust. □

## 6    Conclusions and Open Problems

In this paper we have shown how it is possible to repair tiling errors (as defined in Section 3) on very simple tilesets (the periodic ones). This correction process relies on the fact that we can surround errors by correct zones which we are sure appear in a valid tiling of the plane; while this is true for periodic tilesets we remark that there exists some other tilesets for which it is also true, it would be interesting to obtain a characterization of such tilesets.

We also proved that tilesets that allow countably many tilings are not locally robust; being locally robust is a necessary condition for being able to apply our iterative correction process. While this iterative correction process allows to repair periodic tilesets and even some aperiodic ones [7,8] we believe there exists some tilesets that would be locally robust but on which this iterative correction process will not work.

The recursive inseparability between tilesets that we are able to repair and the ones that are not locally robust shows that there is no simple characterization of error-robustness for tilesets.

## References

1. Ballier, A., Durand, B., Jeandel, E.: Structural aspects of tilings. In: 25th International Symposium on Theoretical Aspects of Computer Science, STACS (2008)
2. Berger, R.: The undecidability of the domino problem. Memoirs of the American Mathematical Society 66 (1966)
3. Börger, E., Grädel, E., Gurevich, Y.: The Classical Decision Problem. In: Perspectives in Mathematical Logic. Springer, Heidelberg (1997)
4. Cervelle, J., Durand, B.: Tilings: recursivity and regularity. Theoretical computer science 310(1-3), 469–477 (2004)
5. Durand, B., Levin, L.A., Shen, A.: Complex Tilings. In: STOC, pp. 732–739 (2001)
6. Durand, B., Romashchenko, A.E.: On stability of computations by cellular automata. In: European Conference on Complex Systems (2005)

7. Durand, B., Romashchenko, A.E., Shen, A.: Fixed point and aperiodic tilings. In: Ito, M., Toyama, M. (eds.) DLT 2008. LNCS, vol. 5257, pp. 276–288. Springer, Heidelberg (2008)

8. Durand, B., Romashchenko, A.E., Shen, A.: High complexity tilings with sparse errors. In: Albers, S., Marchetti-Spaccamela, A., Matias, Y., Nikoletseas, S.E., Thomas, W. (eds.) ICALP 2009. LNCS, vol. 5555, pp. 403–414. Springer, Heidelberg (2009)

9. Gacs, P.: Reliable cellular automata with self-organization. J. of Stat.Phys., 103 (2001)

10. Gurevich, Y., Koriakov, I.: A remark on Berger's paper on the domino problem. Siberian Journal of Mathematics 13, 459–463 (1972) (in Russian)

11. Hanf, W.P.: Nonrecursive tilings of the plane. i. J. Symb. Log. 39(2), 283–285 (1974)

12. Lind, D.A.: Multidimensional Symbolic Dynamics. In: Symbolic dynamics and its applications. In: Proceedings of Symposia in Applied Mathematics, vol. 11, pp. 61–80 (2004)

13. Lind, D.A., Marcus, B.: An Introduction to Symbolic Dynamics and Coding. Cambridge University Press, New York (1995)

14. Myers, D.: Nonrecursive tilings of the plane. ii. J. Symb. Log. 39(2), 286–294 (1974)

15. Ollinger, N.: Two-by-two substitution systems and the undecidability of the domino problem. In: Beckmann, A., Dimitracopoulos, C., Löwe, B. (eds.) CiE 2008. LNCS, vol. 5028, pp. 476–485. Springer, Heidelberg (2008)

16. Radin, C.: Global order from local sources. Bulletin of the American Mathematical Society 25, 335–364 (1991)

17. Radin, C.: $\mathbb{Z}^n$ versus $\mathbb{Z}$ actions for systems of finite type. Contemporary Mathematics 135, 339–342 (1992)

18. Reif, J.H., Sahu, S., Yin, P.: Compact error-resilient computational DNA tiling assemblies. In: Ferretti, C., Mauri, G., Zandron, C. (eds.) DNA 2004. LNCS, vol. 3384, pp. 293–307. Springer, Heidelberg (2005)

19. Robinson, R.M.: Undecidability and nonperiodicity for tilings of the plane. Inventiones Mathematicae 12, 177–209 (1971)

20. Wang, H.: Proving theorems by pattern recognition II. Bell System Technical Journal 40, 1–41 (1961)

21. Wang, H.: Dominoes and the ∀∃∀ case of the decision problem. In: Mathematical theory of Automata, pp. 23 55 (1963)

22. Winfree, E., Bekbolatov, R.: Proofreading tile sets: Error correction for algorithmic self-assembly. In: Chen, J., Reif, J.H. (eds.) DNA 2003. LNCS, vol. 2943, pp. 126–144. Springer, Heidelberg (2004)

# Visiting a Sequence of Points with a Bevel-Tip Needle[*]

Steven Bitner[1], Yam K. Cheung[1], Atlas F. Cook IV[2], Ovidiu Daescu[1],
Anastasia Kurdia[1], and Carola Wenk[2]

[1] University of Texas at Dallas, Department of Computer Science,
Richardson, TX, USA
{sbitner,ykcheung,daescu,akurdia}@utdallas.edu
[2] University of Texas at San Antonio, Department of Computer Science,
San Antonio, TX, USA
{acook,carola}@cs.utsa.edu

**Abstract.** Many surgical procedures could benefit from guiding a bevel-tip needle along circular arcs to multiple treatment points in a patient. At each treatment
point, the needle can inject a radioactive pellet into a cancerous region or extract a tissue sample. Our main result is an algorithm to steer a bevel-tip needle
through a sequence of treatment points in the plane while minimizing the number of times that the needle must be reoriented. This algorithm is related to [6]
and takes quadratic time when consecutive points in the sequence are sufficiently
separated. We can also guide a needle through an arbitrary sequence of points in
the plane by accounting for a lack of optimal substructure.

**Keywords:** Needle Steering, Link Distance, Brachytherapy, Biopsy.

## 1 Introduction

Many surgical procedures could benefit from guiding a bevel-tip needle through a sequence of treatment points in a patient. For example, brachytherapy procedures per-
manently implant radioactive seeds that irradiate the surrounding cancerous tissue, and
biopsy procedures extract tissue samples to test for cancer.

A bevel-tip needle has an asymmetric tip that cuts through tissue along a circular
arc. By rotating the needle during its insertion into soft tissue, the needle can be steered
along a sequence of circular arcs so that it reaches a treatment point while avoiding
bones and vital organs. Although unpredictable deflections can occur due to tissue het-
erogeneity, these deflections can be accounted for by taking snapshots of the needle's
position during surgery [3].

Although current procedures typically create a new puncture for each treatment
point, our work permits multiple treatment points to be visited with a single puncture.
Since each reorientation of the needle inherently complicates the path for the physician,
we minimize the number of arcs that are required for the needle to visit a sequence of
treatment points.

[*] This work has been supported by NSF CAREER CCF-0643597, CCF-0635013, and the 2009
University of Texas at San Antonio Presidential Dissertation Fellowship.

A. López-Ortiz (Ed.): LATIN 2010, LNCS 6034, pp. 492–502, 2010.

## 1.1  Previous Work

The path traveled by a bevel-tip needle is theoretically equivalent to a restricted type of Dubins path [7] that can move along circular arcs but cannot move straight. Needle paths also have similarities to touring problems [6], curvature-constrained shortest path problems [1,4,7], and link distance problems [5,9,10]. Since curvature-constrained paths are NP-Hard to compute in the plane with polygonal obstacles [1], recent work has explored curvature-constrained paths in convex polygons [1] and narrow corridors [4].

Alterovitz et al. [3] consider steering a needle through a plane with polygonal obstacles to a single treatment point. They calculate an optimal entry point for a needle and assume that no rotations are performed during the procedure. Subsequent work in the plane by Alterovitz et al. [2] permits the needle to rotate during its insertion into soft tissue. Their dynamic programming approach uses a Markov model to maximize the probability of successfully reaching a single treatment point without bumping into an obstacle.

Several studies have steered a needle in three dimensions. Duindam et al. [8] guide a needle to a single treatment point in a three dimensional setting without obstacles. Xu et al. [11] use a probabilistic roadmap to reach a single treatment point in a three-dimensional space that contains spherical obstacles.

Related work also exists in the realm of robotic motion. Robots typically have a much easier time moving straight than turning, so the "length" of a path is typically measured by the number of turns on the path instead of the Euclidean length of the path [9]. Steering a needle is similar in the sense that reorienting the needle during the insertion process complicates the path for the physician.

## 1.2  Our Results

We always assume that the needle travels on circular arcs with a fixed radius $r$. Note that this radius can be controlled to some extent by varying the stiffness of the needle. Although deflections can in practice lead to deviations from an ideal path, the location of the needle can be periodically sampled during a procedure to account for these deflections.

Our main result is an algorithm to steer a needle through an (ordered) sequence of treatment points in a plane without obstacles. When consecutive points in this sequence are sufficiently separated, optimal substructure holds, and we can iteratively propagate all optimal paths through the sequence. Although optimal substructure need not hold for an arbitrary sequence of points, we can still propagate all optimal paths through the sequence by accounting for some locally suboptimal paths. We also discuss how to steer a needle between two given points in $\mathbb{R}^d$.

To our knowledge, no previous algorithm exists to steer a needle through *multiple* treatment points. However, another work has previously explored Euclidean shortest paths that tour a sequence of polygons [6].

## 2  Needle Steering in $\mathbb{R}^d$

A bevel-tip needle has an asymmetric tip that cuts through soft tissue along a circular arc. Each rotation of the needle during its insertion into tissue causes the needle to start moving along a new circular arc that is tangent to the previous circular arc [2]. A *needle arc* is a directed circular arc that lies on a *needle circle* with radius $r$. Figure 1a illustrates that a *needle path* $\pi_N(s, t)$ is a connected sequence of needle arcs that connects two points $s$ and $t$. The *length* $d_N(s, t)$ of a needle path equals the smallest possible number of needle arcs that can be used to connect $s$ and $t$. Consecutive needle arcs of $\pi_N(s, t)$ lie on tangent needle circles, and consecutive needle arcs always have opposite orientations (e.g., if the current arc travels clockwise, then the next arc must travel counter-clockwise). A *needle vector* is a vector that is tangent to the needle's current position on a needle arc.

A shortest path map can quickly return a needle path $\pi_N(s, t)$ from some fixed source point $s \in \mathbb{R}^d$ to any treatment point $t \in \mathbb{R}^d$. Assuming that the initial needle vector can be chosen arbitrarily, all points within Euclidean distance $2r$ of $s$ can be reached with one needle arc. More generally, for any integer $j \geq 1$ all points $\{t \in \mathbb{R}^d \mid (j - 1) \cdot 2r < \|s - t\| \leq j \cdot 2r\}$ can be reached by at least one needle path that is composed of $j$ needle arcs (see Figure 1a).[1] Thus, the length of a needle path from $s$ to $t$ is $d_N(s, t) = \lceil \frac{\|s-t\|}{2r} \rceil$. The needle arcs of $\pi_N(s, t)$ can be constructed by stacking a sequence of needle circles with centers on the line segment $\overline{st}$. Note that the final circle on $\pi_N(s, t)$ should be tangent to $t$ and may not have its center on $\overline{st}$. We can now define a shortest path map $\text{SPM}_N(s)$ as a partition of $\mathbb{R}^d$ into an interior-disjoint set of layers $\mathcal{L}^1, ..., \mathcal{L}^M$ such that all points $t \in \mathcal{L}^j$ have $d_N(s, t) = j$. Such a shortest path map is an arrangement of concentric hyperspheres that are centered at $s$ and have radii $j \cdot 2r$ for all integers $j \geq 1$ (see Figure 1a).

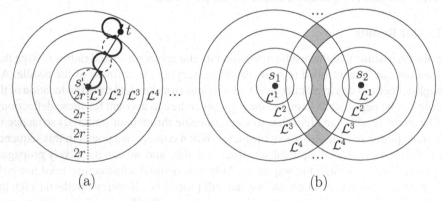

**Fig. 1.** (a) The shortest path map $\text{SPM}_N(s)$ is an arrangement of concentric hyperspheres. A needle path $\pi_N(s, t)$ with four needle arcs is shown. (b) The shaded bisector of $\text{SPM}_N(s_1)$ and $\text{SPM}_N(s_2)$ is a superset of the Euclidean bisector for $s_1, s_2$.

Given multiple candidate start positions $\{s_1, ..., s_n\}$ for the needle, we may wish to quickly determine a nearest starting point to a given query point $t \in \mathbb{R}^d$. Such a task can be efficiently accomplished using a traditional Euclidean Voronoi diagram. The *bisector* for any two points $s_i, s_j$ is the set of all points $\{t \in \mathbb{R}^d \mid d_N(s_i, t) = d_N(s_j, t)\}$ (see Figure 1b). This bisector is always a superset of the Euclidean bisector of $s_i, s_j$

---

[1] The Euclidean distance between points $s$ and $t$ is denoted $\|s - t\|$.

because being equidistant with respect to Euclidean distance implies being equidistant with respect to the length of a needle path.

## 3   Visiting a Sequence of Well-Separated Points in the Plane

This section shows how to compute a needle path $\pi_N(p_1, ..., p_n)$ in the plane that starts at a point $p_1$, visits a sequence of points $p_2, ..., p_{n-1}$, and ends at a point $p_n$. A needle path is said to *visit* a sequence of points when these points appear in order along the path. Let $d_N(p_1, ..., p_n)$ be the number of needle arcs on $\pi_N(p_1, ..., p_n)$. Assume for now that the needle vector is allowed to be arbitrary when the needle visits each point $p_i$. We call a sequence of points $p_1, \ldots, p_n$ *moderately-separated* whenever $||p_i - p_{i-1}|| \geq 2r$ for all $2 \leq i \leq n$ and *well-separated* whenever $||p_i - p_{i-1}|| \geq 4r$ for all $2 \leq i \leq n$. Given an optimal path $\pi_N(p_1, ..., p_n)$, *optimal substructure* guarantees that all possible subpaths $\pi_N(p_i, ..., p_j)$ for $1 \leq i \leq j \leq n$ must also be optimal.

**Lemma 1.** *Optimal substructure need not hold for a needle path that visits an arbitrary sequence of points. Optimal substructure does hold for moderately-separated points.*

*Proof.* Figure 2a illustrates a needle path $\pi_N(p_1, p_2, p_3)$ that does not have optimal substructure. Although a locally optimal path from $p_1$ to $p_2$ is possible with one needle arc, all possible extensions of such a path can require a total of three needle arcs to reach $p_3$. By contrast, a globally optimal path can reach $p_3$ with just two needle arcs if a locally suboptimal path with two needle arcs is used to connect $p_1$ to $p_2$.

To see that optimal substructure does hold for moderately-separated points, consider two paths to $p_i$. Let $c_o$ be a needle circle that intersects $p_i$ and lies on an (optimal) needle path $\pi_N(p_1, ..., p_i)$ with length $d_N(p_1, ..., p_i) = j$. Let $c_s$ be a needle circle that intersects $p_i$ and lies on a suboptimal path with length at least $j+1$. Figure 2b illustrates that the shaded set of all points reachable from $c_o$ with one additional needle arc will always contain all points on the suboptimal circle $c_s$ that lie outside the disk $d_o$ that is bounded by $c_o$. This implies that for any point $p_{i+1} \notin d_o$, there always exists a path $\pi_N(p_1, ..., p_{i+1})$ that contains $c_o$ and has the same length as any path $\pi'_N(p_1, ..., p_{i+1})$ that contains $c_s$. Since moderately-separated points guarantee that $p_{i+1} \notin d_o$, optimal substructure holds for moderately-separated points.    □

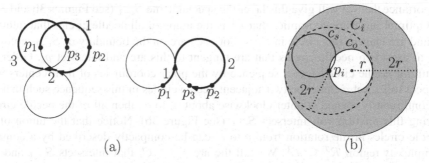

**Fig. 2.** (a) Optimal substructure need not hold for a needle path that visits an arbitrary sequence of points. (b) Optimal substructure does hold for moderately-separated points.

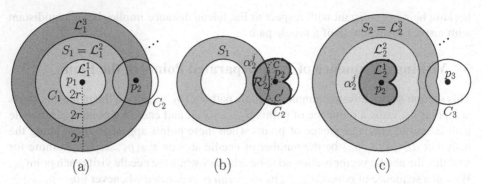

**Fig. 3.** (a) The sequence of layers $\mathcal{L}_1^1, ..., \mathcal{L}_1^M$ for $p_1$ can always be described by the shortest path map $\mathrm{SPM}_N(p_1)$. We refer to $S_1$ as the first layer in this sequence that intersects $C_2$. (b,c) Layer $\mathcal{L}_2^1$ is the set of all needle circles that intersect both $p_2$ and $S_1$. Layer $S_2$ is the first layer in the sequence $\mathcal{L}_2^1, ..., \mathcal{L}_2^M$ that intersects $C_3$.

For each point $p_i$, we can now define a partition of the plane such that each point $t \in \mathbb{R}^2$ has an associated distance $d_N(p_1, ..., p_i, t)$. Such a partition can be described by a sequence of pairwise disjoint layers $\mathcal{L}_i^1, ..., \mathcal{L}_i^M$ such that all points $t \in \mathcal{L}_i^j$ have associated distance $d_N(p_1, ..., p_i) + j - 1$ (see Figure 3). Note that all needle circles composing $\mathcal{L}_i^1$ must necessarily intersect $p_i$, so the boundary of $\mathcal{L}_i^1$ is a connected sequence of needle arcs.

Although the number of layers needed to partition the plane can be unbounded, the optimal substructure of moderately-separated points ensures that it is always possible to efficiently construct any layer $\mathcal{L}_i^j$ directly from $\mathcal{L}_i^1$ because all needle paths $\pi_N(p_1, ..., p_i, t)$ must pass through $\mathcal{L}_i^1$. In particular, any layer $\mathcal{L}_i^{j \geq 2}$ is the set of all points $\{t \in \mathbb{R}^d \mid (j - 2) \cdot 2r < \min_{s \in \mathcal{L}_i^1} \|s - t\| \leq (j - 1) \cdot 2r\}$. The boundary separating $\mathcal{L}_i^{j \geq 2}$ and $\mathcal{L}_i^{j+1}$ can be computed by additively *enlarging* the radius of arcs on the boundary of $\mathcal{L}_i^1$ by $(j - 1) \cdot 2r$ while keeping the center of each arc fixed.

We now show how to compute $\mathcal{L}_i^1$ from $\mathcal{L}_{i-1}^1$ for a sequence of moderately-separated points. Let $C_i$ be the circle with radius $2r$ that is centered at the point $p_i$. Notice that each needle circle in $\mathcal{L}_i^1$ must intersect both $p_i$ and some unique point of $C_i$. For our purposes, the first layer in the sequence $\mathcal{L}_{i-1}^1, ..., \mathcal{L}_{i-1}^M$ that intersects $C_i$ will be of such importance that we will give this layer the special name $S_{i-1}$ (see Figures 3b and 3c).

Optimal substructure implies that $\mathcal{L}_i^1$ is the union of all needle circles that intersect $p_i$ and are tangent to a circle in $S_{i-1}$. For each arc on the boundary of $S_{i-1}$, compute the at most two needle circles that are tangent to this arc and intersect $p_i$. Order the resulting needle circles into a sequence by the polar coordinates of their centers with respect to $p_i$. Let $c$ and $c'$ be two adjacent needle circles in this sequence such that if $c$ is continuously rotated counter-clockwise about $p_i$ to $c'$, then all of the needle circles during this rotation will intersect $S_{i-1}$ (see Figure 3b). Notice that the union of all needle circles in the rotation from $c$ to $c'$ can be compactly described by a constant complexity region $\mathcal{R}_i^j \subseteq \mathcal{L}_i^1$. We call the arc $\alpha_i^j \subseteq C_i$ that intersects $S_{i-1}$ and lies between $c$ and $c'$ a *generating arc* because $\alpha_i^j$ encodes all of the information necessary to generate the region $\mathcal{R}_i^j$.

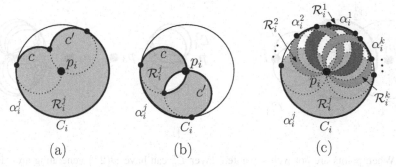

**Fig. 4.** (a) Region $\mathcal{R}_i^j \subseteq \mathcal{L}_i^1$ is bounded by $\alpha_i^j \subseteq C_i$ and two circular arcs. (b) Whenever a generating arc $\alpha_i^j$ covers less than half of $C_i$, the points in the interior of $c \cap c'$ are contained in layer $\mathcal{L}_i^2$. (c) In general, layer $\mathcal{L}_i^1$ is the union of a sequence of regions $\mathcal{R}_i^1, ..., \mathcal{R}_i^k$. These regions can be computed from the associated sequence of generating arcs $\alpha_i^1, ..., \alpha_i^k \subseteq C_i$.

Each region $\mathcal{R}_i^j$ is always bounded by its associated generating arc $\alpha_i^j \subseteq C_i$ and arcs of $c$ and $c'$ (see Figure 4a). Whenever a generating arc $\alpha_i^j$ covers less than half of $C_i$, the points in the interior of $c \cap c'$ are contained in layer $\mathcal{L}_i^2$ (see Figure 4b). Optimal substructure ensures that all layers $\mathcal{L}_i^{j \geq 3}$ will be connected.

Notice that a layer $\mathcal{L}_i^1$ can have multiple associated generating arcs. Consequently, $\mathcal{L}_i^1$ can be defined as the union of a sequence of regions $\mathcal{R}_i^1, ..., \mathcal{R}_i^k$ (see Figure 4c). In the remainder of this section, we will avoid computing an explicit union of these regions and instead work directly with the individual generating arcs and regions that compose a layer.

**Lemma 2.** *A sequence of points $p_1, ..., p_n$ that are not well-separated can define a layer $\mathcal{L}_n^1$ with $\Theta(2^n)$ generating arcs.*

*Proof.* The lower bound can be realized by placing $p_1, ..., p_n$ on a line such that $2r < ||p_i - p_{i-1}|| < 4r$ for all $2 \leq i \leq n$. As illustrated in Figure 5a, layer $\mathcal{L}_1^1$ is a disk. Place $p_2$ to the right of $p_1$ so that layer $\mathcal{L}_2^1$ is defined by one generating arc $\alpha_2^1 \subset C_2$ that represents all needle circles that intersect $p_2$ and are tangent to some needle circle in $\mathcal{L}_1^1$. Place $p_3$ to the right of $p_2$ such that only two discrete needle circles intersect $p_3$ and are tangent to some needle circle in $\mathcal{L}_2^1$. Each of these needle circles defines a (degenerate) generating arc, and the union of these two needle circles is $\mathcal{L}_3^1$. Place $p_4$ to the left of $p_3$ such that four discrete needle circles intersect $p_4$ and are tangent to some needle circle in $\mathcal{L}_3^1$. The union of these four needle circles is $\mathcal{L}_4^1$ (see Figure 5b). Place $p_5$ to the right of $p_4$ such that eight discrete needle circles define $\mathcal{L}_5^1$ (see Figure 5c). Continuing in this fashion for every $i \geq 3$, we can place $p_{2i}$ to the left of $p_{2i-1}$ and $p_{2i+1}$ to the right of $p_{2i}$ such that layer $\mathcal{L}_n^1$ is composed of $2^{n-2}$ circles. Thus, $\mathcal{L}_n^1$ can have $\Omega(2^n)$ generating arcs.

We now show that layer $\mathcal{L}_n^1$ has $O(2^n)$ generating arcs. Assume that $\mathcal{L}_{i-1}^1$ is defined by $k$ generating arcs. By definition, each region associated with these generating arcs is bounded by an arc on $C_{i-1}$ and at most two additional arcs that can lie on unique circles (see Figure 4a). This implies that the arcs bounding any fixed layer $\mathcal{L}_{i-1}^j$ lie on at most

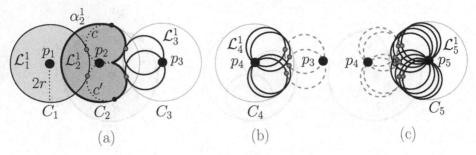

**Fig. 5.** When points are not well-separated, layer $\mathcal{L}_n^1$ can have $\Theta(2^n)$ generating arcs. Part (a) shows the disk defining $\mathcal{L}_1^1$, the generating arc $\alpha_2^1$ defining $\mathcal{L}_2^1$, and the two circles defining $\mathcal{L}_3^1$. Parts (b) and (c), respectively, illustrate the four circles defining $\mathcal{L}_4^1$ and the eight circles defining $\mathcal{L}_5^1$.

$2k + 1$ unique circles. Now consider the circle $C_i$. Since any pair of circles intersect at most twice, $C_i$ intersects each of the $2k+1$ unique circles for $S_{i-1}$ at most twice. Since each generating arc of $\mathcal{L}_i^1$ can always be charged to two unique intersections of $C_i$ with these circles, $\mathcal{L}_i^1$ is defined by at most $2k + 1$ generating arcs. An inductive argument now ensures that $\mathcal{L}_n^1$ has $O(2^n)$ generating arcs.                                  □

Exponential behavior occurred above because $C_i$ was able to pass through $\mathcal{L}_{i-1}^1$ many times. However, if we assume that $\|p_i - p_{i-1}\| \geq 4r$, then $C_i$ intersects $\mathcal{L}_{i-1}^1$ in at most one point. This property is used in the following lemma to show that when the points $p_1, ..., p_n$ are well-separated, each layer $\mathcal{L}_i^1$ is defined by $O(i)$ generating arcs.

**Lemma 3.** *Given both a sequence of well-separated points $p_1, ..., p_n$ and the sequence of generating arcs defining layer $\mathcal{L}_{i-1}^1$, there exist $O(i)$ generating arcs for any layer $\mathcal{L}_i^1, ..., \mathcal{L}_i^M$. These $O(i)$ generating arcs can be constructed in $O(i)$ total time.*

*Proof.* Assume that $\mathcal{L}_{i-1}^1$ is defined by a sequence of $k$ sorted, pairwise disjoint generating arcs $\alpha_{i-1}^1, ..., \alpha_{i-1}^k \subseteq C_{i-1}$. Recall that $S_{i-1}$ denotes the first layer in the sequence $\mathcal{L}_{i-1}^1, ..., \mathcal{L}_{i-1}^M$ that intersects $C_i$. If $S_{i-1} = \mathcal{L}_{i-1}^1$, then the well-separated property ensures that at most one circle defines $\mathcal{L}_{i-1}^1$. Otherwise, $S_{i-1}$ equals some layer $\mathcal{L}_{i-1}^{j \geq 2}$, and the well-separated property ensures that $C_i$ only intersects the connected component of $S_{i-1}$ that lies outside the disk bounded by $C_{i-1}$. Such a connected component has two closed boundaries, and we let $\mathcal{B}$ be the boundary that is farthest from $p_{i-1}$. Observe that $\mathcal{B}$ is a closed sequence of arcs (see Figure 6). Every third arc of $\mathcal{B}$ is an additively enlarged version of a generating arc such that all additively enlarged generating arcs lie on one fixed circle. Adjacent additively enlarged generating arcs in $\mathcal{B}$ are always connected by two arcs that intersect in one concave vertex. Hence, there are at most $k$ concave vertices on $\mathcal{B}$.

We now bound the number of generating arcs that can define $\mathcal{L}_i^1$ by counting the number of times that $C_i$ can intersect $\mathcal{B}$. There are two types of intersections to consider. First, $C_i$ can intersect $\mathcal{B}$ at most twice over all additively enlarged generating arcs because these arcs all lie on a single circle. Second, $C_i$ can intersect $\mathcal{B}$ at most twice

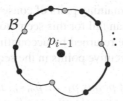

**Fig. 6.** An arbitrary circle $C_i$ intersects the boundary $\mathcal{B} \subseteq S_{i-1}$ at most twice for each of the gray concave vertices on $\mathcal{B}$ and at most twice over all points on the thick additively enlarged generating arcs

for each of the $k$ concave vertices on $\mathcal{B}$. Since a generating arc is always defined by *two* unique intersections, the at most $2k + 2$ intersections of $C_i$ with $\mathcal{B}$ ensure that $\mathcal{L}_i^1$ is defined by at most $k + 1$ generating arcs. An inductive argument can now easily show that $O(i)$ generating arcs define $\mathcal{L}_i^1$. These generating arcs can be constructed in sorted order on $C_i$ in $O(i)$ total time by traversing the boundary $\mathcal{B}$ in sorted order.     □

We can now compute a needle path $\pi_N(p_1, ..., p_n)$ that passes through a sequence of well-separated points.

**Theorem 1.** *Given a sequence $p_1, ..., p_n$ of well-separated points, a needle path $\pi_N$ $(p_1, ..., p_n)$ can be computed in $O(n^2 + K)$ time and space, where $K$ is the complexity of the returned path.*

*Proof.* The generating arcs for $\mathcal{L}_i^1$ and $S_i$ can be constructed from the generating arcs of $\mathcal{L}_{i-1}^1$ in $O(i)$ time by Lemma 3. Hence, the generating arcs for $S_1, ..., S_{n-1}$ and $\mathcal{L}_n^1$ can be constructed in $O(n^2)$ total time.

To compute $\pi_N(p_1, ..., p_n)$, pick any circle $c_n \subseteq \mathcal{L}_n^1$. This circle intersects some point $q \in S_{n-1}$. To connect $q \in S_{n-1}$ to $p_{n-1}$, draw a line segment from $p_{n-1}$ to $q$ and find the last intersection point $q'$ of this line segment with the boundary of $\mathcal{L}_{n-1}^1$. Draw the circle $c_{n-1}$ that touches $p_{n-1}$ and $q'$. To connect the two circles $c_{n-1}$ and $c_n$, identify the closest pair of points $r \in c_{n-1}, r' \in c_n$ (with respect to Euclidean distance) and create a chain of tangent needle circles along the line segment from $r$ to $r'$. Note that the needle circle touching $r'$ may need to be rotated slightly about the previous circle in the chain to ensure that this circle is tangent to $c_n$. We have now found an optimal path from $p_n$ to $p_{n-1}$, and this process can be iteratively repeated from $c_{n-1}$ to construct the remainder of $\pi_N(p_1, ..., p_n)$.     □

Note that if we require the needle to have a specific needle vector when it visits each $p_i$, then each layer $\mathcal{L}_i^1$ consists of the at most two needle circles that are tangent to this vector. For this scenario, our layers technique can return a needle path in only $O(n+K)$ time and space, where $K$ is the complexity of the returned path.

## 4 Visiting a Sequence of Arbitrary Points in the Plane

This section describes a fixed-parameter tractable algorithm that computes a needle path for a sequence of points $p_1, ..., p_{m+n}$ such that any $m$ pairs of consecutive points are positioned arbitrarily and the remaining pairs of consecutive points are well-separated. Note that optimal substructure can fail for this scenario as depicted by Figure 2a. However, an optimal path can still be returned by accounting for locally suboptimal needle circles that touch multiple consecutive points in the sequence.

**Lemma 4.** *Given a sequence of points $p_1, ..., p_{m+n}$, the generating arcs for layer $\mathcal{L}_i^1$ can be constructed in time proportional to the number of generating arcs defining $\mathcal{L}_{i-1}^1$ and $\mathcal{L}_{i-2}^1$.*

*Proof.* If $||p_i - p_{i-1}|| \geq 2r$, then optimal substructure holds by Lemma 1, and we can compute the generating arcs for $\mathcal{L}_i^1$ from the generating arcs of $\mathcal{L}_{i-1}^1$ as in the previous section. Now assume that $||p_i - p_{i-1}|| < 2r$, and let $\gamma = d_N(p_1, ..., p_{i-1})$. Our first step is to determine the value of $d_N(p_1, ..., p_i)$.

If $p_i \in \mathcal{L}_{i-1}^1$, then we can immediately return $d_N(p_1, ..., p_i) = \gamma$. If there exists any fixed needle circle $c \in \mathcal{L}_{i-1}^1$ that does not surround $p_i$, then the precondition $||p_i - p_{i-1}|| < 2r$ ensures that $d_N(p_1, ..., p_i) = \gamma + 1$. The only other way to obtain $d_N(p_1, ..., p_i) = \gamma + 1$ is for a locally suboptimal path with total length $\gamma + 1$ to end in a needle circle that touches both $p_i$ and $p_{i-1}$. Such paths can be accounted for by determining whether the layer $\mathcal{L}_{i-2}^j$ with associated distance $\gamma + 1$ contains a needle circle that touches both $p_i$ and $p_{i-1}$. Otherwise, the precondition $||p_i - p_{i-1}|| < 2r$ ensures that $d_N(p_1, ..., p_i) = \gamma + 2$. This means that we can compute $d_N(p_1, ..., p_i)$ in time proportional to the number of generating arcs defining $\mathcal{L}_{i-1}^1$ and $\mathcal{L}_{i-2}^1$.

We now use the value of $d_N(p_1, ..., p_i)$ to compute $\mathcal{L}_i^1$. If $d_N(p_1, ..., p_i) = \gamma$, then $\mathcal{L}_i^1$ is composed of the at most two circles that touch both $p_i$ and $p_{i-1}$ and are contained in $\mathcal{L}_{i-1}^1$ (see Figure 7a).

If $d_N(p_1, ..., p_i) = \gamma + 1$, then we can partially compute $\mathcal{L}_i^1$ as in Lemma 3 using locally optimal paths from $\mathcal{L}_{i-1}^1$. In addition, we must also account for locally suboptimal paths by testing whether the layer $\mathcal{L}_{i-2}^j$ with associated distance $\gamma + 1$ contains either of the at most two needle circles that touch both $p_i$ and $p_{i-1}$.

If $d_N(p_1, ..., p_i) = \gamma + 2$, then $\mathcal{L}_i^1$ is the disk centered at $p_i$ with radius $2r$. This follows from $||p_i - p_{i-1}|| < 2r$ because it will always be possible to connect any needle circle $c \in \mathcal{L}_{i-1}^1$ to an arbitrary needle circle $c'$ that touches $p_i$ by choosing one needle circle that it tangent to both $c$ and $c'$ (see Figure 7b). Hence, all of the operations

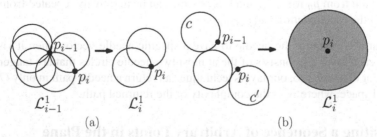

**Fig. 7.** (a) If $d_N(p_1, ..., p_i) = \gamma$, then there are at most two circles that define $\mathcal{L}_i^1$. (b) If $||p_i - p_{i-1}|| < 2r$ and $d_N(p_1, ..., p_i) = \gamma + 2$, then at most one extra circle is needed to connect an arbitrary circle $c \in \mathcal{L}_{i-1}^1$ to an arbitrary circle $c'$ that touches $p_i$.

needed to construct the generating arcs for $\mathcal{L}_i^1$ take time proportional to the number of generating arcs defining $\mathcal{L}_{i-1}^1$ and $\mathcal{L}_{i-2}^1$.    □

**Theorem 2.** *A needle path* $\pi_N(p_1, ..., p_{m+n})$ *can be computed in* $O(2^m n + n^2 + K)$ *time and space, where* $K$ *is the complexity of the returned path.*

*Proof.* Given a sequence of $m$ arbitrary points, layer $\mathcal{L}_m^1$ can have $\Theta(2^m)$ generating arcs by Lemma 2. Lemma 3 ensures that each of the well-separated pairs of consecutive points adds only a constant number of additional generating arcs to the current layer. Consequently, the worst-case time to compute the generating arcs for $\mathcal{L}_1^1, ..., \mathcal{L}_{m+n}^1$ is on the order of $\Sigma_{i=1}^n (2^m + i) \in O(2^m n + n^2)$. Once these generating arcs have been computed, it is a simple matter to construct $\pi_N(p_1, ..., p_{m+n})$ in the same manner as Theorem 1.    □

## 5  Conclusion

We developed two algorithms to guide a bevel-tip needle through a sequence of treatment points in the plane. Such paths are potentially useful for biopsy and brachytherapy procedures because they reduce the number of times that a physician is required to insert and withdraw a needle during a medical procedure. We are currently extending our technique to visit multiple edges instead of multiple points. It would also be interesting to extend our algorithm to points in three dimensions while avoiding an exponential runtime.

## Acknowledgment

We thank Carlos Esquivel, Sotirios Stathakis, and the *Cancer Therapy and Research Center at the University of Texas Health Science Center at San Antonio* for a demonstration of needle steering hardware and clinical procedures.

## References

1. Agarwal, P.K., Biedl, T., Lazard, S., Robbins, S., Suri, S., Whitesides, S.: Curvature-constrained shortest paths in a convex polygon. In: 14th Symposium on Computational Geometry (SoCG), pp. 392–401 (1998)
2. Alterovitz, R., Branicky, M., Goldberg, K.: Motion planning under uncertainty for image-guided medical needle steering. International Journal of Robotics Research 27(1361) (2008)
3. Alterovitz, R., Goldberg, K., Okamura, A.: Planning for steerable bevel-tip needle insertion through 2d soft tissue with obstacles. In: IEEE International Conference on Robotics and Automation, pp. 1640–1645 (2005)
4. Bereg, S., Kirkpatrick, D.: Curvature-bounded traversals of narrow corridors. In: 21st Symposium on Computational Geometry (SoCG), pp. 278–287 (2005)
5. Cook IV, A.F., Wenk, C.: Link distance and shortest path problems in the plane. In: Goldberg, A.V., Zhou, Y. (eds.) AAIM 2009. LNCS, vol. 5564, pp. 140–151. Springer, Heidelberg (2009)

6. Dror, M., Efrat, A., Lubiw, A., Mitchell, J.S.B.: Touring a sequence of polygons. In: 35th ACM Symposium on Theory of Computing (STOC), pp. 473–482 (2003)
7. Dubins, L.E.: On curves of minimal length with a constraint on average curvature, and with prescribed initial and terminal positions and tangents. American Journal of Mathematics 79(3), 497–516 (1957)
8. Duindam, V., Xu, J., Alterovitz, R., Sastry, S., Goldberg, K.: 3d motion planning algorithms for steerable needles using inverse kinematics. In: Eighth International Workshop on Algorithmic Foundations of Robotics, WAFR (2008)
9. Maheshwari, A., Sack, J.-R., Djidjev, H.N.: Link distance problems. In: Handbook of Computational Geometry (1999)
10. Mitchell, J.S.B., Rote, G., Woeginger, G.J.: Minimum-link paths among obstacles in the plane. In: 6th Symposium on Computational Geometry (SoCG), pp. 63–72 (1990)
11. Xu, J., Duindam, V., Alterovitz, R., Goldberg, K.: Motion planning for steerable needles in 3d environments with obstacles using rapidly-exploring random trees and backchaining. In: IEEE Conference on Automation Science and Engineering, CASE (2008)

# Euclidean Prize-Collecting Steiner Forest[*]

MohammadHossein Bateni[1,**] and MohammadTaghi Hajiaghayi[2]

[1] Princeton University, Princeton NJ 08540, USA
mbateni@cs.princeton.edu
[2] AT&T Labs—Research, Florham Park, NJ 07932, USA
hajiagha@research.att.com

**Abstract.** In this paper, we consider *Steiner forest* and its generalizations, *prize-collecting Steiner forest* and *k-Steiner forest*, when the vertices of the input graph are points in the Euclidean plane and the lengths are Euclidean distances. First, we present a simpler analysis of the polynomial-time approximation scheme (PTAS) of Borradaile et al. [11] for the *Euclidean Steiner forest* problem. This is done by proving a new structural property and modifying the dynamic programming by adding a new piece of information to each dynamic programming state. Next we develop a PTAS for a well-motivated case, i.e., the multiplicative case, of prize-collecting and budgeted Steiner forest. The ideas used in the algorithm may have applications in design of a broad class of bicriteria PTASs. At the end, we demonstrate why PTASs for these problems can be hard in the general Euclidean case (and thus for PTASs we cannot go beyond the multiplicative case).

## 1   Introduction

Prize-collecting Steiner problems are well-known network design problems with several applications in expanding telecommunications networks (see e.g. [22,29]), cost sharing, and Lagrangian relaxation techniques (see e.g. [21,14]). The most general version of these problems is called the *prize-collecting Steiner forest (PCSF)* problem[1], in which, given a graph $G = (V, E)$, a set of (commodity) pairs $\mathcal{D} = \{(s_1, t_1), (s_2, t_2), \dots\}$, a non-negative cost function $c : E \to \mathbf{Q}^{\geq 0}$, and finally a non-negative penalty function $\pi : \mathcal{D} \to \mathbf{Q}^{\geq 0}$, our goal is a minimum-cost way of buying a set of edges and paying the penalty for those pairs which are not connected via bought edges. When all penalties are $\infty$, the problem is the classic APX-hard *Steiner forest* problem for which the best approximation factor is $2 - \frac{2}{n}$ ($n$ is the number of vertices of the graph) due to Goemans and Williamson [17]. When all sinks are identical in the PCSF problem, it is the classic prize-collecting Steiner tree problem. Bienstock, Goemans, Simchi-Levi, and Williamson [8] first considered this problem (based on a problem earlier proposed by Balas [3]) for which they gave a 3-approximation algorithm. The current best approximation algorithm for this problem is a recent 1.992-approximation algorithm of Archer, Bateni, Hajiaghayi, and Karloff [1] improving upon

---

[*] Refer to the full version [5] for the omitted proofs and further discussion.
[**] The author was supported by a Gordon Wu fellowship as well as NSF ITR grants CCF-0205594, CCF-0426582 and NSF CCF 0832797, NSF CAREER award CCF-0237113, MSPA-MCS award 0528414, NSF expeditions award 0832797.
[1] It is sometimes called *prize-collecting generalized Steiner tree (PCGST)* in the literature.

A. López-Ortiz (Ed.): LATIN 2010, LNCS 6034, pp. 503–514, 2010.
© Springer-Verlag Berlin Heidelberg 2010

a primal-dual $\left(2 - \frac{1}{n-1}\right)$-approximation algorithm of Goemans and Williamson [17]. When in addition all penalties are $\infty$, the problem is the classic *Steiner tree* problem, which is known to be APX-hard [7] and for which the best known approximation factor is 1.55 [28].

There are several 3-approximation algorithms for the *prize-collecting Steiner forest* problem using LP rounding, primal-dual, or iterative rounding methods which are first initiated by Hajiaghayi and Jain [19] (see [8,20]). Currently the best approximation factor for this problem is a randomized 2.54-approximation algorithm [19]. The approach of Hajiaghayi and Jain has been generalized by Sharma, Swamy, and Williamson [30] for network design problems where violating arbitrary 0-1 connectivity constraints are allowed in exchange for a very general penalty function.

Lots of attention has been paid to budgeted versions of Steiner problems as well. In the *k-Steiner forest* (or just *k*-forest for abbreviation), given a graph $G = (V, E)$ and a set of (commodity) pairs $\mathcal{D}$, the goal is to find a minimum-cost forest that connects at least $k$ pairs of $\mathcal{D}$. The best current approximation factor for this problem is in $O(\min\{\sqrt{k}, \sqrt{n}\})$ [18]. On the other hand, Hajiaghayi and Jain [19] could transform notorious *dense k-subgraph* to this problem, for which the current best approximation factor is $O(n^{1/3-\epsilon})$ [15]. The special case in which we have a root $r$ and $\mathcal{D}$ consists of all pairs $(r, v)$ for $v \in V(G) - \{r\}$ is the well-known NP-hard $k$-MST problem. The first non-trivial approximation algorithm for the $k$-MST problem was given by Ravi et al. [27], who achieved an approximation ratio of $O(\sqrt{k})$. Later this approximation ratio is improved to a constant by Blum et al. [9]. Currently the best approximation factor for this problem is 2 due to Garg [16].

In this paper, we consider *Euclidean prize-collecting Steiner forest* and *Euclidean k-forest* in which the vertices of the input graph are points in the Euclidean plane (or low-dimensional Euclidean space) and the lengths are Euclidean distances. For the *Euclidean Steiner tree* problem, Arora [2] and Mitchell [26] gave polynomial-time approximation schemes (PTASs). Recently Borradaile, Klein and Kenyon-Mathieu [11] claim a PTAS for the more general problem of *Euclidean Steiner forest* .

## 1.1   Problem Definition

Motivated by the settings in which the demand of each pair is the product of the weight of the origin vertex and the weight of the destination vertex in the pair and thus in a sense contributions of each vertex to all adjacent pairs are the same (e.g., see *product multi-commodity flow* in Leighton and Rao [24] or [10,23], and its applications in wireless networks [25] or routing [12,13]), we consider the following multiplicative version of prize-collecting Steiner forest for the Euclidean case.

In the *Multiplicative prize-collecting Steiner forest (MPCSF)* problem, given an undirected graph $G(V, E)$ with non-negative edge lengths $c_e$ for each edge $e \in E$, and also given weights $\phi(v)$ for each vertex $v \in V$, our goal is to find a forest $F$ which minimizes the cost

$$\sum_{e \in F} c_e + \sum_{\substack{u,v \in V : \, u \text{ and } v \text{ are not connected via } F}} \phi(u)\phi(v).$$

Indeed, this is an instance of PCSF in which each ordered vertex pair $(u, v)$ forms a request with penalty $\phi(u)\phi(v)$.[2] We may be asked to *collect a certain prize S*, in which case the goal is to find the forest $F$ of minimum cost for which

$$\sum_{u,v \in V:\, u \text{ and } v \text{ are connected via } F} \phi(u)\phi(v) \geq S.$$

Let us call this problem $S$-MPCSF. We show that this is a generalization of the $k$-MST problem and thus currently there is no approximation better than 2 for this problem, either. When working on the Euclidean case, the input does not include any Steiner vertices, as all the points of the plane are potential Steiner points.

A bicriteria $(\alpha, \beta)$-approximate solution for the the $S$-MPCSF problem is one whose cost is at most $\alpha$OPT, yet collects a prize of at least $\beta S$. Our main contribution in this paper is a bicriteria $(1+\epsilon, 1-\epsilon')$-approximation algorithm that runs in time exponential in $1/\epsilon$ but polynomial in $n$ and $1/\epsilon'$. We then use this algorithm to obtain a PTAS for MPCSF.

## 1.2  Our Contribution

First of all, we present a simpler analysis for the algorithm of Borradaile et al. [11] for the *Euclidean Steiner forest* problem and reprove the following theorem.

**Theorem 1.** *For any constant $\epsilon > 0$, there is an algorithm that runs in polynomial time and approximates the Euclidean Steiner forest problem within $1 + \epsilon$ of the optimal solution.*

This is done by modifying the dynamic programming (DP) algorithm so that instead of storing paths enclosing the *zones* in the algorithm by Borradaile et al., we use a bitmap to identify a zone. The modification results in simplification of the structural property required for the proof of correctness (See Section 3). We prove this structural property in Theorem 6. The proof has some ideas similar to [11], but we present a simpler charging scheme that has a universal treatment throughout. Details of the DP algorithm are deferred to the full version of the paper. We have recently come to know that similar simplifications have been independently discovered by the authors of [11], too.

Next we extend the algorithm for Euclidean $S$-MPCSF and MPCSF problems in Section 4.

**Theorem 2.** *For any $\epsilon, \epsilon' > 0$, there is a bicriteria $(1 + \epsilon, 1 - \epsilon')$-approximation algorithm for the Euclidean S-MPCSF problem, that runs in time polynomial in $n$, $1/\epsilon'$ and exponential in $1/\epsilon$.*

Notice that $\epsilon'$ need not be a constant. In particular, if all weights are polynomially bounded integers, we can find in polynomial time a $(1 + \epsilon)$-approximate solution that collects a prize of at least $S$; this can be done by picking $\epsilon'$ to be sufficiently small ($\epsilon'^{-1}$ is still polynomial). Next we present a PTAS for *Euclidean MPCSF*.

---

[2] We can change the definition to unordered pairs whose treatment requires only a slight modifications of the algorithms. Currently, each unordered pair $(u, v)$ has a prize of $2\phi(u)\phi(v)$ if $u \neq v$.

**Theorem 3.** *For any constant* $\epsilon$, *there is a* $(1 + \epsilon)$-*approximation algorithm for the Euclidean MPCSF problem, that runs in polynomial time.*

We also study the case of asymmetric prizes for vertices in which each vertex $v$ has two types of weights (type one and type two) and the prize for an ordered pair $(u, v)$ is the product of the first type weight of $u$, i.e., $\phi^s(u)$, and the second type weight of $v$, i.e., $\phi^t(v)$. This case is especially interesting because it generalizes the multiplicative prize-collecting problem when we have two disjoint sets $S_1$ and $S_2$ and we pay the multiplicative penalty only when two vertices, one in $S_1$ and the other one in $S_2$, are not connected (by letting for each vertex in $S_1$ the first type weight be its actual weight and the second type weight be zero and for each vertex in $S_2$ the first type weight be zero and the second type weight be its actual weight.) After hinting on the arising complications, we show how we can extend our algorithms for this case as well.

**Theorem 4.** *For any* $\epsilon, \epsilon' > 0$, *there is a bicriteria* $(1 + \epsilon, 1 - \epsilon')$-*approximation algorithm for the Euclidean Asymmetric S-MPCSF problem, that runs in time polynomial in* $n, 1/\epsilon'$ *and exponential in* $1/\epsilon$. *In addition, for any constant* $\epsilon$, *there is a* $(1 + \epsilon)$-*approximation algorithm for the Euclidean Asymmetric MPCSF problem, that runs in polynomial time.*

Indeed, the algorithms in Theorem 4 can be extended to the case in which there are a constant number of different types of weights for each vertex generalizing the case in which we have a constant number of disjoint sets and we pay the multiplicative penalty when two vertices from two different sets are not connected. Notice that the case of two disjoint sets already generalizes the *prize-collecting Steiner tree* problem (by considering $S_1 = \{r\}$ and $S_2 = V - \{r\}$) whose best approximation guarantee is currently 1.992.

At the end, we present in Section 5 why PCSF and $k$-forest problems can be APX-hard in the general case (and thus for PTASs we cannot go beyond the multiplicative case). We conclude with some open problems in Section 6. All the omitted proofs appear in the appendix.

### 1.3  Our Techniques for the Prize-Collecting Version

Here, we summarize our techniques for the multiplicative prize collecting Steiner forest algorithms; see Section 4. In all those algorithms, we store in each DP state extra parameters, including the sum of the weights, as well as the multiplicative prize already collected in each component. These parameters enable us to carry out the DP update procedure. Interestingly, the sum and collected prize parameters have their own precision units.

In the asymmetric version, a major issue is that no fixed unit is good for all sum parameters. Some may be small, yet have significant effect when multiplied by others. To remedy this, we use variable units, reminiscent of the floating-point storage formats (mantissa and exponent). To the best of our knowledge, Bateni and Hajiaghayi [4] were the first to take advantage of this idea in the context of (polynomial time) approximation schemes. The basic idea is that a certain parameter in the description of DP states has a large (not polynomial) range, however, as the value grows, we can afford to sacrifice more on the precision. Thus, we store two (polynomial) integer numbers, say $(i, x)$, where $i$ denotes a variable unit, and $x$ is the coefficient: the actual number is then recovered by $x \cdot u_i$. The conversion between these representations is not lossless, but the aggregate error can be bounded satisfactorily.

In Section 4.3 we consider the problem where the objective is a linear function of penalties paid and the cost of the forest built. The challenging case is when the cost of the optimal forest is very small compared to the penalties paid. In this case, we identify a set of vertices with large penalties and argue they have to be connected in the optimal solution. Then, with a novel trick we show how to ignore them in the beginning, and take them into account only after the DP is carried out.

## 2   Preliminaries

Let $n = |V|$ be the total number of terminals and let OPT be the total length of the optimal solution. A *bitmap* is a matrix with 0-1 entries. Two bitmaps of the same dimensions are called *disjoint* if and only if they do not have value one at the same entry. Consider two partitions $\mathcal{P} = \{P_1, P_2, \ldots, P_{|\mathcal{P}|}\}$ and $\mathcal{P}' = \{P'_1, P'_2, \ldots, P'_{|\mathcal{P}'|}\}$ over the same ground set. Then, $\mathcal{P}$ is said to be a *refinement* of $\mathcal{P}'$ if and only if any set of $\mathcal{P}$ is a subset of a set in $\mathcal{P}'$, namely $\forall P \in \mathcal{P}, \exists P' \in \mathcal{P}' : P \subseteq P'$.

By standard perturbation and scaling techniques, we can assume the following conditions hold incurring a cost increase of $O(\epsilon \text{OPT})$; see [2,11] for example.

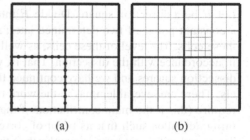

(I) The diameter of the set $V$ is at most $d' = n^2 \epsilon^{-1} \text{OPT}$.

(II) All the vertices of $V$ and the Steiner points have coordinates $(2i + 1, 2j + 1)$ where $i$ and $j$ are integers.

For simplicity of exposition, we ignore the above increase in cost. As we are going to obtain a PTAS, this increase will be absorbed in the future cost increases. We have a grid consisting of vertical and hor-

**Fig. 1.** (a) An example of a dissection square with depth 3, and depiction of portals for a sample dissection square with $m = 8$; (b) the $\gamma \times \gamma$ grid of cells inside a sample dissection square with $\gamma = 4$.

izontal lines with equations $x = 2i$ and $y = 2j$ where $i$ and $j$ are integers. Let $\mathcal{L}$ denote the set of lines in the grid. We let $L$ be the smallest power of two greater than or equal to $2d'$ and perform a dissection on the randomly shifted bounding box of size $L \times L$; see Figure 1(a).

For each dissection square $R$ and each side $S$ of $R$, designate $m + 1$ equally spaced points along $S$ (including the corners) as *portals* of $R$ where $m$ is the smallest power of 2 greater than $4\epsilon^{-1} \log L$. So the square $R$ has $4m$ portals.

There is a notion of *level* associated with each dissection square, line, or side of a square. The bounding box has level zero, and level of each other dissection square is one more than the level of its parent dissection square. The level of a line $\ell$ is the minimum level of a square $R$ a side of which falls on the line $\ell$. Thus, the first two lines dividing the bounding box have level one. If a side $S$ of a square $R$ falls on a line $\ell$, we define level$(S)$ = level$(\ell)$. So level$(S) \leq$ level$(R)$. The thickness of the lines in Figure 1 denotes their level: the thicker the line, the lower is its level.

For a (possibly infinite) set of geometric points $X$, let comp$(X)$ denote the number of connected components of $X$; we will use the shorthand "component" in this paper. With slight abuse of notation, $\ell \in \mathcal{L}$ is used to refer to the set of points[3] on $\ell$. In addition, we use $\mathcal{L}$ to denote the union of points on the lines in $\mathcal{L}$. Similarly, we use $R$ to denote the set of all points on or inside the square $R$. The set of points on (the boundary of) the square $R$ is referred to by $\partial R$. The total length of all line segments in $F$ is denoted by length$(F)$.

The following theorem is mentioned in [11] in a stronger form. We only need its first half whose proof follows from [2].

**Theorem 5.** *[11] There is a solution $F$ having expected length at most $(1 + \frac{1}{4}\epsilon)$OPT such that each dissection square $R$ satisfies the following two properties: for each side $S$ of $R$, $F \cap S$ has at most $\rho = O(\epsilon^{-1})$ non-corner components[4] (boundary components property); and each component of $F \cap \partial R$ contains a portal of $R$ (portal property).*

## 3 Structural Theorem

Let $R$ be a dissection square. Divide $R$ into a regular $\gamma \times \gamma$ grid of *cells*, where $\gamma$ is a constant power of two determined later; see Figure 1(b). We say $R$ is the *owner* of these cells. The level of these cells, as well as the new lines they introduce, is defined in accordance with the dissection. That is, we assign them levels as if they are normal dissection squares and we have continued the dissection procedure for $\log \gamma$ more levels. There are several lemmas in the work of [11] to prove the structural property they require (this is the main contribution of that work). We modify the dynamic programming definition such that its proof of correctness needs a simpler structural property. The proof of this property is simpler than that in the aforementioned paper.

**Theorem 6.** *There is a solution $F$ having expected length at most $(1 + \frac{1}{2}\epsilon)$OPT such that each dissection square $R$ satisfies the locality property: if the terminals $t_1$ and $t_2$ are inside a cell $C$ of $R$ and are connected to $\partial R$ via $F$, then they are connected in $F \cap R$.*

The proof has ideas similar to [11, Theorem 3.2, and Lemmas 3.3, 3.4, 3.5 and 3.9]. We first mention and prove a lemma we need in order to prove Theorem 6. The lemma more or less appears in [2,11].

**Lemma 1.** *For the forest $F$ output by Theorem 5, comp$(F \cap \mathcal{L}) \leq$ length$(F)$.*

We can now prove the main structural result. A side $S$ of a square $R$ is called *private* if it does not lie on a side of the parent square $R'$ of $R$. Observe that out of any two opposite sides of a dissection square, exactly one is private.

---

[3] Not necessarily terminals.

[4] Non-corner components are those not including any corners of squares. Note that each square can have at most four corner components.

*Proof (Theorem 6).* We start with a solution $F$ satisfying Theorem 5. The final solution is produced by iteratively finding the smallest cell $C$ owned by a square $R$ that violates the locality property, and adding $\sigma(C, F)$ to $F$, where $\sigma(C, F)$ is defined as the union of the private sides of $C$ and any side of $C$ having non-empty intersection with $F$. We claim the locality property is realized after finitely many such additions. If after adding $\sigma(C, F)$ to $F$, the cell $C$ still violates the locality property, there has to be exactly two opposite sides of the cell having non-empty intersection with $F$; otherwise, the $\sigma(C, F)$ is clearly connected. However, in case of the opposite sides, one middle side will be a private side of $C$ and hence included as well. We omit the proof that the conditions of Theorem 5 still hold.

Finally we show that the additional length is not large. Let $F^* = F \cap \mathcal{L}$, and let $\mathcal{G} = \{(x, y) : x = 2i, y = 2j\}$ be the set of all grid points. We will charge the additions to the connected components of $F^* - \mathcal{G}$. Notice that

$$\text{comp}(F^* - \mathcal{G}) \leq \text{comp}(F^*) + 3|F^* \cap \mathcal{G}| \tag{1}$$

$$\leq \text{comp}(F^*) + 3 \cdot (\text{length}(F^*) + \text{comp}(F^*)) \tag{2}$$

$$= 4\,\text{comp}(F^*) + 3\,\text{length}(F)$$

$$\leq 7\,\text{length}(F), \qquad\qquad\qquad \text{by Lemma 1.} \tag{3}$$

Inequality (1) holds because removal of each grid point on $F^*$ increases the number of components by at most three. To obtain (2), notice that in any connected component of $F^*$, the distance between any two points of $F^* \cap \mathcal{G}$ is at least 2. Hence, if there are more than one such points, there cannot be more than length$(F^*)$ ones.

We charge this addition to a connected component of $(\partial R \cap F) - \mathcal{G}$, in such a way that each connected component is charged to at most twice: once from each side. For simplicity, we duplicate each connected component of $(F \cap \ell) - \mathcal{G}$: they correspond to squares from either side of $\ell$. For any dissection square $R$, let $\mathcal{C}_R$ refer to the connected components of $F \cap R$ that reach $\partial R$. Further, let $\mathcal{K}_R$ be the set of connected components of $(F \cap \partial R) - \mathcal{G}$. When $\sigma(C, F)$ is added where $R$ is the owner of $C$, there are $k \geq 2$ components $c_1, \ldots, c_k \in \mathcal{C}_R$ that become connected. Any element of $\mathcal{K}_R$ connected via $F \cap R$ to a component $c \in \mathcal{C}_R$ is said to be an *interface* of $c$. The addition will be charged to a *free* interface of some $c \in \mathcal{C}_R$ with maximum level. This element will no longer be free for the rest of the procedure. We argue this procedure successfully charges all the additions to appropriate border components. To this end, we prove the following stronger claim via induction on the number of additions performed; proof omitted here. We call a dissection square $R$ *violated* if the locality property does not hold for a cell $C$ owned by $R$.

*Claim.* At all times during the execution of this procedure, any component $c \in \mathcal{C}_R$ has a free interface, for each violated square $R$. As a result, any addition can be charged to a free component.

Simple calculation (omitted in this version) shows that the expected increase in length is at most $\frac{112}{\gamma}$ length$(F)$. We pick $\gamma$ to be the smallest power of two larger than $112(1 + \epsilon) \cdot 2\epsilon^{-1}$ to finish the proof.

Therefore, with probability $1/2$, we have length$(F) \leq (1 + \epsilon)\text{OPT}$. In the entire argument, no attempt was made to optimize the parameters.

### 3.1 Highlights of the New Ideas

Here, we point out the differences between our work and the previous work of [11]. Borradaile et al. use closed paths to identify the connected zones of the dissection square. These paths consist of vertical and horizontal lines and all the break-points are the corners of the cells. As part of their structural property, they prove that they can guarantee a solution in which these zones can be identified via paths whose total length is at most a constant $\eta$ times the perimeter of the square $R$. Then each path is represented by a chain of $\{1, 2, 3\}$ of length at most $O(\eta\gamma)$: the three values are used to denote moving one unit forward, or turning to the left or right. This results in a storage of $3^{O(\eta\gamma)}$ which is a constant parameter. Instead, we use a bitmap of size $\gamma \times \gamma$ to address this issue. Each zone is represented by a bitmap that has an entry one in the cells of the zone. The bound that we obtain, $2^{\gamma^2}$, may be slightly worse than the previous work, however, a simpler structural property, namely the locality property, suffices as the proof of correctness. Borradaile et al. in contrast need a bound on the total length of the zone boundaries, as noted above.

In addition to the simplification made due to this change, both to the proof and the treatment of the dynamic programming, we simplify the proof further. Borradaile et al. charge the additions of $\sigma(C, F)$ to three different structures, and the argument is described and analyzed separately for each. We manage to perform a universal treatment and charging all the additions to the simplest of the three structures in their work. But this can be done only after showing $F^* - \mathcal{G}$ has a limited number of components. The proof is simple yet elegant—a weaker claim is proved in [11], but even the statement of the claim is hard to read.

## 4 Multiplicative Prizes

We first tackle the *S-multiplicative prize-collecting Steiner forest* problem. Then, we will take a look at its asymmetric generalization. Finally, we show how the *multiplicative prize-collecting Steiner forest* problem can be reduced to *S*-MPCSF.

### 4.1 Collecting a Fixed Prize

Suppose we are given $S$, the amount of prize we should collect. Let OPT be the minimum cost of a forest $F$ that collects a prize of at least $S$, and suppose $Q \subseteq \mathcal{D}$ is the set of terminal pairs connected via $F$. We show how to find a forest with cost at most $(1 + \epsilon)$OPT that collects a prize of at least $(1 - \epsilon')S$. By the structural property, we know that there is a solution $F'$ connecting the same set of terminal pairs $Q$ whose cost is at most $(1 + \epsilon)$OPT, yet it satisfies the conditions of Theorems 5 and 6. Round all the vertex weights down to the next integer multiple of $\theta = \epsilon'\sqrt{S}/2n$. In a connected component of $F'$ of total weight $A_i$ that lost a weight $a_i$ due to rounding, the lost prize is $A_i^2 - (A_i - a_i)^2 \leq 2a_iA_i \leq 2a_i\sqrt{S}$, because the total weight of the component is at most $\sqrt{S}$. Thus, $F'$ collects at least $S - 2n\theta\sqrt{S} \leq (1 - \epsilon')S$ from the rounded weights.

Each dynamic programming state consists of a dissection square $R$, a set of components $\mathcal{K}$, and a new parameter $\Pi$ which denotes the total prize collected inside $R$ by connecting the terminal pairs. Each element of $\mathcal{K}$—corresponding to a connected component in the subsolution—now has the form $\kappa = (P, \Sigma)$ where $P$ denotes the portals of $\kappa$, and $\Sigma$ is the total sum of the weights in $\kappa$. The DP is carried out in a fashion similar to that of [2]. The values of $\Sigma$ and $\Pi$ are easy to determine for the base cases. It is not difficult to update them, either. Whenever two components $\kappa_1 = (P_1, \Sigma_1)$ and $\kappa_2 = (P_2, \Sigma_2)$ merge in the DP, the sum $\Sigma$ for the new component is simply $\Sigma_1 + \Sigma_2$. Besides, the merge increases the $\Pi$ value of the DP state by $2\Sigma_1\Sigma_2$.

To start the algorithm, we need to guarantee the instance satisfies the conditions at the beginning of Section 2. This is discussed in the full version of the paper.

### 4.2 The Asymmetric Prizes

The basic idea is to store two parameters $\Sigma^s$ and $\Sigma^t$ for each component of $\mathcal{K}$. These parameters store the total weight of the first and second type in the component, namely $\sum_i \phi_i^s$ and $\sum_i \phi_i^t$, respectively. The difficulty is that to collect a prize of $A = A^s A^t$ in a component, only one of the parameters $A^s$ or $A^t$ needs to be large. In particular, we cannot do a rounding with a precision like $c'\sqrt{A}/n$. It may even happen that $A^s$ is large in one component, whereas we have a large $A^t$ in another. In fact, we cannot store the values of the $\Sigma^s$ or $\Sigma^t$ as multiples of a fixed unit. To get around the problem, $\Sigma^s$ is stored as a pair $(v, x)$, where $v$ is a vertex of the graph and $x$ is an integer. Together they show that $\Sigma^s$ is $x \cdot \epsilon_1 \phi^s(v)/n^2$; the value of $\epsilon_1$ will be chosen later, and $v$ is supposed to be the vertex of largest type-one weight present in the component. A similar provision is made for $\Sigma^t$. Finally, the value of $\Pi$ is stored as a multiple of $\epsilon_2 A/n$; we will shortly pick the value of $\epsilon_2$.

Whenever $\Sigma_1^s = (v_1, x_1)$ and $\Sigma_2^s = (v_1, x_1)$ are added to give $\Sigma^s = (v, x)$, we do the calculation as follows: let $v$ be the vertex $v_1$ or $v_2$ that has the larger $\phi^s$ value, and then $x = \left\lfloor \frac{x_1 \phi^s(v_1)/n^2 + x_2 \phi^s(v_2)/n^2}{\epsilon_1 \phi^s(v)/n^2} \right\rfloor$.

### 4.3 The Prize-Collecting Version: Trade-off between Penalty and Forest Cost

In the prize-collecting variant, we pay for the cost of the forest, and for the prizes not collected. If the total weight is $\Delta$, the prize not collected is $\Delta^2$ minus the collected prize. One difficulty here is to determine the correct range for the collected prize so that we can use the algorithm of Section 4.1. The trivial range is zero to $\Delta^2$. However, the rounding precision we pick for the penalties should also take into account the cost of the forest. If the cost of the intended solution is much smaller than $\Delta^2$, we cannot simply go with rounding errors like $\epsilon\Delta/n$. Otherwise, the error caused due to rounding the penalties will be too large compared to the solution value.

The trick is to find an estimate of the solution value, and then consider two cases depending on how the cost compares to the total penalty. Using a 3-approximation algorithm, we obtain a solution of value $\omega$. We are guaranteed that OPT $\geq \omega/3$. If $\Delta^2 \leq \omega/3$, the optimum solution is to collect no prize at all. Otherwise, assume $\Delta^2 > \omega/3$. To beat the solution of value $\omega$, we should collect a prize of at least $\Delta^2 - \omega$.

We first consider the simpler case when $\omega/\Delta^2 > 1/n^2$: For an $\epsilon' > 0$ whose precise value will be fixed below, we use the algorithm of Section 4.1 to find a bicriteria $(1 + \epsilon/2, 1 - \epsilon')$-approximate solution for collecting a prize $S$; this is done for any $S$ which is a multiple of $\epsilon'\Delta^2$ in range $[(1-\epsilon')\Delta^2 - \omega, \Delta^2]$. We select the best one after adding the uncollected prize to each of these solutions. Suppose the optimal solution OPT collects a prize $S'$. Let $\text{OPT}_f = \text{OPT} - (\Delta^2 - S')$ be the length of the forest. Round $S'$ down to the next multiple of $\epsilon'\Delta^2$, say $S$. Fed with prize value $S$, the algorithm finds a solution that collects a prize of at least $(1 - \epsilon')S$ with forest cost at most $(1 + \epsilon/2)\text{OPT}_f$.

*Claim.* The total cost of this solution is at most $(1 + \epsilon)\text{OPT}$ if $\epsilon' = \frac{1}{6n^2}\min\left(\frac{\epsilon}{3}, 1\right)$.

The other case, i.e., $\omega/\Delta^2 \leq 1/n^2$, is more challenging. Notice that in order to carry out the same procedure in this case, $\epsilon'$ may not be bounded by $1/\text{poly}(n)$ and thus the running time may not be polynomial. The solution, however, has to collect almost all the prize. Thus, one of the connected components includes almost all the vertex weights. We set aside a subset $\mathcal{B}$ of vertices of large weight. The vertices of $\mathcal{B}$ have to be connected in the solution, or else the paid penalty will be too large. Then, dynamic programming proceeds by ignoring the effect of these vertices and only keeping tabs on how many vertices from $\mathcal{B}$ exist in each component. At the end, we only take into account the solutions that gather *all* the vertices of $\mathcal{B}$ in one component and compute the actual cost of those solutions and pick the best one. In the following, we provide the details of our method and prove its correctness.

Let $\mathcal{B}$ be the set of all vertices whose weight is larger than $n\omega/\Delta$.

**Lemma 2.** *All the vertices of $\mathcal{B}$ are connected in the optimal solution.*

Next, we round up all the weights to the next multiple of $\theta = \epsilon'\omega/\Delta$ for vertices not in $\mathcal{B}$. Define $\text{OPT}'$ as the optimal solution of the resulting instance. Let $\text{OPT}_f$ be the length of the forest in OPT, and define $\text{OPT}'_f$ similarly. Let $\text{OPT}_\pi$ and $\text{OPT}'_\pi$ denote the penalty paid by OPT and OPT', respectively. Assume that $\epsilon' \leq 1$.

**Lemma 3.** $\text{OPT}'_\pi \leq \text{OPT}_\pi + 12n\epsilon'\text{OPT}$.

Suppose we use a dynamic programming approach similar to the previous subsections to find the approximately minimum forest length for any specified collected prize amount; in particular, we obtain a bicriteria $(1 + \epsilon/2, 1 - \epsilon')$-approximate solution. During this process, we ignore the weights associated with vertices in $\mathcal{B}$. Consider a DP state $\chi = (\mathcal{K}, \Pi)$ corresponding to a dissection square $R$. Each component $\kappa \in \mathcal{K}$ looks like $(P, \Sigma, \mu)$: the new piece of information, $\mu$, is an integer number denoting the number of vertices of $\mathcal{B}$ inside $\kappa$. Extending the previous algorithm to populate the new DP table is simple. Finally, we look at all the configurations $\chi$ for the bounding box such that the $\mu$ value of one component is exactly $|\mathcal{B}|$ whereas it is zero for all other components. This guarantees that all elements of $\mathcal{B}$ are inside the former component and hence we can add up the penalties involving those vertices. Let $\mathcal{K} = \{\kappa_1, \kappa_2, \ldots, \kappa_q\}$ where $\kappa_i = (P_i, \Sigma_i)$, and let $\kappa_1$ be the component containing $\mathcal{B}$. The additional cost due to vertices of $\mathcal{B}$ is $\left(\sum_{v \in \mathcal{B}} \phi(v)\right) \cdot \left(\sum_{i=2}^q \Sigma_i\right)$. Finally, we report the best solution corresponding to these configurations.

## 5  Evidence for Hardness

So far PTASs for geometric problems in Euclidean plane including ours and those of Arora [2] and Mitchell [26] can be easily generalized for Euclidean $d$-dimensional space, for any constant $d > 2$. However we can prove the following theorem on the hardness of the problem for Euclidean $d$-dimensional space.

**Theorem 7.** *If notorious densest $k$-subgraph is hard to approximate within a factor $O(n^{\frac{1}{d}})$ for some constant $d$, then for any $d' > 2d+1$, the $k$-forest problem in Euclidean $d'$-dimensional space is hard to approximate within a factor $O(n^{\frac{1}{2d} - \frac{1}{d'-1}})$.*

Note as mentioned above that, despite extensive study, the current best approximation factor for notorious densest $k$-subgraph is $O(n^{1/3-\epsilon})$ [15] and thus we do not expect to have any PTAS for $k$-forest in 8-dimensional Euclidean space.

Unlike the general cases of these problems, as far as PTASs for the case of Euclidean spaces are concerned, it seems *k-forest* and *prize-collecting Steiner forest* problems are essentially equivalent. Indeed we can prove that any PTAS for *k-forest* results in a PTAS for *prize-collecting Steiner forest*, and we believe that any DP algorithm giving a PTAS for PCSF computes along its way the optimal solution to different $k$-forest instances.

Thus based on the evidences above, we do believe *Euclidean k-forest* and *Euclidean prize-collecting Steiner forest* have no PTASs in their general forms.

## 6    Conclusion

Besides presenting a simpler and correct analysis of the PTAS for the *Euclidean Steiner forest problem*, we showed how the approach can be generalized to solve multiplicative prize-collecting problems. Very recently, Bateni, Hajiaghayi and Marx [6] generalized our results and presented a PTAS for *Steiner forest* on graphs of bounded treewidth, and used it to obtain a PTAS for planar and bounded-genus graphs, thereby settling a long-open problem. Finally, obtaining any improvement over the approximation factor 2.54 in [19] for multiplicative prize-collecting Steiner forest in general graphs seems very interesting.

## References

1. Archer, A., Bateni, M., Hajiaghayi, M., Karloff, H.: Improved approximation algorithms for prize-collecting Steiner tree and TSP. In: FOCS (2009)
2. Arora, S.: Polynomial time approximation schemes for Euclidean traveling salesman and other geometric problems. J. ACM 45, 753–782 (1998)
3. Balas, E.: The prize collecting traveling salesman problem. Networks 19, 621–636 (1989)
4. Bateni, M., Hajiaghayi, M.: Assignment problem in content distribution networks: unsplittable hard-capacitated facility location. In: SODA, pp. 805–814 (2009)
5. Bateni, M., Hajiaghayi, M.: Euclidean prize-collecting Steiner forest, CoRR, abs/0912.1137 (2009)
6. Bateni, M., Hajiaghayi, M., Marx, D.: Approximation schemes for Steiner forest on planar graphs and graphs of bounded treewidth, CoRR, abs/0911.5143 (2009)
7. Bern, M., Plassmann, P.: The Steiner problem with edge lengths 1 and 2. Inf. Proc. Lett. 32, 171–176 (1989)
8. Bienstock, D., Goemans, M.X., Simchi-Levi, D., Williamson, D.: A note on the prize collecting traveling salesman problem. Math. Prog. 59, 413–420 (1993)

9. Blum, A., Ravi, R., Vempala, S.: A constant-factor approximation algorithm for the k-MST problem. J. Comput. Syst. Sci. 58, 101–108 (1999)
10. Bonsma, P.: Sparsest cuts and concurrent flows in product graphs. Discrete Appl. Math. 136, 173–182 (2004)
11. Borradaile, G., Klein, P.N., Mathieu, C.: A polynomial-time approximation scheme for Euclidean Steiner forest. In: FOCS, pp. 115–124 (2008),
http://www.math.uwaterloo.ca/~glencora/downloads/
Steiner-forest-FOCS-update.pdf
12. Chekuri, C., Khanna, S., Shepherd, F.B.: Edge-disjoint paths in planar graphs. In: FOCS, pp. 71–80 (2004)
13. Chekuri, C., Khanna, S., Shepherd, F.B.: Multicommodity flow, well-linked terminals, and routing problems. In: STOC, pp. 183–192. ACM, New York (2005)
14. Chudak, F.A., Roughgarden, T., Williamson, D.P.: Approximate $k$-MSTs and $k$-Steiner trees via the primal-dual method and lagrangean relaxation. In: Aardal, K., Gerards, B. (eds.) IPCO 2001. LNCS, vol. 2081, pp. 60–70. Springer, Heidelberg (2001)
15. Feige, U., Kortsarz, G., Peleg, D.: The dense $k$-subgraph problem. Algorithmica 29, 410–421 (2001)
16. Garg, N.: Saving an epsilon: a 2-approximation for the $k$-MST problem in graphs. In: STOC, pp. 396–402 (2005)
17. Goemans, M.X., Williamson, D.P.: A general approximation technique for constrained forest problems. SIAM J. Comput. 24, 296–317 (1995)
18. Gupta, A., Hajiaghayi, M., Nagarajan, V., Ravi, R.: Dial a ride from $k$-forest. In: Arge, L., Hoffmann, M., Welzl, E. (eds.) ESA 2007. LNCS, vol. 4698, pp. 241–252. Springer, Heidelberg (2007)
19. Hajiaghayi, M., Jain, K.: The prize-collecting generalized Steiner tree problem via a new approach of primal-dual schema. In: SODA, pp. 631–640 (2006)
20. Hajiaghayi, M., Nasri, A.: Prize-collecting Steiner networks via iterative rounding. In: LATIN (to appear, 2010)
21. Jain, K., Vazirani, V.V.: Approximation algorithms for metric facility location and k-median problems using the primal-dual schema and Lagrangian relaxation. J. ACM 48, 274–296 (2001)
22. Johnson, D.S., Minkoff, M., Phillips, S.: The prize collecting Steiner tree problem: theory and practice. In: SODA, pp. 760–769 (2000)
23. Kolman, P., Scheideler, C.: Improved bounds for the unsplittable flow problem. In: SODA, pp. 184–193. Society for Industrial and Applied Mathematics, Philadelphia (2002)
24. Leighton, T., Rao, S.: Multicommodity max-flow min-cut theorems and their use in designing approximation algorithms. J. ACM 46, 787–832 (1999)
25. Madan, R., Shah, D., Leveque, O.: Product multicommodity flow in wireless networks. IEEE Trans. Info. Theory 54, 1460–1476 (2008)
26. Mitchell, J.C.: Guillotine subdivisions approximate polygonal subdivisions: A simple polynomial-time approximation scheme for geometric TSP, k-MST, and related problems. SIAM J. Comput. 28, 1298–1309 (1995)
27. Ravi, R., Sundaram, R., Marathe, M.V., Rosenkrantz, D.J., Ravi, S.S.: Spanning trees - short or small. SIAM J. Discrete Math. 9, 178–200 (1996)
28. Robins, G., Zelikovsky, A.: Tighter bounds for graph Steiner tree approximation. SIAM J. Discrete Math. 19, 122–134 (2005)
29. Salman, F.S., Cheriyan, J., Ravi, R., Subramanian, S.: Approximating the single-sink link-installation problem in network design. SIAM J. Opt. 11, 595–610 (2000)
30. Sharma, Y., Swamy, C., Williamson, D.P.: Approximation algorithms for prize collecting forest problems with submodular penalty functions. In: SODA, pp. 1275–1284 (2007)

# Prize-Collecting Steiner Networks
# via Iterative Rounding

MohammadTaghi Hajiaghayi[1] and Arefeh A. Nasri[2]

[1] AT&T Labs– Research, Florham Park, NJ 07932, U.S.A.
hajiagha@research.att.com
[2] Rutgers University, New Brunswick, NJ 08901, U.S.A.
aanasri@gmail.com

**Abstract.** In this paper we design an iterative rounding approach for the classic *prize-collecting Steiner forest problem* and more generally the *prize-collecting survivable Steiner network design* problem. We show as an structural result that in each iteration of our algorithm there is an LP variable in a basic feasible solution which is at least one-third-integral resulting a 3-approximation algorithm for this problem. In addition, we show this factor 3 in our structural result is indeed tight for prize-collecting Steiner forest and thus prize-collecting survivable Steiner network design. This especially answers negatively the previous belief that one might be able to obtain an approximation factor better than 3 for these problems using a natural iterative rounding approach. Our structural result is extending the celebrated iterative rounding approach of Jain [13] by using several new ideas some from more complicated linear algebra. The approach of this paper can be also applied to get a constant factor (bicriteria-)approximation algorithm for degree constrained prize-collecting network design problems.

We emphasize that though in theory we can prove existence of only an LP variable of at least one-third-integral, in practice very often in each iteration there exists a variable of integral or almost integral which results in a much better approximation factor than provable factor 3 in this paper (see patent application [11]). This is indeed the advantage of our algorithm in this paper over previous approximation algorithms for prize-collecting Steiner forest with the same or slightly better provable approximation factors.

## 1  Introduction

Consider a mailing company that wishes to ship packages overnight between several pairs of cities. To this end, this company can build connecting carriers between cities such that at the end by scheduling the carriers, the company is able to ship the packets overnight between pairs of connected cities. Assume the cost of connecting city $i$ to city $j$ is $c_{ij}$ and the costs are symmetric. In addition, the company has the choice of leasing other companies for some pairs $(i, j)$ of cities with cost $\pi_{ij}$ so that without any worry the leased company do the shipment between cities $i$ and $j$ overnight. The goal is to build some carriers and lease some other companies such that the company do the shipments overnight with minimum total cost.

The above network design problem which has also several applications in expanding telecommunications and transportation networks (see e.g. [15,20]), and cost sharing and Lagrangian relaxation techniques (see e.g. [14,6]) is called the *prize-collecting*

A. López-Ortiz (Ed.): LATIN 2010, LNCS 6034, pp. 515–526, 2010.
© Springer-Verlag Berlin Heidelberg 2010

*Steiner forest (PCSF)* problem[1]. In this problem, given a graph $G = (V, E)$, a set of (commodity) pairs $\mathcal{P} = \{(s_1, t_1), (s_1, t_1), \ldots, (s_\ell, t_\ell)\}$, a non-negative cost function $c : E \rightarrow \mathbf{Q}_+$, and finally a non-negative penalty function $\pi : \mathcal{P} \rightarrow \mathbf{Q}_+$, our goal is a minimum-cost way of buying a set of edges and paying the penalty for those pairs which are not connected via bought edges. When all sinks are identical in the PCSF problem, it is the classic prize-collecting Steiner tree problem. Bienstock, Goemans, Simchi-Levi, and Williamson [5] first considered this problem (based on a problem earlier proposed by Balas [2]) for which they gave a 3-approximation algorithm. The current best approximation algorithm for this problem is a primal-dual $2 - \frac{1}{n-1}$ approximation algorithm ($n$ is the number of vertices of the graph) due to Goemans and Williamson [7]. The general form of the PCSF problem first has been formulated by Hajiaghayi and Jain [12]. They showed how by a primal-dual algorithm to a novel integer programming formulation of the problem with doubly-exponential variables, we can obtain a 3-approximation algorithm for the problem (see also [10]). In addition, they show that the factor 3 in the analysis of their algorithm is tight. However they show how a direct randomized LP rounding algorithm with approximation factor 2.54 can be obtained for this problem. Their approach has been generalized by Sharma, Swamy, and Williamson [21] for network design problems where violated arbitrary 0-1 connectivity constraints are allowed in exchange for a very general penalty function. The work of Hajiaghayi and Jain has also motivated a game-theoretic version of the problem considered by Gupta et al. [8].

In this paper, we also consider a generalized version of prize-collecting Steiner forest, called *prize-collecting survivable Steiner network design*, in which we are also given connectivity requirements $r_{uv}$ for all pairs of vertices $u$ and $v$ and a non-increasing marginal penalty function for $u$ and $v$ in case we cannot satisfy all $r_{uv}$. Our goal is to find a minimum way of constructing a network (graph) in which we connect $u$ and $v$ with $r'_{uv} \leq r_{uv}$ edge-disjoint paths and paying the marginal penalty for $r_{uv} - r'_{uv}$ violated connectivity between $u$ and $v$. When all penalties are $\infty$, the problem is the classic survivable Steiner network design problem. For this problem, Jain [13] using the method of iterative rounding obtains a 2-approximation algorithm improving on a long line of earlier research that applied primal-dual methods to this problem.

In this paper, for the first time, we are using the iterative rounding approach for prize-collecting versions of Steiner forest and more generally survivable Steiner network design. To the best of our knowledge, so far this method of iterative rounding has not been used for any prize-collecting problem. After several years since Jain's work, the method of iterative rounding has been revived recently to obtain the best possible bicriteria $(1, B_v + 1)$-approximation algorithm for minimum bounded-degree spanning trees [23] ($B_v$ is the degree bound on vertex $v$) and minimum-bounded degree variants of other problems such as arborescence, Steiner forest and survivable Steiner network design [18,3,19]. The approach of iterative rounding in this paper can be extended further for other prize-collecting problems such as prize-collecting survivable network design with degree constraints $B_v$ on each vertex (i.e., in our solution we should buy at most $B_v$ edges attached to each vertex $v$) to get factor 3 (bicriteria-)approximation algorithms.

---

[1] In the literature, they also called this problem *prize-collecting generalized Steiner tree (PCGST)*.

## 1.1  Our Results

In this paper, we are extending our current knowledge of iterative rounding approaches to prize-collecting Steiner forest and more generally survivable Steiner network design. For the sake of presentation, after introducing the novelty of our approach by stating it precisely for prize-collecting Steiner forest, then we show how it can be extended for prize-collecting survivable Steiner network design. Note that as mentioned in the introduction, so far the only approach to obtain a constant factor approximation algorithm for the survivable Steiner network design, a special case of the prize-collecting survivable Steiner network design problem in which all penalties are ∞, is the method of iterative rounding. Other approaches such as primal-dual methods do not consider the global structure of the network enough to be used for this problem.

We first show as an structural result that in a natural LP for prize-collecting Steiner forest, either a variable corresponding to an edge or a variable corresponding to a penalty for a pair is at least one-third-integral in any basic feasible solution (see Section 3). Indeed we also show this variable of one-third-integral is best that one can hope in a basic feasible solution (see Section 5). This one-third-integral bound obtains a 3-approximation for this problem via much stronger structural results (see Section 2).

There are several novelties in our approach of iterative rounding for the PCSF problem mostly coming from linear algebra. First, so far in all iterative rounding approaches the main constraint is that the fractional value of a cut corresponding to a set $S$ is at least a submodular function of $S$. This has been relaxed in our setting where the fractional value of a cut is also a (not necessarily submodular) function of a penalty associated with a commodity pair separated by this cut. Second, in all previous iterative rounding approaches (in which indeed the heart is obtaining a laminar family using linear algebra, first introduced by Jain [13]) the linear dependence between constraints is a simple addition with all coefficients having absolute values ones (see Theorem 3, Part 5). We show a more complicated fractional dependence between constraints which is crucial to our results. Third, our approach of constructing a laminar family is more complicated than previous approaches when we replace a constraint with one of five (instead of two in previous approaches) constraints (see Theorem 4). Last but not least, obtaining a variable of at least one-third-integral in previous approaches (see e.g. Jain [13]) is relatively easy, however in our case it is much more complicated and needs new ideas from linear algebra (see Theorem 5). Subsequent and separate to our work Konemann et al. [22] obtain the same iterative algorithm as ours for PCSF with some proofs simplified.

After presenting our one-third-integral result for the PCSF problem (which results in a 3-approximation), we show how we can generalize this approach to obtain a variable of at least one-third-integral (and thus a 3-approximation algorithm) for the minimum prize-collecting survivable Steiner network design problem. We briefly discuss the case in which we also have degree constraints on bought edges.

Finally we should emphasize that though in theory we can prove existence of only an LP variable of at least one-third-integral, in practice very often in each iteration there exists a variable of integral or almost integral which results in a much better approximation factor than provable factor 3 in this paper (see AT&T patent application [11] on this regard). This is indeed the advantage of our algorithm in this paper over previous approximation algorithms for prize-collecting Steiner forest with the same or slightly better provable approximation factors.

## 2    Iterative Rounding Approximation Algorithm

The traditional LP relaxation for the PCSF problem which can be solved using Ellipsoid algorithm[2] is as follows:

$$\text{OPT} = \quad \text{minimize} \sum_{e \in E} c_e x_e \quad + \sum_{(i,j) \in \mathcal{P}} \pi_{ij} z_{ij} \tag{1}$$

$$\text{subject to} \quad \sum_{e \in \delta(S)} x_e + z_{ij} \geq 1 \quad \forall S \subset V, (i,j) \in \mathcal{P}, S \odot (i,j) \tag{2}$$

$$x_e \geq 0 \qquad\qquad \forall e \in E \tag{3}$$

$$z_{ij} \geq 0 \qquad\qquad \forall (i,j) \in \mathcal{P} \tag{4}$$

Here for a set $S \subset V$, we denote $|\{i,j\} \cap S| = 1$ by $S \odot (i,j)$.

Let $x^*, z^*$ be an optimal basic feasible solution for LP 1. For $0 < \alpha \leq 1$, let $E_\alpha$ be the set of edges whose value in $x^*$ is at least $\alpha$ and let $\mathcal{P}_\alpha$ be the set of edges whose value in $z^*$ is at least $\alpha$. We define $G_{res} = E - E_\alpha$ and $\mathcal{P}_{res} = \mathcal{P} - \mathcal{P}_\alpha$. Now we consider the following LP, called the *residual LP*, in which we fix all values in edges in $E_\alpha$ and pairs in $\mathcal{P}_\alpha$ to be 1.

$$\text{OPT}_{res} = \quad \text{minimize} \sum_{e \in E} c_e x_e \quad + \sum_{(i,j) \in \mathcal{P}_{res}} \pi_{ij} z_{ij} \tag{5}$$

$$\text{subject to} \quad \sum_{e \in \delta(S)} x_e + z_{ij} \geq 1 \quad \forall S \subset V, (i,j) \in \mathcal{P}_{res}, \tag{6}$$

$$S \odot (i,j), \delta(S) \cap E_\alpha = \emptyset$$

$$x_e \geq 0 \qquad\qquad \forall e \in E_{res} \tag{7}$$

$$z_{ij} \geq 0 \qquad\qquad \forall (i,j) \in \mathcal{P}_{res} \tag{8}$$

Note that in the above LP by contracting edges in $E_\alpha$ and ignoring pairs in $\mathcal{P}_\alpha$, indeed we can always work with an LP similar to that for OPT. Our approximation algorithm for the PCSF problem based on this LP is as follows.

Algorithm PCSF-ALG which is based on the the following theorem is as follows: First we find an optimal basic feasible solution $x^*, z^*$ to LP 1. Then we pay all the penalties of pairs $(i,j)$ whose $z_{ij}^* \geq \alpha$ and remove them from further consideration. We include all edges $e$ whose $x_e^* \geq \alpha$ in the solution and contract them and remove multiple edges by keeping only an edge $e$ with minimum $c_e$ among them. We solve the residual problem recursively.

---

[2] Indeed we can also write the corresponding standard flow-based LP rather than the cut-based LP here, and then use other LP-solver algorithms for a polynomial number of variables and constraints.

**Theorem 1.** *In any basic feasible solution for LP 1, for at least one edge $e \in E$, $x_e \geq \frac{1}{3}$, or for at least one pair $(i,j) \in \mathcal{P}$, $z_{ij} \geq \frac{1}{3}$.*

We prove Theorem 1 in Section 3.

**Theorem 2.** *If $x^I$, $z^I$ is an integral solution to the LP 5 with value at most $\frac{1}{\alpha}\mathrm{OPT}_{res}$, then $E_{x_e=1} \cup E_\alpha$, $P_{z_{ij}=1} \cup \mathcal{P}_\alpha$ is feasible solution for LP 1 with value at most $\frac{1}{\alpha}\mathrm{OPT}$.*

The proof of Theorem 2 is standard and hence omitted. By combining Theorems 1 and 2 we obtain the following conclusion:

**Corollary 1.** *There is an iterative rounding 3-approximation algorithm for PCSF.*

## 3   One-Third-Integrality Result

In this section, we prove Theorem 1. Let $x, z$ be a basic feasible solution. If for an edge $e$, $x_e = 1$ or for a pair $(i', j')$, $z_{ij} = 1$, then the theorem follows. Also, if for an edge $e$, $x_e = 0$, then we can assume that the edge was never there before. This assumption does not increase the cost of the optimum fractional solution $x_e$. Thus we can assume that $0 < x_e < 1$ and $0 \leq z_e < 1$ for all $e \in E$ and $(i,j) \in \mathcal{P}$.

Let $\mathcal{M}(S, ii')$ be the row of the constraint matrix corresponding to a set $S \subset V$ and pair $(i, i') \in \mathcal{P}$. Let $x(A, B)$ be the sum of all $x_e$'s, where $e$ has one end in $A$ and the other end in $B$. We represent $x(A, \overline{A})$ by $x(A)$, for ease of notation. We say a set $A$ is tight with pair $(i, i')$ if $A \odot (i, i')$ and $x(A) + z_{ii'} = 1$.

**Theorem 3.** *If $A$ is tight with $(i, i')$ and $B$ is tight with $(j, j')$ then at least one of the following holds:*

1. *$A - B$ is tight with $(i, i')$, $B - A$ is tight with $(j, j')$ and $\mathcal{M}(A, ii') + \mathcal{M}(B, jj') = \mathcal{M}(A - B, ii') + \mathcal{M}(B - A, jj')$.*
2. *$A - B$ is tight with $(j, j')$, $B - A$ is tight with $(i, i')$ and $\mathcal{M}(A, ii') + \mathcal{M}(B, jj') = \mathcal{M}(A - B, jj') + \mathcal{M}(B - A, ii')$.*
3. *$A \cap B$ is tight with $(i, i')$, $A \cup B$ is tight with $(j, j')$ and $\mathcal{M}(A, ii') + \mathcal{M}(B, jj') = \mathcal{M}(A \cap B, ii') + \mathcal{M}(A \cup B, jj')$.*
4. *$A \cap B$ is tight with $(j, j')$, $A \cup B$ is tight with $(i, i')$ and $\mathcal{M}(A, ii') + \mathcal{M}(B, jj') = \mathcal{M}(A \cap B, jj') + \mathcal{M}(A \cup B, ii')$.*
5. *$A - B$ is tight with $(i, i')$, $B - A$ is tight with $(i', i)$, $A \cap B$ is tight with $(j, j')$, $A \cup B$ is tight with $(j, j')$ and $2\mathcal{M}(A, ii') + 2\mathcal{M}(B, jj') = \mathcal{M}(A - B, ii') + \mathcal{M}(B - A, ii') + \mathcal{M}(A \cap B, jj') + \mathcal{M}(A \cup B, jj')$.*

*Proof.* The proof is by case analysis. For the ease of notation, if a set $A$ is tight with pair $(i, i')$, we assume $i \in A$ (and thus $i' \notin A$).

We consider two cases $i \in A - B$ and $i \in A \cap B$. Without loss of generality, we assume in the latter case $j \in A \cap B$ also (otherwise we consider $j$ instead of $i$ in our arguments). Because of tightness we have:

$$x(A) = x(A - B, B - A) + x(A - B, \overline{A \cup B}) + x(A \cap B, B - A) + x(A \cap B, \overline{A \cup B}) = 1 - z_{ii'}$$

$$x(B) = x(B - A, A - B) + x(B - A, \overline{A \cup B}) + x(A \cap B, A - B) + x(A \cap B, \overline{A \cup B}) = 1 - z_{jj'}$$

Let's first start with the case in which $i \in A \cap B$ (and thus $j \in A \cap B$). In this case $i' \in \overline{A \cup B}$ and $j' \in \overline{A \cup B}$. Because of the feasibility:

$$x(A \cap B) = x(A \cap B, A - B) + x(A \cap B, B - A) + x(A \cap B, \overline{A \cup B}) \geq 1 - z_{ii'}$$

$$x(A \cup B) = x(A - B, \overline{A \cup B}) + x(A \cap B, \overline{A \cup B}) + x(B - A, \overline{A \cup B}) \geq 1 - z_{jj'}$$

Since $x(.,.) \geq 0$, by summing up the two inequalities above and using the equalities for $x(A)$ and $x(B)$, we conclude that the inequalities should be tight, i.e., $x(A \cap B) = 1 - z_{ii'}$ and $x(A \cup B) = 1 - z_{jj'}$ and in addition $x(A - B, B - A) = 0$, i.e., $\mathcal{M}(A, ii') + \mathcal{M}(B, jj') = \mathcal{M}(A \cap B, ii') + \mathcal{M}(A \cup B, jj')$. Thus we are in the case 3 of the statement of the theorem.

Now assume that $i \in A - B$ and $j \in B - A$. Then independent of the place of $i', j'$, by the feasibility of the solution we have:

$$x(A - B) = x(A - B, A \cap B) + x(A - B, B - A) + x(A - B, \overline{A \cup B}) \geq 1 - z_{ii'}$$

$$x(B - A) = x(B - A, A - B) + x(B - A, A \cap B) + x(B - A, \overline{A \cup B}) \geq 1 - z_{jj'}$$

Since $x(.,.) \geq 0$, by summing up the two inequalities above and using the equalities for $x(A)$ and $x(B)$, we conclude that the inequalities should be tight, i.e., $x(A - B) = 1 - z_{ii'}$ and $x(B - A) = 1 - z_{jj'}$ and in addition $x(A \cap B, \overline{A \cup B}) = 0$, i.e., $\mathcal{M}(A, ii') + \mathcal{M}(B, jj') = \mathcal{M}(A - B, ii') + \mathcal{M}(B - A, jj')$. Thus we are in the case 1 of the statement of the theorem.

Finally we consider the case in which $i \in A - B$ and $j \in A \cap B$ (and thus $j' \in \overline{A \cup B}$).

Now if $i' \in \overline{A \cup B}$, then by the feasibility of the solution we have:

$$x(A \cap B) = x(A \cap B, A - B) + x(A \cap B, B - A) + x(A \cap B, \overline{A \cup B}) \geq 1 - z_{jj'}$$

$$x(A \cup B) = x(A - B, \overline{A \cup B}) + x(A \cap B, \overline{A \cup B}) + x(B - A, \overline{A \cup B}) \geq 1 - z_{ii'}$$

Since $x(.,.) \geq 0$, by summing up the two inequalities above and using the equalities for $x(A)$ and $x(B)$, we conclude that the inequalities should be tight, i.e., $x(A \cap B) = 1 - z_{jj'}$ and $x(A \cup B) = 1 - z_{ii'}$ and in addition $x(A - B, B - A) = 0$, i.e., $\mathcal{M}(A, ii') + \mathcal{M}(B, jj') = \mathcal{M}(A \cap B, jj') + \mathcal{M}(A \cup B, ii')$. Thus we are in the case 4 of the statement of the theorem.

Finally if $i' \in B - A$ then, because of feasibility we have

$$x(A - B) = x(A - B, A \cap B) + x(A - B, B - A) + x(A - B, \overline{A \cup B}) \geq 1 - z_{ii'}$$

$$x(A \cap B) = x(A \cap B, A - B) + x(A \cap B, B - A) + x(A \cap B, \overline{A \cup B}) \geq 1 - z_{jj'}$$

$$x(B - A) = x(B - A, A - B) + x(B - A, A \cap B) + x(B - A, \overline{A \cup B}) \geq 1 - z_{jj'}$$

$$x(A \cup B) = x(A - B, \overline{A \cup B}) + x(A \cap B, \overline{A \cup B}) + x(B - A, \overline{A \cup B}) \geq 1 - z_{ii'}$$

Since $x(.,.) \geq 0$, by summing up the four inequalities above and and use the equalities for $2x(A)$ and $2x(B)$, we conclude that all inequalities should be tight, and in addition $x(A - B, B - A) = 0$ and $x(A \cap B, \overline{A \cup B}) = 0$, i.e., and $2\mathcal{M}(A, ii') + 2\mathcal{M}(B, jj') = \mathcal{M}(A - B, ii') + \mathcal{M}(B - A, jj') + \mathcal{M}(A \cap B, ii') + \mathcal{M}(A \cup B, jj')$. So the case 5 of the statement of the theorem holds.    □

Note that especially Case 5 in Theorem 3 is novel to our extension of iterative rounding methods.

Let $\mathcal{T}$ be the set of all tight constraints. For any set of tight constraints $\mathcal{F}$, we denote the vector space spanned by the vectors $M(S, ii')$, where $S \subset V$ and $(i, i') \in \mathcal{P}$, by $Span(\mathcal{F})$. We say two sets $A$ and $B$ *cross* if none of the sets $A - B$, $B - A$ and $A \cap B$ is empty. We say a family of tight constraints is *laminar* if no two sets corresponding to two constraints in it cross.

The proof of the following theorem is similar to that of Jain [13] and hence omitted.

**Theorem 4.** *For any maximal laminar family $\mathcal{L}$ of tight constraints, $Span(\mathcal{L}) = Span(\mathcal{T})$.*

Since $x, z$ is a basic feasible solution, the dimension of $Span(\mathcal{T})$ is $|E(G)| + |\mathcal{P}|$. Since $Span(\mathcal{L}) = Span(\mathcal{T})$, it is possible to choose a basis for $Span(\mathcal{T})$ from the vectors in $\{M(S, ii)\} \in \mathcal{L}$. Let $\mathcal{B} \subseteq \mathcal{L}$ forms a basis for $Span(\mathcal{T})$. Hence we have the following theorem.

**Corollary 2.** *There exists a laminar family, $\mathcal{B}$, of tight constraints satisfying 1) $|\mathcal{B}| = |E(G)| + |\mathcal{P}|$; 2) The vectors in $\mathcal{B}$ are independent; and 3) All constraints in $\mathcal{B}$ are tight.*

Note that in our laminar family if a set $S$ is tight with both $(i, i')$ and $(j, j')$ in two different constraints, since $z_{ii'} = z_{jj'}$, we can remove variable $z_{jj'}$ and just use $z_{ii'}$ instead. Since we removed one variable and one constraint, still we have a basic feasible solution which is laminar. By this reduction, we always can make sure that each set is tight with only one pair. Thus a tight set uniquely determines the tight pair and we use a tight constraint and a tight set interchangeably in our discussion below.

Now we are ready to prove Theorem 1.

**Theorem 5.** *In any basic feasible solution for LP 1, for at least one edge $e \in E$, $x_e \geq \frac{1}{3}$, or for at least one pair $(i, j) \in \mathcal{P}$, $z_{ij} \geq \frac{1}{3}$.*

*Proof.* We are giving a token to each end-point of an edge (and thus two tokens for an edge) and two tokens to all $z$ variables (notice that some $z$ variables are used for more than one commodity pairs as discussed above). Now, we will distribute the tokens such that for every set in the laminar family gets at least two tokens and every root at least four tokens unless the corresponding cut has exactly three edges. (note that each cut has at least three edges since the value of each variable is less than $\frac{1}{3}$) in which the root gets at least three token. This contradict the equality $|V(F)| = |E(G)| + |\mathcal{P}|$ where $F$ is the rooted forest of laminar sets in the laminar family. The subtree of $F$ rooted at $R$ consists of $R$ and all its descendants. We will prove this result by the induction on every rooted subtree of $F$.

Consider a subtree rooted at $R$. Since all $x_e$ and $z_{ij}$ are at most $\frac{1}{3}$, if $R$ is a leaf node, it has at least three edges crossing it and thus gets at least three tokens (and more than 3 tokens if the degree is more than 3). This means the induction is correct for a leaf node, as the basis of the induction.

If $R$ has four or more children, by the induction hypothesis each child has at least three tokens and each of their descendants gets at least two tokens. We re-assign one extra token from each child to the node $R$. Thus $R$ has at least four tokens and the induction hypothesis is correct in this case.

If $R$ has three children, if there is a *private vertex* $u$ to $R$, i.e., a vertex which is in $R$ but not in any of its children, then we are done (since all $x_e$ values are fractional, the degree of $u$ is at least two and thus can contributes at least two extra tokens toward $R$). Also if one of the children has at least four edges in its corresponding cut, by the induction hypothesis it has at least two extra tokens to contribute toward those of $R$ and we are done.

Next, if $R$ has exactly three children each with exactly three edges in its corresponding cut, then by parity $R$ has an odd number of edges in its corresponding cut. If $R$ has edges in the its cut then the three extra token by its children suffices. If $R$ has seven or nine edges in the cut, then at least one of its children has all three edges in the cut and the corresponding pair is not satisfied. But this means all other edges than those of this cut should be zero which is contradiction to fractional value assumption. Now if $R$ tight with $z_{pp'}$ has exactly five edges in the cut, it should be the case that two children $C_1$ tight with $z_{ii'}$ and $C_3$ tight with $z_{kk'}$ have two edges in the cut and $C_2$ tight with $z_{jj'}$ has one edges in the cut. Note that in this case $z_{pp'} > \min\{z_{ii'}, z_{jj'}, z_{kk'}\}$ then at least for one of $z_{ii'}$, $z_{jj'}$, and $z_{kk'}$ all pairs should be inside $R$ for the first time and thus we have at least two extra tokens towards the requirement of $R$ and we are done. It also means that $p$ should be inside the child $C$ with $\min\{z_{ii'}, z_{jj'}, z_{kk'}\}$ and it should be equal to its corresponding $z$ value (otherwise child $C$ violates the condition for $z_{pp'}$). Assume that $z_{pp'} = z_{jj'}$. In this case, it is easy to see that since $C_2$ is tight with three edges and with five edges, the sum of $x$ variables of $C_1$ and $C_3$ in the cut $R$ is equal to the sum of $x$ variables of $C_1$ and $C_3$ to $C_2$. But it means at least for one of $C_1$ and $C_3$, the edge $e$ to $C_2$ has $x_e \geq x_{e'} + x_{e''}$ where $e'$ and $e''$ are the edges in the cut $R$. But since $x_e + x_{e'} + x_{e''} > \frac{2}{3}$ (due to the fact that all $z$ variables are less than $\frac{1}{3}$), $x_e \geq \frac{1}{3}$ which is a contradiction. If $z_{pp} = z_{ii'} \leq z_{kk'}$ where $z_{ii'} < z_{jj'}$. In this case the edge from $C_1$ to $C_2$ should has an $x$ value equal to that those edges of $C_2$ and $C_3$ in the cut $R$. It means the total $x$ value of two edge of $C_3$ in the cut is less than $\frac{1}{3}$ which is a contradiction, since the third edge has $x$ value at least $\frac{1}{3}$.

Now we consider the case in which $R$ is tight with $z_{pp'}$ and has two children . If there is a private vertex $u$ to $R$ we have at least four tokens to satisfy $R$ (two from $u$ and one from each of its children). If both of these children have degree at least four, then we have four extra tokens for $R$ (two from each child). Then at least one of two children, namely $C_1$ tight with $z_{ii'}$, has exactly three edges in its corresponding cut. The other child $C_2$ tight with $z_{jj'}$ has at least three edges in the cut. Note that in this case $z_{pp'} > \min\{z_{ii'}, z_{jj'}\}$ then at least for one of $z_{ii'}$ and $z_{jj'}$ all pairs should be inside $R$ for the first time and thus we have at least two extra tokens towards the requirement of $R$ and we are done. It also means that $p$ should be inside the child $C$ with $\min\{z_{ii'}, z_{jj'}, z_{kk'}\}$ and it should be equal to its corresponding $z$ value (otherwise child $C$ violates the condition for $z_{pp'}$). First assume that $z_{pp'} = z_{ii'} \leq z_{jj'}$. In this case it is not possible that all three edges of $C_1$ are in the cut $R$, since then all edges of $C_2$ are in the cut $R$ and they are zero (since the cut $R$ is already tight with $z_{pp'} = z_{ii'}$). If $C_1$ has two edges in the cut, since $z_{ii'} \leq z_{jj'}$, it means sum of the $x$ values of the edges in the cut corresponding to $R$, which has one edges from $C_1$ and the rest are the edges of $C_2$ in the $R$ cut, should be at least the sum of $x$ values of the cut corresponding to $C_2$. But these means the $x$ value of the edge of $C_1$ in the $R$ cut is at least the sum of $x$ values of the two edges from $C_1$ to $C_2$. Since value of all three edges in $C_1$ is at least

$\frac{2}{3}$, it means the $x$ value of the edge of $C_1$ in the $R$ cut is at least $\frac{1}{3}$, a contradiction. In case $C_1$ does not have any edges in the cut $R$, then all edges should go $C_2$ which means $z_{ii'}$ should be tight with a proper subset of edges of $C_2$ though we know that $x$ values of all edges of $C_2$ is at most $1 - z_{jj}$, a contradiction. we know even all edges of minus those edges should be tight in $R$ with the same $z_{jj'}$ which means all edges between $C_1$ and $C_2$ should be zero which is a contradiction.

Next assume that $z_{pp'} = z_{jj'} \leq z_{ii'}$. In this case it is not possible that all edges of $C_2$ and thus $C_1$ are in the cut since all edges of $C_1$ should have zero $x$ value. In case if one edge of $C_1$ or two edges of $C_1$ are in the cut $R$, then $x$ value of one edge of $C_1$ is equal to the $x$ value of two edges of $C_1$ which means that edge should have $x$ value at least $\frac{1}{3}$ which is a contradiction. In case $C_1$ does not have any edges in the cut $R$, then $C_2$ minus those edges should be tight in $R$ with the same $z_{jj'}$ which means all edges between $C_1$ and $C_2$ should be zero which is a contradiction.

It only remains the case in which $R$ has only one child $C$. In this case if $R$ and $C$ are both tight with respect to $z_{ii'}$ then since $R$ and $C$ are independent there is a vertex $u \in R - C$. However, if $R$ is tight with $z_{ii'}$ and $C$ is tight with $z_{jj'}$ since $z_{ii'} \neq z_{jj'}$, these two cuts should be different and thus again there is vertex $u \in R - C$. Since all $x_e$ values are fractional the degree of $u$ is at least two and thus $u$ gets at least two tokens. Without loss of generality assume $u$ is the node with maximum degree. If $u$ has degree at least three then we can assign at least these three private tokens of $u$ and at least one extra token of $C$ to $R$ to have the induction hypothesis satisfied. In case $u$ has degree two and $C$ has at least four edges in the cut, then we have at least two tokens from $u$ and two extra tokens from $C$ to assign at least four tokens to $R$ and satisfy the hypothesis.

The only remaining case when $R$ has only one child is when $u$ has degree two and $C$ has an odd number of edges in its cut. However in this case because of parity, $R$ should have an odd number of edges in its cut (note that in this case, we may have some other vertices than $u$ of degree two in $R - C$.) If this odd number is three then two tokens of $u$ and one extra token of $C$ satisfies the required number of tokens for $R$. If there is a vertex other than $u$ in $R_C$ it has also two extra tokens and we are done. The only case is that $u$ has degree two, $C$ has three edges and all these five edges are in the cut corresponding to $R$. It means in this case $R$ should be tight with $z_{ii'}$ and $C$ should be tight with $z_{jj'}$ where $z_{ii'} < z_{jj'}$ (otherwise the edges from $u$ in the cut should zero which is a contradiction to the fractional values for $x_e$s). Here $i \neq u$ otherwise, $u$ has degree three and thus three extra tokens and we have at least four tokens for $R$. It means $i \in C$ which is again a contradiction since the current cut for $C$ violates the cut condition for $i$ in the LP.    □

Finally, it is worth mentioning though we guarantee that during the course of the algorithm, we can get a variable which is only one-third-integral, in the first iteration always we can find an integral $z$ variable. Below there is a more general proposition regarding this issue.

**Proposition 1.** *If there is a set $S$ of fractional variables which contains exactly one variable from each tight constraint in our laminar family, our solution cannot be a basic optimum solution. In particular, there is no basic optimum solution in which all constraints are tight with fractional $z$ variables.*

*Proof.* The second statement follows immediately by taking set $S$ in the first statement to be the set of all fractional $z$ variables. The first statement follows from the fact that we can always increase (decrease) each variable $w$ in set $S$ by $\varepsilon(1-w)$, for a very small $\varepsilon > 0$, and decrease (increase) each other variable $u$ by $\varepsilon u$ (increase/decrease is depending on which option does not increase the objective function). It is easy to see in this way we can always get another feasible solution which makes all our current constraints in the laminar family tight and whose value is not larger than that of optimum.    □

## 4    Prize-Collecting Survivable Steiner Network Design

In this section, we show how we can generalize our approach of iterative rounding to obtain a 3-approximation algorithm for the prize-collecting survivable Steiner network design problem. In this problem, we are given connectivity requirements $r_{uv}$ for all pairs of vertices $u$ and $v$ and a non-increasing marginal penalty function $\pi_{uv}(.)$ for $u$ and $v$. Our goal is to find a minimum way of constructing a graph in which we connect $u$ and $v$ with $r'_{uv} \leq r_{uv}$ edge-disjoint paths and paying the marginal penalty $\pi_{uv}(r'_{uv}+1) + \pi_{uv}(r'_{uv}+2) + \cdots + \pi_{uv}(r_{uv})$ for violating the connectivity between $u$ and $v$ to the amount of $r_{uv} - r'_{uv}$.

Let us first start with the following natural LP.

$$\text{OPT} = \quad \text{minimize} \sum_{e \in E} c_e x_e \quad + \sum_{i,j \in \mathcal{P}} \sum_{k=1}^{r_{ij}} \pi_{ij}(k) z_{ij}^k \tag{9}$$

$$\text{subject to} \sum_{e \in \delta(S)} x_e + \sum_{k=1}^{r_{ij}} z_{ij}^k \geq r_{ij} \quad \forall S \subset V, (i,j) \in \mathcal{P}, S \odot (i,j) \tag{10}$$

$$x_e \geq 0 \qquad \qquad \forall e \in E \tag{11}$$

$$z_{ij}^k \geq 0 \qquad \qquad \forall (i,j) \in \mathcal{P}, 0 \leq k \leq r_{ij} \tag{12}$$

First, it is easy to see that since $\pi_{uv}$'s are non-increasing without loss of generality we can assume $0 < z_{uv}^k$ only if $z_{uv}^{k+1} = 1$ for $1 \leq k < r_{uv}$. Now the algorithm indeed is very similar to PCSF-ALG in Figure 1, except for an edge $e$ with $x_e^* \geq \frac{1}{3}$, we do not contract that edge (indeed the contraction was only due to simplicity in PCSF-ALG). Instead we choose edge $e$ to be in our solution and consider it like an edge of $x_e^* = 1$ value in the rest of the rounding. We repeat this process until we satisfy all the commodity pairs either by connecting or paying enough penalty. The argument follows almost the same as the argument for PCSF with the change of connectivity $r_{ij}$ instead of 1 in our arguments in Theorem 3. Note that since $0 < z_{uv}^k$ only if $z_{uv}^{k+1} = 1$ for $1 \leq k < r_{uv}$, we can assume that each constraint is tight with only one variable $z_{uv}^k$, $1 \leq k \leq r_{uv}$ (all $z_{uv}^k = 1$ can be rounded to one and removed from further consideration in the LP without costing any extra penalties with respect to the optimum solution of the LP in Theorem 2). Thus as a result we have the following theorem.

**Theorem 6.** *There is an iterative rounding 3-approximation algorithm for the prize-collecting survivable Steiner network design problem.*

Finally, it is worth mentioning that by combining the technique of this paper in obtaining a one-third-integral variable and that of Lau et. al [18] (which essentially use the work of Jain for survivable network design as a block-box), it is not hard to get $(3, 3B_v + 3)$-approximation algorithm for the *prize-collecting survivable network design with bounded-degree constraints* $B_v$, where the cost of the returned solution is at most three times the cost of an optimum solution satisfying the degree bounds and the degree of each vertex is at most $3B_v + 3$.

## 5  Conclusions and Tight Example

In this paper, we presented a new approach of iterative rounding for prize-collecting problems which generalizes the use of iterative rounding when we do not have necessarily submodular functions. In addition, we used more linear dependence between constraints instead of just some simple additions with all coefficients one. The replacement of one of four sets instead of two sets in our laminar family is another extension to previous iterative rounding approaches (e.g. see [13]). In addition, next we show that indeed our approach of iterative rounding for getting a 3-approximation algorithm is tight even for prize-collecting Steiner forest, i.e., there is an instance with a basic feasible solution in which all $x$ and $z$ variables, except one zero $z$ variable[3], are $\frac{1}{3}$.

**Tight example:** Consider a complete bipartite graph $K_{3,2} = (\{v_1, v_2, v_3\} \cup \{v_4, v_5\}, E)$ with (penalty) pairs $\mathcal{P} = \{(v_1, v_3), (v_2, v_3), (v_4, v_5)\}$. Assume all edges in $E$ and penalties in $\mathcal{P}$ are ones. Consider a basic feasible solution in which all $x$ and $z$ variables are $\frac{1}{3}$, except $z_{4,5} = 0$. The cost of this fractional solution is $\frac{8}{3}$ which is less than the optimum integral solution 3 for this example. Also, it is easy to check that sets $\{v_1\}$ with $(v_1, v_3)$, $\{v_2\}$ with $(v_2, v_3)$, $\{v_3\}$ with $(v_1, v_3)$, $\{v_3\}$ with $(v_2, v_3)$, $\{v_4\}$ with $(v_4, v_5)$, $\{v_5\}$ with $(v_4, v_5)$, $\{v_1, v_4\}$ with $(v_4, v_5)$, $\{v_2, v_5\}$ with $(v_4, v_5)$ form a laminar family of tight constraints. These eight tight constraints in addition of tight constraint $z_{4,5} = 0$ form nine tight independent constraints of the aforementioned basic feasible solution. In this case, by fixing $z_{4,5} = 0$ and omitting variable $z_{4,5}$, we end up with exactly the same instance in which all variables are $\frac{1}{3}$. This shows that $\frac{1}{3}$ in our Theorem 5 is indeed tight.

Finally, we do believe that our iterative rounding approach might be applicable for other problems such as *multicommodity connected facility location (MCFL)* and *multicommodity rent-or-buy (MRoB)* (see e.g. [1,4,9,16,17]) to obtain simpler approximation algorithms with better factors than those currently exist.

**Acknowledgement.** The first author would like to thank Philip Klein and Mohammad-Hossein Bateni for several fruitful discussions and reading an early draft of this paper. Thanks especially goes to Howard Karloff whose program generated an example whose simplified version is the tight example in Section 5.

---

[3] Note that always there exists a $z$ variable with an integral value when we solve the original LP according to Proposition 1.

# References

1. Awerbuch, B., Azar, Y.: Buy-at-bulk network design. In: FOCS 1997, p. 542 (1997)
2. Balas, E.: The prize collecting traveling salesman problem. Networks 19, 621–636 (1989)
3. Bansal, N., Khandekar, R., Nagarajan, V.: Additive guarantees for degree bounded directed network design. In: STOC 2008, pp. 769–778 (2008)
4. Becchetti, L., Konemann, J., Leonardi, S., Pal, M.: Sharing the cost more efficiently: Improved approximation for multicommodity rent-or-buy. In: SODA 2005, pp. 375–384 (2005)
5. Bienstock, D., Goemans, M.X., Simchi-Levi, D., Williamson, D.: A note on the prize collecting traveling salesman problem. Math. Prog. 59, 413–420 (1993)
6. Chudak, F.A., Roughgarden, T., Williamson, D.P.: Approximate k-MSTs and k-Steiner trees via the primal-dual method and lagrangean relaxation. In: Aardal, K., Gerards, B. (eds.) IPCO 2001. LNCS, vol. 2081, pp. 60–70. Springer, Heidelberg (2001)
7. Goemans, M.X., Williamson, D.P.: A general approximation technique for constrained forest problems. SIAM J. Comput. 24, 296–317 (1995)
8. Gupta, A., Könemann, J., Leonardi, S., Ravi, R., Schäfer, G.: An efficient cost-sharing mechanism for the prize-collecting steiner forest problem. In: SODA 2007, pp. 1153–1162 (2007)
9. Gupta, A., Kumar, A., Pal, M., Roughgarden, T.: Approximation via cost-sharing: a simple approximation algorithm for the multicommodity rent-or-buy problem. In: FOCS 2003, p. 606 (2003)
10. Gutner, S.: Elementary approximation algorithms for prize collecting Steiner tree problems. Inf. Process. Lett. 107, 39–44 (2008)
11. M. Hajiaghayi, Designing minimum total cost networks using iterative rounding approximation methods. Pending patent with USPTO of application number 12/315,657 (July 2008)
12. Hajiaghayi, M.T., Jain, K.: The prize-collecting generalized Steiner tree problem via a new approach of primal-dual schema. In: SODA 2006, pp. 631–640 (2006)
13. Jain, K.: A factor 2 approximation algorithm for the generalized Steiner network problem. Combinatorica 21, 39–60 (2001)
14. Jain, K., Vazirani, V.V.: Approximation algorithms for metric facility location and k-median problems using the primal-dual schema and Lagrangian relaxation. J. ACM 48, 274–296 (2001)
15. Johnson, D.S., Minkoff, M., Phillips, S.: The prize collecting Steiner tree problem: theory and practice. In: SODA 2000, pp. 760–769 (2000)
16. Karger, D.R., Minkoff, M.: Building steiner trees with incomplete global knowledge. In: FOCS 2001, p. 613 (2001)
17. Kumar, A., Gupta, A., Roughgarden, T.: A constant-factor approximation algorithm for the multicommodity. In: FOCS 2002, p. 333 (2002)
18. Lau, L.C., Naor, J.S., Salavatipour, M.R., Singh, M.: Survivable network design with degree or order constraints. In: STOC 2007, pp. 651–660 (2007)
19. Lau, L.C., Singh, M.: Additive approximation for bounded degree survivable network design. In: STOC 2008, pp. 759–768 (2008)
20. Salman, F.S., Cheriyan, J., Ravi, R., Subramanian, S.: Approximating the single-sink link-installation problem in network design. SIAM J. Optim. 11, 595–610 (2000)
21. Sharma, Y., Swamy, C., Williamson, D.P.: Approximation algorithms for prize collecting forest problems with submodular penalty functions. In: SODA 2007, pp. 1275–1284 (2007)
22. Konemann, J., Grandoni, F., Rothvoss, T., Qian, J., Schaefer, G., Swamy, C., Williamson, D.P.: An iterated rounding 3-approximation algorithm for prize-collecting Steiner forest (2009) (unpublished manuscript)
23. Singh, M., Lau, L.C.: Approximating minimum bounded degree spanning trees to within one of optimal. In: STOC 2007, pp. 661–670 (2007)

# Kernelization through Tidying
## A Case Study Based on s-Plex Cluster Vertex Deletion*

René van Bevern, Hannes Moser, and Rolf Niedermeier

Institut für Informatik, Friedrich-Schiller-Universität Jena,
Ernst-Abbe-Platz 2, D-07743 Jena, Germany
{rene.bevern,hannes.moser,rolf.niedermeier}@uni-jena.de

**Abstract.** We introduce the NP-hard graph-based data clustering problem s-PLEX CLUSTER VERTEX DELETION, where the task is to delete at most $k$ vertices from a graph so that the connected components of the resulting graph are $s$-plexes. In an $s$-plex, every vertex has an edge to all but at most $s - 1$ other vertices; cliques are 1-plexes. We propose a new method for kernelizing a large class of vertex deletion problems and illustrate it by developing an $O(k^2 s^3)$-vertex problem kernel for s-PLEX CLUSTER VERTEX DELETION that can be computed in $O(ksn^2)$ time, where $n$ is the number of graph vertices. The corresponding "kernelization through tidying" exploits polynomial-time approximation results.

## 1 Introduction

The contributions of this work are two-fold. On the one hand, we introduce a vertex deletion problem in the field of graph-based data clustering. On the other hand, we propose a novel method to derive (typically polynomial-size) problem kernels for NP-hard vertex deletion problems whose goal graphs can be characterized by forbidden induced subgraphs. More specifically, using "kernelization through tidying", we provide a quadratic-vertex problem kernel for the NP-hard s-PLEX CLUSTER VERTEX DELETION problem, for constant $s \geq 1$.

**s-Plex Cluster Vertex Deletion.** Many vertex deletion problems in graphs can be considered as "graph cleaning procedures", see Marx and Schlotter [11]. This view particularly applies to graph-based data clustering, where the graph vertices represent data items and there is an edge between two vertices iff the two items are similar enough [14]. Then, a cluster graph is a graph where every connected component forms a cluster, a dense subgraph such as a clique in the most extreme case. Due to faulty data or outliers, the given graph may not be a cluster graph and it needs to be cleaned in order to become a cluster graph. A recent example for this is the NP-hard CLUSTER VERTEX DELETION problem [9], where the task is to delete as few vertices as possible such that the resulting graph is a disjoint union of cliques. In contrast, in the also NP-hard s-PLEX CLUSTER VERTEX DELETION problem we replace cliques with $s$-plexes (where $s$ is typically a small constant):

---

* Supported by the DFG, project AREG, NI 369/9.

s-PLEX CLUSTER VERTEX DELETION (s-PCVD)
**Input:** An undirected graph $G = (V, E)$ and an integer $k \geq 0$.
**Question:** Is there a vertex set $S \subseteq V$ with $|S| \leq k$ such that $G[V \setminus S]$
is a disjoint union of s-plexes?

Herein, an *s-plex* is a graph where every vertex has an edge to all but at most $s-1$
other vertices [13]. Subsequently, we refer to the solution set $S$ as *s-plex cluster
graph vertex deletion set (s-pvd set)*. The concept of s-plexes has recently received
considerable interest in various fields, see, e.g., [2,6,7,12]. The point of replacing
cliques with s-plexes in the context of cluster graph generation is that s-plexes
better allow to balance the number of vertex deletions against the sizes and
densities of the resulting clusters. Note that too many vertex deletions from the
input graph may too strongly change the original data whereas asking for cliques
as clusters often seems overly restrictive [5,13]. Summarizing, s-PCVD blends
and extends previous studies on CLUSTER VERTEX DELETION [9] (which is the
same as 1-PCVD) and on s-PLEX EDITING [7], where in the latter problem one
adds and deletes as few *edges* as possible to transform a graph into an s-plex
cluster graph.

**Problem kernelization.** Data reduction and problem kernelization is a core
tool of parameterized algorithmics [8]. Herein, viewing the underlying problem
as a decision problem, the goal is, given any problem instance $x$ (a graph in
our case) with a parameter $k$ (the number of vertex deletions in our case), to
transform it in polynomial time into a new instance $x'$ with parameter $k'$ such
that $|x'|$ is bounded by a function in $k$ (ideally, a polynomial in $k$), $k' \leq k$, and
$(x, k)$ is a yes-instance iff $(x', k')$ is a yes-instance. We call $(x', k')$ the *problem
kernel*. It is desirable to get the problem kernel size $|x'|$ as small as possible.
By means of a case study based on s-PCVD, we will present a method to develop small problem kernels for vertex deletion problems where the goal graph
can be characterized by a set of forbidden induced subgraphs. For instance, if
the goal graph shall be a disjoint union of cliques, then it is characterized by
forbidding induced $P_3$'s [14], that is, induced paths containing three vertices.
A more complex characterization has been developed for graphs that are disjoint unions of s-plexes [7]. We term our data reduction approach "kernelization
through tidying"—it uses a polynomial-time constant-factor approximation to
"tidy up" the graph to make data reduction rules applicable.

**Discussion of results.** Complementing and extending results for CLUSTER
VERTEX DELETION [9], we prove an $O(k^2 s^3)$-vertex problem kernel for s-PCVD,
which can be computed in $O(ksn^2)$ time. Note that the related edge modification problem s-PLEX EDITING has an $O(ks^2)$-vertex problem kernel which can
be computed in $O(n^4)$ time [7]. We emphasize that the underlying kernelization
algorithms are completely different in both cases and that vertex deletion is a
"more powerful" operation than edge modification, so a larger problem kernel in
the case of s-PCVD does not come unexpectedly. Our main conceptual contribution is the "kernelization through tidying" method outlined in Section 2. There
is related work by Kratsch [10] that provides polynomial-size problem kernels for
constant-factor approximable problems contained in the classes MIN $F^+\Pi_1$ and

MAX NP. The $s$-PCVD problem is contained in MIN $F^+\Pi_1$. Applying Kratsch's more general method to $s$-PCVD would lead to an $k^{O(s)}$-vertex kernel. By way of contrast, in our $O(k^2 s^3)$-vertex bound, the value of $s$ does not influence the degree of the polynomial in $k$, a significant advantage. Other related work is due to Abu-Khzam [1], who considers problem kernels for HITTING SET problems. Again, translating our problem instances into HITTING SET instances (which can be done in a straightforward way) and applying a kernelization for HITTING SET would yield problem instances which are size-bounded by polynomials whose degree depends on $s$.

Due to space limits, most proofs are deferred to a full version of this paper.

**Notation.** We only consider *undirected* graphs $G = (V, E)$, where $V$ is the set of vertices and $E$ is the set of edges. Throughout this work, we use $n := |V|$ and $m := |E|$. The *open neighborhood* $N(v)$ of a vertex $v \in V$ is the set of vertices that are adjacent to $v$. For a vertex set $U \subseteq V$, we define $N(U) := \bigcup_{v \in U} N(v) \setminus U$. We call a vertex $v \in V$ *adjacent* to $V' \subseteq V$ if $v$ has a neighbor in $V'$. Analogously, we call $U \subseteq V$ *adjacent* to a vertex set $W \subseteq V$ with $W \cap U = \emptyset$ if $N(U) \cap W \neq \emptyset$. We call two vertices $v$ and $w$ *connected* in $G$ if there exists a path from $v$ to $w$ in $G$. For a set of vertices $V' \subseteq V$, the *induced subgraph* $G[V']$ is the graph over the vertex set $V'$ with the edge set $\{\{v, w\} \in E \mid v, w \in V'\}$. For $V' \subseteq V$, we use $G - V'$ as an abbreviation for $G[V \setminus V']$. For a set $\mathcal{F}$ of graphs, we call a graph $\mathcal{F}$-free if it does not contain any graph from $\mathcal{F}$ as an induced subgraph.

# 2    Kernelization through Tidying

For a set $\mathcal{F}$ of forbidden induced subgraphs (FISGs) we outline a general kernelization method for the $\mathcal{F}$-FREE VERTEX DELETION problem. Here, the task is, given an undirected graph $G = (V, E)$ and an integer $k \geq 0$, to decide whether the graph can be made $\mathcal{F}$-free by deleting at most $k$ vertices. If $\mathcal{F}$ is finite (as we have for $s$-PCVD, $s$ being a constant), then $\mathcal{F}$-FREE VERTEX DELETION is clearly fixed-parameter tractable with respect to the parameter $k$, as directly follows from Cai's [4] general result. Moreover, one may observe that its minimization version is in MIN $F^+\Pi_1$; therefore, using a technique due to Kratsch [10], $\mathcal{F}$-FREE VERTEX DELETION admits a problem kernel containing $O(k^h)$ vertices, where $h$ is the maximum number of vertices of a FISG in $\mathcal{F}$. We present an alternative technique to kernelize such vertex deletion problems. While the technique of Kratsch [10] is more general, our approach seems to be useful to obtain smaller problem kernels. For example, the FISGs for $s$-PCVD consist of at most $s + 1 + T_s$ vertices [7], where $T_s$ is the maximum integer satisfying $T_s \cdot (T_s + 1) \leq s$; therefore, Kratsch's technique [10] yields an $O(k^{s+1+T_s})$-vertex problem kernel. In contrast, we obtain an $O(k^2 s^3)$-vertex problem kernel using a novel method. Our approach comprises the following three main steps.

**Approximation Step.** This step is to compute an approximate solution $X$ for $\mathcal{F}$-FREE VERTEX DELETION in polynomial time; let $a$ be the corresponding approximation factor. Obviously, the *residual graph* $G - X$ is $\mathcal{F}$-free. Since $X$

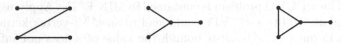

**Fig. 1.** Minimal forbidden induced subgraphs (FISGs) for 2-plex cluster graphs

is a factor-$a$ approximate solution, we can abort with returning "no-instance" if $|X| > ak$. Hence, otherwise, we proceed knowing that $|X| \leq ak$. It remains to bound the number of vertices in the residual graph $G - X$. To this end, we use the property that $G - X$ is $\mathcal{F}$-free.

This property can be difficult to exploit; however, it is always possible to efficiently find a small vertex set $T \subseteq V \setminus X$ (called *tidying set*) such that in $G - T$ (called the *tidy subgraph*), for each $v \in X$, deleting all vertices in $X \setminus \{v\}$ results in an $\mathcal{F}$-free graph (called the *local tidiness property of $G - T$*). As we will see, this additional property helps in finding suitable data reduction rules to shrink the size of $G - X$. The subsequent *Tidying Step* finds such a tidying set $T$.

**Tidying Step.** In this step, polynomial-time data reduction is employed and the tidying set $T$ is computed. Roughly speaking, the data reduction ensures that the tidying set does not become too big.

First, we describe the data reduction. Compute for each $v \in X$ a maximal set $\mathcal{F}(v)$ of FISGs that pairwise intersect exactly in $v$. If $|\mathcal{F}(v)| > k$, then a "high-degree data reduction rule" deletes $v$ from both $G$ and $X$ and decreases the parameter $k$ by one. The correctness of this rule is easy to verify.

Second, we describe how to compute the tidying set $T$, show that its size is bounded, and argue that $G - T$ fulfills the local tidiness property. We define the *tidying set* of a vertex $v \in X$ as $T(v) := \bigcup_{F \in \mathcal{F}(v)} V(F) \setminus X$ (that is, all vertices in $V \setminus X$ that are in a FISG of $\mathcal{F}(v)$). The *tidying set* of the whole graph is defined as $T := \bigcup_{v \in X} T(v)$. Since after the high-degree data reduction rule $|\mathcal{F}(v)| \leq k$, we know that $|T(v)| \leq hk$, where $h$ is the maximum number of vertices of a FISG in $\mathcal{F}$; hence, $|T| \leq hak^2$. The local tidiness property of $G - T$ follows directly from the maximality of $\mathcal{F}(v)$. The local tidiness property can be exploited in the final *Shrinking Step*.

**Shrinking Step.** This is the most unspecified step, which has to be developed using specific properties of the studied vertex deletion problem. In this step, the task is to shrink the tidy subgraph $G - T$ using problem-specific data reduction rules that exploit the local tidiness property. Depending on the strength of these data reduction rules, the total problem kernel size is as follows: as we have seen, the factor-$a$ approximate solution $X$ has size at most $ak$. The tidying set $T$ based on size-at-most-$h$ FISGs has size at most $hak^2$. If we shrink the tidy subgraph $G - T$ to at most $f(k)$ vertices, then we obtain a problem kernel with $O(ak + hak^2 + f(k))$ vertices.

In the next section, we present a case study of kernelization through tidying using 2-PCVD. After that, we generalize the approach to $s$-PCVD with $s > 2$ (Section 4). Finally, we show how to significantly speed up the kernelization.

# 3     A Problem Kernel for 2-Plex Cluster Vertex Deletion

We specify the steps outlined in Section 2 to obtain a problem kernel for 2-PCVD.

*Approximation Step.* Following the tidying kernelization method, greedily compute a factor-4 approximate 2-plex cluster graph vertex deletion set (*2-pvd set*) $X$ using the FISG characterization of 2-plex cluster graphs [7] (simply find one of the three four-vertex FISGs (see Figure 1), add its vertices to $X$, and remove its vertices from the graph, until the remaining graph is a 2-plex cluster graph). If $|X| > 4k$, then simply return "no-instance". Therefore, in the following one may assume that $|X| \leq 4k$. It remains to bound the number of vertices in the residual graph $G - X$.

Using the linear-time algorithm by Guo et al. [7] to find a FISG for 2-plex cluster graphs, we can compute $X$ in $O(k \cdot (n + m))$ time, by either applying the linear-time algorithm at most $k + 1$ times or returning "no-instance".

*Tidying Step.* Let $X$ be the factor-4 approximate 2-pvd set computed in the Approximation Step. For each $v \in X$, greedily compute a maximal set $\mathcal{F}(v)$ of FISGs pairwise intersecting exactly in $v$. Since the FISGs for 2-PCVD contain four vertices, this can be done in $O(n^3)$ time for each $v \in X$ and therefore in $O(|X| \cdot n^3) = O(k \cdot n^3)$ time in total.[1]

**Reduction Rule 1.** *If there exists a vertex $v \in X$ such that $|\mathcal{F}(v)| > k$, then delete $v$ from $G$ and $X$ and decrement $k$ by one.*

**Lemma 1.** *Rule 1 is correct and can be exhaustively applied in $O(n + m)$ time.*

Additionally, we apply a simple and obviously correct data reduction rule in $O(n + m)$ time:

**Reduction Rule 2.** *Delete connected components from $G$ that form 2-plexes.*

In the following, we assume that $G$ is reduced with respect to Rules 1 and 2. Moreover, $X$ is a 2-pvd set of size at most $4k$, and a maximal set of FISGs $\mathcal{F}(v)$ that pairwise intersect exactly in $v$ shall be computed for each $v \in X$. The tidying set is $T(v) := \bigcup_{F \in \mathcal{F}(v)} V(F) \setminus X$ for each $v \in X$ and the tidying set for the whole graph is $T := \bigcup_{v \in X} T(v)$. Since $|\mathcal{F}(v)| \leq k$, $|X| \leq 4k$, and the FISGs have at most four vertices, we can conclude that $|T| \leq 12k^2$. It remains to bound the number of vertices in $G - (X \cup T)$; more specifically, we bound the number of vertices in the tidy subgraph $G - T$.

*Shrinking Step.* Since the set $\mathcal{F}(v)$ of FISGs is maximal, the tidy subgraph $G - T$ fulfills the local tidiness property, that is, for each $v \in X$, deleting $X \setminus \{v\}$ from $G - T$ results in an $\mathcal{F}$-free graph.

---

[1] In Section 4, we present a slightly modified version of our kernelization algorithm that can be performed in $O(kn^2)$ time.

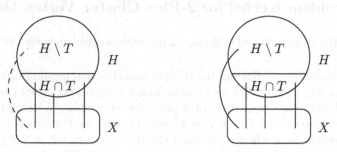

(a) An $X$-separated set $H \in \mathcal{H}(X)$   (b) A non-$X$-separated set $H \in \mathcal{H}(X)$

**Fig. 2.** Diagrams illustrating an $X$-separated and a non-$X$-separated set $H \in \mathcal{H}(X)$. The solid lines between $X$ and $H$ denote possible edges, and the dashed line between $X$ and $H \setminus T$ illustrates that there is no edge between these two sets.

**Definition 1.** *Let $\{G_1, \ldots, G_l\}$ be the set of connected components of $G - X$ and let $\mathcal{H}(X) := \{V(G_1), \ldots, V(G_l)\}$ be the collection of vertex sets of the connected components of $G - X$.*

Since $X$ is a 2-pvd set, each element of $\mathcal{H}(X)$ induces a 2-plex in $G$.

The local tidiness property helps in finding useful structural information; in particular, observe that each vertex $v \in X$ can be adjacent to vertices of arbitrarily many clusters in $G - X$, but in the tidy subgraph $G - T$, each vertex $v \in X$ is adjacent to vertices of at most two clusters in $G - (T \cup X)$; otherwise, if a vertex $v \in X$ were adjacent to at least three clusters in $G - (T \cup X)$, then there would be a FISG (more precisely, a $K_{1,3}$) that contains $v$ and three vertices from $G - (T \cup X)$, contradicting the local tidiness property of $G - T$. With such kind of observations, we can show that the local tidiness property implies the following two properties for each $v \in X$. These properties will later be exploited by the data reduction rules and the corresponding proof of the kernel size:

**Property 1:** There are at most two sets $H \in \mathcal{H}(X)$ with $H \setminus T$ adjacent to $v$.
**Property 2:** If there is a set $H \in \mathcal{H}(X)$ such that $H \setminus T$ is adjacent to $v$, then $v$
  is nonadjacent to at most one vertex in $H \setminus T$.

**Lemma 2.** *The local tidiness property of $G - T$ implies Properties 1 and 2.*

Recall that $|X| \leq 4k$ and that $|T| \leq 12k^2$; it remains to reduce the size of the tidy subgraph $G - T$. To this end, we distinguish between two types of sets in $\mathcal{H}(X)$, namely $X$-separated and non-$X$-separated sets:

**Definition 2.** *A vertex set $H \in \mathcal{H}(X)$ is $X$-separated if $H \setminus T$ is nonadjacent to $X$. A connected component of $G - X$ is $X$-separated if its vertex set is $X$-separated.*

See Figure 2 for an illustration. The remainder of this section is mainly devoted to data reduction rules that shrink the size of large sets in $\mathcal{H}(X)$. We deal with $X$-separated sets and non-$X$-separated sets separately. The intuitive idea to bound the number of vertices in $X$-separated sets is as follows. We use the fact that $T \cap H$ is a separator to show that if there are significantly more vertices in $H \setminus T$ than in $H \cap T$, then some vertices in $H \setminus T$ can be deleted from the graph. Together with the size bound for $T$, one can then obtain a bound on the number of vertices in $X$-separated sets. The intuitive idea to bound the number of vertices in non-$X$-separated sets is to use Properties 1 and 2. Roughly speaking, Property 1 guarantees that in $G-T$, each vertex from $X$ is adjacent to at most two sets from $\mathcal{H}(X)$; if there is a large non-$X$-separated set $H \in \mathcal{H}(X)$, then most of its vertices must have the same neighbors in $X$ due to Property 2. This observation can be used to show that some vertices in $H \setminus T$ can be deleted from the graph.

In the following, we exhibit the corresponding technical details. First, we shrink the size of the $X$-separated sets.

**Bounding the size of $X$-separated connected components.** The following data reduction rule decreases the number of vertices in large $X$-separated sets in $\mathcal{H}(X)$. Recall that $G$ is reduced with respect to Rules 1 and 2.

**Reduction Rule 3.** *If there exists a vertex set $H \in \mathcal{H}(X)$ that is $X$-separated by $T$ such that $|H \setminus T| > |H \cap T| + 1$, then choose an arbitrary vertex from $H \setminus T$ and delete it from $G$.*

**Lemma 3.** *Rule 3 is correct and can be exhaustively applied in $O(n + m)$ time.*

*Proof.* First, we show the correctness. Let $u \in H$ be the vertex that is deleted by Rule 3. If $(G, k)$ is a yes-instance, then obviously $(G - \{u\}, k)$ is a yes-instance as well. Now suppose that $(G - \{u\}, k)$ is a yes-instance. Let $S$ be a 2-pvd set of size at most $k$ for $G - \{u\}$. If $S$ is a 2-pvd set for $G$, then $(G, k)$ is obviously a yes-instance. Otherwise, $u$ must be contained in a FISG $F$ in $G - S$; we show that in this case one can use $S$ to construct a 2-pvd set $S'$ of size at most $k$ for $G$. Since $H$ induces a 2-plex and since $H$ is $X$-separated, $F$ must contain a vertex $v \in X$ and a vertex $w \in H \cap T$. By the preconditions of Rule 3 and because $|H \cap T| \geq 1$, we have $|H \setminus \{u\}| = |H \setminus (T \cup \{u\})| + |H \cap T| \geq 3$. Thus, all vertices in $H \setminus \{u\}$ are connected to $v$ and the at least $|H \cap T| + 1$ vertices in $H \setminus (T \cup \{u\})$ are nonadjacent to $v$, because $H$ is $X$-separated. Since $v, w \notin S$, the 2-pvd set $S$ for $G - \{u\}$ must contain all but at most one vertex from $H \setminus (T \cup \{u\})$, thus $|S \cap (H \setminus T)| \geq |H \cap T|$. Hence $S' := (S \setminus (H \setminus T)) \cup (H \cap T)$ is a 2-pvd set for $G - \{u\}$ of size at most $|S| \leq k$. Since $(H \cap T) \subseteq S'$, $u \in H \setminus T$, and $H$ is $X$-separated, it follows that $S'$ is also a 2-pvd set of size at most $k$ for $G$, thus $(G, k)$ is a yes-instance.

For the running time, consider that we can construct $\mathcal{H}(X)$ in $O(n + m)$ time. If for a $H \in \mathcal{H}(X)$ we find that all vertices in $H \setminus T$ are not adjacent to $X$, then apply Rule 3 to $H$ until $|H \setminus T| \leq |H \cap T| + 1$; deleting vertices from a graph takes $O(n + m)$ time in total.  □

With Rule 3, the number of vertices in $H \setminus T$ is bounded from above by $|H \cap T| + 1$ for each $X$-separated set $H \in \mathcal{H}(X)$. It remains to shrink the size of non-$X$-separated sets.

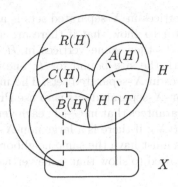

**Fig. 3.** Illustration of the sets $A(H)$, $B(H)$, $C(H)$, and $R(H)$. Solid lines indicate edges and dashed lines "non-edges". Note that the sets $A(H)$, $B(H)$, $C(H)$, and $H \cap T$ might have pairwise non-empty intersections (which is not relevant for our arguments); to keep the figure simple, they are drawn without intersections.

**Bounding the size of non-$X$-separated connected components.** Our goal is to find vertices in a non-$X$-separated set $H \in \mathcal{H}(X)$ that can safely be deleted by a data reduction rule. To this end, we need the following definitions. We call a vertex set $Z \subseteq V$ an $X$-*module* if any two vertices $u, v \in Z$ satisfy $N(u) \cap X = N(v) \cap X$. We define the following set of "candidate" vertices for deletion.

**Definition 3.** *Let $H \in \mathcal{H}(X)$. Then, a redundant subset $R \subseteq H$ is an $X$-module in which every vertex $u \in R$ is adjacent to every vertex in $H \setminus R$.*

With this definition, one can state the following data reduction rule; we describe later how a redundant subset of $H$ can be computed.

**Reduction Rule 4.** *Let $R \subseteq H$ be a redundant subset of some $H \in \mathcal{H}(X)$. If $|R| > k + 3$, then choose an arbitrary vertex from $R$ and delete it from $G$.*

**Lemma 4.** *Rule 4 is correct.*

We next show how to efficiently find a redundant subset $R \subseteq H$. See Figure 3 for an illustration of the following definitions. Let $A(H)$ be the set of vertices in $H$ that are nonadjacent to at least one vertex in $H \cap T$; let $B(H)$ be the set of vertices in $H$ that are nonadjacent to at least one vertex from $X$ that has some neighbor in $H \setminus T$; finally, let $C(H)$ be the set of vertices in $H$ that are nonadjacent to some vertex in $B(H)$. Let $\bar{R}(H) := A(H) \cup B(H) \cup C(H)$ and $R(H) := H \setminus (\bar{R}(H) \cup T)$. Intuitively, $\bar{R}(H)$ contains vertices in $H$ that violate Definition 3, guaranteeing that $R(H)$ contains vertices that satisfy Definition 3.

**Lemma 5.** *For each $H \in \mathcal{H}(X)$, the set $R(H) \subseteq H$ is redundant and the set $\bar{R}(H)$ contains at most $|H \cap T| + 2|N(H \setminus T) \cap X|$ vertices.*

*Proof.* To prove that $R(H)$ is redundant, one has to show that $R(H)$ is an $X$-module (that is, $N(u) \cap X = N(v) \cap X$ for all $u, v \in R(H)$) and that each vertex in $R(H)$ is adjacent to all vertices in $H \setminus R(H)$ (see Definition 3). We first show $N(u) \cap X = N(v) \cap X$ for all $u, v \in R(H)$. Assume that $w \in N(u) \cap X$. This implies $w \in N(H \setminus T) \cap X$. If $w \notin N(v) \cap X$, then $v \in B(H)$, contradicting $v \in R(H)$; as a consequence, $N(u) \cap X \subseteq N(v) \cap X$. The inclusion $N(v) \cap X \subseteq N(u) \cap X$ can be shown analogously.

We now show that each vertex in $R(H)$ is adjacent to all vertices in $H \setminus R(H)$. Every vertex in $R(H)$ is (by definition) adjacent to all vertices in $H \cap T$ and $B(H)$. Because each vertex in $A(H) \cup C(H)$ is nonadjacent to a vertex in $B(H) \cup (H \cap T)$ and because $H$ induces a 2-plex, each vertex in $R(H)$ is adjacent to all vertices in $A(H) \cup C(H)$. Thus, each vertex in $R(H)$ is adjacent to every vertex in $H \setminus R(H)$.

Because $H$ induces a 2-plex, $|A(H)| \leq |H \cap T|$ and $|C(H)| \leq |B(H)|$. For each vertex $v \in X$, Property 2 states that there is at most one vertex in $H \setminus T$ that is nonadjacent to $v$. Thus, $|B(H)| \leq |N(H \setminus T) \cap X|$. As a consequence, $|\bar{R}(H)| = |A(H) \cup B(H) \cup C(H)| \leq |H \cap T| + 2|N(H \setminus T) \cap X|$.    □

**Lemma 6.** *For all $H \in \mathcal{H}(X)$, the set $R(H)$ can be computed in $O(n^2)$ time.*

By Lemma 6, we can compute in $O(n^2)$ time the sets $R(H)$ and shrink them using Rule 4 so that $|R(H)| \leq k + 3$. The number of vertices in $H \setminus (T \cup R(H))$ is upper-bounded by $|H \cap T| + 2|N(H \setminus T) \cap X|$ due to Lemma 5. This shows the following proposition:

**Proposition 1.** *The exhaustive application of Rule 4 takes $O(n^2)$ time. After that, for each non-X-separated set $H \in \mathcal{H}(X)$, $|H \setminus T|$ has size at most $|H \cap T| + 2|N(H \setminus T) \cap X| + k + 3$.*

Now we have all ingredients to show our main result.

**Theorem 1.** *2-PLEX CLUSTER VERTEX DELETION admits a problem kernel of at most $52k^2 + 32k$ vertices, which can be computed in $O(k \cdot n^3)$ time.*

Using a more intricate analysis, without introducing further data reduction rules, we can improve the upper bound from Theorem 1 to $40k^2 + 24k$ vertices [3].

## 4    Generalization to $s$-Plex Cluster Vertex Deletion

In this section, we generalize the kernelization approach for 2-PCVD (Section 3) to $s$-PCVD with $s > 2$, focusing the presentation on the main differences. In addition, in the last part of this section, we discuss how to speed up the kernelization algorithm.

As in Section 3, we start with the Approximation Step. To greedily compute an approximate solution $X$, we employ the algorithm by Guo et al. [7], which finds a $T_s$-vertex FISG if the given graph contains one; here, $T_s$ is the largest integer satisfying $T_s \cdot (T_s + 1) \leq s$. If $X$ contains more than $k \cdot (s + 1 + T_s)$ vertices, then return "no-instance". Therefore, in the following, assume that $|X| \in O(sk)$.

For the Tidying Step, for each $v \in X$, compute a maximal set $\mathcal{F}(v)$ of FISGs that pairwise intersect exactly in $v$. A simple algorithm (trying all subgraphs of $G - \{v\}$ with at most $s + T_s$ vertices) takes $O(|X| \cdot sn^{s+T_s}) = O(s^2 k \cdot n^{s+T_s})$ time in total (that is, polynomial time). Then, apply Rule 1, that is, if there exists a vertex $v \in X$ such that $|\mathcal{F}(v)| > k$, then delete $v$ from $G$ and $X$ and decrement $k$ by one. After that, apply a data reduction rule that deletes connected components from $G$ that are $s$-plexes (cf. Rule 2). The tidying set is $T(v) := \bigcup_{F \in \mathcal{F}(v)} V(F) \setminus X$ for each $v \in X$ and the tidying set for the whole graph is $T := \bigcup_{v \in X} T(v)$. Since $|\mathcal{F}(v)| \leq k$, $|X| = O(sk)$, and the FISGs have at most $O(s)$ vertices, it follows that $|T| = O(s^2 k^2)$. It remains to describe the Shrinking Step, which decreases the number of vertices in the tidy subgraph $G - T$. For $G = (V, E)$, let $\mathcal{H}(X) := \{H \subseteq V \mid H$ induces a connected component in $G - X\}$. Analogous to the case $s = 2$, we can show that the local tidiness property implies the following two properties for each $v \in X$:

**Property 1:** There are at most $s$ sets $H \in \mathcal{H}(X)$ with $H \setminus T$ adjacent to $v$.

**Property 2:** If there is a set $H \in \mathcal{H}(X)$ such that $H \setminus T$ is adjacent to $v$, then $v$ is nonadjacent to at most $2s - 3$ vertices in $H \setminus T$.

We again distinguish between $X$-separated sets $H \in \mathcal{H}(X)$ and non-$X$-separated sets $H \in \mathcal{H}(X)$. The main difference compared to the case $s = 2$ is that the proof of the upper bound on the size of the non-$X$-separated sets is technically more demanding.

**Reduction Rule 5 (Generalization of Rule 3).** *If there exists a vertex set $H \in \mathcal{H}(X)$ that is $X$-separated by $T$ such that $|H \setminus T| \geq |H \cap T| + 2s - 2$, then choose an arbitrary vertex from $H \setminus T$ and delete it from $G$.*

Next, we deal with non-$X$-separated sets in $\mathcal{H}(X)$. To this end, we need the following definition.

**Definition 4 (Generalization of Definition 3).** *Let $H \in \mathcal{H}(X)$. Then, we call a subset $R \subseteq H$ redundant if there is an $X$-module $Z$ with $R \subseteq Z \subseteq H$ that contains all vertices from $H$ that are nonadjacent to a vertex in $R$.*

The main difference to Definition 3 for $s = 2$ is that one cannot guarantee that each vertex in a redundant subset $R$ is adjacent to all vertices in $H \setminus R$.

**Reduction Rule 6 (Generalization of Rule 4).** *Let $R \subseteq H$ be a redundant subset of some $H \in \mathcal{H}(X)$. If $|R| \geq k + 2s$, then choose an arbitrary vertex from $R$ and delete it from $G$.*

**Lemma 7.** *Rule 6 is correct.*

To find a redundant subset $R(H) \subseteq H$, one can use the same definitions of $R(H)$ and $\bar{R}(H)$ as for the case $s = 2$; however, the proof that $R(H)$ is a redundant subset becomes slightly more involved.

**Lemma 8 (Generalization of Lemma 5).** *For each $H \in \mathcal{H}(X)$, the set $R(H) \subseteq H$ is redundant and the set $\bar{R}(H)$ contains $O(s \cdot |H \cap T| + s^2 \cdot |N(H \setminus T) \cap X|)$ vertices.*

Now, we are ready to prove the problem kernel.

**Theorem 2.** $s$-PCVD *admits a problem kernel of* $O(k^2 s^3)$ *vertices, which can be computed in* $O(s^2 k \cdot n^{s+T_s})$ *time, where* $T_s$ *is the maximum integer satisfying* $T_s \cdot (T_s + 1) \leq s$.

*Speeding up the Kernelization Algorithms.* So far, we focused on the kernel size rather than the running time of the kernelization. The bottleneck of the kernelization result given by Theorem 2 is the running time of the simple algorithm that finds for each $v \in X$ a maximal set $\mathcal{F}(v)$ of FISGs that pairwise intersect exactly in $v$. The maximality of $\mathcal{F}(v)$ was used to prove that the tidying set $T$ fulfills Properties 1 and 2 (of Section 4). We obtain a fast kernelization if we do not demand that $\mathcal{F}(v)$ is maximal; rather, we show that we can compute a set $T(v)$ of bounded size such that Properties 1 and 2 are still fulfilled.

**Lemma 9.** *Let* $X$ *be an* $s$-*pvd set. Then, for all* $v \in X$, *a vertex set* $T(v)$ *with* $|T(v)| \leq 2sk$ *and satisfying the following properties can be found in* $O(|X| \cdot n^2)$ *time:*

1. *For each vertex* $v \in X$, *there are at most* $s$ *sets* $H \in \mathcal{H}(X)$ *such that* $H \setminus T(v)$ *is adjacent to* $v$.
2. *If there is a vertex* $v \in X$ *and a set* $H \in \mathcal{H}(X)$ *such that* $H \setminus T(v)$ *is adjacent to* $v$, *then* $v$ *is nonadjacent to at most* $2s - 3$ *vertices in* $H \setminus T(v)$.

Using Lemma 9 instead of the expensive computation of $T(v)$ in Section 4 and a few additional tricks [3], the following result can be shown.

**Theorem 3.** $s$-PCVD *admits a problem kernel of* $O(k^2 s^3)$ *vertices, which can be computed in* $O(ksn^2)$ *time.*

## 5    Conclusion

Our results are based on linking kernelization with polynomial-time approximation, dealing with vertex deletion problems whose goal graphs are characterized by forbidden induced subgraphs. This is a rich class of graphs, among others containing various cluster graphs. When applicable, our method may allow for significantly smaller problem kernel sizes than the more general method by Kratsch [10].

As to future work, it would be desirable to start a general study under which conditions fixed-parameter tractable vertex deletion problems possess polynomial-size kernels.

## References

1. Abu-Khzam, F.N.: A kernelization algorithm for $d$-Hitting Set. J. Comput. System Sci. (2009) (Available electronically)
2. Balasundaram, B., Butenko, S., Hicks, I.V.: Clique relaxations in social network analysis: The maximum $k$-plex problem. Oper. Res. (2009) (Avaiable electronically)
3. van Bevern, R.: A quadratic-vertex problem kernel for s-plex cluster vertex deletion. Studienarbeit, Friedrich-Schiller-Universität Jena, Germany (2009)

4. Cai, L.: Fixed-parameter tractability of graph modification problems for hereditary properties. Inf. Process. Lett. 58(4), 171–176 (1996)
5. Chesler, E.J., Lu, L., Shou, S., Qu, Y., Gu, J., Wang, J., Hsu, H.C., Mountz, J.D., Baldwin, N.E., Langston, M.A., Threadgill, D.W., Manly, K.F., Williams, R.W.: Complex trait analysis of gene expression uncovers polygenic and pleiotropic networks that modulate nervous system function. Nat. Genet. 37(3), 233–242 (2005)
6. Cook, V.J., Sun, S.J., Tapia, J., Muth, S.Q., Argüello, D.F., Lewis, B.L., Rothenberg, R.B., McElroy, P.D.: The Network Analysis Project Team. Transmission network analysis in tuberculosis contact investigations. J. Infect. Dis. 196, 1517–1527 (2007)
7. Guo, J., Komusiewicz, C., Niedermeier, R., Uhlmann, J.: A more relaxed model for graph-based data clustering: s-plex editing. In: Goldberg, A.V., Zhou, Y. (eds.) AAIM 2009. LNCS, vol. 5564, pp. 226–239. Springer, Heidelberg (2009)
8. Guo, J., Niedermeier, R.: Invitation to data reduction and problem kernelization. SIGACT News 38(1), 31–45 (2007)
9. Hüffner, F., Komusiewicz, C., Moser, H., Niedermeier, R.: Fixed-parameter algorithms for cluster vertex deletion. Theory Comput. Syst. (2009) (Available electronically)
10. Kratsch, S.: Polynomial kernelizations for MIN $F^+\Pi_1$ and MAX NP. In: Proc. 26th STACS, pp. 601–612. IBFI Dagstuhl, Germany (2009)
11. Marx, D., Schlotter, I.: Parameterized graph cleaning problems. Discrete Appl. Math., (2009) (Available electronically)
12. Memon, N., Kristoffersen, K.C., Hicks, D.L., Larsen, H.L.: Detecting critical regions in covert networks: A case study of 9/11 terrorists network. In: Proc. 2nd ARES, pp. 861–870. IEEE Computer Society Press, Los Alamitos (2007)
13. Seidman, S.B., Foster, B.L.: A graph-theoretic generalization of the clique concept. J. Math. Sociol. 6, 139–154 (1978)
14. Shamir, R., Sharan, R., Tsur, D.: Cluster graph modification problems. Discrete Appl. Math. 144(1-2), 173–182 (2004)

# Gradual Sub-lattice Reduction and a New Complexity for Factoring Polynomials

Mark van Hoeij[1,*] and Andrew Novocin[2]

[1] Florida State University, 208 Love Building Tallahassee, FL 32306-4510
hoeij@math.fsu.edu, http://www.math.fsu.edu/~hoeij
[2] LIP/INRIA/ENS, 46 allée d'Italie, F-69364 Lyon Cedex 07, France
Andrew.Novocin@ens-lyon.fr, http://andy.novocin.com/pro

**Abstract.** We present a lattice algorithm specifically designed for some classical applications of lattice reduction. The applications are for lattice bases with a generalized knapsack-type structure, where the target vectors are boundably short. For such applications, the complexity of the algorithm improves traditional lattice reduction by replacing some dependence on the bit-length of the input vectors by some dependence on the bound for the output vectors. If the bit-length of the target vectors is unrelated to the bit-length of the input, then our algorithm is only linear in the bit-length of the input entries, which is an improvement over the quadratic complexity floating-point LLL algorithms. To illustrate the usefulness of this algorithm we show that a direct application to factoring univariate polynomials over the integers leads to the first complexity bound improvement since 1984. A second application is algebraic number reconstruction, where a new complexity bound is obtained as well.

## 1 Introduction

Lattice reduction algorithms are essential tools in computational number theory and cryptography. A lattice is a discrete subset of $\mathbb{R}^n$ that is also a $\mathbb{Z}$-module. The goal of lattice reduction is to find a 'nice' basis for a lattice, one which is near orthogonal and composed of short vectors. Since the publication of the 1982 Lenstra, Lenstra, Lovász [15] lattice reduction algorithm many applications have been discovered, such as polynomial factorization [15,11] and attacking several important public-key cryptosystems including knapsack cryptosystems [23], RSA under certain settings [7], and DSA and some signature schemes in particular settings [12]. One of the important features of the LLL algorithm was that it could approximate the shortest vector of a lattice in polynomial time. This is valuable because finding the exact shortest vector in a lattice is provably NP-hard [1,18]. Given a basis $\mathbf{b}_1, \ldots, \mathbf{b}_d \in \mathbb{R}^n$ which satisfies $\| \mathbf{b}_i \| \leq X \ \forall i$, the LLL algorithm has a running time of $\mathcal{O}(d^5 n \log^3 X)$ using classical arithmetic. Recently there has been a resurgence of lattice reduction work thanks to

---

* Supported by NSF 0728853.

A. López-Ortiz (Ed.): LATIN 2010, LNCS 6034, pp. 539–553, 2010.

Nguyen and Stehlé's $L^2$ algorithm [20,21] which performs lattice reduction in $\mathcal{O}(d^4 n \log X [d + \log X])$ CPU operations. The primary result of $L^2$ was that the dependence on $\log X$ is only quadratic allowing for improvement on applications using large input vectors.

**The main result:** Many applications of LLL (see the applications section below) involve finding a vector in a lattice whose norm is known to be small in advance. In such cases it can be more efficient to reduce a basis of a sub-lattice which contains all targeted vectors than reducing a basis of the entire lattice. In this paper we target short vectors in specific types of input lattice bases which we call knapsack-type bases. The new algorithm introduces a search parameter $B$ which the user provides. This parameter is used to bound the norms of targeted short vectors. To be precise:

The *rows* of the following matrices represent a knapsack-type basis

$$\begin{pmatrix} 0 \cdots 0 & 0 & \cdots & P_N \\ 0 \cdots 0 & 0 & \cdots & 0 \\ 0 \cdots 0 & P_1 & \cdots & 0 \\ \hline 1 \cdots 0 & x_{1,1} & \cdots & x_{1,N} \\ \vdots \ddots \vdots & \vdots & & \vdots \\ 0 \cdots 1 & x_{r,1} & \cdots & x_{r,N} \end{pmatrix} \quad \text{or} \quad \begin{pmatrix} 1 \cdots 0 & x_{1,1} & \cdots & x_{1,N} \\ \vdots \ddots \vdots & \vdots & & \vdots \\ 0 \cdots 1 & x_{r,1} & \cdots & x_{r,N} \end{pmatrix}.$$

The specifications of our algorithm are as follows. It takes as input a knapsack-type basis $\mathbf{b}_1, \ldots, \mathbf{b}_d \in \mathbb{Z}^n$ of a lattice $L$ with $\| \mathbf{b}_i \| \leq X \ \forall i$ and a search parameter $B$; it returns a *reduced basis generating a* sub-lattice $L' \subseteq L$ such that if $\mathbf{v} \in L$ and $\| \mathbf{v} \| \leq B$ then $\mathbf{v} \in L'$.

Our algorithm has the following complexity bounds for various input:

| No $P_i$ | $\mathcal{O}(d^2(n + d^2)(d + \log B)[\log X + n(d + \log B)])$ |
|---|---|
| No restriction on $P_i$ | $\mathcal{O}(d^4(d + \log B)[\log X + d(d + \log B)])$ |
| Many $P_i$ large w.r.t. $B$ | $\mathcal{O}(dr^3(r + \log B)[\log X + d(r + \log B)])$ |

These complexity bounds have several distinct parameters, so a comparison with other algorithms is a bit subtle. The most significant parameter to explore is $B$, the search parameter. If one selects $B = X$ then our algorithm will return a reduced basis of $L' = L$ in $\mathcal{O}(d^2 n(n + d^2)[d^2 + \log^2 X])$. This is an interesting result because our algorithm, like the original LLL and the $L^2$ algorithms, uses switches and size-reductions of the vectors to arrive at a reduced basis. The fact that we return a reduced basis with a complexity so similar to $L^2$ implies that there are alternative orderings on the switches which lead to similar performance.

When using a smaller value of $B$ than $X$ the algorithm will return either:

– A reduced basis of a sub-lattice $L'$ which contains all vectors of norm $\leq B$. This sub-lattice may be different than the sub-lattice, $L''$, generated by all vectors of norm $\leq B$, and we do have $L'' \subseteq L' \subseteq L$. Also, because the basis of $L'$ is reduced, we have an approximation of the shortest non-zero vector of $L$.

- The empty set, in which case the algorithm has proved that no non-zero vector of norm $\leq B$ exists in $L$.

We offer the following complexity comparison with $L^2$ [20] for some values of $B$ on square input lattices (with $P_j$'s). When a column has a non-zero $P_j$ we can reduce the $x_{i,j}$ modulo $P_j$. Thus, without loss of generality, we may assume that $P_j$ is the largest element in its column. Note that $r = d - N$.

| $L^2$ | $\mathcal{O}(d^6 \log X + d^5 \log^2 X)$ |
|---|---|
| $B = \mathcal{O}(X)$ | $\mathcal{O}(d^7 + d^5 \log^2 X)$ |
| $B = \mathcal{O}(X^{1/d})$ | $\mathcal{O}(d^2 r^5 + r^3 \log^2 X)$ |
| $B = 2^{\mathcal{O}(d)}$ | $\mathcal{O}(d^4 r^3 + d^2 r^3 \log X)$ |

It should be noted that [20] explores running times of $L^2$ on knapsack lattices with $N = 1$ (such lattice bases are used in [9]). In this case, $L^2$ will have complexity $\mathcal{O}(d^5 \log X + d^4 \log^2 X)$.

**Our approach:** We reduce the basis gradually, using many separate calls to another lattice reduction algorithm. To get the above complexity results we chose H-LLL [19] but there are many suitable lattice reduction algorithms we could use instead such as [13,15,20,24]. For more details on why we made this decision see the discussion in section 5.

There are three important features to our approach. First, we approach the problem column by column. Beginning with the $r \times r$ identity and with each iteration of the algorithm we expand our scope to include one more column of the $x_{i,j}$. Next, within each column iteration, we reduce the new entries bit by bit, starting with a reduction using only the most significant bits, then gradual including more and more bits of data. Third, we allow for the removal of vectors which have become too large. This allows us to always work on small entries, but restricts us to a sub-lattice.

The proof of the algorithm's complexity is essentially a study of two quantities, the product of the Gram-Schmidt lengths of the current vectors which we call the active determinant and an energy function which we call progress. We amortize all of the lattice reduction costs using progress, and we bound the number of iterations and number of vectors using the active determinant. Neither of these quantities is impacted by the choice of lattice reduction algorithm.

**Applications of the algorithm:** As evidence for the usefulness of this new approach we show two new complexity results based on applications of the main algorithm. The first result is a new complexity for the classical problem of factoring polynomials in $\mathbb{Z}[x]$. If the polynomial has degree $N$, coefficients smaller than $\log(A)$, and when reduced modulo a prime $p$ has $r$ irreducible factors then we prove a complexity of $\mathcal{O}(N^3 r^4 + N^2 r^4 \log A)$ for the lattice reduction costs using classical arithmetic. One must also add the cost of multi-factor Hensel lifting which is $\mathcal{O}(N^6 + N^4 \log^2 A)$ ignoring the small terms $\log(r)$ and $\log^2 p$ (see [8] for details). This is the first improvement over the Schönhage bound given in 1984 [25] of $\mathcal{O}(N^8 + N^5 \log^3 A)$.

The second new complexity result comes in the problem of reconstructing a minimal polynomial from a complex approximation of the algebraic number. In this application we know $\mathcal{O}(d^2 + d \log H)$ bits of an approximation of some complex root of an unknown polynomial $h(x)$ with degree $d$ and with maximal coefficient of absolute value $\leq H$. Then our algorithm can be used to find the coefficients of $h(x)$ in $\mathcal{O}(d^7 + d^5 \log^2 H)$ CPU operations.

Other problems of common interest which might be impacted by our algorithm include integer relation finding (where $N = 1$) and simultaneous Diophantine approximation of several real numbers [10,6] (where $r = 1$).

**Notations:** All costs are given for the bit-complexity model. A standard row vector will be denoted $\mathbf{v}$, $\mathbf{v}[i]$ represents the $i^{\text{th}}$ entry of $\mathbf{v}$, $\mathbf{v}[i, \ldots, j]$ a vector consisting of all entries of $\mathbf{v}$ from the $i^{\text{th}}$ entry to the $j^{\text{th}}$ entry, and $\mathbf{v}[-1]$ the final entry of $\mathbf{v}$. Also we will use $\|\mathbf{w}\|_\infty$ as the max-norm or the largest absolute value of an entry in the vector $\mathbf{w}$, $\| \mathbf{w} \| := \sqrt{\sum (\mathbf{w}[i])^2}$ which we call the norm of $\mathbf{w}$, and $\mathbf{w}^T$ as the transpose of $\mathbf{w}$. The scalar product will be denoted $\mathbf{v} \cdot \mathbf{w} := \sum \mathbf{v}[i] \cdot \mathbf{w}[i]$. For a matrix $M$ we will use $M[1, \ldots, k]$ to denote the first $k$ columns of $M$. The $n$ by $n$ identity matrix will be denoted $I_{n \times n}$. For a real number $x$ we use $\lceil x \rceil$ and $\lfloor x \rfloor$ to denote the closest integer $\geq x$ and $\leq x$ respectively.

**Road map:** In section 2 we give a brief introduction to lattice reduction algorithms. In section 3 we present the central algorithm of the paper and prove its correctness. In section 4 we prove several important features by studying quasi-invariants we call the active determinant and progress. In this section we treat lattice reduction as a black-box algorithm. In section 5 we prove the overall complexity and other important claims about the new algorithm by fixing a choice for a standard lattice reduction algorithm. In section 6 we offer new complexity results for factoring polynomials in $\mathbb{Z}[x]$ and algebraic number reconstruction.

## 2    Background on Lattice Reduction

The purpose of this section is to present some facts from [15] that will be needed throughout the paper. For a more general treatment of lattice reduction see [17].

A lattice, $L$, is a discrete subset of $\mathbb{R}^n$ that is also a $\mathbb{Z}$-module. Let $\mathbf{b}_1, \ldots, \mathbf{b}_d \in L$ be a basis of $L$ and denote $\mathbf{b}_1^*, \ldots, \mathbf{b}_d^* \in \mathbb{R}^n$ as the Gram-Schmidt orthogonalization over $\mathbb{R}$ of $\mathbf{b}_1, \ldots, \mathbf{b}_d$. Let $\delta \in (1/4, 1]$ and $\eta \in [1/2, \sqrt{\delta})$. Let $l_i = \log_{1/\delta} \| \mathbf{b}_i^* \|^2$, and denote $\mu_{i,j} = \frac{\mathbf{b}_i \cdot \mathbf{b}_j^*}{\mathbf{b}_j^* \cdot \mathbf{b}_j^*}$. Note that $\mathbf{b}_i, \mathbf{b}_i^*, l_i, \mu_{i,j}$ will change throughout the algorithm sketched below.

**Definition 1.** $\mathbf{b}_1, \ldots, \mathbf{b}_d$ *is* LLL-reduced *if* $\| \mathbf{b}_i^* \|^2 \leq \frac{1}{\delta - \mu_{i+1,i}^2} \| \mathbf{b}_{i+1}^* \|^2$ *for* $1 \leq i < d$ *and* $|\mu_{i,j}| \leq \eta$ *for* $1 \leq j < i \leq d$.

In the original paper the values for $(\delta, \eta)$ were chosen as $(3/4, 1/2)$ so that $\frac{1}{\delta - \eta^2}$ would simply be 2.

**Algorithm 1. (Rough sketch of LLL-type algorithms)**
Input: *A basis* $\mathbf{b}_1, \ldots, \mathbf{b}_d$ *of a lattice* $L$.
Output: *An LLL-reduced basis of* $L$.

$A$ - $\kappa := 2$
$B$ - **while** $\kappa \leq d$ **do:**
    $1$ - (Gram-Schmidt over $\mathbb{Z}$). *By subtracting suitable $\mathbb{Z}$-linear combinations of* $\mathbf{b}_1, \ldots, \mathbf{b}_{\kappa-1}$ *from* $\mathbf{b}_\kappa$ *make sure that* $|\mu_{i,\kappa}| \leq \eta$ *for* $i < \kappa$.
    $2$ - (LLL Switch). *If interchanging* $\mathbf{b}_{\kappa-1}$ *and* $\mathbf{b}_\kappa$ *will decrease* $l_{\kappa-1}$ *by at least 1 then do so.*
    $3$ - (Repeat). *If not switched* $\kappa := \kappa + 1$, *if switched* $\kappa = max(\kappa - 1, 2)$.

That the above algorithm terminates, and that the output is LLL-reduced was shown in [15]. Step B1 has no effect on the $l_i$. In step B2 the only $l_i$ that change are $l_{\kappa-1}$ and $l_\kappa$. The following lemmas present some standard facts which we will need.

**Lemma 1.** *An LLL switch can not increase* $\max(l_1, \ldots, l_d)$, *nor can it decrease* $\min(l_1, \ldots, l_d)$.

**Lemma 2.** *If* $\| \mathbf{b}_d^* \| > B$ *then any vector in* $L$ *with norm* $\leq B$ *is a $\mathbb{Z}$-linear combination of* $\mathbf{b}_1, \ldots, \mathbf{b}_{d-1}$.

In other words, if the current basis of the lattice is $\mathbf{b}_1, \ldots, \mathbf{b}_d$ and if the last vector has sufficiently large G-S length then, provided the user is only interested in elements of $L$ with norm $\leq B$, the last basis element can be removed.

Lemma 2 follows from the proof of [15, Eq. (1.11)], and is true regardless of whether $\mathbf{b}_1, \ldots, \mathbf{b}_d$ is LLL-reduced or not. However, if one chooses an arbitrary basis $\mathbf{b}_1, \ldots, \mathbf{b}_d$ of some lattice $L$, then it is unlikely that the last vector has large G-S length (after all, $\| \mathbf{b}_d^* \|$ is the norm of $\mathbf{b}_d$ reduced modulo $\mathbf{b}_1, \ldots, \mathbf{b}_{d-1}$ over $\mathbb{R}$). The effect of LLL reduction is to move G-S length towards later vectors.

## 3 Main Algorithm

In this section we present the central algorithm of the paper and a proof of its correctness. Our algorithm is a kind of wrapper for other standard lattice reduction algorithms. We try to present it as independently as possible of the choice of lattice reduction algorithm. In order to be general we must first outline the features that we require of the chosen lattice reduction algorithm. Our first requirement is that the output satisfy the following slightly weakened version of LLL-reduction.

**Definition 2.** *Let* $L \subseteq \mathbb{R}^n$ *be a lattice and* $\mathbf{b}_1, \ldots, \mathbf{b}_s \in L$ *be $\mathbb{R}$-linearly independent. We call* $\mathbf{b}_1, \ldots, \mathbf{b}_s$ *an $\alpha$-reduced basis of $L$ if 1,2, and 3a hold, and an $(\alpha, B)$-reduced sequence (basis of a sub-lattice) if 1,2, and 3b hold:*

1. $\| \mathbf{b}_i^* \| \leq \alpha \| \mathbf{b}_{i+1}^* \|$ *for* $i = 1 \ldots s - 1$.
2. $\| \mathbf{b}_i^* \| \leq \| \mathbf{b}_i \| \leq \alpha^{i-1} \| \mathbf{b}_i^* \|$ *for* $i = 1 \ldots s$.
3. *(a)* $L = \mathbb{Z} \mathbf{b}_1 + \cdots + \mathbb{Z} \mathbf{b}_s$.
   *(b)* $\| \mathbf{b}_s^* \| \leq B$ *and for every* $\mathbf{v} \in L$ *with* $\| \mathbf{v} \| \leq B$ *we have* $\mathbf{v} \in \mathbb{Z} \mathbf{b}_1 + \cdots + \mathbb{Z} \mathbf{b}_s$.

The original LLL algorithm from [15] returns output with $\alpha = \sqrt{2}$, $L^2$ from [20] with $\alpha = \sqrt{\frac{1}{\delta - \eta^2}}$ for appropriate choices of $(\delta, \eta)$, and H-LLL from [19] reduced with $\alpha = \frac{\theta \eta + \sqrt{(1 + \theta^2) \delta - \eta^2}}{\delta - \eta^2}$ for appropriate $(\delta, \eta, \theta)$. We may now also make a useful observation about an $(\alpha, B)$-reduced sequence.

**Lemma 3.** *If the vectors* $\mathbf{b}_1, \ldots, \mathbf{b}_s$ *form an* $(\alpha, B)$-*reduced sequence and we let* $\mathbf{b}_1^*, \ldots, \mathbf{b}_s^*$ *represent the GSO, then the following properties are true:*

- $\| \mathbf{b}_i^* \| \leq \alpha^{s-i} B$ *for all* $i$.
- $\| \mathbf{b}_i \| \leq \alpha^{s-1} B$ *for all* $i$.

We use the concept of $\alpha$-reduction as a means of making proofs which are largely independent of which lattice reduction algorithm a user might choose. For a basis which is $\alpha$-reduced, a small value of $\alpha$ implies a strong reduction. In our algorithm we use the variable $\alpha$ as the worst-case guarantee of reduction quality. We make our proofs (specifically Lemma 8 and Theorem 3) assuming an $\alpha \geq \sqrt{4/3}$. This value is chosen because [15,20,19] cannot guarantee a stronger reduction. An $(\alpha, B)$-reduced bases is typically made from an $\alpha$-reduced basis by removing trailing vectors with large G-S length. The introduction of $(\alpha, B)$-reduction does not require creating new lattice reduction algorithms, just the minor adjustment of detecting and removing vectors above a given G-S length.

### Algorithm 2. LLL_with_removals
**Input:** $\mathbf{b}_1, \ldots, \mathbf{b}_s \in \mathbb{R}^n$ *and* $B \in \mathbb{R}$.
**Output:** $\mathbf{b}'_1, \ldots, \mathbf{b}'_{s'} \in \mathbb{R}^n$ $(\alpha, B)$-*reduced,* $s' \leq s$.
**Procedure:** *Use any lattice reduction procedure which returns an* $\alpha$-*reduced basis and follows Assumption 1. However, when it is discovered that the final vector has G-S length provably* $> B$ *remove that final vector (deal with it no further).*

**Assumption 1.** *The lattice reduction algorithm chosen for LLL_with_removals must use switches of consecutive vectors during its reduction process. These switches must have the following properties:*

1. *There exists a number* $\gamma > 1$ *such that every switch of vectors* $\mathbf{b}_i$ *and* $\mathbf{b}_{i+1}$ *increases* $\| \mathbf{b}_{i+1}^* \|^2$ *by a factor provably* $\geq \gamma$.
2. *The quantity* $max\{ \| \mathbf{b}_i^* \|, \| \mathbf{b}_{i+1}^* \| \}$ *cannot be increased by switching* $\mathbf{b}_i$ *and* $\mathbf{b}_{i+1}$.
3. *No steps other than switches can affect G-S norms* $\| \mathbf{b}_1^* \|, \ldots, \| \mathbf{b}_s^* \|$.

Assumption 1 is not very strong as [15,20,19,24,27] and the sketch in Algorithm 1 all conform to these assumptions. We do not allow for the extreme case where $\gamma = 1$, although running times have been studied in [2,16]. It should also be noted that in the floating point lattice reduction algorithms $\| \mathbf{b}_s^* \|$ is only known approximately. In this case one must only remove vectors whose approximate G-S length is sufficiently large to ensure that the exact G-S length is $\geq B$.

The format of the input matrices was given in section 1. A search parameter $B$ is given to bound the norm of the target vectors. The algorithm performs its best when $B$ is small compared to the bit-length of the entries in the input matrix, although $B$ need not be small for the algorithm to work.

**Definition 3.** *We say the $P_j$ are large enough if:*

$$|P_j| \geq 2\alpha^{4r+4k+2}B^2 \text{ for all but } k = \mathcal{O}(r) \text{ values of } j. \tag{1}$$

Note that if $N = \mathcal{O}(r)$ then the $P_j$ are trivially large enough. However, for applications where $N$ is potentially much larger than $r$ this becomes a non-trivial condition. In this case having $B$ close to $X$ means that the $P_j$'s are not large enough.

In the following algorithm we will gradually reduce the input basis. This will be done one column at a time, similar to the experiments in [3,6]. The current basis vectors are denoted $\mathbf{b}_i$ and we will use $M$ to represent the matrix whose rows are the $\mathbf{b}_i$. We will use the notation $\mathbf{x}_j$ to represent the column vector $(x_{1,j}, \ldots, x_{r,j})^T$.

The matrix $M$ will begin as $I_{r \times r}$, and we will adjoin $\mathbf{x}_1$ and a new row $(\mathbf{0}, P_1)$ if appropriate. Each time we add a column $\mathbf{x}_j$ we will need to calculate the effects of prior lattice reductions on the new $\mathbf{x}_j$. We use $\mathbf{y}_j$ to represent a new column of entries which will be adjoined to $M$. In fact $\mathbf{y}_j = M[1,\ldots,r] \cdot \mathbf{x}_j$. Before adjoining the entries we also scale them by a power of 2, to have smaller absolute values. This keeps the entries in $M$ at a uniform absolute value. The central loop of the algorithm is the process of gradually using more and more bits of $\mathbf{y}_j$ until every entry in $M$ is again an integer. No rounding is performed: we use rational arithmetic on the last column of each row. Throughout the algorithm the number of rows of $M$ will be changing. We let $s$ be the current number of rows of $M$. If (1) is satisfied for some $k = \mathcal{O}(r)$ then we can actually bound $s$ by $2r + 2k + 1$. We use $c$ as an apriori upper bound on $s$, either $c := 2r + 2k + 1$ or $c := r + N$. The algorithm has better performance when $c$ is small. We let $L$ represent the lattice generated by the rows of $A$.

### Algorithm 3. Gradual_LLL

*Input: A search parameter, $B \geq \sqrt{5} \in \mathbb{Q}$, an integer knapsack-type matrix, $A$, and an $\alpha \geq \sqrt{4/3}$.*

*Output: An $(\alpha, B)$-reduced basis $\mathbf{b}_1, \ldots, \mathbf{b}_s$ of a sub-lattice $L'$ in $L$ with the property that if $\mathbf{v} \in L$ and $\| \mathbf{v} \| \leq B$ then $\mathbf{v} \in L'$.*

**The Main Algorithm:**

1 - **if** (1) holds set $c := \min(2r + 2k + 1, r + N)$
2 - $s := r; M := I_{r \times r}$
3 - **for** $j = 1 \ldots N$ **do:**
    a - $\mathbf{y}_j := M[1, \ldots, r] \cdot \mathbf{x}_j$; $\ell := \lfloor \log_2(\max\{|P_j|, \|\mathbf{y}_j\|_\infty, 2\}) \rfloor$
    b - $M := \begin{bmatrix} 0 & P_j/2^\ell \\ M & \mathbf{y}_j/2^\ell \end{bmatrix}$; if $P_j \neq 0$ then $s := s + 1$ **else** remove zero row
    c - **while** $(\ell \neq 0)$ **do:**
        i - $\mathbf{y}_j := 2^\ell \cdot M \cdot [0, \cdots, 0, 1]^T$; $\ell := \max\{0, \lceil \log_2(\frac{\|\mathbf{y}_j\|_\infty}{\alpha^{2c} B^2}) \rceil\}$
        ii - $M := [M[1, \ldots, r + j - 1] | \mathbf{y}_j/2^\ell]$
        iii - Call LLL_with_removals on $M$ and set $M$ to output; adjust $s$
4 - **return** $M$

First we will prove the correctness of the algorithm. We need to show that the Gram-Schmidt lengths are never decreased by scaling the final entry or adding a new entry.

**Lemma 4.** *Let $\mathbf{b}_1, \ldots, \mathbf{b}_s \in \mathbb{R}^n$ be the basis of a lattice and $\mathbf{b}_1^*, \ldots, \mathbf{b}_s^*$ its GSO. Let $\sigma : \mathbb{R}^n \to \mathbb{R}^n$ scale up the last entry by some factor $\beta > 1$, then we have $\| \mathbf{b}_i^* \| \leq \| \sigma(\mathbf{b}_i)^* \|$. In other words, scaling the final entry of each vector by the same scalar $\beta > 1$ cannot decrease $\|\mathbf{b}_i^*\|$ for any $i$.*

**Lemma 5.** *Let $\mathbf{b}_1, \ldots, \mathbf{b}_s \in \mathbb{R}^n$ and let $\mathbf{b}_1^*, \ldots, \mathbf{b}_s^* \in \mathbb{R}^n$ be their GSO. The act of adjoining an $(n + 1)^{st}$ entry to each vector and re-evaluating the GSO cannot decrease $\| \mathbf{b}_i^* \|$ for any $i$ (assuming that the new entry is in $\mathbb{R}$).*

The proofs of these lemmas are quite similar and can be found in the appendix. Now we are ready to prove the first theorem, asserting the correctness of algorithm 3's output.

**Theorem 1.** *Algorithm 3 correctly returns an $\alpha$-reduced basis of a sub-lattice, $L'$, in $L$ such that if $\mathbf{v} \in L$ and $\| \mathbf{v} \| \leq B$ then $\mathbf{v} \in L'$.*

*Proof.* When the algorithm terminates all entries are unscaled and each vector in the output is inside of $L$ as it is a linear combination of the original input vectors. Thus the output is a basis of a sub-lattice $L'$ inside $L$. Further, the algorithm terminates after a final call to step 3(c)iii so returns an $(\alpha, B)$-reduced sequence.

Now we show that if $\mathbf{v} \in L$ and $\| \mathbf{v} \| \leq B$ then $\mathbf{v} \in L'$. The removed vectors correspond to vectors $\tilde{\mathbf{b}}_i \in L$ that, by lemmas 4 and 5, have G-S length at least as large as those of $\mathbf{b}_i$. The claim then follows from lemmas 1 and 2.

## 4    Two Invariants of the Algorithm

Here we present the important proofs about the set-up of our algorithm. All proofs in this section and the next allow for a black-box lattice reduction algorithm up to satisfying assumption 1. Each proof in this section involves the study of an invariant. The two invariants which we use are:

- The Active Determinant, $AD(M)$, which is the product of the G-S lengths of the active vectors. This remains constant under standard lattice reduction algorithms, and allows us to bound many features of the proofs.
- The Progress, $PF = \sum_{i=1}^{s}(i-1)\log \|\mathbf{b}_i^*\|^2 + n_{\mathrm{rm}}r\log(4\alpha^{4c}B^4)$, where $n_{\mathrm{rm}}$ is the total number of vectors which have been removed so far. This function is an energy function which never decreases, and is increased by $\geq 1$ for each switch made in the lattice reduction algorithm.

## A Study of the Active Determinant

**Definition 4.** *We call the active determinant of the vectors* $\mathbf{b}_1,\dots,\mathbf{b}_s$ *the product of their Gram-Schmidt lengths. For notation we use, $AD$ or $AD(\{\mathbf{b}_i\}) := \prod_{i=1}^{s}\|\mathbf{b}_i^*\|$. For a matrix $M$ with the $i^{th}$ row denoted by $M[i]$, we use $AD$ or $AD(M) = AD(\{M[1],\dots,M[s]\})$.*

For an $(\alpha, B)$-reduced sequence we can nicely bound the AD. We have such a sequence after each execution of step 3(c)iii.

**Lemma 6.** *If $\mathbf{b}_1,\dots,\mathbf{b}_s$ are an $(\alpha, B)$-reduced sequence then $AD \leq (\alpha^{s-1}B^2)^{s/2}$.*

We now want to attack two problems, bounding the norm of each vector just before lattice reduction, and bounding the number of vectors throughout the algorithm.

**Lemma 7.** *If $s \leq c$ then just before step 3(c)iii we have $\|\mathbf{b}_i\|^2 \leq 2\alpha^{4c}B^4$ for $i = 1\dots s$.*

The full details of this proof can be found in the appendix. The following theorem holds trivially when there is no condition on the $P_j$ or if $N = 0$. When $N > r$ and $B$ is at least a bit smaller than $X$ we can show that not all of the extra vectors stay in the lattice. In other words, if there is enough of a difference between $B$ and $X$ then the sub-lattice aspect of the algorithm begins to allow for some slight additional savings. Here the primary result of this theorem is allowing $\mathcal{O}(r)$ vectors with a relatively weak condition on the $P_j$.

**Theorem 2.** *Throughout the algorithm we have $s \leq c$.*

*Proof.* If $c = r + N$ then $s \leq c$ is vacuously true. So assume $c = 2(r + k) + 1$ and all but $k = \mathcal{O}(r)$ of the $P_j$ satisfy $|P_j| \geq 2\alpha^{4r+4k+2}B^2$. When the algorithm begins, $AD = 1$ and $s = r$. For $s$ to increase step 3 must finish without removing a vector. If this happens during iteration $j$ then the AD has increased by a factor $|P_j|$. The LLL-switches inside of step 3(c)iii do not alter the AD by Assumption 1. Each vector which is removed during step 3(c)iii has G-S length $\leq 2\alpha^{4r+4k+2}B^2$ by Lemmas 7 and 1. After iteration $j$ we have $n_{\mathrm{rm}} = r + j - s$ as the total number of removed vectors. All but $k$ of the $P_i$ have larger norm than any removed vector. Therefore the smallest $AD$ can be after iteration $j$ is $\geq (2\alpha^{(4r+4k+2)}B^2)^{j-k-n_{\mathrm{rm}}}$. Rearranging we get $AD \geq (2\alpha^{4r+4k+2}B^2)^{s-r-k}$. This contradicts Lemma 6 when $s$ reaches $2r + 2k$ for the first time because $(2\alpha^{4r+4k+2}B^2)^{r+k} \geq (\alpha^{2r+2k-1}B^2)^{r+k}$.

**Corollary 1.** *Throughout the algorithm we have* $\| \mathbf{b}_i^* \| \leq 2\alpha^{2c}B^2$.

We also use the active determinant to bound the number of iterations of the main loop, i.e. step 3c. First we show in the appendix that AD is increased by every scaling which does not end the main loop.

**Lemma 8.** *Every execution of step 3(c)ii either increases the AD by a factor* $\geq \frac{\alpha^c B}{2}$ *or sets* $\ell = 0$.

Now we are ready to prove that the number of iterations of the main loop is $\mathcal{O}(r + N)$. This is important because it means that, although we look at all of the information in the lattice, the number of times we have to call lattice reduction is unrelated to $\log X$.

**Theorem 3.** *The number of iterations of step 3c is* $\mathcal{O}(r + N)$.

The strategy of this proof is to show that each succesful scaling increases the active determinant and to bound the number of iterations using Lemma 6 and Corollary 1. For space constraints this proof is provided in the appendix.

**A Study of the Progress Function.** We will now amortize the costs of lattice reduction over each of the $\mathcal{O}(r + N)$ calls to step 3(c)iii. We do this by counting switches, using Progress $PF$ (defined below). In order to mimic the proof from [15] for our algorithm we introduce a type of Energy function which we can use over many calls to LLL (not only a single call).

**Definition 5.** *Let* $\mathbf{b}_1, \ldots, \mathbf{b}_s$ *be the current basis at any point in our algorithm, let* $\mathbf{b}_1^*, \ldots, \mathbf{b}_s^*$ *be their GSO, and* $l_i := \log_\gamma \| \mathbf{b}_i^* \|^2$ *for all* $i = 1 \ldots s$. *We let* $n_{rm}$ *be the number of vectors which have been removed so far in the algorithm. Then we define the progress function* $PF$ *to be:*

$$PF := 0 \cdot l_0 + \cdots + (s-1) \cdot l_s + n_{rm} \cdot c \cdot \log_\gamma (4\alpha^{4c}B^4).$$

This function is designed to effectively bound the largest number of switches which can have occurred so far. To prove that it serves this purpose we must prove the following lemma:

**Lemma 9.** *After step 2 Progress $PF$ has value 0. No step in our algorithm can cause the progress $PF$ to decrease. Further, every switch which takes place in step 3(c)iii must increase $PF$ by at least 1.*

**Theorem 4.** *Throughout our algorithm the total number of switches used by all calls to step 3(c)iii is* $\mathcal{O}((r + N)c(c + \log B))$ *with* $P_j$ *and* $\mathcal{O}(c^2(c + \log B))$ *with no* $P_j$.

*Proof.* Since Lemma 9 shows us that $PF$ never decreases and every switch increases $PF$ by at least 1, then the number of switches is bounded by $PF$.

However $PF$ is bounded by Lemma 7 which bounds $l_i \leq \log_\gamma (\alpha^{4r} B^4)$, Theorem 2 which bounds $s \leq c$, and the fact that we cannot remove more vectors than are given which implies $n_{\mathrm{rm}} \leq r + N$. Further we can see that $(s-1)l_s \leq (c-1)\log_\gamma (4\alpha^{4c} B^4)$ so $PF$ is maximized by making $n_{\mathrm{rm}} = (r+N)$ (or $c$ if no vectors added) and $s = 0$. In which case we have number of switches $\leq$ $PF \leq (r+N)(c-1)(\log_\gamma (4\alpha^{4c} B^4)) = \mathcal{O}((r+N)c(c+\log B))$. Also if there are no $P_j$, we can replace $r + N$ by $c$.

# 5    Complexity Bound of Main Algorithm

In this section we wish to prove a bound for the overall bit-complexity of algorithm 3. The complexity bound must rely on the complexity bound of the lattice reduction algorithm we choose for step 3(c)iii. The results in the previous sections have not relied on this choice. We will present our complexity bound using the H-LLL algorithm from [19]. We choose H-LLL for this result because of its favorable complexity bound and because the analysis of our necessary adaptations is relatively simple. See [19] for more details on H-LLL.

We make some minor adjustments to the H-LLL algorithm and its analysis. The changes to the algorithm are the following:

– We have a single non-integer entry in each vector of bit-length $\mathcal{O}(c + \log X)$.
– Whenever the final vector has G-S length sufficiently larger than $B$, it is removed. This has no impact on the complexity analysis.

We use $\tau$ as the number of switches used in a single call to H-LLL. This allows the analysis of progress $PF$ to be applied directly. The following theorem is an adaptation of the main theorem in [19] adapted to reflect our adjustments.

**Theorem 5.** *If a single call to step 3(c)iii, with H-LLL [19] as the chosen variation of LLL, uses $\tau$ switches then the CPU cost is bounded by $\mathcal{O}((\tau + c + \log B)c^2[(r+N)(c+\log B) + \log X])$ bit-operations.*

Now we are ready to complete the complexity analysis of the our algorithm.

**Theorem 6.** *The cost of executing algorithm 3 with H-LLL [19] as the variant of LLL in step 3(c)iii is*

$$\mathcal{O}((r+N)c^3(c+\log B)[\log X + (r+N)(c+\log B)])$$

*CPU operations, where $B$ is a search parameter chosen by the user, $|A[i,j]| \leq X$ for all $i, j$, and $c = r + N$ or $c = \mathcal{O}(r)$ (see definition 3 for details). If there are no $P_j$'s then the cost is*

$$\mathcal{O}((r+N+c^2)(c+\log B)c^2[\log X + (r+N)(c+\log B)]).$$

*Proof.* Steps 2, 3b, 3(c)i, and 3(c)ii have negligible costs in comparison to the rest of the algorithm. Step 3a is called $N$ times, each call performs $s$ inner products. While each inner product performs $r$ multiplications each of the form

$\mathbf{b}_i[m] \cdot x_{m,j}$ appealing to Corollary 1 we bound the cost of each multiplication by $\mathcal{O}((c + \log B) \log X)$. Since Theorem 2 gives $s \leq c$ we know that the total cost of all calls to step 3a is $\mathcal{O}(Ncr(c + \log B) \log X)$. Let $k = \mathcal{O}(r + N)$ be the number of iterations of the main loop. Let $\tau_i$ be the number of LLL switches used in the $i^{\text{th}}$ iteration. Theorem 5 gives the cost of the $i^{\text{th}}$ call to step 3(c)iii as $= \mathcal{O}((\tau_i + c + \log B)c^2[(r + N)(c + \log B) + \log X])$. Theorem 4 implies that $\tau_1 + \cdots + \tau_k = \mathcal{O}((r + N)c(c + \log B))$ (or $\mathcal{O}(c^2(c + \log B))$ when there are no $P_j$'s). The total cost of all calls to step 3(c)iii is then $\mathcal{O}([k(c + \log B) + \tau_1 + \cdots + \tau_k]c^2[(r + N)(c + \log B) + \log X])$. The term $[k(c + \log B) + \tau_1 + \cdots + \tau_k]$ can be replaced by $\mathcal{O}((r + N)c(c + \log B))$ (if no $P_j$ then $\mathcal{O}((r + N + c^2)(c + \log B)))$. The complete cost of is now $\mathcal{O}(Nrc(c + \log B) \log X + (r + N)c^3[c + \log B](\log X + (r + N)(c + \log B)])$. The first term is absorbed by the cost of the second term, proving the theorem. If there are no $P_j$ then we get $\mathcal{O}((r + N + c^2)(c + \log B)c^2[\log X + (r + N)(c + \log B)])$.

# 6   New Complexities for Applications of Main Algorithm

Our algorithm has been designed for some applications of lattice reduction. In this section we justify the importance of this algorithm by directly applying it to two classical applications of lattice reduction.

**New Complexity Bound for Factoring in $\mathbb{Z}[x]$.** In [4] it is shown that the problem of factoring a polynomial, $f \in \mathbb{Z}[x]$, can be accomplished by the reduction of a large knapsack-type lattice. In this subsection we merely apply our algorithm to the lattice suggested in [4].

**Reminders from [4].** Let $f \in \mathbb{Z}[x]$ be a polynomial of degree $N$. Let $A$ be a bound on the absolute value of the coefficients of $f$. Let $p$ be a prime such that $f \equiv l_f f_1 \cdots f_r \bmod p^a$ a separable irreducible factorization of $f$ in the $p$-adics lifted to precision $a$, the $f_i$ are monic, and $l_f$ is the leading coefficient of $f$. For our purposes we choose $B := \sqrt{r + 1}$.

We will make some minor changes to the All-Coefficients matrix defined in [4] to produce a matrix that looks like:

$$\begin{pmatrix} & & & & & & p^{a-b_N} \\ & & & & & \cdot^{\cdot^{\cdot}} & \\ & & & p^{a-b_1} & & & \\ 1 & & & x_{1,1} & \cdots & & x_{1,N} \\ & \ddots & & \vdots & \ddots & & \vdots \\ & & 1 & x_{r,1} & \cdots & & x_{r,N} \end{pmatrix}.$$

Here $x_{i,j}$ is the $j^{\text{th}}$ coefficient of $f_i' \cdot f/f_i$ mods $p^a$ divided by $p^{b_j}$ and $p^{b_j}$ represents $\sqrt{N}$ times a bound on the $j^{\text{th}}$ coefficient of $g' \cdot f/g$ for any true factor $g \in \mathbb{Z}[x]$ of $f$. In this way the target vectors will be quite small. An empty spot in this matrix represents a zero entry. This matrix has $p^{a-b_j} > 2^{N^2 + N \log(A)} > 2\alpha^{4r+2}B^2$ for all $j$. An $(\alpha, B)$-reduction of this matrix will solve the recombination problem by a similar argument to the one presented in [4] and refined in [22]. Now we look at the computational complexity of making and reducing this matrix which gives the new result for factoring inside $\mathbb{Z}[x]$.

**Theorem 7.** *Using algorithm 3 on the All-Coefficients matrix above provides a complete irreducible factorization of a polynomial $f$ of degree $N$, coefficients of bit-length $\leq \log A$, and $r$ irreducible factors when reduced modulo a prime $p$ in*

$$\mathcal{O}(N^2 r^4 [N + \log A])$$

*CPU operations. The cost of creating the All-Coefficients matrix adds $\mathcal{O}(N^4 [N^2 + \log^2 A])$ CPU operations using classical arithmetic (suppressing small factors $\log r$ and $\log^2 p$) to the complexity bound.*

The following chart gives a complexity bound comparison of our algorithm with the factorization algorithm presented by Schönhage in [25] we estimate both bounds using classical arithmetic and fast FFT-based arithmetic [5]. We also suppress all $\log N$, $\log r$, $\log p$, and $\log \log A$ terms.

| Classical Gradual LLL | $\mathcal{O}(N^3 r^4 + N^2 r^4 \log A + N^6 + N^4 \log^2 A)$ |
|---|---|
| Classical Schönhage | $\mathcal{O}(N^8 + N^5 \log^3 A)$ |
| Fast Gradual LLL | $\mathcal{O}(N^3 r^3 + N^2 r^3 \log A)$ |
| Fast Schönhage | $\mathcal{O}(N^6 + N^4 \log^2 A)$ |

The Schönhage algorithm is not widely implemented because of its impracticality. For most polynomials, $r$ is much smaller than $N$. Our main algorithm will reduce the All-Coefficients matrix with a competitive practical running time, but constructing the matrix itself will require more Hensel lifting than seems necessary in practice. In [22] a similar switch-complexity bound to section 4 is given on a more practical factoring algorithm.

**Algebraic Number Reconstruction.** The problem of finding a minimal polynomial from an approximation of a complex root was attacked in [14] using lattice reduction techniques using knapsack-type bases. For an extensive treatment see [17].

**Theorem 8.** *Suppose we know $\mathcal{O}(d^2 + d \log H)$ bits of precision of a complex root $\alpha$ of an unknown irreducible polynomial, $h(x)$, where the degree of $h$ is $d$ and its maximal coefficient has absolute value $\leq H$. Algorithm 3 can be used to find $h(x)$ in $\mathcal{O}(d^7 + d^5 \log^2 H)$ CPU operations.*

This new complexity is an improvement over the $L^2$ algorithm which would use $\mathcal{O}(d^9 + d^7 \log^2 H)$ CPU operations to reduce the same lattice. Although, one can prove a better switch-complexity with a two-column knapsack matrix by using [10, Lem. 2] to bound the determinant of the lattice as $\mathcal{O}(X^2)$ and thus the potential function from [15] is $\mathcal{O}(X^{2d})$, leading to a switch complexity of $\mathcal{O}(d \log X)$ (posed as an open question in [26, sec. 5.3]). Using this argument the complexity for $L^2$ is reduced to $\mathcal{O}(d^8 + d^6 \log^2 H)$.

**Acknowledgements.** We thank Damien Stehlé, Nicolas Brisebarre, and Valérie Berthé for many helpful discussions. Also Ivan Morel for introducing us to H-LLL. This work was partially funded by the LaRedA project of the Agence Nationale de la Recherche, it was also supported in part by a grant from the National Science Foundation. It was initiated while the second author was hosted by the Laboratoire d'Informatique de Robotique et de Microélectronique de Montpellier (LIRMM).

# References

1. Ajtai, M.: The shortest vector problem in L2 is NP-hard for randomized reductions. In: STOC 1998, pp. 10–19 (1998)
2. Akhavi, A.: The optimal LLL algorithm is still polynomial in fixed dimension. Theor. Comp. Sci. 297(1-3), 3–23 (2003)
3. Belabas, K.: A relative van Hoeij algorithm over number fields. J. Symb. Comp. 37, 641–668 (2004)
4. Belabas, K., van Hoeij, M., Klüners, J., Steel, A.: Factoring polynomials over global fields, preprint arXiv:math/0409510v1 (preprint) (2004)
5. Bernstein, D.: Multiprecision Multiplication for Mathematicians. Accepted by Advances in Applied Mathematics (2001), http://cr.yp.to/papers.html#m3
6. Bright, C.: Vector Rational Number Reconstruction, Masters Thesis, University of Waterloo (2009)
7. Coppersmith, D.: Small solutions to polynomial equations, and low exponent RSA vulnerabilities. J. of Cryptology 10, 233–260 (1997)
8. von zur Gathen, J., Gerhard, J.: Modern Computer Algebra. Cambridge University Press, Cambridge (1999)
9. Goldstein, D., Mayer, A.: On the equidistribution of Hecke points. Forum Mathematicum 15, 165–189 (2003)
10. Hanrot, G.: LLL: a tool for effective diophantine approximation, LLL+25 2007, pp. 81–118 (2007)
11. van Hoeij, M.: Factoring polynomials and the knapsack problem. J. Num. The. 95, 167–189 (2002)
12. Howgrave-Graham, N.A., Smart, N.P.: Lattice attacks on digital signature schemes. Design, Codes and Cryptography 23, 283–290 (2001)
13. Kaltofen, E.: On the complexity of finding short vectors in integer lattices. In: van Hulzen, J.A. (ed.) ISSAC 1983 and EUROCAL 1983. LNCS, vol. 162, pp. 235–244. Springer, Heidelberg (1983)
14. Kannan, R., Lenstra, A.K., Lovász, L.: Polynomial Factorization and Nonrandomness of Bits of Algebraic and Some Transcendental Numbers. In: STOC 1984, pp. 191–200 (1984)
15. Lenstra, A.K., Lenstra Jr., H.W., Lovász, L.: Factoring polynomials with rational coefficients. Math. Ann. 261, 515–534 (1982)
16. Lenstra, H.W.: Flags and lattice basis reduction. In: Euro. Cong. Math., vol. 1. Verlag, Basel (2001)

17. Lovász, L.: An Algorithmic Theory of Numbers, Graphs and Convexity. SIAM, Philadelphia (1986)
18. Micciancio, D.: The Shortest Vector Problem is NP-hard to approximate to within some constant. SIAM Journal on Computing 30(6), 2008–2035 (2001)
19. Morel, I., Stehlé, D., Villard, G.: H-LLL: Using Householder Inside LLL. In: ISSAC 2009, pp. 271–278 (2009)
20. Nguyen, P., Stehlé, D.: Floating-point LLL revisited. In: Cramer, R. (ed.) EURO-CRYPT 2005. LNCS, vol. 3494, pp. 215–233. Springer, Heidelberg (2005)
21. Nguyen, P., Stehlé, D.: An LLL Algorithm with Quadratic Complexity. J. Cmp. 39(3), 874–903 (2009)
22. Novocin, A.: Factoring Univariate Polynomials over the Rationals, Ph.D. Flor. St. Univ. (2008)
23. Odlyzko, A.: The rise and fall of knapsack cryptosystems. In: Proc. of Symposia in Applied Mathematics, A.M.S., Crytology and Computational Number Theory, vol. 42, pp. 75–88 (1990)
24. Schnorr, C.P.: A more efficient algorithm for lattice basis reduction. J. of Algo. 9, 47–62 (1988)
25. Schönhage, A.: Factorization of univariate integer polynomials by Diophantine ap-proximation and an improved basis reduction algorithm. In: Paredaens, J. (ed.) ICALP 1984. LNCS, vol. 172, pp. 436–447. Springer, Heidelberg (1984)
26. Stehlé, D.: Floating-point LLL: Theoretical and Practical Aspects, LLL+25, pp. 33–61 (2007)
27. Storjohann, A.: Faster algorithms for integer lattice basis reduction, Technical re-port 1996, ETH Zürich (1996)

# The Power of Fair Pricing Mechanisms

Christine Chung[1], Katrina Ligett[2], Kirk Pruhs[3], and Aaron L. Roth[4]

[1] Department of Computer Science, Connecticut College, New London, CT
cchung@conncoll.edu
[2] Department of Computer Science, Cornell University, Ithaca, NY
katrina@cs.cornell.edu
[3] Department of Computer Science, University of Pittsburgh, PA
kirk@cs.pitt.edu
[4] Department of Computer Science, Carnegie Mellon University, Pittsburgh, PA
alroth@cs.cmu.edu

**Abstract.** We explore the revenue capabilities of truthful, monotone ("fair") allocation and pricing functions for resource-constrained auction mechanisms within a general framework that encompasses unlimited supply auctions, knapsack auctions, and auctions with general non-decreasing convex production cost functions. We study and compare the revenue obtainable in each fair pricing scheme to the profit obtained by the ideal omniscient multi-price auction. We show (1) for capacitated knapsack auctions, no constant pricing scheme can achieve any approximation to the optimal profit, but proportional pricing is as powerful as general monotone pricing, and (2) for auction settings with arbitrary bounded non-decreasing convex production cost functions, we present a proportional pricing mechanism which achieves a poly-logarithmic approximation. Unlike existing approaches, all of our mechanisms have fair (monotone) prices, and all of our competitive analysis is with respect to the optimal profit extraction.

## 1 Introduction

Practical experience [1,2,3] demonstrates that any store charging non-*monotone* prices (that is, charging some buyer $i$ more than buyer $j$ despite the fact that buyer $i$ receives strictly less of the good than $j$) risks public outrage and accusations of unfair practices. There are of course very simple auction pricing schemes that are monotone: for example, *constant pricing*, in which each bidder is quoted the same price regardless of the quantity of the good she receives, and *proportional pricing* in which each bidder is quoted a price proportional to her demand. Given that fairness may thus in many situations be considered a first-order mechanism design constraint, even at the expense of short-term profit maximization, it is natural to ask, "are clever implementations of these simple monotone pricing schemes capable of maximizing profit?"

We answer this question in the affirmative in a broad class of auctions in which bidders demand different quantities of a given resource, for example, server capacity, bandwidth, or electricity. We consider natural subclasses of this class of

A. López-Ortiz (Ed.): LATIN 2010, LNCS 6034, pp. 554–564, 2010.

auctions: unlimited supply, limited supply ("knapsack auctions"), and a more general setting in which the cost to the mechanism may be some arbitrary non-decreasing convex function of the supply sold. This last model we propose generalizes the first two, and models any way in which the auctioneer may incur decreasing marginal utility as the production of the good being sold increases (for example if increased demand for raw materials increases the producer's per unit cost for these materials).

In general, no truthful auction can acquire the value $h$ from the highest bidder (see, for example, [4]), and at best can hope to compete with $\mathbf{OPT} - h$. In the unlimited supply setting, the constant pricing mechanism from [5] is $O(\log n)$ competitive with $\mathbf{OPT} - h$, where $\mathbf{OPT}$ is the sum of all bidders' valuations, not just the optimal profit obtainable by any constant price mechanism, and $h$ is the highest bid. Our findings are as follows:

- In the limited supply (knapsack) setting, we show that no constant pricing mechanism can achieve an approximation factor of $o(n)$ with $\mathbf{OPT} - o(n)h$. However, we give a mechanism that uses proportional pricing that, aside from an extra profit loss of $h$, achieves an $O(\log S)$ approximation to $\mathbf{OPT} - 2h$, where $S$ is the knapsack size constraint.
- In a general setting in which the mechanism incurs some non-decreasing convex cost as a function $C$ of the supply it sells, we give a proportional pricing mechanism that achieves (again aside from an extra loss of $h$) a polylogarithmic approximation to $\mathbf{REV} - 3h - 2C(S^*)$, where $\mathbf{REV}$ is the revenue obtained, and $C(S^*)$ is the cost incurred by the auctioneer in the optimal solution maximizing $\mathbf{REV} - C(S^*)$. (Here we assume the cost function is polynomially bounded.)

In each of these settings, there is essentially a log lower bound on the profit competitiveness for any monotone pricing mechanism[5]. Additionally, in the generalized auction setting, we show that proportional pricing is strictly weaker than monotone pricing (independent of truthfulness). We give an instance that show that no proportional pricing scheme can achieve profit within any finite factor of $\mathbf{REV} - O(1)C(S^*)$. This perhaps makes it more surprising that a truthful proportional pricing mechanism can be close to optimally competitive with $\mathbf{OPT} - h$. Overall, our results show that there exist proportional pricing schemes that compete with full profit extraction essentially as effectively as the best possible monotone pricing scheme.

## 1.1 Related Work

The framework of competitive analysis in the setting of auction design was introduced by Goldberg et al. [6]. In the digital goods setting, where each bidder demands one unit of the resource, and the supply of the resource is unlimited, [6,7] give randomized truthful mechanisms that are competitive with the optimal constant price profit. In the "knapsack auction" setting, where each bidder may have a different demand and there is a fixed limited supply, [4] gives a randomized truthful mechanism that achieves a profit of $\alpha\mathbf{OPT}_{mono} - \gamma h \lg \lg \lg n$, where $n$ is the number of players, $\mathbf{OPT}_{mono}$ is the optimal monotone pricing profit, $h$ is the maximum valuation of any bidder, and $\alpha$ and $\gamma$ are constants. It is important to note that the mechanisms given in [6] and [4] use random sampling techniques and are not monotone. That is, they can quote customers different prices for identical orders. This can in some sense be justified by the result in [5] that shows that, for either the setting of digital goods or knapsack auctions, no truthful, fair pricing mechanism can be can be $o(\log n / \log \log n)$ competitive with the optimal constant price profit. That is, there is no mechanism that achieves all the properties of (1) truthfulness, (2) fairness, and (3) constant competitiveness with respect to profit.

Goldberg and Hartline [5] go on to show that if the fairness requirement is relaxed (and the auction is allowed to give non-envy-free outcomes with small probability), auctions competitive with the optimal constant price can be found. Guruswami et al. [8] show that in two auction settings closely related to ours, simply computing fair prices that maximize profit, without the requirement of truthfulness, is APX-hard. Very recently, Babaioff et al. considered an auction setting in which the supply arrives online, and generalized the definition of fairness to their setting, also showing an $\Omega(\log n / \log \log n)$ lower bound for even welfare maximization in this online setting [9].

Intuitively, the papers [6,4,7] (among others) study the the profit competitiveness that can be achieved if one gives up on fairness. In contrast, the goal of this work is to understand the pricing techniques that are needed to maximize profit in a variety of settings, if montonicity and truthfulness are objectives we are unwilling to sacrifice.

## 1.2   The Problem

In this work, we consider single-round, sealed-bid auctions with a set $N = [n]$ of single-minded bidders. Each bidder $i$ has a public size, or demand, $x_i$ and a private valuation $v_i$. We write $\mathbf{x} = (x_1, \ldots, x_n)$ and $\mathbf{v} = (v_1, \ldots, v_n)$. We will assume that smallest size $x_i$ of any bidder is 1. Let $X = \sum_{i=1}^{n} x_i$. The demand of a player must be satisfied completely or not at all; we do not allow fractional allocations.

**Definition 1.** *In a single-round, sealed-bid auction, each bidder $i$ submits a bid $b_i$, which is the most she is willing to pay if she wins. We write $\mathbf{b} = \{b_1, \ldots, b_n\}$. Given $\mathbf{b}$ and $\mathbf{x}$, the mechanism returns prices $\mathbf{p} = (p_1, \ldots, p_n)$ and an indicator vector $\mathbf{w} = (w_1, \ldots, w_n)$. If $w_i = 1$, we say player $i$ wins; otherwise, we say she loses. Player $i$ pays $p_i$ if she is a winner and 0 if she is not. The mechanism is valid if and only if every winning bidder has $b_i \geq p_i$, and every losing bidder has $p_i \geq b_i$. The profit achieved by the mechanism depends on the capacity constraints under consideration (discussed below).*

Note that the agents are indistinguishable to the auction mechanism, except for their size. We will only present truthful mechanisms, and so throughout the paper, we will assume $\mathbf{b} = \mathbf{v}$.

**Definition 2.** *A deterministic auction is truthful if, for any* $\mathbf{x}$*, for all bidders* $i \in N$*, for any choice of* $\mathbf{v}_{-i}$*, bidder* $i$*'s profit is maximized by bidding her true value* $v_i$*. We say that a randomized auction is truthful if it is a probability distribution over deterministic truthful auctions.*

We require that our pricing schemes be fair, meaning monotone: we cannot charge some player more than another player if her demand is lower.

**Definition 3.** *A deterministic auction's pricing is* monotone *if for any* $\mathbf{v}$ *and* $\mathbf{x}$*, it assigns prices* $\mathbf{p}$ *such that* $p_i \geq p_j$ *whenever* $x_i \geq x_j$*, for any* $i$ *and* $j$ *in* $N$*. A randomized auction's pricing is* monotone *if it is a distribution over deterministic monotone auctions.*

One example of of a valid monotone pricing scheme is *constant pricing*, where each player is offered the same price $p = p_1 = p_2 = \ldots = p_n$, every player $i$ with $v_i > p$ is a winner, and no player $i$ with $v_i < p$ is a winner. Another monotone scheme, *proportional pricing*, fixes some value $c$ and charges each player $i$ a price $p_i = c \cdot x_i$; every player $i$ with $v_i > p_i$ is a winner, and no player $i$ with $v_i < p_i$ is a winner.

In all cases, we assume that the bidders are trying to maximize profit, that they know the mechanism being used, and that they don't collude. Our goal is to study truthful, monotone pricing mechanisms that maximize our profit, in a variety of capacity constraint settings:

- In the *unlimited supply* setting, there is no limit on the total size of the bidders we can accept. A mechanism's profit here is just $\mathbf{p} \cdot \mathbf{w}$.
- In the *knapsack* setting, there is a hard limit $S$ on the total size of the winning bidders. Here, the mechanism's profit is $\mathbf{p} \cdot \mathbf{w}$ if $\mathbf{x} \cdot \mathbf{w} \leq S$, and $-\infty$ otherwise.
- In the *general cost* setting, there is some non-decreasing convex cost function $C$ of the size of the winning bidder set, and the profit of the auctioneer is the difference between the sum of the prices paid by the winning bidders and the cost of the size of the winning set. Here, we define the mechanism's profit to be $\mathbf{p} \cdot \mathbf{w} - C \left( \sum_{i \in \mathbf{w}} x_i \right)$.

Note that the unlimited supply problem is an instance of the knapsack problem with $S > X$. The knapsack problem is an instance of the general cost setting, with a cost function that takes value 0 for $x < C$ and jumps to $\infty$ at $C$.

In all three cases, we compare our schemes with the optimal multiple-price omniscient allocation that is not constrained to be truthful nor envy-free. In the unlimited supply case, $\mathbf{OPT} = \sum_{i=1}^{n} v_i$. In the knapsack setting,

$$\mathbf{OPT} = \max_{B \subseteq 2^N \mid \sum_{i \in B} x_i \leq S} \sum_{i \in B} v_i.$$

In the general cost setting,

$$\mathbf{OPT} = \max_{B \subseteq 2^N} \left( \left( \sum_{i \in B} v_i \right) - C \left( \sum_{i \in B} x_i \right) \right).$$

Setting $B$ to be the set of winners in an optimal general-cost solution, we will write $\mathbf{REV} = \sum_{i \in B} v_i$ for the revenue of the optimal solution. As mentioned above, in general no truthful algorithm can achieve better than $\mathbf{OPT} - h$, where $h$ is the value of the highest bid, so this will be our performance benchmark.

## 1.3   Unlimited Supply Auctions

In the unlimited supply setting, the auctioneer has an unlimited number of items to sell, at zero marginal cost (equivalently, if each item has some constant marginal cost, we may simply subtract this cost from the valuations of the bidders). Goldberg and Hartline gave a simple randomized mechanism that achieves a $\Theta(\log n)$ approximation to the profit obtained by the best constant price $\mathbf{OPT}_c$,[1] and showed that this is almost optimal [5]. It is also known that $\mathbf{OPT}_c$ can differ by a $\Theta(\log n)$ factor from $\mathbf{OPT}$.[2] That is, if the profit obtained by the mechanism below is $\mathbf{OPT}_c/\alpha$, and $\mathbf{OPT}_c = \mathbf{OPT}/\beta$, it is known that $\alpha$ and $\beta$ can both take values as large as $\Theta(\log n)$, but no larger. This immediately shows that RANDOMPRICE (given below) is an $O(\log^2 n)$ approximation to $\mathbf{OPT}$. In fact, for any instance, $\alpha \cdot \beta = O(\log n)$. In other words, RANDOMPRICE gives a $\Theta(\log n)$ approximation to $\mathbf{OPT}$.

RANDOMPRICE$(\mathbf{v}, \mathbf{x})$
1   Choose $i \in \{1, 2, \ldots, \log n\}$ uniformly at random.
2   Let $g = 2^i$. Sell items to the $g - 1$ highest bidders at price $v_g$
    (where $v_1 \geq v_2 \geq \ldots \geq v_n$).

**Theorem 1 (implicit in [5]).** *For any set of bidder values, let $P$ be the expected profit obtained by* RANDOMPRICE. *Let $\alpha$ be such that $P = \mathbf{OPT}_c/\alpha$, and let $\beta$ be such that $\mathbf{OPT}_c = (\mathbf{OPT} - h)/\beta$. Then $\alpha \cdot \beta = O(\log n)$. Equivalently, $P \geq (\mathbf{OPT} - h)/O(\log n)$.*

We note that although constant pricing is sufficient to obtain an $O(\log n)$ approximation, this mechanism is almost optimal over the set of all *monotone* pricing mechanisms. The following lower bound is implicit in the lower bound proved by Goldberg and Hartline in the context of digital goods auctions:

**Theorem 2 (Goldberg and Hartline [5]).** *In the uncapacitated setting, no truthful mechanism using a monotone pricing scheme can achieve profit within a factor of $o(\log n / \log \log n)$ of $\mathbf{OPT} - c \cdot h$ for any constant $c$.*

## 2   Knapsack Auctions

Knapsack auctions were first studied by Aggarwal and Hartline [4], and model auctions for items for which there is a strict limit on supply: we are given a set

---

[1] Actually, they show something slightly weaker, defining $\mathbf{OPT}_c$ to be the optimal constant price when the mechanism is required to sell at least 2 items.

[2] Consider $n$ bidders with valuations $v_1, \ldots, v_n$ with $v_i = 1/i$. $\mathbf{OPT} = H(n)$, but the best constant price obtains profit $i \cdot v_i = 1$ for all $i$. [4].

of bidders with demands and valuations, and can only sell to a set of bidders whose total demand is smaller than our knapsack capacity $S$.

When supply is unlimited, we have seen that constant pricing is as powerful as monotone pricing in the sense that both can achieve within a $O(\log n)$ factor of **OPT** $- h$, but no better. In this section, we show that in the knapsack case, when supply is limited, no valid constant pricing scheme can achieve within any finite factor of **OPT** $- o(n)h$. However, we show that proportional pricing is as powerful as monotone pricing in the sense that both can achieve $O(\log S)$ competitiveness with **OPT** $- h$, but no better. Our result is also optimal over the set of proportional pricing schemes, even those that are not truthful: Aggarwal and Hartline give an example in which the optimal proportional pricing is an $\tilde{\Omega}(\log S)$ factor off from **OPT** $- h$ [4].[3]

**Theorem 3.** *In the knapsack setting, no mechanism which uses a valid constant pricing scheme can guarantee any approximation to* **OPT** $- o(n)h$.

*Proof.* Consider an instance with knapsack capacity $S = 1/\epsilon$, $1/\epsilon$ bidders $i$ with size $x_i = 1$ and value $v_i = 1$, and one bidder $i^*$ of size $x_{i^*} = 1/\epsilon$ and value $v_{i^*} = 1 + \epsilon$. Clearly, **OPT** $= 1/\epsilon$, which results from selling to each of the $1/\epsilon$ bidders $i \neq i^*$, and **OPT** $- o(n)h = 1/\epsilon - o(1/\epsilon) = \Theta(1/\epsilon)$. However, no constant price $p \leq 1$ results in a valid allocation, since $\sum_{i:v_i \leq 1} x_i = 2/\epsilon > S$. Therefore, **OPT**$_c = 1 + \epsilon$, and results from selling only to bidder $i^*$. Therefore, (**OPT** $- o(n)h)/$**OPT**$_c = \Theta((1/(1+\epsilon)\epsilon)$, which can be unboundedly large.

We can also extend the lower bound for general monotone mechanisms from the unlimited supply setting, simply by choosing the total demand of the lower bound instance to be strictly less than the knapsack size.

**Theorem 4 (Goldberg and Hartline [5]).** *In the knapsack setting, no truthful mechanism using any monotone pricing scheme can guarantee an $o(\log S/\log\log S)$ approximation to* **OPT** $- c \cdot h$ *for any constant $c$.*

Next, we give a truthful mechanism which achieves an $O(\log S)$ approximation to **OPT** $- 2h$ by using a valid proportional pricing scheme. This pricing scheme is similar to the "RANDOM single price" algorithm proposed in [10] for an online, combinatorial auction setting. In [10] they show their algorithm is $O(\log(S) + \log(n))$ competitive with **OPT**. Below we give a bound for our algorithm only in terms of $S$.

---

[3] Aggarwal and Hartline show that proportional pricing cannot in general approximate monotone pricing within a factor of $o(n)$, and in their lower bound instance, use bidders with exponentially large demand (also showing that proportional pricing cannot in general approximate monotone pricing to within an $\tilde{\Omega}(\log T)$ factor, where $T$ is the total demand of all players). Our result implies that there always exists a proportional pricing that approximates **OPT** $- h$ (and not just the optimal monotone pricing) to within an $O(\log S)$ factor.

PROPORTIONALKNAPSACK($\mathbf{v}, \mathbf{x}, S$)

1   Let $\pi$ be an ordering of bidders in non-increasing order of density $d_i = v_i / x_i$.
    Denote by $H$ the largest prefix of bidders that is satisfiable with supply $S$;
    note $\sum_{i \in H} x_i \leq S$. Let $X_i := \sum_{j=1}^{i} x_{\pi(j)}$; $X_0 = 0$.
    Let $g$ be a function mapping points in the knapsack $x \in [0, S]$ to bidders, in
    order of $\pi$ as follows: $g(x) = i$ if $x \in [X_{i-1}, X_i)$
2   Choose $s \in \{0, \ldots, \lfloor \log(S) \rfloor\}$ uniformly at random.
    Consider the bidder $\pi(i^*) = g(2^s - 1)$ who corresponds to the point $2^s - 1$
    in the knapsack; let $d^* = d_{\pi(i^*)}$ be its density.
3   Sell items to bidders $\pi(1), \ldots, \pi(i^* - 1)$ at prices proportional to $d^*$.

**Proposition 1.** PROPORTIONALKNAPSACK *is truthful and produces a valid proportional pricing.*

*Proof.* We observe that fixing any realization of the random coin flips, no winning bidder can become a losing bidder by raising his bid, since a bidder can only increase his rank in $\pi$ by raising his bid. Our pricing scheme is truthful because the price we charge a player is independent of her bid, and all losing players have values at below the price they would be offered if they raised their bid to a winning level. Validity follows since winning bidders are charged proportional to a rate that is at most their own density, and losing bidders are charged proportional to a rate that is at least their own density.

**Theorem 5.** PROPORTIONALKNAPSACK *achieves expected profit at least* $(\mathbf{OPT} - 2h)/O(\log S) - h$.

*Proof.* Let $\mathbf{OPT}(H)$ refer to the value of the optimal solution if the set of bidders were comprised only of those in the set $H$. First observe that $\mathbf{OPT}(H) \geq \mathbf{OPT} - h$. To see this, note that $\mathbf{OPT}$ can take value at most the value of the corresponding knapsack problem. By taking the largest density prefix that fits in our knapsack, we are preserving the value of the optimal solution to the fractional knapsack problem, minus at most the value of a single bidder (the first bidder according to $\pi$ not included in $H$). Since we wish to be competitive with $\mathbf{OPT} - h$, for the rest of the argument, we may restrict our attention to $H$ and assume we are in the unlimited supply setting (since the available supply is larger than the total remaining demand).

   We bound $\mathbf{OPT}(H)$ by considering the bidders in decreasing order of density $\pi(1), \ldots, \pi(|H|)$, and bounding the density of the optimal knapsack solution. Let $f(x)$ denote the density of the bidder occupying position $x$ in the knapsack. We have

$$\mathbf{OPT}(H) \leq \int_0^S f(x) dx \leq \sum_{i=0}^{\lfloor \log(S) \rfloor} f(2^i - 1) 2^i,$$

where the inequality follows since we have ordered the bidders such that their density is non-increasing. Similarly, we may bound the expected profit $P$ obtained by our mechanism:

$$P = \frac{1}{\lfloor \log(S) \rfloor + 1} \left( \sum_{i=0}^{\lfloor \log(S) \rfloor} \left( f(2^i) 2^i - h \right) \right),$$

where we lose the $h$ term since we cannot sell to the bidder from whom we've sampled the sale density $d^*$. Thus,

$$\mathbf{OPT}(H) \leq h + 2 \left( (P + h)(\lfloor \log(S) \rfloor + 1) \right).$$

Recalling that $\mathbf{OPT}(H) \geq \mathbf{OPT} - h$, we get

$$P \geq \frac{\mathbf{OPT} - 2h}{2(\lfloor \log(S) \rfloor + 1)} - h.$$

## 3 General Convex Cost Auctions

In this section, we propose a general setting in which the mechanism incurs a cost, expressed as a function of the amount of supply sold. In the previous section, we showed that in the bounded supply setting, proportional pricing was sufficient to get essentially as good an approximation to $\mathbf{OPT} - h$ as was possible using any monotone pricing scheme. Here, we show that, in general, there is an unboundedly large gap between the profit attainable with proportional pricing and the profit obtainable by monotone pricing, even if we require the monotone pricing scheme to pay a higher cost. Note that this lower bound discusses pricing schemes, not mechanisms. We will show that surprisingly, if we wish to compete with $\mathbf{OPT} - h$ (which any truthful mechanism must do), then proportional pricing *is* sufficient.

**Theorem 6.** *In the general non-decreasing convex cost setting, for any value $d$, there exists a set of bidders and a convex cost function such that the optimal profit is obtained by selling supply $S^*$, yielding profit $\mathbf{OPT} = \mathbf{REV} - C(S^*)$, but no proportional pricing scheme is able to achieve any approximation to $\mathbf{REV} - d \cdot C(S^*)$.*

*Proof.* Consider an instance with a quadratic cost function: $C(x) = x^2$. There are two bidders: One bidder has size 1 and value $d + 2$ (this bidder has density $d + 2$). Let $k$ refer to the size of the second bidder, and set his value to $(d + 2)k + 1$ (this bidder has higher density, $d + 2 + 1/k$). The optimal monotone pricing sells only to the first bidder, and gets profit $d + 1$. Note in this case, even $\mathbf{REV} - d \cdot C(1) = 1 \geq 0$. However, for proportional pricing, it is impossible to sell to the first bidder without selling to the second, since the second bidder is denser. If we sell to the both, however, we get at most profit $(d(k+1) + 1 - (k)^2)$, which is negative for large enough $k$. If we sell to only the denser bidder, we get a most profit $(d+2)k + 1 - k^2$, again negative for large enough $k$. Thus, the best proportional pricing sells no items, and gets profit 0.

Fiat et al. [7] also consider a setting in the presence of a cost function and give a mechanism that is competitive only with the optimal revenue in some class, minus a multiplicative factor times the cost function, and conjecture that this is necessary. Below, we demonstrate a proportional pricing mechanism that is polylog competitive with $\mathbf{REV} - 3h - (1 + \epsilon)C(S^*)$, for any constant $\epsilon$.

Given a non-decreasing convex cost function $C$, the profit $P$ the algorithm obtains when assigning prices $\mathbf{p}$ and allocation $\mathbf{w}$ to players $(\mathbf{v}, \mathbf{x})$ is $P = \mathbf{p} \cdot \mathbf{w} - C\left(\sum_{i \in \mathbf{w}} x_i\right)$. We assume $C$ is continuous and that $C(0) = 0$.

We will write $\mathbf{OPT}_S(\mathbf{v}, \mathbf{x})$ for the maximum profit extraction obtainable from $\mathbf{v}$ and $\mathbf{x}$ with a knapsack restriction $S$ in place of the cost function. Let $S^*$ be the total size of the set of winning bidders under the optimal (non-truthful) allocation:

$$S^* = \sum_{i \in B^*} x_i \text{ where } B^* = \operatorname*{argmax}_{B \subseteq 2^N} \left(\sum_{i \in B} v_i - C\left(\sum_{i \in B} x_i\right)\right).$$

Our algorithm first attempts to guess $S^*$. We then estimate values for each bidder based on their stated valuations minus their contribution to the estimated solution size, and run our proportional-pricing mechanism on these estimated values. We then adjust the resulting prices and output this with the resulting allocation. In what follows, we assume that $h \geq 1$; if this is not the case, the algorithm can be easily adapted to begin guessing the value of $c$ at $h$ instead of 1, losing an additional $\log 1/h$ factor.

GENERALAUCTION$(\mathbf{v}, \mathbf{x}, C)$

1   Select cost $c$ at random from among $\{1, 2, 2^2, \ldots, 2^{\lceil \log C(X) \rceil}\}$.[a]
2   Set $S$ to be the largest value such that $C(S) = c$, or $\infty$ if no such value exists.
3   $v_i' := v_i - \frac{C(S)x_i}{S}$ and $\mathbf{v}' := (v_1', \ldots, v_n')$
4   $(\mathbf{p}', \mathbf{w}') := \text{PROPORTIONALKNAPSACK}(\mathbf{v}', \mathbf{x}, S)$.
5   $p_i := p_i' + \frac{C(S)x_i}{S}$
6   Return $(\mathbf{p}, \mathbf{w}')$

---

[a] Note that if $h \geq 1$, we lose at most an additive $h$ by assuming $S^*$ is such that $c \geq 1$. Also note that it is possible to get a slightly stronger result by excluding values of $S$ such that $S > \sum_{i=1}^n v_i$, although we omit this here.

**Lemma 1.** *Suppose that $2C(S^*) \geq C(S) \geq C(S^*) \geq 1$. The optimal profit obtainable in an $S$-capacitated knapsack, given values $\mathbf{v}'$, is at least the optimal revenue minus twice the cost at that supply, on the original instance:* $\mathbf{OPT}_S(\mathbf{v}', \mathbf{x}) \geq \mathbf{OPT}_{S^*}(\mathbf{v}, \mathbf{x}) - 2C(S^*)$.

*Proof.* We observe $\mathbf{OPT}_S(\mathbf{v}', \mathbf{x}) \geq \mathbf{OPT}_S(\mathbf{v}, \mathbf{x}) - C(S)$, since at worst, the optimal knapsack solution given $\mathbf{v}'$ selects the exact same winners as the optimal knapsack solution given $\mathbf{v}$. Now note that $S^* \leq S$ and $2C(S^*) \geq C(S)$, and so $\mathbf{OPT}_S(\mathbf{v}, \mathbf{x}) - C(S) \geq \mathbf{OPT}_{S^*}(\mathbf{v}, \mathbf{x}) - 2C(S^*)$.

**Theorem 7.** *The* GENERALAUCTION *algorithm obtains expected profit at least*

$$\frac{\textbf{REV} - 2C(S^*) - 3h}{O(\log(X)\log(C(X)))} - \frac{h}{\log(C(X))}.$$

*Note that the denominator is* $O(\log^2(X))$ *when* $C$ *is polynomially bounded.*[4]

*Proof.* Suppose that $2C(S^*) \geq C(S) \geq C(S^*) \geq 1$. Note that by definitions of **REV**, $S^*$, and $\textbf{OPT}_{S^*}(\mathbf{v}, \mathbf{x})$, we have $\textbf{REV} \leq \textbf{OPT}_{S^*}(\mathbf{v}, \mathbf{x})$. Hence, the optimal profit obtainable in an $S$-capacitated knapsack under $(\mathbf{v}', \mathbf{x})$, as shown above, is

$$\textbf{OPT}_S(\mathbf{v}', \mathbf{x}) \geq \textbf{REV} - 2C(S^*).$$

Then by the approximation ratio of the knapsack algorithm, PROPORTIONALK-NAPSACK returns a solution of value

$$\frac{\textbf{OPT}_S(\mathbf{v}', \mathbf{x}) - 2h}{2\lfloor \log(S) \rfloor} - h \geq \frac{\textbf{REV} - 2C(S^*) - 2h}{2\lfloor \log(S) \rfloor} - h.$$

If $C(S^*) < 1$, this becomes at worst

$$\frac{\textbf{REV} - 2C(S^*) - 3h}{2\lfloor \log(S) \rfloor} - h$$

for $h \geq 1$. The additional profit obtained by prices $\mathbf{p}$ over prices $\mathbf{p}'$ in allocation $\mathbf{w}$ is $\frac{C(S)S'}{S}$, where $S'$ is the size of the solution selected by PROPORTIONALK-NAPSACK. The cost imposed by the cost function is $C(S')$. Thus, in this case, the profit GENERALAUCTION obtains is at least

$$\frac{\textbf{REV} - 2C(S^*) - 3h}{2\lfloor \log(S) \rfloor} - h + \left( C(S)\frac{S'}{S} - C(S') \right).$$

Since the cost function $C$ is non-decreasing and convex, this second term is non-negative. Since $2C(S^*) \geq C(S) \geq C(S^*)$ holds with probability $O(1/\log(C(X)))$, and $X = \sum_{i=1}^n x_i$ is an upper bound on $S$, this completes the proof.

**Proposition 2.** GENERALAUCTION *is truthful.*

*Proof.* This is immediate, since it is a distribution over truthful mechanisms. (Specifically, it is a distribution over instances of PROPORTIONALKNAPSACK in which the prices have been modified by a bid-independent function, which preserves truthfulness).

**Proposition 3.** GENERALAUCTION *is a valid mechanism.*

*Proof.* Suppose player $i$ is a winner. Then, by the validity of PROPORTIONALK-NAPSACK, $v_i' \geq p_i'$. Thus,

---

[4] This can be improved to $(\textbf{REV} - (1 + \epsilon)C(S^*) - 3h)/(O(\log(X)\log(C(X)))) - h/\log(C(X))$ for arbitrary constant $\epsilon$ simply by selecting $c$ from among $\{1, (1 + \epsilon), (1 + \epsilon)^2, \ldots\}$.

$$v_i = v_i' + \frac{C(S)x_i}{S} \geq p_i' + \frac{C(S)x_i}{S} = p_i.$$

Now, suppose player $i$ loses. Then, by the validity of the knapsack mechanism, $p_i' \geq v_i'$. Thus,

$$p_i = p_i' + \frac{C(S)x_i}{S} \geq v_i' + \frac{C(S)x_i}{S} = v_i.$$

**Proposition 4.** GENERALAUCTION *produces a proportional pricing.*

*Proof.* PROPORTIONALKNAPSACK produces a proportional pricing scheme, and the prices returned by GENERALAUCTION increase the proportional factor by $\frac{C(S)}{S}$.

**Acknowlegments.** We would like to thank several anonymous referees for thoughtful and helpful comments on earlier versions of this paper.

# References

1. Streitfeld, D.: On the web, price tags blur: What you pay could depend on who you are (2000), http://www.washingtonpost.com/ac2/wp-dyn/A15159-2000Sep25
2. Wolverton, T.: Amazon backs away from test prices (2000), http://news.cnet.com/2100-1017-245631.html
3. Ramasastry, A.: Web sites change prices based on customers' habits (2005), http://www.cnn.com/2005/LAW/06/24/ramasastry.website.prices/
4. Aggarwal, G., Hartline, J.: Knapsack Auctions. In: Proceedings of the Symposium on Discrete Algorithms (2006)
5. Goldberg, A., Hartline, J.: Envy-free auctions for digital goods. In: Proceedings of the 4th ACM conference on Electronic commerce, pp. 29–35. ACM, New York (2003)
6. Goldberg, A., Hartline, J., Karlin, A., Saks, M., Wright, A.: Competitive auctions. Games and Economic Behavior 55, 242–269 (2006)
7. Fiat, A., Goldberg, A., Hartline, J., Karlin, A.: Competitive generalized auctions. In: Proceedings of the thiry-fourth annual ACM symposium on Theory of Computing, pp. 72–81. ACM, New York (2002)
8. Guruswami, V., Hartline, J.D., Karlin, A.R., Kempe, D., Kenyon, C., McSherry, F.: On profit-maximizing envy-free pricing. In: SODA 2005: Proceedings of the sixteenth annual ACM-SIAM symposium on Discrete algorithms, pp. 1164–1173. Society for Industrial and Applied Mathematics, Philadelphia (2005)
9. Babaioff, M., Blumrosen, L., Roth, A.: Auctions with Online Supply (2009)
10. Balcan, M.F., Blum, A., Mansour, Y.: Item pricing for revenue maximization. In: EC 2008: Proceedings of the 9th ACM conference on Electronic commerce (2008)

# Quasi-Proportional Mechanisms: Prior-Free Revenue Maximization

Vahab Mirrokni[1], S. Muthukrishnan[1], and Uri Nadav[2,*]

[1] Google Research, New York, NY
mirrokni@google.com, muthu@google.com
[2] Tel-aviv University
uri.nadav@gmail.com

**Abstract.** Inspired by Internet ad auction applications, we study the problem of allocating a single item via an auction when bidders place very different values on the item. We formulate this as the problem of *prior-free* auction and focus on designing a simple mechanism that always allocates the item. Rather than designing sophisticated pricing methods like prior literature, we design better allocation methods. In particular, we propose *quasi-proportional* allocation methods in which the probability that an item is allocated to a bidder depends (quasi-proportionally) on the bids.

We prove that corresponding games for both all-pay and winners-pay quasi-proportional mechanisms admit pure Nash equilibria and this equilibrium is unique. We also give an algorithm to compute this equilibrium in polynomial time. Further, we show that the revenue of the auctioneer is promisingly high compared to the ultimate, i.e., the highest value of any of the bidders, and show bounds on the revenue of equilibria both analytically, as well as using experiments for specific quasi-proportional functions. This is the first known revenue analysis for these natural mechanisms (including the special case of proportional mechanism which is common in network resource allocation problems).

## 1 Introduction

Consider the following motivating example. There is a single item (in our case, an ad slot) to be sold by auction. We have two bidders $A$ and $B$, $A$ with valuation $b_A = 100$ and $B$ with valuation $b_B = 1$. Who should we *allocate* the item and what is the *price* we charge? In the equilibrium of the *first price* auction, $A$ wins by bidding $1 + \epsilon$. We (the auctioneer) get *revenue* of $1 + \epsilon$ for some small $\epsilon > 0$. In the *second price* auction, $A$ wins and pays $b_B + \epsilon = 1 + \epsilon$ for some $\epsilon > 0$ and the revenue is $1 + \epsilon$, equivalent to the first price revenue. So, neither generates revenue anywhere close to the maximum valuation of $\max\{b_A, b_B\} = 100$. Is there a mechanism that will extract revenue close to the maximum valuation of bidders in equilibrium? What is the formal way to address this situation where valuations are vastly different? In this paper, we look at this problem in

---

* Part of this work was done when the author was at Google.

A. López-Ortiz (Ed.): LATIN 2010, LNCS 6034, pp. 565–576, 2010.
© Springer-Verlag Berlin Heidelberg 2010

a general setting of prior-free auction design, and study revenue maximization. Further, we propose a class of natural allocations and analyze them for revenue and equilibrium properties under different pricing methods.

Our motivation arises from allocation of ad slots on the Internet. Consider the example of sponsored search where when a user enters a phrase in a search engine, an auction is run among advertisers who target that phrase to determine which ads will be shown to the user. There are several instances where the underlying *value* is vastly different for the different participating advertisers. For example, the phrase "shoes" may be targeted by both high end as well as low end shoe retailers and may have vastly different values, budgets or margins in their business. Thus their bids will likely be vastly different. In another example, we have display advertising, where users who visit certain web sites are shown "display" ads like images, banners or even video. Then, depending on the history of the user — e.g., someone who is new to the website versus one who has been previously — different display advertisers value the user significantly differently, and therefore their bid values will be vastly different. In both these motivating scenarios, there are other issues to model and this paper is not a study of these applications, but rather, a study of a fundamental abstract problem inherent in these applications.

*Prior-free Auctions and Revenue Maximization.* Revenue Maximization is a central issue in mechanism design and has been studied extensively. A standard way for maximizing revenue is to derive some value profile from the bids, calculate bidder-specific reserve price, and run a second price auction [16,2,18]. In the example above, say both buyers' *value* comes from some random distribution. Then, if we know this distribution, we can calculate a *reserve price* $r$ using this distribution, and run a second-price auction with this reserve price $r$, i.e, allocate the item to the highest bidder $A$ and charge $A$ the $\max\{b_B, r\}$ if $b_A \geq r$ (else, the item remains unsold); here, $b_A$ and $b_B$'s are bids by $A$ and $B$ resp. Many such mechanisms are known; these mechanisms are incentive-compatible (that is, each bidder has no incentive to lie), and even additionally revenue-optimal, perhaps as the number of bidders goes to infinity. Such methods that rely on some assumptions over the values of bidders, i.e, that the values are drawn from some distribution (known or unknown), are called *prior-aware* mechanisms. Prior-aware mechanisms are popular in Economics. Still, from mathematical and practical point, the following questions arise:

**1.** Are there prior-free mechanisms that work independent of the value distributions of bidders?

This question is of inherent interest: what can be accomplished without knowledge of the value distributions. This is also a question that is motivated by practice. In practical applications, a way to use prior-aware mechanisms is to rely on running the same auction many times, and then use the history of bids to "machine learn" the values. Of course in practice the parameters of the auction change (users evolve), there is sparse data (query phrases are rare), advertisers strategize in complex ways and their values change over time (as they learn their own business feedbacks better), or worse, even if the machine learning methods

converge, they provide "approximate" value distributions and we need to understand the mechanisms under approximate distributions. As a result, there are
challenges in applying prior-aware mechanisms in practice and a natural question
is if they can be avoided.

**2. Are there prior-free mechanisms that work without reserve-prices?**
    This question is a more nuanced concern. First, when there is a reserve price,
the item may remain unsold in instances when $b_A < r$. This may not be desirable
in general. For example, in display ads, if an ad slot is unsold, the webpage has
to find a different template without that ad slot or fill in that space with backup
ads. Also, when the item is not sold, the outcome is not *efficient*, since the value
to the advertisers (defined the value of time to the winner) is not maximized.
And in an ever more nuanced note, advertisers do not find it transparent when
the mechanism has bidder-specific reserve prices, and often see it as a bias. This
is more so when each advertiser may get many different bidder-specific reserve
prices corresponding to different search phrases or display ad locations as implied
by the general prior-aware mechanisms above. More discussions on mechanisms
that always assign the item can be found in [14].
    *Prior-free* revenue-maximizing mechanisms have been developed for various
auction settings [7,10,15]. Lower bounds show that prior-free truthful auction
cannot achieve revenue comparable to the revenue-optimal auctions with prior
[7,10,15], and the mechanism in [15] achieves the best possible revenue among
prior-free truthful mechanisms. Still, these mechanisms work by setting reserve
prices, and do not address the second concern above.

**Our Contribution.** We study a simple, practical prior-free mechanism that
always allocates the item. In contrast to the approaches described above that
allocate the item to the highest bidder, but determine nontrivial prices, we focus
on the allocation problem and allocate the item *probabilistically*. Our contributions are as follows.
**1.** We propose a *quasi-proportional* allocation scheme where the probability that
a bidder wins the item depends (quasi-proportionally) on the bids.
    As an example, for two bidders with bids $b_A$ and $b_B$, we allocate the item
to bidder A with probability $\frac{\sqrt{b_A}}{\sqrt{b_A}+\sqrt{b_B}}$, and to B otherwise. More generally in
the presence of $n$ bidders with bid vector $(b_1, \ldots, b_n)$, we consider a continuous and concave function $w$, and set the probability of winning for bidder $i$
to $\frac{w(b_i)}{\sum_{1 \leq j \leq n} w(b_j)}$. Thus the winner of the auction is not necessarily the bidder
with the highest bid. The special case when $w(b_i) = b_i$ is known as the pro-
portional allocation scheme and has been studied previously e.g., in [11,13,9].
We study both payment methods that are common in auction theory, namely,
*all-pay* (where all bidders pay their bid no matter if they win the item or not) as
well as the *winner-pay* (only the winner pays her bid to the auctioneer) methods.
**2.** We study Nash equilibria of quasi-proportional mechanisms.
    **2.1.** We prove that the corresponding games for both all-pay and winners-pay
quasi-proportional mechanisms admit pure Nash equilibria and this equilibrium

is unique. We also give an algorithm to compute this equilibrium in polynomial time.

**2.2.** We show that the revenue of the auctioneer is promisingly high, while not losing much in the efficiency of the allocation. More precisely, we compare the revenue of such mechanisms against the ultimate: $\max_i v_i$, the highest value of any of the bidders, and show bounds on the revenue of equilibria in such mechanisms. For example, consider an auction among two bidders with values $v_A = \alpha$ and $v_B = 1$ respectively. The revenue of equilibria for both first-price and second-price auctions approaches 1. Instead, with quasi-proportional mechanisms, (i) for the all-pay mechanism with function $w(x) = x^\gamma$ where $\gamma \leq 1$, the revenue of equilibrium is $\gamma\alpha^{1-\gamma}$, and (ii) for winners-pay mechanism, where $\alpha \gg 1$, we show that the revenue of all-pay and winners-pay mechanisms with functions $w(x) = x$ and $w(x) = \sqrt{x}$ are $\Omega(\alpha^{\frac{1}{2}})$ and $\Omega(\alpha^{\frac{2}{3}})$ respectively. For the case of more than two bidders, we first show preliminary results for the revenue of various (specific) valuation vectors for the case that the number of buyers tends to $\infty$, and then we present numerical results for the revenue of equilibria for some key example functions such as $w(x) = \sqrt{x}$ and $w(x) = x$. Taken together, these results give a set of analytical and experimental tools to bound the revenue of these mechanisms against the $max_i v_i$ benchmark.

Proportional allocation, a special case of our quasi-proportional allocation, has been studied extensively in literature, in particular for efficiency analysis. But even for this rather natural allocation method, we do not know of any prior work on revenue analysis.

## 2    Preliminaries

Consider a sealed-bid auction of a single item for a set $A = \{1, \ldots, n\}$ of $n$ potential buyers. Let the value of these $n$ buyers for the single item be $v_1 \geq v_2 \geq \cdots \geq v_n$. Throughout this paper, we assume that $v_1 = \alpha \geq 1$, and $v_n = 1$. Consider a concave function $w : R \to R$ (e.g., $w(x) = \sqrt{x}$). Each buyer $i \in A$ bids an amount $b_i$ to get the item. A *quasi-proportional mechanism* allocates the item in a probabilistic manner. In particular, the item is allocated to exactly one of the buyers, and the probability that buyer $i$ gets the item is $\frac{w(b_i)}{\sum_{j \in A} w(b_j)}$. For a bid vector $(b_1, \ldots, b_n)$, let $b_{-i}$ be the bid vector excluding the bid of buyer $i$. We study the following two variants of quasi-proportional mechanisms with two payment schemes.

1. **All-pay Quasi-proportional Mechanisms.** The allocation rule in this mechanism is described above. For the payment scheme in this mechanism, each buyer pays her bid (no matter if he receives the item or not). This mechanism is ex-ante individually rational, but not ex-post individually rational. Given the above payment scheme, in the all-pay mechanism, we can write the utility of buyer $i$, as a function of the bids vector as follows:

$$u_i(b) = u_i(b_i, b_{-i}) = v_i \frac{w(b_i)}{\sum_{j \in A} w(b_j)} - b_i.$$

2. **Winners-pay Quasi-proportional Mechanisms.** The allocation rule in this mechanism is described above. For the payment scheme in this mechanism, the buyer who receives the item pays her bid, and the other buyers pay zero. This mechanism is ex-post individually rational. As a result buyer $i$'s utility as a function of the bids is

$$u_i(b) = u_i(b_i, b_{-i}) = \frac{w(b_i)}{\sum_{j \in A} w(b_j)}(v_i - b_i).$$

We are interested in Nash equilibria[1] of the above mechanisms. We consider Nash equilibria of normal-form games with complete information. In the corresponding normal-form game of the quasi-proportional mechanism, the strategy of each buyer $i$ is her bid. Formally, a bid vector $(b_1^*, \ldots, b_n^*)$ is a *Nash equilibrium* if for any buyer $i$ and any bid $b_i'$, we have $u_i(b^*) = u(b_i^*, b_{-i}^*) \geq u_i(b_i', b_{-i}^*)$.

In addition, we study efficiency and revenue of quasi-proportional mechanisms: (i) the *efficiency* of a bid vector $(b_1, \ldots, b_n)$ is the expected valuation of buyers, i.e., $\sum_{i \in A}(v_i \frac{w(b_i)}{\sum_{j \in A} w(b_j)})$, and (ii) the *revenue* of a bid vector $(b_1, \ldots, b_n)$ is the expected revenue for the auctioneer given this bid vector, i.e., $\sum_{i \in A} b_i$, in the all-pay auction, and $\sum_{i \in A}(b_i \frac{w(b_i)}{\sum_{j \in A} w(b_i)})$, in the winner-pay auction.

# 3 All-Pay Quasi-Proportional Mechanism: A Warm-Up Example

To demonstrate the kind of analyses we do, and to develop the intuition, we present a study of revenue properties of an all-pay quasi-proportional mechanism for two buyers for functions $w(t) = t^\gamma$ where $\gamma \in [0, 1]$ is a parameter. Let the bid of the first buyer be $y = b_1$ and the bid of the second buyer $x = b_2$. As mentioned earlier, we assume $v_1 = \alpha$ and $v_2 = 1$ are the valuations of the two buyers. The expected utility of the second buyer is $\frac{x^\gamma}{x^\gamma + y^\gamma} - x$, and the utility of the first buyer is $\alpha \frac{y^\gamma}{x^\gamma + y^\gamma} - y$.

For a fixed $y$, the second buyer's utility is a concave function of his bid, in the region $[0, \infty)$ and similarly, for a fixed $x$, the first buyer's utility is concave in his bid. Hence, in equilibrium, both buyers have their first derivative nullified: $\frac{\partial}{\partial x}\left(\frac{x^\gamma}{x^\gamma + y^\gamma} - x\right) = 0$, and $\frac{\partial}{\partial y}\left(\alpha \frac{y^\gamma}{x^\gamma + y^\gamma} - y\right) = 0$. Thus, we get that

$$\frac{\gamma(x)^{\gamma-1} y^\gamma}{(x^\gamma + y^\gamma)^2} = 1 \quad \text{and} \quad \frac{\alpha\gamma(y)^{\gamma-1} x^\gamma}{(x^\gamma + y^\gamma)^2} = 1$$

From which it follows that in equilibrium $\frac{y}{x} = \alpha$. Now, combining with the second equality, we get that $\frac{\alpha\gamma(\alpha x)^{\gamma-1} x^\gamma}{((1+\alpha)x^\gamma)^2} = 1$ or $\frac{\alpha\gamma(\alpha x)^{\gamma-1}}{(1+\alpha^\gamma)^2 x^\gamma} = 1$, and we get that $x = \frac{\gamma\alpha^\gamma}{(1+\alpha^\gamma)^2}; y = \alpha\frac{\gamma\alpha^\gamma}{(1+\alpha^\gamma)^2}$. Hence,

$$x + y = (1 + \alpha)\frac{\gamma\alpha^\gamma}{(1 + \alpha^\gamma)^2} \xrightarrow{\alpha \to \infty} \gamma\alpha^{1-\gamma}.$$

---

[1] Throughout this paper, we study *pure* Nash equilibria and not mixed NE.

Moreover, as $\frac{y}{x} = \alpha$, the probability that buyer 2 receives the item is $\frac{1}{1+\alpha^\gamma}$, and otherwise buyer 1 gets the item. Thus, the efficiency of this mechanism is $\frac{1+\alpha^{\gamma+1}}{1+\alpha^\gamma}$. In particular, as $\alpha \to \infty$, the efficiency is arbitrarily close to $\alpha$. The most efficient allocation rule is to assign the item to buyer 1, and get efficiency $\alpha$. That completes the analysis and shows that

**Theorem 1.** *The all-pay quasi-proportional mechanism with two buyers guarantees a total revenue of $(1 + \alpha)\frac{\gamma\alpha^\gamma}{(1+\alpha^\gamma)^2}$ and expected efficiency of $\frac{1+\alpha^{\gamma+1}}{1+\alpha^\gamma}$ in equilibrium. In particular, for a large enough $\alpha$, the revenue is $\gamma\alpha^{1-\gamma}$ and efficiency is arbitrarily close to $\alpha$.*

## 4   Equilibrium: Existence and Uniqueness

In this section, we establish the existence and uniqueness of Nash equilibria of both the all-pay and winners-pay quasi-proportional auctions.

**Definition 1 (from [6]).** *A game is* socially concave *if the following holds:*

1. *There exists a strict convex combination of the utility functions which is a concave function. Formally, there exists an $n$-tuple $(\lambda_i)_{i\in A}$ , $\lambda_i > 0$, and $\sum_{i\in A} \lambda_i = 1$, such that $g(x) = \sum_{i\in A} \lambda_i u_i(x)$ is a concave function in $x$.*
2. *The utility function of each buyer $i$, is convex in the actions of the other buyers. I.e., for every $s_i \in S_i$ the function $u_i(s_i, x_{-i})$ is convex in $x_{-i} \in S_{-i}$, where $S_i$ is the strategy space of agent $i$, and $S_{-i} = \prod_{j\in A, j\neq i} S_j$.*

Rosen [17] defined the diagonal concavity property for concave games, and showed that when it holds, the Nash equilibrium of the game is unique. Even Dar et al [6] showed that if one of the properties 1 and 2 holds with strict concavity or convexity, respectively, then the diagonal concavity property holds. Now, we show that a quasi-proportional auction is a socially concave game. The uniqueness of Nash equilibrium would follow as a corollary of [17] and [6].

**Lemma 1.** *Let $\Gamma = (A, \{u_i\}_{i\in A})$ be an all-pay quasi-proportional auction, with utility functions for buyer $i$, $u_i()$ defined as above and assume that the weight function $w()$ is a concave function, and that the strategy of each buyer is restricted to a compact set $[B_{\min}, B_{\max}]$, where $0 < B_{\min} < B_{\max} < \infty$. Then $\Gamma$ is a socially-concave game.*

A similar lemma holds for winner-pay auctions, with weight function of the form $w(x) = x^\gamma$, where $0 < \gamma \leq 1$.

**Lemma 2.** *Let $\Gamma = (A, \{u_i\}_{i\in A})$ be an winner-pay quasi-proportional auction, with utility functions for user $i$, $u_i()$ defined as above and assume that the weight function $w(x) = x^\gamma$, where $0 < \gamma \leq 1$, and that the strategy of each user is restricted to a compact set $[B_{\min}, B_{\max}]$, where $0 < B_{\min} < B_{\max} < \infty$. Then $\Gamma$ is a socially concave game.*

## 5   Revenue of Quasi-Proportional Mechanisms

In section 3, we computed the revenue of all-pay quasi-proportional mechanisms for two buyers, and functions $w(x) = x^\gamma$. In this section, we first observe general properties for the revenue of equilibria of quasi-proportional mechanisms. Then, we focus on two special functions and prove tight bounds on the revenue of the winners-pay mechanisms. The utility function $u_i(b_i, b_{-i})$ for both all-pay and winners-pay mechanisms is a strictly concave function of $b_i$ in the region $[0, \infty]$ (as it is a concave function minus a convex function). As a result, in an all-pay quasi-proportional auction, we have: $\frac{\partial}{\partial b_i} \left( \frac{w(b_i)}{\sum_{i \in A} w(b_i)} v_i - b_i \right) = 0$. For a bid vector, $(b_1, b_2, \ldots, b_n)$, let $\sigma(b) = \sum_{i \in A} w(b_i)$. When clear from context, we let $\sigma = \sigma(b)$. As a result, in equilibrium,

$$\frac{\partial}{\partial b_i} \left( \frac{w(b_i)}{\sigma} v_i - b_i \right) = 0$$

From which we derive:

$$v_i = \frac{\sigma^2}{w'(b_i)(\sigma - w(b_i))} \tag{5.1}$$

Similarly, for winners-pay quasi-proportional mechanisms, the bid of each buyer $i$ satisfies the following:

$$\frac{\partial}{\partial b_i} \left( \frac{w(b_i)(v_i - b_i)}{(\sigma - w(b_i)) + w(b_i)} \right) = 0$$

From which it follows that

$$v_i = b_i + \frac{w(b_i)\sigma}{w'(b_i)(\sigma - w(b_i))} \tag{5.2}$$

We will use equations 5.1 and 5.2 in studying the revenue of the equilibrium for various functions. In both equations 5.1 and 5.2 for increasing concave functions such as $w(x) = \sqrt{x}$, the value of $v_i$ increases as $b_i$ increases, i.e, fixing $b_{-i}$ $v_i$ is monotonically increasing in terms of $b_i$. This observation leads to the following fact: For increasing and concave functions $w$, if $v_1 \geq v_2 \geq \ldots \geq v_n$, in the equilibrium bid vector $(b_1^*, b_2^*, \ldots, b_n^*)$, we have $b_1^* \geq b_2^* \geq \ldots b_n^*$.

### 5.1   Revenue for Winners-Pay: Two Bidders

Here, we study winners-pay proportional mechanism for $w(x) = x$. The utility of bidder $i$ as a function of the bids is

$$u_i(b) = \frac{b_i}{\sum_{j \in A} b_j} (v_i - b_i).$$

Given this utility function, it is easy to see that for $v_i > 0$, in equilibrium $b_i > 0$. Let's fix $b_{-i} \neq 0$. In equilibrium, for every $i$ with bid $b_i > 0$,

$$\frac{\partial}{\partial b_i} u_i(b_i, b_{-i}) = -\frac{b_i(v_i - b_i)}{(\sum_{j \in A} b_j)^2} + \frac{v_i - b_i}{\sum_{j \in A} b_j} - \frac{b_i}{\sum_{j \in A} b_j} = -1 + \frac{(\sum_{j \neq i} b_j)(\sum_{j \neq i} b_j + v_i)}{(\sum_{j \in A} b_j)^2}$$

and we get that in equilibrium,

$$b_i = \sqrt{(\sum_{j \neq i} b_j)(\sum_{j \neq i} b_j + v_i)} - \sum_{j \neq i} b_j, \text{ for every } i \in A. \tag{5.3}$$

The revenue from the proportional mechanism as described above is

$$\sum_{i \in A} \Pr[\text{agent } i \text{ wins}] \cdot b_i = \sum_{i \in A} \frac{b_i}{\sum_{j \in A} b_j} b_i = \frac{\sum_{i \in A} b_i^2}{\sum_{j \in A} b_j} \tag{5.4}$$

Consider a setting of two buyers with values $v_1, v_2$. We can, without loss of generality, assume that $v_2 = 1$.

**Theorem 2.** *In the case of two buyers, the revenue from the winners-pay proportional mechanism is $O(\sqrt{\alpha})$, where $\alpha = \max(v_1, v_2)$. Moreover, for arbitrarily large $\alpha$, the efficiency of this mechanism is arbitrarily close to $\alpha$.*

A similar technique can be used for showing a lower bound on the revenue in quasi-proportional winner-pay auctions, with weight function $w(x) = \sqrt{x}$, which asymptotically yields a higher revenue. The proof is left to the appendix.

**Theorem 3.** *The revenue from the winners-pay mechanism for two bidders, with weight function $w(x) = \sqrt{x}$ is $O(\alpha^{2/3})$, where $\alpha = \max(v_1, v_2)$. Moreover, for arbitrarily large $\alpha$, the efficiency of this mechanism is arbitrarily close to $\alpha$.*

We will give numerical results for revenue of other settings like $w(x) = x^{1/4}$ in Section 6. In a full version of this paper we consider other functions as well, e.g., we prove an upper bound on the revenue of both all-pay and winners-pay mechanisms for $w(x) = \log(x + 1)$, and show that the revenue is not more than $\frac{\alpha}{\log(\alpha)}$.

## 5.2   Revenue for Many Buyers

Here, we analyze the revenue for two special valuation vectors for $n$ bidders, i.e, (i) uniform valuation vector, $v_i = V$, and (ii) valuation vector $v_1 = \alpha$, and for $i \neq 1$, $v_i = 1$ for $i \in A$. The second type of valuation is important as it captures examples in which there is a large gap between the highest valuation and value of other buyers.

**Theorem 4.** *For the uniform valuation vector where $v_i = V$ for all $i \in A$, the revenue in the equilibrium for function $w(x) = x^\gamma$ is $\frac{n-1}{n} \gamma V$ for all-pay mechanism, and is $V(\frac{1}{1+(\frac{n}{n-1})\gamma})$ for winners-pay mechanism. Moreover, the equilibrium revenue for uniform valuation vector for function $w(x) = \log(x+1)$ for both all-pay and winners-pay mechanisms is asymptotically $\frac{V}{\log V}$ as $V, n \to \infty$.*

**Theorem 5.** *For the valuation vector* $(\alpha, 1, 1, \ldots, 1)$*, the revenue in the equilibrium of winners-pay quasi-proportional mechanism converges to a constant as* $n$ *goes to* $\infty$ *for a fixed* $\alpha$*. Moreover the revenue of all-pay quasi-proportional mechanism for function* $w(x) = x^\gamma$ *goes to zero as* $n$ *goes to* $\infty$ *for a fixed* $\alpha$*.*

The above theorem shows some bounds on the revenue for a fixed $\alpha$ and as $n$ tends to $\infty$. It would be interesting to understand the trade-off between the revenue for large $\alpha$ and $n$. In particular, it would be interesting to compute the revenue for a fixed $n$ as $\alpha$ tends to $\infty$.

# 6    An Efficient Algorithm and Numerical Study

In this section, we present an efficient algorithm for computing Nash equilibria of quasi-proportional mechanisms and then using this algorithm, we present a family of plots showing the quality of the mechanisms.

## 6.1    A Polynomial-Time Algorithm for Equilibrium Computation

In [6], Even Dar et. al. describe a natural process that converges to a Nash equilibrium in every socially concave game. This method is useful for computing Nash equilibrium of the all-pay and winner-pay auctions. The process considered is known as *no-regret* dynamics. Informally, a buyer's update process is said to have no-regret, if in the long-run, it attains an average utility which is not significantly worse than that of the best fixed action in hindsight (in the context of auctions, the best fixed bid). Even Dar et. al. show that if every buyer uses an update process with no-regret property, in a repeated socially concave game, the joint average action profile converges to a Nash Equilibrium. Many efficient algorithms for attaining the no-regret property (also known as no-external-regret), exist [20,1,12]. In order to compute a Nash equilibrium of the all-pay auction, and the winner-pay auction, one could simulate the process of running a no-regret algorithm for every buyer that participates in the auction. The rate at which the average vector of bids converges to Nash equilibrium, depends on the vector $\lambda$, which existence is guaranteed in property 1. In particular, there exists no-regret algorithms (e.g., [20]), such that the rate of convergence to Nash equilibrium, for the quasi-proportional mechanisms, is $O(\frac{n}{\sqrt{t}} \frac{\sum_{j \in A} v_j}{v_{\min}})$, (I.e., at time $t$ of the simulation process, the average bids vector is an $\epsilon^t$-Nash equilibrium, where $\epsilon^t = O(\frac{n}{\sqrt{t}} \frac{\sum_{j \in A} v_j}{v_{\min}})$. Algorithm 1 describes the simulation of running simultaneous no-regret for every buyer, where the actual no-regret algorithm used is GIGA [20].

## 6.2    Numerical Revenue Computation

In this section, we present numerical results for the revenue of the all-pay and winners-pay quasi-proportional auctions with different weight functions and different number of buyers. Figures 1-4 describe the revenue as a function of the

**Algorithm 1.** Algorithm for computing NE bids for the quasi-proportional auction.

---

**Input:** a vector $v = \{v_1, v_2, \ldots, v_n\}$.
**Output:** an $\epsilon$-NE, $b_1, \ldots, b_n$.

Set $b^0 \leftarrow (1, 1 \ldots, 1)$
  **for** $t = 1$ **to** $T = O(\frac{n}{\epsilon} \frac{\sum_{j \in A} v_j}{v_{\min}})$ **do**
    **for all** $i \in A$ **do**
      $y_i^t \leftarrow b_i^{t-1} + \frac{1}{\sqrt{t}} \frac{\partial}{\partial b_i} u_i(b)$
      **if** $y_i^t > v_i$ **then**
        $b_i^t \leftarrow v_i$
      **else**
        $b_i^t \leftarrow \max(y_i^t, 0)$
      **end if**
    **end for**
    $t \leftarrow t + 1$
  **end for**
  return $b$

---

highest value for the item, over all the bidders, denoted by $\alpha$. Figure 1 describes the revenue in an all-pay auction with two bidders — one bidder has a 'high' value $\alpha \geq 1$, and the other bidder has a value of 1. We consider two versions of the all-pay auctions. In the first, we used a weight function $w(z) = \sqrt{z}$, and in the second we used a weight function $w(z) = z^{\frac{1}{4}}$. Next, in Figure 2, we consider the same setting as in Figure 1, for the winners-pay auction. The revenue in equilibrium is presented for three different versions of the winners-pay auction: The lowest curve describes the winner pay auction with the linear weight function $w(z) = z$. The middle curve describes the revenue when the weight function is $w(z) = \sqrt{z}$ and the upper curve describes the revenue when the weight function is $w(z) = z^{1/4}$.

In Figures 3, and 4 we study numerically the revenue in a winners-pay auction when the number of bidders varies from $n = 2$ to $n = 5$. The bidders' private values are such that a single bidder has a high value $\alpha \geq 1$, and the other $n - 1$ bidders have a low value of 1. Each curve in Figures 3,4 describes the revenue in equilibrium as a function of $\alpha$, and each different curve corresponds to a different number $n$ of bidders. Figure 3 and 4 differ in the weight function used: in Figure 3 we used $w(z) = z$, and in Figure 4 we used $w(z) = \sqrt{z}$. In Theorem 5, we show that the revenue in a winners-pay auction, with values profile $(\alpha, 1, 1, \ldots, 1)$ asymptotically goes to a constant, as the number of bidders with value 1 tends to $\infty$. It is interesting however to notice that in both Figures 3 and 4, while the number of bidders is kept relatively small, the revenue actually increases with the number of low-value bidders.

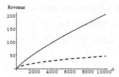

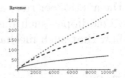

**Fig. 1.** Revenue from equilibrium bids in an all-pay auction with two bidders with values $\alpha$, and 1 respectively. The lower curve describes an all-pay auction with weight function $w(x) = \sqrt{x}$. The upper curve describes an all-pay auction with weight function $w(x) = x^{1/4}$.

**Fig. 2.** Revenue from equilibrium bids in a winners-pay auction with two bidders with values $\alpha$, and 1 respectively. The lower, middle, and upper curves describes a winners-pay auction with weight functions $w(x) = x$, $w(x) = \sqrt{x}$, and $w(x) = x^{1/4}$ respectively.

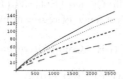

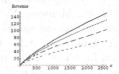

**Fig. 3.** A winners-pay auction with weight $w(x) = x$, and value profile $(\alpha, 1, \ldots, 1)$. The curves from lowest to highest describe the revenue when the number of bidders with value 1 is 1,2,3,4 respectively

**Fig. 4.** A winners-pay auction with weight $w(x) = \sqrt{x}$, and value profile $(\alpha, 1, \ldots, 1)$. The curves from lowest to highest describe the revenue when the number of bidders with value 1 is 1,2,3,4 respectively.

## 7  Concluding Remarks

We study a natural class of quasi-proportional allocation mechanisms. Combined with all-pay or winner-pay methods, this gives a simple prior-free auction mechanism without any reserve prices. Our analytical and experimental study shows the revenue under various quasi-proportional functions in equilibrium, and we showed existence of a unique Nash equilibrium that can also be computed efficiently. We believe quasi-proportional mechanisms will find applications and a deeper understanding of their properties will be useful. An interesting open question is to design an auction for a single item that achieves a total revenue of constant factor of $\alpha = \max_i v_i$ in equilibria. We proved that simple quasi-proportional mechanisms show promising revenue properties in equilibria, however none of our mechanisms achieve a constant approximation factor of $\alpha$ (off by at least facor $\log \alpha$). A main open problem is to design a mechanism for a single item that achieves a constant factor of $\alpha$ in equilibria while not losing much in the efficiency of the allocation. Also as we discussed in Section 5.2, the promising revenue properties of quasi-proportional mechanisms for small number of buyers disappears as the number of buyers tends to $\infty$. An interesting open question is to modify the mechanism to ensure good revenue properties when many buyers are in the system. A simple idea is that for any number of bidders, the auctioneer runs a quasi-proportional mechanism among the highest two bids. One hopes such mechanisms have good revenue properties, however, we can show that such mechanisms may not admit any pure Nash equilibria.

# References

1. Abernethy, J., Hazan, E., Rakhlin, A.: Competing in the Dark: An Efficient Algorithm for Bandit Linear Optimization. In: COLT 2008 (2008)
2. Baliga, S., Vohra, R.: Market Research and Market Design (2003)
3. Baye, M., Kovenock, D., de Vried, C.: The all-pay auction with Complete Information. Economic Theory 8, 291–305
4. Che, Y., Gale, I.: Expected revenue of all-pay auctions and first-price sealed-bid auctions with budget constraints. Economic Letters, 373–379 (1996)
5. Clarke, E.: Multipart pricing of public goods. Public Choice 11, 17–33 (1971)
6. Even Dar, E., Mansour, Y., Nadav, U.: On the convergence of regret minimization dynamics in concave games. In: STOC 2009 (2009)
7. Fiat, A., Goldberg, A.V., Hartline, J.D., Karlin, A.R.: Competitive generalized auctions. In: STOC 2002, pp. 72–81 (2002)
8. Groves, T.: Incentives in teams. Econometrica 41(4), 617–631 (1973)
9. Hajek, B., Gopalakrishnan, G.: Do greedy autonomous systems make for a sensible internet? Presented at the Conference on Stochastic Networks, Stanford University (2002)
10. Hartline, J., Karline, A.: Profit Maximization in Mechanism Design. In: Algorithmic Game Theory (October 2007)
11. Johari, R., Tsitsiklis, J.N.: Efficiency loss in a network resource allocation game. Mathematics of Operations Research 29(3), 407–435 (2004)
12. Kalai, A., Vempala, S.: Efficient algorithms for online decision problems. J. Comput. Syst. Sci. 71(3), 291–307 (2005)
13. Kelly, F.: Charging and rate control for elastic traffic. European Transactions on Telecommunications 8, 33–37 (1997)
14. Liu, D., Chen, J.: Designing online auctions with past performance information. Decision Support Systems 42, 1307–1320 (2006)
15. Lu, P., Teng, S.-H., Yu, C.: Truthful Auctions with Optimal Profit. In: Spirakis, P.G., Mavronicolas, M., Kontogiannis, S.C. (eds.) WINE 2006. LNCS, vol. 4286, pp. 27–36. Springer, Heidelberg (2006)
16. Myerson, R.: Optimal auction design. Mathematics of Operations Research 6, 58–73 (1981)
17. Rosen, J.: Existence and uniqueness of equilibrium points for concave n-person games. Econometrica, 520–534 (1965)
18. Segal, I.: Optimal Pricing Mechanisms with Unknown Demand. American Economic Review 93(3), 509–529 (2003)
19. Vickrey, W.: Counterspeculation, auctions and competitive-sealed tenders. Finance 16(1), 8–37 (1961)
20. Zinkevich, M.: Online convex programming and generalized infinitesimal gradient ascent. In: Twentieth International Conference on Machine Learning (2003)

# Some Observations on Holographic Algorithms

Leslie G. Valiant*

School of Engineering and Applied Sciences
Harvard University
valiant@seas.harvard.edu

**Abstract.** We define the notion of diversity for families of finite functions, and express the limitations of a simple class of holographic algorithms in terms of limitations on diversity. We go on to describe polynomial time holographic algorithms for computing the parity of the following quantities for degree three planar undirected graphs: the number of 3-colorings up to permutation of colors, the number of connected vertex covers, and the number of induced forests or feedback vertex sets. In each case the parity can be computed for any slice of the problem, in particular for colorings where the first color is used a certain number of times, or where the connected vertex cover, feedback set or induced forest has a certain number of nodes. These holographic algorithms use bases of three components, rather than two.

## 1 Introduction

The theory of holographic algorithms is based on a notion of reduction that enables computational problems to be interrelated with unusual fluidity. The theory offers three basic reduction techniques:

(a) *Holographic transformations* that relate pairs of problems by simply taking a different view or *basis*,

(b) *Holographic gadgets* that use internal cancelations custom designed for the problems at hand, and

(c) *Interpolation techniques* for recovering information from the outputs of computations on a set of specially prepared variants of the problem instance at hand.

The overarching open question in the theory is whether this combination of techniques can bridge the gap between classical polynomial algorithms on the one hand, and the class of #P-complete (or NP- or ⊕P-complete) problems as defined by classical reductions, on the other.

In order to further our understanding of this question we introduce here the notion of *diversity* for finite functions, in terms of which some limitations of the simplest kinds of holographic algorithms that we discussed in an earlier paper [V06] can be explored more explicitly. These simplest holographic algorithms are those obtained from what we define as *elementary* reductions. We show that

---

* This work was supported in part by NSF-CCF-04-27129.

A. López-Ortiz (Ed.): LATIN 2010, LNCS 6034, pp. 577–590, 2010.

such algorithms do impose a limitation on the diversity of the functions that can be realized. It remains unresolved, however, whether holographic algorithms that are not bound by the constraints of elementarity, such as those given in later sections of this paper, can evade this diversity limitation.

In the later sections we go on to describe some polynomial time holographic algorithms for three natural problems for undirected graphs of degree three. These compute the parity of the number of solutions of each of the following three problems: feedback vertex sets (or, equivalently, induced forests), connected vertex covers, and vertex 3-colorings up to permutations of colors.

Besides evading the elementarity constraint our algorithms have other features that put them outside the currently better understood regions of holographic theory. For one thing the use of the three element basis **b3** from [V08] puts them outside the collapse theorem of Cai and Lu [CL09], and hence outside any known classification such as [CL07]. (Of course, the possibility has not yet been excluded that #P-complete problems can be solved even within the scope of this collapse theorem or classification.) Second, the results hold for parity rather than counting. For parity problems, or fixed finite fields in general, holographic transformations and interpolation both appear to offer less flexibility than they do for general counting problems. For example, understanding the complexity of the counting problems modulo three, for the structures we analyze here modulo two, appears to remain a challenge.

For brevity of exposition we shall assume familiarity with the basic notions and notations of holographic algorithms as described in ([V04], [V08]).

## 2   Diversity

For a Boolean function $f(x_1, ..., x_m)$ and a subset $S \subseteq X = \{x_1, ..., x_m\}$ of size $n$, we define the *diversity of $S$ in $f$* to be the logarithm to the base two of the number of different functions of the $n$ variables of $S$ that can be obtained by fixing the $m - n$ remaining variables $X - S$ in the $2^{m-n}$ different ways. This is the central concept in Neciporuk's proofs [N66] of lower bounds on formula complexity. He showed that if $X$ has a partition into subsets $S_i$ such that the average diversity of the $S_i$ in $f$ is substantial, then a nonlinear lower bound on the formula size of $f$ follows.

We say that a Boolean function $f(x_1, ..., x_m)$ has *$n$-diversity $D$* if $D$ is the maximum diversity of $S$ in $f$, over all subsets $S \subseteq X = \{x_1, ..., x_m\}$ of size $n$ . Since there are $2^{2^n}$ Boolean functions of $n$ variables, the maximum $n$-diversity of a function is $2^n$.

A Boolean function family $f = \{f_m(x_1, ..., x_m) \mid m = 1, ...\}$ has *diversity $g(n)$* if for each positive integer $n$, $g(n)$ is the maximum $n$-diversity of $f_m$ for any $m \geq n$. Clearly $g(n) \leq 2^n$.

A Boolean function family $f$ has *polynomial diversity* if its diversity $g(n)$ is upper bounded by some polynomial $p(n)$. It has *exponential diversity* if it is lower bounded by $c2^{cn^\kappa}$ for some constants $c, \kappa > 0$. It has *exponential standard diversity* if, for some polynomial $p(n)$, exponential diversity is achieved for all $n$

by $f_m$ with some $m \leq 2^{p(n)}$. It has *polynomial standard diversity* if, for all polynomials $p(n)$, the $n$-diversity achieved for $n$ by $f_m$ with $m \leq 2^{p(n)}$ is polynomial bounded.

Such definitions can also be made for finite fields $F_q$ for families with $f_i$: $\{0, ..., q-1\}^i \mapsto \{0, ..., q-1\}$. In that case the maximum $n$-diversity of a family is $q^n \log_2 q$.

High diversity does not imply high complexity. The Circuit Value problem [L75] $CV_{n,r}(x_1, ..., x_{n+r})$ we shall formulate here as the function that regards its first $n$ inputs as a vector $v$ of $n$ Boolean values, and the remaining bits as a specification of a Boolean circuit $C$ of $n$ inputs with binary gates. An $m$ gate circuit of $n$ inputs can be specified using $r = O((n+m)\log(n+m))$ binary bits. Since all the $2^{2^n}$ Boolean functions of $n$ variables can be realized by a circuit with $O(2^n/n)$ gates [Lu58], they can all be encoded in $CV_{n,r}$ if $r = O(2^n)$. We define CV to have such an encoding of circuits. Hence CV has diversity $2^n$ since $S = (x_1, ..., x_n)$ has diversity $2^n$ in $CV_{n,r}$ for an appropriate $r = O(2^n)$. Clearly CV then also has exponential standard diversity.

Using the following notion of reduction one can deduce that most natural P-, NP- and #P-complete problems have exponential standard diversity. We say that a reduction $\tau$ from $\{CV_{n,r}\}$ to a family of functions $\{Q_i\}$ is *segregating* if in polynomial time $\tau$ maps the pair $v, C$ to a pair of Boolean sequences $(y, z)$ such that (i) for any fixed $n$ and $r$, the lengths of $y$ and of $z$ are uniquely determined, (ii) the length of $y$ is polynomially bounded in terms of $n$, (iii) $y$ depends on $v$ and not $C$, (iv) $z$ depends on $C$ and not $v$, and (v) $Q_i(y, z) = C(v)$. (In short, $y$ encodes $v$, $z$ encodes $C$, and $Q_i$ evaluates $C$ on $v$. The length of $y$ is polynomial in $n$, but the length of $z$ may be exponential in $n$.)

**Proposition 1.** *If CV is reducible to Q by a segregating reduction then Q has exponential standard diversity.*

*Proof.* For a fixed size $n$ consider $CV_{n,r}$ with $r$ exponential in $n$ and large enough that all $2^{2^n}$ Boolean functions of $n$ variables can be expressed. Now consider *one* of the $2^{2^n}$ choices of $C$. Since the reduction, say $\tau$, is segregating, for all $v$ it will map $(v, C)$ to $(y, z)$ for the some fixed value of $z$. For $C$ and $z$ so fixed, as $v$ varies so will $y$, and $Q_i(y, z) = C(v)$. Hence, $Q_i$ will compute on the encoding $y$ of $v$ the same Boolean function as $C$ does on $v$. Hence, fixing $z$ in different ways will make $Q_i$ compute $2^{2^n}$ different functions of $y$. If $S$ is the set of variables that represents $y$ then the diversity of $S$ in $Q_i$ will be $2^n$. Since $\tau$ is segregating, by condition (ii) $|S|$ is polynomially bounded in terms of $n$. It follows that the diversity of $S$ in $Q_i$ will be at least $c2^{c|S|^{1/\kappa}}$ for appropriate positive constants $c$ and $\kappa$. $\qquad \square$.

Now for many NP-complete problems, by tracing through the known reductions, one can derive segregating reductions from CV to them. For example, consider the family $Q$ corresponding to Cook's 3SAT problem. Here $Q_i(x)$ is a 3CNF formula with $i$ clauses and variables from $x_1, ..., x_i$. From a circuit $C$ with inputs $x_1, ..., x_n$, and any vector $v$ of values of $x_1, ..., x_n$, one can construct by now standard methods a polynomial size 3CNF formula that is satisfiable if and only if that circuit $C$ on that input $v$ evaluates to one: The formula will have the first $n$ clauses encode the input with the $j^{th}$ clause being $(x_j)$ or $(x'_j)$ according to whether the $j^{th}$ among the $n$ bits of $v$ is 1 or 0. It will have the remaining clauses encode the gates. This is a segregating polynomial time reduction from CV to 3CNF. Related to the 3CNF satisfiability problem is $\oplus$3CNF, the problem of determining the parity of the number of solutions of a 3CNF formula, and their planar analogs Pl-3CNF and $\oplus$Pl-3CNF. From the above construction we can deduce the following.

**Proposition 2.** *The problems 3CNF, $\oplus$3CNF, Pl-3CNF and $\oplus$Pl-3CNF all have exponential standard diversity.*

*Proof.* The previous paragraph describes a segregating reduction from CV to 3CNF. This establishes the result for 3CNF by virtue of Proposition 1. Since the construction can be made to preserve the number of solutions, the 3CNF formula will have 0 or 1 solutions according to whether the value output by the circuit $C$ is 0 or 1. The result for $\oplus$3CNF therefore also follows. For the planar case one uses additional sets of clauses that act as crossovers and make the formula planar, as described by Lichtenstein [L82]. These can also be made to preserve the number of solutions [HMRS98]. These additional clauses can be viewed as part of the circuit encoding, and then yield a segregating reduction to the planar versions of Pl-3CNF and $\oplus$Pl-3CNF as needed.    □

With this starting point one can ask for each of the known NP-complete problems, such as those of Karp [K72], whether CV is reducible to some natural encodings of them by a segregating reduction. It appears that this is the case for the vast majority, and for those it then follows that some natural encoding of them has exponential standard diversity.

What is the status of the numerous counting problems that are known to be complete in an appropriate counting class, but for which existence is polynomial time computable and not known to be complete for P? Do these counting problems have CV embedded in them equally explicitly? The following shows that in some such cases the embedding is in fact explicit.

**Proposition 3.** *The Permanent modulo $k$ for any prime $k \neq 2$ has exponential standard diversity.*

*Proof.* From the proof in [V79a] one can obtain a segregating reduction from CV via 3CNF to the permanent modulo $k$ for any prime $k$ other than two.    □

On the other hand, if the proof of #P-completeness goes through interpolation (eg [V79b], [J87], [Vad01], [XZZ07], [CLX08]) then exponential diversity does not appear to follow immediately. For example, does counting matchings modulo 3 in some natural encoding of planar graphs [J87], have exponential diversity? There are cases in which the known reductions to the counting problem take particularly circuitous routes through interpolations, raising the possibility that the CV problem is truly disguised, but nevertheless exponential diversity can be deduced from known reductions for the corresponding parity problem. One example of this is planar vertex cover for which known #P completeness proofs [Vad01, XZZ07] are indirect. However, for the subclass of planar 3/2-bipartite graphs (bipartite graphs with degree 2 on one side and 3 on the other) a segregating reduction from 3CNF to this vertex cover problem that preserves the parity of the number of solutions can be derived from the the ⊕P-completeness proof of this problem given in [V06].

**Proposition 4.** *The parity of the number of vertex covers for planar 3/2 graphs has exponential standard diversity.*

Clearly a unary Boolean function family, one that is zero whenever any $x_i = 1$, will have polynomial diversity. Since there exist unary functions of arbitrarily high Turing machine time complexity it follows that polynomial diversity does not imply polynomial time Turing machine complexity. Pavel Pudlak has made the following elegant observation that shows that low diversity can also be possessed by functions that have high complexity in many other senses, such as having exponential circuit complexity or being NP-complete. For any function $f(x)$ on $n$ inputs consider an error correcting code $g: \{0,1\}^n \rightarrow \{0,1\}^r$ that corrects more than $n$ errors and can be computed and inverted efficiently. Let $h : \{0,1\}^r \rightarrow \{0,1\}$ be such that $h(y) = 1$ iff $y = g(x)$ for some $x$ and $f(x) = 1$. Then $h$ has $n$-diversity at most $n + 1$ since for any domain $d$ of $n$ of its input bits, fixing the remaining bits will permit it to have value 1 for at most 1 of the $2^n$ values of $d$, and hence there are at most $2^n + 1$ such different functions possible of the $n$ variables of $d$.

# 3   Elementary Reductions to Matchgrids

We shall now define the notion of an *elementary reduction to matchgrids*. The definition is a generalization of the one given in [V06] that was specific to reductions from 3CNF. Here we consider a *family f with respect to input domain d*. This means that for each $m$ there is a specified function $f_m$ on Boolean variables $x_1, ..., x_m$, and a specified subset $d_m$ of $n_m$ of these $m$ variables.

   We say that $\tau$ is a $k(m)$-*oracle* reduction if for family member $f_m$ it generates $k(m)$ matchgrids in polynomial time and from their Holants it computes the solution to the original problem also in polynomial time. Note that while many holographic algorithms in the literature are 1-oracle, some multi-oracle matchgrid reductions that use interpolation have been described also ([V04], [V08], [CC07]).

   Suppose that $\tau$ is a polynomial time 1-oracle reduction from $f$ to matchgrids over field $F$. For a function $f_m$, a domain of its variables $d_m$ of size $n = n_m$, and an assignment $z$ to the set $c_m$ of variables that is the complement of $d_m$, let $M(f_m, d_m, z)$ be the set of $2^n$ adjacency matrices of the set of matchgrid images under $\tau$ of the $2^n$ restrictions of $f_m$ when the $n = n_m$ variables specified by $d_m$ are fixed in all possible ways.

Then $\tau$ is a *local boundary reduction for* $(f_m, d_m)$ if for each $m$, $z$ and adjacency matrix in $M(f_m, d_m, z)$ there is a planar embedding such that for every fixed $m$ and $z$

(a) the corresponding $2^n$ embeddings have an identical set of nodes $U_{m,z}$.
(b) in these $2^n$ embeddings all the edges and their weights are identical, except possibly those that have both endpoints within a subset $Z_{m,z} \subseteq U_{m,z}$, which is of size upper bounded by a polynomial $L(n)$ independent of $m$.
(c) the $Z_{m,z}$ nodes have degrees bounded by a constant independent of $n$, $m$ or $z$.
(d) the nodes $Z_{m,z}$ all lie in the infinite outer face of the embedding of the graph induced by $T_{m,z} = U_{m,z} - Z_{m,z}$, and
(e) the edges incident to pairs of vertices in $Z_{m,z}$ can be partitioned into $n$ sets such that each such set $S_i$ corresponds to a variable $x_{k_i}$ in $d_m$, and the weights of $S_i$ are functions of the value of $x_{k_i}$ but are independent of the values of the other $x_j$ in $d_m$ (i. e. those with $j \neq k_i$.)

We shall say that *a reduction* $\tau$ *from a family* $f$ *with respect to input domain* $d$ *of size* $n$ *to matchgrids over* $F$ *is* elementary if it has the four properties of (i) being 1-oracle, (ii) being local boundary, (iii) having the number of members $|F|$ polynomial bounded in $m$, and (iv) having $\mathrm{Holant}(\tau(f(x)))$ determine the value of $f(x)$. We then also say that $\tau$ is an *elementary reduction for* $(f, d)$.

We note that there are no constraints on what the transformation does on different $z$. The intent is that when the Circuit Value problem is embedded, then the different circuits $C$ can be embedded in arbitrarily different ways, but for any one circuit there are constraints on the way the matchgrids can vary as the inputs $v$ to $C$ vary. Also note that the field size is allowed to grow polynomially with $m$, so that it can be exponential in $n$ if $m$ is exponential in $n$.

## 4   Elementary Reductions Compute Functions of Polynomial Diversity

Our negative results are based on the following statement [V06]:

**Theorem 1.** *There is a weighted planar graph* $G$ *having* $r$ *external nodes and* $O(r^4)$ *edges of which all but* $O(r^2)$ *have fixed weight 1, such that for any field* $F$, *any* $r$-*component standard signature that is realized by some matchgrid can be realized by* $G$ *by setting the* $O(r^2)$ *variable-weight edges to appropriate constants.*

*Proof.* For $r \leq 4$ this result is proved in [V08] (Propositions 6.1-6.3). For general $r$ the corresponding result was proved by Cai, Choudhary and Lu (Corollary 4.1 in [CCL09]), but for matchcircuits. The general result then follows from the equivalence of matchgrids and matchcircuits (Lemmas 3.1 and 3.2 in [CC07]). $\square$

**Theorem 2.** *For any family* $f$ *and domain* $d$ *if there is an elementary reduction to matchgrids for* $(f, d)$, *then* $d$ *has polynomial standard diversity in family* $f$.

*Proof.* Suppose that $\tau$ is a polynomial time 1-oracle reduction from $f$ to matchgrids over field $F$. For a function $f_m$ and a domain of its variables $d_m$ of size

$n = n_m$, for each $z$ let $M(f_m, d_m, z)$ be the set of adjacency matrices as defined above and consider the planar embeddings that respect conditions (a)-(e) of the definition of elementarity.

By (a), (b) these embeddings are identical with respect to all the edges that are incident to a $T_{m,z}$ node at laest at one end . We regard the embedding of the nodes $T_{m,z}$ as a matchgrid $H_{m,z}$. By (d) the remaining nodes $Z_{m,z}$ are all mapped into the outer face of $H_{m,z}$. Since, by (b) and (c), $|Z_{m,z}|$ is upper bounded by a polynomial $L(n)$, and the degrees of the $Z_{m,z}$ nodes by a constant, $H_{m,z}$ has $O(L(n))$ external connections, and can be regarded as a matchgrid with $O(L(n))$ external nodes. Now, by Theorem 1, $H_{m,z}$ can be replaced by a matchgrid with $O((L(n))^2)$ variable weight edges. From this we deduce that as $z$ varies, the total number of inequivalent matchgrids $H_{m,z}$ is at most $a_1 = |F|^{O((L(n))^2)}$. (In other words it is single exponential in $n$ however large $m$ may be.)

It remains to complete the estimation of the number of different functions of the original $n$ domain $d_m$ variables that the matchgrids can realize as $z$ varies, by also taking into account the remainder of the matchgrid specification, namely the nodes $|Z_{m,z}|$ and the edges incident to them. We can fix the names of the nodes of $Z_{m,z}$ and the external nodes of $H_{m,z}$, which altogether number $O(L(n))$. Then the number of potential edges that have at least one endpoint in $|Z_{m,z}|$ is at most $A = O((L(n))^2)$. By assumption (e), each choice of $z$ partitions the edges among pairs of $Z_{m,z}$ nodes into $n$ sets, and each such set will have a weight assignment that represents the corresponding domain variable having value zero, and a weight assignment corresponding to value one. The number of partitions is upper bounded by $a_2 = A^n$ clearly. Also, for each such partition, $a_3 = |F|^{2A}$ upper bounds the number of distinct values of the edges between $Z_{m,z}$ nodes that among them represent all combinations of representing 0's and 1's for the $n$ variables of $d_m$. Also, the number of possible weight assignments to edges incident to both $|Z_{m,z}|$ and $|T_{m,z}|$ nodes is upper bounded by $a_4 = |F|^{O(L(n))}$. It follows that the total number of functions of the domain variables that the matchgrids can realize is upper bounded by $a_1 a_2 a_3 a_4$, which itself is upper bounded by $(L(n)|F|)^{O((L(n))^2)}$. Now, by condition (iii) of elementarity, $|F|$ is polynomial bounded in $m$. For standard diversity $m$ is single exponential in a polynomial $p(n)$. It follows that the number of distinct functions is at most $2^{O(q(n))}$ for some polynomial $q$. In other words the standard diversity is at most polynomial in $n$.                                         $\square$

From this result one can deduce for problems known to have high diversity that they do not have elementary reductions to matchgrids. The following is an instance that parallels a result in [V06]:

**Corollary 1.** There is no elementary reduction from $(f, d)$ to matchgrids where $f$ is any one of Pl-3CNF, 3CNF, $\oplus$Pl-3CNF or $\oplus$3CNF, and $d$ specifies a subset of $O(\log m)$ of the clauses for formulae with $m$ clauses.

*Proof.* This follows from Proposition 2 and Theorem 2.                                         $\square$

It is an interesting question whether converse implications also hold. For problems such as #$_7$Pl-Rtw-Mon-3CNF [V06] for which 1-oracle holographic algorithms exist, even for fields whose size does not increase at all with the input size, one would like to determine whether they have polynomial diversity. For planar representations of Boolean functions one can define a notion of *planar diversity* where the domains $d_m$ have to be on the outer face of the embedding. Then the particulars of the algorithm just described for $\#_7$Pl-Rtw-Mon-3CNF do imply polynomial planar diversity for that problem. However, such arguments do not appear to apply to domains that are not on the periphery, or to general diversity.

Multiple oracle calls appear to be very useful in reductions among counting problems. There are multitudes of #P-complete problems that have been proved complete via reductions that involve multiple oracle calls and polynomial interpolation on the results ([V79b], [J87], [Vad01], [XZZ07], [CLX08]). For any one of these problems one can ask whether they have polynomial diversity.

In the opposite direction, one can ask whether algorithms that make multiple oracle calls, each via an elementary reduction, can compute functions of exponential diversity. To formulate specific questions of this kind one would need to define specific classes of such multiple oracle call algorithms. One relevant such class is offered by the algorithms described in Sections 6-8 of this paper. These all have the following form: Given an instance $G$ of the problem, one generates a single matchgrid with weights that are polynomials in $x$ with coefficients from a field $F$. The solution sought is the $j$th least significant bit in the coefficient of $x^i$ of the Holant, where $i, j$ are predetermined integers, and all the coefficients are guaranteed to be integral. There is the further constraint that if this coefficient is nonzero then it has at least $j - 1$ factors of 2. Note that the solutions here are obtained by the multi-oracle reduction that evaluates the matchgrid at enough different values of $x$, and then interpolates for the appropriate coefficient. It is an interesting question to determine whether or not these classes of reductions can evade the polynomial constraint on diversity of elementary reductions.

## 5   The Basis b3

The basis **b3** [V08] has three components $\mathbf{z} = (1, 0)$, $\mathbf{n} = (1, -1)$, $\mathbf{p} = (1, 1)$. It has the useful property that for all $x \in F$, $x\mathbf{z}^3 + \mathbf{n}^3 + \mathbf{p}^3$ is an even ternary signature and therefore, by Proposition 6.2 in [V08], is realizable by a planar matchgate. To verify this it is sufficient to expand $x\mathbf{z}^3 + \mathbf{n}^3 + \mathbf{p}^3$ as:

$$(x, 0, 0, 0, 0, 0, 0, 0) + (1, -1, -1, 1, -1, 1, 1, -1) + (1, 1, 1, 1, 1, 1, 1, 1)$$
$$= (x + 2, 0, 0, 2, 0, 2, 2, 0) = [x + 2, 0, 2, 0].$$

We shall call this signature, and the gate realizing it, $\mathbf{g}_3(x)$. The analogous two-output signature $\mathbf{g}_2(x)$ is also even, and therefore realizable by virtue of Proposition 6.1 in [V08], since

$$x\mathbf{z}^2 + \mathbf{n}^2 + \mathbf{p}^2 = (x, 0, 0, 0) + (1, -1, -1, 1) + (1, 1, 1, 1) = [x + 2, 0, 2],$$

as is also the one output signature $\mathbf{g}_1(x) = x\mathbf{z} + \mathbf{n} + \mathbf{p} = [x + 2, 0]$

For each of the three parity problems that we define in the sections that follow, we shall consider planar graphs of $n$ vertices all of maximum degree three. Our constructions do not require that the graph be *cubic* in the sense that every node has degree exactly three.

For each problem we shall construct for any such graph $G$ a family of match-grids $\Omega(G, x)$ indexed by $x$, using a fixed binary recognizer $\mathbf{r}$ for the edges, and the above mentioned generators $\mathbf{g}_1(x)$, $\mathbf{g}_2(x)$ and $\mathbf{g}_3(x)$, for the nodes of degrees one, two and three, respectively. Then for each problem, Holant($\Omega(G, x)$) can be viewed as a polynomial in $x$ of degree at most $n$. If we evaluate Holant($\Omega(G, x)$) for one $G$ and $n + 1$ distinct values of $x$, and interpolate for the coefficients, then the coefficient of $x^i$ will be the sum of the contributions to the Holant of the states in which exactly $i$ of the generators are generating $\mathbf{z}$'s, and the remainder $\mathbf{n}$'s or $\mathbf{p}$'s.

Alternatively, we shall sometimes substitute $\mathbf{g}_1(s)$, $\mathbf{g}_2(t)$ and $\mathbf{g}_3(x)$, with different indeterminates $s, t, x$. Then after evaluating at $O(n^3)$ distinct points, we can interpolate to obtain the coefficient of $s^i t^j x^l$, which gives the contribution to the Holant of states where among the $\mathbf{z}$ generators, exactly $i$ have degree one, $j$ degree two, and $l$ degree three.

We now describe the binary recognizers that we use for the edges. Each of these recognizers is a simple chain, of one, two or three edges, with the end nodes serving as the two external nodes. In our notation below $*$ denotes a node, and $*(w)*$ denotes an edge of weight $w$ between two nodes. The following can be verified by inspection.

**Proposition 5.** *The values of the following three recognizers are as follows when $(a, b)$ is input from the left, and $(c, d)$ from the right:*
**r1:** $*(1)*$ *has value* $ac + bd$,
**r2:** $*(1)*(1)*$ *has value* $ad + bc$, *and*
**r3:** $*(1)*(-1)*$ *has value* $ad - bc$.

# 6    Holographic Algorithm for the Parity of the Number of Induced Forests or Feedback Vertex Sets

The Minimum Feedback Vertex Set problem for undirected graphs is defined as follows: Given an undirected graph $G$ and an integer $k$ the question is to determine whether there is a set of $k$ vertices whose removal leaves a forest (i.e. a graph with no cycles.) There is a substantial literature on this existence problem. The directed version of this problem was proved NP-complete by Karp [K72]. The undirected version we study here was proved NP-complete by Garey and Johnson [GJ79]. Subsequently it was shown to be NP-complete even for planar graphs of degree four by Speckenmeyer [S83]. For cubic (i.e regular degree three) graphs a polynomial algorithm was given by Li and Liu [LL99]. (This last result is also implied by the polynomial time algorithm of Ueno *et al.* [UKG88] for the Minimum Connected Vertex Cover problem (defined in the next section) for cubic graphs, in conjunction with the result of Speckenmeyer that for any cubic graph on $n$ vertices MCVC-MFVS $= n/2 - 1$, where MCVC and MFVS denote the sizes of the minimum connected vertex cover and the minimum feedback vertex set.)

Here we are interested not in the existence problem but in the parity of the number of solutions, not only for forests of the largest size but for forests of every size, and not only for regular graphs of degree three, but for all graphs of maximum degree three. Thus the probem we address, $\oplus mFVS$, is the following: Given a degree $m$ undirected graph $G$ and an integer $k$, determine the parity of the number of sets of $k$ nodes that induce a forest in $G$.

**Theorem 3.** *There is a deterministic polynomial time algorithm for* $\oplus 3FVS$.

*Proof.* We place $\mathbf{g}_3(\mathbf{x})$, $\mathbf{g}_2(\mathbf{x})$ and $\mathbf{g}_1(\mathbf{x})$ generators at vertices of degree three, two and one respectively. We place a recognizer $\mathbf{r1}$ on each edge. Then, by Proposition 5, the value of each recognizer as a function of the nine possible combinations of what the adjacent nodes generate are as follows: $\mathbf{zz} \to 1$; $\mathbf{zp} \to 1$; $\mathbf{zn} \to 1$; $\mathbf{pz} \to 1$; $\mathbf{nz} \to 1$; $\mathbf{pp} \to 2$; $\mathbf{nn} \to 2$; $\mathbf{pn} \to 0$; $\mathbf{np} \to 0$.

We regard each state $\sigma$ (i.e. each combination of states of all the generators) of the matchgrid as a two-coloring, where one color, Z, corresponds to the nodes generating $\mathbf{z}$'s, and the other, Y, those generating $\mathbf{n}$'s and $\mathbf{p}$'s. For each such state we define $\#YY(\sigma)$ to be the number of edges joining a pair of nodes both colored Y, and $\#Ycomponents(\sigma)$ to be the number of connected components induced in $G$ by the removal of the Z nodes and the edges adjacent to them. Then the Holant will be the sum over all such Z/Y 2-colorings of $G$ of the value $V = 2^{\#Ycomponents(\sigma)+\#YY(\sigma)}$, since each connected component has one of two states (all $\mathbf{n}$ or all $\mathbf{p}$), and each edge in such a component contributes a further factor of two. If $G$ has $n$ nodes and the number of Z nodes is fixed as $n - k$, then the minimum number of divisors of 2 in $V$ is $2^{n-(n-k)} = 2^k$, and is achieved if and only if the YY edges induce a forest in $G$. (Note that in any graph with $k$ nodes the sum of the number of edges and the number of connected components is at least $k$, the minimum being achieved only if the graph is a forest.) Hence, if one divides the coefficient of $x^{n-k}$ in Holant($\Omega(G, x)$) by $2^k$, then the parity of that number is the desired solution to $\oplus 3FVS(G)$.  $\square$

# 7   Holographic Algorithm for the Parity of the Number of Connected Vertex Covers

The Minimum Connected Vertex Cover problem is the following. Given an undirected graph $G$ determine the size of the smallest set of nodes that (i) is a vertex cover, and (ii) induces a connected subgraph of $G$.

The existence problem was shown NP-complete for degree four planar graphs by Garey and Johnson [GJ77]. Fernau and Manlove [FN06] showed that this result holds even in the bipartite case. For cubic graphs it was shown to be polynomial time computable by Ueno, Kajitani and Gotoh [UKG88].

Here we are interested in the following parity problem $\oplus mCVC$. Given an undirected planar graph $G$ of maximum degree $m$ and an integer $k$, the problem is to compute the parity of the number of connected vertex covers of $G$ of $k$ vertices.

**Theorem 4.** *There is a deterministic polynomial time algorithm for $\oplus 3CVC$.*

*Proof.* We place $\mathbf{g_3(x)}$, $\mathbf{g_2(t)}$, and $\mathbf{g_1(s)}$ generators at vertices of degree three, two and one respectively. We place a recognizer $\mathbf{r2}$ on each edge. Then, by Proposition 5, the value of each recognizer as a function of the nine possible combinations of what the adjacent nodes generate are as follows: $\mathbf{zz} \to 0$; $\mathbf{zp} \to 1$; $\mathbf{zn} \to -1$; $\mathbf{pz} \to 1$; $\mathbf{nz} \to -1$; $\mathbf{pp} \to 2$; $\mathbf{nn} \to -2$; $\mathbf{pn} \to 0$; $\mathbf{np} \to 0$.

As before, we regard each state $\sigma$ of the matchgrid as a two coloring, where one color, Z, corresponds to the nodes generating $\mathbf{z}$'s, and the other, Y, those generating $\mathbf{n}$'s and $\mathbf{p}$'s. For each such state we define $\#YY(\sigma)$ to be the number of edges joining a pair of nodes both colored Y, and $\#Y\text{components}(\sigma)$ to be the number of connected components induced in $G$ by these YY edges. Now the Holant will be the sum over some such Z/Y 2-colorings of $G$ in which the nodes colored Z form an independent set, of $V = \pm 2^{\#Ycomponents(\sigma)+\#YY(\sigma)}$. This will follow by a similar argument to that used in Theorem 3, except now the Z nodes form an independent set since the value of $\mathbf{r2}$ for $\mathbf{zz}$ input is zero, and we need to analyze potential cancelations.

To derive this value of $V$ we first note that if the graph has $n$ nodes and is cubic, then for a state in which the Z nodes form an independent set of size $n-k$, it will be the case that $\#YY(\sigma) = 3n/2 - 3(n-k) = 3(k-n/2)$. For each Y/Z coloring and for any connected component induced by the Y colored edges in $G$, there will be two valid states, corresponding to the Y-colored nodes having all $\mathbf{p}$ or all $\mathbf{n}$ states. When one changes all the Y nodes from $\mathbf{p}$ to $\mathbf{n}$ then the values of all the recognizers in $G$ will change sign. Hence if the nodes in this component have an even number of edges incident to them then these contributions to the Holant will have the same sign, and otherwise will cancel. Hence the minimum (nonzero) number of divisors of 2 in $V$ is $2^{3(k-n/2)+1}$, and is achieved if and only if the YY edges induce one connected component in $G$ and $G$ has an even number of edges. Hence, if one divides the coefficient of $x^{n-k}$ in $\text{Holant}(\Omega(G,x))$ by $2^{3(k-n/2)+1}$, then the parity of that number is the desired solution to $\oplus 3CVC(G)$.

If the graph is not regular, then by interpolation we can find the coefficient of $s^i t^j x^l$ in $\text{Holant}(\Omega(G,s,t,x))$ for all $i,j,l$. For any specific combination of $i,j,l$ the value of $\#YY(\sigma)$ is $|E| - i - 2j - 3l$, where $|E|$ is the total number of edges in $G$. Hence we can compute the parity of the number of solutions for any combination $i,j,l$, and hence for all the combinations with $i+j+l = n-k$. We shall derive the parity of the number of solutions corresponding to such $Z$ sets by dividing the appropriate coefficient by $2^{|E|-i-2j-3l+1}$ rather than by $2^{3(k-n/2)+1}$ as used in the regular case.

So far we have assumed that the number of edges in $G$ is even. To treat the alternative case we choose an arbitrary edge and replace $\mathbf{r2}$ by $\mathbf{r1}$ on it. This ensures that when switching between all $\mathbf{p}$ and all $\mathbf{n}$ states the sign will not change on this one edge, and hence not for the product of all of these odd number of edges. It only remains to ensure that the Y nodes still form a vertex cover, and for this it is necessary to preclude that the endpoints of the chosen edge be both in state Z. This can be done by multiplying the $x$ term in these two generators by a new indeterminate $w$, and, by interpolation, computing and adding the coefficients of $w^0$ and $w^1$ (while ignoring that of $w^2$.)    □

# 8 Holographic Algorithm for the Parity of the Number of Vertex Colorings

A 3-Vertex Coloring of a graph $G$ is an assignment of a color from a palette of 3 colors to each vertex so that no pair of adjacent vertices has the same color. Clearly the set of all such proper colorings can be partitioned into equivalence classes of 3! colorings, so that the members of each class differ only by a permutation of the colors. Here we are interested in the following two closely related problems. The problem $\oplus mCol$: for an undirected planar graph $G$ of maximum degree $m$ determine the parity of the number of equivalence classes of 3-colorings of $G$. The problem $\oplus mFCol$: for an undirected planar graph $G$ of maximum degree $m$ and an integer $k$ determine the parity of the number of 3-colorings that are invariant under permutations of the remaining two colors when exactly $k$ nodes are given the first color. We note that the corresponding counting problems for 3-colorability of degree three graphs are #P-complete [BDGJ99].

**Theorem 5.** [B04] *For some constant $m$, $\oplus mCol$ is $\oplus P$-complete.*

**Theorem 6.** *There is a deterministic polynomial time algorithm for $\oplus 3FCol$ and for $\oplus 3Col$.*

*Proof.* We place $\mathbf{g_3(x)}$, $\mathbf{g_2(t)}$, and $\mathbf{g_1(s)}$ generators at vertices of degree three, two and one respectively, and $\mathbf{r3}$ recognizers on each edge. The $\mathbf{r3}$ recognizers for $ad - bc$ are not symmetric, and can be placed in arbitrary orientation without influencing our result. By Proposition 5 the value of each recognizer as a function of the nine possible combinations of what the adjacent nodes generate are as follows: $\mathbf{zz} \to 0$; $\mathbf{zp} \to 1$; $\mathbf{zn} \to$ -1; $\mathbf{pz} \to$ -1; $\mathbf{nz} \to 1$; $\mathbf{pp} \to 0$; $\mathbf{nn} \to 0$; $\mathbf{pn} \to$ -2; $\mathbf{np} \to 2$.

Again we regard each state $\sigma$ of the matchgrid as a two coloring, where one color, Z, corresponds to the nodes generating $\mathbf{z}$'s, and the other, Y, those generating $\mathbf{n}$'s and $\mathbf{p}$'s. For each such state we define $\#YY(\sigma)$ to be the number of edges joining a pair of nodes both colored Y, and $\#Y$components$(\sigma)$ to be the number of connected components induced in $G$ by these YY edges. Then the Holant will be the sum over some such Z/Y 2-colorings of $G$ in which the nodes colored Z form an independent set and those colored Y form a bipartite graph, of the values $V = \pm 2^{\#Ycomponents(\sigma) + \#YY(\sigma)}$.

To see this we first assume that the graph is cubic and has an even number of edges. If the graph has $n$ nodes, then for a state in which the Z nodes form an independent set of size $k$, then $\#YY(\sigma) = 3n/2 - 3k$. We note that for each Y/Z coloring the YY edges will form a set of connected bipartite components in $G$. In each component there will be two valid states, corresponding to which of the two parts is in $\mathbf{p}$ or $\mathbf{n}$ state. When one swaps $\mathbf{p}$ and $\mathbf{n}$ all the values of all the recognizers will change sign. Hence if there are an even number of edges incident to the nodes in one such component, then the contributions to the Holant will have the same sign for the two states. Hence the minimum number of divisors of 2 in $V$ is $2^{3n/2-3k+1}$, and is achieved if and only if the YY edges induce one connected bipartite component in $G$. Hence, if one divides the coefficient of $x^k$ in $\text{Holant}(\Omega(G, x))$ by $2^{3n/2-3k+1}$, then the parity of that number is the parity of the number of solutions to $\oplus 3\text{FCol}(G)$.

Graphs that are not regular or have an odd number of edges can be treated exactly as in Theorem 4. $\qquad\Box$

For completeness we mention the following application:

**Theorem 7.** *The parity of the number of edge 3-colorings of planar 3/2 bipartite graphs can be computed in polynomial time.*

*Proof.* The line graph of such a graph is a planar degree three graph, and hence the result follows immediately from Theorem 6. Note that we do not require the 3/2-graph to be a regular 3/2 graph. $\qquad\Box$

# References

[B03]    Barbanchon, R.: Reductions fines entre problèmes NP-complets, PhD Thesis, Université de Caen Basse-Normandie (2003)

[B04]    Barbanchon, R.: On unique graph 3-colorability and parsimonious reductions in the plane. Theoretical Computer Science 319(1-3), 455–482 (2004)

[BDGJ99]  Bubley, R., Dyer, M., Greenhill, C., Jerrum, M.: On approximately counting colourings of small degree graphs. SIAM J. Comput. 29, 387–400 (1999)

[CC07]   Cai, J.-Y., Choudhary, V.: Some Results on Matchgates and Holographic Algorithms. International Journal of Software and Informatics 1(1), 3–36 (2007)

[CCL09]  Cai, J.-Y., Choudhary, V., Lu, P.: On the Theory of Matchgate Computations. Theory of Computing Systems 45(1), 108–132 (2009)

[CL07]   Cai, J.-Y., Lu, P.: Holographic algorithms: from art to science. In: STOC 2007, pp. 401–410 (2007)

[CL09]   Cai, J.-Y., Lu, P.: Holographic algorithms: The power of dimensionality resolved. Theor. Comput. Sci. 410(18), 1618–1628 (2009)

[CLX08]  Cai, J.-Y., Lu, P., Xia, M.: Holographic Algorithms by Fibonacci Gates and Holographic Reductions for Hardness. In: FOCS, pp. 644–653 (2008)

[C71]    Cook, S.A.: The complexity of theorem proving procedures. In: Proceedings of the 3rd ACM STOC, pp. 151–158 (1971)

[EGM09]  Escoffier, B., Gourves, L., Monnot, J.: Complexity and approximation results for the connected vertex cover problem in graphs and hypergraphs. Journal of Discrete Algorithms (2009)

[FM06]   Fernau, H., Manlove, D.: Vertex and edge covers with clustering properties: Complexity and algorithms. Journal of Discrete Algorithms (2009)

[GJ77]   Garey, M.R., Johnson, D.S.: The rectilinear steiner tree problem is NP complete. SIAM Journal of Applied Mathematics 32, 826–834 (1977)

[GJ79]   Garey, M.R., Johnson, D.S.: Computer and Intractability: A Guide to the Theory of NP-Completeness. W.H. Freeman, New York (1979)

590     L.G. Valiant

[HMRS98]   Hunt III, H.B., Marathe, M.V., Radhakrishnan, V., Stearns, R.E.: The
           Complexity of Planar Counting Problems. SIAM J. Comput. 27(4), 1142–
           1167 (1998)
[J87]      Jerrum, M.R.: Two-dimensional monomer-dimer systems are computa-
           tionally intractable. J. Statist. Phys. 48(1-2), 121–134 (1987)
[K72]      Karp, R.M.: Reducibility among combinatorial problems. In: Miller, R.E.,
           Thatcher, J.W. (eds.) Complexity of Computer Computations, pp. 85–
           104. Plenum Press, New York (1972)
[L75]      Ladner, R.E.: The Circuit Value Problem is Log Space Complete for P.
           SIGACT NEWS 7(1), 18–20 (1975)
[LL99]     Li, D.M., Liu, Y.P.: A polynomial algorithm for finding the minimum
           feedback vertex set of a 3-regular simple graph. Acta Math. Sci. 19(4),
           375–381 (1999)
[Li82]     Lichtenstein, D.: Planar formulae and their uses. SIAM J. Comput. 11,
           329–343 (1982)
[Lu58]     Lupanov, O.B.: A method of circuit synthesis. Izv. VUZ Radiofiz 1, 120–
           140 (1958)
[N66]      Neciporuk, E.I.: A Boolean Function. Sov. Math. Dokl. 7, 999–1000
           (1966)
[S83]      Speckenmeyer, E.: Untersuchungen zum Feedback Vertex Set Problem in
           ungerichteten Graphen. PhD Thesis, Universität Paderborn (1983)
[S88]      Speckenmeyer, E.: On feedback vertex sets and nonseparating indepen-
           dent sets in cubic graphs. Journal of Graph Theory 12(3), 405–412 (1988)
[UKG88]    Ueno, S., Kajitani, Y., Gotoh, S.: On the nonseparating independent
           set problem and feedback set problem for graphs with no vertex degree
           exceeding three. Discrete Mathematics 72, 355–360 (1988)
[V79a]     Valiant, L.G.: The complexity of computing the permanent. Theoretical
           Computer Science 8, 189–201 (1979)
[V79b]     Valiant, L.G.: The complexity of enumeration and reliability problems.
           SIAM J. Computing 8(3), 410–421 (1979)
[V04]      Valiant, L.G.: Holographic algorithms (extended abstract). In: Proc. 45th
           Annual IEEE Symposium on Foundations of Computer Science, October
           17-19, pp. 306–315. IEEE Press, Los Alamitos (2004)
[V05]      Valiant, L.G.: Completeness for parity problems. In: Wang, L. (ed.) CO-
           COON 2005. LNCS, vol. 3595, pp. 1–9. Springer, Heidelberg (2005)
[V06]      Valiant, L.G.: Accidental algorithms. In: Proc. 47th Annual IEEE Sym-
           posium on Foundations of Computer Science, October 22-24, pp. 509–517.
           IEEE Press, Los Alamitos (2006)
[V08]      Valiant, L.G.: Holographic algorithms. SIAM J. on Computing 37(5),
           1565–1594 (2008); Earlier version: Electronic Colloquium on Computa-
           tional Complexity, Report TR05-099 (2005)
[XZZ07]    Xia, M., Zhang, P., Zhao, W.: Computational complexity of counting
           problems on 3-regular planar graphs. Theor. Comput. Sci. 384(1), 111–
           125 (2007)

# The Interval Constrained 3-Coloring Problem

Jaroslaw Byrka, Andreas Karrenbauer*, and Laura Sanità**

Institute of Mathematics
EPFL, Lausanne, Switzerland
{Jaroslaw.Byrka,Andreas.Karrenbauer,Laura.Sanita}@epfl.ch

**Abstract.** In this paper, we settle the open complexity status of interval constrained coloring with a fixed number of colors. We prove that the problem is already NP-complete if the number of different colors is 3. Previously, it has only been known that it is NP-complete, if the number of colors is part of the input and that the problem is solvable in polynomial time, if the number of colors is at most 2. We also show that it is hard to satisfy almost all of the constraints for a feasible instance. This implies APX-hardness of maximizing the number of simultaneously satisfiable intervals.

## 1 Introduction

In the interval constrained 3-coloring problem, we are given a set $\mathcal{I}$ of intervals defined on $[n] := \{1, \ldots, n\}$ and a *requirement* function $r : \mathcal{I} \to \mathbb{Z}^3_{\geq 0}$, which maps each interval to a triple of non-negative integers. The objective is to determine a coloring $\chi : [n] \to \{1, 2, 3\}$ such that each interval gets the proper colors as specified by the requirements, i.e. $\sum_{i \in I} e_{\chi(i)} = r(I)$ where $e_1, e_2, e_3$ are the three unit vectors of $\mathbb{Z}^3$.

This problem is motivated by an application in biochemistry to investigate the tertiary structure of proteins as shown in the following illustration. More precisely, in Hydrogen-Deuterium-Exchange (HDX) experiments proteins are put into a solvent of heavy water ($D_2O$) for a certain time after which the amount of residual hydrogen atoms, that have exchanged with deuterium atoms, is measured [1]. Doing this experiment for several timesteps, one can determine the exchange rate of the residues. These exchange rates indicate the solvent accessibility of the residues and hence they provide information about the spatial structure of the protein. Mass spectroscopy is one of the methods for measuring these exchange rates. To this end, the proteins are digested, i.e. cut into parts which can be considered as intervals of the protein chain, and the mass uptake of each interval is measured. But thereby only bulk information about each interval can be obtained. Since there is not only one protein in the solvent but millions and they are not always cut in the same manner, we have this bulk information on overlapping fragments. That is, we are given the number of slow, medium, and fast exchanging residues for each of these intervals and our goal is to find a feasible assignment of these three exchange rates to residues such that for each interval the numbers match with the bulk information.

* Supported by the Deutsche Forschungsgemeinschaft (DFG) within Priority Programme 1307 "Algorithm Engineering".
** Supported by Swiss National Science Foundation within the project "Robust Network Design".

A. López-Ortiz (Ed.): LATIN 2010, LNCS 6034, pp. 591–602, 2010.

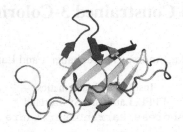

**Fig. 1.** Coloring of the residues of a protein chain according to their exchange rates

Though the interval constrained 3-coloring problem is motivated by a particular application, its mathematical abstraction appears quite simple and ostensibly more general. In terms of integer linear programming, the problem can be equivalently formulated as follows. Given a matrix $A \in \{0,1\}^{m \times n}$ with the *row-wise consecutive-ones property* and three vectors $b_{1,2,3} \in \mathbb{Z}_{\geq 0}^m$, the constraints

$$\begin{pmatrix} A & 0 & 0 \\ 0 & A & 0 \\ 0 & 0 & A \\ I & I & I \end{pmatrix} \cdot \begin{pmatrix} x_1 \\ x_2 \\ x_3 \end{pmatrix} = \begin{pmatrix} b_1 \\ b_2 \\ b_3 \\ 1 \end{pmatrix} \qquad (1)$$

have a binary solution, i.e. $x_{1,2,3} \in \{0,1\}^n$, if and only if the corresponding interval constrained 3-coloring problem has a feasible solution. We may assume w.l.o.g. that the requirements are consistent with the interval lengths, i.e. $A \cdot 1 = b_1 + b_2 + b_3$, since otherwise we can easily reject the instance as infeasible. Hence, we could treat $x_3$ as slack variables and reformulate the constraints as

$$Ax_1 = b_1, \qquad Ax_2 = b_2, \qquad x_1 + x_2 \leq 1. \qquad (2)$$

It is known that if the matrix $A$ has the *column-wise* consecutive ones property (instead of *row-wise*), then there is a reduction from the two-commodity integral flow problem, which has been proven to be NP-complete in [2]. However, the NP-completeness w.r.t. row-wise consecutive ones matrices has been an open problem in a series of papers as outlined in the following subsection.

## 1.1   Related Work

The problem of assigning exchange rates to single residues has first been considered in [3]. In that paper, the authors presented a branch-and-bound framework for solving the corresponding coloring problem with $k$ color classes. They showed that there is a combinatorial polynomial time algorithm for the case of $k = 2$. Moreover, they asked the question about the complexity for $k > 2$. In [4], the problem has been called *interval constrained coloring*. It has been shown that the problem is NP-hard if the parameter $k$ is part of the input. Moreover, approximation algorithms have been presented that allow violations of the requirements. That is, a quasi-polynomial time algorithm that computes a solution in which all constraints are $(1 + \varepsilon)$-satisfied and a plynomial time rounding scheme, which satisfies every requirement within $\pm 1$, based on a technique introduced in [5]. The latter implies that if the LP relaxation of (1) is feasible, then there is a coloring satisfying at least $\frac{5}{16}$ of the requirements. APX-hardness of finding the maximum number of simultaneously satisfiable intervals has been shown in [6] for $k \geq 2$ provided that intervals may be counted with multiplicities. But still, the question about the complexity of the decision problem for fixed $k \geq 3$ has been left open. In [7], several fixed parameter tractability results have been given. However, the authors state that they do not know whether the problem is tractable for fixed $k$.

## 1.2  Our Contribution

In this paper, we prove the hardness of the interval constrained $k$-coloring problem for fixed parameter $k$. In fact, we completely settle the complexity status of the problem, since we show that already the interval constrained 3-coloring problem is NP-hard by a reduction from 3-SAT. This hardness result holds more generally for any problem that can be formulated like (1). Moreover, we even show the stronger result, that it is still difficult to satisfy almost all of the constraints for a feasible instance. More precisely, we prove that there is a constant $\epsilon > 0$ such that it is NP-hard to distinguish between instances where all constraints can be satisfied and those where only a $(1 - \epsilon)$ fraction of constraints can be simultaneously satisfied. To this end, we extend our reduction using expander graphs. This gap hardness result implies APX-hardness of the problem of maximizing the number of satisfied constraints. It is important to note that our construction does neither rely on multiple copies of intervals nor on inconsistent requirements for an interval, i.e. in our construction for every interval $(i, j)$ we have unique requirements that sum up to the length of the interval.

## 2  NP-Hardness

**Theorem 1.** *It is NP-hard to decide whether there exists a feasible coloring $\chi$ for an instance $(\mathcal{I}, r)$ of the interval constrained 3-coloring problem.*

*Proof.* The proof is by reduction from the 3-SAT problem.

Suppose to be given an instance of the 3-SAT problem, defined by $q$ clauses $C_1, \ldots, C_q$ and $p$ variables $x_1, \ldots, x_p$. Each clause $C_i$ $(i = 1, \ldots, q)$ contains 3 literals, namely $y_1(i), y_2(i), y_3(i)$. Each literal $y_h(i)$ $(i = 1, \ldots, q$ and $h = 1, 2, 3)$ *refers to* a variable $x_j$, that means, it is equal to either $x_j$ or $\bar{x}_j$ for some $j$ in $1, \ldots, p$. A truth assignment for the variables $x_1, \ldots, x_p$ satisfies the 3-SAT instance if and only if, for each clause, at least one literal takes the value *true*.

We now construct an instance of the interval constrained 3-coloring problem. For each clause $C_i$ we introduce a sequence of consecutive nodes. This sequence is, in its turn, the union of three subsequences, one for each of the three literals (see Fig. 2).

In the following, for the clarity of presentation, we drop the index $i$, if it is clear from the context. We denote color 1 by RED, color 2 by BLACK and color 3 by WHITE.

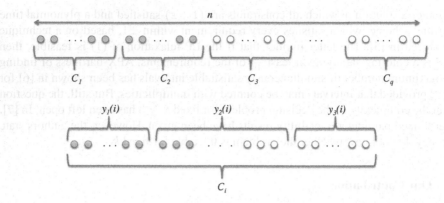

**Fig. 2.** The sequence of nodes in an instance of the interval constrained 3-coloring problem

*Literal $y_1(i)$.* The subsequence representing literal $y_1$ is composed of 8 nodes. Among them, there are three special nodes, namely $t_1$, $f_1$ and $a_1$, that play a key role since they encode the information about the truth value of the literal and of the variable $x_j$ it refers to. The basic idea is to achieve the following two goals: 1) given a feasible coloring, if $\chi(t_1)$ is BLACK, we want to be able to construct a truth assignment setting $x_j$ to *true*, while if $\chi(f_1)$ is BLACK, we want to be able to construct a truth assignment setting the variable $x_j$ to *false*; 2) given a feasible coloring, if $\chi(a_1)$ is RED, we want to be able to construct a truth assignment where $y_1$ is *true*.

To achieve the first goal, we will impose the following property:

*Property 1.* In any feasible coloring, exactly one among $t_1$ and $f_1$ will be BLACK.

To achieve the second goal, and being consistent with the first one, we must have the property that:

*Property 2.* In any feasible coloring, if $\chi(a_1) = RED$, then $\chi(t_1) = BLACK$ if $y_1 = x_j$, while $\chi(f_1) = BLACK$ if $y_1 = \bar{x}_j$.

To guarantee properties (1) and (2), we introduce a suitable set $\mathcal{I}(y_1)$ of six intervals[1], shown in Fig. 3a.

The requirement function for such intervals changes whether $y_1 = x_j$ or $y_1 = \bar{x}_j$. If $y_1 = x_j$, we let $r(I_1) = (1,1,1); r(I_2) = (1,1,1); r(I_3) = (1,0,1); r(I_4) = (1,1,2); r(I_5) = (0,1,0); r(I_6) = (2,3,3)$. For any feasible coloring there are only three possible outcomes for such sequence, reported in Fig. 3b. Observe that the properties (1) and (2) are enforced.

Now suppose that $y_1 = \bar{x}_j$: then we switch the requirement function with respect to WHITE and BLACK, i.e. define it as follows: $r(I_1) = (1,1,1); r(I_2) = (1,1,1); r(I_3) = (1,1,0); r(I_4) = (1,2,1); r(I_5) = (0,0,1); r(I_6) = (2,3,3)$. Trivially, the possible outcomes for such sequence are exactly the ones in Fig. 3b but exchanging the BLACK and WHITE colors.

---

[1] In principle, interval $I_5$ and the node it contains are not needed. However, this allows to have the same number of WHITE and BLACK colored nodes for the sake of exposition.

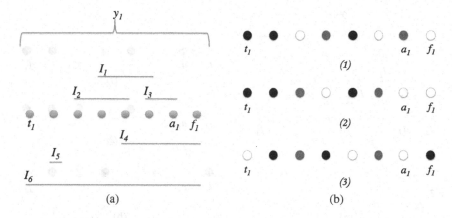

**Fig. 3.** Literal $y_1$. The picture on the right shows the three feasible colorings. On a black and white printout red color appears as grey.

*Literal $y_3(i)$.* The sequence of nodes representing literal $y_3$ is similar to the one representing $y_1$. We still have a sequence of 8 nodes, and three special nodes $t_3$, $f_3$ and $a_3$. As before, we let $t_3$ and $f_3$ encode the truth value of the variable $x_j$ that is referred to by $y_3$, while $a_3$ encodes the truth value of the literal $y_3$ itself. Therefore, we introduce a set $\mathcal{I}(y_3)$ of intervals in order to enforce the following properties:

*Property 3.* In any feasible coloring, exactly one among $t_3$ and $f_3$ will receive color BLACK.

*Property 4.* In any feasible coloring, if $\chi(a_3) = RED$, then $\chi(t_3) = BLACK$ if $y_3 = x_j$, while $\chi(f_3) = BLACK$ if $y_3 = \bar{x}_j$.

Fig. 4a shows the nodes and the six intervals that belong to $\mathcal{I}(y_3)$: observe that the sequence is similar to the one representing $y_1$, but the position of node $a_3$ and the intervals are now "mirrored". If $y_3 = \bar{x}_j$, we let $r(I_1) = (1,1,1)$; $r(I_2) = (1,1,1)$; $r(I_3) = (1,0,1)$; $r(I_4) = (1,1,2)$; $r(I_5) = (0,1,0)$; $r(I_6) = (2,3,3)$. Fig. 4b reports the three possible outcomes for such sequence in a feasible coloring. Note that properties (3) and (4) hold.

Now suppose that $y_3 = x_j$: once again, we switch the requirement function with respect to WHITE and BLACK.

*Literal $y_2(i)$.* The sequence of nodes representing literal $y_2$ is slightly more complicated. It is composed of 36 nodes, and among them there are 4 special nodes, namely $t_2$, $f_2$, $a_2^\ell$ and $a_2^r$ (see Fig. 5). Still, we let $t_2$ and $f_2$ encode the truth value of the variable $x_j$ that is referred to by $y_2$, while $a_2^\ell$ and $a_2^r$ encode the truth value of the literal. Similarly to the previous cases, we want to achieve the following goals: 1) given a feasible coloring, if $\chi(t_2)$ is BLACK, we want to be able to construct a truth assignment setting the variable $x_j$ to *true*, while if $\chi(f_2)$ is BLACK, we want to be able to construct a truth assignment setting the variable $x_j$ to *false*; 2) given a feasible coloring, if

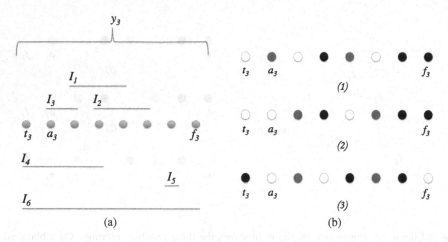

**Fig. 4.** Literal $y_3$

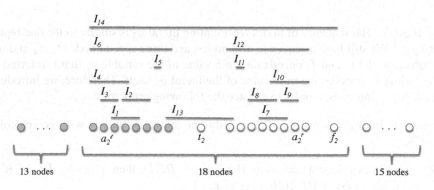

**Fig. 5.** Literal $y_2$

$\chi(a_2^\ell) = \chi(a_2^r) = RED$, we want to be able to construct a truth assignment where the literal $y_2$ is *true*. We are therefore interested in the following properties:

*Property 5.* In any feasible coloring, exactly one among $t_2$ and $f_2$ will receive color BLACK.

*Property 6.* In any feasible coloring, if $\chi(a_2^\ell) = RED$ and $\chi(a_2^r) = RED$, then $\chi(t_2) = BLACK$ if $y_2 = x_j$, and $\chi(f_2) = BLACK$ if $y_2 = \bar{x}_j$.

In this case, we introduce a set $\mathcal{I}(y_2)$ of 14 suitable intervals, shown in Fig. 5. The requirements for the case $y_2 = \bar{x}_j$ are given in the following table.

|  | $I_1$ | $I_2$ | $I_3$ | $I_4$ | $I_5$ | $I_6$ | $I_7$ | $I_8$ | $I_9$ | $I_{10}$ | $I_{11}$ | $I_{12}$ | $I_{13}$ | $I_{14}$ |
|---|---|---|---|---|---|---|---|---|---|---|---|---|---|---|
| *RED* | 1 | 1 | 1 | 1 | 0 | 2 | 1 | 1 | 1 | 1 | 0 | 2 | 0 | 4 |
| *BLACK* | 1 | 1 | 0 | 1 | 1 | 3 | 1 | 1 | 0 | 1 | 1 | 3 | 2 | 7 |
| *WHITE* | 1 | 1 | 1 | 2 | 0 | 3 | 1 | 1 | 1 | 2 | 0 | 3 | 1 | 7 |

Observe that the set of intervals $\{I_1, \ldots, I_6\}$ is defined exactly as the set $\mathcal{I}(y_3)$, therefore the possible outcomes for the sequence of 8 nodes covered by such intervals are as in Fig. 4b. Similarly, the set of intervals $\{I_7, \ldots, I_{12}\}$ is defined exactly as the set $\mathcal{I}(y_1)$, therefore the possible outcomes for the sequence of 8 nodes covered by such intervals are as in Fig. 3b. Combining $r(I_6)$ and $r(I_{12})$ with $r(I_{14})$, it follows that in any feasible coloring $\chi$, exactly one node among $t_2$ and $f_2$ has WHITE (resp. BLACK) color, enforcing Property (5). Still, note that if $\chi(a_2^\ell) = RED$ and $\chi(a_2^r) = RED$, then both the leftmost node and the rightmost node covered by interval $I_{13}$ have color BLACK, therefore $t_2$ must have color WHITE otherwise $r(I_{13})$ is violated. Together with Property (5), this enforces Property (6).

In case $y_2 = x_j$, once again we switch the requirement function with respect to WHITE and BLACK.

It remains to describe the role played by the first 13 nodes and the last 15 nodes of the sequence, that so far we did not consider. We are going to do it in the next paragraph.

*Intervals encoding truth values of literals.* For each clause $C_i$, we add another set $\mathcal{I}(C_i)$ of intervals, in order to link the nodes encoding the truth values of its three literals. The main goal we pursue is the following: given a feasible coloring, we want to be able to construct a truth assignment such that at least one of the three literals is *true*. To achieve this, already having properties (2), (4) and (6), we only need the following property:

*Property 7.* For any feasible coloring, if $\chi(a_1) \neq RED$ and $\chi(a_3) \neq RED$ , then $\chi(a_2^\ell) = \chi(a_2^r) = RED$.

Fig. 6 shows the six intervals that belong to $\mathcal{I}(C_i)$. The requirement function is: $r(I_1) = (1, 2, 2)$; $r(I_2) = (1, 2, 2)$; $r(I_3) = (1, 6, 6)$; $r(I_4) = (1, 3, 3)$; $r(I_5) = (1, 2, 2)$; $r(I_6) = (1, 7, 7)$. We now show that Property (7) holds. Suppose $\chi$ is a feasible coloring, and let $v_1, \ldots, v_{13}$ be the first 13 nodes of the sequence introduced for literal $y_2$. By construction, if $\chi(a_1) \neq RED$, then there is a node $v_j : \chi(v_j) = RED$ and $j \in \{1, 2, 3\}$, otherwise $r(I_1)$ is violated. Similarly, if $\chi(a_2^\ell) \neq RED$, then there is a node $v_j : \chi(v_j) = RED$ and $j \in \{11, 12, 13\}$, otherwise $r(I_2)$ is violated. On the other hand, this subsequence contains exactly one node with RED color, otherwise $r(I_3)$ is violated. It follows that at least one among $a_1$ and $a_2^\ell$ has RED color. The same conclusions can be stated for nodes $a_2^r$ and $a_3$. Putting all together, it follows that the Property (7) holds.

*Intervals encoding truth value of variables (later also called: variable intervals).* Our last set of intervals will force different nodes to take the same color, if they encode the truth value of the same variable. In particular, we aim at having the following property:

*Property 8.* In any feasible coloring, $\chi(t_h(i)) = \chi(t_k(i'))$ if both literals $y_h(i)$ and $y_k(i')$ refer to the same variable $x_j$.

To achieve this, for each pair of such literals we add a big interval $I(y_h(i), y_k(i'))$ from $f_k(i')$ to $t_h(i)$ (assuming $i' < i$ without loss of generality). Note that, by construction, there is a subset of intervals that partitions all the internal nodes covered by the interval. That means, we know exactly the number of such nodes that must be colored with color

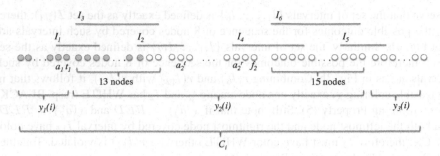

**Fig. 6.** Set of intervals $\mathcal{I}(C_i)$

RED, BLACK and WHITE (say $z_1, z_2, z_3$ respectively). Then, we let the requirement function be $r(I(y_h(i), y_k(i'))) = (z_1, z_2 + 1, z_3 + 1)$. Under these assumptions, if $\chi$ is a feasible coloring then $\chi(t_h(i)) \neq \chi(f_k(i'))$, and in particular one node will have WHITE color and the other one BLACK color. Combining this with properties (1),(3) and (5), the result follows.

Notice that such an interval constrained 3-coloring instance can clearly be constructed in polynomial time. Now we discuss the following claim in more details.

*Claim.* There exists a truth assignment satisfying the 3-SAT instance if and only if there exists a feasible coloring $\chi$ for the interval constrained 3-coloring instance.

First, suppose there exists a feasible coloring. We construct a truth assignment as follows. We set a variable $x_j$ to $true$ if $\chi(t_h(i)) = BLACK$, and to $false$ otherwise, where $y_h(i)$ is any literal referring to $x_j$. Note that, by Property (8), the resulting truth value does not depend on the literal we take. Still, combining Property (7) with properties (2),(4) and (6), we conclude that, for each clause, at least one literal will be $true$. By construction, we therefore end up with a truth assignment satisfying the 3-SAT instance. The result follows.

   Now suppose that there is a truth assignment satisfying the 3-SAT instance. The basic idea, is to construct a coloring $\chi$ such that the following property holds for all literals:

*Property 9.* $\chi(t_h(i)) =$ BLACK (resp. WHITE) if and only if $y_h(i)$ refers to a $true$-variable (resp. $false$-variable).

Consider the sequence of nodes representing literal $y_1(i)$, and suppose $y_1(i) = x_j$ for some $j$. We color such nodes as in Fig. 3b-*(1)* if the literal is $true$ in the truth assignment, and as in Fig. 3b-*(3)* otherwise. If $y_1(i) = \bar{x}_j$, switch BLACK and WHITE colors, in both previous cases. Now focus on the sequence of nodes representing literal $y_3(i)$. If $y_3(i) = \bar{x}_j$ for some $j$, we color such nodes as in Fig. 4b-*(1)* if the literal is $true$, and as in Fig. 4b-*(3)* otherwise. If $y_3(i) = x_j$, switch BLACK and WHITE colors, in both previous cases. Finally, consider the sequence of nodes representing literal $y_2(i)$. Suppose $y_2(i) = \bar{x}_j$. We color the 18 nodes in the middle of the sequence as in Fig. 7-*(1)* if $y_2(i)$ is $true$, as in Fig. 7-*(2)* if both $y_2(i)$ and $y_1(i)$ are $false$, and as in Fig. 7-*(3)* otherwise. Once again, if $y_2(i) = x_j$, we switch BLACK and WHITE

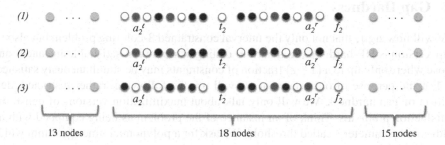

**Fig. 7.** Coloring of nodes representing literal $y_2$

colors, in all the previous three cases. Notice that, by construction, Property (9) holds, and all requirements for the intervals in $\mathcal{I}(y_h(i))$ ($i = 1, \ldots, q$ and $h = 1, 2, 3$) are not violated.

Now we show how to color the first 13 nodes ($v_1, \ldots, v_{13}$) and the last 15 nodes ($w_1, \ldots, w_{15}$) of the sequence representing literal $y_2(i)$, in such a way that the requirements of the intervals $I_1, \ldots, I_6$ in $\mathcal{I}(C_i)$ are not violated ($i = 1, \ldots, q$). Note that, by construction, at least one node among $a_1$ and $a_2^\ell$ is colored with RED. In fact, if $y_1(i)$ is *true* then $\chi(a_1) = $ RED, while if $y_1(i) = false$ then $a_2^\ell$ is colored with RED. Similarly, at least one node among $a_3$ and $a_2^r$ is colored with RED, since $\chi(a_2^r) \neq RED$ only if both literals $y_1(i)$ and $y_2(i)$ are *false*: then, necessarily $y_3(i)$ is *true*, and therefore $\chi(a_3) = RED$. Let us focus on the nodes $v_1, \ldots, v_{13}$, and let $u$ be the node in between $v_{13}$ and $a_2^\ell$. In the following, we refer to WHITE as the *opposite* color of BLACK and vice versa. As we already discuss, we can have only two cases:

Case 1: $\chi(a_1) = \chi(a_2^\ell) = RED$. We color $v_1$ with the opposite color of $f_1$, and the nodes $v_2$ and $v_3$ with BLACK and WHITE. Note that $r(I_1)$ is not violated. We then color $v_4, v_5, v_6$ with the opposite color of $v_1, v_2, v_3$ respectively. Similarly, we color $v_{13}$ with the opposite color of $u$. Then, we color $v_{12}$ and $v_{11}$ with BLACK and WHITE, so that $r(I_2)$ is not violated. Once again, we assign to $v_{10}, v_9, v_8$ the opposite color of $v_{13}, v_{12}, v_{11}$ respectively. Finally, we let $\chi(v_7) = RED$. Note that $r(I_3)$ is not violated.

Case 2: $\chi(a_1) \neq RED$ and $\chi(a_2^\ell) = RED$, or vice versa. Suppose $\chi(a_1) \neq RED$ (the other case is similar). Both nodes $a_1$ and $f_1$ can have only BLACK or WHITE colors. Then, we can color $v_1$ and $v_2$ with the opposite color of $a_1$ and $f_1$ respectively, and $v_3$ with color RED, so that $r(I_1)$ is not violated. Still, we color $v_4$ and $v_5$ with the opposite color of $v_1$ and $v_2$. Finally, we color $v_6$ and $v_7$ with BLACK and WHITE. To the remaining nodes $v_8, \ldots, v_{13}$ we assign the same colors as in Case 1. One checks that requirements of intervals $I_2$ and $I_3$ are not violated.

One can prove in a similar manner that nodes ($w_1, \ldots, w_{15}$) can be properly colored, without violating the requirements of intervals $I_4, I_5, I_6$.

Finally, since Property (9) holds, it is easy to see that, for each couple of literals $y_h(i), y_k(i')$, the requirement $r(I(y_h(i), y_k(i')))$ is also not violated. The result then follows.                                                                                                    □

## 3   Gap Hardness

We will now argue that not only the interval constrained 3-coloring problem but also its gap version is NP-hard, i.e., it is hard to distinguish between satisfiable instances and those where only up to a $(1 - \epsilon)$ fraction of constraints may be simultaneously satisfied.

For the purpose of our argument we will use the following, rather restricted, definition of gap hardness. We will only talk about maximization versions of constraint satisfaction problems. Think of an instance of the problem as being equipped with an additional parameter $t$ called threshold. We ask for a polynomial time algorithm which given the instance answers:

- "YES" if all the constraints can be satisfied,
- "NO" if there is no solution satisfying more than $t$ constraints.

Note that for instances, where more than $t$ but not all constraints can be simultaneously satisfied, any answer is acceptable. We will now restrict our attention to the case where the threshold is a fixed fraction of the total amount of constraints in the instance. We call problem A to be *gap NP-hard* if there exists a positive $\epsilon$ such that there is no polynomial time algorithm to separate feasible instances from those where only at most a $(1 - \epsilon)$ fraction of the constraint can be simultaneously satisfied unless $P = NP$.

Observe that gap NP-hardness implies APX-hardness, but not vice versa. For example the linear ordering problem (also known as max-subdag) is APX-hard [8], but is not gap NP-hard, since feasible instances may be found by topological sorting.

Let us first note that the 3-SAT problem, which we used in the reduction from the previous section, has the gap hardness property. It is the essence of the famous PCP theorems that problems with such gap hardness exist. For a proof of the gap hardness of 3-SAT see [9].

Before we show how to modify our reduction to prove gap hardness of the interval constraint coloring problem, we need to introduce the notion of *expander graphs*. For brevity we will only give the following extract from [9].

**Definition 1.** *Let $G = (V, E)$ be a d-regular graph. Let $E(S, \overline{S}) = |(S \times \overline{S}) \cap E|$ equal the number of edges from a subset $S \subseteq V$ to its complement. The edge expansion of $G$ is defined as*

$$h(G) = \min_{S:|S| \leq |V|/2} \frac{E(S, \overline{S})}{|S|}.$$

**Lemma 1.** *There exists $d_0 \in \mathbb{Z}$ and $h_0 > 0$, such that there is a polynomial-time constructible family $\{X_n\}_{n \in \mathbb{Z}}$ of $d_0$-regular graphs $X_n$ on $n$ vertices with $h(X_n) \geq h_0$. (Such graphs are called expanders).*

Let us now give a "gap preserving" reduction from gap 3-SAT to gap interval constrained 3-coloring. Consider the reduction from the previous section. Observe that the amount of intervals in each literal gadget, and therefore also in each clause gadget, is constant. The remaining intervals are the variable intervals. While it is sufficient for the NP-hardness proof to connect occurrences of the same variable in a "clique" fashion with variable intervals, it produces a potentially quadratic number of intervals. Alternatively, one could connect these occurrences in a "path" fashion, but it would give too little connectivity for the gap reduction. The path-like connection has the desired property of using only linear amount of intervals, since each occurrence of a variable is linked with at most two other ones. We aim at providing more connectivity while not increasing the amount of intervals too much. A perfect tool to achieve this goal is a family of expander graphs.

Consider the instance of the interval coloring problem obtained by the reduction from the previous section, but without any variable intervals yet. Consider literal gadgets corresponding to occurrences of a particular variable $x$. Think of these occurrences as of vertices of a graph $G$. Take an expander graph $X_{|V(G)|}$ and connect two occurrences of $x$ if the corresponding vertices in the expander are connected. For each such connection use a pair of intervals. These intervals should be the original variable interval and an interval that is one element shorter on each of the sides. We will call this pair of intervals a variable link. Repeat this procedure for each of the variables.

Observe that the number of variable links that we added is linear since all the used expander graphs are $d_0$-regular. By contrast to the simple path-like connection, we now have the property, that different occurrences of the same variable have high edge connectivity. This can be turned into high penalty for inconsistent valuations of literals in an imperfect solution.

**Theorem 2.** *Constrained interval 3-coloring is gap NP-hard.*

*Proof.* We will argue that the above described reduction is a gap-preserving reduction from the gap 3-SAT problem to the gap interval 3-coloring problem. We need to prove that there exists a positive $\epsilon$ such that feasible instances are hard to separate from those less than $(1 - \epsilon)$ satisfiable.

Let $\epsilon_0$ be the constant in the gap hardness of gap 3-SAT. We need to show two properties: that the "yes" instances of the gap 3-SAT problem are mapped to "YES" instances of our problem, and also that the "NO" instances are mapped to "NO" instances.

The first property is simple, already in the NP-hardness proof in the previous section it was shown that feasible instances are mapped by our reduction into feasible ones. To show the second property, we will take the reverse direction and argue that an almost feasible solution to the coloring instance can be transformed into an almost feasible solution to the SAT instance.

Suppose we are given a coloring $\chi$ that violates at most $\epsilon$ fraction of the constraints. Suppose the original 3-SAT instance has $q$ clauses, then our interval coloring instance has at most $c \cdot q$ intervals for some constant $c$. The number of unsatisfied intervals in the coloring $\chi$ is then at most $\epsilon q c$.

We will say that a clause is *broken* if at least one of the intervals encoding it is not satisfied by $\chi$. We will say that a variable link is broken if one of its intervals is not satisfied or one of the clauses it connects is broken. An unsatisfied variable link interval contributes a single broken link; an unsatisfied interval within a clause breaks at most $3d_0$ intervals connected to the clause. Therefore, there is at most $3d_0 \epsilon q c$ broken variable links in total.

Recall that each variable link that is not broken connects occurrences of the same variable in two different not broken clauses. Moreover, by the construction of the variable link, these two occurrences display the same logical value of the variable.

Consider the truth assignment $\phi$ obtained as follows. For each variable consider its occurrences in the not broken clauses. Each occurrence associates a logical value to the variable. Take for this variable the value that is displayed in the bigger set of not broken clauses, break ties arbitrarily.

We will now argue, that $\phi$ satisfies a big fraction of clauses. Call a clause *bad* if it is not broken, but it contains a literal such that in the coloring $\chi$ this literal was active, but $\phi$ evaluates this literal to false. Observe that if a clause is neither broken nor bad, then it is satisfied by $\phi$. It remains to bound the amount of bad clauses.

Consider the clauses that become bad from the choice of a value that $\phi$ assigns to a particular variable $x$. Let $b_x$ be the number of such clauses. By the connectivity property of expanders, the amount of variable links connecting these occurrences of $x$ with other occurrences is at least $h_0 b_x$. As we observed above, all these variable links are broken. Since there are in total at most $3d_0 \epsilon qc$ broken links, we obtain that there is at most $\frac{3}{h_0} d_0 \epsilon qc$ bad clauses. Hence, there are at most $(\frac{3}{h_0} d_0 + 1)\epsilon qc$ clauses that are either bad or broken and they cover all the clauses not satisfied by $\phi$.

It remains to fix $\epsilon = \frac{h_0}{(3d_0 + h_0)c} \epsilon_0$ to obtain the property, that more than $\epsilon_0$ unsatisfiable instances of 3-SAT are mapped to more than $\epsilon$ unsatisfiable instances of the constrained interval 3-coloring problem.    □

## Acknowledgement

We thank Steven Kelk for valuable discussion.

## References

1. Lam, T., Lanman, J., Emmett, M., Hendrickson, C., Marshall, A.G., Prevelige, P.: Mapping of protein:protein contact surfaces by hydrogen/deuterium exchange, followed by on-line high-performance liquid chromatography-electrospray ionization fourier-transform ion-cyclotron-resonance mass analysis. Journal of Chromatography A 982(1), 85–95 (2002)
2. Even, S., Itai, A., Shamir, A.: On the complexity of timetable and multicommodity flow problems. SIAM Journal on Computing 5(4), 691–703 (1976)
3. Althaus, E., Canzar, S., Emmett, M.R., Karrenbauer, A., Marshall, A.G., Meyer-Baese, A., Zhang, H.: Computing of H/D-Exchange Speeds of Single Residues form Data of Peptic Fragments. In: Proceedings of the 23rd Annual ACM Symposium on Applied Computing 2008, pp. 1273–1277 (2008)
4. Althaus, E., Canzar, S., Elbassioni, K., Karrenbauer, A., Mestre, J.: Approximating the interval constrained coloring problem. In: Gudmundsson, J. (ed.) SWAT 2008. LNCS, vol. 5124, pp. 210–221. Springer, Heidelberg (2008)
5. Gandhi, R., Khuller, S., Parthasarathy, S., Srinivasan, A.: Dependent rounding and its applications to approximation algorithms. J. ACM 53(3), 324–360 (2006)
6. Canzar, S.: Lagrangian Relaxation - Solving NP-hard problems in Computational Biology via Combinatorial Optimization. PhD thesis, Universität des Saarlandes (2008)
7. Komusiewicz, C., Niedermeier, R., Uhlmann, J.: Deconstructing Intractability - A Case Study for Interval Constrained Coloring. In: Kucherov, G., Ukkonen, E. (eds.) CPM 2009. LNCS, vol. 5577, pp. 207–220. Springer, Heidelberg (2009)
8. Papadimitriou, C.H., Yannakakis, M.: Optimization, approximation, and complexity classes. J. Comput. Syst. Sci. 43(3), 425–440 (1991)
9. Dinur, I.: The pcp theorem by gap amplification. J. ACM 54(3), 12 (2007)

# Counting Hexagonal Patches and Independent Sets in Circle Graphs

Paul Bonsma[1,*] and Felix Breuer[2]

[1] Technische Universität Berlin, Institut für Mathematik,
Sekr. MA 5-1, Straße des 17. Juni 136, 10623 Berlin, Germany
`bonsma@math.tu-berlin.de`
[2] Freie Universität Berlin, Institut für Mathematik,
Arnimallee 3, 14195 Berlin, Germany
`felix.breuer@fu-berlin.de`

**Abstract.** A hexagonal patch is a plane graph in which inner faces have length 6, inner vertices have degree 3, and boundary vertices have degree 2 or 3. We consider the following counting problem: given a sequence of twos and threes, how many hexagonal patches exist with this degree sequence along the outer face? This problem is motivated by the enumeration of benzenoid hydrocarbons and fullerenes in computational chemistry. We give the first polynomial time algorithm for this problem. We show that it can be reduced to counting maximum independent sets in circle graphs, and give a simple and fast algorithm for this problem.

**Keywords:** graph algorithms, computational complexity, counting problem, planar graph, circle graph, fullerene, hexagonal patch, fusene, polyhex.

## 1 Introduction

Since their discovery in 1985, *fullerenes* have become the subject of intense research within chemistry and related sciences. Fullerenes are molecules consisting only of carbon atoms, where every atom is connected to three other atoms, arranged in a spherical structure consisting of hexagons and pentagons. In graph theoretical terms, they can be modelled by 3-regular plane graphs with only 5-faces and 6-faces. One important topic is the enumeration of fullerenes [8]. Informally speaking, this requires solving the following subproblem: when given a partial fullerene, in how many ways can this be completed to a fullerene? This problem is also important for the study of how fullerenes are formed [13]. We now define this more precisely. A plane graph $G$ is a graph together with an embedding in the plane without edge crossings. The unbounded face is called the *outer face* and the other faces are called *inner faces*. The boundary of the outer face is simply called the *boundary* of $G$. A *fullerene patch* is a 2-connected plane graph in which all inner faces have length 5 or 6, boundary vertices have degree 2 or 3, and non-boundary vertices have degree 3. It is a *hexagonal patch* if all inner faces have

* Both authors are supported by the Graduate School "Methods for Discrete Structures" in Berlin, DFG grant GRK 1408.

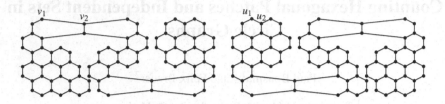

**Fig. 1.** Two different hexagonal patches with the same boundary code

length 6. Hexagonal patches are also known as *fusenes* [19], *hexagonal systems* [10], *polyhexes* [18] and $(6,3)$-*polycycles* [12] in the literature. A sequence $x_0, \ldots, x_{k-1}$ of twos and threes is a *boundary code* of a hexagonal patch $G$ if there is a way to label the boundary vertices of $G$ with $v_0, \ldots, v_{k-1}$ such that $v_0, \ldots, v_{k-1}, v_0$ is a boundary cycle of $G$, and the degree $d(v_i) = x_i$ for all $i$. It can be checked that, when given two fullerene patches, it depends only on their boundary codes whether they can be combined into a fullerene by identifying their boundaries (and embedding the resulting planar graph appropriately). Therefore we study the following counting problem (*Fullerene Patch*): *Given is a sequence $S$ of twos and threes of length $k$. How many fullerene patches exist with boundary code $S$?* This problem is also known as the *PentHex Puzzle* in the literature. It is well-known and easily verified using Euler's formula that the boundary code of a fullerene patch satifies $d_2 - d_3 = 6 - f_5$, where $d_i$ is the number of boundary vertices with degree $i$, and $f_5$ is the number of inner faces of length 5. We define the parameters $d_2(S)$ and $d_3(S)$ also for sequences $S$ of twos and threes, as expected. Until now, the only algorithms known for this problem were (super)exponential time branching algorithms [6] and/or algorithms for special cases of the problem [9,8,10]. In [3] we gave a polynomial Turing reduction from the problem on instances $S$ with $d_2(S) > d_3(S)$ to instances $S'$ with $d_2(S') - d_3(S') = 6$. When $d_2(S') - d_3(S') = 6$, any fullerene patch with boundary code $S'$ is a *hexagonal patch*. When restricted to such sequences, the counting problem is called *Hexagonal Patch*. This is the problem we consider in this paper.

A result by Guo, Hansen and Zheng [19] shows that even this restricted problem is not as easy as was first expected: in Figure 1 their example is shown which shows that also in this case, different patches may exist with the same boundary code. This can be verified by comparing the degree of $v_1$ with $u_1$, $v_2$ with $u_2$, etc. Our drawing of this graph is taken from [7]. (In [19] it is also shown that although multiple solutions may exist, they all have the same size.) Guo et al [19] and Graver [18] give conditions for when solutions are unique, if they exist. Deza et al [10] give an algorithm for deciding whether at least one solution exist. The complexity of their algorithm is however superexponential. In addition they give a polynomial time algorithm for a very restricted case (see Section 3). Brinkmann and Coppens [6] generalize this algorithm to the problem Fullerene Patch, and gave a practical implementation. However, even the question whether it can be decided in polynomial time if at least one hexagonal patch exists with a given boundary code remained open.

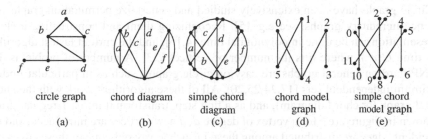

|  circle graph  |  chord diagram  |  simple chord diagram  |  chord model graph  |  simple chord model graph  |
| :---: | :---: | :---: | :---: | :---: |
|  (a)  |  (b)  |  (c)  |  (d)  |  (e)  |

**Fig. 2.** A circle graph, (simple) chord diagram and (simple) chord model graph

*In this paper we show that the counting problem Hexagonal Patch can be solved in time $O(k^3)$, where $k$ is the length of the input sequence.* This is surprising since the number of solutions may be exponential in $k$, which can be shown by generalizing the example from Figure 1. Therefore, we can only return the *number* of solutions in polynomial time, and not return a *list* of all corresponding patches. The algorithm can however be extended to subsequently generate $n$ different solutions (if this many exist) in time $O(k^2n)$. Details will follow in the full version of this paper. Combined with the result in [3], this gives a polynomial time algorithm for many instances of Fullerene Patch, and combined with [8] this allows for much more efficient enumeration of fullerenes. The problem Hexagonal Patch is also interesting by itself, since these graphs model benzenoid hydrocarbons and graphite fragments (see e.g. [19] and the references therein). In addition, our results also have implications outside of computational chemistry, since our algorithm is based on the following idea: with a few intermediate steps, we transform the problem Hexagonal Patch to the problem of counting maximum independent sets in circle graphs. A *circle graph* $G$ is the intersection graph of chords of a circle. Algorithms are known for the optimization problem of finding maximum independent sets in circle graphs [17], but counting problems on circle graphs have not been studied to our knowledge.

*In this paper we give a simple dynamic programming algorithm for counting independent sets in circle graphs.* In addition this algorithm improves the complexity for the optimization problem. Circle graphs can be represented as follows (see Figure 2(a),(b)): Every vertex of $G$ is associated with a *chord* of a circle drawn in the plane, which is a straight line segment between two points on the circle, such that two vertices are adjacent if and only if the two chords overlap (possibly only in a common end). We will represent chord diagrams with graphs as follows (see Figure 2(d)). Number the points on the circle that are ends of chords with $0, \ldots, k-1$, in order around the circle, and view these as vertices. View a chord from $i$ to $j$ as an edge $ij$. Call the resulting graph $G'$ the *chord model graph*. Note that (maximum) independent sets of the circle graph correspond bijectively to *(maximum) planar matchings* or (M)PMs of $G'$, which are (maximum) matchings $M$ that do not contain edges $ij$ and $xy$ with $i < x < j < y$. Hence counting MPMs in $G'$ is polynomially equivalent to counting maximum independent sets in circle graphs.

Circle graphs have been extensively studied and generalize permutation graphs and distance hereditary graphs, see e.g. [5]. Recognizing them and constructing a chord representation can be done in polynomial time [4,16], and the current fastest algorithm uses time $O(n^2)$, where $n$ is the number of vertices [23]. A number of problems that are NP-hard on general graphs are easy on circle graphs, such as in particular finding maximum independent sets [17,24,25,20]. All of these algorithms work with the chord model graph (or chord diagram), and as a first step, transform it into a 1-regular graph as shown in Figure 2(e): for a vertex of degree $d$, $d$ new vertices are introduced, and the $d$ incident edges are distributed among these in such a way only one of these edges can appear in a PM of $G$. This does not change the size and number of MPMs. The resulting graph $G$ has $2m$ vertices and $m$ edges, and is called the *simple chord model graph*. We assume the vertices are numbered $0, \ldots, 2m - 1$, in the proper order. Then the *length* of an edge $ij \in E(G)$ is $|j - i|$. The current fastest algorithm [25] has complexity $O(l)$, where $l$ is the sum of all edge lengths of the simple chord model graph obtained this way. Clearly this is at most $O(m^2)$, and in many cases better. However, when *dense* chord model graphs are given on $n$ vertices and $m \in \Omega(n^2)$ edges, this algorithm may need $\Omega(n^4)$ steps. Our transformation from Hexagonal Patch yields a chord model graph $G'$, which in fact may be dense.

We give a simple algorithm with complexity $O(nm)$, which not only determines the size of a MPM, but also counts the number of MPMs of the chord model graph. This improvement in time complexity is possible by working with arbitrary degrees, and not using the simple chord model graph, in contrast to all previous algorithms for this problem [25].

Finally, our results have possible applications to the problem of counting immersions of disks that extend immersions of circles. This is discussed in Section 7. The outline of the paper is as follows. In Section 2 we give definitions, and a precise formulation of the problem. In Section 3 we define locally injective homomorphisms to the hexagonal lattice (the brickwall) as a way of representing problem instances and solutions and reduce the counting problem to a problem on walks in the brickwall. In Section 4 we reduce that problem to that of counting *proper assignment sets* of the walk, which is in fact the problem of counting MPMs in chord model graphs. In Section 5 we present our algorithm for counting MPMs, and in Section 6 we give a summary of our algorithm for Hexagonal Patch. We end in Section 7 with a discussion. Due to space constraints most proofs are omitted.

## 2    Preliminaries

For basic graph theoretic notions not defined here we refer to [11]. A *walk* of *length* $k$ in a (simple) graph $G$ is a sequence of $k + 1$ vertices $v_0, \ldots, v_k$ such that $v_i$ and $v_{i+1}$ are adjacent in $G$ for all $i \in \{0, \ldots, k - 1\}$. $v_1, \ldots, v_{k-1}$ are the *internal vertices* and $v_0, v_k$ the *end vertices* of the walk. The walk is *closed* if $v_0 = v_k$. Throughout this paper we will in addition assume that $v_{i-1} \neq v_{i+1}$ for all $i \in \{1, \ldots, k - 1\}$, and if the walk is closed, $v_1 \neq v_{k-1}$ (i.e. we will assume walks *do not turn back*). If $v_i \neq v_j$ for all $i \neq j$ then the walk is a *path*. If the walk is closed and $v_i \neq v_j$ for all distinct $i, j \in \{0, \ldots, k - 1\}$ then it is also called a *cycle*. A cycle of length $k$ is also called a *k-cycle*. For a walk

$W = v_0, \ldots, v_k$, $W_x$ denotes $v_x$. If $W$ is a closed walk, then $W_x$ denotes $v_{x \bmod k}$. We will also talk about the *vertices* and *edges* of a walk, which are defined as expected. In a slight abuse of terminology, the graph consisting of these vertices and edges will also be called a walk (or path or cycle if applicable).

Let $H$ be a hexagonal patch, and $B$ be a boundary cycle of $H$ of length $k$. Let $X = x_0, \ldots, x_{k-1}$ be a sequence of twos and threes. We say that the tuple $(H, B)$ is a *solution for the boundary code* $X$ if $d(B_i) = x_i$ for all $i \in \{0, \ldots, k-1\}$. Two solutions $(H, B)$ and $(H', B')$ are considered *equivalent* if there is an isomorphism $\psi$ from $H$ to $H'$ such that $\psi(B_i) = B'_i$ for all $i$. Formally, when we ask for the number of *different* pairs $(H, B)$ that satisfy some property, we want to know how many equivalence classes contain a pair $(H, B)$ satisfying this property. The counting problem Hexagonal Patch is now defined as follows: given a sequence $X$, how many different solutions $(H, B)$ to $X$ exist?

## 3   From Boundary Codes to Walks in the Brickwall

An (infinite) 3-regular plane graph where every face has length 6 is called a *brickwall*. It can be shown that the facial cycles are the only 6-cycles of a brickwall, and that all brickwalls are isomorphic. We will use $\mathscr{B}$ to denote the brickwall as drawn in Figure 3. Edges that are horizontal (vertical) in this drawing are called the *horizontal* (*vertical*) edges of $\mathscr{B}$. Paths consisting of horizontal edges are called *horizontal paths*. Two vertices joined by a horizontal path are said to have the same *height*.

The reason that we study brickwalls is because the following mapping of hexagonal patches into them is very useful. Let $H$ be a hexagonal patch. A *locally injective homomophism (LIH)* of $H$ into $\mathscr{B}$ is a mapping of the vertices of $H$ to vertices of $\mathscr{B}$, such that adjacent vertices are mapped to adjacent vertices in $\mathscr{B}$, and such that all neighbors of any vertex in $H$ are mapped to different vertices in $\mathscr{B}$. Since the shortest cycles in $\mathscr{B}$ are of length 6, a LIH into $\mathscr{B}$ maps 6-cycles to 6-cycles. Since the faces of $\mathscr{B}$ are the only 6-cycles in $\mathscr{B}$, we see that a LIH of $H$ into $\mathscr{B}$ also maps inner faces to faces.

Loosely speaking, the idea behind these mappings is as follows. Let $H$ be a hexagonal patch of which we fix a boundary cycle $B$. When we map $H$ with a LIH $\phi$ into $\mathscr{B}$, then the boundary $B$ is mapped to some walk $W$ in $\mathscr{B}$. But now it can be shown that this walk $W$ is only determined by the choice of the initial vertices and the boundary code of $H$. Hence instead of asking how many hexagonal patches exist with a certain boundary code, we may ask how many patches exist that can be mapped properly to the brickwall, such that the boundary coincides with the walk that is deduced from the boundary code. Below we will go into more detail.

The technique of mapping patches to brickwalls is not new, and is actually considered folklore to some extent [10]. For instance, Deza et al [10] observe that Hexagonal Patch

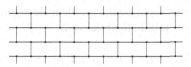

**Fig. 3.** The brickwall $\mathscr{B}$

can be solved in polynomial time if the LIH is bijective, and Graver [18] shows that the problem Hexagonal Patch can only have multiple solutions if there is a brickwall vertex that has at least three preimages in such a LIH. We will however study these mappings more in more detail than has been done before, and develop new concepts, and prove new statements which we feel are of independent interest.

Let $W$ be a walk in a 3-regular plane graph $G$. We say $W$ *makes a right (left) turn at* $i$ when edge $W_iW_{i-1}$ immediately follows edge $W_iW_{i+1}$ in the clockwise (anticlockwise) order around $W_i$. Note that since we assume that walks do not turn back and $G$ is 3-regular, $W$ makes either a left or a right turn at every $i$.

*Walk construction:* Using a given sequence $x_0, \ldots, x_{k-1}$ of twos and threes, we construct a walk $W = v_0, \ldots, v_k$ in $\mathscr{B}$ as follows. For $v_0$ and $v_1$, choose two (arbitrary) adjacent vertices. For $i \geq 1$, choose $v_{i+1}$ such that $W$ makes a left turn at $i$ if $x_i = 3$, and makes a right turn at $i$ if $x_i = 2$.

Let $W$ be a closed walk in $\mathscr{B}$ of length $k$, $H$ be a hexagonal patch, $\phi$ a LIH from $H$ to $\mathscr{B}$ and $B$ a boundary walk of $H$ of length $k$. Then the tuple $(H, \phi, B)$ is said to be a *solution for* $W$ when $\phi(B_i) = W_i$ for all $i$. Two solutions $S = (H, \phi, B)$ and $S' = (H', \phi', B')$ are considered to be *equivalent* if and only if there is an isomorphism $\psi$ from $H$ to $H'$ such that $\psi(B_i) = B'_i$ for all $i$. We say that $\psi$ is an (or demonstrates the) equivalence between $S$ and $S'$. The LIH $\phi$ allows us to use the terminology defined for $\mathscr{B}$ for the graph $H$ as well; we will for instance call edges of $H$ *horizontal* or *vertical* if their images under $\phi$ are horizontal or vertical, respectively.

Let the boundary $B$ of a hexagonal patch $H$ be mapped to the closed walk $W$ in $\mathscr{B}$ by the LIH $\phi$. This is a *clockwise solution* if and only if for every $i$, $d(B_i) = 2$ if $W$ makes a right turn at $i$, and $d(B_i) = 3$ if $W$ makes a left turn at $i$. It is *anticlockwise* when these conditions are reversed. Let $\text{RIGHT}(W)$ and $\text{LEFT}(W)$ denote the number of indices $i \in \{0, \ldots, k-1\}$ such that $W$ makes a right turn or left turn at $i$, respectively. The *turning number* of $W$ is $t(W) = (\text{RIGHT}(W) - \text{LEFT}(W))/6$. Using the fact that for a solution $(H, \phi, B)$, $\phi$ maps faces of $H$ to faces of $\mathscr{B}$, it can be shown that every solution is either clockwise or anticlockwise. Since a hexagonal patch has $d_2 - d_3 = 6$ ($d_i$ is the number of degree $i$ vertices on the boundary), Lemma 1 then follows. Variants of Lemma 2 have been proved in [7,18].

**Lemma 1.** *Let $W$ be a closed walk in $\mathscr{B}$. If $t(W) = 1$, then every solution to $W$ is clockwise. If $t(W) = -1$ then every solution to $W$ is anticlockwise. If $t(W) \notin \{-1, 1\}$, then no solution exists.*

**Lemma 2.** *Let $(H, B)$ be a solution to a boundary code $X$ and let $W$ be a walk in $\mathscr{B}$ that is constructed using $X$. Then there exists a unique LIH $\phi$ such that $(H, \phi, B)$ is a clockwise solution to $W$.*

Because of Lemma 2, we may rephrase the problem Hexagonal Patch in terms of solutions $(H, \phi, B)$ to a closed walk $W$ in the brickwall.

**Theorem 3.** *The number of different (hexagonal) solutions for a boundary code $X$ with $d_2(X) - d_3(X) = 6$ is the same as the number of different clockwise solutions for the walk $W$ in $\mathscr{B}$ that is constructed using $X$.*

*Proof.* For any solution $(H,B)$ for $X$, a unique LIH $\phi$ exists such that $(H,\phi,B)$ is a clockwise solution to $W$ (Lemma 2). For any clockwise solution $(H,\phi,B)$ to $W$, the characterization of clockwise solutions and the construction of $W$ shows that $(H,B)$ is a solution to $X$. (Since $d_2(X) - d_3(X) = 6$ and $t(W) = 1$ by Lemma 1, $W$ turns at 0 as prescribed by $x_0$.) Note that the definitions of equivalence for pairs $(H,B)$ and triples $(H,\phi,B)$ coincide and, in particular, do not depend on $\phi$. □

## 4    From Walks in the Brickwall to Assignment Sets

Throughout Section 4, $W$ denotes a closed walk in $\mathscr{B}$ with length $k$. We first sketch the main idea of this section. Non-boundary vertices and edges of a patch are called *interior*. If we consider a solution $(H,\phi,B)$ to $W$, then, as mentioned above, this defines which edges of $H$ are horizontal and vertical. Now if we start at a boundary vertex $B_i$ of $H$ that is incident with horizontal interior edge of $H$, then we can continue following this horizontal path of $H$ until we end in a different boundary vertex $B_j$. We will say that this solution *assigns $i$ to $j$*. If we only know all assignments defined by the solution this way, we can reconstruct the unique solution. We will deduce properties of such sets of assignments such that there is a solution if and only if these properties are satisfied. The purpose is to show that we may focus on counting such assignment sets instead of solutions to the walk.

For all $i,j$ where $W_i$ and $W_j$ lie on the same height, $H_{i,j}$ denotes the horizontal walk in $\mathscr{B}$ from $W_i$ to $W_j$. Consider an index $i \in \{0,\ldots,k-1\}$ and the vertex $W_i$. Let $u$ be the neighbor of $W_i$ in $\mathscr{B}$ not equal to $W_{i-1}$ or $W_{i+1}$. If $u$ has the same height as $W_i$ and $W$ makes a left turn at $i$, then index $i$ is called a *PA-index*. In Figure 4(b) an example is shown, where vertices corresponding to PA-indices are encircled, and their indices are shown. Note that if $W$ has a clockwise solution $(H,\phi,B)$, then the PA-indices are precisely those indices $i$ such that $B_i$ has degree 3 and the interior edge incident with $B_i$ is horizontal (see Figure 4(a)).

A *possible assignment (PA)* is a pair $\{i,j\}$ of PA-indices with $W_i \neq W_j$ such that $W_i$ and $W_j$ have the same height and $H_{i,j}$ does not contain any of $W_{i-1}, W_{i+1}, W_{j-1}, W_{j+1}$ (note that $H_{i,j}$ has non-zero length). For instance, in Figure 4(b) some PAs are $\{1,28\}$, $\{1,14\}$ and $\{21,32\}$, but $\{1,24\}$ is not.

Let $(H,\phi,B)$ be a clockwise solution to a closed walk $W$ in $\mathscr{B}$. An *assignment path* $P$ is a horizontal path in $H$ from $B_i$ to $B_j$ where $i \neq j$, and all edges and internal vertices of $P$ are interior edges and vertices of $H$. In Figure 4(a) the assignment paths of the given solution are shown in bold.

**Proposition 4.** *If a clockwise solution $(H,\phi,B)$ to $W$ contains an assignment path from $B_i$ to $B_j$, then $\{i,j\}$ is a PA of $W$.*

This motivates the following definition. A clockwise solution $S = (H,\phi,B)$ to a walk $W$ *assigns $i$ to $j$* if there is an assignment path from $B_i$ to $B_j$. For each clockwise solution $S$, we define the set $\mathscr{A}(S) := \{\{i,j\} : \{i,j\}$ is a PA of $W$ and $S$ assigns $i$ to $j\}$. This is the *assignment set* defined by the solution $S$.

**Lemma 5.** *Let $W$ denote a closed walk in $\mathscr{B}$ and let $S,S'$ be clockwise solutions of $W$. If $S$ and $S'$ are equivalent, then $\mathscr{A}(S) = \mathscr{A}(S')$.*

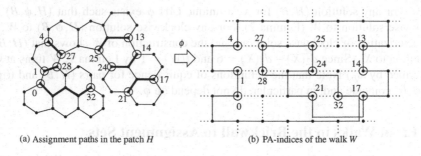

(a) Assignment paths in the patch $H$          (b) PA-indices of the walk $W$

**Fig. 4.** Assignment paths of a solution and PA-indices of a walk

Now we will deduce the properties of a set $\mathscr{A}(S)$. Proposition 6 shows that assignment paths do not share vertices. Combining this with planarity yields Proposition 7.

**Proposition 6.** *Let $(H,\phi,B)$ be a clockwise solution to $W$. Every interior vertex of $H$ and every vertex $B_i$, where $i$ is a PA-index, lies on a unique assignment path.*

**Proposition 7.** *Let $S$ be a solution to a closed walk $W$ that assigns $i$ to $j$. For any $x,y$ with $x < i < y < j$ or $i < x < j < y$, $S$ does not assign $x$ to $y$.*

These two propositions give us properties a set of the form $\mathscr{A}(S)$ for a clockwise solution $S$ necessarily has to have. Given $W$, a set $A$ of possible assignments of $W$ is a *perfect matching* on the set of PA-indices if for every PA-index $i$ of $W$ there is exactly one pair $\{i,j\} \in A$. $A$ is *non-crossing* if there do not exist assignments $\{i,j\},\{x,y\} \in A$ such that $i < x < j < y$. An *assignment set* for $W$ is a set of possible assignments of $W$. It is a *proper assignment set* if it is a non-crossing, perfect matching on the set of PA-indices of $W$. Combining Proposition 4, Proposition 6 and Proposition 7 yields Lemma 8. Lemma 9 states more or less the reverse.

**Lemma 8.** *If $S = (H,\phi,B)$ is a clockwise solution of $W$ then $\mathscr{A}(S)$ is a proper assignment set for $W$.*

**Lemma 9.** *Let $W$ denote a closed walk in $\mathscr{B}$ with $t(W) = 1$, and let $A$ be a proper assignment set of $W$. Then there exists a clockwise solution $S$ of $W$ with $\mathscr{A}(S) = A$.*

It remains to establish the converse of Lemma 5. Suppose we have two solutions $S = (H,\phi,B)$ and $S' = (H',\phi',B')$ with $\mathscr{A}(S) = \mathscr{A}(S')$. Every vertex of $H$ and $H'$ lies on the boundary or on an assignment path (Proposition 6). Therefore we can use the boundary and the assignment paths to define a bijection $\psi : V(H) \to V(H')$. When doing this appropriately, it can be shown that $\psi$ is an equivalence.

**Lemma 10.** *Let $W$ be a closed walk in $\mathscr{B}$, and let $S$ and $S'$ denote clockwise solutions of $W$. If $\mathscr{A}(S) = \mathscr{A}(S')$, then $S$ and $S'$ are equivalent.*

**Theorem 11.** *Let $W$ be a walk in $\mathscr{B}$ with $t(W) = 1$. The number of equivalence classes of solutions to $W$ is the same as the number of different proper assignment sets for $W$.*

*Proof.* The above lemmas show that $S \mapsto \mathscr{A}(S)$ gives a bijection from the set of equivalence classes of clockwise solutions of $W$ to the set of proper assignment sets for $W$, since the following properties are satisfied: (1) $\mathscr{A}$ *is well-defined:* Let $S_1$ and $S_2$ denote clockwise solutions of $W$. If $S_1$ and $S_2$ are equivalent, then $\mathscr{A}(S_1) = \mathscr{A}(S_2)$ (Lemma 5). (2) *The range of $\mathscr{A}$ is correct:* For any clockwise solution $S$ of $W$ the set $\mathscr{A}(S)$ is a proper assignment set for $W$ (Lemma 8). (3) $\mathscr{A}$ *is injective:* Let $S_1$ and $S_2$ denote clockwise solutions of $W$. If $\mathscr{A}(S_1) = \mathscr{A}(S_2)$, then $S_1$ and $S_2$ are equivalent (Lemma 10). (4) $\mathscr{A}$ *is surjective:* For any proper assignment set $A$ for $W$, there exists a clockwise solution $S$ of $W$ with $\mathscr{A}(S) = A$ (Lemma 9). □

It follows that for solving the Hexagonal Patch problem, we may focus on counting proper assignment sets for the walk $W$ (assuming $t(W) = 1$).

## 5   Counting Maximum Planar Matchings

In this section we will observe that the remaining algorithmic problem is that of counting independent sets in circle graphs, and present a fast algorithm for this problem. We use the closed walk $W$ in $\mathscr{B}$ to construct a graph $G^W$ with vertex set $V = \{0, \ldots, n-1\}$, where $n$ is the number of PA-indices of $W$. Let $p_0, \ldots, p_{n-1}$ be all PA-indices of $W$, numbered according to their order in $W$. Then the edge set of $G^W$ will be $E = \{ij \mid \{p_i, p_j\}$ is a PA of $W\}$. The following lemma is now easily observed.

**Lemma 12.** *Let $G^W$ be the graph as constructed above from the walk $W$. The number of proper assignment sets for $W$ is equal to the number of perfect PMs in $G^W$.*

Now we will present an algorithm for counting MPMs of a graph $G$ with $V(G) = \{0, \ldots, n-1\}$. As mentioned in the introduction, this is equivalent to counting maximum independent sets in a circle graph $H$, where $G$ is the chord model graph of $H$. We will present this algorithm for the general case where $G$ has edge weights: $w_{ij}$ denotes the edge weight of $ij$, and a PM $M$ is *maximum* if $\sum_{e \in M} w_e$ is maximum. For $i, j \in V(G)$ with $i \le j$, let $G_{i,j} = G[\{i, \ldots, j\}]$. For $i, j \in V(G)$ with $i \le j$, let $S_{i,j}$ denote the size of a MPM in $G_{i,j}$. In particular, $S_{0,n-1}$ is the size of a MPM in $G$. If $i > j$ or $\{i, j\} \not\subseteq V(G)$, we define $S_{i,j} = 0$. We now give a subroutine $S(i, j)$ for calculating $S_{i,j}$, which considers the sizes of various PMs for $G_{i,j}$, and returns the size of the largest PM.

A subroutine $S(i, j)$ for calculating $S_{i,j}$:
(1)   $m := S_{i+1,j}$
(2)   **For** $v \in N(i)$ with $i + 1 \le v \le j$:
(3)      $m := \max\{m, w_{iv} + S_{i+1,v-1} + S_{v+1,j}\}$
(4)   Return $m$

**Lemma 13.** *Let $G$ be a graph with $V(G) = \{0, \ldots, n-1\}$ and $i, j \in V(G)$. If the values $S_{x,y}$ are known for all $x, y$ with $y - x < j - i$, then the subroutine $S(i, j)$ computes $S_{i,j}$ in time $O(d(i))$.*

Let $N_{i,j}$ denote the number of MPMs in $G_{i,j}$ if $i, j \in V(G)$ and $i \le j$, and let $N_{i,j} = 1$ if $i > j$ or $\{i, j\} \not\subseteq V(G)$. Below is a similar subroutine $N(i, j)$ for calculating $N_{i,j}$, which considers various PMs for $G_{i,j}$, checks whether they are maximum by comparing the size with $S_{i,j}$, and keeps track of the number of MPMs using the variable $N$.

A subroutine $N(i, j)$ for calculating $N_{i,j}$:
(1)  **If** $S_{i,j} = S_{i+1,j}$ **then** $N := N_{i+1,j}$ **else** $N := 0$
(2)  **For** $v \in N(i)$ with $i+1 \le v \le j$:
(3)      **If** $S_{i,j} = w_{iv} + S_{i+1,v-1} + S_{v+1,j}$ **then** $N := N + N_{i+1,v-1} \times N_{v+1,j}$
(4)  Return $N$

**Lemma 14.** *Let $G$ be a graph with $V(G) = \{0, \ldots, n-1\}$ and $i, j \in V(G)$. If the values $S_{x,y}$ and $N_{x,y}$ are known for all $x, y$ with $y - x < j - i$, and $S_{i,j}$ is known, then the subroutine $N(i, j)$ computes $N_{i,j}$ in time $O(d(i))$.*

**Theorem 15.** *Let $G$ be a graph with $V(G) = \{0, \ldots, n-1\}$ on $m$ edges. The size and number of MPMs of $G$ can be computed in time $O(nm)$.*

*Proof.* For $d = 0$ to $n-1$, we consider all $i, j \in \{0, \ldots, n-1\}$ with $j - i = d$, and calculate $S_{i,j}$ and $N_{i,j}$ using the above subroutines. This way, for every value of $d$, every vertex of $G$ is considered at most once in the role of $i$. For this choice of $i$, calculating $S_{i,j}$ and $N_{i,j}$ takes time $O(d(i))$ (Lemma 13, Lemma 14). Hence for one value of $d$ this procedure takes time $O(\sum_{i \in V(G)} d(i)) = O(m)$.  □

We remark that Valiente's algorithm [25] for simple (1-regular) chord model graphs can also be extended by using Subroutine $N(i, j)$ to calculate $N_{i,j}$ in constant time, immediately any time after a value $S_{i,j}$ is calculated. This then yields time complexity $O(l)$ and space complexity $O(n)$ (see Section 1). In some cases it may be better to transform to a simple chord model graph and use Valiente's algorithm.

## 6  Summary of the Algorithm

We now summarize how counting the number of hexagonal patches that satisfy a given boundary code $X$ of length $k$ can be done in time $O(k^3)$. W.l.o.g. $d_2(X) - d_3(X) = 6$. First use $X$ to construct a walk $W$ in $\mathscr{B}$ of length $k$, as shown in Section 3. Theorem 3 shows that we may now focus on counting clockwise solutions to $W$. If $W$ is not closed it clearly has no solution. Since $d_2(X) - d_3(X) = 6$ we may now assume $t(W) = 1$. Then Theorem 11 shows we may focus on counting proper assignment sets for $W$. Now construct $G^W$ as shown in Section 5. $G^W$ has $n$ vertices where $n < k$ is the number of PA-indices of $W$ (and $O(n^2)$ edges). By Lemma 12, the number of proper assignment sets for $W$ is equal to the number of MPMs of $G^W$, provided that $G^W$ has a perfect PM. This number and property can be determined in time $O(n^3) \in O(k^3)$ (Theorem 15).

## 7  Discussion

Our first question is whether the complexity of $O(k^3)$ can be improved. Secondly, considering the motivation from benzenoid hydrocarbons, it is interesting to study whether a patch exists that can be realized in $\mathbb{R}^3$ using *regular* hexagons with 'minimal distortion'. This is the problem from Section 3, but requires in addition giving a consistent linear order $\prec$ ('depth') for all vertices mapped to the same vertex of $\mathscr{B}$. That is, if patch $H$ is mapped to $\mathscr{B}$ by LIH $\phi$, $uv, xy \in E(H)$, $\phi(u) = \phi(x)$ and $\phi(v) = \phi(y)$, then

$u \prec x$ should hold if and only if $v \prec y$. Surprisingly these problems are not equivalent. It may also be interesting to study generalizations such as to surfaces of higher genus.

After we presented an early version of this work [2], Jack Graver pointed us to a similar well-studied problem in topology. Let $S^1$ denote the unit circle and $D^2$ the unit disk in $\mathbb{R}^2$. An *immersion* is a continuous function $f : A \to B$ such that for every $x$ in $A$ there is a neighborhood $N$ of $x$ such that $f|_N$ is a homeomorphism. (A *curve* when $A = S^1$, $B = \mathbb{R}^2$.) An immersion $c : S^1 \to \mathbb{R}^2$ of the circle into the plane is *normal* if $c$ has only finitely many double-points and $c$ crosses itself at each of these. Two immersions $d, d'$ are *equivalent* if there exists a homeomorphism $\phi : \mathbb{R}^2 \to \mathbb{R}^2$ such that $d \circ \phi = d'$. Now the *Immersion Extension* problem is this: given an immersion $c : S^1 \to \mathbb{R}^2$, how many immersions $d : D^2 \to \mathbb{R}^2$ exist that extend $c$? Note that this problem is not combinatorial, therefore it makes no sense to study its computational complexity. One can turn it into a combinatorial problem by restricting the input to *piecewise linear (PL)* curves $c : S^1 \to \mathbb{R}^2$.

When viewing the walk constructed in Section 3 as a curve, there are obvious similarities between the Hexagonal Patch problem and the Immersion Extension problem. However, to our knowledge it is an open problem to prove that these problems are in fact equivalent. The ideas introduced here may be helpful for giving such a proof. Establishing this would provide insight to both problems, since the Immersion Extension problem is well-studied – at least on normal curves – see e.g. [1,14,22]. Interestingly, Blank [1,15] reduces the Immersion Extension problem problem to a combinatorial problem that is essentially the same as counting MPMs in simple chord model graphs. He does not address the complexity of this problem. Shor and Van Wyk [22] were the first to study the complexity of the combinatorial Immersion Extension problem on normal curves. They give an $O(n^3 \log n)$ algorithm where $n$ is the number of pieces of the PL curve $c$. Assuming the equivalence of the Immersion Extension problem and the Hexagonal Patch problem, this would give an alternative algorithm for Hexagonal Patch; note that there are methods for transforming general PL curves to equivalent normal PL curves [21]. Since our algorithm does not need such a step, it is not only faster but also much easier to implement (see also [21]). However, the question of equivalence of these problems is still interesting because many generalizations of the Immersion Extension problem have been studied [14]. Finally, we believe that in fact our method can be adapted to give a simple and fast algorithm for the combinatorial Immersion Extension problem that does not require the assumption that the given curve is normal, but that is beyond the scope of this paper.

**Acknowledgement.** We thank Gunnar Brinkmann for introducing us to this subject and his suggestions, Hajo Broersma for the discussions on this topic, and Jack Graver for referring us to the literature on immersions.

# References

1. Blank, S.J.: Extending immersions of the circle. PhD thesis, Brandeis University (1967)
2. Bonsma, P., Breuer, F.: Finding fullerene patches in polynomial time I: Counting hexagonal patches (2008), http://arxiv.org/abs/0808.3881

3. Bonsma, P., Breuer, F.: Finding fullerene patches in polynomial time. In: Dong, Y., Du, D.-Z., Ibarra, O.H. (eds.) ISAAC 2009. LNCS, vol. 5878, pp. 750–759. Springer, Heidelberg (2009)

4. Bouchet, A.: Reducing prime graphs and recognizing circle graphs. Combinatorica 7(3), 243–254 (1987)

5. Brandstädt, A., Le, V.B., Spinrad, J.P.: Graph classes, a survey. SIAM, Philadelphia (1999)

6. Brinkmann, G., Coppens, B.: An efficient algorithm for the generation of planar polycyclic hydrocarbons with a given boundary. MATCH Commun. Math. Comput. Chem. (2009)

7. Brinkmann, G., Delgado-Friedrichs, O., von Nathusius, U.: Numbers of faces and boundary encodings of patches. In: Graphs and discovery. DIMACS Ser. Discrete Math. Theoret. Comput. Sci, vol. 69, pp. 27–38. Amer. Math. Soc., Providence (2005)

8. Brinkmann, G., Dress, A.W.M.: A constructive enumeration of fullerenes. J. Algorithms 23(2), 345–358 (1997)

9. Brinkmann, G., Nathusius, U.v., Palser, A.H.R.: A constructive enumeration of nanotube caps. Discrete Appl. Math. 116(1-2), 55–71 (2002)

10. Deza, M., Fowler, P.W., Grishukhin, V.: Allowed boundary sequences for fused polycyclic patches and related algorithmic problems. J. Chem. Inf. Comput. Sci. 41, 300–308 (2001)

11. Diestel, R.: Graph theory, 3rd edn. Springer, Berlin (2005)

12. Dutour Sikirić, M., Deza, M., Shtogrin, M.: Filling of a given boundary by $p$-gons and related problems. Discrete Appl. Math. 156, 1518–1535 (2008)

13. Endo, M., Kroto, H.W.: Formation of carbon nanofibers. J. Phys. Chem. 96, 6941–6944 (1992)

14. Eppstein, D., Mumford, E.: Self-overlapping curves revisited. In: SODA 2009, pp. 160–169. SIAM, Philadelphia (2009)

15. Francis, G.K.: Extensions to the disk of properly nested plane immersions of the circle. Michigan Math. J. 17(4), 377–383 (1970)

16. Gabor, C.P., Supowit, K.J., Hsu, W.L.: Recognizing circle graphs in polynomial time. J. ACM 36(3), 435–473 (1989)

17. Gavril, F.: Algorithms for a maximum clique and a maximum independent set of a circle graph. Networks 3(3), 261–273 (1973)

18. Graver, J.E.: The $(m, k)$-patch boundary code problem. MATCH Commun. Math. Comput. Chem. 48, 189–196 (2003)

19. Guo, X., Hansen, P., Zheng, M.: Boundary uniqueness of fusenes. Discrete Appl. Math. 118, 209–222 (2002)

20. Nash, N., Lelait, S., Gregg, D.: Efficiently implementing maximum independent set algorithms on circle graphs. ACM J. Exp. Algorithmics 13 (2009)

21. Seidel, R.: The nature and meaning of perturbations in geometric computing. Discrete Comput. Geom. 19(1), 1–17 (1998)

22. Shor, P.W., Van Wyk, C.J.: Detecting and decomposing self-overlapping curves. Comput. Geom. 2, 31–50 (1992)

23. Spinrad, J.P.: Recognition of circle graphs. J. Algorithms 16(2), 264–282 (1994)

24. Supowit, K.J.: Finding a maximum planar subset of a set of nets in a channel. IEEE T. Comput. Aid. D. 6(1), 93–94 (1987)

25. Valiente, G.: A new simple algorithm for the maximum-weight independent set problem on circle graphs. In: Ibaraki, T., Katoh, N., Ono, H. (eds.) ISAAC 2003. LNCS, vol. 2906, pp. 129–137. Springer, Heidelberg (2003)

# Approximating Maximum Diameter-Bounded Subgraphs

Yuichi Asahiro[1], Eiji Miyano[2], and Kazuaki Samizo[2]

[1] Department of Information Science, Kyushu Sangyo University,
Fukuoka 813-8503, Japan
asahiro@is.kyusan-u.ac.jp
[2] Department of Systems Design and Informatics, Kyushu Institute of Technology,
Fukuoka 820-8502, Japan
{miyano@,samizo@theory.}ces.kyutech.ac.jp

**Abstract.** The paper studies the maximum diameter-bounded subgraph problem (MAXDBS for short) which is defined as follows: Given an $n$-vertex graph $G$ and a fixed integer $d \geq 1$, the goal is to find its largest subgraph of the diameter $d$. If $d = 1$, the problem is identical to the maximum clique problem and thus it is $\mathcal{NP}$-hard to approximate MAXDBS to within a factor of $n^{1-\varepsilon}$ for any $\varepsilon > 0$. Also, it is known to be $\mathcal{NP}$-hard to approximate MAXDBS to within a factor of $n^{1/3-\varepsilon}$ for any $\varepsilon > 0$ and a fixed $d \geq 2$. In this paper, we first strengthen the hardness result; we prove that, for any $\varepsilon > 0$ and a fixed $d \geq 2$, it is $\mathcal{NP}$-hard to approximate MAXDBS to within a factor of $n^{1/2-\varepsilon}$. Then, we show that a simple polynomial-time algorithm achieves an approximation ratio of $n^{1/2}$ for any even $d \geq 2$, and an approximation ratio of $n^{2/3}$ for any odd $d \geq 3$. Furthermore, we investigate the (in)tractability and the (in)approximability of MAXDBS on subclasses of graphs, including chordal graphs, split graphs, interval graphs, and $k$-partite graphs.

## 1 Introduction

MAXCLIQUE is one of the central problems in graph theory and combinatorial optimization and thus many researches are devoted to it [4]: A clique $Q$ in a graph is a set of pairwise adjacent vertices, i.e., the diameter of the subgraph induced by $Q$ is one. Given a graph $G$, the goal of MAXCLIQUE is to find a clique of maximum cardinality in $G$. The decision version of MAXCLIQUE was one of Karp's original 21 problems shown $\mathcal{NP}$-complete in [11]. The currently best polynomial-time approximation algorithm is only guaranteed to find a clique within a factor of $n(\log \log n)^2/(\log n)^3$ of optimal for $n$-vertex graphs [6]. On the negative side, under the assumption that $\mathcal{NP} \neq \mathcal{ZPP}$, Bellare, Goldreich and Sudan [3] proved that MAXCLIQUE cannot be efficiently approximated within a factor of $n^{1/3-\varepsilon}$ and then Håstad [10] showed a stronger hardness ratio of $n^{1-\varepsilon}$ for any $\varepsilon > 0$. Note that under a weaker assumption that $\mathcal{P} \neq \mathcal{NP}$, Håstad's hardness ratio becomes $n^{1/2-\varepsilon}$. The strongest hardness ratio so far is by Zuckerman [15] who proved that, assuming only $\mathcal{P} \neq \mathcal{NP}$, MAXCLIQUE cannot be approximated within a factor of $n^{1-\varepsilon}$ for any $\varepsilon > 0$ in polynomial time.

A. López-Ortiz (Ed.): LATIN 2010, LNCS 6034, pp. 615–626, 2010.
© Springer-Verlag Berlin Heidelberg 2010

In this paper, we consider a natural generalization of MAXCLIQUE, named the maximum diameter-bounded subgraph problem (MAXDBS for short), which is defined as follows: Given a graph and an integer $d \geq 1$, the goal is to find its largest subgraph of the diameter $d$. If $d = 1$, the problem is identical to MAX-CLIQUE. Thus, under the assumption that $\mathcal{P} \neq \mathcal{NP}$, for any $\varepsilon > 0$, MAXDBS cannot be efficiently approximated within a factor of $n^{1-\varepsilon}$ if $d = 1$. Furthermore, it can be shown that, for any $\varepsilon > 0$ and a fixed $d \geq 2$, it is $\mathcal{NP}$-hard to approximate MAXDBS to within a factor of $n^{1/3-\varepsilon}$ by using the gap preserving reduction provided by Marinček and Mohar [12] (although they assumed that $\mathcal{NP} \neq \mathcal{ZPP}$ in their original proof). In this paper we first strengthen the hardness result for $d \geq 2$; we prove that, for any $\varepsilon > 0$ and a fixed $d \geq 2$, it is $\mathcal{NP}$-hard to approximate MAXDBS to within a factor of $n^{1/2-\varepsilon}$ for general graphs. Then, we present a simple polynomial-time algorithm which achieves an approximation ratio of $n^{1/2}$ for any even $d \geq 2$, and an approximation ratio of $n^{2/3}$ for any odd $d \geq 3$. As far as the authors know, there has been no result on the approximability of MAXDBS for $d \geq 2$ so far.

MAXCLIQUE (and also MAXDBS) is very difficult even to approximate. Fortunately, however, it is known [8] that if the input graph is restricted to planar graphs, chordal graphs, or graphs with bounded degrees, then MAX-CLIQUE is solvable in polynomial time. This tractability suggests that if we restrict the set of instances to such subclasses of graphs, MAXDBS for a fixed $d \geq 2$ might be also solvable efficiently. From this point of view, this paper investigates MAXDBS, namely, our work focuses on the (in)tractability and the (in)approximability of MAXDBS on subclasses of graphs, including chordal graphs, split graphs, interval graphs, and $k$-partite graphs. The split graphs and the interval graphs are subclasses of the chordal graphs. The definitions of these graphs are presented in Section 2.

The following is a list of our main results shown in the paper, the tractability and the (in)approximability for the graph classes:

(i) For general graphs, MAXDBS can be approximated within a factor of $n^{1/2}$ for even $d$'s and $n^{2/3}$ for odd $d$'s. [Section 3.1]

(ii) For any $\varepsilon > 0$ it is $\mathcal{NP}$-hard to approximate MAXDBS to within a factor of $n^{1/2-\varepsilon}$ for general graphs. [Section 3.2]

(iii) For odd $d$'s, MAXDBS can be solved in polynomial time for chordal graphs and also for split graphs. [Sections 4 and 5]

(iv) For $d = 2$, MAXDBS can be approximated within a factor of $n^{1/3}$ for split graphs. [Section 5]

(v) For any $\varepsilon > 0$ and even $d$'s, it remains $\mathcal{NP}$-hard to approximate MAXDBS to within a factor of $n^{1/3-\varepsilon}$ for chordal graphs and also for split graphs. [Sections 4 and 5]

(vi) MAXDBS can be solved in polynomial time for interval graphs. [Section 6]

(vii) For any $\varepsilon > 0$ it remains $\mathcal{NP}$-hard to approximate MAXDBS to within a factor of $n^{1/3-\varepsilon}$ for bipartite graphs with $d \geq 3$, and also for $k$-partite graphs ($k \geq 3$) with $d \geq 2$. [Section 7]

One can see that the complexity of MAXDBS depends on the parity of $d$, especially, if the set of input graphs is limited to chordal graphs.

## 2    Problem and Algorithms

**Notation.** Let $G = (V, E)$ be a connected undirected graph, where $V$ and $E$ denote the set of vertices and the set of edges, respectively. $V(G)$ and $E(G)$ also denote the vertex set and the edge set of $G$, respectively. We denote an edge with endpoints $u$ and $v$ by $(u, v)$. The maximum degree among all vertices in a graph $G$ is denoted by $deg_{\max}(G)$. For a vertex $v$, the set of vertices adjacent to $v$ in $G$ is denoted by $N_G(v)$, and $N_G^+(v)$ denotes $N_G(v) \cup \{v\}$.

A graph $G_S$ is a subgraph of a graph $G$ if $V(G_S) \subseteq V(G)$ and $E(G_S) \subseteq E(G)$. For a subset of vertices $U \subseteq V$, let $G[U]$ be the subgraph induced by $U$. For a subgraph $G_S = (V_S, E_S)$ of $G$, if $E_S = V_S \times V_S$, then $G_S$ (or $G[V_S]$) and $V_S$ are called a *clique* and a *clique set*, respectively. If $G_S$ is the maximum clique in $G$, then $V_S$ is called the *maximum clique set*. Also, If $(V_S \times V_S) \cap E = \emptyset$, then $G_S$ and $V_S$ are called a *stable graph* and a *stable set*, respectively.

For a positive integer $d \geq 1$ and a graph $G$, the $d$-th power of $G$, denoted by $G^d = (V(G), E^d)$, is the graph formed from $V(G)$, where all pairs of vertices $u, v \in G$ such that $dist_G(u, v) \leq d$ are connected by an edge $(u, v)$. Note that $E(G) \subseteq E^d$, i.e., the original edges in $E(G)$ are retained.

A path $P$ of length $\ell$ from a vertex $v_0$ to a vertex $v_\ell$ is represented as a sequence of vertices such that $P = \langle v_0, v_1, \cdots, v_\ell \rangle$. A cycle $C$ of length $\ell$ is similarly written as $C = \langle v_0, v_1, \cdots, v_{\ell-1}, v_0 \rangle$. In this paper, we deal with simple paths and simple cycles only, that is, $v_i \neq v_j$ for any $v_i$ and $v_j$ in the sequences of vertices. For a pair of vertices $u$ and $v$, the length of a shortest path from $u$ to $v$, i.e., the distance between $u$ and $v$ is denoted by $dist_G(u, v)$, and the diameter of $G$ is defined as $diam(G) = \max_{u,v \in V} dist_G(u, v)$.

The definitions of graph classes are from [5]: A *chord* of a cycle is an edge between two vertices of the cycle that is not an edge of the cycle. A graph $G$ is *chordal* if each cycle in $G$ of length at least 4 has at least one chord. A graph $G = (V, E)$ is a *split graph* if there is a partition of $V$ into a clique set $V_1$ and a stable set $V_2$ such that $V_1 \cap V_2 = \emptyset$ and $V_1 \cup V_2 = V$. A graph $G = (V, E)$ is an *interval graph* if the following two conditions are satisfied for a collection $\mathcal{I}$ of intervals on the real line: (i) There is a one-to-one correspondence between $V$ and $\mathcal{I}$, and (ii) for a pair of vertices $u, v \in V$ and their corresponding two intervals $I_u, I_v \in \mathcal{I}$, $I_u \cap I_v \neq \emptyset$ if and only if $(u, v) \in E$. We would like to note again that the split graphs and interval graphs are subclasses of the chordal graphs [5]. A graph $G = (V, E)$ is a $k$-*partite graph* if there is a partition of $V$ into $k$ disjoint stable sets $V_1, V_2, \cdots, V_k$, namely, for any $i \neq j$, $V_i \cap V_j = \emptyset$, $V = \bigcup_{i=1}^{k} V_i$, and each $V_i$ is a stable set. Note that the class of $k$-partite graphs is a subclass of $(k + 1)$-partite graphs by definition. A graph is a *star* if it is a rooted tree of height one.

For the maximization problems, an algorithm ALG is called a $\sigma$-approximation algorithm and ALG's approximation ratio is $\sigma$ if $OPT(G)/ALG(G) \leq \sigma$ holds for every input $G$, where $ALG(G)$ and $OPT(G)$ are the numbers of vertices of obtained subgraphs by ALG and an optimal algorithm, respectively.

**Problem.** The maximum diameter-bounded subgraph problem (MaxDBS) considered in this paper is formulated as follows:

---

**Problem:** MAXIMUM DIAMETER-BOUNDED SUBGRAPH (MaxDBS)

**Input:** A connected undirected graph $G = (V, E)$ and an integer $d$
  such that $1 \le d \le |V| - 1$

**Output:** A subgraph $G_S$ of $G$ such that $diam(G_S) \le d$ and $|V(G_S)|$
  is the maximum among all such subgraphs.

---

Throughout the paper, we use the following notation: $n = |V|$ and $m = |E|$ for the input graph $G$. If $d = 1$, the problem is identical to MaxClique, and hence MaxDBS is a generalization of MaxClique in terms of the diameter of output subgraphs.

**Algorithms.** Let us start with a construction algorithm PowerOfGraph of the $d$-th power of a graph; given a connected undirected graph $G = (V, E)$ and an integer $d$, PowerOfGraph outputs the $d$-th power $G^d$ of $G$: First, compute $dist_G(u, v)$ for any pair of vertices $u, v \in V$, and then add an edge $(u, v)$ if $dist_G(u, v) \le d$. The correctness of PowerOfGraph is obvious by the definition of the power of a graph, and PowerOfGraph runs in polynomial time.

Let an algorithm which can obtain optimal solutions for MaxClique (or, MaxDBS for $d = 1$) be FindClique. Here, the running time of FindClique might be exponential. By using FindClique as the main procedure, we can design a simple algorithm called ByFindClique for MaxDBS with $d \ge 2$, whose output is optimal. The following is a description of the algorithm ByFindClique.

---

**Algorithm ByFindClique**

**Input:** A connected undirected graph $G = (V, E)$ and an integer $d$
  such that $1 \le d \le |V| - 1$

**Output:** A subgraph $G_S$ of $G$ such that $diam(G_S) \le d$ and $|V(G_S)|$
  is the maximum among all such subgraphs.

**Step 1.** Obtain the $d$-th power $G^d$ of $G$ by PowerOfGraph$(G, d)$.

**Step 2.** Apply FindClique to $G^d$, and then obtain a maximum
  clique $G_Q = (V_Q, E_Q)$ in $G^d$.

**Step 3.** Output $G_S = (V_Q, E \cap E_Q)$.

---

Since any subgraph $H$ of $G$ such that $diam(H) \le d$ transforms into a clique in $G^d$ after applying PowerOfGraph$(G, d)$, ByFindClique can find a maximum subgraph such that its diameter is bounded by $d$. As for the running time, ByFindClique is an exponential time algorithm if the running time of `FindClique` is exponential, since `ByFindClique` uses `FindClique` as a sub-procedure. If `FindClique` is a polynomial time algorithm, so is `ByFindClique`, because all the steps can be executed in polynomial time.

If $d = 2$, we may design a simple approximation algorithm called `FindStar` which finds a vertex $v$ having the maximum degree $deg_{\max}(G)$ in the input graph $G$ and then outputs the subgraph $G[N_G^+(v)]$ induced by the vertex $v$ and its adjacent $deg_{\max}(G)$ vertices. The number of vertices of the output is $deg_{\max}(G) + 1$ in total. `FindStar` is rather simple and runs in linear time. Little surprisingly, however, `FindStar` outputs good solutions as we can see in the following sections. Combining two algorithms `PowerOfGraph` and `FindStar`, we obtain the following algorithm `ByFindStar`, in which the $\lfloor d/2 \rfloor$-th power of the input graph is constructed while the $d$-th power is constructed in `ByFindClique`. The running time of `ByFindStar` is also polynomial in the input size.

---

**Algorithm `ByFindStar`**

**Input:** A connected undirected graph $G = (V, E)$ and an integer $d$ such that $1 \leq d \leq |V| - 1$

**Output:** A subgraph $G_S$ of $G$ such that $diam(G_S) \leq d$ and $|V(G_S)|$ is the maximum among all such subgraphs.

**Step 1.** Obtain the $\lfloor d/2 \rfloor$-th power $G^{\lfloor d/2 \rfloor}$ of the graph $G$ by applying `PowerOfGraph`$(G, \lfloor d/2 \rfloor)$.

**Step 2.** Apply `FindStar` to $G^{\lfloor d/2 \rfloor}$, and then obtain a largest star $G_T = (V_T, E_T)$ in $G^{\lfloor d/2 \rfloor}$.

**Step 3.** Output $G_S = (V_T, E \cap E_T)$.

---

## 3   General Graphs

### 3.1   Approximability

Given an $n$-vertex graph, the currently best polynomial-time approximation algorithm achieves an approximation ratio of $O(n(\log \log n)^2/(\log n)^3)$ for MAX-CLIQUE [6] as mentioned in Section 1. We can adopt this algorithm as `FindClique` to design the approximation version of the algorithm `ByFindClique` for MaxDBS with $d \geq 2$. The obtained algorithm has the same approximation ratio as the one for MAXCLIQUE:

**Proposition 1.** *Given an $n$-vertex graph and a fixed integer $d \geq 2$, we can design a polynomial time $O(n(\log \log n)^2/(\log n)^3)$-approximation algorithm for MAXDBS.* □

Now we examine the approximation ratio of the algorithm `FindStar`.

**Lemma 1.** *Given an $n$-vertex graph and a fixed integer $d \geq 2$, `FindStar` achieves an approximation ratio of $O(n^{1-1/d})$ for MAXDBS.*

*Proof.* The output size $FindStar(G)$ of the algorithm FindStar is $deg_{max}(G)+1$. If $deg_{max}(G) \geq n^{1/d}$, then

$$\frac{OPT(G)}{FindStar(G)} \leq \frac{n}{deg_{max}(G)+1} < n^{1-1/d}$$

since $OPT(G) \leq n$. Conversely, assume that $deg_{max}(G) < n^{1/d}$. Since the number of vertices in the optimal subgraph with the diameter $d$ is at most $1 + deg_{max}(G)+(deg_{max}(G))^2+\cdots+(deg_{max}(G))^d = (deg_{max}(G)^{d+1}-1)/(deg_{max}(G) - 1) = (deg_{max}(G))^d + O(deg_{max}(G)^{d-1})$, the following holds:

$$\frac{OPT(G)}{FindStar(G)} \leq \frac{deg_{max}(G)^d + O(deg_{max}(G)^{d-1})}{deg_{max}(G)+1}$$
$$= O(deg_{max}(G)^{d-1}) = O(n^{1-1/d}). \qquad \square$$

By the above Lemma 1, one can see that, for the case $d = 2$, the approximation ratio of FindStar is $O(n^{1/2})$. Although FindStar is quite simple and its approximation ratio is easy to analyze, the approximation ratio of FindStar is the best possible for the case $d = 2$ in the sense that the lower bound of the approximation ratio is $\Omega(n^{1/2-\varepsilon})$ for any $\varepsilon > 0$ as shown later. Furthermore, utilizing Lemma 1 and PowerOfGraph, we can obtain the following theorem for any even $d$, which is also best possible in the sense that the inapproximability bound for any even $d$ is also $\Omega(n^{1/2-\varepsilon})$ as shown later.

**Theorem 1.** *Given an $n$-vertex graph and a fixed even integer $d \geq 2$, we can design a polynomial time $O(n^{1/2})$-approximation algorithm for MAXDBS.*

*Proof.* We show ByFindStar achieves the approximation ratio of $O(n^{1/2})$ for $d \geq 4$. For any pair of vertices $u$ and $v$, it is easy to see that $dist_G(u, v) \leq d$ if and only if $dist_{G^{d/2}}(u, v) \leq 2$. Hence an optimal set of vertices for MAXDBS with an input graph $G$ and distance $d$ is also an optimal set of vertices for MAXDBS with an input graph $G^{d/2}$ and distance two. Since the approximation ratio of FindStar is $O(n^{1/2})$ for the latter problem by Lemma 1, the approximation ratio of ByFindStar is also $O(n^{1/2})$ for the former problem. $\qquad \square$

Next, take a look at odd $d$'s. By a similar argument to Lemma 1, we obtain the following theorem.

**Theorem 2.** *Given an $n$-vertex graph and a fixed odd integer $d \geq 3$, we can design a polynomial time $O(n^{2/3})$-approximation algorithm for MAXDBS. (The proof is omitted.)* $\qquad \square$

### 3.2 Inapproximability

In this section, we show lower bounds of the approximability of MAXDBS for general graphs. Under the assumption that $\mathcal{P} \neq \mathcal{NP}$, the best approximation lower bound so far is $n^{1-\varepsilon}$ for MAXCLIQUE (MAXDBS with $d = 1$) [15]. We will show the inapproximability of MAXDBS by providing a *gap-preserving reduction*

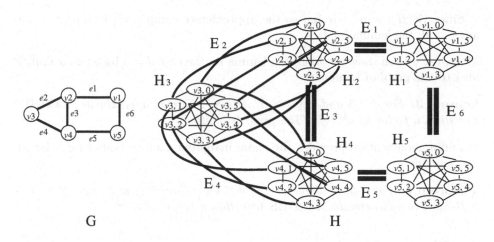

**Fig. 1.** A graph $G$ and the reduced graph $H$ from $G$

from MAXCLIQUE (refer to, e.g., pp.307–308 in [14]). First of all, the following lemma deals with the case $d = 2$:

**Lemma 2.** *For $d = 2$ and any $\varepsilon > 0$, it is $\mathcal{NP}$-hard to approximate MAXDBS to within a factor of $n^{1/2-\varepsilon}$.*

*Proof.* We give a gap-preserving reduction from MAXCLIQUE to MAXDBS. Namely, from an instance $G = (V(G), E(G))$ of MAXCLIQUE, we construct an instance $H = (V(H), E(H))$ of MAXDBS. Let $OPT_1(G)$ (or $OPT_2(H)$, resp.) denote the number of vertices of an optimal solution for $G$ of MAX-CLIQUE (or $H$ of MAXDBS, resp.). Let $V(G) = \{v_1, v_2, \cdots, v_n\}$ of $n$ vertices and $E(G) = \{e_1, e_2, \cdots, e_m\}$ of $m$ edges. Let $g(n)$ be a parameter function of the instance $G$. Then, we can provide the gap-preserving reduction such that (1) if $OPT_1(G) \geq g(n)$, then $OPT_2(H) \geq (n+1) \times g(n)$, and (2) if $OPT_1(G) < \frac{g(n)}{n^{1-\varepsilon_1}}$ for a positive constant $\varepsilon_1$, then $OPT_2(H) < (n+1) \times \frac{g(n)}{n^{1-\varepsilon_1}}$. In the following, only a brief construction of $H$ will be presented due to space limitations.

A graph $H$ consists of (i) $n$ subgraphs, $H_1$ through $H_n$, which are associated with $n$ vertices, $v_1$ through $v_n$, and (ii) $m$ edge sets, $E_1$ through $E_m$, which are associated with $m$ edges, $e_1$ through $e_m$: (i) For $i = 1, 2, \cdots, n$, the subgraph $H_i = (V(H_i), E(H_i))$ includes $n+1$ vertices and $n(n+1)/2$ edges joining them to form a complete graph: $V(H_i) = \{v_{i,0}, v_{i,1}, \cdots, v_{i,n}\}$ and $E(H_i) = \{(v_{i,k}, v_{i,\ell}) \mid k \neq \ell, 0 \leq k, \ell \leq n\}$. In total, there are $n(n+1)$ vertices in $H$. For simplicity, let $V_i$ denote $V(H_i)$. (ii) For each $j = 1, 2, \cdots, m$, $E_j = \{(v_{k,i}, v_{\ell,i}) \mid e_j = (v_k, v_\ell), i = 0, 1, \cdots, n\}$ includes $n+1$ "parallel" edges connecting $H_k$ to $H_\ell$. See Fig. 1 for an example of the input $G$ and its reduced graph $H$, where the sets of six edges $E_1$, $E_3$, $E_5$, and $E_6$ are simplified by replacing the six edges with two bold lines. This reduction can be done in polynomial time.

Since $|V(H)| = n^2 + n$ and so the approximation gap is $\Theta(|V(H)|^{1/2-\varepsilon})$ for any $\varepsilon > 0$, the lemma holds.                                                                  □

Moreover we can show the following lemma for the case $d = 3$ based on a similar idea to the proof of Lemma 2:

**Lemma 3.** *For $d = 3$ and any $\varepsilon > 0$, it is $\mathcal{NP}$-hard to approximate MaxDBS to within a factor of $n^{1/2-\varepsilon}$. (The proof is omitted.)*                     □

A minor modification provides the same inapproximability bound for a larger diameter $d$:

**Theorem 3.** *For a fixed diameter $2 \le d \le O(\sqrt{diam(G)})$ and any $\varepsilon > 0$, it is $\mathcal{NP}$-hard to approximate MaxDBS to within a factor of $n^{1/2-\varepsilon}$.*                     □

## 4   Chordal Graphs

In this section, we restrict the set of input graphs to chordal ones. We will show that the complexity of MaxDBS alternates between tractableness and intractableness according to the parity of $d$, i.e., MaxDBS is in $\mathcal{P}$ for odd $d$'s, but $\mathcal{APX}$-hard for even $d$'s.

**Theorem 4.** *If the input graph $G$ is chordal, then MaxDBS is in $\mathcal{P}$ for every fixed odd $d$ such that $1 \le d \le diam(G)$.*

*Proof.* There exits a polynomial-time algorithm that can find a maximum clique of any chordal graph [8]. Thus MaxDBS is in $\mathcal{P}$ if $d = 1$. As for any odd $d \ge 3$, it is known that the $d$-th power $G^d$ of $G$ is also chordal [1,2]. Hence, utilizing the polynomial-time algorithm [8] as `FindClique`, we can find an optimal solution by `ByFindClique`.                                                                  □

If $d$ is even, the $d$-th power of a chordal graph $G$ is not always chordal, which means that `ByFindClique` may not find an optimal solution in polynomial time. Indeed, we can show the inapproximability of MaxDBS for chordal graphs with $d = 2$.

**Lemma 4.** *For $d = 2$ and any $\varepsilon > 0$, it is $\mathcal{NP}$-hard to approximate MaxDBS to within a factor of $n^{1/3-\varepsilon}$ for chordal graphs.*

*Proof.* From an input graph $G_1 = (V_1, E_1)$ of MaxClique, we construct a graph $G_2 = (V_2, E_2)$ for MaxDBS with $d = 2$. Let $V_1 = \{v_1, v_2, \cdots, v_n\}$ and $E_1 = \{e_1, e_2, \cdots, e_m\}$. First, we construct a vertex set $V_E = \{w_1, w_2, \cdots, w_m\}$ which mimics the edge set $E_1$. Then, another vertex set $V_{\alpha V} = \{u_{1,1}, u_{1,2}, \cdots, u_{1,m}, u_{2,1}, u_{2,2}, \cdots, u_{n,m}\}$ is constructed, which mimics $V_1$. The vertex set $V_2$ of $G_2$ contains these two sets: $V_2 = V_E \cup V_{\alpha V}$. The edge set $E_2$ also consists of the following two edge sets, $E_E$ and $E_{\alpha E}$, and $E_2 = E_E \cup E_{\alpha E}$: $E_E = \{(w_i, w_j) \mid 1 \le i, j \le m, i \ne j\}$, by which $G_2[V_E]$ turns to be a clique. $E_{\alpha E}$ connects between $V_{\alpha V}$ and $V_E$: $E_{\alpha E} = \{(w_\ell, u_{i,h}), (w_\ell, u_{j,h}) \mid e_\ell = (v_i, v_j) \in E_1, 1 \le \ell, h \le m\}$. Fig. 2

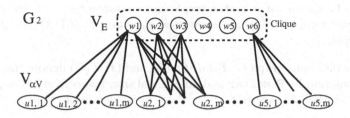

**Fig. 2.** The reduced graph $G_2$ from the graph $G$ in Fig. 1

illustrates the reduced graph $G_2$ from $G$ in Fig. 1. It is clear that this reduction can be done in polynomial time because $|V_2| = |V_E| + |V_{\alpha V}| = m + m \times n$ and $|E_2| = |E_E| + |E_{\alpha E}| = m(m-1)/2 + m \times 2m$. Note that since $G_2[V_E]$ is a clique and $G_2[V_{\alpha V}]$ is a stable graph, $G_2$ is a split graph and is also a chordal graph.

It can be shown that (1) if $OPT_1(G_1) \geq g(n)$, then $OPT_2(G_2) \geq m \times g(n) + m$, and (2) if $OPT_1(G_1) \leq \frac{g(n)}{n^{1-\varepsilon_1}}$ for a constant $\varepsilon_1$, then $OPT_2(G_2) \leq m \times \frac{g(n)}{|V_2|^{1/3-\varepsilon_2}} + m$ for a constant $\varepsilon_2$. The key construction of the reduction is that $dist_{G_2}(u_{i,k}, u_{j,h}) = 2$ if and only if $dist_{G_1}(v_i, v_j) \leq 1$ for any $v_i, v_j \in V_1$ and any $u_{i,k}, u_{j,h} \in V_{\alpha V}$. Details are omitted. □

In case $d \geq 4$, we can arbitrarily enlarge the diameter of the reduced graph $G_2$ and also that of an optimal solution by adding a (redundant) path starting from each vertex in $V_{\alpha V}$; for example, for each vertex $v \in V_{\alpha V}$, adding one vertex and one edge connecting to $v$ enlarges the diameter of $G_2$ and an optimal solution by two. Thus, we obtain the following theorem:

**Theorem 5.** *For any even $d$ such that $2 \leq d \leq O(diam(G))$ and any $\varepsilon > 0$, it is $\mathcal{NP}$-hard to approximate MAXDBS to within a factor of $n^{1/3-\varepsilon}$ for chordal graphs.* □

## 5   Split Graphs

Since the graph $G_2$ constructed in the proof of Lemma 4 is a split graph, we obtain the following corollary straightforwardly. Remind that the diameter of any split graph is at most three, and so our interest here is only the case of $d = 2$.

**Corollary 1.** *It remains $\mathcal{NP}$-hard to approximate MAXDBS to within a factor of $n^{1/3-\varepsilon}$ even for split graphs if $d = 2$.* □

In the following, we show the $O(n^{1/3})$ approximability for split graphs, which is tight in the sense that the lower bound of the approximability is $\Omega(n^{1/3-\varepsilon})$ as shown above. Here again, the simple algorithm `FindStar` is the best approximation algorithm.

**Theorem 6.** *Given an n-vertex graph and an integer $d = 2$, we can design a polynomial time $O(n^{1/3})$-approximation algorithm for* MAXDBS *if the input graph is split.*

*Proof.* For the input graph $G$, $FindStar(G)$ and $OPT(G)$ denote the size of the obtained solution by `FindStar` and an optimal size, respectively. If $deg_{\max}(G) \geq n^{2/3}$, it holds that $FindStar(G) \geq n^{2/3}+1$ by its definition. Since $OPT(G) \leq n$ holds,

$$\frac{OPT(G)}{FindStar(G)} \leq \frac{n}{n^{2/3}+1} < n^{1/3}.$$

Suppose that $deg_{\max}(G) < n^{2/3}$. Let us partition the optimal set $V^*$ of vertices into two disjoint subsets $C$ and $S$ such that $V^* = C \cup S$ and $C \cap S = \emptyset$, where $C$ is a clique set and $S$ is a stable set. Note that we can do this because $G$ is a split graph, and $|C| \leq deg_{\max}(G) + 1 \leq n^{2/3}$. For simplicity, $G^*$ denotes $G[V^*]$. We denote an upper bound of $|N(v) \cap S|$ for any vertex $v \in C$ by $k$, i.e., $k = deg_{\max}(G^*) - (|C| - 1)$.

In the following we obtain an upper bound of $|S|$. Consider a pair of vertices $u, w \in S$. Since $diam(G^*) = 2$ and $S$ is a stable set, there must be two edges $(u, v)$ and $(v, w)$ for some $v \in C$ in order to satisfy the condition $dist_{G^*}(u, w) \leq 2$. Here we say that $(u, v)$ *covers* $w$ for $u$. Since $|N_{G^*}(v) \cap S| \leq k$, $(u, v)$ can cover at most $k$ vertices in $S$ for $u$. This implies that $N(u) \geq \lceil |S|/k \rceil$ holds in order to cover all vertices in $S$ by at least one edge for $u$. Since this is true for all vertices in $S$, we observe that there exist at least $\lceil |S|/k \rceil \cdot |S|$ edges between $C$ and $S$. Since every vertex in $C$ is adjacent to at most $k$ vertices in $S$, the following condition must be satisfied:

$$k \cdot |C| \geq \left\lceil \frac{|S|}{k} \right\rceil \cdot |S| \geq \frac{|S|^2}{k},$$

which implies that $|S| \leq k\sqrt{|C|}$.

Finally, the approximation ratio of the algorithm `FindStar` is estimated as follows:

$$\frac{OPT(G)}{FindStar(G)} \leq \frac{|C| + |S|}{|C| + k} \leq \frac{|C| + k\sqrt{|C|}}{|C| + k} \leq \sqrt{|C|} \leq n^{1/3},$$

where the first inequality comes from the fact that $|C| + k = deg_{\max}(G^*) + 1 \leq deg_{\max}(G) + 1 = FindStar(G)$. □

## 6    Interval Graphs

Interval graphs are very popular and useful in graph theory and operations research; for example, several resource allocation problems can be naturally represented by using interval graphs. For interval graphs, MAXDBS is tractable for all $d$'s:

**Theorem 7.** *If the input graph is an interval graph,* MAXDBS *is in* $\mathcal{P}$ *for a fixed* $d \geq 1$.

*Proof.* Recall that the class of interval graphs is a subclass of chordal graphs. For chordal graphs, MAXCLIQUE can be solved in polynomial time [8], and hence MAXDBS is also in $\mathcal{P}$ for $d = 1$. As for the case $d \geq 2$, it is known that the $d$-th power $G^d$ of $G$ remains an interval graph [1]. Hence, we can obtain an optimal solution by `ByFindClique` with the polynomial-time algorithm in [8]. $\square$

## 7   $k$-Partite Graphs

In this section, we mention the complexity of MAXDBS for $k$-partite graphs. First of all, it is known [13] that MAXCLIQUE is $\mathcal{NP}$-hard even for $k$-partite graphs. However, if $k$ is constant, MAXCLIQUE is solvable in polynomial time since the size of the maximum clique is at most $k$. As for MAXDBS on bipartite graphs, it is still solvable in polynomial time if $d = 2$, by using a polynomial time algorithm which finds the maximum complete bipartite subgraph in a given bipartite graph [7,9]. On the other hand, we can show the $\mathcal{NP}$-hardness and the inapproximability of MAXDBS for $k$-partite graphs for $k \geq 3$ even if $d = 2$ by providing a gap-preserving reduction from MAXCLIQUE.

**Lemma 5.** *For any* $\varepsilon > 0$, *it is* $\mathcal{NP}$-hard *to approximate* MAXDBS *to within a factor of* $n^{1/3-\varepsilon}$ *if* $d = 2$ *and the input graph is restricted to 3-partite graphs. (The proof is omitted.)* $\square$

Since the class of $k$-partite graphs is a subclass of $(k + 1)$-partite graphs, the following theorem for $k$-partite graphs $(k \geq 3)$ holds from Lemma 5:

**Theorem 8.** *For any* $\varepsilon > 0$ *and* $d = 2$, *it is* $\mathcal{NP}$-hard *to approximate* MAXDBS *to within a factor of* $n^{1/3-\varepsilon}$, *even if the input graph is restricted to $k$-partite graphs for* $k \geq 3$. $\square$

It can be also shown by a small modification that the same inapproximability holds for $k$-partite graphs having larger diameter.

**Corollary 2.** *For $k$-partite graphs, it is* $\mathcal{NP}$-hard *to approximate* MAXDBS *to within a factor of* $n^{1/3-\varepsilon}$ *for any* $\varepsilon > 0$, $k \geq 3$ *and* $2 \leq d \leq diam(G) - 1$. *(The proof is omitted.)* $\square$

Furthermore, for a fixed $d \geq 3$, we can obtain the same approximation lower bound even for bipartite graphs by using a similar, but different gap-preserving reduction:

**Theorem 9.** *For any* $\varepsilon > 0$ *and a fixed* $3 \leq d \leq diam(G) - 1$, *it is* $\mathcal{NP}$-hard *to approximate* MAXDBS *to within a factor of* $n^{1/3-\varepsilon}$ *for bipartite graphs.* $\square$

# 8    Conclusion

In this paper we have presented several approximability and inapproximability results for MAxDBS, with respect to graph classes. There still, however, exists a gap between the $n^{2/3}$-approximability and the $n^{1/2}$-inapproximability for general graphs for odd $d$'s. Also, currently we have not obtained tight bounds of the approximability for chordal and $k$-partite graphs. As a further research, we can consider an extended problem whose input is a weighted graph.

## Acknowledgments

This work is partially supported by KAKENHI, 18700015 and 20500017.

## References

1. Agnarsson, G., Greenlaw, R., Halldórsson, M.M.: On powers of chordal graphs and their colorings. Congr. Numer. 144, 41–65 (2000)
2. Balakrishnan, R., Paulraja, P.: Powers of Chordal Graphs. Australian J. Mathematics, Series A 35, 211–217 (1983)
3. Bellare, M., Goldreich, O., Sudan, M.: Free bits, PCPs and non-approximability - Towards tight results. SIAM J. Computing 27(3), 804–915 (1998)
4. Bomze, I.M., Budinich, M., Pardalos, P.M., Pelillo, M.: The maximum clique problem. In: Handbook of Combinatorial Optimization, pp. 1–74. Kluwer Academic, Dordrecht (1999)
5. Brandstädt, A., Le, V.B., Spinrad, J.P.: Graph Classes: A Survey. SIAM, Philadelphia (1999)
6. Feige, U.: Approximating maximum clique by removing subgraphs. SIAM Journal on Discrete Mathematics 18, 219–225 (2004)
7. Garey, M.R., Johnson, D.S.: Computers and Intractability (1979)
8. Gavril, F.: Algorithms for minimum coloring, maximum clique, minimum covering by cliques, and maximum independent set of a chordal graph. SIAM Journal on Computing 1(2), 180–187 (1972)
9. Goerdt, A., Lanka, A.: An approximation hardness result for bipartite clique. Electronic Colloquium on Computational Complexity, Report No. 48 (2004)
10. Håstad, J.: Clique is hard to approximate within $n^{1-\varepsilon}$. Acta Mathematics 182(1), 105–142 (1999)
11. Karp, R.M.: Reducibility among combinatorial problems. In: Complexity of Computer Computations, pp. 85–103 (1972)
12. Marinček, J., Mohar, B.: On approximating the maximum diameter ratio of graphs. Discrete Math. 244, 323–330 (2002)
13. Schröder, B.S.W.: Algorithms for the fixed point property. Theoretical Computer Science 217, 301–358 (1999)
14. Vazirani, V.V.: Approximation Algorithms. Springer, Heidelberg (2003)
15. Zuckerman, D.: Linear degree extractors and the inapproximability of max clique and chromatic number. Theory of Computing 3, 103–128 (2007)

# Largest Induced Acyclic Tournament in Random Digraphs: A 2-Point Concentration

Kunal Dutta and C.R. Subramanian

The Institute of Mathematical Sciences, Chennai 600 113, India
{kdutta,crs}@imsc.res.in

**Abstract.** Given a simple directed graph $D = (V, A)$, let the size of the largest induced acyclic tournament be denoted by $mat(D)$. Let $D \in \mathcal{D}(n, p)$ be a *random* instance, obtained by choosing each of the $\binom{n}{2}$ possible undirected edges independently with probability $2p$ and then orienting each chosen edge in one of two possible directions with probability $1/2$. We show that for such a random instance, $mat(D)$ is asymptotically almost surely one of only 2 possible values, namely either $b^*$ or $b^* + 1$, where $b^* = \lfloor 2(\log_{p^{-1}} n) + 0.5 \rfloor$. It is then shown that almost surely any maximal induced acyclic tournament is of a size which is at least nearly half of any optimal solution. We also analyze a polynomial time heuristic and show that almost surely it produces a solution whose size is at least $\log_{p^{-1}} n + \Theta(\sqrt{\log_{p^{-1}} n})$. Our results also carry over to a related model in which each possible directed arc is chosen independently with probability $p$. An immediate corollary is that (the size of a) minimum feedback vertex set can be approximated within a ratio of $1 + o(1)$ for random tournaments.

**Keywords:** Random graphs, acyclic tournaments, analysis of algorithms, 2-point concentration, feedback vertex set.

## 1 Introduction

By a simple directed graph, we mean a directed graph having no self-loops, parallel arcs, or 2-cycles. Throughout the paper, we assume, w.l.o.g., that $V = \{1, \ldots, n\}$. Given a directed graph $D = (V, A)$, we want to find the maximum size of an induced acyclic tournament in $D$, denoted by $mat(D)$. A tournament is a simple directed graph whose underlying undirected graph is a complete graph. A tournament is acyclic if and only if it is transitive. By $\text{MAT}(D, k)$, we denote the following computational problem : Given a simple directed graph $D = (V, A)$ and $k$, determine if $mat(D) \geq k$. When restricted to the class of tournaments, this problem is the complement of the Feedback Vertex Set (FVS) problem. The FVS problem is to determine if the removal of a given number of vertices results in an acyclic subgraph. It comes up in various applications in computer science, such as in proving partial correctness of programs [5], in deadlock recovery in operating systems [4], and in VLSI design. It has been widely studied by the approximation and parameterized algorithms communities [10, 12].

A. López-Ortiz (Ed.): LATIN 2010, LNCS 6034, pp. 627–637, 2010.

The MAT$(D, k)$ problem is known to be NP-complete [6], even if $D$ is restricted to be a tournament [13]. Also, it is NP-complete even when $D$ is restricted to be a directed acyclic graph $(DAG)$, as shown in Theorem 1 stated below.

**Theorem 1.** *MAT$(D, k)$ is NP-complete even when $D$ is restricted to be acyclic.*

*Proof.* We reduce the NP-complete Maximum Clique problem MC$(G, k)$ to the MAT$(D, k)$ problem as follows. Given an instance $(G = (V, E), k)$ of the first problem, compute an instance $(G' = (V, A), k)$ in polynomial time where

$$A = \{(u, v) : uv \in E, u < v\}.$$

Clearly, $G'$ is a DAG and it is easy to see that a set $V' \subseteq V$ induces a clique in $G$ if and only if $V'$ induces an acyclic tournament in $G'$. This establishes that MAT$(D, k)$ is NP-hard even if $D$ is restricted to be a DAG.    ∎

In fact, even the approximation version is known to be hard [9] when the input is an arbitrary digraph: a polynomial-time approximation algorithm with an approximation ratio of $O(n^\epsilon)$ is not possible unless $P = NP$.

However, the *average* case version of the problem - finding $mat(D)$ for a random digraph $D$ - might offer some hope. We study the following model of a simple random digraph introduced in [14]. In what follows, $p \leq 0.5$ is a real number.

**Model $\mathcal{D}(n, p)$:** Let the vertex set be $V = \{1, 2, ..., n\}$. Choose each *undirected edge* joining distinct elements of $V$ independently with probability $2p$. For each chosen $\{u, v\}$, independently orient it in one of the two directions $\{u \to v, v \to u\}$ in $D$ with equal probability $= 1/2$. The resulting directed graph is an orientation of a simple graph, i.e., there are no 2-cycles.

Subramanian [14] first studied the related problem of determining $mas(D)$, the size of a largest induced DAG in a random digraph $D = (V, E)$, and later Spencer and Subramanian [15] obtained the following result.

**Theorem 2.** *[15] Let $D \in \mathcal{D}(n, p)$ and $w = np$. There is a sufficiently large constant $W$ such that : If $p$ satisfies $w \geq W$, then, asymptotically almost surely (referred to as a.a.s),*

$$mas(D) \in \left[ \left( \frac{2}{\ln q} \right) (\ln w - \ln \ln w - O(1)), \left( \frac{2}{\ln q} \right) (\ln w + 3e) \right]$$

*where $q = (1 - p)^{-1}$.*

Thus, $mas(D)$ is concentrated in an integer band of width $O \left( \frac{\ln \ln w}{\ln q} \right)$ which can become large for small values of $p$. However, if we focus on more restricted subgraphs, namely, induced acyclic tournaments, then the optimum size can be shown to be one of two consecutive values a.a.s. In other words, we obtain a 2-point concentration for $mat(D)$. This is our main result in this paper. Further, under certain assumptions on $p(n)$, $mat(D)$ is shown to be a unique value almost surely.

**Theorem 3.** *Let $\{\mathcal{D}(n, p) | n \geq 1, p = p(n)\}$ be an infinite sequence of probability distributions. Let $w = w(n)$ be an arbitrarily slow-growing function of $n$. Let $D \in \mathcal{D}(n, p)$. Then, asymptotically almost surely:*

*(i) Suppose $p \geq 1/n$. Then, $mat(D)$ is either $b^*$ or $b^* + 1$ where*

$$b^* = \lfloor d - 1/2 \rfloor; \quad d = 2\log_{p^{-1}} n + 1 = \frac{2(\ln n)}{\ln p^{-1}} + 1.$$

*(ii) $mat(D) \in \{2, 3\}$ if $1/(wn) \leq p < 1/n$.*
*(iii) $mat(D) = 2$ if $wn^{-2} \leq p < 1/(wn)$.*
*(iv) $mat(D) \leq 2$ if $p < wn^{-2}$.*

With some assumptions about $p = p(n)$, one can also prove a stronger one-point concentration on $mat(D)$ for all large values of $n$. The proof of the following theorem is omitted from this abstract.

**Theorem 4.** *Let $\mathcal{D}(n, p)$, $d$ be as defined in in Theorem 3. Let $w = w(n)$ be an arbitrarily slow-growing function of $n$. If $p \geq 1/n$ is such that $d$ satisfies $\frac{w}{\ln n} \leq \lceil d \rceil - d \leq 1 - \frac{w}{\ln n}$ for all large values of $n$, then a.a.s. $mat(D) = \lfloor d \rfloor$.*

The proof of Theorem 1 suggests a correspondence between cliques in arbitrary undirected graphs and acyclic tournaments in specific orientations of these graphs. When random graphs are compared to random digraphs instead, this relation can be seen more clearly. $\mathcal{G}(n, p)$ denotes the standard random model for simple undirected graphs on $V$ defined by including each of the possible $\binom{n}{2}$ edges independently with probability $p$. The clique number $\omega(G)$ (maximum size of a clique in $G$) has been well-studied for this model and very tight concentration results (see [3, 2, 8]) have been obtained for this number. We relate the two quantities $mat(D)$ ($D \in \mathcal{D}(n, p)$) and $\omega(G)$ ($G \in \mathcal{G}(n, p)$) in the following lemma (proof omitted due to lack of space). A similar relationship was established relating $mas(D)$ and $\alpha(G)$ (maximum size of an independent set in $G$) in [14].

**Lemma 1.** *For any integer $b > 0$, $D \in \mathcal{D}(n, p)$ and $G \in \mathcal{G}(n, p)$,*

$$Pr[mat(D) \geq b] \geq Pr[\omega(G) \geq b].$$

**Note:** Recall that we first draw an undirected $G \in \mathcal{G}(n, 2p)$ and then choose uniformly randomly an orientation of $E(G)$. Hence, for any fixed $A \subseteq V$ of size $b$ with $b = \omega(1)$,

$$\mathbf{Pr}(D[A] \text{ is an acy. tourn. } | G[A] \text{ induces a clique}) = \frac{b!}{2^{\binom{b}{2}}} = o(1).$$

However, there are so many cliques of size $b$ in $G$ that one of them manages to induce an acyclic tournament.

**Outline:** The remaining sections are organized as follows: In Section 2, we provide the proofs of Theorems 3 and 4. The proofs are based on the Second Moment Method. In Section 3, we show (Theorem 5) that almost surely *every* maximal induced acyclic tournament is of size which is at least nearly half of the optimal size. Hence any greedy heuristic obtains a solution whose approximation factor is almost surely $2 + O((\ln \ln n)/(\ln n))$. This is similar to the case of the clique number $\omega(G(n,p))$, but unlike the case of $\omega(G(n,p))$, we obtain explicit closed-form expression for $mat(D)$ for every $p$.

In Section 4, we study another heuristic which combines greedy and brute-force approaches as follows. We first apply the greedy heuristic to get a partial solution whose size is nearly $\log_{p^{-1}} n - c\sqrt{\log_{p^{-1}} n}$ for some arbitrary constant $c$. Amongst the remaining vertices, let $C$ be the set of vertices such that each vertex in $C$ can be individually "safely" added to the partial solution. Then, in the subgraph induced by $C$ we find an optimal solution by brute-force and combine it with the partial solution. It is shown in Theorem 6 that this modified approach produces a solution whose size is at least $\log_{p^{-1}} n + c\sqrt{\log_{p^{-1}} n}$. This results in an additive improvement of $\Theta(\sqrt{\log_{p^{-1}} n})$ over the simple greedy approach. The improvement is mainly due to the fact we stop using greedy heuristic at a point where it is possible to apply brute-force efficiently. This approach is similar to (and was motivated by) the one used in [7] for finding large independent sets in $\mathcal{G}(n, 1/2)$.

Each of the concentration and algorithmic results mentioned before also carry over (with some slight changes) to a related random model $\mathcal{D}_2(n,p)$ where we allow 2-cycles to be present and each of the potential arcs is chosen independently. These are presented in Section 5. In Section 6, we show how our results lead to a $(1 + o(1))$ - approximation for the feedback vertex set problem in random tournaments. Finally, in Section 7, we conclude with a summary and some open problems.

## 2   Analysis of $\mathcal{D}(n,p)$

Each of the following two easy-to-verify observations play a role in the analysis. Their proofs are omitted here.

**Fact 1.** *A DAG $H = (U, A)$ has at most one (directed) hamilton path.*

Let $U$ be any fixed subset of $V$ of size $b$.

**Proposition 1.** $\Pr[D[U]$ *is an acyclic tournament* $] = b! \, p^{\binom{b}{2}}$

Before we proceed further, we introduce some notations which play an important role in the analysis. Define $\delta = \lceil d \rceil - d$. Then, clearly

$$b^* = \begin{cases} d - 2 + \delta \text{ if } \delta > 1/2; \\ d - 1 + \delta \text{ if } \delta \le 1/2. \end{cases}$$

For a given $b$, let $m = \binom{n}{b}$ and let $\{A_1, \ldots, A_m\}$ denote the set of all $b$-sized subsets of $V$. For $i \in [m]$, let $X_i$ denote the random variable that indicates whether $D[A_i]$ induces an acyclic tournament or not. Let $X = X(b) = X(n, b)$ denote the number of induced acyclic tournaments of size $b$ in $D$. Since there are $\binom{n}{b}$ sets of size $b$, it follows by Linearity of Expectation that

$$E[X(n,b)] = \sum_i E[X_i] = \binom{n}{b} b! \, p^{\binom{b}{2}}.$$

We are only interested in the behavior of $E[X(n,b)]$ for $b \in [1, b^* + 2]$. From the definition of $b^*$, it follows that $b^* + 2 \le \lceil d \rceil + 1 \le \frac{2(\ln n)}{\ln p^{-1}} + 3 \le 3(\ln n)$ for sufficiently large $n$ since $p \le 1/2$. As a result,

$$[1 - o(1)] \cdot f(n, p, b)^b \le E[X(n, b)] \le f(n, p, b)^b \quad \ldots\ldots \text{ (A)}$$

$$\text{where} \quad f(n, p, b) = n \, p^{(b-1)/2}.$$

Now, for the sake of analysis, we extend the definition of $f(n, p, b)$ for the case of non-negative reals also. Setting $f(n, p, b) = 1$ and solving for $b$, we see that

$$f(n, p, b) > 1 \text{ if } b < d; \quad f(n, p, d) = 1; \quad f(n, p, b) < 1 \text{ if } b > d.$$

## 2.1 Proof of $mat(D) \le b^* + 1$

First, we focus on proving the upper bound of Theorem 3. This is done by proving that

$$\mathbf{Pr}(X(b^* + 2) > 0) \le E[X(b^* + 2)] = o(1).$$

This is established by using the upper bound of $(A)$ with $b = b^*$. The details are omitted in this abstract.

## 2.2 Proof of $mat(D) \ge b^*$

Next, we focus on proving the lower bound of Theorem 3. For this, we first show that $E[X(b^*)] \to \infty$ as $n \to \infty$. The proof is omitted for lack of space.

For the sake of notational simplicity, we use $X$ to denote $X(b^*)$ and use $b$ to denote $b^*$ for the rest of this section. Now, we need to show that $X > 0$ with high probability. We use the well-known Second Moment Method to establish this. Let $Var(X)$ denote the variance of $X$. From Chebyshev's Inequality, it follows that

$$Pr[X = 0] \le Var(X)/(E[X])^2 \tag{1}$$

It follows from standard arguments (see [1]) that

$$Var(X) \le E[X] + \sum_{i \neq j} E[X_i X_j] \tag{2}$$

where the second sum is over ordered pairs $(i, j)$ such that $2 \le |A_i \cap A_j| < b$. For $(i, j)$ with $|A_i \cap A_j| = l$, we have

$$E[X_i X_j] = E[X_i]E[X_j | X_i = 1]$$
$$= b! p^{\binom{b}{2}} \cdot (b!/l!) \cdot p^{\binom{b}{2} - \binom{l}{2}} \tag{3}$$

where the last equality follows from Proposition 1. Also, for any fixed $i$, the number of $b$-sized subsets $A_j$ such that $|A_i \cap A_j| = l$ is exactly $\binom{b}{l}\binom{n-b}{b-l}$. As a result, it is verified that

$$\sum_{i \neq j} E[X_i X_j] = \sum_i \sum_{j \,:\, 2 \leq |A_i \cap A_j| \leq b-1} E[X_i X_j] \leq E[X]^2 \cdot M \qquad (4)$$

$$\text{where } M = M(n,p,b) = \left( \sum_{2 \leq l \leq b-1} \frac{(b)_l}{(l!)^2} \binom{n-b}{b-l} \binom{n}{b}^{-1} \cdot p^{-\binom{l}{2}} \right)$$

Combining (1), (2) and (4), we notice that

$$\mathbf{Pr}(X = 0) \leq E[X]^{-1} + M = o(1) \qquad (5)$$

provided $M = o(1)$ since it has already been shown that $E[X] \to \infty$. Thus, we only need to show that $M = o(1)$.

Before that, we make the following proposition whose proof is skipped.

**Proposition 2.** $np^{(b-1)/2} \geq 1$.

Write $M = \sum_{2 \leq l \leq b-1} F_l$ where

$$F_l = \frac{(b)_l}{(l!)^2} \cdot \binom{n-b}{b-l} \cdot \binom{n}{b}^{-1} \cdot p^{-\binom{l}{2}} \leq \binom{b}{l}^2 \cdot \frac{1}{(n)_l} \cdot p^{-\binom{l}{2}}$$

$$\leq \binom{b}{l}^2 \cdot \frac{1 + o(1)}{n^l} \cdot p^{-\binom{l}{2}} \leq 2 \binom{b}{l}^2 \cdot \left( \frac{1}{n \cdot p^{(l-1)/2}} \right)^l$$

$$\leq 2 \binom{b}{l}^2 \left( \frac{p^{(b-l)/2}}{np^{(b-1)/2}} \right)^l \leq 2 \binom{b}{l}^2 \left( p^{(b-l)/2} \right)^l \quad \text{(by Proposition 2)} . \qquad (6)$$

**Case $l < b/2$:** In this case,

$$(b-l)/2 > b/4 \geq (d-2+\delta)/4 \geq \frac{\ln n}{2(\ln p^{-1})} - \frac{1}{4}.$$

Hence, $p^{(b-l)/2} \leq \dfrac{p^{-1/4}}{n^{0.5}} = O\left( n^{-1/4} \right)$ since $p \geq 1/n$.

As a result, $F_l = O\left( \dfrac{b^{2l}}{n^{l/4}} \right) = O\left( n^{-l/8} \right) = O(n^{-1/4})$

where we have used the fact that $b = O(\ln n)$ and $l \geq 2$. This is true for every $p \geq 1/n$. It then follows that

$$\sum_{2 \leq l < b/2} F_l = b(O(n^{1/4})) = O(b/n^{1/4}) = o(1) \quad \ldots \ldots \quad (B)$$

**Case $l \geq b/2$ and $p \leq (\ln n)^{-1}$:** From (6), it follows that

$$F_l \leq 2 \left( 4e^2 \cdot p^{(b-l)/2} \right)^l \leq 2 \left( 4e^2 \cdot (\ln n)^{-1/2} \right)^l \Rightarrow$$

$$\sum_{l \geq b/2} F_l \leq 2 \left( \sum_{l \geq b/2} \left( 4e^2 \cdot (\ln n)^{-1/2} \right)^l \right) = o(1) \quad \ldots \ldots \quad (C)$$

$(B)$ and $(C)$ together prove that $M = \sum_l F_l = o(1)$ for $p \leq (\ln n)^{-1}$.

**Case** $b/2 \leq l < b - 12$ **and** $p \geq (\ln n)^{-1}$: We have

$$F_l \leq 2\left((4e^2) \cdot p^6\right)^l \leq 2\left(4e^2 \cdot 2^{-6}\right)^l \leq 2 \cdot 2^{-l}.$$

Also, if $p \geq (\ln n)^{-1}$, then it follows that $b \geq d - 2 = \frac{2(\ln n)}{\ln p^{-1}} - 1 \geq \frac{2(\ln n)}{\ln \ln n} - 1$.
Hence

$$\sum_{b/2 \leq l < b-12} F_l \leq 2\left(\sum_{l \geq b/2} 2^{-l}\right) = o(1) \quad \ldots\ldots \quad (D)$$

**Case** $l \geq b - 12$ **and** $p \geq (\ln n)^{-1}$: We have

$$\sum_{l \geq b-12} F_l \leq 24 \binom{b}{12}^2 \cdot 2^{-(b-12)} \leq 24(3(\ln n))^{24} \cdot 2^{-\frac{\ln n}{\ln \ln n}} = o(1) \quad \ldots \quad (E)$$

$(B)$, $(D)$ and $(E)$ together prove that $M = \sum_l F_l = o(1)$ for $p \geq (\ln n)^{-1}$.

We have thus completely established that $M = o(1)$ for all $p \geq 1/n$, thereby establishing that $\mathbf{Pr}(M = 0) = o(1)$. Hence, almost surely, $mat(D) \in \{b^*, b^*+1\}$ for the stated range of $p$. The proofs of the remaining parts of Theorem 3 are easier and are omitted. This completes the proof of Theorem 3.

## 3    Finding an Induced Acyclic Tournament

In this section, we study a simple greedy heuristic for finding a large induced acyclic tournament inside a random digraph. We study the greedy heuristic GRDMAT($D$) (described below), and show that almost surely, it produces an acyclic tournament of size within a constant factor ($\geq 1/2$) of the optimal. The algorithm is simple:

GRDMAT($D = (V, E)$)

1. $A := \emptyset$.
2. **while** $\exists\, u \in V \setminus A$ such that $D[A \cup \{u\}]$ induces an acyclic tournament **do**
3.     Add $u$ to $A$.    (* ties are broken arbitrarily *)
4. **end**
5. Return $D[A]$ and halt.

It is easy to inductively verify that GRDMAT($D$) always outputs a maximal acyclic tournament. The following theorem proves a lower bound on the size of any maximal solution. Its proof is omitted for lack of space.

**Theorem 5.** *Given $D \in \mathcal{D}(n,p)$ with $p \geq 1/n$ and any $\omega = \omega(n)$ such that $\omega(n) \to \infty$ as $n \to \infty$, with probability $1 - o(1)$, every maximal induced acyclic tournament is of size at least $d = \lfloor \delta \log_{p^{-1}} n \rfloor$, where $\delta = 1 - \frac{\ln(\ln n + \omega)}{\ln n}$.*

# 4   Another Efficient Heuristic with Improved Guarantee

We present below another efficient heuristic which will be analyzed and be shown to have an additive improvement of $\Theta\left(\sqrt{\frac{\ln n}{\ln p^{-1}}}\right)$ over the guarantee given (in Section 3) on the size of any maximal solution. It is similar to a heuristic presented in [7] for finding large independent sets in $G \in \mathcal{G}(n, 1/2)$. We show that for every fixed $c > 0$, one can find in polynomial time an acyclic tournament of size at least $\lfloor \log_{p^{-1}} n + c\sqrt{\log_{p^{-1}} n} \rfloor$.

For a given partial solution $A$ (obtained using the greedy method), let $C$ be the set of the vertices in the remaining graph such that each vertex $u \in C$ can be individually added as a sink vertex (vertex of zero outdegree) to $A$. Any acyclic tournament in $C$ together with $A$ forms a larger acyclic tournament. The idea is therefore to construct greedily a solution $A$ of size $g(n, p, c) = \lceil \log_{p^{-1}} n - c\sqrt{\log_{p^{-1}} n} \rceil$ and then add an optimal solution (found by an exhaustive search) in the subgraph induced by $C$. We will show that an exhaustive search can be done in polynomial time and yields (a.s.) a solution of size $2c\sqrt{\log_{p^{-1}} n}$. As a result, we finally get a solution of the stated size. The algorithm is described below.

ACYTOUR($D = (V, E), c$)

1. Choose and fix a linear ordering $\sigma$ of $V$.
2. $c' = 1.2c$; $A = \emptyset$; $B = V$.
3. **while** $B \neq \emptyset$ and $|A| < g(n/2, p, c')$ **do**
4.      Let $u$ be the $\sigma$-smallest vertex in $B$.
5.      **If** $D[A \cup \{u\}]$ is an acyclic tournament **then** add $u$ to $A$.
6.      remove $u$ from $B$. **endwhile**
7. **if** $|A| < g(n/2, p, c')$ **or** $|B| < n/2$, **then** Return FAIL and halt.
8. $C = \{u \in B : (v, u) \in E, \forall v \in A\}$; $r = p^{-1}$; $\mu = |B|p^{|A|}$.
9. **if** $|C| \notin [(0.9)\mu, (1.1)\mu]$ **then** Return FAIL.
10. **for each** $X \subset C$ : $|X| = \lfloor 2c'\sqrt{\log_r n/2} \rfloor - 1$ **do**
11.          **if** $D[X]$ is an acyclic tournament **then** Return $D[A \cup X]$ and halt.
    **endfor**
12. Return FAIL.

We analyze the above algorithm and obtain the following result.

**Theorem 6.** *Let $D \in \mathcal{D}(n, p)$. For every sufficiently large constant $c \geq 1$ : if $p$ is such that $n^{-1/c^2} \leq p \leq 0.5$, then, with probability $1 - o(1)$, ACYTOUR(D) will output an induced acyclic tournament of size at least $b' = \lfloor (1 + \epsilon') \log_{p^{-1}} n \rfloor$, where $\epsilon' = c/\sqrt{\log_{p^{-1}} n}$.*

*Proof.* Recall our assumption that $c$ is sufficiently large.

**Correctness:** Note that $D[A]$ is always an acyclic tournament. Also, each $u \in C$ is such that $D[A \cup \{u\}]$ is an acyclic tournament with $u$ as the unique sink vertex (having zero out-degree). Hence, any acyclic tournament $D[X]$ present as a subgraph in $D[C]$ can be added to $A$ so that $D[A \cup X]$ is also an acyclic tournament.

**Time Complexity:** It is easy to see that the running time is polynomial except for the *for* loop of lines 10 and 11. The maximum number of iterations of the **for** loop is at most

$$\left( \frac{(1.05) \cdot r^{c'\sqrt{\log_r n}}}{\lfloor 2c'\sqrt{\log_r n} \rfloor} \right) = O\left( r^{4c'^2(\log_r n)} \right) = O\left( n^{O(1)} \right).$$

Since each iteration takes polynomial time, the algorithm finishes in polynomial time always.

**Analysis:** It is possible to analyze ACYTOUR($D$) and prove (details omitted for lack of space) that it outputs a solution of size at least $(1 + \epsilon') \log_{p^{-1}} n$ with probability $1 - o(1)$.  ∎

*Remark 1.* In Theorem 6, we assume that $p \geq n^{-1/c^2}$. This is because if $p \leq n^{-1/c^2}$, then $mat(D) \leq \lceil 2c^2 + 1 \rceil$ almost surely and hence even a provably optimal solution can be found in polynomial time a.a.s.

## 5    $mat(D)$ for Non-simple Random Digraphs

We also consider another model introduced in [15] which does not force the random digraph to be simple and allows cycles of length 2.

**Model $D \in \mathcal{D}_2(n, p)$:** Choose each *directed* edge $u \to v$ joining distinct elements of $V$ independently with probability $p$.

Note that if $D \in \mathcal{D}_2(n, p)$ and $D' \in \mathcal{D}_2(n, 1 - p)$, then for every $b$,

$$\mathbf{Pr}(mat(D) = b) = \mathbf{Pr}(mat(D') = b).$$

Hence, for the rest of this section, without loss of generality, we assume that $p \leq 0.5$ and use $q$ to denote $1 - p$.

Using similar arguments, we can obtain the following analogues of Theorems 3, 4, 5, 6 and Lemma 1.

**Theorem 7.** *Let $D \in \mathcal{D}_2(n, p)$ with $p \geq 1/n$. Define*

$$d = 2 \log_{(pq)^{-1}} n + 1 = \frac{2(\ln n)}{\ln (pq)^{-1}} + 1; \quad b^* = \lfloor d - 1/2 \rfloor.$$

*Then, almost surely as $n \to \infty$, $mat(D)$ is either $b^*$ or $b^* + 1$.*

**Theorem 8.** *Let $D \in \mathcal{D}_2(n, p)$. Let $\omega = \omega(n)$ be any slowly growing function. If $p = p(n)$, $p \geq 1/n$, is such that $d$ (defined in Theorem 7) satisfies $\frac{\omega}{\ln n} \leq \lceil d \rceil - d \leq 0.5$ for all large values of $n$, then $mat(D)$ is a.a.s equal to $b^*$.*

**Lemma 2.** *For any integer $b > 0$, $D \in \mathcal{D}_2(n, p)$ and $G \in \mathcal{G}(n, pq)$,*

$$Pr[mat(D) \geq b] \geq Pr[\omega(G) \geq b].$$

**Theorem 9.** *Given $D \in \mathcal{D}_2(n, p)$ with $p \geq 1/n$ and any $\omega = \omega(n)$ such that $\omega(n) \to \infty$ as $n \to \infty$, with probability $1 - o(1)$, every maximal induced acyclic tournament is of size at least $d = \lfloor \delta \log_{(pq)^{-1}} n \rfloor$, where $\delta = 1 - \frac{\ln(\ln n + \omega)}{\ln n}$.*

**Theorem 10.** *Let $D \in \mathcal{D}_2(n, p)$. For every sufficiently large constant $c \geq 1$, if $p \leq 0.5$ is such that $n^{-1/c^2} \leq pq \leq 0.25$, then, with probability $1 - o(1)$, ACYTOUR(D) will output an induced acyclic tournament of size at least $b' = \lfloor (1 + \epsilon') \log_{(pq)^{-1}} n \rfloor$, where $\epsilon' = c/\sqrt{\log_{(pq)^{-1}} n}$.*

*Remark 2.* However, in the case of $\mathcal{D}_2(n, p)$, we need to slightly modify the description of ACYTOUR(D) as follows : In the definition of $C$ (Line 8), we also need to require that $(u, v) \notin E$ for each $v \in A$.

## 6    Remarks on Approximating Minimum FVS

Since the vertices left on removing an induced acyclic subtournament form a FVS, we notice that we obtain a 2-point concentration for the minimum FVS in random tournaments also. This follows from Theorems 3 and 7. While the presently best known algorithm in [10] has a worst-case approximation ratio of 2.5, we notice that the 2-approximate algorithms GRDMAT and ACYTOUR yield an approximate solution (for minimum FVS) with an approximation ratio of $1 + \frac{mat(D)}{2*opt(minfvs(D))} = 1 + O((\log_{p^{-1}} n)/n)$ for random tournaments.

## 7    Summary

We studied the problem of determining the size of the largest induced acyclic tournament $mat(D)$ in a random digraph $D = (V, E)$. We showed that asymptotically almost surely $mat(D)$ takes one of only two possible values. The result is valid for all ranges of the arc probability $p$. The value of $mat(D)$ also has a closed form expression (for all ranges of $p$) which is simpler than the one for $\omega(G), G \in \mathcal{G}(n, p)$).

We then showed that every maximal acyclic tournament is of size which is at least nearly half of the optimal size. As a result, we have a simple greedy heuristic whose approximation ratio is bounded by $2 + O((\ln \ln n)/(\ln n))$. We also considered and analyzed another efficient heuristic whose approximation ratio was shown to be $2 - O(1/\sqrt{\log_{p^{-1}} n})$.

Although we proved only a 2-point concentration for $mat(D)$, we believe that for every *fixed* $p$, actually $mat(D)$ is just one value a.s. for every $n \in S$, where $S$ is a subset of positive integers with density 1. This is presently being verified. An interesting and natural open problem that comes to mind is the following.

**Open Problem:** Let $p$, $0 < p \leq 0.5$, be a constant. Design a polynomial time algorithm which, given $D \in \mathcal{D}(n, p)$, almost surely finds an induced acyclic tournament of size at least $(1 + \epsilon) \log_{p^{-1}} n$ for some positive constant $\epsilon$.

Solving this problem could turn out to be as hard as designing an efficient algorithm which finds, given $G \in \mathcal{G}(n, 1/2)$, a clique of size $(1 + \epsilon) \log_2 n$, which has remained open for more than three decades.

Unlike the case of $mat(D)$, the gap between lower and upper bounds on $mas(D)$ obtained in [14, 15] is not very sharp. However, further progress has been made on shortening this gap and the results are being written up.

# References

[1] Alon, N., Spencer, J.H.: The Probabilistic Method. Wiley International, Chichester (2001)
[2] Bollobás, B.: Random Graphs, 2nd edn. Cambdrige University Press, Cambdrige (2001)
[3] Bollobás, B., Erdös, P.: Cliques in random graphs. Math. Proc. Camb. Phil. Soc. 80, 419–427 (1988)
[4] Rosen, B.K.: Robust linear algorithms for cutsets. J. Algorithms 3, 205–217 (1982)
[5] Floyd, R.W.: Assigning meaning to programs. In: Proc. Symp. Appl. Math., vol. 19, pp. 19–32 (1967)
[6] Garey, M.R., Johnson, D.S.: Computers and Intractability: A Guide To The Theory of NP-Completeness. W.H. Freeman, San Francisco (1978)
[7] Krivelevich, M., Sudakov, B.: Coloring random graphs. Information Processing Letters 67, 71–74 (1998)
[8] Janson, S., Łuczak, T., Ruciński, A.: Random Graphs. John Wiley & Sons, Inc., Chichester (2000)
[9] Lund, C., Yannakakis, M.: The Approximation of Maximum Subgraph Problems. In: Lingas, A., Carlsson, S., Karlsson, R. (eds.) ICALP 1993. LNCS, vol. 700, pp. 40–51. Springer, Heidelberg (1993)
[10] Cai, M., Deng, X., Zang, W.: An Approximation Algorithm for Feedback Vertex Set in Tournaments. SIAM Journal of Computing 30(6), 1993–2007 (2001)
[11] Motwani, R., Raghavan, P.: Randomized Algorithms. Cambridge University Press, Cambridge (1995)
[12] Raman, V., Saurabh, S.: Parameterized algorithms for feedback set problems and their duals in tournaments. Theoretical Computer Science 351(3), 446–458 (2006)
[13] Speckenmeyer, E.: On feedback problems in digraphs. In: Nagl, M. (ed.) WG 1989. LNCS, vol. 411, pp. 218–231. Springer, Heidelberg (1990)
[14] Subramanian, C.R.: Finding induced acyclic subgraphs in random digraphs. The Electronic Journal of Combinatorics 10, R46 (2003)
[15] Spencer, J.H., Subramanian, C.R.: On the size of induced acyclic subgraphs in random digraphs. Disc. Math. and Theoret. Comp. Sci. 10(2), 47–54 (2008)

# The Complexity of Counting Eulerian Tours in 4-Regular Graphs*

Qi Ge and Daniel Štefankovič

Department of Computer Science
University of Rochester
{qge,stefanko}@cs.rochester.edu

**Abstract.** We investigate the complexity of counting Eulerian tours (#ET) and its variations from two perspectives—the complexity of exact counting and the complexity w.r.t. approximation-preserving reductions (AP-reductions [DGGJ04]). We prove that #ET is #P-complete even for planar 4-regular graphs.

A closely related problem is that of counting A-trails (#A-TRAILS) in graphs with rotational embedding schemes (so called maps). Kotzig [Kot68] showed that #A-TRAILS can be computed in polynomial time for 4-regular plane graphs (embedding in the plane is equivalent to giving a rotational embedding scheme). We show that for 4-regular maps the problem is #P-hard. Moreover, we show that from the approximation viewpoint #A-TRAILS in 4-regular maps captures the essence of #ET, that is, we give an AP-reduction from #ET in general graphs to #A-TRAILS in 4-regular maps. The reduction uses a fast mixing result for a card shuffling problem [Wil04].

## 1 Introduction

An Eulerian tour in a graph is a tour which travels each edge exactly once. The problem of counting Eulerian tours (#ET) of a graph is one of a few recognized counting problems (see, e.g., [Vaz01], p. 339). The exact counting in general graphs is #P-complete [BW05], and thus there is no polynomial-time algorithm for it unless P=NP. For the approximate counting one wants to have a fully polynomial randomized approximation scheme (FPRAS), that is, an algorithm which on every instance $x$ of the problem and error parameter $\varepsilon > 0$, will output a value within a factor $\exp(\pm\varepsilon)$ of $f(x)$ with probability at least $2/3$ and in time polynomial in the length of the encoding of $x$ and $1/\varepsilon$, where $f(x)$ is the value we want to compute. The existence of an FPRAS for #ET is an open problem [TV01, Jer, Vaz01].

A closely related problem to #ET is the problem of counting A-trails (#A-TRAILS) in graphs with rotational embedding schemes (called maps, see Section 2 for a definition). A-trails were studied in the context of decision problems (for example, it is NP-complete to decide whether a given plane graph has an A-trail [BM87, AF95]; on the other hand for 4-regular maps the problem is in

---

* Research supported, in part, by NSF grant CCF-0910584.

A. López-Ortiz (Ed.): LATIN 2010, LNCS 6034, pp. 638–649, 2010.

P [Dvo04]), as well as counting problems (for example, Kotzig [Kot68] showed that #A-TRAILS can be computed in polynomial time for 4-regular plane graphs, reducing the problem to counting of spanning trees).

In this paper, we investigate the complexity of #ET in 4-regular graphs and its variations from two perspectives. First, the complexity of exact counting is considered. We prove that #ET in 4-regular graphs (even in 4-regular planar graphs) is #P-complete. We also prove that #A-TRAILS in 4-regular maps is #P-complete (recall that the problem can be solved in polynomial time for 4-regular *plane* graphs).

The second perspective is the complexity w.r.t. the AP-reductions proposed by Dyer, Goldberg, Greenhill and Jerrum [DGGJ04]. We give an AP-reduction from #ET in general graphs to #A-TRAILS in 4-regular maps. Thus we show that if there is an FPRAS for #A-TRAILS in 4-regular maps, then there is also an FPRAS for #ET in general graphs. The existence of AP-reduction from #ET in general graphs to #ET in 4-regular graphs is left open.

## 2  Definitions and Terminology

For the definitions of cyclic orderings, A-trails, and mixed graphs, we follow [Fle90]. Let $G = (V, E)$ be a graph. For a vertex $v \in V$ of degree $d > 0$, let $K(v) = \{e_1, \ldots, e_d\}$ be the set of edges adjacent to $v$ in $G$. The *cyclic ordering* $O^+(v)$ of the edges adjacent to $v$ is a $d$-tuple $(e_{\sigma(1)}, \ldots, e_{\sigma(d)})$, where $\sigma$ is a permutation in $S_d$. We say $e_{\sigma(i)}$ and $e_{\sigma(i+1)}$ are cyclicly-adjacent in $O^+(v)$, for $1 \le i \le d$, where we set $\sigma(d + 1) := \sigma(1)$. The set $O^+(G) = \{O^+(v) | v \in V\}$ is called a *rotational embedding scheme* of $G$. For a plane graph $G = (V, E)$, if $O^+(v)$ is not specified, we usually set $O^+(v)$ to be the clockwise order of the half-edges adjacent to $v$ for each $v \in V$.

Let $G = (V, E)$ be a graph with a rotational embedding scheme $O^+(G)$. An Eulerian tour $v_0, e_1, v_1, e_2, \ldots, e_\ell, v_\ell = v_0$ is called an *A-trail* if $e_i$ and $e_{i+1}$ are cyclicly-adjacent in $O^+(v_i)$, for each $1 \le i \le \ell$, where we set $e_{\ell+1} := e_1$.

Let $G = (V, E, E')$ be a mixed graph, that is, $E$ is the set of edges and $E'$ is the set of half-edges (which are incident with only one vertex in $V$). Let $|E'| = 2d$ where $d$ is a positive integer and assume that the half-edges in $E'$ are labelled by numbers from 1 to $2d$. A *route* $r(a, b)$ is a trail (no repeated edges, repeated vertices allowed) in $G$ that starts with half-edge $a$ and ends with half-edge $b$. A collection of $d$ routes is called *valid* if every edge and every half-edge is travelled exactly once.

We say that a valid set of routes is of the *type* $\{\{a_1, b_1\}, \ldots, \{a_d, b_d\}\}$ if it contains routes connecting $a_i$ to $b_i$ for $i \in [d]$. We use $VR(\{a_1, b_1\}, \ldots, \{a_d, b_d\})$ to denote the set of valid sets of routes of type $\{\{a_1, b_1\}, \ldots, \{a_d, b_d\}\}$ in $G$.

We will use the following concepts from Markov chains to construct the gadget in Section 4 (see, e. g., [Jer03] for more detail). Given two probability distributions $\pi$ and $\pi'$ on finite set $\Omega$, the *total variation distance* between $\pi$ and $\pi'$ is defined as

$$\|\pi - \pi'\|_{TV} = \frac{1}{2} \sum_{\omega \in \Omega} |\pi(\omega) - \pi'(\omega)| = \max_{A \subseteq \Omega} |\pi(A) - \pi'(A)|.$$

Given a finite ergodic Markov chain with transition matrix $P$ and stationary distribution $\pi$, the *mixing time* from initial state $x$, denoted as $\tau_x(\varepsilon)$, is defined as

$$\tau_x(\varepsilon) = \min\{t : \|P^t(x, \cdot) - \pi\|_{TV} \le \varepsilon\},$$

and the mixing time of the chain $\tau(\varepsilon)$ is defined as

$$\tau(\varepsilon) = \max_{x \in \Omega}\{\tau_x(\varepsilon)\}.$$

# 3    The Complexity of Exact Counting

## 3.1    Basic Gadgets

We describe two basic gadgets and their properties which will be used as a basis for larger gadgets in the subsequent sections.

The first gadget, which is called the $(X, Y, Y)$ node, is shown in Figure 1(a), and it is represented by the symbol shown in Figure 1(b). There are $k$ internal vertices in the gadget, and the labels 0, 1, 2 and 3 are four half-edges of the $(X, Y, Y)$ node which are the only connections from the outside.

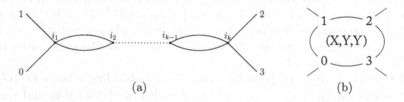

(a)                                                                (b)

**Fig. 1.** An $(X, Y, Y)$ node and its symbol. (a): an $(X, Y, Y)$ node consisting of $k$ internal vertices; (b): symbol representing the $(X, Y, Y)$ node.

By elementary counting we obtain the following fact.

**Lemma 1.** *The $(X, Y, Y)$ node with parameter $k$ has three different types of valid sets of routes and these satisfy*

$$|VR(\{0, 1\}, \{2, 3\})| = k2^{k-1},$$
$$|VR(\{0, 2\}, \{1, 3\})| = |VR(\{0, 3\}, \{1, 2\})| = 2^{k-1}.$$

*The gadget has $k$ vertices.*

The second gadget, which is called the $(0, X, Y)$ node, is shown in Figure 2(a), and it is represented by the symbol shown in Figure 2(b). Let $p$ be any odd prime. In the construction of the $(0, X, Y)$ node we use $p$ copies of $(X, Y, Y)$

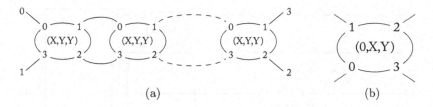

**Fig. 2.** A $(0, X, Y)$ node and its symbol. (a): a $(0, X, Y)$ node consisting of $p$ copies of $(X, Y, Y)$ nodes; (b): symbol representing the $(0, X, Y)$ node.

nodes as basic components, and each $(X, Y, Y)$ node has the same parameter $k$. As illustrated, half-edges are connected between two consecutive $(X, Y, Y)$ nodes. The four labels 0, 1, 2 and 3 at four corners in Figure 2(a) are the four half-edges of the $(0, X, Y)$ node, and they are the only connections from the outside.

By elementary counting, binomial expansion, Fermat's little theorem, and the fact that 2 has a multiplicative inverse mod $p$, we obtain the following:

**Lemma 2.** *Let $p$ be an odd prime and let $k$ be an integer. The $(0, X, Y)$ node with parameters $p$ and $k$ has three different types of valid sets of routes and these satisfy*

$$|VR(\{0, 1\}, \{2, 3\})| = pA(A + B)^{p-1} \equiv 0 \mod p, \tag{1}$$

$$|VR(\{0, 2\}, \{1, 3\})| = \frac{(A + B)^p - (B - A)^p}{2} \equiv A \mod p, \tag{2}$$

$$|VR(\{0, 3\}, \{1, 2\})| = \frac{(A + B)^p + (B - A)^p}{2} \equiv B \mod p, \tag{3}$$

*where $A = 2^{k-1}$ and $B = k2^{k-1}$. The gadget has $kp$ vertices.*

### 3.2   #ET in 4-Regular Graphs Is #P-Complete

Next, we will give a reduction from #ET in general Eulerian graphs to #ET in 4-regular graphs.

**Theorem 1.** *#ET in general Eulerian graphs is polynomial time Turing reducible to #ET in 4-regular graphs.*

The proof of Theorem 1 is postponed to the end of this section.

We use the gadget, which we will call $Q$, illustrated in Figure 3 to prove the Theorem. The gadget is constructed in a recursive way. The $d$ labels $1, \ldots, d$ on the left are called input half-edges of the gadget, and the $d$ labels on the right are called output half-edges. Given a prime $p$ and a positive integer $d$, the gadget consists of $d - 1$ copies of $(0, X, Y)$ nodes with different parameters and one recursive part represented by a rectangle with $d - 1$ input half-edges and $d - 1$ output ones. For $1 \le i \le d - 1$, the $i$-th $(0, X, Y)$ node from left has parameters

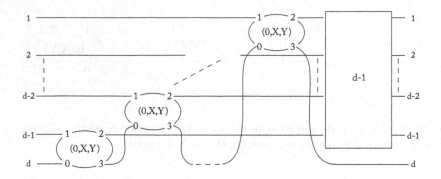

**Fig. 3.** Gadget $Q$ with $d$ input half-edges and $d$ output half-edges

$p$ and $i$. Half-edge 0 of the $i$-th $(0, X, Y)$ node is connected to half-edge 3 of the $(i - 1)$-st $(0, X, Y)$ node except that for the 1st $(0, X, Y)$ node half-edge 0 is the $d$-th input half-edge of the gadget. Half-edge 1 of the $i$-th $(0, X, Y)$ node is the $(d - i)$-th input half-edge of the gadget. Half-edge 2 of the $i$-th $(0, X, Y)$ node is connected to the $(d - i)$-th input half-edge of the rectangle. Half-edge 3 of the $(d - 1)$-st $(0, X, Y)$ node is the $d$-th output half-edge of the gadget. For $1 \leq j \leq d - 1$, the $j$-th output half-edge of the rectangle is the $j$-th output half-edge of the gadget. From the constructions of $(X, Y, Y)$ nodes and $(0, X, Y)$ nodes, the total size of the $d - 1$ copies of $(0, X, Y)$ nodes is $O(pd^2)$. Thus, the size of the gadget is $O(pd^3)$.

**Lemma 3.** *Consider the gadget $Q$ with parameters $d$ and $p$. Let $\sigma$ be a permutation in $S_d$. Then*

$$|VR(\sigma)| := |VR(\{IN_1, OUT_{\sigma(1)}\}, \dots, \{IN_d, OUT_{\sigma(d)}\})| \equiv R_d \mod p, \qquad (4)$$

*where $R_d \equiv \prod_{i=1}^{d-1}(2^{i(i-1)/2} i!)$.*

*Moreover, any type $\tau$ which connects two IN (or two OUT) half-edges satisfies*

$$|VR(\tau)| \equiv 0 \mod p. \qquad (5)$$

*Proof.* The proof is by induction on $d$, the base case $d = 1$ is trivial. Suppose the statement is true for gadget $Q$ with $(d - 1)$ input half-edges, that is, $|VR(\varrho)| \equiv R_{d-1} \mod p$ for every $\varrho \in S_{d-1}$.

Now, consider gadget $Q$ with $d$ input half-edges. For $1 \leq j \leq d - 1$, we cut the gadget by a vertical line just after the $j$-th $(0, X, Y)$ node and only consider the part of the gadget to the left of the line, we will call this partial gadget $Q_j$.

*Claim.* Let $A_s$ be the set of permutations in $S_d$ which map $s$ to $d$. In the partial gadget $Q_j$ we have that for $s \in \{d - j, \dots, d\}$ have

$$\sum_{\sigma \in A_s} |VR_{Q_j}(\sigma)| \equiv j! 2^{j(j-1)/2} \mod p,$$

where the subscript $Q_j$ is used to indicate that we count routes in gadget $Q_j$.

*Proof of the Claim.* We prove the claim by induction on $j$, the base case $j = 1$ is trivial.

Now assume that the claim is true for $j - 1$, that is, for all $s \in \{d - j + 1, \ldots, d\}$ in gadget $Q_{j-1}$ we have

$$\sum_{\sigma \in A_s} |VR_{Q_{j-1}}(\sigma)| \equiv (j - 1)! 2^{(j-1)(j-2)/2} \mod p.$$

The $j$-th $(0, X, Y)$ node takes $(d - j)$-th input half-edge of the gadget and the half-edge 3 of the $(j - 1)$-st $(0, X, Y)$ node, and has parameters $p$ and $j$.

The type of the $j$-th $(0, X, Y)$ node is $\{\{0, 2\}, \{1, 3\}\}$ if and only if the resulting permutation in $Q_j$ is in $A_{d-j}$. Thus we have

$$\sum_{\sigma \in A_{d-j}} |VR_{Q_j}(\sigma)| \equiv 2^{j-1} \prod_{k=1}^{j-1} (2^{k-1}(k + 1)) \equiv j! 2^{j(j-1)/2} \mod p,$$

where the first term is the number of choices (modulo $p$) in the $j$-th $(0, X, Y)$ node to make it $\{\{0, 2\}, \{1, 3\}\}$ and the $k$-th term in the product is the number of choices (modulo $p$) in the $k$-th $(0, X, Y)$ node to make it either $\{\{0, 2\}, \{1, 3\}\}$ or $\{\{0, 3\}, \{1, 2\}\}$.

If the type inside the $j$-th $(0, X, Y)$ node is $\{\{0, 3\}, \{1, 2\}\}$ then the resulting permutation is in $A_s$ for $s \in \{d - j + 1, \ldots, d\}$. Thus

$$\sum_{\sigma \in A_s} |VR_{Q_j}(\sigma)| \equiv j 2^{j-1} \sum_{\sigma \in A_s} |VR_{Q_{j-1}}(\sigma)| \equiv j 2^{j-1} (j - 1)! 2^{(j-1)(j-2)/2}$$

$$\equiv j! 2^{j(j-1)/2} \mod p,$$

where $j 2^{j-1}$ is the number of choices (modulo $p$) in the $j$-th $(0, X, Y)$ node to make it $\{\{0, 3\}, \{1, 2\}\}$. □

Now we continue with the proof of the Lemma 3.

Let $\sigma$ be a permutation in $S_d$. Let $l = \sigma^{-1}(d)$. In order for $\sigma$ to be realized by gadget $Q$ we have to have $l$ mapped to $d$ by $Q_{d-1}$ and the permutation realized by the recursive gadget of size $d - 1$ must "cancel" the permutation of $Q_{d-1}$. By the claim there are $(d - 1)! 2^{(d-1)(d-2)/2}$ (modulo $p$) choices in $Q_{d-1}$ which map $l$ to $d$ and by the inductive hypothesis there are $R_{d-1}$ (modulo $p$) choices in the recursive gadget of size $d - 1$ that give the unique permutation that "cancels" the permutation of $Q_{d-1}$. Thus

$$|VR(\sigma)| \equiv R_d \equiv (d - 1)! 2^{(d-1)(d-2)/2} R_{d-1} \mod p,$$

finishing the proof of (4).

To see (5) note that the number of valid sets of routes which contain route starting and ending at both input half-edges or both output half-edges is 0 modulo $p$. This is because the number of valid set of routes of type $\{\{0, 1\}, \{2, 3\}\}$ inside the $(0, X, Y)$ node is 0 modulo $p$. □

*Proof (of Theorem 1).* The reduction is now a standard application of the Chinese remainder theorem. Given an Eulerian graph $G = (V, E)$, we can, w.l.o.g., assume that the degree of vertices of $G$ is at least 4 (vertices of degree 2 can be removed by contracting edges). The number of Eulerian tours of a graph on $n$ vertices is bounded by $n^{n^2}$ (the number of pairings in a vertex of degree $d$ is $d!/(2^{d/2}(d/2)!) \leq n^n$).

We choose $n^2$ primes $p_1, \ldots, p_{n^2} > n$ such that $\prod_{i=1}^{n^2} p_i > n^{n^2}$ and each $p_i$ is bounded by $O(n^3)$ (see, e.g., [BS96], p.296). For each $p_i$, we construct graph $G_i$ by replacing each vertex $v$ of degree $d > 4$ with $Q$ gadget with $d$ input and $d$ output half-edges where the $(2j - 1)$-st and $2j$-th output half-edge are connected (for $j = 1, \ldots, d/2$), and the input half-edges are used to replace half-edges emanating from $v$ (that is, they are connected to the input half-edges of other gadgets according to the edge incidence at $v$). Note that $G_i$ is a 4-regular graph. Since $p_i = O(n^3)$, the construction of $G_i$ can be done in time polynomial in $n$. Having $G_i$, we make a query to the oracle and obtain the number $T_i$ of Eulerian tours in $G_i$. Let $T$ be the number of Eulerian tours in $G$. Then

$$T_i \equiv T \prod_{d=6}^{n} \left( \left( \frac{d}{2} \right)! 2^{d/2} R_d \right)^{n_d} \mod p_i, \tag{6}$$

where $n_d$ is the number of vertices of degree $d$ in $G$.

Since $T_i$ is of length polynomial in $n$, we can compute $T_i \mod p_i$ for each $i$ and thus $T \mod p_i$ (since on the right hand side of (6) $T$ is multiplied by a term that has an inverse modulo $p_i$). By the Chinese remainder theorem, we can compute $T$ in time polynomial in $n$ (see, e.g., [BS96], p.106). □

### 3.3  #ET in 4-Regular Planar Graphs Is #P-Complete

First, it's easy to see that #ET in 4-regular planar graphs is in #P. We will give a reduction from #ET in 4-regular graphs to #ET in 4-regular planar graphs.

**Theorem 2.** *#ET in 4-regular graphs is polynomial time Turing reducible to #ET in 4-regular planar graphs.*

*Proof.* Given a 4-regular graph $G = (V, E)$, we first draw $G$ in the plane. We allow the edges to cross other edges, but i) edges do not cross vertices, ii) each crossing involves 2 edges. The embedding can be found in polynomial time.

Let $p$ be an odd prime, we will construct a graph $G_p$ from the embedded graph as follows. Let $e, e'$ be two edges in $G$ which cross in the plane as shown in Figure 4(a), we split $e$ (and $e'$) into two half-edges $e_1, e_2$ ($e_1', e_2'$, respectively). As illustrated in Figure 4(b), a $(0, X, Y)$ node with parameters $p$ and $k = p$ is added, and $e_1, e_1', e_2, e_2'$ are connected to the half-edges 0,1,2,3 of the $(0, X, Y)$ node, respectively.

Let $G_p$ be the graph after replacing all crossings by $(0, X, Y)$ nodes. We have that $G_p$ is planar since $(X, Y, Y)$ nodes and $(0, X, Y)$ nodes are all planar. The construction can be done in time polynomial in $p$ and the size of $G$ (since the

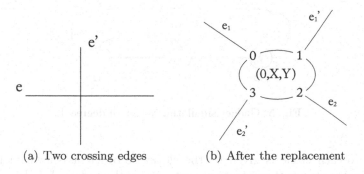

(a) Two crossing edges          (b) After the replacement

**Fig. 4.** To replace a crossover point by a $(0, X, Y)$ node with parameters $p$ and $k = p$

number of crossover points is at most $O(|E|^2)$ and the size of each $(0, X, Y)$ node is $O(p^2)$).

In the reduction, we choose $n = |V|$ primes $p_1, p_2, \ldots, p_n$ such that $p_i = O(n^2)$ for $i \in [n]$ and $\prod_{i=1}^{n} p_i \geq 3^n$, where $3^n$ is an upper bound for the number of Eulerian tours in $G$ (the number of pairings in each vertex is 3). For each $p_i$, we construct a graph $G_{p_i}$ from the embedded graph as described above with $p = p_i$. Let $T$ be the number of Eulerian tours in $G$ and $T_i$ be the number of Eulerian tours in $G_{p_i}$, we have

$$T \equiv T_i \quad \mathrm{mod}\ p_i. \tag{7}$$

Equation (7) follows from the fact that the number of Eulerian tours in which the set of routes within any $(0, X, Y)$ node is not of type $\{\{0, 2\}, \{1, 3\}\}$ is zero (modulo $p_i$) (since in (2) we have $A \equiv 1 \mod p_i$ and in (3) we have $B \equiv 0 \mod p_i$). We can make a query to the oracle to obtain the number $T_i$. By the Chinese remainder theorem, we can compute $T$ in time polynomial in $n$.    □

### 3.4 #A-TRAILS in 4-Regular Graphs with Rotational Embedding Schemes Is #P-Complete

In this section, we consider #A-TRAILS in graphs with rotational embedding schemes (maps). We prove that #A-TRAILS in 4-regular maps is #P-complete by a simple reduction from #ET in 4-regular graphs.

First, it's not hard to verify that #A-TRAILS in 4-regular maps is in #P.

**Theorem 3.** *#ET in 4-regular graphs is polynomial time Turing reducible to #A-TRAILS in 4-regular maps.*

*Proof.* Given a 4-regular graph $G = (V, E)$, for each vertex $v$ of $G$, we use the gadget shown in Figure 5 to replace $v$.

The gadget consists of three vertices which are represented by circles in Figure 5. The labels 0, 1, 2 and 3 are the four half-edges which are used to replace half-edges emanating from $v$. The cyclic ordering of the 4 (half-)edges

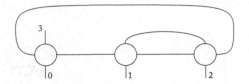

**Fig. 5.** Gadget simulating vertex of degree 4

incident to each circle is given by the clockwise order, as shown in Figure 5. There are three types of valid sets of routes inside the gadget, $VR(\{0,1\},\{2,3\})$, $VR(\{0,2\},\{1,3\})$ and $VR(\{0,3\},\{1,2\})$. By enumeration, we have the size of each of the three sets is 2.

Let $G'$ be the 4-regular map obtained by replacing each vertex $v$ by the gadget. Let $T$ be the number of Eulerian tours in $G$, we have the number of A-trails in $G'$ is $2^{|V|}T$.                                                                              □

Note that Kotzig [Kot68] gave a one-to-one correspondence between the A-trails in any 4-regular plane graph $G$ (the embedding in the plane gives the rotational embedding scheme) and the spanning trees in a plane graph $G'$, where $G$ is the medial graph of $G'$. By the Kirchhoff's theorem (c.f. [Jer03]), the number of spanning trees of any graph can be computed in polynomial time. Thus #A-TRAILS in 4-regular plane graphs can be computed in polynomial time.

## 4    The Complexity of Approximate Counting

In this section, we show that #ET in general graphs is AP-reducible to #A-TRAILS in 4-regular maps. AP-reductions were introduced by Dyer, Goldberg, Greenhill and Jerrum [DGGJ04] for the purpose of comparing the complexity of two counting problems in terms of approximation (given two counting problems $f, g$, if $f$ is AP-reducible to $g$ and there is an FPRAS for $g$, then there is also an FPRAS for $f$).

In the AP-reduction from #ET to #A-TRAILS in 4-regular maps, we use the idea of simulating the pairings in a vertex by a gadget as what we did in the construction of the $Q$ gadget. The difference is that the new gadget works in an approximate way, that is, instead of having the number of valid sets of routes to be the same for each of the types, the numbers can be different but within a small multiplicative factor. The analysis of the gadget uses a fast mixing result for a card shuffling problem.

We use the gadget illustrated in Figure 6. The circles represent the vertices in the map. Let $d$ be an even number. The gadget has $d$ input half-edges on left and $d$ output half-edges (Figure 6 demonstrates the case of $d = 6$). There are $T$ layers in the gadget which are numbered from 1 to $T$ from left to right. In an odd layer $t$, the $(2i-1)$-st and the $2i$-th output half-edges of layer $t-1$ are connected to a vertex of degree 4, for $i \in [d/2]$. In an even layer $t$, the $2i$-th and

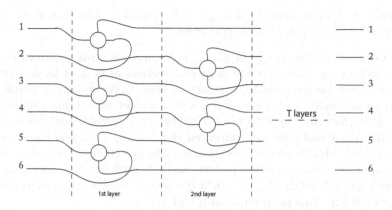

**Fig. 6.** Construction of the gadget for a vertex of degree 6

the $(2i+1)$-st output half-edges of layer $t-1$ are connected to a vertex of degree 4, for $i \in [d/2 - 1]$. In Figure 6, we illustrate the first two layers each of which is in two consecutive vertical dashed lines. The cyclic ordering of each vertex is given by the clockwise ordering (in the drawing in Figure 6), and so we have that the two half-edges in each vertex which are connected to half-edges of the previous layer are not cyclicly-adjacent.

Note that a valid route in the gadget always connects an input half-edge to an output half-edge. Thus a valid set of routes always realizes some permutation $\sigma$ connecting input half-edge $i$ to output half-edge $\sigma(i)$.

In order to prove that $|VR(\sigma)|$ is almost the same for each permutation $\sigma \in S_d$, we show that for $T = \Theta(d^2 \log d \log(d!/\varepsilon))$ we have $|VR(\sigma)| / \sum_{\varrho \in S_d} |VR(\varrho)| \in [(1-\varepsilon)/d!, (1+\varepsilon)/d!]$ for each permutation $\sigma \in S_d$. The gadget can be interpreted as a process of a Markov chain for shuffling $d$ cards. The simplest such chain proceeds by applying adjacent transpositions. The states of the chain are all the permutations in $S_d$. In each time step, let $\sigma \in S_d$ be the current state, we choose $i \in \{1, \ldots, d-1\}$ uniformly at random, and then switch $\sigma(i)$ and $\sigma(i+1)$ with probability $1/2$ and stay the same with probability $1/2$. For our gadget, it can be viewed as an even/odd sweeping Markov chain on $d$ cards [Wil04]. The ratio $|VR(\sigma)| / \sum_{\varrho \in S_d} |VR(\varrho)|$ is exactly the probability of being $\sigma$ at time $T$ when the initial state of the even/odd sweeping Markov chain is the identity permutation. By the analysis in [Wil04], we can relate $T$ with the ratio as follows.

**Lemma 4 ([Wil04]).** *Let $T$ be the number of layers of the gadget with $d$ input half-edges and $d$ output half-edges as shown in Figure 6, and let $\mu, \lambda$ be two distributions on $S_d$ such that $\mu(\sigma) = |VR(\sigma)| / \sum_{\varrho \in S_d} |VR(\varrho)|$ and $\lambda(\sigma) = 1/d!$ ($\lambda$ is the uniform distribution on $S_d$). For*

$$T = O(d^2 \log d \log(d!/\varepsilon)),$$

*then $\|\mu - \lambda\|_{TV} \leq \varepsilon/d!$, and thus $(1-\varepsilon)/d! \leq \mu(\sigma) \leq (1+\varepsilon)/d!$.*

**Theorem 4.** *If there is an FPRAS for* #A-TRAILS *in 4-regular maps, then we have an FPRAS for* #ET *in general graphs.*

*Proof.* Given an Eulerian graph $G = (V, E)$ and an error parameter $\varepsilon > 0$, we can, w.l.o.g., assume that the degree of vertices of $G$ is at least 4 (vertices of degree 2 can be removed by contracting edges). We construct graph $G'$ by replacing each vertex $v$ of degree $d > 2$ with a gadget with $d$ input half-edges, $d$ output half-edges and $T_d = \Theta(d^2 \log d \log(4d! n/\varepsilon))$ layers where the $(2i - 1)$-st and $2i$-th output half-edge are connected (for $1 \le i \le d/2$), and the input half-edges are used to replace half-edges emanating from $v$ (that is, they are connected to the input half-edges of other gadgets according to the edge incidence at $v$). We have that $G'$ has $O(n^2 T_n) = O(n^4 \log n(n \log n + \log(1/\varepsilon)))$ vertices and can be constructed in time polynomial in $n$ and $1/\varepsilon$.

Let $\mathcal{A}$ be an FPRAS for #A-TRAILS in 4-regular maps by the assumption of the theorem, we run $\mathcal{A}$ on $G'$ with error parameter $\varepsilon/2$. Let $\mathcal{A}(G', \varepsilon/2)$ be the output of $\mathcal{A}$ and $N_A$ be the number of A-trails in $G'$, we have $\mathcal{A}(G', \varepsilon/2) \in [e^{-\varepsilon/2} N_A, e^{\varepsilon/2} N_A]$ with probability at least 2/3. This process can be done in time polynomial in the size of $G'$ and $1/\varepsilon$, which is polynomial in $n$ and $1/\varepsilon$.

Let $D_d$ be the number of vertices in the gadget of $d$ input half-edges and $d$ output half-edges, and let $R_d = 2^{D_d} 2^{d/2} (d/2)!/d!$ and $R = \prod_{d=4}^n R_d^{n_d}$ where $n_d$ is the number of vertices of degree $d$ in $G$. Our algorithm $\mathcal{B}$ will output

$$\mathcal{B}(G, \varepsilon) = \mathcal{A}(G', \varepsilon/2)/R. \tag{8}$$

We next prove that $\mathcal{B}$ is an FPRAS for #ET in general graphs. For every Eulerian tour in $G$, the type of the pairing in each vertex in $G$ is fixed. Note that each pairing corresponds to $(d/2)! 2^{d/2}$ permutations in a gadget with $d$ input half-edges and $d$ output half-edges. By Lemma 4, we have $(1 - \varepsilon/(4n)) 2^{D_d}/d! \le |VR(\sigma)| \le (1 + \varepsilon/(4n)) 2^{D_d}/d!$ for each $\sigma \in S_d$ where $VR(\sigma)$ is counted in a gadget with $d$ input half-edges and $d$ output half-edges. Thus, the number of A-trails in $G'$ which correspond to the same Eulerian tour in $G$ is in $[(1 - \varepsilon/(4n))^n R, (1 + \varepsilon/(4n))^n R]$. Let $N_E$ be the number of Eulerian tours in $G$, we have $N_A \in [(1 - \varepsilon/(4n))^n R N_E, (1 + \varepsilon/(4n))^n R N_E]$, and thus for $\varepsilon \le 2n$, $N_A/R \in [e^{-\varepsilon/2} N_E, e^{\varepsilon/4} N_E]$ (the case when $\varepsilon > 2n$ is trivial, $\mathcal{B}$ can just output $3^n$). Since $\mathcal{A}(G', \varepsilon/2) \in [e^{-\varepsilon/2} N_A, e^{\varepsilon/2} N_A]$ with probability at least 2/3, then by (8), we have $\mathcal{B}(G, \varepsilon) \in [e^{-\varepsilon} N_E, e^{\varepsilon} N_E]$ with probability at least 2/3. This completes the proof. $\qquad\square$

# References

[AF95]    Andersen, L.D., Fleischner, H.: The NP-completeness of finding A-trails in Eulerian graphs and of finding spanning trees in hypergraphs. Discrete Appl. Math. 59(3), 203–214 (1995)

[BM87]    Bent, S.W., Manber, U.: On nonintersecting Eulerian circuits. Discrete Appl. Math. 18(1), 87–94 (1987)

[BS96]    Bach, E., Shallit, J.: Algorithmic number theory. Foundations of Computing Series, vol. 1. MIT Press, Cambridge (1996); Efficient algorithms

[BW05]      Brightwell, G., Winkler, P.: Counting eulerian circuits is #p-complete.
            In: ALENEX/ANALCO, pp. 259–262 (2005)
[DGGJ04]    Dyer, M., Goldberg, L.A., Greenhill, C., Jerrum, M.: The relative com-
            plexity of approximate counting problems. Algorithmica 38(3), 471–500
            (2004); Approximation algorithms
[Dvo04]     Dvořák, Z.: Eulerian tours in graphs with forbidden transitions and
            bounded degree. KAM-DIMATIA (669) (2004)
[Fle90]     Fleischner, H.: Eulerian graphs and related topics. Part 1. vol. 1. Annals
            of Discrete Mathematics, vol. 45. North-Holland Publishing Co., Amster-
            dam (1990)
[Jer]       Jerrum, M.:
            http://www.ams.org/mathscinet/search/publdoc.html?pg1=ISSI&
            s1=191567&r=2&mx-pid=1822924
[Jer03]     Jerrum, M.: Counting, sampling and integrating: algorithms and com-
            plexity. Lectures in Mathematics. ETH Zürich. Birkhäuser Verlag, Basel
            (2003)
[Kot68]     Kotzig, A.: Eulerian lines in finite 4-valent graphs and their transforma-
            tions. In: Theory of Graphs (Proc. Colloq., Tihany, 1966), pp. 219–230.
            Academic Press, New York (1968)
[TV01]      Tetali, P., Vempala, S.: Random sampling of Euler tours. Algorith-
            mica 30(3), 376–385 (2001); Approximation algorithms for combinatorial
            optimization problems
[Vaz01]     Vazirani, V.V.: Approximation algorithms. Springer, Berlin (2001)
[Wil04]     Wilson, D.B.: Mixing times of Lozenge tiling and card shuffling Markov
            chains. Ann. Appl. Probab. 14(1), 274–325 (2004)

# Efficient Edge Domination on Hole-Free Graphs in Polynomial Time

Andreas Brandstädt, Christian Hundt, and Ragnar Nevries

Institut für Informatik, Universität Rostock, D-18051 Rostock, Germany
{Andreas.Brandstaedt,Christian.Hundt,Ragnar.Nevries}@uni-rostock.de

**Abstract.** This paper deals with the *Efficient Edge Domination* Problem (*EED*, for short), also known as *Dominating Induced Matching Problem*. For an undirected graph $G = (V, E)$ EED asks for an induced matching $M \subseteq E$ that simultaneously dominates all edges of $G$. Thus, the distance between edges of $M$ is at least two and every edge in $E$ is adjacent to an edge of $M$. EED is related to parallel resource allocation problems, encoding theory and network routing. The problem is NP-complete even for restricted classes like planar bipartite and bipartite graphs with maximum degree three. However, the complexity has been open for chordal bipartite graphs.

This paper shows that EED can be solved in polynomial time on hole-free graphs. Moreover, it provides even linear time for chordal bipartite graphs.

Finally, we strengthen the NP-completeness result to planar bipartite graphs of maximum degree three.

**Keywords:** efficient edge domination; dominating induced matching; chordal bipartite graphs; weakly chordal graphs; hole-free graphs; linear time algorithm; polynomial time algorithm.

## 1 Introduction

Let $G = (V, E)$ be a simple undirected graph with vertex set $V$ and edge set $E$. A subset $M$ of $E$ is an *induced matching* if every pair of edges $e, e'$ in $M$ has distance at least two. That means that $e \cap e' = \emptyset$ and there is no edge $xy \in E$ with $x \in e$ and $y \in e'$. A subset $M \subseteq E$ is called *dominating edge set* if every edge $e \in E \setminus M$ shares one endpoint with an edge $e' \in M$, i.e., if $e \cap e' \neq \emptyset$. The *Efficient Edge Domination Problem* (*EED* for short) is to decide whether a given graph has a *dominating induced matching* (*d.i.m.* for short). Therefore, EED is also known as the *Dominating Induced Matching Problem*. Moreover, in literature d.i.m.s are sometimes called *efficient edge packings*. A brief history of EED as well as some applications in the fields of resource allocation, encoding theory and networking routing are presented in [8] and [11].

Grinstead et al. [8] show that EED is NP-complete in general. It remains hard for bipartite graphs [13] and even for some of their very special subclasses. In

A. López-Ortiz (Ed.): LATIN 2010, LNCS 6034, pp. 650–661, 2010.

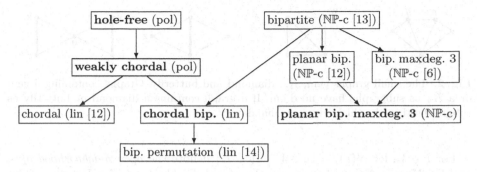

**Fig. 1.** An inclusion diagram for graph classes related to this work. The time complexity of MEED is given in brackets. The graph classes for which new complexity results are given in this paper are printed boldly.

particular, [12] shows the intractability of EED for planar bipartite graphs and [6] for bipartite graphs with maximum degree three.

On the other hand, EED can be solved efficiently on various specific graph classes such as series-parallel graphs [8], bipartite permutation graphs [14] and claw-free graphs [6]. Since EED can be stated in monadic second order logic it is also polynomial time solvable on any graph class with bounded clique-width [6]. The complexity of EED for weakly chordal graphs is stated as an open problem in [12]. However, it has neither been known whether EED is tractable on the subclass of chordal bipartite graphs nor if it is NP-complete for the superclass of hole-free graphs.

In this paper we consider the generalized optimization version of EED, the *Minimum Efficient Edge Domination Problem (MEED)*, which asks for a d.i.m. $M$ in $G = (V, E)$ of minimum weight with respect to some given weight function $\omega : E \to \mathbb{R}$. Hence, we try to find $M$ in such a way that it minimizes $\sum_{e \in M} \omega(e)$. Clearly, MEED is computationally harder than EED.

In this paper, we show that MEED can be solved in linear time on chordal bipartite graphs. This extends the result in [14], since bipartite permutation graphs are chordal bipartite. Furthermore, we show that MEED can be solved in polynomial time on hole-free graphs. In this way we answer the open question posed in [12]. Finally we generalize the hardness results of [6] and [12] by showing that EED is NP-complete for planar bipartite graphs even with maximum degree three. See Figure 1 for a small overview.

## 2 Basic Notions

Let $G = (V, E)$ be a finite simple (i.e., without loops and multiple edges) undirected graph with vertex set $V$ and edge set $E$. The *complement graph* $\overline{G} = (V, \overline{E})$ of $G$ contains for all $x, y \in V$ the edge $xy \in \overline{E}$ iff $xy \notin E$.

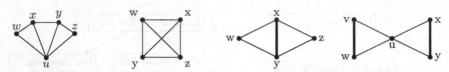

**Fig. 2.** The small graphs gem, $K_4$, diamond and butterfly. Graphs containing a gem or a $K_4$ as subgraphs have no d.i.m. If a graph contains a diamond or butterfly as subgraphs then the thick edges are mandatory.

For $v \in V$, let $N(v) := \{u \in V \mid uv \in E\}$ denote the *open neighborhood of $v$*, and let $N[v] := N(v) \cup \{v\}$ denote the *closed neighborhood of $v$*. For $uv \in E$ let $N(uv) := N(u) \cup N(v) \setminus \{u, v\}$ and $N[uv] := N[u] \cup N[v]$. For $U \subseteq V$, let $G[U]$ denote the induced subgraph of $G$ with vertex set $U$, hence, the graph which contains exactly the edges $xy \in E$ with both vertices $x$ and $y$ in $U$.

A path $P_k$ (resp. cycle $C_k$) is a sequence of distinct vertices, say $v_1, \ldots, v_k$, which are connected by edges $v_i v_{i+1}$, $1 \leq i \leq k - 1$ (and $v_k v_1$, resp.). We say that such a path has length $k - 1$ and such a cycle has length $k$. For two vertices $x, y \in V$, let $dist_G(x, y)$ denote the *distance between $x$ and $y$ in $G$*, i.e., the length of a shortest path between $x$ and $y$ in $G$. The *distance of two edges* $e, e' \in E$ is the length of a shortest path between $e$ and $e'$, i.e., $dist_G(e, e') = \min\{dist_G(u, v) \mid u \in e, v \in e'\}$. In particular, this means that $dist_G(e, e') = 0$ if and only if $e \cap e' \neq \emptyset$. A *connected component* of $G$ is a maximal vertex subset $U \subseteq V$ such that all pairs of vertices of $U$ are connected by paths in $G[U]$. A *biconnected component* of $G$ is a maximal vertex subset $U \subseteq V$ such that all pairs of vertices of $U$ are connected by at least two paths in $G[U]$ that share no other vertices than start and end vertex.

See Figure 2 and [5,7] for important graphs and graph classes used in this paper but not defined in the folling listing: A graph is a *block graph* if all its biconnected components $U$ induce complete graphs (cliques) $G[U]$, i.e., $xy \in E$ for all $x, y \in U$. For a set $\mathcal{F}$ of graphs, a graph $G$ is called *$\mathcal{F}$-free* if $G$ contains no induced subgraph from $\mathcal{F}$. A graph is *chordal* if it is $C_k$-free for all $k \geq 4$. A bipartite graph $B = (X, Y, E)$ is *chordal bipartite* if $B$ is $C_{2k}$-free for all $k \geq 3$.

A graph $G(u, v)$ is *generalized series-parallel* if it has two designated vertices $u$ and $v$ and is either a single edge $uv$ or can be built out of two smaller generalized series-parallel graphs $G_1(u_1, v_1)$ and $G_2(u_2, v_2)$ with the following compositions:

*Series-1:* Combine $G_1$ and $G_2$ by identifying $v_1$ with $u_2$ to obtain $G(u_1, v_2)$.
*Series-2:* Combine $G_1$ and $G_2$ by identifying $v_1$ with $u_2$ to obtain $G(u_1, v_1)$.
*Parallel:* Combine $G_1$ and $G_2$ by identifying $u_1$ with $u_2$ and $v_1$ with $v_2$ to obtain $G(u_1, v_1)$.

The paper deals with *dominating induced matchings (d.i.m.s)* as defined in the previous section. If an edge $e$ of $G$ is part of a d.i.m. $M$ we call it an edge *matched by $M$*. Furthermore, we call a vertex $v \in V$ *matched by $M$* if it is endpoint of an edge matched by $M$. If an edge $e \in E$ is matched by every d.i.m. of $G$, we call it *mandatory in $G$*.

## 3   Basic Properties of Graphs Having Dominating Induced Matchings

The following general observations are useful when dealing with d.i.m.s.

**Observation 1.** *Let $G$ be a graph having a d.i.m. $M$. Then the following holds:*

(i) *$M$ contains exactly one edge of every triangle of $G$.*
(ii) *$M$ contains no edge of any (not necessarily induced) cycle of length four in $G$.*

As a simple consequence we obtain:

**Corollary 2.** *If a graph $G$ contains a triangle such that two of its edges are in a cycle of length four (not necessarily the same), the third edge of the triangle is mandatory in $G$. If a graph $G$ contains a triangle such that all its three edges are in cycles of length four then $G$ has no d.i.m.*

In a reduction described later in this paper, we want to remove induced diamonds and butterflies by replacing mandatory edges. So we need:

**Observation 3.** *In a graph $G$, in every induced diamond, its mid-edge $xy$ is mandatory in $G$, and for every induced butterfly in $G$, its two peripheral edges $vw$ and $xy$ are mandatory in $G$ (see Figure 2).*

For every d.i.m. $M$ of a graph $G$, the matched edges must have distance at least two. This must be true in particular for mandatory edges, so we obtain:

**Corollary 4.** *If a graph $G$ has a d.i.m. $M$, then $G$ is $(K_4, gem)$-free.*

## 4   Dominating Induced Matchings in Chordal Bipartite Graphs

MEED is known to be solvable in linear time on generalized series-parallel graphs by [12]. The aim of this section is a linear time reduction from a chordal bipartite graph $G$ into a $K_4$-free block graph $G'$, such that solving MEED on $G'$ is equivalent to solving MEED on $G$. Since $K_4$-free block graphs are generalized series-parallel, this gives a linear time algorithm for MEED on chordal bipartite graphs.

Therefore, we substitute every biconnected component of $G$ by a gadget to obtain $G'$. The gadget must ensure that $G'$ behaves like $G$ with respect to d.i.m.s.

The reduction from $G = (V, E)$ to $G' = (V', E')$ works as follows: Let $G$ be a chordal bipartite graph and let $U_1, \ldots, U_k$ be the biconnected components of $G$. Since $G$ is bipartite, the biconnected components are bipartite. So let $U_i = X_i \cup Y_i$ be a bipartition for every $1 \leq i \leq k$. Note that all edges of a biconnected component in a chordal bipartite graph $G$ are part of a $C_4$, because $G$ contains no other induced cycles. By Observation 1, no edge of $G[U_i]$ is matched in any d.i.m. Hence, all edges of $G[U_i]$ must be dominated from the outside. The following lemma demonstrates the only two ways of how the edges of a biconnected component can be dominated:

**Lemma 5.** *Let $G = (X, Y, E)$ be a chordal bipartite graph and $U_i = X_i \cup Y_i$ a biconnected component of $G$. Then for every d.i.m. $M$ of $G$, either all vertices in $X_i$ and none of $Y_i$ are matched or vice versa.*

*Proof.* For every d.i.m. $M$ and every biconnected component $U_i = X_i \cup Y_i$ of $G$ consider an edge $xy \in E$ with $x \in X_i$ and $y \in Y_i$. By Observation 1, $xy \notin M$. Thus, $x$ or $y$ must be matched by $M$. If $M$ matches both $x$ and $y$ then there are two edges $xx' \in M$ and $yy' \in M$ with distance one and $M$ is not an induced matching, a contradiction.

Hence $M$ matches $x$ if and only if $M$ does not match $y$. Suppose that $M$ matches a vertex $x' \in X_i$ and a vertex $y' \in Y_i$. Then there is an induced path $P$ in $U$ between $x'$ and $y'$ such that $P$ has an odd number $2k + 1$, $k \geq 1$, of edges. Then a simple inductive argument shows that there is a shorter odd path with matched endpoints in both $X_i$ and $Y_i$. Thus, either $M$ matches all vertices in $X_i$ and none of $Y_i$ or $M$ matches all vertices in $Y_i$ and none of $X_i$.     □

To simulate this property of biconnected components, we modify every biconnected component using a certain gadget that also allows exactly two different d.i.m.s. It consists of a triangle $(x_i, y_i, d_i)$ and a pending edge $d_i p_i$. The reduction replaces the edges of $G[U_i]$ for every biconnected component $U_i$ by the gadget $D_i = (\{d_i, x_i, y_i, p_i\}, \{d_i x_i, d_i y_i, d_i p_i, x_i y_i\})$. From Observation 1 we know that exactly one edge of every triangle must be matched in every d.i.m. The pending edge $d_i p_i$ forces to match either $d_i x_i$ or $d_i y_i$, because if $x_i y_i$ were matched, there would be no way to dominate $d_i p_i$. Now, we add edges $u x_i$ for all $u \in X_i$ and edges $u y_i$ for all $u \in Y_i$. This retains the property of the Lemma 5 for all vertices of $U_i$ in $G'$ by the reduction, i.e. either all vertices in $X_i$ and none in $Y_i$ are matched or vice versa. The edge weights for $G'$ are inherited from $G$ and new edges get weight zero: $\omega'(e) = \omega(e)$ for all $e \in E \cap E'$ and $\omega'(e) = 0$ for all $e \in E' \setminus E$. See an example for the reduction of one single biconnected component in Figure 3. By construction, the resulting graph $G'$ is a $K_4$-free block graph.

It remains to show that finding a d.i.m. in $G$ is equivalent to finding one in $G'$.

**Lemma 6.** *A chordal bipartite graph $G$ has a d.i.m. $M$ with weight $\omega(M) = W$ if and only if the graph $G'$ has a d.i.m. $M'$ with weight $\omega'(M') = W$.*

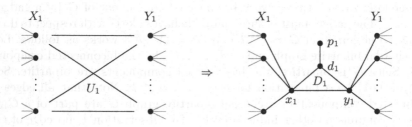

**Fig. 3.** The biconnected component $U_1$ of $G$ (left) is substituted by the gadget $D_1$ and additional edges in $G'$ (right)

*Proof.* Let $G = (V, E)$ be a chordal bipartite graph with biconnected components $U_1, \ldots, U_k$ and their bipartitions $U_i = X_i \cup Y_i$. Let $M$ be a d.i.m. of $G$. We define $M'$ as follows:

$$M' := M \cup M^D$$

where $M^D$ fulfills the following conditions for all $1 \leq i \leq k$:

1. Let $d_i y_i \in M^D$ if and only if all vertices of $X_i$ are matched by $M$.
2. Let $d_i x_i \in M^D$ if and only if all vertices of $Y_i$ are matched by $M$.

Since $M^D \subseteq E'$ by definition of $G'$, $M \cap U_i = \emptyset$ for all $1 \leq i \leq k$ by Observation 1 and $M \subseteq E$ it is true $M' \subseteq E'$. We have to show that $M'$ is an induced matching in $G'$ and dominates all edges of $G'$.

Assume there are two edges $uv, u'v' \in M'$ with distance smaller than two in $G'$. If $uv \in M$ and $u'v' \in M$, then they have distance smaller than two also in $G$ and $M$ was not an induced matching. If $uv \in M$ and $u'v' \in M^D$, then there are two matched vertices $u \in X_i$ and $x_i$ ($u \in Y_i$ and $y_i$ respectively) for some $1 \leq i \leq k$. Then by definition of $M^D$, $d_i y_i \in M^D$ ($d_i x_i \in M^D$ respectively) and then there exists a matched vertex $u' \in Y_i$ ($u' \in X_i$ respectively). But by Lemma 5, then $M$ was not a d.i.m. of $G$, a contradiction. This is also the case if $uv \in M^D$ and $u'v' \in M^D$.

Since $M$ dominates all edges in $E$ and $M \subseteq M'$, all edges in $E' \cap E$ are dominated by $M'$. By Lemma 5, either $d_i x_i \in M'$ or $d_i y_i \in M'$ for all $1 \leq i \leq k$ and hence, $d_i p_i$ and $x_i y_i$ are dominated by $M'$. Assume there is an edge in $E'$ that is not dominated by $M'$, it must have the form $u x_i$ with $u \in X_i$ or $u y_i$ with $u \in Y_i$ for some $1 \leq i \leq k$. If $u \in X_i$, then $d_i x_i \notin M'$, because otherwise $u x_i$ would be dominated and $d_i y_i \notin M'$ by definition of $M^D$, because $u$ is not matched. Analogously, if $u \in Y_i$. But if both $d_i x_i$ and $d_i y_i$ are not in $M'$, then by Lemma 5, $M$ is not a d.i.m. of $G$, a contradiction.

Now let $M'$ be a d.i.m. of $G'$. We define $M$ as $M := M' \cap E$, so it is an edge subset of $G$. Note that in every d.i.m. $M'$ of $G'$ either $d_i x_i \in M'$ or $d_i y_i \in M'$ for all $1 \leq i \leq k$, because exactly one edge of the triangle $(d_i, x_i, y_i)$ must be in $M'$ by Observation 1 and it cannot be $x_i y_i \in M'$, because then there would be no way to dominate $d_i p_i$. Furthermore, if $d_i x_i \in M'$ for some $1 \leq i \leq k$, then all vertices in $Y_i$ are matched by $M'$, because otherwise an edge between $y_i$ and $Y_i$ is not dominated, and analogously for $d_i y_i \in M'$. Assume there are two edges $uv \in M$ and $u'v' \in M$ with a distance smaller two. Since $E \setminus E'$ contains only $C_4$-edges and $M'$ is an induced matching, $uv$ and $u'v'$ have both one endpoint in $X_i$ or $Y_i$ for some $1 \leq i \leq k$. Without loss of generality, let $u \in X_i$ and $u' \in Y_i$. But then $M'$ was not an induced matching, because either $u$ is too close to $x_i$ or $u'$ is too close to $y_i$, a contradiction.

Since edges in $M' \setminus E$ cannot dominate edges in $E$ by definition of $G'$, $M$ already dominates all edges in $E \cap E'$. Assume there is a not dominated edge $uu' \in U_i$ for some $1 \leq i \leq k$ and without loss of generality let $u \in X_i$ and $u' \in Y_i$. Then $M'$ was not a d.i.m., since either $u x_i$ or $u' y_i$ is not dominated in $G'$.

It is easy to see that $M$ and $M'$ have the same weight in both cases, since it is true $M \subseteq M'$ and all edges in $M' \setminus M$ have weight zero by definition of $G'$.    $\square$

Now we can formulate the following theorem.

**Theorem 1.** *MEED is solvable in linear time on chordal bipartite graphs.*

*Proof.* Let $G$ be a chordal bipartite graph. The problem can be solved by constructing $G'$ of $G$ as above and then solving the problem on $G'$. The correctness results directly from Lemma 6.

To find all biconnected components of $G$, we can use Tarjan's algorithm (see, e.g., [15]). The algorithm can easily be modified to give a bipartition for each biconnected component in a single depth-first search run as well. The reduction from $G$ to $G'$ adds a constant number of new vertices for every biconnected component of $G$. Since the number of biconnected components is at most the number of vertices of $G$, the number of vertices of $G'$ is linear in the number of vertices of $G$. The reduction modifies also the number of edges; for every vertex of $G$ there is at most one new edge in $G'$, so the increase is linear, too. Since $K_4$-free block graphs are generalized series-parallel graphs, we can use the algorithm given in [12] to solve MEED on $G'$ in linear time.                                □

## 5   Dominating Induced Matchings in Hole-Free Graphs

We use a reduction from hole-free graphs to $K_4$-free block graphs, that is similar to the reduction described in the previous section. Since hole-free graphs can contain more complex biconnected components than chordal bipartite graphs, the reduction is more sophisticated, needs two steps and requires polynomial time.

The first step simplifies the biconnected components of the given hole-free graph $G$ such that the resulting graph $G'$ contains only triangles of a simpler structure. This reduction uses a graph operation $f_1$ that is repeatedly applied until the result does not change any more, we say $G' = f_1^*(G)$.

Then, the second reduction step from $G'$ to $G''$ performs a gadget based substitution of biconnected components. It also uses a graph operation $f_2$ that is repeatedly applied such that $G'' = f_2^*(G')$. After that, $G''$ is a $K_4$-free block graph. Again, the reductions have to guarantee, that solving MEED on $G$ is equivalent to solving MEED on $G''$. This is done by showing that applying $f_1$ and $f_2$ keeps this equivalence.

Let $G$ be a hole-free graph. Since graphs containing a gem or a $K_4$ are not of interest by Observation 4 and because finding such induced subgraph can be done in polynomial time, we can assume that $G$ is $(K_4, gem)$-free in the rest of this section.

The first reduction step from $G$ to $G'$ works as follows: We define a graph operation $f_1$ that replaces certain kinds of triangle edges. In particular, $f_1$ concerns the mandatory edges in diamonds, butterflies (see Observation 3) and other triangles described in Corollary 2. Let $uv \in E$ be such a mandatory edge in $G = (V, E)$. Then $f(G)$ is defined as follows: Remove vertices $u$ and $v$ from $G$ and add a $P_4$ with new vertices $p_1, p_2, p_3, p_4$ and edges $p_i p_{i+1}$, $i = 1, 2, 3$. Add additional edges such that $N(p_2) = N(uv) \cup \{p_1, p_3\}$. Let $\omega(p_1 p_2) = \omega(p_3 p_4) = 0$ and $\omega(p_2 p_3) = \omega(uv)$. Moreover, let $\omega(p_2 x) = 0$ for all $x \in N(uv)$.

**Lemma 7.** *Let $G$ be a hole-free graph. Then $G$ has a d.i.m. $M$ with weight $\omega(M) = W$ if and only if $f_1(G)$ has a d.i.m. $M_{f_1}$ with weight $\omega_{f_1}(M_{f_1}) = W$. Furthermore, $f_1(G)$ is hole-free.*

The graph $G' = f_1^*(G)$ results from applying $f_1$ repeatedly for all mandatory edges given by Observation 3 and Corollary 2. Since $f_1$ reduces the number of edges that are in a cycle, the number of iterations is smaller than $|E(G)|$. From Lemma 7 follows:

**Corollary 8.** *$G$ has a d.i.m. $M$ with weight $\omega(M) = W$ if and only if $G'$ has a d.i.m. $M'$ with $\omega'(M') = W$.*

**Observation 9.** *Let $G$ be a hole-free graph and let $G'$ be defined as above. If $G$ has a d.i.m., then for every triangle in $G'$, exactly one of the following conditions holds:*

(1) *No edge of the triangle is in another cycle.*
(2) *Exactly one edge of the triangle is in one or more cycles of length four.*

If $G$ does not fulfill the condition of Observation 9, it has no d.i.m. This can obviously be checked in polynomial time. Consequently, from now on we assume that $G$ fulfills this condition.

In the reduction step from $G'$ to $G''$, we use a gadget transformation $f_2$ to generate a $K_4$-free block graph $G''$. The graph $f_2(G)$ is defined as follows: Let $U$ be a biconnected component of $G$ with more than three vertices. Hence, $G[U]$ contains either at least one triangle of the second form of Observation 9 or it is triangle-free. In both cases, we replace all edges of $G[U]$ by the gadget $D = (\{d, x, y, p\}, \{dx, dy, dp, xy\})$, known from the previous section.

In the easy case, where $G[U]$ is triangle-free, that is bipartite, let $U = X \cup Y$ be a bipartition of $U$. Then the graph $f_2(G)$ results from $G$ by adding edges $ux$ for all $u \in X$ and $uy$ for all $u \in Y$.

In the other case, the embedding of the gadget $D$ is more sophisticated due to the presence of triangles in the biconnected components. Let $T_1 = (t_1^X, t_1^Y, t_1), \ldots, T_\ell = (t_\ell^X, t_\ell^Y, t_\ell)$ be all triangles of $G[U]$ and let $T \subset U$ contain all triangle vertices of $G[U]$. Note that all triangles are disjoint because the first reduction step eliminates all diamonds and butterflies in $G'$. Every triangle $T_i$ has exactly one edge being part of an induced cycle of length four. Without loss of generality let this be the edge $t_i^X t_i^Y$. Let $U' = U \setminus \{t_1, \ldots, t_\ell\}$. Then $G[U']$ is obviously bipartite. Let $U' = X' \cup Y'$ be a bipartition of $U'$. Without loss of generality let $t_i^X \in X'$ and $t_i^Y \in Y'$ for all $1 \leq i \leq \ell$. Let $\mathcal{N}(v) = N(v) \setminus U$ denote all neighbors of $v$ outside of $U$. Then $f_2(G)$ results from $G$ by the following steps:

- Remove all vertices of $T$ from $G$.
- Add an edge $ux$ for all $u \in X' \setminus T$.
- Add an edge $uy$ for all $u \in Y' \setminus T$.
- Add edges such that $\mathcal{N}(t_i) \subseteq N(d)$ for all $1 \leq i \leq \ell$.
- Add edges such that for all $1 \leq i \leq \ell$:

$$\mathcal{N}(t_i^Y) \subseteq N(x) \quad \text{and} \quad \mathcal{N}(t_i^X) \subseteq N(y).$$

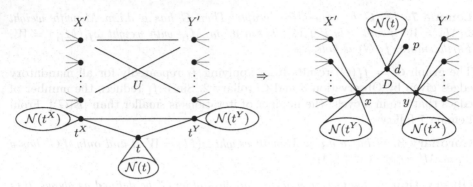

**Fig. 4.** The biconnected component $U$ of a graph $G$ (left) is substituted by the gadget $D$ and appropriate edges in $f_2(G)$ (right)

Furthermore let

$$w^{\star\star}(yd) = \sum_{1 \le i \le \ell} w'(t_i^X t_i) \quad \text{and} \quad w^{\star\star}(xd) = \sum_{1 \le i \le \ell} w'(t_i^Y t_i).$$

All other new edges get weight zero. See Figure 4 for an example of this transformation.

**Lemma 10.** *Let $G$ be a (hole,diamond,butterfly)-free graph that fulfills the condition given in Observation 9. Then $G$ has a d.i.m. $M$ with weight $\omega(M) = W$ if and only if $f_2(G)$ has a d.i.m. $M_{f_2}$ with weight $\omega_{f_2}(M_{f_2}) = W$.*

The graph $G'' = f_2^*(G')$ results by repeatedly applying $f_2$. Since $f_2$ reduces the number of biconnected components of more than three vertices, the number of iterations is polynomial. The resulting graph $G''$ is a $K_4$-free block graph, since all biconnected components of more than three vertices get substituted by a gadget containing a single triangle.

**Corollary 11.** *$G'$ has a d.i.m. $M'$ with weight $\omega'(M') = W$ if and only if $G''$ has a d.i.m. $M''$ with $\omega''(M'') = W$.*

**Theorem 2.** *MEED is solvable in polynomial time on hole-free graphs.*

## 6   Dominating Induced Matchings in Planar Bipartite Graphs with Maximum Degree Three

The previous sections show quite clearly that restricting the length of induced cycles makes MEED less difficult. On the other hand, EED becomes $\mathbb{NP}$-complete for planar bipartite graphs [12]. A closer examination of the proof in [12] shows that one can additionally restrict to graphs of maximum degree four. For maximum degree two however EED is trivial again and thus, it would be interesting to know the complexity of EED on the same problem with max-degree *three*.

**Theorem 3.** *EED is* $\mathbb{NP}$-*complete on planar bipartite graphs of maximum degree three (3-PBEED).*

The basic ideas for the polynomial time reduction of the $\mathbb{NP}$-complete Planar Monotone One-In-Three-3SAT Problem [9] to 3-PBEED are as follows: We construct a planar bipartite graph $G(F)$ (of max-degree three) from any planar monotone 3CNF-formula $F$ such that $G(F)$ has a dominating induced matching $M$ if and only if $F$ has a satisfying assignment with exactly one true variable in each clause. Monotone means, that all clauses of $F$ contain only positive literals (i.e., variables). Moreover, if $X$ are the variables and $Y$ the clauses of $F$ let $I(F) = (X, Y, E)$ be $F$'s incidence graph, where $x \in X$ and $y \in Y$ are connected by an edge $xy$ iff $x$ occurs in $y$. Then $F$ is called planar if $I(F)$ is planar.

The basic idea is to construct $G(F)$ from $I(F)$ by replacing the vertices $x$ of $X$ and $y$ of $Y$ by specific planar gadgets $V(x)$ and $C(y)$. This modification preserves the planarity for $G(F)$.

Every d.i.m. $M$ of $G(F)$ represents an assignment $t : X \to \{0,1\}$ of truth values. In fact, $M$ can match the edges of each $V(x)$ in only two ways. We can interpret this two states as the truth value of $x$. Particularly, $V(x)$ contains an edge $e_x^0$ such that $t(x) = 1$ iff $e_x^0$ is in $M$. Moreover, $V(x)$ clones the matching state of $e_x^0$ to fan out edges $e_x^1, \ldots, e_x^p$. The number $p$ is chosen to be a power of four with $p \geq |Y|$. Beside preserving the bipartite structure for $V(x)$ this makes sure that there will be enough copies of $x$'s truth value.

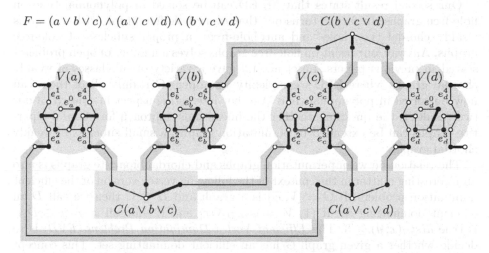

**Fig. 5.** An example of how a formula $F$ is encoded into a graph $G(F)$. The planar incidence graph $I(F)$ can be found grayed-out in the background with roundly variable vertices and rectangular clause vertices. Obviously, $G(F)$ follows the structure of $I(F)$ by replacing vertices with gadgets. The vertex coloring shows the bipartition of $G(F)$. Thick edges in $G(F)$ give a d.i.m. to encode $a = b = d = 0$ and $c = 1$, a solution for Planar Monotone One-In-Three-3SAT.

The gadgets $C(y)$ guarantee that $M$ encodes exactly one true literal for each clause $y$. If $y$ contains the variables $a$, $b$, $c$ then $C(y)$ connects to free output edges $e_a^i$, $e_b^j$ and $e_c^k$ of $V(a)$, $V(b)$ and $V(c)$.

Finally, every $V(x)$ is a binary tree of at most $6p - 4 \leq 24|Y|$ edges and every $C(y)$ is only a claw of 6 edges. Consequently, $G(F)$ has max-degree three and can be constructed in polynomial time. Figure 5 gives an intuitive example.

# 7   Conclusion

In this paper we studied complexity aspects of the Minimum Efficient Edge Domination Problem. We generalized various results of [14,12]. This section draws a wider picture on the relation and impact of our findings.

The central concept of our polynomial time results is a reduction of MEED on chordal bipartite and hole-free graphs to the same problem on $K_4$-free block graphs. Such graphs have bounded clique-width which implies bounded treewidth. Particularly, their treewidth is at most two. Because EED can be expressed in terms of monadic second order logic this gives in fact immediately the tractability of MEED on these classes (see the corresponding discussion in [4]).

In case of chordal bipartite graphs we have shown that MEED can be solved even in linear time. However, it should be clarified that the recognition of chordal bipartite graphs in linear time is a long-standing open problem. Consequently, the input graph for the given linear time algorithm for MEED on chordal bipartite graphs must be chordal bipartite.

Our second result states that MEED can be solved in polynomial time on hole-free graphs. In fact it turns out that every hole-free graph with a d.i.m. is weakly chordal (i.e., hole- and anti-hole-free), a proper subclass of hole-free graphs. Anyway, our result on hole-free graphs solves a number of open problems stated in previous papers. Lu et al. [12] gave a variety of subclasses of weakly chordal graphs where EED's complexity was open. Certainly all of them can now be solved in polynomial time. Another interesting aspect in context of our two results is the question whether the linear time approach for chordal bipartite graphs can be extended to permutation graphs, a small subclass of weakly chordal graphs.

The relation between permutation graphs and chordal bipartite graphs is also an interesting matter in the context of the following vertex version of the efficient domination problem: If $G = (V, E)$ is a graph and $D \subseteq V$ then we call $D$ an efficient dominating set if (1) $V = \cup_{x \in D} N[x]$ and (2) for all $x \neq y \in D$ it is true $dist_G(x, y) \geq 3$. The *Efficient Vertex Domination Problem* (*EVD*) is to decide whether a given graph $G$ has an efficient dominating set. This concept was independently invented in [1], [2] and [3]. EVD is NP-complete on chordal bipartite graphs [13] which nicely contrasts with our result for EED. On the other hand EVD is efficiently solvable on permutation graphs [10] just as EED.

Our third result provides NP-completeness of EED on planar bipartite graphs with maximum degree three. We can conclude that neither degree constraints, restrictions on $G$'s coloring nor $G$'s planarity influence the complexity of this problem significantly. However, the two polynomial time results motivate the question for a precise answer on how the complexity is influenced by the maximum length of induced cycles. We leave this as a challenge for future work.

## References

1. Bange, D., Barkauskas, A., Slater, P.: Efficient near-domination of graphs. Congr. Numer. 58, 83–92 (1987)
2. Bange, D., Barkauskas, A., Slater, P.: Efficient domination sets in graphs. In: Ringeisen, R.D., Roberts, F.S. (eds.) Applications of Discrete Mathematics, pp. 189–199. SIAM, Philadelphia (1988)
3. Biggs, N.: Perfect codes in graphs. J. Combin. Theory Ser. B 15, 289–296 (1973)
4. Bodlaender, H.L.: A tourist guide through treewidth. Acta Cybernetica 11, 1–23 (1993)
5. Brandstädt, A., Le, V.B., Spinrad, J.P.: Graph Classes: A Survey. SIAM Monographs on Discrete Math. Appl., vol. 3. SIAM, Philadelphia (1999)
6. Cardozo, D.M., Lozin, V.V.: Dominating induced matchings. In: Lipshteyn, M., Levit, V.E., McConnell, R.M. (eds.) Graph Theory, Computational Intelligence and Thought. LNCS, vol. 5420, pp. 77–86. Springer, Heidelberg (2009)
7. Golumbic, M.C.: Algorithmic Graph Theory and Perfect Graphs. Academic Press, London (1980)
8. Grinstead, D.L., Slater, P.L., Sherwani, N.A., Holmes, N.D.: Efficient edge domination problems in graphs. Information Processing Letters 48, 221–228 (1993)
9. Laroche, P.: Planar monotone 1-in-3 satisfiability is NP-complete. In: ASMICS Workshop on Tilings, Deuxième Journées Polyominos et pavages, Ecole Normale Supérieure de Lyon (1992)
10. Liang, Y.D., Lu, C.L., Tang, C.Y.: Efficient domination on permutation graphs and trapezoid graphs. In: Jiang, T., Lee, D.T. (eds.) COCOON 1997. LNCS, vol. 1276, pp. 232–241. Springer, Heidelberg (1997)
11. Livingston, M., Stout, Q.: Distributing resources in hypercube computers. In: Proceedings 3rd Conf. on Hypercube Concurrent Computers and Applications, pp. 222–231 (1988)
12. Lu, C.L., Ko, M.-T., Tang, C.Y.: Perfect edge domination and efficient edge domination in graphs. Discrete Applied Math. 119(3), 227–250 (2002)
13. Lu, C.L., Tang, C.Y.: Efficient domination in bipartite graphs (1997) (manuscript)
14. Lu, C.L., Tang, C.Y.: Solving the weighted efficient edge domination problem on bipartite permutation graphs. Discrete Applied Math. 87, 203–211 (1998)
15. Tarjan, R.E.: Depth-First Search and Linear Graph Algorithms. SIAM J. Comput. 1(2), 146–160 (1972)

# Computational Complexity of the Hamiltonian Cycle Problem in Dense Hypergraphs

Marek Karpiński[1,*], Andrzej Ruciński[2,**], and Edyta Szymańska[2,***]

[1] Department of Computer Science, University of Bonn
marek@cs.uni-bonn.de
[2] Faculty of Mathematics and Computer Science,
Adam Mickiewicz University, Poznań
{rucinski,edka}@amu.edu.pl

**Abstract.** We study the computational complexity of deciding the existence of a Hamiltonian Cycle in some dense classes of k-uniform hypergraphs. Those problems turned out to be, along with the hypergraph Perfect Matching problems, exceedingly hard, and there is a renewed algorithmic interest in them. In this paper we design a polynomial time algorithm for the Hamiltonian Cycle problem for k-uniform hypergraphs with density at least $\frac{1}{2}+\epsilon$, $\epsilon > 0$. In doing so, we depend on a new method of constructing Hamiltonian cycles from (purely) existential statements which could be of independent interest. On the other hand, we establish NP-completeness of that problem for density at least $\frac{1}{k} - \epsilon$. Our results seem to be the first complexity theoretic results for the Dirac-type dense hypergraph classes.

## 1 Introduction

We address the problem of deciding the existence and construction of a Hamiltonian Cycle in some dense classes of hypergraphs. The corresponding problem is being well understood for dense graphs (cf., e.g., [7] and [19]), as well as random graphs (cf., e.g., [2],[4], [5], and [10]). However, the computational status of the problem for hypergraphs was widely open and has become a challenging issue recently.

In this paper we shed some light on the computational complexity of that problem for $k$-uniform hypergraphs. For any $\epsilon > 0$, we design the first polynomial time algorithm for the Hamiltonian Cycle problem for $k$-uniform hypergraphs with Dirac-type density at least $1/2 + \epsilon$. We prove also a complementary intractability result for $k$-uniform hypergraphs with density at least $1/k - \epsilon$. The techniques used in this paper could be also of independent interest.

We consider $k$-*uniform hypergraphs*, that is, hypergraphs $H$ whose edges are $k$-element subsets of $V := V(H)$. We refer to $k$-uniform hypergraphs as $k$-*graphs*.

* Research supported by DFG grants and the Hausdorff Center grant EXC59-1.
** Research supported by grant N201 036 32/2546.
*** Research supported by grant N206 017 32/2452.

A. López-Ortiz (Ed.): LATIN 2010, LNCS 6034, pp. 662–673, 2010.

For $k$-graphs with $k \geq 3$, a cycle may be defined in many ways (see, e.g., [3], [13] and [14]). Here by *a cycle* of length $l \geq k + 1$ we mean a $k$-graph whose vertices can be ordered cyclically $v_1, \ldots, v_l$ in such a way that for each $i = 1, \ldots, l$, the set $\{v_i, v_{i+1}, \ldots, v_{i+k-1}\}$ is an edge, where for $h > l$ we set $v_h = v_{h-l}$. Such cycles are sometimes called *tight*. A *Hamiltonian cycle* in a $k$-graph $H$ is a spanning cycle in $H$. A $k$-graph containing a Hamiltonian cycle is called *Hamiltonian*.

For a $k$-graph $H$ and a set of $k-1$ vertices $S$, let $N_H(S)$ be the set of vertices $v$ of $H$ such that $S \cup \{v\} \in H$. We define the degree of $S$ as $deg_H(S) = |N_S(H)|$, and write $deg_H(S, T)$ for the degree restricted to the subset $T \subseteq V$, that is, $deg_H(S, T) = |N_S(H) \cap T|$. We define $\delta(H) = \min_S deg_H(S)$ and refer to it as the $(k-1)$-*wise, collective minimum degree of* $H$, or simply, *minimum co-degree*. The ratio $\delta(H)/|V(H)|$ is sometimes called a *Dirac-type density of* $H$.

We denote by **HAM**$(k, c)$ the problem of deciding the existence of a Hamiltonian cycle in a $k$-graph with minimum co-degree $\delta(H)$ satisfying $\delta(H) \geq c|V(H)|$.

For graphs, that is, for $k = 2$, one of the classic theorems of Graph Theory by Dirac [8] states that if the minimum degree in an $n$-vertex graph is at least $n/2$, $n \geq 3$, then the graph is Hamiltonian. Hence, the problem **HAM**$(2, 1/2)$ is trivial. Complementing this result, it was shown in [7] that **HAM**$(2, c)$ is NP-complete for any $c < \frac{1}{2}$.

Turning to genuine hypergraphs $(k \geq 3)$, it was recently shown in [16] that for all $k \geq 3$, $c > \frac{1}{2}$, and sufficiently large $n$, every $k$-graph $H$ with $|V(H)| = n$ and $\delta(H) \geq cn$ contains a Hamiltonian cycle. Hence, again, **HAM**$(k, c)$ is trivial for all $c > \frac{1}{2}$. In the case in which $c = \frac{1}{2}$ Rödl et al.[17] proved that the same holds for 3-graphs and the problem remains open for $k \geq 4$.

Our main contribution are two complementary results on **HAM**$(k, c)$.

**Theorem 1.** *For all $k \geq 3$ and $c < \frac{1}{k}$ the problem* **HAM**$(k, c)$ *is NP-complete.*

Interestingly, Theorem 1 leaves a similar hardness gap of $(\frac{1}{k}, \frac{1}{2})$ as for the problem of deciding the existence of a perfect matching in a $k$-graph with $\delta(H) \geq c|V(H)|$ (see [18] and [12]). Note that, in view of [7], this gap collapses for graphs. In Section 2, Theorem 1 is proved by a reduction from **HAM**$(2, c)$, $c < \frac{1}{2}$.

In the second part of this paper, we strengthen the above mentioned result from [16], by designing a polynomial time algorithm for the search version of **HAM**$(k, c)$.

**Theorem 2.** *For all $k \geq 3$ and $c > \frac{1}{2}$ there exists a polynomial time algorithm, called* HamCycle, *which finds a Hamiltonian cycle in every $k$-graph with $\delta(H) \geq c|V(H)|$.*

In view of [17], we believe that also the proof from there can be turned into a polynomial time algorithm extending Theorem 2 to $c = \frac{1}{2}$ for $k = 3$.

Our construction is based on the existential proofs from [16] and [17]. In short, the idea is as follows. First, procedure AbsorbingPath constructs a special, relatively short path $A$ in $H$, called *absorbing*. Next, procedure AlmostHamCycle finds an almost Hamiltonian cycle $C$ containing $A$. Finally, the remaining vertices are absorbed by $A$ into $C$ to form a Hamiltonian cycle. Along the way, two probabilistic lemmas from [16] are derandomized using the Erdős-Selfridge method of conditional expectations [1].

## 2    The Reduction

In this section we prove Theorem 1. We will show that for all $k \geq 3$ and all $\epsilon > 0$, the problem $\mathbf{HAM}(k-1, \frac{1}{k-1} - \epsilon')$, where $\epsilon' = \frac{k}{k-1}\epsilon$ reduces to $\mathbf{HAM}(k, \frac{1}{k} - \epsilon)$. This, together with the known fact proved in [7] that $\mathbf{HAM}(2, c)$ is NP-complete for all $c < \frac{1}{2}$, shows that also $\mathbf{HAM}(k, c)$ is NP-complete for all $c < \frac{1}{k}$.

Let $H$ be a $(k-1)$-graph on $(k-1)n$ vertices with $\delta(H) \geq (\frac{1}{k-1}-\epsilon')(k-1)n$. We construct a (gadget) $k$-graph $G$ as follows. Let $V(G) = A \cup B$ where $A = V(H)$ and $B$ is disjoint from $A$ with $|B| = n$. The edge set $E(G)$ is union of three sets:

$$E(G) = E_{\leq k-3} \cup E_H \cup E_k,$$

where for $i = 0, \ldots, k$, $E_i$ consists of all $k$-element subsets of $V(G)$ which intersect $A$ in precisely $i$ vertices, $E_{\leq k-3} = \bigcup_{i \leq k-3} E_i$ and $E_H$ consists of all $k$-element subsets of $V(G)$ whose intersection with $A$ is an edge of $H$ (see Figure 1). Let us check first that $\delta(G) \geq (\frac{1}{k}-\epsilon)kn$. We assume, as we can, that $\epsilon kn \geq 2$. Let $S \in \binom{V(G)}{k-1}$. If $|S \cap A| = k-3$ (and so $|S \cap B| = 2$) then $deg_G(S) = |B|-2 = n-2$. If $|S \cap A| \leq k-4$ then $deg_G(S) = |V(G)| - (k-1) = kn - k + 1$. If $S \subset A$, then $deg_G(S) \geq |A| - (k-1) \geq n$, regardless whether $S \in E(H)$ or not. Finally, if $|S \cap A| = k-2$, we know by the assumption on $\delta(H)$ that there are at least

$$\left(\frac{1}{k-1} - \frac{k}{k-1}\epsilon\right)(k-1)n = \left(\frac{1}{k} - \epsilon\right)kn$$

vertices $v \in A \setminus S$ such that $(S \cap A) \cup \{v\} \in E(H)$, and hence, $S \cup \{v\} \in E_H \subset E(G)$.

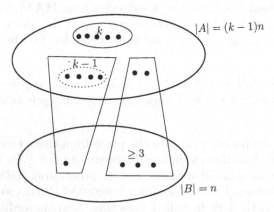

**Fig. 1.** The gagdet. The dotted oval represents an edge of $H$.

It remains to show that $H$ has a Hamiltonian cycle if and only if $G$ does. Let $v_1 v_2 \ldots v_{(k-1)n}$ be a Hamiltonian cycle in $H$ and let us order the vertices of $B$ arbitrarily, say $B = \{w_1, \ldots, w_n\}$. Then the sequence

$$v_1 \ldots v_{k-1} w_1 v_k \ldots v_{2k-2} w_2 \ldots w_{n-1} v_{(k-1)(n-1)+1} \ldots v_{(k-1)n} w_n \qquad (1)$$

forms a Hamiltonian cycle in $G$. Indeed, every $k$ consecutive (cyclically) vertices of that string contain exactly one vertex of $B$ and an edge of $H$, and thus, by the definition of $E_H$, form an edge of $G$.

Conversely, let $G$ have a Hamiltonian cycle $F$. Note first that $E(F) \not\subseteq E_{\leq k-3}$, because each edge of $E_{\leq k-3}$ contains at most $k-3$ vertices of $A$ and each vertex of $F$ is contained in precisely $k$ edges of $F$. Hence, $F$ could cover only at most

$$kn \times (k-3) \times \frac{1}{k} = (k-3)n < |A|$$

vertices of $A$. But then $E(F) \cap E_{\leq k-3} = \emptyset$, because, due to the lack of edges of $E_{k-2}$ in $G$, the cycle cannot traverse from an edge of $E_{\leq k-3}$ to any edge in $E_H \cup E_k$.

Secondly, no edge of $E_k$ can be in $F$ either. Indeed, since the edges of $F$ covering $B$ are all in $E_H \subseteq E_{k-1}$, each vertex of $B$ has to be immediately preceded in $F$ by exactly $k-1$ vertices of $A$, making no room for any edge of $E_k$ in $F$. So, $F$ looks exactly like in (1). Note that every $k$ consecutive (cyclically) vertices of that string form an edge of $G$ and contain exactly one vertex of $B$ and a set $S$ of $k-1$ vertices of $A$. Thus, by the definition of $G$, this set $S$ must be an edge of $H$. Hence, the sequence $v_1 v_2 \ldots v_{(k-1)n}$ forms a Hamiltonian cycle in $H$.    □

# 3   Subroutines

In this section we describe several subroutines which will be used by the main algorithm, HAMCYCLE. We begin with procedures constructing tight paths in a dense hypergraph. Wherever convenient, we will identify a sequence of distinct vertices $(v_1, v_2, \ldots)$ with the set of its elements $\{v_1, v_2, \ldots\}$.

## 3.1   Paths

A *path* is a $k$-graph $P$, whose vertices can be ordered $v_1, \ldots, v_l$, where $l = |V(P)|$, in such a way that for each $i = 1, \ldots, l - k + 1$, we have $\{v_i, v_{i+1}, \cdots, v_{i+k-1}\} \in P$. We say that $P$ *connects* the sequences $(v_1, v_2, \ldots v_{k-1})$ and $(v_l, \ldots, v_{l-k+2})$, which will be called *the ends* of $P$. A path on $l$ vertices (and thus with $l - k + 1$ edges) will be said to have *length* $l$.

Our algorithm will frequently use the following subroutine. Let $\gamma > 0$. By Lemma 4 in [16], for sufficiently large $n$, every $k$-graph $H$ with $n$ vertices and $\delta(H) \geq (\frac{1}{2} + \gamma)n$ contains a path of length at most $2k/\gamma^2$ between any pair of $(k-1)$-element sequences of distinct vertices. Thus, an exhaustive search of all

$O(n^{2k/\gamma^2})$ sequences of distinct $2k/\gamma^2$ vertices would certainly find such a path. However, for better complexity, a BFS-type search can be applied.

---

### SUBROUTINE CONNECT

**In:** $k$-graph $H$ with $\delta(H) \geq (\frac{1}{2} + \gamma)n$ and two disjoint $(k-1)$-element sequences of distinct vertices, $\mathbf{u}$ and $\mathbf{v}$
**Out:** Path $P$ in $H$ with ends $\mathbf{u}$ and $\mathbf{v}$ of length at most $2k/\gamma^2$.

---

CONNECT begins its BFS search at $\mathbf{u}$ and moves on by one vertex at a time until the reverse of $\mathbf{v}$ is found. Throughout it maintains a record of the path by which the current end has been reached, and uses this record to verify that the new vertex added is distinct from all previous on the current path. Each step corresponds to traversing one edge of $H$ in a particular order and no edge is traversed twice in the same order. Hence, the time complexity of CONNECT is $O(n^k)$.

In fact, we will rather need a restricted version of CONNECT, where the connecting path is supposed to use, except for the ends $\mathbf{u}$ and $\mathbf{v}$, only the vertices from a specified „transfer set" $T$.

---

### SUBROUTINE CONNECTVIA

**In:** $k$-graph $H$, a subset $T \subset V$ such that for every $S \in \binom{V}{k-1}$ we have $deg_H(S,T) \geq (\frac{1}{2} + \gamma)|T|$, and two $(k-1)$-element sequences of distinct vertices from $V \setminus T$, $\mathbf{u} = (u_1, u_2, \ldots u_{k-1})$ and $\mathbf{v} = (v_1, v_2, \ldots v_{k-1})$
**Out:** Path $P$ in $H$ with ends $\mathbf{u}$ and $\mathbf{v}$ of length at most $2k/\gamma^2 + 2(k-1)$ and such that $V(P) \setminus (\mathbf{u} \cup \mathbf{v}) \subset T$

---

In its first $2k - 2$ steps CONNECTVIA moves from $\mathbf{u}$ and $\mathbf{v}$ to, resp., $\mathbf{u}'$ and $\mathbf{v}'$, where all vertices of $\mathbf{u}'$ and $\mathbf{v}'$ are in the set $T$. Then it invokes CONNECT with $H[T]$, $\mathbf{u}'$, and $\mathbf{v}'$ as inputs.

Another subroutine finds a long path in any dense $k$-graph. It is an algorithmic generalization of Claim 6.1 from [17]. For a $k$-graph $F$ denote by $\delta_{>0}(F)$ the minimum of $deg_F(S)$ taken over all $S \in \binom{V(F)}{k-1}$ with $deg_F(S) > 0$.

---

### SUBROUTINE LONGPATH

**In:** $k$-graph $F$ with $l$ vertices and $m > 0$ edges
**Out:** Path $P$ in $F$ of length at least $d := m/\binom{l}{k-1}$.

1. $V(F) := V$, $F' := F$
2. Find a set $S \in \binom{V}{k-1}$ for which $deg_{F'}(S) = \delta_{>0}(F')$;
3. If $\delta_{>0}(F') < d$, then $F' := F' \setminus \{e \in F' : e \supset S\}$ and go to Step 2;
4. Greedily find a maximal path $P$ in $F'$;
5. Return $P$.

---

Observe that at the outset of Step 3 we have $deg_{F'}(S) = 0$ and so, every set $S$ is selected in Step 2 at most once. Note also that once we get to Step 4, we have

$F' \neq \emptyset$ and $\delta_{>0}(F') \geq d$. Hence, any maximal path in $F'$ has length at least $d$. The time complexity of LONGPATH is $O(l^{k-1} + m)$.

## 3.2 Derandomization

At the heart of our algorithm lies the following procedure based on a simple probabilistic fact. Let $\tau > 0$, $\beta > \tau$, and $m, N$, and $r \leq N$, be positive integers. Set $\rho := 2mr/N^2$.

---

**ALGORITHM SELECTSUBSET**

**In:** Graph $G = (U \cup W, E)$ such that

- $|U| = M$ and $e(G[U]) = 0$
- $|W| = N$ and $e(G[W]) = m$
- $\min_{u \in U} deg_G(u) \geq \beta N$,

and an integer $r$, $1 \leq r \leq N$
**Out:** Independent set $R \subset W$ with $(1 - \rho)r \leq |R| \leq r$ and $\min_{u \in U} deg_G(u, R) \geq (\beta - \tau - \rho)r$.

1. Set $U = \{u_1, \ldots, u_M\}$, $W = \{w_1, \ldots, w_N\}$;
2. $R' := \emptyset$;
3. For $k = 1$ to $r$ do:
   (a) For $i = 1$ to $M$ and $j = 1$ to $N$ do:

   $$d'_{i,j} := deg_G(u_i, R' \cup \{w_j\}) \quad \text{and} \quad d''_{i,j} := deg_G(u_i, W \setminus (R' \cup \{w_j\})).$$

   (b) For $j = 1$ to $N$ do:

   $$e_j^{(0)} := e(G[R' \cup \{w_j\}]), e_j^{(1)} := e(G[R', W \setminus R']), e_j^{(2)} := e(G[W \setminus (R' \cup \{w_j\})]).$$

   (c) Find $w_{j_k} \in W \setminus R'$ such that, with $y := 2m(r/N)^2$,

   $$\sum_{i=1}^{M} \sum_{d \leq (\beta - \tau)r - d'_{i,j_k}} \frac{\binom{d''_{i,j_k}}{d}\binom{N-k-d''_{i,j_k}}{r-k-d}}{\binom{N-k}{r-k}}$$

   $$+ \frac{1}{y}\left(e_{j_k}^{(0)} + e_{j_k}^{(1)} \frac{r-k}{N-k} + e_{j_k}^{(2)} \frac{(r-k)(r-k-1)}{(N-k)(N-k-1)}\right) < 1. \tag{2}$$

   (d) $R' := R' \cup \{w_{j_k}\}$.
4. Remove one vertex from each edge of $G[R']$ and call the resulting set $R$.
5. Return $R$.

---

**Lemma 1.** *If* $\log M = o(r)$, *then* SELECTSUBSET *finds the desired set* $R$ *in time* $O(M \times poly(N))$.

In the proof we use the following probabilistic fact, which together with Markov's inequality implies the existence of the required set $R$. Algorithm SELECTSUBSET derandomizes this fact.

**Fact 3.** *Let $G$ be the graph given as an input of* SELECTSUBSET. *Further, let $R'$ be a random subset of $W$ chosen uniformly from $\binom{W}{r}$, let $X$ be the number of vertices $u \in U$ with $deg_G(u, R') \leq (\beta - \tau)r$ and $Y = e(G[R'])$. Then*

$$EX = o(1) \qquad and \qquad EY \leq m \left(\frac{r}{N}\right)^2.$$

*Proof.* First observe that $X = \sum_{u \in U} I_u$, where $I_u$ is the indicator of the event $\{deg_G(u, R') \leq (\beta - \tau)r\}$. Note also that $P(I_u = 1) = P(Z_u \leq (\beta - \tau)r)$, where $Z_u$ is a hypergeometric random variable with parameters $N, deg_G(u), r$ and that, by the properties of $G$, the expectation of $Z_u$ is $rdeg_G(u)/N \geq \beta r$. Thus, by a Chernoff bound for hypergeometric distributions (see, e.g., [11], Theorem 2.10, formula (2.6)), $EX \leq Me^{-\Theta(r)} = o(1)$. Finally, by the linearity of expectation, $EY = m\binom{N-2}{r-2}/\binom{N}{r} \leq m(r/N)^2$. ☐

*Proof of Lemma 1:* By Fact 3 and the definition of $y$, $EX + \frac{EY}{y} \leq o(1) + \frac{1}{2} < 1$. We can view the selection of $R'$ as a result of a random process $w_{j_1}, \ldots, w_{j_r}$, where in step $k$, a vertex $w_{j_k} \in W$ is randomly selected without repetitions.

Let $\alpha_j = E(X|j_1 = j) + \frac{EY}{y}E(Y|j_1 = j)$. Then, by the law of total probability, $EX + \frac{EY}{y} = \frac{1}{N}\sum_{j=1}^{N} \alpha_j$, and so, there exists an index $j$ such that $\alpha_j < 1$. Take that index as $j_1$. Repeat until the whole set $R'$ is selected. Then, $E(X|R') + \frac{1}{y}E(Y|R') = X(R') + \frac{1}{y}Y(R') < 1$, which implies that $X(R') = 0$ and $Y(R') < y$. This proves that $R'$, and consequently $R$, have the desired properties.

Note that the conditional expectations $E(X|j_1, \ldots, j_k)$ and $E(Y|j_1, \ldots, j_k)$ correspond to the quantities appearing in the expression (2) given in Step 3(c) of the algorithm. ☐

## 4   The Algorithm

In this section we prove Theorem 2 by giving the main algorithm HAMCYCLE. It will be based on two major procedures, ABSORBINGPATH and ALMOSTHAM-CYCLE which we will describe first.

In order to formulate our main procedures, we need a few definitions from [16]. We choose $0 < \epsilon < c - \frac{1}{2}$ small enough.

Given a vertex $v$ we say that a $(2k - 2)$-element sequence of vertices $\mathbf{x} = (x_1, \ldots, x_{2k-2})$ is $v$-*absorbing* in $H$ if for every $i = 1, \ldots, k - 1$ we have $\{x_i, x_{i+1}, \ldots, x_{i+k-1}\} \in H$ (that is, $\mathbf{x}$ spans a path in $H$) and for every $i = 1, \ldots, k$ we also have edges $\{x_i, x_{i+1}, \ldots, x_{i+k-2}, v\} \in H$. Note that, if $\mathbf{x}$ is actually a segment of a path $P$ and $v$ is not a vertex of $P$, then the segment $\mathbf{x}$ of $P$ can be replaced by the new segment $(x_1, \ldots, x_{k-1}, v, x_k, \ldots, x_{2k-2})$, absorbing $v$ onto $P$.

A path $A$ in $H$ is called *absorbing* if $|V(A)| \leq 8k\epsilon^{k-1}n$ and for every $v \in V$ there are at least

$$q := 2^{k-4}\epsilon^{2k}n \tag{3}$$

disjoint $v$-absorbing sequences, each of which is a segment of $A$. Note that if $A$ is an absorbing path in $H$, then for every subset $U \subset V \setminus V(A)$ of size $|U| \leq q$ there is a path $A_U$ in $H$ with $V(A_U) = V(A) \cup U$ and such that $A_U$ has the same ends as $A$.

The idea behind our algorithm is the same as the idea of the existential proofs in [16] and [17], and can be summarized as follows.

- Find an absorbing path $A$ in $H$.
- Find a cycle $C$ in $H$ containing $A$ as well as all but. at most $q$ vertices of $V(H) \setminus V(A)$.
- Extend $C$ to a Hamiltonian cycle of $H$ using the absorbing property of $A$ with respect to $U = V(H) \setminus V(C)$.

To build an absorbing path we use PROCEDURE ABSORBINGPATH and to build the long cycle – PROCEDURE ALMOSTHAMCYCLE, both described below.

We are now ready to give our main algorithm which finds a Hamiltonian cycle in every $k$-graph with $\delta(H) \geq cn$, for $c > \frac{1}{2}$.

---

ALGORITHM HAMCYCLE

**In:** $n$-vertex $k$-graph $H$ with $\delta(H) \geq cn, c > \frac{1}{2}$
**Out:** Hamiltonian cycle $C$ in $H$

1. Fix a sufficiently small $0 < \epsilon < c - \frac{1}{2}$;
2. Apply ABSORBINGPATH to $H$ obtaining an absorbing path $A$;
3. Apply ALMOSTHAMCYCLE to $H$ with $P_0 = A$, obtaining a cycle $C$ in $H$ of length at least $n - q$ which contains $A$;
4. For each vertex $v \in V \setminus V(C)$ do:
   (a) Find a $v$-absorbing sequence $\mathbf{x} = (x_1, \ldots, x_{2k-2})$ which is a segment of $C$;
   (b) Replace $(x_1, \ldots, x_{2k-2})$ by $(x_1, \ldots, x_{k-1}, v, x_k, \ldots, x_{2k-2})$ and call the new cycle $C$;
5. Return $C$.

---

It remains to describe how the two procedures used by HAMCYCLE work. By Claim 3.2 in [16] we know that for every vertex $v \in V(H)$ there are at least $2^{k-2}\gamma^{k-1}$ $v$-absorbing sequences in $H$. In [16] a random selection of $(2k - 2)$-sequences was chosen and proved to contain enough $v$-absorbing sequences for every $v$. Here we derandomize this step by invoking SELECTSUBSET.

---

PROCEDURE ABSORBINGPATH

**In:** $n$-vertex $k$-graph $H$ with $\delta(H) \geq (\frac{1}{2} + \epsilon)n$
**Out:** Absorbing path $A$ in $H$

1. Build an auxiliary graph $G = (U \cup W, E)$, where $U = V(H)$, $W$ is the set of all $(2k - 2)$-element sequence of vertices $\mathbf{x} = (x_1, \ldots, x_{2k-2})$ in $H$, and $E$ consists of all pairs $v \in U, \mathbf{x} \in W$ such that $\mathbf{x}$ is $v$-absorbing in $H$, as well as of all pairs $\mathbf{x}, \mathbf{x}' \in W$ which share at least one element;

2. Apply SELECTSUBSET to $G$ with $r = \epsilon^{k+1}n$, $\rho = 8(k-1)^2\epsilon^{k+1}$, and $\tau = \beta/2$, to obtain a family $\mathcal{F}$ of $s \leq r$ vertex-disjoint sequences and such that for each vertex $v$ of $H$ the number of $v$-absorbing sequences in $\mathcal{F}$ is at least $2^{k-4}\epsilon^{2k}n$;

3. Use repeatedly CONNECTVIA to connect all sequences of $\mathcal{F}$ into one path $A$.

---

Note that in the above application of SELECTSUBSET, $M = n$, $N = (n)_{(2k-2)} \sim n^{2k-2}$, $m \leq (2k-2)^2 n^{4k-5}$, and $\beta = 2^{k-2}\gamma^{k-1}$. Thus, SELECTSUBSET does find a family $\mathcal{F}$ as described in Step 2. As the final path $A$ contains all elements of $\mathcal{F}$ as disjoint segments, the absorbing property of $A$ follows.

Our second major procedure constructs in $H$ an almost Hamiltonian cycle containing any given, not too long path. In [16] this has been done by applying a weak regularity lemma to $H$ and finding in the cluster $k$-graph an almost perfect matching. Then, applying repeatedly the existential analog of LONGPATH to the dense and regular clusters, a collection of finitely many paths covering almost all vertices of $H$ was found. These paths were then connected into a cycle by applying CONNECTVIA with a preselected reservoir set $R$.

That proof can be turned into an algorithm by recalling the algorithmic version of the weak hypergraph regularity lemma from [6]. We, however, prefer to follow the more elementary approach from [17], generalizing it to $k$-graphs without any effort.

In fact, the single difficulty in both these approaches was the same: to derandomize the selection of a reservoir set $R$, a small subset of vertices which reflects the property of the entire hypergraph and can be used to connect paths during the whole procedure. This step is now derandomized by using algorithm SELECTSUBSET (Steps 1 and 2 of procedure ALMOSTHAMCYCLE).

Once we have $R$ which is disjoint from $P_0$, we keep extending $P_0$ in $H - R$ by little increments until it reaches the desired length. Initially, we extend $P_0$ greedily (Step 3), using the fact that $\delta(H - R) > n/2$. After reaching the length of $n/2$, in every step we look at $L := V \setminus (V(P) \cup R)$, where $P$ is the current path, and consider two cases.

If $H[L]$ is dense we apply LONGPATH to find a long path $P'$ in $|H[L]|$ and connect it via $R$ using CONNECTVIA with the transfer set $R$ (Step 5(c)).

If $H[L]$ is sparse then many edges of $H$ have $k-1$ vertices in $L$ and one in $P$. By averaging, there must be a constant length segment $I$ of $P$ with many such edges incident to $I$, and, again by averaging, a subset $J \subset I$ with $|J| \geq \frac{4}{3k}|I|$ and whose *every* vertex is hit by *the same* set $H_0$ of $(k-1)$-tuples from $L$ (Step 5(d)(i)). Next a $(k-1)$-partite $(k-1)$-clique $K$ is found in $H_0$ and trivially extended, by adding $J$, to a $k$-partite $k$-clique $K'$. Clique $K'$ contains a spanning Hamiltonian path $Q$ whose length is $\frac{4}{3}|I|$. We then cut $I$ out of $P$ and reconnect the two remaining subpaths, $P_1$ and $P_2$, with $Q$, obtaining a path longer by $\frac{1}{3}|I|$ (Step 5(d)(ii)-(vi)). Finally, when $P$ has grown long enough, we connect the two ends of $P$ to form the desired cycle. All connections are via $R$ using CONNECTVIA.

For details of the case $k = 3$ we refer to [17]. Since the general case has not appeared in the literature yet, we provide here a detailed pseudo-code, followed by a formal proof of the most crucial steps.

Let $D$ be a large integer, say

$$D \gg n/q = 2^{4-k}\epsilon^{-2k},$$

where $q$ is given by (3).

---

PROCEDURE ALMOSTHAMCYCLE

**In:** $n$-vertex $k$-graph $H$ with $\delta(H) \geq (\frac{1}{2} + \epsilon)n$ and a path $P_0$ in $H$ of length at most $\frac{1}{3}\epsilon n$
**Out:** Cycle $C$ in $H$ such that $P_0 \subset C$ and $|V(C)| \geq n - q$.

1. Build an auxiliary graph $G = (U \cup W, E)$, where $U = \binom{V}{k-1}$, $W = V \setminus V(P_0)$, and $E$ consists of all pairs $S \in U, v \in W$ such that $S \cup \{v\} \in H$ ;
2. Apply SELECTSUBSET to $G$ with $r = \frac{1}{2}q$, and $\tau = \epsilon/6$, to obtain a set $R \subset V \setminus V(P_0)$ of size $|R| = r$ with the property that $deg_H(S, R) \geq \frac{1}{2}(1+\epsilon)r$ for all $S \in \binom{V}{k-1}$.
3. Extend greedily $P_0$ (at one end only) to a path $P$ in $H - R$ of length at least $n/2$;
4. Let $\mathbf{x}$ be the common end of $P_0$ and $P$;
5. While $|V(P)| < n - q$ do:
   (a) let $\mathbf{y}$ be the end of $P$ other than $\mathbf{x}$;
   (b) $L := V \setminus (V(P) \cup R)$, $l := |L|$;
   (c) If $|H[L]| > D\binom{l}{k-1}$ then do:
      i. Apply LONGPATH to $H[L]$ obtaining a path $P'$ of length at least $D$, disjoint from $P$.
      ii. Apply CONNECTVIA with $\gamma = \frac{1}{3}\epsilon$ and $T = R$, obtaining a path $Q$ of length at most $20k/\epsilon^2$ from $\mathbf{y}$ to $\mathbf{x}'$, and thus connecting paths $P$ and $P'$ into a new path $PQP'$;
      iii. $P := PQP'$, $R := R \setminus V(Q)$;
   (d) If $|H[L]| \leq D\binom{l}{k-1}$ then do:
      i. Find (by exhaustive search) a segment (that is, a set of consecutive vertices) $I \subset V(P) \setminus (V(P_0) \cup \mathbf{y})$, a subset $J \subset I$, and a $(k-1)$-graph $H_0 \in \binom{I}{k-1}$ such that $|I| = D$, $|J| = \frac{4}{3k}D$, $|H_0| \geq 2^{-D}(\frac{1}{2} - \frac{4}{3k})\binom{l}{k-1}$, and for every $e \in H_0$ and every $v \in J$ we have $e \cup \{v\} \in H$;
      ii. Find (by exhaustive search) a $(k-1)$-partite, complete $(k-1)$-graph $K$ in $H_0$ with all partition classes of size $|J|$;
      iii. Let $K'$ be the $k$-partite, complete $k$-graph spanned in $H$ by the partition classes of $K$ and $J$;
      iv. Take any Hamiltonian path $Q$ in $K'$ with ends $\mathbf{z}$ and $\mathbf{z}'$;
      v. Remove $I$ from $P$ obtaining two disjoint paths $P_1 \supset P_0$ and $P_2$;
      vi. Apply CONNECTVIA with $\gamma = \frac{1}{3}\epsilon$ and $T = R$, to connect $P_1$, $Q$, and $P_2$ together (see Figure 2); call the resulting path $P$;

6. Apply CONNECTVIA with $\gamma = \frac{1}{3}\epsilon$ and $T = R$, to the ends $\mathbf{x}$ and $\mathbf{y}$ of $P$, obtaining a path $Q$ of length at most $20k/\epsilon^2$ from $\mathbf{x}$ to $\mathbf{y}$, and thus creating a cycle $C = PQ$ of length at least $n - q$;
7. Return $C$.

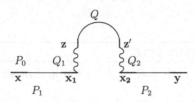

**Fig. 2.** Illustration to Step 4(d) of procedure ALMOSTHAMCYCLE

**Fact 4.** ALMOSTHAMCYCLE *constructs a cycle $C$ in $H$ such that $P_0 \subset C$ and* $|V(C)| \geq n - q$.

*Proof.* The graph $G$ constructed in Step 1 has parameters $M = \binom{n}{k-1}$, $(1 - 8k\epsilon^{k-1})n \leq N \leq n$, $m = 0$, and $\beta \geq \frac{1}{2} + \frac{2}{3}\epsilon$, and so SELECTSUBSET does find a set $R$ as described in Step 2.

Now we prove that the sets $I$ and $J$, and a $(k-1)$-graph $H_0$ searched for in Step 5(d)(i) do exist. By estimating the sum $\sum_{S \in \binom{L}{k-1}} \deg_H(S)$ in two ways we derive the inequality

$$(\tfrac{1}{2} + \gamma)n\binom{l}{k-1} \leq kD\binom{l}{k-1} + |R \cup V(P_0) \cup \mathbf{y}|\binom{l}{k-1} + N,$$

where $N$ counts the number of edges of $H$ with $k-1$ vertices in $L$ and one vertex in $V(P) \setminus (V(P_0) \cup \mathbf{y})$. Since $|R \cup V(P_0) \cup \mathbf{y}| \leq \frac{3}{4}\gamma n$, this yields that

$$N \geq [(\tfrac{1}{2} + \tfrac{1}{4}\gamma)n - O(1)]\binom{l}{k-1}.$$

Let $N_i$ be the number of edges of $H$ counted by $N$, with one vertex in the $i$-th $D$-element segment $I_i$ of $V(P) \setminus (V(P_0) \cup \mathbf{y})$. Then, with $s := |V(P) \setminus (V(P_0) \cup \mathbf{y})|$ we have

$$\sum_{i=1}^{s-D+1} N_i \geq ND - O(1)\binom{l}{k-1} \geq (\tfrac{1}{2} + \tfrac{1}{5}\gamma)n\binom{l}{k-1}D,$$

so, by averaging, there exists $i$ such that $N_i \geq \frac{1}{2}\binom{l}{k-1}D$. Let $H_i$ be the $(k-1)$-graph of all $S \in \binom{L}{k-1}$ with at least $\frac{4}{3k}D$ neighbors in $I := I_i$. Then $|H_i| \geq (\frac{1}{2} - \frac{4}{3k})\binom{l}{k-1}$. For each $J \subset I$, $|J| \geq \frac{4}{3k}D$, let $H_J$ be the set of those edges of $H_i$ whose $H$-neighborhood in $I$ is exactly $J$. By averaging there exists a set $J$ such that $|H_J| \geq 2^{-D}|H_i|$.

The existence of a $(k-1)$-partite, complete $(k-1)$-graph $K$ in $H_0$ with all partition classes of size $|J|$ searched for in Step 5(d)(ii) follows by an old result of Erdős [9], see also [15], Lemma 8. (Recall that $|L| > \frac{1}{2}q$.) Note that the initial path $P_0$ has stayed intact throughout the entire procedure and so, it is contained in $C$.     □

Finally, note that the time complexity of ALMOSTHAMCYCLE is $O(poly(n))$. This completes the proof of Theorem 2.

## References

1. Alon, N., Spencer, J.H.: The Probabilistic Method, 3rd edn. Wiley, Chichester (2008)
2. Angluin, D., Valiant, L.: Fast probabilistic algorithms for Hamiltonian circuits and matchings. J. Comput. System Sci. 18, 155–193 (1979)
3. Bermond, J.C., et al.: Hypergraphes hamiltoniens. Prob. Comb. Theorie Graph Orsay 260, 39–43 (1976)
4. Bollobas, B., Fenner, T.I., Frieze, A.: An algorithm for finding hamiltonian cycles in random graphs. In: Proceedings of the 17th Annual ACM Symposium on Theory of Computing, pp. 430–439 (1985)
5. Broder, A., Frieze, A., Shamir, E.: Finding hidden Hamilton cycles. Random Structures Algorithms 5, 395–410 (1994)
6. Czygrinow, A., Rödl, V.: An algorithmic regularity lemma for hypergraphs. SIAM Journal on Computing 30, 1041–1066 (2000)
7. Dalhaus, E., Hajnal, P., Karpiński, M.: On the parallel complexity of Hamiltonian cycle and matching problem on dense graphs. J. Algorithms 15, 367–384 (1993)
8. Dirac, G.A.: Some theorems of abstract graphs. Proc. London Math. Soc. 3, 69–81 (1952)
9. Erdős, P.: On extremal problems of graphs and generalized graphs, Israel. J. Math. 2, 183–190 (1964)
10. Frieze, A., Reed, B.: Probabilistic Analysis of Algorithms. In: Probabilistic Methods for Algorithmic Discrete Mathematics, pp. 36–92. Springer, Heidelberg (1998)
11. Janson, S., Łuczak, T., Ruciński, A.: Random Graphs. John Wiley and Sons, New York (2000)
12. Karpiński, M., Ruciński, A., Szymańska, E.: The Complexity of Perfect Matching Problems on Dense Hypergraphs. In: Dong, Y., Du, D.-Z., Ibarra, O. (eds.) ISAAC 2009. LNCS, vol. 5878, pp. 626–636. Springer, Heidelberg (2009)
13. Katona, G.Y., Kierstead, H.A.: Hamiltonian chains in hypergraphs. J. Graph Theory 30, 205–212 (1999)
14. Kühn, D., Osthus, D.: Loose Hamilton cycles in 3-uniform hypergraphs of high minimum degree. JCT B 96(6), 767–821 (2006)
15. Rödl, V., Ruciński, A., Schacht, M.: Ramsey Properties of Random k-Partite, k-Uniform Hypergraphs. SIAM J. of Discrete Math. 21(2), 442–460 (2007)
16. Rödl, V., Ruciński, A., Szemerédi, E.: An approximate Dirac-type theorem for k-uniform hypergraphs. Combinatorica 28(2), 229–260 (2008)
17. Rödl, V., Ruciński, A., Szemerédi, E.: Dirac-type conditions for hamiltonian paths and cycles in 3-uniform hypergraphs. The Mittag-Leffler preprint series (Spring 2009) (submitted)
18. Szymańska, E.: The Complexity of Almost Perfect Matchings in Uniform Hypergraphs with High Codegree. In: Fiala, J. (ed.) IWOCA 2009. LNCS, vol. 5874, pp. 438–449. Springer, Heidelberg (2009)
19. Särközy, G.: A fast parallel algorithm for finding Hamiltonian cycles in dense graphs. Discrete Mathematics 309, 1611–1622 (2009)

# Rank Selection in Multidimensional Data*

Amalia Duch, Rosa M. Jiménez, and Conrado Martínez

Departament de Llenguatges i Sistemes Informàtics
Universitat Politècnica de Catalunya
Barcelona, Spain
{duch,jimenez,conrado}@lsi.upc.edu

**Abstract.** Suppose we have a set of $K$-dimensional records stored in
a general purpose spatial index like a $K$-d tree. The index efficiently
supports insertions, ordinary exact searches, orthogonal range searches,
nearest neighbor searches, etc. Here we consider whether we can also efficiently support search by rank, that is, to locate the $i$-th smallest element
along the $j$-th coordinate. We answer this question in the affirmative
by developing a simple algorithm with expected cost $\mathcal{O}(n^{\alpha(1/K)} \log n)$,
where $n$ is the size of the $K$-d tree and $\alpha(1/K) < 1$ for any $K \geq 2$.
The only requirement to support the search by rank is that each node in
the $K$-d tree stores the size of the subtree rooted at that node (or some
equivalent information). This is not too space demanding. Furthermore,
it can be used to randomize the update algorithms to provide guarantees on the expected performance of the various operations on $K$-d
trees. Although selection in multidimensional data can be solved more
efficiently than with our algorithm, those solutions will rely on ad-hoc
data structures or superlinear space. Our solution adds to an existing
data structure ($K$-d trees) the capability of search by rank with very
little overhead. The simplicity of the algorithm makes it easy to implement, practical and very flexible; however, its correctness and efficiency
are far from self-evident. Furthermore, it can be easily adapted to other
spatial indexes as well.

## 1 Introduction

Selection is a fundamental computing task: given a collection $A$ of $n$ items drawn
from a totally ordered domain, and a *rank* $i$, $1 \leq i \leq n$, the goal is to retrieve the
$i$-th smallest item from $A$. The selection problem can be trivially solved in time
$\mathcal{O}(n \log n)$ by sorting $A$, but it can be solved more efficiently in either expected
linear time [1] or worst-case linear time [2].

Suppose that the collection $A$ is stored in some balanced (or unbalanced)
binary search tree. Then we can dynamically mantain the collection, supporting
both updates and searches in (expected) time $\mathcal{O}(\log n)$, but we can also support
selection in (expected) time $\mathcal{O}(\log n)$ quite easily. We will only need to augment

---

* This research was supported by the Spanish Min. of Science and Technology project
TIN2006-11345 (ALINEX).

A. López-Ortiz (Ed.): LATIN 2010, LNCS 6034, pp. 674–685, 2010.

the data structure so that each node stores the size of the subtree rooted at that node. This is a very modest price to pay. In fact, the information about subtree sizes can be used advantageously to balance the tree, either probabilistically [3] or deterministically [4]. Hence, we might argue that adding the capability of searching or deleting items by rank comes at no cost.

When dealing with multidimensional point data, we also face frequently the need to sort the data according to one of the coordinates or to find the $i$-th smallest element along some given coordinate. Those problems can be solved like in the unidimensional case in time $\mathcal{O}(n \log n)$ and $\mathcal{O}(n)$, respectively. But it is natural to question if we can do better when the collection of multidimensional points is stored in a data structure like a $K$-d tree [5] or a quadtree [6] (see also [7,8] for background on multidimensional data structures).

Here we show that we can select the $i$-th point along a given coordinate $j$, $0 \le j < K$, in expected sublinear time, when the collection of $K$-dimensional points is stored in a $K$-d tree. More specifically, for a collection of $n$ points, we can find the answer in expected time $\mathcal{O}(n^{\alpha(1/K)} \log n)$, where $\alpha(x)$ is a function that depends on the type of $K$-d tree we use. Furthermore, the exponent $\alpha(x) < 1$ for all $x \in (0, 1)$, with $\alpha(x) \to 1$ as $x \to 0$ (that is, when $1/x = K \to \infty$). Although better performance for rank search in multidimensional data can be obtained (using more than linear space, for instance), we stress here that our solution adds efficient rank search to general purpose multidimensional data structures like $K$-d trees or quadtrees, with only a modest increase of space, namely, storing the size of the subtree rooted at each node. Thus the total space comsumption remains linear in $n$. Like in the case of ordinary "unidimensional" binary search trees, the information about subtree sizes can be used to randomize the insertion and deletion in $K$-d trees, thus guaranteeing the expected time bounds of several operations like ordinary search, partial match search, orthogonal range search and nearest neighbor search, even when the dynamic updates are not random [9,10].

Section 2 briefly summarizes the standard $K$-d trees and several of its variants, the probabilistic model that will be used in the sequel, and recalls a few important previous results, e. g., the expected cost of partial match search in $K$-d trees. Then we describe in Sect. 3 the main contribution of this paper, the algorithm to find the $i$-th smallest element of a $K$-d tree $T$ along the $j$-th coordinate. The following section, Sect. 4, is devoted to the analysis of the expected cost of the algorithm, and we prove there that this cost is sublinear for any $K$. Sect. 5 reports the results of several experiments that we have conducted. The results match very well the predictions of the theoretical analysis in Sect. 4.

## 2   Preliminaries

A $K$-dimensional search tree $T$ ($K$-d tree, for short) of size $n \ge 0$ stores a set of $n$ $K$-dimensional records, each holding a key $x = (x_0, \ldots, x_{K-1}) \in D$, where $D = D_0 \times \cdots \times D_{K-1}$, and each $D_j$ is a totally ordered domain. The $K$-d tree $T$ is a binary tree such that:

– Either it is empty and $n = 0$, or
– Its root stores a record with key $x$ and has a discriminant $j$, $0 \leq j < K$, and
  the remaining $n - 1$ records are stored in the left and right subtrees of $T$,
  say $L$ and $R$, in such a way that both $L$ and $R$ are $K$-d trees; furthermore,
  for any key $u \in L$, it holds that $u_j \leq x_j$, and for any key $v \in R$, it holds
  that $x_j < v_j$.

We will assume without loss of generality that $D = [0,1]^K$. We will also use the
notation $\langle x, j \rangle$ to refer to a node that contains the key $x$ and the discriminant $j$.

A $K$-d tree of size $n$ induces a partition of the domain $D$ into $n + 1$ regions,
each corresponding to a leaf in the $K$-d tree. The *bounding box* of a node $z$ is
the region of the space associated to the leaf replaced by $z$ when it was inserted
into the tree. Thus, the bounding box of the root $\langle x, j \rangle$ is $[0,1]^K$, the bounding
box of the root of the left subtree is $[0,1] \times \cdots \times [0, x_j] \times \cdots \times [0,1]$, and so on.

Different variants of $K$-d trees have been proposed so far; many differ in the
way the discriminants are assigned to nodes. In the original or *standard* $K$-d
trees by Bentley [5] the root of the tree gets discriminant 0, the nodes in the
first level get discriminant 1, and so on, in a cyclic fashion. Notice that, since
there is a fixed, data-independent rule, to assign discriminants to nodes, there is
no need to explicitly store the discriminants. Much later Duch et al. [9] proposed
*relaxed* $K$-d trees, where each node is assigned a random discriminant, uniform
and independently drawn from $\{0, \ldots, K-1\}$. The *squarish* $K$-d trees of Devroye
et al. [11] try to get a more balanced partition of the space by discriminating
along the coordinate for which the bounding box of the node is more elongated.
We will consider along the paper the two first variants mentioned above, as
representative variants of $K$-d trees.

Because of their definition, the insertion and search algorithms for $K$-d trees
are straightforward, and we will not give here the details. Insertions work iden-
tically in the three variants, except in the way discriminants are assigned to
newly inserted nodes. We also mention here two other algorithms, common
to all variants of $K$-d trees. In *partial match search* we are given a pattern
$q = (q_0, \ldots, q_{K-1})$ where $q_j \in [0,1]$ or $q_j = \bot$, for $0 \leq j < K$. Coordinates such
that $q_j \neq \bot$ are called *specified*, otherwise they are called *unspecified*; we assume
that the number $s$ of specified coordinates satisfies $0 < s < K$. The goal of the
partial match search is to retrieve all points in the $K$-d tree that match the
pattern $q$, that is, the points $x$ such that $x_j = q_j$ whenever $q_j \neq \bot$. To perform a
partial match the $K$-d tree is recursively explored. First, we check whether the
root matches or not the pattern, to report it in the former case. Then, if the root
discriminates with respect to an unspecified coordinate, we make recursive calls
in both subtrees. Otherwise, if the root containing $x$ discriminates with respect
to a specified coordinate $j$ we continue recursively in the appropriate subtree,
depending on whether $q_j \leq x_j$ or $x_j < q_j$. The other algorithm is *orthogonal
range search*. The input to the algorithm is a $K$-d tree $T$ and a $K$-dimensional
rectangle $Q = [\ell_0, u_0] \times \cdots \times [\ell_{K-1}, u_{K-1}]$, and the goal is to retrieve all the
points in $T$ that lie within $Q$. The algorithm is very similar to partial match
search; the recursion proceeds into one of the subtrees if the root stores $\langle x, j \rangle$

and $x_j < \ell_j$ or $u_j < x_j$; otherwise we have to make recursive calls in both subtrees and check if $x$ does actually fall inside $Q$ or not.

We now turn our attention to the probabilistic model that we will use later in Sect. 4, when analyzing the expected performance of our algorithm. We say that a $K$-d tree built from a given set of $n$ keys is *random* if it is built with identical probability from any of the $n!$ possible input sequences. The discriminants must be assigned according to a fixed rule (standard, squarish $K$-d trees) or at random (relaxed $K$-d trees). As a consequence, a $K$-d tree $T$ of size $n$ is random if and only if it is either empty ($n = 0$), or if its left and right subtrees, $L$ and $R$, are independent random $K$-d trees of sizes $\ell$ and $n - 1 - \ell$, respectively, with

$$\Pr\left[|L| = \ell \mid |T| = n\right] = \frac{1}{n},$$

for any $0 \le \ell < n$.

There is another equivalent, alternative formulation of the probabilistic model above which is also useful. A random $K$-d tree of size $n$ is built by $n$ successive random insertions in a initially empty $K$-d tree. An insertion in a random $K$-d tree of size $n$ is random if it has the same probability to fall in any of the $n + 1$ leaves of the tree. Thus the insertion of $n$ points independently drawn from a continuous distribution in $[0, 1]^K$ into an initially empty $K$-d tree will produce always a random $K$-d tree.

The probabilistic model for random $K$-d trees is equivalent, as far as the shape of trees are concerned, to the probabilistic model of binary search trees. It follows then that the expected cost of insertions and the expected cost of exact searches is $\Theta(\log n)$ (see, for instance, [12]).

On the other hand, the expected cost of a partial match search with $s$ random specified coordinates in a random $K$-d tree of size $n$ is

$$P_n = \beta_q n^{\alpha(s/K)} + \mathcal{O}(1), \qquad 0 < s < K, \tag{1}$$

where $\beta_q$ is a constant that might depend on the alternance of specified and unspecified coordinates in the pattern $q$, and $\alpha(x)$ is a function depending on the type of $K$-d tree that we consider. In all cases, $1 - x \le \alpha(x) \le 1$ for $x \in [0, 1]$ with $\alpha(x) < 1$ if $x > 0$ and $\alpha \to 1$ as $x \to 0$. For squarish $K$-d trees $\alpha(x) = 1 - x$ [11], for relaxed $K$-d trees $\alpha(x) = (\sqrt{9 - 8x} - 1)/2$ [9,13] and for standard $K$-d trees $\alpha(x) = 1 - x + \phi(x)$ [14,15], where $\phi = \phi(x)$ is the unique solution in $[0, 1]$ of $(\phi + 3 - x)^x (\phi + 2 - x)^{1-x} - 2 = 0$.

For instance, for standard $K$-d trees, $\alpha(1/2) \approx 0.561$, $\alpha(1/3) \approx 0.716$ and $\alpha(1/4) \approx 0.790$. For relaxed $K$-d trees, we have $\alpha(1/2) \approx 0.618$, $\alpha(1/3) \approx 0.758$ and $\alpha(1/4) \approx 0.823$.

The expected cost of orthogonal range search comes as a combination of partial match costs [16,17]. The query rectangle $Q$ induces a division of the space into $2^K$ regions, which can be indexed with bitstrings of length $K$. The query rectangle itself is the region $R_{00...0}$. By extending $Q$ along each one of the $K$ coordinates and then substracting $Q$, we obtain $K$ regions $R_{100...0}, R_{010...0}, \ldots, R_{00...01}$. By extending $Q$ along two coordinates and then substracting $Q$ and all regions of the previous step, we obtain the regions $R_{00...011}, \ldots, R_{110...0}$, and so on. Denoting $p_w$ the probability that a point falls in region $R_w$ when the point is drawn from the continuous distribution in $[0,1]^K$ used to build the random $K$-d tree, and the center of the query is also drawn using the same distribution, the expected cost of an orthogonal range search is [17]

$$S_n = p_{00...0} \cdot n + 2p_{11...1} \cdot \log n + \sum_{j=1}^{K-1} \sum_{w:w \text{ has } j \text{ ones}} \beta_w p_w n^{\alpha(j/K)} + \mathcal{O}(1), \quad (2)$$

The probabilities $p_w$ will depend on the dimensions $\Delta_0, \ldots, \Delta_{K-1}$ of the query $Q$ and can be thought of as the "volumes" of the corresponding regions. For instance, if the data points and the center of the queries are uniformly distributed in $[0,1]^K$ then

$$p_w = \left( \prod_{i:w_i=0} \Delta_i \right) \cdot \left( \prod_{i:w_i=1} (1 - \Delta_i) \right).$$

For the particular case where the query hyperrectangle is a slice $Q = [0,1] \times [0,1] \times \cdots \times [\ell_j, u_j] \times [0,1] \times \cdots \times [0,1]$ the expected cost reduces to

$$S_n = p \cdot n + \beta_{000...1...0} \cdot (1 - p) \cdot n^{\alpha(1/K)} + \mathcal{O}(1), \quad (3)$$

since all regions except $R_{00...0} = Q$ and $R_{00...1...0}$ are empty. Here we use $p$ for the probability that a random point falls inside the slice; the first term is thus the expected number of points that fall inside the slice.

## 3   The Algorithm

We present now the algorithm to find the $i$-th smallest point along the coordinate $j$, $0 \le j < K$, in a $K$-d tree $T$. The algorithm has three main steps. In the first step, it does a breadth-first traversal of the tree $T$ using a queue $Q$ of pointers to nodes. This first step can also be easily formulated using a recursive preorder traversal of the tree.

During the first step, at any of its iterations, we have a current subtree $t$ and two values $low$ and $high$ with the guarantee that the $j$-th coordinate of the sought element is between those two values. The purpose of the first step is either to locate the $i$-th point along the $j$-th coordinate—and we would be then done—or to return a reasonably "thin" slice defined by $low$ and $high$ that must contain the sought element. If the sought element is not found during the first phase, the algorithm performs a convential orthogonal range search to find all the points within the slice $[low, high]$. Finally, the third step finds the sought element using a standard selection algorithm applied to the elements returned by the second step.

We give now a detailed description of the first step. If the current subtree $t$ discriminates with respect to $j' \ne j$, then the sought element could be eventually

**Algorithm 1.** The first phase of multidimensional selection

```
kdt kdselect(kdt T, int i, int j) {
    queue<kdt> Q; Q.push(T);
    double low = 0.0;
    double high = 1.0;
    bool found = false;
    kdt t;
    while (not Q.empty() and not found) {
        t = Q.pop(); if (t == NULL) continue;
        if (t -> discr != j) {
            Q.push(t -> left); Q.push(t -> right);
        } else { // t -> discr == j
            double z = t -> key[j];
            if (low <= z and z <= high) {
                int r = below(z, j, T);
                if (i < r) high = z;
                else if (i > r) low = z;
                else found = true;
            }
            if (z <= low) Q.push(t -> right);
            if (z >= high) Q.push(t -> left);
        }
    }
    if (found) return t;
    ...
}
```

found in any of its subtrees; therefore, nothing useful can be inferred and both subtrees of $t$ are enqueued for further processing in later iterations. If, on the other hand, the root of $t$ contains $\langle x, j \rangle$, then we have to consider three possibilities. If $x_j < low$ then none of the elements in the left subtree of $t$ can be the sought element; therefore, we enqueue the right subtree of $t$ only. Similarly, if $high < x_j$ then the right subtree of $t$ can be pruned, and we need only to explore (part of) the left subtree of $t$. Finally, if $low \leq x_j \leq high$ then we compute how many points in the collection, that is, in $T$, have coordinate $j$ less than or equal to $x_j$. This is done using the procedure below. Let $r$ be that number. Then if $i = r$ the root of $t$ is the sought element. If $i < r$ then the sought element might be in the left subtree of $t$ but not in its right subtree, thus we push only the left subtree of $t$ into the queue. Furthermore, the $j$-th coordinate of the sought element must be less than or equal to $x_j$, hence we set $high := x_j$. If $i > r$ then the sought element cannot be in the left subtree of $t$, and we enqueue the right subtree of $t$; additionally, we set $low := x_j$, since the $j$-th coordinate of the $i$-th element must be greater than or equal to $x_j$. This first part of the algorithm is given in Algorithm 1. Figure 1 illustrates a standard $K$-d tree with $K = 2$, the partition of $[0, 1]^2$ that it induces, and the outcome (the shaded slice) of the first

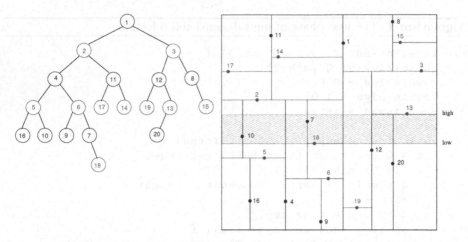

**Fig. 1.** An example of $K$-d tree and the execution of Algorithm 1

step of kdselect when looking for the 11-th smallest element along coordinate 1 (the $y$-axis). Note that the labels of the items in the figure only indicate the order of insertion.

To complete our description of the first step, we now draw our attention to the procedure below. It is a simple variation of partial match (see Algorithm 2). The algorithm uses the sizes of subtrees that we store at each node. Using this stored information is essential to avoid computation and thus to achieve a reasonable expected performance. We assume, for convenience, that each node stores its rank relative to its bounding box and the discriminating coordinate, that is, the size of its left subtree plus one.

If the tree is empty we return 0. Otherwise, if the root of $T$ discriminates with respect to a coordinate different from $j$, we count recursively how many points there are below the given line in both subtrees. We also add one if the root itself meets the condition $x_j \leq z$. If the root of $T$ discriminates with respect to $j$ then we have to continue counting recursively in only one of the subtrees. Note that if $x_j \leq z$ then we count how many points are below $z$ in the right subtree, as all the points in the left subtree and the root itself are below $z$. We avoid making any traversal of the left subtree since its root stores the corresponding size.

## 4   Analysis

For our analysis of kdselect, we will consider that the input $K$-d tree is random, that the given rank $i$ is random, namely, uniformly distributed in $\{1, \ldots, n\}$, and that the given coordinate $j$ is also uniformly chosen from $\{0, \ldots, K-1\}$.

The analysis of the expected performance of kdselect is based upon the following ingredients, which we will later prove formally:

1. The number of visited nodes in the main loop of kdselect is at most the number of nodes that we would visit in an orthogonal range search to locate the points that lie within the slice defined by $[low, high]$.

---

**Algorithm 2.** Below counts how many points in $T$ have coordinate $j$ less than $z$

---

```
int below(double z, int j, kdt T) {
    if (T == NULL) return 0;
    if (T -> discr != j) {
        int c = (T -> key[j] <= z) ? 1 : 0;
        return below(z, j, T -> left) +
               below(z, j, T -> right) + c;
    } else {
        if (z < T -> key[j])
            return below(z, j, T -> left);
        else
            return T -> rank + below(z, j, T -> right);
    }
}
```

---

2. The expected cost of a call to below is that of a partial match query with a single specified coordinate (the $j$-th).

3. The expected number of calls to below is $\mathcal{O}(\log n)$.

4. If the $i$-th smallest element along coordinate $j$ discriminates along coordinate $j$, it will be found during the first step; otherwise, the first step will report the smallest slice $[low, high]$ that contains the $i$-th element along the $j$-th coordinate and no interior point discriminating along coordinate $j$.

5. If the first step of kdselect does not find the sought element, then the expected number of points in $[low, high]$ is $\Theta(1)$.

Before going on with a formal proof of each of the statements above, we discuss now how they affect the overall expected performance of kdselect. The expected cost of the first phase will have two contributions, one coming from the calls to below, the other from the main loop. From the items 2 and 3 above, and since the expected cost of a partial match (1) is $\Theta(n^{\alpha(1/K)})$, it follows that the first contribution is $\mathcal{O}(n^{\alpha(1/K)} \log n)$. For the second contribution, we deduce from items 1 and 5 and (3) that it is $\Theta(n^{\alpha(1/K)})$. In total, the first step of the algorithm has expected cost $\mathcal{O}(n^{\alpha(1/K)} \log n)$. The second and third steps are only necessary if the $i$-th element has not been found (this happens[1] with probability $(K-1)/K$, when it does not discriminate with respect to $j$). The second step is an orthogonal range search for points falling in the slice $[low, high]$ and has expected cost $\Theta(n^{\alpha(1/K)})$. The third and last step is an ordinary selection algorithm applied to the points found in the previous step; since the expected number of points within the slice is $\Theta(1)$ (item 5), this part has expected cost $\Theta(1)$. Summing up everything we conclude with the following theorem.

---

[1] Actually, for variants of $K$-d trees such as standard and relaxed $K$-d trees; in general, for any variant which does not exhibit a bias in the distribution of the coordinates assigned to the discriminants.

**Theorem 1.** *The expected cost to select the i-th smallest element along coordinate j in a K-d tree of size n is $\mathcal{O}(n^{\alpha(1/K)} \log n)$, where $\alpha(x)$ is a function depending on the variant of K-d tree used, such that $1 - x \leq \alpha(x) \leq 1$ for all $x \in [0, 1]$. Furthermore, $\alpha(x) < 1$ for all $x > 0$, and $\alpha(x) \to 1$ as $x \to 0$.*

We now prove the key five statements above. For the first statement, relating the number of iterations of the first phase and the cost of an orthogonal range search, the proof relies in the fact that a node $x$ in a $K$-d tree is visited during an orthogonal range search with query $Q$ if and only if $Q$ and the bounding box of $x$ intersect [16,17]. Let $\ell_0 = 0$, $\ell_1$, ..., $\ell_r$ be the sequence of values assigned to the variable *low* along the execution of Algorithm 1 and similarly, $h_0 = 1$, $h_1$, ..., $h_{r'}$ for the values of *high*. Suppose $t = \langle x, j' \rangle$ is the current node, and that its bounding box intersects $[\ell_r, h_{r'}]$. Suppose also that at that iteration $low = \ell_m$ and $high = h_{m'}$. If $j' \neq j$ both subtrees of $t$ will be visited (if they are non-empty) and their corresponding bounding boxes intersect $[\ell_r, h_{r'}]$. If $j' = j$ and $x_j < low = \ell_m$ then only the right subtree of $t$ will be visited. Since $\ell_m \leq \ell_r$, the bounding box of the right subtree of $t$ does intersect $[\ell_r, h_{r'}]$, whereas the bounding box of the left subtree of $t$ does not. For the case where $high = h_{m'} < x_j$, we have that the left subtree is visited and its bounding box intersects $[\ell_r, h_{r'}]$, and the right subtree is not visited and its bounding box does not intersect $[\ell_r, h_{r'}]$. Finally, if $low \leq x_j \leq high$ we will update either *low* or *high* (or finish because we find the sought element). If we have not yet finished, the new current $[low, high]$ contains the final $[low, high]$ slice, that is, $[\ell_r, h_{r'}]$, and we apply the same reasoning as above. The basis of this inductive proof is provided by the root of the tree, whose bounding box $[0, 1]^K$ obviously intersects the slice $[\ell_r, h_{r'}]$.

The second statement, that the cost of below is that of a partial match is also very easy to prove. The algorithm below behaves exactly as a partial match with a query pattern $q = (\perp, \perp, \ldots, z, \perp \ldots)$, where only the $j$-th coordinate of $q$ is specified.

For the third statement, where we claim that the expected number of calls to below is $\mathcal{O}(\log n)$ we reason as follows. Consider the points whose $j$-th coordinate is smaller than or equal to that of the element for which we set the final value of *low*. Of those, only the points whose bounding box intersects $[low, high]$ will be visited. Furthermore, only a fraction $1/K$ (on average, see the remarks in the footnote of the previous page) of them discriminate with respect to $j$, so eventually a call to below will be made when visiting them. Since the $K$-d tree is random, the sequence of $j$-th coordinates of these points will form a random permutation of $[1, \ldots, N]$, where $N \leq n$ is the number of points discriminating with respect to $j$, whose bounding box intersects $[low, high]$ and such that its $j$-th coordinate is less than or equal to *low*. Each call to below to update the value of *low* corresponds to a left-to-right maxima in that permutation, and it is well-known (see for instance [12]) that the expected number of left-to-right maxima in a random permutation of size $N$ is $\Theta(\log N)$. Analogously, each call to below to update the value of *high* corresponds to a left-to-right minima in the random permutation induced by the sequence of $j$-th coordinates larger or equal to $high$, for visited points discriminating with respect to $j$.

The fourth statement says that if the sought element discriminates with respect to the given coordinate $j$, then it will be found; otherwise, the first phase of the algorithm will terminate returning the slice $[low, high]$ that contains the sought element. The last part follows by design of the algorithm: the invariant of the iteration guarantees that the sought element lies within the slice $[low, high]$. On the other hand, the bounding box of the sought elements always intersects $[low, high]$ so, if it discriminates with respect to $j$, sooner or later it will be visited and its rank will be computed using below.

The last statement establishes that the expected number of points within the slice is $\Theta(1)$, when the sought element is not found by the first step of kdselect. By item 4, the sought element does not discriminate with respect to $j$. Moreover, $low$ and $high$ are the $j$-th coordinates of two points, say $u$ and $v$, that discriminate with respect to $j$; all points properly falling within $[low, high]$ have been visited but do not discriminate with respect to $j$. The expected number of points in the slice is the number of points that we see when we start from the $j$-th coordinate of the sought element and go towards $low = u_j$, plus the number of points that we see when we go towards $high = v_j$. These points that we count must not discriminate with respect to $j$. Since the probability that a point does not discriminate with respect to $j$ is $(K-1)/K$, the expected number of points within the slice in either direction is $K$, including the two points discriminating with respect to $j$ that define the boundaries of the slice. In total, the expected number of points is $2K + 1$.

## 5    Experiments

To corroborate the analysis of kdselect we have performed a preliminary set of experiments in two diferent variants of $K$-d trees (standard and relaxed).

For every dimension going from $K = 2$ to $K = 4$, we generate $M$ $K$-d trees of size $n$, with $n$ going from 1000 to 50000 with a step of 1000 elements. In each tree we look for the $i$-th element (with $i$ going from 1 to $n$ with a step of $n/100$) in each of the $K$ possible coordinates (going from 0 to $K - 1$). For each tree we count the total number of visited nodes in the main loop of kdselect, the number of calls to function below and the number of points lying in interval $[low, high]$ and take the corresponding averages.

Figure 2 contains the experimental results regarding the total number of visited nodes in the main loop of kdselect. In particular, we plot the ratio of the number of visited nodes to $n^{\alpha(1/K)}$, so the figures exhibit the convergence of the ratio to a constant factor as $n$ grows.

The number of calls to below can be found in Fig. 3; we actually plot the number of calls to below minus $2 \log n$, which converges to a constant factor depending on $K$. Our analysis of the previous section can be refined to show that the expected number of calls to below when the rank is chosen at random is $2 \log n + \mathcal{O}(1)$, as the experiment corroborates.

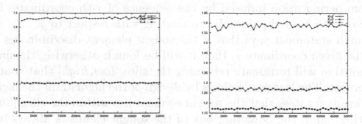

**Fig. 2.** Number of visited nodes in the main loop of Algorithm 1 divided by $n^{\alpha(1/K)}$, for standard (left) and relaxed (right) $K$-d trees

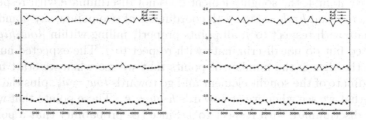

**Fig. 3.** Number of calls made to Algorithm 2 (below) minus $2 \log n$ in standard (left) and relaxed (right) $K$-d trees

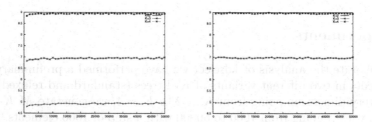

**Fig. 4.** Number of points within interval $[low, high]$ in in standard (left) and relaxed (right) $K$-d trees

Finally, Fig. 4 shows the number of points contained in the interval $[low, high]$. The experiments confirm very well the predicted value $2K + 1$, that does not depend on the variant of $K$-d trees considered.

# References

1. Hoare, C.A.R.: FIND (Algorithm 65). Comm. ACM 4, 321–322 (1961)
2. Blum, M., Floyd, R., Pratt, V., Rivest, R., Tarjan, R.: Time bounds for selection. J. Comp. Syst. Sci. 7, 448–461 (1973)

3. Martínez, C., Roura, S.: Randomized binary search trees. J. Assoc. Comput. Mach. 45(2), 288–323 (1998)
4. Roura, S.: A new method for balancing binary search trees. In: Orejas, F., Spirakis, P.G., van Leeuwen, J. (eds.) ICALP 2001. LNCS, vol. 2076, pp. 469–480. Springer, Heidelberg (2001)
5. Bentley, J.L.: Multidimensional binary search trees used for associative retrieval. Comm. ACM 18(9), 509–517 (1975)
6. Bentley, J., Finkel, R.: Quad trees: A data structure for retrieval on composite keys. Acta Informatica 4, 1–9 (1974)
7. Samet, H.: The Design and Analysis of Spatial Data Structures. Addison-Wesley, Reading (1990)
8. Gaede, V., Günther, O.: Multidimensional access methods. ACM Computing Surveys 30(2), 170–231 (1998)
9. Duch, A., Estivill-Castro, V., Martínez, C.: Randomized $k$-dimensional binary search trees. In: Chwa, K.-Y., Ibarra, O.H. (eds.) ISAAC 1998. LNCS, vol. 1533, pp. 199–208. Springer, Heidelberg (1998)
10. Duch, A., Martínez, C.: Updating relaxed $k$-d trees. ACM Trans. on Algorithms (2008) (accepted for publication)
11. Devroye, L., Jabbour, J., Zamora-Cura, C.: Squarish $k$-d trees. SIAM J. Comput. 30, 1678–1700 (2000)
12. Knuth, D.E.: The Art of Computer Programming: Sorting and Searching, 2nd edn., vol. 3. Addison-Wesley, Reading (1998)
13. Martínez, C., Panholzer, A., Prodinger, H.: Partial match queries in relaxed multidimensional search trees. Algorithmica 29(1-2), 181–204 (2001)
14. Flajolet, P., Puech, C.: Partial match retrieval of multidimensional data. J. Assoc. Comput. Mach. 33(2), 371–407 (1986)
15. Chern, H.H., Hwang, H.K.: Partial match queries in random $k$-d trees. SIAM J. Comput. 35(6), 1440–1466 (2006)
16. Chanzy, P., Devroye, L., Zamora-Cura, C.: Analysis of range search for random $k$-d trees. Acta Informatica 37, 355–383 (2001)
17. Duch, A., Martínez, C.: On the average performance of orthogonal range search in multidimensional data structures. J. Algorithms 44(1), 226–245 (2002)

# Layered Working-Set Trees*

Prosenjit Bose, Karim Douïeb, Vida Dujmović, and John Howat

School of Computer Science, Carleton University
1125 Colonel By Dr., Ottawa, Ontario, Canada, K1S 5B6
{jit,karim,vida,jhowat}@cg.scs.carleton.ca

**Abstract.** The *working-set bound* [Sleator and Tarjan, J. ACM, 1985] roughly states that searching for an element is fast if the element was accessed recently. Binary search trees, such as splay trees, can achieve this property in the amortized sense, while data structures that are not binary search trees are known to have this property in the worst case. We close this gap and present a binary search tree called a *layered working-set tree* that guarantees the working-set property in the worst case. The *unified bound* [Bădoiu et al., TCS, 2007] roughly states that searching for an element is fast if it is near (in terms of rank distance) to a recently accessed element. We show how layered working-set trees can be used to achieve the unified bound to within a small additive term in the amortized sense while maintaining in the worst case an access time that is both logarithmic and within a small multiplicative factor of the working-set bound.

## 1 Introduction

Let $S$ be a set of keys from a totally ordered universe and let $X$ be a sequence of elements from $S$. Typically, one is required to store elements of $S$ in some data structure $D$ such that accessing the elements of $S$ using $D$ in the order defined by $X$ is "fast." Here, "fast" can be defined in many different ways, some focusing on worst case access times and others on amortized access times. For example, the search times of splay trees [9] can be stated in terms of the rank difference between the current and previous elements of $X$; this is the *dynamic finger property* [4,5].

If $x$ is the $i$-th element of $X$, we say that $x$ is accessed at time $i$ in $X$. The working-set number of $x$ at time $i$, denoted $w_i(x)$, is the number of distinct elements accessed since the last time $x$ was accessed or inserted, or $|D|$ if $x$ is either not in $D$ or has not been accessed by time $i$.

The *working-set property* states the time to access $x$ at time $i$ is $O(\lg w_i(x))$.[1] Splay trees were shown by Sleator and Tarjan [9] to have the working-set property in the amortized sense. One drawback of splay trees, however, is that most of the access bounds hold only in an amortized sense. While the amortized cost of a query can be stated in terms of its rank difference between successive queries or

---

* This research was partially supported by NSERC and MRI.
[1] In this paper, $\lg x$ is defined to be $\log_2(x + 2)$.

A. López-Ortiz (Ed.): LATIN 2010, LNCS 6034, pp. 686–696, 2010.

the number of distinct queries since a query was last made, any particular operation can take $\Theta(n)$ time. In order to address this situation, attention has turned to finding data structures that maintain the distribution-sensitive properties of splay trees but guarantee good performance in the worst case.

The data structure of Bădoiu et al. [2], called the working-set structure, guarantees this property in the worst case. However, this data structure departs from the binary search tree model and is instead a collection of binary search trees and queues.

Bădoiu et al. [2] also describe a data structure called the unified structure that achieves the *unified property*, which states that searching for $x$ at time $i$ takes time $O(\min_{y \in S} \lg(w_i(y) + d(x, y)))$ where $d(x, y)$ is the rank difference between $x$ and $y$. Again, this data structure is not a binary search tree. The skip-splay algorithm of Derryberry and Sleator [7] fits into the binary search tree model and comes within a small additive term of the unified bound in an amortized sense.

*Our Results.* We present a binary search tree that is capable of searching for a query $x$ in worst-case time $O(\lg w_i(x))$ and performs insertions and deletions in worst-case time $O(\lg n)$, where $n$ is the number of keys stored by the tree at the time of the access. This fills in the gap between binary search trees that offer these query times only in an amortized sense and data structures which guarantee these query times in the worst-case but do not fit in the binary search tree model. We have also shown how to use this binary search tree to achieve the unified bound to within a small additive term in the amortized sense while maintaining in the worst case an access time that is both logarithmic and within a small multiplicative factor of the working-set bound. Due to space constraints, proofs have been omitted; full proofs can be found in the technical report [3].

## 1.1   The Working-Set Structure

The working-set structure of Bădoiu et al. [2] maintains a dynamic set under the operations INSERT, DELETE and SEARCH. Denote by $S_i \subseteq S$ the set of keys stored in the data structure at time $i$.

The structure is composed of $t = O(\lg \lg |S_i|)$ balanced binary search trees $T_1, T_2, \ldots, T_t$ and $t$ doubly linked lists $Q_1, Q_2, \ldots, Q_t$. For any $1 \le j \le t$, the contents of $T_j$ and $Q_j$ are identical, and pointers (in both directions) are maintained between their common elements. Every element in the set $S_i$ is contained in exactly one tree and in its corresponding list. For $j < t$, the size of $T_j$ and $Q_j$ is $2^{2^j}$, whereas the size of $T_t$ and $Q_t$ is $|S_i| - \sum_{j=1}^{t-1} 2^{2^j} \le 2^{2^t}$.

The working-set structure achieves its stated query time of $O(\lg w_i(x))$ by ensuring that an element $x$ with working-set number $w_i(x)$ is stored in a tree $T_j$ with $j \le \lceil \lg \lg w_i(x) \rceil$. Every list $Q_j$ orders the elements of $T_j$ by the time of their last access, starting with the youngest (most recently accessed) and ending with the oldest (least recently accessed).

Operations in the working-set structure are facilitated by an operation called a *shift*. A shift is performed between two trees $T_j$ and $T_k$. Assume $j < k$, since the other case is symmetric. To perform a shift, we begin at $T_j$. We look in $Q_j$

to determine the oldest element and remove it from $Q_j$ and delete it from $T_j$. We then insert it into $T_{j+1}$ and $Q_{j+1}$ (as the youngest element) and repeat the process by shifting from $j+1$ to $k$. This process continues until we attempt to shift from one tree to itself. Observe that a shift causes the size of $T_j$ to decrease by one and the size of $T_k$ to increase by one (although the contents of intermediate trees change).

To search for an element $x$, we search sequentially in $T_1, T_2, \ldots$ until we find $x$ or search all of the trees and fail to find $x$. If $x \notin T_j$ for any $j$, then we will search every tree at a total cost of $O(\lg |S_i|)$ and then report that $x$ is not in the structure. Otherwise, assume $x \in T_j$. We delete $x$ from $T_j$ and $Q_j$ and insert it in $T_1$ and place it at the front of $Q_1$. We then perform a shift from 1 to $j$ to restore the sizes of the trees and lists. The time required for a search is dominated by the search time in $T_j$. Observe that if $x \in T_j$ and $j > 1$, then it must have been removed as the oldest element from $Q_{j-1}$, and so $w_i(x) \geq 2^{2^{j-1}}$. Thus, the search time is $O(\lg w_i(x))$.

Insertions are performed by inserting the element into $T_1$ and $Q_1$ (as the youngest element) and shifting from 1 to $t$ (possibly creating a new tree) at total cost $O(\lg |S_i|)$. Deletions are performed by first searching for the element to be deleted. Once found, say in $T_j$, it is removed from $T_j$ and $Q_j$ and a shift from $t$ to $j$ is performed at total cost $O(\lg |S_i|)$. If the last tree becomes empty, it can be removed.

## 2    The Binary Search Tree

In this section, we describe a binary search tree that has the working-set property in the worst case.

*Model.* Recall the binary search tree model of Wilber [11]. Each node of the tree stores the key associated with it and has a pointer to its left and right children and its parent. The keys stored in the tree are from a totally ordered universe and are stored such that at any node, all of the keys in the left subtree are less than that stored in the node and all of the keys in the right subtree are greater than that stored at the node. Furthermore, each node may keep a constant[2] amount of additional information called *fields*, but no additional pointers may be stored. To perform an access to a key, we are given a pointer initialized to the root of the tree. An access consists of moving this pointer from a node to one of its adjacent nodes (through the parent pointer or one of the children pointers) until the pointer reaches the desired key. Along the way, we are allowed to update the fields and pointers in any nodes that the pointer reached. The access cost is the number of nodes reached by the pointer.

### 2.1    Tree Decomposition

Our binary search tree $T$ will adapt the working-set structure described in the previous section to the binary search tree model. At a high level, our binary

---

[2] By standard convention, $O(\lg |S_i|)$ bits are considered to be "constant."

search tree layers the trees $T_1, T_2, \ldots, T_t$ of the working-set structure together to form $T$, and then augments nodes with enough information to recover which is the oldest in each tree at any given time.

Consider a labelling of $T$ where each node $x \in T$ has a label from $\{1, 2, \ldots, t\}$ such that no node has an ancestor with a label greater than its own label. This labelling partitions the nodes of $T$. We say that the nodes with label $j \in \{1, 2, \ldots, t\}$ form a *layer* $L_j$. Observe that layer $L_j$ can be connected to any layer $L_k$ with $k > j$. A layer $L_j$ will play the same role as $T_j$ in the working-set structure. Like $T_j$, $L_j$ contains exactly $2^{2^j}$ elements for $j < t$, and $L_t$ contains the remaining elements. Unlike $T_j$, $L_j$ is typically a collection of subtrees of $T$. We refer to a subtree of a layer $L_j$ as a *layer-subtree*. Every node $x \in T$ stores as a field the value $j$ such that $x \in L_j$ which we denote by layer[$x$]. We also record the total number of layers $t$ and the size of $L_t$ at the root as fields of each node.

Each layer-subtree $T'_j \in L_j$ is maintained independently as a tree that guarantees that each node of $T'_j$ has depth in $T'_j$ at most $O(\lg |T'_j|) = O(\lg |L_j|)$ (*e.g.*, a red-black tree [1,8]). Balance criteria are applied only to the elements within a single layer-subtree.

Our first observation concerns the depth of a node in a given layer.

**Lemma 1.** *The depth of a node $x \in L_j$ is $O(2^j)$.*

The main obstacle in creating our tree comes from the fact that the core operations are performed on subtrees rather than trees, as is the case for the working-set structure. Consequently, standard red-black tree operations cannot be used for the operations spanning more than one layer as described in Section 2.3. We break the operations into those restricted to one layer, those spanning two neighbouring layers, and finally those performed on the tree as a whole. These operations are described in the following sections.

Another difficulty arises from having to implement the queues of the working-set structure in the binary search tree model. The queues are needed to determine the oldest element in a layer at any given time. We encode the linked lists in our tree as follows. Each node $x \in L_j$ stores the key of the node inserted into $L_j$ directly before and after it. This information is stored in the fields older[$x$] and younger[$x$], respectively. We also store a key value in the field nextlayer[$x$]. If $x$ is the oldest element in layer $L_j$, then no element was inserted before it and so we set older[$x$] = nil. In this case, we use nextlayer[$x$] to store the key of the oldest element in layer $L_{j+1}$. Similarly, if $x$ is the youngest element in layer $L_j$, then no element was inserted after it and so we set younger[$x$] = nil and use nextlayer[$x$] to store the key of the youngest element in layer $L_{j+1}$. If $x$ is neither the youngest nor the oldest element in $L_j$, then we have nextlayer[$x$] = nil.

Before we describe how operations are performed on this binary search tree, we must make a brief note on storage. Each node $x$ stores three pointers (parent and children) and a key. The root also maintains the number of trees $t$ and the size of $L_t$. In addition, we must store balance information (one bit for red-black trees) and three additional key values (exactly one of which is nil): older[$x$], younger[$x$] and nextlayer[$x$]. If keys are assumed to be of size $O(\lg n)$, then our structure fits the binary search tree model described earlier. Note that we are storing *key values*, not pointers. Given a key value stored at a node, we do not have a pointer to it, so we must search for it by performing a standard search from the root of the tree. If keys have size $\omega(\lg n)$, it is true that we use more than $O(\lg n)$ additional space per node. However, since every node would then store a key of size $\omega(\lg n)$, we are only increasing the size of a node by a constant factor.

## 2.2   Intra-layer Operations

Within a layer, we need notions of restoring balance after insertions and deletions and of splitting and joining. We will state these operations and the required time bounds, and then show how red-black trees [1,8] can be used to fulfill this role. Any binary search trees that meets the requirements of each operation could also be used. Layer-subtrees must also ensure that their operations do not leave the layer-subtree; this can be done by checking the layer number of a node before visiting it.

Intra-layer operations rearrange layer-subtrees in some way. Observe that layer-subtrees hanging off a given node are maintained even after rearranging the layer-subtree, since the roots of such layer-subtrees can be viewed as the results of unsuccessful searches. Therefore, when describing these operations, we need not concern ourselves with explicitly maintaining layer-subtrees below the current one.

Consider a node $x$ in a layer-subtree $T_j'$ of $L_j$.

INSERT-FIXUP$(x)$. This operation is responsible for ensuring that each node of $T_j'$ has depth $O\big(\lg |T_j'|\big)$ after the node $x$ has been inserted into the layer-subtree. For red-black trees, this operation is precisely the RB-INSERT-FIXUP operation presented by Cormen et al. [6, Section 13.3]. Although the version presented there does not handle colouring $x$, it is straightforward to modify it to do so.

DELETE-FIXUP$(x)$. This operation is responsible for ensuring that each node of $T_j'$ has depth $O\big(\lg |T_j'|\big)$ after a deletion in the layer-subtree. The exact node $x$ given to the operation is implementation dependent. For red-black trees, this operation is precisely the RB-DELETE-FIXUP operation presented by Cormen et al. [6, Section 13.4]. In this case, the node $x$ is the child of the node spliced out by the deletion algorithm; we will elaborate on this when describing the layer operations in Section 2.3.

SPLIT$(x)$. This operation will cause the node $x \in T_j'$ to be moved to the root of $T_j'$. The rest of the layer-subtree will be split between the left and right side of $x$ such that each side is independently balanced and thus guarantee depth $O\big(\lg |T_j'|\big)$ of their respective nodes; this may mean that the layer-subtree is no longer balanced as a whole. For red-black trees, this operation is described by Tarjan [10, Chapter 4], except we do not destroy the original trees, but rather stop when $x$ is the root of the layer-subtree.

JOIN($x$). This operation is the inverse of SPLIT($x$): given a node $x \in T'_j$, we will restructure $T'_j$ to consist of $x$ at the root of the $T'_j$ and the remaining elements in subtrees rooted at the children of $x$ such that all nodes in the layer-subtree have depth $O(\lg |T'_j|)$. For red-black trees, this operation is described by Cormen et al. [6, Problem 13-2].

**Lemma 2.** *When red-black trees are used as layer-subtrees, the operations* INSERT-FIXUP($x$), DELETE-FIXUP($x$), SPLIT($x$) *and* JOIN($x$) *can be implemented to take worst-case time* $O(2^j)$ *for a node* $x \in L_j$.

## 2.3    Inter-layer Operations

The operations performed between layers correspond to the queue and shift operations of the working-set structure. The four operations performed on layers are YOUNGESTINLAYER($L_j$) and OLDESTINLAYER($L_j$) for a layer $L_j$ and MOVEUP($x$) and MOVEDOWN($x$) for a node $x$. Only the operation MOVEDOWN($x$) will require knowledge of the implementation of the layer-subtrees; the remaining operations simply make use of the operations defined in Section 2.2.

YOUNGESTINLAYER($L_j$). This operation returns the key of the youngest node in layer $L_j$. We first examine all elements in $L_1$ (of which there are $O(1)$). Once we find the element that is the youngest (by looking for the element for which younger[$x$] = nil), say $x_1$, we go back to the root and search for nextlayer[$x_1$], which will bring us to the youngest element in $L_2$, say $x_2$. We then go back to the root and search for nextlayer[$x_2$], and so on. This repeats until we find the youngest element in $L_j$, as desired.

OLDESTINLAYER($L_j$). The process is the same as the previous operation, except our initial search in $L_1$ is for the oldest element.

MOVEUP($x$). This operation will move $x$ from its current layer $L_j$ to the next higher layer $L_{j-1}$. To accomplish this, we first split $x$ to the root of its layer-subtree using SPLIT($x$). We remove $x$ from $L_j$ by setting layer[$x$] = $j-1$. We now must restore balance properties. Observe that, by the definition of split, both of the layer-subtrees rooted at the children of $x$ are balanced. Therefore, we only need to ensure the balance properties of $L_{j-1}$. Since we have just inserted $x$ into the layer $L_{j-1}$, this can be done by performing the intra-layer operation INSERT-FIXUP($x$). Finally, we remove $x$ from the implicit queue structure of $L_j$ and place it in the implicit queue structure of $L_{j-1}$.

To do this, we look at both older[$x$] and younger[$x$]. If they are both non-nil, then we go to the root and perform searches for older[$x$] and younger[$x$], setting younger[older[$x$]] = younger[$x$] and older[younger[$x$]] = older[$x$]. Otherwise, if only younger[$x$] is nil, then we conclude that $x$ is the youngest in its former layer. After removing it from that layer, older[$x$] will be the new youngest element in that layer, so we go to the root search for older[$x$] and set younger[older[$x$]] = nil. Since older[$x$] is the youngest element in that layer, we also copy nextlayer[$x$] into nextlayer[older[$x$]]. We must also update the key stored by the youngest element in the next higher layer. In order to do this, we run YOUNGESTINLAYER($L_{j-1}$) to find this element, say $y$, and set nextlayer[$y$] = older[$x$]. The case for when only older[$x$] is nil is symmetric: the new oldest element in the layer is younger[$x$], so we update older[younger[$x$]] = nil, we copy nextlayer[$x$] into nextlayer[younger[$x$]], and update the pointer to the oldest element in this layer that is stored in $L_{j-1}$ in the same was as we did for the youngest.

We now must insert $x$ into the implicit queue structure of layer $L_{j-1}$. To do this, we search for the youngest node in $L_{j-1}$, say $y$. We then set older[$x$] = $y$, younger[$x$] = nil and younger[$y$] = $x$. We then go to the next layer $L_{j-2}$ and update its pointer to the youngest element in this layer the same way we did before.

MOVEDOWN($x$). This operation will move $x$ from its current layer $L_j$ to the next lower layer $L_{j+1}$. We describe how to perform this operation for red-black trees. Let $p$ denote the predecessor of $x$ in $L_j$. If $x$ does not have a predecessor in $L_j$, set $p = x$. Similarly, let $s$ denote the successor of $x$ in $L_j$, and if $x$ does not have a successor in $L_j$, set $s = x$. Our first goal is to move $x$ such that it becomes a leaf of its layer-subtree. If $x$ is not already a leaf in $L_j$, then $x$ has at least one child in its layer-subtree. To make it a leaf of it layer-subtree, we *splice* out the node $s$ by making the parent of $s$ point to the right child of $s$ instead of $s$ itself. Note that this is well-defined since $s$ has no left child in $L_j$ as it is the smallest element greater than $x$. We then move $s$ to the location of $x$. Finally, we make $x$ a child of $p$ and make the new children of $x$ the old children of $p$ and $s$.

Observe that we now have that $x$ is a leaf of its layer-subtree. The layer-subtree is configured exactly as if we had deleted $x$ using the deletion operation described by Cormen et al. [6, Section 13.4]. Therefore, we can perform DELETE-FIXUP($s'$), where $s'$ is the (only) child of $s$, to restore the balance properties of the nodes of the layer-subtree. Thus, $s'$ is exactly the child of the node spliced out by the deletion ($s$), as required by the operation of Cormen et al. [6, Section 13.4].

To complete the movement to the next layer, we change the layer number of $x$ and execute JOIN($x$) to create a single balanced layer-subtree from $x$ and its children.[3] We then update the implicit queue structure as we did before. Observe that once $x$ has been removed from its original layer-subtree, layer-subtree balance has been restored because no node on that path was changed.

**Lemma 3.** *For a layer $L_j$ and a node $x \in L_j$, YOUNGESTINLAYER($L_j$), OLDESTINLAYER($L_j$), MOVEUP($x$) and MOVEDOWN($x$) take worst-case time $O(2^j)$.*

## 2.4   Tree Operations

We are now ready to describe how to perform the usual dictionary operations SEARCH($x$), INSERT($x$) and DELETE($x$) on the tree as a whole.

---

[3] Note that if these children have larger layer numbers than the new layer number for $x$, nothing is performed and $x$ becomes the lone element in its (new) layer-subtree; this follows from the fact that JOIN($x$) only joins nodes that are in the same layer.

SEARCH($x$). To perform a search for $x$, we begin by performing the usual method of searching in a binary search tree. Once we have found $x \in L_j$, we execute MOVEUP($x$) a total of $j - 1$ times to bring $x$ into $L_1$. We then restore the sizes of the layers as was done in the working-set structure. We run OLDESTINLAYER($L_1$) to find the oldest element $y_1$ in layer $L_1$ and then run MOVEDOWN($y_1$). We then run OLDESTINLAYER($L_2$) to find the oldest element $y_2$ in layer $L_2$ and run MOVEDOWN($y_2$). This process of moving elements down layer-by-layer continues until we reach a layer $L_k$ such that $|L_k| < 2^{2^k}$.[4] Note that efficiency can be improved by remembering the oldest elements of previous layers instead of finding the oldest element in each of $L_1, \ldots, L_j$ when running OLDESTINLAYER($L_j$). Such an improvement does not alter the asymptotic running time, however.

INSERT($x$). To insert $x$ into the tree, we first examine the index $t$ and size $|L_t|$ of the deepest layer, which we have stored at the root. If $|L_t| = 2^{2^t}$, then we increment $t$ and set $|L_t| = 1$. Otherwise, if $|L_t| < 2^{2^t}$, we simply increment $|L_t|$. We now insert $x$ into the tree (ignoring layers for now) using the usual algorithm where $x$ is placed in the tree as a leaf. We set layer$[x] = t + 1$ (*i.e.*, a temporary layer larger than any other) and update the implicit queue structure for $L_t$ (and the youngest and oldest elements of $L_{t-1}$) as we did before. Finally, we run SEARCH($x$) to bring $x$ to $L_1$. Note that since SEARCH($x$) stops moving down elements once the first non-full layer is reached, we do not place another element in layer $t + 1$. Thus, this layer is now empty and we update the youngest and oldest elements in layer $t$ to indicate that there is no layer below.

DELETE($x$). To delete $x$ from the tree, we look at the total number $t$ of layers in the tree that is stored at the root. We then locate $x \in T_j$ and perform MOVEDOWN($x$) a total of $t - j + 1$ times. This will cause $x$ to be moved to a new (temporary) layer that is guaranteed to have no other nodes in it. Therefore, $x$ must be a leaf of the tree, and we can simply remove it by setting the corresponding child pointer of its parent to nil. As was the case for insertion, this temporary layer is now empty and so we update the youngest and oldest elements in layer $t$ to indicate that there is no layer below. We then perform $t - j + 1$ MOVEUP($y$) operations for the youngest element $y$ of each layer from $t$ to $j$ to restore the sizes of the layers. At this point, it could be the case that $|L_t| = 0$. If this happens, we decrement the number of layers $t$ which is stored at the root, and update the youngest and oldest elements in the new deepest layer to indicate that there is no layer below.

**Theorem 1.** *There exists a binary search tree that performs accesses in* $O(\lg w_i(x))$ *worst-case time and insertions and deletions in* $O(\lg n)$ *worst-case time.*

---

[4] Note that for an ordinary search, we have $k = j$. However, thinking of the algorithm this way gives us a clean way to describe insertions.

# 3   Skip-Splay and the Unified Bound

In this section, we show how to use layered working-set trees in the skip-splay structure of Derryberry and Sleator [7] to achieve the unified bound to within a small multiplicative factor. The unified bound [2] requires that the time to search for an element $x$ at time $i$ is $\text{UB}(x) = O(\min_{y \in S_i} \lg(w_i(y) + d(x, y)))$ where $w_i(y)$ is the working-set number of $y$ at time $i$ (as in Section 1) and $d(x, y)$ is defined as the rank distance between $x$ and $y$. This property implies both the working-set and the dynamic finger properties. Informally, the unified bound states that an access is fast if the current access is close in term of rank distance to some element that has been accessed recently. Bădoiu et al. [2] introduced a data structure achieving the unified bound in the amortized sense. This structure does not fit into the binary search tree model, but the splay tree [9], which does fit into this model, is conjectured to achieve the unified bound [2].

Recently, Derryberry and Sleator [7] developed the first binary search tree that guarantees an access time close to the unified bound. Their binary search tree, called *skip-splay*, performs an access to an element $x$ in $O(\text{UB}(x) + \lg \lg n)$ amortized time. Insertions and deletions are not supported. In the remainder of this section, we briefly describe skip-splay and then show how to modify it using the layered working-set tree presented in Section 2 in order to achieve a new bound in the binary search tree model.

The skip-splay algorithm works in the following way. Assume for simplicity that the tree $T$ stores the set $\{1, 2, \ldots, n\}$ where $n = 2^{2^{k-1}} - 1$ for some integer $k \geq 0$ and that $T$ is initially perfectly balanced. Nodes of height $2^i$ (where the leaves of $T$ have height 1) for $i \in \{0, 1, \ldots, k-1\}$ are marked as the root of a subtree. Such nodes partition $T$ into a set of splay trees called *auxiliary trees*. Each auxiliary tree is maintained as an independent splay tree. Observe that the $i$-th auxiliary tree encountered on a path from the root to a leaf in $T$ has size $2^{\lg_2 n / 2^i} = n^{1/2^i}$. Define $\text{aux}[x]$ to be the auxiliary tree containing the node $x$.

To access an element $x$, we perform a standard binary search in $T$ to locate $x$. We then perform a series of splay operations on some of the auxiliary trees of $T$. We begin by splaying $x$ to the root of $\text{aux}[x]$ using the usual splay algorithm. If $x$ is now the root of $T$, the operation is complete. Otherwise, we *skip* to the new parent of $x$, say $y$, and splay $y$ to the root of $\text{aux}[y]$. This process is repeated until we reach the root of $T$.

As suggested by Derryberry and Sleator [7], instead of using splay trees to maintain the auxiliary trees, we could use any data structure that satisfies the working-set property. By using layered working-set trees as auxiliary trees and by doubling each access, we obtain

**Theorem 2.** *There exists a binary search tree that performs an access to* $x_i$ *in worst-case time* $O(\min\{\lg n, (\lg \lg n) \lg w_i(x_i)\})$ *and amortized time* $O(\text{UB}(x_i) + \lg \lg n)$.

# 4    Conclusion and Open Problems

We have given the first binary search tree that guarantees the working-set property in the worst-case. We have also shown how to combine this binary search tree with the skip-splay algorithm of Derryberry and Sleator [7] to achieve the unified bound to within a small additive term in the amortized sense while maintaining in the worst case an access time that is both logarithmic and within a small multiplicative factor of the working-set bound. Several directions remain for future research.

For layered working-set trees, it seems that by forcing the working-set property to hold in the worst case, we sacrifice good performance on some other access sequences. Is it the case that a binary search tree that has the working-set property in the worst case cannot achieve other properties of splay trees? For example, what kind of scanning bound can we achieve if we require the working-set property in the worst case? It would also be interesting to bound the number of rotations performed per access. Can we guarantee at most $O(\lg \lg w_i(x_i))$ rotations to access $x_i$? Red-black trees guarantee $O(1)$ rotations per update, for instance. For the results on the unified bound, the most obvious improvement would be to remove the $\lg \lg n$ term, as posed by Derryberry and Sleator [7], or removing the $\lg \lg n$ factor from the worst-case access cost.

*Acknowledgements.* We thank Jonathan Derryberry and Daniel Sleator for sending us a preliminary version of their skip-splay paper [7] and Stefan Langerman for stimulating discussions.

# References

1. Bayer, R.: Symmetric binary B-trees: data structures and maintenance algorithms. Acta Informatica 1, 290–306 (1972)
2. Bădoiu, M., Cole, R., Demaine, E.D., Iacono, J.: A unified access bound on comparison-based dynamic dictionaries. Theoretical Computer Science 382(2), 86–96 (2007)
3. Bose, P., Douïeb, K., Dujmovic, V., Howat, J.: Layered working-set trees. CoRR, abs/0907.2071 (2009)
4. Cole, R.: On the dynamic finger conjecture for splay trees. Part II: the proof. SIAM Journal on Computing 30(1), 44–85 (1990)
5. Cole, R., Mishra, B., Schmidt, J., Siegel, A.: On the dynamic finger conjecture for splay trees. Part I: splay sorting log $n$-block sequences. SIAM Journal on Computing 30(1), 1–43 (1990)
6. Cormen, T.H., Leiserson, C.E., Rivest, R.L., Stein, C.: Introduction to Algorithms, 2nd edn. MIT Press, Cambridge (2001)
7. Derryberry, J.C., Sleator, D.D.: Skip-splay: Toward achieving the unified bound in the BST model. In: WADS 2009: Proceedings of the 16th Annual International Workshop on Algorithms and Data Structures, pp. 194–205 (2009)
8. Guibas, L.J., Sedgewick, R.: A dichromatic framework for balanced trees. In: FOCS 1978: Proceedings of the 19th Annual IEEE Symposium on Foundations of Computer Science, pp. 8–21 (1978)

9. Sleator, D.D., Tarjan, R.E.: Self-adjusting binary search trees. J. ACM 32(3), 652–686 (1985)
10. Tarjan, R.E.: Data structures and network algorithms. Society for Industrial and Applied Mathematics, Philadelphia (1983)
11. Wilber, R.: Lower bounds for accessing binary search trees with rotations. SIAM Journal on Computing 18(1), 56–67 (1989)

# Lightweight Data Indexing and Compression in External Memory[*]

Paolo Ferragina[1], Travis Gagie[2], and Giovanni Manzini[3]

[1] Dipartimento di Informatica, Università di Pisa, Italy
[2] Departamento de Ciencias de Computación, Universidad de Chile, Chile
[3] Dipartimento di Informatica, Università del Piemonte Orientale, Italy
ferragina@di.unipi.it, travis.gagie@gmail.com, manzini@unipmn.it

**Abstract.** In this paper we describe algorithms for computing the BWT and for building (compressed) indexes in external memory. The innovative feature of our algorithms is that they are lightweight in the sense that, for an input of size $n$, they use only $n$ bits of disk working space while all previous approaches use $\Theta(n \log n)$ bits of disk working space. Moreover, our algorithms access disk data only via sequential scans, thus they take full advantage of modern disk features that make sequential disk accesses much faster than random accesses.

We also present a scan-based algorithm for inverting the BWT that uses $\Theta(n)$ bits of working space, and a lightweight *internal-memory* algorithm for computing the BWT which is the fastest in the literature when the available working space is $o(n)$ bits.

Finally, we prove *lower* bounds on the complexity of computing and inverting the BWT via sequential scans in terms of the classic product: internal-memory space $\times$ number of passes over the disk data, showing that our algorithms are within an $O(\log n)$ factor of the optimal.

## 1 Introduction

Full-text indexes are data structures that index a text string $T[1, n]$ to support subsequent searches for arbitrarily long patterns like substrings, regexp, errors, *etc.*, and have many applications in computational biology and data mining. Recent years have seen a renewed interest in these data structures since it has been proved that full-text indexes can be compressed up to the $k$-th order empirical entropy of the input text $T$, and searched without being fully decompressed [22]. Clearly, data compression and indexing are mandatory when the data to be processed and/or transmitted has large size. But larger data means more memory levels involved in their storage and hence, more costly memory references. It is already known how to design an optimal external-memory (uncompressed) full-text index [7], and some results on external memory compressed indexes have recently appeared in the literature [3,14]. However, whichever is the index chosen

---

[*] The first author has been partially supported by Yahoo! Research and FIRB Linguistica 2006. The second author has been partially funded by Millennium Institute for Cell Dynamics and Biotechnology (ICDB), Grant ICM P05-001-F, Mideplan, Chile.

A. López-Ortiz (Ed.): LATIN 2010, LNCS 6034, pp. 697–710, 2010.
© Springer-Verlag Berlin Heidelberg 2010

(compressed or uncompressed), to use it one must first *build* it! The sheer size of data available nowadays for mining and search applications has turned the construction/compression phase into a bottleneck that can even prevent these indexing and compression tools from being used in large-scale applications.

Recent research [11,16,17,21,23] has highlighted that a major issue in the construction of such data structures is the large amount of *working space* usually needed for the construction. Here working space is defined as the space required by an algorithm in addition to the space required for the input (the text to be indexed/compressed) and the output (the index or the compressed file). If the data to be indexed is too large to fit in main memory one must resort to external memory construction algorithms. Such algorithms are known (see e.g. [5,18]), but they all use $\Theta(n \log n)$ bits of working space. We found (see Section 3) that this working space can be up to 500 times larger than the final size of the compressed output that, for typical data, is three to five times smaller than the original input and is anyway $O(n)$ bits in the worst case.

Given these premises, the first issue we address in this paper is the design of construction algorithms for full-text indexes which work on a disk-memory system and are *lightweight* in that their working space is as small as possible. The second issue we address concerns the way our algorithms fetch/write data onto disk: we design them to access disk data only via *sequential* scans. This approach is motivated by the well known fact that sequential I/Os are much faster than random I/Os. Sequential access to data has the additional advantage of using modern caching architectures optimally, making the algorithm cache-oblivious. In this paper we investigate the problems of building (compressed) full-text indexes and compressing data using only sequential scans (i.e. streaming-like). We provide nearly matching *upper* and *lower* bounds for them in terms of the product "internal-memory space × passes over the disk data".

In the following we consider the classical I/O model [24]: a fast internal memory with $M$ words (i.e. $\Theta(M \log n)$ bits) and $O(1)$ disks of unbounded capacity. Disks are organized in pages consisting of $B$ consecutive words (i.e. $\Theta(B \log n)$ bits overall). Since our algorithms access disk data only by sequential scans, we analyze them counting the number of disk passes as in the streaming models: From that number is straightforward also to derive the cost in terms of the number of I/Os (disk page accesses).

Our first contribution is a lightweight algorithm for computing the BWT — a basic ingredient of both compressors and compressed indexes — in $O(n/M)$ passes and $n$ bits of disk working space. Note that the total space usage of the algorithm is $\Theta(n)$ bits and therefore proportional to the size of the input. Since at each pass we scan $\Theta(n)$ bits of disk data, each pass scans $\Theta(n/(B \log n))$ pages and the overall I/O complexity is $O(n^2/(MB \log n))$. We have implemented a prototype of this algorithm (available from people.unipmn.it/manzini/bwtdisk). The prototype takes advantage of the sequential disk access by storing all files (input, output, and intermediate) in compressed form, thus further reducing the disk usage and the total I/Os. Our tests show that our tool is the fastest currently available for the computation of the BWT in external memory, and that its disk working space is much smaller than the size of the input.

The second contribution of the paper is to show that from our algorithm we can derive: **(1)** a lightweight *internal-memory* algorithm for computing the BWT, which is the fastest in the literature when the amount of available working space is $o(n)$ bits (Theorem 2), and **(2)** lightweight algorithms for computing: the suffix array, the $\Psi$ array, and a sampling of the suffix array, which are important ingredients of (compressed) indexes (see Theorems 3, 4, and 5).

Another contribution is a lightweight algorithm to invert the BWT which uses $O(n/M)$ passes with one disk or $O(\log^2 n)$ passes with two disks, and $\Theta(n)$ bits of working space (Theorem 6). This result is based on different techniques than the ones used for our construction algorithms.

Finally, we try to assess to what extent we can improve our scan-based algorithms for computing/inverting the BWT with only one disk. In this setting, lower bounds are often established considering the product "internal-memory space $\times$ passes" [20]. For our BWT construction and inversion algorithms such product is $O(n \log n)$ bits; by strengthening a lower bound from [12], we prove that we cannot reduce it to $o(n)$ bits with a scan-based algorithm using a single disk (Theorem 7). Hence our algorithms are within an $O(\log n)$ factor of the optimal. We note that our lower bound is "best possible" because, if we have $\Omega(n)$ bits of memory, then we can read the input into internal memory with one pass over the disk and then compute the BWT there. Due to space limitations some proofs will be omitted, the reader can find them in [8].

**Related results.** As we mentioned above, the problem of the lightweight computation of (compressed) indexes in internal memory has recently received much attention [11,15,16,17,21,23]. However, all the proposed algorithms perform many random memory-accesses so they cannot be easily transformed into external memory algorithms. To our knowledge no lightweight algorithms specific for external memory are known. The construction of most full-text indexes reduces to suffix-array construction, which in turn needs $\log n$ recursive sorting-levels [6]. In external memory this sort-based approach takes $O\left(\frac{n}{B} \log_{M/B} \frac{n}{B}\right)$ I/Os [7] and is faster than our algorithms when $M = O\left(n/\left(\log n \log_{M/B} \frac{n}{B}\right)\right)$. However, this approach is not lightweight since it uses $\Theta(n \log n)$ bits of disk working space.

## 2  Notation

We briefly recall some definitions related to compressed full-text indexes; for further details see [22]. Let $T[1, n]$ denote a text drawn from a constant size alphabet $\Sigma$. As is usual, we assume that $T[n]$ is a character not appearing elsewhere in $T$ and is lexicographically smaller than all other characters. Given two strings $s, t$ we write $s \prec t$ to denote that $s$ precedes $t$ lexicographically. The suffix array $\mathsf{sa}[1, n]$ is the permutation of $[1, n]$ giving the lexicographic order of the suffixes of $T$, that is $T[\mathsf{sa}[i], n] \prec T[\mathsf{sa}[i + 1], n]$ for $i = 1, \ldots, n - 1$. The inverse of the $\mathsf{sa}$ is the $\mathsf{pos}$ array, such that $\mathsf{pos}[i]$ is the rank of suffix $T[i, n]$

in the suffix array. This way, $\mathsf{sa}[\mathsf{pos}[i]] = i$. We denote by $\mathsf{pos}_d$ the set of $(n/d)$ values $\mathsf{pos}[d], \mathsf{pos}[2d], \ldots, \mathsf{pos}[n]$ that indicate the distribution of the positions of the $d$-spaced suffixes within $\mathsf{sa}$.

The Burrows-Wheeler transform is an array of characters $\mathsf{bwt}[1, n]$ defined as $\mathsf{bwt}[i] = T[(\mathsf{sa}[i] - 1) \bmod n]$. The array $\varPsi[1, n]$ is the permutation of $[1, n]$ such that $\mathsf{sa}[\varPsi(i)] = \mathsf{sa}[i] + 1 \bmod n$. The value $\varPsi[i]$ is the lexicographic rank of the suffix which is one character shorter than the suffix of rank $i$. The basic ingredients of most compressed indexes are either the $\mathsf{bwt}$ or the $\varPsi$ array, optionally combined with the set $\mathsf{pos}_d$ for some $d = \varOmega(\log n)$. In this paper we describe external memory lightweight algorithms for the computation of all these three basic ingredients.

## 3   Lightweight Scan-Based BWT Construction

In this section we describe the algorithm $\mathsf{bwt}$-$\mathsf{disk}$ for the computation of the $\mathsf{bwt}$ of a text $T[1, n]$ when $n$ is so large that the computation cannot be done in internal memory. Our algorithm is lightweight in the sense that it uses only $M$ words of RAM and $n$ bits of disk space — in addition to the disk space used for the input $T[1, n]$ and the output $\mathsf{bwt}(T[1, n])$. Our algorithm is scan-based in the sense that all data on disk is accessed by sequential scans only. Note that in the description below our algorithm scans the input file right-to-left: in the actual implementation we scan the input rightward which means that we compute the $\mathsf{bwt}$ of $T$ reversed. The $\mathsf{bwt}$-$\mathsf{disk}$ algorithm is an evolution of a disk-based construction algorithm for suffix arrays first proposed in [13] and improved in [4]. However, our algorithm constructs the $\mathsf{bwt}$ *directly* without passing through the $\mathsf{sa}$ and uses some new ideas to reduce the working space from $\varTheta(n \log n)$ to $n$ bits.

The algorithm $\mathsf{bwt}$-$\mathsf{disk}$ logically partitions the input text $T[1, n]$ into blocks of size $m = \varTheta(M)$ characters each, i.e. $T = T_{n/m} T_{n/m-1} \cdots T_2 T_1$, and computes incrementally the $\mathsf{bwt}$ of $T$ via $n/m$ passes, one per block of $T$. Text blocks are examined right to left so that at pass $h + 1$ we compute and store on disk $\mathsf{bwt}(T_{h+1} \cdots T_1)$ given $\mathsf{bwt}(T_h \cdots T_1)$. The fundamental observation is that going from $\mathsf{bwt}(T_h \cdots T_1)$ to $\mathsf{bwt}(T_{h+1} \cdots T_1)$ requires only that we insert the characters of $T_{h+1}$ in $\mathsf{bwt}(T_h \cdots T_1)$. In other words, adding $T_{h+1}$ does not modify the relative order of the characters already in $\mathsf{bwt}(T_h \cdots T_1)$.

At the beginning of pass $h + 1$, in addition to the $\mathsf{bwt}$ of $T_h \cdots T_1$ we assume we have on disk a bit array, called $\mathsf{gt}$, such that $\mathsf{gt}[i] = 1$ if and only if the suffix $T[i, n]$ starting in $T_h \cdots T_1$ is greater than the suffix $T_h \cdots T_1$ (hence at pass $h+1$ this array takes exactly $hm - 1$ bits). For simplicity of exposition, we denote by $\mathsf{gt}_h[1, m - 1]$ the part of the array $\mathsf{gt}$ referring to the text suffixes which start in $T_h$: namely, it is $\mathsf{gt}_h[i] = 1$ iff the suffix starting at $T_h[1 + i]$ is lexicographically greater than the suffix starting at $T_h[1]$, for $i = 1, \ldots, m - 1$ (note that all these suffixes extend past $T_h$ up to the last character of $T$).

The pseudo-code of the generic $(h + 1)$-th pass is given in Figure 1. Step 1 reads into internal memory the substring $t[1, 2m] = T_{h+1} T_h$ and the binary

1. Compute in internal memory the array $\mathsf{sa}_{int}[1, m]$ which contains the lexicographic ordering of the suffixes starting in $T_{h+1}$ and extending up to $T[n]$ (the end of $T$). This step uses $T_{h+1}, T_h$ and the first $m - 1$ entries of gt. Let us call the suffixes starting in $T_{h+1}$ *new* suffixes, and the ones starting in $T_h \cdots T_1$ *old* suffixes.
2. Compute in internal memory the array $\mathsf{bwt}_{int}[1, m]$ defined as $\mathsf{bwt}_{int}[i] = T_{h+1}[\mathsf{sa}_{int}[i] - 1]$, for $i = 1, \ldots, m$. If $\mathsf{sa}_{int}[i] = 1$ set $\mathsf{bwt}_{int}[i] = \#$ where $\#$ is a character not appearing in $T$.
3. Using $\mathsf{bwt}_{int}$ and scanning both $T_h T_{h-1} \cdots T_1$ and gt, compute how many old suffixes fall between two lexicographically consecutive new suffixes. At the same time update gt so that it contains the correct information for the extended string $T_{h+1} T_h \cdots T_1$.
4. Merge $\mathsf{bwt}_{ext}$ and $\mathsf{bwt}_{int}$ so that at the end of the step $\mathsf{bwt}_{ext}$ contains the bwt of $T_{h+1} T_h \cdots T_1$.

**Fig. 1.** Pass $h + 1$ of the bwt-disk algorithm. At the beginning of pass $h + 1$, we assume that $\mathsf{bwt}_{ext}$ contains the bwt of $T_h T_{h-1} \cdots T_1$, and the bit array gt is defined as described in the text. Both arrays are stored on disk.

array $\mathsf{gt}_h[1, m - 1]$. Then we build $\mathsf{sa}_{int}$ by lexicographically sorting the suffixes starting in $T_{h+1}$ and possibly extending up to $T[n]$ (the last character of $T$). Observe that, given two such suffixes starting at positions $i$ and $j$ of $T_{h+1}$, with $i < j$, we can compare them lexicographically by comparing the strings $t[i, m]$ and $t[j, j + m - i]$, which have the same length and are completely contained in $t[1, 2m]$ (thus, they are in internal memory). If these strings differ we are done; otherwise, the order between the above two suffixes is determined by the order of the suffixes starting at $t[m + 1] \equiv T_h[1]$ and $t[j + m - i + 1] \equiv T_h[1 + j - i]$. This order is given by the bit stored in $\mathsf{gt}_h[j - i]$, also available in internal memory. This argument shows that $t[1, 2m]$ and $\mathsf{gt}_h$ contain all the information we need to build $\mathsf{sa}_{int}$ working in internal memory. The actual computation of $\mathsf{sa}_{int}$ is done in $O(m)$ time as follows. First we compute the rank $r_{m+1}$ of the suffix starting at $t[m + 1] \equiv T_h[1]$ among all suffixes starting in $T_{h+1}$; that is, we compute for how many indices $i$ with $1 \le i \le m$ the suffix starting at $t[i]$ is smaller than the suffix starting at $t[m + 1]$ (both extending up to $T[n]$). This can be done in $O(m)$ time using the above observation and Lemma 5 in [17]. At this point the problem of building $\mathsf{sa}_{int}$ is equivalent to the problem of building the suffix array of the string $t[1, m]\$$, where $\$$ is a special end-of-string character that has rank precisely $r_{m+1}$ (instead of being lexicographically smaller than all other suffixes, as usual). Thus, we can compute $\mathsf{sa}_{int}$ in $O(m)$ time and $O(m \log m)$ bits of space with a straightforward modification of the algorithm DC3 [18].

At Step 2 we build the array $\mathsf{bwt}_{int}$ which is a sort of bwt of the string $T_{h+1}$: it is not a *real* bwt because it refers to suffixes which are not confined to $T_{h+1}$ but start in this string and extend up to $T[n]$. The crucial point of the algorithm is then to compute some additional information that allows us to *merge* $\mathsf{bwt}_{int}$ and $\mathsf{bwt}_{ext}$ I/O-efficiently. This additional information consists of a counter array $\mathsf{gap}[0, m]$ which stores in $\mathsf{gap}[j]$ the number of (old) suffixes of the string $T_h \cdots T_1$

which lie lexicographically between the two new suffixes— i.e. $\mathsf{sa}_{int}[j-1]$ and $\mathsf{sa}_{int}[j]$— starting in $T_{h+1}$. Note that the gap array was used also in [4]. However in [4] gap is computed in $O(n \log M)$ time using $\Theta(n \log n)$ extra bits; here we compute gap in $O(n)$ time using only the $n$ extra bits of gt. The following lemma is the key to this improvement.

**Lemma 1.** *For any character* $c \in \Sigma$, *let* $C[c]$ *denote the number of charac-ters in* $\mathsf{bwt}_{int}$ *that are smaller than* $c$, *and let* $\mathsf{Rank}(c, i)$ *denote the number of occurrences of* $c$ *in the prefix* $\mathsf{bwt}_{int}[1, i]$. *Assume that the old suffix* $T[k, n]$ *is lexicographically larger than precisely* $i$ *new suffixes, that is,*

$$T[\mathsf{sa}_{int}[i], n] \prec T[k, n] \prec T[\mathsf{sa}_{int}[i+1], n].$$

*Now fix* $c = T[k-1]$. *Then, the old suffix* $T[k-1, n] = cT[k, n]$ *is lexicograph-ically larger than precisely* $j$ *new suffixes, that is,* $T[\mathsf{sa}_{int}[j], n] \prec T[k-1, n] \prec T[\mathsf{sa}_{int}[j+1], n]$, *where*

$$j = \begin{cases} C[c] + \mathsf{Rank}(c, i) & \text{if } c \neq T_{h+1}[m]; \\ C[c] + \mathsf{Rank}(c, i) + \mathsf{gt}[k] & \text{if } c = T_{h+1}[m]. \end{cases}$$

Step 3 uses the above lemma to compute the array gap with a single right-to-left scan of the two arrays $T_h \cdots T_1$ and gt. Step 3 takes $O(n)$ time because we can build a $o(m)$-bit data structure supporting $O(1)$ time Rank queries over $\mathsf{bwt}_{int}$ [22]. Finally, Step 4 uses gap to create the new array $\mathsf{bwt}_{ext}$ by merging $\mathsf{bwt}_{int}$ with the current $\mathsf{bwt}_{ext}$. The idea is very simple: for $i = 0, \ldots, m-1$ we copy $\mathsf{gap}[i]$ old values in $\mathsf{bwt}_{ext}$ followed by the value $\mathsf{bwt}_{int}[i+1]$.

At Step 3 we also compute the content of gt for the next pass: namely, $\mathsf{gt}[k] = 1$ iff $T_{h+1} \cdots T_1 \prec T[k, n]$. We know the lexicographic relation between $T_{h+1} \cdots T_1$ and all new suffixes since it does exist $r_1$ such that $T[\mathsf{sa}_{int}[r_1], n] = T_{h+1} \cdots T_1$ (the latter is a new suffix, indeed). The relation between $T_{h+1} \cdots T_1$ and any old suffix $T[k, n]$ is available during the construction of gap: when we find that $T[k, n]$ is larger than $i$ new suffixes of $\mathsf{sa}_{int}$, we know that $T_{h+1} \cdots T_1 \prec T[k, n]$ iff $r_1 \leq i$. So we can write the correct value for $\mathsf{gt}[k]$ to disk.

Our algorithm uses $O(m \log m)$ bits of internal memory. Hence, if the internal memory consists of $M$ words, we can take $m = \Theta(M)$ and establish the following:

**Theorem 1.** *We can compute the* bwt *of a text* $T[1, n]$ *in* $O(n/M)$ *passes over* $\Theta(n)$ *bits of disk data, using* $n$ *bits of disk working space. The total number of I/Os is* $O(n^2/(MB \log n))$ *and the CPU time is* $O(n^2/M)$.

**Single-disk implementation.** In the bwt-disk algorithm, and in its derivatives described below, we scan $T$ and the gt array in parallel so we need at least *two disks*. However, in view of the lower bounds in Section 6, which hold for a single disk, it is important to point out that our algorithm (and its derivatives) can work via sequential scans using *only one* disk. This is possible by interleaving $T$ and the gt array in a single file. At pass $h$ we interleave $m$ new bits within the segment $T_h$ (so that the portion $T_{n/m} \cdots T_{h+1}$ is shifted by $m$ bits). These new bits together with the bits already interleaved in $T_{h-1} \cdots T_1$ allow us to store the portion of the gt array that is needed at the next pass. Note also that the merging of bwt$_{ext}$ and bwt$_{int}$ at Step 4 can be done on a single disk. This requires that, at the beginning of the algorithm, we reserve on disk the space for the full output ($n$ characters), and that we fill this space right-to-left (that is, at the end of pass $h$ bwt($T_h \cdots T_1$) is stored in the rightmost $mh$ characters of the reserved space).

**Working with compressed files.** Accessing files only by sequential scans makes it possible to store them on disk in compressed form. This is not particularly significant from a theoretical point of view — in the worst case the compressed files still take $\Theta(n)$ bits — but is a significant advantage in practice. If the input file $T[1, n]$ is large, it is likely that it will be given to us in compressed form. If the compression format allows for the scanning of a file without full decompression (as, for example, gzip, bzip, and ppm) our algorithm is able to work on the compressed input without additional overhead. An algorithm that accesses the input non-sequentially would require the additional space for an uncompressed image of $T[1, n]$. The same considerations apply to the output file bwt($T$) and the intermediate files bwt($T_h \cdots T_1$). Since they are bwt's of (suffixes of) $T$ they are likely to be highly compressible, so it is very convenient to be able to store them in compressed form: this makes our algorithm even more "lightweight". It goes without saying that using compressed files also yields a reduction of the I/O transfer so this is advantageous also in terms of running time (see experimental results below).

Note that the use of compressed files is straightforward if we use two disks: in this way we can store $T$ and gt separately and at Step 4 we can store on two different disks the compressed images of bwt($T_h \cdots T_1$) and bwt($T_{h+1} \cdots T_1$). The use of compression in the single disk version is trickier and requires the use of ad-hoc compressors.

**Experimental results.** To test how bwt-disk works in practice, we have implemented a prototype in C (see [8] for the details of our implementation). The main modification wrt the description of Fig. 1 is that, instead of storing the entire array gt on disk, we maintain a "reduced" version in RAM. In fact, Step 1 uses gt$_h[1, m-1]$ which can be stored in RAM. At Step 3 we need the entire gt to lexicographically compare all suffixes $T[k, n]$ of $T_h \cdots T_1$ with $T_h \cdots T_1$ itself (see proof of Lemma 1 [8]). Instead of storing the whole gt, we keep in internal memory the length-$\ell$ prefix of $T_h \cdots T_1$, call it $\alpha_h$, and the entries gt$[k]$ such that $\alpha_h$ is a prefix of $T[k, n]$. Unless $T$ is a very pathological string, this "reduced" version of gt is much more succinct: by setting $\ell = 1024$ we were able to store it in internal memory in just 128KB. Using this "reduced" version, the comparison between $T[k, n]$ and $T_h \cdots T_1$, can be done by comparing $T[k, n]$ with $\alpha_h$. If these two strings are different, we are done; otherwise, $\alpha_h$ is a prefix of $T[k, n]$ and thus the bit gt$[k]$ is available and provides the result of that suffix comparison. Hence, by using standard string-matching techniques, it is possible to compare all suffixes $T[k, n]$ with $T_h \cdots T_1$ in $O(n + \ell)$ time overall. Our implementation can work with a block size $m$ of up to 4GB and uses $8m$ bytes of internal memory for the storage (and computation) of sa$_{int}$, bwt$_{int}$, and the gap array. We ran

| File Name | Description |
|---|---|
| Proteins | Sequence of bare protein sequences from the Pizza&Chili corpus [10]. |
| Swissprot | Annotated Swiss-prot Protein knowledge base. |
| Genome | Human genome filtered to have a string over the alphabet A,C,G,T,N. |
| Gutenberg | Concatenation of English texts from Project Gutenberg. |
| Random2 | Two concatenated copies of a random string of length 2GB. |
| Mice&Men | Concatenation of the Mouse (mm9) and Human (hg18) genomes. |
| Html | Collection of html pages crawled from the UK-domain in 2006-07 [1]. |

| Input file | | bwt-disk space | | bwt-disk time | | | | DC3 | |
|---|---|---|---|---|---|---|---|---|---|
| name | size | output | total | step 1 | step 3 | step 4 | total | time | w. space |
| Proteins | 1.10 | 0.29 | 1.02 | 0.49 | 0.59 | 0.15 | 1.45 | 6.31 | 30.62 |
| SwissProt | 1.88 | 0.08 | 0.33 | 0.36 | 1.16 | 0.10 | 1.88 | 6.67 | 30.68 |
| Genome | 2.86 | 0.22 | 0.69 | 0.50 | 2.32 | 0.55 | 3.72 | 6.88 | 30.68 |
| Gutenberg | 3.05 | 0.18 | 0.74 | 0.85 | 2.19 | 0.36 | 3.76 | 7.14 | 30.58 |
| Random2 | 4.00 | 0.56 | 2.00 | 0.80 | 3.28 | 2.32 | 6.90 | 7.48 | 30.66 |
| Mice&Men-4 | 4.00 | 0.22 | 0.70 | 0.52 | 3.37 | 0.80 | 5.06 | 7.22 | 30.66 |
| Html-4 | 4.00 | 0.06 | 0.33 | 0.58 | 2.55 | 0.19 | 3.60 | 7.49 | 30.66 |
| Mice&Men | 5.43 | 0.22 | 0.70 | 0.52 | 4.06 | 1.14 | 6.10 | — | — |
| Html | 8.00 | 0.05 | 0.32 | 0.49 | 4.90 | 0.33 | 6.01 | — | — |

**Fig. 2.** Dataset (top) and experimental results (bottom). Since DC3 cannot handle files larger than 4GB we considered also the files Mice&Men and Html truncated at 4GB (indicated by the suffix -4). In the bottom table, column 2 reports the size (in Giga-bytes) of the uncompressed input file: the values in all other columns are normalized with respect to this size. Column 3 reports the size of the compressed bwt which is also an upper bound to the working space of bwt-disk(see text). Column 4 reports the total (working + input + output) disk space used by bwt-disk. Columns 5–9 report running (wallclock) times in microseconds per input byte. The last column reports the size of DC3 working space (again normalized with respect to the size of the input file).

our experiments with $m = 400$MB on a Linux box with a 2.5Ghz AMD Phenom 9850 Quad Core processor (only one CPU was used for our tests) and 3.7GB of RAM. On the same machine we also tested the best competitor of our algorithm. Since all other known approaches for computing the BWT in external memory compute the suffix array first, we tested the DC3 tool [5] which is the current best algorithm for computing the suffix array in external memory. We ran DC3 using two disks for the storage of temporary files and setting the ram_usage parameter to 1500MB. With these settings the peak heap memory usage reported by memusage was between 3.2 and 3.3 Gigabytes for both bwt-disk and DC3.

In our implementation we store the files in compressed form: the input $T$ is gzip-compressed, whereas the partial (and final) bwt's are compressed by Rle followed by *range coding*: according to the experiments in [9] this combination offers the best compression/speed tradeoff for compressing the BWT. Our cur-rent implementation uses a single disk. Since at Step 4 we scan simultaneously two partial bwt's (say bwt($T_h \cdots T_1$) and bwt($T_{h+1} \cdots T_1$)) in that step the disk head has to move between the two files and the algorithm is not "scan-only". We plan to support the use of two disks to remove this inefficiency in a future version.

Our algorithm stores on disk only the compressed input and at most two compressed partial bwt's. Hence, the working space (the space used in addition to the input and the output) has the size of a single compressed partial bwt: in Fig. 2 we bound it with the size of the final compressed bwt. The results in Fig. 2 show that our algorithm is indeed lightweight: for all files the working space is (much) smaller than the size of the input text uncompressed; for most files even the *total* space usage is less than the size of the uncompressed input. The algorithm DC3 uses consistently a working space of more than 30 times the size of the uncompressed input. Comparing columns 3 and 9 we see that, for all files except Random2, DC3 working space is more than 100 times the size of the compressed bwt; for the file Html-4, which is highly compressible, DC3 working space is more than 500 times the size of the compressed bwt! (recall that bwt-disk working space is at most the size of the compressed bwt). By comparing the running times (columns 8 and 9 in Fig. 2) we see that bwt-disk is always faster than DC3 (recall that DC3 only computes the suffix array so we are ignoring the additional cost of computing the BWT from the suffix array). The results show that the more compressible is the input, the faster is bwt-disk, while DC3's running time is less sensitive to the content of the input file. Another interesting data is the total I/O volume of the two algorithms (measured as the ratio between total I/Os and input size and not reported in Fig. 2). According to [5] for files up to 4GB for DC3 such ratio is between 200 and 300. For bwt-disk such ratio is less than 6 for all files except Random2 for which the ratio is 14.76.

The asymptotic analysis predicts that, if $M \ll n$, as the size of the input grows, our algorithm will eventually become slower than DC3 (our algorithm is designed to be lightweight, not to be fast!). However, the above results show that the use of compressed files and avoiding the construction of the suffix array make our algorithm, not only lightweight, but also faster than the available alternatives on real world inputs.

## 4   Other Lightweight Scan-Based Algorithms

**Internal Memory Lightweight BWT construction.** Our bwt-disk algorithm can be turned into a lightweight *internal memory* algorithm with interesting time-space tradeoffs. For example, setting $M = n/\log n$ we get an internal memory algorithm that runs in $O(n \log n)$ time and uses $2n$ bits of working space: $n$ bits for the gt array and $n$ bits for the $M$ words that play the same role as the internal memory in bwt-disk. Setting $M = n/\log^{1+\epsilon} n$, with $\epsilon > 0$, the running time becomes $O(n \log^{1+\epsilon} n)$ and the working space is reduced to $n + o(n)$ bits. This algorithm still accesses the text and the partial bwt's by sequential scans, hence it takes full advantage of the very fast caches available on modern CPU's.

We can further reduce the working space by replacing the $n$ bits of the gt array with a $o(n)$-bit data structure supporting $O(1)$-time Rank queries over bwt($T_h \cdots T_1$). This data structure can provide in constant time the lexicographic rank of each suffix of $T_h \cdots T_1$ (in right-to-left order, see [22]) and therefore can emulate, without asymptotic slowdown, the scanning of gt.

If we no longer need the input text $T$, we can write the (partial) bwt's over the already processed portion of text. That is, at the end of pass $h$, we store bwt($T_h \cdots T_1$) in the space originally used for $T_h \cdots T_1$. The right-to-left scan of $T_h \cdots T_1$ required at Step 3 can be emulated, without any asymptotic slowdown, using the same data structure used to replace gt (see again [22]). Note that overwriting $T$ roughly doubles the size of the largest input that can be processed with a given amount of internal memory. Summing up, we have:

**Theorem 2.** *For any $\epsilon > 0$, we can compute the BWT in internal memory in $O(n \log^{1+\epsilon} n)$ time, using $o(n)$ bits of working space. The BWT can be stored in the space originally containing the input text.*

The only internal-memory BWT construction algorithm that can use such a small working space is [17] which—when restricted to using $o(n)$ bits of working space—runs in $\omega(n \log^2 n)$ time. Note, however, that the algorithm [17] has the advantage of working also for non constant alphabets and can use as little as $\Theta(n \log n / \sqrt{v})$ bits of working space with $v = O(n^{2/3})$, running in $O(n \log n + vn)$ worst case time. The algorithms in [16,21] build directly a compressed suffix array but, at least in their original formulation, they use $\Omega(n)$ bits of working space. The algorithm in [23] build a compressed suffix array of a collections of texts. For a collection of $p = \Theta(\log n)$ texts of size $n/p$ the algorithm in [23] runs in $O(n \log n)$ time using $O(n)$ bits of working space, storing the output in compressed form and overwriting the input.

**Lightweight SA construction.** We can transform our bwt-disk algorithm into a lightweight algorithm for computing the Suffix Array. The key observation is that the values stored in bwt$_{ext}$ are never used in subsequent computations. Therefore, to compute the sa we can simply replace bwt$_{ext}$ with an array sa$_{ext}$ containing the sa entries (that is, at the end of pass $h$ sa$_{ext}$ contains sa($T_h \cdots T_1$)). The only change in the algorithm is that, after the computation of the gap array, at Step 4 we update sa$_{ext}$ as follows: we copy gap[$i$] old sa$_{ext}$ entries followed by sa$_{int}$[$i + 1$], for $i = 0, \ldots, m - 1$. Summing up, we have the following result.

**Theorem 3.** *We can compute the suffix array in $O(n/M)$ passes over $\Theta(n \log n)$ bits of disk data, using $n$ bits of disk working space. The total number of I/Os is $O(n^2/(MB))$ and the CPU time is $O(n^2/M)$.*

**Lightweight Computation of the $\Psi$ Array.** We use the same framework as above and maintain an array $\Psi_{ext}$ that, at the end of pass $h$, contains the $\Psi$ values for the string $T_h \cdots T_1$. Since the value $\Psi[j]$ refers to the suffix of lexicographic rank $j$, at Step 4 $\Psi_{ext}$ values are computed using the same scheme used for BWT and suffix array entries: for $i = 0, \ldots, m - 1$, we first update gap[$i$] values in $\Psi_{ext}$ referring to old suffixes and then compute and write the $\Psi$ value referring to $T[\text{sa}_{int}[i + 1], n]$. We can compute $\Psi$ values for the new suffixes using information available in internal memory, while for old suffixes we make use of the relationship $\Psi_{h+1}[j] = \Psi_h[j] + k_j$ where $k_j$ is the largest integer such that $\mathsf{gap}[0] + \mathsf{gap}[1] + \cdots + \mathsf{gap}[k_j] < \Psi_h[j]$ (details in the full paper). Since each value $k_j$ can be computed in $O(\log m)$ time with a binary search over the array whose $i$-th element is $\mathsf{gap}[0] + \cdots + \mathsf{gap}[i]$, we have the following result.

**Lemma 2.** *We can compute $\Psi$ in $O(n/M)$ passes over $\Theta(n \log n)$ bits of disk data, using $n$ bits of working space. The CPU time is $O((n^2 \log M)/M)$.*

To reduce the amount of processed data, we observe that although $\Psi$ values are in the range $[1, n]$, it is well known [22] that the sequence $\Psi[1], \Psi[2] - \Psi[1], \Psi[3] - \Psi[2], \ldots, \Psi[n] - \Psi[n-1]$, can be represented in $\Theta(n)$ bits. Thus, by storing an appropriate encoding of the differences $\Psi[i] - \Psi[i-1]$ we can obtain an algorithm that works over a total of $O(n)$ bits.

**Theorem 4.** *We can compute the array $\Psi$ in $O(n/M)$ passes over $\Theta(n)$ bits of disk data, using $n$ bits of disk working space. The total number of I/Os is $O(n^2/(MB \log n))$ and the CPU time is $O((n^2 \log M)/M)$.*

**Lightweight Computation of $\mathsf{pos}_d$.** To compute the set $\mathsf{pos}_d$ with a sampling step $d = \Omega(\log n)$, we modify our bwt-disk algorithm as follows. At the end of pass $h$, instead of $\mathsf{bwt}_{ext} = \mathsf{bwt}(T_h \cdots T_1)$ we store on disk the pairs $\langle i_1, j_1 \rangle, \langle i_2, j_2 \rangle, \ldots \langle i_k, j_k \rangle$ such that $\mathsf{sa}_h[i_\ell] = j_\ell$ is a multiple of $d$ (here $\mathsf{sa}_h = \mathsf{sa}(T_h \cdots T_1)$). These pairs are sorted according to their first component and essentially represent $\mathsf{pos}_d(T_h \cdots T_1)$. The update of this set of pairs at pass $h + 1$ is straightforward: the second component does not change, whereas the value $i_\ell$ must be increased by the number of new suffixes which are lexicographically smaller than $i_\ell$ old suffixes. This can be done via a sequential scan of the already computed set of pairs and of the $\mathsf{gap}$ array. Since the set $\mathsf{pos}_d$ contains $n/d = O(n/\log n)$ pairs, we have:

**Theorem 5.** *We can compute $\mathsf{pos}_d$ in $O(n/M)$ passes over $\Theta(n)$ bits of disk data, using $n$ bits of disk working space. The number of I/Os is $O(n^2/(MB \log n))$ and the CPU time is $O(n^2/M)$.*

## 5   Lightweight Scan-Based BWT Inversion

The standard algorithm for inverting the BWT is based on the fact that the "successor" of character $\mathsf{bwt}[i]$ in $T$ is $\mathsf{bwt}[\Psi[i]]$. Since we can set up a pointer from position $i$ to position $\Psi[i]$ for $i = 1, \ldots, n$ in linear time, to retrieve $T$ we essentially need to solve a *list ranking* problem in which we have to restore a sequence given the first element and a pointer to each element's successor. The naïve algorithm for list ranking — follow each pointer in turn — is optimal when the permuted sequence and its pointers fit in memory, but very slow when they do not. List ranking in external memory has been extensively studied, and Chiang *et al.* [2] showed how to reduce this problem to sorting a set of

$n$ items (recursively), each of size $\Theta(\log n)$ bits. If we invert bwt by turning it into an instance of the list-ranking problem and solve that by using Chiang *et al.*'s algorithm, then we end up with a solution requiring $\Theta(n \log n)$ bits of disk space. The following Theorem establishes that, using Chiang *et al.*'s algorithm as a subroutine, we can invert bwt using a sorting primitive now applied on $O(n/\log n)$ items, for a total of $O(n)$ bits of disk space. In the full paper we will also show how we can similarly recover $T$ from the array $\Psi$ still using $O(n)$ bits of total disk space, and how to take advantage of the $\text{pos}_d$ array.

Our algorithm for BWT-inversion works in $O(\log n)$ rounds, each working on two files. The first file contains a set $\mathcal{S}$ of $n/\log n$ substrings of $T$. Each substring is prefaced by a header, which specifies (i) the position in bwt of the substring's first character, (ii) the position in bwt of the successor in $T$ of the substring's last character, (iii) eventually, the character whose index is in (ii) and, (iv) eventually, the substring's position in a certain partial order that we will define later. These substrings are non-overlapping and their length increases as the algorithm proceeds with its rounds. The second file contains the bwt plus an $n$-bit array bwtMark which marks the characters of bwt already appended to some substring of $\mathcal{S}$. The overall space taken by both files is $O(n)$ bits.

The main idea underlying our algorithm is to cover $T$ by the substrings of $\mathcal{S}$, avoiding their overlapping. The substrings of $\mathcal{S}$ consist initially of the characters which occupy the first $n/\log n$ positions of bwt; then, they are extended one character after the other along the $O(\log n)$ rounds, always taking care that they do not overlap. If, at some round, $c$ of those substrings become adjacent in $T$, they are merged to form one single, longer substring which is then inserted in $\mathcal{S}$, and those $c$ constituting substrings are deleted. In each round, we use Chiang *et al.*'s list-ranking algorithm on the headers both to detect when substrings become adjacent and to determine the order in which we should merge adjacent substrings. Our algorithm preserves the condition $|\mathcal{S}| = n/\log n$, by selecting $(c-1)$ new substrings which are inserted in $\mathcal{S}$ and consist of one single character not already belonging to any substring of $\mathcal{S}$. This is easily done by scanning bwt and bwtMark and taking the first $(c-1)$ characters of bwt which result *unmarked* in bwtMark. Keep in mind that whenever a character is appended to a substring, its corresponding bit in bwtMark is set to 1.

**Theorem 6.** *We can invert the BWT in $O(n/M)$ passes on one disk and in $O(\log^2 n)$ passes on two or more disks. If we allow random (i.e. non sequential) disk accesses we can invert the BWT in $O(\frac{n}{B} \log_{M/B} \frac{n}{B \log n})$ I/Os. For all algorithms the total disk usage is $\Theta(n)$ bits.*

## 6   Lower Bounds

Our scan-based algorithms to compute or invert the bwt have a product "memory's size × number of passes" which is $O(n \log n)$ bits. We prove in this section that we cannot reduce them to $o(n)$ bits via any algorithm that uses only *one single disk* (accessed sequentially). Hence our algorithms are an $O(\log n)$-factor from the optimal. We note that our lower bound is best-possible because, if we have $\Omega(n)$ bits of memory, then we can read the input into internal memory with one pass over the disk and then compute the BWT there using, e.g., Theorem 2.

In a recent paper [12] we observed that, if the repeated substring is larger than the product of the size of the memory and the number of passes, then an algorithm that uses multiple passes but only one disk still cannot take full advantage of the string's periodicity. Using properties of De Bruijn sequences we proved that, with polylogarithmic memory and polylogarithmic passes over one disk, we cannot achieve entropy-only bounds and, therefore, we also cannot compute the BWT. In that paper, however, we were mostly concerned with low-entropy bounds, and only considered the BWT as a means to achieve them. Our new lower bound for the BWT alone is stronger, with a simple and direct proof. Our previous lower bound was based on a technical lemma that we can restate as follows:

**Lemma 3.** *Consider an invertible function from strings to strings and a machine that computes (or inverts) that function using only one disk. We can compute any substring of an input string given*

1. *for each pass, the machine's memory configurations when it reaches and leaves the part of the disk that initially (resp., eventually) holds that substring,*
2. *the eventual (resp., initial) contents of that part of the disk.*

Our new lower bound is based on the same lemma but, instead of combining it with properties of De Bruijn sequences, we now combine it with a property of the BWT itself, demonstrated by Mantaci, Restivo and Sciortino [19]: it turns periodic strings with relatively short periods into strings consisting of relatively few runs.

**Lemma 4.** *If $T$ is periodic and its minimum period $r$ divides $n$, then $\text{bwt}(T)$ consists of $r$ runs, each of length $n/r$ and containing only one distinct character.*

Lemma 3 implies that, if the initial contents of some part of the disk are much more complex than its eventual contents (or vice versa), then the product of the memory's size and the number of passes must be at least linear in the initial (resp., eventual) contents' complexity. To see why, consider that we can compute the initial contents from the eventual contents (or vice versa) and two memory configurations for each pass; therefore, the product of the memory's size and the number of passes must be at least the difference between the complexities. Lemma 4 implies that, if $T$ is periodic, then short substrings of $\text{bwt}(T)$ are simple. Combining these ideas in a fairly obvious way gives us our lower bound.

**Theorem 7.** *In the worst case, we can neither compute nor invert the BWT using only one disk when the product of the memory's size in bits and the number of passes is $o(n)$.*

# References

1. Boldi, P., Codenotti, B., Santini, M., Vigna, S.: Ubicrawler: A scalable fully distributed web crawler. Software: Practice & Experience 34(8), 711–726 (2004)
2. Chiang, Y., Goodrich, M., Grove, E., Tamassia, R., Vengroff, D., Vitter, J.: External-memory graph algorithms. In: ACM-SIAM SODA, pp. 139–149 (1995)

3. Chien, Y.-F., Hon, W.-K., Shah, R., Vitter, J.S.: Geometric Burrows-Wheeler transform: Linking range searching and text indexing. In: IEEE DCC (2008)
4. Crauser, A., Ferragina, P.: A theoretical and experimental study on the construction of suffix arrays in external memory. Algorithmica 32(1), 1–35 (2002)
5. Dementiev, R., Kärkkäinen, J., Mehnert, J., Sanders, P.: Better external memory suffix array construction. ACM Journal of Experimental Algorithmics 12 (2008)
6. Farach-Colton, M., Ferragina, P., Muthukrishnan, S.: On the sorting-complexity of suffix tree construction. Journal of the ACM 47(6), 987–1011 (2000)
7. Ferragina, P.: String search in external memory: Data structures and algorithms. In: Aluru, S. (ed.) Handbook of Computational Molecular Biology (2005)
8. Ferragina, P., Gagie, T., Manzini, G.: Lightweight data indexing and compression in external memory. CoRR, abs/0909.4341 (2009)
9. Ferragina, P., Giancarlo, R., Manzini, G.: The engineering of a compression boosting library. In: Azar, Y., Erlebach, T. (eds.) ESA 2006. LNCS, vol. 4168, pp. 756–767. Springer, Heidelberg (2006)
10. Ferragina, P., Navarro, G.: The Pizza & Chili corpus home page (2007), http://pizzachili.dcc.uchile.cl/, pizzachili.di.unipi.it
11. Franceschini, G., Muthukrishnan, S.: In-place suffix sorting. In: Arge, L., Cachin, C., Jurdziński, T., Tarlecki, A. (eds.) ICALP 2007. LNCS, vol. 4596, pp. 533–545. Springer, Heidelberg (2007)
12. Gagie, T.: On the value of multiple read/write streams for data compression. In: Kucherov, G., Ukkonen, E. (eds.) CPM 2009. LNCS, vol. 5577, pp. 68–77. Springer, Heidelberg (2009)
13. Gonnet, G.H., Baeza-Yates, R.A., Snider, T.: New indices for text: PAT trees and PAT arrays. In: Frakes, B., Baeza-Yates, R.A. (eds.) Information Retrieval: Data Structures and Algorithms, ch. 5, pp. 66–82 (1992)
14. González, R., Navarro, G.: A compressed text index on secondary memory. In: Proceedings IWOCA 2007, pp. 80–91. College Publications, UK (2007)
15. Hon, W., Sadakane, K., Sung, W.: Breaking a time-and-space barrier in constructing full-text indices. SIAM J. Comput. 38, 2162–2178 (2009)
16. Hon, W., Lam, T., Sadakane, K., Sung, W., Yiu, S.: A space and time efficient algorithm for constructing compressed suffix arrays. Algorithmica 48, 23–36 (2007)
17. Kärkkäinen, J.: Fast BWT in small space by blockwise suffix sorting. Theoretical Computer Science 387, 249–257 (2007)
18. Kärkkäinen, J., Sanders, P., Burkhardt, S.: Linear work suffix array construction. Journal of the ACM 53(6), 918–936 (2006)
19. Mantaci, S., Restivo, A., Sciortino, M.: Burrows-Wheeler transform and Sturmian words. Information Processing Letters 86(5), 241–246 (2003)
20. Munro, J., Paterson, M.: Selection and sorting with limited storage. Theor. Comput. Sci. 12, 315–323 (1980)
21. Na, J., Park, K.: Alphabet-independent linear-time construction of compressed suffix arrays using $o(n \log n)$-bit working space. TCS 386, 127–136 (2007)
22. Navarro, G., Mäkinen, V.: Compressed full-text indexes. ACM Computing Surveys 39(1) (2007)
23. Sirén, J.: Compressed suffix arrays for massive data. In: Hyyro, H. (ed.) SPIRE 2009. LNCS, vol. 5721, pp. 63–74. Springer, Heidelberg (2009)
24. Vitter, J.: Algorithms and Data Structures for External Memory. Foundations and Trends in Theoretical Computer Science, vol. 2(4). NOW (2008)

# Author Index